Heart Development

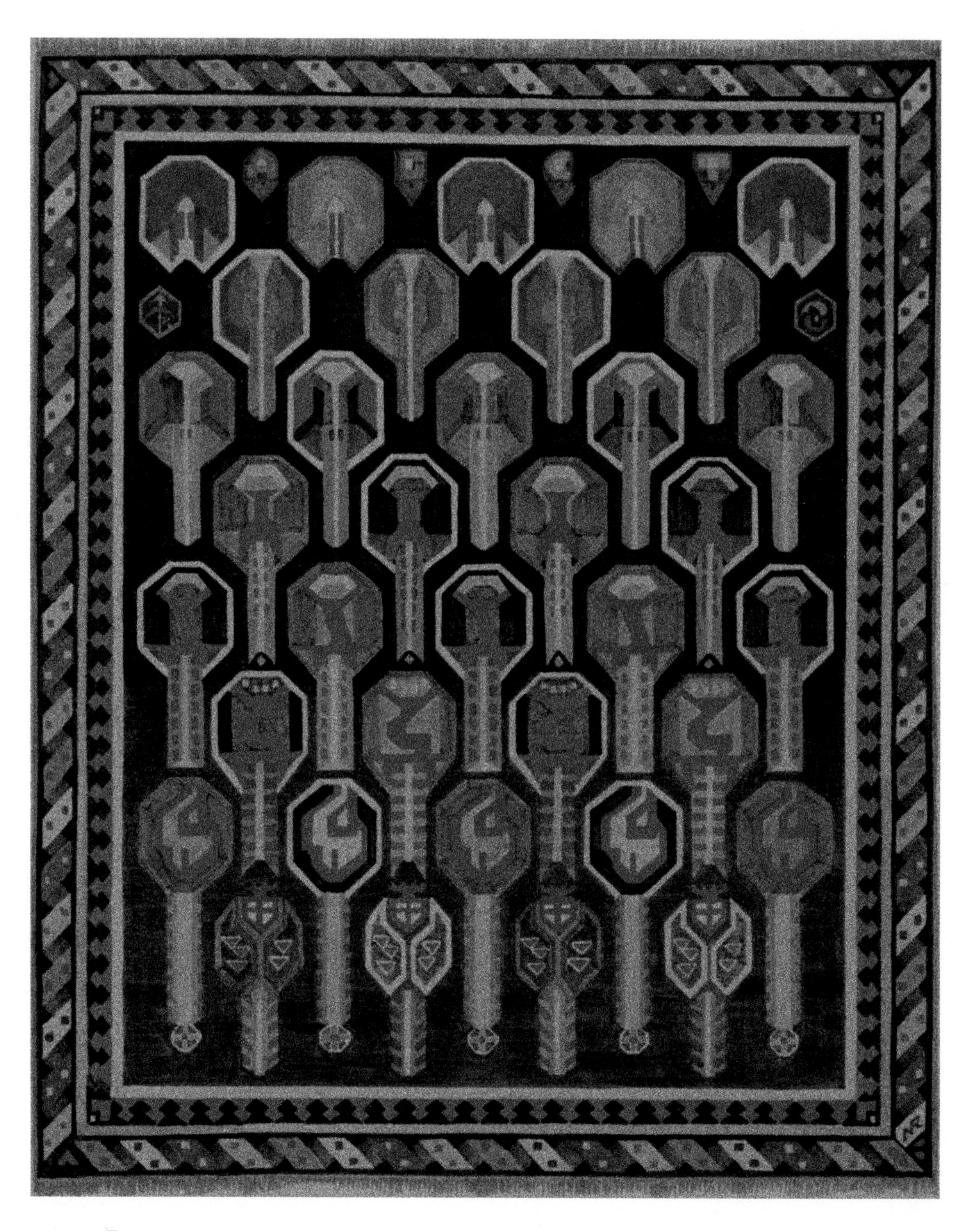

HEARTCARPET—Inspired by a 19th century Afshar rug from Persia. The Afshar were a nomadic people of Turkic descent, reaching Persia via Central Asia and the Caucasus in the 12th and 13th centuries and settling throughout much of the western part of the country. Their carpets were made on horizontal looms in the villages of the Kerman province. The diagonally staggered columns of geometric polychrome medallions, or botehs, are typical of Afshar weavings, as is the dark blue background. Fields of cardiac tissue are depicted in red. Progressively later stages of heart development are depicted running down the carpet, followed on the diagonal in two different colorways. The bottom motif, representing the fully developed cardiopulmonary system, is a direct rendition of the double boteh on the original carpet, as is the elaborate border which includes a reciprocating trefoil, here representing the vascular–tissue interface, and a typical barberpole design, rendered as a DNA double helix.—NR

Heart Development

Edited by

Richard P. Harvey
The Victor Chang Cardiac Research Institute
St. Vincent's Hospital
Sydney, Australia

Nadia Rosenthal
Cardiovascular Research Hospital
Massachusetts General Hospital–East
Consultant in Molecular Medicine
New England Journal of Medicine
Boston, Massachusetts

Academic Press
San Diego • London • Boston • New York • Sydney • Tokyo • Toronto

This book is printed on acid-free paper. 

Academic Press
a division of Harcourt Brace & Company
525 B Street, Suite 1900, San Diego, California 92101-4495, USA
http://www.apnet.com

Academic Press
24-28 Oval Road, London NW1 7DX, UK
http://www.hbuk.co.uk/ap/

Library of Congress Catalog Card Number: 98-86437

International Standard Book Number: 0-12-329860-1

PRINTED IN CANADA
98 99 00 01 02 03 FR 9 8 7 6 5 4 3 2 1

In her heart she carried an instinct
like that of a migrant species.
She would find the tundra,
the deeps, she would journey home.

James Salter
Light Years (1975)

Contents

III Genetic Dissection of Heart Development

5 Genetic Determination of *Drosophila* Heart Development

Rolf Bodmer and Manfred Frasch

6 Mutations Affecting Cardiac Development in Zebrafish

Jonathan Alexander and Didier Y. R. Stainier

7 Transcriptional Control and Pattern Formation in the Developing Vertebrate Heart: Studies on NK-2 Class Homeodomain Factors

Richard P. Harvey, Christine Biben, and David A. Elliott

8 Control of Cardiac Development by the MEF2 Family of Transcription Factors

Brian L. Black and Eric N. Olson

9 Segmental Regulation of Cardiac Development by the Basic Helix–Loop–Helix Transcription Factors dHAND and eHAND

Deepak Srivastava

IV Normal and Abnormal Morphogenesis

10 Mechanisms of Segmentation, Septation, and Remodeling of the Tubular Heart: Endocardial Cushion Fate and Cardiac Looping

Corey H. Mjaatvedt, Hideshi Yamamura, Andy Wessels, Anne Ramsdell, Debi Turner, and Roger R. Markwald

11 Contribution of Neural Crest to Heart and Vessel Morphology

Margaret L. Kirby

12 Development of the Conduction System of the Vertebrate Heart

Antoon F. M. Moorman and Wouter H. Lamers

13 Retinoids in Heart Development

Steven W. Kubalak and Henry M. Sucov

14 Molecular Mechanisms of Vascular Development

Ondine Cleaver and Paul A. Krieg

V Genetic Control of Muscle Gene Expression

15 The MLC-2 Paradigm for Ventricular Heart Chamber Specification, Maturation, and Morphogenesis

Vân Thi Bich Nguyêñ-Trân, Ju Chen, Pilar Ruiz-Lozano, and Kenneth Randall Chien

Contributors

Numbers in parentheses indicate the pages on which the authors' contributions begin.

Jonathan Alexander (91)
Department of Biochemistry and Biophysics, Programs in Developmental Biology and Human Genetics, University of California, San Francisco, California 94143

Robert H. Anderson (447)
Section of Pediatrics, Royal Brompton Campus, Imperial College School of Medicine, National Heart and Lung Institute, London SW3 6LY, United Kingdom

Narasimhaswamy S. Belaguli (273)
Department of Cell Biology, Baylor College of Medicine, Houston, Texas 77030

Brian L. Black (131)
Department of Molecular Biology and Oncology, University of Texas Southwestern Medical Center, Dallas, Texas 75235

Christine Biben (111)
Victor Chang Cardiac Research Institute, St. Vincent's Hospital, Sydney 2010, Australia

Rolf Bodmer (65)
Department of Biology, University of Michigan, Ann Arbor, Michigan 48109

Nigel A. Brown (447)
Department of Anatomy and Developmental Biology, St. George's Hospital Medical School, University of London, London SW17 0RE, United Kingdom

Margaret Buckingham (333)
Department of Molecular Biology, CNRS URA 1947, Pasteur Institute, 75724 Paris Cedex 15, France

Marcia L. Budarf (463)
Division of Human Genetics and Molecular Biology, The Children's Hospital of Philadelphia, and the Department of Pediatrics, University of Pennsylvania School of Medicine, Philadelphia, Pennsylvania 19104

Brett Casey (479)
Department of Pathology, Baylor College of Medicine, and Texas Children's Hospital, Houston, Texas 77030

Ju Chen (255)
Department of Medicine, Center for Molecular Genetics, and American Heart Association—Bugher Foundation Center for Molecular Biology, University of California School of Medicine at San Diego, La Jolla, California 92093

Kenneth R. Chien (255)
Department of Medicine, Center for Molecular Genetics, and American Heart Association—Bugher Foundation Center for Molecular Biology, University of California School of Medicine at San Diego, La Jolla, California 92093

Ondine Cleaver (221)
Department of Zoology and Institute for Cellular and Molecular Biology, University of Texas at Austin, Austin, Texas 78712

David Elliott (111)
Victor Chang Cardiac Research Institute, St. Vincent's Hospital, Sydney 2010, Australia

Beverly S. Emanuel (463)
Division of Human Genetics and Molecular Biology, The Children's Hospital of Philadelphia, and the Department of Pediatrics, University of Pennsylvania School of Medicine, Philadelphia, Pennsylvania 19104

Diego Franco (333)
Department of Anatomy and Embryology, Academic Medical Center, University of Amsterdam, 1105 AZ Amsterdam, The Netherlands

Manfred Frasch (65)
Brookdale Center for Developmental and Molecular Biology, Mount Sinai School of Medicine, New York, New York 10029

Richard P. Harvey (111)
Victor Chang Cardiac Research Institute, St. Vincent's Hospital, Sydney 2010, and University of New South Wales, Kensington 2033, Australia

Robert G. Kelly (333)
Department of Molecular Biology, CNRS URA 1947, Pasteur Institute, 75724 Paris Cedex 15, France

Margaret L. Kirby (179)
Developmental Biology Program, Institute of Molecular Medicine and Genetics, Medical College of Georgia, Augusta, Georgia 30912

Kenjiro Kosaki (479)
Department of Pediatrics, Keio University School of Medicine, Shinjuku, Tokyo 160, Japan

Paul A. Krieg (221)
Department of Zoology and Institute for Cellular and Molecular Biology, University of Texas at Austin, Austin, Texas 78712

Steven W. Kubalak (209)
Department of Cell Biology and Anatomy, Cardiovascular Developmental Biology Center, Medical University of South Carolina, Charleston, South Carolina 29425

Wouter H. Lamers (195)
Department of Anatomy and Embryology, The Cardiovascular Research Institute Amsterdam, Academic Medical Center, University of Amsterdam, 1105 AZ Amsterdam, The Netherlands

Sarah B. Larkin (307)
Department of Anatomy, University of California at San Francisco, San Francisco, California 94143

Andrew B. Lassar (51)
Department of Biological Chemistry and Molecular Pharmacology, Harvard Medical School, Boston, Massachusetts 02115

Jeffrey M. Leiden (291)
Departments of Medicine and Pathology, University of Chicago, Chicago, Illinois 60637

Li Ming Leong (37)
National Institute for Medical Research, London NW7 1AA, United Kingdom

W. Robb MacLellan (405)
Department of Medicine, Molecular Cardiology Unit, Houston Veteran's Affairs Medical Center, Baylor College of Medicine, Houston, Texas 77030

Kumud Majumder (391)
Department of Cell Biology, Baylor College of Medicine, Houston, Texas 77030

Roger R. Markwald (159)
Department of Cell Biology and Anatomy, Medical University of South Carolina, Charleston, South Carolina 29425

Michael J. McGrew (493)
Developmental Biology Institute of Marseille, 13288 Marseille Cedex 9, France

Takashi Mikawa (19)
Department of Cell Biology, Cornell University Medical College, New York, New York 10021

Corey H. Mjaatvedt (159)
Department of Cell Biology and Anatomy, Medical University of South Carolina, Charleston, South Carolina 29425

Timothy J. Mohun (37)
National Institute for Medical Research, London NW7 1AA, United Kingdom

Antoon F. M. Moorman (195, 333)
Department of Anatomy and Embryology, The Cardiovascular Research Institute Amsterdam, Academic Medical Center, University of Amsterdam, 1105 AZ Amsterdam, The Netherlands

Vân Thi Bich Nguyêñ-Trân (255)
Department of Medicine, Center for Molecular Genetics, and American Heart Association—Bugher Foundation Center for Molecular Biology, University of California School of Medicine at San Diego, La Jolla, California 92093

Eric N. Olson (131)
Department of Molecular Biology and Oncology, University of Texas Southwestern Medical Center, Dallas, Texas 75235

Charles P. Ordahl (307)
Department of Anatomy and Cardiovascular Research Institute, University of California, San Francisco, San Francisco, California 94143

Paul A. Overbeek (391)
Department of Cell Biology, Baylor College of Medicine, Houston, Texas 77030

Michael S. Parmacek (291)
Departments of Medicine and Pathology, University of Chicago, Chicago, Illinois 60637

Harris R. Perlman (429)
Division of Cardiovascular Research, St. Elizabeth's Medical Center, and Program in Cell, Molecular, and Developmental Biology, Sackler School of Biomedical Studies, Tufts University School of Medicine, Boston, Massachusetts 02135

Olivier Pourquie (493)
Developmental Biology Institute of Marseille, 13288 Marseille Cedex 9, France

Anne Ramsdell (159)
Department of Cell Biology and Anatomy, Medical University of South Carolina, Charleston, South Carolina 29425

James M. Reecy (273)
Department of Cell Biology, Baylor College of Medicine, Houston,Texas 77030

Nadia Rosenthal (493)
Cardiovascular Research Center, Massachusetts General Hospital–East, Boston, Massachusetts 02129

Pilar Ruiz-Lozano (255)
Department of Medicine, Center for Molecular Genetics, and American Heart Association—Bugher Foundation Center for Molecular Biology, University of California School of Medicine at San Diego, La Jolla, California 92093

Peter J. Scambler (463)
Molecular Medicine Unit, Institute of Child Health, University College London, London WC1N 1EH, United Kingdom

Michael D. Schneider (405)
Departments of Medicine, Cell Biology, and Molecular Physiology and Biophysics, Molecular Cardiology Unit, Baylor College of Medicine, Houston, Texas 77030

Gary C. Schoenwolf (3)
Department of Neurobiology and Anatomy, University of Utah School of Medicine, Salt Lake City, Utah 84132

Thomas M. Schultheiss (51)
Molecular Medicine Unit, Beth Israel Deaconess Medical Center, Harvard Medical School, Boston, Massachusetts 02215

Robert J. Schwartz (273)
Department of Cell Biology, Baylor College of Medicine, Houston, Texas 77030

Roy C. Smith (429)
Division of Cardiovascular Research, St. Elizabeth's Medical Center, and Program in Cell, Molecular, and Developmental Biology, Sackler School of Biomedical Studies, Tufts University School of Medicine, Boston, Massachusetts 02135

Deepak Srivastava (143)
Departments of Pediatric Cardiology, and Molecular Biology and Oncology, University of Texas Southwestern Medical Center, Dallas, Texas 75235

Didier Y. R. Stainier (91)
Department of Biochemistry and Biophysics, Programs in Developmental Biology and Human Genetics, University of California, San Francisco, California 94143

Frank E. Stockdale (357)
Department of Medicine, Stanford University School of Medicine, Stanford, California 94305

Henry M. Sucov (209)
Department of Cell and Neurobiology, Institute for Genetic Medicine, University of Southern California School of Medicine, Los Angeles, California 90033

Patrick P. L. Tam (3)
Embryology Unit, Children's Medical Research Institute, Wentworthville, New South Wales 2145, Australia

Debi Turner (159)
Department of Cell Biology and Anatomy, Medical University of South Carolina, Charleston, South Carolina 29425

Kenneth Walsh (429)
Division of Cardiovascular Research, St. Elizabeth's Medical Center, and Program in Cell, Molecular, and Developmental Biology, Sackler School of Biomedical Studies, Tufts University School of Medicine, Boston, Massachusetts 02135

Gang Feng Wang (357)
Department of Medicine, Stanford University School of Medicine, Stanford, California 94305

Andy Wessels (159)
Department of Cell Biology and Anatomy, Medical University of South Carolina, Charleston, South Carolina 29425

José Xavier-Neto (493)
Laboratório de Biologia Molecular, Instituto do Coração, São Paulo SP 05403, Brazil, and Cardiovascular Research Center, Massachusetts General Hospital–East, Charlestown, Massachusetts 02129

Hideshi Yamamura (159)
Department of Cell Biology and Anatomy, Medical University of South Carolina, Charleston, South Carolina 29425

H. Joseph Yost (373)
Departments of Oncological Sciences and Pediatrics, Huntsman Cancer Institute Center for Children, University of Utah, Salt Lake City, Utah 84112

Preface

This book is about how hearts develop. The heart is an organ of key medical, developmental, and evolutionary significance, and it is a subject that has captured the imaginations of embryologists and clinicians over the centuries. More recently, we have begun to dissect the genetic pathways and cellular interactions that lead to heart formation and specialization of its components. Here, the most basic questions in developmental biology come to the fore: where do heart precursor cells arise; when do they become different from their neighbors; how do they activate genetic programs that specify cell type and organ morphology; how do defects in those programs manifest as congenital abnormalities? Today's biologists have at their disposal an unprecedented richness of experimental paradigms and techniques with which to explore the control circuitry of the developing embryo, and these are being applied in elegant and ingenious ways to the study of the heart.

Our present focus on the developing heart is driven, in part, by a clinical urgency. Cardiovascular disease is one of our greatest killers. Recent progress in understanding the genetic and molecular bases of cardiovascular development, function, and disease is leading to a redefinition of its pathophysiology, and innovative diagnostic and prognostic tools are emerging. But we are only at the beginning of revolutions in biopharmaceuticals, gene therapy, and organ culture. Tracking the development of the cardiovascular system in intricate detail will offer broader dimensions of understanding within which to explore novel interventional strategies for treatment of disease.

Interest in the embryonic heart is also driven by a fascination with developmental processes. The heart is the first organ to form and function during mammalian embryogenesis and serves to focus our attention on the issues of lineage, inductive interactions between cells, and morphogenesis, and how these conspire to craft a complex organ. The surprising similarities in the cellular and molecular mechanisms of cardiac development among different organisms, even those as unrelated as mammals and fruitflies, have inspired an immediate interest in each experimental system, and a deeper questioning and understanding of the evolution of the body and its component parts.

But there is a less tangible side to our fascination, an element of aesthetic awe. The formation of the cardiovascular system is extraordinarily beautiful, whether we are witnessing the improbably early heartbeat in a chick embryo or the patterns of genes expressed in heart progenitor cells. It is this sense of beauty and universality that was the real inspiration for this book, which we hope comes through in both content and design.

We express our gratitude to Craig Panner at Academic Press, who shared our original concept for the project and whose continued enthusiasm and unconditional support have been essential for its fruition. We also thank Kathy Stern, our artist, whose creativity, curiosity, and patience carried us through the seemingly endless adjustments necessary to realize the visions of each contributor and whose talent is evident in the gorgeous images that enrich their words. Our laboratories deserve acknowledgment for the hours of our attention they lost to this book and for their invaluable editorial assistance in the final harried stages of production. Finally, we thank the authors themselves, friends and colleagues all of them, who have lent their time, inspiration, and knowledge to help us capture the current collective excitement that is the hallmark of this fast-moving field. It has been a rewarding experience to work with each of them, and they have our deepest gratitude.

Nadia Rosenthal

Richard P. Harvey

Introduction

Dialogues in Cardiovascular Development

The formation of the heart and vasculature during embryogenesis is a miraculous event. Our fundamental understanding of this critical process is based on decades of meticulous study by anatomists. Analyses of the sequential evolution of structures through distinct developmental stages have provided us with a series of snapshots of how components grow, differentiate, and sometimes die. Detailed anatomical catalogues cannot, however, distinguish initiating stimuli from responding events. Hence, cardiovascular embryogenesis often appears rather like a stage play without the dialogue; the observer can only wonder why events are linked and how sequences are orchestrated. This book provides some of the answers.

By intertwining descriptions of modern molecular technologies with classical embryology, chapters of this text provide the reader with an understanding of the dynamic forces that shape normal and aberrant cardiovascular development. The complexity of these processes cannot be underestimated. Nevertheless, analyses of multiple vertebrate species help to delineate the fundamental principles of cardiac induction, diversification, morphogenesis, and maturation that transcend species specificity.

For experimentalists, the studies detailed herein mark the beginning of a new era in molecular developmental biology. Targeted gene ablation and random mutagenic screens have provided an initial scaffold of transcription factors that orchestrate cardiac development. Future investigations will undoubtedly expand this data set and assemble the participants into formidable cascades. Fate mapping studies are making use of a range of molecular and cellular markers: viruses, vital dyes, genetically tagged allografts, and normal expression patterns.

These tools allow lineage detection at increasing resolution, often challenging existing paradigms. Their further application should delineate the origin of specialized subpopulations within the myocardium. Another exciting area is the identification of the genes and cellular mechanisms that signal laterality to the primitive heart tube. These studies will give us insights into how physiological heart looping is directed and how endocardial cushions and great vessels are integrated.

For clinicians, these studies herald the promise of new insights into normal cardiac development and the pathogenesis of congenital heart disease. Most certainly, genes and pathways identified in lower vertebrates should continue to provide clues that are relevant to human cardiovascular embryogenesis. For while some species-specific events must occur to account for the obvious differences in the mature hearts of humans, mice, and flies, increasing evidence indicates conservation of genetic pathways involved in heart formation. The rapid pace of investigations into molecular cardiovascular development in lower vertebrates will expedite study of equivalent human pathways, and the participants in these processes should provide candidate molecules that may be perturbed in congenital heart disease. We can expect not only that the etiologies of extraordinarily complex human disorders will be defined, but ultimately that these data will improve techniques for correcting congenital heart defects. Because cardiac malformations occur in 5 to 8 of 1000 live births and are associated with significant mortality in neonatal life and beyond, achievement of these lofty goals would have substantial merit to society.

Unlike the final scene in the theater, this book does not resolve all questions in cardiac development. Rather, it sets the stage for the rebirth of cardiovascular embryology and promises an accelerated and exciting dialogue between experimentalist and clinician, developmental biologist and cardiologist. Let the play begin!

Christine Seidman
Department of Genetics
Harvard Medical School

I

Origins and Early Morphogenesis

1

Cardiac Fate Maps: Lineage Allocation, Morphogenetic Movement, and Cell Commitment

Patrick P. L. Tam* and Gary C. Schoenwolf†

**Embryology Unit, Children's Medical Research Institute, Wentworthville, New South Wales 2145, Australia*
†Department of Neurobiology and Anatomy, University of Utah School of Medicine, Salt Lake City, Utah 84132

I. Prelude to Germ Layer Formation

The freshly laid egg of the chick consists of an enormous mass of yolk on which floats a flat, disc-shaped blastoderm (Fig. 1). The diameter of the blastoderm at laying is about 3.5 mm, whereas that of the yolk is about 35 mm. At the time the egg is laid, the blastoderm consists of about 10,000 cells and is subdivided into an outer area opaca, firmly attached to the underlying yolk, and a central, translucent area pellucida, separated from the underlying yolk by a fluid-filled cavity called the subgerminal cavity. The area opaca forms only extraembryonic tissues, whereas the area pellucida forms the embryo in addition to contributing to the extraembryonic membranes. Moreover, at the time of laying, a sheet of cells, called the hypoblast, is forming at the caudal end of the area pellucida. The hypoblast continues to expand anteriorly and ultimately underlies the entire area pellucida. However, its position in the area pellucida is only temporary, and as the endoderm forms during gastrulation the hypoblast is displaced rostrally to an extraembryonic site. With formation of the hypoblast, the area pellucida becomes two layered. The outer layer is called the epiblast and the inner layer is the hypoblast (Fig. 1).

Early postimplantation development of the mouse embryo is preoccupied with the expansion of the cell population descended from the inner cell mass of the blastocyst. About 15 cells are found in the inner cell mass of the implanting blastocyst (at 3.5 days postcoitum). The other population of about 60 cells forms a

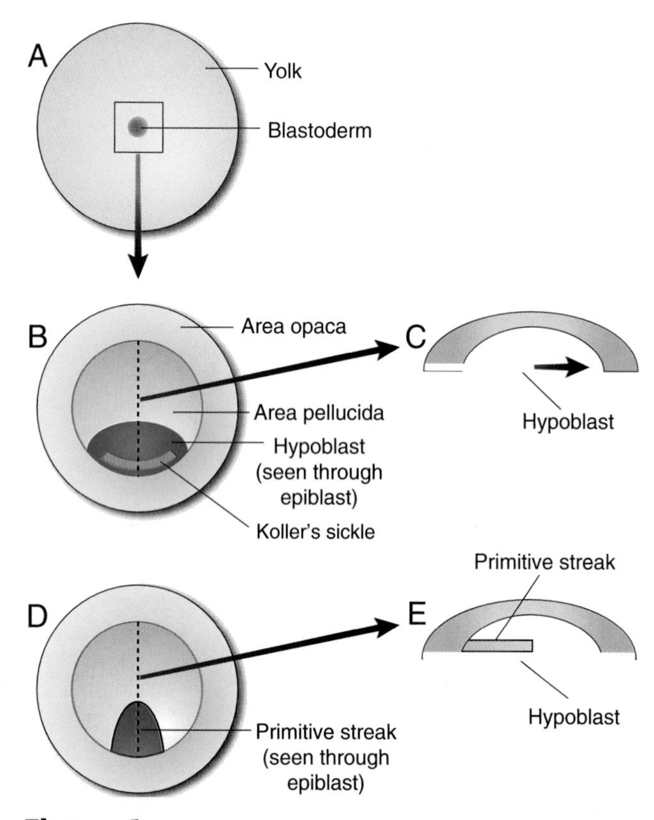

Figure 1 Development of the avian blastoderm from the time the egg is laid (A) until formation of the primitive streak (D). The blastoderm at the time the egg is laid floats on top of a large mass of yolk (A). Both the blastoderm and the yolk are enclosed in the noncellular vitelline membranes (not shown). In B and D, which show dorsal views of the blastoderm (rostral is at the top and caudal is at the bottom of each of these two illustrations), note that the blastoderm is subdivided into an outer area opaca and an inner area pellucida. The hypoblast is beginning its formation at the caudal end of the area pellucida at the stage shown in B (hypoblast is seen through the epiblast) and has completed its formation at the stage shown in D. C and E show sections cut down the midline axis (caudal to the left and rostral to the right). The hypoblast in C is formed only caudally; it is growing rostrally (i.e., in the direction of the arrow) at this stage. The hypoblast in E is completely formed. In addition, the primitive streak has just formed at the caudal end; it is extending rostrally (i.e., in the direction of the arrow in C) at this stage. At the tip of the primitive streak, epiblast cells condense into a structure called the Hensen's node (not shown).

thin epithelial layer called the trophectoderm that encases the inner cell mass and a blastocyst cavity (Fig. 2). A significant departure of the early development of the mouse from that of the chick is the intervention of implantation, whereby the mouse embryo adheres by an intimate contact with the uterine tissue. Implantation, which happens at 3.5 days postcoitum, involves the extensive proliferation of the trophoblasts and the invasion of these cells into the uterine stroma to establish the first embryo–maternal connection. Following implantation, the inner cell mass population increases to about 120 cells during the first 2 days after implantation. Initially, cells are clustered in the inner cell mass, but by 5.5 days postcoitum, they are organized into a pseudostratified epithelium, known as the epiblast. As a result of the physical constraint imposed on the embryo by the confine of the uterine cavity, the epiblast is restrained from growing into a flat discoid shape like that of the chick embryo. Instead, the embryo grows into a cylinder inside the cavity bounded by the trophoblast and the primitive endoderm (Fig. 2). To conform with this rather peculiar configuration, the epiblast epithelium and a layer of primitive endoderm are shaped into a double-layered cup, with the epiblast inside the endoderm (Fig. 2). Shortly before gastrulation starts at 6.5 days postcoitum, there are about 660 cells in the epiblast and 300 cells in the primitive endoderm. The cup-shaped epiblast measures about 200 μm in height and 120 μm in width. The epiblast of the mouse conceptus is therefore only about one-tenth the size of the area pellucida of the chick embryo shortly before gastrulation.

Despite the marked difference in shape, size, and number of cells, the embryos of the chick and the mouse are similar in many respects. At the onset of gastrulation, both embryos are composed of two cell layers, the epiblast and the hypoblast (in the chick) or the primitive endoderm (in the mouse). The epiblast in both is the primary source of all embryonic and some extraembryonic tissues. Although the chick blastoderm contains many more cells than does the mouse epiblast and primitive endoderm combined, a major proportion of cells in the chick is allocated to the area opaca, which is destined to form extraembryonic tissues, some of which must grow to fully enclose the large yolk mass. In both embryos, the formation of the germ layers (ectoderm, mesoderm, and endoderm) is accomplished by the recruitment of cells from the epiblast to the primitive streak where they undergo ingression and are incorporated into three new tissue layers (Bellairs, 1986). This critical phase of embryonic development is known as gastrulation.

Analysis of gene activity during the gastrulation of chick and mouse embryos has revealed remarkable conservation in molecular control of cell differentiation and tissue patterning between these species. This finding, when considered in conjunction with the similar construction of chick and mouse early embryos, suggests that the morphogenetic processes of gastrulation will share many common features. The immediate outcome of gastrulation is the formation of the definitive germ layers and the organization of a blueprint of the embryonic body. The establishment of this blueprint is accomplished by the generation of diverse tissue lineages from a pluripotent progenitor cell population and the placement of the different cell types in their appropriate locations in the embryo. This chapter provides an overview of the morphogenetic events that culminate in the regionalization of the progenitors of the heart tis-

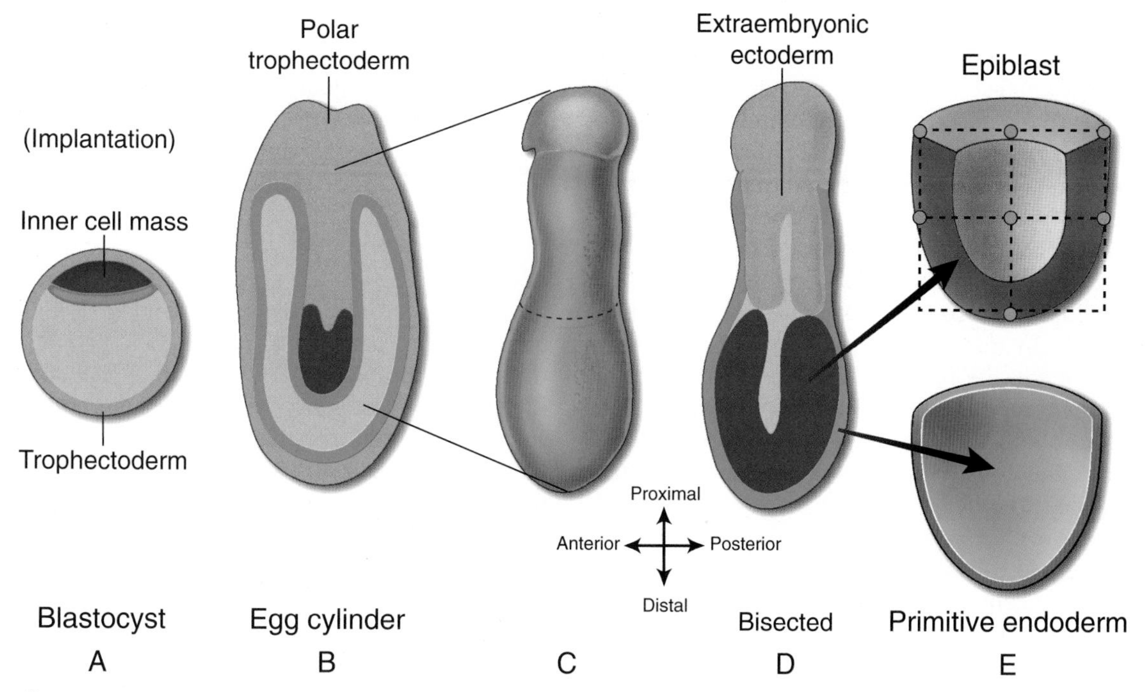

Figure 2 Early development of the mouse embryo from the blastocyst (A) to the pre-primitive-streak (egg cylinder) stage (B). Implantation of the blastocyst occurs on the fourth day postcoitum. The inner cell mass (composed of the epiblast and the primitive endoderm) and the polar trophectoderm grow into the blastocyst cavity to form a cylindrical embryo (B). The embryo has asymmetrical curvatures on the anterior and the posterior sides of the cylinder which seem to define the prospective anterior–posterior axis (C). Shown in the bisected view (D), the cylindrical embryo is made of the extraembryonic ectoderm, the epiblast, and the primitive endoderm. The epiblast and the primitive endoderm are shown separately in E. A grid system is mapped visually on the epiblast to provide the Cartesian reference points for positioning the cell population in the epiblast during fate mapping. The orientation of the anterior–posterior and proximal–distal axes is shown by the cross-arrows.

sues and the specification of the cardiac cell lineages during gastrulation.

II. Experimental Strategy of Fate Mapping

A. A Critical Prerequisite: Reliable Tracking of Cells

1. Locating the Cell Population for Fate Mapping

In fate-mapping experiments, single cells or groups of cells under examination have to be located as precisely as possible in the embryo. The position of the cells in the embryo may be determined by their distance from a recognizable morphological landmark. In the gastrulating mouse embryo, this can be the primitive streak, the allantois, or the junction of the amnion and the epiblast. Because fate-mapping experiments will be repeated using different embryos, it is important to ensure that only embryos of similar developmental stages are used for each fate-mapping experiment. Often, due to the variation in the dimensions of the embryos, even for those at comparable developmental milestones, an absolute measure of position is unable to pinpoint consistently an identical group of cells in different embryos. It is therefore more practical to determine the location of cells by their relative position along the proximal–distal and the anterior–posterior axes (Figs. 1 and 2). In essence, the position of a group of cells is determined by their Cartesian coordinates in a grid system that maps onto the embryo (Fig. 2). The number of coordinate points in the grid is fixed and the dimension of the grid is adjusted in proportion to the size of the embryos. In the chick embryo, the common landmarks are the posterior marginal zone, Koller's sickle, the primitive streak, and the Hensen's node (Fig. 1). The position of the cells is often determined by their distance from these landmarks.

2. Methods for Labeling Cells

The clonal descendants of the cell(s) whose fate is being mapped must be identified by some means so that their distribution and differentiation can be monitored during embryonic development. This can be achieved by marking the cell(s) with an exogenous label. For both chick and mouse cells, the most commonly used labels are the lipophilic carbocyanine dyes [DiI 1,1′-dioctadecyl-3,3,3′,3′-tetramethylindocarbocyanine perchlorate], DiO (3′,3′-dioctadecyloxacarbocyanine perchlorate), and their derivatives, available from Molecular Probes Inc. (Eugene, OR). A group of cells is marked by injecting a small (10–20 pl) volume into the extracellular space around the cells (Darnell and Schoenwolf, 1997). The dyes are taken up by the lipid component of the cell membrane and are distributed to daughter cells that inherit a portion of the membrane of the parent cell during cell division. When viewed under fluorescence illumination, the labeled cells in the embryo glow in bright colors. More refined techniques are used to label single cells by microinjecting the dye into the cytoplasm. Active enzyme (e.g., horseradish peroxidase) or a fluorescent compound (e.g., rhodamine dextran) can be introduced into single cells by iontophoresis (Beddington and Lawson, 1990). However, uncertainty can exist regarding whether only single cells are always injected. Moreover, the efficacy of introduced label is often compromised by the uneven passage of the marker to the daughter cells and the dilution of the label after each cell division. The duration of time that the progeny of cells can be followed in an experiment is therefore limited by the longevity of the label.

An alternative approach to trace the clonal descendants of cells is to employ transgenes as inheritable cell markers. A transgene containing the *Escherichia coli lacZ* sequence that encodes the enzyme β-galactosidase is commonly used. It is important that the transgene is expressed ubiquitously in all progeny of the parent cells irrespective of the types of tissues to which they may differentiate. In our fate-mapping experiments we have used the promoter of a mouse housekeeping gene, which encodes the enzyme 3′-hydroxyl- 5′-methylglutaryl coenzyme A reductase catalyzing a rate-limiting step in the cholesterol biosynthetic pathway, to drive unbiased and ubiquitous expression of the *lacZ* gene in all cell lineages (Tam and Tan, 1992). Cells expressing the *lacZ* gene can be identified by histochemical detection of β-galactosidase activity using X-gal and other colored galactosidase substrates. With this technique, the dilution and the uneven distribution of the label are no longer a problem because the active marker enzyme is always produced in the transgenic cells and their progeny throughout the experiment.

An additional method is used to track cells in avian embryos: the construction of quail/chick transplantation chimeras. In these experiments, donor cells obtained from quail embryos are transplanted in place of cells in host chick embryos. Quail cells later can be identified in chimeras by using a quail-specific antibody or by staining DNA to reveal the endogenous quail-specific nucleolar marker—a clump of heterochromatin associated with the nucleolus (Le Douarin, 1976). Alternatively, cells from donor embryos labeled with fluorescent dyes (e.g., 5- [and -6]-carboxyfluorescein diacetate succinimidyl ester (Molecular Probes Inc.) can be used in place of quail cells. Such cells can be followed over time with time-lapse videomicroscopy, and their fluorescent, transient label can be converted subsequently into a permanent label using antibodies coupled to peroxidase (Garton and Schoenwolf, 1996; Darnell and Schoenwolf, 1997).

The developmental fate of a cell(s) under examination is assessed by the types of embryonic tissues that have been colonized by its descendants. Additionally, the cells are examined for evidence of proper differentiation by the acquisition of histological features and the expression of appropriate molecular and biochemical markers that are characteristic of the tissues that have been colonized. The analysis of the results of fate mapping therefore necessitates the preservation of the cell label during histological preparation of the specimens and the ability to simultaneously detect the cell label and the lineage-specific marker. In this respect, all types of markers described previously have been used successfully for studies that required codetection of markers (Tam and Zhou, 1996; Tam *et al.,* 1997; Garcia-Martinez *et al.,* 1997).

B. Techniques of Fate Mapping

The major technical impediments for studying the differentiation of cell types in the mouse embryo during the immediate postimplantation period are the small size of the gastrula, the relatively small size of its cells compared to those of lower vertebrates at this stage, and that the embryo is developing inside an inaccessible uterine environment. Significant advances have been made in the past decade to culture the mouse embryo through the stages of gastrulation and early organogenesis, stages during which the cultured embryo grows and undergoes morphogenesis similar to that achieved *in utero* (Sturm and Tam, 1993). Embryos maintained *in vitro* are readily amenable to experimental procedures that are essential to fate mapping, such as cell marking, cell transplantation, and continuous scrutiny of cell movement and tissue differentiation.

Chick embryos at gastrula stages are larger than mouse embryos but their cells are still relatively small, and although embryos develop outside of the hen, they still develop within a calcified shell. At gastrula stages, the presence of the shell makes embryos essentially inaccessible because breaching the shell to gain access to embryos *in ovo* distorts the blastoderm, causing further development of the embryo *in ovo* to occur abnormally. Thus, as with mouse embryos, whole-embryo culture is used. With 1 day of culture, embryos removed from freshly laid eggs initiate and complete gastrulation. With 2 days they initiate and virtually complete neurulation, establishing a neural tube along their entire length. Moreover, during this 2-day period of culture prospective cardiac cells are recruited from the superficial epiblast into the primitive streak, ingress and migrate into the interior of the embryo, and are assembled into paired heart tubes, which fuse into a single tube in the ventral midline. This primitive heart contains multiple layers, exhibits rostrocaudal regionalization, and initiates looping during the 2-day period of culture. Thus, several crucial events of early embryonic development can be studied in whole-embryo culture.

To date, mapping the fate of cells in the embryo can be accomplished by two experimental approaches. First, individual cells or groups of cells at a defined position in the embryo are labeled by microinjection. The embryo is then examined at various intervals of culture to assess where the labeled cells are found and into which tissues they have differentiated. Second, the developmental fate of cells can be studied following cell transplantation from a donor to a host embryo. Cells used for transplantation are isolated from donor embryos that have been labeled with cell markers, express a transgene, or, in the case of avian embryos, are derived from quail embryos. These cells are transplanted in groups as small as 5–10 to an equivalent site in the host embryo. By performing such orthotopic transplantations, the donor cells are placed among cells that are likely of the same fate. The differentiation of the transplanted cells in the host embryo will therefore reflect the normal fate of the cells at a specific position in the embryo. The results obtained from these two types of experiments are generally consistent. The results obtained after labeling or transplanting a group of cells may, however, only reveal the collective fate of the population. It is not possible to distinguish if individual cells of that population are displaying a multitude of cell fates (i.e., pluripotent) or whether they are restricted in their potency so that each cell differs from the others and has a unique fate (i.e., determined). This issue can best be resolved by analyzing the differentiation of clonal descendants following the transplanting or marking of single cells.

III. Fate Maps of Cardiac Tissues

A. The Emergence of a Basic Body Plan

The appearance of the primitive streak at gastrulation provides a morphological indication of the orientation of the embryonic axis. The primitive streak marks the caudal pole of the rostrocaudal axis. However, in the chick, a hint of the orientation of the rostrocaudal embryonic axis is given by the formation of a fold of tissues called the Koller's sickle near the posterior marginal zone of the blastoderm, which marks the future caudal aspect of the embryo (Fig. 1). In the mouse, a similar morphological landmark for the caudal aspect of the embryo is not found. The cylindrical embryo often shows some asymmetry in its curvature (Fig. 2) but a direct association of this asymmetry to any embryonic axis has not been demonstrated. There is now, however, evidence from the localized activity of several genes, such as *Evx1, Fgf8,* and *Hesx1 (Rpx)* (Dush and Martin, 1992; Crossley and Martin, 1995; Hermesz *et al.,* 1996; Thomas and Beddington, 1996), and the regionalization of cell fate in the epiblast (Lawson *et al.,* 1991) that the embryonic axes may have been established before gastrulation.

Some molecular indication of the left–right-handedness of chick and mouse embryos can be discerned in the asymmetrical expression of a number of genes at late gastrulation. These include the genes that are expressed predominantly on one side of the primitive streak and the node, which is a condensed group of cells at the anterior end of the primitive streak (e.g., mouse, nodal; chick, *nodal*-related gene [cNR1], sonic hedgehog, HNF3β, *Snail*-related zinc finger gene, and activin receptor IIA; King and Brown, 1995; Levin *et al.,* 1995; Stern *et al.,* 1995; Collignon *et al.,* 1996; Isaac *et al.,* 1997). There is, however, a close correlation between the sided expression of the *nodal* gene in the perinodal tissue and the expression of the *nodal, lefty,* and *ehand* genes in the lateral mesoderm that is associated with the sinoatrial tissues of the heart during cardiac morphogenesis (Collignon *et al.,* 1996; Meno *et al.,* 1996; Srivastava *et al.,* 1995). In the precardiac mesoderm of the stage 5 chick embryo, there is an asymmetrical distribution of two extracellular matrix proteins. The heart-specific lectin-associated matrix protein is more abundant in the mesoderm on the left side of the heart field but the fibrillin-related protein is more abundant on the right side (Smith *et al.,* 1997). It is not known whether this early asymmetry in the distribution of extracellular matrix protein plays a role in the development of heart that displays handedness in looping. The precardiac mesoderm on the two sides of the embryo has been shown to display differential sensitivity to the induction of abnormal heart looping by retinoic acid (RA). Application of RA to the right heart field results in randomized looping of the heart tube, but similar treatment of the left heart field only leads to reversed looping at high RA doses. The alteration of the heart looping pattern is accompanied by changes in the distribution of the extracellular matrix proteins. There is, however, no consistent alteration in the sided expression of *Shh* and *nodal* in the RA-treated embryos that show randomized heart looping (Smith *et al.,* 1997). Nevertheless, the discovery of an asymmetrical distribution of matrix protein in the precardiac mesoderm raises the possibility that there is an early determination of the laterality of the heart primordium during gastrulation.

The discovery of the rostrocaudal polarity and the handedness of the embryo at the beginning of gastrulation indicates that a preliminary blueprint for body pattern may have been established in the morphologically homogeneous epiblast and endoderm layer of the embryo. Studies on cell movement in the epiblast and primitive streak have revealed that there is a predictable site-specific pattern of cell movement. We do not know for certain if the movement of different cell populations in the germ layers is related to the fate of these cells. This issue could be addressed by studying whether there is a regionalization of developmental fate of cells in the germ layers such that a consistent pattern of allocation of cells to various lineages is evident during embryogenesis. The correlation of the developmental fate of cells originating from different sites in the early embryo allows the construction of topographical maps showing the approximate localization of the progenitors of tissue lineages that collectively compose the whole embryo. These are called fate maps and they show pictorially the position of the various precursor populations in the embryo (Figs. 3 and 4). A series of fate maps can be constructed for different developmental stages (Figs. 4 and 5). A correlative study of sequential fate maps is especially informative in projecting the morphogenetic movements of cells of a particular lineage during development and in assessing the neighboring relationship of tissues that may be involved in inductive interactions during differentiation. Also, comparisons among animal species provide information on the degree of universality of morphogenetic movements.

Clonal analyses and fate-mapping studies have shown that cells in the epiblast of the early- primitive-streak-stage (early PS) embryo are not restricted in their lineage potency. Descendants of single epiblast cells can be found in diverse lineages that classically have been regarded as belonging to different germ layers. This strongly implies that there is little restriction of the lineage potency of cells in the epiblast before ingression through the primitive streak (Lawson *et al.*, 1991; Tam *et al.*, 1997). The plasticity or the lack of determination of cell fate has also been demonstrated by cell transplantation experiments. Epiblast cells that

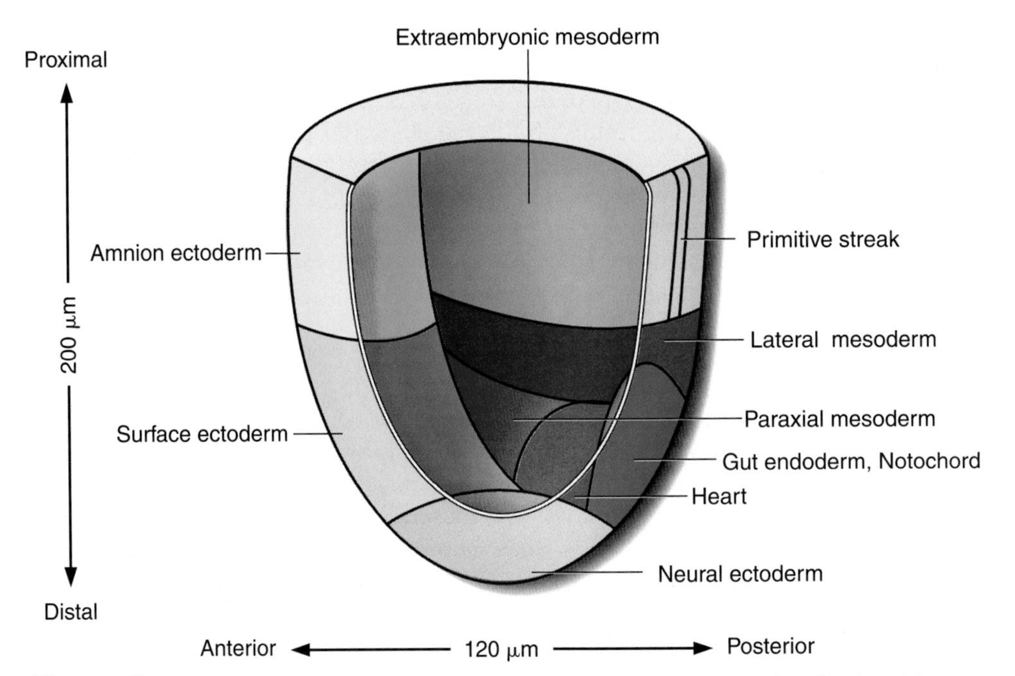

Figure 3 A fate map of the epiblast of the early mouse gastrula showing the regionalization of the precursor cell populations of the various major tissue lineages. The map is shown on the right half of the epiblast viewed from within the proamniotic cavity. The proximal one-third of the epiblast (about 220 cells) is allocated to the extraembryonic mesoderm and the amnion. Precursors of the embryonic tissues are localized to the distal two-thirds of the epiblast. Approximately 50 cells are allocated to the heart mesoderm, which is found in the posterior–lateral region of the epiblast immediately proximal to the neuroectoderm. Other embryonic tissue precursors include those of the surface ectoderm, paraxial mesoderm, lateral plate mesoderm, gut endoderm, and notochord. The arrows show the orientation of the embryonic axes and the dimension of the epiblast.

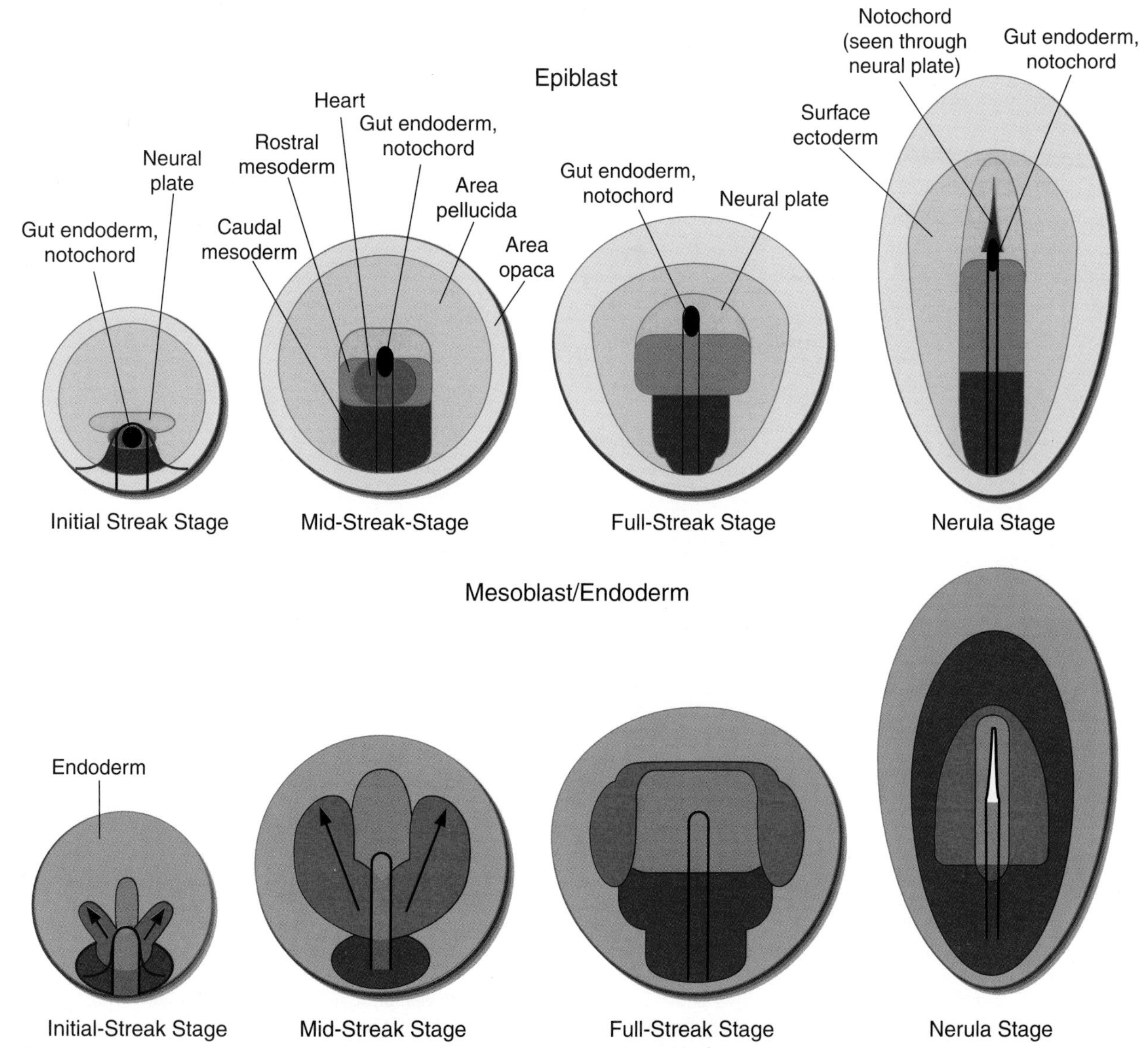

Figure 4 Fate maps of the avian blastoderm showing the epiblast (top) and the combined mesoblast and endoderm layers (bottom) at the initial-streak, midstreak, full-streak, and neurula stages. The maps show the dorsal view of the blastoderm (see Fig. 1). The heart precursors are found initially in the posterior part of the blastoderm adjacent to the middle segment of the primitive streak. The heart mesoderm ingresses through the primitive streak during the initial- to midstreak stage and moves anterolaterally (arrows) during the expansion of the mesoderm. By the full-streak stage, the heart mesoderm is found in a crescent anterior and lateral to the rostral mesoderm.

have been transplanted to different sites in the epiblast will differentiate with the cells in their new neighborhood and not according to the fate of the cell population from which they come. In order words, the epiblast cells may acquire a different fate after transplantation to a new site (Tam and Zhou, 1996; Schoenwolf and Alvarez, 1991; Garcia-Martinez *et al.,* 1997).

Despite this plasticity of cell potency, distinct regionalization of cell fate can be shown by fate mapping. Fate maps depicting the location of the progenitors of major tissue lineages in the epiblast have been constructed (Figs. 3 and 4). In these fate maps, cells in different regions of the embryo are assigned a specific fate according to the probability that descendants of these cells are allocated to tissues of a particular part of the body. For clarity and simplicity, distinct borders are drawn for domains of epiblast that are occupied by the progenitors of specific tissue types. Clonal analysis and cell transplantation studies suggest, however, that there is considerable overlap in the distribution of the progenitor cells for the different lineages. The stylized fate maps therefore, show the most probable localization of the different precursors in the epiblast at the beginning of gastrulation.

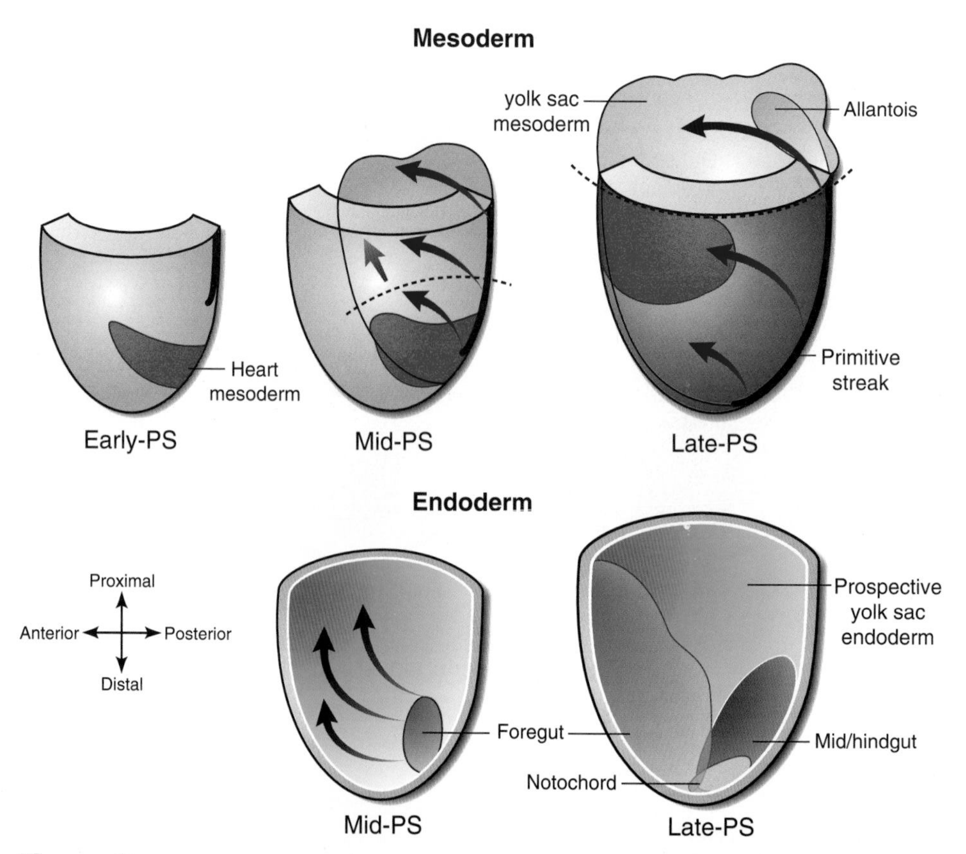

Figure 5 Tissue movement in the mesoderm and the endoderm during gastrulation. The precursor of the heart mesoderm is initially found in the lateral posterior epiblast at the early primitive streak (Early PS) stage. Epiblast cells destined for the heart ingress through the primitive streak during early gastrulation, and recruitment to heart mesoderm continues for some time after the mid-PS stage. The heart mesoderm is displaced anteriorly and proximally to the cardiogenic region, in concert with the expansion of the mesodermal layer (straight solid arrow). The movement of the heart mesoderm matches both spatially and temporally with that of the definitive endoderm that becomes the foregut endoderm. This finding strongly suggests that a neighbor relationship has been established at the inception of the germ layers between the two tissues that may be involved in the inductive interaction underlying heart morphogenesis. The curved arrows show the movement of the ingressed mesodermal and endodermal cells away from the primitive streak (the solid line on the posterior side of the egg cylinder). The red areas in the epiblast and the mesoderm indicate heart precursors (the mesoderm diagrams). The dashed line marks the boundary between extraembryonic and embryonic mesoderm. The orientation of the proximal–distal and the anterior–posterior axes is shown by the crossed arrows.

B. Precursors of Different Cardiac Cell Types Are Colocalized in the Epiblast

The heart is one of the first structures to form during organogenesis (DeRuiter *et al.,* 1992; Icardo, 1996). Fate maps of chick and mouse show the lateral epiblast as the major source of mesoderm (Lawson *et al.,* 1991; Garcia-Martinez *et al.,* 1993; Parameswaran and Tam, 1995). Broadly, the lateral epiblast can be divided into two mesodermal domains: extraembryonic mesoderm (proximal in the mouse and caudal in the chick) and embryonic mesoderm (distal in the mouse and rostral in the chick, Figs. 3 and 4). The heart precursors are localized together with the cranial mesoderm in the embryonic mesoderm domain. They are located posterior to the precursor cells of the neural plate, close to the site that later becomes the rostral part of the primitive streak. Further analysis of the lineage composition of the heart precursor cells in the mouse epiblast has shown that descendants of this population contribute predominantly to the myocardium and to a lesser extent to the endocardium and the pericardium (Tam *et al.,* 1997). These findings suggest that the precursors for these three cardiac tissues are localized to the same regions of the epiblast. Similarly in chick, transplantation of segments of the primitive streak reveals that endocardium and myocardium for each of the subdivisions (e.g., ventricles and atria) of the heart arise in concert

(Garcia-Martinez and Schoenwolf, 1993). However, the transplantation of groups of cells in these two fate-mapping experiments precludes the possibility of revealing whether cardiac lineages share a common progenitor.

It has been suggested that in the zebrafish blastula, the endocardium and the myocardium may be derived from common progenitors and that these two lineages may segregate later during development (Stainier *et al.*, 1993; Lee *et al.*, 1994). The mutation of the *cloche* gene in zebrafish, however, results in the selective deletion of the endocardial cells and does not affect the myocardial lineage (Stainier *et al.*, 1995). This could mean that these two cell lineages are derived from independent progenitors, but it is also possible that the mutation only impacts cell differentiation after the segregation of these lineages has occurred.

C. The Early Association of the Heart Mesoderm and the Foregut Endoderm

At the onset of gastrulation in the mouse (6.5 days), the primitive streak first forms at the most proximal site on the posterior midline of the epiblast (Hashimoto and Nakatsuji, 1989) and is adjacent to the precursors of the extraembryonic mesoderm. The primitive streak extends distally during gastrulation (Tam *et al.*, 1993) and reaches the domain occupied by the heart and cranial mesoderm at about 7.0 days (mid-primitive streak stage). The strategic location of the mesodermal precursors in the epiblast thus predicts that the first cell types that are recruited from epiblast cells to the new germ layer will be the extraembryonic mesoderm, followed by the heart and cranial mesoderm about 12 hr after the onset of gastrulation.

Analysis of the fate of cells in the mouse primitive streak at the mid-primitive streak stage reveals that different cell types are ingressing through different segments of the primitive streak. Because of the technical difficulty in delineating finer regions of the primitive streak, it is not possible to distinguish whether cells destined for different cranial–caudal portions of the heart tube are ingressing through specific segments of the primitive streak during gastrulation. Some heart progenitor cells, however, are still found in the epiblast near the anterior end of the primitive streak at this stage (Parameswaran and Tam, 1995) and presumably will be ingressing later than during the mid-primitive streak stage. It is possible that different segments of the heart tube are ingressing sequentially instead of simultaneously through the primitive streak during gastrulation. In the chick, heart cells ingress through the rostral-middle portion of the primitive streak (Figs. 4 and 6). Ingression of heart precursors begins at the early primitive streak stage and is virtually completed by the fully elongated primitive-streak stage, with the exception of some cells that continue to ingress over the next few hours through the rostral end of the primitive streak; these cells contribute to the most rostral end of the straight heart tube (bulbus cordis; Fig. 6). Furthermore, heart precursor cells ingress in the chick in roughly rostral-to-caudal sequence. That is, cells contributing to the rostral part of the straight heart tube ingress through the most rostral part of the primitive streak, whereas those that contribute to the caudal part of the straight heart tube ingress through more caudal levels of the primitive streak (Fig. 6).

An immediate outcome of the ingression of cells at the primitive streak is the formation of the mesodermal layer. This layer expands as more cells join the mesoderm from the primitive streak and as a result of the proliferation of cells already incorporated into the mesoderm. In the mouse at about 12 hr after the onset of gastrulation (7.0 days postcoitum), the mesodermal layer has expanded to cover about two-thirds of the posterior undersurface of the epiblast. Fate-mapping studies performed by cell transplantation reveal that, consistent with the sequential recruitment of first the extraembryonic mesoderm and then the embryonic mesoderm, the majority of cells in the proximal part of the mesoderm are destined for the mesodermal tissues of the extraembryonic membranes. The heart and cranial mesoderm occupy the most distal part of the mesodermal layer and are localized near the anterior segment of the extending primitive streak (Fig. 5; Parameswaran and Tam, 1995).

Results of fate-mapping (Parameswaran and Tam, 1995) and cell-tracking experiments (Tam *et al.*, 1997) in the mouse show that the mesodermal cells destined for the heart are displaced in a distal to anterior–proximal direction along the lateral aspect of mid- to late-PS embryos (Fig. 5). This movement of the heart mesoderm in the mouse gastrula is comparable to that seen in *Xenopus* (Keller and Tibbetts, 1989) and in the chick (Fig. 4; Garcia-Martinez and Schoenwolf, 1992; Schoenwolf *et al.*, 1992; Psychoyos and Stern, 1996). By the late-PS stage of the mouse, the prospective heart mesoderm is found in the anterior proximal regions of the mesodermal layer underneath the cephalic neural plate (Parameswaran and Tam, 1995). The localization of the heart precursors in the mesoderm of 7.0- to 7.5-day mouse embryos coincides with the distribution of the transcripts of the endothelium-specific *GATA4* and *Flt1* genes (Heikinheimo *et al.*, 1994; Yamaguchi *et al.*, 1993; see Chapter 17) and the *Nkx2.5 (Tinman)* gene which is expressed in the precursor of the myocardium (Lints *et al.*, 1993; reviewed by Lyons, 1994, 1996; see Chapters 5 and 7). Similar localization of heart precursors in the

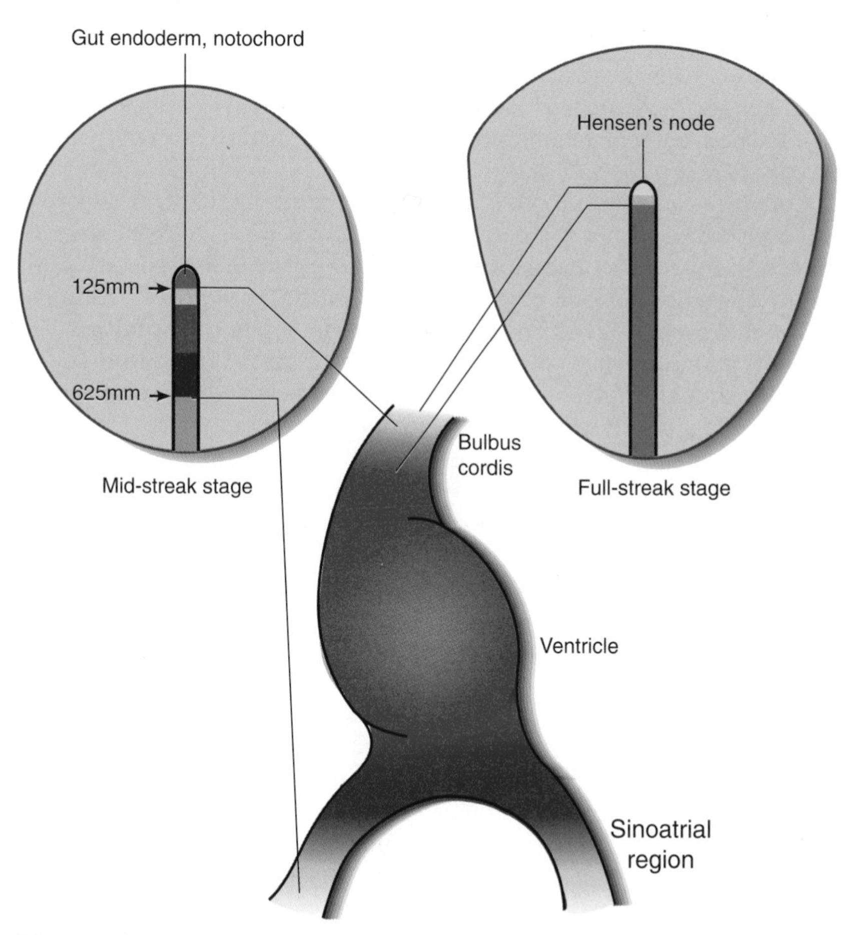

Figure 6 Primitive streak origin of the rostrocaudal subdivisions of the heart in the avian blastoderm. The heart precursors are mapped to a segment of the primitive streak between 125 and 625 mm caudal to the Hensen's node in the midstreak embryo. Cells that are allocated to the three major heart segments (bulbus cordis, ventricle, and the sinoatrial region) are mapped in the corresponding rostrocaudal order in the primitive streak as in the heart (Garcia-Martinez and Schoenwolf, 1993). The Hensen's node does not contribute any cells to the heart. Ingression of the heart precursors finishes by the full-streak stage and there is no further contribution by the primitive streak cells to the heart.

mesoderm and *Nkx2.5* gene expression occur in the chick (Schultheiss *et al.,* 1995; see Chapter 4).

It is interesting to note that a generally similar direction of cellular displacement occurs in the endoderm of the chick and mouse embryos at gastrulation (Rosenquist, 1971; Lawson *et al.,* 1986; Lawson and Pedersen, 1987). The progenitor cells of gut endoderm have been mapped to the posterior epiblast next to the heart mesoderm and the cranial mesoderm (Fig. 5). These progenitor cells ingress through the anterior primitive streak and are recruited to the endodermal layer at early- to mid-PS stages, about the same time as when the heart mesoderm ingresses through the primitive streak. The newly recruited definitive endoderm cells intercalate between the preexisting primitive endoderm cells and are integrated into the epithelium. As more epiblast cells are recruited to the endoderm layer, this new population of cells progressively displaces the primitive endoderm in a proximal–anterior direction over to the extraembryonic yolk sac (Fig. 5). The early waves of definitive endoderm contribute mostly to the foregut (Lawson *et al.,* 1986; Lawson and Pedersen, 1987; Tam and Beddington, 1992). The spatial overlap of the heart mesoderm and the foregut endoderm, therefore, strongly suggests that a neighbor relationship may be established early at the inception of these two tissues. Similarly, in chick, the heart mesoderm, after its ingression, is found next to the rostral endoderm (Rosenquist, 1971), and in *Xenopus,* the heart mesoderm is found next to the pharyngeal endoderm of the late blastula (Keller, 1976; see Chapter 3). During gastrulation, endoderm and heart mesoderm move together anteriorly, along the lateral sides of the neural plate (Fig. 5; Rosenquist, 1971).

IV. Patterning and Cell Commitment to a Heart Fate

A. Role of Cell Ingression and Movement in Fate Specification

Cell populations in different regions of the primitive streak display a diverse cell fate (Schoenwolf *et al.*, 1992; Tam and Beddington, 1987; Smith *et al.*, 1994), although considerable plasticity of cell fate is still evident (Garcia-Martinez and Schoenwolf, 1992; Inagaki *et al.*, 1993). It is not known what constitutes the signal(s) for lineage specification or restriction. The analysis of gene activity in the primitive streak has revealed some regionalization of transcription that is consistent with the allocation of the ingressing cells to various mesodermal lineages (Sasaki and Hogan, 1993; Tam and Trainor, 1994). This concept has been applied to the interpretation of the inappropriate allocation of cells to different mesodermal derivatives in mutant mouse embryos that have lost FGFR1 and HNF3β function (Ang and Rossant, 1994; Yamaguchi *et al.*, 1994; Sasaki and Hogan, 1993).

Experiments in the mouse involving the reciprocal transplantation of the posterior epiblast and newly ingressed mesoderm reveal that the ingression of cells through the primitive streak leads to a restriction of lineage potency (Fig. 7). Cells taken directly from the posterior epiblast to the mesoderm, and therefore not subject to an influence of the primitive streak, retain full epiblast potency and colonize a range of host tissues concordant with their expected cell fate and not with that of the mesoderm at the host site. When ingressed cells in the distal mesoderm of the mid-PS embryo are transplanted to the epiblast of the early PS embryo, the transplanted cells can ingress for a second time through the primitive streak and contribute to a wider variety of mesodermal tissues than that expected from their original fate. The transplanted cells differ from the native epiblast cells and do not colonize the lateral plate mesoderm. Results of these two transplantation experiments indicate that the process of ingression may result in some restriction of lineage potency. Previous ingression through the primitive streak, however, does not seem to reduce the ability of the mesodermal cells to reenact the morphogenetic movement of gastrulation (Tam *et al.*, 1997).

Similar plasticity of cell fate has been found in the avian embryos. Prospective neural plate, when transplanted to the primitive streak, ingresses through the streak and forms mesoderm. Thus, the ability of cells to ingress is not determined by their prospective fate. Moreover, primitive streak cells when transplanted to the prospective neural plate can respond to neural-inductive signals. The transplanted tissues epithelialize and form a neural plate showing the molecular and histological features that are characteristic to the brain region into which they have integrated (Garcia-Martinez *et al.*, 1997). Thus, although ingression can lead to restriction of lineage potency, a remarkable degree of flexibility in cell differentiation can still exist.

Other studies in the avian embryo have shown that the commitment of embryonic cells to myocardial fate occurs immediately following ingression of cells through the primitive streak. Heterotopic transplantations of primitive streak segments (Fig. 8; Inagaki *et al.*, 1993) between prospective cardiogenic and noncardiogenic levels and heterotopic transplantation of prospective mesodermal areas of the epiblast (Garcia-Martinez *et al.*, 1997) reveal that prospective mesodermal cells are largely uncommitted at the time of their ingression, with the exception of cells that form the notochord (Garcia-Martinez and Schoenwolf, 1992; Selleck and Stern, 1992). Thus, prospective heart mesoderm can readily form somites and lateral plate mesoderm, and prospective somites and lateral plate mesoderm can readily form heart, including both endocardial and myocardial layers. However, just a few hours later in development, which is shortly after cells ingress through the primitive streak, prospective myocardial cells become committed. When these cells are explanted so that they are no longer subject to further interaction with other embryonic cells, they will differentiate autonomously into cardiomyocytes (Antin *et al.*, 1994; Sugi and Lough, 1994). Moreover, rostrocaudal patterning of the heart occurs shortly thereafter. When the rostrocaudal position of the heart precursor cells is altered by reversing the orientation of a segment of the primitive streak, cells that are destined for specific segments of the heart can still retain their morphogenetic characteristics during subsequent development (Orts-Llorca and Collado, 1967, 1969; Satin *et al.*, 1988).

B. Inductive Interactions

Studies performed in amphibian and chick embryos have shown that the developmental fate of heart precursors may be specified early during gastrulation through inductive interactions (Sater and Jacobson, 1989; Yutzey *et al.*, 1995; Tonegawa *et al.*, 1996). The cardiogenic potency of these cells is initially labile and progressively becomes fixed through the cumulative effect of ongoing inductive tissue interactions (Sater and Jacobson, 1990). Results of experiments that involve the ablation, recombination, and transplantation of the germ layers of amphibian and avian embryos have shown that some inductive signals emanating from the endoderm may be required for the specification of cell

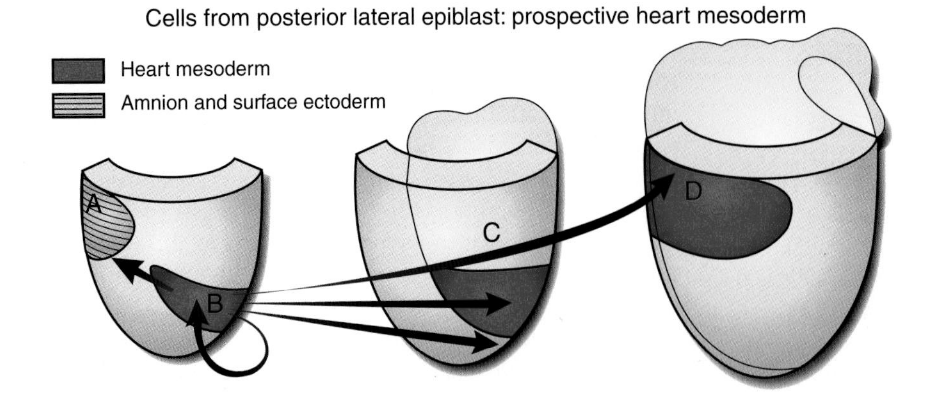

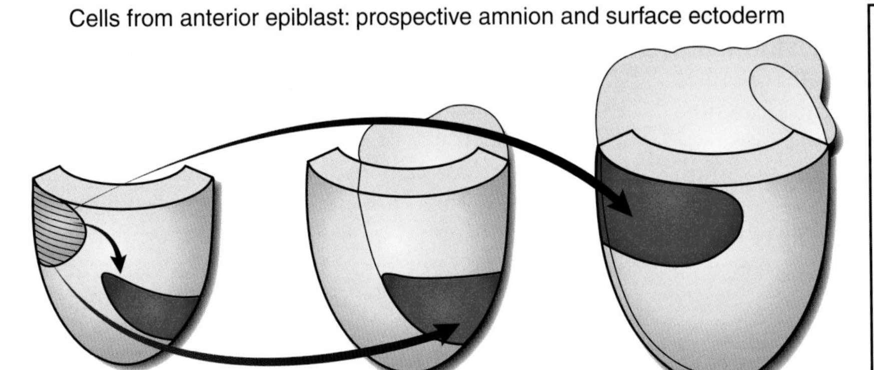

Cell from	Grafted to	Myocardial differentiation
A	B	Yes
	C	Yes
	D	Yes
B	A	No
	B	Yes
	C	Yes
	D	Yes
C	B	Yes/No
	C	Yes
	D	Yes

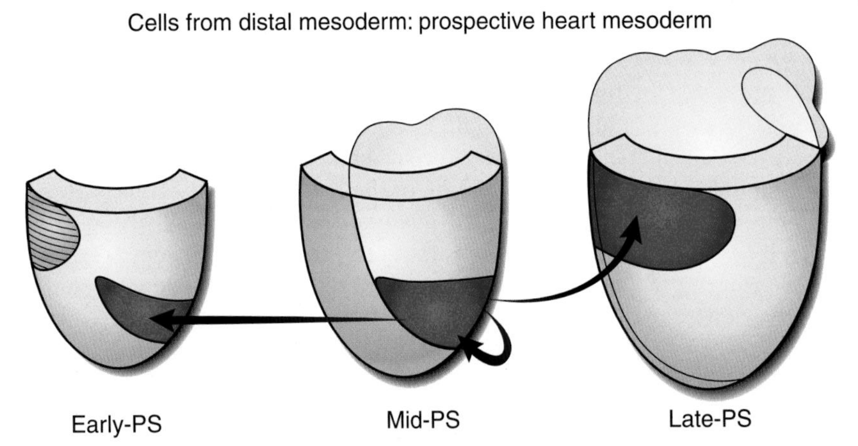

Figure 7 The experimental strategies of orthotopic and heterotopic transplantations to test the cardiogenic potency of the presumptive heart precursors in the epiblast, the newly formed mesoderm, and the nonheart epiblast. Embryonic cells for the potency tests are isolated from the anterior epiblast (middle), the posterior epiblast (top), and the distal region of the nascent mesoderm (bottom). For testing the developmental fate of the epiblast cells and the recently ingressed mesodermal cells, they are transplanted to posterior epiblast (Site B) of the early primitive streak (PS) stage embryo, distal mesoderm (Site C) of mid-PS embryo, the anterior–proximal mesoderm (Site D) of late-PS embryo, which are the sites where heart precursors are normally found during gastrulation. Posterior epiblast cells that contain heart precursors are also transplanted to the anterior epiblast (Site A) of early PS embryo to test if the cardiogenic potency could be maintained in the ectopic site. Results of these transplantation studies show that the allocation of cells to the heart lineage is independent of the act of ingression or mesodermal movement during gastrulation. Nonheart precursor cells can be induced to differentiate into heart tissues when they are transplanted directly to the cardiogenic field of the late-PS embryo, suggesting that the necessary signals for heart differentiation are given to the cells only after they have reached their destination at the conclusion of gastrulation.

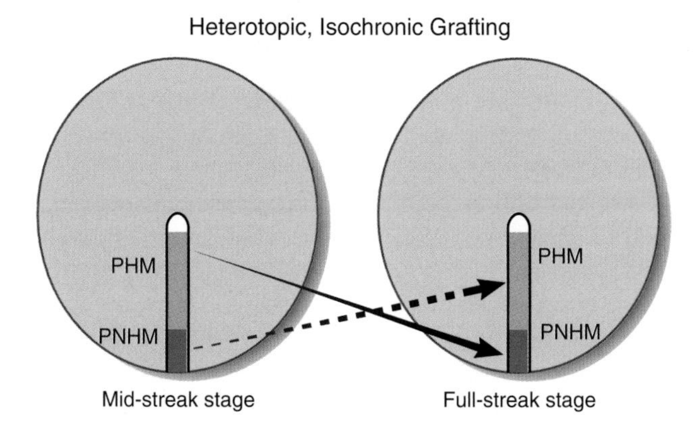

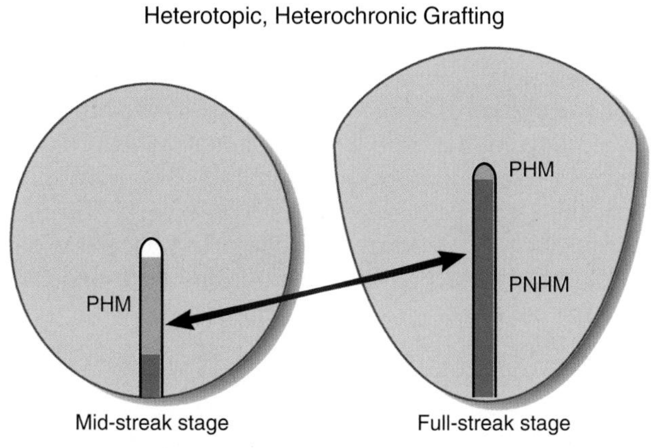

Figure 8 Experiments utilizing heterotopic, isochronic grafting (top) and heterotopic, heterochronic grafting (bottom) test the state of commitment of prospective heart mesoderm (PHM) and prospective nonheart mesoderm (PNHM) at mid- and full-streak stages of avian blastoderms. Results of these transplantations show that the cells in the primitive streak are not committed to any particular fate and the allocation of cells to the embryonic heart is dependent on the site of ingression of the cells in the primitive streak.

fate, the maintenance of tissue competence, and the induction to terminal differentiation of cardiogenic mesoderm. However, there are conflicting views on how critical the role of the endoderm is to these cardiogenic processes in different vertebrate embryos (Gannon and Bader, 1995; reviewed by Jacobson and Sater, 1988).

A series of transplantation studies in mouse (Fig. 7) have shown that the descendants of cells taken from two different regions of the epiblast can contribute to the heart of the host embryo. The graft-derived cells also express the appropriate tropomyosin isoform when they colonize the myocardium of the host (Tam *et al.,* 1997). The only condition for heart cell differentiation is the placement of the epiblast cells at the sites in the germ layers where the heart precursor cells of host embryos are normally found. Interestingly, epiblast cells can acquire a myocardial fate after direct transplantation to the heart field of the late-PS embryo (Fig. 7). This observation provides the most compelling evidence that the necessary information for specifying heart differentiation is given to the cells only after they have reached their destination and not while they are traveling to the heart field. This information may be derived from either the inductive signals from the endoderm (Sugi and Lough, 1995) or the cellular interactions with the host heart mesoderm community (Gonzalez-Sanchez and Bader, 1990). Results from transplantation studies do not preclude the possibility that some steps of specification of the heart mesoderm may have taken place during the initial phases of gastrulation. In the avian embryo, non-heart precursor cells isolated from stage 4 posterior primitive streak which normally contribute to haematopoietic cells are induced to differentiate into cardiac myocytes when they are cocultured with the endoderm from the anterior lateral region of the stages 4 and 5 embryo (Schultheiss *et al.,* 1995). Interestingly, although the mesoderm in the more central region of the embryo does not differentiate to heart tissues, the mesoendoderm tissues have been shown to acquire cardiac-inducing activity *in vitro* when they are free of other germ-layer tissues (Schultheiss *et al.,* 1995). It is suggested that the inductive signal(s) for heart differentiation is derived from the anterior endoderm and is normally absent from the central mesoendoderm, probably as a result of the suppressive activity emanating from the node or its derivatives (Yuan *et al.,* 1995a,b; see Chapter 4).

The specific requirement of endoderm interaction in cardiac differentiation has been disputed, however (Jacobson and Sater, 1988). Studies on cardiac differentiation, assayed by the expression of heart-specific transcripts, in coexplants of anterior ectoderm and mesoderm of stages 6–9 chick embryos suggest that the initial steps of myocardial differentiation may occur in the absence of the endoderm. The endoderm of early gastrula (stages 3 and 4) is also not required for the patterning of tissues in the embryonic heart (Inagaki *et al.,* 1993) but may be necessary for establishing the diversity of myocardial cell types (Yutzey *et al.,* 1995). Furthermore, the proper assembly of sarcomeric proteins and the initiation of contractile activity require the support of the anterior endoderm at least until stage 6 (Antin *et al.,* 1994; Sugi and Lough, 1994; Gannon and Bader, 1995). These findings raise the possibility that the specification of heart mesoderm is initiated early during gastrulation by the inductive activity of the endoderm (Sugi and Lough, 1994). At a later stage, the inductive activity of the anterior endoderm is dispensable because the assembly of sarcomeres and the differentiation of the atrial myocytes can occur in the absence of endoderm (Gannon and Bader, 1995; Yutzey *et al.,* 1995). In *Xenopus,* the heart mesoderm seems to be specified at early gastrulation, and its differentiation does not require any interaction with the superficial pharyngeal endoderm (Sater and Jacobson, 1989). The anterior endoderm is needed at the proneurula stage (stage 20) only to maintain the cardiogenic competence of the lateral mesoderm (Sater and Jacobson, 1990). Heart differentiation in these explants was assayed by the appearance of foci of contractile cells and not by the expression of heart-specific molecules. Recently, the deep endoderm at the dorsal marginal zone of the early *Xenopus* stage 10–10.5 gastrula has been shown to be involved in the induction of heart mesoderm (Nascone and Mercola, 1995). The nature of the endoderm-derived inducing molecules is not known (Tonegawa *et al.,* 1996), but members of the transforming growth factor-β, (BMP), bone morphogenetic proteins and fibroblast growth factor families of growth factors or some endoderm-derived secreted factors are strong candidates (Kokan-Moore *et al.,* 1991; Sugi and Lough, 1995; Lough *et al.,* 1996; Bouwmeester *et al.,* 1996; Zhu *et al.,* 1996; Schultheiss *et al.,* 1997). In chick, the prospective heart mesoderm is sandwiched between ectoderm and endoderm, both of which express BMP (Schultheiss *et al.,* 1997; see Chapter 4). Consequently, anterior ectoderm may also be involved in heart induction in this organism. If an inductive interaction between the endoderm (and/or ectoderm) and the mesoderm is critical for heart differentiation in the mouse embryo, the early spatial association of the heart mesoderm and the gut endoderm during gastrulation would be important for initiating and perpetuating such tissue interactions.

C. Diversification of Cell Lineage

Recent studies in avian embryos on cell lineage allocation, utilizing replication-incompetent retrovirus carrying a reporter gene, have provided insight into whether single progenitor cells can contribute to multiple lineages during cardiogenesis. Transfection of prospective cardiogenic cells shortly after their ingression through the primitive streak (i.e., at stage 4) reveals clones consisting of only one cell type. In about 95% of the clones, only myocardial cells were identified, whereas in about 5% of the clones only endocardial cells were present. Clones consisting of both types were never found, suggesting that at the time of commitment to a cardiogenic fate, two lineages are already established. Similarly, within the myocyte population, the three main lineages—atrial myocytes, ventricular myocytes, and cells of the cardiac conduction system—arise individually from separate progenitors (Cohen-Gould and Mikawa, 1996; Mikawa and Fischman, 1996; See Chapter 2). Thus, the current evidence is consistent with a restriction of cardiac lineage at early gastrulation.

V. Summary

Results of fate-mapping experiments and transplantation studies that test the developmental potency of early mesodermal cells have provided new insight into the process of lineage specification in gastrulating chick and mouse embryos. Specifically, cellular ingression and cell movement in the mesoderm through the primitive streak are associated with a restriction in mesodermal potency. During normal development, such a morphogenetic repertoire may be essential for establishing an early association of the heart mesoderm and the gut endoderm. This would ensure that tissues involved in inductive interactions are properly placed during cardiac differentiation. We have shown in the mouse and chick that different epiblast cell populations have equivalent developmental potencies, which enable them to respond to the cardiogenic signals in the gastrulating embryo following heterotopic transplantation. It is plausible that the initial morphogenetic cell movement associated with germ layer formation is not critical for the specification of heart mesoderm. The necessary signal(s) for myocardial specification may be found only after the cells have reached the cardiogenic field.

Acknowledgments

We thank Peter Rowe, Richard Harvey, Bruce Davidson and Gabriel Quinlan for reading the manuscript. Research works in our laboratories are supported by the National Health and Medical Research Council of Australia, Human Frontier of Science Program, the National Institute of Health, and Mr. James Fairfax.

References

Ang, S.-L., and Rossant, J. (1994). *HNF-3β* is essential for node and notochord formation in mouse development. *Cell* (*Cambridge, Mass.*) **78,** 561–574.

Antin, P., Taylor, R. G., and Yatskievych, T. (1994). Precardiac mesoderm is specified during gastrulation in quail. *Dev. Dyn.* **200,** 144–154.

Beddington, R. S. P., and Lawson, K. A. (1990). Clonal analysis of cell lineages. *In* "Postimplantation Mammalian Embryos" (A. J. Copp and D. L. Cockroft, eds.), pp. 267–292. IRL Press, Oxford.

Bellairs, R. (1986). The primitive streak. *Anat. Embryol.* **174,** 1–14.

Bouwmeester, T., Kim, S. H., Sasai, Y., Lu, B., and De Robertis, E. M. (1996). *Cerberus,* a head-inducing secreted factor expressed in the anterior endoderm of the Spemann's organizer. *Nature* (*London*) **382,** 595–601.

Cohen-Gould, L., and Mikawa, T. (1996). The fate diversity of mesodermal cells within the heart field during chick early embryogenesis. *Dev. Biol.* **177,** 265–273.

Collignon, J., Varlet, I., and Robertson, E. J. (1996). Relationship between asymmetric *nodal* expression and the direction of embryonic turning. *Nature* (*London*) **381,** 155–158.

Crossley, P., and Martin, G. (1995). The mouse *Fgf8* gene encodes a family of polypeptides that is expressed in regions that direct outgrowth and patterning in the developing embryo. *Development* (*Cambridge, UK*) **121,** 439–451.

Darnell, D. K., and Schoenwolf, G. C. (1997). Modern techniques for cell labelling in avian and murine embryos. *In* "Molecular and Cellular Methods in Developmental Toxicology" (G. P. Daston, ed), pp. 231–268. CRC Press, Boca Raton, FL

DeRuiter, M. C., Poelmann, R. E., Vander Plas-deVries, I., Menthink, M. M. T., and Gittenberger-de Groot, A. C. (1992). The development of myocardium and endocardium in mouse embryos. Fusion of two heart tubes? *Anat. Embryol.* **185,** 461–473.

Dush, M. K., and Martin, G. R. (1992). Analysis of mouse *Evx* genes, *Evx-1* displays graded expression in the primitive streak. *Dev. Biol.* **151,** 273–287.

Gannon, M., and Bader, D. (1995). Initiation of cardiac differentiation occurs in the absence of anterior endoderm. *Development (Cambridge, UK)* **121,** 2439–2450.

Garcia-Martinez, V., and Schoenwolf, G. C. (1992). Positional control of mesoderm movement and fate during avian gastrulation and neurulation. *Dev. Dyn.* **193,** 249–256.

Garcia-Martinez, V., and Schoenwolf, G. C. (1993). Primitive-streak origin of the cardiovascular system in avian embryos. *Dev. Biol.* **159,** 706–719.

Garcia-Martinez, V., Alvarez, I. S., and Schoenwolf, G. C. (1993). Locations of the ectodermal and non-ectodermal subdivisions of the epiblast at stages 3 and 4 of avian gastrulation and neurulation. *J. Exp. Zool.* **267,** 431–446.

Garcia-Martinez, V., Darnell, D. K., Sosic, D., Olson, E. N., and Schoenwolf, G. C. (1997). State of commitment of prospective neural plate and prospective mesoderm in late gastrula/early neurula stages of avian embryos. *Dev. Biol.* **181,** 102–115.

Garton, H. J. L., and Schoenwolf, G. C. (1996). Improving the efficacy of fluorescent labeling for histological tracking of cells in early mammalian and avian embryos. *Anat. Rec.* **244,** 112–117.

Gonzalez-Sanchez, A., and Bader, D. (1990). In vitro analysis of cardiac progenitor cell differentiation. *Dev. Biol.* **139,** 197–209.

Hashimoto, K., and Nakatsuji, N. (1989). Formation of the primitive streak and mesoderm cells in mouse embryos—detailed scanning electron microscopical study. *Dev. Growth Differ.* **31,** 209–218.

Heikinheimo, M., Scandrett, J. M., and Wilson, D. B. (1994). Localization of transcription factor GATA-4 to regions of the mouse embryo involved in cardiac development. *Dev. Biol.* **164,** 361–373.

Hermesz, E., Mackem, S., and Mahon, K. A. (1996). *Rpx,* A novel anterior-restricted homeobox gene progressively activated in the prechordal plate, anterior neural plate and Rathkes's pouch of the mouse embryo. *Development (Cambridge, UK)* **122,** 41–52.

Icardo, J. M. (1996). Developmental biology of the vertebrate heart. *J. Exp. Zool.* **275,** 144–161.

Inagaki, T., Garcia-Martinez, V., and Schoenwolf, G. C. (1993). Regulative ability of the prospective cardiogenic and vasculogenic areas of the primitive streak during avian gastrulation. *Dev. Dyn.* **197,** 57–68.

Isaac, A., Sargent, M. G., and Cooke, J. (1997). Control of vertebrate left-right asymmetry by a *Snail*-related zinc finger gene. *Science* **275,** 1301–1304.

Jacobson, A. G., and Sater, A. K. (1988). Features of embryonic induction. *Development (Cambridge, UK)* **104,** 341–360.

Keller, R., and Tibbetts, P. (1989). Mediolateral cell intercalation in the dorsal, axial mesoderm of *Xenopus laevis. Dev Biol.* **131,** 539–549.

Keller, R. E. (1976). Vital dye mapping of the gastrula and neurula of *Xenopus laevis.* II. Prospective areas and morphogenetic movements in the deep layer. *Dev. Biol.* **51,** 118–137.

King, T., and Brown, N. A. (1995). The embryo's one-sided genes. *Curr. Biol.* **5,** 1364–1366.

Kokan-Moore, N. P., Bolender, D. L., and Lough, J. (1991). Secretion of inhibin βA by endoderm cultured from early embryonic chicken. *Dev Biol.* **146,** 242–245.

Lawson, K. A., and Pedersen, R. A. (1987). Cell fate, morphogenetic movement and population kinetics of embryonic endoderm at the time of germ layer formation in the mouse. *Development (Cambridge, UK)* **101,** 627–652.

Lawson, K. A., Meneses, J. J., and Pedersen, R. A. (1986). Cell fate and cell lineage in the endoderm of the presomite mouse embryo, studied with an intracellular tracer. *Dev. Biol.* **115,** 325–339.

Lawson, K. A., Meneses, J. J., and Pedersen, R. A. (1991). Clonal analysis of epiblast fate during germ layer formation in the mouse embryo. *Development (Cambridge, UK)* **113,** 891–911.

Le Douarin, N. (1976). Cell migration in early vertebrate development studied in interspecific chimaeras. *Ciba Found. Symp.* **40,** 71–97.

Lee, R. K. K., Stainier, D. Y. R., Weinstein, B. M., and Fishman, M. C. (1994). Cardiovascular development in the zebrafish. II. Endocardial progenitors are sequestered within the heart field. *Development (Cambridge, UK)* **120,** 3361–3366.

Levin, M., Johnson, R. L., Stern, C. D., Kuehn, M., and Tabin, C. (1995). A molecular pathway determining left-right asymmetry in chick embryogenesis. *Cell (Cambridge, Mass.)* **82,** 803–814.

Lints, T. J., Parson, L. M., Hartley, L., Lyons, I., and Harvey, R. P. (1993). *Nkx-2.5,* a novel murine homeobox gene expressed in early heart progenitor cells and their myogenic descendants. *Development (Cambridge, UK)* **119,** 419–431.

Lough, J., Barron, M., Brogley, M., Sugi, Y., Bolender, D. L., and Zhu, X. (1996). Combined BMP-2 and FGF-4, but neither factor alone, induces cardiogenesis in non-precardiac embryonic mesoderm. *Dev. Biol.* **178,** 198–202.

Lyons, G. E. (1994). Insitu analysis of the cardiac muscle gene program, during embryogenesis. *Trends Cardiovasc. Med.* **4,** 70–77.

Lyons, G. E. (1996). Vertebrate heart development. *Curr. Opin. Genet. Dev.* **6,** 454–460.

Meno, C., Saijoh, Y., Fujii, H., Ikeda, M., Yokoyama, T., Yokoyama, M., Toyoda, Y., and Hamada, H. (1996). Left-right asymmetric expression of the TGFβ-family member *lefty* in mouse embryos. *Nature (London)* **381,** 151–155.

Mikawa, T., and Fischman, D. A. (1996). The polyclonal origin of myocyte lineages. *Annu. Rev. Physiol.* **58,** 509–521.

Nascone, N., and Mercola, M. (1995). An inductive role for the endoderm in *Xenopus* cardiogenesis. *Development (Cambridge, UK)* **121,** 515–523.

Orts-Llorca, F., and Collado, J. J. (1967). Determination of heart polarity (arterio venous axis) in the chicken embryo. *Wilhelm Roux, Arch. Entwicklungsmech. Org.* **158,** 147–163.

Orts-Llorca, F., and Collado, J. J. (1969). The development of heterologous grafts, labeled with thymidine-^{3}H in the cardiac area of the chick blastoderm. *Dev. Biol.* **19,** 213–227.

Parameswaran, M., and Tam, P. P. L. (1995). Regionalisation of cell fate and morphogenetic movement of the mesoderm during mouse gastrulation. *Dev Genet.* **17,** 16–28.

Psychoyos, D., and Stern, C. D. (1996). Fates and migratory routes of primitive streak cells in the chick embryo. *Development (Cambridge, UK)* **122,** 1523–1534.

Rosenquist, G. C. (1971). The location of the pregut endoderm in the chick embryo at the primitive streak stage as determined by radioautographic mapping. *Dev. Biol.* **26,** 323–335.

Sasaki, H., and Hogan, B. L. M. (1993). Differential expression of multiple fork head related genes during gastrulation and axial pattern formation in the mouse embryo. *Development (Cambridge, UK)* **118,** 47–59.

Sater, A. K., and Jacobson, A. G. (1989). The specification of heart mesoderm occurs during gastrulation in *Xenopus laevis. Development (Cambridge, UK)* **105,** 821–830.

Sater, A. K., and Jacobson, A. G. (1990). The restiction of the heart morphogenetic field in *Xenopus laevis. Dev. Biol.* **140,** 328–336.

Satin, J., Fujii, S., and De Haan, R. L. (1988). Development of cardiac beat rate in early chick embryos is regulated by regional cues. *Dev. Biol.* **129,** 103–113.

Schoenwolf, G. C., and Alvarez, I. S. (1991). Specification of neurepithelium and surface epithelium in avian transplantation chimeras. *Development (Cambridge, UK)* **112,** 713–722.

Schoenwolf, G. C., Garcia-Martinez, V., and Dias, M. S. (1992). Mesoderm movement and fate during avian gastrulation and neurulation. *Dev. Dyn.* **193,** 235–248.

Schultheiss, T. M., Xydas, S., and Lassar, A. B. (1995). Induction of avian cardiac myogenesis by anterior endoderm. *Development (Cambridge, UK)* **121,** 4203–4214.

Schultheiss, T. M., Burch, J. B. E., and Lassar, A. B. (1997). A role for bone morphogenetic proteins in the induction of cardiac myogenesis. *Genes Dev.* **11,** 451–462.

Selleck, M. A. J., and Stern, C. D. (1992). Commitment of mesoderm cells in Hensen's node of the chick embryo to notochord and somite. *Development (Cambridge, UK)* **114,** 403–415.

Smith, J. L., Gesteland, K. M., and Schoenwolf, G. C. (1994). Prospective fate map of the mouse primitive streak at 7.5 days of gestation. *Dev. Dyn.* **201,** 279–289.

Smith, S. M., Dickman, E. D., Thompson, R. P., Sinning, A. R., Wunsch, A. M., and Markwald, R. R. (1997). Retinoic acid directs cardiac laterality and the expression of early markers of precardiac asymmetry. *Dev. Biol.* **182,** 162–171.

Srivastava, D., Cserjesi, P., and Olson, E. N. (1995). A subclass of bHLH proteins required for cardiogenesis. *Science* **270,** 1995–1999.

Stainier, D. Y. R., Lee, R. K., and Fishman, M. C. (1993). Cardiovascular development in zebrafish. I. Myocardial fate map and heart formation. *Development (Cambridge, UK)* **119,** 31–40.

Stainier, D. Y. R., Weinstein, B. M., Detrich, H. W., III, Zon, L. I., and Fishman, M. C. (1995). *Cloche,* an early acting zebrafish gene, is required by both the endothelial and haematopoietic lineages. *Development (Cambridge, UK)* **121,** 3141–3150.

Stern, C. D., Yu, R. T., Kakizuka, A., Kintner, C. R., Mathews, L. S., Vale, W. V., Evans, R. M., and Umesono, K. (1995). Activin and its receptors during gastrulation and the later phases of mesoderm development in the chick embryo. *Dev. Biol.* **172,** 192–205.

Sturm, K. S., and Tam, P. P. L. (1993). Isolation and culture of whole post-implantation embryos and germ layer derivatives.. In "Methods in Enzymology (P. M. Wassarman and M. L. DePamphillis, eds.), Vol. 225, pp. 164–189. Academic Press, San Diego, CA.

Sugi, Y., and Lough, J. (1994). Anterior endoderm is a specific effector of terminal cardiac myocyte differentiation of cells from the embryonic heart forming region. *Dev. Dyn.* **200,** 155–162.

Sugi, Y., and Lough, J. (1995). Activin-A and FGF-2 mimic the inductive effects of anterior endoderm on terminal cardiac myogenesis in vitro. *Dev. Biol.* **168,** 567–574.

Tam, P. P. L., and Beddington, R. S. P. (1987). The formation of mesodermal tissues in the mouse embryo during gastrulation and early organogenesis. *Development (Cambridge, UK)* **99,** 109–126.

Tam, P. P. L., and Beddington, R. S. P. (1992). Establishment and organization of germ layers in the gastrulating mouse embryo. *Ciba Found. Symp.* **165,** 27–49.

Tam, P. P. L., and Tan, S. S. (1992). The somitogenetic potential of cells in the primitive streak and the tail bud of the organogenesis-stage mouse embryo. *Development (Cambridge, UK)* **115,** 703–715.

Tam, P. P. L., and Trainor, P. A. (1994). Specification and segmentation of the paraxial mesoderm. *Anat. Embryol.* **189,** 275–306.

Tam, P. P. L., and Zhou, S. X. (1996). The allocation of epiblast cells to ectodermal and germ-line lineage is influenced by the position of the cells in the gastrulating mouse embryo. *Dev. Biol.* **178,** 124–132.

Tam, P. P. L., Williams, E. A., and Chan, W. Y. (1993). Gastrulation in the mouse embryo: Ultrastructural and molecular aspects of germ layer morphogenesis. *Microsc. Res. Tech.* **26,** 301–328.

Tam, P. P. L., Parameswaran, M., Kinder, S. J., and Weinberger, R. P. (1997). The allocation of epiblast cells to the embryonic heart and other mesodermal lineages: The role of ingression and tissue movement during gastrulation. *Development (Cambridge, UK)* **124,** 1631–1642.

Thomas, P., and Beddington, R. S. P. (1996). Anterior primitive endoderm may be responsible for patterning the anterior neural plate in the mouse embryo. *Curr. Biol.* **6,** 1487–1496.

Tonegawa, A., Moriya, M., Tada, M., Nishimoto, S., Katagiri, C., and Ueno, N. (1996). Heart formative factor(s) is localized in the anterior endoderm of early *Xenopus neurula. Roux's Arch. Dev. Biol.* **205,** 282–289.

Yamaguchi, T. P., Dumont, D. J., Conlon, R. A., Breitman, M. L., and Rossant, J. (1993). *flk-1,* and *flt*-related receptor tyrosinase kinase is an early marker for endothelial cell precursors. *Development (Cambridge, UK)* **118,** 489–498.

Yamaguchi, T. P., Harpal, K., Henkemeyer, M., and Rossant, J. (1994). *fgfr-1* is required for embryonic growth and mesodermal patterning during mouse gastrulation. *Genes. Dev.* **8,** 3032–3044.

Yuan, S., Darnell, D. K., and Schoenwolf, G. C. (1995a). Identification of inducing, responding, and suppressing regions in an experimental model of notochord formation in avian embryos. *Dev. Biol.* **172,** 567–584.

Yuan, S., Darnell, D. K., and Schoenwolf, G. C. (1995b). Mesodermal patterning during avian gastrulation and neurulation, experimental induction of notochord from nonnotochordal precursor cells. *Dev. Genet.* **17,** 38–54.

Yutzey, K. E., Gannon, M. and Bader, D. (1995). Diversification of cardiomyogenic cell lineages in vitro. *Dev. Biol.* **170,** 531–541.

Zhu, X., Sasse, J., McAllister, D., and Lough, J. (1996). Evidence that fibroblast growth factors 1 and 4 participate in regulation of cardiogenesis. *Dev. Dyn.* **207,** 429–438.

2

Cardiac Lineages

Takashi Mikawa

Department of Cell Biology, Cornell University Medical College, New York, New York 10021

I. Introduction

In vertebrates, the heart is first established as a single tube consisting of two epithelial layers: the inner endocardium and the outer myocardium (Manasek, 1968; see Chapters 1, 6, and 10). The primitive tubular heart partitions further into two chambers—atrial and ventricular—confined by atrioventricular septa (see Chapters 10 and 19). Myocardial contractions begin during this double-walled stage of heart formation; initial pulsations are generated at the right myocardium and spread posteriorly to anteriorly over the whole myocardium. Except for the pacemaker cells, the cardiac conduction system has not yet developed at this stage. Coronary vascular cells, connective tissue cells, and autonomic neurons are not present in the tubular stage heart (Manasek, 1968).

The two-chambered heart is the mature form in the primitive vertebrates such as fish (see Chapter 6). In contrast, the heart of higher vertebrates, including birds and mammals, further undergoes a series of morphogenetic steps: looping (see Chapters 21–22); septation generating the four compartments characteristic of the adult heart (see Chapter 10); trabeculation and thickening of the ventricular walls; and the cranial shift of atrial chambers, assuming their adult position rostrodorsal to the ventricles. The coordinated contraction of the topologically repositioned atrial and ventricular chambers depends on developing and patterning the cardiac conduction system (see Chapter 12). The survival of thickened ventricular myocardium relies on the coronary vessel system. These two specialized tissue systems are absent from hearts of primitive vertebrates. To understand the molecular mechanisms that govern sequentially programmed development and integration of these subcardiac components during heart formation in higher vertebrates, it is essential to define their lineage relationships and migration patterns within the embryo.

II. Origin of Endocardial and Myocardial Cell Lineages

Only three cell types, endocardial endothelia, ventricular myocytes, and atrial myocytes, constitute the tubular heart when it begins rhythmic contraction (Manasek, 1968). As in the adult heart, the primitive heart tube consists of a greater proportion of myocardial cells than endocardial cells (Fig. 1) with the ratio of ~18:1 (Cohen-Gould and Mikawa, 1996). Fate map studies in chicken embryos have established that cardiogenic mesoderm rostrolaterally flanking Hensen's node contains precursor cells of all three early cardiac lineages (Rawles, 1943; Rosenquist and De Haan, 1966; Garcia-Martinez and Schoenwolf, 1993). However, to date it remains unexplained how more myocardial cells than endocardial cells segregate from the cardiogenic mesoderm, partly due to controversy regarding our understanding of the lineage relationship between the myocardial and endocardial lineages. It is uncertain when and how these two cardiac lineages are established.

A. Common vs Separate Origin of Endocardial and Myocyte Cell Lineages

During tubular heart formation, the cardiogenic mesoderm first becomes an epithelial layer (Fig. 1) in which all cells express N-cadherin, a calcium-dependent cell adhesion molecules (Takeichi, 1991), by Hamburger–Hamilton (HH) stages (Hamburger and Hamilton, 1951) 6 and 7 (Linask, 1992). The epithelialized cardiogenic mesoderm then generates two subpopulations; the majority of cells, maintaining N-cadherin expression, remain as epithelia and later differentiate into myocytes, whereas the minor population, downregulating N-cadherin expression, segregates from the original epithelial layer as a progenitor population of the endocardial endothelia (Manasek, 1968; Linask and Lash, 1993). The epithelioid myocardial cells begin expression of muscle-specific genes and proteins substantially before initiation of the heart beat (Han *et al.,* 1992; Yutzey *et al.,* 1994), whereas endocardial cells begin expression of an endothelial marker at their segregation from the epithelialized cardiogenic mesoderm (Linask and Lash, 1993; Sugi and Markwald, 1996).

There is confusion in the understanding of the genesis of endocardial and myocardial cell lineages. Based on morphological heterogeneity within cardiogenic mesoderm, it was originally predicted that two lineages are already separated when their progenitors migrate to the heart field (Pardanaud *et al.,* 1987a,b; Coffin and Poole, 1988; DeRuiter *et al.,* 1992). Based on expression patterns of myocardial and endocardial cell markers in cardiogenic mesoderm (Linask and Lash, 1993) and in an immortalized myogenic cell line, QCE-6 (Eisenberg and Bader, 1995), a recent model predicts that mesodermal cells in the heart field commonly produce both cell lineages. QCE-6 cells can be induced to express myocyte or endothelial phenotypes in response to several different growth factors.

If individual cells of cardiogenic mesoderm com-

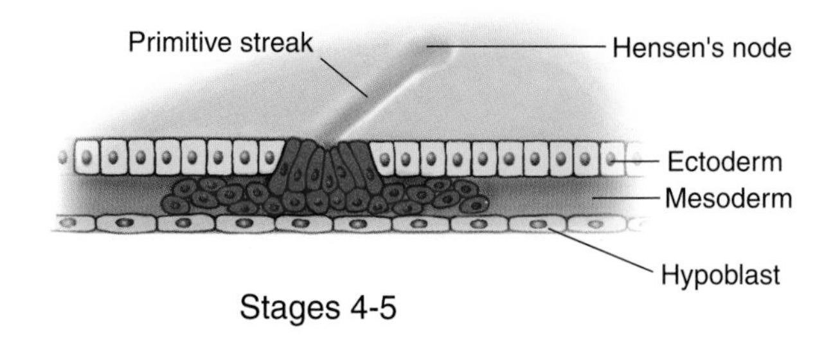

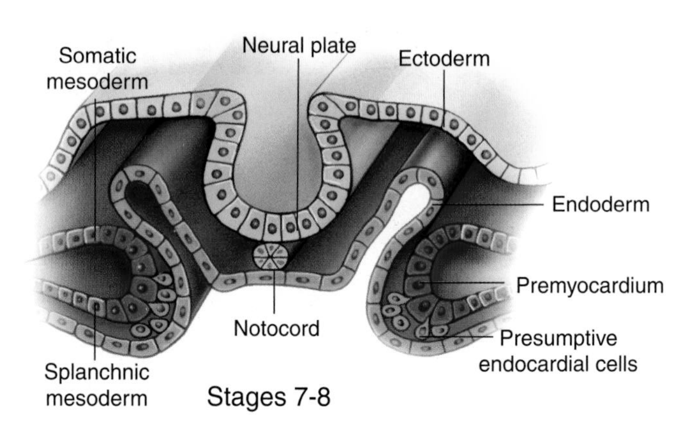

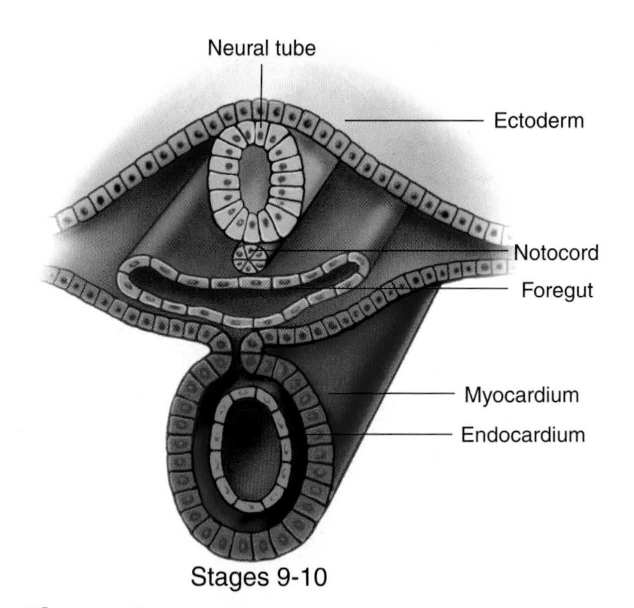

Figure 1 Morphogenetic processes in the early heart development. Gastrulating cardiac mesodermal cells (red) at HH-stage 4 (top panel), segregation of endocardial precursor cells (green) from epithelioid presumptive myocardium (red) during lateral body fold at the neurula stage (second panel), and subsequent tubular heart formation by fusion of bilateral heart primordia, respectively.

monly generate both myocardial and endocardial cells, the mechanism inducing these distinct cell types could be uncovered by addressing the diversification process of cardiogenic mesodermal cells into either cell lineage. This type of approach would not be appropriate, however, if these two lineages were already separated prior to their migration to the heart field. Thus, understanding the fate of individual cells within the heart field is crucial for addressing mechanisms by which myocardial and endocardial lineages of the heart are established. However, morphological analysis, fate maps based on migration patterns of groups of cells, and expression of marker genes do not provide sufficient evidence to determine whether the endocardial and myocardial lineages segregate at the time of gastrulation or if both lineages arise from a common progenitor present in the cardiogenic mesoderm. We have recently addressed this question by fate-mapping analysis of individual progenitor cells within the heart field.

B. Retroviral Cell Lineage Studies of Cardiogenic Mesoderm

The value of replication-defective retrovirus-mediated genetic tags for analyzing cell lineages was first demonstrated in pioneering studies on the central nervous system and the eye (Sanes, 1989; Cepko, 1988). Retroviruses stably integrate their genetic material into the infected host cell and this genetic information is inherited by every descendant of that initially infected cell. If horizontal transmission from primary infected cells is inhibited, retroviral-mediated gene transfer can be one of the most reliable methods for restricting the effects of transgenes to the desired cell type and time point in the developing heart. Several replication-defective variants of retroviruses have been engineered as a vector suitable for cell lineage studies in the avian embryo by replacing viral structural genes with reporter genes, such as β-galactosidase (reviewed in Mikawa *et al.*, 1996). The simplicity of the retroviral cell-tagging protocol provides a reliable way to study the cell lineage of higher vertebrates *in vivo,* resolving some debate on previous fate map data using *in vitro* explants.

Retroviral cell lineage studies in the chicken embryo revealed that individual cells in the heart field give rise to a clone consisting only of one cell type, either endocardial or myocardial cells (Cohen-Gould and Mikawa, 1996). No mesodermal cells generate clones containing both these cell types. Importantly, ~95% of the mesoderm-derived clones are localized in the myocardium only, whereas ~5% of them are found in endocardium only. The results do not support the model that cells in the heart field commonly generate both lineages and rather indicate that the heart field mesoderm consists of at least two distinct subpopulations with substantially more premyocardial cells than preendocardial cells. This uneven sorting process within the heart field mesoderm can be explained by two new models illustrated in Fig. 2 (Cohen-Gould and Mikawa, 1996).

According to the first model, in which cardiogenic mesoderm is equipotential, there must be a mechanism which locally defines the region inducing the endocardial or myocardial cell lineage within the heart field. It is known that bone morphogenetic protein from underlying endoderm is involved in specification of anterior mesoderm into the myocyte lineage (Schultheiss *et al.*, 1995) and vascular endothelial growth factor is necessary for induction of the endothelial lineage (Shalaby *et al.*, 1995; Ferrara *et al.*, 1996; Fong *et al.*, 1995). However, there is no evidence that either of these ligands or receptors are expressed locally within the heart field (Flamme *et al.*, 1994).

According to the second model, in which the heart field consists of two subpopulations already restricted to the myocardial or endocardial lineage, the uneven sorting of cardiogenic mesodermal cells into either lineages can be explained by the presence of more premyocardial progenitors than preendocardial progenitors within the heart field. There are several lines of evidence consistent with the model. Endothelial cell commitment occurs before and independent of gastrulation, whereas myocyte commitment has not yet occurred at the prestreak stage (von Kirschhoffer *et al.*, 1994). Single-cell marking and tracing studies in zebrafish (Lee *et al.*, 1994) have identified a blastomere population which generates only endocardial or myocardial cells. Thus, separation of these two lineages can occur at blastula stage prior to formation of mesoderm. The endocardial progenitor population is sequestered from myocyte progenitors within the heart field of zebrafish blastula (Lee *et al.*, 1994), as observed in the chicken cardiogenic mesoderm (Cohen-Gould and Mikawa, 1996). Recent fate-mapping analysis of the mouse blastula has revealed that both myocardial and endocardial cells arise from two small sites bilateral to the anteroposterior axis in the distal region of the epiblast (Tam *et al.*, 1997; see Chapter 1). It remains to be seen whether the myocardial and endocardial precursors are already segregated within the heart field of the mouse embryo.

Several transcriptional factors, including members of *Tbx, GATA, HAND, Nkx2.5,* and *MEF2* gene families (see Sections III, V), have been identified in the heart-forming region. Although all these factors are necessary for survival and/or proliferation of already differentiated myocytes, their roles in induction of the myocyte lineage remain uncertain. The previous cell lineage data

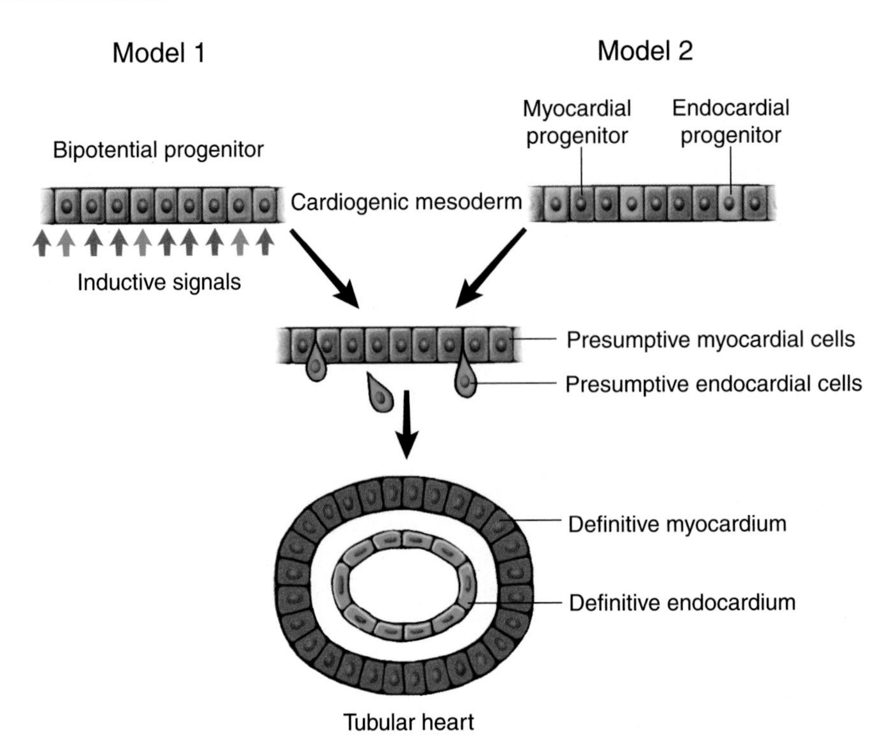

Figure 2 Models for genesis of myocyte and endocardial endothelial lineages (modified from Cohen-Gould and Mikawa, 1996). Model 1 proposes that cardiogenic mesoderm cells are bipotential and enter either endothelial or myocyte lineage by receiving distinct inductive signals. Model 2 illustrates that 1) prior to or at the initiation of gastrulation, each cardiac lineage in the epiblast or primitive streak cells is established; 2) during gastrulation, precursors of both lineages co-migrate to the cardiogenic mesoderm area; and 3) responding to instructive signals from endoderm and/or ectoderm, the precursors undergo terminal differentiation and morphogenetic processes.

would provide a foundation for addressing their potential roles in trasncriptional regulation underlying the establishment of cardiac lineages.

C. Origin of Atrial and Ventricular Myocyte Lineages

Before the tubular heart begins its contraction, cells in the caudal region of the presumptive myocardium are distinguished as atrial myocytes from those in the rostral region differentiating as ventricular myocytes, based on electrophysiological properties (Kamino *et al.*, 1981) and the expression of contractile protein genes (O'Brien *et al.*, 1993; Yutzey *et al.*, 1994). Fate-mapping studies have shown that the more rostral cells of the primitive streak form the more rostral regions of the tubular heart and the more caudal cells form the caudal, inflow regions (Rosenquist and De Haan, 1966; Garcia-Martinez and Schoenwolf, 1993; see Chapter 1). Although this spatial relationship is not absolute (Garcia-Martinez and Schoenwolf, 1993), the results suggest the atrial and ventricular lineages may segregate at or prior to gastrulation.

This hypothesis has been tested by cell lineage studies using single-cell tagging procedures in chicken (Mikawa *et al.*, 1992a) and zebrafish (Stainier *et al.*, 1993) embryos. Retroviral-mediated genetic marking and subsequent fate analyses of individual cells present in the cardiogenic mesoderm have proved that individual cells in the rostral cardiogenic mesoderm only enter the ventricular myocyte lineage, whereas cells in the caudal region differentiate into the atrial myocyte lineage (Mikawa *et al.*, 1996). Clone-based fate maps in zebrafish embryos showed that atrial and ventricular myocyte lineages separate at the midblastula stage (Stainier *et al.*, 1993; see Chapter 6). Thus, these two myocyte lineages are already established when mesodermal cells migrate into the cardiogenic area; the ventricular myocytes arise from a subpopulation residing in the rostral regions of the cardiogenic mesodermal plates, whereas the atrial myocyte lineage originates from cells present in the caudal regions of cardiogenic mesoderm. Indeed, explants from the caudal region of cardiogenic mesoderm can differentiate into atrial but not ventricular myocytes in culture (Yutzey *et al.*, 1995).

Molecular signals inducing these two myocyte lineages are currently unknown. However, the caudal region of cardiogenic mesoderm has the potential to change its beat rate from atrial type to that of prospective ventricular myocytes, if placed in the rostral heart-forming region (Satin *et al.,* 1988). Ectopic retinoic acid treatment of cardiogenic mesoderm induces the expression of atrial myosin in presumptive ventricular myocytes (Yutzey *et al.,* 1994). Such plasticity can be seen only before atrial and ventricular myocyte phenotypes become apparent. These observations suggest that myogenic progenitors within the cardiogenic mesoderm remain bipotential until terminal differentiation into either atrial or ventricular phenotypes becomes evident. Therefore, it is likely that the terminal differentiation of either atrial or ventricular lineages is defined by positionally delineated extracellular signal(s). Limited migratory activities of myocyte progenitors within the epithelioid presumptive myocardium (Mikawa *et al.,* 1992a,b) may play a role in stabilizing the terminal differentiation process.

D. Fate of Epithelioid Myocytes during Myocardial Wall Morphogenesis

Soon after myocyte lineages complete their terminal differentiation at the tubular heart stage, the epithelioid, beating myocytes then divide in a plane perpendicular to the heart wall, delaminate, and migrate toward the endocardium, generating a number of protrusions or trabeculae (Manasek, 1968). The trabeculated pattern increases endocardial surface area to facilitate diffusion of oxygen and nutrients into the avascular ventricular myocardium prior to coronary vessel system development. Subsequent coalescence of trabeculae gives rise to the thickened myocardium in which all myocytes are tightly connected with intercalated discs. Although the morphological transitions that occur during myocardium formation have been extensively characterized, the mechanism underlying asymmetrical muscle growth of the left and right ventricles and genesis of muscular septa and papillary muscles remains poorly understood.

Retroviral cell lineage procedures have defined the fate of individual myocytes during trabeculation and subsequent thickening and multilayering of the ventricular myocardial wall (Mikawa *et al.,* 1992a,b; Mikawa, 1995). Myocardial wall morphogenesis can be divided into four major steps (Fig. 3). (i) During tubular heart formation, individual myocyte precursors become epithelialized; (ii) the differentiated, epithelioid myocytes generate a series of progeny which undergo an epithelial–mesenchymal transformation and migrate more vertically than horizontally, creating ridge-like protrusions (trabeculae); (iii) gradients of myocyte proliferation are evident across the myocardial wall, greater at peripheral than deeper layers, resulting in the formation of the cone- or wedge-shaped sectors which span the entire thickness of the myocardium. The clonally related sectors serve as the fundamental growth units of the myocardial wall, and (iv) two-dimensional arrays of these cone-shaped growth units give rise to the three-dimensional ovoid structure of the ventricular walls. The interventricular septum consists of cones of more axially elongated dimensions than those in the lateral walls. Thus, the thickened and multilayered ventricular myocardium is defined by the locally regulated migration and proliferation of progeny derived from individual epithelioid parental myocytes.

Some molecular signalings involved in induction of trabeculation and/or proliferation of myocytes have been identified. Signaling of neuregulin, a peptide secreted by endocardial endothelia, through its cognate tyrosin kinase receptors (erbB2 and erbB4) expressed by epithelioid myocytes is required for initiating trabeculation: Mutation of neureglin (Meyer and Birchmeier, 1995), erbB2 (Lee *et al.,* 1995), or erbB4 (Gassmann *et al.,* 1995) results in loss of trabeculation. Signaling of fibroblast growth factor (FGF) through its tyrosin kinase receptor (FGFR1) serves as a potent mitogen during myocardial wall thickening (Mima *et al.,* 1995; Mikawa, 1995). Recent studies of the transforming growth factor-β (TGF-β) null mice revealed that there is a significant thickening of the ventricular wall and a loss of ventricular chamber volume, both resulting from myocyte hyperplasia (Letterio *et al.,* 1994). The results suggest that TGF-β signaling may serve as a supressor of cardiomyocyte proliferation.

III. Origin of Coronary Vascular Cell Lineages

The embryonic heart of higher vertebrates is avascular until the myocardium becomes thickened through both myocyte hyperplasia and fusion of trabeculae (Rychter and Ostadal, 1971). In the chicken embryo, coronary vascularization begins at E6 with formation of a capillary plexus and venous sinusoids (Bogers *et al.,* 1989; Waldo *et al.,* 1990). The subsequent anastomosis with coronary arteries establishes the closed coronary vessel network by E14 (Rychter and Ostadal, 1971). Only two of several connections to the aorta persist and differentiate as definitive coronary orifices by E18 (Hood and Rosenquist, 1992). Currently, the mechanisms governing the development and patterning of coronary vessel network are not well understood since until recently the ontogeny and lineage of cells comprising coronary vessels were controversial.

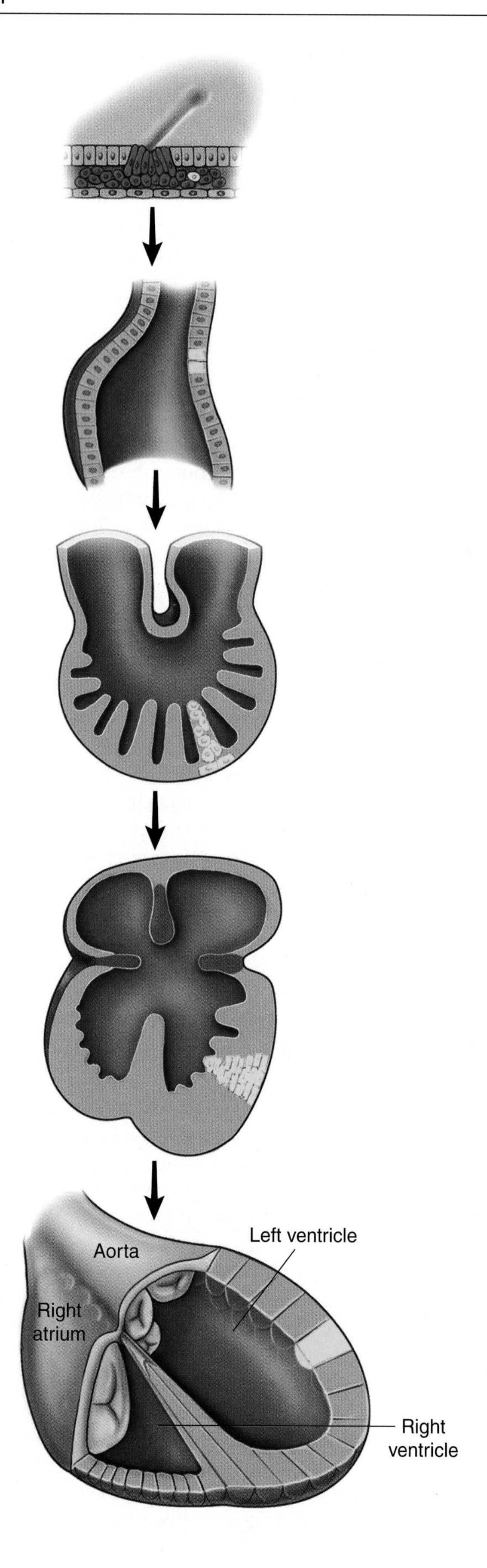

Precise characterization of the origin of cell populations in the coronary arterial system appears to be important for not only clarifying embryological questions but also providing a foundation for understanding clinically related problems, such as acute atherosclerosis. Chicken/quail chimera studies of noncoronary vessels first identified two embryonic origins, neural crest and local mesoderm, as progenitors for vascular smooth muscle (Le Lièvre and Le Douarin, 1975). Recent studies have demonstrated that there is a high degree of heterogeneity in the response to various cytokines within vascular smooth muscle. Although the basis of this variation is not thoroughly understood, it has been suggested that cell lineage (e.g., neural crest vs mesenchyme) may be one of its determinants (Topouzis *et al.,* 1992; Schwartz and Liaw, 1993). More specifically, blood vessel ontogeny may be a factor in the varying susceptibility of different components of the vascular system to atherosclerosis (Topouzis *et al.,* 1992; Hood and Rosenquist, 1992).

A. Angiogenesis vs Vasculogenesis

There are two distinct processes by which embryonic blood vessels develop: "angiogenesis" by outgrowth or branching of preformed vessels and "vasculogenesis" by fusion of locally formed endothelial vesicles (Noden, 1990; Fig. 4). The origin of the coronary vessel system was controversial (Waldo *et al.,* 1990; Bogers *et al.,* 1989); it was uncertain whether the coronary vessel network was established by an outgrowth from the root of aorta (angiogenesis) or by *in situ* fusion of angioblasts (vasculogenesis).

To identify the timing of coronary stem cell entry into the embryonic heart, cells of the heart tube were tagged through infection with a replication-defective virus encoding β-gal at various stages *in ovo.* The fate analysis of the tagged cells revealed that coronary vascular precursors, the endothelial cells, vascular smooth muscle cells, and perivascular connective tissue cells enter the heart on the third day of embryogenesis; none of these precursors enter the heart before this stage (Mikawa and Fischman, 1992). Importantly, each precursor generates a daughter population which consists of only one cell type and forms a colony with a sharp boundary within the coronary vessels (Fig. 5; Mikawa and Fischman, 1992). Thus, the three lineages making up

Figure 3 The main morphogenetic events in the myocyte lineage during formation of four chambered heart (modified from Mikawa, 1995). A clonal myocyte population was highlighted in yellow. The endoderm is omitted from the diagram.

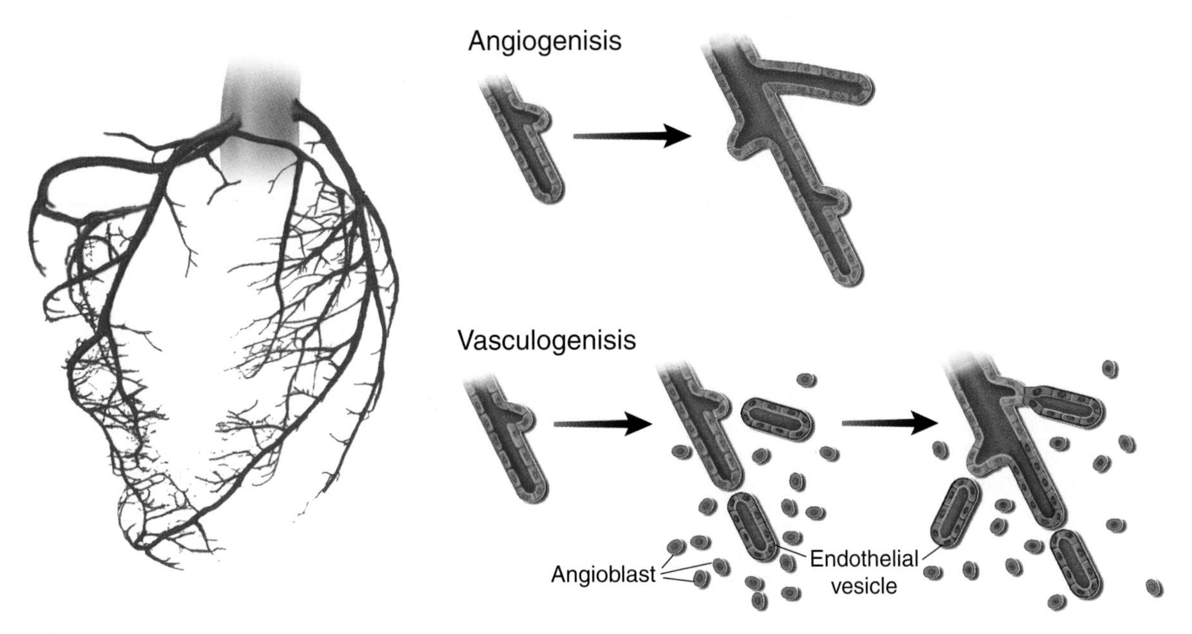

Figure 4 Two potential mechanisms, angiogenesis and vasculogenesis, for establishing coronary arterial network (red in left diagram) connected to aorta (pink).

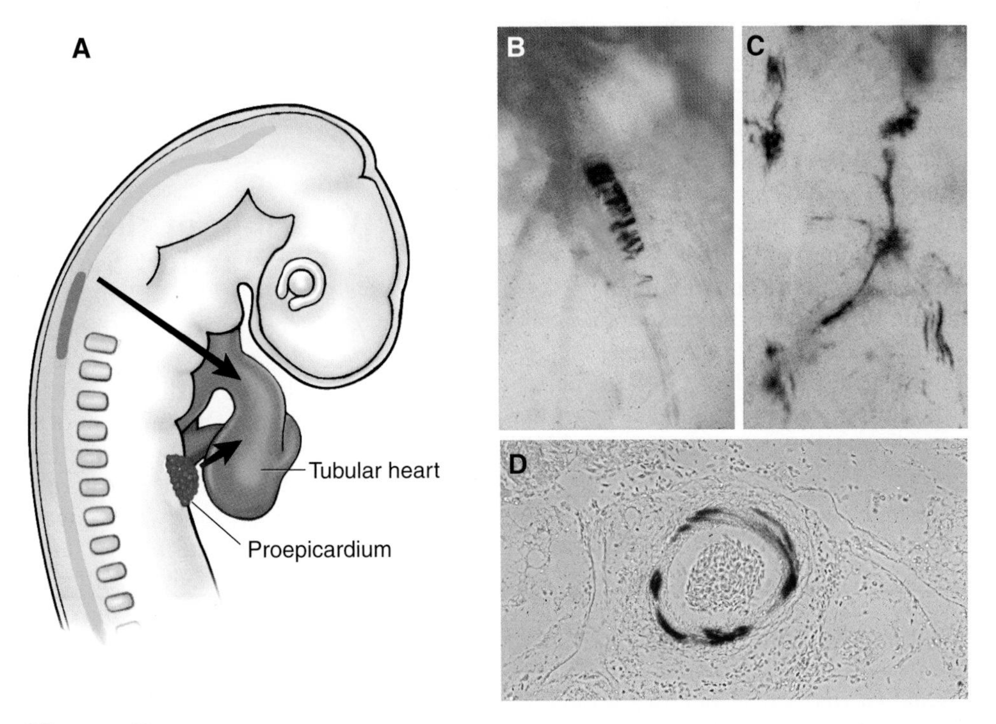

Figure 5 (A) Origin and migrate pathway of cardiac neural crest and coronary vessel precursors. (B) A β-gal+ colony of coronary arterial smooth muscle cells of E18 heart which received single cell infection in proepicardium at E3. (C) β-gal+ neuronal elements in E18 heart which received infection of cardiac neural crest at E2. (D) Transverse section of coronary artery containing β-gal+ smooth muscle cells.

the coronary vasculature appear to be distinct when they migrate into the tubular heart. The distribution of clonally related vascular cells restricted to a segment of arterial vessels indicates vasculogenesis, rather than angiogenesis, as the mechanism of coronary vessel formation.

B. Neural Crest vs Proepicardial Origin of Coronary Vasculatures

Studies using chick/quail chimeras have demonstrated that vascular smooth muscle cells in the proximal great vessels, including the thoracic aorta and aortic arches, are derived from the neural crest (Le Lièvre and Le Douarin, 1975; d'Amico-Martel and Noden, 1983; Kirby and Stewart, 1983; Hood and Rosenquist, 1992; see Chapter 11). In the heart, however, it has been difficult to distinguish between chicken and quail vascular cells because the chromatin pattern of quail nuclei is modified during differentiation of the vessels (Le Lièvre and Le Douarin, 1975). Thus, it is uncertain if the neural crest provides a progenitor population for coronary vasculatures. A genetic tag (β-gal) was introduced into neural crest cells by retroviral infection (Epstein *et al.,* 1994) and their contribution to both aortic and coronary vessels during chicken cardiogenesis was examined (Gourdie *et al.,* 1995; Noden *et al.,* 1995). The results demonstrate that smooth muscle cells in the tunica media of the aortic and pulmonary trunks, but not those of the coronary arteries, derive from the neural crest.

In addition to the neural crest, there is another migratory cell population that enters the heart tube; the epicardial mantle begins to envelop the myocardium on E3 (Ho and Shimada, 1988; Hiruma and Hirakow, 1989) at the same time that coronary precursor cells first appear in the tubular heart (Mikawa and Fischman, 1992). The epicardial mantle originates from mesothelial cell clusters on the right side of the external surface of the sinus venosus (Fig. 5). The protrusions of the proepicardium contact the dorsal wall of the tubular heart in the region of the atrioventricular junction at E3 (HH stage 17) and subsequently form a cellular monolayer which gradually covers the heart in a well-characterized progression (Hiruma and Hirakow, 1989).

Chicken/quail chimera studies have demonstrated that the proepicardium contains endothelial cells which can differentiate into the coronary endothelia if implanted (Poelmann *et al.,* 1993). Retroviral genetic tagging studies have shown that single putative vasculogenic cells of the proepicardium differentiate into solitary vessel-associated clusters consisting of only one cell type—endothelial, smooth muscle, or perivascular connective tissue cells (Mikawa and Gourdie, 1996). In no case were vessels labeled along their entire length. The segmental distribution of clonally related daughter cells was identical to that characterized in earlier clonal analyses of developing coronary arteries (Fig. 5; Mikawa and Fischman, 1992). These results proved unequivocally that the coronary vasculature is derived from mesodermal cells in the dorsal mesocardium (the proepicardium), and these results do not support the hypothesis of a neural crest origin as suggested by earlier investigators. Importantly, three distinct lineages of the coronary vasculature—the smooth muscle, endothelial, and connective tissue cells—are already segregated at the time of viral tagging in the proepicardium prior to migration into the heart tube.

C. A Model of Coronary Vasculogenesis

The data previously described indicate that the coronary vessel network is established by the vasculogenic mechanism and not by outgrowth from the aorta. The vasculogenic steps in coronary vessel morphogenesis are summarized in Fig. 6 (reviewed in Mikawa and Gourdie, 1996). First, independent endothelial and smooth muscle precursors migrate from the proepicardium to the tubular heart during formation of the epicardial mantle. Second, the endothelial cells differentiate to form sinusoidal vesicles. These endothelial vesicles fuse, eventually forming capillary channels. Third, once the closed vessel network is established and hooked to the aorta, intracardiac smooth muscle cell progenitors migrate to defined segments of the endothelial channels and differentiate to form the spiral segments.

IV. Origin of the Cardiac Conduction System

The rhythmic contraction of the heart in higher vertebrates is coordinated by electrical impulses generated and transmitted by the cardiac conduction system. Conduction disturbance by dysfunction of this essential tissue is a direct cause of arrhythmia, leading to sudden death. Regeneration and/or repair of cardiac conduction tissue after heart injury or congenital disease has not been considered plausible; to date little is known about mechanisms that regulate differentiation and patterning of this essential tissue. This is mainly due to the long-standing controversy concerning the ontogeny of the conduction system (see Chapter 12). The debate concerns (i) whether cells of the conduction system arise from the myocyte or neural crest lineage and (ii) whether the branched network of the entire conduction system is established by outgrowth from the common progenitor or by *in situ* linkage of subcomponents with independent origins. These questions have recently

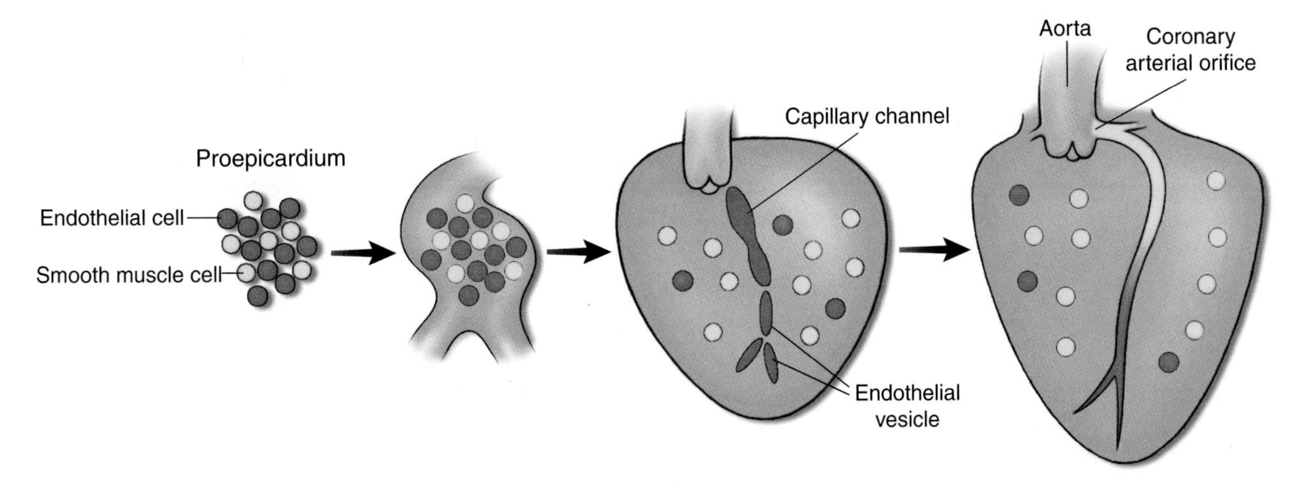

Figure 6 Model representing the differentiation of coronary arteries from migratory endothelial and smooth muscle progenitors (from Mikawa and Gourdie, 1996). Independent endothelial (red) and smooth muscle (yellow) precursors migrate from the proepicardium to the tubular heart during formation of the epicardial mantle. The endothelial cells first form sinusoidal sacs or channels. These sinusoids fuse along certain preferred axes, eventually forming capillary channels. Intracardiac smooth muscle cell progenitors migrate to defined segments of the endothelial channels and differentiate to form the spiral segments observed following retrovirus tagging.

been settled, in large part by retroviral cell lineage studies in chicken embryonic hearts, as discussed in the following sections.

A. Topological Changes of Impulse Conduction Pathway during Heart Development

In contrast to invertebrate hearts, which stochastically reverse the direction of the contractile wave (see Chapter 5), the vertebrate hearts unidirectionally pump blood by the precisely timed contractile sequence of atrial and ventricular chambers. After formation of the tubular heart, but before it begins contraction, epithelioid myocytes become electrically active. Action potentials are first detected in the posterior inflow tract (the presumptive sinus venosus and atrium). These impulses propagate to the rostral end of the heart through gap junctions between the presumptive myocytes (Kamino *et al.,* 1981). This caudal-to-rostral pattern of electrical propagation is sufficient for producing a caudal-to-rostral contractile wave in the tubular heart (Fig. 7). However, once the four-chambered heart is established, electrical impulses from atria need to be transmitted to the apex of the ventricle, avoiding direct propagation to the basal part of the ventricle (Fig. 7). This topological shift of the pulse-transmission site in the ventricle depends on the development and patterning of a specialized tissue called the "cardiac conduction system" (Tawara, 1906).

This system consists of four subcomponents (Fig. 7); the sinoatrial node, the atrioventricular node, the atrioventricular bundle, and Purkinje fibers. Pacemaking action potentials generated at the sinoatrial node spread through atrial myocytes, initiating contraction of both atrial chambers. Unlike in tubular heart, the impulse from the atrial myocardium does not propagate directly into the ventricular myocardium, but instead converges on the atrioventricular node, in which after a brief delay the action potential is propagated rapidly through the highly coupled cells of the atrioventricular bundle. Finally, activation is spread into ventricular muscle via the subendocardial and intramural network of Purkinje fibers (Fig. 8), thereby synchronizing contraction of the ventricular chambers of the heart (reviewed in Viragh and Challice, 1982; Lamers *et al.,* 1991). To address mechanisms governing the differentiation and patterning of the cardiac conduction system, it is essential to determine the origin and lineage relationships of its cellular components.

B. Neural Crest vs Myocyte Origin of the Conduction System

Cells in the cardiac conduction system are characterized by a diameter considerably greater than that of ordinary cardiac muscle cells, a reduced number of myofibrils, and large accumulations of glycogen. In addition to their anatomical properties, they can be distinguished from contractile myocytes by their unique gene expression. Conduction cells contain a unique set of ion channels for pacemaker activity (Cavalie et al., 1983; Callewaert *et al.,* 1986; Hagiwara *et al.,* 1988; DiFrancesco, 1995) and unique connexins for gap-junctional electrical couplings between conduction cells (van Kempen *et al.,* 1991; Gourdie *et al.,* 1992, 1993;

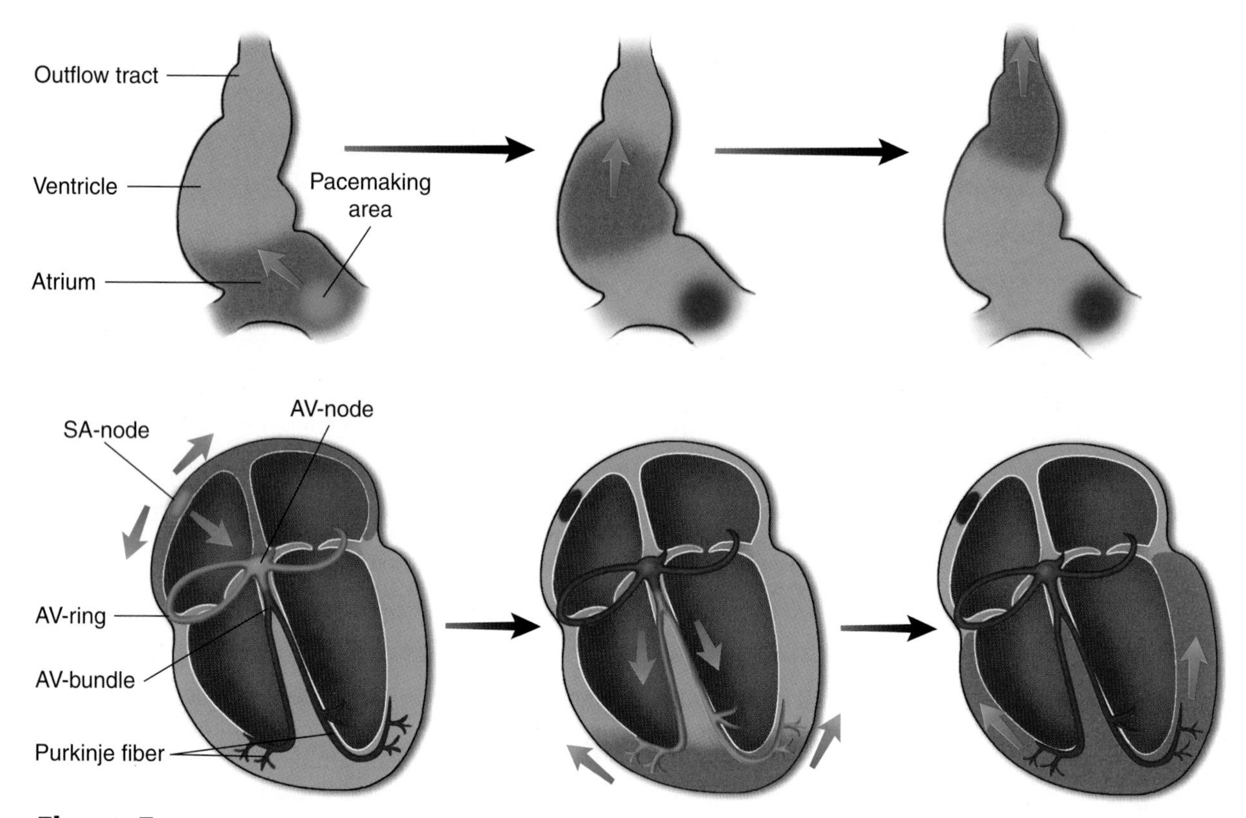

Figure 7 Impulse-conducting pathways between two-chambered tubular heart and four-chambered heart. Arrows: direction of propagation of action potentials and contractile wave. Red: action potential-positive myocytes. Dark blue: cells of the cardiac conduction system. Light blue: action potential-positive conduction cells.

Bastide *et al.*, 1993; Kanter *et al.*, 1993; Gros *et al.*, 1994). Importantly, conduction cells express genes that are usually restricted to neurons, e.g., neurofilament proteins, and brain-associated glycoproteins (Sartore *et al.*, 1978; Gonzalez-Sanchez and Bader, 1985; reviewed in Gorza *et al.*, 1994). Both the coexpression of neural and muscle genes and the migration of neural crest cells into the heart (see Chapter 11) have proven puzzling in the determination of the origin of the cardiac conduction system and led to the suggestion of two possible origins: myogenic (Patten and Kramer, 1933; Patten, 1956) and neural crest (Gorza *et al.*, 1988, 1994; Vitadello *et al.*, 1990).

C. Matters of Ingrowth or Outgrowth

The development of the conduction system in embryonic hearts has been studied in a number of species, including human (Wessels *et al.*, 1992), rat (Gourdie *et al.*, 1992), and avian (Vassal-Adams, 1982; Gourdie *et al.*, 1993; Chan-Thomas *et al.*, 1993): In all cases, a ring-like structure (Wessels *et al.*, 1992) at the atrioventricular junction has been mapped as the initiation site of conduction system formation. The primary ring can be identified at E2 in the chicken (HH stage 15+) by probing the *Msx-2* expression (Chan-Thomas *et al.*, 1993). *Msx-2* expression extends to the proximal conduction system but does not occur in Purkinje fibers (Chan-Thomas *et al.*, 1993). The Purkinje fibers cannot be identified immunohistologically until E10 (Gourdie *et al.*, 1995). Thus, differentiation of the conduction system first occurs proximally and secondarily extends to the Purkinje fibers.

Based on the proximal–distal wave of conduction system development, it was proposed that the "primary conduction ring" contains precursor cells which enable the progeny to differentiate into the entire conduction system (Wessels *et al.*, 1992; Lamers *et al.*, 1991; Chan-Thomas *et al.*, 1993). Neither morphology nor gene expression is adequate to address directly questions concerning the ontogeny of the conduction system and the lineage relationships between its proximal and distal subcomponents. A clear resolution of these lineage relationships has recently been obtained in retroviral cell

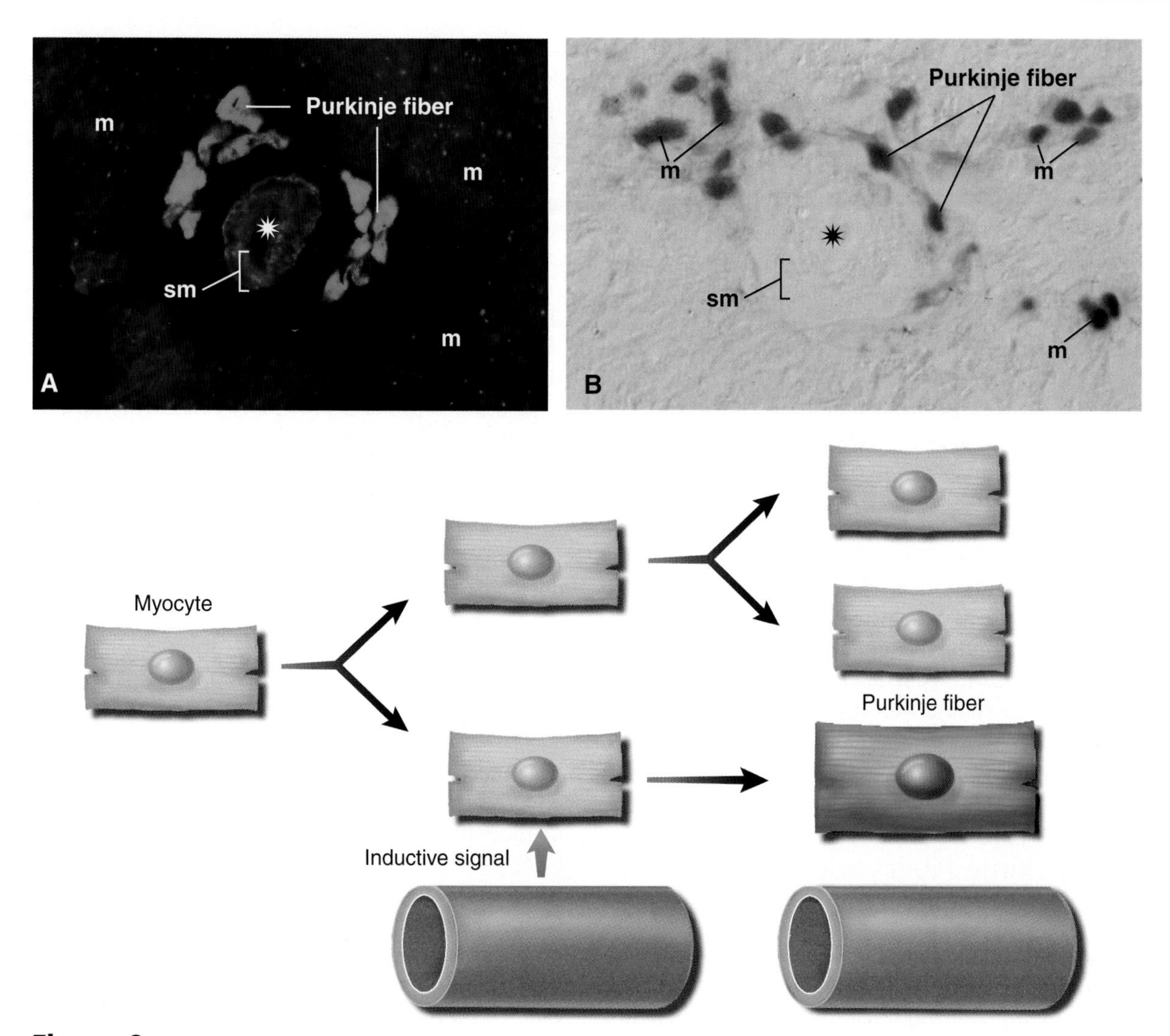

Figure 8 Purkinje fiber differentiation within the myocyte lineage. (A) Periatrial localization of Purkinje fibers in the avian heart. Purkinje fibers (green) and atrial bed (red). (B) A subpopulation of clonally related myocytes expressing nuclear directed β-galactosidase (blue-stained nuclei) differentiate into periarterial Purkinje fibers (arrows). (C) Proposed model of induction of Purkinje fibers within the myocyte lineage.

lineage studies on the Purkinje fiber network of the chicken embryonic heart (Gourdie *et al.,* 1995).

D. Purkinje Fibers Arise from Contractile Ventricular Myocytes

In the embryonic heart, the differentiating Purkinje fibers can first be detected as cells highly expressing a member of the connexin family, Cx-42, along the growing coronary arteries on E10 (Gourdie *et al.,* 1995). It is known that all cells in the epithelioid myocardium of the tubular heart differentiate into contractile myocytes by E2 (Manasek, 1968), whereas cardiac neural crest cells begin migration from the embryonic hindbrain at E2 or E3 and enter the heart at E4 (reviewed in Kirby, 1993). Therefore, if Purkinje fibers are of neural crest origin, their precursors must be absent before E3. On the other hand, if the conduction cells are of myogenic origin, the precursor cells must be present in the beating tubular heart at E3.

Retroviral-mediated genetic marking of single differentiated and contractile myocytes in the E3 tubular heart and subsequent inspection of clonal populations have revealed that a subset of clonally related myocytes differentiates into conducting Purkinje fibers, invariably in close spatial association with forming coronary arterial blood vessels. Importantly, infection of cardiac neural crest did not produce β-gal-positive Purkinje fibers, and in no case did a clone containing both Purkinje fibers and cells of more proximal components of the conduction system such as the atrioventricular node and bundles, or sinoatrial component (Gourdie *et al.*, 1995). These data indicate that pulse generating and conduction cells are derived by localized recruitment of differentiated, beating myocytes specifically along the developing coronary arterial bed (Fig. 8).

The lineage data do not support a hypothesis that the branched network of the entire conduction system is established by outgrowth from the primary ring. It has also been shown that cells in the primary ring cease their proliferation very early in development (Thompson *et al.*, 1990). Both data suggest that Purkinje fibers have a different parental lineage from that of the proximal conduction system (Gourdie *et al.*, 1995). The definite mapping of Purkinje fiber progenitor cells to a myocyte lineage, and not to neural crest, allows us to study the mechanism of Purkinje fiber differentiation by analyzing the process by which heart cells are converted from a contractile to a conductive lineage. The lineage data of independent origins for central and peripheral elements of the conduction system raise the question of how the central and peripheral components link *in situ* to form an integrated conduction network.

E. Potential Mechanisms Inducing Purkinje Fibers within the Myocyte Lineage

As described in section II,C, the coronary vasculature does not arise by outgrowth from the root of the aorta but rather it migrates into the tubular heart from extracardiac mesenchyme (Mikawa and Fischman, 1992; Poelmann *et al.*, 1993; Mikawa and Gourdie, 1996) along with a proepicardial sheet (Hiruma and Hirakow, 1989; Ho and Shimada, 1988). In the avian, entry of these coronary precursors to the heart begins at E3 (Mikawa and Fischman, 1992; Poelmann *et al.*, 1993). Following inward migration, vasculogenic cells first form discontinuous endothelial channels, and subsequent fusion between endothelial channels by E6 and connection to the aorta establishes the closed coronary vessel network by E14 (Bogers *et al.*, 1989; Waldo *et al.*, 1990; Mikawa and Fischman, 1992; Mikawa and Gourdie, 1996). Coincident with this early vasculogenic process, recruitment of Purkinje fibers begins exclusively in myocyte subpopulations juxtaposed to developing coronary arteries but not veins (Gourdie *et al.*, 1995). The close spatiotemporal relationship between Purkinje fiber differentiation and coronary blood vessel development suggests that an inductive role of coronary vasculature (i.e., coronary arteries) may recruit contractile myocytes to form Purkinje fibers. If coronary arteries play a role in the recruitment of Purkinje fibers from contractile myocytes, the vessel network may be a key factor in defining the branching pattern of the peripheral conduction system (Gourdie *et al.*, 1995, 1998; Mikawa and Fischman, 1996).

As discussed in Chapter 11, ablation of the cardiac neural crest alters the pattern of the coronary arterial tree (Hood and Rosenquist, 1992), neural crest derivatives are necessary for the survival of branches of the coronary artery system (Waldo *et al.*, 1994). Thus, there is evidence that neural crest-derived cells affect the development of coronary arteries. Although cell lineage studies prove a myogenic origin for the Purkinje fibers, neural crest-derived cells may indirectly contribute to their differentiation and network patterning through regulation of coronary arterial development. It remains to be seen if the induction and patterning of Purkinje fibers are in concert with the development of coronary arteries and the differentiation of cardiac neural crest cells.

V. Concluding Remarks

Molecular and cell biological approaches have uncovered the potentiality or plasticity of embryonic cells in their lineage commitment and terminal differentiation, whereas genetics has been powerful in identifying genes or gene networks involved in these processes. Clone-based cell lineage data, together with fate mapping on a group of cells, define the timing and site at which embryos really induce the cardiac lineages and their differentiation and patterning.

Except for neural crest derivatives, the lineage relationships of all cell types in the heart have been identified (Fig. 9). The cell lineage data lead to the following questions: (i) When and where is the endocardial cell lineage established? (ii) What extracellular factors restrict bipotential myogenic precursors to either atrial or ventricular myocytes within the heart field in a spatially defined pattern? (iii) How is asymmetrical mus-cle growth locally regulated in the genesis of muscular septa and papillary muscles? (iv) How are the branching patterns of coronary arteries defined? What is the role of the cardiac neural crest in this process? (v) Which signals and cell types of the developing coronary arteries are responsible for induction and patterning of Purkinje fibers within the myocyte lineage? and (vi) What mechanisms govern linkage of the Purkinje fiber network to the proximal conduction system? Answers to these questions will contribute significantly to our understanding of the development and integrated function of the vertebrate heart.

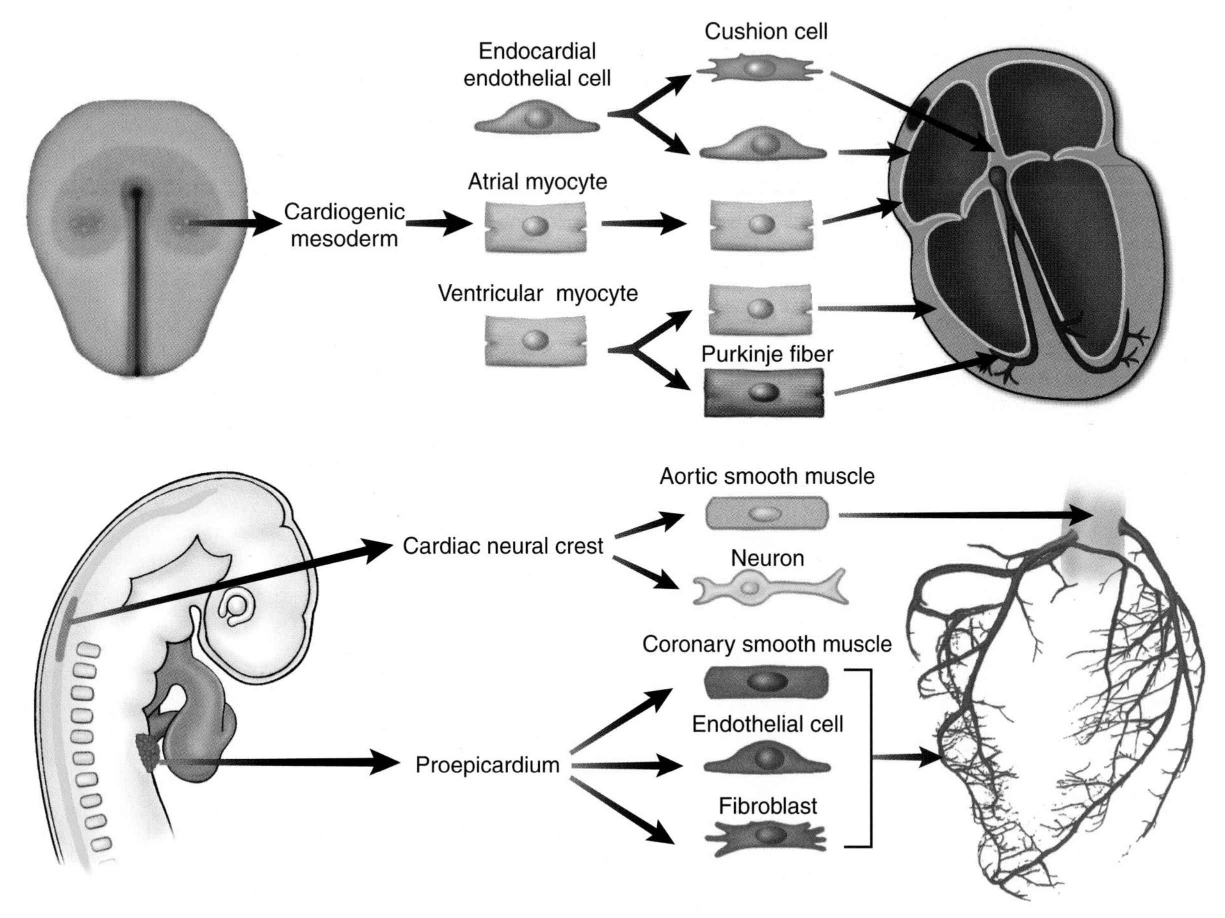

Figure 9 The origin and lineage relationships of cardiac cell types. Each cardiac cell type is established by lineage diversification of embryonic cells which arise from one of three distinct origins: cardiogenic mesoderm, neural crest, or proepicardium. These data define the chronology and distribution for the development of all cell lineages in the avian heart.

References

Bastide, B., Neyses, L., Ganten, D., Paul, M., Willecke, K., and Traub, O. (1993). Gap junction protein connexin40 is preferentially expressed in vascular endothelium and conductive bundles of rat myocardium and is increased under hypertensive conditions. *Circ. Res.* **73,** 1138–1149.

Bogers, A. J. J. C., de Groot, A. C., Poelmann, R. E., and Huysmans, H. A. (1989). Development of the origin of the coronary arteries, a matter of ingrowth or outgrowth? *Anat. Embryol.* **180,** 437–441.

Callewaert, G., Vereecke, J., and Carmeliet, E. (1986). Existence of a calcium-dependent potassium channel in the membrane of cow cardiac Purkinje cells. *Pfluegers Arch.* **406,** 424–426.

Cavalie, A., Ochi, R., Pelzer, D., and Trautwein, W. (1983). Elementary currents through Ca2+ channels in guinea pig myocytes. *Pfluegers Arch.* **398,** 284–297.

Cepko, C. (1988). Retrovirus vectors and their applications in neurobiology. *Neuron* **1,** 345–353.

Chan-Thomas, P. S., Thompson, R. P., Robert, B. Y., Yacoub, M. H., and Barton, P. J. R. (1993). Expression of homeobox genes Msx-1 (Hox-7) and Msx-2 (Hox-8) during cardiac development in the chick. *Dev. Dyn.* **197,** 203–216.

Coffin, J. D., and Poole, T. J. (1988). Embryonic vascular development: Immunohistochemical identification of the origin and subsequent morphogenesis of the major vessel primordia in quail embryos. *Development (Cambridge, UK)* **102,** 735–748.

Cohen-Gould, L., and Mikawa, T. (1996). The fate diversity of mesodermal cells within the heart field during chicken early embryogenesis. *Dev. Biol.* **177,** 265–273.

d'Amico-Martel, A., and Noden, D. (1983) Contributions of placodal and neural crest cells to avian cranial peripheral ganglia. *Am. J. Anat.* **166,** 445–468.

Davies, F. (1930). The conducting system of the bird's heart. *J. Anat.* **64,** 129–146.

DeRuiter, M. C., Poelmann, R. E., Vander Plas-de Vries, I., Mentink, M. M. T., and Gittenberger-de Groot, A. C. (1992). The development of the myocardium and endocardium in mouse embryos. *Anat. Embryol.* **185,** 461–473.

DiFrancesco, D. (1995). The onset and autonomic regulation of cardiac pacemaker activity: Relevance of the f current. *Cardiovasc. Res.* **29,** 449–456.

Eisenberg, C. A., and Bader, D. (1995). QCE-6: A clonal cell line with cardiac myogenic and endothelial cell potentials. *Dev. Biol.* **167,** 469–481.

Epstein, M. L., Mikawa, T., Brown, A. M. C., and McFarlin, D. R. (1994). Mapping the origin of the avian enteric nervous system with a retroviral marker. *Dev. Dyn.* **201,** 236–244.

Ferrara, N., Carver-Moore, K., Chen, H., Dowd, M., Lu, L., O'Shea,

K. S., Powell-Braxton, L., Hillan, K. J., and Moore, M. W. (1996). Heterozygous embryonic lethality induced by targeted inactivation of the VEGF gene. *Nature (London)* **380,** 439–442.

Flamme, I., Breier, G., and Risau, W. (1994). Vascular endothelial growth factor (VEGF) and VEGF receptor 2 (flk-1) are expressed during vasculogenesis and vascular differentiation in the quail embryo. *Dev. Biol.* **169,** 699–712.

Fong, G. H., Rossant, J., Gertsenstein, M., and Breitman, M. L. (1995). Role of the Flt-1 receptor tyrosine kinase in regulating the assembly of vascular endothelium. *Nature (London)* **376,** 66–70.

Garcia-Martinez, V., and Schoenwolf, G. C. (1993). Primitive-streak origin of the cardiovascular system in avian embryos. *Dev. Biol.* **159,** 706–719.

Gassmann, M., Casagranda, F., Orioli, D., Simon, H., Lai, C., Klein, R., and Lemke, G. (1995). Aberrant neural and cardiac development in mice lacking the ErbB4 neuregulin receptor. *Nature (London)* **378,** 390–394.

Gonzalez-Sanchez, A., and Bader, D. (1985). Characterization of a myosin heavy chain in the conductive system of the adult and developing chicken heart. *J. Cell Biol.* **100,** 270–275.

Gorza, L., Schiaffino, S., and Vitadello, M. (1988). Heart conduction system: A neural crest deivative. *Brain Res.* **457,** 360–366.

Gorza, L., Vettore, S., and Vitadello, M. (1994). Molecular and cellular diversity of heart conduction system myocytes. *Trends in Cardiovasc. Med.* **4,** 153–159.

Gourdie, R. G., Green, C. R., Severs, N. J., and Thompson, R. P. (1992). Immunolabelling patterns of gap junction connexins in the developing and mature rat heart. *Anat. Embryol.* **185,** 363–378.

Gourdie, R. G., Green, C. R., Severs, N. J., Anderson, R. H., and Thompson, R. P. (1993). Evidence for a distinct gap-junctional phenotype in ventricular conduction tissues of the developing and mature avian heart. *Circ. Res.* **72,** 278–289.

Gourdie, R. G., Mima, T., Thompson, R. P., and Mikawa, T. (1995). Terminal diversification of the myocyte lineage generates Purkinje fibers of the cardiac conduction system. *Development (Cambridge, UK)* **121,** 1423–1431.

Gourdie, R. G., Wei, Y., Kim, D., Klatt, S. C., and Mikawa, T. (1998). Endothelin-induced conversion of heart muscle cells into impulse-conducting Purkinje fibers. *Proc. Natl. Acad. Sci. U.S.A.* **95,** 6815–6818.

Gros, D., Jarry-Guichard, T., Ten-Velde, I., de-Maziere, A., van Kempen, M. J., Davoust, J., Briand, J. P., Moorman, A. F., and Jongsma, J. J. (1994). Restricted distribution of connexin40, a gap junctional protein, in mammalian heart. *Circ. Res.* **74,** 839–851.

Hagiwara, N., Irisawa, H., and Kameyama, M. (1988). Contribution of two types of calcium currents to the pacemaker potentials of rabbit sino-atrial node cells. *J. Physiol. (London)* **395,** 233–253.

Hamburger, V., and Hamilton, H. L. (1951). A series of normal stages in the development of the chick embryo. *J. Morphol.* **88,** 49–92.

Han, Y., Dennis, J. E., Cohen-Gould, L., Bader, D. M., and Fischman, D. A. (1992). Expression of sarcomeric myosin in the presumptive myocardium of chicken embryos occurs within six hours of myocyte commitment. *Dev. Dyn.* **193,** 257–265.

Hiruma, T., and Hirakow, R. (1989). Epicardial formation in embryonic chick heart: Computer-aided reconstruction, scanning, and transmission electron microscopic studies. *Am. J. Anat.* **184,** 129–138.

Ho, E., and Shimada, Y. (1988). Formation of the epicardium studied with the scanning electron microscope. *Dev. Biol.* **66,** 579–585.

Hood, L. A., and Rosenquist, T. H. (1992). Coronary artery development in the chick: Origin and development of smooth muscle cells, and effects of neural crest ablation. *Anat. Rec.* **234,** 291–300.

Kamino, K., Hirota, A., and Fujii, S. (1981). Localization of pacemaking activity in early embryonic heart monitored using voltage-sensitive dye. *Nature (London)* **290,** 595–597.

Kanter, H. L., Laing, J. G., Beau, S. L., Beyer, E. C., and Saffitz, J. E. (1993). Distinct patterns of connexin expression in canine Purkinje fibers and ventricular muscle. *Circ. Res.* **72,** 1124–1131.

Kirby, M. L. (1993). Cellular and molecular contributions of the cardiac neural crest to cardiovascular development. *Trends Cardiovasc. Med.* **3,** 18–23.

Kirby, M. L., and Stewart, D. E. (1983). Neural crest origin of cardiac ganglion cells in the chick embryo: Identification and extirpation. *Dev. Biol.* **97,** 433–443.

Lamers, W. H., De Jong, F., De Groot, I. J. M., and Moorman, A. F. M. (1991). The Developmant of the avian conduction system, a review. *Eur. J. Morphol.* **29,** 233–253.

Lee, K., Simon, H., Chen, H., Bates, B., Hung, M., and Hauser, C. (1995). Requirement for neuregulin receptor erbB2 in neural and cardiac development. *Nature (London)* **378,** 394–398.

Lee, R. R. K., Stainier, D. Y. R., Weinstein, B. M., and Fishman, M. C. (1994). Cardiovascular development in the zebrafish. II. Endocardial progenitors are sequestered within the heart field. *Development (Cambridge, UK)* **120,** 3361–3366.

Le Lièvre, C. S., and Le Douarin, N. M. (1975). Mesenchymal derivatives of the neural crest: Analysis of chimeric quail and chick embryos. *J. Embryol. Exp. Morphol.* **34,** 125–154.

Letterio, J. J., Geiser, A. G., Kulkarni, A. B., Roche, N. S., Sporn, M. B., and Roberts, A. B. (1994). Maternal rescue of transforming growth factor-beta 1 null mice. *Science* **264,** 1936–1938.

Linask, K. K. (1992). N-cadherin localization in early heart development and polar expression of Na^+, K^+-ATPase, and integrin during pericardial coelom formation and epithelialization of the differentiating myocardium. *Dev. Biol.* **151,** 213–224.

Linask, K. K., and Lash, J. W. (1993). Early heart development: Dynamics of endocardial cell sorting suggests a common origin with cardiomyocytes. *Dev. Dyn.* **195,** 62–66.

Manasek, F. J. (1968). Embryonic development of the heart: A light and electron microscopic study of myocardial development in the early chick embryo. *J. Morphol.* **125,** 329–366.

Meyer, D., and Birchmeier, C. (1995). Multiple essential functions of neuregulin in development. *Nature (London)* **378,** 386–390.

Mikawa, T. (1995). Retroviral targeting of FGF and FGFR in cardiomyocytes and coronary vascular cells during heart development. *Ann. N.Y. Acd. Sci.* **752,** 506–516.

Mikawa, T., and Fischman, D. A. (1992). Retroviral analysis of cardiac morphogenesis: Discontinuous formation of coronary vessels. *Proc. Natl. Acad. Sci. U.S.A.* **89,** 9504–9508.

Mikawa, T., and Fischman, D. A. (1996). The polyclonal origin of myocyte lineages. *Annu. Rev. Physiol.* **58,** 509–521.

Mikawa, T., and Gourdie, R. G. (1996). Pericardial mesoderm generates a population of coronary smooth muscle cells migrating into the heart along with ingrowth of the epicardial organ. *Dev. Biol.* **173,** 221–232.

Mikawa, T., Borisov, A., Brown, A. M. C., and Fischman, D. A. (1992a). Clonal analysis of cardiac morphogenesis in the chicken embryo using a replication-defective retrovirus: I. Formation of the ventricular myocardium. *Dev. Dyn.* **193,** 11–23.

Mikawa, T., Cohen-Gould, L., and Fischman, D. A. (1992b). Clonal analysis of cardiac morphogenesis in the chicken embryo using a replication-defective retrovirus. III: Polyclonal origin of adjacent ventricular myocytes. *Dev. Dyn.* **195,** 133–141.

Mikawa, T., Hyer, J., Itoh, N., and Wei, Y. (1996). Retroviral vectors to study cardiovascular development. *Trends Cardiovasc. Med.* **6,** 79–86.

Mima, T., Ueno, H., Fischman, D. A., Williams, L. T., and Mikawa, T. (1995). FGF-receptor is required for *in vivo* cardiac myocyte proliferation at early embryonic stages of heart development. *Proc. Natl. Acad. Sci. U.S.A.* **92,** 467–471.

Noden, D. M. (1990). Origins and assembly of avian embryonic blood vessels. *Ann. N.Y. Acad. Sci.* **588,** 236–249.

Noden, D. M., Poelmann, R. E., and Gittenberger-de Groot, A. C. (1995). Cell origins and tissue boundaries during outflow tract development. *Trends Cardiovasc. Med.* **5,** 69–75.

O'Brien, T. X., Lee, K. J., and Chien, K. R. (1993). Positional specification of ventricular myosin light chain 2 expression in the primitive murine heart tube. *Proc. Natl. Acad. Sci. U.S.A.* **90,** 5157–5161.

Pardanaud, L., Altmann, C., Kitos, P., Dieterlen-Lievre, F., and Buck, C. A. (1987a). Vasculogenesis in the early quail blastodisc as studied with a monoclonal antibody recognizing endothelial cells. *Development (Cambridge, UK)* **100,** 339–349.

Pardanaud, L., Buck, C., and Dieterlen-Lievre, D. (1987b). Early germ cell segregation and distribution in the quail blastodisc. *Cell Differ.* **22,** 47–60.

Patten, B. M. (1956). The development of the sinoventricular conduction system. *Univ. Mich. Med. Bull.* **22,** 1–21.

Patten, B. M., and Kramer, T. C. (1933). The initiation of contraction in the embryonic chick heart. *Am. J. Anat.* **53,** 349–375.

Poelmann, R. E., Gittenberger-de Groot, A. C., Mentink, M. T., Bokenkamp, R., and Hogers, B. (1993). Development of the cardiac coronary vascular endothelium, studied with anti-endothelial antibodies, in chicken-quail chimeras. *Circ. Res.* **73,** 559–568.

Rawles, M. E. (1943). The heart forming regions of the early chick blastoderm. *Physiol. Zool.* **16,** 22–42.

Rosenquist, G. C., and DeHaan, R. L. (1966). Migration of precardiac cells in the chick embryo: a radioautographic study. *Carnegie Inst. Washington Publ.* **625,** (*Contrib. to Embryol*). **38,** 111–121.

Rychter, Z., and Ostadal, R. (1971). Mechanism of development of the coronary arteries in chick embryo. *Folia. Morphol. (Prague)* **16,** 113–124

Sanes, J. R. (1989). Analyzing cell lineage with a recombinant retrovirus. *Trends Neurosci.* **12,** 21–28.

Sartore, S., Pierobon-Bormioli, S., and Schiaffino, S. (1978). Immunohistochemical evidence for myosin polymorphism in the chicken heart. *Nature (London)* **274,** 82–83.

Satin, J., Fujii, S., and De Haan, R. L. (1988). Development of cardiac beat rate in early chick embryos is regulated by regional cues. *Dev. Biol.* **129,** 103–113.

Schultheiss, T. M., Xydas, S., and Lassar, A. B. (1995). Induction of avian cardiac myogenesis by anterior endoderm. *Development (Cambridge, UK)* **121,** 4203–4214.

Schwartz, S. M., and Liaw, L. (1993). Growth control and morphogenesis in the development and pathology of arteries. *J. Cardiovasc. Pharmacol.* 21, S31–S49.

Shalaby, F., Rossant, J., Yamaguchi, T. P., Gertsenstein, M., Wu, X. F., Breitman, M. L., and Schuh, A. C. (1995). Failure of blood-island-formation and vasculogenesis in Flk-1-deficient mice. *Nature (London)* **376,** 62–66.

Stainier, D. Y. R., Lee, R. K., and Fishman, M. C. (1993). Cardiovascular development in the zebrafish I. Mocardial fate and heart tube formation. *Development (Cambridge, UK)* **119,** 31–40.

Sugi, Y., and Markwald, R. R. (1996). Formation and early morphogenesis of endocardial endothelial precursor cells and the role of endoderm. *Dev. Biol.* **175,** 66–83.

Takeichi, M. (1991). Cadherin cell adhesion receptors as a morphogenetic regulator. *Science* **251,** 1451–1455.

Tam, P. P., Parameswaran, M., Kinder, S. J., and Weinberg, R. P. (1997). The allocation of epiblast cells to the embryonic heart and other mesodermal lineages: The role of ingression and tissue movement during gastrulation. *Development (Cambridge, UK)* **124,** 1631–1642.

Tawara, S. (1906). Das reizleitungssystem des Säugetierherzens. Gustv Fischer, Jena.

Thompson, R. P., Lindroth, J. R., and Wong, Y. -M. M. (1990). Regional differences in DNA-synthetic activity in the preseptation myocardium of the chick. *In* "Developmental Cardiology; Morphogenesis and Function" (E. Clark and A. Takao, eds.), pp. 219–234, Futura Publ. Co., Mt. Kisco, NY.

Topouzis, S., Catravas, J. D., Ryan, J. W., and Rosenquist, T. H. (1992). Influence of vascular smooth muscle heterogeneity on angiotensin converting enzyme activity in chicken embryonic aorta and in endothelial cells in culture. *Circ. Res.* **71,** 923–931.

van Kempen, M. J., Fromaget, C., Gros. D., Moorman, A. F., and Lamers, W. H. (1991). Spatial distribution of connexin43, the major cardiac gap junction protein, in the developing and adult rat heart. *Circ. Res.* **68,** 1638–1651.

Vassal-Adams, P. R. (1982). The development of the atrioventricular bundle and its branches in the avian heart. *J. Anat.* **134,** 169–183.

Viragh, Sz., and Challice, C. E. (1982). The development of the conduction system in the mouse embryo heart. IV. Differentiation of the atrioventricular conduction system. *Dev. Biol.* **89,** 25–40.

Vitadello, M., Matteoli, M., and Gorza, L. (1990). Neurofilament proteins are co-expressed with desmin in heart conduction system myocytes. *J. Cell Sci.* **97,** 11–21.

von Kirschhofer, K., Grim, M., Christ, B., and Wachtler, F. (1994). Emergence of myogenic and endothelial cell lineages in avian embryos. *Dev. Biol.* **163,** 270–278.

Waldo, K. L., Willner, W., and Kirby, M. L. (1990). Origin of the proximal coronary artery stems and a review of ventricular vascularization in the chick embryo. *Am. J. Anat.* **188,** 109–120.

Waldo, K. L., Kumiski, D. H., and Kirby, M. L. (1994). Association of the cardiac neural crest with development of the coronary arteries in the chick embryo. *Anat. Rec.* **239,** 315–331.

Wessels, A., Vermeulen, J. L. M., Verbeek, F. J., Viragh, Sz., Kalman, F., Lamers, W. H., and Moorman, A. F. M. (1992). Spatial distribution of "tissue-specific" antigens in the developing human heart. *Anat. Rec.* **232,** 97–111.

Yutzey, K. E., Rhee, J. T., and Bader, D. (1994). Expression of the atrial-specific myosin heavy chain AMHC1 and the establishment of anteroposterior polarity in the developing chicken heart. *Development (Cambridge, UK)* **120,** 871–883.

Yutzey, K., Gannon, M., and Bader, D. (1995). Diversification of cardiomyogenic cell lineages in vitro. *Dev. Biol.* **170,** 531–541.

II

Cardiac Induction

3

Heart Formation and the Heart Field in Amphibian Embryos

Timothy J. Mohun and Li Ming Leong

National Institute for Medical Research, London NW7 1AA, United Kingdom

I. Introduction

With the resurgence of interest in the mechanisms regulating vertebrate heart development, it is timely to review the results of embryological studies that have examined this process in amphibian embryos. Beginning almost a century ago, this work belongs to an era in which experimental analysis was restricted to observations of morphology or function, without the aid of biochemical or molecular tools. Conclusions were expressed in a language of "potentialities," "potencies," and "tendencies" which appears unfamiliar today (Huxley and De Beer, 1934; Weiss, 1939; Needham, 1942). Nevertheless, these studies, along with those of limb formation and lens induction, provide the basis for our current understanding of organogenesis (Jacobson and Sater, 1988). As the application of molecular techniques expands our knowledge of the cardiogenic program, a central aim remains the identification of molecular signals that regulate cardiac differentiation and morphogenesis. By identifying tissue interactions essential for heart formation, the early studies provide important clues toward this goal.

II. The Heart Field

A recurrent term in older work is that of the "heart field," a term that sits uneasily with modern molecular studies. Its context is a model of amphibian development derived from explant culture and graft transplantation experiments. These suggested that the early neurula embryo contained a series of overlapping regions or "organ fields" (Weiss, 1939: Needham, 1942). Each field extended beyond the limits of the corresponding organ, contributing as well to adjacent tissues. Each was established through inductive interactions during gastrulation and orderly development of the embryo depended on their positions and relative strengths. The field concept therefore combined a descriptive (or geographical) element with the idea that regions of the embryo were spatially organized in a state of dynamic equilibrium. Attempts to codify a definition of organ fields emphasized two properties that were evident from embryological experiments; the regulative behavior of fields and their transitory character (Weiss, 1939). In response to subdivision or addition of tissue, an organ field would reorganize to encompass its new size, but this organizing ability declined during development.

Amphibian heart formation provided a clear example of such behavior and investigators used a combination of grafting and tissue explanation to study the capacity of heart field tissue to undergo morphological differentiation (Copenhaver, 1955). These studies provided a detailed description of cardiac morphogenesis and also identified a number of tissue interactions that influenced this process. From a modern perspective, the limitations of this work lie not in the imprecision of the heart field concept itself but rather in the inability of early investigators to distinguish between cellular differentiation and morphogenesis. In reviewing these studies, a further difficulty is the use of several different amphibian species, each with its own staging system and rate of development. Urodele embryos (particularly *Triturus* and *Ambystoma*) were commonly used, whereas in recent years the anuran *Xenopus* has predominated as the laboratory amphibian. To simplify matters and facilitate comparisons, we have therefore avoided use of stage numbers and resorted instead to less precise descriptive designations of developmental stage.

A. Heart Formation in the Urodele, *Ambystoma*

The most detailed description of the origins and formation of the amphibian heart have come from studies of urodele embryos. Using a combination of histological observation and vital dye mapping, these studies have identified regions of the early gastrula embryo destined to form the heart tube and traced the location of heart precursors through subsequent relocation that characterizes amphibian cardiogenesis. The classical fate maps for urodele embryos were provided by the studies of Vogt (1929) and subsequently ammended by Pasteels (1942) for the axolotl, *Ambystoma.* Reinvestigations in the modern era (Nakamura *et al.,* 1978; Cleine and Slack, 1985) largely confirm the classical fate map of the urodele early gastrula embryo. Using these data, the following summary of heart formation in the axolotl is drawn from the studies of Copenhaver (1926, 1930, 1939a,b, 1955), Fales (1946), and Wilens (1955) (Fig. 1).

The prospective heart region is not identified in the urodele fate maps but has been inferred to comprise a pair of mesodermal rudiments in the deep rather than superficial marginal zone, immediately lateral to the blastopore lip (Jacobson and Sater, 1988). In this location, the heart precursors lie at the anterior edge of the involuting mesoderm, immediately adjacent to anterior endoderm. Explant assays show that anterior endoderm is capable of differentiation into liver, stomach, and esophageal endoderm (Holtfreter, 1939), and fate mapping indicates that it forms the pharyngeal endoderm in tailbud embryos (Cleine and Slack, 1985). As the mesodermal mantle advances anteriorly during gastrulation, the presumptive heart regions remain in apposition to the pharyngeal endoderm and come to lie at the anterior of the embryo, lateral to the future hindbrain region of the neural plate and beneath the otic ectoderm (placode). This arrangement is disrupted during neural tube formation as the overlying ectoderm moves medially while the two heart rudiments simultaneously move toward the ventral midline. By the tailbud stage, the heart rudiments lie on the ventral side of the embryo adjacent to endoderm of the anterior archenteron floor, whereas the endoderm with which they were previously associated shows little relative movement and forms the dorsolateral walls of the foregut (Wilens, 1955).

As the anterior, lateral mesoderm approaches the ventral midline, it divides into a thick, somatopleural layer, separated from a thin splanchnopleure. In the heart region, the former will give rise to the ventral lining of the pericardium, whereas the latter forms the endocardium and myocardium. Prior to fusion of the two heart rudiments, a population of cells separates from the medial borders of the splanchnopleure and spread between the endoderm and the splanchnic mesoderm. These become the endocardial cells of the heart and their mesodermal origin is confirmed by vital dye-mapping experiments (Wilens, 1955). The pericardial cavity between the two mesodermal layers becomes enlarged and the ventral borders of the splanchnopleure fuse to form the ventral mesocardium. The central mass of endocardial cells then becomes organized into a tube structure which is progressively enclosed by dorsal fusion of the myocardial layers. Fusion occurs in a cephalocaudal direction and is accompanied by the onset of contractions, which commence in the anterior region of the heart tube but extend to regions in which dorsal fusion of the myocardium has not yet occurred (Copenhaver, 1939a).

The simple linear heart tube at this stage encompasses the bulbus, ventricular, atrial, and sinus regions in

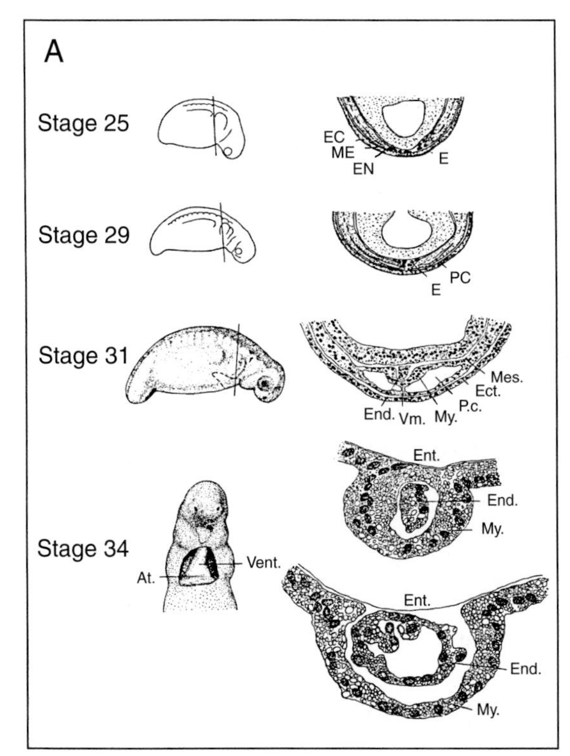

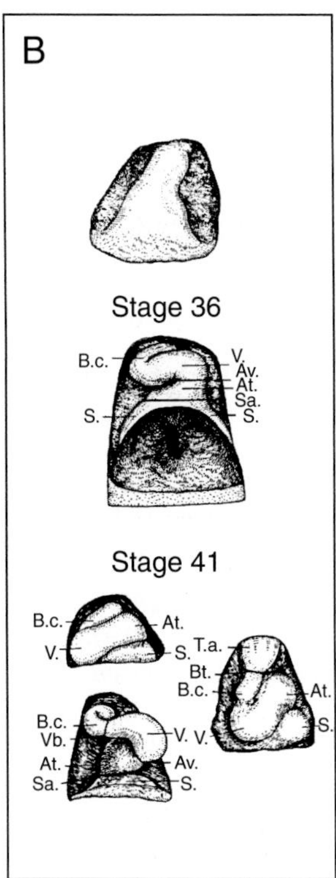

Figure 1 (A) Sections through the heart-forming region from the appearance of endocardial precursors, just prior to ventral fusion of the heart mesoderm (stage 25), to the formation of a linear heart tube (stage 34). Cross sections through the ventricular (upper) and atrial (lower) regions of the heart tube are shown. E, endocardial rudiment; EC and Ect, ectoderm; EN and Ent, endoderm; End, endocardium; ME and Mes, mesoderm; My, myocardium; Pc, pericardial cavity; Vm, ventral myocardium. (B) Heart tube looping in the axolotl. The linear tube (stage 34), which initially shows a clear leftward inclination, loops to the right in the future ventricular region (stage 36). By stage 41, the sinus venosus (S), atrium (At), ventricle (V), bulbus cordis (Bc), and truncus arteriosus (Ta) can be distinguished, separated by the sinoatrial (Sa), atrioventricular (Av), and bulbotruncus (Bt) junctions. This is seen most clearly if the rightward loop of the ventricular region is lifted to the left (stage 41) [drawings of the heart-forming region in embryos of the axolotl, *Ambystoma punctatum,* by Copenhaver (1926, 1939a,b), based on a compilation shown in Rugh (1962)].

linear cephalocaudal sequence which become progressively more delineated. Contractions are initially restricted to the bulbus and ventricular regions comprising the cephalic or anterior two-thirds of the tube. Atrial contractions commence immediately prior to the onset of circulation. At its caudal (posterior) end, the endocardium broadens and separates into left and right halves of the sinus venosus rudiment which extend around the anterior border of the liver, joining the vitelline veins and Cuverian ducts. Contractions in the sinus venosus are detected last, after circulation has commenced.

Left and right heart rudiments make an equal contribution to the heart tube, as revealed by the ordered restriction of vital dye staining applied to one rudiment at the onset of neurulation and maintained in the same half of the heart tube in the tailbud embryo (Wilens, 1955). However, after circulation has commenced, dye label from the left side of the tube shifts rightward across the midline. This is most pronounced in the ventricular region, which is almost entirely stained by tissue from the left rudiment. The explanation for this lies in the onset of heart looping; the ventricular region of the linear tube first shifts to the right and rotates to move the left side of the ventricle over to the right. Simultaneous expansion of the ventricular region results in a characteristic S-shaped heart. The sinus lies posterior to the atrium, which connects by a broad opening at its posterior end to the ventricle. The ventricle lies ventral and largely posterior to the atrium, on the right side of the embryo. At its anterior end, the ventricle opens to the conus, which extends anterior and leftward to the bulbus, from which the aortic arches arise (Fales, 1946).

At this stage of development, the heart is uniformly thin walled and is relatively broad in cross section but short along the anteroposterior axis. In subsequent tadpole stages, the sinus endocardium becomes completely surrounded by myocardial tissue and forms a definite chamber, the separate chambers become more distinct as the sinoatrial, atrioventricular, and ventriculobulbus junctions become delineated, and the heart extends along the anteroposterior axis. As looping continues, heart size increases threefold and the ventricle wall becomes thickened and trabeculated. Shortly after the onset of feeding, the tadpole heart has achieved its final form (Copenhaver, 1939a; Fales, 1946).

B. Heart Formation in the Anuran, *Xenopus*

The origins and location of heart-forming tissue in the early *Xenopus* embryo are well established from modern fate-mapping studies (Keller, 1975, 1976). While the precise topographies of early gastrula are somewhat different between *Ambystoma* and *Xenopus,* reflecting differences in gastrulation, the relative arrangement of heart precursors and adjacent tissue is broadly similar.

In *Xenopus,* unlike other amphibians, mesoderm originates entirely from within the deep layer of the marginal zone during gastrulation. Fate mapping with vital dyes has established that the pair of heart precursors lie at the anterior edge of the involuted mesoderm in the early gastrula, flanked medially by the chordamesoderm (presumptive notochord) and laterally by somitic mesoderm. They therefore lie beneath the suprablastoporal, superficial layer (which forms the endodermal archenteron roof), and the preinvolution mesoderm of the deep layer which will occupy more posterior positions in the neurula embryo (Keller, 1976). A similar location is suggested by explant studies in which regions of marginal zone were tested for their ability to give heart differentiation in culture (Sater and Jacobson, 1989; Nascone and Mercola, 1995). Projected onto the surface of the early gastrula, these regions lie above the dorsal lip, minimally estimated as spanning 15° on either side of a dorsomedial 60° segment, generally considered to define the Spemann "organizer." In this location, the heart precursors are intimately associated with two regions of endodermal tissue which they retain throughout gastrulation: a region of deep endoderm that is driven anteriorly by gastrulation movements and the dorsolateral bottle cells which are pulled along by the involuting mesoderm. The precise fate of deep endoderm is unclear, but it probably contributes to pharyngeal and branchial regions (Sater and Jacobson, 1989), whereas the dorsal and dorsolateral bottle cells give rise to endoderm of the pharynx (Keller, 1981; Gerhart and Keller, 1986; Sater and Jacobson, 1989). Throughout gastrulation the heart mesoderm appears as a loose crawling population rather than a cohesive sheet (Gerhart and Keller, 1986). At the onset of neurulation, it has come to lie adjacent to the anterolateral edges of the neural plate. By the early tailbud stage, the heart mesoderm has migrated to the ventral midline. Terminal differentiation commences shortly afterwards, as judged by cardiac muscle-specific gene expression (Logan and Mohun, 1993; Chambers *et al.,* 1994; Drysdale *et al.,* 1994), followed shortly by heart tube formation and contraction. Subsequent steps in heart tube looping and chamber formation have been detailed by Nieuwkoop and Faber (1956) and are similar to the course of axolotl heart formation (Fig. 2). By feeding stage, the myocardium of the ventricle has become trabeculated and the single atrium has divided, yielding the three-chambered heart characteristic of amphibians.

III. Specification of the Amphibian Heart Field

Explants of amphibian embryos will survive and continue to develop without the need for complex nutrients or growth factors in the culture medium. This convenient property has been exploited throughout this century to study heart formation. Early studies [most notably of Ekman (1921, 1924), Stohr (1924), and Copenhaver (1926)] demonstrated that explants of the ventral, heart-forming region from tailbud embryos from urodele and anuran species could differentiate into heart structures. Differentiation occurred within ectodermal jackets, usually derived from the original explanted fragment, producing ectodermal vesicles within which a beating heart tube with distinct chambers and looped morphology could be observed. Similar results were also obtained with explants from younger embryos, although these showed a reduced frequency of heart formation. Inclusion of underlying endoderm invariably improved the extent of explant differentiation, as judged by a variety of criteria such as size, chamber formation, degree of looping, or length of culture before beating.

In principle, such results could indicate a progressive specification of cardiac fate in the mesodermal heart rudiments during neurula and tailbud stages. However, since the explants frequently included other tissue associated with the heart precursors, the results could also reflect the presence of heart-inducing capacity of adjacent tissue. Transplantation experiments by Copenhaver (1926) showed that at the tailbud stage, heterotopic grafts of heart mesoderm with its associated ectoderm formed hearts when moved to more posterior, ventral regions, demonstrating that the endoderm normally underlying the heart mesoderm in urodele embryos was unnecessary for its differentiation. Taken together with the explant studies, this result suggested that the heart mesoderm of the tailbud embryo was al-

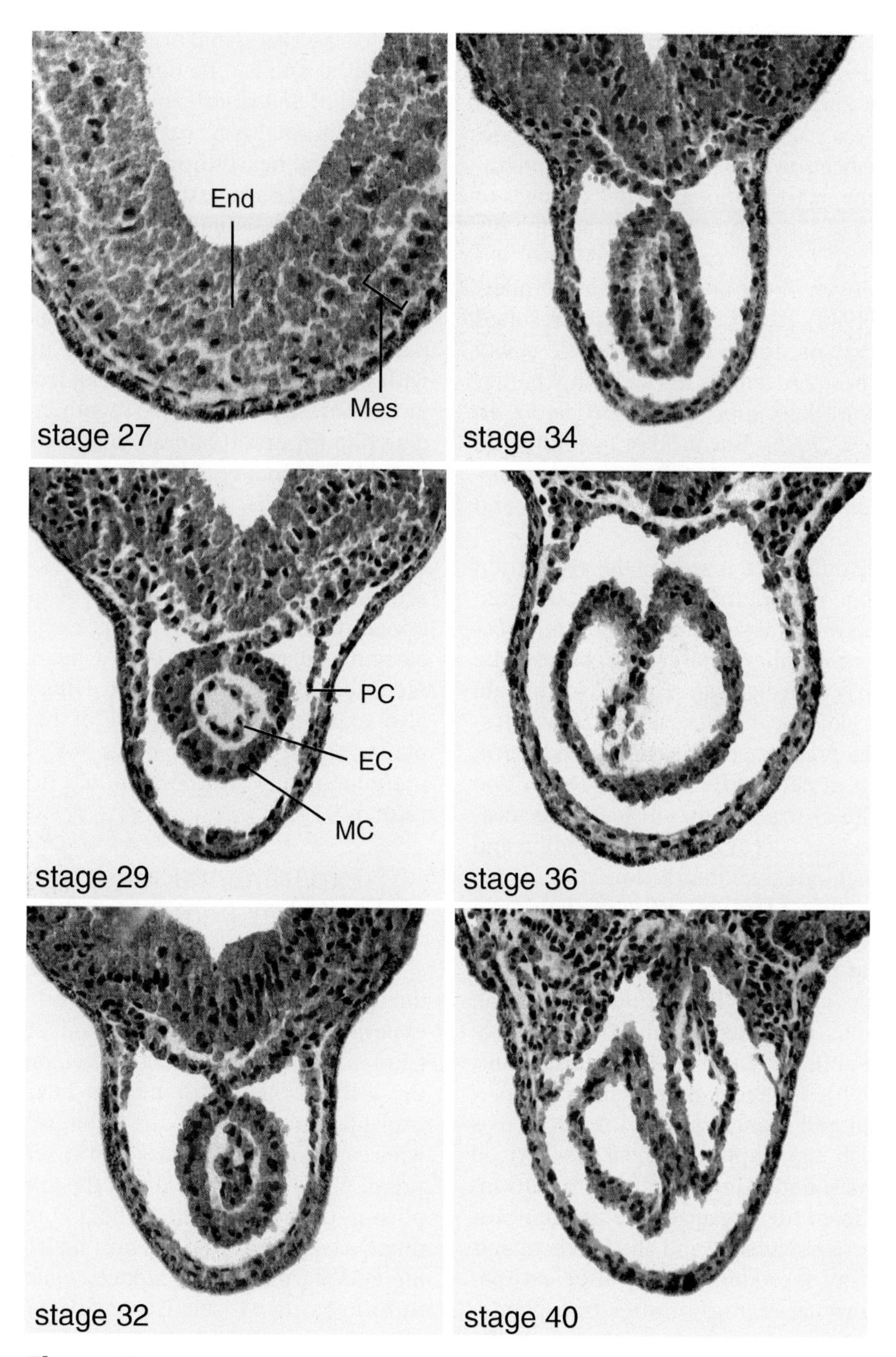

Figure 2 Transverse sections through the anterior ventral region of *Xenopus laevis* embryos. At tailbud stage 27, the bilateral mesodermal precursors of the heart can be seen between the endoderm and ectoderm layers prior to fusion. By stage 29, a linear heart tube has formed, comprising myocardial (MC) and endocardial (EC) layers enclosed within the pericardium (PC). The dorsal mesocardium is more evident in later sections (stages 32 and 34). By stage 36, the heart tube adopts a tight S shape as looping commences, and transverse sections therefore pass through the heart tube region twice. As looping continues, the future atrial and ventricular regions of the tube are more clearly separated (stage 40). (Note that the endocardial layer is poorly resolved in looping stages.)

ready specified to undergo cardiac differentiation. A similar capacity for self-differentiation was found in a small proportion of endoderm-free mesoderm explants from neurula embryos as early as the neural plate stage, indicating that specification may precede ventral migration and fusion of the heart rudiments. After changes to the salt composition of the culture medium, beating tissue was also obtained from a small proportion of explants at this stage, even in the absence of any ectodermal jacket (Bacon, 1945). Better differentiation resulted if the tissue was implanted into the gastrocoele cavity of tailbud embryo hosts. Attempts to establish whether heart mesoderm of urodele embryos was specified even earlier (Copenhaver, 1926; Bacon, 1945) foundered on the difficulty of removing all endoderm from gastrula explants (Chuang and Tseng, 1957; Jacobson and Duncan, 1968).

In *Xenopus,* little difference is seen in the proportion of neurula and tailbud mesoderm explants that undergo heart differentiation in the absence of endoderm. Specification of each heart rudiment therefore appears to be complete by the early neurula stage. Explants from late gastrulae will also yield beating tissue at a lower frequency, although the presence of pharyngeal endoderm improves the extent of heart differentiation (Sater and Jacobson, 1989). The critical events of heart specification in *Xenopus* therefore occur during gastrulation and comprise two distinct inductive interactions.

Embryos lacking the organizer region of the dorsal marginal zone during early gastrula stages fail to form heart tissue, despite the presence of presumptive cardiac mesoderm. In contrast, early gastrulae lacking the heart-forming regions but retaining the organizer region retained some ability to form beating hearts (Sater and Jacobson, 1990b). Classical organizer grafts pioneered by Spemann and Mangold (1924) result in axis duplication in which the duplicated heart is derived from host ventral mesoderm. Together, these results indicate that signals from the organizer are essential for specification of heart precursors and that such signals can respecify adjacent mesoderm either after extirpation of the heart-forming regions or after transplantation into the ventral side of the gastrula (Sater and Jacobson, 1990b). Signaling by the organizer, however, is required only early during gastrulation since its removal in mid- and late gastrula stages has no effect on heart formation.

Large explants of dorsal marginal zone from early gastrulae of *Xenopus* yield beating tissue with only low frequency despite including both organizer and prospective heart mesoderm tissue (Sater and Jacobson, 1989). The explanation for the result lies in the existence of a second signal which is also required before heart mesoderm becomes specified (Nascone and Mercola, 1995). This signal originates from the anterior deep endoderm and can be demonstrated in two ways. First, removal of anterior deep endoderm from early gastrula dorsal marginal zone explants dramatically reduces the frequency of heart formation. Second, respecification of ventral mesoderm explants by culture in association with organizer tissue is successful only in the presence of anterior deep endoderm. These experiments lead to a model of heart specification for *Xenopus* in which signals from the organizer, responsible for dorsoventral patterning of the newly forming mesoderm, combine with inductive signals emanating from immediately adjacent anterior endoderm. Together, they specify a cardiac fate for anterolateral regions of dorsal mesoderm. A similar model could also apply to urodeles, with the qualification that endodermal signaling appears more protracted though development and specification is not complete until much later in development. This would account for the low frequency of heart formation obtained with heart mesoderm explants from urodele early neurulae (Bacon, 1945; Chuang and Tseng, 1957; Jacobson, 1960; Jacobson and Duncan, 1968). It would also explain the association of heart formation and pharynx differentiation noted for explants of gastrula marginal zone cultured with anterior endoderm tissue (Amano, 1958).

C. Experimental Induction of Hearts in Other Regions of Mesoderm

There are several reports describing experimental induction of hearts from mesoderm that might not have experienced both organizer and endodermal signals. Ekman (1925) found that *Bombinator* mesoderm from the gill region contributed to heart formation when transplanted between the bilateral heart primordia, whereas more caudal mesoderm remained unincorporated. While it is possible that the graft tissue included a portion of the donor heart field, it may also be that inductive signaling from the host heart fields, the underlying endoderm, or both induced cells of the graft tissue to form cardiac tissue. Bacon (1945) reported a similar outcome using *Ambystoma* early gastrula donors to provide grafts of presumptive mesoderm from the marginal zone. Grafts were chosen to avoid the regions of presumptive heart mesoderm but may have included endodermal precursors. They nevertheless gave rise to hearts when used to replace the entire fused heart field of host embryos. Newly ingressed dorsolateral mesoderm (presumptive somite or pronophros) from late gastrulae was also tested but proved to be unaffected by the transplantation. Instead, such grafts differentiated independently to form somitic muscle or pronephric tubules in their ectopic location.

The induction of heart tissue from regions of marginal zone that exclude the presumptive heart mesoderm has also been reported by Amano (1958). Explants of marginal zone from the early gastrula of *Triturus* could be induced to form heart tissue by blastoporal presumptive endoderm, irrespective of their original location in the embryo. Indeed, the frequency of cardiac differentiation was greatest with explants of the ventral marginal zone. It is difficult to reconcile these examples of heart induction with the simple two-signal model proposed from studies of *Xenopus*. At the least, they deserve reexamination and they serve to emphasize the provisional character of the current model.

IV. Regulative Properties of the Heart Field

The amphibian heart field provides a dramatic example of regulative behavior by embryonic tissue in response to pertubations. Extirpation of the fused heart rudiment in early tailbud embryos of both urodeles and anurans has long been known to result in regulative replacement by adjacent lateral mesodermal tissue and the formation of either single or bilateral ectopic hearts (Ekman, 1921, 1925; Copenhaver, 1926). Indeed, only by removing both the ventral heart rudiment and a considerable portion of adjacent anterior lateral tissue (e.g., by replacing it with posterior ventral graft tissue) can heart formation be prevented (Copenhaver, 1926).

Left and right sides of the heart field each have the capacity for heart differentiation. Hence, if the two heart rudiments are prevented from fusing on the ventral midline (either by a physical block or by heterotopic graft tissue), a pair of hearts will result (Ekman, 1925; Copenhaver, 1926). Even after fusion of the left and right primordia, physical splitting of the field can result in the formation of multiple hearts. In one such study, Ekman (1927) obtained as many as five hearts by repeated subdivision of a single heart field. Heart-forming ability therefore extends throughout the heart field, encompassing tissue that would not undergo cardiac differentiation in normal development. Only relatively small regions of the field are necessary to obtain an entire heart tube.

Selective extirpation of either left or right sides of the field prior to fusion reveals a notable asymmetry in the ability of the remaining side to differentiate. While left sides develop into reasonably well-formed hearts, right sides differentiate only poorly (Copenhaver, 1926). Does this reflect inherent differences between the tissue comprising the left and right sides of the field or is it due to the sidedness of their position? Heart mesoderm from the left side of urodele embryos has been reported to differentiate more completely in explant culture than that from the right side (Goertller, 1928), suggesting that differences may be inherent to the tissues. Similar differences have also been reported in the development of left and right twinned hearts, produced when fusion of the two fields is prevented. However, embryos in which the left field is replaced by a second right field prior to fusion (or vice versa) subsequently form a normal, ventrally located heart (Copenhaver, 1926), suggesting that left and right sides possess equivalent heart-forming potency. Differences in the development of twinned hearts have also been attributed to mechanical effects of the field-splitting procedure rather than differences inherent to each side of the heart field (Fales, 1946).

The regulative behavior of the heart field is not only observed in response to reductions or divisions in the heart field but also has been found after experimental enlargement of the field. Thus, if the anterior ventral mesoderm of a urodele tailbud embryo is grafted on top of the same tissue in a second, a single heart usually results (Copenhaver, 1926). While this is initially larger than usual, it subsequently becomes more normal in size.

A. Polarity of the Heart Field

In chick embryos, lineage labeling studies have shown that ventricular and atrial precursors lie in distinct regions of the lateral heart fields and their anteroposterior (AP) pattern corresponds to that of the fused, linear heart tube. While equivalent fate-mapping studies are probably not feasible with amphibian embryos, early grafting experiments do provide evidence of a similar AP patterning in the heart field. Thus, 180° rotation of the fused heart field about its AP axis reverses the subsequent orientation of the heart tube (Stohr, 1925; Copenhaver, 1926; Ekman, 1929) and the same result can be obtained in some species prior to ventral fusion. Similarly, heterotopic grafting of the fused heart field to more posterior ventral locations yields ectopic hearts which develop according to the AP axis of the graft rather than that of the host embryo (Copenhaver, 1926). Finally, in experiments to enlarge the heart field by superimposing a second heart field on that of the host, reversal of the graft field resulted in independent differentiation of a second heart rather than incorporation into the host heart.

These experiments demonstrate AP polarity of the heart field before fusion of the left and right heart rudiments but do not indicate when this polarity is established (Copenhaver, 1955). The relatively narrow width of the prospective heart rudiments during neurula stages and the absence of early cardiac lineage markers currently preclude any experimental approaches to this problem other than the use of graft reversals.

The existence of AP polarity prior to differentiation does not preclude regulation along this axis of the heart field, either before or even after the heart tube has formed. In urodeles, complete hearts can develop, despite extirpation of the anterior half of the fused heart field. They can also be obtained from heart fields comprising two anterior or two posterior halves, indicating that regulative ability is present in both regions. Even after the linear heart tube has formed, relatively normal hearts develop after local ablation of regions within the heart tube (Copenhaver, 1926). In this case, regeneration of the ablated tissue presumably results from local proliferation within the heart tube rather than regulative reprogramming of noncardiac tissue. Nevertheless, it demonstrates that the signals which establish or maintain AP patterning of the heart field continue during heart tube differentiation.

Recent molecular studies have identified differences in gene expression between dorsal and ventral sides of the heart tube (Biben and Harvey, 1997; see Chapters 7 and 9) and these may well be important in the process of heart looping (see Chapter 22). However, from embryological studies there is little evidence for prior dorsoventral polarity in the amphibian heart field (Copenhaver, 1955).

V. Cardiac Differentiation and the Heart Field

Anterior lateral mesoderm flanking the heart rudiment contributes to a variety of tissues during normal development, including mesenchyme, mesothelium, and pronephros. Its capacity to form heart tissue through regulative reprogramming prompts a question central to our understanding of heart differentiation: What signals restrict heart formation to the most ventral portion of the heart field during normal development? The relatively broad size of the heart field that emerges from gastrulation presumably reflects the extent of initial inductive signaling specifying a cardiac fate to newly formed mesoderm. Consistent with this, expression of the tinman homologs, XNkx2.3 and XNkx2.5, is detected in broad domains of the early neurula that coincide with the bilateral heart fields (see Chapter 7). However, since much of this tissue does not normally undergo cardiac differentiation, it must be exposed to other signals within the embryo that divert it to other fates. These cannot originate from endoderm or ectoderm immediately adjacent to mesoderm of the heart field because such tissue can be included in explants without inhibiting heart differentiation. On the contrary, a stimulatory role for anterior endoderm is well established and there is some evidence for a similar but nonspecific effect of epidermis (Jacobson, 1960; Jacobson and Duncan, 1968).

Normal heart formation is restricted to the most ventral region of the heart field, and after removal of the heart rudiment, ectopic heart differentiation is also generally restricted to the most ventral edge of the lateral mesoderm. This suggests that inhibitory signals may be dorsoventral in character. One possibility is that they originate from the heart rudiment itself, with ablation of this tissue thereby relieving adjacent mesoderm from its inhibitory influence. However, in explant culture, combinations of anterior ventral and lateral mesoderm from *Xenopus* tailbud embryos both make substantial contributions to the resulting heart tube (T. Mohun and M. Logan, unpublished observations), and there is little direct evidence in support of this proposal.

An alternative model is that inhibitory signals originate from dorsal, axial structures of the embryo and there is some evidence from urodele embryos to support this view. In explant cultures, inclusion of neural plate or neural fold tissue suppresses the capacity of either whole embryo mesoderm or anterior lateral mesoderm to form beating hearts (Chuang and Tseng, 1957). This effect can be obtained with anterior or posterior neural tissue, although the former has a more potent effect (Jacobson, 1960). Conflicting results have been reported for the effect of notochord in such assays (see Jacobson, 1961).

While these experiments clearly demonstrate an inhibitory role for neural tissue, it seems unlikely that this is sufficient to account for ventrally restricted cardiac differentiation within the heart field. In heart ablation experiments, ectopic hearts tend to form at the ventralmost edge of the lateral mesoderm, but if wound closure is poor, this constitutes a much more dorsal position than normal. Furthermore, grafting of neural plate tissue into the presumptive heart region of urodele early tailbud embryos does not prevent heart formation on either side of the graft (Jacobson, 1961). These observations point to the presence of other influences in the embryo which can counteract the inhibitory effect of neural tissue. Anterior endoderm has long been considered a source of such an influence.

A. The Influence of Endoderm on Heart Morphogenesis

Explants of the heart field from neurula embryos differentiate relatively poorly in the absence of underlying endoderm, as judged by their morphology, the length of culture before heartbeat, or the proportion showing beating structures (Stohr, 1924; Chuang and Tseng, 1957; Jacobson, 1960; Jacobson and Duncan, 1968; Sater and Jacobson, 1990a). In the case of urodeles, such results could reflect the relatively prolonged period of endodermal signaling required for cardiac specification. However, even with explants from tailbud embryos, the "qualitative" nature of differentiation is improved by the presence of endoderm. Similar effects can be seen with explants from *Xenopus* embryos, in which the role of deep anterior endoderm in heart specification is restricted to the period of gastrulation. This suggests that the endoderm has a second and more protracted role in which it acts as a stimulatory influence on the progress of heart morphogenesis. Recent studies of avian cardiogenesis point to a similar conclusion: an endodermal influence is critical for myofibrillogenesis and the acquisition of contractility (Nascone and Mercola, 1996; see Chapter 4).

While much of the evidence for the "formative influence" of the endoderm has come from the use of cultured explants, support for this view has also come from embryo ablation experiments. Several studies have shown that if the entire endoderm is removed from early embryos, beating hearts fail to form (Balinsky, 1939; Nieuwkoop, 1946; Chuang and Tseng, 1957; Jacobson, 1960, 1961; Jacobson and Duncan, 1968; Nascone and Mercola, 1995). In *Xenopus,* the effectiveness of this operation is restricted to early gastrula stages (Nascone and Mercola, 1995). In urodeles the proportion of endodermless embryos producing a beating heart increases with the stage at which the ablation is performed, and by the onset of ventral fusion, the operation has no effect on heart formation (Jacobson, 1961).

The interpretation of these experiments is complicated by two factors. First, ablation of the endoderm during a critical phase of heart mesoderm specification may be expected to block heart formation since no heart field will be established. This accounts for the absence of hearts after removal of endoderm at the gastrula stage (Jacobson and Duncan, 1968; Nascone and Mercola, 1995). It may also explain the same outcome with urodele neurula embryos. Second, the absence of hearts after endoderm ablation in older urodele embryos may be due, at least in part, to the inhibitory influence of th neural tissue in the embryo. Thus, simultaneous removal of the entire neural plate and folds along with the endoderm restores heart differentiation (Mangold, 1954; Jacobson, 1961). In urodeles, therefore, the combined results of explant and embryo ablation experiments indicate that the heart fields are subject to antagonistic influences of the endoderm and neural tissue (Chuang and Tseng, 1957).

Endodermless embryos have also been used in an assay to test the stimulatory effect of different regions of urodele endoderm (Jacobson, 1961; Jacobson and Duncan, 1968). These experiments show that the ability to stimulate heart morphogenesis is restricted to a broad anterior domain (Jacobson, 1961; Jacobson and Duncan, 1968). Explant assays have provided a more precise localization, indicating that the most powerful effects are associated with endoderm directly underlying the heart fields at the neural plate stage (Fullilove, 1970). In *Xenopus,* the stimulatory activity is also restricted to anterior endoderm and can be conferred on more posterior ventral endoderm by ectopic expression of anterior endoderm RNA (Tonegawa *et al.,* 1996).

B. The Axolotl *c/c* Mutation

In the axolotl, *Ambystoma mexicanum,* a recessive lethal mutation (*c*) causes incomplete differentiation of the myocardium. The mutation follows simple Mendelian laws and in homozygotes, the embryonic heart fails to beat and the tadpoles become edemic and microcephalic (Humphrey, 1972). Mutant heart tissue contains apparently normal levels of most contractile proteins with the exception of tropomyosin (Erginel-Unaltuna *et al.,* 1995), indicating that much of the normal cardiac differentiation program is unimpaired. However, the sarcomeric proteins are not organized into sarcomeric arrays (Hill and Lemanski, 1979; Erginel-Unaltuna and Lemanski, 1994; Lemanski *et al.,* 1997).

Such a phenotype indicates the the *c* mutation affects myofibrillogenesis during tailbud stages rather than the earlier events of heart mesoderm specification or cardiac differentiation (Smith and Armstrong, 1991). The absence of beating tissue in the c/c mutant could result from a failure in the endodermal signaling that stimulates heart morphogenesis, or it could result from failure of the mesoderm to respond. A number of studies have therefore attempted to establish which tissue is affected in the c/c mutant.

Transplantion of mutant heart primordium into normal hosts at the tailbud stage rescues beating in the mutant organ (Humphrey, 1972). In reciprocal experiments, wild-type primordia are unable to form beating tissue in a mutant embryo. In explant culture, the rescuing activity of wild-type endoderm is restricted to anterior tissue (Lemanski *et al.,* 1979), as is its ability to stimulate morphogenesis in normal heart mesoderm. Myofibrillar organization in *cc* heart mesoderm can also be obtained with media conditioned by anterior endoderm (Davis and Lemanski, 1987) or total RNA extracted from the tissue (LaFrance *et al.,* 1993). These suggest that the *c* mutation affects some aspect of signaling by the endoderm (Lemanski *et al.,* 1995).

However, explants of wild-type heart field from early neurula (i.e., prior to specification) can be induced to beat by coculture with either wild-type or c/c endoderm. In this assay, mutant endoderm therefore retains inducing capacity. In addition, explants of c/c heart field from the same early stage cannot be rescued by wild-type anterior endodermal RNA, unlike the older c/c heart rudiments used in earlier experiments (Smith and Armstrong, 1991). To reconcile such conflicting results, Armstrong and co-workers have proposed that specification of heart mesoderm results in the appearance of both an activator and an inhibitor, in opposing gradients throughout the heart field (Smith and Armstrong, 1991, 1993; Holloway *et al.,* 1994). Myofibrillogenesis is promoted by a high ratio of activator to inhibitor, whereas a low ratio in lateral regions prevents them from contributing to the final heart. In this model, the *c* mutation results in overproduction of inhibitor and/or underproduction of activator.

Whether the lesion blocking heart morphogenesis in the c/c mutant affects the endoderm or mesoderm, it is worth noting that processes other than myofibrillogenesis are affected. Thus, mutant embryos rescued by transplantation of wild-type heart mesoderm remain unable to feed and also suffer from heart valve defects (Smith and Armstrong, 1993).

VI. Restriction of the Heart Field during Development

The ability of anterior lateral mesoderm to undergo regulative redirection to a cardiac fate does not persist beyond heart tube formation (Ekman, 1921; Copenhaver, 1926). During tailbud stages, the heart field therefore becomes progressively smaller until it comprises the ventral mesoderm of the heart rudiment. This restriction has been carefully documented in *Xenopus* (Sater and Jacobson, 1990a) and is complete by the onset of terminal differentiation (Logan and Mohun, 1993; Chambers *et al.,* 1994; Drysdale *et al.,* 1994). Since cells of the anterior lateral mesoderm do not move more ventrally during tailbud stages, restriction of the heart field therefore reflects changes within this tissue rather than its relocation (Sater and Jacobson, 1990a).

What is the nature of these changes? In the absence of molecular markers, heart-forming ability of explants has always been assessed using morphological criteria, with restriction of the heart field being identified as a decline in the capacity of lateral mesoderm explants to form a heart tube. Since this is the outcome of a complex multistep process, requiring both terminal differentiation and morphogenetic changes, its failure could be caused in a number of different ways. One possibility is that it reflects a loss of the capacity by anterior lateral mesoderm to initiate terminal cardiac differentiation. However, restriction of the heart field does not appear to be accompanied by a comparable restriction in the expression domains of the tinman homologs, XNkx2.3 and XNkx2.5 (Tonissen *et al.,* 1994; Evans *et al.,* 1995). Furthermore, lateral explants that have lost their heart-forming capacity still activate expression of cardiac muscle-specific markers (T. Mohun and M. Logan, unpublished observations). These results suggest that, as in avian embryos, heart morphogenesis in amphibians can be distinguished from prior steps of cardiac differentiation. Restriction of the heart field may therefore result from a failure in the capacity of anterior endoderm to stimulate heart morphogenesis, through either a loss of signaling capacity by the endoderm or a loss of responsiveness by anterolateral mesoderm. In heterochronic explant combinations, heart-forming ability of older lateral mesoderm remains poor even in the presence of younger anterolateral endoderm (Sater and Jacobson, 1990a). This is consistent with a loss of competence by the lateral mesoderm to respond to stimulatory, endodermal signals.

VII. Conclusions

The embryological studies we have reviewed point to three main conclusions. First, the initial steps in establishing cardiac precursors occur during gastrulation, when regions of anterior mesoderm receive signals from the organizer and from adjacent anterior endoderm. It seems likely, although unproven, that the former defines the anterodorsal character of the heart mesoderm, whereas the latter imparts cardiac specificity. Molecular candidates for the organizer signal(s) include members of the fibroblast growth factor and transforming growth factor-β (TGF-β) families, along with several other proteins expressed in the early gastrula embryo that have mesoderm-inducing or patterning activity (Smith, 1995; Heasman, 1997). Much less is known about the nature of the anterior endodermal signal, but recent studies suggest two possible components. In *Xenopus,* the secreted protein, Cerberus, is localized in the anterior endoderm in the gastrula embryo and can induce expression of tinman homologs in explants of presumptive ectoderm (Bouwmeester *et al.,* 1996). The homeobox protein, Hex (Newman *et al.,* 1997) is also localized to this region and can induce ectopic Cerberus expression in the embryo (M. Jones, personal communication).

The combined effect of organizer and endoderm signaling can be mimicked in explants of presumptive ectoderm by high doses of the TGF-β, activin A (Logan and Mohun, 1993), and in the embryo these signals together specify the bilateral heart fields. The operational nature of this term is graphically illustrated by the fact that only part of each heart field actually forms the heart, whereas the entire region will undergo cardiac differentiation in explant culture. At the molecular level, the bilateral heart fields are marked by expression of the tinman homologs. While these proteins probably function as regulators of the cardiac-specific transcription (see Chapter 7), ectopic expression outside of the heart field does not cause secondary sites of cardiac differentiation (Cleaver *et al.,* 1996). For this reason, their expression in the heart field is probably insufficient to account for cardiac specification. Their role in regulating cardiac-specific gene expression probably depends on interaction with other transcription factors and these must be present throughout the heart field to allow its differentiation in explant culture.

A second conclusion is that signals from anterior endoderm play an important role in facilitating heart morphogenesis (see Chapter 4). The distinction between this and the prior role of the endoderm in cardiac specification has sometimes become blurred. In part, this has occurred because heart specification is a relatively protracted process in the urodele embryos favored for embryological study. It is also a consequence of the limited morphological assays that were originally available for scoring heart differentiation. For the same reason, it is unclear whether the signals stimulating myofibrillogenesis are necessary prior to the onset of cardiac differentiation. Nor do we know if heart morphogenesis requires their maintenance after terminal differentiation has commenced. A progressive restriction occurs in the heart-forming capacity of tissue flanking the prospective heart rudiment, probably as a result of loss of responsiveness to endodermal signaling. This, however, should be distinguished from any change in the ability of anterolateral mesoderm to initiate cardiac differentiation in explant culture, which has yet to be studied.

Finally, embryo manipulations have demonstrated that the heart field is initially patterned in a dynamic manner. Both cardiac differentiation and morphogenesis are restricted to the anteroventral region of the field within the embryo, even after alteration of the field size. The patterning that underlies this regulative behavior originates at least in part from outside of the heart field since explants of the tissue are freed from such constraints. Inhibition from neural tissue plays some part in this patterning and a role for similar signals within the mesoderm has also been proposed. In a similar manner, the heart field is polarized along both anteroposterior and dorsoventral axes, but for a time, both can be redefined after tissue reversal. Differentiation of cardiac tissue from within the heart field is therefore regulated by a number of signals from different tissues in the embryo. Identifying their molecular basis will be central to understanding cardiogenesis.

References

Amano, H. (1958). Some experiments on heart development. *Doshisha Kogaku Kaishi* **8,** 203–207.

Bacon, R. L. (1945). Self-differentiation and induction in the heart of *Amblystoma. J. Exp. Zool.* **91,** 87–121.

Balinsky, B. I. (1939). Experiments on total extirpation of the whole endoderm in *Triton* embryos. *C. Rend. Acad. Sci. URSS* **23,** 196–198.

Biben, C., and Harvey, R. P. (1997). Homeodomain factor Nkx2-5 controls left/right asymmetric expression of bHLH gene eHand during murine heart development. *Genes Dev.* **11,** 1357–1369.

Bouwmeester, T., Kim, S., Sasai, Y., Lu, B., and De Robertis, E. M. (1996). Cerberus is a head-inducing secreted factor expressed in the anterior endoderm of Spemann's organizer. *Nature (London)* **382,** 595–601.

Chambers, A. E., Logan, M., Kotecha, S., Towers, N., Sparrow, D., and Mohun, T. J. (1994). The RSRF/MEF2 protein SL1 regulates cardiac muscle specific transcription of a myosin light-chain gene in *Xenopus* embryos. *Genes Dev.* **8,** 1324–1334.

Chuang, H. H., and Tseng, M. P. (1957). An experimental analysis of the determination and differentiation of the mesodermal structures of neurula in urodeles. *Sci. Sin.* **6,** 669–708.

Cleaver, O. B., Patterson, K. D., and Krieg, P. A. (1996). Overexpression of the tinman-related genes XNkx-2.5 and XNkx-2.3 in *Xenopus* embryos results in myocardial hyperplasia. *Development (Cambridge, UK)* **122,** 3549–3556.

Cleine, J. C., and Slack, J. M. W. (1985). Normal fates and states of specification of different regions in the axolotl gastrula. *J. Embryol. Exp. Morph.* **86,** 247–269.

Copenhaver, W. M. (1926). Experiments on the development of the heart of *Amblystoma punctatum. J. Exp. Zool.* **43,** 321–371.

Copenhaver, W. M. (1930). Results of heteroplastic transplantation of anterior and posterior parts of the heart rudiment in *Amblystoma* embryos. *J. Exp. Zool.* **55,** 293–318.

Copenhaver, W. M. (1939a). Initiation of beat and intrinsic contraction rates in different parts of the *Amblystoma* heart. *J. Exp. Zool.* **80,** 193–224.

Copenhaver, W. M. (1939b). Some observations on the growth and function of heteroplastic heart grafts. *J. Exp. Zool.* **82,** 239–272.

Copenhaver, W. M. (1955). Heart, blood vessels, blood, and entodermal derivatives. *In* "Analysis of Development" (B. H. Wilier, P. Weiss, and V. Hamburger, eds.), pp. 440–461. Saunders, Philadelphia.

Davis, L. A., and Lemanski, L. F. (1987). Induction of myofibrillogenesis in cardiac lethal mutant axolotl hearts rescued by RNA derived from normal endoderm. *Development (Cambridge, UK)* **99,** 145–154.

Drysdale, T. A., Tonissen, K. F., Patterson, K. D., Crawford, M. J., and Krieg, P. A. (1994). Cardiac troponin I is a heart-specific marker in the *Xenopus* embryo: Expression during abnormal heart morphogenesis. *Dev. Biol.* **165,** 432–441.

Ekman, G. (1921). Experimentelle beitrage zur entwicklung des Bombinator-herzens. *Oevers. Fin. Vetensk. Soci. Foerh. A* **63,** 1–37.

Ekman, G. (1924). Neue experimentelle Beitrage zur fruhesten entwicklung des amphibienherzens. *Commentat. Biol.* **1**(9), 1–37.

Ekman, G. (1925). Experimentelle beitrage zur herzentwicklung der amphibien. *Wilhelm Roux' Arch. Entwicklungsmech. Org.* **106,** 320–352.

Ekman, G. (1927). Einige experimentelle beitrage zur fruhesten hertzenwicklung bei *Rana fusca. Ann. Acad. Sci. Fenn., Ser. A* **27,** 1–26.

Ekman, G. (1929). Experimentelle untersuchungen uber die fruheste Herzentwicklung bei *Rana fusca. Wilhelm Roux' Arch. Entwicklungsmech. Org.* **116,** 327–347.

Erginel-Unaltuna, N., and Lemanski, L. F. (1994). Immunofluorescent studies on titin and myosin in developing hearts of normal and cardiac mutant axolotls. *J. Morph.* **222,** 19–32.

Erginel-Unaltuna, N., Dube, D. K., Salsbury, K. G., and Lemanski, L. F. (1995). Confocal microscopy of a newly identified protein associated with heart development in the Mexican axolotl. *Cell. Mol. Biol. Res.* **41,** 117–130.

Evans, S. M., Yan, W., Murillo, M. P., Ponce, J., and Papalopulu, N. (1995). Tinman, a *Drosophila* homeobox gene required for heart and visceral mesoderm specification, may be represented by a family of genes in vertebrates: XNkx-2.3, a second vertebrate homologue of tinman. *Development (Cambridge, UK)* **121,** 3889–3899.

Fales, D. (1946). A study of double hearts produced experimentally in embryos of *Amblystoma* punctatum. *J. Exp. Zool.* **101,** 281–298.

Fullilove, S. (1970). Heart induction: Distribution of active factors in newt endoderm. *J. Exp. Zool.* **175,** 323–326.

Gerhart, J., and Keller, R. (1986). Region-specific cell activities in amphibian gastrulation. *Annu. Rev. Cell Biol.* **2,** 201–229.

Goertller, K. (1928). Die bedeutung der ventrolateralen mesodermbezirke fur die herzenlage der amphibienkeime. *Anat. Anz., Ergaenzungsh.* **66,** 132–139.

Heasman, J. (1997). Patterning the *Xenopus* blastula. *Development (Cambridge, UK)* **124,** 4179–4191.

Hill, C. S., and Lemanski, L. F. (1979). Morphological studies on cardiac lethal mutant salamander hearts in organ cultures. *J. Exp. Zool.* **209,** 1–20.

Holloway, D. M., Harrison. L. G., and Armstrong, J. B. (1994). Computations of post-inductive dynamics in axolotl heart formation. *Dev. Dyn.* **200,** 242–256.

Holtfreter, J. (1939). Studien zur entwicklung der gestaltungsfaktoren in der orgenentwicklung der amphibien. *Wilhelm Roux' Arch. Entwicklungsmech. Org.* **139,** 227–273.

Humphrey, R. R. (1972). Genetic and experimental studies on a mutant gene (*c*) determining absence of heart action in embryos of the Mexican axolotl (*Ambystoma mexicanum*). *Dev. Biol.* **27,** 138–154.

Huxley, J. S., and De Beer, G. R. (1934). "The Elements of Experimental Embryology." Cambridge University Press, Cambridge, UK.

Jacobson, A. G. (1960). Influences of ectoderm and endoderm on heart differentiation in the newt. *Dev. Biol.* **2,** 138–154.

Jacobson, A. G. (1961). Heart differentiation in the newt. *J. Exp. Zool.* **146,** 139–151.

Jacobson, A. G., and Duncan, J. T. (1968). Heart induction in salamanders. *J. Exp. Zool.* **167,** 79–103.

Jacobson, A. G., and Sater, A. K. (1988). Features of embryonic induction. *Development (Cambridge, UK)* **104,** 341–359.

Keller, R. E. (1981). An experimental analysis of the role of bottle cells and the deep marginal zone in gastrulation of *Xenopus laevis. J. Exp. Zool.* **216,** 81–101.

Keller, R. E. (1975). Vital dye mapping of the gastrula and neurula of *Xenopus laevis.* I. Prospective areas and morphogenetic movements of the superficial layer. *Dev. Biol.* **42,** 222–241.

Keller, R. E. (1976). Vital dye mapping of the gastrula and neurula of *Xenopus laevis.* II. Prospective areas and morphogenetic movements of the deep layer. *Dev. Biol.* **51,** 118–137.

Lafrance, S. M., Fransen, M. E., Erginel-Unatuna, N., Dube, D. K., Robertson, D. R., Stefanu, C., Ray, T. K., and Lemanski, L. F. (1993). RNA from normal anterior endoderm-conditioned medium stimulates myofibrillogenesis in developing mutant axolotl hearts. *Cell. Mol. Biol. Res.* **39,** 547–560.

Lemanski, L. F., Paulson, D. J., and Hill, C. S. (1979). Normal anterior endoderm corrects the heart defect in cardiac mutant salamanders (*Ambystoma mexicanum*). *Science* **204,** 860–862.

Lemanski, L. F., Lafrance, S. M., Erginel-Unaltuna, N., Luque, E. A., Ward, S. M., Fransen, M. E., Mangiacapra, F. J., Nakatsugawa, M., Lemanski, S. L., Capone, R. B., Goggins, K. J., Nash, B P., Bhatia, R., Dube, A., Gaur, A., Zajdel, R. W., Zhu, Y. Z., Spinner, B. J., Pietras, K. M., Lemanski, S. F., Kovacs, C. P., Vanarsdale, X., Lemanski, J. L., and Dube, D. K. (1995). The cardiac mutant-gene-*c* in Axolotls-cellular, developmental and molecular studies. *Cell. Mol. Biol. Res.* **41,** 293–305.

Lemanski, S. F., Kovacs, C. P., and Lemanski, L. F. (1997). Analysis of the three-dimensional distributions of alpha-actinin, ankyrin and filamin in developing hearts of normal and cardiac mutant axolotls (*Ambystoma mexicanum*). *Anat. Embryol.* **195,** 155–163.

Logan, M., and Mohun, T. (1993). Induction of cardiac muscle differentiation in isolated animal pole explants of *Xenopus laevis* embryos. *Development (Cambridge, UK)* **118,** 865–875.

Mangold, O. (1954). Entwicklung und differenzierung der presumptiven epidermis und ihres unterlagernden entomesoderms aus der neurula von *Triton alpestris* alsilsolat. *Wilhelm Roux' Arch. Entwicklungsmech, Org.* **147,** 131–170.

Nakamura, O., Hayashi, Y., and Asashima, M. (1978). *In* "Organizer, A Milestone of a Half Century from Spemann" (O. Nakamura and S. Toivonen, eds.), pp. 1–47. Elsevier, Amsterdam.

Nascone, N., and Mercola, M. (1995). An inductive role for the endoderm in *Xenopus* cardiogenesis. *Development (Cambridge, UK)* **121,** 515–523.

Nascone, N., and Mercole, M. (1996). Endoderm and cardiogenesis. *Trends Cardiovasc. Med.* **6,** 211–216.

Needham, J. (1942). "Biochemistry and Morphogenesis." Cambridge University Press, Cambridge, UK.

Newman, C. S., Chia, F., and Krieg, P. A. (1997). The XHex homeobox gene is expressed during development of the vascular endothelium: Overexpression leads to an increase in vascular endothelial cell number. *Mech. Dev.* **66,** 83–93.

Nieuwkoop, P. (1946). Experimental investigations on the origin and determination of the germ cells and on the development of the lateral plates and germ ridges in urodeles. *Arch. Neerl. Zool.* **8,** 1–205.

Nieuwkoop, P., and Faber, J. (1956). "Normal Table of *Xenopus laevis* (Daudin)." North-Holland, Amsterdam.

Pasteels, J. (1942). New observations concerning the maps of presumptive areas of the young amphibian gastrula. (*Amblystoma* and *Disglossus*). *J. Exp. Zool.* **89,** 255–281.

Rugh, R. (1962). "Experimental Embryology." Burgess, Minneapolis, MN.

Sater, A. K., and Jacobson, A. G. (1989). The specification of heart mesoderm occurs during gastrulation in *Xenopus laevis. Development (Cambridge, UK)* **105,** 821–830.

Sater, A. K., and Jacobson, A. G. (1990a). The restriction of the heart morphogenetic field in *Xenopus laevis. Dev. Biol.* **140,** 328–336.

Sater, A. K., and Jacobson, A. G. (1990b). The role of the dorsal lip in the induction of heart mesoderm in *Xenopus laevis. Development (Cambridge, UK)* **108,** 461–470.

Smith, J. C. (1995). Mesoderm-inducing factors and mesodermal patterning. *Curr. Opin. Cell Biol.* **7,** 856–861.

Smith, S. C., and Armstrong, J. B. (1991). Heart development in normal and cardiac-lethal mutant axolotls: A model for the control of vertebrate cardiogenesis. *Differentiation (Berlin)* **47,** 129–134.

Smith, S. C., and Armstrong, J. B. (1993). Reaction-diffusion control of heart development: Evidence for activation and inhibition in precardiac mesoderm. *Dev. Biol.* **160,** 535–542.

Spemann, H., and Mangold, H. (1924). Ueber induktion von embryonalenlagen durch implantation artfremder organisatoren. *Arch. Mikrosk. Anat. Entwicklungsmech.* **100,** 599–638.

Stohr, P., Jr. (1924). Experimentelle studien an embryonalen amphibienherze. I. Uber explanation embryonaler amphibienherzen. *Arch. Mikrosk. Anat. Entwicklungsmech.* **102,** 426–451.

Stohr, P., Jr. (1925). Experimentelle studien an embryonalen amphibienherzen. III. Ueber die entsthehung der herzform *Wilhelm Roux' Arch. Entwicklungsmech.Org.* **106,** 409–455.

Tonegawa, A., Moriya, M., Tada, M., Nishimatsu, S., Katagiri, C., and Ueno, N. (1996). Heart formative factor(s) is localised in the anterior endoderm of early *Xenopus* neurula. *Roux's Arch. Dev. Biol.* **205,** 282–289.

Tonissen, K. F., Drysdale, T. A., Lints, T. J., Harvey, R. P., and Krieg, P. A. (1994). XNkx-2.5, a *Xenopus* gene related to Nkx-2.5 and tinman: Evidence for a conserved role in cardiac development. *Dev. Biol.* **162,** 325–328.

Vogt, W. (1929). Gesteltungsanalyse am amphibienleim mit ortlicher vitalfarbung. *Wilhelm Roux' Arch. Entwicklungsmech. Org.* **120,** 384–706.

Weiss, P. (1939). "Principles of Development." Holt, New York.

Wilens, S. (1955). The migration of heart mesoderm and associated areas in *Amblystoma punctatum. J. Exp. Zool.* **129,** 576–606.

4

Vertebrate Heart Induction

Thomas M. Schultheiss[1] and Andrew B. Lassar

Department of Biological Chemistry and Molecular Pharmacology, Harvard Medical School, Boston, Massachusetts 02115

The main focus of this chapter is to review the current state of knowledge concerning the tissue interactions that regulate heart determination in vertebrates. Before we can discuss cardiac induction itself, however, we must review what is known about the timing of events during cardiogenesis and the degree of commitment of cells to the cardiac lineage at various stages in their life histories, because it is these events and changes in commitment that models of inductive interactions must attempt to explain. Following a review of these basic aspects of cardiac development, we proceed to consider positive and negative tissue interactions that regulate cardiogenesis and finally to consider recent data in which the molecular basis of some of these tissue interactions is beginning to be understood.

The method throughout is comparative, with a given phenomenon examined in several species in order to extract fundamental principles as well as to understand how the cardiac developmental program can be modified in different organisms. It is also hoped that by discussing cardiac development in several vertebrate species, we will encourage workers on one species to try to integrate their results with the findings of those who work in other experimental systems. Primary attention is given to birds and amphibians since most vertebrate cardiac experimental embryology has been undertaken in these groups, but some mention is made of other organisms. Although progress has been made, we are far from a complete understanding of the molecular basis of cardiac induction. One of the goals of this review is to point out particular tissue interactions which have been well characterized at the tissue level but whose molecular basis is not understood, with the hope that such a discussion will stimulate further study of these embryological events.

[1]Current address: Molecular Medicine Unit, Beth Israel Hospital, RW 663, Harvard Medical School, Boston, MA 02215.

I. Fate Mapping

A. Chick

The determination of which embryonic cells contribute to the chick heart has been the subject of study for over 50 years and has been approached using increasingly precise methods (Fig. 1). The heart precursors are among the first embryonic cells to gastrulate. Prior to stage 3, precardiac cells are found in the epiblast lateral to the midportion of the primitive streak (Rosenquist and De Haan, 1966; Hatada and Stern, 1994). At the onset of gastrulation, at stage 3, heart precursors are located throughout a broad zone in the primitive streak, absent from only the most anterior (where Henson's node precursors are found) and posterior portions of the streak (Garcia-Martinez and Schoenwolf, 1993). The precursors then leave the streak and enter the nascent mesodermal layer, where they migrate in an anterior/lateral direction. Since they are among the first cells to gastrulate, the heart precursors are always found at the anterior leading edge of the migrating mesodermal sheet. By stage 6 (early neurula), the precursors have come to occupy a crescent in the anterior lateral region of the embryo (Rosenquist, 1966; Rosenquist and De Haan, 1966). Starting at stage 6, with

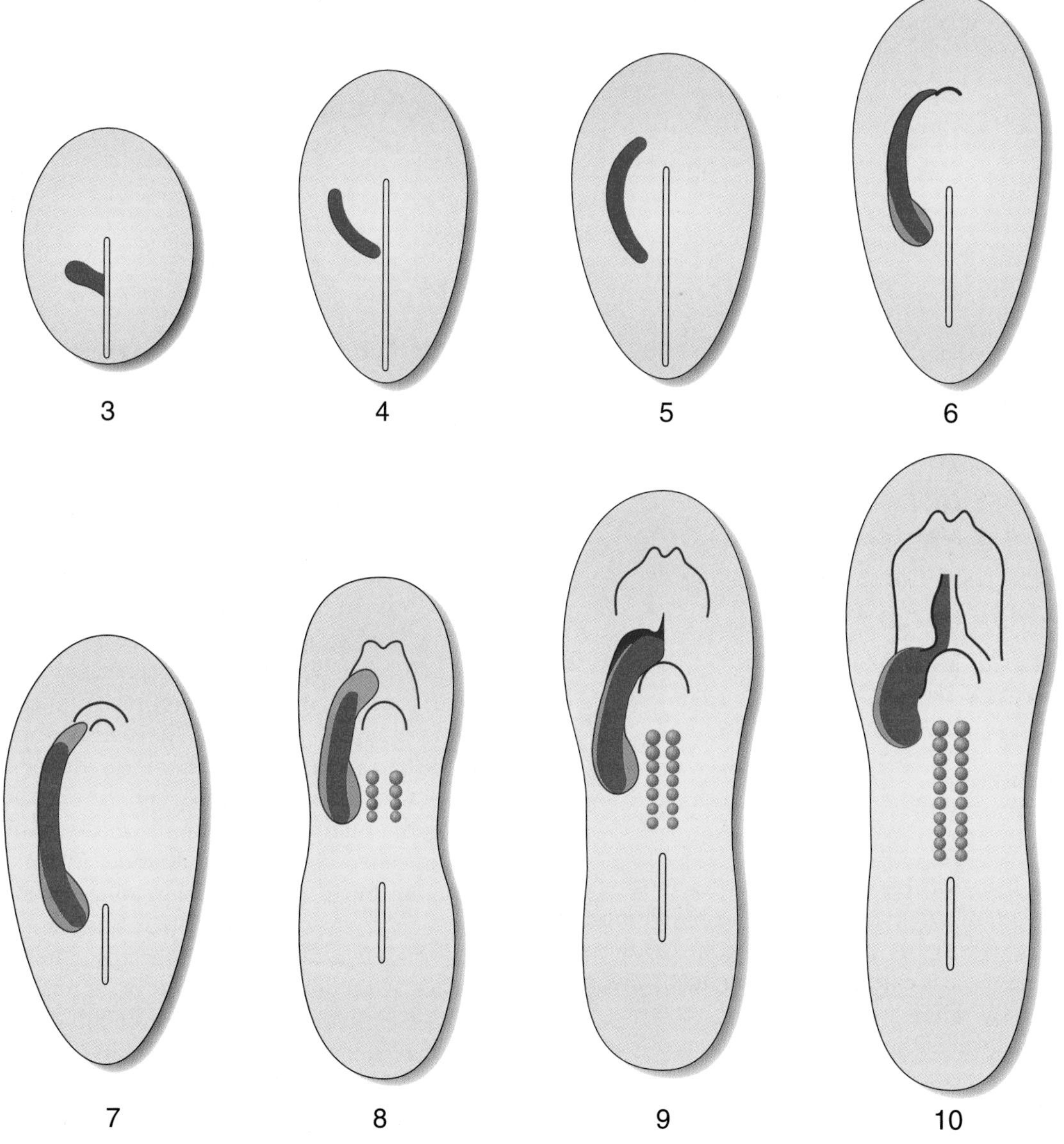

Figure 1 Cardiac fate map of the chick embryo: The location of cells that will contribute to the heart, from stages 3 to 10. Prospective heart and differentiated heart cells are indicated by shading (adapted from Rosenquist and De Haan, 1966, with additional information from Garcia-Martinez and Schoenwolf, 1993).

the formation of the head fold and the anterior intestinal portal, the lateral regions of the anterior embryo begin to approach each other to form the foregut. As a part of this process, the heart precursors in the lateral plate are brought together on the ventral surface of the forming foregut. By stage 10, the precursor regions from the two sides of the embryo have fused to form the tubular heart, which begins beating at this time (Rosenquist and DeHaan, 1966).

B. Amphibians

Amphibian heart development has classically been followed in urodeles and, recently, in the anuran *Xenopus laevis* (see Chapter 3). In urodeles, classical studies have localized heart precursors to the lateral edges of the blastopore (Vogt, 1929; Holtfreter, 1938). These regions gastrulate over the lateral blastopore lip and, by early neurula stages (stage 14), are found in the anterior mesoderm lateral to the hindbrain (Wilens, 1955). Subsequently, the bilateral heart primordia move ventrally toward the anterior ventral midline, where they fuse (Wilens, 1955). Although the *Xenopus* heart fate map is similar to that of the urodeles that have been studied, there are some differences. At the onset of gastrulation, the *Xenopus* heart primordia are not located on the surface of the marginal zone, as in urodeles, but rather in the deep marginal zone at the lateral edge of the blastopore (Keller, 1976). As a result, the *Xenopus* heart primordia never actually migrate over the lip of the blastopore because they are already in the interior of the embryo at the onset of gastrulation. Nevertheless, after the onset of gastrulation they undergo a similar set of movements to those in urodeles and end up in the anterior ventral midline by tailbud stages (Keller, 1976).

A comparison of the cardiac fate maps of avians and amphibians points to fundamental similarities as well as to some interesting differences. Gastrulation of the precardiac mesoderm takes place at homologous positions: the mid-primitive streak in birds and the lateral blastopore in amphibians. In both classes of organism, the cardiac primordia migrate anteriorly at the anterior edge of the spreading lateral mesoderm and then move ventrally to their definitive location in the anterior ventral midline. However, while the order of events is quite similar in birds and amphibians, the duration of each stage is quite variable. Migration of the primordia to the anterior ventral midline is relatively delayed in amphibians when compared to birds. Thus, in birds, cells approach the anterior ventral midline and begin terminal differentiation as the first somites are forming; in amphibians, the somites are well formed before the heart tube begins to form.

II. Acquisition of a Cardiac Fate: The Timing of Specification and Commitment

Fate maps establish, at each stage of development, which embryonic cells will contribute to the heart, but fate maps do not provide information regarding the state of commitment of these cells to the cardiac lineage. One source of information on the state of specification of cells comes from experiments in which tissues from within the cardiac fate map are removed from the embryo at various stages of development and placed into foreign environments. The ability of the prospective cardiac cells to differentiate into heart tissue is then evaluated. Traditionally, cells are considered "specified" if they can differentiate into heart when placed into a "neutral" environment, such as tissue culture medium, and are considered "determined" (a higher level of commitment) if they can differentiate into heart even when placed into "antagonistic" environments, such as noncardiogenic regions of the embryo (Slack, 1991).

A. Chick

Prior to gastrulation, when chick precardiac cells still reside in the epiblast (i.e., prior to stage 3), they will not differentiate into cardiac muscle (as defined by undergoing rhythmic beating or expressing sarcomeric myofibrillar proteins) when placed into tissue culture (Holtzer *et al.*, 1990; Yatskievych *et al.*, 1997). Hence, current evidence suggests that heart tissue is not specified prior to gastrulation. In contrast, if precardiac tissue is explanted after it has moved into the primitive streak, at stages 3 and 4, it will then undergo cardiac differentiation (Gonzalez-Sanchez and Bader, 1990; Holtzer *et al.*, 1990; Montgomery *et al.*, 1994). It must be kept in mind, however, that this stage 3 or 4 mid-primitive streak tissue is heterogeneous; in addition to heart precursors, it also contains the precursors of other tissues, including endoderm, which, as discussed later, has cardiac-inducing properties.

Two other types of assays have been performed on precardiac cells residing in the primitive streak. If, instead of being cultured as a mass, the cells are instead dispersed and grown at low density, then very little if any cardiac differentiation is observed (Montgomery *et al.*, 1994), suggesting that precardiac cells in the primitive streak are not specified at the individual cell level (note that it cannot be determined from these experiments where the cells are blocked in the differentiation process). Finally, if the precardiac region of the stage 3 primitive streak is transplanted to the posterior, noncardiogenic region of a similarly staged embryo, the transplanted tissue does not undergo cardiogenesis, implying that this tissue is not yet determined to undergo cardiogenesis (Inagaki *et al.,* 1993). Taken together, the data suggest that, when residing in the primitive streak, precardiac cells themselves do not have a high degree of commitment to the cardiac lineage. However, when cultured in a manner that allows them to interact with their neighbors, cardiac differentiation will take place, suggesting that at this stage cell–cell interactions are necessary for cardiogenesis.

Once chick precardiac cells enter the mesodermal layer, they acquire a greater capacity for autonomous differentiation. Initially, as precardiac cells migrate away from the primitive streak, they are part of an endomesodermal layer, and the mesodermal precardiac cells cannot be cleanly separated from prospective endoderm. If precardiac cells from this stage (stage 4) are cultured as single cells, a small percentage will express sarcomeric myosin heavy chain (Gonzalez-Sanchez and Bader, 1990), indicating that stage 4 precardiac mesoderm is specified to some degree. By stage 4+ or 5, the mesodermal and endodermal layers become distinct. At this stage, the precardiac mesoderm will differentiate into cardiac muscle if placed in tissue culture, either as a group of cells or, to a lesser degree, as single cells (Gonzalez-Sanchez and Bader, 1990; Antin *et al.,* 1994; Montgomery *et al.,* 1994; Gannon and Bader, 1995). Thus, by stage 4+ or 5, the precardiac mesoderm can be considered "specified." Interestingly, the region of the stage 5 embryo that can give rise to heart if placed in culture (the "specification map") is broader than the region of the embryo that actually gives rise to heart *in vivo* (the "fate map") (Rawles, 1943; Rosenquist and De Haan, 1966) (Fig. 2). This is a common feature of embryogenesis, and we shall return to it several times in the following discussion.

Orts-Llorca and colleagues performed classical "determination" assays and found that stage 5 precardiac tissue will give rise to heart when transplanted to posterior regions of the embryo (Orts-Llorca and Collado, 1969); however, these transplants involved both mesoderm and endoderm, so the state of determination of the precardiac mesoderm itself could not be evaluated. In a different type of determination assay, it has been found that various general inhibitors of differentiation, including BrdU (Chacko and Joseph, 1974; Montgomery *et al.,* 1994), will inhibit the differentiation of cultured precardiac mesoderm if treatment is initiated at or before stage 5, suggesting that at stage 5 the precardiac cells still lack a certain degree of commitment to a cardiac fate. Interestingly, as discussed later, differentiation of precardiac cells from this stage can also be prevented in a more physiological manner by treatment with noggin, an inhibitor of bone morphogenetic protein (BMP) activity (Schultheiss *et al.,* 1997). By stage 6,

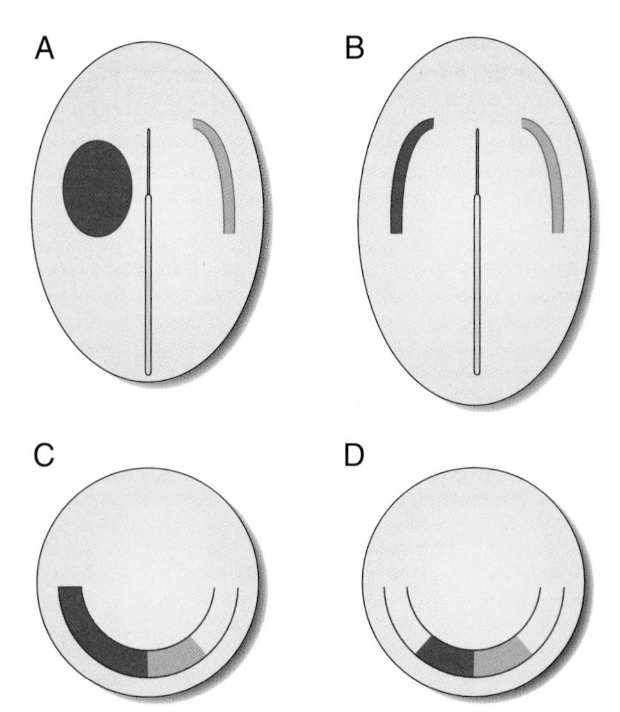

Figure 2 Comparison of the specification map and the fate map of the heart in representative avian and amphibian species. (A, B) Diagram of stage 5 (A) or stage 6 (B) chick embryos, with the heart specification map indicated on the left in red and the heart fate map indicated on the right in pink. Anterior is up. (C, D) Scheme of a cross section through the cardiac region of an amphibian (*Xenopus*) embryo at stage 22 (C) or stage 26 (D), with specification and fate maps indicated as in A and B. Ventral is down. In both classes of vertebrates, at early stages the specification map appears to be broader than the fate map, whereas at later stages the specification map has narrowed to coincide more closely with the actual fate map. See text for details and references.

differentiation of the precardiac mesoderm is resistant to inhibition by BrdU, (Chacko and Joseph, 1974; Montgomery *et al.,* 1994) and noggin (Schultheiss and Lassar, 1997), suggesting that, by stage 6, the precardiac mesoderm has attained an additional level of commitment. Shortly thereafter, at stage 7, markers of terminal differentiation, such as myosin heavy chain, can be detected in the cardiac mesoderm (Han *et al.,* 1992).

B. Amphibians

The timing of cardiac specification has been studied extensively in amphibians. Because the results have been somewhat different in urodeles and anurans, the two situations will be considered separately. Because this topic has been the subject of several reviews (Jacobson, 1961; Jacobson and Sater, 1988; Nascone and Mercola, 1996; see Chapter 3), only the basic outlines will be presented here for the purposes of comparison with the situation in birds.

Studies in the California newt *Taricha torosa* indicate

that prior to stage 15 (early neural fold stage), cardiac differentiation (as assessed by beating) does not occur in heart mesoderm placed into tissue culture (Jacobson, 1960). Low levels of differentiation are seen in stages 15–18 (neural fold to neural groove stages), with reliable differentiation occurring only if explants are taken at stage 19 or later (Jacobson, 1960). Thus, heart specification appears to take place later in the newt than in the chicken. However, account must be taken of the differing culture conditions and the different means of detection of cardiac differentiation. Avian tissues require richer media for survival, and it is possible that some components of this media have the effect of supporting the differentiation or survival of cardiac tissue. In addition, the avian studies have typically examined molecular markers in addition to beating and are therefore more sensitive than most of the assays used in the urodele studies. However, even if beating is used as the criterion for cardiac differentiation, chick heart specification can be observed in explants taken from stage 5 embryos (late gastrula) (Gonzalez-Sanchez and Bader, 1990; Antin *et al.,* 1994), significantly earlier than in urodeles.

Experiments using the anuran *X. laevis* reveal that cardiac specification takes place earlier in this species than in the urodeles that have been studied. In *Xenopus,* stage 12.5 (late gastrula) is the earliest stage in which the three germ layers are distinct. Explantation of the cardiac primordia at this stage results in robust cardiac differentiation (Sater and Jacobson, 1989), implying that specification of the cardiac mesoderm in *Xenopus* occurs by late gastrula, as in chick . Indeed, cardiac primordia from as early as stage 10.5 (early gastrula) have been found to differentiate into heart autonomously in culture (Nascone and Mercola, 1995), although it should be noted that explants from this early stage are quite heterogeneous and give rise to other tissues besides heart. It should also be noted that, unlike the chick experiments, all the amphibian studies were performed in the presence of overlying ectoderm. It is thought that the ectoderm plays a largely neutral role and predominantly augments the general health of the explants, although a nonspecific cardiac-promoting role has been reported (Jacobson and Duncan, 1968).

It is also of interest that in amphibians, as in the chick, there are developmental stages in which the specification map is broader than the fate map, i.e., there are cells which will give rise to heart when placed into tissue culture (i.e., they are specified) but which do not normally give rise to heart *in vivo* (Fig. 2). Thus, in stage 22 *Xenopus* embryos and urodele species of similar developmental stages, both ventral and lateral regions of the anterior mesoderm can give rise to heart when cultured *in vitro,* but only the ventral portion will actually form heart *in vivo* (Sater and Jacobson, 1990a). As discussed later, one possibility is that inhibitory signals from neural tissues play a role in this restriction of the cardiac field.

In brief, the timing of the acquisition of cardiac commitment in *Xenopus* appears to be quite similar to that in the chick. In both species, the early gastrula primordia can differentiate into heart but do so in concert with other nearby tissues; as soon as the germ layers are clearly separable from each other, by late gastrula, the cardiac primordia will differentiate autonomously and with relative independence from other tissues. In the urodeles examined, acquisition of specification is a slower process, which is not completed until neural tube stages.

III. Timing of Differentiation: Molecular Markers

Another source of information regarding the state of differentiation of cells in the cardiac fate map comes from studies of the expression of molecular markers of cardiogenesis. Such studies complement the functional studies reviewed in the previous section and can provide suggestions of the molecular basis for the state of commitment of cells to the cardiac lineage. The following description is limited to an overview of the expression of various cardiac markers with respect to how they relate to the acquisition of commitment to a cardiac fate; more thorough descriptions of the expression patterns and presumed function of these molecules are provided in other chapters of this book (see Sections III and V).

There are currently no known molecular markers for avian precardiac cells while they are still in the primitive streak nor for amphibian heart precursors in the lateral blastoporal lip. Perhaps the earliest marker that is relatively specific for cells in the cardiac fate map is the homeobox gene *Nkx-2.5* (Bodmer *et al.,* 1990; Komuro and Izumo, 1993; Lints *et al.,* 1993; Tonissen *et al.,* 1994; Schultheiss *et al.,* 1995; Lee *et al.,* 1996; see Chapter 7). In the chick, *Nkx-2.5* is first detectable by *in situ* hybridization in precardiac cells when they are located in the anterior lateral plate at the early neurula stages 5 and 6 (Schultheiss *et al.,* 1995), whereas in amphibians, expression of this gene is detected in the anterior lateral heart precursors at stage 15 (also early neurula) (Tonissen *et al.,* 1994). In both species (and in mouse and zebrafish), expression of *Nkx-2.5* is not restricted to the precardiac cells: it is also expressed in neighboring cells of the endoderm (both chick and *Xenopus*), mesoderm (*Xenopus*), and ectoderm (chick).

The degree of commitment to a cardiac fate in cells which have begun to express *Nkx-2.5* is still somewhat unclear. By combining data from several studies, it seems reasonable to conclude that, in both chick and *Xenopus,* some cells which express *Nkx-2.5* are already specified to become heart since cells from these stages will differentiate into heart if placed into tissue culture (Jacobson, 1960, 1961; Sater and Jacobson, 1989; Gonzalez-Sanchez and Bader, 1990; Antin *et al.,* 1994; Montgomery *et al.,* 1994; Tonissen *et al.,* 1994; Nascone and Mercola, 1995; Schultheiss *et al.,* 1995). It is more difficult to determine the relationship of *Nkx-2.5* expression to commitment to undergo cardiogenesis. In the chick, *Nkx-2.5* begins to be expressed in the precardiac region at stages 5 to 6 (Schultheiss *et al.,* 1995), which is when cells have been found to become committed to a cardiac fate (Orts-Llorca and Collado, 1969; Gonzalez-Sanchez and Bader, 1990; Antin *et al.,* 1994; Montgomery *et al.,* 1994). However, it is not clear how *Nkx-2.5* expression and cardiac commitment are related on an individual cell level. In *Xenopus,* on the other hand, it appears that some of the cells in the "cardiac field" which express *Nkx-2.5* will not actually contribute to the heart *in vivo* (such cells subsequently appear to downregulate *Nkx-2.5* expression) (Tonissen *et al.,* 1994); thus, even among cells which are specified to undergo cardiogenesis, *Nkx-2.5* expression does not imply irreversible commitment to a cardiac fate.

Whether *Nkx-2.5* expression merely correlates with cardiac commitment or actually plays a role in committing cells to a cardiac fate is still unclear. Recent studies in *Xenopus* (Cleaver *et al.,* 1996) and zebrafish (Chen and Fishman, 1996) have found that ectopic expression of *Nkx-2.5* can lead to ectopic expression of cardiac genes, pointing to a role for *Nkx-2.5* in cardiac cell type determination. However, the ectopic cardiogenesis is modest and, in *Xenopus,* is limited to tissues surrounding the endogenous heart tissue (Cleaver *et al.,* 1996), suggesting that other factors are required for cardiac specification.

Transcription factors of the GATA-4/5/6 class are also expressed in early heart precursors and the developing heart, but again they are not specific for heart precursors: Prior to definitive heart formation they are also expressed in neighboring cells which do not contribute to the heart proper (Laverriere *et al.,* 1994; Jiang and Evans, 1996; Schultheiss *et al.,* 1997; see Chapter 17). One of the earliest currently known markers that maps to heart precursors but not to immediately neighboring noncardiac cells is the transcription factor MEF-2C (see Chapter 8). In the mouse it is expressed in precardiac mesoderm beginning at about Day7.5 (Edmondson *et al.,* 1994), and in the chick it is expressed in the precardiac mesoderm of the anterior lateral plate and anterior intestinal portal beginning at about stage 7 (T. M. Schultheiss, unpublished data). At about the same time as the expression of MEF-2C, one can begin to detect expression of the first myocardial structural proteins, including muscle-specific actins, myosins, and associated proteins (Han *et al.,* 1992; Edmondson *et al.,* 1994).

IV. Induction of Cardiogenesis

We now discuss the central topic of this article: the role of tissues and signaling molecules in regulating which cells adopt a cardiac fate during embryogenesis. Induction experiments typically involve combining a "responding" and an "inducing" tissue and determining the ability of the latter to induce a response in the former. The responder must, of course, not undergo the particular process spontaneously in the absence of any inducer. To study induction of cardiogenesis, two types of responder tissues can and have been used. Responder tissues have either been taken from within the fate map of the heart but at a time prior to specification, or they have been taken from outside of the fate map of the heart, i.e., from tissues that normally will give rise to noncardiac tissues. As shall be shown, the source of responding tissue influences the conclusions that can be drawn regarding the nature of the tissue interactions.

A. Chick and Mouse

For studying very early inductive events in cardiogenesis in the chick, using responding tissues from within the cardiac fate map is not a problem since, if taken prior to gastrulation, these responding tissues will not spontaneously differentiate into heart (Yatskievych *et al.,* 1997). Few experiments have been reported using these early, pregastrulation stages as responders. Antin and coworkers found that the pregastrula epiblast will give rise to heart tissue if cultured with the hypoblast with which it is normally in contact (Yatskievych *et al.,* 1997). Since several other tissues besides heart are induced in these experiments, the hypoblast appears to provide an early, general signal(s) required for induction of a variety of mesodermal tissues including heart.

As discussed previously, when chick precardiac cells are in the primitive streak, at stage 3, they have not yet acquired a high degree of commitment to the cardiac lineage. However, these cells are intimately intermixed with cells that have cardiac-inducing properties so that if the cardiac region of the primitive streak is cultured alone it will spontaneously differentiate into beating heart tissue (Montgomery *et al.,* 1994). In other words, in stages 3 and 4, when significant events in cardiac specification are taking place, the precardiac tissue is not experimentally separable from surrounding tissues (most important, the prospective endoderm) and hence is not amenable for use in induction assays. At stage 4+, when they are experimentally separable from prospective endoderm, the mesodermal cardiac precursors are already specified to a significant extent (Gonzalez-Sanchez and Bader, 1990; Antin *et al.,* 1994; Gannon and Bader, 1995). Therefore, in order to study these important stages (stages 3 and 4) in cardiac induction, researchers have had to utilize responding tissues that are outside of the specification map of the heart.

The groundwork for these studies was laid by the commitment studies of Schoenwolf and colleagues (Inagaki *et al.,* 1993). They found that the posterior-most portion of the stage 3 primitive streak, which normally gives rise to blood and extraembryonic membranes, will contribute to the heart if transplanted into the cardiac region of the primitive streak (it should be mentioned that, although these cells ended up in the heart, whether they actually differentiated into heart tissue was not definitively established). Here, then, was a tissue that was not in the heart fate map but could contribute to the heart if placed in the proper environment. In a series of induction experiments, we used this posterior primitive streak (PPS) as a responder tissue, and cultured it in combination with various potential inducers in order to identify regions of the embryo with cardiac-inducing properties (Schultheiss *et al.,* 1995). Responder PPS tissue was taken from chickens, whereas inducing tissues were taken from quail, so that one could identify the source of any heart tissue that emerged in culture using species-specific reverse-transcriptase polymerase chain reaction. Using such a culture system, we found that the anterior endoderm in stages 4 and 5 embryos (mid- to late gastrula) could induce cardiogenesis from the posterior primitive streak. Induction of cardiogenesis was accompanied by repression of erythropoiesis in the posterior streak tissue. At these stages, cardiac-inducing activity was restricted to the anterior endoderm and was not found in the posterior endoderm or in the precardiac mesoderm itself. As we have seen, when the precardiac cells are in the primitive streak they are not committed to a cardiac fate. When they leave the primitive streak, they migrate into the anterior lateral plate mesoderm, where they are in close contact with the anterior endoderm, which has cardiac-inducing properties. It is therefore plausible that signals from the anterior endoderm play a role in inducing cardiogenesis in the anterior mesodermal cells with which they come into contact. A working model based on this data suggests that large portions of the primitive streak are competent to becomes heart, and that the portion of the streak which actually does give rise to heart is determined by which subset of gastrulating mesodermal cells come into contact with the anterior endoderm (Fig. 3A).

Two additional features of cardiac induction by ante-

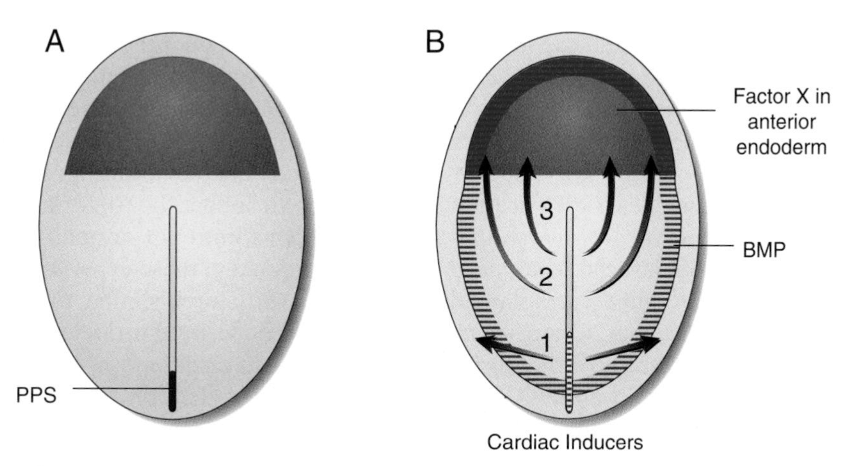

Figure 3 (A) Localization of cardiac-inducing activities in chick embryos. The red-shaded region indicates the area of stages 4 or 5 endoderm that can induce cardiogenesis from the posterior primitive streak (PPS, indicated in dark shading), whose normal fate is to form blood and extraembryonic membranes. (B) A two-signal model of heart induction in the chick. One signal is transmitted by BMPs, which are present in the peripheral regions of the gastrula embryo and also in the posterior primitive streak (purple hatching). The other activity is a factor or factors present in the anterior endoderm, here designated "factor X" (red shading). Cardiogenesis occurs where the two activities overlap, in the anterior lateral region. The numbered arrows indicate migration paths of mesodermal cells from the primitive streak. Normally only cells which follow path 2 encounter both inducing signals and hence undergo cardiogenesis. Cells that follow path 1 encounter only BMP signaling and form blood and extraembryonic membranes, but will form heart if experimentally exposed to anterior endoderm (Schultheiss *et al.,* 1995). Cells that follow path 3 encounter only factor X and differentiate as paraxial head mesoderm (Rosenquist, 1966), although they can form heart if exposed to BMP signaling (adapted from Schultheiss *et al.,* 1997, © Cold Spring Harbor Laboratory Press).

rior endoderm in the chick should be noted. First, at stage 5, both lateral and medial portions of the anterior endoderm have cardiac-inducing properties (Fig. 3A) (Schultheiss *et al.,* 1995), whereas only the anterior lateral mesoderm actually differentiates into heart *in vivo* (Fig. 1) (Rosenquist and De Haan, 1966). In other words, the region of anterior endoderm that has cardiac-inducing properties is broader than the region of overlying mesoderm that actually gives rise to heart. Several explanations are possible for this discrepancy, including inhibitory influences from medial tissues and the requirement for additional signals present in lateral regions which are not necessary for induction of the posterior streak in our experimental system but which are required *in vivo*. Both of these explanations appear to be correct, and we shall return to this issue later when we discuss BMP signaling in cardiogenesis. Here, it shall suffice to remark that, at stage 5, the region of the anterior endoderm that has cardiac-inducing properties coincides more with the cardiac specification map than the fate map (Fig. 2). This suggests that one role of the anterior endoderm is to create a "cardiac field" in the overlying anterior mesoderm, and that other events restrict that field to the region of the embryo that actually becomes heart.

The second issue that should be mentioned is that, while our data indicate that anterior endoderm displays cardiac-inducing properties at stages 4+ and 5, it is possible that anterior endoderm has heart-inducing activity prior to this stage, when endodermal and mesodermal precursors are in intimate contact within the stage 3 primitive streak (Rosenquist, 1966; Garcia-Martinez and Schoenwolf, 1993).

Less experimental data regarding heart specification are available in the mouse. However, recent studies have found that noncardiogenic epiblast can be induced to undergo cardiac differentiation when transplanted into the cardiogenic region of the heart (Tam *et al.,* 1997). This situation is highly reminiscent of the observations in chick (Inagaki *et al.,* 1993; Schultheiss *et al.,* 1995), and suggests that the mechanisms of cardiogenesis in chick and mouse may be quite similar.

B. Amphibians

In urodeles, the cardiac primordia are associated with prospective dorsolateral foregut endoderm (which will contribute to the esophagus, stomach, and duodenum) from the time these tissues lie in the lateral blastopore lip through the neurula stages (Jacobson, 1961). Several investigators, mainly Jacobson and colleagues, have found that this dorsolateral endoderm strongly promotes heart formation from gastrula and neurula stage precardiac mesoderm (Jacobson, 1960, 1961; Jacobson and Duncan, 1968; Jacobson and Sater, 1988). Since the experiments were performed using cardiac primordia as the responding tissue, these data strongly support a role for the anterior endoderm in heart formation *in vivo*.

Following neurula stages, the cardiac primordia migrate ventrally and become associated with the ventral foregut endoderm (which will contribute to the liver and the floor of the pharynx) (Jacobson, 1961). Bacon (1945) has shown that this ventral endoderm also has cardiac-inducing properties since it can induce cardiogenesis from regions of the gastrula embryo which normally do not form heart. Thus, in these urodele species, as in the chick, both dorsolateral and ventral anterior endoderm have cardiac-inducing properties. The combined evidence points to the importance of the anterior endoderm in vertebrate heart induction.

Recent experiments carried out on *Xenopus* cast a somewhat different light on the situation. As discussed previously (see Chapter 3), in *Xenopus* the heart is specified earlier than it is in the urodeles that have been studied. Specification events take place during gastrula stages when the cardiac primordia are not clearly separable from surrounding tissues, making interpretation of experiments somewhat more difficult (Sater and Jacobson, 1989). Sater and Jacobson (1990b) have demonstrated a requirement for the dorsal lip (i.e., the organizer) in the induction of heart from lateral lip cardiac primordia. Subsequently, Nascone and Mercola (1995) showed that endoderm lying deep to the cardiac primordia at gastrula stages also promoted cardiogenesis, either in the presence or absence of organizer tissue, and that this deep endoderm could act in concert with organizer to induce heart formation in ventral mesoderm which would not normally form heart. The deep endoderm used in these experiments is fated to give rise to foregut structures, similar to the fate of the cardiac-inducing endoderm of urodeles (Nascone and Mercola, 1995). Thus, in chick and in two subclasses of amphibians, anterior endoderm is seen to play an important role in heart induction. It is currently unclear whether the organizer plays a role in heart induction in avians or urodeles. Since heart can be induced from noncardiogenic tissues by endoderm alone in chick or urodele embryos, organizer activity is apparently not required for heart induction in these species (Bacon, 1945; Schultheiss *et al.,* 1995). However, because these experiments were performed with anterior endoderm isolated from more developmentally advanced stages than were employed in the *Xenopus* experiments (Nascone and Mercola, 1995), one potential unifying hypothesis is that the organizer may pattern the endoderm and render it cardiogenic. Under this scenario, gastrula stage endoderm would still require organizer to induce cardiogenesis (as in the *Xenopus* experiments), whereas at later stages the endoderm would have already been patterned by the organizer and could therefore induce heart by itself (as in the urodele and chick studies).

V. Inhibition of Cardiogenesis

Tissue identity is regulated not only by positive, inducing signals but also by inhibitory signals from adjacent tissues. Negative tissue interactions that affect cardiogenesis have been most extensively studied in amphibians (see Chapter 3). The best characterized cardiac inhibitory interaction is the effect of neural tissue on cardiogenesis. When explants are made containing cardiac primordia as well as surrounding tissues, there is a strong negative correlation between the presence of neural tissue and the presence of cardiac tissue (Jacobson, 1960, 1961; Jacobson and Duncan, 1968; Jacobson and Sater, 1988). When both are present, the cardiac tissue is generally found as far from the neural tissue as possible. Jacobson and colleagues have reported a general antagonism between positive inducing effects of anterior endoderm and inhibitory effects of neural tissue (Jacobson, 1960, 1961; Jacobson and Duncan, 1968). Neurula stage cardiogenic mesoderm can undergo cardiogenesis if cultured alone but not if combined with neural plate. However, adding anterior endoderm to the neural plate/cardiac mesoderm explants restores the ability of the cardiac primordia to differentiate into heart. Anterior neural tissue has stronger cardiac inhibitory properties than posterior tissue, and strong inhibitory activity is found in the neural crest (Jacobson, 1960, 1961; Jacobson and Duncan, 1968). It is interesting that the portion of the amphibian heart field which does not actually become heart *in vivo* lies nearest to neural tissue. Recent studies have implicated neural tissue in this restriction of the cardiac field in *Xenopus* (M. Mercola, personal communication), and cardiac inhibitory properties of neural tissues have also been reported in the chick (Climent *et al.*, 1995).

VI. Molecular Mediators of Cardiac Induction

Recently, progress has been made in studying the molecular level of vertebrate cardiac induction. Not only do these studies begin to explain cardiac induction in molecular terms but also molecular studies have necessitated the reformulation of some of our ideas about the process of cardiac induction. The molecular analysis of cardiac induction is only in an early phase, and many phenomena are not well understood at a molecular level. Therefore, we shall concentrate on only a few of the more well-studied areas.

A. Bone Morphogenetic Proteins

As emphasized in other chapters in this book, there is a surprising degree of similarity between heart development in vertebrates and that in invertebrates. In particular, the homeobox gene *tinman* is expressed in the *Drosophila* heart and is required for heart formation in that species (Bodmer *et al.*, 1990; Bodmer, 1993). In several vertebrate species, genes related to *tinman* are expressed in the heart (Komuro and Izumo, 1993; Lints *et al.*, 1993; Tonissen *et al.*, 1994; Schultheiss *et al.*, 1995; Lee *et al.*, 1996), and in the mouse, mutations in *Nkx-2.5* result in abnormal heart development (I. Lyons *et al.*, 1995; see Chapter 7). In *Drosophila*, the regulation of *tinman* has been the subject of considerable study (see Chapter 5). Initially, *tinman* is expressed in all mesoderm, under the control of the transcription factor twist (Bodmer *et al.*, 1990; Azpiazu and Frasch, 1993). Subsequently, as twist levels decline, *tinman* is downregulated in all but the most dorsal mesoderm (Staehling-Hampton *et al.*, 1994; Frasch, 1995). This dorsal mesoderm is in contact with ectoderm that expresses the transforming growth factor-β (TGF-β) family signaling molecule dpp, and Frasch (1995) has shown that ectopic expression of dpp in more ventral regions of the ectoderm results in expansion of *tinman* into more ventral regions of the mesoderm. In other words, maintenance of *tinman* in the dorsal mesoderm appears to be under the control, either directly or indirectly, of dpp.

In order to gain insight into the molecular basis of vertebrate cardiac induction, we conducted experiments to explore whether such regulatory mechanisms were conserved between vertebrates and flies (Schultheiss *et al.*, 1997). In the chick, the *tinman* homolog *Nkx-2.5* is expressed in the precardiac mesoderm beginning at about stage 5, and is the earliest known marker of the precardiac cells (Schultheiss *et al.*, 1995). We conducted a degenerate PCR screen for TGF-β family members expressed in the stage 5 anterior lateral plate. Several members of the BMP family were detected in this screen, including BMP-2 and BMP-4, the two closest relatives of *Drosophila* dpp (Schultheiss *et al.*, 1997). By *in situ* hybridization, *BMP-2* is expressed in the endoderm, and *BMP-4* in the ectoderm, immediately surrounding the precardiac mesoderm. Beads carrying recombinant BMP-2 or BMP-4 could induce ectopic *Nkx-2.5* expression when placed into noncardiogenic regions of the embryo *in vivo* and could induce complete cardiac differentiation (including beating) when combined in tissue culture with regions of the embryo that are not normally cardiogenic. Interestingly, only certain areas of the mesoderm were competent to undergo cardiogenesis in response to BMP administration, namely, the anterior mesoderm that lies medial to the heart-forming region. The normal fate of this mesoderm is to form skeletal muscle, cartilage, and mesenchyme of the head. Not only can BMP signaling induce ectopic cardiogenesis in this anterior medial mesoderm but also administration of the BMP-antagonist noggin to stage 4 mesoendoderm from the precardiac region completely blocks differentiation of this tissue into heart, implying that BMP signaling is a crucial component of heart specification *in vivo* (Schultheiss *et al.,* 1997).

How should these BMP data be integrated with the experimental data on cardiac induction discussed previously? First, it should be appreciated that BMP signaling is not identical with the anterior endodermal cardiac inducer. While the anterior endodermal activity can induce heart from PPS tissue (Schultheiss *et al.,* 1995), BMP signaling cannot elicit cardiogenesis in this tissue (indeed, BMP-2 and -4 are expressed at high levels in the PPS, which is fated to give rise to blood and extraembryonic membranes) (Schultheiss *et al.,* 1997). Instead, BMP signaling appears to determine which cells, from within a competent field in the anterior part of the embryo, will become heart. The establishment of this competent field appears to be a function of the anterior endodermal inducer since the region of mesoderm which is competent to become heart in response to BMP signaling (Schultheiss *et al.,* 1997) is precisely that region of mesoderm which is in contact with this anterior endoderm (Schultheiss *et al.,* 1995). We thus arrive at a two-signal model for heart induction. In one step, currently unknown signals from the anterior endoderm create a heart field in the overlying anterior mesoderm. BMP signaling selects which portion of this heart field will actually differentiate into heart. In other words, heart development occurs where anterior endoderm and BMP signaling domains overlap (Fig. 3B).

While the previous data indicate a role for BMP signaling in chick heart development, less data are available regarding the role of BMP signaling in cardiogenesis in other vertebrate species. In the mouse, BMP-2 is expressed in the cardiogenic regions of the embryo (K. M. Lyons *et al.,* 1995). Knockouts of mouse BMP-2 produce cardiac defects ranging from the absence of heart to various heart malformations (Zhang and Bradley, 1996), consistent with a role for BMP signaling in murine cardiogenesis. BMP-4 knockouts generally die prior to gastrulation (Winnier *et al.,* 1995), and thus cardiogenesis in these mutants has not been investigated. The zebrafish mutant *swirl* has recently been found to be a mutant in BMP-2 (Kishimoto *et al.,* 1997). *Swirl* fish have severe defects in cardiogenesis, with most mutants failing to express even the early cardiac markers *Nkx-2.5* and *Nkx-2.7* (Kishimoto *et al.,* 1997). While much work remains to be done, it appears that BMP signaling is a general requirement for cardiogenesis in both zebrafish and mouse (see Chapter 6). It will be important to determine the precise roles of BMP signaling in these species.

There are no direct data addressing the role of BMP signaling in amphibian cardiogenesis. However, it is interesting to recall that in the investigations of cardiac inhibition in amphibians discussed previously, it was found that at neurula stages the heart specification map is significantly broader than the region of the embryo that will actually give rise to heart (Fig. 2). This narrowing down of the heart specification map was attributed to cardiac inhibitory properties of neural tissues (Jacobson, 1960, 1961). In the chick, the cardiac specification map at stage 5 is also broader than the cardiac fate map (Fig. 2), and the portion of the specification map that actually differentiates into heart is that portion which comes into contact with BMP signals (Fig. 3) (Schultheiss *et al.,* 1997). It is of interest that, in amphibians, at the time when the heart field is being narrowed down to its ventral-most portion, the cardiac primordia are coming into contact with the pharyngeal endoderm, where BMPs are expressed (Clement *et al.,* 1995). It will thus be important to determine whether BMP signaling is involved in this later phase of cardiac differentiation in amphibians. If so, it would point to a general role for BMP signaling in selecting which subdomain from within a cardiac field actually becomes heart. It would also suggest that these two components of cardiac induction (anterior endoderm and BMP signaling), which occur concurrently in chicks, may be separated in time in other organisms. Indeed, given the cardiac inhibitory properties of the BMP antagonist noggin, it is even possible that the cardiac inhibitory properties of neural tissue are attributable either to noggin or to other BMP antagonists produced in the neural plate.

B. Other Molecules

Other molecules have been implicated in vertebrate cardiac induction, but their precise roles are less well-understood than the role of BMP signaling.

The TGF-β signaling molecule activin has been implicated in early events in cardiogenesis in both amphibians and birds. Heart tissue can be produced from *Xenopus* animal caps if several caps are aggregated together and the aggregate is treated with high levels of activin (Logan and Mohun, 1993; see Chapter 3). In birds, treatment of pregastrula epiblast with activin also induces cardiac tissue (Yatskievych *et al.,* 1997). In both of these cases, activin induces many types of mesoderm besides heart (and also probably endoderm), and secondary interactions undoubtedly occur between these induced tissues. Thus, the role of an activin-like substance in cardiogenesis is most likely due to the role of such a substance in mesoderm induction in general.

No strong candidate has yet emerged for the other anterior endodermal signal which synergizes with BMP signaling to induce cardiogenesis. In the chick, it has been reported that a combination of FGF-4 and BMP-2 signaling can induce cardiogenesis from posterior regions of the stage 4 embryo, suggesting that FGF signaling may play such a role (Lough *et al.,* 1996). However, there is no evidence of an FGF that is expressed specifically in the anterior endoderm, as one would expect for an endogenous cardiac inducer. Since the FGF-4 levels used in those experiments were quite high (100 ng/ml), it is possible that FGF-4 is mimicking an endogenous, still unknown signal. In *Drosophila,* wingless signaling has been found to synergize with dpp signaling to specify heart precursors in the dorsal mesoderm (Wu *et al.,* 1995). Thus far, no data have been presented that Wnt signaling can induce cardiogenesis in vertebrates, although this possibility is still under investigation in a number of laboratories, including our own. Recently, the *Xenopus* gene *cerberus* has been described which has features that make it a possible candidate for a cardiac-promoting signal in the anterior endoderm (Bouwmeester *et al.,* 1996). *Cerberus* is expressed in the anterior endoderm and, when ectopically expressed in ventral blastomeres of the 32-cell embryo, can lead to formation of an ectopic head, including an ectopic heart (see Chapter 7). Investigations of a specific role for *cerberus* in heart induction in *Xenopus* and other species are currently being performed.

VII. Summary

A two-step model of vertebrate cardiac induction is emerging from studies in avian and amphibian species. Signals from anterior endoderm induce a "cardiac field" in the overlying mesoderm. Only a subdomain of this field actually becomes heart *in vivo*. The restriction of the heart field is affected both by BMP signals from the surrounding ventral/lateral tissues which promote cardiogenesis and by signals from dorsal/medial structures, including neural tissues, which inhibit heart formation. The end result is specification of heart tissue in the anterior ventral mesoderm. Future studies will be directed at elucidating the identity of the cardiac-inducing activity in the anterior endoderm and the cardiac inhibitory signals produced by neural tissues, and at understanding in greater depth how BMP signaling activates the cardiac program.

Acknowledgements

The authors thank Kyu-Ho Lee, Matha Marvin, and Mark Mercola for making valuable comments on the manuscript. This work was done during the tenure of an established investigatorship from the American Heart Association to A. B. L. T. M. S. is a Howard Hughes Medical Institute Physician Postdoctoral Fellow.

References

Antin, P. B., Taylor, R. G., and Yatskievych, T. (1994). Precardiac mesoderm is specified during gastrulation in quail. *Dev. Dyn.* **200,** 144–154.

Azpiazu, N., and Frasch, M. (1993). Tinman and bagpipe: Two homeo box genes that determine cell fates in the dorsal mesoderm of Drosophila. *Genes Dev.* **7,** 1325–1340.

Bacon, R. L. (1945). Self-differentiation and induction in the heart of Amblystoma. *J. Exp. Zool.* **98,** 87–125.

Bodmer, R. (1993). The gene tinman is required for specification of the heart and visceral muscles in Drosophila. *Development (Cambridge, UK)* **118,** 719–729.

Bodmer, R., Jan, L. Y., and Jan, Y. N. (1990). A new homeobox-containing gene, *msh-2,* is transiently expressed early during mesoderm formation in *Drosophila. Development (Cambridge, UK)* **110,** 661–669.

Bouwmeester, T., Kim, S., Sasai, Y., Lu, B., and De Robertis, E. M. (1996). Cerberus is a head-inducing secreted factor expressed in the anterior endoderm of Spemann's organizer. *Nature (London)* **382,** 595–601.

Chacko, S. and Joseph, X. (1974). The effect of 5-bromodeoxyuridine (BrdU) on cardiac muscle differentiation. *Dev. Biol.* **40,** 340–354.

Chen, J. N., and Fishman, M. C. (1996). Zebrafish tinman homolog demarcates the heart field and initiates myocardial differentiation. *Development (Cambridge, UK)* **122,** 3809–3816.

Cleaver, O. B., Patterson, K. D., and Krieg, P. A. (1996). Overexpression of the tinman-related genes XNkx-2.5 and XNkx-2.3 in *Xenopus* embryos results in myocardial hyperplasia. *Development (Cambridge, UK)* **122,** 3549–3556.

Clement, J. H., Fettes, P., Knochel, S., Lef, J., and Knochel, W. (1995). Bone morphogenetic protein 2 in the early development of *Xenopus laevis. Mech. Dev.* **52,** 357–370.

Climent, S., Sarasa, M., Villar, J. M., and Murillo-Ferrol, N. L. (1995). Neurogenic cells inhibit the differentiation of cardiogenic cells. *Dev. Biol.* **171,** 130–48.

Edmondson, D. G., Lyons, G. E., Martin, J. F., and Olson, E. N. (1994). Mef2 gene expression marks the cardiac and skeletal muscle lineages during mouse embryogenesis. *Development (Cambridge, UK)* **120,** 1251–1263.

Frasch, M. (1995). Induction of visceral and cardiac mesoderm by ectodermal Dpp in the early Drosophila embryo. *Nature (London)* **374,** 464–467.

Gannon, M., and Bader, D. (1995). Initiation of cardiac differentiation occurs in the absence of anterior endoderm. *Development (Cambridge, UK)* **121,** 2439–2450.

Garcia-Martinez, V., and Schoenwolf, G. C. (1993). Primitive streak origin of the cardiovascular system in avian embryos. *Dev. Biol.* **159,** 706–719.

Gonzalez-Sanchez, A., and Bader, D. (1990). In vitro analysis of cardiac progenitor cell differentiation. *Dev. Biol.* **139,** 197–209.

Han, Y., Dennis, J. E., Cohen-Gould, L., Bader, D. M., and Fischman, D. A. (1992). Expression of sarcomeric myosin in the presumptive myocardium of chicken embryos occurs within six hours of myocyte commitment. *Dev. Dyn.* **193,** 257–265.

Hatada, Y., and Stern, C. (1994). A fate map of the epiblast of the early chick embryo. *Development (Cambridge, UK)* **120,** 2879–2889.

Holtfreter, J. (1938). Differenzierungspotenzen isolierter Teile der Urodelengastrula. *Wilhelm Roux' Arch. Entwicklungsmech. Org.* **138,** 522–656.

Holtzer, H., Schultheiss, T., Dilullo, C., Choi, J., Costa, M., Lu, M., and Holtzer, S. (1990). Autonomous expression of the differentiation programs of cells in the cardiac and skeletal myogenic lineages. *Ann. N. Y. Acad. Sci.* **599,** 158–169.

Inagaki, T., Garcia-Martinez, V., and Schoenwolf, G. C. (1993). Regulative ability of the prospective cardiogenic and vasculogenic areas of the primitive streak during avian gastrulation. *Dev. Dyn.* **197,** 57–68.

Jacobson, A. G. (1960). Influences of ectoderm and endoderm on heart differentiation in the newt. *Dev. Biol.* **2,** 138–154.

Jacobson, A. G. (1961). Heart determination in the newt. *J. Exp. Zool.* **146,** 139–152.

Jacobson, A. G., and Duncan, J. T. (1968). Heart induction in salamanders. *J. Exp. Zool.* **167,** 79–103.

Jacobson, A. G., and Sater, A. K. (1988). Features of embryonic induction. *Development (Cambridge, UK)* **104,** 341–359.

Jiang, Y., and Evans, T. (1996). The *Xenopus* GATA-4/5/6 genes are associated with cardiac specification and can regulate cardiac-specific transcription during embryogenesis. *Dev. Biol.* **174,** 258–270.

Keller, R. E. (1976). Vital dye mapping of the gastrula and neurula of *Xenopus laevis* II: Prospective areas and morphogenetic movements of the depp layer. *Dev. Biol.* **51,** 118–137.

Kishimoto, Y., Lee, K.-H., Zon, L., Hammerschmidt, M., and Schulte-Merker, S. (1997). The molecular nature of zebrafish *swirl:* BMP-2 function is essential during early dorsoventral patterning. *Development (Cambridge, UK)* **124,** 4457–4466.

Komuro, I. and Izumo, S. (1993). Csx: A murine homeobox-containing gene specifically expressed in the developing heart. *Proc. Natl. Acad. Sci. U.S.A.* **90,** 8145–8149.

Laverriere, A. C., MacNeill, C., Mueller, C., Poelmann, R. E., Burch, J. B. E., and Evans, T. (1994). GATA-4/5/6, a subfamily of three transcription factors transcribed in developing heart and gut. *J. Biol. Chem.* **269,** 23177–23184.

Lee, K.-H., Xu, Q., and Breitbart, R. E. (1996). A new tinman-related gene, nkx-2.7, anticipates the expression of nkx-2.5 and nkx-2.3 in zebrafish heart and pharyngeal endoderm. *Dev. Biol.* **180,** 722–731.

Lints, T. J., Parsons, L. M., Hartley, L., Lyons, I., and Harvey, R. P. (1993). Nkx-2.5: A novel murine homeobox gene expressed in early heart progenitor cells and their myogenic descendants. *Development (Cambridge, UK)* **119,** 419–431.

Logan, M., and Mohun, T. (1993). Induction of cardiac muscle differentiation in isolated animal pole explants of *Xenopus laevis* embryos. *Development (Cambridge, UK)* **118,** 865–875.

Lough, J., Barron, M., Brogley, M., Sugi, Y., Bolender, D. L., and Zhu, X. L. (1996). Combined BMP-2 and FGF-4, but neither factor alone, induces cardiogenesis in non-precardiac embryonic mesoderm. *Dev. Biol.* **178,** 198–202.

Lyons, I., Parsons, L. M., Hartley, L., Li, R., Andrews, J. E., Robb, L., and Harvey, R. P. (1995). Myogenic and morphogenetic defects in the heart tubes of murine embryos lacking the homeobox gene Nkx-2.5. *Genes Dev.* **9,** 1654–1666.

Lyons, K. M., Hogan, B. L., and Robertson, E. J. (1995). Colocalization of BMP 7 and BMP 2 RNAs suggests that these factors cooperatively mediate tissue interactions during murine development. *Mech. Dev.* **50,** 71–83.

Montgomery, M. O., Litvin, J., Gonzalez-Sanchez, A., and Bader, D. (1994). Staging of commitment and differentiation of avian cardiac myocytes. *Dev. Biol.* **164,** 63–71.

Nascone, N., and Mercola, M. (1995). An inductive role for the endoderm in *Xenopus* cardiogenesis. *Development (Cambridge, UK)* **121,** 515–523.

Nascone, N., and Mercola, M. (1996). Endoderm and cardiogenesis—new insights. *Trends Cardiovasc. Med.* **6,** 211–216.

Orts-Llorca, F., and Collado, J. J. (1969). The development of heterologous grafts of the cardiac area (labelled with thymidine-^{3}H) to the caudal area of the chick blastoderm. *Dev. Biol.* **19,** 213–227.

Rawles, M. E. (1943). The Heart-forming areas of the early chick blastoderm. *Phys. Zool.* **41,** 22–42.

Rosenquist, G. C. (1966). A radioautographic study of labelled grafts in the chick blastoderm. *Carnegie Inst. Washington Contrib. Embryol.* **38,** 71–110.

Rosenquist, G. C., and De Haan, R. L. (1966). Migration of precardiac cells in the chick embryo: A radioautographic study. *Carnegie Inst. Washington Contrib. Embryol.* **38,** 111–121.

Sater, A. K., and Jacobson, A. G. (1989). The specification of heart mesoderm occurs during gastrulation in *Xenopus laevis. Development (Cambridge, UK)* **105,** 821–830.

Sater, A. K., and Jacobson, A. G. (1990a). The restriction of the heart morphogenetic field in *Xenopus laevis. Dev. Biol.* **140,** 328–336.

Sater, A. K., and Jacobson, A. G. (1990b). The role of the dorsal lip in the induction of heart mesoderm in *Xenopus laevis. Development (Cambridge, UK)* **108,** 461–470.

Schultheiss, T. M., and Lassar, A. B. (1997). Induction of chick cardiac myogenesis by bone morphogenetic proteins. *Cold Spring Harbor Symp. Quant. Biol.* **57,** 413–419.

Schultheiss, T. M., Xydas, S., and Lassar, A. B. (1995). Induction of avian cardiac myogenesis by anterior endoderm. *Development (Cambridge, UK)* **121,** 4203–4214.

Schultheiss, T. M., Burch, J. B. E., and Lassar, A. B. (1997). A role for bone morphogenetic proteins in the induction of cardiac myogenesis. *Genes Dev.* **11,** 451–462.

Slack, J. M. W. (1991). "From Egg to Embryo." Cambridge University Press, Cambridge, UK.

Staehling-Hampton, K., Hoffmann, F. M., Baylies, M. K., Rushton, E., and Bate, M. (1994). dpp induces mesodermal gene expression in Drosophila. *Nature (London)* **372,** 783–786.

Tam, P. P. L., Parameswaran, M., Kinder, S. J., and Weinberger, R. P. (1997). The allocation of epiblast cells to the embryonic heart and other mesodermal lineages: the role of ingression and tissue movement during gastrulation. *Development (Cambridge, UK)* **124,** 1631–1642.

Tonissen, K. F., Drysdale, T. A., Lints, T. J., Harvey, R. P., and Krieg, P. A. (1994). *XNkx-2.5,* a *Xenopus* gene related to *Nkx-2.5* and *tinman:* Evidence for a conserved role in cardiac development. *Dev. Biol.* **162,** 325–328.

Vogt, W. (1929). Gestaltungsanalyse am Amphibienkeim mit ortlicher Vitalfarbung. II. Gastrulation und Mesodermbildung bei Urodelen und Anuren. *Wilhelm Roux' Arch. Entwicklungsmech. Org.* **120,** 384–706.

Wilens, S. (1955). The migration of heart mesoderm and associated areas in amblystoma punctatum. *J. Exp. Zool.* **129,** 579–605.

Winnier, G., Blessing, M., Labosky, P. A., and Hogan, B. L. (1995). Bone morphogenetic protein-4 is required for mesoderm formation and patterning in the mouse. *Genes Dev.* **9,** 2105–2116.

Wu, X., Golden, K., and Bodmer, R. (1995). Heart development in drosophila requires the segment polarity gene *wingless. Dev. Biol.* **169,** 619–628.

Yatskievych, T. A., Ladd, A. N., and Antin, P. B. (1997). Induction of cardiac myogenesis in avian pregastrula epiblast: The role of the hypoblast and activin. *Development (Cambridge, UK)* **124,** 2561–2170.

Zhang, H., and Bradley, A. (1996). Mice deficient for BMP2 are nonviable and have defects in amnion/chorion and cardiac development. *Development (Cambridge, UK)* **122,** 2977–2986.

III

Genetic Dissection of Heart Development

5

Genetic Determination of *Drosophila* Heart Development

Rolf Bodmer* and Manfred Frasch†

* *Department of Biology, University of Michigan, Ann Arbor, Michigan 48109*
† *Brookdale Center for Developmental and Molecular Biology, Mount Sinai School of Medicine, New York, NewYork 10029*

I. Introduction

The *Drosophila* heart is a linear tube that is reminiscent of the primitive heart tube in vertebrate embryos. In both invertebrates and vertebrates, the heart originates from embryologically equivalent, bilaterally symmetrical groups of mesodermal cells. Recently, some of the genetic mechanisms of heart specification have been elucidated in *Drosophila*. Here, we summarize the functions and interactions of genes that subdivide the early mesoderm and orchestrate the specification and initial differentiation of the *Drosophila* heart. Interestingly, many of these genes have vertebrate counterparts with apparently conserved cardiogenic functions, suggesting that basic molecular control mechanisms of heart development are conserved. The emerging picture of an interplay between mesoderm-intrinsic transcription factors and inductive signals from the ectoderm provides a framework that can now be used as a prototype to explore the genetic basis of cardiogenesis in vertebrates.

II. Morphology and Morphogenesis of the *Drosophila* Heart

A. Heart Structure

The heart of *Drosophila* has a simple tubular structure and is probably the only structural component of the fly's circulatory system (Poulson, 1950; Rizki, 1978; Bodmer, 1995). Its major function apparently is to pump various types of blood cells of the hemolymph through the body cavity in an open circulatory system. Because insects have elaborate tracheal systems, the heart is unlikely to be critical as a provider of oxygen. The heart of *Drosophila* is located at the dorsal midline, which differs from vertebrates in which the heart lies ventral to the gut (Fig. 1). Because the *Drosophila* heart looks more like a pulsating blood vessel, it is also called a dorsal vessel.

In the larva, the heart extends along most of the body axis and is suspended at each segment border beneath the dorsal epidermis by the skeletal alary muscles (Fig. 1A). The heart consists of two major cell types: the inner, contractile muscle cells (the "cardial" or "myocardial" cells) are aligned in two rows flanked on each side by an outer row of pericardial cells (Figs. 1A and 1B). The two rows of cardial cells form a central cavity, generating the lumen of the heart (Rizki, 1978; Rugendorff *et al.,* 1994; Bodmer, 1995). As do many other muscles, the cardial cells contain numerous muscle-specific structural proteins, such as actins, tropomyosins, myosin heavy chain, and β3-tubulin. The cardial cells also express muscle-specific transcription factors, such as myocyte-specific enhancer-binding factor 2 (MEF2) (Fig. 1D; see Section VI,A). The subcellular arrangement of myofilaments has been described by Rugendorff *et al.* (1994).

The pericardial cells are loosely associated with the cardial cells (Figs. 1 and 1B). Their function is not well-defined. They do not contain muscle-specific proteins as do the cardial cells and are not contractile. In the thoracic region, the pericardial cells are arranged in two bilaterally symmetrical clusters, termed lymph glands, which serve as blood-forming organs during larval stages (Rizki, 1978). In the embryo, the blood cells derive from the head mesoderm (Tepass *et al.,* 1994), which is distinct from the mesodermal origin of the cardial and pericardial cells (see Section V,D). Anterior to the lymph glands, surrounding the anterior-most part of the heart, is the ring gland, an endocrine organ which is unlikely to be of mesodermal origin. There are at least two different populations of pericardial cells; one has larger nuclei and expresses the gene *even-skipped,* and

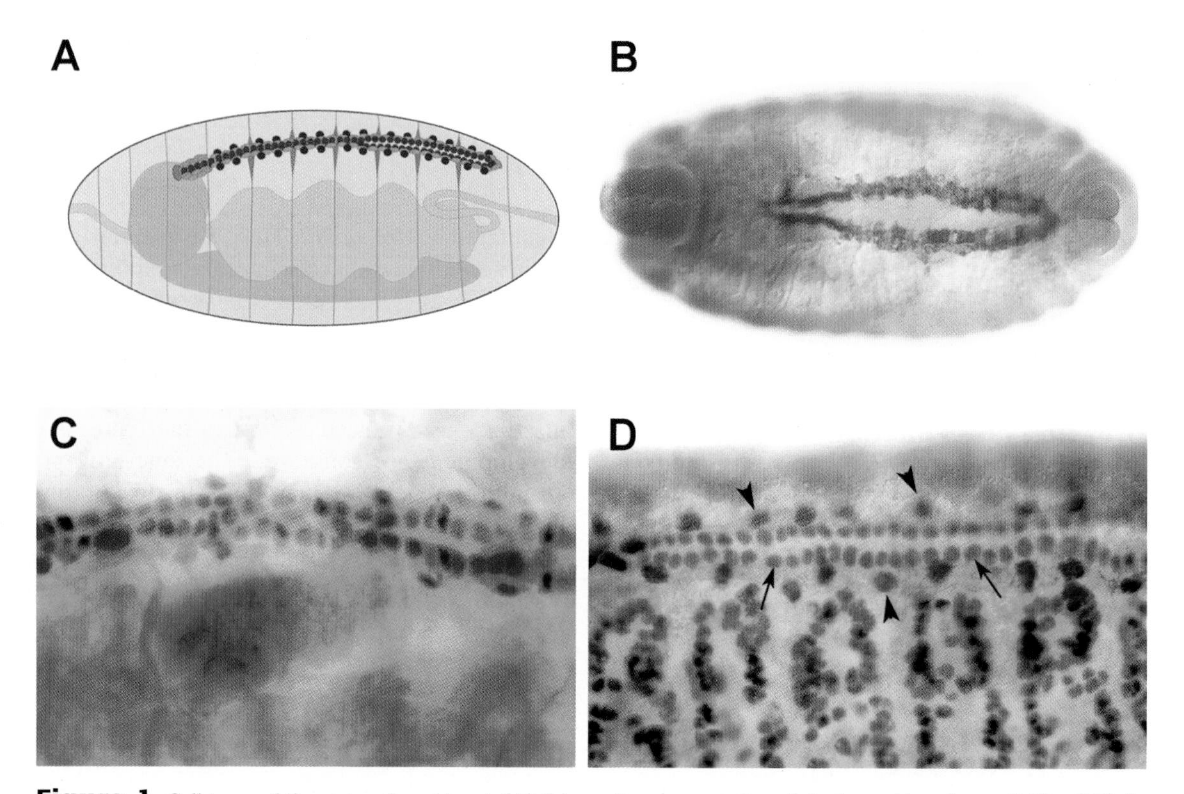

Figure 1 Cell types of the mature larval heart: (A) Schematic representation of the heart (dorsal vessel). The CNS (in blue) is located ventrally. The gut is in pink in the middle. Anterior as in all other figures is to the right (drawing courtesy of K. Golden). (B) Dorsal view of an embryo double labeled for a pericardial (in gray) and cardial (in brown) cell marker (micrograph courtesy of T. V. Venkatesh). (C) Section of the heart double labeled for Tinman protein (brown) and reporter gene expression under the control of the *seven-up* gene (blue-gray) (micrograph courtesy of K. Golden). (D) Section of the heart double labeled for Mef2 (in blue-gray, arrows) in the cardial cells and Eve (in brown, arrowheads) in a subset of pericardial cells (the EPCs) (micrograph courtesy of M.-T. Su).

another expresses *ladybird* (Figs. 1A and 1D; see Section VI,D).

During midembryonic stages of heart development, before the heart tube has formed, no morphological differences are detected between heart precursors along the anterior–posterior axis (Fig. 1B). After heart tube formation, the anterior portion of the heart (also called "aorta") is narrower than the posterior portion (the heart proper). In later larval stages, a valve separates the aorta and the wider portion of the heart. Posterior to this aortic valve in the heart wall, there is a set of small valves, the ostia (Rizki, 1978). The aortic valve controls the pumping of hemolymph and blood cells through the embryo in coordination with the openings and closings of the ostia. Thus, during the rhythmic contractions of the heart, the ostia close and the aortic valve opens, causing the hemolymph to be pumped out through the aorta. When the heart expands again, the aortic valve closes and the ostia open to allow entry of the body fluid.

B. Embryology of Heart Development

In the trunk region of the *Drosophila* embryo, the mesoderm gives rise to four major derivatives: The skeletal (or somatic) muscles of the body wall are arranged in a segmentally repeated pattern of syncytial myofibers, the visceral muscles surround the gut, the fat body is located between the somatic and visceral muscles, and the heart is located most dorsally. At the beginning of gastrulation, the ventral-third portion of the blastoderm embryo, the presumptive mesoderm, forms a furrow along the ventral midline and invaginates into the interior of the embryo (Fig. 2A). Shortly thereafter, the mesodermal cell mass flattens and spreads dorsally to form a monolayer of cells in close apposition to the ectoderm (Figs. 2B and 2C). After the mesoderm has reached the dorsal ectoderm, the first morphologically visible subdivision of mesodermal lineages occurs. At this stage, the dorsal half of the mesoderm segregates into an inner and outer mesodermal layer: The inner layer contributes to the visceral gut musculature (dorsally) and to the fat body (ventrally); the outer layer gives rise to the heart and dorsal body wall muscles (Fig. 2D; Dunin Borkowski *et al.,* 1995). A portion of the mesodermal cells which have migrated most dorsally on either side of the embryo (in the trunk region) becomes specified as heart progenitor cells (Figs. 2D and 2E). The cardiac mesoderm then subdivides into two rows of bilateral symmetrical cells, the future cardial and pericardial cells (Figs. 1B and 2F). In the context of the dorsal extension of the entire germ band, the heart precursors from either side of the embryo move toward each other and form the linear heart at the dorsal midline (Fig. 3, left). The assembly of the contractile heart tube occurs in a highly ordered fashion, in that the left- and right-hand cardial precursors align perfectly with each other (Figs. 1C and 1D), and form a lumen between them. Thus, the structure and formation of the *Drosophila* heart appears to be very simple and is easily amenable to histological and genetic studies.

A prerequisite for the usefulness of the *Drosophila* heart as an experimental system for studying heart development in general is the existence of considerable similarities between the *Drosophila* and vertebrate heart. Superficially, however, the vertebrate heart looks very different from the *Drosophila* heart in that it consists of multiple chambers with numerous specialized cell types; it is looped and connected to an elaborate circulatory system. In contrast, the *Drosophila* heart is a linear structure composed of only a few cell types. Nevertheless, when comparing the very early developmental events between these species, several interesting similarities become apparent (Fig. 3; for review, see Bodmer, 1995; Harvey, 1996; Olson and Srivastava, 1996): Similar to the *Drosophila* heart, the vertebrate heart is also formed from bilaterally symmetrical rows of mesodermal cells, and these appear to be of equivalent embryological origin in that they have migrated most distally from the point of invagination during gastrulation (see Chapter 1). Since the dorsal–ventral axis is reversed between vertebrates and invertebrates (François and Bier, 1995; De Robertis and Sasai, 1995), the cardiac precursors appear dorsally in *Drosophila* and laterally and then ventrally in vertebrates (anterior lateral plate mesoderm). As in *Drosophila,* the bilateral heart primordia of vertebrates also fuse together at the midline and initially form a linear heart tube. In addition, there is an emerging body of evidence that heart development initiates through molecular and cellular mechanisms that have been largely conserved during the evolution of invertebrates and vertebrates (Scott, 1994; Bodmer, 1995; Harvey, 1996). The most compelling evidence stems from studies comparing the gene *tinman,* which is required for heart formation in *Drosophila,* with the vertebrate gene family of *Nkx2-5* and related genes (see Section IV,C). These genes are similar in amino acid sequence and cardiac pattern of expression and they also appear to be functionally equivalent to *tinman.* When the vertebrate *tinman*-like genes are expressed in transgenic *Drosophila,* they are capable of substituting for the *Drosophila tinman* gene and rescue some of the abnormalities caused by the lack of *tinman* function (M. Park and R. Bodmer,

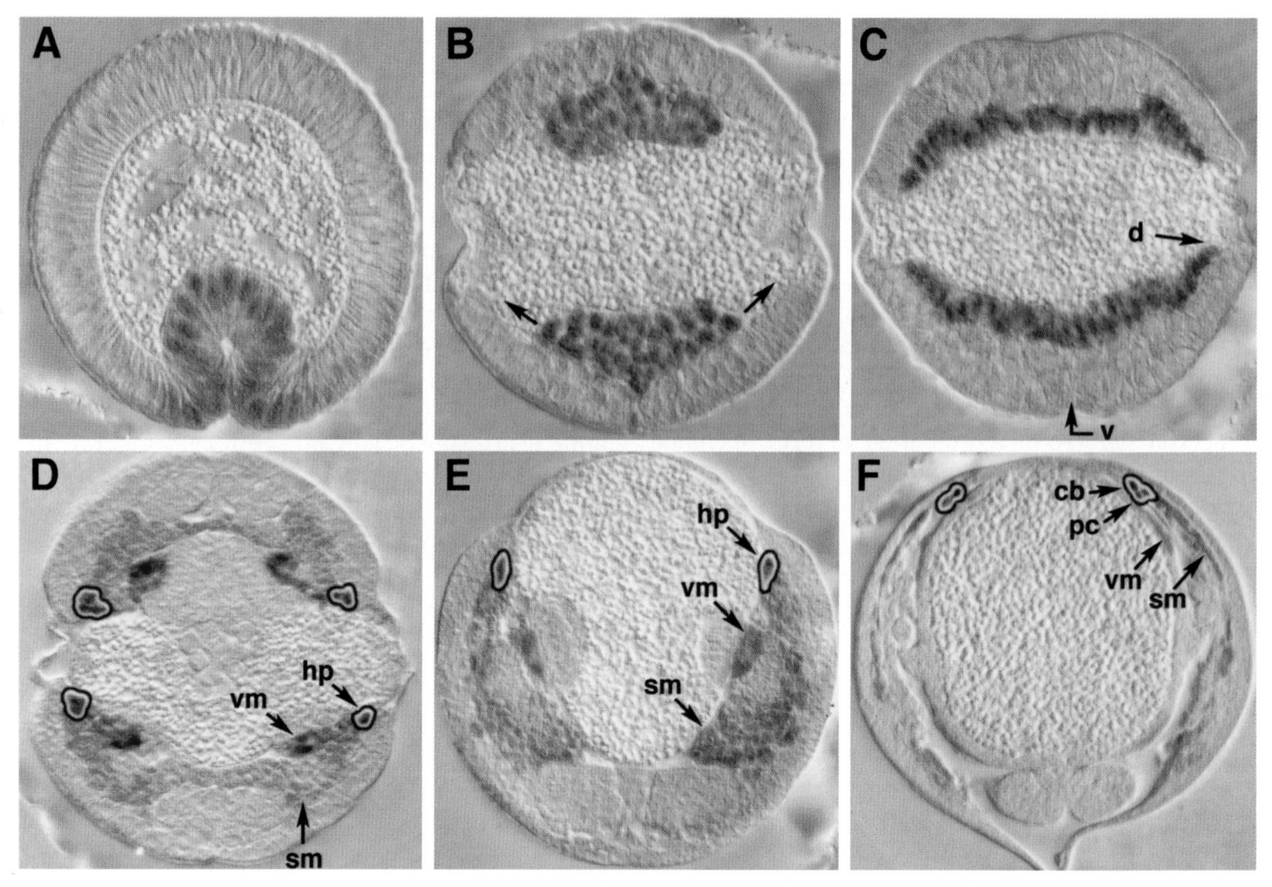

Figure 2 Mesoderm formation and differentiation of mesodermal tissues. (A–F) Cross sections through embryos at progressively advanced developmental stages [ventral sides are down; dorsal (d) and ventral (v) borders of the germ bands in elongation stages (B–D) are as indicated in C]. Embryos in A–C were stained for Twist protein and those in D–F for both Mef2 (light brown) and Tinman (dark brown). (A) Stage 7 (3 hr after fertilization): The mesoderm invaginates on the ventral side of the gastrulating embryo. (B) Stage 8 (3.5 hr): The mesoderm spreads dorsally (arrows). (C) Stage 9 (4 hr): The mesoderm has reached the dorsal border of the ectoderm. (D) Late stage 11 (7 hr): Heart precursors (hp) and visceral mesoderm (vm) have segregated from the somatic mesoderm (sm). Fat body primordia (between vm and sm) are unstained. (E) Stage 13 (10 hr): Both mesoderm and ectoderm extend toward the dorsal side. (F) Stage 15 (11.5 hr): Mesodermal primordia differentiate into somatic and visceral muscles (sm and vm) and into the cardioblasts (cb) and pericardial cells (pc) of the future dorsal vessel.

unpublished observations). Taken together, there seem to be sufficient molecular and embryological similarities between *Drosophila* and vertebrates, and thus *Drosophila* can serve as an excellent model for the early steps of cardiogenesis.

III. Genetic Control of the Formation and Dorsal Expansion of the Mesoderm

A. *twist* and *snail*

In *Drosophila,* the formation of the mesoderm in ventral regions of blastoderm embryos is largely controlled autonomously, i.e., by mechanisms that act within the nuclei or cells that acquire mesodermal fates. This is in contrast to vertebrate embryos in which mesoderm formation is thought to depend on inductive processes. *Drosophila* mesoderm formation is initiated by a nuclear gradient of the maternally provided, NK-κB-related morphogen Dorsal[1] (reviewed in Rusch and Levine, 1996). Peak levels of nuclear Dorsal protein are present along the ventral midline of blastoderm embryos and are required to activate two zygotic genes, *twist* and *snail,* that are essential for mesoderm formation. *twist* encodes a basic helix–loop–helix protein and

[1]According to general practice in the *Drosophila* literature, we write names of genes in italics and lowercase (e.g., *dorsal*) and names of the corresponding gene products in plain text with the first letter in uppercase (e.g., Dorsal).

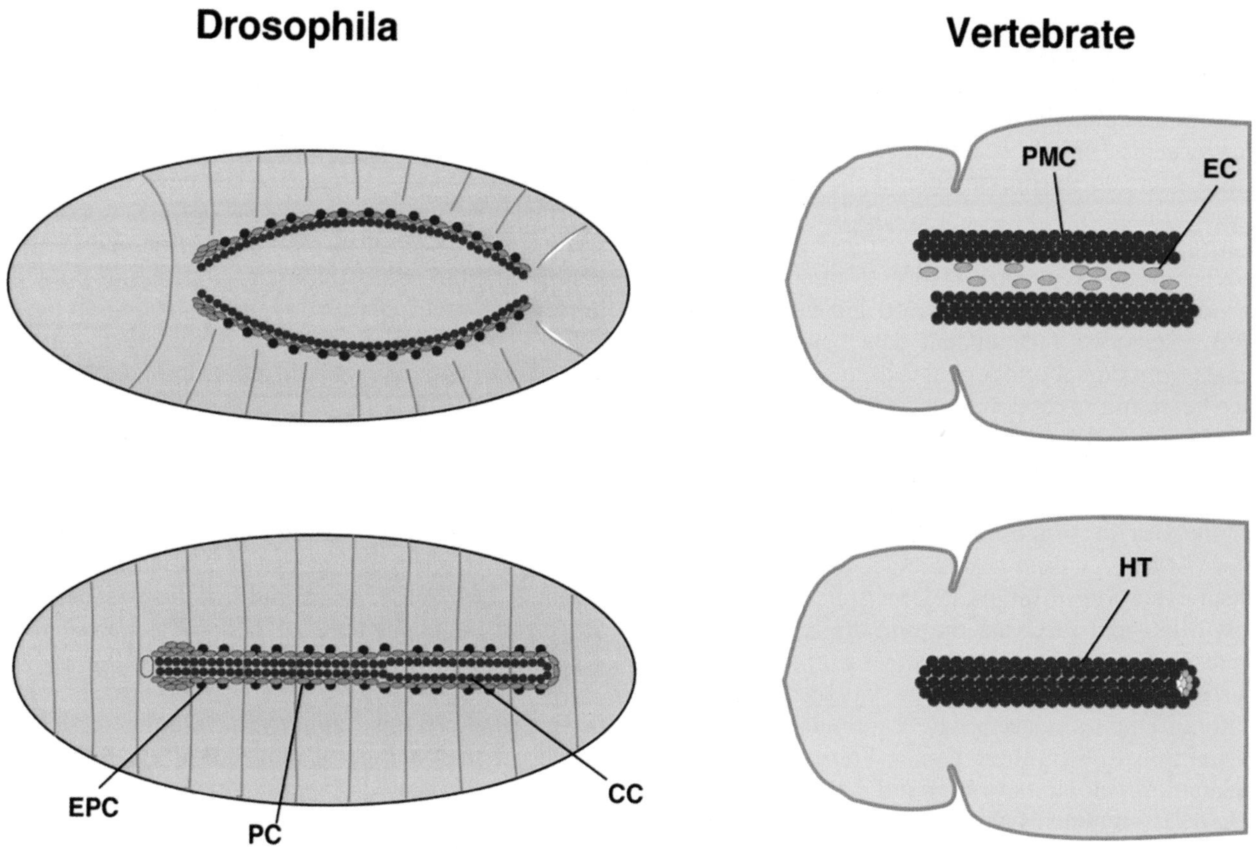

Figure 3 Comparison of heart tube formation in *Drosophila* and vertebrates. (drawings courtesy of K. Golden). CC, cardial cells; EC, endocardial cells; EPC, Eve pericardial cells; HT, heart; PC, pericardial cells; PMC, premyocardial cells.

snail a zinc finger-containing protein, thus indicating that both gene products act as transcription factors (Boulay *et al.,* 1987; Thisse *et al.,* 1988). Mutations of either of the two genes cause virtually identical phenotypes which consist of the complete lack of invagination and mesoderm differentiation (Simpson, 1983; Grau *et al.,* 1984). However, there is genetic evidence suggesting that the two genes play different roles in mesoderm development. *snail* appears to function mainly to repress nonmesodermal genes in the prospective mesoderm and thus has a permissive role in mesoderm formation. By contrast, *twist* is thought to have a key role in activating downstream genes that are required for the processes of invagination, patterning, and differentiation of the mesoderm (Kosman *et al.,* 1991; Leptin, 1991). Candidates for target genes of *twist* at blastoderm stage include *folded gastrulation,* which codes for a secreted molecule controlling aspects of mesoderm invagination (Costa *et al.,* 1994); *heartless,* which encodes a fibroblast growth factor (FGF) receptor homolog that is involved in mesoderm migration; the homeobox gene *tinman* that is crucial for mesoderm patterning and heart formation (see Section IV.); *mef2,* which encodes a MADS domain transcription factor that functions in later aspects of muscle and heart differentiation (see Section VI,A); and *zfh-1,* a zinc finger- and homeobox-containing gene which is required for the differentiation of a subset of heart cells (see Section VI,B).

B. *heartless*

The *heartless* gene is expressed uniformly in the early mesoderm in a *twist*-dependent manner (Shishido *et al.,* 1993). In *heartless* mutants, the formation of the dorsal vessel and midgut visceral mesoderm is severely reduced. However, the primary phenotype of *heartless* mutants appears to be the inability of the invaginated mesoderm mass to spread toward the dorsal ectoderm. Instead, the mesoderm flattens only partially, and only small portions of it fortuitously reach more dorsal positions (Beiman *et al.,* 1996; Gisselbrecht *et al.,* 1996; Shishido *et al.,* 1997). These observations suggest that the extension of the mesoderm toward dorsal areas of the germ band is triggered by FGF signaling. A candidate for the FGF ligand involved in this process has not been found and its source is unknown. Interestingly, an antibody that specifically recognizes activated MAP kinase molecules detects high activities, presumably due to activated Heartless receptors, in the dorsolateral edges of the spreading mesoderm of wild-type embryos (Gabay *et al.,* 1997). This indicates that the ligand is released from dorsal areas, most likely being secreted from dorsal ectodermal or amnioserosa cells, and serves to attract mesodermal cells toward these areas.

Why does a reduction of dorsal spreading of the mesoderm cause defects in heart and visceral mesoderm formation? Independent data have shown that signals from dorsal ectodermal cells are required to induce heart and visceral mesoderm formation in the underlying cells of the dorsal mesoderm (see Section V). As discussed later, the major inducing signal in this process has been identified as Dpp, a secreted protein homologous to vertebrate bone morphogenetic proteins (BMPs). In *heartless* mutants, disruption of the dorsal extension of the mesoderm appears to prevent most cells from receiving the dorsally restricted Dpp signal. Thus, the loss of heart and visceral mesoderm in *heartless* mutant embryos can be explained by the inability of Dpp to induce a significant number of mesodermal cells to form these tissues. Consistent with this explanation, ectopic expression of Dpp in ventral areas of *heartless* mutants is able to restore the formation of both tissues (Beiman *et al.,* 1996; Gisselbrecht *et al.,* 1996).

IV. Expression and Function of *tinman* in Mesoderm Patterning and Heart Development

A. The Function of *tinman*

In contrast to the mesoderm determinant *twist,* which is required for the establishment of all mesoderm, and *heartless,* which is needed for the flattening and dorsal extension of the mesoderm, the gene *tinman* is crucial for the subsequent subdivision of mesoderm into visceral, cardiac, and somatic mesoderm (Azpiazu and Frasch, 1993; Bodmer, 1993). Although *tinman* is expressed initially in the entire mesoderm of the trunk (Figs. 4A and 4B), as are *heartless* and *twist* (Figs. 2A–2C), *tinman* is not required for gastrulation nor for dorsal migration of the mesoderm. The major role of *tinman* is in the initial specification of tissues in dorsal portions of the mesoderm. Indeed, in *tinman* mutants, all derivatives from the dorsal mesoderm appear to be absent. The heart and midgut musculature are not formed (Fig. 5), and the absence of the dorsal muscle markers *eve* and *msh* and of dorsal muscle-specific enhancer/*lacZ* expression in *tinman* mutants indicates that dorsal body wall muscles (including muscles 1, 2, 9, and 10) also fail to be specified (Azpiazu and Frasch, 1993; Bodmer, 1993; D'Alessio and Frasch, 1996; Yin and Frasch, 1998). In the place of these muscles, lateral muscle fibers appear enlarged and extend dorsally.

By contrast, most tissues and cell types derived from ventral and ventrolateral portions of the mesoderm develop in the absence of *tinman* function and thus the majority of the somatic muscles and the fat body are present. However, *tinman* does have functions in the development of ventrally derived tissues, albeit more selective ones compared to the dorsal mesoderm. For example, *tinman* is required for the specification of a distinct subset of ventrolateral muscle founders and the development of the corresponding muscles to which they give rise (Azpiazu and Frasch, 1993). In addition, *tinman* is required for the formation of a set of mesodermally derived glial cells, called DM cells, that are positioned along the ventral midline of the mesoderm and are involved in guidance of peripheral nerves (Gorczyka *et al.,* 1994).

The direct downstream targets of *tinman* during these early functions of *tinman* have not been defined. Based on their temporal and spatial expression patterns, as well as on the absence of their expression in *tinman* mutants, several genes have been proposed as candidate targets for *tinman.* These include the homeobox genes *even-skipped* and *ladybird* in heart progenitors (Azpiazu and Frasch, 1993; Bodmer, 1993; Jagla *et al.,* 1997), *bagpipe* in visceral mesoderm primordia (Azpiazu and Frasch, 1993), *msh* in muscle progenitors, and *buttonless* in DM cell precursors (Yin and Frasch, 1998). In addition, there is evidence for autoregulatory functions of *tinman* (R. Bodmer and M. Frasch, unpublished observations). Molecular and genetic analysis will be required to demonstrate that *tinman* binds to essential enhancer elements from these genes.

B. The Expression of *tinman*

tinman seems to be a crucial link between the determination of mesoderm by *twist* and later events that lead to the differentiation of mesodermal sublineages. Indeed, *tinman* is expressed only minutes after Twist protein is present in the mesodermal anlagen at blastoderm stage and this expression depends on *twist* function (Figs. 4A and 4B; see Section V,D,). Upon completion of the dorsal spreading of the mesoderm, prior to its subdivision into somatic and visceral components, *tinman* expression becomes restricted to the dorsal half of the mesodermal monolayer, the "dorsal mesoderm" (Fig. 4C; Bodmer *et al.,* 1990). The restriction of *tinman* expression to the dorsal mesoderm is likely to be functionally important since this is where *tinman* ex-

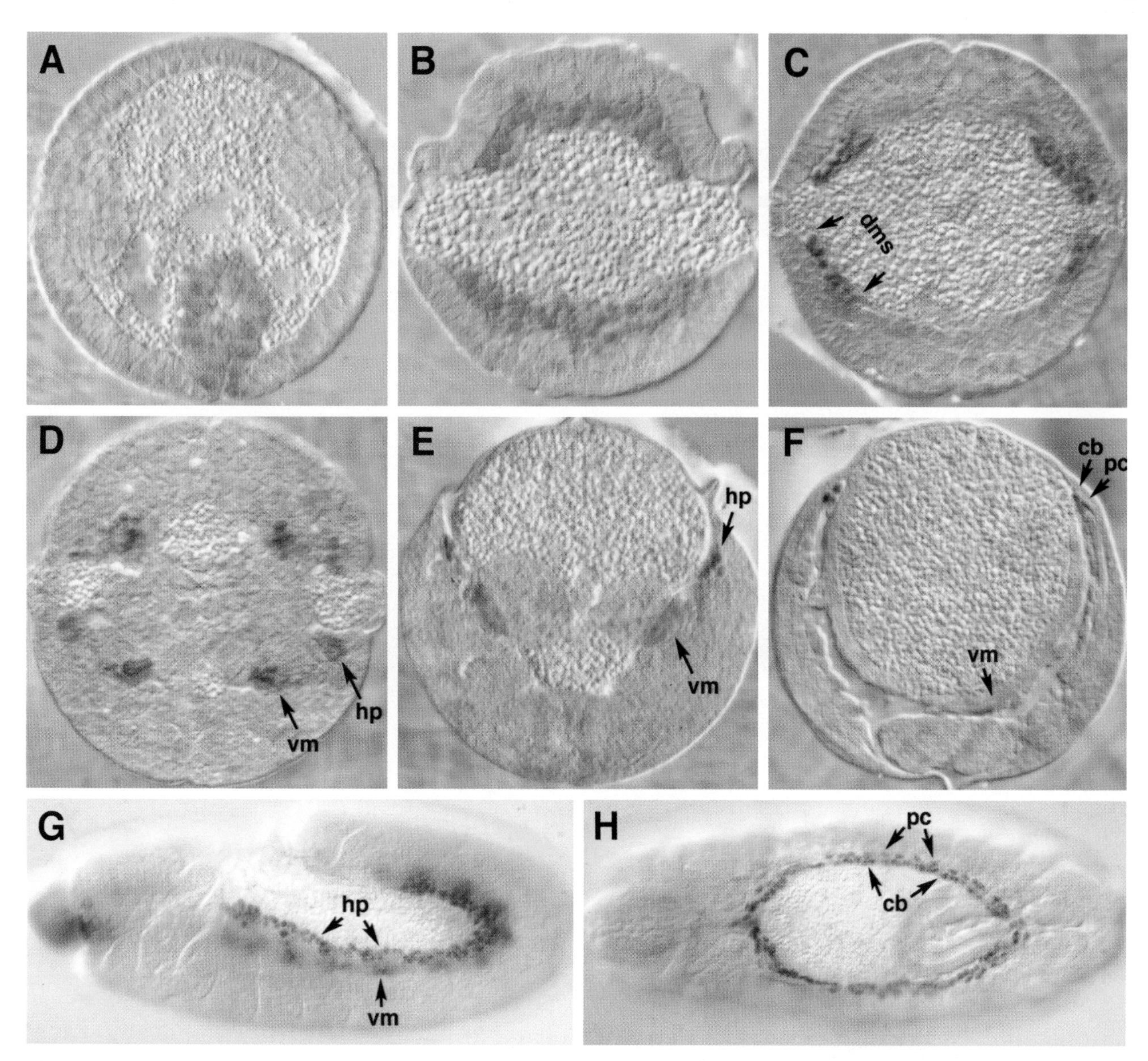

Figure 4 Expression of Tinman protein during embryogenesis. The developmental stages shown in A–F are similar to those shown in Figs. 2A–2F. During gastrulation (A) and dorsal spreading of the mesoderm (B), Tinman is expressed ubiquitously in the trunk mesoderm. On completion of the dorsal spreading, Tinman expression becomes restricted to the dorsal mesoderm (dsm). At later stages (D–F), Tinman is observed in heart precursors (hp), which segregate into a dorsal row of cardioblasts (cb) and ventrally adjacent rows of pericardial cells (pc). In addition, *tinman* mRNA is transiently expressed in the visceral mesoderm (vm), where detectable protein (but not mRNA) levels are maintained until late stages (see F). (G) Tinman protein expression in the hp and vm of an early stage 12 embryo (side view), and (H) Tinman in cb and pc of a stage 14 embryo (dorsal view).

erts its major functions during the specification of cardiac and visceral mesoderm. By contrast, the *tinman*-independent fat body and somatic muscles originate from more ventral regions. The more selective activities of *tinman* in specification of certain somatic muscles and glial cells are most likely required during its early expression in the entire trunk mesoderm.

After the mesoderm has morphologically subdivided and just before the retraction of the germ band, *tinman* RNA expression ceases in forming visceral mesoderm and becomes exclusively restricted to the cardiac precursors, where its expression persists after the formation of the heart tube (Bodmer *et al.,* 1990). Tinman protein, which otherwise mimics faithfully the RNA pattern, persists for some time in the developing visceral and gonadal mesoderm (Figs. 4 D–H; Jagla *et al.,* 1997; Yin *et al.,* 1997; K. Ocorr, R. Bodmer, and M. Frasch, unpublished observations). Thus, it is con-

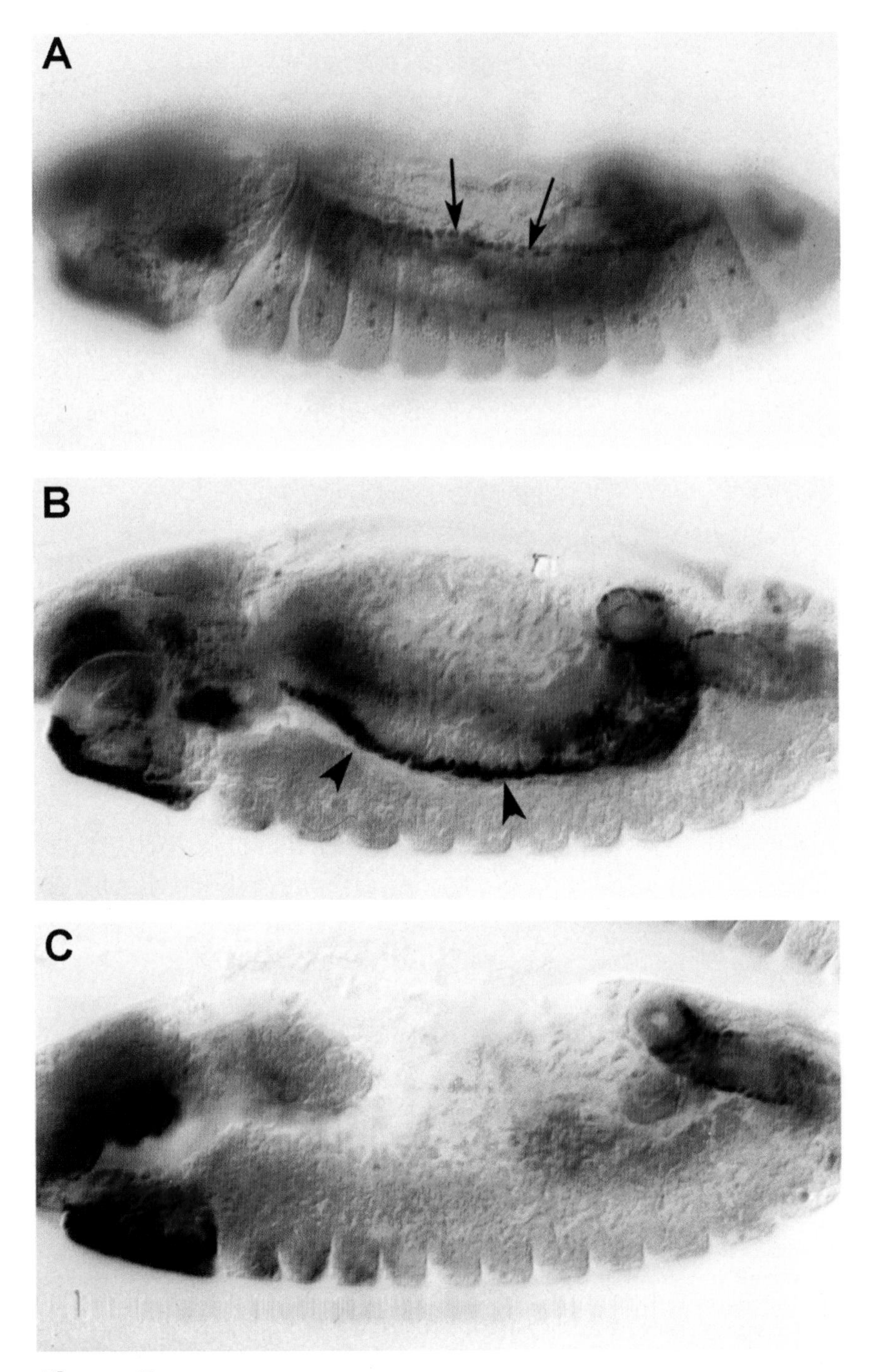

Figure 5 Disco expression in *wt* (A, B) and *tin* mutant embryos (C). (A) Cardial cells of the heart are indicated by arrows. (B) Visceral mesoderm is indicated by arrowheads. (C) No heart or visceral mesoderm formation in *tinman* mutants (micrographs courtesy of K. Golden).

ceivable that *tinman* may also be required for the correct differentiation and morphogenesis of the heart (and perhaps of the visceral muscles and gonads) in addition to its requirement for the initial specification of dorsal mesodermal fates. As the heart tube forms, Tinman protein is present in a segmentally repeated pattern in that four of an average of six cardial and pericardial cells in each hemisegment express Tinman protein, whereas the remaining two cells lack expression (Fig. 1C; Jagla *et al.,* 1997).

C. Functional Equivalence of *tinman* and Vertebrate *tinman*-Related Genes

Thus far, six vertebrate genes have been isolated from many divergent species and have been postulated to be structurally and possibly functionally related to *Drosophila tinman: Nkx2-3, Nkx2-5, Nkx2-6, Nkx2-7, Nkx2-8,* and *Nkx2-9* (note that not all six genes have been found in a single species; for review, see Chapters 7 and 16; Bodmer, 1995; Harvey, 1996; Newman and Krieg, 1998). Interestingly, many of them are expressed in the early precardiac mesoderm. These observations suggest that a set of *tinman*-related homeobox genes may be crucial for cardiac myogenesis in vertebrates. Elucidation of heart development in *Drosophila* may provide important clues concerning the molecular basis of cardiogenesis in vertebrates, just as the discovery of the *Drosophila Hox* genes revealed striking similarities between the developmental mechanisms by which the anterioposterior body axis is elaborated in vertebrates and many invertebrates (Krumlauf, 1994).

tinman, bagpipe, and their vertebrate relatives encode a homeodomain of the NK type that has been shown to preferentially bind to DNA sites containing sequences with a CAAG core (Chen and Schwartz, 1995; Damante *et al.*, 1996). In addition, the products of this gene family contain a conserved stretch of 10–12 amino acids (TN domain) close to the amino terminus, which has homology to the *engrailed* repressor element (Smith and Jaynes, 1996). Bagipe and all Tinman-related gene products in vertebrates, but not Tinman itself, also contain a conserved hydrophobic domain carboxy terminal to the homeodomain [the NK2-specific domain (NK2-SD)]. It is likely that these genes have arisen from a common ancestor by gene duplication. This idea is supported by the fact that *tinman* and *bagpipe* are closely linked in the genome (Kim and Nirenberg, 1989; Azpiazu and Frasch, 1993). Since *tinman* does not have a NK2-SD and the six vertebrate "*tinman*-like" genes are as similar to *bagpipe* as they are to *tinman,* it has been proposed that *tinman* has diverged considerably from its other relatives (Bodmer *et al.*, 1997). Although this hypothesis must be confirmed, several lines of circumstantial evidence corroborate this proposal: The *bona fide* vertebrate relatives of *bagpipe* are much more closely related (in the homeodomain) to their fly counterpart than the putative vertebrate *tinman*-like genes are to fly *tinman.* The majority of the *tinman*-related genes in a variety of species are expressed in the early heart primordium, whereas the *bap*-related genes are not (Newman and Krieg, 1998). In addition, one pair of linked genes has been found in the mouse that may have arisen from an ancestral *tinman/bagpipe* pair (Komuro and Izumo, 1993). Moreover, some of the vertebrate *tinman*-like genes can substitute (at least partially) for *tinman* function when expressed in *Drosophila,* but *bagpipe* cannot.

The homeodomains of the vertebrate *tinman*-like genes share 80% or more sequence identity but are only about 65% identical to *Drosophila tinman.* Thus, these genes must have arisen relatively recently from a common (vertebrate) ancestor, which is likely to be related to *tinman.* Similar to the *Hox* genes, the myogenic genes, or the MEF2 genes, the *tinman*-related genes may have retained partially overlapping functions in heart development. In all species reported to date, *Nkx2-5,* which is the best characterized member of this gene family, is prominently expressed in a bilaterally symmetrical group or a crescent of lateral plate mesoderm at the time when the cardiac progenitors are thought to become committed as a result of an inductive signal from the anterior endoderm (Schultheiss *et al.,* 1995; Nascone and Mercola, 1995). A targeted knockout mutation in *Nkx2-5* results in severe abnormalities in heart development, indicating a vital role of this gene in vertebrate cardiogenesis (Lyons *et al.,* 1995). In contrast to eliminating *tinman* function in *Drosophila,* which causes a total lack of heart progenitors, the *Nkx2-5* mutant mice do form a linear (and beating) heart tube, possibly because other *tinman*-like genes partially substitute for *Nkx2-5* function.

To test for possible functional equivalence between the vertebrate and *Drosophila tinman* genes, heterologous expression was used to examine whether the vertebrate genes can rescue the *tinman* mutant phenotype of fly embryos (Park *et al.,* 1998a). Expression of a chimeric *tinman* gene product, in which the 3′ portion including the homeobox was replaced by the mouse *Nkx2-5* homeobox and NK2-SD, restores heart and visceral mesoderm development, suggesting that the *tinman* and *Nkx2-5* homeodomains are functionally interchangeable despite their limited homology (65%). Therefore, the *in vivo* target specificity of these homeodomains appears to be sufficiently similar, which is consistent with them binding the same consensus sequence *in vitro* (Chen and Schwartz, 1995; Gajewski *et al.,* 1997). The addition of a NK2-SD to *Drosophila tinman* does not appear to interfere substantially with its function. On the other hand, full-length *Nkx2-5, Nkx2-3,* and *Nkx2-7* genes from several vertebrate species in *tinman* mutant fly embryos are able to rescue only the visceral mesoderm but not the heart development. This is in contrast to *bap,* which has no rescue activity of *tinman* mutants (Park *et al.,* 1998a). This suggests that this N-terminal sequence from the *Drosophila* Tinman protein has an important role in heart development but not in visceral mesoderm specification. These findings clearly demonstrate that *tinman* and its vertebrate relatives are similar in their structure and expression patterns and suggest that they are also equivalent in many functional aspects.

V. A Combination of Inductive Signals Is Necessary for Cardiac Specification

A. The Signaling Factors Encoded by *Decapentaplegic* and *Wingless*

Immediately after gastrulation, the fate of mesodermal cells appears to be largely uncommitted (Beer *et al.*, 1987). Consistent with this, expression of *tinman* is ubiquitous in the trunk mesoderm at this time, and only later does its expression become confined to the dorsal portion of the mesoderm, where it is thought to have its major function. Experiments in embryos of other insects suggest that a likely source of patterning information for mesodermal subdivisions is the ectoderm, with which the invaginated mesoderm is in close contact (Seidel *et al.*, 1940). Results from more recent experiments, which involved blocking gastrulation by treatment with colchicine and thus preventing an apposition of ectoderm and mesoderm, suggest that the ectoderm has an instructive role in patterning the mesoderm (Baker and Schubiger, 1995). Similar conclusions were obtained in genetic experiments in which the mesoderm was prevented from reaching the dorsal ectoderm, which resulted in the preferential loss of heart and visceral mesoderm [e.g., in *heartless* mutants (see Section III,B) and in embryos genetically engineered to have fewer mesodermal cells; Maggert *et al.*, 1995]. Recently, two secreted factors that are primarily expressed in the ectoderm have been shown to serve as inductive signals for patterning the mesoderm, in particular the formation of the cardiac and visceral mesoderm. One of these molecules is the product of the gene *decapentaplegic* (*dpp*), a member of the bone morphogenetic protein subgroup of the transforming growth factor-β (TGF-β) superfamily proteins (Kingsley, 1994). The other secreted molecule is encoded by *wingless* (*wg*), a homolog of the vertebrate *wnt-1* gene (Nusse and Varmus, 1992; Perrimon, 1994; Klingensmith and Nusse, 1994; Venkatesh and Bodmer, 1995).

1. Signaling by *dpp*

Members of the TGF-β superfamily, widely conserved in different organisms, mediate key events in growth and development and exhibit diverse activities such as specification of the dorsoventral body axis in invertebrates, control of sexual development, and bone and cartilage formation in mammals (Kingsley, 1994). TGF-β proteins have also been shown to induce mesoderm in amphibian embryos (Smith and Howard, 1992). In *Drosophila, dpp* is expressed throughout the dorsal half of the embryo beginning at blastoderm stage, in which it is required for the development of dorsal and lateral derivatives of the epidermis as part of the zygotic components that specify the embryonic dorsal–ventral axis of ectodermal cuticular structures (Spencer *et al.*, 1982; Ferguson and Anderson, 1992; François *et al.*, 1994). Much later in embryogenesis, *dpp* is also expressed in the visceral mesoderm and acts as an inductive signal across germ layers (Bienz, 1994). *dpp* also has a inductive role in axis specification of the imaginal disc primordia of adult appendages (Spencer *et al.*, 1982; Nellen *et al.*, 1996; Lecuit *et al.*, 1996; see also review by Blair, 1995). Recent evidence suggests that ectodermally expressed *dpp* also has a direct role in patterning the mesoderm.

After gastrulation, when *tinman* expression becomes restricted to the dorsal mesoderm, its ventral limits coincide with the ventral limits of *dpp* in the dorsal ectoderm directly above (Figs. 6A and 6D). Mesodermal cells that do not "contact" *dpp* expressing ectodermal cells lose *tinman* expression shortly after the mesoderm has reached its dorsal margin. This led to the hypothesis that *dpp* may encode an inductive ectodermal signal necessary for patterning the mesoderm (Frasch, 1995; Staehling-Hampton *et al.*, 1994). Indeed, in *dpp* mutant embryos *tinman* expression fades shortly after gastrulation, indicating that *dpp* signaling from the dorsal ectoderm is needed to maintain *tinman* expression in the dorsal mesoderm (in contrast to suppressing it ventrally). The early pan-mesodermal *twist*-dependent expression of *tinman* is not affected in *dpp* mutants. The absence of *dpp*-dependent *tinman* expression in the dorsal mesoderm results in a failure to develop cardiac mesoderm. In addition, loss of *tinman* expression in the dorsal mesoderm of *dpp* mutants results in the absence of *bap* expression, and visceral mesoderm is not formed.

Further evidence that *dpp* encodes a direct signal for mesodermal patterning stems from experiments in which *dpp* is ectopically expressed. *dpp* expression in the entire ectoderm maintains *tinman* and activates *bap* expression in the ventral mesoderm at a stage when the expression of both genes is normally confined to the dorsal mesoderm. This indicates that *dpp* is not only necessary but also sufficient for dorsal mesodermal cell fates, at least with respect to *tinman* and *bagpipe* expression and visceral mesoderm formation (Frasch, 1995; Staehling-Hampton *et al.*, 1994). The *dpp*-dependent patterning cannot exclusively be mediated by maintaining *tinman* expression in the dorsal mesoderm since ubiquitous *tinman* expression alone at the time of its dorsal restriction is insufficient to confer dorsal mesodermal fates to ventral mesoderm. This suggests that

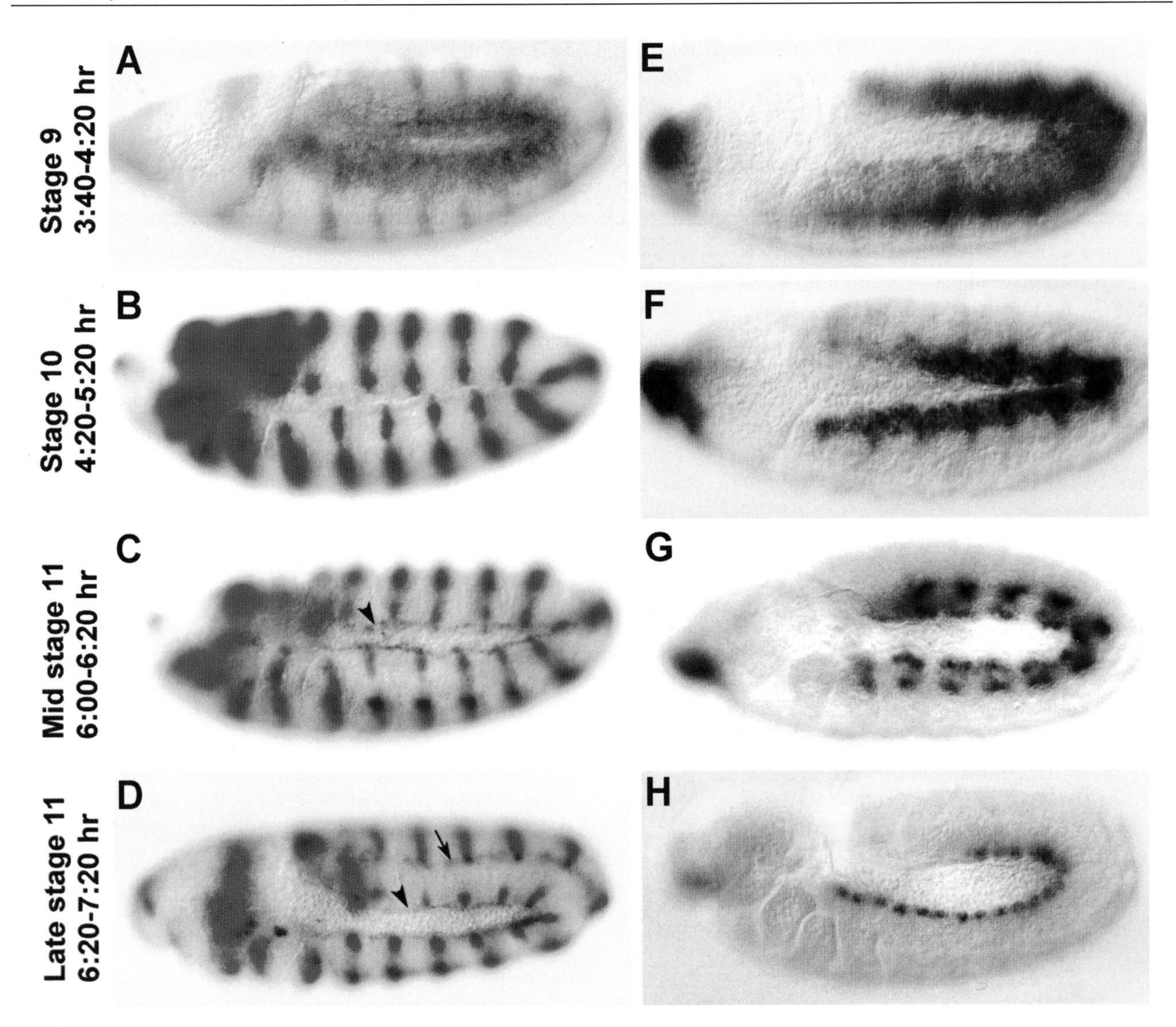

Figure 6 Wild-type expression patterns of ectodermal *wg* (turquoise) and *dpp* (purple) (A–D) and mesodermal *tinman* (E–H). *dpp* and *tinman* are visualized by *in situ* hybridization and *wg* by X-gal staining to a *wg-lacZ* marker. (A) Stage 9 *wg* and *dpp* patterns. *dpp* is expressed in a broad band in the dorsal ectoderm, and *wg* is expressed in 15 continuous circumferential stripes. (B) Stage 10 *wg* and *dpp* patterns: The broad band of *dpp* expression decreases during late stage 9 and is further reduced during stage 10. During stage 10 the pattern of *wg* also changes. The *wg* continuous stripes become interrupted laterally and persist only in the dorsal and ventral regions. This interrupted pattern continues throughout embryogenesis. (C) Mid-stage 11 *wg* and *dpp* patterns. *dpp* reappears in a fine line of expression at the extreme dorsal ectodermal edge (arrowhead), and *wg* expression persists in laterally interrupted stripes. (D) Late stage 11 *wg* and *dpp* patterns: The dorsal line of *dpp* expression persists (arrowhead), and a second stripe of *dpp* appears at a lateral position, just dorsal to the CNS (arrow). *wg* expression is unchanged. (E) At stage 9 *tin* is expressed ubiquitously in the trunk mesoderm. (F) At stage 10 *tin* is restricted to the dorsal mesoderm (by *dpp;* see text). (G) At mid-stage 11 *tin* is further restricted to segmental regions corresponding to the *wg* stripes, in two clusters. The dorsal cluster corresponds to heart precursors and is dependent on *wg* signaling. (H) At late stage 11 *tin* expression is confined to the cardial and pericardial cells of the heart (micrographs courtesy of W. Lockwood).

additional patterning factors are induced or activated in response to *dpp* signaling. It is not known whether these factors are induced independently of *tinman,* or whether *tinman* and Dpp signals are required together for their induction. In summary, it is strongly suggested that *dpp* is indeed producing an ectodermal signal to specify dorsal mesoderm, which is in part accomplished by maintaining *tinman* expression. This appears to generate one of the first subdivisions of the mesoderm, leading to a distinction between "dorsal" and "ventral" mesoderm that enter separate developmental pathways.

If the ectodermal Dpp signal is directly instructing the underlying mesoderm, one would predict that Dpp receptors are present in the mesoderm. Indeed, the putative *dpp* receptors, *punt* and *thickveins,* which encode serine/threonine transmembrane kinases, are present and required in the mesoderm for *dpp*-dependent signaling (Ruberte *et al.,* 1995; Letsou *et al.,* 1995; for review, see Wharton, 1995). Mutation of *thickveins* and ectopic mesodermal expression of constitutively active Thickveins receptors have similar effects on *tinman* expression as mutation and overexpression of *dpp,* respectively (Yin and Frasch, 1998). This indicates that the Dpp signals are transmitted through Thickveins to the dorsal mesoderm. Three presumed downstream components of the Dpp signaling pathway have been identified: *Mothers against dpp* (*Mad;* Sekelsky *et al.,* 1995) and *Medea* (Raftery *et al.,* 1995), both of which code for cytoplasmic proteins likely to be translocated to the nucleus (as suggested by studies of vertebrate homologs of *Mad;* reviewed in Heldin *et al.,* 1997); and *schnurri,* which codes for a zinc finger-containing transcription factor (Arora *et al.,* 1995; Grieder *et al.,* 1995; Staehling-Hampton *et al.,* 1995). Because in *schnurri* mutant embryos there is only a mild reduction of the dorsal *tinman* domains and heart formation is not completely disrupted, it is possible that *schnurri* has a stimulating rather than an essential function in Dpp-mediated dorsal mesoderm induction (Yin and Frasch, 1998). By contrast, embryos lacking both the maternal and zygotic activities of *Medea* display the same mesodermal phenotypes as *dpp* mutant embryos, indicating that *Medea* is an essential component in the *dpp* pathway between *thickveins* and *tinman* (X. Xu, E. Ferguson, and M. Frasch, unpublished results). It will be interesting to determine if *schnurri* and *Mad/Medea* are directly involved in regulating *tinman* (and *bagpipe*) expression in the dorsal mesoderm.

2. Signaling by *wingless*

Although *dpp* is required to maintain *tinman* expression in the dorsal mesoderm, additional spatial cues are likely to play a role in distinguishing cardiac versus visceral versus somatic muscle cell fates within the dorsal mesoderm. A second signaling molecule which functions in mesodermal patterning has been identified and is encoded by *wg,* a *Drosophila* homolog of the mouse oncoprotein Wnt-1, which is a secreted glycoprotein (van den Heuvel *et al.,* 1989; reviewed in Nusse and Varmus, 1992). Like members of the TGF-β family, Wnt family proteins are also highly conserved in different species and play key roles in developmental processes. They are involved in morphogenesis of the central nervous system (CNS) (McMahon and Bradley, 1990; Thomas and Capecchi, 1990), mesodermal patterning (Takada *et al.,* 1994), and cell proliferation (Roelink *et al.,* 1990). They are also implicated in embryonic axis specification and mesodermal patterning in *Xenopus* embryos (McMahon and Moon, 1989).

In *Drosophila, wg* is required for a variety of inductive signaling events during both embryonic and imaginal development (e.g., Basler and Struhl, 1994; Ng *et al.,* 1996; for reviews see Peifer and Bejsovec, 1992; Perrimon, 1994; Klingensmith and Nusse, 1994; Blair, 1995). In the early embryo, *wg* is expressed in 15 transverse stripes in the trunk region (Baker, 1987) and plays a central role in determining the anterior–posterior segmental polarity of the ectoderm. In addition, it is also required for head patterning (Schmidt-Ott and Technau, 1992), for the specification of subsets of CNS neuroblasts (Chu-LaGraff and Doe, 1993), for patterning of the midgut (Bienz, 1994), and for growth of Malphigian tubules (Skaer and Martinez Arias, 1992). In the developing leg, *wg* is required for both dorsal–ventral and anterior–posterior axial patterning (Campbell and Tomlinson, 1995), and in the developing wing it is required for the growth of the wing blade as well as for patterning the bristles in the wing margin (Williams *et al.,* 1993; Couso *et al.,* 1994; Axelrod *et al.,* 1996). Thus, as is the case for *dpp, wg* plays a key signaling role in a plethora of developmental processes.

Recent evidence suggests that *wg* is also directly involved in heart formation (Wu *et al.,* 1995; Park *et al.,* 1996). In contrast to a loss of *dpp* function, which causes a failure of visceral and cardiac mesoderm to form, *wg* mutants do form visceral and some somatic mesoderm (albeit abnormally patterned; Volk and Vijay-Raghaven, 1994; Baylies *et al.,* 1995; Azpiazu *et al.,* 1996; Ranganayakulu *et al.,* 1996), but the heart and its precursors are entirely missing. Thus, it appears that *wg* is involved in further subdividing the dorsal mesoderm and is necessary for specifying cardiac cell fates. Given the requirement for *wg* in ectodermal segmentation, a temperature-sensitive *wg* allele was instrumental in addressing the question of timing and specificity for *wg* in cardiac mesoderm formation. Elimination of *wg* function shortly after gastrulation, at a time when *tinman* becomes restricted to the dorsal mesoderm, results in the selective loss of heart progenitor cells with little effect on segmental patterning of the cuticle or other mesodermal derivatives (Wu *et al.,* 1995). As expected, *tinman* expression is only lost in the presumptive cardiac precursors and is normal in the early mesoderm and during its subsequent dorsal restriction. Although *wg* is expressed at its highest level in the ectoderm, transplantation experiments with syncytial blastoderm nuclei between wild-type and *wg* mutant embryos have shown that the germ layer-specific origin of the *wg* activity is not of crucial importance and cardiogenic *wg* activity can be provided either by the ectoderm or by the mesoderm (Lawrence *et al.,* 1995).

Although *wg* affects heart formation specifically, it is important to determine whether the Wg pathway provides a direct cardiogenic signal or whether it exerts its effect via regulating the expression or activity of other segmentation genes. Indeed, another secreted factor encoded by *hedgehog* (*hh*) is also required for heart precursor formation during the same time period as *wg* (Park *et al.*, 1996). A further complication is that maintenance of *wg* expression after gastrulation requires *hh* function, and *vice versa* (Klingensmith and Nusse, 1994). This complex situation has been resolved by genetic epistasis experiments, which showed that *wg* activity is absolutely necessary for cardiogenesis and cannot be substituted by *hh* or other segmentation genes that regulate *wg* function (Park *et al.*, 1996).

Since the intracellular transducers downstream of the Wg signal are provided maternally and are present ubiquitously, developmental patterns generated by Wg signaling appear to be solely due to the striped zygotic expression of *wg* itself (Noordermeer *et al.*, 1994; Siegfried *et al.*, 1994; Peifer *et al.*, 1994; Perrimon, 1994; Klingensmith and Nusse, 1994). One of these transducers, Disheveled (Dsh), which has a GLGF/DHR motif, is thought to be activated upon Wg signaling immediately downstream of a recently identified Wg receptor, Dfz2 (Woods and Bryant, 1991; Klingensmith *et al.*, 1994; Noordermeer *et al.*, 1994; Siegfried *et al.*, 1994; Bhanot *et al.*, 1996). However, when Dsh is overexpressed in transgenic flies it appears to constitutively activate the Wg pathway in the absence of *wg* (Yanagawa *et al.*, 1995). This property of *dsh* overexpression was exploited to show that *dsh* is epistatic to *wg* and required autonomously within the mesoderm for heart development (Park *et al.*, 1996; M. Park and R. Bodmer, unpublished). *dsh* is thought to then negatively regulate Shaggy/Zeste-white3 (Sgg/Zw3), a serine/threonine kinase closely related to GSK-3β (Bourouis *et al.*, 1990, Siegfried *et al.*, 1990, 1992). *sgg/zw3*, in turn, negatively regulates the activity of Armadillo (Arm), a β-catenin homolog (Riggleman *et al.*, 1990; Oda *et al.*, 1993), causing translocation of Arm protein from its inactive membrane-bound state to the cytoplasm (Peifer *et al.*, 1994; Yanagawa *et al.*, 1995). In vertebrate systems, cytoplasmic β-catenin has been shown to translocate to the nucleus upon binding to a HMG box-containing transcription factor (Behrens *et al.*, 1996; Molenaar *et al.*, 1996; Miller and Moon, 1996). It is likely that activated Arm protein interacts with a similar factor in *Drosophila*, which is encoded by the *pangolin/dTCF* gene (Brunner *et al.*, 1997; van de Wetering *et al.*, 1997). Genetic studies involving elimination of both the maternal and the zygotic activities of these signal transducers in the Wingless pathway confirmed that they are all required for normal heart development (Park *et al.*, 1996, 1998b). However, *sgg/zw3* appears to have an additional, earlier function during the Dpp-induced maintenance of dorsal *tinman* expression, which is reduced in sgg null embryos (Park *et al.*, 1998b). Taken together these experiments clearly indicate that Wg is a direct signal for specifying cardiac cell fates. It remains to be shown whether the Wingless signaling pathway, through the activity of Pangolin/dTCF, acts directly to control *tinman* expression in heart precursors, or whether it activates the transcription of unknown intermediates in the precardiac mesoderm that are in turn required to activate *tinman* expression.

B. Heart Development in the Context of Segmental Patterning Events in the Mesoderm

Prior to the determination of cardiac precursors, the ectoderm is already subdivided into segmentally repeated units, and each unit is further partitioned into an anterior compartment (A compartment) and a posterior compartment (P compartment). The expression of *wg* is restricted to striped domains in each of the A compartments. Because of the limited diffusion range of the secreted Wg protein, this would suggest that Wg induces cardiac precursors at positions below the ectodermal A compartments but not below the P compartments. Histological analysis has indeed indicated that the heart anlagen arise as 11 segmental clusters of cells on either side that only subsequently merge to form continuous structures along the anterioposterior axis (Dunin Borkowski *et al.*, 1995). The analysis of early markers for heart precursors, including *even-skipped* (*eve*) and *ladybird* (*lb*), confirmed that induction of heart precursors is restricted to areas below the *wingless* stripes, i.e., below the ectodermal A compartments (Jagla *et al.*, 1997). Within these areas, heart induction occurs near the dorsal margins of the *wingless* stripes. Importantly, the results of several studies have shown that *wingless* functions also in more ventral areas of the mesoderm, where it is required for the development of somatic muscles and their progenitor cells (Baylies *et al.*, 1995; Ranganayakulu *et al.*, 1996; D'Alessio and Frasch, 1996). Thus, it appears that the development of the entire mesodermal areas below each of the *wg* stripes is under the influence of *wg*, although the response to the Wg signal differs along the dorsoventral extent of these areas. The *sloppy paired* (*slp*) gene pair, which encodes *forkhead* domain proteins and is expressed in stripes that are in register with the *wingless* stripes, is similarly required for heart and somatic muscle formation (Park *et al.*, 1996; Riechmann *et al.*, 1997). At least in part, this role of *slp* is thought to be due to the known requirement of *slp* for the activation of *wg* expression (Cadigan *et al.*, 1994).

The mesodermal domains between the *wg* and *slp* stripes, which are located below the ectodermal P compartments, have been shown to be governed by different genetic inputs and give rise to distinct mesodermal derivatives. Notably, the dorsal portions of these domains form visceral mesoderm (midgut musculature), and the lateral portions develop fat body (Azpiazu *et al.,* 1996; Riechmann *et al.,* 1997). Thus, similar to the heart anlagen, the anlagen of the visceral mesoderm and fat body are specified as metamerically repeated clusters of cells which only later merge into continuous structures along the anteroposterior embryo axis. This conclusion has been derived from the analysis of the expression domains of genes that are involved in the specification of these tissues, in particular *bagpipe,* which is required for visceral mesoderm specification, and *serpent,* a GATA factor encoding gene that is required for fat body specification. *bagpipe* is expressed in 11 metameric domains that are positioned below the ectodermal P compartments in the dorsal mesoderm. These and other observations showed that the dorsal mesoderm is composed of alternating domains of cardiac/dorsal somatic muscle progenitors and visceral muscle progenitors along the anteroposterior axis. Whereas the formation of cardiac and somatic muscle precursors is positively influenced by *wg* and *slp,* expression of *bagpipe* and thus formation of visceral mesoderm is negatively regulated by these two genes. Conversely, *bagpipe* is positively controlled by *hh,* which is expressed in striped ectodermal domains neighboring the *wg* stripes and encodes a secreted molecule related to vertebrate Shh. Another important regulator of *bagpipe* is the pair-rule gene *eve,* which encodes a homeodomain protein and is involved in establishing segmental subdivisions in the early embryo and only later becomes expressed in a subset of dorsal mesodermal cells. In the absence of *eve, bagpipe* expression is missing and the visceral mesoderm fails to develop. This function of *eve* is probably partially mediated by *hh,* which requires *eve* for its normal expression. However, both the stronger phenotype of *eve* mutants with respect to *bagpipe* and the observation that the function of *eve* for *bagpipe* activation is required in the mesoderm, but not in the ectoderm, suggest that additional *eve* downstream genes are essential to activate *bagpipe.* It has been proposed that this gene(s), which has not been identified, may be expressed in striped domains in the mesoderm that are located below the ectodermal P compartments (Azpiazu *et al.,* 1996).

Based on these and additional observations, it has been proposed that after gastrulation, the mesoderm becomes subdivided into segmentally repeated units, each of which consists of two separate domains (Dunin Borkowski *et al.,* 1995; Azpiazu *et al.,* 1996; Riechmann *et al.,* 1997).[2] The domains that are located below the ectodermal P compartments are subject to influences from the striped regulators *eve* and *hh;* for these reasons, they have been termed "P domains" or "*eve* domains." The P (*eve*) domains form visceral mesoderm (dorsally), fat body (laterally), and presumably some somatic muscles (ventrally). The requirement of *eve* for heart formation is likely through its influence on other segmentation genes (such as *hh* and *wg*).[3] By contrast, the development of the metameric domains that are located below the A compartments of the ectoderm depends largely on the striped regulators *wg* and *slp.* Hence, these domains have been termed "A domains" or "*slp* domains." The A (*slp*) domains form cardiac precursors (at their dorsal borders), dorsal somatic muscles (in the remaining dorsal mesoderm), and the majority of the lateral and ventral somatic muscles further ventrally. These anteroposterior subdivisions are also reflected in the expression pattern of *twist,* which becomes periodically modulated prior to the segregation of mesodermal tissues (Baylies and Bate, 1996). Under the control of *slp* (but not *wg*), high levels of *twist* are maintained in the A (*slp*) domains, which appears to be important for normal somatic muscle development in these areas. Conversely, *twist* expression is reduced in the P (*eve*) domains at this stage, which appears to be a prerequisite for visceral mesoderm specification by *bagpipe.* While a number of key regulators of these anteroposterior subdivisions, including *eve, slp, wg, hh,* and *twist,* have been defined, additional gene functions have to be elucidated to obtain a complete understanding of the genetic and molecular mechanisms involved in this patterning process.

---

[2]The nomenclature of mesodermal segmentation is under some debate. Here we follow the nomenclature proposed by Azpiazu *et al.* (1996) which is based on the notion that the earliest periodic subdivisions in the ectoderm are laid down in a parasegmental register. Parasegments are shifted in register by one-half of a segment compared to the segments that emerge from them later in development. The nomenclature by Dunin Borkowski *et al.* (1995) uses the segmental register for early stages. In their description, the correlation of metameric units in the mesoderm with ectodermal A and P compartments is reversed and slightly out of register (for a more extensive discussion, see Riechmann *et al.,* 1997).

[3]Note that the initial expression of *eve* occurs in seven parasegmental stripes, i.e., in a pair-rule pattern, whereas the functions of *eve* in mesoderm patterning are required later and in every parasegment. This suggests that *eve* acts through *eve*-dependent downstream genes that are expressed in stripes in every parasegment. The primary intermediates of *eve,* including *hedgehog,* function in the ectodermal P compartments and mesodermal P domains (also called "eve domains"). However, since *hedgehog* is required to maintain *wingless* expression in the ectodermal A compartments, *eve* is more indirectly also required for heart specification in the A domains. By contrast, *slp,* which is expressed in the A compartments, appears to be required more directly for *wingless* activation (hence the alternative term "*slp* domains").

The mutant phenotypes of genes affecting heart development, as described in sections III,B–V,B, are schematically summarized in Fig. 7.

C. A Combinatorial Model for Specifying the Precardiac Mesoderm

In addition to the genes involved in the initial specification of the mesoderm, we have discussed in some detail the function of a number of genes that play a role in specification of individual cell types within the mesoderm. Based on their functions and expression patterns, these genes can be divided into three basic groups. The first group contains genes that pattern the mesoderm along the dorsoventral axis. The primary members of this group, *dpp* and *tinman,* are needed for specifying the dorsal mesoderm, which includes the formation of the heart, the visceral muscles, and dorsal somatic muscles (Azpiazu and Frasch, 1993; Bodmer, 1993; Staehling-Hampton *et al.,* 1994; Frasch, 1995). The second group comprises genes that act in the anteroposterior patterning of the mesoderm and subdivide it into segmental and subsegmental units. These genes include *wg* and *slp,* which are required for the formation of the heart (and most skeletal muscles) but not for specifying visceral muscles (Wu *et al.,* 1995; Park *et al.,* 1996; Riechmann *et al.,* 1997). Other members of this group are *eve* and *hh,* which are required for the formation of visceral muscles but not (at least not directly) for the formation of the heart and the majority of somatic muscles (Azpiazu *et al.,* 1996). Finally, genes of the third group are expressed at defined dorsoventral and anteroposterior positions in the mesoderm and control the tissue specification of the cells in which they are expressed. The best studied example for a gene in this third group is *bagpipe,* which is activated in the P domains of the dorsal mesoderm and specifies visceral mesoderm identities in these areas (Azpiazu and Frasch, 1993).

The expression patterns of genes from the first and second groups intersect in every segment. We propose that it is at these intersections where the genes of the third group are activated and serve to specify individual tissues. For example, *bagpipe,* activation requires intersections of the dorsal domains of *dpp* and *tinman* with the transverse stripes of *hh* (apparently together with mesodermal stripes of an unknown *eve*-dependent regulator). Likewise, specification of precardiac and dorsal somatic mesoderm requires intersections of the dorsal domains of *dpp* and *tinman* with the transverse stripes of *wg* and *slp.* It is important to note that, in these quadrants, an additional dorsoventral subdivision appears to occur, which restricts precardiac mesoderm formation to their dorsal-most areas. This indicates the existence of an additional, unidentified cue which is active in narrow domains along the dorsal margins of the germ band and acts in combination with *dpp, tin, wg,* and *slp* to determine precardiac mesoderm. This interpretation of combinatorial events of tissue specification in the dorsal mesoderm is consistent with the results from ectopic expression experiments. For example, ectopic expression of *dpp* in the ventral ectoderm and of *dpp* or activated Tkv in the entire mesoderm results in an expansion of *bagpipe* expression and visceral mesoderm formation toward the ventral midline (Staehling-Hampton *et al.,* 1994; Frasch, 1995; Yin and Frasch, 1998; W. Lockwood and R. Bodmer, unpublished observations). By contrast, heart formation is not expanded to a similar extent under these conditions, presumably because an additionally required cue is still confined to the dorsal-most areas of the germ band. Interestingly, however, a small area of ventral mesoderm does form heart precursors upon ectopic Dpp signaling, perhaps suggesting that this (hypothetical) cue has a second domain of activity in ventral embryonic regions (Yin and Frasch, 1998; W. Lockwood and R. Bodmer, unpublished observations). The ventral mesoderm in the vicinity of the CNS is therefore poised to form a second ventral heart, but in normal embryos it is unable to do so because of the absence of ventral Dpp signals at early postgastrulation stages (Fig. 8).

In summary, the convergence of *wg, dpp, tinman* (Fig. 6), and probably an additional unidentified cue appears to be a pivotal event in specification of the heart. Since the intersection of wg and *dpp* also plays a role in imaginal disc development and development in other parts of the embryo, *wg* and *dpp* are not instructive in terms of the specific tissue type that is formed. Rather, it appears that the patterning information they provide is used to direct the subdivision of a primordial tissue, including the mesoderm, into different subfates, such as cardiac versus visceral or somatic mesoderm. The mesoderm-specific (and at this point, dorsal mesoderm-specific) expression of *tinman* appears to target the converging *wg* and *dpp* signals to the mesoderm, thus eliciting distinct responses that result in the formation of cardiac and dorsal somatic mesoderm.

D. Molecular Mechanisms of *tinman* Regulation

As described previously, the functions of *tinman* in mesoderm patterning and heart development are reflected in its dynamic mesodermal expression pattern, which can be subdivided into three distinct phases: early, *twist*-dependent pan-mesodermal expression; *dpp*-dependent dorsal mesodermal expression after mesoderm migration; and heart-specific expression during mesoderm differentiation. Recent studies have

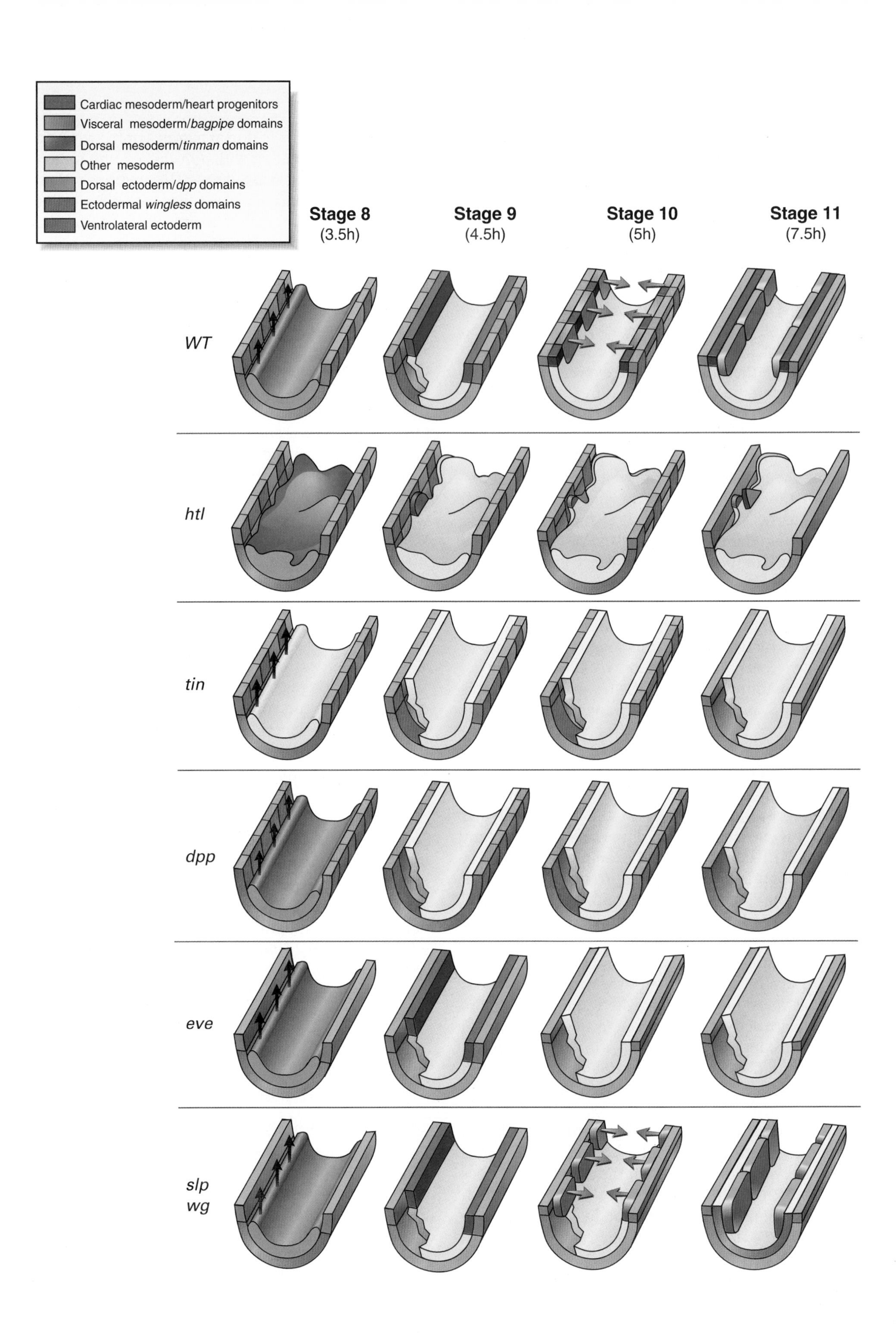
Cardiac mesoderm/heart progenitors
Visceral mesoderm/*bagpipe* domains
Dorsal mesoderm/*tinman* domains
Other mesoderm
Dorsal ectoderm/*dpp* domains
Ectodermal *wingless* domains
Ventrolateral ectoderm
Stage 8 (3.5h)
Stage 9 (4.5h)
Stage 10 (5h)
Stage 11 (7.5h)
WT
htl
tin
dpp
eve
slp
wg

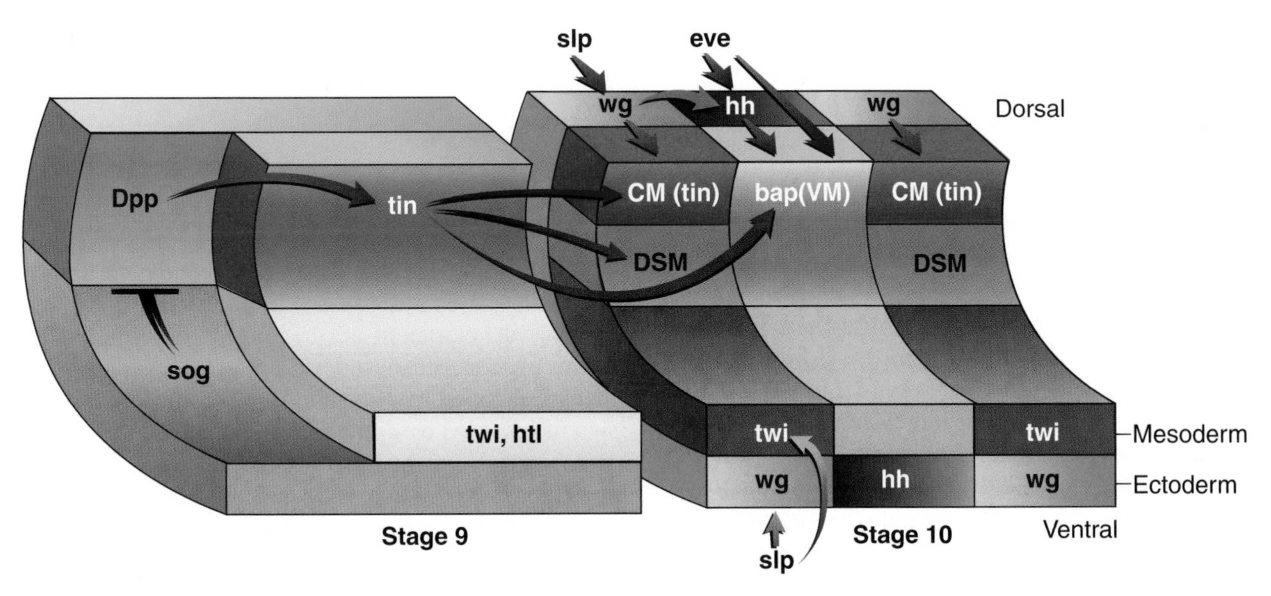

Figure 8 Summary of genetic interactions during early mesoderm patterning. Schematically shown are the left halves of the germ band at stage 9 (to the left) and stage 10 (to the right). At stage 9, Dpp induces the maintenance of *tinman* (*tin*) expression in the dorsal ectoderm, thus triggering a dorsoventral subdivision of the mesodermal layer. The activity of Dpp itself is limited to the dorsal ectoderm by *short gastrulation* (*sog*), which encodes an inhibitor of Dpp. During stage 10, *tin* activity is modulated by gene products that are expressed in transverse stripes. This results in anteroposterior subdivisions of the dorsal mesoderm and, thus, in differential determination of the primordia of the cardiac mesoderm (CM), visceral mesoderm (VM), and dorsal somatic mesoderm (DSM) (see text).

identified minimal enhancer elements from the *tinman* locus that are sufficient to drive individual aspects of this pattern of *tinman* expression. These studies have defined three discrete mesodermal enhancer elements: The first driving broad expression in the early mesoderm, the second driving expression at stage 11 in the dorsal mesoderm, and the third being active after stage 11 exclusively in the cardial cells (Yin *et al.,* 1997; T. V. Venkatesh and R. Bodmer, unpublished observations). This shows that the tinman cis-regulatory elements are organized in a modular fashion, and the DNA sequences that respond to different genetic inputs can be physically separated and act independently from one another. Unexpectedly, none of these elements is located in 5' flanking regions of *tinman* since the early mesodermal element (180 bp) is found in the first intron, the cardioblast element (300 bp) in proximal 3' flanking regions, and the dorsal mesodermal element (350 bp) about 1.5 kb downstream of the *tinman* coding region. To date, the early mesodermal element, which is expected to be a target of *twist,* has been analyzed in most detail. This element contains three E-box sequences (CATGTG or CATATG) that bind Twist *in vitro* and are essential for its activity *in vivo,* thus suggesting that Twist binding to these E-box sequences is responsible for activating *tinman* in the early mesoderm. The *in vivo* relevance of this interaction is further supported by the observation that ectodermally expressed Twist can ectopically activate enhancer elements containing these E boxes in the ectoderm. Moreover, a 29-bp element containing two of the three E boxes is sufficient for mesodermal activation, and all three E boxes are conserved in a corresponding element from the *tinman* gene of a distantly related species, *D. virilis.* It is interesting to note that, unlike *twist,* both the *tinman* gene and reporter genes driven by the 180-bp early *tinman* enhancer are not expressed in the entire mesoderm. Rather, expression is excluded from a small area of the mesoderm in the embryonic head which is fated to develop into hemocytes. This area of the mesoderm expresses a GATA factor-encoding gene, *serpent,* that is essential for blood cell differentiation (Rehorn *et al.,* 1996). It was found that the early 180-bp element of tinman includes a repressor element (52 bp) that functions to abrogate activation by Twist in the head mesoderm. The SP1-related zinc finger-encoding gene *buttonhead* (*bth*) has been identified as an important regulator of this process. *bth,* which is

Figure 7 Summary of early mesoderm development and mesodermal phenotypes in embryos mutant for *heartless* (*htl*), *tinman* (*tin*), *decapentaplegic* (*dpp*), *even-skipped* (*eve*), *sloppy paired* (*slp*), and *wingless* (*wg*). Schematic drawings of wild-type embryos (WT) are shown in the top row.

expressed in a circumferential stripe during early head development (Wimmer *et al.,* 1991), is required for the exclusion of *tinman* expression (and reporter gene expression driven by the 180-bp early enhancer) from the head mesoderm and is needed for the activation of *serpent* in the same area. The absence of *bth* activity results in a strong reduction and apparently disrupted differentiation of hemocytes, which is presumably due to the loss of *srp* expression and ectopic *tinman* expression in the head mesoderm (Yin *et al.,* 1997). These observations suggest that Twist-binding to the E boxes of the early *tinman* enhancer has the potential to activate *tinman* in the entire mesoderm, but closely linked repressor binding sites that are direct or indirect targets of Bth restrict Twist activity, and thus *tinman* expression, to the trunk mesoderm. This is the area in which *tinman* subsequently functions to allow heart formation and the development of visceral and certain somatic muscles.

The identification of a discrete dorsal mesodermal enhancer element demonstrates that the "maintenance" of *tinman* expression in the dorsal mesoderm is in fact due to a second event of transcriptional activation during the time when *twist*-dependent activation ceases. This element is likely to serve as a "Dpp response element" that receives inputs from the Dpp signal transduction cascade to activate *tinman* expression. Indeed, its activity is abolished in *dpp* mutant embryos, while it expands ventrally in embryos containing mesodermally expressed, constitutively active versions of Tkv (M. Frasch, unpublished). In addition to *dpp,* this enhancer element requires *tinman* for its full activity (X. Xu and M. Frasch, unpublished observations). Thus, one function of early expressed *tinman* appears to involve an autoregulatory activity that enhances the inputs from *dpp* during its second phase of expression.

Since the heart enhancer of *tinman* is only active in cardioblasts, it appears that *tinman* expression in cardioblasts and pericardial cells is controlled by different (although perhaps overlapping) genetic inputs (X. Xu and M. Frasch, unpublished observations; T. V. Venkatesh and R. Bodmer, unpublished observations). It will be important to establish whether the *wingless* signaling cascade is targeting any of these enhancer sequences directly, for example, by stimulating the binding of the downstream factor Pangolin/dTCF to them. The expression of *tinman*-related genes from vertebrates in the precardiac mesoderm and developing heart, which depends on the activities of BMPs (Schultheiss *et al.,* 1997), is comparable to the expression of *tinman* during its second and third phases of expression (Bodmer, 1995; Harvey, 1996; Bodmer *et al.,* 1997). Therefore, it will be interesting to determine whether these vertebrate genes contain enhancers that are related to the *dpp*-response element or the cardioblast element of *Drosophila tinman.*

VI. Specification of Cardiac Cell Types

As discussed previously, the molecular mechanisms of the early events in mesoderm differentiation leading to heart development are beginning to be understood. In this process, the mesoderm-endogenous transcription factors encoded by *twist* and *tinman,* and the inductive signals encoded by *dpp* and *wg,* have been shown to be most crucial. Further distinctions into different heart-specific cell types, such as cardial and pericardial cells or subsets thereof, may be generated by additional external cues, by cell–cell interactions, or by asymmetric cell divisions. These next stages of heart formation, morphogenesis, and differentiation, are less well understood. To date, only a few genes have been shown to be involved in the differentiation of specific cardiac cell types. Best characterized among these is *mef2,* a MADS box containing muscle differentiation gene required for the myogenic differentiation of the cardial cells of the heart tube (Fig. 1D and 10; Bour *et al.,* 1995; Lilly *et al.,* 1995; Ranganayakulu *et al.,* 1995). Recent evidence shows that the homeobox genes *zfh-1, eve,* and *ladybird* are involved in the formation and differentiation of a subset of heart cells (M.-T. Su and R. Bodmer, unpublished; Jagla *et al.,* 1997). In addition, the lineage genes, *numb* and *Notch,* which specify alternative cell fates during asymmetric cell divisions in the nervous system (Uemura *et al.,* 1989; Rhyu *et al.,* 1994; Guo *et al.,* 1996; Spana *et al.,* 1995; Spana and Doe, 1996), are also involved in cell fate decisions of cardiac lineages (Park *et al.,* 1998c). For all these genes, vertebrate homologs have been identified, and some of them have already been shown to be present in the developing heart (Verdi *et al.,* 1996; Zhong *et al.,* 1996; Takagi *et al.,* 1997). It is thus tempting to speculate that some of the molecular mechanisms of vertebrate cardiac differentiation and cell type specification may again share a common basis with *Drosophila,* as appears to be the case for the initial cardiac specification involving *tinman/Nkx* and *dpp*/BMP2,4.

A. The MADS Box Gene *mef2*

In *Drosophila, mef2* is a member of the MADS box family of transcription factor genes first characterized in yeast. MEF2 proteins are distinguished from other MADS box-containing genes by an adjacent 29 amino acid sequence called the MEF2 domain. One of the four vertebrate MEF2 has been shown to interact with bHLH myogenic factors of the MyoD class (Molkentin *et al.*, 1995; see Chapter 8).

The expression of the vertebrate *mef2* genes correlates with the development of skeletal, cardial, and smooth muscle cells (Edmondson *et al.*, 1994), and suggests a role for this gene in all muscle lineages (Olson *et al.*, 1995; Olson and Srivastava, 1996). In contrast to the *mef2* genes, the early muscle-specific genes of the *MyoD* class are not expressed in the cardiac muscle progenitors (Rudnicki and Jaenisch, 1995). As is *tinman, mef2* is first expressed ubiquitously in the early mesoderm, under the control of *twist,* then undergoes dynamic changes in its expression (Nguyen *et al.*, 1994; Lilly *et al.*, 1994; Taylor *et al.*, 1995). *mef2* expression finally persists in all progenitor cells of contractile muscle cells and thus also in the cardial cells of the heart (Fig. 1D).

In *mef2* mutants, the formation of cardiac, skeletal, and visceral muscle progenitors per se does not seem to be affected, suggesting that *mef2* is not required for the initial specification of the major mesodermal subtypes. Rather, the major requirement for *mef2* appears to be in muscle differentiation and muscle-specific gene expression (Bour *et al.*, 1995; Lilly *et al.*, 1995; Ranganayakulu *et al.*, 1995; Lin *et al.*, 1996, 1997). Most strikingly, myocytes of the somatic musculature do not fuse into polynucleated muscle fiber syncytia, although marker gene expression in the putative founder cells of individual muscles is not affected. In contrast, some of the muscle structural genes required for muscle function are not expressed in *mef2* mutants: *Drosophila* muscle myosin heavy chain (DMM) and tropomyosin I are absent, whereas β3-tubulin which is expressed earlier than DMM is still present (Fig. 9). Since the contractile cardial cells are mononucleate, it is perhaps not surprising that heart morphogenesis per se is unaffected in the absence of *mef2* function. Nevertheless, the requirement of *mef2* for muscle-specific gene expression in the heart suggests that *mef2* is a crucial differentiation factor not only for the body wall muscles but also for the heart of *Drosophila.*

tinman is expressed in two clusters of heart precursors per hemisegment early on and later in both cardial and pericardial cells (Fig. 1C; Bodmer *et al.*, 1990). In contrast, *mef2* appears to be present in only one cluster per hemisegment and later only in cardial cells (Fig. 1D; Nguyen *et al.*, 1994). This is consistent with the hypothesis that the cardial and pericardial lineages separate very early in cardiac development (see Section VI,D). There is evidence that *mef2* expression in cardial cells is directly activated by *tinman.* A *mef2* enhancer element has been identified that is specifically active in cardial cells and contains Tinman binding sites that are essential for its activity (Gajewski *et al.*, 1997). Since this enhancer is inactive in other mesodermal areas where *tinman* is expressed during earlier stages, additional binding factors must contribute to its restricted activity.

B. The Homeobox Genes *zfh-1* and *even-skipped*

zfh-1, which contains nine zinc fingers in addition to its divergent homeodomain (Fortini *et al.*, 1991), is expressed initially in all mesoderm, similar to *tinman* and *mef2.* Later, *zfh-1* is present in a variety of tissues, including the cardial and pericardial cells of the forming heart (Lai *et al.*, 1991; Fig. 3). After formation of the heart tube, cardiac expression of *zfh-1* becomes restricted primarily to the pericardial cells. The early panmesodermal *zfh-1* expression depends on *twist* but not on *tinman,* although later cardiac expression of *zfh-1* is dependent on *tinman. zfh-1* mutant embryos show only weak abnormalities, with variable penetrance, in the body wall muscle pattern and possibly in heart muscle morphology (Lai *et al.*, 1993). The formation of cardial and pericardial cells is not noticeably affected by loss of *zfh-1* function, except for the Eve pericardial cells (EPCs) (see Fig. 1; M.-T. Su and R. Bodmer, unpublished). This suggests that *zfh-1,* in contrast to *tinman,* is not essential for the initial subdivision of the mesoderm, despite the similarities in their early mesodermal expression pattern.

Although expressed earlier as a pair-rule gene, Eve is also present later in segmentally arranged, mesodermal clusters of three or four cells at the dorsal edge of the mesoderm (Frasch *et al.*, 1987). After germ band retraction, two of these cells in each hemisegment differentiate into a distinct subset of EPCs (also termed e-PCs) with large nuclei (Fig. 1). The Eve-positive clusters also contain a founder for one of the dorsal body wall muscles, DA1 (Bate, 1993). In *zfh-1* mutants, the initial Eve clusters are formed normally, whereas later the EPCs but not the DA1 founders fail to differentiate (Fig. 10; M.-T. Su and R. Bodmer, unpublished). Moreover, *eve* is also required for the correct development of the EPCs since inactivation of *eve* function (using a temperature-sensitive allele of *eve*) at the time when it is normally expressed in the EPC precursors causes most EPCs to be missing or misplaced (M.-T. Su and R. Bodmer, unpublished). As in *zfh-1* mutants, other aspects of heart development are little affected in these conditional *eve* mutant embryos. Thus, both *zfh-1* and *eve* seem to be needed for cardiac progenitor diversification by regulating EPC differentiation.

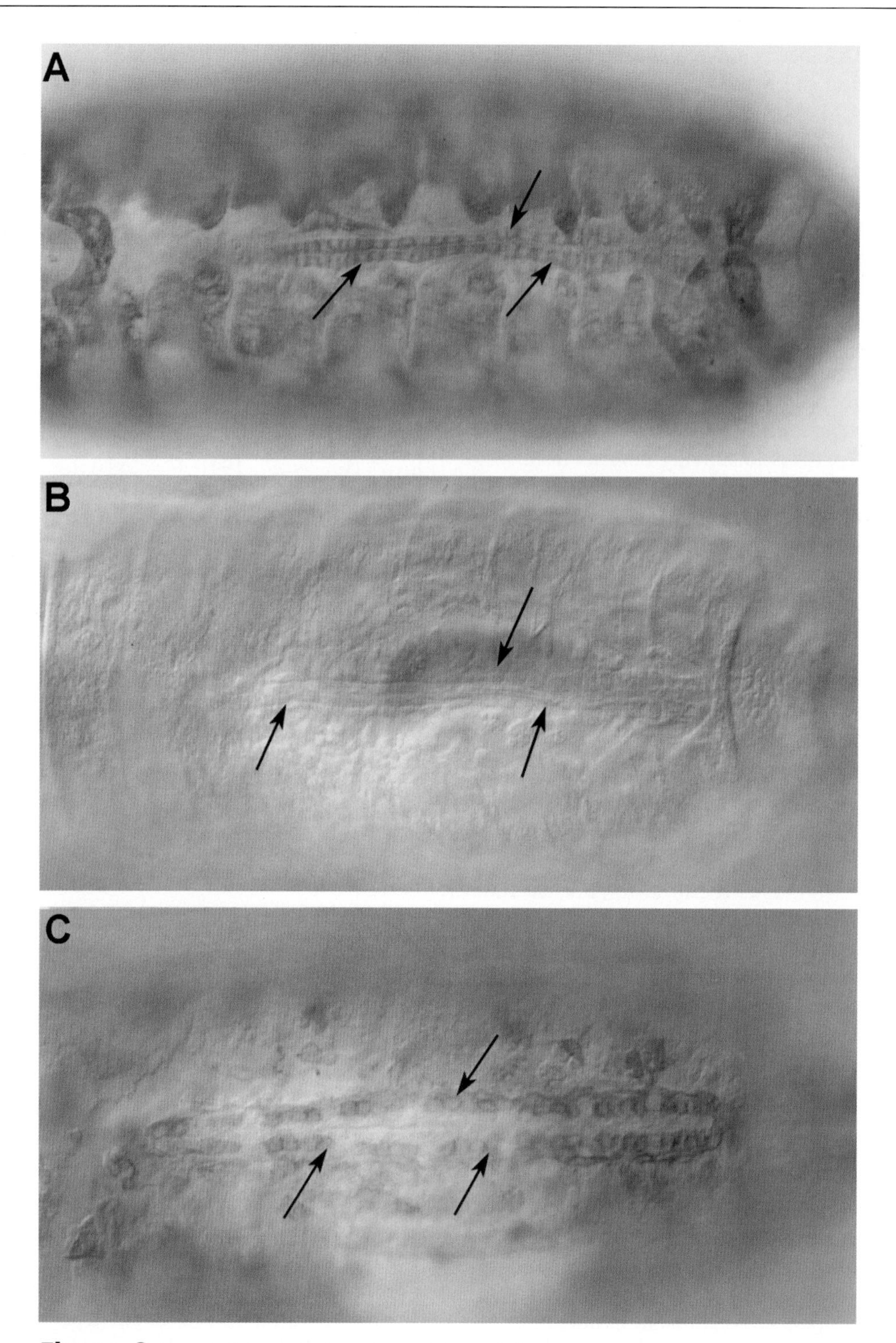

Figure 9 Dorsal view of the heart in a wild-type (A) and *mef2* mutant embryo (B) showing muscle-specific myosin expression. Note the lack of myosin in the cardial cells of the heart in the mutant (indicated by arrows). (C) β3-tubulin expression in the heart of *mef2* mutant embryo (indicated by arrows) reveals that heart morphogenesis as such is unaffected at this stage (micrographs courtesy of B. Bour, S. Abmayr, and H. Nguyen).

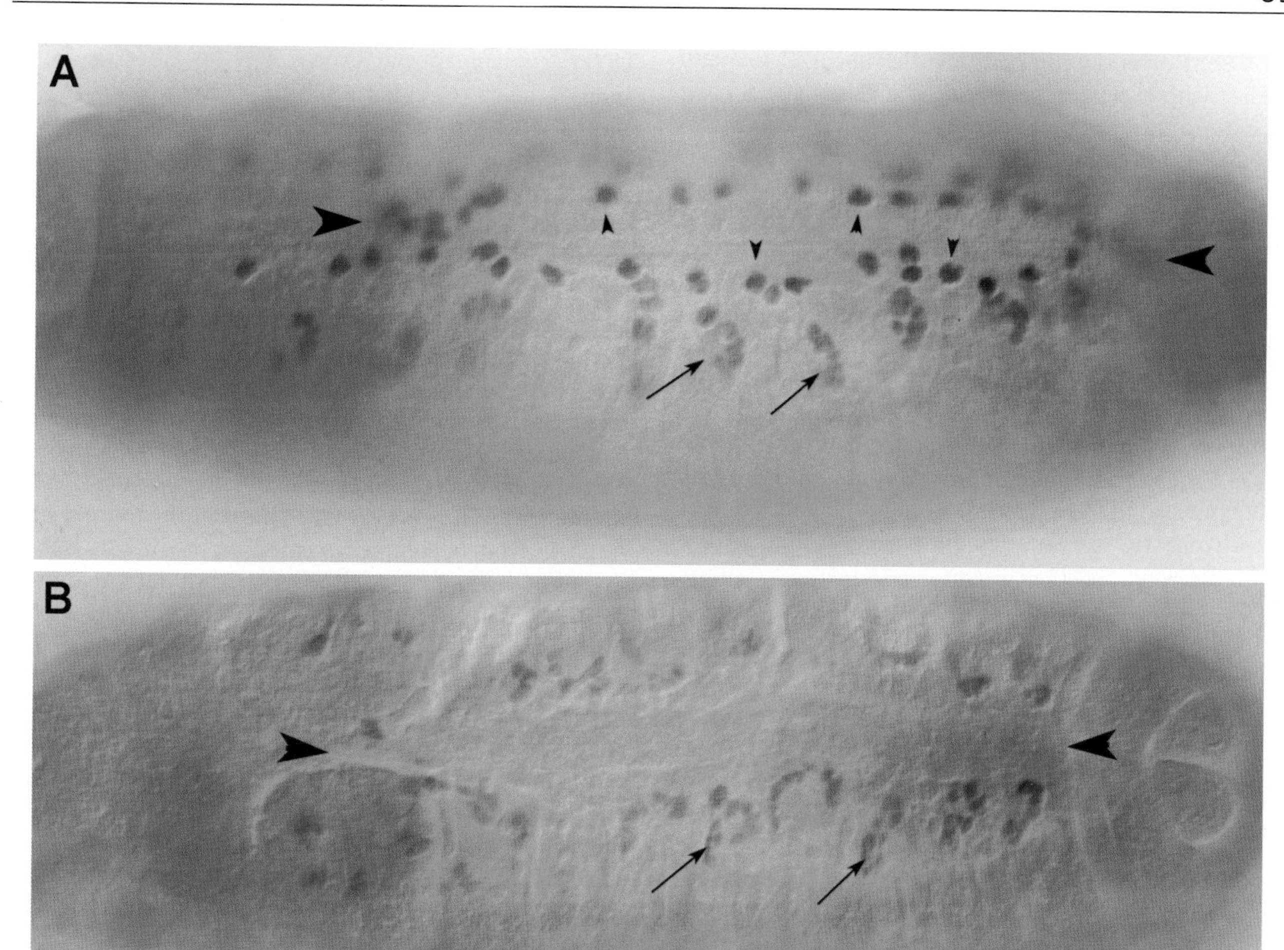

Figure 10 Eve expression in pericardial cells (EPCs; small arrowheads) and dorsal muscle DA1 (arrows) showing a wild-type (A) and mutant (B) embryo. The absence of EPCs but not of most of the other heart cells (note the presence of a dorsal vessel between arrowheads) in the mutant is characteristic for loss of function of the genes *zfh-1, spdo* or overexpression of *numb* (see text) (micrographs courtesy of M.-T. Su).

C. The Lineage Genes, *numb* and *Notch*

In the nervous system of *Drosophila,* the lineages of neuroblasts and of sensory organ precursors appear to be fixed, giving rise to a defined set of progeny. The specification of the cell fates of individual daughter cells depends on asymmetric cell divisions mediated by the phosphotyrosine-binding (PTB) domain-containing gene product of *numb,* which is inherited by one daughter cell but not the other (Uemura *et al.,* 1989; Rhyu *et al.,* 1994). In the Numb-containing daughter cell the transmembrane receptor Notch is thought to be inhibited from signaling, thus allowing the specification of a different cell fate from its sibling where Notch signaling proceeds unimpaired (Guo *et al.,* 1996; Spana and Doe; 1996).

Although many lineage aspects of mesoderm development seem to be less fixed than those of the nervous system (Beer *et al.,* 1987), recent results from clonal and genetic analysis support the view that lineage mechanisms are indeed an important aspect of mesoderm development. Clonal analysis using cell transplantation between early gastrula embryos (R. Bodmer and Gerd Technau, unpublished; see Beer *et al.,* 1987, for methodology) or random labeling of postblastoderm cells with the Flp/FRT site-specific recombinase system from yeast (R. Bodmer and R. Brewster, unpublished; for methodology, see Struhl and Basler, 1993; Brewster and Bodmer, 1995) demonstrated that all the obtained clones exclusively contained either pericardial and or cardial cells. The absence of mixed clones indicates an early split of pericardial and cardial lineages. Genetic studies with the neurogenic genes such as Notch have implicated this group of genes in cardial versus pericardial cell differentiation (Hartenstein *et al.,* 1992), although it is not clear whether they act at the level of specification or at the level of maintaining cell fates that had been specified earlier. Lineage genes, including *numb,* appear to also function in specifying mesodermal cell fates, similar to their previously described functions in neuronal development. Not only the identities of body wall muscles but also a subset of pericardial cells of the heart (the EPCs) depend on these genes. In *numb* mutants the number of EPCs is doubled at the expense of DA1 muscles and in embryos with ubiquitous *numb* overexpression DA1 muscles form but the EPCs do not (Park *et al.,* 1998c). These findings suggest that cardiac and body wall muscle cells can derive from a common lineage and that alternative cell fate decisions may play a role in cardiac development (and other mesodermal situations; Ruiz Gomez and Bate, 1997) that is as important as that in neurodevelopment.

D. Ladybird Genes

As described in Section V, the *Drosophila* heart is derived from 11 segmentally repeated anlagen that subsequently merge with one another to form a continuous vessel along the dorsal midline. There is indication that the segmental character of the dorsal vessel is maintained even after completion of this morphogenetic process. For example, *tinman* is expressed in only four of the six cardioblasts per hemisegment, and several enhancer trap lines also express *lacZ* also in defined subsets of cardioblasts in each segment (Fig. 1C; Hartenstein and Jan, 1992). The expression of the *ladybird* genes (*lbe* and *lbl*), a gene pair of the homeobox gene family, represents another interesting example for the segmental organization of the dorsal vessel. In late stage embryos, *lb* expression is observed in two of the four *tinman*-expressing cardioblasts in each hemisegment and, in addition, in two pericardial cells (Jagla *et al.,* 1997). These pericardial cells are distinct from those expressing *even-skipped* (e-PCs) and have been termed l-PCs. The analysis of lb expression during earlier stages has provided additional insights into the the development and regulation of the diversification of heart cells. At early stage 12, *lb* expression is activated in segmental clusters of about four cells each in the segregating heart anlagen. At the same stage, *even-skipped* expression is observed in clusters of three cells each (two pericardial precursors and one somatic muscle progenitor), which are located posteriorly adjacent to the *lb* clusters. During the subsequent processes of heart segregation, cell rearrangements which involve a 90° clockwise rotation of heart precursors within each segment move the *lb*-expressing cells to the dorsal side and the eve-expressing cells ventrally to them. These observations indicate that the cardioblasts and pericardial cells are originally specified by anteroposterior patterning processes and are subsequently aligned by morphogenetic movements into rows of dorsal cardial and ventrally adjacent pericardial cells. Experiments with temperature-sensitive alleles of *wingless* and *hedgehog* indicate that these two segment polarity genes are among the regulators of this diversification process. The data obtained from these experiments suggest that *wingless* is required for the specification of *lb*-expressing cells, whereas *hedgehog* is needed to prevent the more posterior cells from becoming "*lb* cells." Apparently, these activities of *wg* and *hh* occur after their initial function in mesoderm segmentation and precardiac mesoderm specification (see Section V). Additional specific markers for the cells that neither express *lb* nor *eve* at early stages of heart development will be essential for a full understanding of these processes of heart cell diversification and morphogenesis. In addition, *lb* mutations are required to obtain information on the significance of the diversification into different types of pericardial and cardial cells within each segment. The results from ectopic *lb* expression and homozygous *lb*-deficiency embryos strongly indicate that *lb* does have roles in specifying distinct cell fates within the cardiac mesoderm (Jagla *et al.,* 1997). Although closely related homeobox genes have been identified in a number of vertebrate species, none have been reported to be expressed in embryonic heart tissues (Jagla *et al.,* 1995).

References

Arora, K., Dai, H., Kazuko, S. G., Jamal, J., O'Connor, M. B., Letsou, A., and Warrior, R. (1995). The *Drosophila schnurri* gene acts in the Dpp/TGF signaling pathway and encodes a transcription factor homologous to the human MBP family. *Cell (Cambridge, Mass.)* **8,** 781–790.

Axelrod, J. D., Matsuno, K., Manoukian, A., Artavanis-Tsakonas, A., and Perrimon, N. (1996). Interaction between inductive and lateral intercellular signalling mediated by Dishevelled and Notch. *Science* **271,** 1826–1832.

Azpiazu, N., and Frasch, M. (1993). *Tinman* and *bagpipe:* Two homeobox genes that determine cell fates in the dorsal mesoderm of *Drosophila. Genes Dev.* **7,** 1325–1340.

Azpiazu, N., Lawrence, P., Vincent, J.-P., and Frasch, M. (1996). Segmentation and specification of the *Drosophila* mesoderm. *Genes Dev.* **10,** 3183–3194.

Baker, N., (1987). Molecular cloning of sequences from *wingless,* a segment polarity gene in *Drosophila:* The spatial distribution of a transcript in embryos. *EMBO J.* **6,** 1765–1773.

Baker, R., and Schubiger, G. (1995). Ectoderm induces muscle-specific gene expression in *Drosophila* embryos. *Development (Cambridge, UK)* **121,** 1387–1398.

Basler, K., and Struhl, G. (1994). Compartment boundaries and the control of *Drosophila* limb pattern by *hedgehog* protein. *Nature (London)* **368,** 208–214.

Bate, M. (1993). The mesoderm and its derivatives. *In* "The Development of *Drosophila melanogaster*" (M. Bate and A. Martinez-Arias, eds.), pp. 1013-1090. Cold Spring Harbor Lab. Press, Cold Spring Harbor, NY.

Baylies, M. K., and Bate, M. (1996). *Twist,* a myogenic switch in *Drosophila. Science* **272,** 1481–1484.

Baylies, M. K., Martinez Arias, A., and Bate, M. (1995). *Wingless* is required for the formation of a subset of muscle founder cells during *Drosophila* embryogenesis. *Development (Cambridge, UK)* **121,** 3829–3837.

Beer, J., Technau, G., and Camps-Ortega, J. A. (1987). Lineage analysis of transplanted individual cells in embryos of *Drosophila melanogaster.* IV. Commitment and proliferative capabilities of mesodermal cells. *Roux's Arch. Dev. Biol.* **196,** 222–230.

Behrens, J., von Kries, J. P., Kuhl, M., Bruhn, L., Wedlich, D., Grosschedl, R., and Birchmeier, W. (1996). Functional interaction of β-catenin with the transcription factor LEF-1. *Nature (London)* **382,** 638–642.

Beiman, M., Shilo, B., and Volk, T. (1996). Heartless, a *Drosophila* FGF receptor homolog, is essential for cell migration and establishment of several mesodermal lineages. *Genes Dev.* **10,** 2993–3002.

Bhanot, P., Brink, M., Samos, C. H., Hsieh, J.-C., Wang, Y., Macke, J. P., Andrew, D., Nathans, J., and Nusse, R. (1996). A new member of the *frizzled* family from *Drosophila* functions as a wingless receptor. *Nature (London)* **382,** 225–230.

Bienz, M. (1994). Homeotic genes and positional signalling in the *Drosophila* viscera. *Trends Genet.* **10,** 22–26.

Blair, S. S. (1995). Compartments and appendage development in *Drosophila. BioEssays* **17,** 299–309.

Bodmer, R. (1993). The gene *tinman* is required for specification of the heart and visceral muscles in *Drosophila. Development (Cambridge, UK)* **118,** 719–729.

Bodmer, R. (1995). Heart development in *Drosophila* and its relationship to vertebrate systems. *Trends Cardiovasc. Med.* **5,** 21–27.

Bodmer, R., Jan, L. Y., and Jan, Y. N. (1990). A new homeobox-containing gene, *msh-2 (tinman)*, is transiently expressed early during mesoderm formation in *Drosophila. Development (Cambridge, UK)* **110,** 661–669.

Bodmer, R., Golden, K., Lockwood, W. B., Ocorr, K. A., Park, M., Su, M.-T., and Venkatesh, T. V. (1997). Heart development in *Drosophila. In* "Advances in Developmental Biology" (P. Wassarman, ed.), Vol. 5, pp. 201–236. JAI Press, Greenwich, CT.

Boulay, J. L., Dennedeld, C., and Alberga, A. (1987). The *Drosophila* developmental gene *snail* encodes a protein with nucleic acid binding fingers. *Nature (London)* **330,** 395–398.

Bour, B. A., O'Brien, M. A., Lockwood, W. L., Goldstein, E. S., Bodmer, R., Taghert, P. H., Abmayr, S. M., and Nguyen, H. T. (1995). *Drosophila* MEF2, a transcription factor that is essential for myogenesis. *Genes Dev.* **9,** 730–741.

Bourouis, M., Moore, P., Ruel, L., Grau, Y., Heitzler, P., and Simpson, P. (1990). An early embryonic product of the gene *shaggy* encodes a serine/threonine protein kinase related to the *CDC28/cdc2*$^{+}$ subfamily. *EMBO J.* **9,** 2877–2884.

Brewster, R., and Bodmer, R. (1995). Origin and specification of type II sensory neurons in *Drosophila. Development (Cambridge, UK)* **121,** 2923–2936.

Brunner, E., Peter, O., Schweizer, L., and Basler, K. (1997). *Pangolin* encodes a Lef-1 homologue that acts downstream of Armadillo to transduce the Wingless signal in *Drosophila. Nature (London)* **385,** 829–833.

Cadigan, K., Grossniklaus, U., and Gehring, W. (1994). Localized expression of *sloppy paired* protein maintains the polarity of *Drosophila* parasegments. *Development (Cambridge, UK)* **8,** 899–913.

Campbell, G., and Tomlinson, A. (1995). Initiation of the proximodistal axis in insect legs. *Development (Cambridge, UK)* **121,** 619–628.

Chen, C. Y., and Schwartz, R. J. (1995). Identification of novel DNA binding targets and regulatory domains of a murine *tinman* homeodomain factor, *nkx*-2.5. *J. Biol. Chem.* **270,** 15628–15633.

Chu-LaGraff, Q., and Doe, C. Q. (1993). Neuroblast specification and formation regulated by wingless in the *Drosophila* CNS. *Science* **261,** 1594–1597.

Costa, M., Wilson, E., and Wieschaus, E. (1994). A putative cell signal encoded by the *folded gastrulation* gene coordinates cell shape changes during *Drosophila* gastrulation. *Cell (Cambridge, Mass.)* **76,** 1075–1089.

Couso, J. P., Bishop, S. A., and Martinez Arias, A. (1994). The wingless signaling pathway and the patterning of the wing margin in *Drosophila. Development (Cambridge, UK)* **120,** 621–636.

D'Alessio, M., and Frasch, M. (1996). *msh* may play a conserved role in dorsoventral patterning of the neuroectoderm and mesoderm. *Mech. Dev.* **58,** 217–231.

Damante, G., Pellizzari, L., Esposito, G., Fogolari, F., Viglino, P., Fabbro, D., Tell, G., Formisano, S., and Di Lauro, R. (1996). A molecular code dictates sequence-specific DNA recognition by homeodomains. *EMBO J.* **15,** 4992–5000.

De Robertis, E. M., and Sasai, Y. (1996). A common plan for dorsoventral patterning in Bilateria. *(London)* **380,** 37–40.

Dunin Borkowski, O. M., Brown, N. H., and Bate, M. (1995). Anterior-posterior subdivision and the diversification of the mesoderm in *Drosophila. Development (Cambridge, UK)* **121,** 4183–4193.

Edmondson, D., Lyons, G., Martin, J., and Olson, E. (1994). Mef2 gene expression marks the cardiac and skeletal muscle lineages during mouse embryogenesis. *Development (Cambridge, UK)* **120,** 1251–63.

Ferguson, E. L., and Anderson, K. V. (1992). Decapentaplegic acts as a morphogen to organize dorsal-ventral pattern in the *Drosophila* embryo. *Cell (Cambridge, Mass.)* **71,** 451–461.

Fortini, M., Lai, Z., and Rubin, G. (1991). The *Drosophila zfh-1* and *zfh-2* genes encode novel proteins containing both zinc-finger and homeodomain motifs. *Mech. Dev.* **34,** 113–122.

François, V., and Bier, E. (1995). *Xenopus chordin* and *Drosophila short gastrulation* genes encode homologous proteins functioning in dorsal-ventral axis formation. *Cell (Cambridge, Mass.)* **80,** 19–20.

Francois, V., Solloway, M., O'Neil, J. W., Emery, J., and Bier, E. (1994). Dorsal-ventral patterning of the *Drosophila* embryo depends on a putative negative growth factor encoded by the *short gastrulation* gene. *Genes Dev.* **8,** 2602–2616.

Frasch, M. (1995). Induction of visceral and cardiac mesoderm by ectodermal Dpp in the early *Drosophila* embryo. *Nature (London)* **374,** 464–467.

Frasch, M., Hoey, T., Rushlow, C., Doyle, H., and Levine, M. (1987). Characterization and localization of the even-skipped protein of *Drosophila. EMBO J.* **6,** 749–759.

Gabay, L., Seger, R., and Shilo, B.-Z. (1997). MAP kinase in situ activation atlas during *Drosophila* embryogenesis. *Development (Cambridge, UK)* **124,** 3535–3541.

Gajewski, K., Kim, Y., Lee, Y., Olson, E., and Schulz, R. (1997). *D-mef2* is a target for Tinman activation during *Drosophila* heart development. *EMBO J.* **16,** 515–522.

Gisselbrecht, S., Skeath, J., Doe, C., and Michelson, A. (1996). *Heartless* encodes a fibroblast growth factor receptor (DFR1/DFGF-R2) involved in the directional migration of early mesodermal cells in the *Drosophila* embryo. *Genes Dev.* **10,** 3003–3017.

Gorczyka, M. G., Phyllis, R. W., and Budnik, V. (1994). The role of *tinman,* a mesodermal cell fate gene, in axon pathfinding during the development of the transverse nerve in *Drosophila. Development (Cambridge, UK)* **120,** 2143–2152.

Grau, Y., Carteret, C., and Simpson, P. (1984). Mutations and chromosomal rearrangements affecting the expression of *snail,* a gene involved in embryonic patterning in *Drosophila melanogaster. Genetics* **108,** 347–360.

Grieder, N. C., Nellen, D., Burke, R., Basler, K., and Affolter, M. (1995). *Schnurri* is required for *Drosophila* Dpp signalling and encodes a zinc finger protein similar to mammalian transcription factor PRDII-BFI. *Cell (Cambridge, Mass.)* **81,** 791–800.

Guo, M., Jan, L. Y,. and Jan, Y. N. (1996). Control of daughter cell fates during asymmetric cell division: Interaction of Numb and Notch. *Neuron* **17,** 27–24.

Hartenstein, A. Y., Rugendorff, A., Tepass, U., and Hartenstein, V. (1992). The function of the neurogenic genes during epithelial development in the *Drosophila* embryo. *Development (Cambridge, UK)* **116,** 1203–1220.

Hartenstein, V., and Jan, Y. N. (1992). Studying *Drosophila* embryogenesis with *P-lacZ* enhancer trap lines. *Roux's Arch. Dev. Biol.* **201,** 194–220.

Harvey, R. P. (1996). NK-2 homeobox genes and heart development. *Dev. Biol.* **178,** 203–216.

Heldin, C.-H., Miyazono, K., and ten Dijke, P. (1997). TGF-b signalling from cell membrane to nucleus through SMAD proteins. *Nature (London)* **390,** 465–471.

Jagla, K., Dolle, P., Mattei, M., Jagla, T., Schuhbaur, B., Dretzen, G., Bellard, F., and Bellard, M. (1995). Mouse *Lbx1* and human *LBX1* define a novel mammalian homeobox gene family related to the *Drosophila lady bird* genes. *Mech. Dev.* **53,** 345–356.

Jagla, K., Frasch, M., Jagla, T., Dretzen, G., Bellard, F., and Bellard, M. (1997). *Ladybird,* a new component of the cardiogenic pathway in *Drosophila* required for diversification of heart precursors. *Development (Cambridge, UK)* **124,** 3471–3479.

Kim, Y., and Nirenberg, M. (1989). *Drosophila* NK-homeobox genes. *Proc. Natl. Acad. Sci. U.S.A.* **86,** 7716–7720.

Kingsley, D. (1994). The TGF-β superfamily: New members, new receptors, and new genetic tests of function in different organisms. *Genes Dev.* **8,** 133–146.

Klingensmith, J., and Nusse, R. (1994). Signalling by wingless in *Drosophila. Dev. Biol.* **166,** 396–414.

Klingensmith, J., Nusse, R., and Perrimon, N. (1994). The *Drosophila* segment polarity gene *dishevelled* encodes a novel protein required for the response to the wingless signal. *Genes Dev.* **8,** 118–130.

Komuro, I., and Izumo, S. (1993). *csx,* a murine homeobox-containing gene specifically expressed in the developing heart. *Proc. Natl. Acad. Sci. U.S.A.* **90,** 8145–1949.

Kosman, D., Yp, Y. T., Levine, M., and Arora, K. (1991). Establishment of the mesoderm-neuroectoderm boundary in the *Drosophila* embryo. *Science* **254,** 118–122.

Krumlauf, R. (1994). *Hox* genes in vertebrate development. *Cell (Cambridge, Mass.)* **78,** 191–201.

Lai, Z., Fortini, M. E., and Rubin, G. M. (1991). The embryonic expression patterns of *zfh-1* and *zfh-2,* two *Drosophila* genes encoding novel zinc-finger homeodomain proteins. *Mech. Dev.* **34,** 123–134.

Lai, Z., Rushton, E., Bate, M., and Rubin, G. M. (1993). Loss of function of the *Drosophila zfh-1* gene results in abnormal development of mesodermally derived tissues. *Proc. Natl. Acad. Sci. U.S.A.* **90,** 4122–4126.

Lawrence, P. A., Bodmer, R., and Vincent, J. P. (1995). Segmental patterning of heart precursors in *Drosophila. Development (Cambridge, UK)* **121,** 4303–4308.

Lecuit, T., Brook, W. J., Ng, M., Calleja, M., Sun, H., and Cohen, S. M. (1996). Two distinct mechanisms for long-range patterning by Decapentaplegic in the *Drosophila* wing. *Nature (London)* **381,** 387–393.

Leptin, M. (1991). *Twist* and *snail* as postive and negative regulators during *Drosophila* mesoderm development. *Genes Dev.* **5,** 1568–1576.

Letsou, A., Arora, K., Wrana, J. L., Simin, K., Twombly, V., Jamal, J., Staehing-Hampton, K., Hoffman, F. M., Gelbart, W. M., Massague, J., and O'Connor, M. B. (1995). *Drosophila* Dpp signaling is mediated by the *punt* gene product: A dual ligand-binding type II receptor of the TGFb receptor family. *Cell (Cambridge, Mass.)* **80,** 899–908.

Lilly, B., Galewsky, S., Firulli, A. B., Schulz, R. A., Olson, E. N. (1994). D-MEF2: A MADS Box transcription expressed in differentiating mesoderm and muscle cell lineages during *Drosophila* embryogenesis. *Proc. Natl. Acad. Sci. U.S.A.* **91,** 5662–5666.

Lilly, B., Zhao, B., Ranganayakulu, G., Patterson, B. M., Schulz, R. A., and Olson, E. N. (1995). Requirement of MADS domain transcription factor D-MEF2 for muscle formation in *Drosophila. Science* **267,** 688–693.

Lin, M., Nguyen, H., Dybala, C., and Storti, R. (1996). Myocyte-specific enhancer factor 2 acts cooperatively with a muscle activator region to regulate *Drosophila* tropomyosin gene muscle expression. *Proc Natl Acad Sci U.S.A.* **93,** 4623–4628.

Lin, M., Bour, B., Abmayr, S., and Storti, R. (1997). Ectopic expression of MEF2 in the epidermis induces epidermal expression of muscle genes and abnormal muscle development in *Drosophila. Dev. Biol.* **182,** 240–255.

Lyons, I., Parsons, L. M., Hartley, L., Li, R., Andrews, J. E., Robb, L., and Harvey, R. P. (1995). Myogenic and morphogenetic defects in the heart tubes of murine embryos lacking the homeobox gene *Nkx2-5. Genes Dev.* **9,** 1654–1666.

Maggert, K., Levine, M., and Frasch, M. (1995). The somatic-visceral subdivision of the embryonic mesoderm is initiated by dorsal gradient thresholds in *Drosophila. Development (Cambridge, UK)* **121,** 2107–2116.

McMahon, A. P., and Bradley, A. (1990). The *Wnt* (int1) protooncogene is required for development of a large region of the mouse brain. *Cell (Cambridge, Mass.)* **62,** 1073–1085.

McMahon, A. P., and Moon, R. T. (1989). Ectopic expression of the proto-oncogene *intl* in *Xenopus* embryos leads to duplication of the embryonic axis. *Cell (Cambridge, Mass.)* **58,** 1075–1084.

Miller, J. R., and Moon, R. T. (1996). Signal transduction through β-catenin and specification of cell fate during embryogenesis. *Genes Dev.* **10,** 2527–2539.

Molenaar, M., van de Wetering, M., Oosterwegel, M., Peterson-Maduro, J., Godsave, S., Korinek, V., Roose, J., Destree, O., and Cevers, H. (1996). Xtcf3 transcription factor mediates β-catenin-indiced axis formation in *Xenopus* embryos. *Cell (Cambridge, Mass.)* **86,** 391–399.

Molkentin, J., Black, B., Martin, J., and Olson, E. (1995). Cooperative activation of muscle gene expression by MEF2 and myogenic bHLH proteins. *Cell (Cambridge, Mass.)* **83,** 1125–1136.

Nascone, N., and Mercola, M. (1995). An inductive role for the endoderm in *Xenopus* cardiogenesis. *Development (Cambridge, UK)* **118,** 877–892.

Nellen, D., Burke, R., Struhl, G., and Basler, K. (1996). Direct and long-range action of a DPP morphogen gradient. *Cell (Cambridge, Mass.)* **85,** 357–368.

Newman, C. S., and Krieg, P. A. (1998). *Tinman*-related genes expressed during heart development in *Xenopus. Dev. Genet.* **22,** 230–238.

Ng, M., Diaz-Benjumea, F. J., Vincent, J.-P., Wu, J., and Cohen, M. (1996). Specification of the wing by localized expression of *wingless* protein. *Nature (London)* **381,** 316–318.

Nguyen, H. T., Bodmer, R., Abmayr, S., McDermott, J. C., and Spoerel, N.A. (1994). D-mef2: A new *Drosophila* mesoderm-specific MADS box-containing gene with a bi-modal expression profile during embryogenesis. *Proc. Natl. Acad. Sci. U.S.A.* **91,** 7520–7524.

Noordermeer, J., Klingensmith, J., Perrimon, N., and Nusse, R. (1994). *Dishevelled* and *armadillo* act in the *wingless* signalling pathway in *Drosophila. Nature (London)* **367,** 80–83.

Nusse, R., and Varmus, H. E. (1992). Wnt genes. *Cell (Cambridge, Mass.)* **69,** 1073–1087.

Oda, H., Uemura, T., Shiomi, K., Nagafuchi, A., Tsukita, S., and Takeichi, M. (1993). Identification of a *Drosophila* homologue of α-catenin and its association with the armadillo protein. *J. Cell Biol.* **121,** 1133–1140.

Olson, E. N., and Srivastava, D. (1996). Molecular pathways controlling heart development. *Science* **272,** 671–676.

Olson, E. N., Perry, M., and Schulz, R. A. (1995). Regulation of muscle differentiation by the MEF2 family of MADS box transcription factors. *Dev. Biol.* **172,** 2–14.

Park, M., Wu, X., Golden, K., Axelrod, J., and Bodmer, R. (1996). The Wingless signalling pathway is directly involved in *Drosophila* development. *Dev. Biol.* **177,** 104–116.

Park, M., Lewis, C., Turbay, D., Chung, A., Chen, J.-N., Evans, S., Breitbart, R. E., Fishman, M. C., Izumo, S., Bodmer, R. (1998a). *Proc. Natl. Acad. Sci. U.S.A.* (in press).

Park, M., Venkatesh, T. V., and Bodmer, R. (1998b). A dual role for the *zeste-white3/shaggy*-encoded kinase in mesoderm and heart development of *Drosophila. Dev. Genet.* **22,** 201–211.

Park, M., Yaich, L. E., and Bodmer, R. (1998c). Mesodermal cell fate decisions in *Drosophila* are under the control of the lineage genes *numb*, *Notch* and *sanpodo*. *Mech. Develop.* (in press).

Peifer, M., and Bejsovec, A. (1992). Knowing your neighbor: Cell interactions determine intrasegmental pattering in *Drosophila. Trends Genet.* **8,** 243–249.

Peifer, M., Sweeton, D., Casey, M., and Wieschaus, E. (1994). *Wingless* signal and *zeste-white 3 kinase* trigger opposing changes in the intracellular distribution of armadillo. *Development (Cambridge, UK)* **120,** 369–380.

Perrimon, N. (1994). The genetic basis of patterned baldness in *Drosophila. Cell (Cambridge, Mass.)* **76,** 781–784.

Poulson, D. F. (1950). Histogenesis, organogenesis and differentiation in the embryo of *Drosophila melanogaster. In* "Biology of *Drosophila*" (M. Demerec, ed.), pp. 168–274. Cold Spring Harbor Lab. Press, Cold Spring Harbor, New York.

Raftery, L., Twombly, V., Wharton, K., and Gelbart, W. (1995). Genetic screens to identify elements of the *decapentaplegic* signaling pathwayin *Drosophila. Genetics* **139,** 241–254.

Ranganayakulu, G., Zhao, B., Dokidis, A., Molkentin, J. D.,Olson, E. N., and Schulz, R. A. (1995). A series of mutations in the D-MEF2 transcription factor reveals multiple functions in larval and adult myogenesis in *Drosophila. Dev. Biol.* **171,** 169–181

Ranganayakulu, G., Schulz, R. A., and Olson, E. N. (1996). Wingless signaling induces *nautilus* expression in the ventral mesoderm of the *Drosophila* embryo. *Dev. Biol.* **176,** 143–148.

Rehorn, K.-P., Thelen, H., Michelson, A., and Reuter, R. (1996). A molecular aspect of hematopoiesis and endoderm development common to vertebrates and *Drosophila. Development (Cambridge, UK)* **122,** 4023–4031.

Rhyu, M. S., Jan, L. Y., and Jan, Y. N. (1994). Asymmetric distribution of Numb protein during division of the sensory organ precursor cell confers distinct fates to daughter cells. *Cell (Cambridge, Mass.)* **76,** 477–491.

Riechmann, V., Irion, U., Wilson, R., Grosskortenhaus, A., and Leptin, M. (1997). Control of cell fates and segmentation in the *Drosophila* mesoderm. *Development (Cambridge, UK)* **124,** 2915–2922.

Riggleman, B., Schedl, P., and Wieschaus, E. (1990). Spatial expression of the *Drosophila* segment polarity gene *armadillo* is post-transcriptionally regulated by *wingless. Cell (Cambridge, Mass.)* **63,** 549–560.

Rizki, T. M. (1978). The circulatory system and associated cells and tissues. *In* "The Genetics and Biology of *Drosophilia*" (M. Ashburner and T. R. F. Wright, eds.), pp. 397–452. Academic Press, London and New York.

Roelink, H., Wagenaar, E., Lopes da Silva, S., and Nusse, R. (1990). Wnt-3, a gene activated by proviral insertion in mouse mammary tumors is homologs to int-1/Wnt-1 and normally expressed in mouse embryos and adult brain. *Proc. Natl. Acad. Sci. U.S.A.* **87,** 4519–4523.

Ruberte, E., Marty, T., Nellen, D., Affolter, M., and Basler, K. (1995). An absolute requirement for both the type II and type I receptors, Punt and Thick Veins, for Dpp signaling in vivo. *Cell (Cambridge, Mass.)* **80,** 889–897.

Rudnicki, M., and Jaenisch, R. (1995). The MyoD family of transcription factors and skeletal myogenesis. *BioEssays* **17,** 203–209.

Rugendorff, A., Younossi-Hartenstein, A., and Hartenstein, V. (1994). Embryonic origin and differentiation of the *Drosophila* heart. *Roux's Arch Devl. Biol.* **203,** 266–280.

Ruiz Gomez, M., and Bate, M. (1997). Segregation of myogenic lineages in *Drosophila* requires *numb*. *Development (Cambridge, UK)* **124,** 4857–4866.

Rusch, J., and Levine, M. (1996). Threshold responses to the dorsal regulatory gradient and the subdivision of primary tissue territories in the *Drosophila* embryo. *Curr. Opin. Genet. Dev.* **6,** 416–423.

Schmidt-Ott, U., and Technau, G. M. (1992). Expression of *en* and *wg* in the embryonic head and haltere of *Drosophila* indicates a refolded band of seven segment remants. *Development (Cambridge, UK)* **116,** 111–125.

Schultheiss, T. M., Xydas, S., and Lassar, A. B. (1995). Induction of avian cardiac myogenesis by anterior endoderm. *Development (Cambridge, UK)* **121,** 4203–4214.

Schultheiss, T. M., Burch, J. B., and Lassar, A. B. (1997). A role for bone morphogenetic proteins in the induction of cardiac myogenesis. *Genes Dev.* **11,** 451–462.

Scott, M. P. (1994). Intimations of a creature. *Cell (Cambridge, Mass.)* **79,** 1121–1124.

Seidel, F., Bock, E., and Krause, G. (1940). Die Organisation des Insekteneies. *Naturwissenschaften* **28,** 433–446.

Sekelsky, J. J., Newfeld, S. J., Raftery, L. A., Chartoff, E. H., and Gelbart, W. M. (1995). Genetic characterization and cloning of *mothers against dpp,* a gene required for *decapentaplegic* function in *Drosophila* melanogaster. *Genetics* **139,** 1347–1358.

Shishido, E., Higashijima, S.-I., Emori, Y., and Saigo, K. (1993). Two FGF-receptor homologs of *Drosophila:* One is expressed in mesodermal primordium in early embryos. *Development (Cambridge, UK)* **117,** 751–761.

Shishido, E., Ono, N., Kojima, T., and Saigo, K. (1997). Requirements of DFR1/Heartless, a mesoderm-specific *Drosophila* FGF-receptor, for the formation of heart, visceral and somatic muscles, and ensheathing of longitudinal axon tracts in CNS. *Development (Cambridge, UK)* **124,** 2119–2128.

Siegfried, E., Perkins, L. A., Capaci, T. M., and Perrimon, N. (1990). Putative protein kinase product of the *Drosophila* segment-polarity gene *zeste-white 3. Nature (London)* **345,** 825–829.

Siegfried, E., Chou, T. B., and Perrimon, N. (1992). *Wingless* signaling acts through *zeste-white 3,* the *Drosophila* homolog of glycogen synthase kinase-3, to regulate engrailed and establish cell fate. *Cell (Cambridge, Mass.)* **71,** 1167–1179.

Siegfried, E., Wilder, E. L., and Perrimon, N. (1994). Components of *wingless* signaling in *Drosophila. Nature (London)* **367,** 76–79.

Simpson, P. (1983). Maternal-zygotic gene interactions involving the dorsal-ventral axis in *Drosophila* embryos. *Genetics* **105,** 615–632.

Skaer, H., and Martinez Arias, A. (1992). The *wingless* product is required for cell proliferation in the Malphigian tubule anlagen of *Drosophila melanogaster. Development* (*Cambridge, UK*) **116,** 745–754.

Smith, J. C., and Howard, J. E. (1992). Mesoderm inducing factors and the control of gastrulation. *Development* (*Cambridge, UK*) Suppl., pp. 127–136.

Smith, T. S., and Jaynes, J. B. (1996). A conserved region of *engrailed,* shared among all en, Nk1, Nk2- and msh-class homeoproteins, mediates active transcriptional represssion *in vivo. Development* (*Cambridge, UK*) **122,** 3141–3150.

Spana, E. P. , and Doe, C. Q. (1996). Numb antagonizes Notch signalling to specify sibling neuron cell fates. *Neuron* **17,** 21–26.

Spana, E. P., Kopczynski, C., Goodman, C. S., and Doe, C. Q. (1995). Asymmetric localization of Numb autonomously determines sibling neuron identity in the *Drosophila* CNS. *Development* (*Cambridge, UK*) **121,** 3489–3494.

Spencer, F. A., Hoffmann, F. M., and Gelbert, W. M. (1982). *Decapentaplegic:* A gene complex affecting morphogenesis in *Drosophila melanogaster. Cell* (*Cambridge, Mass.*) **28,** 451–461.

Staehling-Hampton, K., Hoffmann, F. M., Baylies, M. K., Rushton, E., and Bate, M. (1994). *dpp* induces mesodermal gene expression in *Drosophila. Nature* (*London*) **372,** 783–786.

Staehling-Hampton, K., Laughon, A. S., and Hoffmann, F. M. (1995). A *Drosophila* protein related to the human zinc finger transcription factor PRDII/MBPi/-HIV-EP1 is required for *dpp* signalling. *Development* (*Cambridge, UK*) **121,** 3393–3403.

Struhl, G., and Basler, K. (1993). Organizing activity of *wingless* protein in *Drosophila. Cell* (*Cambridge, Mass.*) **72,** 527–540.

Takagi, T., Kondoh, H., and Higashi, Y. (1997). *deltaEF1* null mutant mice exhibit multiple skeletal defects in craniofacial, vertebral column, rib and limb formation. *Dev. Biol.* **186,** 346–351.

Takada, S., Stark, K. L., Shea, M. J., Vassileva, G., McMahon, J. A., and McMahon, A. P. (1994). *Wnt-3A* regulates somite and tailbud formation in the mouse embryos. *Genes Dev.* **8,** 174–189.

Taylor, M. V., Meatty, K. E., Unter, H. K., and Baylies, M. K. (1995). *Drosophila* MEF2 is regulated by *twist* and is expressed in both the primordia and differentiated cells of the embryonic somatic, visceral and heart musculature. *Mech. Dev.* **50,** 29–41.

Tepass, U., Fessler, L. I., Aziz, A., and Hartenstein, V. (1994). Embryonic origin of hemocytes and their relationship to cell death in *Drosophila. Development* (*Cambridge, UK*) **120,** 1829.

Thisse, B., Stoetzel, C., Gorositiza-Thisse, C., and Perrin-Schmitt, F. (1988). Sequence of the *twist* gene and nuclear localization of its protein in endomesodermal cells of early *Drosophila* embryos. *EMBO J.* **7,** 2175–2183.

Thomas, K. R., and Capecchi, M. R. (1990). Targeted disruption of the murine int1 protooncogene resulting in severe abnormalities in midbrain and cerebellar development. *Nature* (*London*) **346,** 847–850.

Uemura, T., Shepherd, S., Ackerman, L., Jan, L. Y., and Jan, Y. N. (1989). *numb,* a gene required in determination of cell fate during sensory organ formation in *Drosophila* embryos. *Cell* (*Cambridge, Mass.*) **58,** 349–360.

van den Heuvel, M., Nusse, R., Johnston, P., and Lawrence, P. A. (1989). Distribution of the *wingless* gene product in *Drosophila* embryos: A protein involved in cell-cell communication. *Cell* (*Cambridge, Mass.*) **59,** 739–749.

van de Wetering, M., Cavallo, R., Dooijes, D., van Beest, M., van Es, J., Loureiro, J., Ypma, A., Hursh, D., Jones, T., Bejsovec, A., Peifer, M., Mortin, M., and Clevers, H. (1997). Armadillo coactivates transcription driven by the product of the *Drosophila* segment polarity gene dTCF. *Cell* (*Cambridge, Mass.*) **88,** 789–799.

Venkatesh, T. V., and Bodmer, R. (1995). How many signals does it take? *BioEssays* **17,** 754–757.

Verdi, J. M., Schmandt, R., Bashirullah, A., Jacob, S., Salvino, R., Craig, C. G., Program, A. E., Lipshitz ,H. D., and McGlade, C. J. (1996). Mammalian NUMB is an evolutionarily conserved signaling adapter protein that specifies cell fate. *Curr. Biol.* **6,** 1134–1145.

Volk, T., and VijayRaghavan, K. (1994). A central role for epidermal segment border cells in the induction of muscle patterning in the *Drosophila* embryo. *Development* (*Cambridge, UK*) **120,** 59–70.

Wharton, K. A. (1995). How many receptors does it take? *BioEssays* **17,** 13–16.

Williams, J. A., Paddock, S. W., and Carroll, S. B. (1993). Pattern formation in a secondary field: A hierachy of regulatory genes subdivides the developing *Drosophila* wing disc into discrete subregions. *Development* (*Cambridge, Mass.*) **117,** 571–584.

Wimmer, E., Jäckle, H., Pfeifle, C., and Cohen, S. (1991). A *Drosophila* homologue of human Sp1 is a head-specific segmentation gene. *Nature* (*London*) **366,** 690–694.

Woods, D. F., and Bryant, P. J. (1991). The discs-large tumor suppressor gene of *Drosophila* encodes a guanylate kinase homolog localized at septate junctions. *Cell* (*Cambridge, Mass.*) **66,** 451–464.

Wu, X., Golden, K., and Bodmer, R. (1995). Heart development in *Drosophila* requires the segment polarity gene wingless. *Dev. Biol.* **169,** 619–628.

Yanagawa, S., Leeuwen, F., Wodarz, A., Klingensmith, J., and Nusse, R. (1995). The Dishevelled protein is modified by *wingless* signaling in *Drosophila. Genes Dev.* **9,** 1087–1097.

Yin, Z., and Frasch, M. (1998). Regulation and function of *tinman* during dorsal mesoderm induction and heart specification in *Drosophila. Dev. Genet.* **22,** 187–200.

Yin, Z., Xu, X.-L., and Frasch, M. (1997). Regulation of the Twist target gene *tinman* by modular *cis*-regulatory elements during early mesoderm development. *Development* (*Cambridge, UK*) **124,** 4871–4982.

Zhong, W., Feder, J. N., Jiang, M. M., Jan, L. Y., and Jan, Y. N. (1996). Asymmetric localization of a mammalian Numb homolog during mouse cortical neurogenesis. *Neuron* **17,** 43–53.

6

Mutations Affecting Cardiac Development in Zebrafish

Jonathan Alexander and Didier Y. R. Stainier

Department of Biochemistry and Biophysics, Programs in Developmental Biology and Human Genetics, University of California, San Francisco, San Francisco, California 94143

I. Introduction

The zebrafish *Danio rerio* represents the newest addition to the pantheon of model organisms used to study vertebrate development. Originally introduced by George Streisinger and colleagues at the University of Oregon in the late 1970s (Streisinger *et al.,* 1981), the zebrafish is distinguished from the other vertebrate model systems (principally the frog, chick, and mouse) by the opportunity it provides to apply classical genetics to problems in vertebrate development (Driever *et al.,* 1994; Kimmel, 1989; Mullins *et al.,* 1994). Classical genetics refers to the search for mutations that perturb specific biological events without foreknowledge of the gene or genes involved. The power of this approach has been amply demonstrated by previous work in invertebrate organisms such as *Drosophila* and *Caenorhabditis elegans* (Horvitz, 1988; St. Johnston and Nüsslein-Volhard, 1992). Similarly important contributions to our understanding of vertebrate development are beginning to emerge from work on the zebrafish.

Tremendous attention has been focused on the zebrafish in the wake of the recent completion of two large-scale screens for mutations affecting numerous aspects of embryonic development (Driever *et al.,* 1996; Haffter *et al.,* 1996). Among these newly identified mutations are a number that disrupt heart formation (Table I) or function (Table II) (Chen *et al.,* 1996; Stainier *et al.,* 1996). Although studies of the various cardiac mutants have only just begun, taken as a whole the mutant phenotypes demonstrate that the individual developmental events necessary to form the heart can to some degree be uncoupled. Indeed, despite disruption of an early step, for example, formation of a single definitive heart tube, later events such as chamber formation and myocardial maturation appear to proceed for the most part normally. As studies progress, both on existing mutants and on others that undoubtedly will be identified in new screens directed specifically toward the heart (Alexander *et al.,* 1998), we can expect to gain

Table I Mutations Affecting Cardiac Morphogenesis in the Zebrafish *Danio rerio*

Genetic loci	Alleles	Phenotype	Other phenotypes	Reference
Induction of precardiac mesoderm				
one-eyed pinhead (oep)	m134, tz257	Reduced heart tissue, cardia bifida	Brain, eye, prechordal plate, body shape	c, d, e
Formation of the definitive heart tube				
miles apart (mil)	m93, te273	Cardia bifida	Tail	a, b
bonnie and clyde (bon)	m425	Cardia bifida		a
casanova (cas)	ta56	Cardia bifida	Motility	b
faust (fau)	tm236a	Cardia bifida		b
natter (nat)	ta219, tl43c	Cardia bifida		b
heart and soul (has)	m129, m567, m781	Small compact heart	Brain, eye, body shape	a, c
Endocardial differentiation				
cloche (clo)	m39, m378	No endocardium		a, f
Chamber formation				
pandora (pan)	m313	Reduced or absent ventricle	Eye, ear, somite, body shape	a, j, k
lonely atrium (loa)	tu29d	Absent ventricle		b
Heart looping				
no tail (ntl)	b160, b195, m147, m550, tb244e, tc41, ts260	Randomized heart looping	Notochord, tail	g, i, l, m, n
floating head (flh)	nl, b327, m614, tm229, tk241	Randomized heart looping	Notochord, somites	h, i, l, m, n
dino (din)	tm84, tt350	Randomized heart looping	Gastrulation	l, n, p, q
lost-a-fin (laf)	tm110b	Randomized heart looping	Gastrulation	n, p
piggytail (pgy)	dty40, dti216c, tc227a, tm124, tx223	Randomized heart looping	Gastrulation	n, p
spadetail (spt)	b104, tm41, tq5	Randomized heart looping	Gastrulation	n, o, q
snailhouse (snh)	ty68a	Randomized heart looping	Gastrulation	n, p
momo (mom)	th211	Randomized heart looping	Notochord	l, n
cyclops (cyc)	b16, b213, b229, m101, m294, tf219, te262c	Randomized heart looping	Floor plate	e, n, r
iguana (igu)	tm79a, ts294e	Randomized heart looping	Floor plate	e, n
schmalspur (sur)	ty68b	Randomized heart looping	Floor plate	e, n
schmalhans (smh)	tn222a	Randomized heart looping	Floor plate	e, n
curly up (cup)	ty30, tp85a, tc321, tg266d	Randomized heart looping	Curly tail	e, n
locke (lok)	tm138a, tj8, ts277, tl215, to237b	Randomized heart looping	Curly tail, kidney	b, e, n
(−)	tg238a	Randomized heart looping	Curly tail	n
(−)	tj2a	Randomized heart looping	Curly tail	n
(−)	tm243b	Randomized heart looping	Curly tail	n
(−)	tm317b	Randomized heart looping	Curly tail	n
(−)	tn20b	Randomized heart looping	Curly tail	n
(−)	tw29b	Randomized heart looping	Curly tail	n
Valve formation				
jekyll (jek)	m151, m310	No valves	Branchial arches, jaw	a
(−)	m27	No valves		a

Note. The b and n alleles were identified in Eugene, Oregon; the m alleles were identified in Boston; the t alleles were identified in Tübingen. a, Stainier *et al.*, 1996; b, Chen *et al.*, 1996; c, Schier *et al.*, 1996; d, Schier *et al.*, 1997; e, Brand *et al.*, 1996; f, Stainier *et al.*, 1995; g, Halpern *et al.*, 1993; h, Talbot *et al.*, 1995; i, Danos and Yost, 1996; j, Malicki *et al.*, 1996a; k, Malicki *et al.*, 1996b; l, Odenthal *et al.*, 1996; m, Stemple *et al.*, 1996; n, Chen *et al.*, 1997; o, Kimmel *et al.*, 1989; p, Mullins *et al.*, 1996; q, Hammerschmidt *et al.*, 1996; r, Hatta *et al.*, 1991.

a full description of the genes required for each individual event in heart development, as well as to understand how these events are coordinated.

In this chapter we introduce the zebrafish as a genetic model system, describing the characteristics that make it amenable to classical genetic analyses, and briefly cover the basics of zebrafish embryonic development. We then describe zebrafish heart development in more detail, in particular noting similarities to other vertebrate species. Finally, we proceed stepwise through the process of heart development, discussing briefly what is known about particular events (more detailed discussions of these events can be found in various other chapters within this book) and describing mutants

Table II Mutations Affecting Cardiac Function in the Zebrafish *Danio rerio*

Genetic loci	Alleles	Phenotype	Other phenotypes	References
Heart does not beat				
silent heart (sih)	b109, tc300b	No heart beat		b
throbless (thr)	b212, to241a, tm123a	No heart beat	Severely reduced motility	b
(−)	tw212e	No heart beat	Brain	b
viper (vip)	ta52e	No heart beat		b, c
Reduced contractility of both chambers				
pickwick (pik)	m171, m186, m242, m740	Both chambers beat weakly		a
lazy susan (laz)	m647	Both chambers beat weakly		a
pipe dream (ppd)	m301	Both chambers beat weakly		a
beach bum (bem)	m281	Both chambers beat weakly		a
pipe line (ppl)	m340	Both chambers beat weakly		a
weak beat (web)	tc288b	Both chambers beat weakly		b
quiet heart (quh)	tq268	Both chambers beat weakly		b
stretched (str)	tu255b	Both chambers beat weakly		b
herzschlag (hel)	tg287	Both chambers beat weakly	Immotile	b
heart attack (hat)	te313	Both chambers beat weakly	Immotile	b
slop (slp)	tq235c	Both chambers beat weakly	Reduced motility, reduced muscle striations	b
jam (jam)	tr254a	Both chambers beat weakly	Reduced motility, reduced muscle striations	b
slinky (sky)	ts254	Both chambers beat weakly	Reduced motility, reduced muscle striations	b
(−)	tk34b	Both chambers beat weakly	Immotile	b
Reduced ventricular contractility				
sisyphus (sis)	m351, m644	Weak ventricle		a
main squeeze (msq)	m347	Weak ventricle	Body shape, pigment	a
hal (hal)	m235	Weak ventricle		a
weiches herz (whz)	m245	Weak ventricle		a
low octane (loc)	m543	Weak ventricle		a
dead beat (ded)	m582	Weak ventricle	Body shape	a
pipe heart (pip)	tq286e	Weak ventricle		b
tango (tan)	tg299c, to216b	Weak ventricle		b
Reduced atrial contractility				
weak atrium (wea)	m58, m229, m448, tw220a	Silent atrium		a, b
Heart rhythm and conduction				
tremblor (tre)	m116, m139, m158, m276, m736, tc318d, te381b	Fibrillation		a, b
polka (plk)	tg290, ty105b	Isolated cell contractions		b
legong (leg)	tj201	Weak beat initially; later becomes fibrillating		b
slip jig (sli)	tm117c	Weak beat initially; later becomes fibrillating		b
island beat (isl)	m231, m379, m458	Isolated twitches		a
reggae (reg)	m230	Spasmodic beat		a
silent partner (sil)	m656	Silent ventricle		a
ginger (gin)	m47, m155, m739	Ventricle becomes silent		a
tell tale heart (tel)	m225	Nearly silent heart		a
breakdance (bre)	tb218	2:1 ratio of atrial: ventricular beating		b
hiphop (hip)	tx218	3:1 ratio of atrial: ventricular beating		b
Heart rate				
slow mo (smo)	m51	Slow beat		a, d
Large heart				
santa (san)	m775, ty219c	Large heart, no valves		a, b
valentine (vtn)	m201	Large heart, no valves		a
heart of glass (heg)	m552	Large heart, no valves		a

Note. The b alleles were identified in Eugene, Oregon; the m alleles were identified in Boston; the t alleles were identified in Tübingen. a, Stainier *et al.*, 1996; b, Chen *et al.*, 1996; c, Jiang *et al.*, 1996; d, Baker *et al.*, 1997.

whose study may shed light on these various processes. We conclude by speculating about the contributions that future work in zebrafish will likely make to the field of heart development.

II. The Zebrafish as a Genetic Model System

A fortuitous ensemble of characteristics makes the zebrafish suitable for classical genetic analyses (Fig. 1) (Westerfield, 1993). Adult zebrafish are small (typically 3 to 4 cm in length), are easy and inexpensive to maintain in large numbers, and reach sexual maturity in approximately 3 months. Methods exist to generate mutations of various types, including point mutations (Mullins and Nüsslein-Volhard, 1993; Solnica-Krezel *et al.,* 1994), deletions (Walker and Streisinger, 1983), and insertional mutations (Gaiano *et al.,* 1996), which can be bred to homozygosity in order to determine their phenotypic consequences. Additionally, the ability to produce haploid and diploid embryos of either gynogenetic or androgenetic origin allows one to force mutations to hemi- (present in only a single dose) or homozygosity in a single generation (Kimmel, 1989; Streisinger *et al.,* 1981). Single adult zebrafish mating pairs yield hundreds of embryos at weekly intervals, and the oviparous nature of the process allows for continual observation and manipulation of the embryos throughout development. The striking transparency of the zebrafish embryo greatly facilitates morphological observation of internal structures *in vivo,* without the need for fixation or stains. Finally, zebrafish development proceeds quite rapidly; for example, the heart begins beating at approximately 22 hr postfertilization (hpf), and by 36 hpf has completed chamber formation and looping morphogenesis (Stainier and Fishman, 1994). This constellation of traits occurs in no other vertebrate model system; the zebrafish therefore provides a unique tool for the study of vertebrate development.

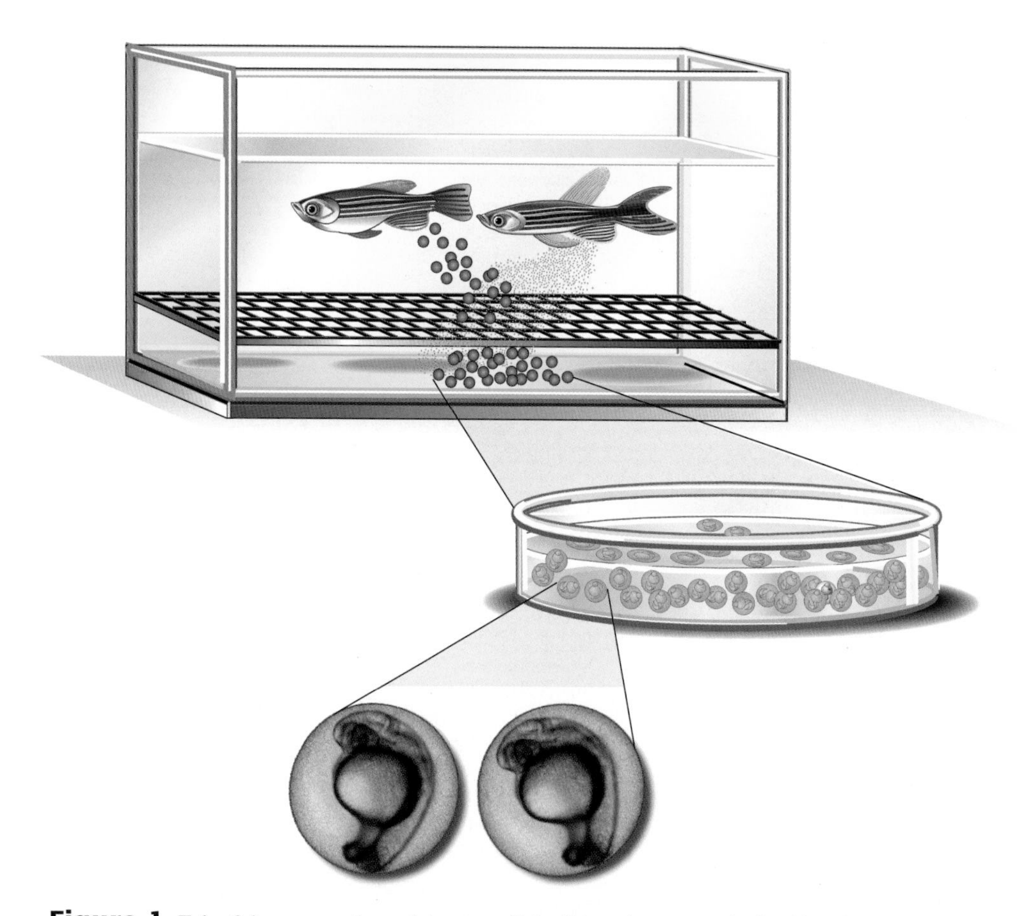

Figure 1 Zebrafish as a genetic model system. Zebrafish embryos are obtained by placing an adult male and female together in a breeding cage in the evening. The following morning, after the lights come on ("sunrise"), the male and female release sperm and eggs into the water, respectively; fertilization occurs externally. The eggs fall through a mesh screen positioned near the bottom of the breeding trap that prevents the adults from eating the eggs. The eggs are collected and raised in petri dishes in a temperature-controlled incubator and are accessible for observation and manipulation throughout embryonic development. The embryos shown are approximately 24 hours old.

III. Zebrafish Embryonic Development

The timing of particular events in zebrafish development is generally given in terms of standard developmental time, which reflects the hours postfertilization at 28.5°C (Kimmel *et al.,* 1995); development proceeds faster or slower at higher and lower temperatures, respectively, a fact which can be experimentally useful. For the purposes of this discussion, all times given represent standard developmental times.

The mating behavior of zebrafish follows a diurnal rhythm, usually initiating soon after "sunrise" (i.e., when the lights come on) (Westerfield, 1993). Male and female gametes are released into the water and the activated sperm must quickly encounter and fertilize an egg. Upon fertilization of the egg the process of cytoplasmic streaming begins (Kimmel *et al.,* 1995). This coordinated cytoplasmic movement segregates the non-yolky cytoplasm toward the animal pole to form the blastodisc, from which the embryo itself arises (Fig. 2A). The yolk cytoplasm is contained within a single large yolk cell upon and around which the embryo develops; this mode of development differs from amphibian development, in which each individual cell contains yolk granules. Starting at 40 min postfertilization the zygote undergoes a series of rapid and synchronous cell divisions, driven entirely by maternal components deposited in the egg, that results in an embryo of approximately 500 cells by 2.75 hpf (Kimmel and Law, 1985a).

The blastula period follows (2.75–5.3 hpf) (Fig. 2B), marked by three critical events. First, the midblastula transition begins during the tenth cell cycle; at this time cell cycle length increases and the occurrence of cell division throughout the embryo becomes markedly asynchronous (Kane and Kimmel, 1993). Importantly, zygotic gene expression first initiates at the midblastula transition. At nearly the same time the yolk syncytial layer forms, as the vegetal-most blastomeres collapse and release their nuclei and cytoplasm into the yolk cell (Kimmel and Law, 1985b). Some very limited evidence suggests that the dorsal region of the yolk syncytial layer may function as the teleost equivalent to the *Xenopus* Nieuwkoop Center, inducing the formation of the dorsal organizer (Long, 1983; Mizuno *et al.,* 1996). Epiboly, the morphogenetic movement by which the cellular blastoderm thins out and spreads over the yolk cell, like a ski cap being pulled over one's head, also begins during the blastula period (Solnica-Krezel and Driever, 1994).

Gastrulation begins at 50% epiboly (Fig. 2C) (5.3 hpf) with the involution of the vegetal-most blastodermal cells, generating a single layer of mesendodermal cells, the hypoblast, beneath the existent epiblast (Kimmel *et al.,* 1995). Some cells in the zebrafish embryo also form mesendoderm via ingression (Shih and Fraser, 1995), entering the hypoblast directly from positions several cell diameters away from the margin; neither the relative importance of ingression versus involution nor the developmental significance of ingression have been determined. Cells also begin to converge dorsally, producing at 6 hpf (Fig. 2D) a site of increased thickening, the embryonic shield, that marks the dorsal side of the embryo and functions as the zebrafish equivalent to the dorsal organizer in *Xenopus* (Ho, 1992). Gastrulation continues in concert with the progression of epiboly until the yolk cell has been entirely engulfed, at which point the notochord has begun to form and the tail bud is just appearing (Fig. 2E) (10 hpf) (Kimmel *et al.,* 1995).

During the next period of development, the segmentation period (Fig. 2F) (10–24 hpf), somites form, the tail elongates, and the rudiments of many organs first appear, including the brain, eyes, otic vesicles, kidneys, and heart, so that by 24 hpf the zebrafish embryo has reached the vertebrate phylotypic stage (Fig. 2G) (Kimmel *et al.,* 1995). Pharyngeal arch, pigment cell, and fin development are prominent hallmarks of the pharyngula period (Fig. 2H) (24–48 hpf). Gill slits form and the jaw and pectoral fins develop rapidly during the following 24 hr, which is also the time when the embryo hatches from its chorion. As growth proceeds, the swim bladder inflates and the larva begins to swim actively and soon thereafter initiates feeding (usually around 5 days postfertilization) (Kimmel *et al.,* 1995). Depending on the conditions under which they are raised, larvae reach sexual maturity in approximately 3 months.

IV. Development of the Zebrafish Heart

Fate-mapping studies have revealed the early embryonic location of the cells that give rise to the zebrafish heart. At early to midblastula stages (3 or 4 hpf) cells located throughout the ventrolateral region of the embryo (from 90 to 270° longitude, where the future dorsal axis is 0° longitude) can later populate both myocardial and endocardial lineages (Fig. 3) (Lee *et al.,* 1994; Stainier *et al.,* 1993). Nearer to the onset of gastrulation (40% epiboly, approximately 5 hpf) both the myocardial and endocardial progenitors are located more laterally (spanning areas from 60 to 150° and 210 to 300° longitude) and close to the margin between the blastoderm and the yolk cell (Warga, 1996). Chen and Fishman (1996) have reported seeing expression of the heart-specific homeobox gene *nkx2.5* at this stage in a gradient that is strongest ventrally and diminishes in the lateral regions of the embryo. They propose that this pattern of expression represents the embryonic "heart field," those cells capable of giving rise to the heart, and that the level of *nkx2.5* expression in a given cell deter-

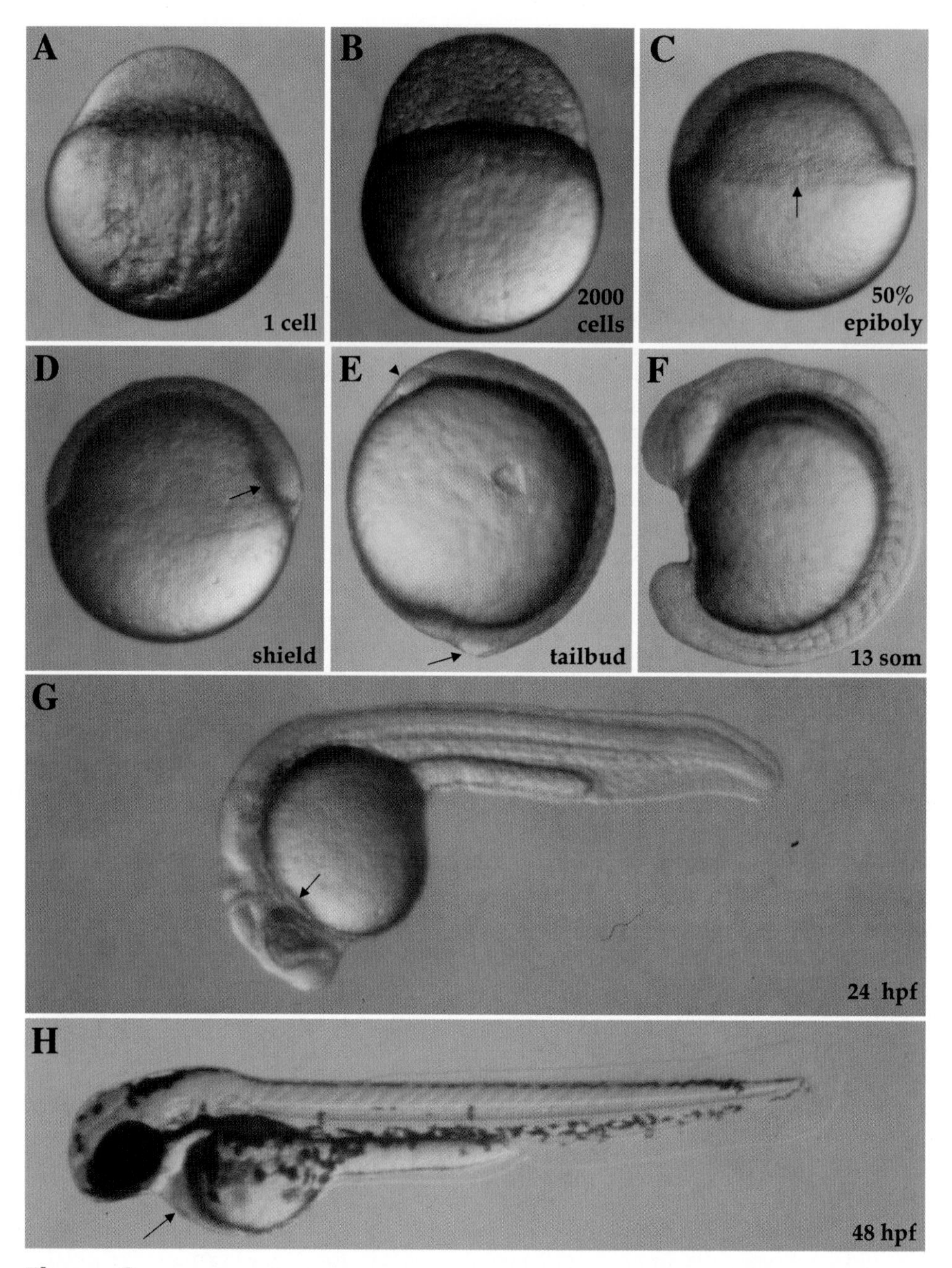

Figure 2 Zebrafish embryonic development. Zebrafish embryos at various stages of development: (A) 1-cell stage (0.2 hpf); (B) 2000-cell stage (3.3 hpf), (C) 50% epiboly (5.3 hpf) (arrow indicates the blastoderm margin); (D) shield stage (6 hpf) (arrow marks the embryonic shield); (E) tailbud stage (10 hpf) (arrow marks the tailbud; arrowhead marks the forming left optic vesicle); (F) 13-somite stage (15.5 hpf); (G) pharyngula stage (24 hpf) (arrow indicates the location of the heart); (H) 48 hpf (arrow indicates the location of the heart, which is filled with red blood cells). Embryos in A–F are shown with the animal pole to the top; dorsal is to the right in C–F but cannot be determined in A and B. Embryos in G and H are shown with anterior to the left and dorsal to the top. The chorion has been removed in order to view the embryos more clearly. [From Kimmel, C. B., Ballard, W. W., Kimmel, S. R., Ullmann, B., and Schilling, T. F. (1995). Stages of embryonic development of the zebrafish. *Dev. Dyn.* **203,** 253–310. Copyright © 1995. Reprinted by permission of Wiley-Liss, Inc. a subsidiary of John Wiley & Sons, Inc.]

mines the likelihood that the cell will in fact contribute to the heart. However, these data are not entirely consistent with the fate-mapping data described previously and their significance remains to be determined.

Following their involution, the bilateral populations of heart progenitors, which now constitute the anterior-most lateral plate mesoderm, migrate to their destination alongside the embryonic axis at the level of the future hindbrain (Fig. 3). Examination of the expression of *nkx2.5* (Lee *et al.,* 1996), the earliest specific marker

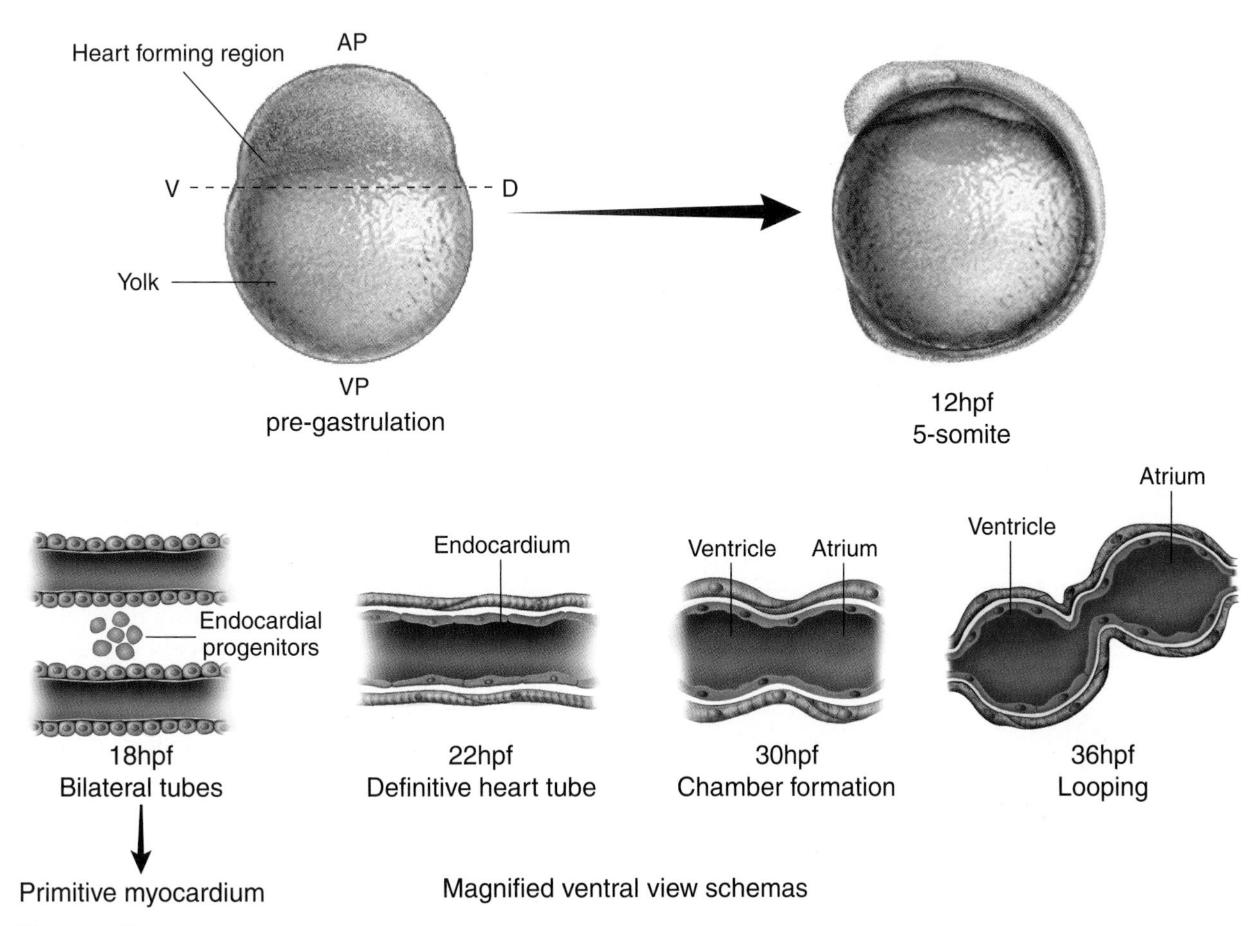

Figure 3 Zebrafish heart development. Prior to gastrulation, at approximately the 2000-cell stage (3.3 hpf), the progenitors of the zebrafish heart are located throughout the ventral and lateral regions of the blastula. Following involution these cells converge toward the embryonic axis, reaching their destination at the level of the future hindbrain by the 5-somite stage (approximately 12 hpf). These cells then form two primitive myocardial tubes that flank the endocardial progenitors (18-somite stage, 18 hpf). The primitive myocardial tubes have fused to form the definitive heart tube by the 26-somite stage (22 hpf), enclosing the endocardial progenitors in the process. Subsequently, visibly distinct atrial and ventricular chambers form (by approximately 30 hpf) and the heart undergoes looping morphogenesis (by approximately 36 hpf). AP, animal pole; VP, vegetal pole; D, dorsal; V, ventral.

of precardiac mesoderm in all vertebrates, by RNA *in situ* hybridization reveals the location and arrangement of these cells at different times in development: At the 5-somite stage (12 hpf) the myocardial progenitors appear as oval-shaped groups of loosely packed cells located on either side of the embryonic axis (Fig. 4A); over the next few hours these cells pack together more tightly, extend rostrocaudally, and by the 15-somite stage (16 hpf) have assembled into two primitive myocardial tubes (Figs. 3 and 4B) (Stainier and Fishman, 1992).

Soon thereafter, these primitive myocardial tubes begin to move closer to each other and by the 26-somite stage (approximately 22 hpf) have fused to form the definitive heart tube (Fig. 3) (Stainier *et al.,* 1993). The primitive heart tubes do not, however, fuse along their long axis as might be expected. Rather, their fusion creates a cone-shaped structure whose base rests on the yolk and whose apex is continuous with the ventral aorta (Fig. 4C) (Stainier *et al.,* 1993). Thus, the long axis of the definitive heart tube is initially orthogonal to that of the primitive heart tubes and to the anterior–posterior axis of the embryo itself. Later in development, as the head straightens out and the yolk is resorbed, the anterior–posterior axis of the heart is brought into alignment with that of the embryo (Fig. 4D). How exactly fusion of the primitive myocardial tubes is achieved, and which cells within them give rise to which portions of the definitive heart tube, remains unclear.

The endocardium appears to arise from anterior groups of endothelial progenitors that populate the head region; these cells can be identified by the 7-somite stage (12 hpf) due to their expression of the endothelial-specific receptor tyrosine kinase gene *flk-1* (Fig. 4E) (Liao *et al.,* 1997). It appears that cells from the caudalmost portion of these anterior angioblast populations

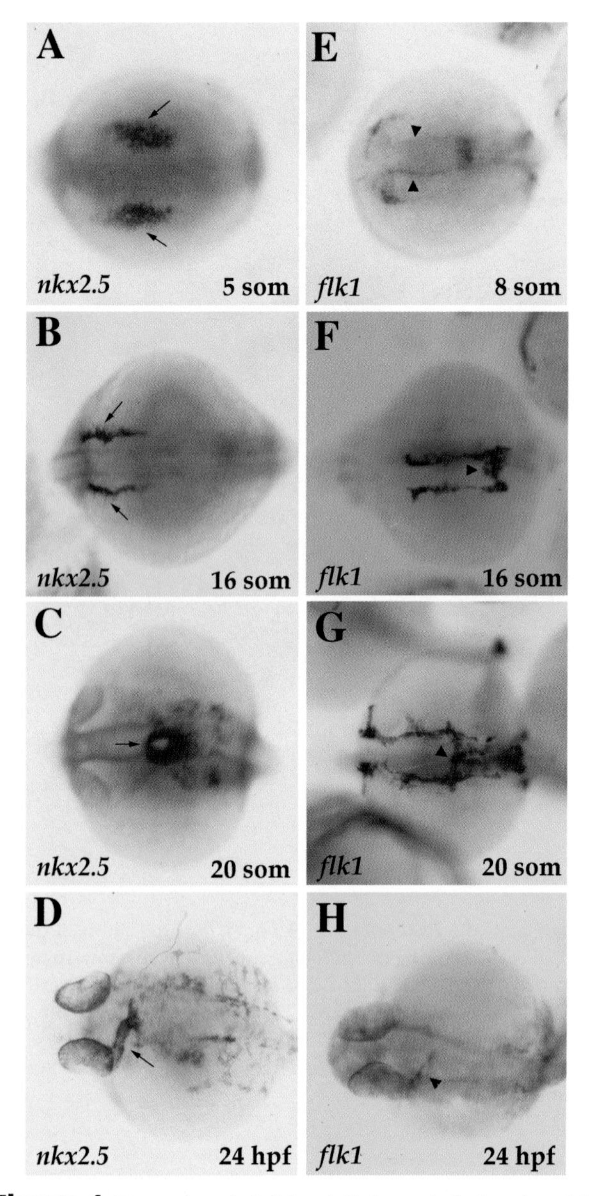

Figure 4 Expression of *nkx2.5* and *flk-1* reveals the location of the myocardial and endocardial progenitors, respectively. Expression of the myocardial-specific homeobox gene *nkx2.5* [arrows in A–D mark the myocardial progenitors (A, 5-somite stage), the primitive myocardial tubes (B, 16-somite stage), the cone-shaped heart primordium (C, 20-somite stage), and the definitive heart tube (D, 24 hpf)]. The endocardium derives from *flk-1*-expressing cranial endothelial progenitors (E, 8-somite stage, arrowheads). Cells from the caudal-most region of these bilateral populations migrate medially (F, 16-somite stage, arrowhead) and form the endocardium [arrowhead in G, (20-somite stage) and H (24 hpf)]. Embryos were examined by RNA *in situ* hybridization with riboprobes directed against *nkx2.5* and *flk-1*. Embryos are viewed dorsally with anterior to the left, though not always rotated to the same extent; for example, the embryo in F is viewed more dorsoanteriorly to reveal better how the endocardial cells derive from the caudal region of the cranial endothelial progenitor populations.

begin to migrate medially at approximately 16 hpf (Fig. 4F) and by 19 hpf occupy a position between the bilateral primitive myocardial tubes (Figs. 3 and 4G). Cells in this location, the so-called *portion moyenne du mesoblaste* (Swaen and Brachet, 1901), later give rise to the endocardium (Stainier *et al.,* 1993). The process of fusion described previously serves to enclose these endocardial progenitors within the definitive heart tube (Figs. 3 and 4H), where they undergo further differentiation into mature endocardium, as assessed by the expression of later endothelial markers such as the receptor tyrosine kinase gene *tie-1* (Liao *et al.,* 1997).

Subsequent development of the heart proceeds rapidly (Stainier and Fishman, 1994). Occasional contraction of cells within the primitive myocardial tubes can be seen even prior to their fusion into the definitive heart tube. Regular and coordinated beating initiates shortly after fusion, starting slowly (approximately 25 beats/min at 22 hpf) and increasing with time (about 90 beats/min at 24 hpf). Circulation also begins at 24 hpf. The morphological demarcation of the atrium and ventricle is apparent by approximately 30 hpf (Fig. 3), although the expression of chamber-specific myosin heavy-chain isoforms indicates the existence of distinct atrial and ventricular tissues several hours earlier (Stainier and Fishman, 1992). By 36 hpf the heart has completed the rightward looping characteristic of vertebrate cardiac development (Fig. 3), positioning the atrium to the left of the more medially situated ventricle. Lastly, endocardial cells in the regions of the chamber boundaries appear to undergo an epithelial to mesenchymal transition, forming the cardiac cushions (at approximately 48 hpf) that later give rise to the heart valves.

The adult zebrafish heart is a two-chambered organ that contains a single atrium and a single ventricle, as well as a recognizable sinus venosus and bulbus arteriosus at its extreme posterior and anterior ends, respectively (Stainier and Fishman, 1994). There is no separate pulmonary circulation in the zebrafish; blood leaving the ventricle is oxygenated upon passage through the gills and traverses the vasculature before returning to the atrium. The three- and four-chambered hearts characteristic of higher vertebrates, which enable separation of the pulmonary and systemic circulations, are generated by septation and further morphogenetic remodeling of an embryonic two-chambered heart that looks very much like the zebrafish heart.

Viewed as a whole, the zebrafish heart develops in a manner quite similar to that of other vertebrate hearts (Fishman and Chien, 1997). The same fundamental events appear to occur in all vertebrates, although they may proceed somewhat differently due to the morphological constraints inherent to the development of a particular organism. For the purposes of this discussion, we have artificially divided zebrafish heart development into seven events (in approximate chronological order) (i) induction of precardiac mesoderm, (ii) definitive heart tube formation, (iii) endocardial cell differentiation, (iv) onset of cardiac function, (v) chamber formation, (vi) looping, and (vii) valve formation. In the following sections, we consider each of these events separately, briefly reviewing what is known from work in other systems and then describing zebrafish mutants that may be relevant to the particular event under consideration.

A. Induction of Precardiac Mesoderm

Studies of heart induction date to the early years of this century (DeHaan, 1965). Nonetheless, definitive answers regarding which tissues induce the formation of the precardiac mesoderm and what the heart-inducing molecules are remain elusive. Substantial classical embryological data, gathered principally from studies using urodele amphibian and avian species, implicate the anterior endoderm as the heart inducer (Jacobson and Sater, 1988; see Chapters 3 and 4). Recent work provides support for this model. In the chick the anterior endoderm is capable of inducing cardiac-specific gene expression in mesoderm not normally fated to form heart (Schultheiss *et al.,* 1995), and bone morphogenetic protein (BMP)-2 secreted by the anterior endoderm appears to be required to maintain *cNkx2.5* expression in the precardiac mesoderm (Schultheiss *et al.,* 1997). Also, the endoderm has been shown to play an important role in *Xenopus* heart induction (Nascone and Mercola, 1995), perhaps in conjunction with direct or indirect signals from the dorsal organizer (Sater and Jacobson, 1990).

While these results are consistent with the idea that the endoderm sends important signals to the precardiac mesoderm, these signals may not be truly inductive but instead may act to sustain differentiation of the precardiac mesoderm (Sugi and Lough, 1994). Indeed, experiments in the chick suggest that one role of the endoderm is to cause cardiac contractile proteins to assemble into organized myofibrils; when the anterior endoderm is removed after Hamburger–Hamilton stage 6, gene expression in the precardiac mesoderm appears to be unaffected but myofibrils fail to form and the tissue does not initiate contractions (Gannon and Bader, 1995). Furthermore, mice homozygous for a targeted mutation in the *HNF-3β* gene show significant endodermal defects, with foregut morphogenesis most severely affected, yet nonetheless develop beating hearts (Ang and Rossant, 1994; Weinstein *et al.,* 1994). Thus, in the mouse, it is also not clear whether the anterior definitive endoderm is strictly required for induction of the precardiac mesoderm.

An alternative source of heart-inducing signals is the head organizer, consisting of those cells in the gastrula that induce and pattern the head and anterior neural structures (Lemaire and Kodjabachian, 1996). Tissues thought to function as the head organizer are the yolk syncytial layer in zebrafish, the deep endoderm in frog, the hypoblast in chick, and the anterior visceral endoderm in mouse (Long, 1983; Mizuno *et al.,* 1996; Lemaire and Kodjabachian, 1996; Thomas and Beddington, 1996; Varlet *et al.,* 1997). Hints that these cells may also provide heart-inducing signals have recently emerged. First, the heart progenitors are among the earliest mesodermal cells to gastrulate (Tam and Quinlan, 1996) and therefore are likely well removed from the margin or primitive streak by the time the canonical trunk, or dorsal, organizer (e.g., the shield in zebrafish, the dorsal blastopore lip in *Xenopus,* or the node in chick and mouse) has formed. Second, as mentioned previously, the heart forms in *HNF-3β* mutant mice despite severe defects in the anterior endoderm; these mice show molecular evidence of anterior–posterior neurectodermal patterning, suggesting that the head organizer is intact (Ang and Rossant, 1994; Weinstein *et al.,* 1994). Third, the hypoblast in chick (Yatskievych *et al.,* 1997) and the deep dorsal endoderm (Nascone and Mercola, 1995), though not the superficial pharyngeal endoderm (Sater and Jacobson, 1989), in *Xenopus* have been shown to induce cardiac-specific gene expression in unspecified mesoderm. In fact, the *Xenopus* deep dorsal endoderm expresses *cerberus,* a recently discovered secreted factor that when overexpressed induces ectopic heads and hearts in embryos and *XNkx2.5* expression in animal caps (Bouwmeester *et al.,* 1996). Furthermore, the mouse anterior visceral endoderm also expresses a homolog of *cerberus* (R. Harvey, personal communication). Taking these data together it seems plausible that the initial heart-inducing signals may come from the head organizer, whereas the definitive endoderm may provide later signals that reinforce and extend the initial induction of the precardiac mesoderm.

The zebrafish mutant *one-eyed pinhead (oep)* may provide additional support for the hypothesis that the head organizer functions in heart induction. The most obvious phenotypic defect in *oep* embryos is severe cyclopia that results from the absence of the prechordal plate, the anterior-most axial mesoderm in vertebrates (Hammerschmidt *et al.,* 1996; Schier *et al.,* 1996). The *oep* gene product likely functions to maintain prechordal plate cell fate because *oep* mutants initiate expression of the prechordal plate-specific homeobox gene *goosecoid* normally but fail to maintain *goosecoid* expression (Schier *et al.,* 1997). Interestingly, *oep* mutants also entirely lack mature endodermal derivatives (Schier *et al.,* 1997). Although morphologically detectable only at later stages (older than 48 hpf), this endodermal defect in fact manifests itself as early as the onset of gastrulation. In wild-type zebrafish embryos *axial,* a homolog of the mouse *HNF-3β* gene, is expressed in endodermal progenitors scattered throughout the entire margin of the early gastrula; this *axial* expression is absent in *oep* mutants (Schier *et al.,* 1997; J. Alexander and D. Y. R. Stainier, unpublished results), indicating a severe defect in, or perhaps a complete absence of, endodermal cells from a very early time in development. Nonetheless, the vast majority of *oep* embryos do have hearts; these hearts are reduced in size and most display cardia bifida (Fig. 5C) (Schier *et al.,* 1997; J. Alexander and D. Y. R. Stainier, unpublished results)—bilateral hearts that result from the failure of the primitive myocardial tubes to fuse—but they beat and express several markers of myocardial differentiation (J. Alexander and D. Y. R. Stainier, unpublished results). *oep* thus provides another example in which heart tissue forms even though the endoderm is greatly abnormal or absent. Furthermore, similar to the *HNF-3β* mutant mice described previously, *oep* mutants show normal anterior–posterior neurectodermal patterning, demonstrating that they possess a functional head organizer (Schier *et al.,* 1996). The data regarding heart and endoderm differentiation in *oep* mutants are therefore consistent with the suggestion that the head organizer plays an important role in heart induction. On the other hand, it remains possible that in the absence of the *oep* gene product endodermal cells that do not express *axial* provide sufficient "endodermal signaling" to induce formation of precardiac mesoderm. Resolution of this question will require the isolation of additional early endodermal markers. Also, further studies will determine whether the heart defects in *oep* mutants result from cell-autonomous, nonautonomous, or a combination of these actions of the *oep* gene product.

Our laboratory has recently completed a screen for mutations that affect cardiac induction (Alexander *et al.,* 1998). This screen used expression of the cardiac-specific homeobox gene *nkx2.5* as an assay for the occurrence of cardiac induction. Analysis of the mutants discovered in this screen should help to clarify the roles played by the potential heart-inducing tissues as well as the ways in which the prospective precardiac mesoderm responds to signals from these tissues.

It is noteworthy that among the hundreds of zebrafish mutations thus far identified in a number of different screens, none have been found to eliminate heart tissue completely. Such "heartless" mutations may occur only very rarely so that their discovery simply awaits additional screening. Alternatively, the substantial functional overlap that appears to be the rule in vertebrate genomes may make it essentially impossible for such a profound phenotype to result from the disruption of a single gene.

B. Definitive Heart Tube Formation

In all vertebrates the heart arises from bilaterally paired regions of anterior lateral plate mesoderm that migrate medially and subsequently fuse to form the definitive heart tube (DeHaan, 1965). Achieving proper definitive heart tube formation likely requires complex regulation of myocardial progenitor cell adhesive and migratory properties, appropriate interactions with various components of the extracellular matrix that serve as substrates for migration, and topologically correct execution of fusion. Additionally, acquisition of the cellular properties that facilitate formation of the definitive heart tube represents a significant step in myocardial differentiation that must occur at the developmentally appropriate time.

Very little is known about the specific molecules that control and execute the complex morphogenetic events that form the definitive heart tube. Mice homozygous for a null mutation in *Gata4,* which encodes a zinc finger-containing transcription factor expressed early in the precardiac mesoderm as well as in other tissues, exhibit cardia bifida, indicating a requirement for this gene in the formation of the definitive heart tube (Kuo *et al.,* 1997; Molkentin *et al.,* 1997; see Chapter 17). In these *Gata4* mutant mice, however, the lateral-to-ventral and rostral-to-caudal foldings that generate the anterior intestinal portal and initiate foregut development during mouse embryogenesis do not occur. Recent mouse chimera experiments in fact demonstrate that the presence of *Gata4*$^{+/+}$ cells solely in the primitive endoderm and portions of the foregut and hindgut is sufficient to rescue the cardia bifida and other morphogenetic defects (Narita *et al.,* 1997). Thus, the cardia bifida seen in *Gata4* mutant mice likely results not from

Figure 5 Zebrafish cardiac mutants. Wild-type and various mutant zebrafish hearts: A and B, wild-type (note: the ventricle is obscured by the atrium in A); C, *one-eyed pinhead* (arrows indicate the bilateral hearts; the abnormal head can also be seen (compare to B); D, *miles apart* (arrows indicate the bilateral hearts); (E) *heart and soul* (arrow indicates the heart); F, *cloche;* G, *weak atrium;* H, *pandora* (arrow indicates the stalk of ventricular or outflow tract tissue); I, wild-type (arrows indicate cardiac valves); J, *jekyll.* A–H are Nomarski optics images of living embryos; I and J are histological sections. A and E–J are lateral views; B–D are ventral views. In all panels anterior is to the left. Embryos in B–D are at 36 hpf; embryos in A and E–H are at 48 hpf; and embryos in I and J are at 96 hpf. A, atrium; V, ventricle; WT, wild-type.

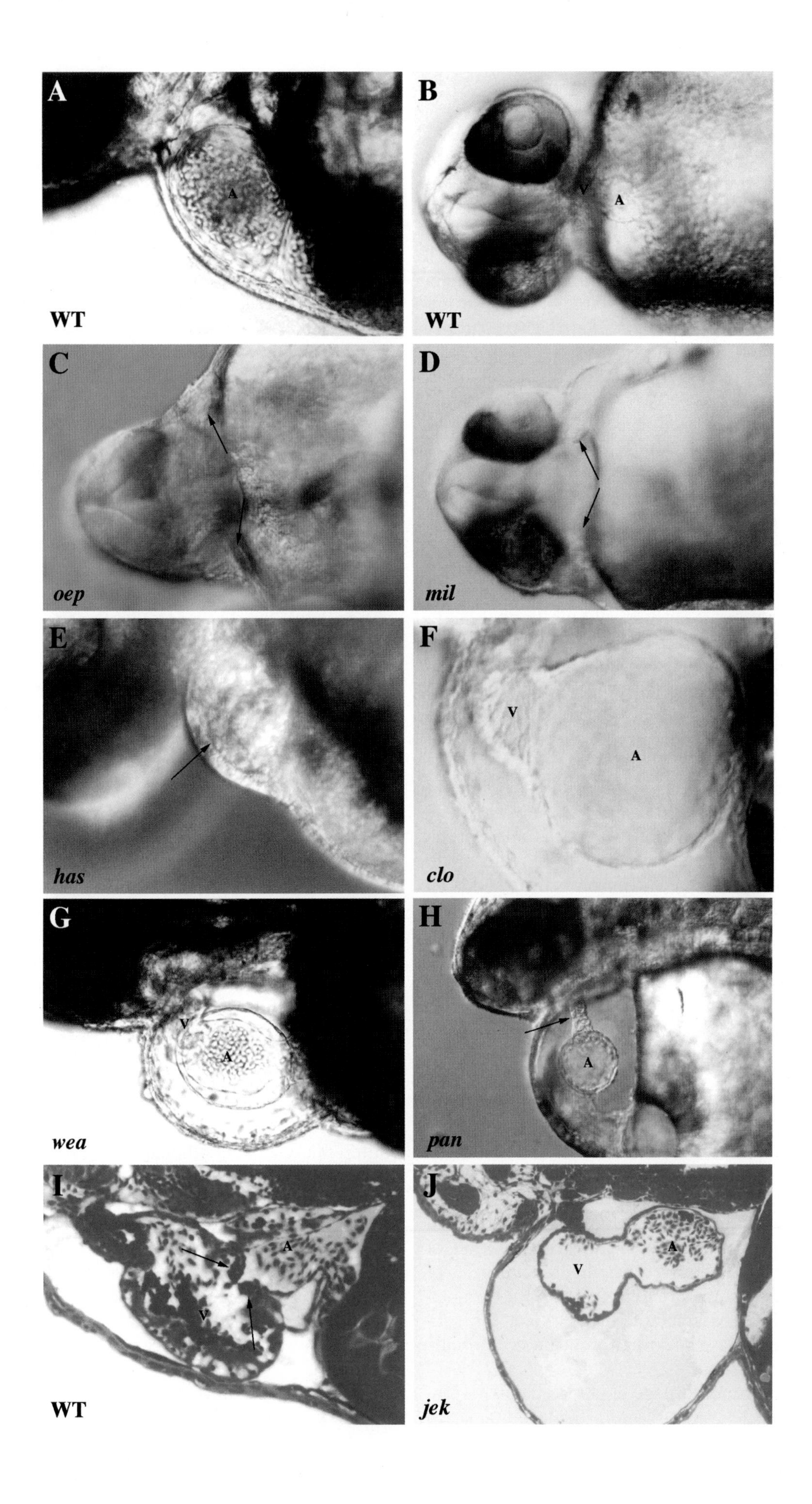
A
A
WT
B
V
A
WT
C
oep
D
mil
E
has
F
V
A
clo
G
V
A
wea
H
A
pan
I
A
V
WT
J
A
V
jek

an intrinsic defect in the myocardial cells but rather from morphogenetic problems due to an absence of Gata4 function in either the primitive or definitive endoderm (or both). Targeted disruption of the mouse fibronectin gene also results in cardia bifida in a small percentage of the mutant embryos (George *et al.,* 1993), as does treament of chick embryos with anti-fibronectin antibodies (Linask and Lash, 1988), indicating an important role for the extracellular matrix in the proper formation of the definitive heart tube.

Fusion of the bilateral primitive myocardial tubes in zebrafish appears to be very sensitive to mutational disruption; at least six mutants display cardia bifida: *miles apart* (*mil*), *bonnie and clyde* (*bon*), *faust* (*fau*), *casanova* (*cas*), *natter* (*nat*), and *oep* (Fig. 5D) (Chen *et al.,* 1996; Schier *et al.,* 1997; Stainier *et al.,* 1996). The *oep* mutant was described previously; whether cardia bifida in this mutant results from a defect in the myocardial cells themselves or is secondary to the endoderm defect remains to be determined. The other five mutations appear to affect the heart fairly specifically. In addition to cardia bifida, *mil* embryos develop blisters at the distal end of the tail; *cas* embryos exhibit gradually reduced motility during the second day of development and by 48 hpf are immotile. *bon* and *fau* embryos show no other obviously detectable morphological defects besides cardia bifida.

A wide variety of defects could explain the failure of the primitive heart tubes to fuse in these mutants. Myocardial progenitors may not express the appropriate cell adhesion molecule(s) necessary to migrate medially. The tail blisters seen in *mil* embryos suggest that an adhesion defect may indeed be present in this mutant. Alternatively, cardia bifida could result from a cell motility defect; this would seem to require that the motility of myocardial progenitors be somehow different than that of other cell types, given how morphologically normal the mutants otherwise appear. A more general delay or block in myocardial differentiation could also explain the failure of the primitive heart tubes to migrate and fuse in these mutants. Such a defect could result cell autonomously, or it could be that the mutation affects the production of a signal by another tissue, for example, the anterior endoderm, that induces the myocardial progenitors to migrate. In all these mutants the bilaterally situated hearts beat, indicating that functional maturation of the myocardium has occurred; whether other aspects of terminal differentiation are defective has not been examined. Finally, these mutations may affect the substrate on which the myocardial progenitors normally migrate.

Clearly there is no shortage of potential defects that could underlie the various cardia bifida mutants. To date, however, the actual molecular defect that results in cardia bifida has not been determined for any of these mutants. In light of the effect of its disruption in the mouse, zebrafish *gata4* is certainly a candidate gene for several of these mutations (e.g., *bon, cas,* and *fau*); however, it is not clear whether an anterior intestinal portal-like structure exists in the zebrafish. Given our paucity of knowledge regarding the process of heart tube fusion, it is certain that a more detailed understanding of these numerous zebrafish cardia bifida mutants, at both the cellular and molecular level, will prove extremely valuable.

One additional mutant, *heart and soul* (*has*), may also be defective in definitive heart tube formation. In *has* mutants the heart is small and appears to be composed of compact myocardial tissue rather than being a hollow tube (Fig. 5E) (Stainier *et al.,* 1996). It is not known when exactly heart development first goes awry in *has*. The abnormal *has* heart may, however, result from a failure to achieve topologically correct fusion of the primitive heart tubes so that the atrium and ventricle lie not in series but instead are wrapped about each other (Fishman and Chien, 1997; S. A. Horne and D. Y. R. Stainier, unpublished results). *has* mutants also display a curved body axis and retinal defects, suggesting that the *has* gene product functions in a variety of tissues.

C. Endocardial Cell Differentiation

The endocardium represents a specialized subset of endothelial cells that lines the inside of the heart and is continuous with the endothelium of the cardiac inflow and outflow tracts. These cells ensure the smooth flow of blood through the heart, contribute importantly to formation of the cardiac valves, and also play a role in modulating the proliferation and contractility of the adult myocardium (Shah and Lewis, 1993; Gassmann *et al.,* 1995; Kramer *et al.,* 1996; Lee *et al.,* 1995; Meyer and Birchmeier, 1995).

Endothelial cell differentiation occurs via three distinct mechanisms: vasculogenesis type I refers to the differentiation of endothelial cells *in situ,* such as occurs in the trunk vessels; vasculogenesis type II describes endothelial cell differentiation following their migration from a distant location; and angiogenesis is the process whereby new vessels form by sprouting from preexisting ones. The endocardium forms via type II vasculogenesis; in the zebrafish the endocardial progenitors migrate medially from bilateral positions starting at approximately 16 hpf (15-somite stage) (Liao *et al.,* 1997).

Recent studies have implicated several receptor tyrosine kinases in endothelial cell differentiation. In mouse the angiopoietin receptor Tek (also known as Tie-2) and the orphan receptor Tie (also known as Tie-1) are required for vascular formation including angiogenesis and the integrity and survival of endothelial cells (Dumont *et al.,* 1994; Puri *et al.,* 1995; Sato *et al.,* 1995). Flt-1, a receptor for vascular endothelial growth factor (VEGF), appears to regulate the assembly of vascular endothelium (Fong *et al.,* 1995).

Because of the close proximity in which endothelial and hematopoietic precursors develop in vertebrates, it has been proposed that a common progenitor, the hemangioblast, gives rise to both endothelial and hematopoietic cells. Alternatively, this close spatial relationship may belie an inductive and/or trophic interaction between endothelium and hematopoietic progenitors. Targeted mutation of murine Flk-1, a second VEGF receptor, results in embryos that lack both endothelial and hematopoietic lineages, suggesting that Flk-1 may indeed function in a common hemangioblast progenitor population (Shalaby *et al.,* 1995). Mouse chimera experiments demonstrate that homozygous Flk-1-mutant cells cannot form endothelium, nor do they contribute to either the primitive or the definitive hematopoietic populations (Shalaby *et al.,* 1997). Embryoid bodies derived from homozygous Flk-1-mutant cells also do not form endothelium; however, they do generate erythroid progenitors (Shalaby *et al.,* 1997). These data suggest that Flk-1 may not play a role in hematopoietic differentiation per se but rather function in the migration of cells to the sites of both primitive and definitive hematopoiesis. Other *in vitro* studies provide evidence that the survival of both hematopoietic and endothelial progenitors requires Flk-1 function (Eichmann *et al.,* 1997). Clearly, the question of whether the hemangioblast actually exists remains unanswered; careful lineage analysis, possible in the zebrafish, may provide some conclusive data.

Studies of the zebrafish mutation *cloche* (*clo*) may also help to resolve this issue. *clo* mutant embryos develop a bell-shaped heart, with a small ventricle and enlarged atrium, that entirely lacks endocardium (Fig. 5F) (Stainier *et al.,* 1995). The *clo* heart exhibits diminished contractility, suggesting that endocardial signaling may modulate differentiation and/or function of the embryonic as well as the adult myocardium. As expected, given the absence of endocardium, the *clo* mutant heart also lacks cardiac cushions, the valve precursors. *clo* acts cell-autonomously in endocardial cells; wild-type cells transplanted into a *clo* mutant host are able to form endocardium (Stainier *et al.,* 1995). Further studies demonstrate that *clo* acts upstream of a *flk-1* homolog in all endothelial cells of the embryo, making *clo* the earliest gene known to function in endothelial cell development (Liao *et al.,* 1997).

Interestingly, *clo* mutant embryos also display a severe blood defect: Early expression of the hematopoietic transcription factor genes *gata-1* and *gata-2* is not detectable; the intermediate cell mass, from which the first hematopoietic cells arise in zebrafish, is greatly depleted of cells; and few if any erythrocytes are present at later embryonic stages (Stainier *et al.,* 1995). The intriguing concurrence of endothelial and hematopoietic defects in *clo* mutants again raises the issue of whether a common hemangioblast progenitor population exists and is affected by this mutation. Lineage data in the zebrafish demonstrate that single cells in the ventral region of the zebrafish blastula can indeed give rise to both endothelium and blood cells (Lee *et al.,* 1994). However, at least two distinct populations of endothelial progenitors appear to exist in the zebrafish embryo prior to gastrulation; cells destined to give rise to trunk and tail endothelium are concentrated in the ventral marginal zone, whereas those that form the head endothelium are scattered throughout the marginal zone (Warga, 1996). Thus, at least some endothelial cells must already be distinct from the blood lineage prior to gastrulation. An equally tenable explanation for the *clo* blood defect is that it results secondarily from a lack of endothelial-derived signals that normally promote hematopoietic differentiation or support the survival of hematopoietic progenitors. The observation that at later stages *clo* mutants exhibit some *gata-1* expression in the same posterior region where *flk-1* is also expressed can be interpreted to support either hypothesis (Liao *et al.,* 1997). Experiments to test the cell autonomy of the *clo* blood defect or the cloning of the *clo* locus and analysis of its expression pattern may help to resolve this issue. One must also consider the possibility that only either the primitive or definitive blood cell populations derive from hemangioblasts.

The endocardial cells become enclosed within the definitive heart tube as a result of the fusion of the bilateral primitive myocardial tubes. The fate of the endocardial progenitors in the numerous mutants defective in the formation of the definitive heart tube (described previously) has not been determined. It will be quite interesting to determine how endocardial differentiation and morphogenesis are affected when the primitive heart tubes either fail to fuse or fuse abnormally, especially should differences exist between mutants with very similar myocardial phenotypes (e.g., cardia bifida).

D. Onset of Cardiac Function

Perhaps more than any other characteristic, its spontaneous contractility distinguishes cardiac from skeletal or smooth muscle. This property is inherent to differentiated cardiomyocytes and its acquisition may depend on signals from the anterior endoderm (Gannon and Bader, 1995; Sugi and Lough, 1994). The intrinsic rate at which a given cardiomyocyte contracts increases in an anterior (slow) to posterior (fast) gradient along the heart tube (DeHaan, 1965). Because the myocardial cells are electrically coupled via gap junctions, those cells with the fastest intrinsic rate of beating, located in the sinoatrial node, serve as the heart's pacemaker.

Cardiac muscle contracts in essentially the same way as other muscle types; membrane depolarization leads to increased cytoplasmic calcium, which binds to the troponin–tropomyosin complex, permitting the interaction and subsequent antiparallel sliding of actin and myosin filaments. The same basic components form the myocardial and skeletal muscle contractile apparatuses, although cardiac-specific versions of many proteins are used in the heart (e.g., α-myosin heavy chain; Bisaha and Bader, 1991).

Many zebrafish mutants that exhibit abnormal cardiac function have been identified (Table II). *silent heart* (*sih*) mutant embryos develop a morphologically normal heart that never initiates contractions; skeletal and smooth muscles function normally in *sih* embryos (Chen *et al.,* 1996). Calcium flux in *sih* hearts is normal, indicating that the *sih* mutation uncouples excitation from contraction (D. Y. R. Stainier and M. C. Fishman, unpublished data). Two other mutants also develop hearts that do not beat: *throbless* (also known as *still heart*) embryos additionally show a strong motility defect and their skeletal muscle structure is severely disrupted, suggesting that this mutation affects a common component of cardiac and skeletal muscle; mutation of the *viper* gene results in brain abnormalities in addition to a heart that does not beat (Chen *et al.,* 1996). In addition to identifying gene products specifically required for myocardial function, these mutants may provide insights into the different gene-regulatory networks that act in cardiac and skeletal muscle. Also, these mutants are of great practical value because they reveal the secondary defects that result from an absence of circulation and thereby enable one to distinguish more easily between the primary and secondary effects of other cardiac mutations.

Numerous mutations affect the contractility of the cardiac chambers. Both chambers exhibit diminished contractility in *weak beat* and *pickwick* mutants. The *main squeeze* and *pipe heart* mutations appear to affect the ventricle specifically, whereas only the atrium is silent in *weak atrium* mutants (Fig. 5G) (Chen *et al.,* 1996; Stainier *et al.,* 1996). In each of these and the other "reduced contractility" mutants, the resulting morphology of the heart is variable. For example, the atrium and ventricle appear dilated in *pickwick* mutants, whereas several other mutations that affect the contractility of both chambers, such as *lazy susan* and *pipe dream,* result in localized contractures and aneurysms of the chamber wall (Stainier *et al.,* 1996). Knowledge of the genes that when mutant give rise to these different phenotypes will help to elucidate the molecular basis of functional differences between the atrium and ventricle.

Other mutations cause either abnormal rate or rhythmicity of the heartbeat. The *tremblor* mutation results in disorganized cardiac contractions that may represent fibrillation (Chen *et al.,* 1996; Stainier *et al.,* 1996). *slow mo* (*smo*) mutant hearts beat at a significantly slower rate than wild-type hearts; this is one of the few nonlethal cardiac mutants (Stainier *et al.,* 1996). Electrophysiological studies have demonstrated that *smo* cardiomyocytes have a reduction in I_h, a hyperpolarization-activated inward current, indicating that I_h likely plays an important role in normal cardiac pacemaker function (Baker *et al.,* 1997). Mutants such as *reggae* and *hiphop* exhibit sinoatrial conduction abnormalities or various degrees of atrioventricular block, respectively (Chen *et al.,* 1996; Stainier *et al.,* 1996). Identification of the genes responsible for these various phenotypes will complement pharmacological and genetic studies aimed at understanding the mechanism and regulation of the rate and rhythmicity of the heart beat.

Lastly, mutations in three loci, *santa, valentine,* and *heart of glass,* result in hearts that are abnormally large (Chen *et al.,* 1996; Stainier *et al.,* 1996). Cardiac contractility in these mutants appears essentially normal, indicating that the hearts are not simply dilated secondary to poor function. In these cases defects in structural components of the cardiomyocytes may impair the heart's ability to maintain the appropriate size under pressure from the returning systemic circulation without markedly diminishing its contractility.

The various zebrafish cardiac function mutants provide unique opportunities for understanding the molecular basis of numerous aspects of cardiac function and physiology. Some of these mutants may also represent cases in which abnormal cardiac function results from a failure to complete myocardial differentiation. Resolving which mutants fall into this class will require substantial additional phenotypic characterization but may prove very revealing regarding late steps in myocardial differentiation.

E. Chamber Formation

Anterior–posterior (A–P) patterning of the heart manifests itself morphologically in the formation of distinct ventricular (anterior) and atrial (posterior) chambers. Differences in gene expression exist between the two chambers even prior to their morphological demarcation (Bisaha and Bader, 1991; Kubalak *et al.,* 1994; O'Brien *et al.,* 1993; Stainier and Fishman, 1992; Yutzey *et al.,* 1994; see Chapters 3 and 4), yet little is known about how these differences arise. Lineage analyses suggest that separate populations of cells in the pregastrula embryo may contribute to the atrium and ventricle (Gonzalez-Sanchez and Bader, 1984; Stainier *et al.,* 1993; Yutzey and Bader, 1995). Also, in chick embryos the location of cells within the definitive heart tube reflects the level at which they ingressed through the primitive streak during gastrulation (DeHaan, 1965), indicating that the relative A–P orientation of the myocardial progenitor pool is maintained from a very early point in development. However, it is not known when commitment to one or the other chamber actually occurs, nor what signals direct myocardial progenitors to these different fates.

Even less is known about the molecules involved in cardiac A–P patterning. Retinoid signaling pathways can influence the A–P patterning of the heart; low doses of retinoic acid affect the heart in a relatively specific fashion, causing the transformation of anterior (ventricular) to posterior (atrial) tissue in both chick and zebrafish embryos (Stainier and Fishman, 1992; Yutzey and Bader, 1995; see Chapter 13). Still, the importance of endogenous retinoids for the correct A–P patterning of the heart as well as the identity of other molecules involved in this process remain largely unknown.

Two zebrafish mutants, *pandora* (*pan*) and *lonely atrium* (*loa*), may represent defects in the A–P patterning of the heart (Chen *et al.,* 1996; Stainier *et al.,* 1996). In both cases the mutant hearts appear to consist principally of an atrium connected to a stalk of tissue that may either be a ventricular rudiment or the cardiac outflow tract (Fig. 5H). While *loa* seems to function specifically in the heart, *pan* mutants also exhibit defects in the eye, ear, and somites (Malicki *et al.,* 1996a,b). The *pan* eye defect is particularly interesting; at 72 hpf *pan* mutant embryos specifically lack the ventral retina (Malicki *et al.,* 1996a). Recent work indicates that retinoic acid plays a role in the development of the zebrafish ventral retina (Marsh-Armstrong *et al.,* 1994). In light of the potential importance of retinoids in cardiac A–P patterning, it is tempting to speculate that *pan* encodes a component of some retinoid signaling pathway that functions to establish or maintain A–P pattern in the heart and dorsal–ventral pattern in the retina. Clearly, further studies will be required to test this hypothesis.

Determining the fate of the ventricular cardiomyocyte progenitors in *pan* and *loa* mutants will be particularly important for understanding the basis of their phenotypes. These mutations may cause an anterior (ventricular) to posterior (atrial) transformation; alternatively, in the absence of the *pan* or *loa* gene products prospective ventricular cardiomyocytes may not receive or respond to specific factors needed for their specification, proliferation, differentiation, or survival. Knowing which of these or other possibilities occurs in each case should provide insight into the mechanisms by which cardiac A–P patterning is generated.

Our lab has recently completed a screen that used expression of chamber-specific myosin heavy-chain isoforms (Stainier and Fishman, 1992) to detect additional mutants with abnormal cardiac A–P pattern (Alexander *et al.,* 1998). The many new mutants discovered in this screen provide the opportunity to initiate a truly thorough genetic and molecular analysis of cardiac chamber formation. It is notable that in the large majority of these cardiac A–P patterning mutants, as well as in *pan* and *loa* embryos, the ventricle is more affected than the atrium. This observation hints at the possibility that cardiac A–P patterning may represent primarily the induction of anterior (ventricular) fates (Alexander *et al.,* 1998).

As noted in the preceding section describing the cardiac function mutants, several mutations specifically affect the function of either the atrium or the ventricle. These mutants may well be defective in executing particular aspects of one or the other chamber-specific differentiation programs. Their study will therefore likely reveal much about how cardiac A–P pattern is translated into the distinct molecular and physiological properties of the ventricle and atrium.

F. Looping

Rightward looping of the heart tube is conserved throughout the vertebrate subphylum, reflecting its profound importance for establishing the spatial relationships necessary to form properly the cardiac chambers, valves, and vascular connections. As the first overt manifestation of left–right asymmetry in the vertebrate embryo, heart looping has come under intensive study. Yet much remains to be learned about both the intrinsic and the extrinsic regulation of this process.

The rightward direction in which the heart loops appears to result from the biasing of an intrinsic cardiac looping program by interactions with an axial left–right system (Biben and Harvey, 1997; see Chapter 7). Numerous genes have been described to be expressed in the upper trunk region in left–right asymmetric patterns. These include *activin receptor IIa, Sonic hedgehog* (*Shh*), and the transforming growth factor-β (TGF-β) superfamily members *nodal* and *lefty* (Collignon *et al.,* 1996; Levin *et al.,* 1995; Lowe *et al.,* 1996; Meno *et al.,* 1996). Manipulation of *Shh* expression can reverse heart situs (Levin *et al.,* 1995). Also, the expression patterns of *nodal* and *lefty* are altered vis-a-vis the left–right axis in *iv* and *inv* mutant mice, which show 50 and 100% situs inversus, respectively (Collignon *et al.*, 1996; Lowe *et al.*, 1996; Meno *et al.*, 1996). Few of the cells that express *nodal* and *lefty* appear to contribute to the heart, however, suggesting that these molecules may act inductively to steer the direction of heart looping.

How the heart actually executes looping is not known. Mechanisms proposed to drive heart looping include left–right differences in the proliferation, contractility, or shape of myocardial cells (DeHaan, 1965), although no strong evidence exists that supports any of these models. It has been shown recently that in mouse embryos the caudal heart tube undergoes a leftward displacement prior to looping morphogenesis (Biben and Harvey, 1997); how this movement is achieved is also not clear. However, some understanding of the intrinsic regulators of looping has been gained. The hearts of mice homozygous for mutations in either *Nkx2-5* or the MADS box transcription factor *Mef2c* fail to loop (Lin *et al.*, 1997; Lyons *et al.*, 1995), although at least in the *Nkx2-5* mutants leftward cardiac displacement does occur (Biben and Harvey, 1997). In the chick simultaneous inhibition with antisense oligonucleotides designed against dHand and eHand, two related basic helix–loop–helix proteins, prevents looping morphogenesis (Srivastava *et al.*, 1995); a mutation in the mouse *dHand* gene also blocks looping (Srivastava *et al.*, 1997; see Chapter 9). Interestingly, in normal mouse embryos *eHand* is expressed more strongly on the left side of the heart; this expression is reversed in *inv* mutant mice and is absent in *Nkx2-5* mutant mice (Biben and Harvey, 1997; see Chapter 7). Expression of both *dHand* and *eHand* is abnormal in *Mef2c* mutant mice (Lin *et al.*, 1997). Thus, it seems that *eHand* and *dHand* lie downstream of *Nkx2-5* and *Mef2c* in an intrinsic myocardial pathway that regulates heart looping.

The analysis of two zebrafish notochord mutants supports a role for the axial mesoderm in influencing the direction of heart looping (Danos and Yost, 1996). In *no tail* (*ntl*) mutants mature notochord cells do not form (Halpern *et al.*, 1993), whereas in *floating head* (*flh*) mutants notochordal cell fate cannot be maintained so that notochord progenitors are respecified as muscle (Halpern *et al.*, 1995; Talbot *et al.*, 1995); the direction of heart looping is randomized in both mutants (Danos and Yost, 1996). Some notochord signaling potential does remain in *ntl* and *flh* embryos, as evidenced by the presence of floor plate cells in the ventral spinal cord (Halpern *et al.*, 1993; Talbot *et al.*, 1995). The signal(s) that imparts left–right information to the heart, however, must not be produced in these mutants, not be present at the appropriate time to direct heart looping to the right, or not be restricted to the correct side in the absence of an important boundary (i.e., the notochord).

A recent study examined heart looping in a large collection of zebrafish mutants (Chen *et al.*, 1997). The results indicate that, similar to the mouse, an initial displacement or "jogging" of the heart tube predicts the subsequent direction of looping; leftward displacement, which occurs in the vast majority of wild-type embryos, is followed by rightward looping and vice versa. In a number of mutants, many of which affect either the notochord or the ventral spinal cord or display a so-called "curly tail" phenotype, heart jogging is either randomized or does not occur. Heart displacement and looping are correlated in the "randomized jogging" mutants, whereas in the "no jogging" mutants looping is randomized. This report further describes asymmetric cardiac expression of the TGF-β superfamily member *bmp4* at the 22-somite stage (20 hpf), soon after fusion of the primitive myocardial tubes has occurred. In the hearts of wild-type embryos *bmp4* expression is greater on the left side, correlating with the later direction of heart tube jogging. Cardiac *bmp4* expression can be either left or right predominant in the randomized jogging mutants but remains symmetric in the no jogging mutants. The authors perturb *bmp4* expression or signaling by injection of RNA encoding either Shh or a dominant-negative BMP receptor construct, respectively; *shh* overexpression appears to cause symmetric cardiac *bmp4* expression and randomizes heart looping, whereas the dominant-negative BMP receptor also randomizes the direction of heart looping. Together these data suggest that the axial left–right system influences the direction of heart displacement and consequent looping by generating asymmetric *bmp4* expression in the heart. Additionally, these mutants with randomized heart looping provide a unique opportunity to determine the components of the left–right signaling system through the molecular cloning of the affected loci.

G. Valve Formation

The mature heart valves function to prevent retrograde blood flow from occurring as the heart beats. The valves derive from the cardiac cushions, which form as a result of the invasion by endocardial cells of the cardiac jelly or matrix at specific sites in the atrioventricular canal and outflow tract (Markwald *et al.*, 1996; see Chapter 10). The endocardium in the presumptive cushion regions exhibits distinct gene expression patterns. For example, in the chick heart these cells express the homeobox gene *Msx-2* (Chan-Thomas *et al.*, 1993) and increased levels of *TGF-β1* (Akhurst *et al.*, 1990), whereas in the mouse *desert hedgehog* is expressed in the prospective cushion endocardium (Bitgood and McMahon, 1995). Additionally, appropriate expression by the cushion progenitors of cell adhesion molecules

(Crossin and Hoffman, 1991) and secreted proteases (McGuire and Alexander, 1993) may also play an important role in cardiac cushion formation.

The myocardium in the regions overlying the prospective cardiac cushions is also specialized, and signals from these myocardial cells appear to induce the epithelial-to-mesenchymal transition of the invading endocardial cells leading to cardiac cushion formation. The molecules that induce this endocardial invasion are not known but candidates, based on their expression patterns, include various TGF-βs and BMPs, among others (Bitgood and McMahon, 1995; Jones *et al.,* 1991; Potts *et al.,* 1991, 1992).

Two zebrafish mutants specifically lack cardiac valves. In the hearts of both *jekyll* and *m27* (a name will be assigned to this locus pending completion of complementation testing) mutants, the cardiac cushions fail to appear and later valve formation does not occur (Fig. 5J) (Stainier *et al.,* 1996). Whether the affected gene in each case functions in the prospective cushion endocardium or instead is required for production of a signal by the overlying myocardium remains to be determined. This knowledge, combined with the data from other species mentioned previously, will help to guide the testing of candidate genes.

In four other zebrafish mutants the heart also lacks valves. The *clo* mutant described earlier entirely lacks endocardium and consequently does not form cardiac cushions or valves (Stainier *et al.,* 1995). In the large heart mutants *santa, valentine,* and *heart of glass* the endocardium is present but again neither cushions nor valves form (Chen *et al.,* 1996; Stainier *et al.,* 1996). The lack of valves in these mutants may result from a failure of the myocardium to induce the appropriate endocardial cells to undergo the required epithelial-to-mesenchymal transition. It is equally possible, however, that valves fail to form in these mutants due to an intrinsic defect in the endocardium. Cell transplantation studies that test the cell autonomy of these valve defects will be necessary to distinguish between these possibilities.

V. Conclusions

The zebrafish heart mutants described in this review offer tremendous opportunities to further our understanding of essentially all aspects of early cardiac development. As detailed morphological and molecular analyses of the mutant phenotypes emerge, a more precise classification of the various mutants will become possible. This classification will enable these loci to be ordered into genetic pathways controlling specific aspects of heart development, such as definitive heart tube formation or looping, and may also indicate those loci critical for the coordination of these diverse developmental events. The generation of allelic series of varying phenotypic strength will aid epistasis testing and may allow additional genes important for heart development to be identified through screens for phenotypic suppressors or enhancers. Screens that use molecular markers to monitor specific events in cardiac development will further add to the collection of zebrafish heart mutants (Alexander *et al.,* 1998); such screens are likely to be particularly useful because of their ability to detect phenotypes found infrequently or not at all in the large-scale morphological screens. Ultimately, the isolation of the genes affected in the zebrafish cardiac mutants will reveal the molecules that control vertebrate heart development. The embryological manipulations and detailed observation possible in the zebrafish will allow us to extend this molecular knowledge, revealing how these genes regulate the behavior of the cells that form the heart and the rest of the cardiovascular system.

While we are just beginning to tap into the potential of the zebrafish, interest and enthusiasm for this system are clearly on the upswing. Given the rapid progress and increasing sophistication of cellular, molecular, and genetic studies on the zebrafish, we expect that the field of heart development will soon begin to mine rich rewards from the many zebrafish cardiac mutants.

Acknowledgments

We thank R. Harvey and S. Horne for communicating results prior to publication. C. Kimmel graciously provided the slides for Fig. 2, and A. Gamarnik helped invaluably in the production of Fig. 5. We are grateful to D. Yelon, E. Walsh, L. Parker, E. Kupperman, S. Horne, C.-Y. Ho, and R. Harvey for their valuable comments, discussions, and suggestions regarding the manuscript. J. A. is a member of the Medical Scientist Training Program at UCSF and an American Heart Association predoctoral fellow. D. Y. R. S. is a Basil O'Connor Scholar and a Packard Foundation Fellow. Work from our laboratory cited in this chapter has been supported by the National Institutes of Health and the American Heart Association.

References

Akhurst, R. J., Lehnert, S. A., Faissner, A., and Duffie, E. (1990). TGF-β in murine morphogenetic processes: The early embryo and cardiogenesis. *Development (Cambridge, UK)* **108,** 645–656.

Alexander, J., Stainier, D. Y. R., and Yelon, D. (1998). Screening mosaic F1 females for mutations affecting zebrafish heart induction and patterning. *Dev. Genet.* **22,** 288–299.

Ang, S. L., and Rossant, J. (1994). *HNF-3β* is essential for node and notochord formation in mouse development. *Cell (Cambridge, Mass.)* **78,** 561–574.

Baker, K., Warren, K. S., Yellen, G., and Fishman, M. C. (1997). Defective "pacemaker" current (Ih) in a zebrafish mutant with a slow heart rate. *Proc. Natl. Acad. Sci. U. S. A.* **94,** 4554–4559.

Biben, C., and Harvey, R. P. (1997). Homeodomain factor Nkx2-5 controls left/right asymmetric expression of bHLH gene eHand during murine heart development. *Genes Dev.* **11,** 1357–1369.

Bisaha, J. G., and Bader, D. (1991). Identification and characterization of a ventricular-specific avian myosin heavy chain, VMHC1: Expression in differentiating cardiac and skeletal muscle. *Dev. Biol.* **148,** 355–364.

Bitgood, M. J., and McMahon, A. P. (1995). Hedgehog and Bmp genes are coexpressed at many diverse sites of cell-cell interaction in the mouse embryo. *Dev. Biol.* **172,** 126–138.

Bouwmeester, T., Kim, S., Sasai, Y., Lu, B., and De Robertis, E. M. (1996). Cerberus is a head-inducing secreted factor expressed in the anterior endoderm of Spemann's organizer. *Nature (London)* **382,** 595–601.

Brand, M., Heisenberg, C. P., Warga, R. M., Pelegri, F., Karlstrom, R. O., Beuchle, D., Picker, A., Jiang, Y. J., Furutani-Seiki, M., van Eeden, F. J., Granato, M., Haffter, P., Hammerschmidt, M., Kane, D. A., Kelsh, R. N., Mullins, M. C., Odenthal, J., and Nüsslein-Volhard, C. (1996). Mutations affecting development of the midline and general body shape during zebrafish embryogenesis. *Development (Cambridge, UK)* **123,** 129–142.

Chan-Thomas, P. S., Thompson, R. P., Robert, B., Yacoub, M. H., and Barton, P. J. (1993). Expression of homeobox genes *Msx-1* (*Hox-7*) and *Msx-2* (*Hox-8*) during cardiac development in the chick. *Dev. Dyn.* **197,** 203–216.

Chen, J.-N., and Fishman, M. C. (1996). Zebrafish *tinman* homolog demarcates the heart field and initiates myocardial differentiation. *Development (Cambridge, UK)* **122,** 3809–3816.

Chen, J.-N., Haffter, P., Odenthal, J., Vogelsang, E., Brand, M., van Eeden, F. J., Furutani-Seiki, M., Granato, M., Hammerschmidt, M., Heisenberg, C. P., Jiang, Y.-J., Kane, D. A., Kelsh, R. N., Mullins, M. C., and Nüsslein-Volhard, C. (1996). Mutations affecting the cardiovascular system and other internal organs in zebrafish. *Development (Cambridge, UK)* **123,** 293–302.

Chen, J.-N., van Eeden, F. J. M., Warren, K. S., Chin, A. S., Nüsslein-Volhard, C., Haffter, P., and Fishman, M. C. (1997). Left-right pattern of cardiac *BMP4* may drive asymmetry of the heart in zebrafish. *Development (Cambridge, UK)* **124,** 4373–4382.

Collignon, J., Varlet, I., and Robertson, E. J. (1996). Relationship between asymmetric nodal expression and the direction of embryonic turning. *Nature (London)* **381,** 155–158.

Crossin, K. L., and Hoffman, S. (1991). Expression of adhesion molecules during the formation and differentiation of the avian endocardial cushion tissue. *Dev. Biol.* **145,** 277–286.

Danos, M. C., and Yost, H. J. (1996). Role of notochord in specification of cardiac left-right orientation in zebrafish and *Xenopus*. *Dev. Biol.* **177,** 96–103.

DeHaan, R. L. (1965). Morphogenesis of the vertebrate heart. *In* "Organogenesis" (R. L. DeHaan and H. Ursprung, eds.), pp. 377–419. Holt, Rinehart & Winston, New York.

Driever, W., Stemple, D., Schier, A., and Solnica-Krezel, L. (1994). Zebrafish: Genetic tools for studying vertebrate development. *Trends Genet.* **10,** 152–159.

Driever, W., Solnica-Krezel, L., Schier, A. F., Neuhauss, S. C., Malicki, J., Stemple, D. L., Stainier, D. Y. R., Zwartkruis, F., Abdelilah, S., Rangini, Z., Belak, J., and Boggs, C. (1996). A genetic screen for mutations affecting embryogenesis in zebrafish. *Development (Cambridge, UK)* **123,** 37–46.

Dumont, D. J., Gradwohl, G., Fong, G. H., Puri, M. C., Gertsenstein, M., Auerbach, A., and Breitman, M. L. (1994). Dominant-negative and targeted null mutations in the endothelial receptor tyrosine kinase, tek, reveal a critical role in vasculogenesis of the embryo. *Genes Dev.* **8,** 1897–1909.

Eichmann, A., Corbel, C., Nataf, V., Vaigot, P., Bréant, C., and Le Douarin, N. M. (1997). Ligand-dependent development of the endothelial and hemopoietic lineages from embryonic mesodermal cells expressing vascular endothelial growth factor receptor 2. *Proc. Natl. Acad. Sci. U. S. A.* **94,** 5141–5146.

Fishman, M. C., and Chien, K. R. (1997). Fashioning the vertebrate heart: Earliest embryonic decisions. *Development (Cambridge, UK)* **124,** 2099–2117.

Fong, G. H., Rossant, J., Gertsenstein, M., and Breitman, M. L. (1995). Role of the Flt-1 receptor tyrosine kinase in regulating the assembly of vascular endothelium. *Nature (London)* **376,** 66–70.

Gaiano, N., Amsterdam, A., Kawakami, K., Allende, M., Becker, T., and Hopkins, N. (1996). Insertional mutagenesis and rapid cloning of essential genes in zebrafish. *Nature (London)* **383,** 829–832.

Gannon, M., and Bader, D. (1995). Initiation of cardiac differentiation occurs in the absence of anterior endoderm. *Development (Cambridge, UK)* **121,** 2439–2450.

Gassmann, M., Casagranda, F., Orioli, D., Simon, H., Lai, C., Klein, R., and Lemke, G. (1995). Aberrant neural and cardiac development in mice lacking the ErbB4 neuregulin receptor. *Nature (London)* **378,** 390–394.

George, E. L., Georges-Labouesse, E. N., Patel-King, R. S., Rayburn, H., and Hynes, R. O. (1993). Defects in mesoderm, neural tube and vascular development in mouse embryos lacking fibronectin. *Development (Cambridge, UK)* **119,** 1079–1091.

Gonzalez-Sanchez, A., and Bader, D. (1984). Immunochemical analysis of myosin heavy chains in the developing chicken heart. *Dev. Biol.* **103,** 151–158.

Haffter, P., Granato, M., Brand, M., Mullins, M. C., Hammerschmidt, M., Kane, D. A., Odenthal, J., van Eeden, F. J., Jiang, Y. J., Heisenberg, C. P., Kelsh, R. N., Furutani-Seiki, M., Vogelsang, E., Beuchle, D., Schach, U., Fabian, C., and Nüsslein-Volhard, C. (1996). The identification of genes with unique and essential functions in the development of the zebrafish, *Danio rerio*. *Development (Cambridge, UK)* **123,** 1–36.

Halpern, M. E., Ho, R. K., Walker, C., and Kimmel, C. B. (1993). Induction of muscle pioneers and floor plate is distinguished by the zebrafish *no tail* mutation. *Cell (Cambridge, Mass.)* **75,** 99–111.

Halpern, M. E., Thisse, C., Ho, R. K., Thisse, B., Riggleman, B., Trevarrow, B., Weinberg, E. S., Postlethwait, J. H., and Kimmel, C. B. (1995). Cell-autonomous shift from axial to paraxial mesodermal development in zebrafish *floating head* mutants. *Development (Cambridge, UK)* **121,** 4257–4264.

Hammerschmidt, M., Pelegri, F., Mullins, M. C., Kane, D. A., Brand, M., van Eeden, F. J., Furutani-Seiki, M., Granato, M., Haffter, P., Heisenberg, C. P., Jiang, Y. J., Kelsh, R. N., Odenthal, J., Warga, R. M., and Nüsslein-Volhard, C. (1996). Mutations affecting morphogenesis during gastrulation and tail formation in the zebrafish, *Danio rerio*. *Development (Cambridge, UK)* **123,** 143–151.

Hatta, K., Kimmel, C. B., Ho, R. K., and Walker, C. (1991). The *cyclops* mutation blocks specification of the floor plate of the zebrafish central nervous system. *Nature (London)* **350,** 339–341.

Ho, R. K. (1992). Axis formation in the embryo of the zebrafish, *Brachydanio rerio*. *Semin. Dev. Biol.* **3,** 53–64.

Horvitz, H. R. (1988). Genetics of cell lineage. *In* "The Nematode Caenorhabditis Elegans" (W. B. Wood, ed.), pp. 157–190. Cold Spring Harbor Lab., Plainview, NY.

Jacobson, A. G., and Sater, A. K. (1988). Features of embryonic induction. *Development (Cambridge, UK)* **104,** 341–359.

Jiang, Y. J., Brand, M., Heisenberg, C. P., Beuchle, D., Furutani-Seiki, M., Kelsh, R. N., Warga, R. M., Granato, M., Haffter, P., Hammerschmidt, M., Kane, D. A., Mullins, M. C., Odenthal, J., van Eeden, F. J., and Nüsslein-Volhard, C. (1996). Mutations affecting neurogenesis and brain morphology in the zebrafish, *Danio rerio*. *Development (Cambridge, UK)* **123,** 205–216.

Jones, C. M., Lyons, K. M., and Hogan, B. L. (1991). Involvement of Bone Morphogenetic Protein-4 (BMP-4) and Vgr-1 in morphogenesis and neurogenesis in the mouse. *Development (Cambridge, UK)* **111,** 531–542.

Kane, D. A., and Kimmel, C. B. (1993). The zebrafish midblastula transition. *Development* (*Cambridge, UK*) **119,** 447–456.

Kimmel, C. B. (1989). Genetics and early development of zebrafish. *Trends Genet.* **5,** 283–288.

Kimmel, C. B., and Law, R. D. (1985a). Cell lineage of zebrafish blastomeres. I. Cleavage pattern and cytoplasmic bridges between cells. *Dev. Biol.* **108,** 78–85.

Kimmel, C. B., and Law, R. D. (1985b). Cell lineage of zebrafish blastomeres. II. Formation of the yolk syncytial layer. *Dev. Biol.* **108,** 86–93.

Kimmel, C. B., Kane, D. A., Walker, C., Warga, R. M., and Rothman, M. B. (1989). A mutation that changes cell movement and cell fate in the zebrafish embryo. *Nature* (*London*) **337,** 358–362.

Kimmel, C. B., Ballard, W. W., Kimmel, S. R., Ullmann, B., and Schilling, T. F. (1995). Stages of embryonic development of the zebrafish. *Dev. Dyn.* **203,** 253–310.

Kramer, R., Bucay, N., Kane, D. J., Martin, L. E., Tarpley, J. E., and Theill, L. E. (1996). Neuregulins with an Ig-like domain are essential for mouse myocardial and neuronal development. *Proc. Natl. Acad. Sci. U. S. A.* **93,** 4833–4838.

Kubalak, S. W., Miller-Hance, W. C., O'Brien, T. X., Dyson, E., and Chien, K. R. (1994). Chamber specification of atrial myosin light chain-2 expression precedes septation during murine cardiogenesis. *J. Biol. Chem.* **269,** 16961–16970.

Kuo, C. T., Morrisey, E. E., Anandappa, R., Sigrist, K., Lu, M. M., Parmacek, M. S., Soudais, C., and Leiden, J. M. (1997). GATA4 transcription factor is required for ventral morphogenesis and heart tube formation. *Genes Dev.* **11,** 1048–1060.

Lee, K. F., Simon, H., Chen, H., Bates, B., Hung, M. C., and Hauser, C. (1995). Requirement for neuregulin receptor erbB2 in neural and cardiac development. *Nature* (*London*) **378,** 394–398.

Lee, K. H., Xu, Q., and Breitbart, R. E. (1996). A new tinman-related gene, *nkx2.7,* anticipates the expression of *nkx2.5* and *nkx2.3* in zebrafish heart and pharyngeal endoderm. *Dev. Biol.* **180,** 722–731.

Lee, R. K., Stainier, D. Y., Weinstein, B. M., and Fishman, M. C. (1994). Cardiovascular development in the zebrafish. II. Endocardial progenitors are sequestered within the heart field. *Development* (*Cambridge, UK*) **120,** 3361–3366.

Lemaire, P., and Kodjabachian, L. (1996). The vertebrate organizer: Structure and molecules. *Trends Genet.* **12,** 525–531.

Levin, M., Johnson, R. L., Stern, C. D., Kuehn, M., and Tabin, C. (1995). A molecular pathway determining left-right asymmetry in chick embryogenesis. *Cell* (*Cambridge, Mass.*) **82,** 803–814.

Liao, W., Bisgrove, B. W., Sawyer, H., Hug, B., Bell, B., Peters, K., Grunwald, D. J., and Stainier, D. Y. R. (1997). The zebrafish gene *cloche* acts upstream of a *flk-1* homologue to regulate endothelial cell differentiation. *Development* (*Cambridge, UK*) **124,** 381–389.

Lin, Q., Schwarz, J., Bucana, C., and Olson, E. N. (1997). Control of mouse cardiac morphogenesis and myogenesis by transcription factor MEF2C. *Science* **276,** 1404–1407.

Linask, K. K., and Lash, J. W. (1988). A role for fibronectin in the migration of avian precardiac cells. I. Dose-dependent effects of fibronectin antibody. *Dev. Biol.* **129,** 315–323.

Long, W. L. (1983). The role of the yolk syncytial layer in determination of the plane of bilateral symmetry in the rainbow trout, *Salmo gairdneri* Richardson. *J. Exp. Zoo.* **228,** 91–97.

Lowe, L. A., Supp, D. M., Sampath, K., Yokoyama, T., Wright, C. V., Potter, S. S., Overbeek, P., and Kuehn, M. R. (1996). Conserved left-right asymmetry of nodal expression and alterations in murine situs inversus. *Nature* (*London*) **381,** 158–161.

Lyons, I., Parsons, L. M., Hartley, L., Li, R., Andrews, J. E., Robb, L., and Harvey, R. P. (1995). Myogenic and morphogenetic defects in the heart tubes of murine embryos lacking the homeo box gene *Nkx2-5*. *Genes Dev*. **9,** 1654–1666.

Malicki, J., Neuhauss, S. C., Schier, A. F., Solnica-Krezel, L., Stemple, D. L., Stainier, D. Y., Abdelilah, S., Zwartkruis, F., Rangini, Z., and Driever, W. (1996a). Mutations affecting development of the zebrafish retina. *Development* (*Cambridge, UK*) **123,** 263–273.

Malicki, J., Schier, A. F., Solnica-Krezel, L., Stemple, D. L., Neuhauss, S. C., Stainier, D. Y., Abdelilah, S., Rangini, Z., Zwartkruis, F., and Driever, W. (1996b). Mutations affecting development of the zebrafish ear. *Development* (*Cambridge, UK*) **123,** 275–283.

Markwald, R., Eisenberg, C., Eisenberg, L., Trusk, T., and Sugi, Y. (1996). Epithelial-mesenchymal transformations in early avian heart development. *Acta Anat.* **156,** 173–186.

Marsh-Armstrong, N., McCaffery, P., Gilbert, W., Dowling, J. E., and Drager, U. C. (1994). Retinoic acid is necessary for development of the ventral retina in zebrafish. *Proc. Natl. Acad. Sci. U.S.A.* **91,** 7286–7290.

McGuire, P. G., and Alexander, S. M. (1993). Inhibition of urokinase synthesis and cell surface binding alters the motile behavior of embryonic endocardial-derived mesenchymal cells in vitro. *Development* (*Cambridge, UK*) **118,** 931–939.

Meno, C., Saijoh, Y., Fujii, H., Ikeda, M., Yokoyama, T., Yokoyama, M., Toyoda, Y., and Hamada, H. (1996). Left-right asymmetric expression of the TGF beta-family member lefty in mouse embryos. *Nature* (*London*) **381,** 151–155.

Meyer, D., and Birchmeier, C. (1995). Multiple essential functions of neuregulin in development. *Nature* (*London*) **378,** 386–390.

Mizuno, T., Yamaha, E., Wakahara, M., Kuroiwa, A., and Takeda, H. (1996). Mesoderm induction in zebrafish. *Nature* (*London*) **383,** 131–132.

Molkentin, J. D., Lin, Q., Duncan, S. A., and Olson, E. N. (1997). Requirement of the transcription factor GATA4 for heart tube formation and ventral morphogenesis. *Genes Dev.* **11,** 1061–1072.

Mullins, M. C., and Nüsslein-Volhard, C. (1993). Mutational approaches to studying embryonic pattern formation in the zebrafish. *Curr. Opin. Genet. Dev.* **3,** 648–654.

Mullins, M. C., Hammerschmidt, M., Haffter, P., and Nüsslein-Volhard, C. (1994). Large-scale mutagenesis in the zebrafish: In search of genes controlling development in a vertebrate. *Curr. Biol.* **4,** 189–202.

Mullins, M. C., Hammerschmidt, M., Kane, D. A., Odenthal, J., Brand, M., van Eeden, F. J. M., Furutani-Seiki, M., Granato, M., Haffter, P., Heisenberg, C. P., Jiang, Y. J., Kelsh, R. N., and Nüsslein-Volhard, C. (1996). Genes establishing dorsoventral pattern formation in the zebrafish embryo: The ventral specifying genes. *Development* (*Cambridge, UK*) **123,** 81–93.

Narita, N., Bielinska, M., and Wilson, D. B. (1997). Wild-type endoderm abrogates the ventral developmental defects associated with *Gata-4* deficiency in the mouse. *Dev. Biol.* **189,** 270–274.

Nascone, N., and Mercola, M. (1995). An inductive role for the endoderm in *Xenopus* cardiogenesis. *Development* (*Cambridge, UK*) **121,** 515–523.

O'Brien, T. X., Lee, K. J., and Chien, K. R. (1993). Positional specification of ventricular myosin light chain 2 expression in the primitive murine heart tube. *Proc. Natl. Acad. Sci. U. S. A.* **90,** 5157–5161.

Odenthal, J., Haffter, P., Vogelsang, E., Brand, M., van Eeden, F. J., Furutani-Seiki, M., Granato, M., Hammerschmidt, M., Heisenberg, C. P., Jiang, Y. J., Kane, D. A., Kelsh, R. N., Mullins, M. C., Warga, R. M., Allende, M. L., Weinberg, E. S., and Nüsslein-Volhard, C. (1996). Mutations affecting the formation of the notochord in the zebrafish, *Danio rerio. Development* (*Cambridge, UK*) **123,** 103–115.

Potts, J. D., Dagle, J. M., Walder, J. A., Weeks, D. L., and Runyan, R. B. (1991). Epithelial-mesenchymal transformation of embryonic cardiac endothelial cells is inhibited by a modified antisense oligodeoxynucleotide to transforming growth factor-β 3. *Proc. Natl. Acad. Sci. U. S. A.* **88,** 1516–1520.

Potts, J. D., Vincent, E. B., Runyan, R. B., and Weeks, D. L. (1992). Sense and antisense TGF-β 3 mRNA levels correlate with cardiac valve induction. *Dev. Dyn.* **193,** 340–345.

Puri, M. C., Rossant, J., Alitalo, K., Bernstein, A., and Partanen, J. (1995). The receptor tyrosine kinase TIE is required for integrity and survival of vascular endothelial cells. *EMBO J.* **14,** 5884–5891.

Sater, A. K., and Jacobson, A. G. (1989). The specification of heart mesoderm occurs during gastrulation in *Xenopus laevis. Development (Cambridge, UK)* **105,** 821–830.

Sater, A. K., and Jacobson, A. G. (1990). The role of the dorsal lip in the induction of heart mesoderm in *Xenopus laevis. Development (Cambridge, UK)* **108,** 461–470.

Sato, T. N., Tozawa, Y., Deutsch, U., Wolburg-Buchholz, K., Fujiwara, Y., Gendron-Maguire, M., Gridley, T., Wolburg, H., Risau, W., and Qin, Y. (1995). Distinct roles of the receptor tyrosine kinases Tie-1 and Tie-2 in blood vessel formation. *Nature (London)* **376,** 70–74.

Schier, A. F., Neuhauss, S. C., Harvey, M., Malicki, J., Solnica-Krezel, L., Stainier, D. Y., Zwartkruis, F., Abdelilah, S., Stemple, D. L., Rangini, Z., Yang, H., and Driever, W. (1996). Mutations affecting the development of the embryonic zebrafish brain. *Development (Cambridge, UK)* **123,** 165–178.

Schier, A. F., Neuhauss, S. C., Helde, K. A., Talbot, W. S., and Driever, W. (1997). The *one-eyed pinhead* gene functions in mesoderm and endoderm formation in zebrafish and interacts with *no tail. Development (Cambridge, UK)* **124,** 327–342.

Schultheiss, T. M., Xydas, S., and Lassar, A. B. (1995). Induction of avian cardiac myogenesis by anterior endoderm. *Development (Cambridge, UK)* **121,** 4203–4214.

Schultheiss, T. M., Burch, J. B., and Lassar, A. B. (1997). A role for bone morphogenetic proteins in the induction of cardiac myogenesis. *Genes Dev.* **11,** 451–462.

Shah, A. M., and Lewis, M. J. (1993). Modulation of myocardial contraction by endocardial and coronary vascular endothelium. *Trends Cardiovasc. Med.* **3,** 98–103.

Shalaby, F., Rossant, J., Yamaguchi, T. P., Gertsenstein, M., Wu, X. F., Breitman, M. L., and Schuh, A. C. (1995). Failure of blood-island formation and vasculogenesis in Flk-1-deficient mice. *Nature (London)* **376,** 62–66.

Shalaby, F., Ho, J., Stanford, W. L., Fischer, K. D., Schuh, A. C., Schwartz, L., Bernstein, A., and Rossant, J. (1997). A requirement for Flk1 in primitive and definitive hematopoiesis and vasculogenesis. *Cell (Cambridge, Mass.)* **89,** 981–990.

Shih, J., and Fraser, S. E. (1995). Distribution of tissue progenitors within the shield region of the zebrafish gastrula. *Development (Cambridge, UK)* **121,** 2755–2765.

Solnica-Krezel, L., and Driever, W. (1994). Microtubule arrays of the zebrafish yolk cell: Organization and function during epiboly. *Development (Cambridge, UK)* **120,** 2443–2455.

Solnica-Krezel, L., Schier, A. F., and Driever, W. (1994). Efficient recovery of ENU-induced mutations from the zebrafish germline. *Genetics* **136,** 1401–1420.

Srivastava, D., Cserjesi, P., and Olson, E. N. (1995). A subclass of bHLH proteins required for cardiac morphogenesis. *Science* **270,** 1995–1999.

Srivastava, D., Thomas, T., Lin, Q., Kirby, M. L., Brown, D., and Olson, E. N. (1997). Regulation of cardiac mesodermal and neural crest development by the bHLH transcription factor, dHAND. *Nat. Genet.* **16,** 154–160.

Stainier, D. Y. R., and Fishman, M. C. (1992). Patterning the zebrafish heart tube: Acquisition of anteroposterior polarity. *Dev. Biol.* **153,** 91–101.

Stainier, D. Y. R., and Fishman, M. C. (1994). The zebrafish as a model system to study cardiovascular development. *Trends Cardiovasc. Med.* **4,** 207–212.

Stainier, D. Y. R., Lee, R. K., and Fishman, M. C. (1993). Cardiovascular development in the zebrafish. I. Myocardial fate map and heart tube formation. *Development (Cambridge, UK)* **119,** 31–40.

Stainier, D. Y. R., Weinstein, B. M., Detrich, H. W., Zon, L. I., and Fishman, M. C. (1995). *Cloche,* an early acting zebrafish gene, is required by both the endothelial and hematopoietic lineages. *Development (Cambridge, UK)* **121,** 3141–3150.

Stainier, D. Y. R., Fouquet, B., Chen, J. N., Warren, K. S., Weinstein, B. M., Meiler, S. E., Mohideen, M. A., Neuhauss, S. C., Solnica-Krezel, L., Schier, A. F., Zwartkruis, F., Stemple, D. L., Malicki, J., Driever, W., and Fishman, M. C. (1996). Mutations affecting the formation and function of the cardiovascular system in the zebrafish embryo. *Development (Cambridge, UK)* **123,** 285–292.

Stemple, D. L., Solnica-Krezel, L., Zwartkruis, F., Neuhauss, S. C., Schier, A. F., Malicki, J., Stainier, D. Y., Abdelilah, S., Rangini, Z., Mountcastle-Shah, E., and Driever, W. (1996). Mutations affecting development of the notochord in zebrafish. *Development (Cambridge, UK)* **123,** 117–128.

St. Johnston, D., and Nüsslein-Volhard, C. (1992). The origin of pattern and polarity in the *Drosophila* embryo. *Cell (Cambridge, Mass.)* **68,** 210–219.

Streisinger, G., Walker, C., Dower, N., Knauber, D., and Singer, F. (1981). Production of clones of homozygous diploid zebra fish (*Brachydanio rerio*). *Nature (London)* **291,** 293–296.

Sugi, Y., and Lough, J. (1994). Anterior endoderm is a specific effector of terminal cardiac myocyte differentiation of cells from the embryonic heart forming region. *Dev. Dyn.* **200,** 155–162.

Swaen, A., and Brachet, A. (1901). Etude sur les premières phases du développement des organes dérivés du mesoblaste chez les poissons Téléostéens. *Arch. Biol.* **18,** 73–190.

Talbot, W. S., Trevarrow, B., Halpern, M. E., Melby, A. E., Farr, G., Postlethwait, J. H., Jowett, T., Kimmel, C. B., and Kimelman, D. (1995). A homeobox gene essential for zebrafish notochord development. *Nature (London)* **378,** 150–157.

Tam, P. P., and Quinlan, G. A. (1996). Mapping vertebrate embryos. *Curr. Biol.* **6,** 104–106.

Thomas, P., and Beddington, R. (1996). Anterior primitive endoderm may be responsible for patterning the anterior neural plate in the mouse embryo. *Curr. Biol.* **6,** 1487–1496.

Varlet, I., Collignon, J., and Robertson, E. J. (1997). *Nodal* expression in the primitive endoderm is required for specification of the anterior axis during mouse gastrulation. *Development (Cambridge, UK)* **124,** 1033–1044.

Walker, C., and Streisinger, G. (1983). Induction of mutations by gamma-rays in pregonial germ cells of zebrafish embryos. *Genetics* **103,** 125–136.

Warga, R. (1996). "Origin and Specification of the Endoderm in the Zebrafish, *Danio rerio,*" pp. 1–87. University of Tuebingen.

Weinstein, D. C., Ruiz i Altaba, A., Chen, W. S., Hoodless, P., Prezioso, V. R., Jessell, T. M., and Darnell, J. E., Jr. (1994). The winged helix transcription factor HNF-3β is required for notochord development in the mouse embryo. *Cell (Cambridge, Mass.)* **78,** 575–588.

Westerfield, M. (1993). "The Zebrafish Book," 3rd ed. University of Oregon Press, Eugene.

Yatskievych, T. A., Ladd, A. N., and Antin, P. B. (1997). Induction of cardiac myogenesis in avian pregastrula epiblast: The role of the hypoblast and activin. *Development (Cambridge, UK)* **124,** 2561–2570.

Yutzey, K. E., and Bader, D. (1995). Diversification of cardiomyogenic cell lineages during early heart development. *Circ. Res.* **77,** 216–219.

Yutzey, K. E., Rhee, J. T., and Bader, D. (1994). Expression of the atrial-specific myosin heavy chain AMHC1 and the establishment of anteroposterior polarity in the developing chicken heart. *Development (Cambridge, UK)* **120,** 871–883.

7

Transcriptional Control and Pattern Formation in the Developing Vertebrate Heart: Studies on NK-2 Class Homeodomain Factors

Richard P. Harvey,[*,†] Christine Biben,[*] and David A. Elliott[*]

**The Victor Chang Cardiac Research Institute, St. Vincent's Hospital, Sydney 2010, Australia*
†University of New South Wales, Kensington 2033, Australia

I. Introduction

Due to a failing heart, 400,000 individuals are admitted into the clinic annually in the United States, at a direct cost of $10.2 billion (Williams, 1995). This burden provides an incentive to seek out the genetic blueprint of the heart and how it relates to disease. This, of course, involves consideration of its developmental history and the mechanisms underlying lineage specification and pattern. Explaining the origin of lineage and pattern within multicellular organisms is one of the greatest challenges for modern biologists. Pattern refers to the activities of the cells of an embryo in space and is the net result of genetic and epigenetic processes determining cellular shape, movement, identity, proliferation, and death. There is enormous diversity in pattern among different organisms, even between homologous structures, such as limbs, in closely related species. The task of understanding the differences, similarities, or common origins seems daunting. In recent decades, however, extensive progress has been made in understanding how lineages and pattern are generated and the molecules involved (Scott, 1997). These studies have given us paradigms that can now be applied to the heart. In this chapter, we summarize what is known about one specific gene family acting in transcriptional control and pattern formation in vertebrate and invertebrate hearts: the NK-2 class of homeodomain factors.

II. Overview of Cardiac Development in Mammals

As described in detail in other chapters of this book, precursors of the mammalian heart can be mapped early in development as tiny populations of cells located just lateral to the primitive streak at gastrulation (Buckingham, 1997; Lawson *et al.,* 1991; Lawson and Pedersen, 1992; Tam *et al.,* 1997; see Chapters 1 and 2). After ingression through the streak, precardiac cells, now part of the nascent mesodermal mantle, migrate toward the anterior of the embryo and enter the heart field (Tam *et al.,* 1997). It is during this process that they encounter signals that identify them as anterior tissue (Belo *et al.,* 1997; Biben *et al.,* 1998b; Bouwmeester and Leyns, 1997; Thomas *et al.,* 1997) and which commit them to the cardiac lineages (Schultheiss *et al.,* 1997; Tam *et al.,* 1997; Yatskievych *et al.,* 1997). The heart field is most likely shaped by positive and negative influences from surrounding tissues, such as notochord (Schultheiss *et al.,* 1997; Serbedzija *et al.,* 1998; see Chapter 3), and, by definition, may be highly regulative, as it is in lower vertebrates (see Chapter 3). Committed myocardial and endocardial precursor cells become arranged in the form of a layered crescent (DeRuiter *et al.,* 1992; Linask and Lash, 1993; Sugi and Markwald, 1996). At late crescent stages, myocardial cells begin to express myofilament proteins (Lints *et al.,* 1993; Lyons, 1994; Lyons *et al.,* 1995), then coalesce at the ventral midline to form a roughly linear tube consisting of an inner endothelial layer, the endocardium, and an outer muscular layer, the myocardium. The transition from crescent to tube occurs in concert with expansion of the headfolds and formation of the foregut pocket and may depend on functions within both mesoderm and endoderm (George *et al.,* 1997; Kuo *et al.,* 1997; Linask and Lash, 1988a,b; Molkentin *et al.,* 1997). Explantation and marker analysis in various systems clearly show that the linear heart tube is a highly regionalized structure with distinct (although evolving) anterior/posterior (A/P), dorsal/ventral (D/V), and left/right (L/R) identities (Biben and Harvey, 1997; Chen *et al.,* 1997; Collignon *et al.,* 1996; Lyons *et al.,* 1995; Ross *et al.,* 1996; Satin *et al.,* 1988; Smith *et al.,* 1997; Tsuda *et al.,* 1996; Yutzey *et al.,* 1994). Rapidly, the linear tube, now fully enclosed within an embryonic space called the pericardial cavity, begins to beat and initiates the process of looping morphogenesis (Harvey, 1998). Guided by information from the embryonic L/R asymmetry pathway (Levin, 1997), the caudal region of the heart tube shifts to the left and the ventricular region loops to the right (Biben and Harvey, 1997; Chen *et al.,* 1997; see Chapters 21 and 22). At this time, individual ventricular chambers become distinct (Firulli *et al.,* 1998; Riley *et al.,* 1998), and looping brings them into the basic pattern that prefigures the adult structure. Extensive differentiation and complex remodeling of the "primary" cardiac tissues occurs during looping. Over a period of a few days, the inner walls of individual primitive chambers and their associated inflow and outflow vessels become aligned and integrated into a compact organ; valve precursors differentiate, first as the endocardial cushions (Webb *et al.,* 1998; see Chapter 10); neural crest cells invade and contribute to the spiral septum (Kirby and Waldo, 1995; see Chapter 11); fibrous tissue is created from the cushions, giving rigidity to the heart and its valves and insulating the atria from the ventricles (Webb *et al.,* 1998); the conduction system develops (Moorman *et al.,* 1998; see Chapter 12); and the epicardium, from which coronary vessels and intramyocardial fibroblasts derive, covers the myocardium like a glove (Dettman *et al.,* 1998; Mikawa and Gourdie, 1996; see Chapter 2). All of these processes involve intricate morphogenetic events, and, in most cases, we are far from understanding their molecular basis.

III. Genetic Regulation of Myocardial Development: A General Concept

The muscles of the vertebrate skeletal system and heart appear similar. Both assemble a striated myofilament and both express regulatory, structural, and metabolic genes in common (see Chapter 28). However, cardiac cells also share similarities with smooth muscle.

What is clear is that the basic helix–loop–helix (bHLH) transcription factors related to *myod*, the so-called master regulators of skeletal muscle development (Olson and Klein, 1994), are not expressed in cardiac or smooth muscle. Specification of lineage and pattern in these systems is apparently driven by different genes. Several families of transcription factors are now known to play critical roles in myocardial development and patterning (Olson and Srivastava, 1996). The MADS box-containing Mef2 and SRF proteins are expressed in all muscle types (Chen *et al.*, 1996; Olson *et al.*, 1995), reflecting their ancient common origin. Others, such as the bHLH Hand factors, the zinc finger GATA-4/5/6 factors, and NK-2 homeodomain factors (Olson and Srivastava, 1996), are expressed in a broad range of tissues including the heart. Even though these transcription factors can activate heart target genes in some circumstances when ectopically expressed (Chen and Schwartz, 1996; Chen and Fishman, 1996; Fu and Izumo, 1995), there is little evidence to suggest that the master regulatory gene concept, as often applied to vertebrate skeletal muscle development, is appropriate for the heart (Evans *et al.*, 1994). It follows that specificity in heart development is generated by the biochemical and genetic interactions between different groups of factors. Much of this specificity may be realized at the promoter and enhancer regions of cardiac-expressed genes (Firulli and Olson, 1997).

IV. Homeobox Genes and Embryonic Patterning

How are the major events in cardiac development controlled and what type of genes are we looking for? Important clues have come from genetic analysis of embryonic patterning in various model systems, particularly the fruit fly, *Drosophila melanogaster*. Key players appear to be members of the homeobox gene family (Manak and Scott, 1994). Throughout this century, mutations that exhibit transformations of one body part into the form of another, so-called "homeotic" changes, have been collected and studied, and a lifetime's work on the mutant phenotypes by E. B. Lewis was celebrated in 1995 as part of a shared Nobel prize. The studies led to the identification of the homeotic or *Hox* genes, encoding transcription factors which carry a DNA-binding domain called the homeodomain or homeobox. *Hox* genes appear to specify the type of structure found along the A/P axis of embryos, controlling the complex morphogenesis of individual body parts (Manak and Scott, 1994). This is most strikingly evident in the formation of segmental structures, such as the cuticle of the fly and the skeleton and rhombomeric hindbrain of vertebrates, but also for less overtly segmented structures such as the gut (Roberts *et al.*, 1995). Molecular data have largely substantiated an early hypothesis that the *Hox* genes were "selector" genes, controlling batteries of subordinate "realizator" genes, encoding functions directly related to cellular differentiation and morphogenesis. Known realizator genes for *Drosophila* Hox proteins encode other transcription factors, secreted signaling molecules, cell cycle regulators, and/or structural proteins (Andrew and Scott, 1992). In one form or another, this developmental paradigm is played out time and time again in the formation of the body parts of all metazoa (Carroll, 1995; Manak and Scott, 1994).

The *Hox* genes are linked within tight clusters. In man, there are four clusters composed of a total of 38 genes, with up to 11 genes per cluster. However, the clustered *Hox* genes coexist with a variety of chromosomally dispersed homeobox-containing genes (Duboule, 1994). Currently, in excess of 100 such genes are known in the mouse. Since these genes appear to be involved variously in lineage specification, cell type diversification, growth, and survival, some may serve as realizator genes for the *Hox* proteins. Homeobox genes are found in organisms from hydra to man; thus, they appear to be ancient components of the genetic circuitry that generates cellular diversity and pattern in multicellular animals.

V. Pattern Formation in the Heart

During the course of vertebrate evolution, the heart has developed enormously in structural sophistication. From simple beginnings as a muscular vessel (Randall and Davie, 1980), multiple chambers, valves, septa, endocardium, trabeculae, conduction system, coronary circulation, and a critical interface with laterality have all been incorporated. These adaptations no doubt arose in a stepwise sequence, after considerable experimentation, with some requiring complex interactions between component tissues, for example, between endocardium and myocardium (trabeculation), myocardium and endocardium (cushion formation), and epicardium and myocardium (formation of Purkinje fibers in the chick) (Eisenberg and Markwald, 1995; Gassmann *et al.*, 1995; Gourdie *et al.*, 1995; Meyer and Birchmeier, 1995). Improvements to cardiovascular output would, in the course of evolution, be expected to confer a considerable selective advantage, and indeed, various features of the vertebrate pattern of unidirectional flow, low-pressure collecting chamber, high-pressure pumping chamber, valves, neuronal control and closed vascular system appear to have been arrived at independently in different invertebrate species (Martin, 1980; Randall and Davie, 1980).

Is there a prepattern upon which these modifications to the mammalian heart and vessels were crafted? The embryonic heart has often been referred to as a segmented structure (Davis, 1927; De La Cruz *et al.,* 1989; Moorman *et al.,* 1998; Stalsberg, 1969; see Chapter 12). For example, the early heart tube has alternating slow and fast conducting regions (Moorman *et al.,* 1998) and endocardial cushions are induced by myocardium in a segment-like periodicity (Eisenberg and Markwald, 1995). While the myocardium is clearly regionalized at the linear tube stage (Biben and Harvey, 1997; Lyons, 1994; Lyons *et al.,* 1995; Ross *et al.,* 1996; Yutzey *et al.,* 1995), explantation experiments suggest that regionalization may begin much earlier, at the crescent stage, in the heart field, or even earlier (Garcia-Martinez and Schoenwolf, 1993; Satin *et al.,* 1988; Stainier *et al.,* 1993; Yutzey *et al.,* 1995). How this regionalization is achieved is a key question in cardiac developmental biology. It is possible that a segmental structure is established, with identity of segments determined by a unifying patterning system, such as one utilizing *Hox* genes. In this regard it is noteworthy that retinoic acid, which can directly influence *Hox* gene expression via its family of ligand-dependent DNA-binding receptors (Marshal *et al.,* 1996), can induce atrial character in the primitive ventricle of the chick heart (Yutzey *et al.,* 1994), a possible homeotic change (see Chapter 13).

From a totally different perspective, however, the heart may have developed unfettered by the logic of a unifying patterning system. A hint in this direction comes from mutagenesis studies in mice and zebrafish which have revealed cardiac phenotypes in which individual segments (e.g., ventricles) or other units (e.g., endocardium) are deleted without more global effects on pattern (Fishman and Olson, 1997). The modular architecture of the *cis*-regulatory regions of heart-expressed genes (Firulli *et al.,* 1997; see Chapter 19) could also be interpreted as evidence for modular heart construction. However, this issue is clearly not resolved: The modules of heart gene regulation may equally be revealing the segmental units upon which the heart is constructed, just as the separate enhancers of the *Drosophila even skipped* gene (*eve*), which collectively control expression in seven cuticular stripes, reveal the units of the patterning system known to underlie cuticular segmentation (Small *et al.,* 1996). It is still possible, however, to envisage a model in which the modules of the heart were assembled during evolution as separate and isolated innovations, built one upon the other in a progressive fashion.

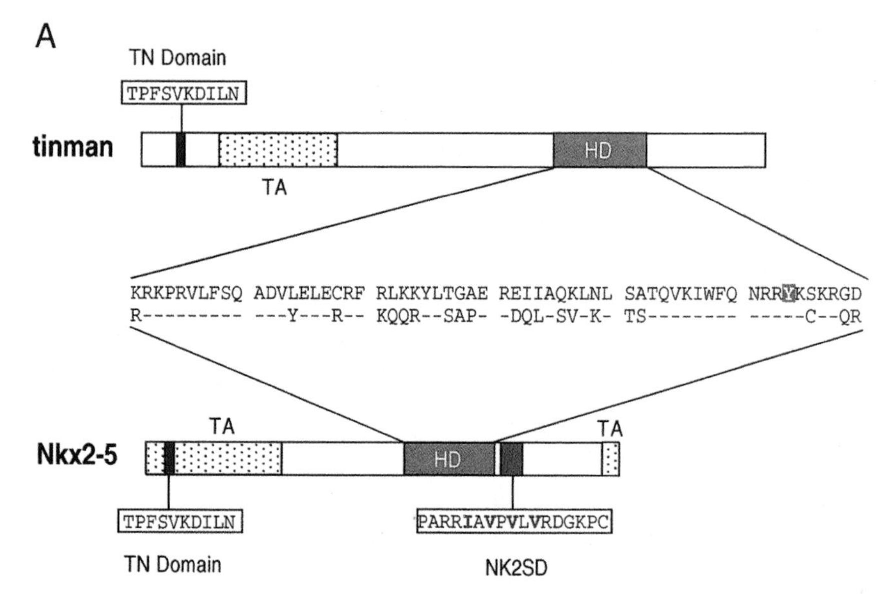

Figure 1 (A) Domain structure of the tinman and Nkx2-5 homeobox proteins. Amino acid sequences of the homeodomain (HD) and conserved TN domain and NK2 specific domain (NK2SD) are shown. Transcriptional transactivation domains (TA), as mapped in a GAL4-dependent heterologous system (Ranganayakulu *et al.,* 1998), are shaded. The Tyr residue at position 54 of the homeodomain that is the defining characteristic of the NK-2 homeodomain class is highlighted in red. Within the NK2SD sequence, the conserved repetitive amino acids of the V/IxVxVxV core are highlighted. (B) Phylogenetic tree analysis of NK-2 homeodomain proteins using the parsimony method and based on comparison of both the homeodomain and the NK2SD sequences (adapted from Jacobs *et al.,* 1998).

VI. The *Drosophila* Heart and the *tinman* Gene

The homeobox gene *tinman* (*tin;* formally *NK4* and *msh-2*) is expressed in the *Drosophila* heart and is essential for its formation (Bodmer *et al.,* 1990). Analysis of the *tinman* gene and its vertebrate relatives has provided important insights into the genetic regulation of cardiac development and has highlighted the possibility that the genetic underpinning of vertebrate heart development may be mirrored in its fly counterpart (Bodmer, 1995; Harvey, 1996; Olson and Srivastava, 1996; Ranganayakulu *et al.,* 1998).

The fly has a mesodermally derived heart-like organ, often called the dorsal vessel (see Chapter 5). It is a relatively simple linear tube composed of an inner muscular (cardial) cell layer surrounding a lumen and an outer pericardial layer (Rizki, 1978; Rugendorff *et al.,* 1994). The fly heart pumps cellular hemolymph around an open body cavity: Hemolymph is drawn in through numerous small openings called ostia and then is pumped cranially by unidirectional peristalsis. While structural detail in the fly heart is minimal, the anterior portion (aorta) is thinner, and a valve separates it from the more caudal "ventricle." In addition, both cardial and pericardial cells are heterogeneous in origin, as assessed by expression of markers (Jagla *et al.,* 1997). Superficially at least, this structure resembles the primitive vertebrate heart before looping, although a better analogy might be with an ancestral vertebrate heart, for example, the pulsatory muscular vessels found in our invertebrate chordate ancestor, amphioxus (Randall and Davie, 1980). Like the vertebrate heart, the fly heart derives from lateral mesodermal progenitor cells that migrate to the midline before forming a tube.

tinman was first isolated by Kim and Nirenberg in 1989 as part of a screen for new homeobox sequences in the fly. *tin* encodes a homeodomain protein of the NK-2 class (Burglin, 1993), distinguished principally by the presence of a tyrosine at position 54 of the homeodomain (Fig. 1A), an amino acid that most

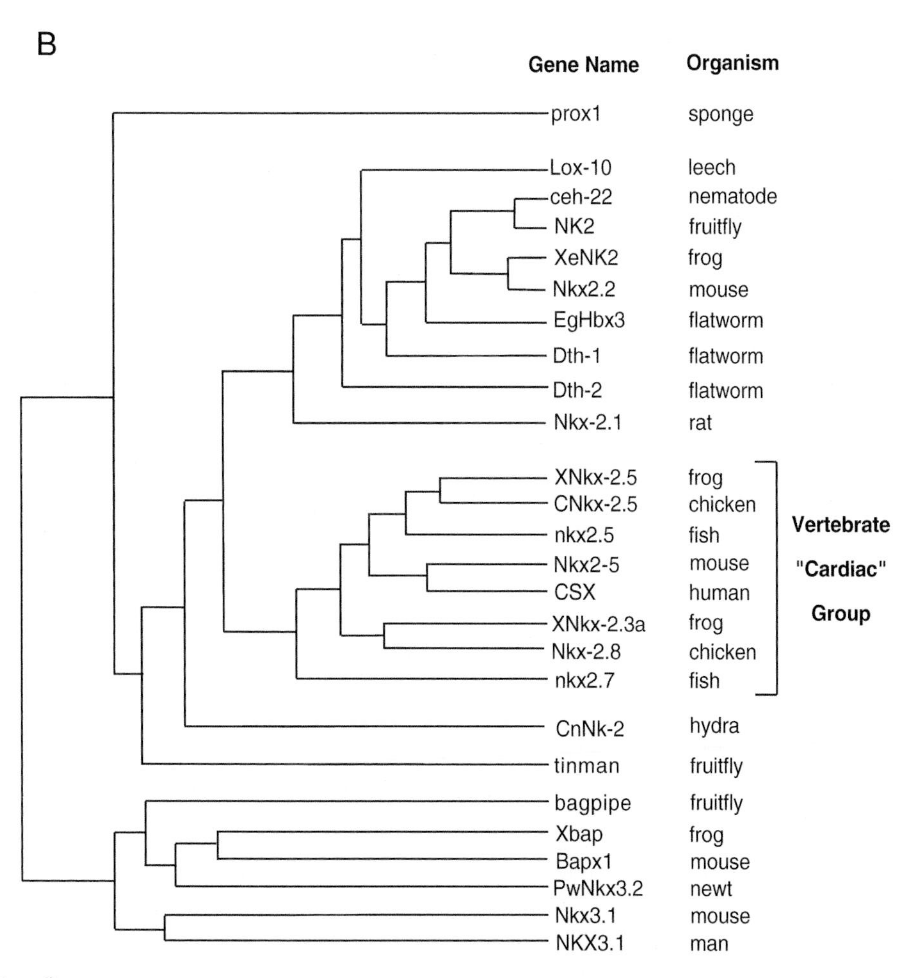

Figure 1 (*Continued*)

likely contributes to its unique DNA binding site specificity (Damante *et al.,* 1994; Tsao *et al.,* 1994; Viglino *et al.,* 1993). The gene is first expressed in cells of the cellular blastoderm fated to become mesoderm, and expression remains during ingression and spreading of those cells at gastrulation. In this early phase, *tinman* is controlled by twist, a bHLH protein which sits at the top of the mesodermal regulatory hierarchy (Yin *et al.,* 1997). *tin* expression then becomes restricted in nascent mesoderm to the dorsal-most cells, those that will yield the precursors of the visceral and cardiac muscles, as well as some dorsal body wall muscles (Azpiazu and Frasch, 1993; Bodmer *et al.,* 1990; Frasch, 1995). This restriction involves maintenance of *tin* expression in dorsal mesoderm by an inductive signal from ectoderm, mediated by dpp, a *Drosophila* member of the bone morphogenetic protein family (Frasch, 1995). At this time, the cardiac and visceral progenitors sort out and the *tinman* expression domain takes on a segmented appearance, with transcripts restricted along the A/P body axis to reiterated cell clusters that correspond to the *tin*-dependent cardiac precursors (Park *et al.,* 1996). These are interdigitated with clusters of visceral progenitors which now no longer express *tin.* It is interesting that all mesodermal derivatives arising from the trunk region of the fly embryo appear to have a segmental origin, before merging later into a homogeneous structure (see Chapter 5). This mode of organogenesis appears in clear contrast to similar processes in vertebrates. After subdivision of cardiac and visceral precursors, *tin* expression remains restricted to the cardiac lineages (Fig. 2A).

Three roles for the tinman protein can currently be defined. The most obvious relates to its expression in the dorsal cells of the mesodermal mantle. In *tin* mutants, dorsal precursor cells are not specified and the cardiac, visceral, and dorsal body wall muscles do not form (Azpiazu and Frasch, 1993; Bodmer, 1993). Thus, *tin* has an essential role in mesodermal patterning. A second role for tin in cytodifferentiation is also evident. While there are a number of candidate target genes for tin in the heart (Bodmer, 1995), only one, *Dmef2,* has been shown to be regulated directly by the tin protein (Gajewski *et al.,* 1997). *Dmef2* is a MADS box transcription factor, essential for differentiation of all muscle types in the fly (Ranganayakulu *et al.,* 1995; see Chapter 8), and its dependence on *tin* for expression in heart shows a clear role for tin in differentiation of heart muscle cells. A third role in specification of some ventral body wall muscles and glial cells has also been identified (Azpiazu and Frasch, 1993; Gorczyka *et al.,* 1994).

VII. Vertebrate Relatives of *tinman* and Evolution of the *NK-2* Gene Family

The discovery and characterization of the *tinman* gene has guided us directly to a novel family of vertebrate *tin* relatives, also *NK-2* homeobox genes, including several expressed in the heart (Harvey, 1996). The first of the heart *NK-2* genes identified was *Nkx2-5/Csx* (Komuro and Izumo, 1993; Lints *et al.,* 1993), and thus far this gene is the only member expressed in cardiac progenitor cells of all vertebrate models examined (Harvey, 1996). Others are expressed in the hearts of some species but not in those of others. For example, *Nkx2-3* is expressed along with *Nkx2-5* in heart progenitors and the heart tube of frog embryos (Evans *et al.,* 1995; Tonissen *et al.,* 1994) but from a later time in heart development in chick embryos (Buchberger *et al.,* 1996) and not at all in the mouse heart (Pabst *et al.,* 1997). In the mouse, two genes, *Nkx2-5* and *Nkx2-6,* are expressed in heart—*Nkx2-5* from an early time in progenitor cells (Lints *et al.,* 1993) and *Nkx2-6* beginning in the caudal region of the late crescent and early heart tube, then later in the outflow tract (Biben *et al.,* 1998a). A recently described chick gene, *Nkx-2.8,* also shows regional expression (Boettger *et al.,* 1997; Brand *et al.,* 1997; Reecy *et al.,* 1997). While the species differences are interesting, and presumably relate to the different degrees of gene and whole genome duplication and divergence in different vertebrates (Garcia-Fernandez and Holland, 1994), the studies raise the issue of redundancy among *Nkx2* genes and whether the regional expression of some genes relates to heart patterning. So far, no striking functional differences have been mapped between cardiac Nkx2 proteins: Frog Nkx2.5 and Nkx2.3, for example, have the same high-affinity DNA-binding site [Mohun, 1997 No. 808] and both can induce large hearts when overexpressed in embryos (Cleaver *et al.,* 1996). However, it has been noted that Nkx2.5 is considerably more efficient than Nkx-2.3 in activating a synthetic promoter carrying a mutant version of its high-affinity binding site, one closely resembling that bound by Hox class homeodomain factors [Mohun, 1997 No. 808]. Thus, in addition to their common roles, individual regional functions for heart *Nkx2* genes may eventually be revealed.

The *NK-2* homeobox genes are ancient in origin. In addition to *tinman* and its vertebrate relatives, members of this class have been isolated from sponges, cnidarians, flatworms, nematodes, and leeches (Fig. 1B). Their presence in sponges indicates that they evolved coincident with or prior to the evolution of Metazoa (Jacobs *et al.,* 1998) and certainly prior to the appearance of hearts. At least two families (or clades) can be distin-

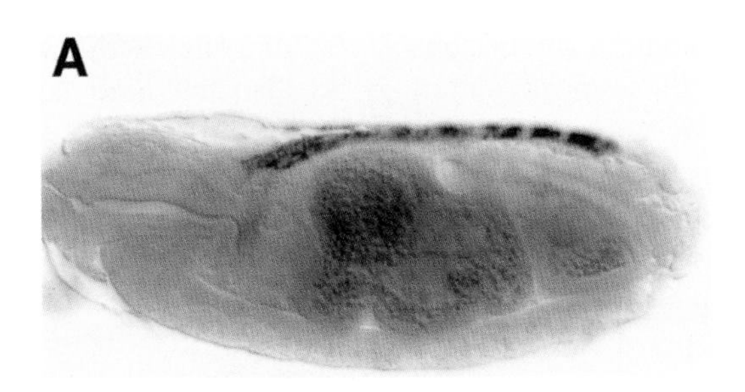

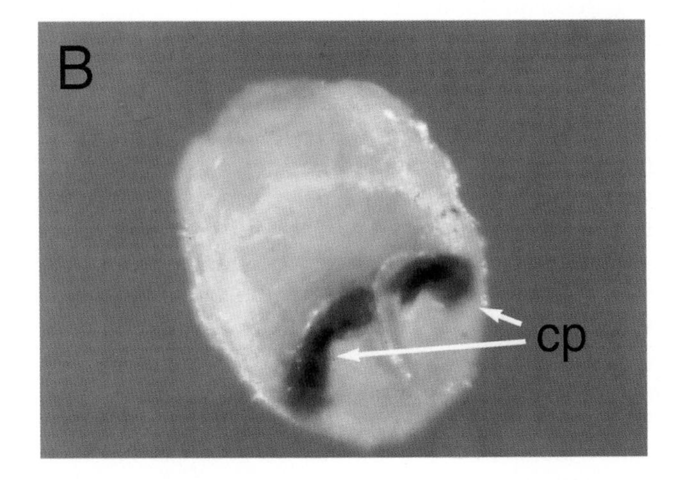

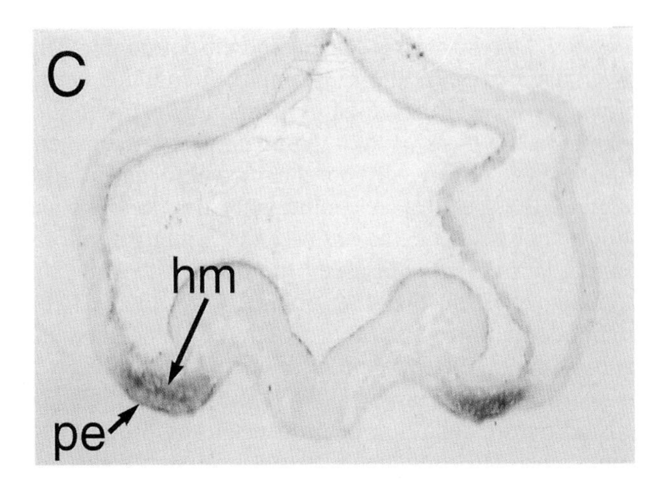

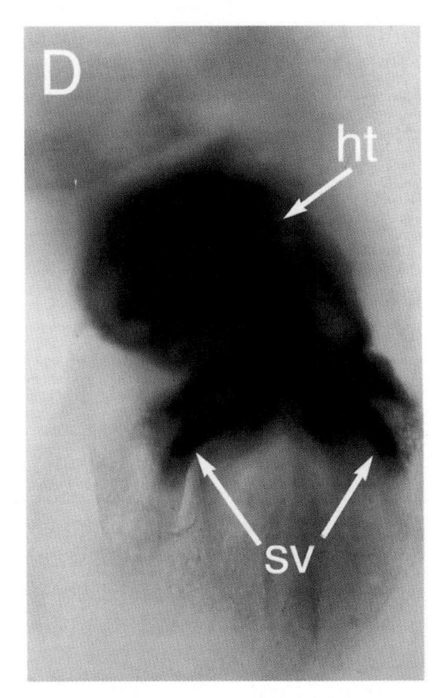

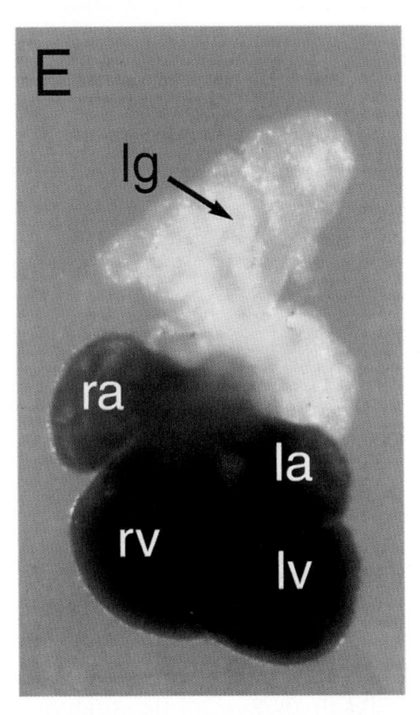

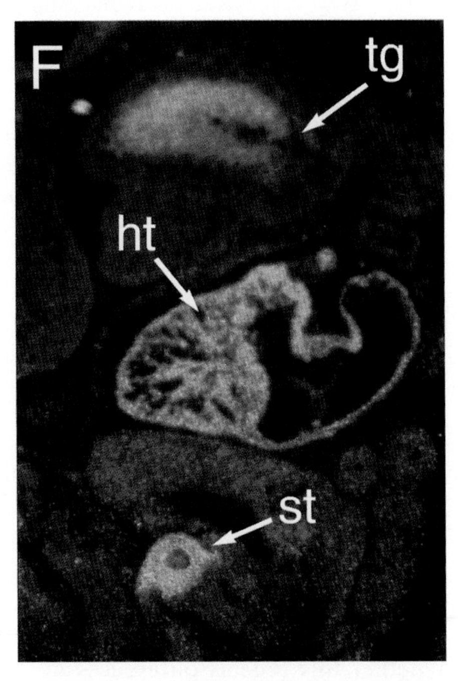

Figure 2 Expression of *tinman*-related genes during embryogenesis. (A) *Drosophila* embryo showing *tinman* expression in the dorsal vessel or heart. (B) E7.5 mouse embryo showing *Nkx2-5* expression in the paired cardiac progenitor cell pools (cp). (C) E8.0 mouse embryo in transverse section at the level of the cardiac crescent. Note *Nkx2-5* expression in the splanchnopleural heart mesoderm (hm) and within adjacent pharyngeal endoderm (pe). (D) E8.5 mouse embryo showing *Nkx2-5* transcription throughout the looping heart tube (ht), including sinus venosa (sv). (E) E12.5 mouse heart. *Nkx2-5* is expressed in all cardiac chambers (ra and la, right and left atria; rv and lv, right and left ventricles, respectively). The attached lungs (lg) do not express the gene. (F) Sagittal section through an E13.0 mouse embryo showing *Nkx2-5* expression in heart (ht), tongue (tg), and stomach (st). [Parts A, B, and D are reprinted with permission from Harvey (1996).]

guished on the basis of homeodomain comparisons (Fig. 1B): for convenient reference, those most related to the *Drosophila* proteins ventral nervous system defective (vnd; formally NK2) and bagpipe (bap; formally NK3). *vnd* is a gene required for nervous system development in the fly (Jiminez *et al.,* 1995), whereas *bap* is a potential tin target gene in visceral mesoderm, essential for formation of this lineage (Azpiazu and Frasch, 1993). Within the *vnd/NK2* group, there are two subgroups. One of these is composed entirely of those genes expressed in vertebrate hearts (the "cardiac" *NK-2* genes; Fig. 1B), whereas the other, containing *vnd /NK2* itself, appears more ancient and contains vertebrate and invertebrate members uniquely expressed in brain. On the basis of homeodomain comparisons, tinman itself cannot easily be classified into one or the other of these clades, as evolutionary tree analysis has shown (Harvey, 1996; Jacobs *et al.,* 1998). Furthermore, tinman lacks a key conserved motif, the NK-2-specific domain (NK2SD; Fig. 1A), found C terminal to the homeodomain in most other examples of the class, including all other *Drosophila* members and all vertebrate members (Harvey, 1996). The NK2SD clearly existed before radiation of protostomes and deuterostomes since *Drosophila* vnd carries a version highly related to the consensus vertebrate sequence (Harvey, 1996). However, NK-2 proteins from flatworms, which lie basal to the branch point, lack the NK2SD (Garcia-Fernandez *et al.,* 1991), and those from hydra and nematode appear to carry a rudimentary version (Grens *et al.,* 1996; Okkema and Fire, 1994). Thus, the evolutionary relationship between tin and its vertebrate relatives seems distant. Alternatively, tin structure may appear atavistic due to the recent loss of the NK2SD and further divergence. Evidence for this comes from functional experiments and from the fact that at least one mouse gene pair related to *tin* and *bap* are linked (Komuro and Izumo, 1993), as they are in the fly (Kim and Nirenberg, 1989).

VIII. Expression of *Nkx2-5*

Mouse *Nkx2-5* is first expressed in late gastrulation embryos at around embryonic day (E) 7.5 (Lints *et al.,* 1993). Unlike *tinman* in *Drosophila, Nkx2-5* is not expressed in early mesoderm at the onset of gastrulation. Whether this reflects differences in the timing of developmental events or differences in function is not totally clear. Expression begins in the classically described, paired cardiac progenitor cell pools (Rawles, 1943; Rosenquist, 1966) upon their arrival at the anterior (Fig. 2B) and continues as these cells form the cardiac crescent. *Mef2C* and *Mef2B,* other transcription factor genes acting in the cardiac regulatory program, are also activated in precardiac cells at the same time or slightly after induction of *Nkx2-5* (Edmondson *et al.,* 1994; Molkentin *et al.,* 1996). The zinc finger factor genes *GATA-4, -5,* and *-6,* however, appear to be expressed earlier but in broader domains that include primitive and definitive endoderm in the cardiogenic region and more caudal mesoderm (Morrisey *et al.,* 1997). Expression of *Nkx2-5* is maintained in the myocardial layer of the forming and looping heart tube (Fig. 2D) and during subsequent heart morphogenesis (Fig. 2E).

Expression of *Nkx2-5* is not cardiac specific. In late crescent stage embryos (E7.75–8), transcripts appear in future foregut endoderm where juxtaposed to the cardiogenic mesoderm (Fig. 2C). At later stages, expression is also seen in branchial arch ectoderm and mesoderm, developing spleen, lingual muscle, and in a small region of stomach mesoderm at its most caudal aspect (Lints *et al.,* 1993; Fig. 2F). Thus, the cardiogenic function of Nkx2-5 may depend on interaction with other factors. The pattern in foregut endoderm is worthy of comment. Expression in this tissue is the one unifying feature of the mammalian *Nkx2* genes (Harvey, 1996; Fig. 1B). All studied members of the "cardiac" subgroup as well as its apparently more ancient sister group are activated in distinct and overlapping patterns in the developing pharynx and/or foregut (Biben *et al.,* 1998a; Lazzaro *et al.,* 1991; Lints *et al.,* 1993; R. P. Harvey and C. Biben, unpublished data). The same may be true for *Nkx2* genes in other vertebrates (Lee and Breitbart, 1996; Reecy *et al.,* 1997). Most of the genes are first expressed in the floor of the developing foregut pocket, before becoming restricted along its A/P axis to regions associated with different organ anlage, for example, thyroid, esophagus, lung, and pancreas. The mediolateral axis also seems important, with expression of some genes restricted to the pharyngeal floor and others to the pharyngeal pouches (Biben *et al.,* 1998a). Reecy *et al.* (1998) have proposed an "Nkx code" for specification of pharyngeal organs (see Chapter 16). This may have some validity since knockout mice lacking *Nkx2-1/TTF-1* do not form thyroid or lung epithelium (Kimura *et al.,* 1996) and those lacking *Nkx2-5* show ectopic expression of the bHLH factor gene *eHand/Hand1* in a distinct caudal segment of the pharynx (Biben and Harvey, 1997). However, whether *Nkx* genes themselves create regionalities through their interactions, and whether a combinatorial NK code determines identity, is still unclear. *Nkx* genes, including the *Nkx6* group, also appear to mark specific A/P (and D/V) regions of the brain (Qiu *et al.,* 1998), and as noted previously *Nkx2* genes are regionally expressed in the hearts of chicken and mouse (Biben *et al.,* 1998a; Boettger *et al.,*

1997; Brand *et al.*, 1997; Buchberger *et al.*, 1996; Reecy *et al.*, 1997).

IX. Expression of *Nkx2* Genes and the Heart Morphogenetic Field

A morphogenetic field is the term used in the classical literature to describe an embryonic region that will give rise to the differentiated cell types of a particular organ or structure if explanted and cultured, and will participate in regulation if the zone definitively fated to form that organ or structure is removed or killed (see Chapter 3). Organ fields have been studied extensively in amphibia and the heart provides an impressive example of the concept (Jacobson and Sater, 1988). An interesting finding is that frog *Nkx2.5* is expressed in a region much broader than that occupied by the definitive heart progenitors, in a pattern resembling that of the morphogenetic field mapped in other amphibia (Evans *et al.*, 1995; Tonissen *et al.*, 1994). Thus, the regulatory ability of the field may be limited to the zone of *Nkx2.5* expression. On the other hand, loss of regulatory ability in the field occurs in advance of loss of *Nkx2.5* expression. In the fish, two *NK-2* genes, *nkx2.5* and *nkx2.7*, are expressed in different but overlapping domains in a region encompassing the heart field (Lee and Breitbart, 1996; Serbedzija *et al.*, 1998). *nkx2.5* expression occurs in those cells definitively fated to the heart but also extends more caudally beyond the heart field into cells that will never participate in heart development, even if definitive progenitors are killed by laser ablation (Serbedzija *et al.*, 1998). The notochord may provide inhibitory signals which keep cells in the caudal region from differentiating as heart (Schultheiss *et al.*, 1997; Serbedzija *et al.*, 1998; see Chapter 4). Cells that participate in regulation have been mapped to the cranial region of the heart field, and these cells express *nkx2.7* but not *nkx2.5*. The findings are consistent with the view that heart potency in the morphogenetic field requires the presence of a cardiac *Nkx2* gene, but that other positive and negative interactions determine the actual extent of the definitive progenitors and of the field itself. Ectopic expression of Nkx2.5 in frog and fish embryos, achieved by microinjection of *Nkx2.5* mRNA into fertilized eggs, leads to larger hearts and myocardial hyperplasia, without ectopic activation of the myogenic program (Chen and Fishman, 1996; Cleaver *et al.*, 1996; Fu and Izumo, 1995). While the mechanism is not known, one possibility is that the boundaries of the heart field, and hence the extent of the interactions within it, are expanded by enforced *Nkx2.5* expression. Induction of myocardial hyperplasia is an interesting and novel activity for Nkx2.5.

X. Known Target Genes

NK-2 genes encode transcriptional activators. The high-affinity DNA-binding site has been determined for several NK-2 proteins (Chen and Schwartz, 1995; Mohun, 1997; Tsao *et al.*, 1994) and the common consensus sequence (T[C/T]AAGTG) corresponds well with known NK-2 protein binding sites within the promoters and/or enhancers of target genes (Durocher *et al.*, 1996; Gajewski *et al.*, 1997; Okkema and Fire, 1994; Ray *et al.*, 1996). Lower affinity sites have also been mapped (Chen *et al.*, 1996; Durocher *et al.*, 1996) and activation from a variant Hox factor consensus site (TAAT) demonstrated (Chen and Schwartz, 1996; Mohun, 1997). However, few genuine target genes have been identified. As mentioned previously, tinman directly regulates a cardiac enhancer within the *Drosophila mef2* gene (Gajewski *et al.*, 1997). Here, two high-affinity tin binding sites are both essential for activation of *Dmef2* reporter genes *in vitro*. This same enhancer is used in combination with additional flanking sequences for activation of *Dmef2* in visceral and some ventral body wall muscle founder cells (Ranganayakulu *et al.*, 1998; see Chapter 8). Interestingly, Nkx2-5 cannot activate this enhancer in embryos, pointing to functional differences between the two proteins. tinman also regulates expression of its own gene (Lee *et al.*, 1997).

Nkx2-5 has been implicated in direct activation of two downstream target genes in the heart—those encoding atrial natriuretic factor (ANF), a vasoactive hormone, and α-cardiac actin (α-CA), a myofilament protein (Chen *et al.*, 1996; Chen and Schwartz, 1996; Durocher *et al.*, 1997). For ANF, an "NK"-binding element containing high- and low-affinity sites lies close to a GATA factor consensus element (Durocher *et al.*, 1996). Both are required for efficient transactivation *in vitro*. In addition, Nkx2-5 and GATA-4 appear to physically interact and can synergistically activate the ANF promoter (Durocher *et al.*, 1997). For α-CA, maximum activity *in vitro* requires multiple serum response elements (SREs) that bind serum response factor (SRF) (Chen *et al.*, 1996; Chen and Schwartz, 1996). Nkx2-5 can bind weakly to these sites as well as facilitate the independent binding of SRF. Nkx2-5 and SRF synergize strongly on promoters containing multiple SREs (see Chapter 16). The results are consistent with the prior notion that SRF-related proteins from yeast to mammals serve as platforms for recruitment of other proteins, including homeodomain and ets-domain factors, into ternary complexes (Grueneberg *et al.*, 1992; Treisman, 1994). Like tin, Nkx2-5 also appears to regulate expression of its own gene (Oka *et al.*, 1997).

XI. Are *tinman* and Vertebrate *Nkx2* Genes Functionally Homologous?

The finding that vertebrate cardiac development utilizes *tinman*-related genes is unlikely to be trivial. However, the functional relationship between *tin* and the vertebrate *NK-2* genes is not necessarily obvious. As described previously, evolutionary tree analysis places tin in a position remote from its vertebrate relatives. Furthermore, while *tin* is essential for formation of the cardiac lineages in flies, *Nkx2-5* is not required for formation of cardiac muscle in the murine system. This issue is an important one since genetic analysis of tin structure and function could potentially teach us much about the mechanism of cardiogenesis in vertebrates. To gain perspective on this issue, we must first ask whether the common ancestor of arthropods and vertebrates had a heart-like organ or a recognizable structure from which the hearts of both lines evolved as independent adaptations (Harvey, 1996; Jacobs *et al.,* 1998). In fact, comparative anatomists have never considered the structurally diverse hearts of the protostome invertebrates to be homologous to those of vertebrates, largely because there is no evidence that the common protostome/deuterostome ancestor possessed a heart at all (Harvey, 1996). However, the central position occupied by *NK-2* genes in the cardiac programs of flies and vertebrates allows us to propose that a common structure for heart formation did exist, perhaps in the form of a muscular vessel (Harvey, 1996), and that this structure was dependent for its formation or maturation on an *NK-2* gene.

The functional relationship between *tinman* and vertebrate *NK-2* genes has recently been examined directly in transspecific assays. Using a GAL4-dependent conditional transgenic system in *Drosophila* (Ranganayakulu *et al.,* 1998), or one dependent on heat shock activation (Bodmer *et al.,* 1997), vertebrate *NK-2* genes have been assessed for their ability to rescue the *tin* mutant phenotype. As previously discussed, *tin* is essential for specification of both cardiac and visceral muscles, and various markers have been used to score for rescue of these muscle types in transgenic flies, including fasciclin III for visceral muscle, Dmef2 for visceral and cardiac muscle, and eve and zfh-1 for pericardial cells (Figs. 4A–D). Using the conditional system (Ranganayakulu *et al.,* 1998), tin and Nkx2-5 can be stably expressed in early mesodermal progenitors, where tin is first active (Gorczyka *et al.,* 1994; Park *et al.,* 1996), and then in all muscle cell types (Fig. 3). The results show that while tin itself rescues both cardiac and visceral markers, Nkx2-5 rescues only visceral muscle (Fig. 4). Interestingly, the ability of vertebrate *NK-2* genes to rescue visceral muscle appears specific to the "cardiac" group (Fig. 1B) since members *Nkx2-5* and *Nkx2-3* can rescue (Bodmer *et al.,* 1997; Ranganayakulu *et al.,* 1998), whereas noncardiac members *Nkx2-1/TTF* and *bap* cannot (Ranganayakulu *et al.,* 1998). In this regard, it is interesting that *Nkx2-5* can substitute for *ceh-22,* a noncardiac *NK-2* gene involved in specification of a pharyngeal muscle type in the nematode (Haun *et al.,* 1998).

The experiments demonstrate a close functional kinship between *tinman* and the vertebrate cardiac *NK-2* genes. Despite the atavistic structure of tinman, it shares functional properties with its vertebrate relatives in the stringent fly rescue assay. However, the experiments also suggest that the genetic mechanisms of cardiogenesis in flies and vertebrates have diverged. The dominant cardiogenic activity of tin has been mapped to within 42 amino acids of its N terminus, a region that has no apparent counterpart in Nkx2-5 (Ranganayakulu *et al.,* 1998). When transferred to the N terminus of Nkx2-5, however, this region confers upon it full cardiogenic activity in the rescue assay. The experiments suggest that an ancestral structure, perhaps a muscular vessel, gave rise to the hearts of arthropods and vertebrates

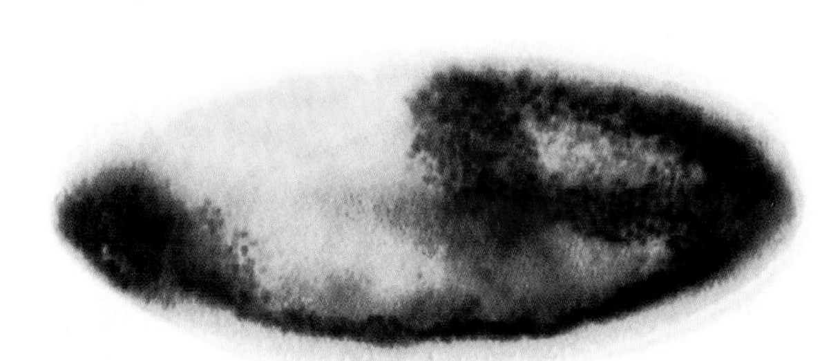

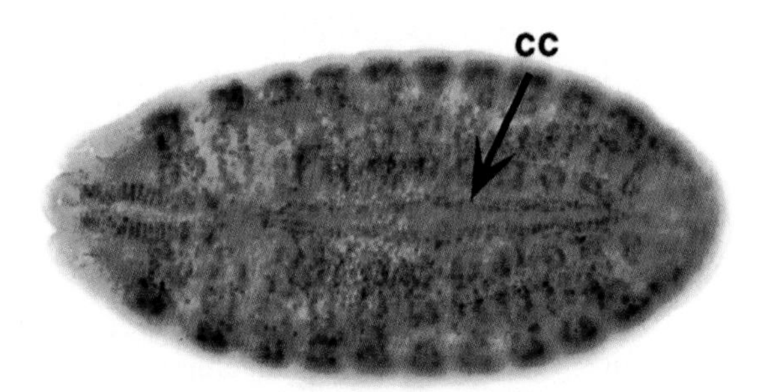

Figure 3 Nkx2-5 protein expression in transgenic *Drosophila* embryos. The conditional GAL4-dependent *Nkx2-5* transgene is driven by the twist and Dmef2 promoters (Ranganayakulu *et al.,* 1998, © Company of Biologists, Ltd.). Expression is seen throughout the early mesoderm at stage 9 (left) and then in all muscle cell lineages at stage 16 (right), including cardial cells of the heart (cc). Embryos were stained for Nkx2-5 using a polyclonal anti-Nkx2-5 antibody.

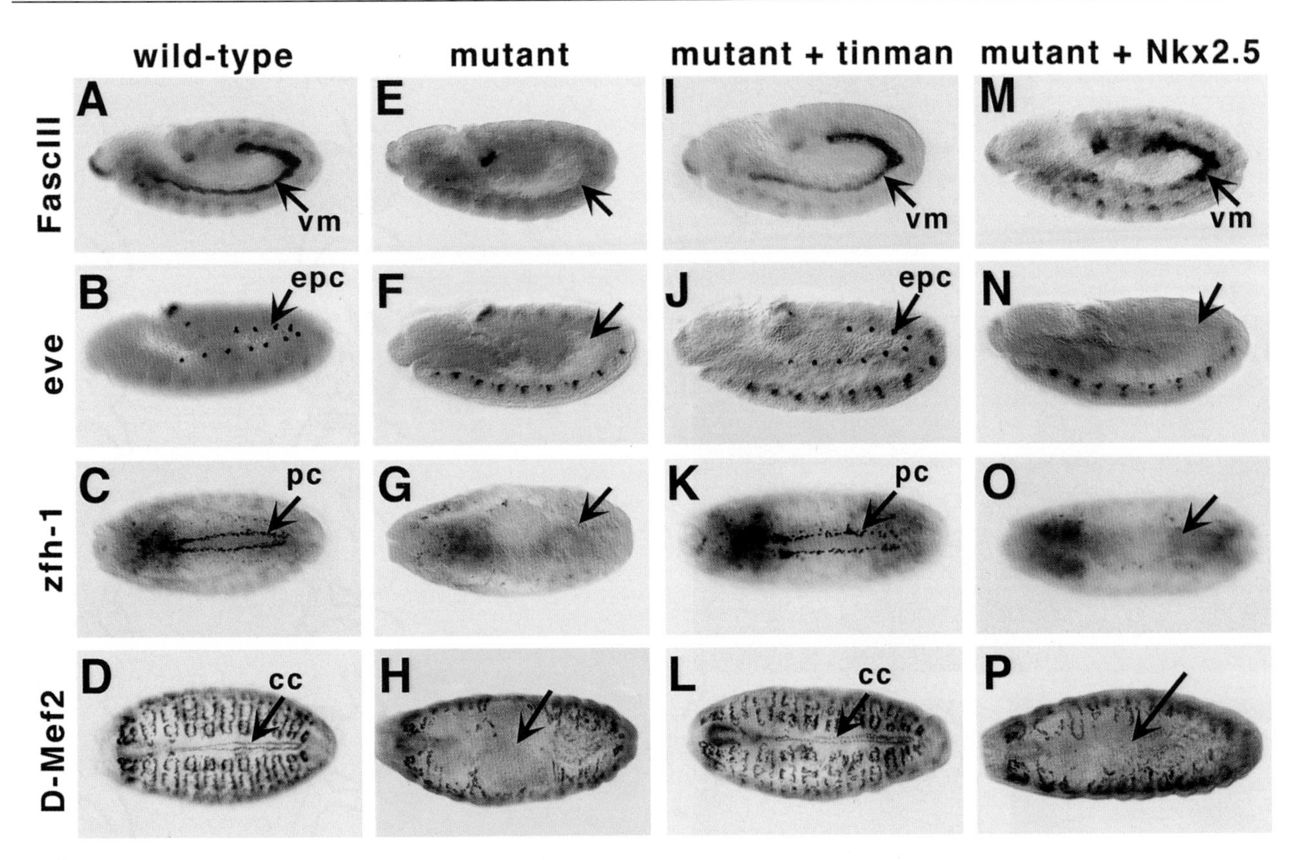

Figure 4 *Nkx2-5* can rescue visceral but not cardiac muscle in *tinman* mutants. A–D, Wild-type embryos; E–H, *tinman* mutant embryos; I–L, *tinman* mutant embryos expressing a *tinman* transgene; M–P, *tinman* mutant embryos expressing an *Nkx2-5* transgene. *Fascilin III* stains visceral muscle, *eve* and *zfh-1* stain pericardial cells of the heart, and *Mef-2* stains all muscle lineages. For all embryos, anterior is to the left. A,B,E,F,I,J,M,N, lateral views C,D,G,H,K,L,O,P, dorsal views. Note that *tinman* rescues visceral and cardiac markers in *tinman* mutant embryos, whereas *Nkx2-5* rescues only visceral markers. vm, visceral mesoderm; epc, eve positive cells; pc, pericardial cells; cc, cardial cells. (Ranganayakulu *et al.,* 1998, © Company of Biologists, Ltd.)

via independent adaptations to an underlying *NK-2*-dependent genetic pathway. The results may also have general relevance to a proposal that the muscular vessels of oligochaete worms, which have often developed further into heart-like organs, are of visceral myogenic origin (Martin, 1980; Stephenson, 1930). While our understanding of cardiogenesis in vertebrates may continue to benefit from genetic analysis of heart formation in flies, it seems that this will relate only to the nature of their ancient common visceral-like program.

XII. *Nkx2-5* Knockout Phenotype

A mutation in the *Nkx2-5* gene has been generated by gene targeting (Lyons *et al.,* 1995). The knockout allele inserts a neomycin drug-resistance cassette into the conserved and essential helix 3 of the homeodomain. Homozygous embryos show growth retardation beginning around E8.5–9 and they die over the next 1 or 2 days, apparently from cardiac insufficiency. A linear heart tube forms and begins to beat but fails to undergo correct looping morphogenesis (Figs. 5A–C). With subsequent development, the tube remains in a largely linear conformation, with dilatation and edema occurring as embryos enter their demise. Dissection of mutant hearts clearly shows the linear relationship between the body of the heart and the outflow tract (Fig. 5B), although some asymmetry in the caudal region is evident (Fig. 5C). Mutant hearts have an endocardial and myocardial layer, with ultrastructural details typical for that stage of development (Lyons *et al.,* 1995). Other features, however, are highly abnormal (Figs. 5D and 5E). Chamber identity is indeterminate, although marker expression suggests some atrial and ventricular character can be developed (Lyons *et al.,* 1995). A sulcus forms in the region of the atrioventricular canal (AVC; Fig. 5B), but further development of this structure is completely blocked and no endocardial cushions form (Figs. 5D and 5E). In the open atrioventricular chamber, trabeculation is poor or absent (Figs. 5D and 5E).

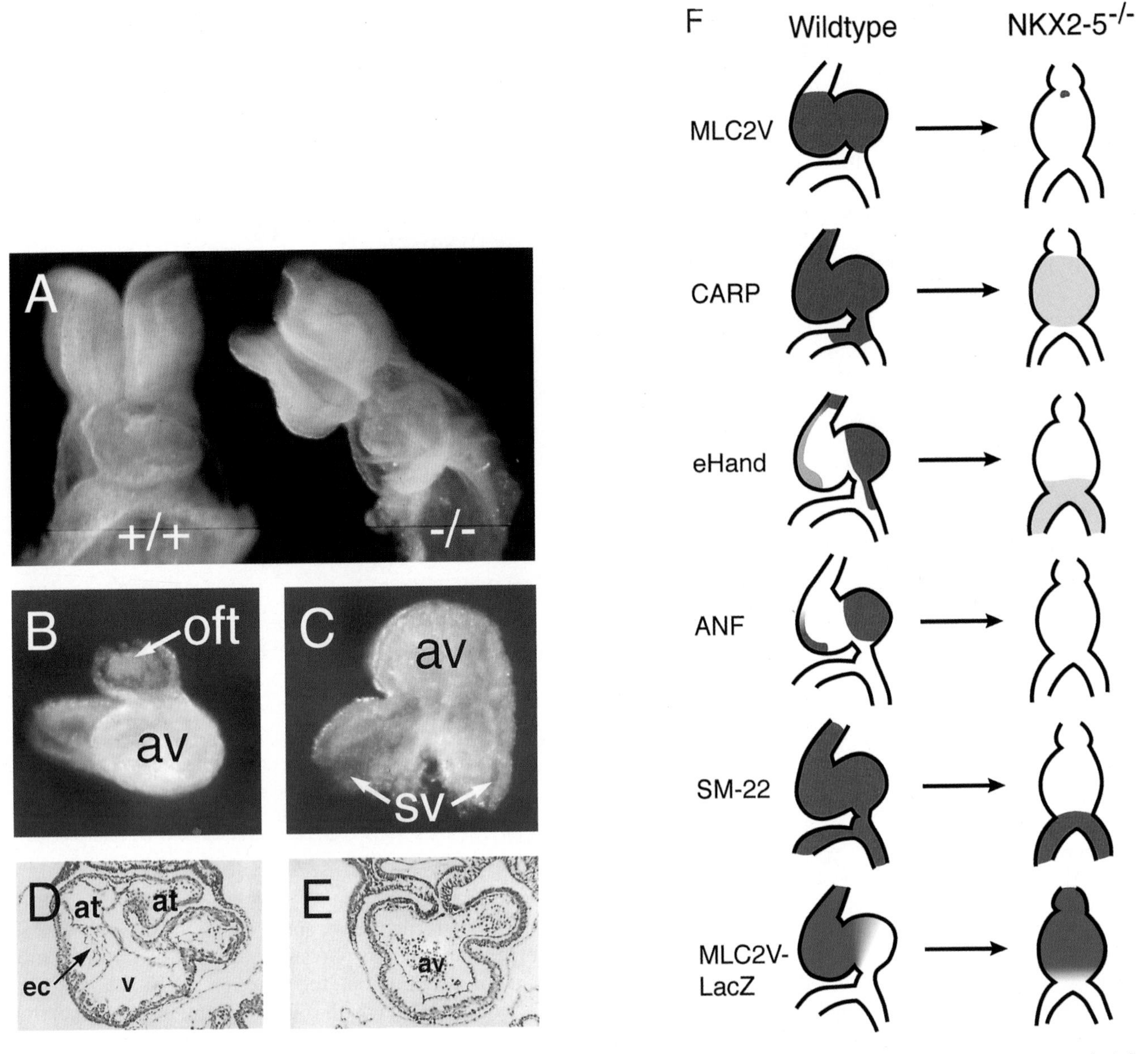

Figure 5 Targeted inactivation of *Nkx2-5* leads to cardiac dysmorphogenesis at E8.5. (A) E8.5 embryos wild type (+/+) or mutant (−/−) for a targeted interruption in the *Nkx2-5* gene. Note the looped heart tube in normal embryos, whereas at the same stage, mutant hearts retain a linear morphology. (B) Dissected heart tube of an E8.5 *Nkx2-5*$^{-/-}$ embryo. Ventral view showing the linear relationship between the open atrioventricular chamber (av) and the outflow tract (oft). (C) Caudal view of the same heart as in B. The sinus venosa (sv) show L–R asymmetry suggesting that *Nkx2-5*$^{-/-}$ hearts can correctly interpret early laterality signals. (D,E) Sections through a heart of E9.5 wild-type (D) and *Nkx2-5* mutant (E) embryos. Note developing endocardial cushions (ec) and trabeculae in the wild type but not in the mutant. at, atrium; av, atrioventricular chamber; v, ventricle. (F) Summary of the expression patterns of genes misexpressed in *Nkx2-5*$^{-/-}$ hearts at E8.5–9.0 (see text). The expression pattern of the gene or transgene is depicted on the looped normal heart (left) and on the mutant heart (right). [Parts A, D, and E reprinted from Lyons *et al.* (1995), and Part F from Biben *et al.* (1997), with permission of Cold Spring Harbor Laboratory Press.]

Although some of the defects may be secondary, the overall impression is that *Nkx2-5* knockout hearts have deranged A/P identity and abnormal chamber formation. The lack of a distinct AVC suggests that the primitive segments of the early heart tube (Eisenberg and Markwald, 1995; Moorman *et al.*, 1998) have not been correctly established.

XIII. Marker Analysis

In situ hybridization analysis has identified a number of genes downregulated in *Nkx2-5* mutant hearts. These include genes encoding the transcription factors *CARP* and *eHand/Hand1,* as well as *myosin light chain 2V* (*MLC2V*), *ANF,* and the *calponin*-related gene *SM-22*

(Biben and Harvey, 1997; Biben *et al.,* 1997; Lyons *et al.,* 1995; Zou *et al.,* 1997). The expression patterns of these genes is affected in a variety of ways which suggest defects not only in myogenesis but also in chamber patterning (Fig. 5F). For example, expression of *MLC2V* and *ANF* is all but obliterated. *CARP* is severely downregulated globally but remains on at a low level in the open atrioventricular chamber. *eHand* is also severely downregulated, remaining on very weakly in the ventrocaudal region. *SM-22* expression is eliminated from the atrioventricular chamber but remains at robust levels in sinus venosa. Thus, Nkx2-5 appears to be differentially required for expression of genes in different heart segments and no region of the heart tube is spared some form of alteration in gene expression. The data support our gross morphological assessment that segmental patterning in the heart is deranged.

While the underlying molecular basis of these defects is currently difficult to assess due to our general ignorance of patterning mechanisms acting in heart development, some progress has been made in understanding the *Nkx2-5* mutant phenotype, and of heart patterning generally, by examination of the expression profiles of the Hand genes (Biben and Harvey, 1997; Srivastava *et al.,* 1997; Thomas *et al.,* 1998).

XIV. *Hand* Gene Expression in Normal and *Nkx2-5*$^{-/-}$ Hearts

eHand and dHand (Hand1 and Hand2; Firulli *et al.,* 1998; Riley *et al.,* 1998) are bHLH transcription factors expressed in the early heart and in other embryonic tissues (Cross *et al.,* 1995; Cserjesi *et al.,* 1995; Hollenberg *et al.,* 1995; Srivastava *et al.,* 1995; see Chapter 9). In the chick, both *Hand* genes appear to be expressed throughout the heart, where they are required for morphogenesis beyond early looping stages (Srivastava *et al.,* 1995). In the mouse, however, the patterns are regional and highly dynamic (Biben and Harvey, 1997; Srivastava *et al.,* 1997; Thomas *et al.,* 1998). The *eHand* pattern is restricted along both the A/P and D/V axes of the heart tube. At linear and early looping stages, expression occurs throughout the sinus venosa and atrial region as well as in most of the presumptive left ventricle (Fig. 6A). However, an important finding is that only the ventral surface of the myocardium expresses the gene, establishing the presence of D/V patterning in the early heart tube (Fig. 6B). By E10.5, expression in sinus venosa and atrium has withdrawn, and the pattern is largely left ventricle specific, although by E11.5 some expression in right ventricle is seen (Figs. 6C and 6D). In the left ventricle, expression is set back from the region that will form the interventricular muscular septum (Figs. 6C and 6D), and when the septum forms it shows no expression (Firulli *et al.,* 1998).

The *dHand/Hand2* expression pattern in the mouse heart differs considerably from that of *eHand/Hand1* (Biben and Harvey, 1997; Srivastava *et al.,* 1997). Transcripts are present across the whole myocardium at the linear tube stage, before becoming restricted to the outer curvature of the right ventricle. Thus, in the well-looped heart, the *Hand* genes have developed a complementarity in their expression patterns, with *eHand* predominant in the left ventricle and *dHand* predominantly in the right (Srivastava *et al.,* 1997).

The working myocardium of the ventricles and atria develop as regional specializations of the primitive myocardium of the early heart tube (see Chapter 12). In a particular temporal sequence, these regions expand and bulge outwards, filling out the shape of the looping heart. In any circumference of the heart tube, one arc will mature into working myocardium, whereas the complementary arc will form part of the "inner curvature," which undergoes extensive remodeling in the end stages of looping (see Chapter 10). The Hand expression patterns are consistent with a role for these genes in defining the character of the working myocardium in the left and right ventricles, an idea that is strengthened by the apparent loss of right ventricle in the hearts of *dHand/Hand2* knockout mice (Srivastava *et al.,* 1997). Sections show that *eHand* is expressed in a ventral arc of the caudal region of the heart tube (Fig. 6B) and then in the outer curvature of the left ventricle (Figs. 6C and 6D; Biben *et al.,* 1997). We believe, to a first approximation, that the ventral arc of *eHand* expression corresponds to the future working myocardium of the looped heart, the dorsal arc forming the inner curvature. The transition between *eHand*-expressing cells in the ventral arc to their eventual location in the outer curvature of the left ventricle may involve torsion of the caudal heart tube during looping and expansion of the whole chamber, perhaps by proliferation. The model implies that the A/V and D/V regionalities set up in the linear heart tube, as revealed by *eHand* expression, provide a major component of the information required to complete the morphogenetic events associated with cardiac looping, supplementing that from the L/R axial system (Harvey, 1998). Conflicting perceptions on whether the *eHand* expression pattern is influenced by the L/R asymmetry pathway (Biben and Harvey, 1997; Sparrow *et al.,* 1997; Thomas *et al.,* 1998) remain to be resolved.

We have found that expression of *eHand/Hand* is severely downregulated in *Nkx2-5* mutant hearts while being unaffected in tissues that do not normally express *Nkx2-5* (Fig. 6E; Biben *et al.,* 1997). As revealed recently, the morphological defects in *Nkx2-5*$^{-/-}$ hearts resemble those seen in the hearts of *eHand/Hand1* mu-

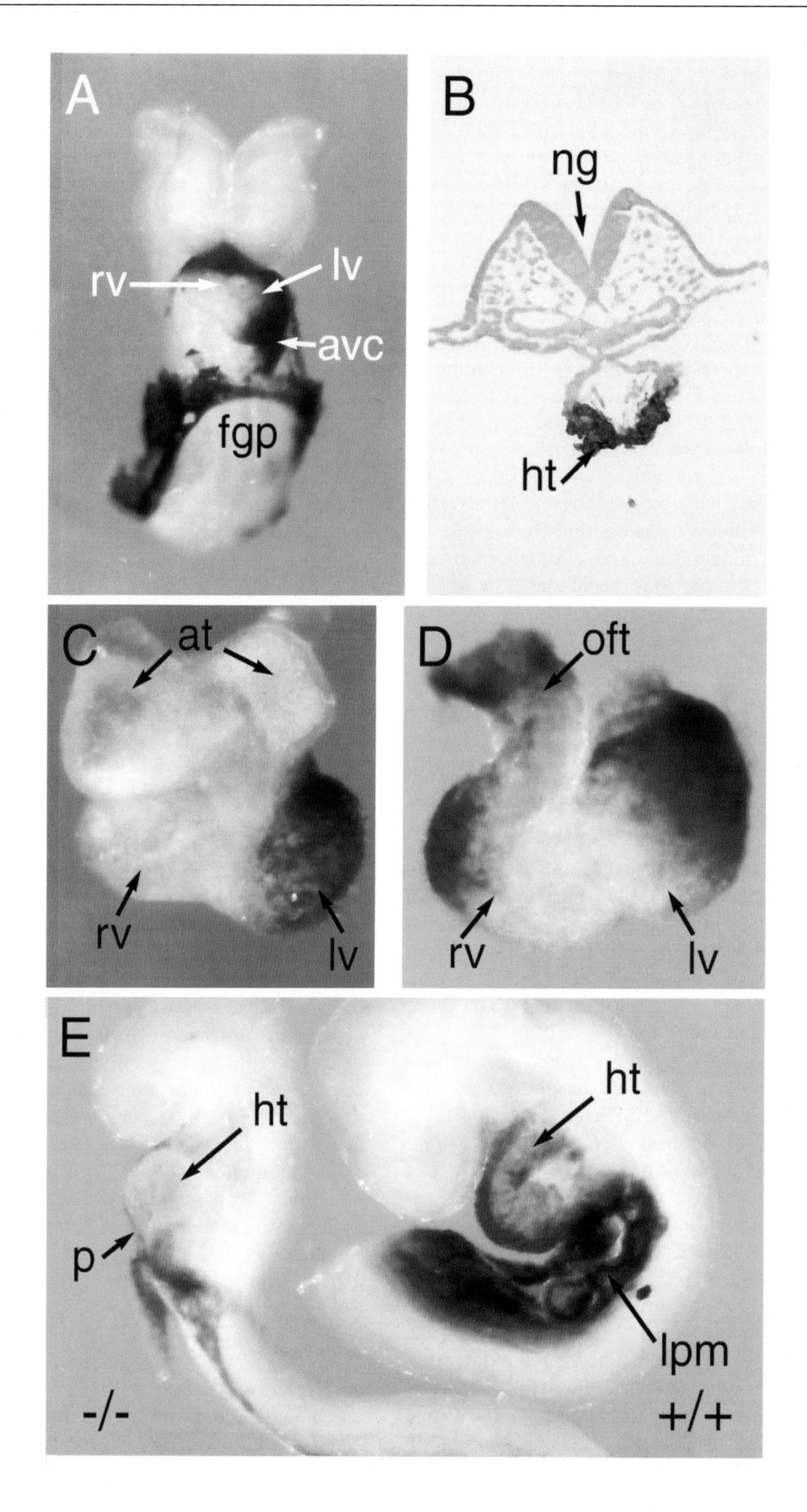
A
rv
lv
avc
fgp
B
ng
ht
C
at
rv
lv
D
oft
rv
lv
E
ht
p
ht
lpm
-/-
+/+

tants (Firulli *et al.*, 1998; Riley *et al.*, 1998). One hypothesis to account for the common defect is that left ventricular cells are severely reduced in both Nkx2-5 and *eHand* mutant hearts due to loss of *eHand* expression. Evidence in favor of this comes from analysis of the expression of a right ventricle-specific transgene marker in the *Nkx2-5*$^{-/-}$ background. A transgenic *LacZ* reporter driven by the proximal region of the *MLC2V* promoter is expressed predominantly in the right ventricle and outflow tract of the forming heart tube (Ross *et al.*, 1996). Interestingly, this transgene is expressed robustly in the *Nkx2-5* mutant background, even though the endogenous *MLC2V* gene is not. Nevertheless, the *LacZ* pattern is revealing in that it almost completely encompasses the open atrioventricular chamber seen in the mutant hearts, suggesting that it has right ventricular character and that the left ventricle is reduced or absent.

These studies identify a key *Nkx2-5*-dependent genetic circuit acting in early morphogenesis of the mammalian heart. The *Hand* expression patterns have provided us a direct visualization of how unique left and right ventricular identities evolve from complex beginnings in the linear heart tube, sequences that must betray something of earlier events. Of some significance is the D/V patterning now evident in the early heart and its possible direct connection to specification of the inner and outer curvatures. This, in turn, links D/V pattern with both formation of a working myocardium and remodeling of the primitive tube into a compact organ. That *Hand* genes appear essential for chamber formation indicates that they are key determinants of pattern in the developing heart. The transcription factor gene, *Msg1,* has an expression pattern in the caudal region of the early mouse heart almost identical to that of *eHand,* indicating another potential link in this pathway (Dunwoodie *et al.,* 1998). How the expression domains of these factors are established and the nature of their specific functions will be intriguing to dissect.

XV. *Nkx2-5* and Human Cardiac Defects

The quest to identify mutations responsibe for familial human diseases is gaining pace. Among the most common cardiac defects are those affecting septation of the atria and/or ventricles. In a rare form of hand–heart anomaly, Holt-Oram syndrome, affected individuals show atrial and/or ventricular septal defects and conduction system abnormalities, along with a range of preaxial radial ray limb deformities. Single nucleotide changes in the T-box transcription factor gene, *TBX5,* have now been shown to underlie both familial and sporadic cases (Basson *et al.,* 1997; Li *et al.,* 1997).

The causal mutations underlying a nonsyndromic, autosomal dominant condition manifesting atrial septal defects (secundum type) with atrioventricular conduction delay, have also been investigated. This condition is associated with a high risk of sudden death and may require surgical correction of atrial defects and/or insertion of a pacemaker. In four families, the mutation mapped to the *NKX2.5* gene (Schott *et al.,* 1998). Of three different mutations identified, one truncates the NKX2.5 protein midway through the homeodomain, thus eliminating helix 3, known to be essential for DNA binding. Another substitutes one amino acid at the N-terminal side of helix 3, and the third truncates the protein immediately after the homeodomain. The two homeodomain mutations may be null, in which case the cardiac defects would appear to derive from a haploinsufficiency for the NKX2.5 protein. Although atrial septal defects have not been recognized in the mouse model, these, like most in humans, may be benign. These human studies add an exciting dimension to investigations of the mouse gene, and reveal a heretofore unrecognized function for *Nkx2-5* in atrial septation and conduction system development.

Acknowledgments

RPH is indebted to the Finley family for their generosity and support. We also thank Veronica Hammond for assistance with the manuscript. This work was supported by Grant G96M 4675 from the Heart Foundation of Australia and Grant RG-308/95 from the Human Frontier Science Program.

References

Andrew, D. J., and Scott, M. P. (1992). Downstream of the homeotic genes. *The New Biologist* **4,** 5–15.

Azpiazu, N., and Frasch, M. (1993). *tinman* and *bagpipe:* Two homeo

Figure 6 *eHand* expression in normal and *Nkx2-5* mutant hearts, as studied by whole mount *in situ* hybridization. (A) E8.3 mouse embryo at an early looping stage showing *eHand* expression restricted to the caudal region of the heart. avc, atrioventricular canal; fgp, foregut pocket; lv, left ventricle; oft, outflow tract; rv, right ventricle. (B) Transverse section of a hybridized E8.3 embryo showing the early looping heart tube. *eHand* expression is localized to the ventral aspect of the heart tube (ht). ng, neural groove. (C) E10.5 heart stained for *eHand.* Transcripts are restricted to the left ventricle. at, atria. (D) E11.5 heart stained for *eHand.* Note expression in the outer curvature of the right ventricle and outflow tract, in addition to left ventricle. (E) *eHand* expression in E10.0 embryos wild type (+/+) or mutant (−/−) for targeted inactivation of the *Nkx2-5* gene. *eHand* is not expressed in the mutant heart (ht) but transcripts are still present in pericardium (p) and lateral plate mesoderm (lpm), in which *Nkx2-5* is not normally expressed. [Reprinted with permission from Biben and Harvey (1997), © 1997 Cold Spring Harbor Laboratory Press.]

box genes that determine cell fates in the dorsal mesoderm of *Drosophila. Genes Dev.* **7,** 1325–1340.

Basson, C. T., Bachinsky, D. R., Lin, R. C., Levi, T., Elkins, J. A., Soults, J., Grayzel, D., Kroumpouzou, E., Traill, T. A., Leblanc-Straceski, J., Renault, B., Kucherlapati, R., Seidman, J. G., and Seidman, C. E. (1997). Mutations in human cause limb and cardiac malformation in Holt-Oram syndrome. *Nature Genet.* **15,** 30–35.

Belo, J. A., Bouwmeester, T., Leyns, L., Kertesz, N., Gallo, M., Follettie, M., and De Robertis, E. M. (1997). Cerberus like is a secreted factor with neuralizing activity expressed in the anterior primitive endoderm of the mouse gastrula. *Mech. of Dev.* 68, 45-57.

Biben, C., and Harvey, R. P. (1997). Homeodomain factor Nkx2-5 controls left–right asymmetric expression of bHLH *eHand* during murine heart development. *Genes Dev.* **11,** 1357–1369.

Biben, C., Palmer, D. A., Elliot, D. A., and Harvey, R. P. (1997). Homeobox genes and heart development. In *Cold Spring Harbor Symposia on Quantitative Biology,* Vol. LXIII, pp. 395–403. Cold Spring Harbor Laboratory Press, Cold Spring Harbor, NY.

Biben, C., Hatzistavrou, T., and Harvey, R. P. (1998a). Expression of *NK-2* class homeobox gene *Nkx2-6* in foregut endoderm and heart. *Mech. Dev.* **73,** 125–127.

Biben, C., Stanley, E., Fabri, L., Kotecha, S., Rhinn, M., Drinkwater, C., Lah, M., Wang, C.-C., Nash, A., Hilton, D., Ang, S.-L., Mohun, T., and Harvey, R. P. (1998b). Murine cerberus homologue mCer-1: A candidate anterior patterning molecule. *Dev. Biol.* **194,** 135–151.

Bodmer, R. (1993). The gene *tinman* is required for specification of the heart and visceral muscles in *Drosophila. Development* 118, 719–729.

Bodmer, R. (1995). Heart development in Drosophila and its relationship to vertebrates. *Trends Cardiovasc. Med.* **5,** 21–28.

Bodmer, R., Jan, L. Y., and Jan, Y. N. (1990). A new homeobox-containing gene, *msh-2,* is transiently expressed early during mesoderm formation in *Drosophila. Development* **110,** 661–669.

Bodmer, R., Golden, K., Lockwood, W. B., Ocorr, K. A., Park, M., Su, M.-T., and Venkatesh, T. V. (1997). Heart development in *Drosophila.* In *Advances in Developmental Biochemistry* (P. Wassarman, Ed.), Vol. 5, pp. 201–236. JAI, Greenwich, CT.

Boettger, T., Stefan, S., and Kessel, M. (1997). The chicken *NKX2-8* gene: A novel member of the NK-2 family. *Dev. Genes Evol.* **207,** 65–70.

Bouwmeester, T., and Leyns, L. (1997). Vertebrate head induction by anterior primitive endoderm. *BioEssays* **19,** 855–862.

Brand, T., Andree, B., Schneider, A., and Arnold, H.-H. (1997). Chicken NKx2.8, a novel homeobox gene expressed during early heart and foregut development. *Mech. of Dev.* **64,** 53–59.

Buchberger, A., Pabst, O., Brand, T., Seidl, K., and Arnold, H.-H. (1996). Chick NKx-2.3 represents a novel family member of vertebrate homologues to Drosophila homeobox gene tinman: Differential expression of cNKx-2.3 and cNKx-2.5 during heart and gut development. *Mech. Dev.* **56,** 151–163.

Buckingham, M. (1997). Fate-mapping pre-cardiac cells in the developing mouse. In *Genetic Control of Heart Development* (E. N. Olson, R. P. Harvey, R. A. Schulz, and J. S. Altman, Eds.). Human Frontier Science Program, Strasbourg.

Burglin, T. R. (1993). A comprehensive classification of homeobox genes. In *Guidebook to the Homeobox Genes* (D. Duboule, Ed.), pp. 25–71. Oxford Univ. Press, Oxford, UK.

Carroll, S. B. (1995). Homeotic genes and the evolution of arthropods and chordates. *Nature* **376,** 479–485.

Chen, C. Y., and Schwartz, R. J. (1995). Identification of novel DNA binding targets and regulatory domains of a murine Tinman homeodomain factor, *Nkx-2.5. J. of Biol. Chem.* **270,** 15628–15633.

Chen, C. Y., and Schwartz, R. J. (1996). Recruitment of the Tinman Homologue Nkx-2.5 by serum response factor activates cardiac α-actin gene transcription. *Mol. Cell. Biol.* **16,** 6372–6384.

Chen, C. Y., Croissant, J., Majesky, M., Topouzis, S., McQuinn, T., Frankovsky, M. J., and Schwartz, R. J. (1996). Activation of the cardiac alpha-actin promoter depends upon serum response factor, Tinman homologue, Nkx-2.5, and intact serum response elements. *Dev. Genet.* **19,** 119–130.

Chen, J.-N., and Fishman, M. C. (1996). Zebrafish *tinman* homolog demarcates the heart field and initiates myocardial differentiation. *Development* **122,** 3809–3816.

Chen, J.-N., van Eeden, J. M., Warren, K. S., Chin, A., Nusslein-Volhard, C., Haffter, P., and Fishman, M. C. (1997). Left–right pattern of cardiac *BMP4* may drive asymmetry of the heart in zebrafish. *Development* **124,** 4373–4382.

Cleaver, O. B., Patterson, K. D., and Krieg, P. A. (1996). Overexpression of the tinman-related genes *XNkx-2.5* and *XNkx-2.3* in *Xenopus* embryos results in myocardial hyperplasia. *Development* **122,** 3549–3556.

Collignon, J., Varlet, I., and Robertson, E. J. (1996). Relationship between asymmetric *nodal* expression and the direction of embryonic turning. *Nature* **381,** 155–158.

Cross, J. C., Flannery, M. L., Blanar, M. A., Steingrimsson, E., Jenkins, N. A., Copeland, N. G., Rutter, W. J., and Werb, Z. (1995). *Htx* encodes a basic helix–loop–helix transcription factor that regulates trophoblast cell development. *Development* **121,** 2513–2523.

Cserjesi, P., Brown, B., Lyons, G. E., and Olson, E. N. (1995). Expression of the novel basic helix–loop–helix gene *eHAND* in neural crest derivatives and extraembryonic membranes during mouse development. *Dev. Biol.* **170,** 664–678.

Damante, G., Fabbro, D., Pellizzari, L., Civitareale, D., Guazzi, S., Polycarpou-Schwartz, M., Cauci, S., Quadrifoglio, F., Formisano, S., and Di Lauro, R. (1994). Sequence-specific DNA recognition by the thyroid transcription factor-1 homeodomain. *NAR* **22,** 3075–3083.

Davis, C. L. (1927). Development of the human heart from its first appearance to the stage found in embryos of twenty paired somites. *Contrib. Embryol.* **19,** 245–284.

De La Cruz, M. V., Sanchez-Gomez, C., and Palomino, M. A. (1989). The primitive cardiac regions in the straight tube heart (stage 9-) and their anatomical expression in the mature heart: An experimental study in the chick embryo. *J. Anat.* **165,** 121–131.

DeRuiter, M. C., Poelmann, R. E., VanderPlas-de Vries, I., Mentink, M. M. T., and Gittenberger-de Groot, A. C. (1992). The development of the myocardium and endocardium in mouse embryos. *Anat. and Embryol.* **185,** 461–473.

Dettman, R. W., Denetclaw Jn., W., Ordahl, C. P., and Bristow, J. (1998). Common epicardial origin of coronary vascular smooth muscle, perivascular fibroblasts, and intermyocardial fibroblasts in the avian heart. *Dev. Biol.* **193,** 169–181.

Duboule, D. (1994). *Guidebook of Homeobox Genes.* IRL, Oxford, UK.

Dunwoodie, S. L., Rodriguez, T. A., and Beddington, R. S. P. (1998). Msgl and mrgl, founding members of a gene family, show distinct patterns of gene expression during mouse embryogenesis. *Mech. Dev.* **72,** 27–40.

Durocher, D., Chen, C.-Y., Ardati, A., Schwartz, R. J., and Nemer, M. (1996). The atrial natriuretic factor promoter is a downstream target for Nkx-2.5 in the myocardium. *Mol. Cell. Biol.* **16,** 4648–4655.

Durocher, D., Charron, F., Warren, R., Schwartz, R. J., and Nemer, M. (1997). The cardiac transcription factors Nkx2-5 and GATA-4 are mutual cofactors. *EMBO J.* **16,** 5687–5696.

Edmondson, D. G., Lyons, G. E., Martin, J. F., and Olson, E. N. (1994). Mef2 gene expression marks the cardiac and skeletal muscle lineages during mouse embryogenesis. *Development* **120,** 1251–1263.

Eisenberg, L. M., and Markwald, R. R. (1995). Molecular regulation of atrioventricular valvuloseptal morphogenesis. *Circ. Res.* **77,** 1–6.

Evans, S. M., Tai, L.-J., Tan, V. P., Newton, C. B., and Chien, K. R. (1994). Heterokaryons of cardiac myocytes and fibroblasts reveal the lack of dominance of the cardiac phenotype. *Mol. Cell. Biol.* **14,** 4269–4279.

Evans, S., Yan, W., Murillo, M. P., Ponce, J., and Papalopulu, N. (1995). *tinman,* a *Drosophila* homeobox gene required for heart and visceral mesoderm specification, may be represented by a family of genes in vertebrates: *XNkx-2.3,* a second homologue of *tinman. Development* **121,** 3889–3899.

Firulli, A. B., and Olson, E. N. (1997). Modular regulation of muscle gene transcription: A mechanism for muscle cell diversity. *Trends Genet.* **13,** 364–369.

Firulli, A. B., McFadden, D. G., Lin, Q., Srivastava, D., and Olson, E. N. (1998). Heart and extra-embryonic mesoderm defects in mouse embryos lacking the bHLH transcription factor Hand1. *Nature Genet.* **18,** 266–270.

Fishman, M. C., and Olson, E. N. (1997). Parsing the heart: genetic modules for organ assembly. *Cell* **91,** 153–156.

Frasch, M. (1995). Induction of visceral and cardiac mesoderm by ectopic Dpp in the early Drosophila embryo. *Nature* **374,** 464–467.

Fu, Y., and Izumo, S. (1995). Cardiac myogenesis: Overexpression of *XCsx2* or *XMEF2A* in whole *Xenopus* embryos induces the precocious expression of *XMHCα* gene. *Roux's Arch. Dev. Biol.* **205,** 198–202.

Gajewski, K., Kim, Y., Lee, Y. M., Olson, E. N., and Schulz, R. A. (1997). *D-mef2* is a target for Tinman activation during *Drosophila* heart development. *EMBO J.* **16,** 515–522.

Garcia-Fernandez, J., and Holland, P. W. H. (1994). Archetypal organisation of the amphioxus Hox gene cluster. *Nature* **370,** 563–566.

Garcia-Martinez, V., and Schoenwolf, G. C. (1993). Primitive-streak origin of the cardiovascular system in avian embryos. *Dev. Biol.* **159,** 706–719.

Garcia-Fernandez, J., Baguna, J., and Salo, E. (1991). Planarian homeobox genes: Cloning, sequence analysis, and expression. *Proc Natl. Acad. Sci. USA* **88,** 7338–7342.

Gassmann, M., Casagranda, F., Orioli, D., Simon, H., Lai, C., Klein, R., and Lemke, G. (1995). Aberrent neural and cardiac development in mice lacking the ErbB4 neuregulin receptor. *Nature* **378,** 390–394.

George, E. L., Baldwin, H. S., and Hynes, R. O. (1997). Fibronectins are essential for heart and blood vessel morphogenesis but are dispensible for initial specification of precursor cells. *Blood* **15,** 3073–3081.

Gorczyka, M. G., Phyllis, R. W., and Budnik, V. (1994). The role of *tinman,* a mesodermal cell fate gene, in axon pathfinding during the development of the transverse nerve in *Drosophila. Development* 120, 2143-2152.

Gourdie, R. G., Mima, T., Thompson, R. P., and Mikawa, T. (1995). Terminal diversification of the myocyte lineage generates Purkinje fibers of the cardiac conduction system. *Development* **121,** 1423–1431.

Grens, A., Gee, L., Fisher, D. A., and Bode, H. R. (1996). *CnNK-2,* an *NK-2* homeodomain gene, has a role in patterning the basal end of the axis in hydra. *Dev. Biol.* **180,** 473–488.

Grueneberg, D. A., Natesan, S., Alexandre, C., and Gilman, M. Z. (1992). Human and Drosophila homeodomain proteins that enhance the DNA-binding activity of serum response factor. *Science* **257,** 1089–1095.

Harvey, R. P. (1996). NK-2 homeobox genes and heart development. *Dev. Biol.* **178,** 203–216.

Harvey, R. P. (1998). Cardiac looping: An uneasy deal with laterality. *Sem. Cell Dev. Biol.* **9,** 101–108.

Haun, C., Alexander, J., Stainier, D. Y., and Okkema, P. G. (1998). Rescue of caenorhabditis elegans pharyngeal development by a vertebrate heart specification gene. *Proc. Natl. Acad. Sci. USA* **95,** 5072–5075.

Hollenberg, S. M., Sternglanz, R., Cheng, P. F., and Weintraub, H. (1995). Identification of a new family of tissue-specific basic helix–loop–helix proteins with a two hybrid system. *Mol. Cell. Biol.* **15,** 3813–3822.

Jacobs, D. K., Lee, S. E., Dawson, M. N., Staton, J. L., and Raskoff, K. A. (1998). The history of development through the evolution of molecules: gene trees, hearts, eyes and dorsoventral inversion. In *Individuals, Population and Species: A Molecular Perspective* (R. DeSalle and B. Schierwater, Eds.). Birkhauser, Basel.

Jacobson, A. G., and Sater, A. K. (1988). Features of embryonic induction. *Development* **104,** 341–359.

Jagla, K., Frasch, M., Jagla, T., Bretzen, G., Bellard, F., and Bellard, F. (1997). *ladybird,* a new component of the cardiogenic pathway in *Drosophila* required for diversification of heart progenitors. *Development* **124,** 3471–3479.

Jiminez, F., Marin-Morris, L. E., Velasco, L., Chu, H., Sierra, J., Rossen, D. R., and White, K. (1995). *vnd,* a gene required for early neurogenesis of Drosophila, encodes a homeodomain protein. *EMBO J.* **14,** 3487–3495.

Kim, Y., and Nirenberg, M. (1989). *Drosophila* NK-homeobox genes. *Proc. Natl. Acad. Sci. USA* **86,** 7716–7720.

Kimura, S., Hara, Y., Pineau, T., Fernandez-Salguero, P., Fox, C. H., Ward, J. M., and Gonzalez, F. J. (1996). The T/ebp null mouse: Thyroid-specific enhancer-binding protein is essential for organogenesis of the thryroid, lung, ventral forebrain, and pituitary. *Genes Dev.* **10,** 60–69.

Kirby, M. L., and Waldo, K. L. (1995). Neural crest and cardiovascular patterning. *Circ. Res.* **77,** 211–215.

Komuro, I., and Izumo, S. (1993). Csx: A murine homeobox-containing gene specifically expressed in the developing heart. *Proc. Natl. Acad. Sci. USA* **90,** 8145–8149.

Kuo, C. T., Morrisey, E. E., Anandappa, R., Sigrist, K., Lu, M. M., Parmacek, M. S., Soudais, C., and Leiden, J. M. (1997). GATA4 transcription factor is required for ventral morphogenesis and heart tube formation. *Genes Dev.* **11,** 1048–1060.

Lawson, K. A., and Pedersen, R. A. (1992). Clonal analysis of cell fate during gastrulation and early neurulation in the mouse. *Ciba Foundation Symp.* **165,** 3–21.

Lawson, K. A., Meneses, J. J., and Pederson, R. A. (1991). Clonal analysis of epiblast during germ layer formation in the mouse embryo. *Development* **113,** 891–911.

Lazzaro, D., Price, M., Felice, M., and Di Lauro, R. (1991). The transcription factor TTF-1 is expressed at the onset of thyroid and lung morphogenesis and in restricted regions of the foetal brain. *Development* **113,** 1093–1104.

Lee, K. H., and Breitbart, R. E. (1996). A new tinman-related gene, nxk2.7, anticipates the expression of nkx2.5 and nkx2.3 in zebrafish heart and pharyngeal endoderm. *Dev. Biol.* **180,** 722–731.

Lee, Y. M., Park, T., Schulz, R. A., and Kim, Y. (1997). Twist-mediated activation of the NK-2 homeobox gene in the visceral mesoderm of Drosophila requires two distinct clusters of E-box regulatory elements. *J. Biol. Chem.* **272,** 17531–17541.

Levin, M. (1997). Left–right asymmetry in vertebrate embryogenesis. *Bioessays* **19,** 287–296.

Li, Q. Y., Newbury-Ecob, R. A., Terrett, J. A., Wilson, D. I., Curtis, A. R. J., Yi, C. H., Gebuhr, T., Bullen, P. J., Robson, S. C., Strachan, T., Bonnet, D., Lyonnet, S., Young, I. D., Raeburn, A., Buckler, A. J., Law, D. J., and Brook, J. D. (1997). Holt-Oram syndrome is caused by mutations in *TBX5,* a member of the Brachyury (T) gene family. *Nature Genet.* **15,** 21–29.

Linask, K. K., and Lash, J. W. (1988a). A role for fibronectin in the migration of avian precardiac cells I. Dose-dependent effects of fibronectin antibody. *Dev. Biol.* **129,** 315–323.

Linask, K. K., and Lash, J. W. (1988b). A role for fibronectin in the migration of avian precardiac cells. II Rotation of the heart-forming region during different stages and its effects. *Dev. Biol.* **129,** 324–329.

Linask, K. K., and Lash, J. W. (1993). Early heart development: Dynamics of endocardial cell sorting suggests a common origin with cardiomyocytes. *Dev. Dyn.* **195,** 62–66.

Lints, T. J., Parsons, L. M., Hartley, L., Lyons, I., and Harvey, R. P. (1993). *Nkx-2.5:* A novel murine homeobox gene expressed in early heart progenitor cells and their myogenic descendants. *Development* **119,** 419–431.

Lyons, G. E. (1994). In situ analysis of the cardiac muscle gene program during embryogenesis. *Trends Cardiovasc. Med.* **4,** 70-77.

Lyons, I., Parsons, L. M., Hartley, L., Li, R., Andrews, J. E., Robb, L., and Harvey, R. P. (1995). Myogenic and morphogenetic defects in the heart tubes of murine embryos lacking the homeobox gene *Nkx2-5. Genes Dev.* **9,** 1654–1666.

Manak, J. R., and Scott, M. P. (1994). A class act: Conservation of homeodomain protein functions. *Development* (Suppl.), 61–71.

Marshal, H., Morrison, A., Studer, M., Popperl, H., and Krumlauf, R. (1996). Retinoids and Hox genes. *FASEB J.* **10,** 969–978.

Martin, A. (1980). Some invertebrate myogenic hearts: The hearts of worms and molluscs. In *Hearts and Heart-Like Organs* (G. H. Bourne, Ed.), Vol. 1, pp. 1–39. Academic Press, New York.

Meyer, D., and Birchmeier, C. (1995). Multiple essential functions of neuregulin in development. *Nature* **378,** 386–390.

Mikawa, T., and Gourdie, R. G. (1996). Pericardial mesoderm generates a population of coronary smooth muscle cells migrating into the heart along with ingrowth of the epicardial organ. *Dev. Biol.* **173,** 221–232.

Mohun, T. (1997). Transcription factors in cardiogenesis in the amphibian, *Xenopus laevis.* In *Genetic Control of Heart Development* (E. N. Olson, R. P. Harvey, R. A. Schulz, and J. S. Altman, Eds.), pp. 76–84. HFSP, Strasbourg.

Molkentin, J. D., Firulli, A. B., Black, B. L., Martin, J. F., Hustad, C. M., Copeland, N., Jenkins, N., Lyons, G., and Olson, E. (1996). MEF2B is a potent transactivator expressed in early myogenic lineages. *Mol. Cell. Biol.* **16,** 3814–3824.

Molkentin, J. D., Lin, Q., Duncan, S. A., and Olson, E. N. (1997). Requirement of the transcription factor GATA4 for heart tube formation and ventral morphogenesis. *Genes Dev.* **11,** 1061–1072.

Moorman, A. F. M., de Jong, F., Denyn, M. M. F. J., and Lamers, W. H. (1998). Development of the cardiac conduction system. *Circ. Res.* **82,** 629–644.

Morrisey, E. E., Ip, H. S., Tang, Z., Lu, M. M., and Parmacek, M. S. (1997). GATA-5: A transcriptional activator expressed in a novel and temporally and spatially-restricted pattern during embryonic development. *Dev. Biol.* **183,** 21–36.

Oka, T., Komuro, I., Shiojima, I., Hiroi, Y., Mizuno, T., Aikawa, R., Akazawa, H., Yamazaki, T., and Yazaki, Y. (1997). Autoregulation of human cardiac homeobox gene CSX1: Mediation by the enhancer element in the first intron. *Heart Vessels* **12** (Suppl.), 10–14.

Okkema, P. G., and Fire, A. (1994). The *Caenorhabditis elegans* NK-2 class homeoprotein CEH-22 is involved in combinatorial activation of gene expression in pharyngeal muscle. *Development* **120,** 2175–2186.

Olson, E., and Klein, W. H. (1994). bHLH factors in muscle development: deadlines and commitment, what to leave in and what to leave out. *Genes Dev.* **8,** 1–8.

Olson, E. N., and Srivastava, D. (1996). Molecular pathways controlling heart development. *Science* **272,** 671–676.

Olson, E. N., Perry, M., and Schulz, R. A. (1995). Regulation of muscle differentiation by the MEF2 family of MADS box transcription factors. *Dev. Biol.* **172,** 2–14.

Pabst, O., Schneider, A., Brand, T., and Arnold, H.-H. (1997). The mouse Nkx2-3 homeodomain gene is expressed in gut mesenchyme during pre- and postnatal mouse development. *Dev. Dyn.* **209,** 29–35.

Park, M., Wu, X., Golden, K., Axelrod, J. D., and Bodmer, R. (1996). The wingless pathway is directly involved in *Drosophila* heart development. *Dev. Biol.* **177,** 104–116.

Qiu, M., Shimamura, K., Sussel, L., Chen, S., and Rubenstein, J. L. R. (1998). Control of anteroposterior and dorsoventral domains of nkx-6.1 gene expression relative to other nkx genes during vertebrate CNS development. *Mech. Dev.* **72,** 77–88.

Randall, D. J., and Davie, P. S. (1980). The hearts of Urochordates and Cephalochordates. In *Hearts and Heart-Like Organs* (G. H. Bourne, Ed.), Vol. 1, pp. 41–59. Academic Press, New York.

Ranganayakulu, G., Zhao, B., Dokidis, A., Molkentin, J. D., Olson, E. N., and Schulz, R. A. (1995). A series of mutations in the D-MEF2 transcription factor reveal multiple functions in larval and adult myogenesis in drosophila. *Dev. Biol.* **171,** 169–181.

Ranganayakulu, G., Elliot, D. A., Harvey, R. P., and Olson, E. N. (1998). Divergent roles for NK-2 class homeobox genes in cardiogenesis in flies and mice. *Development* **125,** 3037–3048.

Rawles, M. E. (1943). The heart forming area of the chick blastoderm. *Physiol. Zool.* **16,** 22–41.

Ray, M. K., Chen, C.-Y., Schwartz, R. J., and DeMayo, F. J. (1996). Transcriptional regulation of a mouse clara cell-specific protein (mCC10) gene by the NKx transcription factor family members thyroid transcription factor 1 and cardiac muscle-specific homeobox protein (CSX). *Mol. Cell. Biol.* **16,** 2056–2064.

Reecy, J. M., Yamada, M., Cummings, K., Sosic, D., Chen, C.-Y., Eichele, G., Olson, E. N., and Schwartz, R. J. (1997). Chicken Nkx-2.8: A novel homeobox gene expressed in early heart progenitor cells and pharyngeal pouch-2 and -3 endoderm. *Dev. Biol.* **188,** 295–311.

Riley, P., Anson-Cartwright, L., and Cross, J. C. (1998). The Hand1 bHLH transcription factor is essential for placentation and cardiac morphogenesis. *Nature Genet.* **18.**

Rizki, T. M. (1978). The circulation system and associated cells and tissues. In *The Genetics and Biology of Drosophila* (M. Ashburner and T. R. F. Wright, Eds.), Vol. 26, pp. 397–452. Academic Press, New York.

Roberts, D. J., Johnson, R. L., Burke, A. C., Nelson, C. E., Morgan, B. A., and Tabin, C. (1995). Sonic hedgehog is an endodermal signal inducing *Bmp-4* and *Hox* genes during induction and regionalization of the chick hindgut. *Development* **121,** 3163–3174.

Rosenquist, G. C. (1966). A radioautographic study of labeled grafts in the chick blastoderm. Development from primitive streak stages to stage 12. *Carnegie Inst. Wash. Contrib. Embryol.* **38,** 71–110.

Ross, R. S., Navankasattusas, S., Harvey, R. P., and Chien, K. R. (1996). An HF-1a/HF-1b/MEF-2 combinatorial element confers cardiac ventricular specificity and establishes an anterior–posterior gradient of expression via an *Nkx2-5* independent pathway. *Development* **122,** 1799–1809.

Rugendorff, A., Younossi-Hartenstein, A., and Hartenstein, V. (1994). Embryonic origin and differentiation of the *Drosophila* heart. *Roux's Arch. Dev. Biol.* **203,** 266–280.

Satin, J., Fujii, S., and DeHaan, R. L. (1988). Development of cardiac beat rate in early chick embryos is regulated by regional cues. *Dev. Biol.* **129,** 103–113.

Schott, J.-J., Benson, D. W., Basson, C. T., Pease, W., Silberach, G. M., Moak, J. P., Maron, B. J., Seidman, C. E., and Seidman, J. G. (1998). Congenital heart disease caused by mutations in the transcription factor *NKX2-5. Science* (in press).

Schultheiss, T. M., Burch, J. B. E., and Lassar, A. B. (1997). A role for bone morphogenetic proteins in the induction of cardiac myogenesis. *Genes Dev.* **11,** 451–462.

Scott, M. P. (1997). Summary: A common language. In *Patter Formation during Development,* Vol. LXII, pp. 555–562. Cold Spring Harbor Laboratory Press, Cold Spring Harbor, NY.

Serbedzija, G. N., Chen, J.-N., and Fishman, M. C. (1998). Regulation in the heart field of zebrafish. *Development* **125,** 1095–1101.

Small, S., Blair, A., and Levine, M. (1996). Regulation of two pair-rule stripes by a single enhancer in the Drosophila embryo. *Dev. Biol.* **175,** 314–324.

Smith, S. M., Dickman, E. D., Thompson, R. P., Sinning, A. R., Wunsch, A. M., and Markwald, R. R. (1997). Retinoic acid directs cardiac laterality and the expression of early markers of precardiac asymmetry. *Dev. Biol.* **182,** 162–171.

Sparrow, D. B., Kotecha, S., Towers, N., and Mohun, T. J. (1997). The *Xenopus eHAND* gene is expressed in the developing cardiovascular system of the embryo and is regulated by BMPs. *Mech. Dev.* **71,** 151–163.

Srivastava, D., Cserjesi, P., and Olson, E. N. (1995). A subclass of bHLH proteins required for cardiac morphogenesis. *Science* **270,** 1995–1999.

Srivastava, D., Thomas, T., Lin, Q., Brown, D., and Olson, E. N. (1997). Regulation of cardiac mesodermal and neural crest development by the bHLH transcription factor, dHAND. *Nature Genet.* **16,** 154–160.

Stainier, D. Y. R., Lee, R. K., and Fishman, M. C. (1993). Cardiovascular development in the zebrafish I. Myocardial fate map and heart tube formation. *Development* **119,** 31–40.

Stalsberg, H. (1969). The origin of heart asymmetry: Right and left contributions to the early chick embryo heart. *Dev. Biol.* **19,** 109–129.

Stephenson, J. (1930). *The Oligoshaeta.* Oxford Univ. Press, London.

Sugi, Y., and Markwald, R. R. (1996). Formation and early morphogenesis of endocardial endothelial precursor cells and the role of endoderm. *Dev. Biol.* **175,** 66–83.

Tam, P. P. L., Parameswaran, M., Kinder, S. J., and Weinberger, R. P. (1997). The allocation of epiblast cells to the embryonic heart and other mesodermal lineages: The role of ingression and tissue movement during gastrulation. *Development* **124,** 1631–1642.

Thomas, P., Brickman, J. M., Popperl, H., Krumlauf, R., and Beddington, R. S. P. (1997). Axis duplication and anterior identity in the mouse embryo. In *Patter Formation during Development,* Vol. LXII, pp. 115–125. Cold Spring Harbor Laboratory Press, Cold Spring Harbor, NY.

Thomas, T., Yamagishi, H., Overbeek, P. A., Olson, E. N., and Srivastave, D. (1998). The bHLH factors, dHAND and eHAND, specify pulmonary and systemic cardiac ventricles independent of left–right sidedness. *Dev. Biol.* **196,** 228–236.

Tonissen, K. F., Drysdale, T. A., Lints, T. J., Harvey, R. P., and Krieg, P. A. (1994). *XNkx2.5,* a *Xenopus* gene related to *Nkx-2.5* and *tinman:* Evidence for a conserved role in cardiac development. *Dev. Biol.* **162,** 325–328.

Treisman, R. (1994). Ternary complex factors: Growth factor regulated transcriptional activators. *Curr. Opin. Genet. Dev.* **4,** 96–101.

Tsao, D. H. H., Gruschus, J. M., Wang, L.-H., Nirenberg, M., and Ferretti, J. A. (1994). Elongation of helix III of the NK-2 homeodomain upon binding to DNA: Secondary structure study by NMR. *Biochemistry* **33,** 15053–15060.

Tsuda, T., Philp, N., Zile, M. H., and Linask, K. K. (1996). Left–right asymmetric localization of flectin in the extracellular matrix during heart looping. *Dev. Biol.* **173,** 39–50.

Viglino, P., Fogolari, F., Formisano, S., Bortolotti, N., Damante, G., Di Lauro, R., and Esposito, G. (1993). Structural study of rat thyroid transcription factor 1 homeodomain (TTF-1 HD) by nuclear magnetic resonance. *FEBS Lett.* **336,** 397–402.

Webb, S., Brown, N. A., and Anderson, R. H. (1998). Formation of the atrioventricular septal structures in the normal mouse. *Circ. Res.* **82,** 645–656.

Williams, R. S. (1995). Boosting cardiac contractility with genes. *N. Engl. J. Med.* **332,** 817–818.

Yatskievych, T. A., Ladd, A. N., and Antin, P. B. (1997). Induction of cardiac myogenesis in avian pregastrula epiblast: The role of the hypoblast and activin. *Development* **124,** 2561–2570.

Yin, Z., Xu, X. L., and Frasch, M. (1997). Regulation of twist target gene tinman by modular cis-regulatory elements during early mesoderm development. *Development* **124,** 4971–4982.

Yutzey, K. E., Rhee, J. T., and Bader, D. (1994). Expression of the atrial-specific myosin heavy chain AMHC-1 and the establishment of anteroposterior polarity in the developing chicken heart. *Development* **120,** 871–883.

Yutzey, K. E., Gannon, M., and Bader, D. (1995). Diversification of cardiomyogenic cell lineages *in vitro. Dev. Biol.* **170,** 531–541.

Zou, Y., Evans, S., Chen, J., Kou, H.-C., Harvey, R. P., and Chien, K. R. (1997). CARP, a cardiac ankyrin repeat protein, is downstream in the *Nkx2-5* homeobox gene pathway. *Development* **124,** 793–804.

8

Control of Cardiac Development by the MEF2 Family of Transcription Factors

Brian L. Black and Eric N. Olson

Department of Molecular Biology and Oncology, The University of Texas Southwestern Medical Center, Dallas, Texas 75235

There has been rapid progress in recent years toward defining the mechanisms that control skeletal muscle development (Molkentin and Olson, 1996; Yun and Wold, 1996; Ludolph and Konieczny, 1995). In contrast, relatively little is known of the mechanisms that control cardiac gene expression and development. Recent studies in fruit flies and mice have demonstrated that members of the myocyte enhancer factor-2 (MEF2) family of transcription factors play multiple roles in cardiac myogenesis and morphogenesis. Here, we discuss the multiple functions of MEF2 factors in heart development and speculate about the potential mechanisms whereby these transcription factors control different sets of target genes at multiple steps in the cardiogenic pathway.

I. Heart Formation during Vertebrate Embryogenesis

The heart is the first organ to form during vertebrate embryogenesis (Fig. 1). In the mouse, heart formation begins at about Embryonic Day 7.5 (E 7.5) when a population of cells within the anterior lateral plate mesoderm become committed to a cardiogenic fate in response to inductive signals from the adjacent endoderm (Olson and Srivastava, 1996; Fishman and Chien, 1997; see Chapters 1–4). These cardiogenic cells, which are localized to a region known as the cardiac crescent, mi-

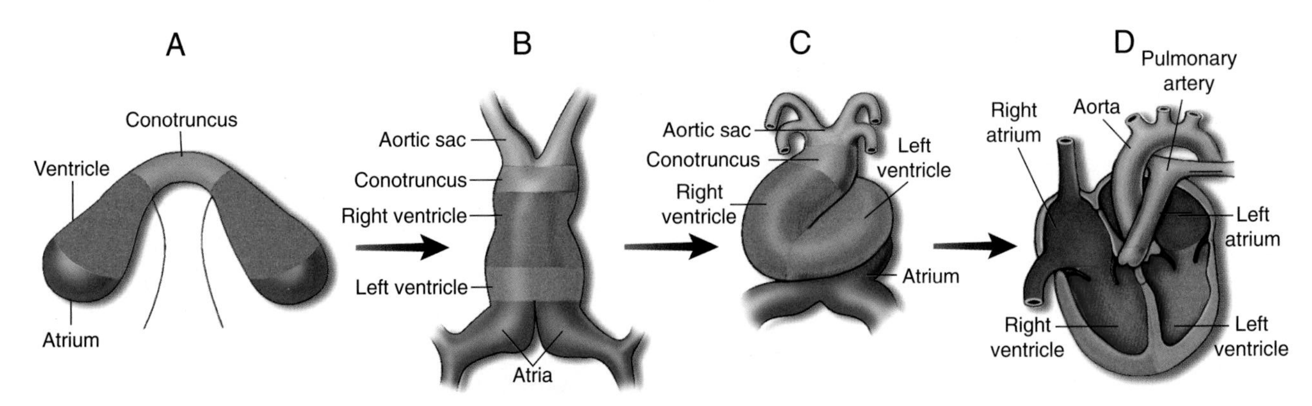

Figure 1 Schematic of the major morphologic events in vertebrate heart formation. Cells from the precardiogenic mesoderm are specified at the embryonic primitive streak stage (A) to form the various structures of the linear heart tube (B). Later, the heart loops rightward (C) and finally septates and forms the various regions and chambers of the looped and mature heart (D).

grate ventromedially to form the linear heart tube. Although the linear heart tube appears homogeneous along its length, it is divided into distinct segments of gene expression that prefigure the eventual atrial and ventricular chambers (Yutzey and Bader, 1995; see Chapters 19 and 20). Little is known of the mechanisms involved in specification of these different populations of cardiac precursors.

The first evidence of left–right asymmetry in the embryo is the rightward looping of the heart tube, which occurs in all vertebrate species (Yost, 1995; Srivastava and Olson, 1997; see Chapters 21 and 22). The process of looping converts anterior-posterior into left–right patterning of the heart tube (Fig. 1) and is essential for the proper orientation of the cardiac chambers and for alignment of the heart with the vascular system. Following looping, the atrial and ventricular chambers become demarcated by the formation of the interventricular septa and valves (see Chapter 10). Neural crest cells, which migrate from rhombomeres 6–8, also play an important role in heart formation by contributing to the outflow tract and the aortic arch arteries (Kirby and Waldo, 1995; see Chapter 11).

II. Cardiogenesis at the Cellular Level

The earliest marker for cells that have acquired cardiogenic potential in the embryo is the homeobox gene *Nkx2.5,* one of several related *NK*-type homeobox genes in vertebrates (Harvey, 1996; see Chapter 7). These genes share high homology with *Drosophila tinman,* which is expressed throughout the early mesoderm of the embryo before becoming restricted specifically to cardioblasts of the heart-like organ known as the dorsal vessel (Apiazu and Frasch, 1993; Bodmer, 1993). The precise functions of Tinman remain to be determined, but the fact that cardioblasts are not specified in *tinman* mutant embryos indicates that it is essential for specification of cardiac cell fate (Apiazu and Frasch, 1993; Bodmer, 1993). However, other factors or signals must also be involved because only a subset of cells that express *tinman* adopt a cardiac fate. The acquisition of cardiac fate in the *Drosophila* embryo has been shown to be dependent on decapentaplegic (Dpp) and wingless (Wg) signaling from adjacent cells (Staehling-Hampton *et al.,* 1994; Frasch, 1995; Lawrence *et al.,* 1995; Wu *et al.,* 1995; Park *et al.,* 1996). Thus, Tinman may act together with these growth factor signals to specify cardiac identity. In addition, as discussed later, an important function of Tinman is to activate the expression of *D-mef2* during *Drosophila* heart development (Gajewski *et al.*, 1997; see Chapter 5).

III. Transcriptional Control by MEF2

Soon after cardiac cells are specified, they begin to express a large array of contractile protein genes. Members of the MEF2 family of MADS box transcription factors play an important role in this aspect of cardiac development. The MADS box is named for the original four transcription factors identified as belonging to the family. MCM1 is a yeast protein which controls mating type; Agamous and Deficiens are flower homeotic gene products which control whorl and petal identity; and serum response factor is a transcription factor which mediates activation of immediate early response genes and muscle genes (Shore and Sharrocks, 1995).

Members of the MEF2 subfamily of MADS box proteins each encode the highly conserved MADS domain within their first 57 amino acids, in addition to a highly conserved MEF2 domain immediately adjacent to the MADS domain (Olson *et al.,* 1995). Together, the MADS and MEF2 domains mediate DNA binding and dimerization of these factors. *Drosophila* encodes a single *mef2* gene product, D-MEF2 (Lilly *et al.,* 1994; Nguyen *et al.,* 1994), whereas vertebrates have four *mef2* genes, *mef2a–d* (Breitbart *et al.,* 1993; Leifer *et al.,* 1993; Martin *et al.,* 1993, 1994; McDermott *et al.,* 1993; Pollock and Treisman, 1991; Yu *et al.,* 1992). The four vertebrate MEF2 factors and the single fly MEF2 protein share a high degree of sequence homology within their MADS and MEF2 domains but are divergent in the carboxyl-terminal transactivation domains (Fig. 2).

MEF2 was originally identified as a muscle-specific binding activity present in differentiated muscle cells (Gossett *et al.,* 1989) and was later found to be important in the transcriptional activation of many cardiac and skeletal muscle genes (Cserjesi and Olson, 1991). MEF2 proteins bind as homo- and heterodimers and to a conserved DNA consensus sequence, C/TTA(A/T)$_4$TAG/A, found in the control regions of many muscle- and heart-specific genes (Gossett *et al.,* 1989; Cserjesi and Olson, 1991; Pollock and Treisman, 1991; Andres *et al.,* 1995). MEF2 proteins bind to this target sequence to directly activate the transcriptional machinery via potent transactivation domains present in their carboxyl termini (Martin *et al.,* 1994; Molkentin *et al.,* 1996a). Numerous cardiac-restricted genes have been identified to contain essential MEF2 sites in their control regions. For example, there is an essential MEF2 site in the promoter of the *desmin* gene that when mutated completely abolishes all cardiac-specific expression of *desmin* (Kuisk *et al.,* 1996). Likewise, MEF2 sites found in the *myosin light chain* (*MLC*)*-2V* and *α-myosin heavy chain* genes are essential for cardiac-specific expression (Molkentin *et al.,* 1994; Navankasattusas *et al.,* 1992).

Each of the vertebrate *mef2* genes produces several alternatively spliced messages which give rise to multiple *mef2* gene products (Fig. 2) (Breitbart *et al.,* 1993; Leifer *et al.,* 1993; Martin *et al.,* 1993, 1994; McDermott *et al.,* 1993; Pollock and Treisman, 1991; Yu *et al.,* 1992). This multitude of different MEF2 proteins is capable of heterodimerization, allowing for a large number of variant DNA-binding heterodimers. Such a diversity of MEF2 proteins in vertebrates is likely to account for the

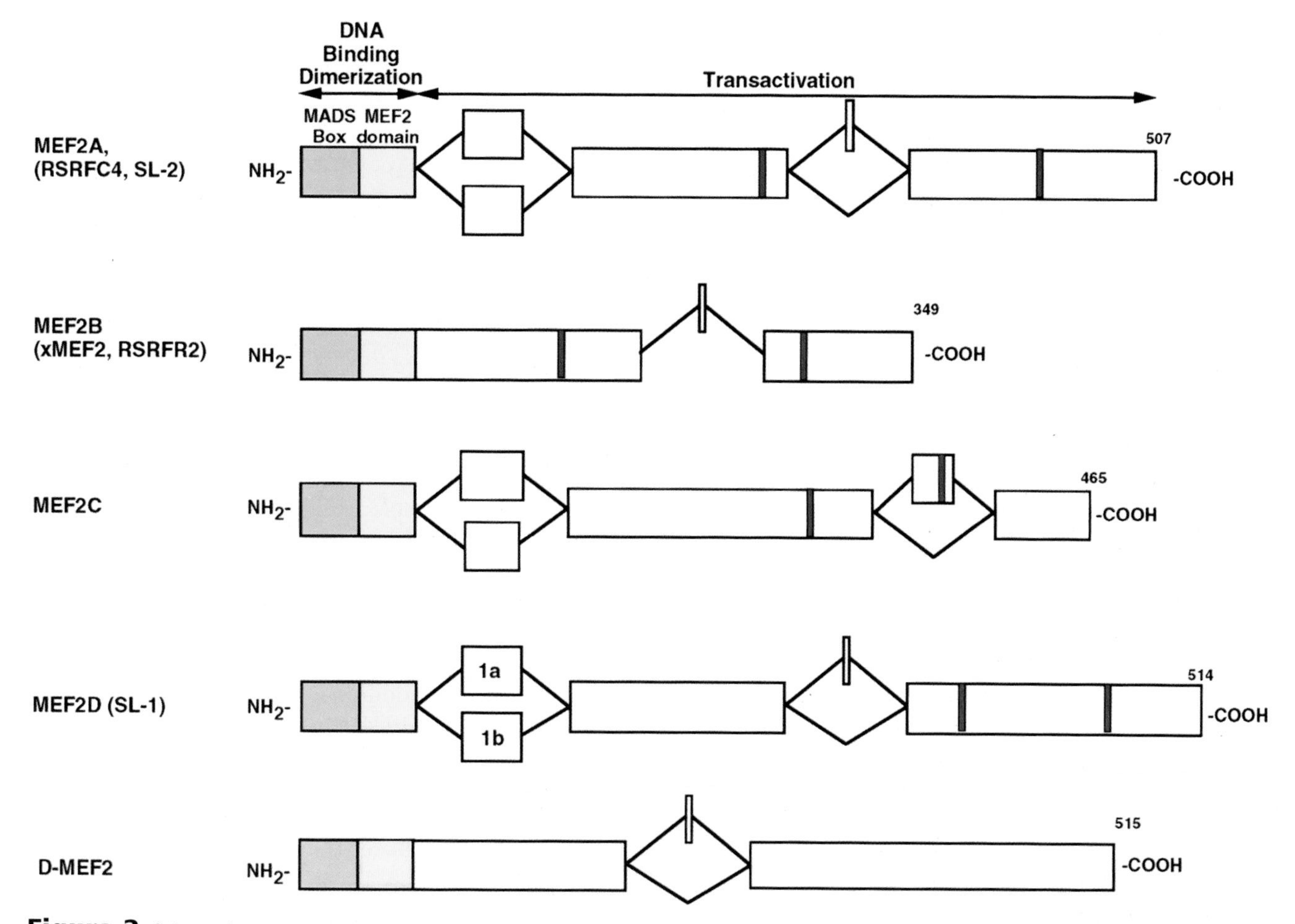

Figure 2 Schematic representation of MEF2 proteins. The four MEF2 proteins (MEF2A–D) in vertebrates and the single MEF2 protein (D-MEF2) in *Drosophila* are depicted. Each MEF2 protein comprises the highly conserved MADS (blue) and MEF2 (yellow) domains which mediate DNA binding, dimerization, and protein–protein interactions with other classes of transcription factors and one or more transactivation domains in the carboxy terminus. Each of the *mef2* genes is alternatively spliced and the boundaries for the exons encoding the MADS and MEF2 domains are conserved among all *mef2* genes.

numerous potential roles played by MEF2 factors in the differentiation of muscle, heart, and neural lineages. In addition to the role of MEF2 factors in tissue-specific differentiation, certain splice variants of MEF2 may be important in cell proliferation. Recently, a ubiquitous isoform of MEF2D was shown to regulate the serum inducibility of the *c-jun* promoter (Han and Prywes, 1995). Likewise, MEF2 protein expression was shown to be up regulated upon proliferation of vascular smooth muscle cells following balloon injury (Firulli *et al.*, 1996). Thus, the presence of four *mef2* genes with numerous splice variants and the possibility of heterodimerization among MEF2 factors creates multiple different *mef2* products with numerous potential roles in the transcriptional control of genes important in a variety of cellular processes.

A. Expression of *mef2* Genes in Vertebrates and Flies

During vertebrate embryogenesis, the four *mef2* genes are expressed throughout developing muscle cell lineages as well as in other cell types (Edmondson *et al.*, 1994; Molkentin *et al.*, 1996b; Subramanian and Nadal-Ginard, 1996; G. E. Lyons *et al.*, 1995; Leifer *et al.*, 1993, 1994). *mef2b* and *-c* are the first members of the family to be expressed in the heart, with transcripts appearing in the precardiogenic mesoderm at about E 7.75 in the mouse (Edmondson *et al.*, 1994; Molkentin *et al.*, 1996b). By E 8.0, *mef2a* and *-d* expression is detected in the newly formed linear heart tube (Edmondson *et al.*, 1994). Thereafter, all four *mef2* genes continue to be expressed throughout the developing heart.

The *mef2* genes are also expressed in developing skeletal and smooth muscle (see Chapters 24 and 28), as well as in endothelial cells that form the vascular template within the yolk sac and embryo. In addition, the four *mef2* genes show highly specific expression patterns throughout the developing brain (Leifer *et al.*, 1994 ; G. E. Lyons *et al.*, 1995). By the late fetal period, transcripts for the different *mef2* genes begin to appear in a variety of tissues, and in the adult they are widely expressed (Edmondson *et al.*, 1994).

Despite the widespread expression of *mef2* mRNAs in adult tissues and in a variety of cell lines, MEF2 protein and DNA-binding activity are highly enriched in muscle, heart, and brain tissues (Gossett *et al.*, 1989; Yu *et al.*, 1992; Suzuki *et al.*, 1995). This disparity between the expression of *mef2* mRNA and MEF2 protein has suggested the existence of post transcriptional mechanisms for MEF2 regulation (Suzuki *et al.*, 1995; Black *et al.*, 1997). Indeed, the 3′ untranslated region of the *mef2a* mRNA has been shown to contain a highly conserved 428-nucleotide region that confers translational repression preferentially in nonmuscle cells when fused to an exogenous reporter gene (Black *et al.*, 1997).

In addition to transcriptional and translational control mechanisms, MEF2 proteins also appear to be regulated posttranslationally. Recent studies have demonstrated that a casein kinase-II phosphorylation site in the MEF2 domain of MEF2C augments DNA binding by MEF2C *in vivo* (Molkentin *et al.*, 1996c). This phosphorylation site is conserved in all MEF2 proteins, which is consistent with a role for phosphorylation in the regulation of MEF2 function. In addition, there are conserved phosphorylation sites in the carboxy-terminal transactivation domains of MEF2 factors. These sites may be modified by phosphorylation to increase their transactivation potential (Han *et al.*, 1997). Furthermore, studies of *myogenin* gene transcription have shown that a MEF2 site in the promoter is required for transcriptional activation *in vivo* (Cheng *et al.*, 1993). When skeletal myoblasts are induced to differentiate, preexisting MEF2 is modified, probably by phosphorylation, such that it can activate *myogenin* transcription through this site (Buchberger et al., 1994).

Thus, numerous studies have shown that MEF2 expression and function are regulated at multiple levels. By utilizing multiple control mechanisms, a cell ensures tight regulation of MEF2 activity which is likely to be important given the role of MEF2 factors in mediating tissue-specific differentiation.

Unlike in vertebrates, there is a single *mef2* gene product in *Drosophila*, D-MEF2, which also is expressed in developing muscle and neural lineages. *D-mef2* is first expressed at gastrulation throughout the mesoderm of *Drosophila* embryos (Lilly *et al.*, 1994; Nguyen *et al.*, 1994). As the embryos develop, *D-mef2* expression becomes restricted to cells of the somatic, visceral, and cardiac musculature, and this expression is maintained throughout embryonic, larval, and adult myogenesis (Lilly *et al.*, 1994; Nguyen *et al.*, 1994). During larval development, *D-mef2* is also expressed in the mushroom bodies of the fly brain (Schulz *et al.*, 1996).

The fly heart, known as the dorsal vessel, is a linear tube which pumps hemolymph throughout the organism. The dorsal vessel consists of a single layer of cardial cells and an additional layer of pericardial cells (Rugendorff *et al.*, 1994). *D-mef2* is expressed in both these cell types prior to and during their differentiation. *D-mef2* expression in the dorsal vessel is controlled by Tinman, which binds to two sites in a distal upstream cardiac-specific enhancer (Gajewski *et al.*, 1997).

B. Genetic Analysis of MEF2 in *Drosophila*

The existence in vertebrate species of four *mef2* genes, which are expressed in overlapping patterns in

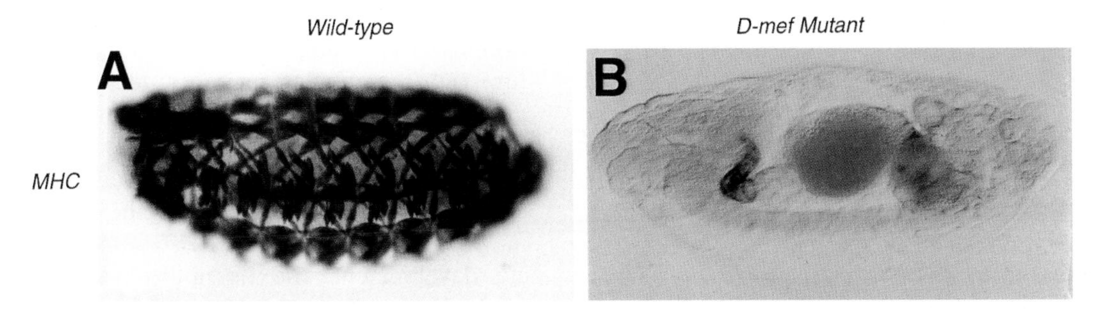

Figure 3 D-MEF2 is required for muscle differentiation in *Drosophila*. Myosin-heavy chain (MHC) expression, as a marker for terminally differentiated muscle, in wild-type (A) and *mef2* mutant (B) *Drosophila* embryos at stage 15. The brown color of the MHC staining pattern in A shows the metameric pattern of somatic muscle fibers. (B) The nearly complete absence of muscle differentiation in embryos lacking D-MEF2.

developing muscle cell lineages, complicates the interpretation of loss-of-function phenotypes because of possible redundancy of functions. However, in *Drosophila*, which contains only a single *mef2* gene, it has been possible to determine the consequences of complete elimination of all MEF2 activity from the embryo. Using P-element insertion and ethylmethanosulfonate mutagenesis, we and others have generated null and hypomorphic alleles of *D-mef2* (Lilly *et al.*, 1995; Bour *et al.*, 1995; Ranganayakulu *et al.*, 1995). In embryos lacking *D-mef2*, myoblasts from all three muscle cell lineages—cardiac, somatic, and visceral—are properly specified and positioned, but they are unable to differentiate (Fig. 3) (Lilly *et al.*, 1995; Bour *et al.*, 1995; Ranganayakulu *et al.*, 1995). This dramatic mutant phenotype provided the first molecular evidence for a commonality in the molecular mechanisms that control differentiation of diverse muscle cell types and demonstrated that D-MEF2 is a central component of all muscle differentiation programs in *Drosophila*.

C. Genetic Analysis of *mef2* Genes in the Mouse

The functions of the vertebrate *mef2* genes are only beginning to be determined through gene inactivation studies in the mouse. The first of the *mef2* genes to be inactivated was *mef2c* (Lin *et al.*, 1997). *mef2c* null embryos appear normal until about E 9.0, when they begin to show retarded growth and pericardial effusion, indicative of cardiac insufficiency. At the linear heart tube stage (E 8.0), the mutants do not show obvious cardiac defects. However, they do not initiate rightward looping and the future right ventricular region fails to form (Fig. 4). Instead, the mutant hearts form a single hypoplastic ventricular chamber fused directly to an enlarged atrial chamber. Because of the absence of a right ventricular region in the mutant, the remaining portion of the ventricular region is displaced to the left.

Normally, as the ventricular chambers develop, the trabeculae form as finger-like projections along the inner myocardial wall, which becomes separated from the endocardium by a layer of loose mesenchyme called the cardiac jelly (Fishman and Chien, 1997). The endocardium is present in *mef2c* mutant embryos, but the trabeculae develop poorly, and both the endocardial cells and cardiomyocytes within the ventricular wall appear disorganized (Lin *et al.*, 1997).

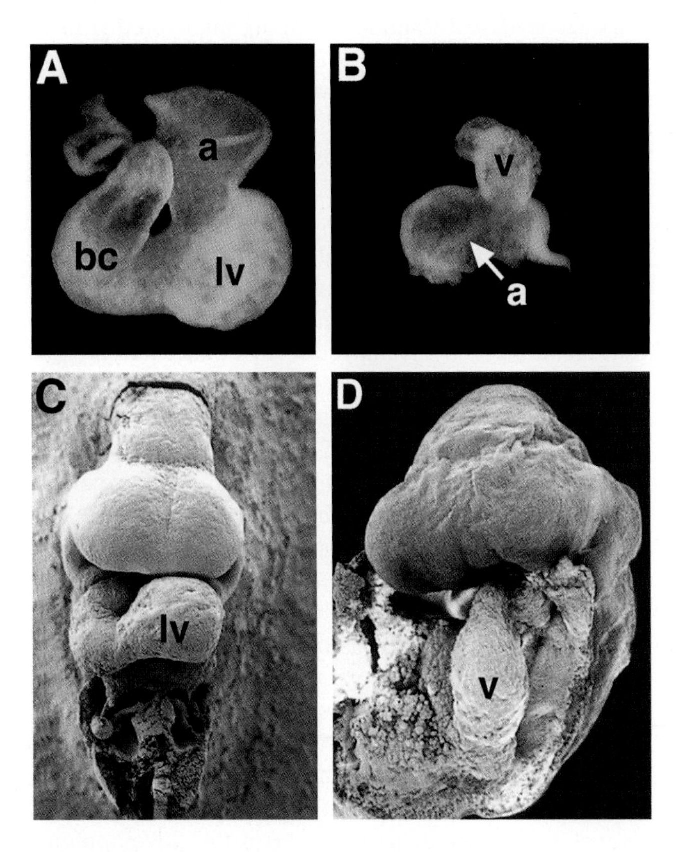

Figure 4 Cardiac defects in MEF2C null embryos. Wild-type (A, C) and *mef2c* mutant (B, D) hearts from E 9.0 embryos. The mutant heart (B, D) fails to undergo rightward looping and there is no evidence of the future right ventricle but rather a single hypoplastic ventricle is fused to an enlarged atrium. a, atrium; bc, bulbus cordis; lv, left ventricle; v, ventricle.

The basic-helix–loop–helix (bHLH) genes *dHAND* and *eHAND* are expressed in complementary patterns within the developing heart (Cserjesi *et al.,* 1995; Srivastava *et al.,* 1995; see Chapter 9). *dHAND* is normally expressed throughout the linear heart tube but becomes restricted to the future right ventricular region during looping (Srivastava *et al.,* 1997). In contrast, *eHAND* is expressed in two specific segments of the heart tube which give rise to the conotruncus and the left ventricle (Cserjesi *et al.,* 1995; Biben and Harvey, 1997; see Chapter 7). In *mef2c* mutants, *dHAND* is expressed in the linear heart tube, but expression is downregulated at the time of looping, concomitant with the failure of the right ventricle to form (Lin *et al.,* 1997). Consistent with the conclusion that the right ventricular region is deleted in the mutant, *eHAND* is expressed contiguously throughout the heart tube without the gap in expression normally observed in the right ventricle (Lin *et al.,* 1997).

In the hearts of *mef2c* mutant embryos, a subset of cardiac contractile protein genes, including *atrial natriuretic factor* (*ANF*), *cardiac α-actin, α-MHC,* and *myosin light chain* (*MLC*)*-1A,* is downregulated, whereas others, such as *MLC2A* and *MLC2V,* are unaffected (Lin *et al.,* 1997). Interestingly, the *MLC2V* gene contains an essential MEF2 site in its promoter (Navankasattusas *et al.,* 1992). The observation that this and other cardiac genes are expressed in the absence of MEF2C suggests that other members of the MEF2 family can support the expression of these genes. *mef2b* is coexpressed with *mef2c* during the early stages of cardiogenesis and would, therefore, be a likely candidate for functional redundancy with MEF2C (Edmondson *et al.,* 1994; Molkentin *et al.,* 1996b). We have recently generated *mef2b* mutant mice, which do not exhibit obvious developmental defects (J. Molkentin and E. Olson, unpublished observations). Whether *mef2b/mef2c* double mutants will show cardiac defects that are more severe than those of *mef2c* mutants remains to be determined.

The finding that only a subset of MEF2-dependent genes are downregulated in *mef2c* mutant embryos also demonstrates that members of the MEF2 family can discriminate between downstream genes. The molecular basis for this discrimination is unclear but might occur through differential protein–protein interactions with other tissue-restricted transcription factors.

The selective deletion of the right ventricular region of the heart tube in *mef2c* mutant embryos suggests that distinct regulatory programs control the development of different regions of the heart. Since MEF2C is expressed homogeneously throughout the heart tube, it may cooperate with a regionally restricted cofactor to control right ventricular development. The identity of this potential cofactor remains to be determined, however, *dHAND* expression becomes localized to the future right ventricular region at the onset of looping morphogenesis. Moreover, mouse embryos homozygous for a *dHAND* null mutation resemble *mef2c* mutants in their failure to form a right ventricle (Srivastava *et al.,* 1997; see Chapter 9). Given the cooperative roles of MEF2 and myogenic bHLH factors in skeletal muscle development, the possibility that MEF2C and dHAND establish a combinatorial code for right ventricular development warrants further investigation (Srivastava and Olson, 1997).

The left ventricular region of the developing heart tube appears to be correctly specified in *mef2c* mutant embryos, based on the expression of *eHAND* in these embryos. While obviously speculative, this could indicate that eHAND does not require MEF2C for function. Whether left ventricular formation represents a default pathway for cardiac development or whether another member of the MEF2 family might substitute for a potential role of MEF2C in this aspect of cardiogenesis remain to be determined.

While we know the identities of many of the cardiac contractile protein genes that are regulated by MEF2 factors during differentiation of individual cardiomyocytes, little is known of the MEF2 target genes that could participate in looping morphogenesis or right ventricular chamber development. There is evidence suggesting that cardiac looping may require asymmetries in expression of cell adhesion molecules, mechanical forces, and cell proliferation across the left–right axis of the heart tube (Taber *et al.,* 1995). Thus, components of these pathways would be promising candidates as essential regulators of the morphogenic events in the MEF2C pathway. In this regard, previous studies have shown that the integrin subunit gene, *α-PS2,* in *Drosophila* is a direct target for transcriptional activation by D-MEF2 in the developing visceral mesoderm and its failure to be expressed in *D-Mef2* mutant embryos accounts for the severe morphologic defects in the developing gut (Ranganayakulu *et al.,* 1995).

IV. Combinatorial Control of Muscle Development by MEF2

Based on the absence of muscle differentiation in *Drosophila* embryos lacking *D-mef2* and the presence of MEF2 binding sites in the control regions of numerous muscle genes (Cserjesi and Olson, 1991; Lilly *et al.,* 1995; Bour *et al.,* 1995; Ranganayakulu *et al.,* 1995), we have concluded that MEF2 regulates muscle gene expression in different muscle cell types. How does MEF2 control the expression of different sets of muscle-specific genes in the cardiac, skeletal, and smooth muscle cell lineages? It is clear that MEF2 proteins can bind DNA and directly activate transcription of cardiac and skeletal muscle genes but this function of MEF2 cannot account fully for the role of MEF2 proteins in myogenic gene activation. Therefore, we have proposed that MEF2 acts in a combinatorial mechanism through direct protein–protein interactions with other lineage-restricted transcription factors to activate each myogenic differentiation program (Molkentin *et al.,* 1995; Molkentin and Olson, 1996). Such a combinatorial model for transcriptional activation by the MEF2 family is consistent with the role of MADS domain proteins in other species (Shore and Sharrocks, 1995). In flowers, the MADS domain proteins, Agamous and Deficiens, mediate their homeotic effects on leaf identity through protein–protein interactions with a variety of different classes of transcription factors. Furthermore, serum response factor (SRF) is known to directly interact with numerous different classes of transcription factors to synergistically activate transcription. In each of these cases, the interaction is mediated through the MADS domains of these factors.

MEF2 factors also appear to activate transcription through a similar mechanism. In the skeletal muscle lineage, MEF2 factors physically interact with members of the MyoD family of bHLH transcription factors to cooperatively activate skeletal muscle-specific transcription (Fig. 5) (Kaushal *et al.,* 1994; Molkentin *et al.,* 1995). Coexpression of MEF2 and myogenic bHLH proteins results in synergistic activation of myogenesis. This activation does not require direct DNA binding by MEF2 because MEF2 can cooperatively activate myogenic transcription when bound to a promoter solely through protein–protein interactions (Molkentin *et al.,* 1995; Black *et al.,* 1998). These interactions are mediated by the DNA-binding and dimerization motifs of these factors (Kaushal *et al.,* 1994; Molkentin *et al.,* 1995). The MEF2 proteins utilize the MADS and MEF2 domains

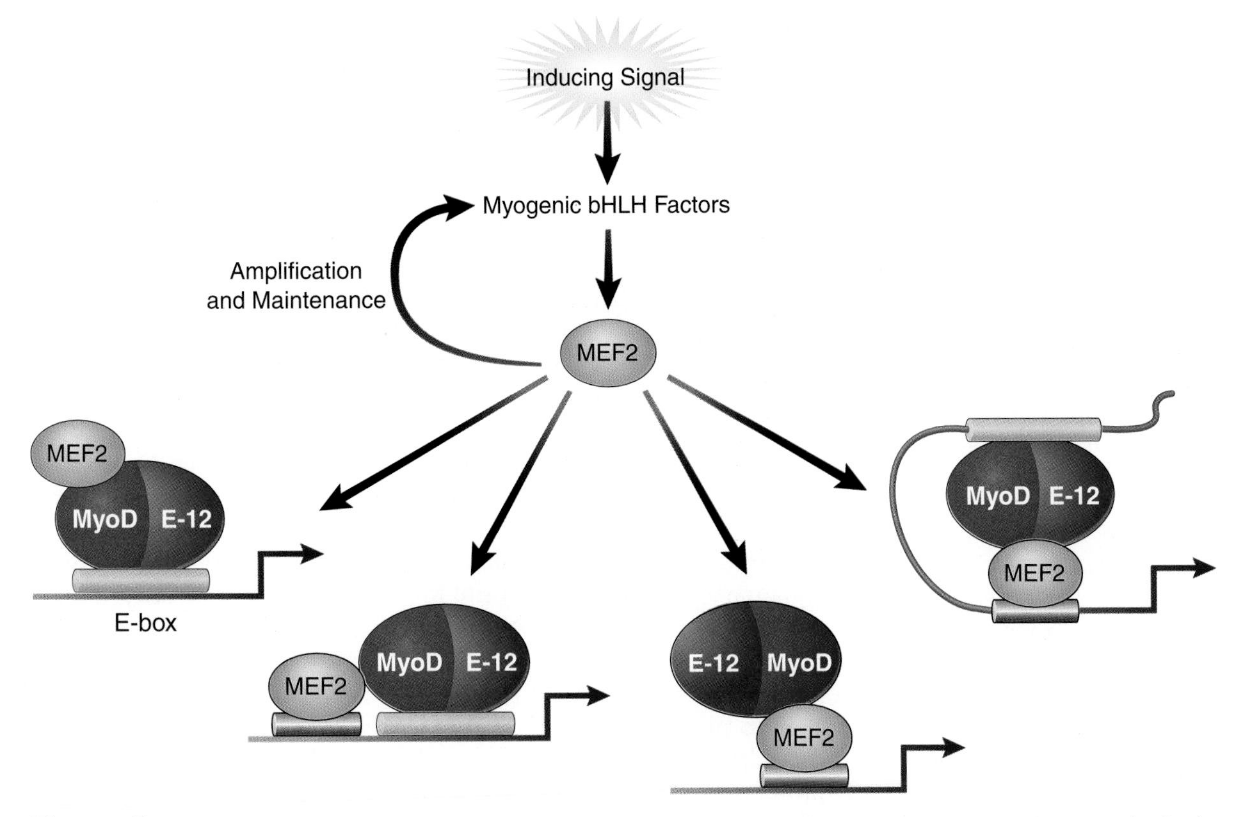

Figure 5 Combinatorial control of myogenesis by MEF2. This figure depicts possible mechanisms for muscle gene activation by MEF2. In skeletal muscle, MEF2 proteins collaborate with myogenic bHLH factors such as MyoD (and possibly other factors) to synergistically activate myogenic transcription. Myogenic bHLH proteins activate E-box-dependent genes that also contain MEF2 sites in their control regions. MEF2 or myogenic bHLH proteins also can function to activate myogenic genes which contain only one factor's binding sites through protein–protein interactions. Interactions between myogenic bHLH proteins and MEF2 factors may also provide a mechanism for linking muscle promoters with distal enhancers through protein–protein interactions. Similar mechanisms for MEF2 function may also operate in cardiac muscle lineages, however, the factors with which MEF2 collaborates in that lineage are unknown.

for interaction and the bHLH factors mediate interaction through the basic domain, which is also responsible for DNA binding.

While transcriptional activation can occur when either factor is bound to DNA, activation of skeletal myogenesis requires activation of transcription to occur through a myogenic bHLH factor bound to DNA (Molkentin *et al.*, 1995). Recent studies indicate that interaction of MEF2 factors and myogenic bHLH proteins alone is not sufficient to activate myogenesis or to activate transcription synergistically (Black *et al.*, 1998). This is apparent from studies of a class of MyoD mutants which support interaction with MEF2 but are incapable of activation of transcription and myogenesis. These studies indicate that initiation of myogenesis requires activation signals by both MEF2 and myogenic bHLH factors to be transmitted to the transcriptional machinery through the myogenic bHLH factor (Black *et al.*, 1998). The paradigm of MEF2 functioning as a coactivator of tissue-specific transcription has also been extended to neural lineages in which MEF2 factors have been shown to cooperatively activate transcription in collaboration with the neural-specific bHLH protein, MASH1 (Black *et al.*, 1996; Mao and Nadal-Ginard, 1996).

No cardiac-specific factors have been shown to interact with MEF2 factors. However, it seems likely that this general model for MEF2 activation of cell type-specific genes is probably functioning in cardiac lineages as well, and several candidate molecules may serve as MEF2 cofactors. The *tinman* homolog, *Nkx2.5*, is expressed in developing cardiac cells, and its gene product has been shown to interact with the MADS domain protein SRF (Chen and Schwartz, 1996). Likewise, members of the GATA family of transcription factors have been shown to interact with SRF (R. J. Schwartz, personal communication; see Chapter 16). Nkx2.5 is required for proper looping of the developing heart (I. Lyons *et al.*, 1995; see Chapter 7) and GATA4 is required for migration of cells in the cardiac lineage (Molkentin *et al.*, 1997; see Chapter 17). These two classes of transcription factors serve as potential candidates for MEF2 interaction. Finally, two bHLH factors, eHAND and dHAND, which are expressed in the developing heart, have been identified (see Chapter 9). While no direct evidence has been obtained to suggest that these factors associate with MEF2 to specify differentiation of the cardiac lineage, bHLH proteins clearly utilize MEF2 factors in the transcriptional activation of the skeletal muscle and neural differentiation programs, suggesting by analogy that similar mechanisms may also be functioning in the cardiac lineage. Furthermore, as noted previously, the cardiac phenotypes of the *mef2c* and *dHAND* knockouts are strikingly similar in that both mutations disrupt the rightward looping of the linear heart tube at similar times during development (Srivastava *et al.*, 1997; Lin *et al.*, 1997). The similarity of these two mutant phenotypes suggests that these factors may cooperate to specify cardiac morphogenesis such that mutation of either gene disrupts the transcriptional program.

V. Regulation of MEF2 Expression

The vertebrate *mef2* genes contain multiple alternatively spliced exons and large introns in their 5' untranslated regions, which has made it difficult to identify regulatory regions that control their expression in different muscle cell types (B. Black, T. Firulli, J. Molkentin, and E. Olson, unpublished observations). However, there has been considerable progress toward defining the regulatory elements that control expression of *mef2* during *Drosophila* embryogenesis. The intron–exon organization of the *D-mef2* gene is similar to that of the vertebrate genes, and within the conserved MADS and MEF2 domains, the introns are located at the identical codons in *D-mef2* and the vertebrate *mef2* genes (Lilly *et al.*, 1995). There is a single intron of about 6 kb in the 5' untranslated region of *D-mef2* (Lilly *et al.*, 1995; Bour *et al.*, 1995).

Within the 12 kb of DNA preceding the *D-mef2* gene, there are at least a dozen independent transcriptional enhancers that direct the expression of the gene in different mesodermal precursor cells and muscle cell types (R. Cripps, B. Zhao, K. Gajewski, R. Schulz, and E. Olson, unpublished observations). During development of the mesoderm into cardiac, skeletal, and visceral muscle, different enhancers are activated independently. There are three cardiac enhancers that show distinct transcriptional activities during development of the dorsal vessel. We have focused primarily on one of these enhancers, which is contained within a 237-bp region located about 5.5 kb upstream of the transcription initiation site for the gene (Gajewski *et al.*, 1997). This enhancer is active in four of six cardioblasts within each hemisegment (Gajewski *et al.*, 1997). Within the enhancer are two identical, convergently oriented binding sites (CTCAAGTGG) for Tinman, separated by about 180 bp (Gajewski *et al.*, 1997). Mutation of either site completely abolishes enhancer activity, suggesting some form of transcriptional cooperativity between Tinman molecules bound at the two sites. Not only is Tinman necessary for activation of this cardiac enhancer but also it is sufficient for expression, at least in certain regions of the embryo. If a reporter gene linked to this cardiac enhancer is introduced into a *Drosophila* strain harboring a heat shock-inducible *tinman* allele, the enhancer can be activated outside the cardiac lineage in response to ectopic *tinman* expression (Gajewski *et al.,* 1997). Under these conditions, the enhancer can be activated in most regions of the embryo, except in a subset of ventral cells. This suggests that activation of the enhancer by Tinman may require a cofactor that is missing from these cells in which the enhancer cannot be activated, or that these cells contain an inhibitor of Tinman function. In this regard, the NK-2 homeodomain protein, which is the product of the *ventral nervous system defective* (*vnd*) locus (Jimenez *et al.,* 1995), is expressed in neurogenic cells in the ventral region of the embryo and binds the same DNA sequence as Tinman (Tsao *et al.,* 1994), making it possible that it could interfere with Tinman function.

Like the Tinman-dependent enhancer, *tinman* is expressed in four of six cardioblasts within each hemisegment (Apiazu and Frasch, 1993; Bodmer, 1993; see Chapter 5). There is another *D-mef2* cardiac enhancer that is active in the remaining two cardioblasts within each hemisegment (R. Cripps, B. Zhao, and E. Olson, unpublished observations; K. Gajewski and B. Schulz, unpublished observations). We do not yet know the identities of the transcription factors that activate this enhancer, but they promise to be extremely interesting because they represent a *tinman*-independent pathway for cardiac gene expression.

VI. An Evolutionarily Conserved Pathway for Cardiogenesis

Based on work from our group and others, it is possible to consider a transcriptional cascade for cardiogenesis in the *Drosophila* embryo. As schematized in Fig. 6, dorsal mesodermal cells express *tinman* and become committed to a cardiogenic fate in response to Wg and Dpp (Staehling-Hampton *et al.,* 1994; Frasch, 1995; Lawrence *et al.,* 1995; Wu *et al.,* 1995; Park *et al.,* 1996). Tinman directly activates *D-mef2* transcription in cardioblasts (Gajewski *et al.,* 1997), probably in collaboration with other cofactors that are as yet unidentified. D-MEF2 then directly activates the transcription of cardiac contractile protein genes, also in collaboration with other cofactors. While this pathway clearly is incomplete, now that key transcription regulators acting at each step and their direct targets have been identified, it should be possible to identify the additional regulators in the pathway.

What have the MEF2 loss-of-function phenotypes in fruit flies and mice revealed about the possible evolutionary similarities and differences in cardiogenesis in these unrelated organisms? In flies, MEF2 is required for activation of contractile protein genes in the dorsal vessel, but it does not appear to play a role in formation or patterning of the heart tube. In mice, MEF2C is required for activation of a subset of contractile protein

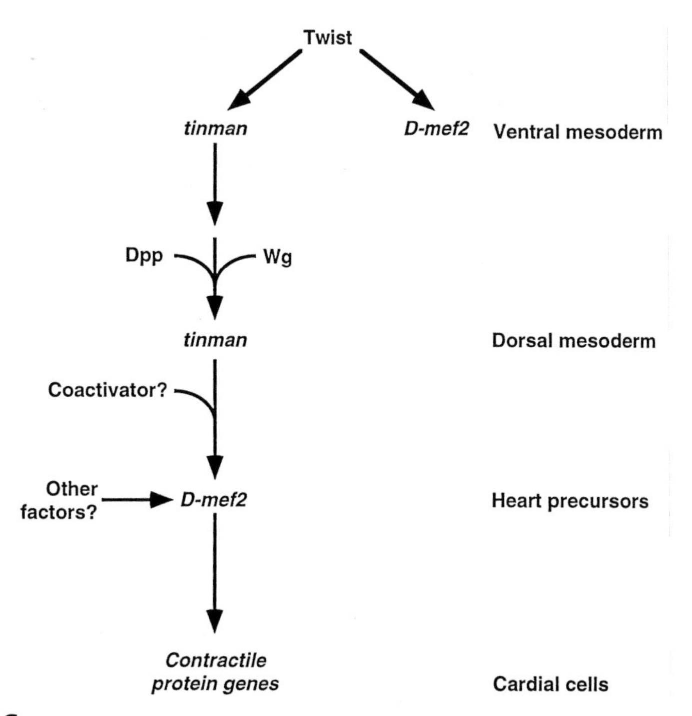

Figure 6 Schematic of a regulatory hierarchy controlling cardiac gene expression in *Drosophila.*

genes, but many are independent of MEF2C (Lin *et al.,* 1997). Thus, it is likely that the early functions of the vertebrate MEF2 factors in muscle gene activation are distributed among multiple family members. An unanticipated conclusion from the mouse mutants is that MEF2C also plays an important role in ventricular morphogenesis and looping of the heart tube—morphogenic events that are unique to vertebrate cardiogenesis (Lin *et al.,* 1997). One interpretation of these findings is that MEF2 factors play evolutionarily conserved roles in the early steps of cardiomyocyte differentiation but are also reemployed during vertebrate evolution to control later aspects of morphogenesis.

VII. Future Questions

MEF2 factors are currently the best understood cardiac regulatory factors with respect to their target genes, mechanisms of action, and regulation. However, many important questions remain. Among these is how MEF2 factors are able to control different programs of myogenesis and morphogenesis. What are the identities of the cofactors with which MEF2 factors cooperate? How are the functions of the different vertebrate MEF2 factors similar and different? The loss-of-function phenotype of MEF2C reveals only the earliest function of the gene in embryogenesis. Does MEF2C play later roles in heart development and function and what are the functions of the other *mef2* genes? How is *mef2* gene expression controlled in the developing heart, as well as in other cell types?

Given the central role of MEF2 factors in several steps in cardiac development, the answers to these questions promise to provide many important insights into the mechanisms that underlie normal heart development as well as the disruption of cardiac regulatory pathways in disease.

Acknowledgments

Work in our laboratory was supported by the National Institutes of Health, the Muscular Dystrophy Association, the American Heart Association, and the Robert A. Welch Foundation. B. L. B. was supported by a postdoctoral fellowship from the American Cancer Society.

References

Andres, V., Cervera, M., and Mahdavi, V. (1995). Determination of the consensus binding site for MEF2 expressed in muscle and brain reveals tissue-specific sequence constraints. *J. Biol. Chem.* **270,** 23246–23249.

Apiazu, N., and Frasch, M. (1993). *Tinman* and *bagpipe:* Two homeo box genes that determine cell fates in the dorsal mesoderm of *Drosophila. Genes Dev.* **7,** 1325–1340.

Biben, C., and Harvey, R. P. (1997). Homeodomain factor Nkx2.5 controls left/right asymmetric expression of bHLH gene *eHAND* during murine heart development. *Genes Dev.* **11,** 1357–1369.

Black, B. L., Ligon, K. L., Zhang, Y., and Olson, E. N. (1996). Cooperative transcriptional activation by the neurogenic bHLH protein MASH1 and members of the MEF2 family. *J. Biol. Chem.* **271,** 26659–26663.

Black, B. L., Lu, J-R., and Olson, E. N. (1997). The MEF2A 3′ untranslated region functions as a *cis*-acting translational repressor. *Mol. Cell. Biol.* **17,** 2756–2763.

Black, B. L., Molkentin, J. D., and Olson, E. N. (1998). Multiple roles for the MyoD basic region in transmission of transcriptional activation signals and interaction with MEF2. *Mol. Cell. Biol.* **18,** 69–77.

Bodmer, R. (1993). The gene *tinman* is required for specification of the heart and visceral muscles in *Drosophila. Development (Cambridge, UK)* **118,** 719–729; published erratum: **119**(3), 969.

Bour, B. A., O'Brien, M. A., Lockwood, W. L., Goldstein, E. S., Bodmer, R., Tagher, P. H., Abmayr, S. M., and Nguyen, H. T. (1995). *Drosophila* MEF2, a transcription factor that is essential for myogenesis. *Genes Dev.* **9,** 730–741.

Breitbart, R., Liang, C., Smoot, L.B., Laheru, D., Mahdavi, V., and Nadal-Ginard, B. (1993). A fourth human MEF-2 transcription factor, hMEF2D, is an early marker of the myogenic lineage. *Development (Cambridge, UK)* **118,** 1095–1106.

Buchberger, Ragge, A. K., and Arnold, H. H. (1994). The *myogenin* gene is activated during myocyte differentiation by pre-existing, not newly synthesized transcription factor MEF-2. *J. Biol. Chem.* **269,** 17289–17296.

Chen, C. Y., and Schwartz, R. J. (1996). Recruitment of the tinman homolog Nkx-2.5 by serum response factor activates cardiac alpha-actin gene transcription. *Mol. Cell. Biol.* **16,** 6372–6384.

Cheng, T. C., Wallace, M. C., Merlie, J. P., and Olson, E. N. (1993). Separable regulatory elements governing myogenin transcription in mouse embryogenesis. *Science* **261,** 215–218.

Cserjesi, P., and Olson, E. N. (1991). Myogenin induces muscle-specific enhancer binding factor MEF-2 independently of other muscle-specific gene products. *Mol. Cell. Biol.* **11,** 4854–4862.

Cserjesi, P., Brown, D., Lyons, G. E., and Olson, E. N. (1995). Expression of the novel basic helix-loop-helix gene *eHAND* in neural crest derivatives and extraembryonic membranes during mouse development. *Dev. Biol.* **170,** 664–678.

Edmondson, D. G., Lyons, G. E., Martin, J. F., and Olson, E. N. (1994). *mef2* gene expression marks the cardiac and skeletal muscle lineages during mouse embryogenesis. *Development (Cambridge, UK)* **120,** 1251–1263.

Firulli, A. B., Miano, J. M., Weizhen, B. I., Johnson, A. D., Casscells, W., Olson, E. N., and Schwarz, J. J. (1996). Myocyte enhancer-binding-factor-2 expression and activity in vascular smooth muscle cells: Association with the activated phenotype. *Circ. Res.* **78,** 196–204.

Fishman, M. C., and Chien, K. R. (1997). Fashioning the vertebrate heart: earliest embryonic decisions. *Development (Cambridge, UK)* **124,** 2099–2117.

Frasch, M. (1995). Induction of visceral and cardiac mesoderm by ectodermal Dpp in the early *Drosophila* embryo. *Nature (London)* **374,** 464–467.

Gajewski, K., Kim, Y., Lee, Y. M., Olson, E. N., and Schulz, R. A. (1997). *D-mef2* is a target for Tinman activation during *Drosophila* heart development. *EMBO J.* **16,** 515–522.

Gossett, L. A., Kelvin, D. J., Sternberg, E. A., and Olson, E. N. (1989). A new myocyte-specific enhancer-binding factor that recognizes a conserved element associated with multiple muscle-specific genes. *Mol. Cell Biol.* **9,** 5022–5033.

Han, J., Jiang, Y., Li, Z., Kravchenko, V. V., and Ulevitch, R. J. (1997). Activation of the transcription factor MEF2C by the MAP kinase p38 in inflammation. *Nature* (*London*) **386,** 296–299.

Han, T.-H., and Prywes, R. (1995). Regulatory role of MEF2D in serum induction of the *c-jun* promoter. *Mol. Cell. Biol.* **15,** 2907–2915.

Harvey, R. P. (1996). NK-2 homeobox genes and heart development. *Dev. Biol.* **178,** 203–216.

Jimenez, F., Martin-Morris, L. E., Velasco, L., Chu, H., Sierra, J., Rosen, D. R., and White, K. (1995). *vnd,* a gene required for early neurogenesis of Drosophila, encodes a homeodomain protein. *EMBO J.* **14,** 3487–3495.

Kaushal, S., Schneider, J. W., Nadal-Ginard, B., and Mahdavi, V. (1994). Activation of the myogenic lineage by MEF2A, a factor that induces and cooperates with MyoD. *Science* **266,** 1236–1240.

Kirby, M. L., and Waldo, K. L. (1995). Neural crest and cardiovascular patterning. *Circ. Res.* **77,** 211–215.

Kuisk, I. R., Li, H., Tran, D., and Capetanaki, Y. (1996). A single MEF2 site governs *desmin* transcription in both heart and skeletal muscle during mouse embryogenesis. *Dev. Biol.* **174,** 1–13.

Lawrence, P. A., Bodmer, R., and Vincent, J. P. (1995). Segmental patterning of heart precursors in Drosophila. *Development* (*Cambridge, UK*) **121,** 4303–4308.

Leifer, D., Krainc, D., Yu, Y.T., McDermott, J. C., Breitbart, R., Heng, J., Neve, R. L., Kosofsky, B., Nadal-Ginard, B., and Lipton, S. A. (1993). MEF2C, a MADS/MEF2-family transcription factor expressed in a laminar distribution in cerebral cortex. *Proc. Natl. Acad. Sci. U.S.A.* **90,** 1546–1550.

Leifer, D., Golden, J., and Kowall, N. W. (1994). Myocyte-specific enhancer binding factor 2C expression in human brain development. *Neuroscience* **63,** 1067–1079.

Lilly, B., S. Galewski, S., Firulli, A. B., Schulz, R. A., and Olson, E. N. (1994). *mef2:* A MADS gene expressed in the differentiating mesoderm and the somatic muscle lineage during *Drosophila* embryogenesis. *Proc. Natl. Acad. Sci. U.S.A.* **91,** 5662–5666.

Lilly, B., Zhao, B., Ranganayakulu, G., Paterson, B. M., Schulz, R. A., and Olson, E. N. (1995). Requirement of MADS domain transcription factor D-MEF2 for muscle formation in *Drosophila. Science* **267,** 688–693.

Lin, Q., Schwarz, J. J., Bucana, C., and Olson, E. N. (1997). Control of mouse cardiac morphogenesis and myogenesis by the myogenic transcription factor MEF2C. *Science* **276,** 1404–1407.

Ludolph, D. C., and Konieczny, S. F. (1995). Transcription factor families: Muscling in on the myogenic program. *FASEB J.* **9,** 1595–1604.

Lyons, G. E., Micales, B. K., Schwarz, J. J., Martin, J. F., and Olson, E. N. (1995). Expression of *mef2* genes in the mouse central nervous system suggests a role in neuronal maturation. *J. Neurosci.* **15,** 5727–5738.

Lyons, I., Parsons, L. M., Hartley, L., Li, R., Andrews, J. E., Robb, L., and Harvey, R. P. (1995). Myogenic and morphogenetic defects in the heart tubes of murine embryos lacking the homeo box gene *Nkx2.5. Genes Dev.* **9,** 1654–1666.

Mao, Z., and Nadal-Ginard, B. (1996). Functional and physical interactions between mammalian achaete-scute homolog 1 and myocyte enhancer factor 2A. *J. Biol. Chem.* **271,** 14371–14375.

Martin, J. F., Schwarz, J. J., and Olson, E. N. (1993). Myocyte enhancer factor (MEF) 2C: A tissue-restricted member of the MEF-2 family of transcription factors. *Proc. Natl. Acad. Sci. U.S.A.* **90,** 5282–5286.

Martin, J. F., Miano, J., Husted, C. M., Copeland, N. G., Jenkins, N. A., and Olson, E. N. (1994). A *mef2* gene that generates a muscle-specific isoform via alternative mRNA splicing. *Mol. Cell Biol.* **14,** 1647–1656.

McDermott, J. C., Cardoso, M. C., Yu, Y. T., Andres, V., Leifer, D., Krainc, D., Lipton, S. A., and Nadal-Ginard, B. (1993). hMEF2C gene encodes skeletal muscle- and brain-specific transcription factors. *Mol. Cell Biol.* **13,** 2564–2577.

Molkentin, J. D., and Olson, E. N. (1996). Combinatorial control of muscle development by basic helix-loop-helix and MADS-box transcription factors. *Proc. Natl. Acad. Sci. U.S.A.* **93,** 9366–9373.

Molkentin, J. D., Kalvakolanu, D., and Markham, B. E. (1994). Transcription factor GATA-4 regulates cardiac muscle-specific expression of the *α-myosin heavy chain* gene. *Mol. Cell. Biol.* **14,** 4947–4957.

Molkentin, J. D., Black, B. L., Martin, J. F., and Olson, E. N. (1995). Cooperative activation of muscle gene expression by MEF2 and myogenic bHLH proteins. *Cell* (*Cambridge, Mass.*) **83,** 1125–1136.

Molkentin, J. D., Black, B. L., Martin, J. F., and Olson, E. N. (1996a). Mutational analysis of the DNA binding, dimerization, and transcriptional activation domains of MEF2C. *Mol. Cell. Biol.* **16,** 2627–2636.

Molkentin, J. D., Firulli, A. B., Black, B. L., Martin, J. F., Husted, C. M., Copeland, N., Jenkins, N., Lyons, G., and Olson, E. N. (1996b). MEF2B is a potent transactivator expressed in early myogenic lineages. *Mol. Cell. Biol.* **16,** 3814–3824.

Molkentin, J. D., Li, L., and Olson, E. N. (1996c). Phosphorylation of the MADS-box transcription factor MEF2C enhances its DNA binding activity. *J. Biol. Chem.* **271,** 17199–17204.

Molkentin, J. D., Lin, Q., Duncan, S. A., and Olson, E. N. (1997). Requirement of the transcription factor GATA4 for heart tube formation and ventral morphogenesis. *Genes Dev.* **11,** 1061–1072.

Navankasattusas, S., Zhu, H., Garcia, A. V., Evans, S. M., and Chien, K. R. (1992). A ubiquitous factor (HF-1a) and a distinct muscle factor (HF-1b/MEF2) form an E-box-independent pathway for cardiac muscle gene expression. *Mol. Cell. Biol.* **12,** 1469–1479.

Nguyen, H. T., Bodmer, R., Abmayr, S. M., McDermott, J. C., and Spoerel, N. A. (1994). *D-mef2:* A Drosophila mesoderm-specific MADS box-containing gene with a biphasic expression profile during embryogenesis. *Proc. Natl. Acad. Sci. U.S.A.* **91,** 7520–7524.

Olson, E. N., and Srivastava, D. (1996). Molecular pathways controlling heart development. *Science* **272,** 671–676.

Olson, E. N., Perry, M., and Schulz, R. A. (1995). Regulation of muscle differentiation by the MEF2 family of MADS box transcription factors. *Dev. Biol.* **172,** 2–14.

Park, M. Y., Wu, X. S., Golden, K., Axelrod, J. D., and Bodmer, R. (1996). The wingless signaling pathway is directly involved in *Drosophila* heart development. *Dev. Biol.* **177,** 104–116.

Pollock, R., and Treisman, R. (1991). Human SRF-related proteins: DNA-binding properties and potential regulatory targets. *Genes Dev.* **5,** 2327–2341.

Ranganayakulu, G., Zhao, B., Dokodis, A., Molkentin, J. D., Olson, E. N., and Schulz, R. A. (1995). A series of mutations in the D-MEF2 transcription factor reveals multiple functions in larval and adult myogenesis in *Drosophila. Dev. Biol.* **171,** 169–181.

Rugendorff, A., Younossi-Hartenstein, A., and Hartenstein, V. (1994). Embryonic origin and differentiation of the *Drosophila* heart. *Roux's Arch. Dev. Biol.* **203,** 266–280.

Schulz, R. A., Chromey, C., Lu, M.-F., Zhao, B., and Olson, E. N. (1996). Expression of the D-MEF2 transcription factor in the *Drosophila* brain suggests a role in neuronal cell differentiation. *Oncogene* **12,** 1827–1831.

Shore, P., and Sharrocks, A. D. (1995). The MADS-box family of transcription factors. *Eur. J. Biochem.* **229,** 1–13.

Srivastava, D., and Olson, E. N. (1997). Knowing in your heart what's right. *Trends Cell Biol.* **7,** 447–453.

Srivastava, D., Cserjesi, P., and Olson, E. N. (1995). A subclass of bHLH proteins required for cardiac morphogenesis. *Science* **270,** 1995–1999.

Srivastava, D., Thomas, T., Lin, Q., Kirby, M. L., Brown, D., and Olson,

E. N. (1997). Regulation of cardiac mesodermal and neural crest development by the bHLH transcription factor, dHAND. *Nat. Genet.* **16,** 154–160.

Staehling-Hampton, K., Hoffman, F. M., Baylies, M. K., Rushton, E., and Bate, M. (1994). dpp induces mesodermal gene expression in *Drosophila. Nature* (*London*) **372,** 783–786.

Subramanian, S. V., and Nadal-Ginard, B. (1996). Early expression of the different isoforms of the myocyte enhancer factor-2 (MEF2) protein in myogenic as well as non-myogenic cell lineages during mouse embryogenesis. *Mech. Dev.* **57,** 103–112.

Suzuki, E., Guo, K., Kolman, M., Yu, Y. T., and Walsh, K. (1995). Serum induction of MEF2/RSRF expression in vascular myocytes is mediated at the level of translation. *Mol. Cell. Biol.* **15,** 3415–3423.

Taber, L. A., Lin, I.-E., and Clark, E. B. (1995). Mechanics of cardiac looping. *Dev. Dyn.* **203,** 42–50.

Tsao, D. H. H., Gruschus, J. M., Wang, L.-H., Nirenberg, M., and Ferretti, J. A. (1994). Elongation of helix III of the NK-2 homeodomain upon binding to DNA: A secondary structure study by NMR. *Biochemistry* **33,** 15053–15060.

Wu, X., Golden, K., and Bodmer, R. (1995). Heart development in *Drosophila* requires the segment polarity gene wingless. *Dev. Biol.* **169,** 619–628.

Yost, H. J. (1995). Vertebrate left-right development. *Cell* (*Cambridge, Mass.*) **82,** 689–692.

Yu, Y. T., Breitbart, R. E., Smoot, L. B., Lee, Y., Mahdavi, V., and Nadal-Ginard, B. (1992). Human myocyte-specific enhancer factor 2 comprises a group of tissue-restricted MADS box transcription factors. *Genes Dev.* **6,** 1783–1798.

Yun, K., and Wold, B. (1996). Skeletal muscle determination and differentiation: Story of a core regulatory network and its context. *Curr. Opin. Cell Biol.* **8,** 877–889.

Yutzey, K. E., and Bader, D. (1995). Diversification of cardiomyogenic cell lineages during early heart development. *Circ. Res.* **77,** 216–219.

9

Segmental Regulation of Cardiac Development by the Basic Helix–Loop–Helix Transcription Factors dHAND and eHAND

Deepak Srivastava

Departments of Pediatric Cardiology, and Molecular Biology and Oncology, University of Texas Southwestern Medical Center, Dallas, Texas 75235

In one of the more interesting forms of malformation the septum of the ventricles is not only incomplete, but is found to deviate from its natural position, one of the ventricles being unduly developed, while the other is atrophied."

—Thomas B. Peacock (1858)

I. Introduction

During human fetal development, the majority of organogenesis is completed during the first trimester of pregnancy after which further maturation and growth predominate. The heart is the first organ to form, with the earliest recognizable cardiac structure evident at 3 weeks of gestation concomitant with the onset of rhythmic heartbeats. By 7 weeks, morphogenesis of the heart is complete with establishment of appropriate venous and arterial connections. The heart is derived from multiple cell lineages and must differentiate into unique regions, each possessing different physiologic, electrical, and anatomic properties. The impact of hemodynamic loads introduces a separate level of complexity in cardiac chamber growth and alignment with respect to the vasculature.

In light of the multiple events necessary for normal cardiogenesis, it is not surprising that defects in development of the heart are the most common of human birth defects. In fact, only a fraction of cardiac developmental defects are seen clinically, with the more severe abnormalities resulting in spontaneous abortions in the first trimester (Hoffman, 1995). Defects in myocardial function result in cardiac insufficiency and embryonic lethality. In contrast, newborns with congenital heart disease typically have normal myocardial function and have a cardiac anatomy which is suitable for fetal circulation but not for a newborn circulatory system, in which separation of the pulmonary (lungs) and systemic (body) circulations is necessary. Such anatomy is usually a result of malalignment, malformation, or arrested development of specific regions of the heart rather than global defects of the heart. This observation suggests that each cardiac chamber and its vascular connections are under distinct regulatory controls during development, although obvious interactions between programs are critical for appropriate cardiogenesis. A segmental approach to understanding the regulation of cardiac development may therefore be useful, just as it has been instrumental in understanding the anatomy of cardiogenesis.

II. Cardiac Morphogenesis

Cardiogenesis involves cellular determination, migration, and differentiation along with a series of critical morphogenetic events (Olson and Srivastava, 1996). Cells from the anterior lateral plate mesoderm give rise to the precardiogenic mesoderm, which is fated to form the heart well before cardiac morphogenesis begins (Yutzey and Bader, 1995). The bilaterally symmetric heart primordia migrate to the midline and fuse to form a single beating heart tube. Although seemingly homogeneous, the straight heart tube is patterned in an anterior–posterior (AP) fashion to form the future regions of the four-chambered heart, including the aortic sac, conotruncus, right ventricle, left ventricle, and atria (DeHaan, 1965; see Chapters 19 and 20) (Fig. 1). The heart tube forms a rightward loop, the direction being conserved in all vertebrates, which begins to establish the left–right (LR) asymmetry of the heart and the spatial orientation of the cardiac chambers (Levin, 1997). Proper looping of the heart tube is necessary for correct alignment of the chambers and outflow tract of the heart and converts the AP polarity to a LR polarity (see Chapters 21 and 22). Further septation leads eventually to a four-chambered heart with two atrioventricular valves in higher organisms (see Chapter 10). Dissecting the molecular pathways controlling development and alignment of the individual chambers will be necessary to determine how individual regions of the heart are malformed.

The other major cell type which contributes to heart formation is a population of migratory neural crest cells known as the cardiac neural crest. Neural crest cells migrate from the neural folds and condense in the cardiac outflow tract (conotruncus) and developing aortic arches (Kirby and Waldo, 1990; see Chapter 11). They are involved in formation of the truncus arteriosus and subsequent septation of the truncus into the aorta and pulmonary artery as well as formation of the conotruncal portion of the ventricular septum (Kirby *et al.*, 1983). After septation of the aorta and pulmonary artery, the vessels rotate in a twisting fashion to achieve their final connection with the left and right ventricles, respectively. Neural crest cells also contribute to the development of the bilaterally symmetric aortic arches which undergo extensive remodeling resulting in formation of the ascending aorta, proximal subclavian, carotid, and pulmonary ar-

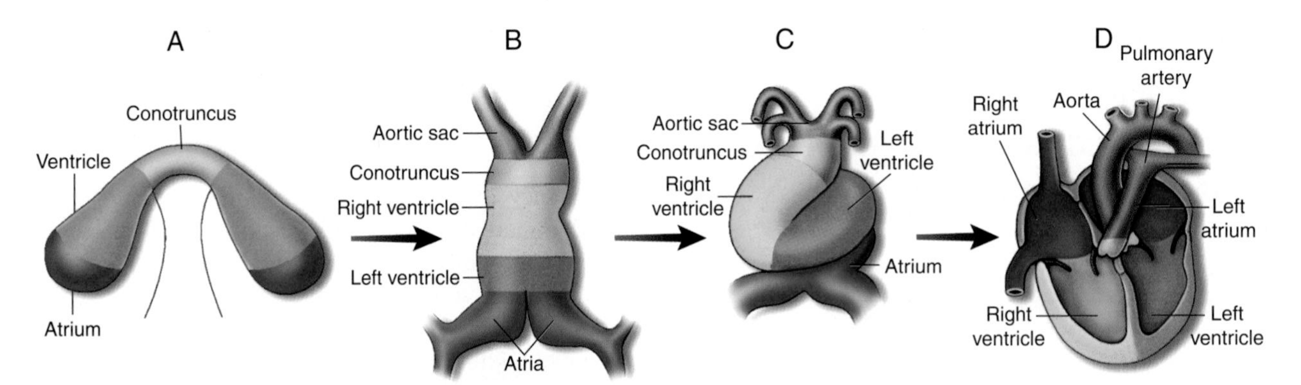

Figure 1 Schematic diagram of cardiogenesis. Bilaterally symmetrical cardiac progenitor cells (A) are prepatterned to form distinct regions of the heart as shown in color-coded fashion. The precardiac mesodermal cells give rise to a linear heart tube (B) which forms a rightward loop (C) and begins to establish the spatial orientation of the four-chambered mature heart (D) (adapted from Srivastava and Olson, 1997).

teries (Kirby and Waldo, 1990). Determining how neural crest cells are instructed to migrate, differentiate, and proliferate is fundamental to understanding the pathogenesis of a variety of conotruncal and aortic arch defects.

III. Molecular Segmentation of the Heart

Analysis of the regulatory regions of some cardiac structural genes in transgenic mice provides molecular evidence for segmental regulation of the developing heart. The *desmin* gene is expressed uniformly in the heart, but unique *cis* elements exist which govern expression in the right ventricle (Kuisk *et al.,* 1996). The smooth muscle marker, SM-22, is also expressed in the developing heart uniformly, but *cis* elements have been identified which regulate predominately right ventricular expression (Li *et al.,* 1996). Similarly, a ventricular isoform of myosin light chain (MLC2V) is expressed in the right and left ventricles, but a stretch of 28 nucleotides in the regulatory region of the gene is responsible for right but not left ventricular expression (Ross *et al.,* 1996; see Chapter 15). Finally, a regulatory sequence in the *myosin light chain 1F* gene has been demonstrated to control the gene's left but not right ventricular expression (Kelly *et al.,* 1995; Franco *et al.,* 1997; see Chapter 19). These findings are consistent with a model in which unique regulatory pathways control development of each chamber of the heart.

Embryologic evidence of segmental formation of the heart has also come from retinoic acid treatment of embryos from multiple species. Frog (Drysdale *et al.,* 1994), zebrafish (Stainier and Fishman, 1992), and chick (Osmond *et al.,* 1991) embryos display anterior truncation of the cardiac tube after exposure to retinoic acid. A lacZ insertion into a locus on chromosome 13, termed *hdf,* results in anterior truncation of a segment of the heart tube, which is further truncated by exposure to retinoic acid (Yamamura *et al.,* 1997). The gene responsible for this phenotype is yet to be identified.

The distinct regulatory patterns of cardiac genes in individual chambers suggest that *trans*-acting factors will exist in a segmental fashion in order to establish the observed pattern. At least four families of transcription factors are expressed in the cardiac primordia fated to form the heart: the myocyte enhancer binding factor-2 (MEF2) factors (Olson *et al.,* 1995; Yu *et al.,* 1992; Pollock and Treisman, 1991), NK homeodomain proteins (Harvey, 1996), the zinc finger containing-GATA factors (Arceci *et al.,* 1993; Ip *et al.,* 1994; Laverriere *et al.,* 1994), and the basic helix–loop–helix (bHLH) proteins dHAND and eHAND (Srivastava *et al.,* 1995; Cserjesi *et al.,* 1995). The MEF2 family of transcription factors are expressed throughout the developing heart tube (Edmondson *et al.,* 1994); gene deletion of one member, MEF2C, in mice results in hypoplastic right and left ventricles (Lin *et al.,* 1997; see Chapter 8). Nkx2.5/Csx, a member of the NK family, is also expressed throughout the heart (Lints *et al.,* 1993; Komuro and Izumo, 1993), but disruption of this gene in mice leads to cardiac insufficiency and failure to complete the looping process (Lyons *et al.,* 1995; see Chapter 7). The cardiac role of GATA-4, -5, and -6, all coexpressed uniformly in the heart, remains unclear although GATA-4 null mice have defects in ventral morphogenesis, including failure to fuse the paired heart tubes in the midline (Molkentin *et al.,* 1997; Kuo *et al.,* 1997; see Chapter 17). The uniform cardiac expression of these factors argues against a role in specific chambers, although interactions with other chamber-specific factors may confer unique spacial properties.

IV. Basic Helix–Loop–Helix Factors in Development

Members of the bHLH family of transcription factors regulate determination and differentiation of skeletal muscle (Olson and Klein, 1994; Weintraub, 1993), neuronal (Jan and Jan, 1993; Lee *et al.,* 1995; Ma *et al.,* 1996) and hematopoietic cells (Zhuang *et al.,* 1994; Shivdasani *et al.,* 1995). The HLH motif mediates dimerization of bHLH proteins, which juxtaposes their basic regions to form a bipartite DNA-binding domain which recognizes the E box consensus sequence (CANNTG) in the control region of downstream target genes (Fig. 2). Ubiquitous bHLH proteins, known as class A bHLH proteins (E proteins) (Murre *et al.,* 1989b; Henthorn *et al.,* 1990; Hu *et al.,* 1992), dimerize preferentially with cell type-specific class B bHLH proteins (Murre *et al.,* 1989a). There are also HLH proteins which lack a basic domain and dimerize with bHLH proteins to form heterodimers which cannot bind DNA. Among this class are members of the Id family (Benezra *et al.,* 1990) and *Drosophila* extramacrochaete (Ellis *et al.,* 1990).

There are four skeletal muscle-specific bHLH proteins (MyoD, myogenin, myf5, and MRF4) which share extensive homology within their bHLH regions (Olson and Klein, 1994; see Chapter 28). The myogenic bHLH proteins have the remarkable ability to independently induce a muscle phenotype when overexpressed in fibroblast cells (Davis *et al.,* 1987). During embryogenesis, the four bHLH factors are expressed in distinct but overlapping patterns in the skeletal, but not cardiac, muscle lineage. Gene knockout experiments have shown that the myogenic bHLH genes comprise a regulatory network which controls myoblast determination and differentiation. Similarly, members of the achaete scute family of bHLH proteins in *Drosophila* (Jan and Jan, 1993) and their mammalian homologs are expressed in neurogenic

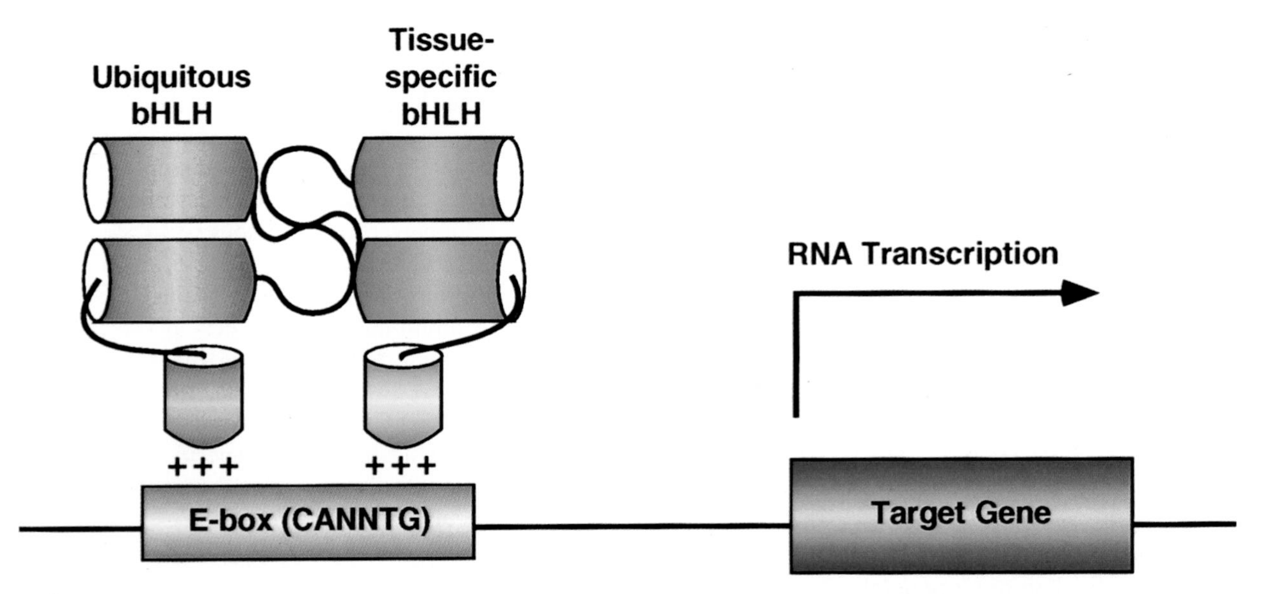

Figure 2 Schematic diagram of basic helix–loop–helix (bHLH) proteins. Tissue-specific bHLH proteins heterodimerize with ubiquitous bHLH factors via the HLH domain. Heterodimerization enables a bipartite basic domain (+++) to interact with consensus Ebox sequences (CANNTG) located in the regulatory region of target genes. Such interaction initiates a cascade of events culminating in regulation of downstream genes.

precursors and their descendants (Johnson *et al.,* 1990), in which they regulate neurogenesis (Guillermot *et al.,* 1993).

V. Basic Helix–Loop–Helix Factors in Heart Development

The bHLH transcription factors dHAND and eHAND (*d*eciduum/*e*xtraembryonic membrane, *h*eart, *a*utonomic nervous system, *n*eural crest-*d*erived cell types) provide an entry to investigate the regulatory programs which might govern cardiac transcription in a segmental fashion. dHAND and eHAND, also referred to as Thing-2/Hed and Thing-1/Hxt, respectively (Hollenberg *et al.,* 1995; Cross *et al.,* 1995), share high homology within their bHLH regions and are encoded by genes with similar intron–exon organization, suggesting that the genes arose by duplication of an ancestral *HAND*-like gene (Fig. 3). The dHAND proteins from mouse, chick, human and frog (D. Srivastava, unpublished observations) and zebrafish (K. Lee, unpublished observations) share >95% amino acid homology, whereas eHAND is much less conserved across species, suggesting that *dHAND* may be the more ancient of the two genes.

In the chick (Srivastava *et al.,* 1995) (Fig. 4) and frog (D. Srivastava, J. Lohr, and J. Yost, unpublished observations), *dHAND* and *eHAND* are coexpressed in a bilaterally symmetric pattern throughout the precardiac mesoderm, linear and looped heart tube, as well as the lateral mesoderm and certain neural crest-derived structures. Antisense experiments in the chick suggest that dHAND and eHAND play redundant roles in cardiac development beyond the stage of cardiac looping (Srivastava *et al.,* 1995). Disruption of *dHAND* and *eHAND* mRNA in combination, but not alone, resulted in arrest of cardiac development just after the heart began to loop in the rightward direction.

In contrast to their apparently homogeneous expression throughout the developing heart in the chick, *dHAND* and *eHAND* exhibit distinct expression patterns during cardiogenesis in the mouse (Fig. 5). Both genes are initially expressed in the precardiac mesoderm. *dHAND* is also expressed throughout the linear heart tube but becomes restricted predominantly to the future right ventricular compartment during cardiac looping (Srivastava *et al.,* 1995, 1997). By contrast, eHAND expression is restricted to the anterior and posterior segments of the straight heart tube, which are fated to form the conotruncus and left ventricle, respectively, but is undetectable in the intervening right ventricle-forming region. The interrupted AP pattern of expression is maintained as the heart loops and becomes a distinct LR cardiac asymmetry by virtue of the morphogenetic movements of cardiac looping with expression of *eHAND* in the left, but not right, ventricle (Srivastava *et al.,* 1997; Biben and Harvey, 1997; see Chapter 7). The spatially distinct expression patterns of *dHAND* and *eHAND* make them candidate genes for controlling the segmental development of the heart tube.

The complementary expression of *dHAND* and *eHAND* in the right and left ventricles, respectively, raises the question of whether they are involved in specification of particular chambers of the heart or in determining the direction of looping of the heart. It is

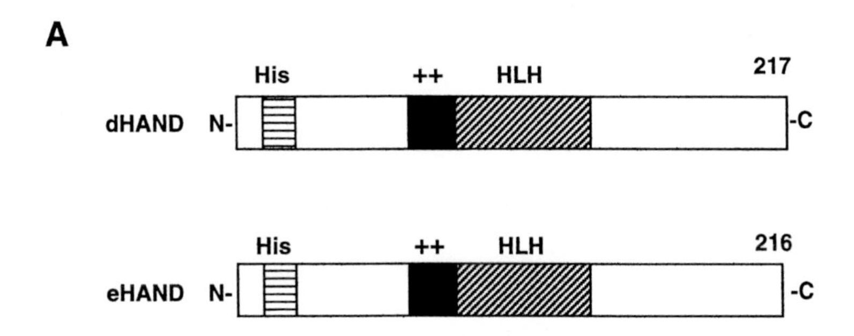

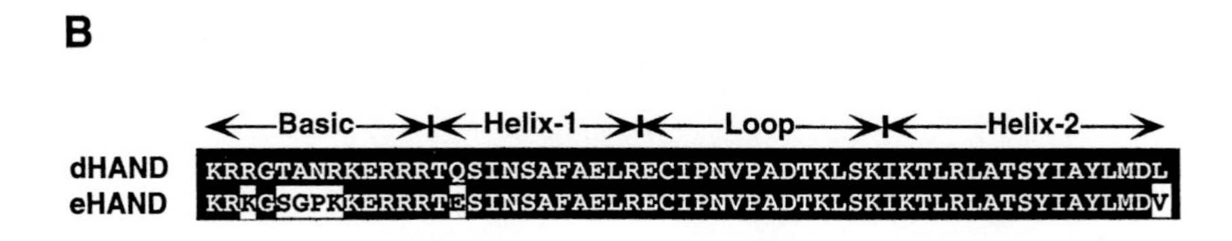

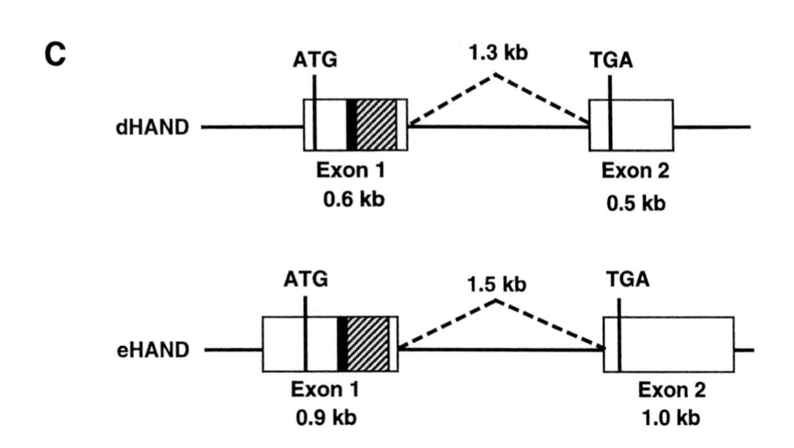

Figure 3 Comparison of the mouse bHLH proteins dHAND and eHAND. (A) dHAND and eHAND share similar protein structure composed of 217 and 216 amino acids, respectively. Both have histidine (His)-rich regions in the amino terminal and a centrally located bHLH region. (B) Comparison of amino acid sequences in the critical bHLH domain reveals a high homology between dHAND and eHAND as shown in shaded boxes. Sequences diverge more in the basic, or DNA-binding, domain (+ +). (C) *dHAND* and *eHAND* share a similar genomic structure, with each having two exons separated by a single intron.

unlikely that they are involved in establishing the direction of looping given the bilaterally symmetric expression in the lateral mesoderm, unlike members of the transforming growth factor-β family (Lowe *et al.,* 1996; Collignon *et al.,* 1996; Meno *et al.,* 1996), whose LR asymmetry of expression controls direction of cardiac looping (Levin, 1997). In addition, the *HAND* genes are expressed symmetrically along the left–right axis of the straight heart tube (Srivastava *et al.,* 1997; Thomas *et al.,* 1998), although Biben and Harvey (1997) suggest that a caudal left–right asymmetry of *eHAND* expression exists. In a mouse model of situs inversus (*inv/inv*) (Yokoyama *et al.,* 1993), *dHAND* and *eHAND* expression in the looped heart tube is reversed along the LR axis, but *dHAND* continues to be expressed in the pulmonary ventricle and eHAND expression persists in the systemic ventricle (Thomas *et al.,* 1998). This suggests that *HAND* gene expression is chamber specific rather than embryonic side specific.

Although no asymmetry is detectable along the LR axis of the linear heart tube, in the mouse the *HAND* genes are expressed on only the ventral and not the dor-

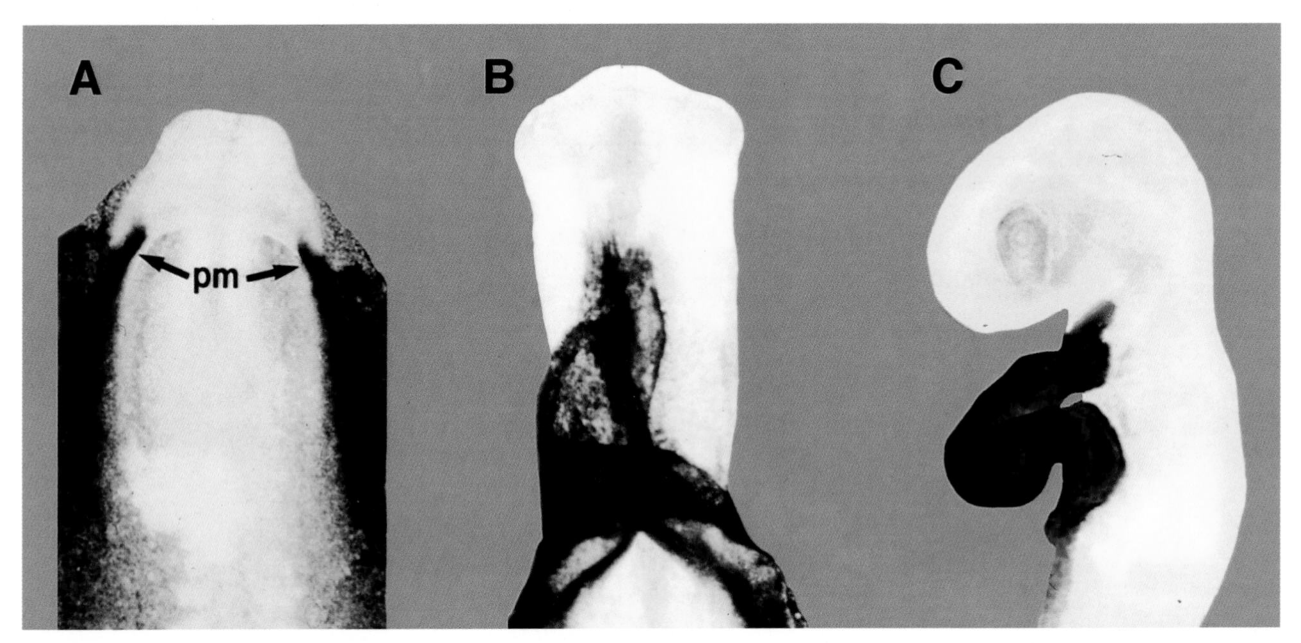

Figure 4 Expression of *dHAND* and *eHAND* transcripts during cardiac development in chick embryos as monitored by whole mount *in situ* hybridization. *dHAND* alone is shown here because *eHAND* is coexpressed in chick embryos. *HAND* transcripts were first detected in the precardiogenic mesoderm (pm) and lateral mesoderm of stage 8^- embryos (A). Expression of *dHAND* and *eHAND* was seen uniformly in the straight heart tube of stage 10 embryos (B) and in the looped heart tube of stage 16 embryos (C).

sal surface of the straight heart tube. After the torsion of looping, it is thought that the ventral surface forms the outer curvature of the looped heart, whereas the dorsal surface becomes the inner curvature. *HAND* gene expression remains nonconcentric and is seen along the outer curvature but not the inner curvature (Srivastava *et al.,* 1997; Biben and Harvey, 1997). It is interesting to speculate that the *HAND* genes may regulate differential cell proliferation or cell death in the outer and inner curvatures, thereby contributing to the process of cardiac looping and remodeling of the inner curvature.

VI. Targeted Gene Deletion in Mice

A. Cardiac Mesodermal Defects

Significant insight into the role of the *HAND* genes in cardiogenesis has come from mouse knockout studies. Heterozygote *dHAND* null mice survive to reproductive age, but homozygous null mice die by Embryonic (E) Day 10.5, apparently from cardiac failure (Srivastava *et al.,* 1997). *dHAND* null embryos begin the process of cardiac looping in the rightward direction but fail to develop the segment of the heart tube that forms the right ventricle (Fig. 6). This is consistent with the predominant expression of *dHAND* in the right ventricle-forming region. The atrial chamber moves dorsally and to the left as it should during cardiac looping in the mutant, suggesting that the process and direction of cardiac looping is initiated correctly but appears abnormal because of growth failure in a specific segment of the looping heart tube. The left-sided ventricle, where *dHAND* is normally expressed at lower levels, forms in the mutant but lacks trabeculations, the finger-like projections of myocardium necessary for increasingly forceful contractions of the heart. The gene encoding the cardiac transcription factor, GATA-4, is downregulated in the left ventricle, a finding which may be related to the absence of trabeculations. Thus, dHAND is necessary for morphogenesis of an entire segment of the developing heart tube and for proper function of other areas where *dHAND* is expressed less robustly.

How might dHAND regulate right ventricular development? We favor the interpretation that dHAND is required for the expansion or specification of a population of cardiogenic precursor cells within the linear heart tube which is destined to form the right ventricular region. At the linear heart tube stage, there are no obvious differences between wild-type embryos and *dHAND* mutants. However, after E8.0, when looping begins and the future right ventricular segment should expand, there is no growth of the corresponding region of the mutant heart tube. Instead, the heart tube develops abruptly leftward and ultimately gives rise to the left-sided ventricle. Because *eHAND* is expressed in the single ventricular chamber of the *dHAND* null heart, we believe that this region of the heart tube is specified to form the morphologic left ventricle and the right ventricle region is deleted (Srivastava *et al.,* 1997).

Antisense experiments in cultured chick embryos and gene knockout experiments in mice both demon-

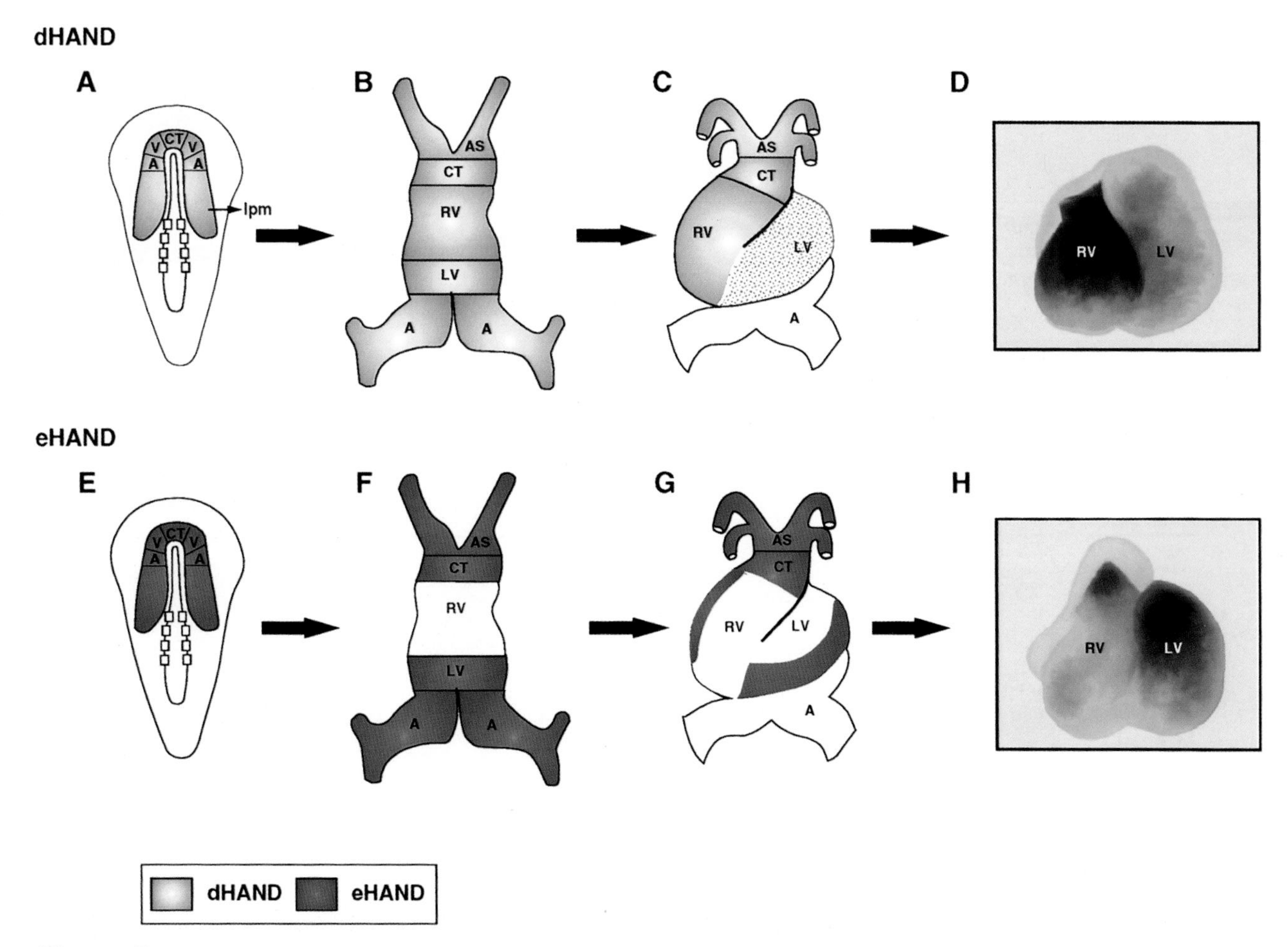

Figure 5 Summary of *dHAND* and *eHAND* expression patterns during mouse cardiogenesis. *dHAND* (blue) and *eHAND* (green) are uniformly expressed in cardiac progenitor cells (A, E) and in the left and right lateral plate mesoderm (lpm) but become restricted to the future right and left ventricle-forming regions, respectively, as the heart tube loops. *dHAND* is expressed throughout the linear heart tube (B) but becomes predominately right sided after looping (C). *eHAND* is expressed in the conotruncus (CT) and left ventricle (LV)-forming regions of the linear heart tube (F); the anterior–posterior interrupted pattern becomes left–right by virtue of cardiac looping (G). Expression is nonconcentric and is along the outer curvature of the heart (G). RNA *in situ* hybridization with isolated E 10.0 mouse hearts shows expression of *dHAND* in the right ventricle (D) and *eHAND* in the left ventricle (H) (frontal views). Both genes are expressed in the aortic sac (AS), which gives rise to the aorta and pulmonary arteries, but are downregulated in the myocardium of the heart once formation is complete. RV, right ventricle; A, atria; LA, left atrium; RA, right atrium (adapted from Srivastava and Olson, 1997).

strate an important role for the *HAND* genes at the time of cardiac looping (Srivastava *et al.,* 1995, 1997), but there are some intriguing differences in the expression patterns of *dHAND* and *eHAND* in chick and mouse embryos that suggest how these genes might function. In chicks, *dHAND* and *eHAND* are coexpressed throughout the heart without segmental restriction and appear to have some degree of functional redundancy. Frogs, which have only three-chambered hearts, also coexpress the *HAND* genes uniformly. By contrast, *eHAND* is not expressed in the right ventricle-forming segment of the mouse heart and would therefore be unable to compensate for loss of *dHAND* expression in this segment. *dHAND*-null mice have a less severe defect in the left ventricle, where *eHAND* is expressed, suggesting some compensation by *eHAND.* This interpretation supports a model in which dHAND and eHAND play similar roles, possibly in proliferation or cell survival of cardiomyocytes, within spatially distinct regions of the heart. Alternatively, dHAND and eHAND may confer the unique physiological properties of the right and left ventricles, respectively, and not be functionally similar in the mouse.

Intriguingly, the defects in cardiac morphogenesis in *dHAND* mutant embryos are similar to, but less severe than, those of embryos lacking the MADS box transcription factor MEF2C (Lin *et al.,* 1997), one of four members of the MEF2 family of myogenic transcription factors. During cardiac development, *MEF2C* is expressed throughout the linear and looping heart tube. In addition to lacking a right ventricle, *MEF2C* null embryos have a severely hypoplastic left ventricle. In the skeletal muscle lineage, MEF2 proteins have been shown to act as cofactors for members of the MyoD

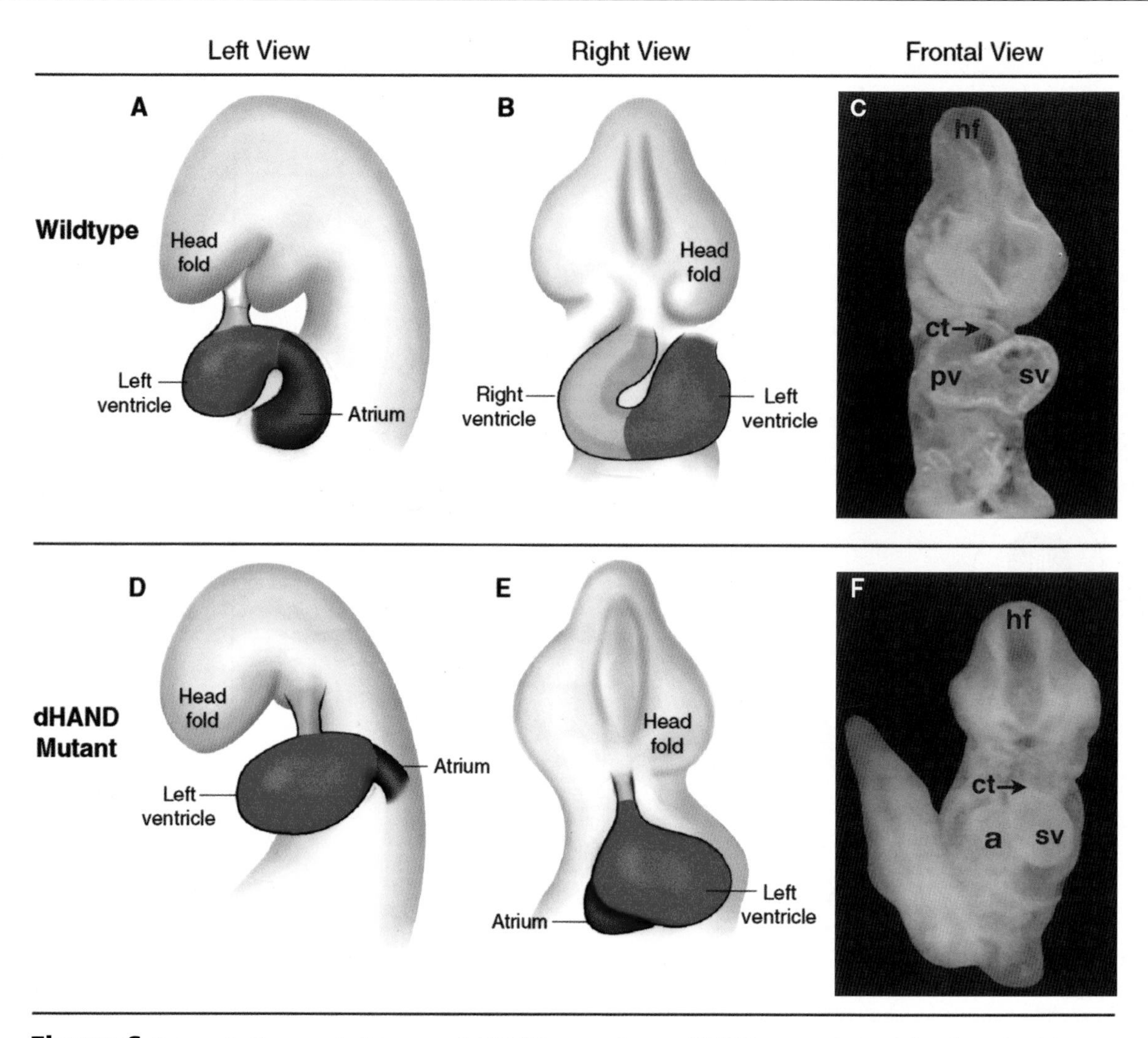

Figure 6 Schematic diagram of wild-type and *dHAND* mutant hearts. *dHAND* mutant hearts fail to form the right ventricle (blue) and have a dilated left ventricle (red) compared with wild type, apparent from lateral (A, D) and frontal (B, E) views as shown schematically. Actual wild-type (C) and *dHAND* null (F) embryos are seen from a frontal view at E 9.5. hf, head fold; rv, right ventricle; lv, left ventricle; a, atria; ct, conotruncus; pv, pulmonary ventricle; sv, systemic ventricle.

family of bHLH proteins (Molkentin *et al.,* 1995; see Chapter 8). Thus, it is tempting to speculate that MEF2C normally cooperates with dHAND to specify right ventricular development. A model in which MEF2C serves as a cofactor for dHAND in the right ventricle and eHAND in the left ventricle would explain the MEF2C null phenotype in which looping does not occur and both ventricles are hypoplastic (Fig. 7).

It is interesting that mice deficient in Nkx2.5 have hearts which fail to develop beyond the straight heart tube stage and also have diminished cardiac expression of the left ventricle marker, *eHAND* (Biben and Harvey, 1997). In contrast, expression of the right ventricle-specific MLC2V-lacZ transgene in an Nkx2.5 null background is seen throughout most of the straight heart tube (Lyons *et al.,* 1995). Together, these data might indicate the absence of a left ventricle in Nkx2.5 null mice, possibly as a result of *eHAND* downregulation. This scenario would assign eHAND an analogous role to dHAND in left ventricular development and would be consistent with its left ventricle-specific expression pattern. Mice homozygous null for the *eHAND* gene appear to have a cardiac defect (Firulli *et al.,* 1998; Riley *et al.,* 1998). However, the precise role of eHAND in cardiogenesis remains unclear secondary to the coincident placental insufficiency and early embryonic lethality of eHAND-null embryos.

B. Cardiac Neural Crest-Derived Defects

dHAND and eHAND are also expressed during development of the aortic sac and the bilaterally symmetric aortic arch arteries (I–VI) which arise from it (Sri-

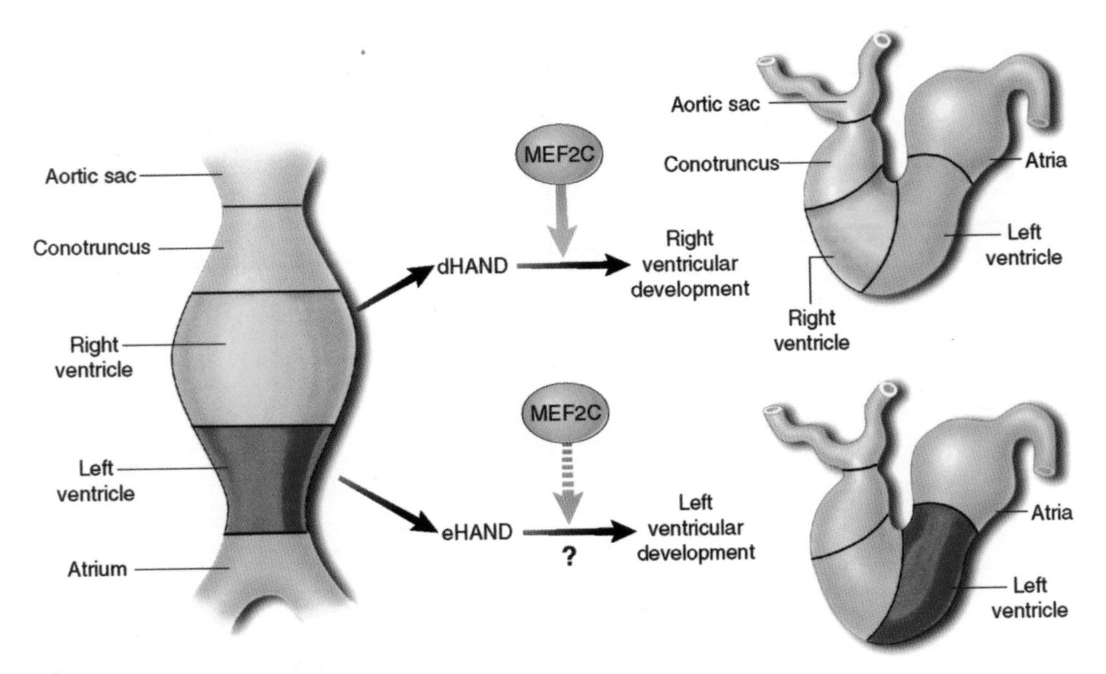

Figure 7 A model for the roles of dHAND and eHAND in ventricular development. dHAND and eHAND may control development of the right and left ventricles, respectively. MEF2C may be an important cofactor for dHAND and eHAND in ventricular development, whereas Nkx2.5 may lie upstream of eHAND in regulation of left ventricle formation.

vastava *et al.,* 1995; Cserjesi *et al.,* 1995). Each aortic arch artery traverses through a branchial arch and is remodeled during development to form the mature aortic arch and proximal pulmonary artery. The normal development and maintenance of these vascular structures is dependent on migration of cardiac neural crest cells, which undergo an ectomesenchymal transformation and give rise to smooth muscle and connective tissue in the great vessels (Kirby and Waldo, 1995). Ablation of the cardiac neural crest in chick embryos results in defects of the cardiac outflow tract, aortic arch, and proximal pulmonary arteries which are similar to those seen in children with congenital heart disease (CHD) (Kirby and Waldo, 1995). Neural crest cells begin populating the aortic sac and aortic arches at E 9 in the mouse (Serbedzija *et al.,* 1992). Mouse embryos lacking the *dHAND* gene form an aortic sac which becomes markedly dilated by E 9.5 (Fig. 8). This progresses until a grossly dilated balloon-like structure is evident, representing the aortic sac at E 10.5.

Why does the aortic sac become dilated in the absence of dHAND? The extent of development of the aortic sac and aortic arch arteries is clarified by India ink injection into the beating hearts of E 9.5 embryos (Srivastava *et al.,* 1997). Whereas the first and second aortic arch arteries are readily apparent following ink injection into wild-type embryos, there are no aortic arch arteries arising from the aortic sac in the mutant (Fig. 8). The lack of aortic arch vessels is likely to have resulted in dilation of the aortic sac and severe cardiac failure as evidenced by collection of a large effusion in the pericardial sac with embryonic demise occurring soon thereafter. These findings suggest that the expression of *dHAND* in the aortic arches is necessary for the persistence of the neural crest-derived aortic arch arteries.

dHAND and eHAND are also implicated in development of other structures derived from the neural crest. Targeted disruption of the signaling peptide, endothelin-1 (ET-1) (Kurihara *et al.,* 1995), and its G protein-coupled receptor, ET_A (M. Yanagisawa, personal communication) results in defects of craniofacial structures derived from the pharyngeal arches which are populated by neural crest cells. ET-1 and ET_A-deficient mice also have defects of the cardiac outflow tract and aortic arch, structures which are populated by the cardiac neural crest. The defects include malalignment of the aorta and pulmonary arteries with the left and right ventricles and anomalous developmental patterning of the aortic arch arteries. The phenotype of such mice is reminiscent of the defects seen in chick embryos after cardiac neural crest ablation, which include tetralogy of Fallot, persistent truncus arteriosus, and interrupted aortic arch. It is believed that the defects in *ET-1*-deficient mice are a result of abnormal differentiation of the circumpharyngeal neural crest, although marker analyses indicate that the neural crest cells migrate appropriately to their respective areas. The

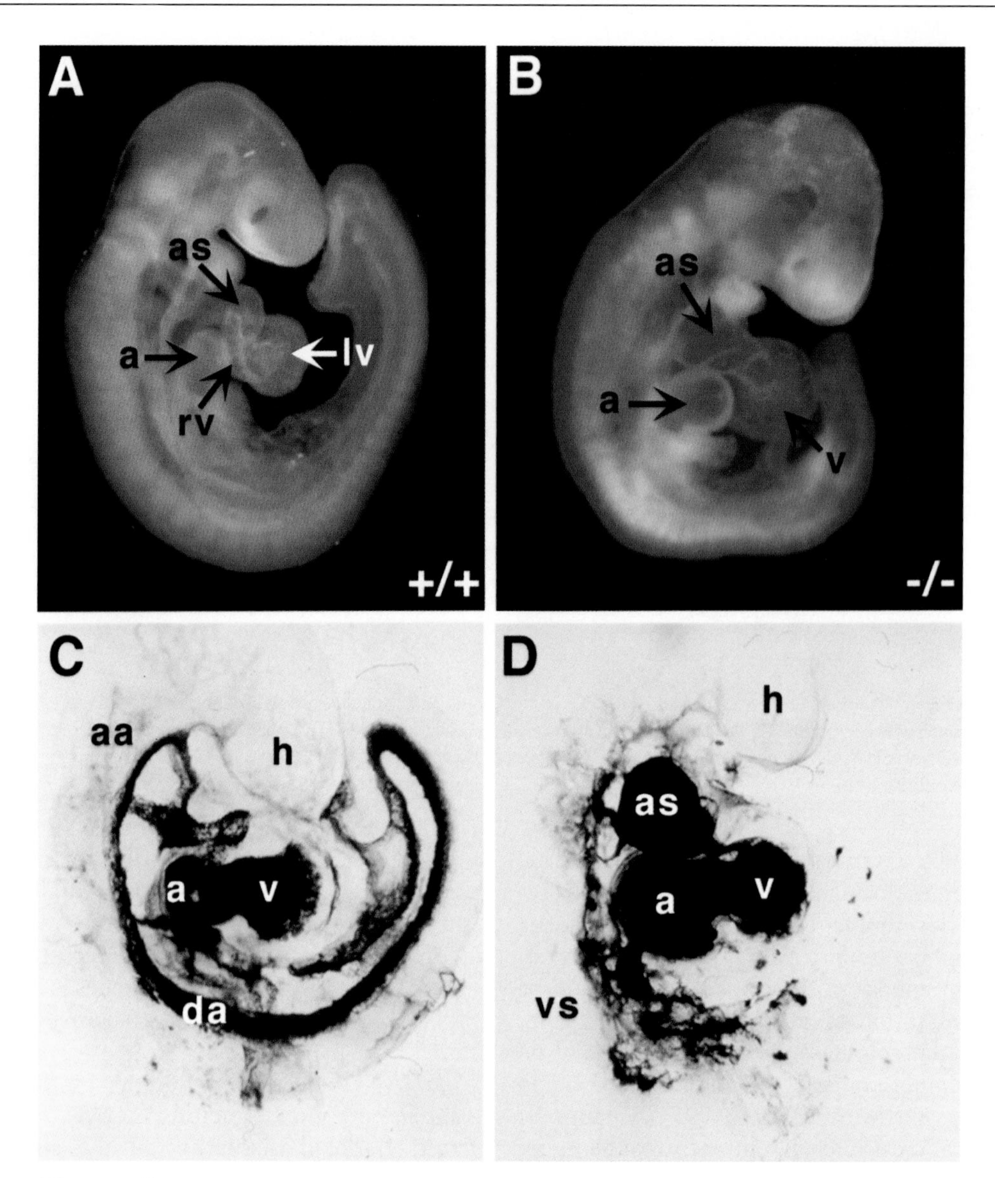

Figure 8 The absence of aortic arch arteries in *dHAND* null embryos. The aortic sac (as) of a *dHAND* null embryo is dilated at E 9.5 (B) in comparison to the wild type (A) as seen in lateral views. India ink injection into the beating hearts of E 9.5 embryos traces development of the vascular system in wild-type (C) and *dHAND* null (D) embryos. The first and second aortic arch arteries (aa) fail to form in the *dHAND* mutant resulting in dilation of the aortic sac and the absence of anterograde blood flow from the heart. Vascular markings are from retrograde flow into the venous system (vs) from the heart. h, head; a, atria; v, ventricle; da, dorsal aorta; rv, right ventricle; lv, left ventricle (adapted from Srivastava *et al.,* 1997).

HAND genes, which are normally expressed in the pharyngeal arches and aortic arch arteries but not in migratory neural crest cells, are dramatically downregulated in the pharyngeal arches and aortic arch arteries of *ET-1* null mice (Thomas *et al.,* 1998b). This observation provides an important link between a signal transduction cascade and a transcriptional response in the development of neural crest-derived cell types. The notion that the *HAND* genes play a role in development of the circumpharyngeal neural crest is further supported by hypoplasia of the first and second pharyngeal arches in mice completely lacking dHAND (Thomas *et al.,* 1998b).

What might the mechanism of action be which results in the observed phenotype in *dHAND*-deficient and *dHAND* null mice? It is possible that dHAND has unique target genes and roles in mesodermal and neural crest-derived cells. The specificity of action may be conferred by the expression of tissue-specific cofactors present in one or the other cell type. How the HAND proteins interact with other factors known to be important in cardiac neural crest development, such as the retinoic acid receptors (Kastner *et al.,* 1994), NF1 (Brannan *et al.,* 1994), and Pax 3 (Epstein *et al.,* 1993), remains to be determined. Whether dHAND is playing a role in cell determination or whether it may be playing a more general role in tissue development, such as controlling proliferation or survival of cells during organogenesis, will be important to determine. Further analysis of *dHAND* null embryos may identify the precise molecular pathway through which the *HAND* genes function during embryogenesis.

VII. Relevance to Human Disease

A. Mesodermal Defects of the Heart

The majority of defects of heart muscle result in embryonic lethality from cardiac insufficiency. However, defects which are specific to a particular chamber can often result in normal *in utero* development since separation of the pulmonary and systemic circulations is unnecessary. Hypoplasias of the right (pulmonary) or left (systemic) ventricle are relatively common congenital heart defects and are some of the most severe types of CHD. Despite aggressive medical and surgical therapy for these conditions, morbidity and mortality remain high among this population.

It has long been thought that hypoplastic right and left heart syndromes were the result of decreased blood flow to one side of the heart resulting in diminished growth stimuli. The decreased blood flow has been thought to be a result of atresia or underdevelopment of the right- or left-sided atrioventricular valve. While this may be true in some cases, the finding that a single gene defect (*dHAND*) can result in hypoplasia of an entire chamber of the heart demands a reconsideration of this hypothesis. The primary defect may actually be in hypoplasia of the right or left ventricle with an obligate secondary decrease in size of the atrioventricular valve. The model of a primary ventricular hypoplasia is supported by the recent demonstration of hypoplastic left heart syndrome in the human fetus early during development rather than during the later growth phase of the ventricles (Yagel *et al.,* 1997).

In light of their restricted expression patterns in the right and left ventricles and the hypoplastic right ventricle seen in *dHAND* null embryos, dHAND and eHAND are obvious candidates for controlling segmental development in the heart and may be responsible for hypoplastic right and left ventricles, respectively, in humans. It will be interesting to determine if these genes are mutated or deleted in individuals with such conditions. Identification of upstream and downstream members of the molecular pathway in which these transcription factors act will be important because they, rather than the *HAND* genes, may be the culprits in many cases.

B. Defects of the Neural Crest

The human syndrome known as CATCH-22 (*c*ardiac defects, *a*bnormal facies, *t*hymic hypoplasia, *c*left palate, and *h*ypocalcemia associated with chromosome *22* microdeletion) is a defect of neural crest-derived cell types (Wilson *et al.,* 1993; Driscoll *et al.,* 1992). The cardiac defects are predominately of the outflow tract (conotruncus) and aortic arch, including persistent truncus arteriosus, tetralogy of Fallot, double-outlet right ventricle, and interrupted aortic arch. Mice deficient in ET-1 and its receptor have a phenotype very similar to CATCH-22 (Kurihara *et al.,* 1995). The downregulation of the *HAND* genes in ET-1 null mice and the defects in aortic arch development in dHAND null mice suggest that the *HAND* genes may be involved in a pathway responsible for normal cardiac and pharyngeal neural crest development. Although the human homologs of *dHAND* and *eHAND* do not map to chromosome 22 (D. Srivastava, unpublished observations), the unidentified gene(s) residing on chromosome 22 may lie upstream or downstream of the *HAND* genes in a molecular pathway controlling neural crest development.

VIII. Summary

Discovery of the bHLH proteins, dHAND and eHAND, provides a point of entry into the molecular pathways which govern both mesodermal and neural crest development of the heart. Moreover, the *HAND* genes provide a transcriptional basis for a segmental model of cardiac development in which chamber-specific regulatory programs operate relatively independently of one another. Finally, through gene knockout experiments of *dHAND* and other cardiac transcription factors, we have begun to assimilate a variety of critical factors in a molecular pathway controlling individual steps of heart formation.

Although significant insights have been gained regarding the biology of the HAND transcription factors, many questions remain. Through what mechanism do dHAND and eHAND regulate right and left ventricle development? What are the target genes which lie downstream of these transcription factors? What factors lie upstream of *dHAND* and *eHAND* and control the chamber-specific expression patterns observed? What other, yet unidentified, transcriptional regulators might confer positional identity of the developing heart tube? Do the *HAND* genes affect neural crest cell migration, differentiation, and/or proliferation? Finally, will mutations in the *HAND* genes or other upstream or downstream genes be the cause of hypoplastic right and left heart syndromes or cardiac outflow tract and aortic arch defects? It is hoped that further understanding of these and other factors will provide answers to the previous and other exciting questions in the future.

Acknowledgments

I thank A. Tizenor and T. Thomas for assistance with the figures and E. N. Olson for invaluable collaboration. I also thank the multiple investigators for their personal communication of unpublished data. The author is supported by grants from the National Institutes of Health (RO1HL57181-01), March of Dimes, and American Heart Association.

References

Arceci, R. J., King, A. A. J., Simon, M. C., Orkin, S. H., and Wilson, D. B. (1993). Mouse GATA-4: A retinoic acid-inducible GATA-binding transcription factor expressed in endodermal derivatives and heart. *Mol. Cell. Biol.* **13,** 2235–2246.

Benezra, R., Davis, R. L., Lockshon, D., Turner, D., and Weintraub, H. (1990). The protein Id: A negative regulator of helix-loop-helix DNA binding proteins. *Cell (Cambridge, Mass.)* **61,** 49–59.

Biben, C., and Harvey, R. P. (1997). Homeodomain factor Nkx2-5 controls left/right asymmetric expression of bHLH gene eHand during murine heart development. *Genes Dev.* **11,** 1357–1369.

Brannan, C. I., Perkins, A. S., Vogel, K. S., Ratner, N., Nordlund, M. L., Reid, S. W., Buchberg, A. M., Jenkins, N. A., Parada, L. F., and Copeland, N. G. (1994). Targeted disruption of the neurofibromatosis type-1 gene leads to developmental abnormalities in heart and various neural crest-derived tissues. *Genes Dev.* **8**(9), 1019–1029.

Collignon, J., Varlet, I., and Robertson, E. J. (1996). Relationship between asymmetric nodal expression and the direction of embryonic turning. *Nature (London)* **381,** 155–158.

Cross, J. C., Flannery, M. L., Blanar, M. A., Steingrimsson, E., Jenkins, N. A., Copeland, N. G., Rutter, W. J., and Werb, Z. (1995). Hxt encodes a basic helix-loop-helix transcription factor that regulates trophoblast cell development. *Development (Cambridge, UK)* **121,** 2513–2523.

Cserjesi, P., Brown, D., Lyons, G. E., and Olson, E. N. (1995). Expression of the novel basic helix-loop-helix gene eHAND in neural crest derivatives and extraembryonic membranes during mouse development. *Dev. Biol.* **170,** 664–678.

Davis, R. L., Weintraub, H., and Lassar, A. B. (1987). Expression of a single transfected cDNA converts fibroblasts to myoblasts. *Cell (Cambridge, Mass.)* **51**(6), 987–1000.

DeHann, R. L. (1965). *In* "Organogenesis" (R. L. DeHaan and H. Ursprung, eds.), pp. 377–419. Holt, Rinehart & Winston, New York.

Driscoll, D. A., Budarf, M.L., and Emanuel, B. S. (1992). A genetic etiology for DiGeorge syndrome: Consistent deletions and microdeletions of 22q11. *Am. J. Hum. Genet.* **50,** 924–933.

Drysdale, T. A., Tonissen, K. F., Patterson, K. D., Crawford, M. J. and Krieg, P. A. (1994). Cardiac troponin I is a heart-specific marker in the Xenopus embryo: Expression during abnormal heart morphogenesis. *Dev. Biol.* **165,** 432–441.

Edmondson, D. G., Lyons, G. E., Martin, J. E., and Olson, E. N. (1994). MEF2 gene expression marks the cardiac and skeletal muscle lineages during mouse embryogenesis. *Development (Cambridge, UK)* **120,** 1251–1263.

Ellis, H. M., Spann, D. R., and Posakony, J. W. (1990). Extramachrochaete, a negative regulator of sensory organ development on Drosophila, defines a new class of helix-loop-helix proteins. *Cell (Cambridge, Mass.)* **61,** 27–38.

Epstein, D. J., Vogan, K. J., Trasler, D. G., and Gros, P. A. (1993). A mutation within intron 3 of the Pax-3 gene produces aberrantly spliced mRNA transcripts in the Splotch mouse mutant. *Proc. Natl. Acad. Sci. U.S.A.* **90,** 532–536.

Firulli, A. B., McFadden, D., Lin, Q., Srivastava, D., and Olson, E. N. (1998). Heart and extraembryonic mesodermal defects in mouse embryos lacking the bHLH transcription factor, HAND1. *Nature Genet.* **18,** 266–270.

Franco, D., Kelly, R., Lamers, W. H., Buckingham, M., and Moorman, A. F. M. (1997). Regionalized transcriptional domains of myosin light chain 3f transgenes in the embryonic mouse heart: Morphogenetic implication. *Dev. Biol.* **188,** 17–33.

Guillermot, F., Lo, L. C., Johnson, J. E., Auerbach, A., Anderson, D. J., and Joyner, A. L. (1993). Mammalian achaete-scute homolog-1 is required for the early development of olfactory and autonomic neurons. *Cell (Cambridge, Mass.)* **75,** 463–476.

Harvey, R. P. (1996). NK-2 homeobox genes and heart development. *Dev. Biol.* **178,** 203–216.

Henthorn, P., Kiledjian, M., and Kadesch, T. (1990). Two distinct transcription factors that bind the immunoglobulin enhancer uE5/KE2 motif. *Science* **247,** 467–470.

Hoffman, J. I. E. (1995). Incidence of congenital heart disease: I. Postnatal incidence. *Pediatr. Cardiol.* **16,** 103–113.

Hollenberg, S. M., Sternglanz, R., Cheng, P. F., and Weintraub, H. (1995). Identification of a new family of tissue-specific basic helix-loop-helix proteins with a two-hybrid system. *Mol. Cell. Biol.* **15,** 3813–3822.

Hu, J. S., Olson, E. N., and Kingston, R. E., (1992). HEB, a helix-loop-helix protein related to E2A and ITF2 that can modulate the DNA-binding ability of myogenic regulatory factors. *Mol. Cell. Biol.* **12,** 1031–1042.

Ip, H. S., Wilson, D. B., Heikinheimo, M., Tang, Z., Ting, C. N., Simon, M. C., Leiden, J. M., and Parmacek, M. S. (1994). The GATA-4 transcription factor transactivates the cardiac muscle-specific troponin C promoter-enhancer in nonmuscle cells. *Mol. Cell. Biol.* **14**(11), 7517–7526.

Jan, Y. N., and Jan, L. Y. (1993). HLH proteins, fly neurogenesis, and vertebrate myogenesis. *Cell (Cambridge, Mass.)* **75,** 827–830.

Johnson, J. E., Birren, S. J., and Anderson, D. J. (1990). Two rat homologues of Drosophila achaete-scute specifically expressed in neuronal precursors. *Nature (London)* **346,** 858–861.

Kastner, P., Grondona, J. M., Mark, M., Gansmuller, A., LeMeur, M., Decimo, D., Vonesch, J. L., Dolle, P., and Chambon, P. (1994). Genetic analysis of RXR alpha developmental function: Convergence of RXR and RAR signaling pathways in heart and eye morphogenesis. *Cell (Cambridge, Mass.)* **78,**(6), 987–1003.

Kelly, R., Alonso, S., Tajbakhsh, S., Coss, G., and Buckingham, M. (1995). Myosin light chain 3F regulatory sequences confer regionalized cardiac and skeletal muscle expression in transgenic mice. *J. Cell Biol.* **129,** 383–396.

Kirby, M. L., and Waldo, K. L., (1990). Role of neural crest in congenital heart disease. *Circulation* **82,** 332–340.

Kirby, M. L., and Waldo, K. L. (1995). Neural crest and cardiovascular patterning. *Circ. Res.* **77**(2), 211–215.

Kirby, M. L., Gale, T. F., and Stewart, D. E., (1983). Neural crest cells constitute tonormal aorticopulmonary septation. *Science* **220,** 1059–1061.

Komuro, I., and Izumo, S. (1993). Csx: A murine homeobox-containing gene specifically expressed in the developing heart. *Proc. Natl. Acad. Sci. U.S.A.* **90,** 8145–8149.

Kuisk, I. R., Li, H., Tran, D., and Capetanaki, Y. (1996). A single MEF2 site governs desmin transcription in both heart and skeletal muscle during mouse embryogenesis. *Dev. Biol.* **174,** 1–13.

Kuo, C. T., Morrisey, E. E., Anandappa, R., Sigrist, K., Lu, M. M., Parmacek, M. S., Soudais, C., and Leiden, J. M. (1997). GATA4 transcription factor is required for ventral morphogenesis and heart tube formation. *Genes Dev.* **11,** 1048–1060.

Kurihara, Y., Kurihara, H., Oda, H., Maemura, K., Nagai, R., Ishikawa, T., and Yazaki, Y. (1995). Aortic arch malformations and ventricular septal defect in mice deficient in endothelin-1. *J. Clin. Invest.* **96**(1), 293–300.

Laverriere, A. C., MacNeill, C., Mueller, C., Poelmann, R. E., Burch, J. B., and Evans, T. (1994). GATA-4/5/6, a subfamily of three transcription factors transcribed in developing heart and gut. *J. Biol. Chem.* **269**(37), 23177–23184.

Lee, J. E., Hollenberg, S. M., Snider, L., Turner, D. L., Lipnick, N., and Weintraub, H. (1995). Conversion of xenopus ectoderm into neurons by NeuroD, a basic helix-loop-helix protein. *Science* **268,** 836–844.

Levin, M. (1997). Left-right asymmetry in vertebrate embryogenesis. *BioEssays* **19**(14), 287–296.

Li, L., Miano, J. M., Mercer, B., and Olson, E. N. (1996). Expression of the SM22α promoter in transgenic mice provides evidence for distinct transcriptional regulatory programs in vascular and visceral smooth muscle cells. *J. Cell Biol.* **132**(5), 849–859.

Lin, Q., Schwartz, J. A., and Olson, E. N. (1997). Control of cardiac morphogenesis and myogenesis by the myogenic transcription factor MEF2C. *Science* **276,** 1404–1407.

Lints, T. J., Parsons, L. M., Hartley, L., Lyons, I., and Harvey, R. P. (1993). Nkx-2.5: A novel murine homeobox gene expressed in early heart progenitor cells and their myogenic descendants. *Development (Cambridge, UK)* **119,** 419–431.

Lowe, L. A., Supp, D. M., Sampath, K., Yokoyama, T., Wright, C. V. E., Potter, S. S., Overbeek, P., and Kuehn, M. R. (1996). Conserved left-right asymmetry of nodal expression and alterations in murine situs inversus. *Nature (London)* **381,** 158–161.

Lyons, I., Parsons, L. M., Hartley, A. L., Li, R., Andrews, J. E., Robb, L., and Harvey, R. P. (1995). Myogenic and morphogenetic defects in the heart tubes of murine embryos lacking the homeo box gene Nkx2-5. *Genes Dev.* **9,** 1654–1666.

Ma, Q., Kintner, C., and Anderson, D. J. (1996). Identification of neurogenin, a vertebrate neuronal determination gene. *Cell (Cambridge, Mass.)* **87**(1), 43–52.

Meno, C., Saijoh, Y., Fujii, H., Ikeda, M., Yokoyama, T., Yokoyama, M., Toyoda, Y., and Hamada, H. (1996). Left-right asymmetric expression of the TGFβ-family member lefty in mouse embryos. *Nature (London)* **381,** 151–155.

Molkentin, J. D., Black, B., Martin, J. F., and Olson, E. N. (1995). Cooperative activation of muscle gene expression by MEF2 and myogenic bHLH proteins. *Cell (Cambridge, Mass.)* **83,** 1125–1136.

Molkentin, J. D., Lin, Q., Duncan, S. A., and Olson, E. N. (1997). Requirement of the transcription factor GATA4 for heart tube formation and ventral morphogenesis. *Genes Dev.* **11,** 1061–1072.

Murre, C., McCaw, P. S., and Baltimore, D. (1989a). A new DNA binding and dimerization motif in immunoglobulin enhancer binding, daughterless, MyoD, and myc proteins. *Cell (Cambridge, Mass.)* **56,** 777–783.

Murre, C., McCaw, P., Vaessin, H., Claudy, M., Jan, L. Y., Jan, Y. N., Cabrera, C. V., Buskin, J. N., Hauschka, S. D., Lassar, A. B., Weintraub, H., and Baltimore, D. (1989b). Interactions between heterologous helix-loop-helix proteins generate complexes that bind specifically to a common DNA sequence. *Cell (Cambridge, Mass.)* **58,** 537–544.

Olson, E. N., and Klein, W. H. (1994). bHLH factors in muscle development: Dead lines and commitments, what to leave in and what to leave out. *Genes Dev.* **8**(1), 1–8.

Olson, E. N., and Srivastava, D. (1996). Molecular pathways controlling heart development. *Science* **272,** 671–676.

Olson, E. N., Perry, M., and Schulz, R. A. (1995). Regulation of muscle differentiation by the MEF2 family of MADS box transcription factors. *Dev. Biol.* **172,** 2–14.

Osmond, M. K., Butler, A. J., Voon, F. C., and Bellairs, R. (1991). The effects of retinoic acid on heart formation in the early chick embryo. *Development (Cambridge, UK)* **113**(4), 1405–1417.

Peacock, T. B. (1858). *In* "Malformations of the Human Heart" (T. B. Peacock, ed.), p. 26. John Churchill, London.

Pollock, R., and Treisman, R. (1991). Human SRF-related proteins: DNA binding properties and potential regulatory targets. *Genes Dev.* **5,** 2327–2341.

Riley, P., Anson-Cartwright, L., and Cross, J. C. (1998). The Hand1 bHLH transcription factor is essential for placentation and cardiac morphogenesis. *Nature Genet.* **18**(3), 271–275.

Ross, R. S., Navankasattusas, S., Harvey, R. P., and Chien, K. R. (1996). An HF-1a/HF-1b/ MEF2 combinatorial element confers cardiac ventricular specificity and establishes an anterior-posterior gradient of expression via an Nkx2.5 independent pathway. *Development (Cambridge, UK)* **122,** 1799–1809.

Serbedzija, G. N., Bronner-Fraser, M., and Fraser, S. E. (1992). Vital dye analysis of cranial neural crest cell migration in the mouse embryo. *Development (Cambridge, UK)* **116,** 297–307.

Shivdasani, R. A., Mayer, E. L., and Orkin, S. H. (1995). Absence of blood formation in mice lacking the T-cell leukaemia oncoprotein tal-1/SCL. *Nature (London)* **373,** 432–434.

Srivastava, D., and Olson, E. N. (1997). Knowing in your heart what's right. *Trends Cell Biol.* **7,** 447–453.

Srivastava, D., Cserjesi, P., and Olson, E. N. (1995). A subclass of bHLH proteins required for cardiogenesis. *Science* **270,** 1995–1999.

Srivastava, D., Thomas, T., Lin, Q., Kirby, M. L., Brown, D., and Olson, E. N. (1997). Regulation of cardiac mesodermal and neural crest development by the bHLH transcription factor, dHAND. *Nat. Genet.* **16**(2), 154–160.

Stainier, D. Y., and Fishman, M. C. (1992). Patterning the zebrafish heart tube: acquisition of anteroposterior polarity. *Dev. Biol.* **153**(1), 91–101.

Thomas, T., Yamagishi, H., Overbeek, P. A., Olson, E. N., and Srivastava, D. (1998a). The bHLH factors, dHAND and eHAND, specify pulmonary and systemic cardiac ventricles independent of left-right sidedness. *Dev. Biol.* **196,** 228–236.

Thomas, T., Kurihara, H., Yamagishi, H., Kurihara, Y., Yazaki, Y., Olson, E. N., and Srivastava, D. (1998b). A signaling cascade involving endothelin-1, dHAND and Msx1 regulates development of neural crest-derived branchial arch mesenchyme. *Development (Cambridge, UK)*, in press.

Weintraub, H. (1993). The MyoD family and myogenesis: Redundancy, networks, and thresholds. *Cell* **75,** 1241–1244.

Wilson, D. I., Burn, J., Scambler, P., and Goodship, J. (1993). DiGeorge syndrome: Part of CATCH-22. *J. Med. Genet.* **30,** 852–856.

Yagel, S., Weissman, A., Rotstein, Z., Manor, M., Hegesh, J., Anteby, E., Lipitz, S., Achiron, R. (1997). Congenital heart defects: Natural course and in utero development. *Circulation* **96,** 550.

Yamamura, H., Zhang, M., Mjaatvedt, C. H., and Markwald, R. R. (1997). A heart segmental defect in the anterior–posterior axis of a transgenic mutant mouse. *Dev. Biol.* **186,** 58–72.

Yokoyama, T., Copeland, N. G., Jenkins, N. A., Montgomery, C. A., Elder, F. F., and Overbeek, P. A. (1993). Reversal of left–right asymmetry: a situs inversus mutation. *Science* **260,** 679–682.

Yu, Y.-T, Breitbart, R. E., Smoot, L. B., Lee, Y., Mahdavi, V., and Nadal-Ginard, B. (1992). Human myocyte-specific enhancer factor 2 comprises a group of tissue-restricted MADS box transcription factors. *Genes Dev.* **6,** 1783–1798.

Yutzey, K., and Bader, D. (1995). Diversification of cardiomyogenic cell lineages during early heart development. *Circ. Res.* **77,** 216–219.

Zhuang, Y., Soriano, P., and Weintraub, H. (1994). The helix-loop-helix gene E2A is required for B cell formation. *Cell (Cambridge, Mass.)* **79,** 875–884.

IV

Normal and Abnormal Morphogenesis

10

Mechanisms of Segmentation, Septation, and Remodeling of the Tubular Heart: Endocardial Cushion Fate and Cardiac Looping

Corey H. Mjaatvedt, Hideshi Yamamura, Andy Wessels, Anne Ramsdell, Debi Turner, and Roger R. Markwald

Department of Cell Biology and Anatomy, Medical University of South Carolina, Charleston, South Carolina 29425

I. Development of Heart Segments

The primary heart tube forms from the progressive fusion of two cardiac primordia that arise from a region of splanchnic mesoderm found on either side of the anterior–posterior embryonic axis and ventral to the endoderm (Rawles, 1943; Rosenquist and DeHaan, 1966; see Chapters 1–4). These paired primordia fuse anteriorly in a progressive fashion to form a series of primitive tubular segments (De la Cruz *et al.,* 1989). Each segment consists of an inner endothelium surrounded by an outer myocardial layer. Between the epithelial layers a complex layer of extracellular matrix is present. The epithelial cell layer and the associated intervening matrix of each segment appear superficially to be homogeneous throughout the anterior–posterior axis of the primary heart tube. Segments are demarcated by constrictions (Fig. 1) that possess differences in contractility, expression of actins (Ruzicka and Schwartz, 1988), specific myosins (Kubalak *et al.,* 1994; O'Brien *et al.,* 1993; Yutzey *et al.,* 1994), and matrix components (Mjaatvedt *et al.,* 1991). By HH stage 13 (Hamburger

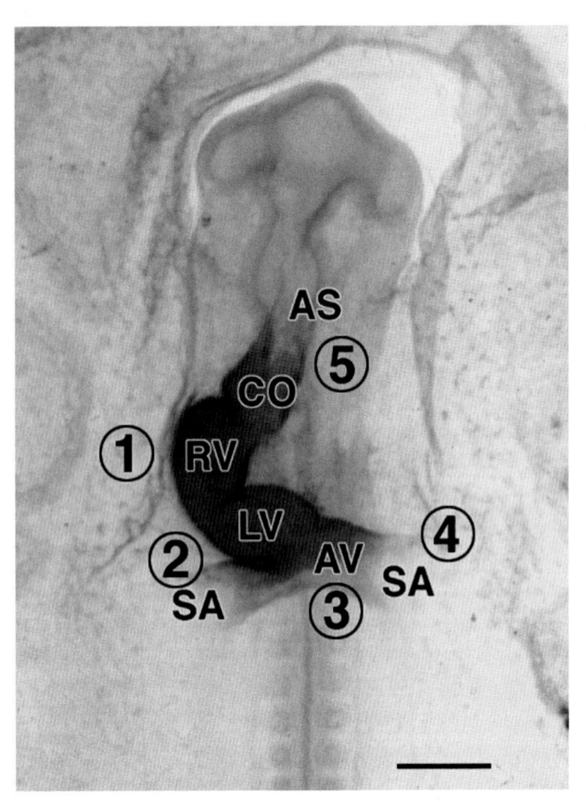

Figure 1 Progressive formation of the five primitive segments of the single heart tube. The heart segments are incorporated into the heart tube over time. The primitive heart segments present by HH stage 12 are shown in a ventral view of an embryonic chick heart stained with the MF20 antibody. These segments include (1) the primitive right ventricle (RV), (2) the primitive left ventricle (LV), (3) the atrioventricular canal (AV) segment (in the process of formation at this stage), (4) the sinoatrial segment (SA) (not yet formed by HH stage 12), and (5) the conus segment (CO) which completes its formation between HH stages 14 and 22. Aortic sac (AS). Scale bar, 500 μM.

and Hamilton, 1951), the chick heart tube consists of four identifiable segments which form in temporal succession as follows (De la Cruz *et al.,* 1977, 1983). The first two segments to form are the primitive right ventricle and primitive left ventricle. Next, the atrioventricular (AV) segment followed by the sinoatrial segment are formed. A fifth segment, the conus segment, develops after the tubular heart is formed (De la Cruz *et al.,* 1977). The atrium forms from incorporation of cells from the sinoatrial region. The derivation of the conus–truncus (outflow tract) has not been clearly determined. It may arise from the previously formed primitive right ventricular segment. Alternatively, the conus and primitive truncus, as suggested by Dr. De la Cruz, may be formed from precursor mesoderm that lies anterior to the right ventricle. The significance of a progressive formation of segments may be that expression of genes in the earliest formed segments may influence the gene expression in later segments as they form along the anterior–posterior axis of the tube. This could account for the variation in segmental fates, particularly regarding their role in septation. Although the cells within endothelial inner and myocardial outer layers appear to be phenotypically similar in all five primitive segments of the heart. The endocardial cushions are formed in only two of the five segments of the heart tube—the AV and conus–truncus. The endocardial and myocardial epithelium found in these segments include subpopulations of cells that are functionally distinct in their ability to form cushion tissue. In the case of the endothelium, the subpopulation competent to form cushion tissue is derived from a cell lineage unique from the endothelial cells present in other segments of the heart tube (Wunsch *et al.,* 1994). As segments are added before the future endocardial cushions begin to form, the single heart tube begins the looping process. Looping begins as a bending of the tube at the junction of the primitive right and left ventricles which functionally divides the tube into inlet and outlet segments. The direction of looping is influenced by action of the laterality genes, e.g., *inv/iv* (Brueckner *et al.,* 1989; Yokoyama *et al.,* 1993), lefty genes (Meno *et al.,* 1996), sonic hedgehog and activin or activin-like ligands that are recognized by the ActIIa receptor (Levin *et al.,* 1995). The heart continues to loop as the ventricular infundibular fold deepens and the endocardial cushions in AV and conus regions expand and then fuse to form the AV and conal septa. The atrial, ventricular, and aorticopulmonary septa also form during this time, the latter by the contribution of invading neural crest cells that divide the aortic sac. In the final stages of looping, the inlet and outlet segments of the heart tube are brought together through a process of "morphogenetic shifts" which bring all five of the septa together in proper alignment to produce a normal four-chambered heart. It is important to note that none of the initial segments found in the single heart tube independently give rise to an equivalent adult heart chamber. The definitive adult heart chambers form by the integration and remodeling of the primitive segments (De la Cruz and Markwald, 1998). The developmental mechanisms by which the endocardial cushions form and contribute to these morphogenetic shifts are discussed later.

II. Progenitors of the Cardiac Epithelia

Cells of the epiblast that are specified to the heart lineage migrate through the primitive streak into the lateral plate mesoderm forming the bilateral heart-forming fields (Garcia-Martinez *et al.,* 1997) (Fig. 2). The heart-forming fields continue their movement ante-

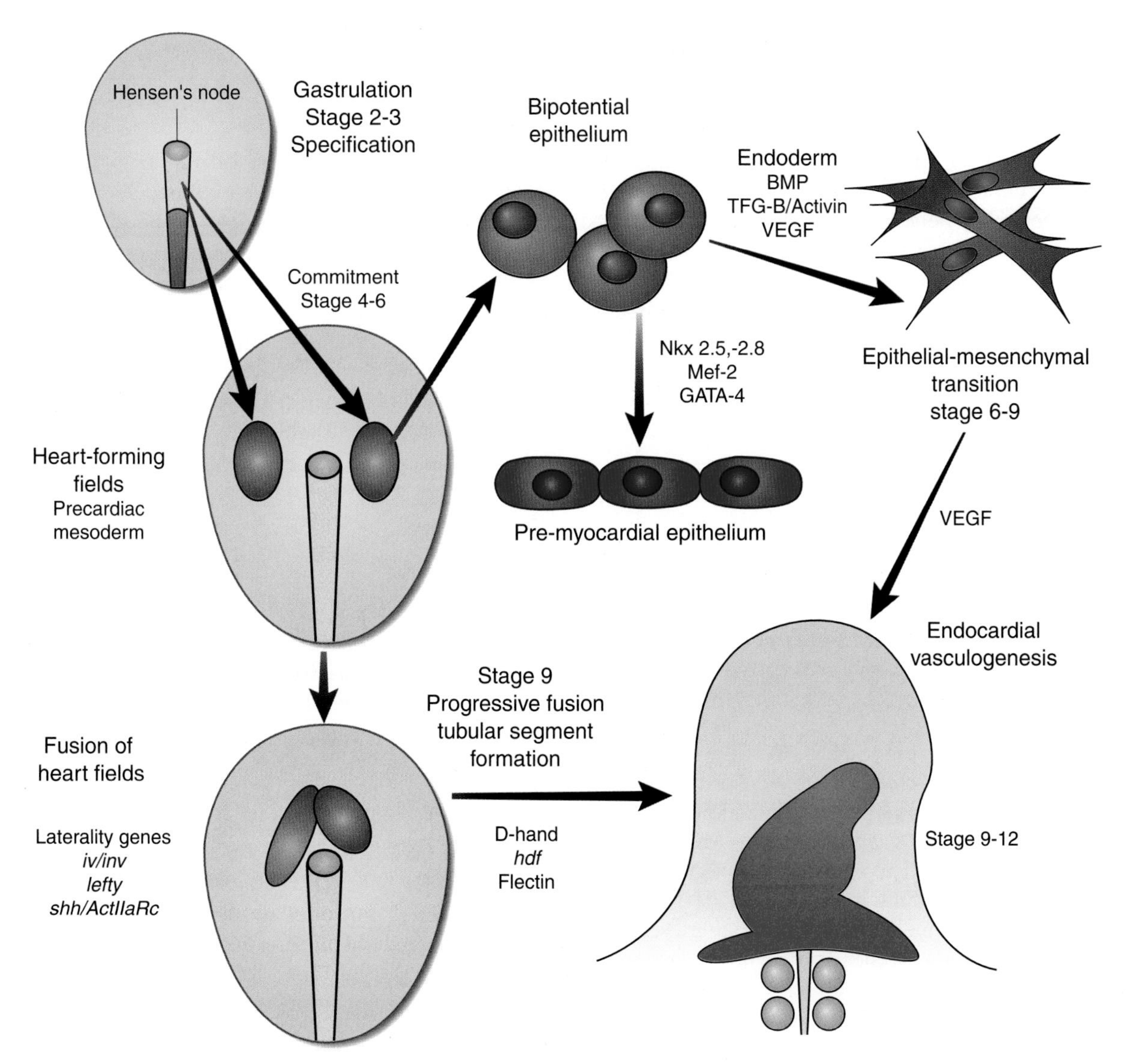

Figure 2 Schematic overview of the early genetic and morphogenic events associated with formation of the heart myocardium and endocardium. At some time prior to or during gastulation (stages 2 or 3), cells invading the primitive streak are specified (reversibly committed) to the heart lineage. Irreversible commitment (determination) of the specified mesodermal cells to the heart lineage occurs by HH stages 4–6 in the chick and appears to involve the action of BMP-2. The precardiac mesoderm of the heart-forming fields is a bipotential epithelium that gives rise to both myocardial and endocardial cells. In response to an inductive interaction with the adjacent endoderm, some cells of the precardiac mesoderm delaminate from the epithelium and form the preendocardial mesenchyme. Extracellular signaling by the endoderm may be mediated by more than one factor, including members of the transforming growth factor-β superfamily (TGF-β and activins), vascular endothelial growth factor (VEGF), and bone morphogenetic protein (BMP) by the adjacent foregut endoderm. Under the influence of VEGF, the preendocardial mesenchyme reestablish cell–cell interactions to form the endocardial epithelium (endocardial vasculogenesis). Cells that do not undergo an epithelial–mesenchymal transition and remain in the epithelial layer of the precardiac mesoderm form the premyocardial epithelium through a morphogenetic pathway involving the genes *Mef2, GATA-4,* and homologs of the *Drosophlia* gene (*tinman* and *nkx2.5*). Fusion of the heart fields begins by HH stage 9 in the chick. The direction of the initial heart tube bending is controlled by expression within the mesoderm of the heart laterality genes that include the *situs inversus* genes *iv/inv, lefty-1* and *-2, sonic hedgehog* (*shh*), and an activin-like ligand that presumably interacts with the activin receptor IIa (*ActIIaRc*). As the cardiac primordia fuse in a progressive fashion during HH stages 9–12, new segments are added to the heart tube. The segmental nature of the heart tube establishes domains of transcriptional activity that regulate the expression of genes such as *flectin, d-* and *e-hand,* and *hdf.*

riorly in the embryo, eventually forming the cardiac crescent located above the anterior intestinal portal. Just prior to partial fusion of the two lateral heart fields, a few contractile and related regulatory genes are expressed to reveal the initial differentiation of the myocyte cell precursors (Kubalak *et al.,* 1994; Lints *et al.,* 1993). It is generally believed that cells forming the myocardium arise from precursors found only in the definitive heart fields (Olson and Srivastava, 1996). Because segments within the heart tube are added over time in a progressive fashion, the myocytes and endothelial cells within different segments will undergo differentiation at progressively different times. One hypothesis to explain this progression is that mesodermal cells located caudally (posterior) to the heart fields are recruited into the cardiomyogenic lineage. As new cells are recruited to the heart fields, the more differentiated cells move anteriorly (see Chapters 3 and 4). This mechanism is consistent with findings that the mesodermal cells immediately posterior and lateral to the heart fields can enter the cardiomyogenic lineage through bone morphogenetic protein (BMP) signaling (Lough *et al.,* 1996) and reportedly the gene *cerberus* (Bouwmeester *et al.,* 1996). Cells specified to the myocardial lineage of the mesodermal epithelium enter a cascade of gene expression that includes the *tinman* homolog *NKx 2.5* (Lints *et al.,* 1993), CARP (Zou *et al.,* 1997), Mef2c (Edmondson *et al.,* 1994; Lin *et al.,* 1997), and GATA-4 (Molkentin *et al.,* 1994; see Section V). The endothelial cell precursors that form the endocardium of the heart derive from mesenchymal cells formed by an epithelial–mesenchymal transition (Fig. 2). Some studies have suggested that the endocardium derives directly from the heart-forming fields of the splanchnic mesoderm (De Haan, 1965; Rosenquist and DeHaan, 1966; Sabin, 1920; Stalsberg and De Haan, 1969; Viragh *et al.,* 1989), whereas other studies have supported the hypothesis that the cardiac endothelial cells arise from regions of splanchnic mesoderm peripheral to the heart fields (Drake and Jacobson, 1988; Manasek, 1968). Evidence that the endocardium consisted of two subpopulations of endothelial cells was also inferred from *in vitro* observations that only endothelial cells in the AV or proximal conus region of the heart could give rise to cushion mesenchyme (Mjaatvedt *et al.,* 1987). Also, in cushion-forming regions, not all the endothelial cells leave the monolayer to invade the underlying cushion swellings (Markwald *et al.,* 1975, 1977). Most of the endothelial cells remain within the endocardial lining during the cellularization of the cushion. One endothelial cell marker, called JB3, has been identified that discriminates between the two endothelial subpopulations present in the cushion monolayer (Wunsch *et al.,* 1994). The JB3 monoclonal antibody recognizes a 350-kDa band on Western blots and may be related to the large extracellular matrix protein fibrillin-2 (Rongish *et al.,* 1998). In the heart, JB3 antigen is only expressed on the subset of endothelial cells competent to transdifferentiate into cushion mesenchyme.

The origin of the JB3-positive ($JB3^+$) endothelial cells has been traced by immunohistochemical analysis of embryos (Wunsch *et al.,* 1994). The first cells to express JB3 are the prestreak epiblast (HH stage 2); by HH stage 3, JB3-expressing cells are located in the rostral primitive streak. From the streak, $JB3^+$ cells migrate to the paired heart-forming fields (HH stages 4 and 5), leaving behind residual "trails" of JB3 antigen. JB3 is among the earliest known lineage markers to be expressed in the epithelium of the heart-forming fields (Smith *et al.,* 1997). At HH stages 6^+ and 7^-, JB3 antigen localizes with the first mesenchymal cells seeded by the precardiac mesodermal epithelium into the endodermal basement membrane. Myocardial precursors do not express JB3.

In quail embryos, QH1 (positive for all endothelial cells) and JB3 immunostaining patterns have been compared. All $JB3^+$ cells are also positive for the QH1 antigen; however, many $QH1^+$ cells are not positive for JB3 antigen. At HH stages 8–10, $JB3^+$ cells become interspersed with $QH1^+$ cells that have migrated into the heart-forming fields during their fusion to form the tubular heart (Sugi and Lough, 1995; Sugi and Markwald, 1996; Wunsch *et al.,* 1994). Thus, within the developing tubular heart (HH stages 11–14), all endocardial cells are $QH1^+$ but only about 10% are $JB3^+$. Initially, $JB3^+$ endocardial cells are present in all segments present at HH stages 11–13, but thereafter they are restricted to the cushion-forming segments. These data showed that there were indeed two populations of endothelial precursor cells, the larger one originating outside the heart fields and the smaller ($JB3^+$) arising from within the heart fields. Significantly, the $JB3^+$ subset constitutes the anlagen of cushion cells.

A. Role of Endoderm during Formation of Cardiac Endothelium

In vitro experiments have suggested the endoderm plays a significant role during the derivation of $JB3/QH1^+$ endothelial cells from the mesodermal heart fields. These studies utilized a primary culture system in which precardiac mesoderm was microdissected free of the associated endoderm and then cocultured on the surface of thick collagen gels with isolated endoderm or endoderm conditioned medium (Sugi and Markwald, 1996). Precardiac mesoderm explanted in this way from HH stage 5 chicken embryos, cultured in the absence of endoderm or endoderm conditioned medium, eventually expressed muscle markers only within the central mass of tissue. The monolayer of cells surrounding the explant on the surface of the collagen gel expressed neither myocardial or endocardial markers. Cells do not invade the collagen gels under these conditions. If, however, the precardiac mesoderm is cocultured with endoderm or its conditioned medium, cells invade into the gel and will express both JB3 and QH1 markers. This is the first evidence that the endoderm induces the heart field to delaminate endocardial precursor cells from the epithelium and thereby segregate lineages committed to myocardial or endocardial fate. Similar results have been obtained using collagen gels and QCE-6 cells as a model for precardiac mesoderm (Eisenberg and Bader, 1995; Markwald *et al.*, 1996).

III. Molecular Interactions of JB3$^+$ Endothelial Cells in the Heart Cushions

As stated previously, the JB3 antigen, at later stages, is only expressed in the heart on the subset of endothelial cells competent to transdifferentiate into cushion mesenchyme. The fate of the JB3/QH1 endothelial cells appears to be tied to the initial formation and ultimate fate of the endocardial cushions.

A. Regulation of Endocardial Cushion Formation

The heart cushions form in only two segments of the heart tube—the AV and conotruncus. Within these two segments the JB3$^+$ endothelial cells undergo a transition to form mesenchyme that invade and populate the underlying expanded layer of extracellular matrix (cardiac jelly). The myocardium in these segments is highly secretory and is responsible for the localized expansion of the cushion by the addition of matrix molecules which include laminin, collagen, fibronectin, and proteogycans. These matrix molecules are organized as a myocardially derived basement membrane that extends across the cardiac jelly and joins the basement membrane of the endothelium (Kitten *et al.*, 1987). The regulatory mechanisms governing the selective myocardial secretion of matrix components into these two cushion-forming heart segments remain virtually unexplored. However, some mouse mutant models have been identified in which the endocardial cushions fail to form. These models include mice deficient for *Nkx2.5* (Lyons *et al.*, 1995), *mef2C* (Lin *et al.*, 1997), or the heart defect (*hdf*) gene (Yamamura *et al.*, 1997). The *Nkx2.5* and *mef2C* genes, deleted by targeted homologous recombination, have been placed early in the hierarchy of muscle lineage formation (see Chapters 7 and 8); however, both probably act too early in development to provide insight into the selective regulation of the genes later required to initiate cushion formation. The *hdf* mouse line, however, may provide a useful model for understanding the selective regulation of cardiac matrix expansion in the AV and outlet regions.

1. The *hdf* Mouse Model

The *hdf* mouse line arose as an insertional mutation of a single gene locus located on chromosome 13. The transgene used to create the mouse line contained a lacZ reported and analysis of lacZ expression in developing embryos revealed its earliest detectable expression within the myocardium during cushion-forming stages of heart development (Fig. 3; Yamamura *et al.*, 1997). In the homozygous mutant the heart cushion swellings are absent and endothelial cells remain in the endocardium in a phenotypically normal monolayer. If the AV region is removed from the heart and grown in culture on the surface of a three-dimensional collagen gel, the endothelial cells will undergo a normal appearing epithelial–mesenchymal transition and form migrating mesenchyme that invades the collagen lattice. These rescue experiments imply that the *hdf* endothelium is intrinsically competent to form cushion mesenchyme and suggest that the defect is an extrinsic factor probably selectively secreted by the myocardium into the cardiac jelly of cushion-forming segments. Recent data have shown that an antibody marker, TC2, that recognizes glycosaminoglycan chains in the normal embryonic mouse heart is not detectable in the *hdf* hearts (Fig. 4; Mjaatvedt *et al.*, 1997). Sequences flanking the transgene have been identified and used to map the *hdf* locus to chromosome 13 between markers D13Bir18 and D13Hum34 by interspecific backcross analysis (Yamamura *et al.*, 1997). Recently, we have mapped the locus onto another mouse DNA-mapping panel [(JAX-BSS: C57BL/6JEi × SPRET/Ei) F1 female × SPRET/Ei male] and shown that it segregates with the marker D13Mit27 (Mjaatvedt *et al.*, 1997). Although no coding loci have been mapped in the JAX-BSS panel at this locus, the dihydrofolate reductase (*Dhfr*) gene lies within 2.4 cm of this site (Fig. 5). Also, a neural specific homeobox gene (*Nsx1*) and a yeast gene designated *D13Xrf249* (homology with *mef2C*) have been mapped approximately 5.0 cm from the *hdf* locus. On an integrated map of chromosome 13, the anomymous marker D13Mit27 has been placed approximately 0.5 cm from the gene *Cspg2* (Seymour *et al.*, 1996). We have not yet mapped the *Cspg2* gene onto the JAX-BSS panel to determine if it lies within the *hdf* locus. Identification of the *hdf* gene and molecular analysis of how the *hdf* locus is regulated should provide insights into understanding the mechanism of how the endocardial cushion

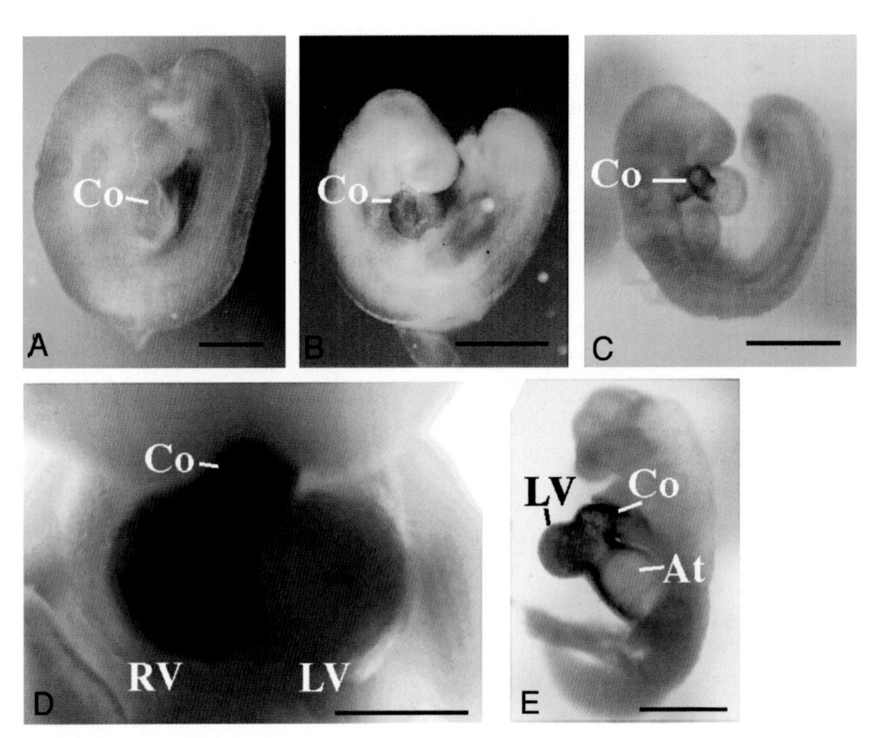

Figure 3 Expression of the *hdf* gene. The *hdf* mouse is an insertional mutant in which endocardial cushions and the right outlet of the heart fail to form. (A) Embryonic Day 9.5 pc wild-type normal mouse littermate. (B) Based on expression of the *lacZ* portion of the transgene, the *hdf* gene is restricted to the right side [conus and right ventricle] at stages of heart looping. (C) Homozygous *hdf* embryos show a similar right-sided expression pattern. (D) Embryonic Day 10.5 pc embryos shows a strong right-sided expression of the *lacZ* transgene in embryos hemizygous for the transgene. (E) Embryos homozygous for the transgene also continue to show predominately right-sided expression of *lacZ*; however, the hearts are severely distorted and embryos fail to survive past this stage. Scale bars; A–C, 500 μM; D, 0.5 mm; E, 1 mm.

swellings form within their specific locations along the single heart tube. The expression distribution of the *hdf* gene, based on the lacZ reporter, is segmentally restricted to the right side of the looped heart tube (Fig. 3). This pattern of expression is very similar to expression patterns observed for members of the BMP family of morphogens, especially BMP-4 (Hogan *et al.,* 1994; Jones *et al.,* 1991). This pattern of coexpression suggests that the *hdf* gene may be coordinately regulated with BMP-4 expression.

B. Induction of the Cushion Mesenchyme

The inductive and permissive molecular mechanisms involved in stimulating the JB3/QH1$^+$ endothelial cells to leave the vascular (endocardial) monolayer and populate the subjacent cushion matrix have been explored with the aid of a three-dimensional collagen culture system that closely simulates the regionally and temporally specific events observed *in vivo* (Eisenberg and Markwald, 1995). The results of these studies are summarized in Fig. 6. Cushion mesenchyme derived from the JB3$^+$ endothelial cells forms only in the segments of the heart tube where the myocardium secretes a particulate form of matrix we termed adherons after a precedent in the literature (Mjaatvedt and Markwald, 1989; Schubert *et al.,* 1986).

1. Identification of ES Antigens

Adherons represent aggregates of matrix proteins including fibronectin and other proteins called ES (EDTA soluble) antigens extracted from the myocardial basement membrane (Kitten *et al.,* 1987). Direct application of isolated matrix aggregates induces competent AV (not ventricular) endothelial cells to seed mesenchyme in culture (Mjaatvedt and Markwald, 1989). Antibodies prepared to ES aggregates confirmed the segmental distribution of adheron particulates in the heart and removed inductive activity from two potent sources of endothelial transforming signals: myocardial conditioned medium (myoCM) and EDTA

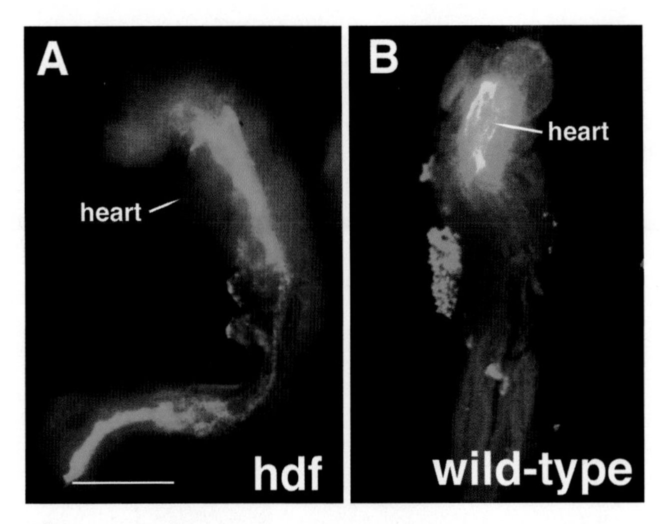

Figure 4 Immunostaining for TC2 antigen in the *hdf* embryonic heart. Whole embryo immunofluorescent images of *hdf* homozygous (A; *hdf*) and normal wild-type (B) embryos (Embryonic Days 8.5–9.0 pc) using the TC2 monoclonal antibody shows intense localization in the wild-type but not the *hdf* homozygous heart tube. The TC2 monoclonal antibody is specific for a glycosaminoglycan chain epitope and was developed by Dr. Tony Capehart (personal communication) using a novel technique for immunogen presentation. Scale bar, 1 mm.

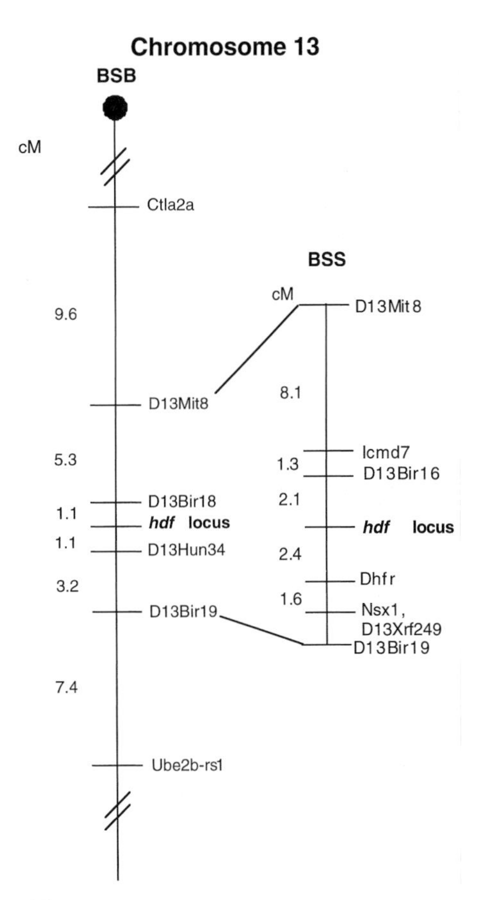

Figure 5 A partial linkage map of mouse chromosome 13 near the *hdf* mouse transgene insertion site. The location of *hdf* locus is shown in relation to flanking markers mapped in this (C57BL/6J × *Mus spretus*) F1 × C57BL/6J backcross (BSB). Recently, we mapped the locus on the (C57BL/6J × *Mus spretus*) F1 × *Mus spretus* backcross (BSS) (shown in the expanded region between D13Mit8 and D13Bir19). Ctla2a on the BSB panel is the gene symbol for cysteine protease Ctla2 α which maps approximately 16 cm proximal and Ube2B is the symbol for ubiquitin-conjugating enzyme and is approximately 11.7 cm distal to the *hdf* locus. Dhfr, dihydrofolate reductase; Nsx1, neural specific homeobox 1; D13Xrf249, yeast gene homologous to the myocyte-enhancing factor 2 (Mef2); *Imcd7* is an anonymous gene from kidney-tubule cell line. All the raw mapping data have been deposited with the Jackson Laboratories mapping panel resource. Genetic distances in cM are shown to the left between markers.

extracts of the cardiac jelly (Mjaatvedt *et al.,* 1991). Similar results were obtained in parallel studies using soybean agglutinin (SBA; Sinning *et al.,* 1996). SBA is a lectin that specifically recognizes ES$^+$ particulate matrix (Sinning *et al.,* 1992). Monoclonal antibodies have also been prepared against SBA-bound proteins called hLAMP antigens (heart lectin-associated myocardial proteins). Like ES antibodies, the hLAMP antibodies, the hLAMP antibodies effectively remove the stimulus for mesenchyme formation from EDTA extracts and myoCM in culture assays. Proteins eluted from hLAMP antibodies are capable of inductive signaling (Sinning *et al.,* 1996).

2. Characterization of ES Proteins

Several proteins are recognized by both the ES and hLAMP antibodies (28-, 46-, 60/63-, 70-, 93-, 130-, and 230-kDa proteins). The 70-kDa protein has been identified as transferrin (Isokawa *et al.,* 1994). Interestingly, transferrin has been shown to activate a migratory phenotype from differentiated monolayers of human endothelial cell cultures (Carlevaro *et al.,* 1997). The 60/63-kDa band showed homology with hepatocyte mitogen (scatter factor) and zymograms have indicated that the 46-kDa band contains a serine protease activity (Y. Nakajima, unpublished results). Antibodies to the ES proteins were used to screen a chick heart cDNA library. Several positive clones were identified, including a novel clone called r2.ES-1.

3. Identification and Role of the ES/130 Protein

Recombinant fusion protein made from a clone, r2.ES-1.R2, was used as immunogen to prepare a polyclonal antibody. The anti-r.2ES-1 antibody recognized a 130-kDa band in EDTA extracts of hearts or myocardial cell-conditioned medium. Designated ES/130, this protein was found in expression studies to be virtually heart specific in HH stages 8–13 (i.e., before and after fusion of heart fields). Thereafter, it is expressed at other sites of known inductive interactions, including the limb, notochord, and somites (Krug *et al.,* 1995;

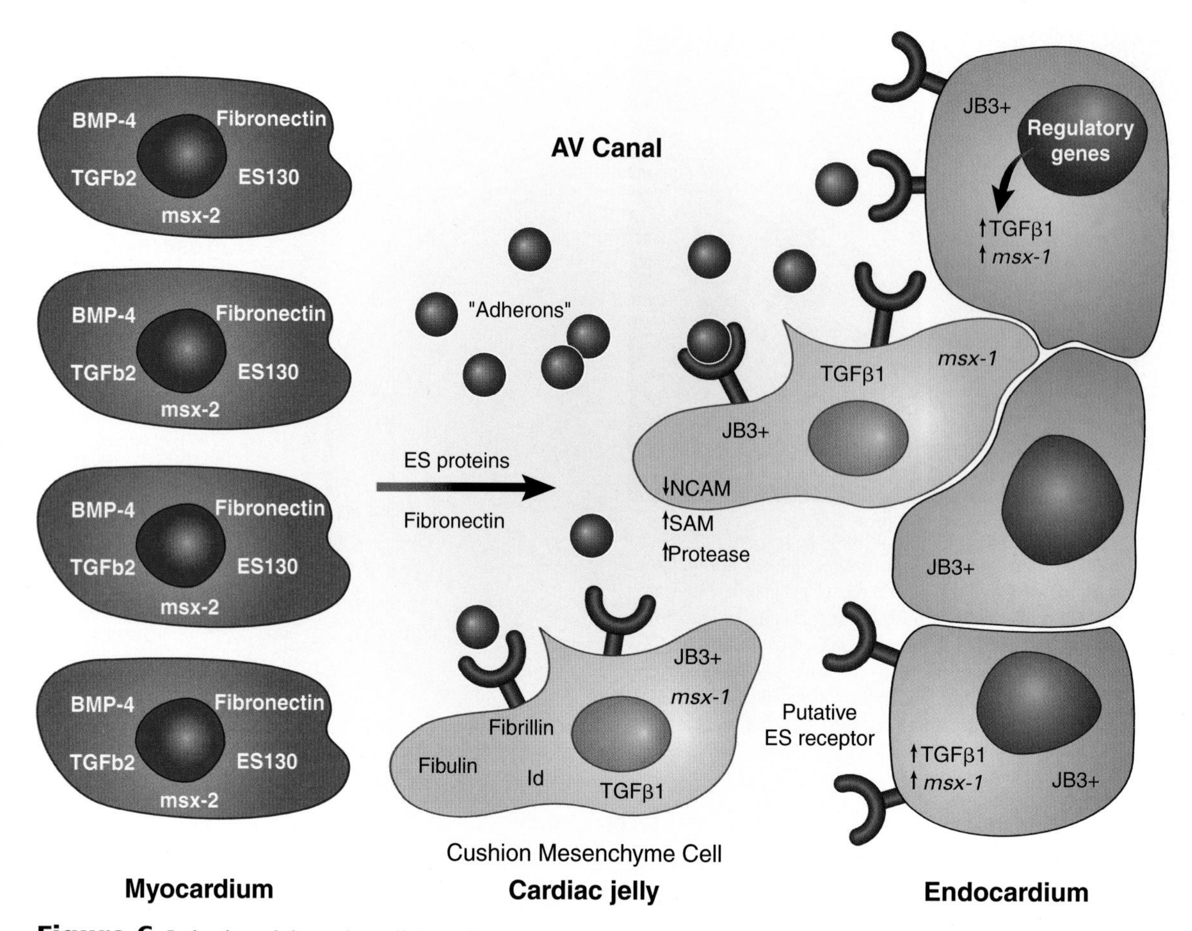

Figure 6 Induction of the endocardial cushion mesenchyme. The endocardial cushions of the atrioventricular (AV) canal and the outlet of the heart form in the extracellular matrix (cardiac jelly) between the myocardium and the endocardium. Two subpopulations of endothelial cells are present within the endocardium monolayer. These are the endocardial cells that are positive for the anti-fibrillin JB3 antibody ($JB3^+$) and those endothelial cells that are negative ($JB3^-$) for this marker. The myocardium induces an epithelial–mesenchymal transition of only the $JB3^+$ cells forming a population of migrating cushion mesenchyme that invade the cardiac jelly. The signaling myocardial cells express the homeobox containing gene *Msx-2,* growth factors TGF-β2 and BMP-4, fibronectin, and other extracellular proteins. The inductive signal can be found within an EDTA soluble fraction of H.H. stage 14 chick hearts (ES proteins) that consists of several proteins, including ES/130 and transferrin. Antibodies to different ES proteins colocalize within the cardiac jelly on multicomponent matrix particles, termed adherons, that transiently appear throughout the cardiac jelly prior to the epithelial–mesenchymal transition of the $JB3^+$ endothelial cells. The ES proteins signal the epithelial–mesenchymal transition presumably via an unknown receptor (ES receptor) on the endothelial cell surface. The competent ($JB3^+$) endocardial cells lose their cell–cell contacts (e.g., NCAM) and upregulate substratum adhesion molecules (SAM) and proteases such as MMP-1 required for migration into the cardiac jelly matrix. Induced endothelial–mesenchymal cells also express fibulin, TGF-β1, and *Msx-1.*

Markwald *et al.,* 1995; Rezaee *et al.,* 1993). Specifically, for the chick heart, ES/130 is expressed first in the myocardium and then in the endocardium–mesenchyme (Krug *et al.,* 1995; Markwald *et al.,* 1995). Similarly, in the limb, ES/130 is first expressed in the apical ridge ectoderm and then in the subjacent mesenchyme, and for the developing neural tube, it is first expressed in the notochord and then in the neural tube's ventral floorplate. This pattern of ES/130 expression suggested to us that the formation of cushion cells might be a progressive inductive event called homogenetic induction (Spemann, 1938): a process in which stimulator cells (such as the myocardium) induce target responder cells (such as the endocardium and cushion mesenchyme), in turn, to become signaling cells. This process might function to sustain or amplify the initial inductive event. This hypothesis predicts that medium conditioned by cushion mesenchymal cells (MCCM) should elicit mesenchyme formation from an endothelial cell monolayer in culture assays. The results of such experiments have shown that the mesenchymal-conditioned medium is an even more effective inducer than that obtained from myocardial cells (Ramsdell and Markwald, 1997).

4. Role of Transforming Growth Factor Superfamily Members

Both ES/130 and transforming growth factor-β (TGF-β3) are upregulated upon induction of the target endothelial–mesenchymal cells and secreted into MCCM. Antibodies to TGF-β3 added to cocultures of myocardium block mesenchyme formation, as do antisense oligos to TGF-β3 (Nakajima *et al.,* 1994; Potts *et al.,* 1991). The inhibitory effect can be overridden by adding TGF-β3 (Nakajima *et al.,* 1998). However, the outcome of culture assays in which anti-TGF antibodies were used to immunoabsorb myoCM vs MCCM indicated that only MCCM lost signaling activity. TGF-β3 directly added to control (unconditioned) medium also initiated mesenchyme formation in HH stages 13^+ and 14 endothelial monolayers but it took 24 hr longer for induced cells to acquire competency to invade the gel than occurs for MCCM-treated or myocardial coculture (Ramsdell and Markwald, 1997). These findings indicate that TGF-β3 expression is a target cell response to the myocardial induction which may function as a homogenetic signal to sustain and amplify cushion formation. ES/130 appears to be required for the myocardial signaling upstream of TGF-β3 activity. Endothelial cells upregulate expression of ES/130 prior to the formation and growth of mesenchyme. Also, those endothelial cells present in a minority of culture assays that do not respond to a inductive stimulus (myocardium, myoCM, MCCM, or TGF-β3) do not express ES/130. Since the extracellular levels of ES/130 are relatively low in MCCM compared with myoCM (Ramsdell *et al.,* 1998), the functional role of ES/130 in the endocardial cushions is unclear. Further characterization of the proteolytic fragments of ES/130 and analysis of any extracellular form of the protein should help to clarify its functional role. One possibility is that endothelial ES/130 is upregulated and functions within the early steps of transformation leading to a local expression of TGF-β3. TGF-β3, in turn, functions to upregulate those proteins shown to be expressed both *in vivo* and *in vitro* during endothelial transformation. Also under consideration is the possibility that ES/130 may be part of the intracellular secretory pathway that operates as the inductive signal moves into the extracellular matrix (E. Krug, personal communication). This hypothetical role is consistent with the localization of ES/130 within secretory cells found at other embryonic sites of induction such as the apical ectodermal ridge and notochord.

Explant studies have shown that the ventricular muscle cannot substitute for the AV myocardium as a signaling source to induce mesenchyme (Mjaatvedt *et al.,* 1987). These experiments suggested that the unidentified myocardial factor that initiates mesenchyme formation would be expressed within the myocardium of cushion-forming segments (e.g., AV canal) but not expressed in non-cushion-forming segments (e.g., atrium or ventricle) prior to mesenchyme formation. Neither TGF-β3 nor ES/130 have a segmentally restricted myocardial pattern of expression that might be expected for the initial signal. However, ES/130 in chick does, in time, become restricted (Ramsdell and Markwald, 1997). In the mouse, two members of the TGF-β superfamily that have been localized to the myocardium in cushion-forming segments prior to mesenchyme formation are BMP-2 and BMP-4 (Lyons *et al.,* 1992). Targeted deletions of these genes have confirmed that they are both important during development (Winnier *et al.,* 1995; Zhang and Bradley, 1996) but have not shed light on their potential role in cushion development. One possibility is to segmentally activate TGF-β3 or ES/130 or to support segmental growth. Recent preliminary studies in mouse (H. Yamamura and C. H. Mjaatvedt, 1998, in preparation) have indicated that BMP can directly induce the endothelial mesenchymal transition when added to isolated monolayers of AV endothelial cells on collagen gels.

5. Response of Endothelial Genes to Induction

Part of the target cell initial response to myocardial signaling is the gradual loss of N-CAM and the upregulation of serine and metalloproteinases (Alexander *et al.,* 1997; Mjaatvedt and Markwald, 1989). Both these genes could facilitate the loss of adhesion prior to matrix invasion. As cell–cell adhesions are lost in the transforming endocardial cells, other genes needed for matrix interactions are upregulated, such as fibronectin (Mjaatvedt *et al.,* 1987), cytotactin and fibulin (Bouchey *et al.,* 1996; Hoffman *et al.,* 1994), hyaluronate synthase (Spicer *et al.,* 1996), and proteoglycans (Funderburg and Markwald, 1986; Hoffman *et al.,* 1994; Little and Rongish, 1995). In addition to changes in the expression of cell–cell adhesions molecules and of matrix associated molecules, morphogens or growth factors, such as TGF-β (Brand and Schneider, 1995; Dickson *et al.,* 1993; Hogan *et al.,* 1994; Huang *et al.,* 1995; Nakajima *et al.,* 1998; Potts *et al.,* 1992; Runyan *et al.,* 1992), insulin-like growth factor (Fig. 7), and fibroblast growth factor (Choy *et al.,* 1996), are upregulated in cells within the zone of epithelial–mesenchymal transition. The expression of these morphogens may mediate the other changes in gene expression already described. For example, TGF-β has been shown in this and other systems to stimulate the production of matrix proteins and associated receptors. Fibroblast and insulin-like growth factors may mediate the proliferative response of mesenchymal cells located in a zone beneath the endothelial cell monolayer.

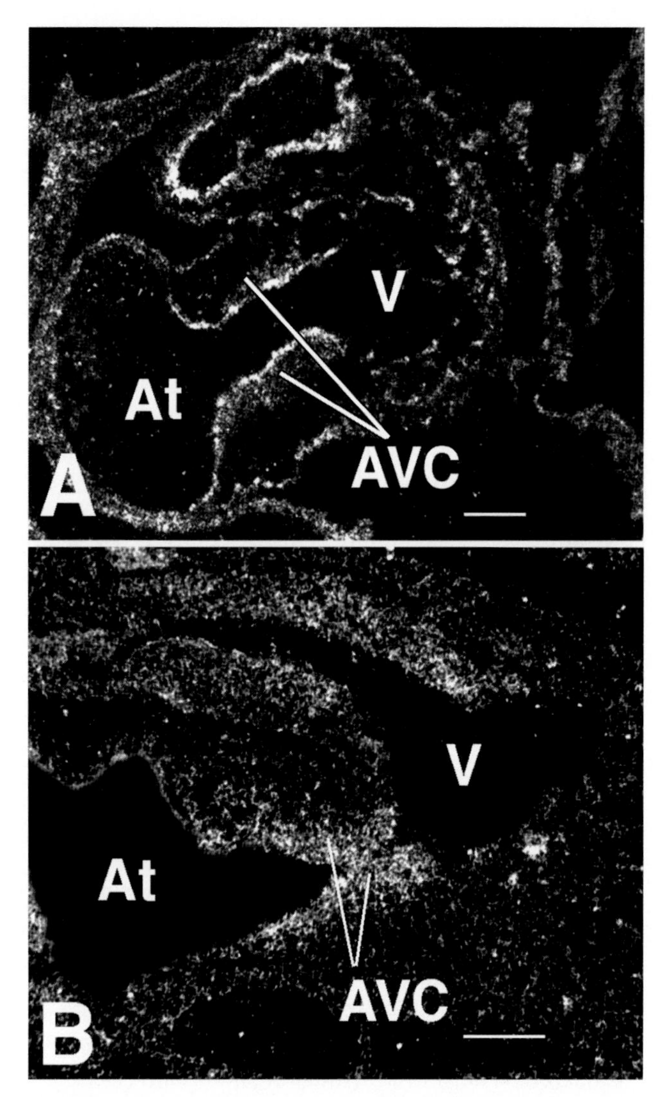

Figure 7 Localization of IGF-1 and H19 mRNA in the embryonic mouse heart. Dark field image of sections through an embryonic mouse heart (Embryonic Days 10.5 pc) hybridized with radiolabeled probes for H19 (A) and IGF-1 mRNA (B). Localization of IGF-1 and H19 mRNA is apparent within the endothelial and mesenchymal cells of the atrioventricular (AV) and outlet (Ot) regions. The density of silver grains is highest in the endocardium and appears to decrease in the mesenchymal cells that have migrated from the endothelial monolayer. In contrast to the AV and Ot endocardium, the myocardium throughout the heart and the ventricular endothelial monolayer shows little or no developed silver grains. Atrium (At), atrioventricular canal (AVC), ventricle (V). Scale bar, 50 μM.

IV. Developmental Fates of the JB3$^+$ Endocardial Cushion Cells

The normal formation of the cellularized cushion is a prerequisite for the septation, valve leaflet formation, and structural alignment of the four-chambered heart. After the initial epithelial to mesenchymal transition, subpopulations of endocardial-derived mesenchyme continue to proliferate and express genes associated with proliferation. However, with the expansion of the mesenchyme, other subpopulations close to the myocardium begin to selectively express genes associated with differentiation. The differential gene expression observed within subpopulations of cushion mesenchyme appears to correspond to the potential developmental fates assumed by endocardial cushion cells.

A. Formation of Valve Leaflets

The mature AV valve leaflets are composed in mammals of approximately equal parts mesenchymal (fibroblastic) and myocardial tissues. Previous work has indicated that the derivation and integration of these two tissue types in the valve leaflets are dependent on the normal maturation of the cushion mesenchyme (Bouchey *et al.*, 1996; De la Cruz *et al.*, 1977; Lamers *et al.*, 1995; Wunsch *et al.*, 1994).

Fusion at the midline of the superior (SAVC) and inferior cushions forms the AV septum, which divides the common inlet into the right (tricuspid) and left (mitral) inlets. Each inlet opening is surrounded by a rim of cushion tissue containing cells derived from the lateral cushions and the newly formed AV septum. Mesenchyme from the AV cushions extend into the ventricle along the inner curvature and into the atrium in a plane associated with the primary atrial septum. The ventricular projections may mediate connections of the valve leaflets to the papillary muscles. Studies of heart defects in mice trisomic for chromosome 16 suggest that the atrial extensions are required for normal atrial septation (Tasaka *et al.*, 1996a). In the tricuspid inlet, mitral inlets, and the conal outlet, the prevalvular mesenchyme proliferates producing bulges that project into each lumen. The cellular distribution and proliferation is not uniform throughout the endocardial cushion outgrowth. Subpopulations of cells immediately adjacent to the lumen divide more rapidly than at sites proximal to the myocardium (R. R. Markwald and C. H. Mjaatvedt, unpublished results). In the zone of proliferating cushion mesenchyme, the pattern of valvular fibroblasts is correlated with the expression of growth and transcription factors. Secreted morphogens or growth factors that are correlated with the proliferative zone of cushions include members of the fibroblast family of growth factors such as FGF-1, FGF-2 (Choy *et al.*, 1996), and FGF-4 (C. Mjaatvedt *et al.*, unpublished data); members of the insulin family of growth factors, i.e., IGF-1 and -2 and the IGF-associated gene H19 (Fig. 7), and *desert hedgehog* and BMP-2 (Bitgood and McMahon, 1995). Homeobox-containing genes (Kern *et al.*, 1995) and other presumed transcription factors expressed within the proliferating zone of the cushion mesenchyme include *Msx1* (Chan-Thomas *et al.*, 1993),

Prx1, Prx2 (Kern *et al.,* 1995; Leussink *et al.,* 1995), *Sox4* (Schilham *et al.,* 1996), and *NF-ATc* (Park *et al.,* 1996). Although the expression pattern and nature of these genes are suggestive of irreplaceable function, functional roles for many of these molecules in heart cushions remain to be determined. Interestingly, mice made deficient for *Msx1* (Satokata and Maas, 1994), *Prx1,* and *Prx2* did not provide clearly defective phenotypes in the cushions. Conversly, the homeobox-containing gene *Sox4* did show abnormalities in semilunar valve formation and the muscular portion of the outlet septum consistent with the gene expression in these regions during development (Schilham *et al.,* 1996). Several of these genes could have a redundant function with each other or with genes not yet detected in the cushions. Alternatively, mice deficient for these genes may have subtle deviations not yet detected but consistent within the context of a different hypothesis.

B. Role for Cushion Mesenchyme in the Completion of Heart Looping

During the first phase of looping, the primary segments are still within a linear array and must be repositioned and remodeled during the second phase of looping for the forming septa to come into proper alignment. During this process of remodeling and completion of looping, the conal septum must be repositioned or shifted to overlie the left and right ventricles. Simultaneously, the right AV canal must be aligned with the primitive right ventricle (Fig. 8). This process of alignment is accomplished by the second phase of the looping process in which the ventricular infundibular fold deepens to bring the inlet and outlet limbs of the heart together. Prior to this alignment the blood flow from both the right and left AV canals fills the left ventricle and then exits through the right ventricle and conus outlet. This primitive pattern of blood flow, if retained after birth, is termed double-outlet right ventricle (DORV), a common heart defect observed from a diverse and seemingly unrelated list of embryonic perturbations in more than one animal species. For example, in avians, any perturbation that affects normal cardiac looping, especially the second phase, often results in a DORV phenotype, including small changes in hemodynamics (Hogers *et al.,* 1997) and treatments with retinoids (Bouman *et al.,* 1995). In mice such perturbations include trisomies of chromosomes 13 or 16, the *iv/iv* mouse (Icardo and Sanchez de Vega, 1991), and targeted gene deletions of the endothelin-1 (Kurihara *et al.,* 1995), retinoid receptors (Dyson *et al.,* 1995), neuregulin and its receptors (Gassmann *et al.,* 1995; Lee *et al.,* 1995; Meyer and Birchmeier, 1995) and TFG-β2 (Sanford *et al.,* 1997).

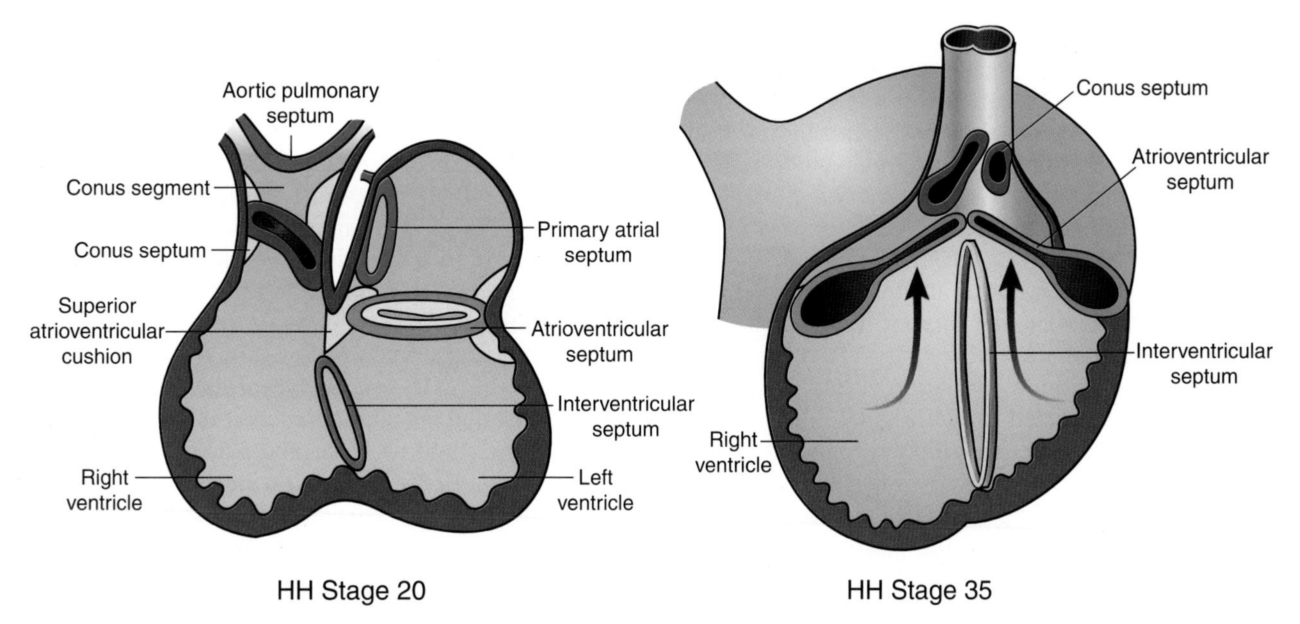

Figure 8 Diagram depicting alignment of the septal ridges. At HH stage 20 the right ventricle receives blood flow only from the left ventricle. The flow of blood exits the RV through a common conotruncal region. Subsequent remodeling of the heart tube results in a shift of the atrioventricular septum (AVS) toward the right to position the AVS over the interventricular septum as seen by HH stage 35. During the AVS shift, the conotruncal region and conotruncal septum are also positioned directly over the AVS. These morphogenetic shifts transform the single inlet and single outlet (double-inlet left ventricle and double-outlet right ventricle) into a four-chambered phenotype in which the right and left ventricles each have a separate atrial inlet and separate outlets for the blood flow (arrows; HH Stage 35).

1. The Trisomy 16 Mouse Model

The trisomy 16 (Ts16) mouse (Gropp *et al.,* 1983) displays a high frequency of specific cardiac defects and provides a useful model for understanding the mechanisms of cushion morphogenesis and their relationship to cardiac looping. A significant degree of synteny exists between human chromosome 21 (Chr21) and the distal region of mouse chromosome 16 (Cabin *et al.,* 1996; Mjaatvedt *et al.,* 1993; Reeves *et al.,* 1986). Also, the Ts16 mouse displays some phenotypic similarities with phenotypes observed in Down's syndrome and has been considered as a partial model of human trisomy 21 (Down's syndrome). The incidence of persistent AV septal defect reported in the Ts16 mouse (Gearhart *et al.,* 1987; Miyabara *et al.,* 1982) is similar to that found for Down's syndrome patients (Ferencz *et al.,* 1989). However, a more detailed analysis of the phenotype shows that the Ts16 hearts possess a more complicated spectrum of defects (Hiltgen *et al.,* 1995; Miyabara *et al.,* 1982; Webb *et al.,* 1996). Besides defects in AV septum, other defects include tricuspid atresia, double-inlet left ventricle (DILV), and conotruncal defects, e.g., DORV, subaortic VSD, infundibular and pulmonary stenosis, and transposition of the great arteries not normally associated with Down's syndrome (Miyabara *et al.,* 1982; Webb *et al.,* 1996).

The 22.q11 region of human chromosome 22, which is deleted in the DiGeorge syndrome (or CATCH 22), is also found on murine chromosome 16 (Wilson *et al.,* 1993). Therefore, the structural defects found in the outlet of the Ts16 heart may result from the overexpression of genes found more proximally on mouse chromosome 16, outside of the Chr21 homologous region. In both the AV and the outlet region of the Ts16 heart there is a lag or delay in the onset of cushion mesenchyme formation. This delay gives rise to a change in the overall morphogenetic shape of the cushion that correlates with the AV septal defect (Hiltgen *et al.,* 1995). The change in cushion shape appears to result from alterations in the Ts16 cell–matrix interactions that reorder the normally randomized collagen fibers.

The delay in cushion formation results in the AV cushions being restricted to the original boundaries of the AV canal. Although not widely recognized, the superior AV cushion extends both ventricularly and atrially. Both these extensions occur in a midsaggital plane. The atrial extension contacts the posterior wall of the atrium at the point where extracardiac mesenchyme, called the spina vestibuli (Asami and Koizumi, 1995), enters the atrial wall. Tasaka *et al.* (1996b) have noted that the spina vestibuli enters the atrium, growing along the atrial extension of the superior cushion. They have suggested that the spina vestibuli brings an infolding of myocardium with it—the future primary atrial septum. Consistent with this hypothesis, a primary atrial septal defect occurs in 100% of trisomic 16 mice and correlates with a hypoplastic or absent atrial extension of the superior cushion and spina vestibuli (Markwald *et al.,* 1997).

In normal mice and chick embryos, an extension of the SAVC also develops along the inner curvature of the looping heart and fuses with a similar ventricular extension of the sinistro-ventral conal cushion (SVCC). The fusion of these two cushions creates a layer of mesenchyme that underlies the progressively deepening inner curvature fold of the ventricular myocardium (Fig. 9). Collectively, this region is sometimes called the ventricular–infundibular fold (VIF). The inner curvature of the heart is established with the onset (first phase) of looping (stage 10^{+}) and anatomically extends from the right AV canal to the right side of the distal conal segment. In the second phase of looping the loop of VIF progressively deepens until HH stage 38.

However, in the Ts16 mouse, the inner curvature was incompletely lined by cushions, i.e., the SAVC component does not fuse with the SVCC (Wessels and Markwald, 1998). This correlated 100% with DORV, in which the conus remained connected to the right ventricle. As shown in Fig. 10, both the aorta and pulmonary trunk were open to the right ventricle. Also, there was a spectrum of impaired development of the right AV inlet ranging from a slight narrowing (tricuspid atresia) to the absence of the tricuspid orifice with DILV.

V. The Cardiac Inner Curvature: Remodeling by Myocardialization

Analysis of the developing Ts16 heart has shown that the absence of a ventricular extension from the SAVC might be linked to the failure of the conus to shift leftward and counterclockwise (resulting in DORV) and, possibly, the failure to create a right AV inlet. Further analysis showed that the muscularization of cushion tissue normally observed in this region was delayed or inhibited in the Ts16 heart. Myocardialization is a term used to describe the phenomenon at the inner curvature in which the conal cushions and part of the SAVC become muscularized through an invasive migration of nonproliferating myocardial cells (Fig. 11; chick HH. stage 28). As a result of this process, the myocardium is removed from the inner curvature, allowing the posterior wall of the conus and anterior wall of the right AV canal to merge, forming a shared wall—the mitroaortic continuity. Myocardialization is the final step (or fate) in the morphogenesis of both the SVCC and superior AV cushion. Results from Ts16 mouse studies indicate that the failure to initiate myocardialization is a

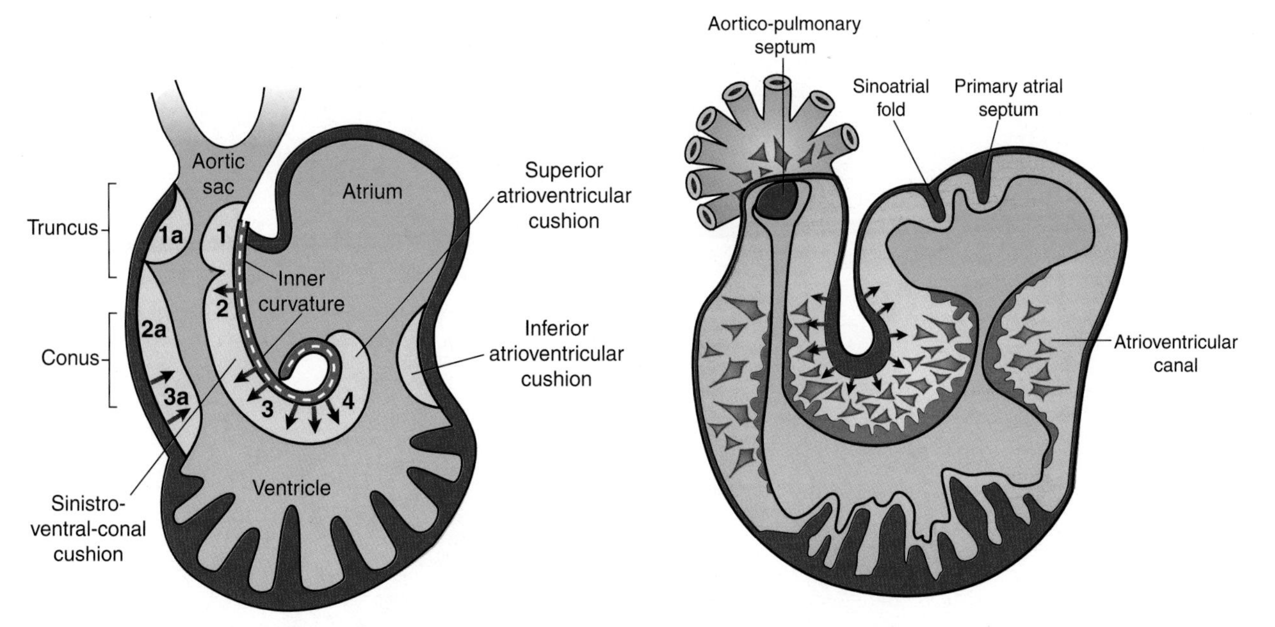

Figure 9 Segmental interaction during remodeling of the inner curvature. (Left) The relative positions (1–5) of the endocardial cushion swellings are shown within the looping heart tube (approximately HH stage 24). The outlow tract of the heart consists of a proximal region defined here as the conus and a more distal part defined as the truncus arteriosus. The myocardium of the heart's inner curvature is completely lined with cushion tissue (2–4) by the fusion of the sinistro-ventralconal cushion (SVCC) and the superior atrioventricular cushion (SAVC). The inner curvature myocardium (dashed line) is removed (arrows) during remodeling to allow proper septal alignments. The inner curvature myocardium is the only region of the heart that comes into close interaction with cushion mesenchyme derived from two separate heart segments (the SAVC and the SVCC). (Right) The myocardial cells invade (arrows) the cellularized cushions that line the inner curvature of the heart tube.

major cause of many of the most common forms of congenital heart disease. The outcome of myocardialization is to literally reposition the myocardium from the surface of the inner curvature into the cushion itself, explaining (i) why both the conal septum and superventricular crest (part of the SVCC) become muscularized, (ii) why heart looping should be considered complete only when the conus segment (anterior limb) of the tubular heart is brought into direct contact with the AV inlet, (iii) why completion of looping is required to establish a mitroaortic continuity and an outlet for the left ventricle, and (iv) how the alignment of the conal, A–P, and ventricular septa is established (De la Cruz and Markwald, 1998).

A. Mechanism to Explain Ts16 Heart Defects

Observations of the Ts16 mouse indicate that the fate of the superior AV cushion is crucial to establishing a normal four-chambered heart, including atrial, AV, and outflow septation. Its expansion and extension onto the inner curvature is necessary to mediate the septal alignments that occur during the completion of looping. The inability to adequately infiltrate the inner curvature with cushion mesenchyme is the central deficiency that gives rise to the Ts16 heart phenotype. The recent observation that the primary problem of Ts16 mouse is a looping problem fits this hypothesis (Webb *et al.*, 1996). Without the SAVC extension, the myocardium of the inner curvature in the Ts16 mouse cannot be completely removed or remodeled (because there is no cushion for it to invade or trigger its remodeling). This leads to the hypothesis that a principal fate of cushion tissues—specifically those that line the inner curvature (SAVC and SVCC)—is to initiate myocardialization.

B. The Morphogenetic Mechanism of Myocardialization

The anatomical relationship between the superior AV cushion, the SVCC, and the inner curvature myocardium suggests that the cushion mesenchyme interacts with the myocardium to promote or induce the removal or redistribution of the associated myocardium. The removal of the myocardium at the inner curvature correlates temporally with the muscularization of the cushions by the formation of migrating myocardial cells that invade the underlying cushion matrix. Although cushion matrix is heavily populated with mesenchyme along the inner curvature by stage HH 24 (Embryonic Day 12 in mouse), myocardialization is not initiated un-

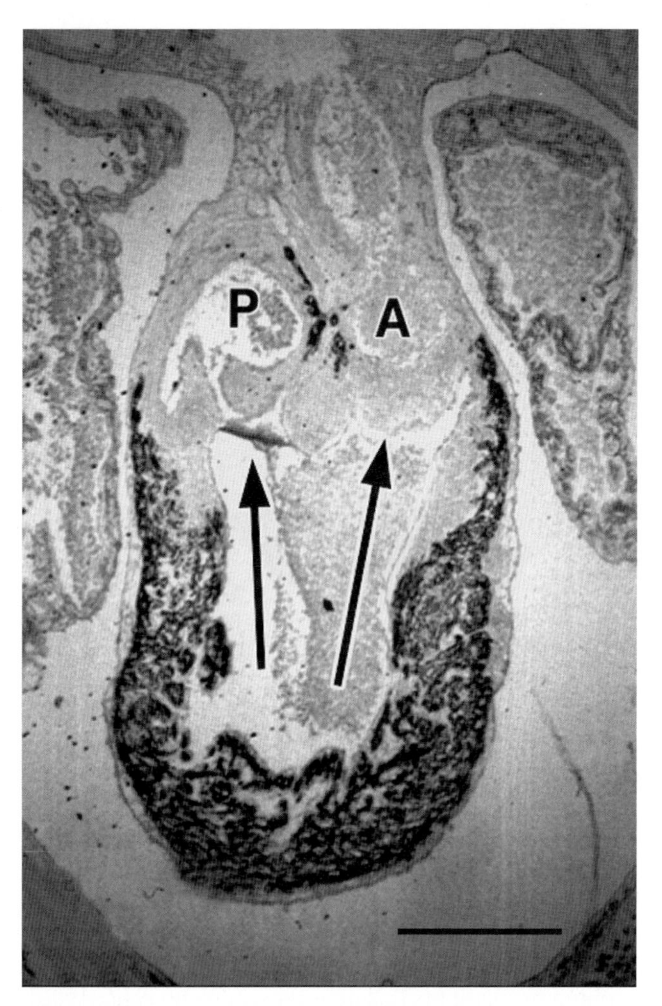

Figure 10 Photomicrograph of a trisomy 16 (Ts16) mouse heart (Embryonic Day 16 pc) showing double-outlet right ventricle. The myocardium is immunostained for MHC. Blood flow (arrows) from the right ventricle (RV) enters both the aortic (A) and pulmonary (P) outflow (double-outlet right ventricle). Scale bar, 500 μM.

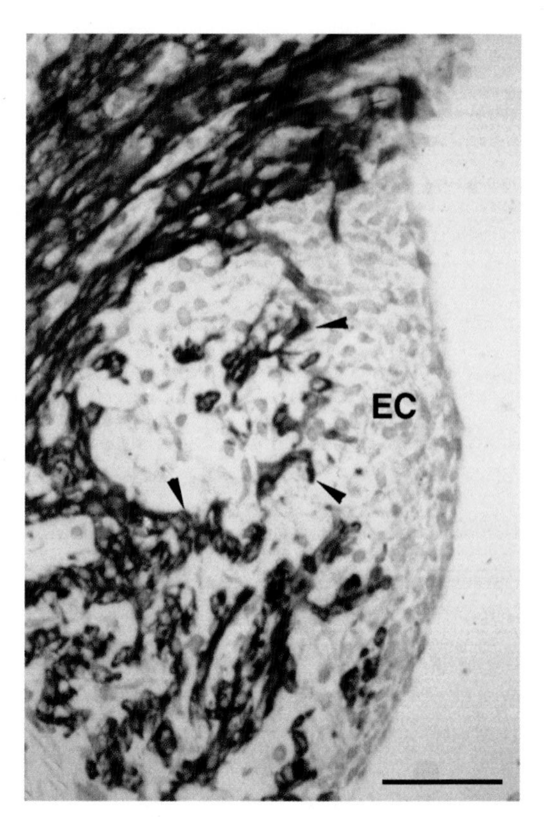

Figure 11 The process of myocardialization. Photomicrograph of the superior endocardial cushion. The myocardium has been immunostained with a myosin heavy chain. By this stage, myocardial cells (arrowheads) can be seen to invade the adjacent cellularized endocardial cushion (EC) in a process termed "myocadialization." Scale bar, 100 μM.

til after this time point is reached. This suggests that differentiation of the cushion mesenchyme may be as important as its initial formation. The upregulation of new sets of genes by the "differentiated" cushion mesenchyme may actually induce the invasion of the myocardial cells. A hypothetical mechanism of how cushion mesenchyme might induce myocardialization is considered in Fig. 12. Cushion mesenchyme at the interface with the myocardium is postulated to be older and further differentiated than newly formed mesenchyme at the lumenal surface and therefore more competent to induce the myocardialization process.

Homebox genes known to be expressed in these putatively differentiated and undifferentiated regions are good candidates for molecular regulatory mechanism. Four nonclustered homeobox genes are known to be expressed in both AV and outflow cushion tissues: *Msx-1* (Chan-Thomas *et al.,* 1993), *Meox-1* (Candia *et al.,* 1992), and *Prx-1* and *Prx-2* (Kern *et al.,* 1995; Leussink *et al.,* 1995). The helix–loop–helix gene, *Id,* is also expressed (Evans and O'Brien, 1993). However, the role of these genes in cushion morphogenesis is not known. The timing of their expression would suggest that their function would follow after the initial transformation of endocardium into mesenchyme. In mice, *Msx-1* and *Meox-1* are expressed reciprocally, with *Meox-1* strongest in the differentiated mesenchyme proximal to the myocardial interface (Barton *et al.,* 1995). Accordingly, we hypothesize that *Msx-1* is antidifferentiative to the remodeling process and *Meox-1* is prodifferentiative to remodeling. This would be consistent with the effect of *Msx-1* on inhibiting myoD expression in myoblasts (Wang and Sassoon, 1995) and the *Msx-1* knockout mouse in which no obvious defects were seen in the heart (Satokata and Maas, 1994). Our hypothesis, if correct, would predict normal (or accelerated) remodeling of the inner curvature if *Msx-1*

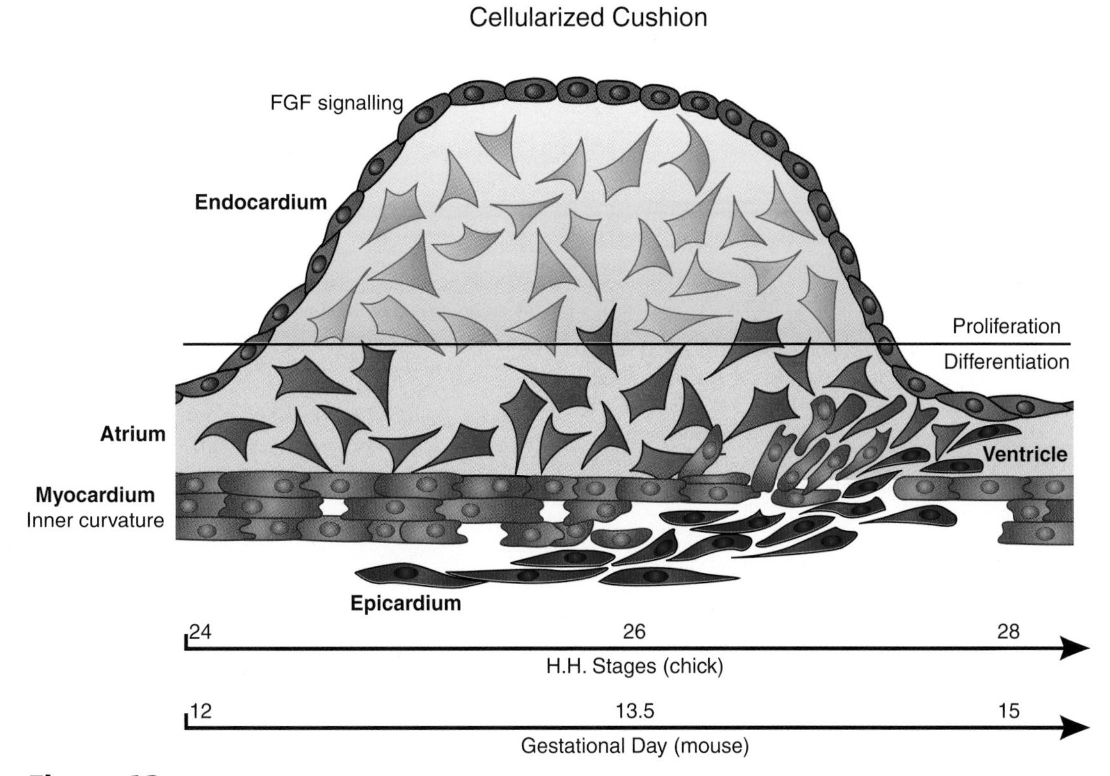

Figure 12 A schematic diagram depicting a proposed model of the fates of AV endocardial cushion along the inner curvature. The cellularized endocardial cushion can be divided into a proliferative zone and a differentiative zone. The proliferating zone is the region of mesenchyme distal to the myocardium and located just beneath the endocardial "ridge" of cushion that extends into the heart lumen. Mesenchyme located in the proliferative zone have been shown to express the genes associated with cell proliferation, such as *FGF, Msx1, Prx1,* and *Id.* The differentiative zone of mesenchyme lies more proximal to the myocardium. The mesenchyme in this region are less proliferative and no longer express *Msx1* and *Id* but express other genes, such as *Meox1* and *Prx1,* that might alter the local matrix environment and facilitate the inner curvature myocardial invasion of the cushion. The ridge hypothesis is consistent with observations that during stages of cushion remodeling (chick HH stages 24–28 and mouse (Embryonic Days 12–15 pc), the distal tip of the cushions becomes elongated and extends into the lumen of the heart-forming septal ridges (conus) or valve leaflets (truncus), whereas the proximal portion of the cushion, particularly along the inner curvature, becomes muscularized during these stages of development.

were inhibitory to this morphogenetic event. Thus, overexpression of *Msx-1* might be more revealing regarding the role of this homeobox gene. Conversely, inhibiting *Meox-1* might be the best approach to determine if the fate of cushions is to promote myocardialization.

VI. Conclusion

The formation and differentiation of the endocardial cushions and the process of heart looping are central events required for the reorganization of the single heart tube into a four-chambered structure. Both these events appear to be regulated by the segmental restriction of gene expression along the anterior–posterior axis of the heart tube. Because of segmental restrictions, we proposed that the endocardial cushions form in only two regions of the tube. The initial rightward bend of the tube separates the first two segments into the initial inlet and outlet of the primitive heart. The completion of looping, which accompanies remodeling, brings the inlet and outlet segments of the tube into their correct anatomical relationship and alignment. The endocardial cushions then have multiple functions: (i) to divide the inlet into a right and left side, (ii) to form the valvular leaflets through the growth and differentiation of the cushion's luminal surface, and (iii) to induce myocardialization which removes the myocardium from the inner curvature of the heart, bringing the internal septa into alignment. Thus, any perturbations that restrict the growth and/or normal differentiation of the cushions could be expected to give rise to a common defective phenotype. This may explain why a wide variety of environmental insults and a diverse assortment of gene mutations can result in a similar spectrum of inlet and outlet heart defects.

References

Alexander, S. M., Jackson, K. J., Bushnell, K. M., and McGuire, P. G. (1997). Spatial and temporal expression of the 72-kDa type IV collagenase (MMP-2) correlates with development and differentiation of valves in the embryonic avian heart. *Dev. Dyn.* **209,** 261–268.

Asami, I., and Koizumi, K. (1995). Development of the atrial septal complex in the human heart: Contribution of the spinal vestibuli. *In* "Developmental Mechanisms of Heart Disease" (E. B. Clark and A. Takao, eds.), pp. 255–260. Futura Publ. Co., Armonk, NY.

Barton, P. J. R., Boheler, K. R., Brand, N. J., and Thomas, P. S. (1995). "Molecular Biology of Cardiac Development and Growth." Springer-Verlag, Heidelberg.

Bitgood, M. J., and McMahon, A. P. (1995). Hedgehog and Bmp genes are coexpressed at many diverse sites of cell-cell interaction in the mouse embryo. *Dev. Biol.* **172,** 126–138.

Bouchey, D., Argraves, W. S., and Little, C. D. (1996). Fibulin-1, vitronectin, and fibronectin expression during avian cardiac valve and septa development. *Anat. Rec.* **244,** 540–551.

Bouman, H. G., Broekhuizen, M. L., Baasten, A. M., Gittenberger-de Groot, A. C., and Wenink, A. C. (1995). Spectrum of looping disturbances in stage 34 chicken hearts after retinoic acid treatment. *Anat. Rec.* **243,** 101–108.

Bouwmeester, T., Kim, S., Sasai, Y., Lu, B., and De Roberts, E. M. (1996). Cerberus is a head-inducing secreted factor expressed in the anterior endoderm of Spemann's organizer. *Nature* (*London*) **382,** 595–601.

Brand, T., and Schneider, M. D. (1995). The TGF beta superfamily in myocardium: Ligands, receptors, transduction, and function. *J. Mol. Cell. Cardiol.* **27,** 5–18.

Brueckner, M., D'Eustachio, P., and Horwich, A. L. (1989). Linkage mapping of a mouse gene, iv, that controls left-right asymmetry of the heart and viscera. *Proc. Natl. Acad. Sci. U.S.A.* **86,** 5035–5038.

Cabin, D., Citron, M., McKee-Johnson, J., Mjaatvedt, A. E., and Reeves, R. (1996). Encyclopedia of the mouse genome V. Mouse chromosome 16. *Mamm. Genome* **6** (Spec. No.), S271–S280.

Candia, A. F., Hu, J., Crosby, J., Lalley, P. A., Noden, D., Nadeau, J. H., and Wright, C. V. (1992). Mox-1 and Mox-2 define a novel homeobox gene subfamily and are differentially expressed during early mesodermal patterning in mouse embryos. *Development* (*Cambridge, UK*) **116,** 1123–1136.

Carlevaro, M. F., Albini, A., Ribatti, D., Chiara, G., Benelli, R., Cermelli, S., Cancedda, R., and Cancedda, F. D. (1997). Transferrin promotes endothelial cell migration and invasion: Implication in cartilage neovascularization. *J. Cell Biol.* **136,** 1375–1384.

Chan-Thomas, P. S., Thompson, R. P., Robert, B., Yacoub, M. H., and Barton, P. J. (1993). Expression of homeobox genes Msx-1 (Hox-7) and Msx-2 (Hox-8) during cardiac development in the chick. *Dev. Dyn.* **197,** 203–216.

Choy, M., Oltjen, S. L., Otani, Y. S., Armstrong, M. T., and Armstrong, P. B. (1996). Fibroblast growth factor-2 stimulates embryonic cardiac mesenchymal cell proliferation. *Dev. Dyn.* **206,** 193–200.

De Haan, R. L. (1965). "Morphogenesis of the Vertebrate Heart." Holt, Rinehart & Winston, New York.

De la Cruz, M. V., and Markwald, R. R. (1998). "Living Morphogenesis of the Heart." Springer-Verlag, Heidelberg.

De la Cruz, M. V., Sanchez Gomez, C., Arteaga, M. M., and Arguello, C. (1977). Experimental study of the development of the truncus and the conus in the chick embryo. *J. Anat.* **123,** 661–686.

De la Cruz, M. V., Gimenez-Ribotta, M., Saravalli, O., and Cayre, R. (1983). The contribution of the inferior endocardial cushion of the atrioventricular canal to cardiac septation and to the development of the atrioventricular valves: Study in the chick embryo. *Am. J. Anat.* **166,** 63–72.

De la Cruz, M. V., Sanchez-Gomez, C., and Palomino, M. A. (1989). The primitive cardiac regions in the straight tube heart (stage 9−) and their anatomical expression in the mature heat: An experimental study in the chick heart. *J. Anat.* **165,** 121–131.

Dickson, M. C., Slager, H. G., Duffie, E., Mummery, C. L., and Akhurst, R. J. (1993). RNA and protein localisations of TGF beta 2 in the early mouse embryo suggest an involvement in cardiac development. *Development* (*Cambridge, UK*) **117,** 625–639.

Drake, C. J., and Jacobson, A. G. (1988). A survey by scanning electron microscopy of the extracellular matrix and endothelial components of the primordial chick heart. *Anat. Rec.* **222,** 391–400.

Dyson, E., Sucov, H. M., Kubalak, S. W., Schmid-Schonbein, G. W., De Lano, F. A., Evans, R. M., Ross, J., Jr., and Chien, K. R. (1995). Atrial-like phenotype is associated with embryonic ventricular failure in retinoid X receptor alpha −/− mice. *Proc. Natl. Acad. Sci. U.S.A.* **92,** 7386–7390.

Edmondson, D. G., Lyons, G. E., Martin, J. F., and Olson, E. N. (1994). Mef2 gene expression marks the cardiac and skeletal muscle lineages during mouse embryogenesis. *Development* (*Cambridge, UK*) **120,** 1251–1263.

Eisenberg, C. A., and Bader, D. (1995). QCE-6: A clonal cell line with cardiac myogenic and endothelial cell potentials. *Dev. Biol.* **167,** 469–481.

Eisenberg, L. M., and Markwald, R. R. (1995). Molecular regulation of atrioventricular valvuloseptal morphogenesis. *Circ. Res.* **77,** 1–6.

Evans, S. M., and O'Brien, T. X. (1993). Expression of the helix-loop-helix factor Id during mouse embryonic development. *Dev. Biol.* **159,** 485–499.

Ferencz, C., Neill, C., Boughman, J., Rubin, J., Brenner, J., and Perry, L. (1989). Congenital cardiovascular malformations associated with chromosome abnormalities: An epidemiologic study [see comments]. *J. Pediatr.* **114,** 79–86.

Funderburg, F. M., and Markwald, R. R. (1986). Conditioning of native substrates by chondroitin sulfate proteoglycans during cardiac mesenchymal cell migration. *J. Cell Biol.* **103,** 2475–2487.

Garcia-Martinez, V., Darnell, D. K., Lopez-Sanchez, C., Sosic, D., Olson, E. N., and Schoenwolf, G. C. (1997). State of commitment of prospective neural plate and prospective mesoderm in late gastrula/early neurula stages of avian embryos. *Dev. Biol.* **181,** 102–115.

Gassmann, M., Casagranda, F., Orioli, D., Simon, H., Lai, C., Klein, R., and Lemke, G. (1995). Aberrant neural and cardiac development in mice lacking the ErbB4 neuregulin receptor [see comments]. *Nature* (*London*) **378,** 390–394.

Gearhart, J., Oster-Granite, M., Reeves, R., and Coyle, J. (1987). Developmental consequences of autosomal aneuploidy in mammals. *Dev. Genet.* **8,** 249–265.

Gropp, A., Winking, H., Herbst, E., and Claussen, C. (1983). Murine trisomy: Developmental profiles of the embryo, and isolation of trisomic cellular systems. *J. Exp. Zool.* **228,** 253–269.

Hamburger, V., and Hamilton, H. L. (1951). A series of normal stages in the development of the chick embryo. 1951 [classical article]. *Dev. Dyn.* **195,** 231–272.

Hiltgen, G., Litke, L., and Markwald, R. R. (1995). Morphogenetic alterations during endocardiac cushion development in trisomy 16 (Down's) mouse. *Pediatr. Cardiol.* **17,** 21–30.

Hoffman, S., Dutton, S. L., Ernst, H., Boackle, M. K., Everman, D., Tourkin, A., and Loike, J. D. (1994). Functional characterization of antiadhesion molecules. *Perspect. Dev. Neurobiol.* **2,** 101–110.

Hogan, B. L., Blessing, M., Winnier, G. E., Suzuki, N., and Jones, C. M. (1994). Growth factors in development: The role of TGF-beta related polypeptide signalling molecules in embryogenesis. *Development (Cambridge, UK) Suppl.,* pp. 53–60.

Hogers, B., De Ruiter, M. C., Gittenberger-de Groot, A. C., and Poelmann, R. E. (1997). Unilateral vitelline vein ligation alters intracardiac blood flow patterns and morphogenesis in the chick embryo. *Circ. Res.* **80,** 473–481.

Huang, J. X., Potts, J. D., Vincent, E. B., Weeks, D. L., and Runyan, R. B. (1995). Mechanisms of cell transformation in the embryonic heart. *Ann. N. Y. Acad. Sci.* **752,** 317–330.

Icardo, J. M., and Sanchez de Vega, M. J. (1991). Spectrum of heart malformations in mice with situs solitus, situs inversus, and associated visceral heterotaxy. *Circulation* **84,** 2547–2558.

Isokawa, K., Rezaee, M., Wunsch, A., Markwald, R. R., and Krug, E. L. (1994). Identification of transferrin as one of multiple EDTA-extractable extracellular proteins involved in early chick heart morphogenesis. *J. Cell Biochem.* **54,** 207–218.

Jones, C. M., Lyons, K. M., and Hogan, B. L. (1991). Involvement of Bone Morphogenetic Protein-4 (BMP-4) and Vgr-1 in morphogenesis and neurogenesis in the mouse. *Development (Cambridge, UK)* **111,** 531–542.

Kern, M., Argao, E., and Potter, S. (1995). Homeobox genes and heart development. *Trends Cardiovasc. Med.* **5,** 47–54.

Kitten, G. T., Markwald, R. R., and Bolender, D. L. (1987). Distribution of basement membrane antigens in cryopreserved early embryonic hearts. *Anat. Rec.* **217,** 379–390.

Krug, E. L., Rezaee, M., Isokawa, K., Turner, D. K., Litke, L. L., Wunsch, A. M., Bain, J. L., Riley, D. A., Capehart, A. A., and Markwald, R. R. (1995). Transformation of cardiac endothelium into cushion mesenchyme is dependent on ES/130: Temporal, spatial, and functional studies in the early chick embryo. *Cell Mol. Biol. Res.* **41,** 263–277.

Kubalak, S. W., Miller-Hance, W. C., O'Brien, T. X., Dyson, E., and Chien, K. R. (1994). Chamber specification of atrial myosin light chain-2 expression precedes septation during murine cardiogenesis. *J. Biol. Chem.* **269,** 16961–16970.

Kurihara, Y., Kurihara, H., Oda, H., Maemura, K., Nagai, R., Ishikawa, T., and Yazaki, Y. (1995). Aortic arch malformations and ventricular septal defect in mice deficient in endothelin-1. *J. Clin. Invest.* **96,** 293–300.

Lamers, W. H., Viragh, S. S., Wessels, A., Moorman, A. F. M., and Anderson, R. H. (1995). Formation of the tricuspid valve in the human heart. *Circ. Res.* **91,** 111–121.

Lee, K., Simon, H., Chen, H., Bates, B., Hung, M., and Hauser, C. (1995). Requirement for neuregulin receptor erbB2 in neural and cardiac development [see comments]. *Nature (London)* **378,** 394–398.

Leussink, B., Brouwer, A., el Khattabi, M., Poelmann, R. E., Gittenberger-de Groot, A. C., and Meijlink, F. (1995). Expression patterns of the paired-related homeobox genes MHox/Prx1 and S8/Prx2 suggest roles in development of the heart and the forebrain. *Mech. Dev.* **52,** 51–64.

Levin, M., Johnson, R., Stern, C., Kuehn, M., and Tabin, C. (1995). A molecular pathway determining left-right asymmetry in chick embryogenesis. *Cell (Cambridge, Mass.)* **82,** 803–814.

Lin, Q., Schwarz, J., Bucana, C., and Olson, E. N. (1997). Control of mouse cardiac morphogenesis and myogenesis by transcription factor MEF2C. *Science* **276,** 1404–1407.

Lints, T. J., Parsons, L. M., Hartley, L., Lyons, I., and Harvey, R. P. (1993). Nkx-2.5: a novel murine homeobox gene expressed in early heart progenitor cells and their myogenic descendants. *Development (Cambridge, UK)* **119,** 419–431; published erratum: **119**(3), 969.

Little, C. D., and Rongish, B. J. (1995). The extracellular matrix during heart development. *Experientia* **51,** 873–882.

Lough, J., Barron, M., Brogley, M., Sugi, Y., Bolender, D. L., and Zhu, X. (1996). Combined BMP-2 and FGF-4, but neither factor alone, induces cardiogenesis in nonprecardiac embryonic mesoderm. *Dev. Biol.* **178,** 198–202.

Lyons, I., Parsons, L. M., Hartley, L., Li, R., Andrews, J. E., Robb, L., and Harvey, R. P. (1995). Myogenic and morphogenetic defects in the heart tubes of murine embryos lacking the homeo box gene Nkx2-5. *Genes Dev.* **9,** 1654–1666.

Lyons, K. M., Jones, C. M., and Hogan, B. L. (1992). The TGF-beta-related DVR gene family in mammalian development. *Ciba Found. Symp.* **165,** 219–230.

Manasek, F. J. (1968). Embryonic development of the heart. I. A light and electron microscopic study of myocardial development in the early chick embryo. *J. Morphol.* **125,** 329–365.

Markwald, R. R., Rezaee, M., Nakajima, Y., Wunsch, A., Isokawa, K., Litke, L., and Krug, E. (1995). "Molecular Basis for the Segmental Pattern of Cardiac Cushion Mesenchyme Formation: Role of ES/130 in the Embryonic Chick Heart." Futura Publ. Co., Armonk, NY.

Markwald, R. R., Eisenberg, C., Eisenberg, L., Trusk, T., and Sugi, Y. (1996). Epithelial-mesenchymal transformations in early avian heart development. *Acta Anat.* **156,** 173–186.

Markwald, R. R., Fitzharris, T. P., and Smith, W. N. (1975). Structural analysis of endocardial cytodifferentiation. *Dev. Biol.* **42,** 160–180.

Markwald, R. R., Fitzharris, T. P., and Manasek, F. J. (1977). Structural development of endocardial cushions. *Am. J. Anat.* **148,** 85–119.

Markwald, R. R., Trusk, T., Gittenberger-de Groot, A., and Poelmann, R. (1997). Cardiac morphogenesis: Formation and septation of the primary heart tube. *In* "Drug Toxicity in Embryonic Development I" (R. Kavlock and G. Daston, eds.), pp. 11–33. Springer-Verlag, Heidelberg.

Meno, C., Saijoh, Y., Fujii, H., Ikeda, M., Yokoyama, T., Yokoyama, M., Toyoda, Y., and Hamada, H. (1996). Left–right asymmetric expression of the TGF beta-family member lefty in mouse embryos. *Nature (London)* **381,** 151–155.

Meyer, D., and Birchmeier, C. (1995). Multiple essential functions of neuregulin in development [see comments]. *Nature (London)* **378,** 386–390.

Miyabara, S., Gropp, A., and Winking, H. (1982). Trisomy 16 in the mouse fetus associated with generalized edema and cardiovascular and urinary tract anomalies. *Teratology* **25,** 369–380.

Mjaatvedt, A. E., Citron, M. P., and Reeves, R. H. (1993). High-resolution mapping of D16led-1, Gart, Gas-4, Cbr, Pcp-4, and Erg on distal mouse chromosome 16. *Genomics* **17,** 382–386.

Mjaatvedt, C. H., and Markwald, R. R. (1989). Induction of an epithelial-mesenchymal transition by an in vivo adheron-like complex. *Dev. Biol.* **136,** 118–128; published erratum: **137**(1), 217 (1990).

Mjaatvedt, C. H., Lepera, R. C., and Markwald, R. R. (1987). Myocardial specificity for initiating endothelial-mesenchymal cell transition in embryonic chick heart correlates with a particulate distribution of fibronectin. *Dev. Biol.* **119,** 59–67.

Mjaatvedt, C. H., Krug, E. L., and Markwald, R. R. (1991). An antiserum (ES1) against a particulate form of extracellular matrix blocks the transition of cardiac endothelium into mesenchyme in culture. *Dev. Biol.* **145,** 219–230.

Mjaatvedt, C. H., Yamamura, H., and Markwald, R. R. (1997). Molecular and pheotypic characterization of the hdf (heart defect) mouse insertional mutation. *Int. Congr. Dev. Biol., 13th,* Snowbird, UT.

Molkentin, J. D., Kalvakolanu, D. V., and Markham, B. E. (1994). Transcription factor GATA-4 regulates cardiac muscle-specific expression of the alpha-myosin heavy-chain gene. *Mol. Cell Biol.* **14,** 4947–4957.

Nakajima, Y., Krug, E. L., and Markwald, R. R. (1994). Myocardial regulation of transforming growth factor-beta expression by outflow tract endothelium in the early embryonic chick heart. *Dev. Biol.* **165,** 615–626.

Nakajima, Y., Yamagishi, T., Nakamura, H., Markwald, R. R., and Krug, E. L. (1998). An autocrine function for transforming growth factor (TGF)-B3 in the transformation of atrioventricular canal endocardium into mesenchyme during chick heart development. *Dev. Biol.* **194,** 99–113.

O'Brien, T. X., Lee, K. J., and Chien, K. R. (1993). Positional specification of ventricular myosin light chain 2 expression in the primitive murine heart tube. *Proc. Natl. Acad. Sci. U.S.A.* **90,** 5157–5161.

Olson, E. N., and Srivastava, D. (1996). Molecular pathways controlling heart development. *Science* **272,** 671–676.

Park, J., Takeuchi, A., and Sharma, S. (1996). Characterization of a new isoform of the NFAT (nuclear factor of activated T cells) gene family member NFATc. *J. Biol. Chem.* **271,** 20914–20921.

Potts, J. D., Dagle, J. M., Walder, J. A., Weeks, D. L., and Runyan, R. B. (1991). Epithelial-mesenchymal transformation of embryonic cardiac endothelial cells is inhibited by a modified antisense oligodeoxynucleotide to transforming growth factor beta 3. *Proc. Natl. Acad. Sci. U.S.A.* **88,** 1516–1520.

Potts, J. D., Vincent, E. B., Runyan, R. B., and Weeks, D. L. (1992). Sense and antisense TGF beta 3 mRNA levels correlate with cardiac valve induction. *Dev. Dyn.* **193,** 340–345.

Ramsdell, A. F., and Markwald, R. R. (1997). Induction of endocardial cushion tissue in the avian heart is regulated, in part, by TGFbeta-3-mediated autocrine signaling. *Dev. Biol.* **188,** 64–74.

Ramsdell, A., Moreno-Rodriguez, R., Wienecke, M., Sugi, Y., Turner, D., Mjaatvedt, C., and Markwald, R. (1998). Identification of an autocrine signaling pathway that amplifies induction of endocardial cushion tissue in the avian heart. *Acta Anat.* In press.

Rawles, M. (1943). The heart forming areas of the early chick blastoderm. *Physiol. Zool.* **16,** 22–42.

Reeves, R., Gearhart, J., and Littlefield, J. (1986). Genetic basis for a mouse model of Down syndrome. *Brain Res. Bull.* **16,** 803–814.

Rezaee, M., Isokawa, K., Halligan, N., Markwald, R. R., and Krug, E. L. (1993). Identification of an extracellular 130-kDa protein involved in early cardiac morphogenesis. *J. Biol. Chem.* **268,** 14404–14411.

Rongish, B. J., Drake, C. J., Argraves, W. S., and Little, C. D. (1998). Identification of the developmental marker, JB3-antigen, as fibrillin-2 and its de novo organization into embryonic microfibrous arrays. *Dev. Dyn.* **212,** 461–471.

Rosenquist, G., and DeHaan, R. (1966). Migration of precardiac cells in the chick embryo: A radiographic study. *Carnegie Inst. Washington, Contrib. Embryol.* **38,** 111–121.

Runyan, R. B., Potts, J. D., and Weeks, D. L. (1992). TGF-beta 3-mediated tissue interaction during embryonic heart development. *Mol. Reprod. Dev.* **32,** 152–159.

Ruzicka, D. L., and Schwartz, R. J. (1988). Sequential activation of alpha-actin genes during avian cardiogenesis: Vascular smooth muscle alpha-actin gene transcripts mark the onset of cardiomyocyte differentiation. *J. Cell Biol.* **107,** 2575–2586.

Sabin, F. R. (1920). Studies on the origin of blood-vessels and of red blood-corpuscles as seen in the living blastoderm of chicks during the second day of incubation. *Carnegie Inst. Washington, Contrib. Embryol.* **9,** 213–262.

Sanford, L. P., Ormsby, I., Gittenberger-de Groot, A. C., Sariola, H., Friedman, R., Boivin, G. P., Cardell, E. L., and Doetschman, T. (1997). TGFbeta2 knockout mice have multiple developmental defects that are non-overlapping with other TGFbeta knockout phenotypes. *Development (Cambridge, UK)* **124,** 2659–2670.

Satokata, I., and Maas, R. (1994). Msx1 deficient mice exhibit cleft palate and abnormalities of craniofacial and tooth development. *Nat. Genet.* **6,** 348–356.

Schilham, M. W., Oosterwegel, M. A., Moerer, P., Ya, J., de Boer, P. A., van de Wetering, M., Verbeek, S., Lamers, W. H., Kruisbeek, A. M., Cumano, A., and Clevers, H. (1996). Defects in cardiac outflow tract formation and pro-B-lymphocyte expansion in mice lacking Sox-4. *Nature (London)* **380,** 711–714.

Schubert, D., La Corbiere, M., and Esch, F. (1986). A chick neural retina adhesion and survival molecule is a retinol-binding protein. *J. Cell Biol.* **102,** 2295–2301.

Seymour, A., Yanak, B., O'Brien, E., Rusiniak, M., Novak, E., Pinto, L., Swank, R., and Gorin, M. (1996). An integrated genetic map of the pearl locus of mouse chromosome 13. *Genome Res.* **6,** 538–544.

Sinning, A. R., Krug, E. L., and Markwald, R. R. (1992). Multiple glycoproteins localize to a particulate form of extracellular matrix in regions of the embryonic heart where endothelial cells transform into mesenchyme. *Anat. Rec.* **232,** 285–292.

Sinning, A. R., Hewitt, C. H., and Markwald, R. R. (1995). A subset of SBA lectin binding proteins isolated from myocardial conditioned medium transforms cardiac endothelium. *Acta Anat.* **154,** 111–119.

Smith, S. M., Dickman, E. D., Thompson, R. P., Sinning, A. R., Wunsch, A. M., and Markwald, R. R. (1997). Retinoic acid directs cardiac laterality and the expression of early markers of precardiac asymmetry. *Dev. Biol.* **182,** 162–171.

Spemann, H. (1938). "Embryonic Development and Induction." Yale University Press, New Haven, CT.

Spicer, A. P., Augustine, M. L., and McDonald, J. A. (1996). Molecular cloning and characterization of a putative mouse hyaluronan synthase. *J. Biol. Chem.* **271,** 23400–23406.

Stalsberg, H., and De Haan, R. L. (1969). The precardiac areas and formation of the tubular heart in the chick embryo. *Dev. Biol.* **19,** 128–159.

Sugi, Y., and Lough, J. (1995). Activin-A and FGF-2 mimic the inductive effects of anterior endoderm on terminal cardiac myogenesis in vitro. *Dev. Biol.* **168,** 567–574.

Sugi, Y., and Markwald, R. R., (1996). Formation and early morphogenesis of endocardial precursor cells and the role of the endoderm. *Dev. Biol.* **175,** 66–83.

Tasaka, H., Krug, E. L., and Markwald, R. R. (1996a). Origin of the pulmonary venous orifice in the mouse and its relation to the morphogenesis of the sinus venosus, extracardiac mesenchyme (spina vestibuli), and atrium. *Anat. Rec.* **246,** 107–113.

Tasaka, H., Krug, E. L., and Markwald, R. R. (1996b). Origin of the pulmonary venous orifice in the mouse and its relationship to the morphogenesis of the sinus venosus, extracardiac mesenchyme (sina vestibuli) and atrium. *Anat. Rec.* **246,** 107–113.

Viragh, S., Szabo, E., and Challice, C. E. (1989). Formation of the primitive myo- and endocardial tubes in the chicken embryo. *J. Mol. Cell. Cardiol.* **21,** 123–137.

Wang, Y., and Sassoon, D. (1995). Ectoderm-mesenchyme and mesenchyme-mesenchyme interactions regulate Msx-1 expression and cellular differentiation in the murine limb bud. *Dev. Biol.* **168,** 374–382.

Webb, S., Anderson, R. H., and Brown, N. A. (1996). Endocardial cushion development and heart loop architecture in the trisomy 16 mouse. *Dev. Dyn.* **206,** 301–309.

Wessels, A., and Markwald, R. R. (1998). In preparation.

Wilson, D., Burn, J., Scrambler, P., and Goodship, J. (1993). DiGeorge syndrome: Part of CATCH 22. *J. Med. Genet.* **30,** 852–856.

Winnier, G., Blessing, M., Labosky, P. A., and Hogan, B. L. (1995). Bone morphogenetic protein-4 is required for mesoderm formation and patterning in the mouse. *Genes Dev.* **9,** 2105–2116.

Wunsch, A. M., Little, C. D., and Marwald, R. R., (1994). Cardiac endothelial heterogeneity defines valvular development as demonstrated by the diverse expression of JB3, an antigen of the endocardial cushion tissue. *Dev. Biol.* **165,** 585–601.

Yamamura, H., and Mjaatvedt, C. H. (1998). In preparation.

Yamamura, H., Zhang, M., Markwald, R. R., and Mjaatvedt, C. (1997). A heart segmental defect in the anterior-posterior axis of a transgenic mutant mouse. *Dev. Biol.* **186,** 58–72.

Yokoyama, T., Copeland, N. G., Jenkins, N. A., Montgomery, C. A., Elder, F. F., and Overbeek, P. A. (1993). Reversal of left-right asymmetry: A situs inversus mutation. *Science* **260,** 679–682.

Yutzey, K. E., Rhee, J. T., and Bader, D. (1994). Expression of the atrial-specific myosin heavy chain AMHC1 and the establishment of anteroposterior polarity in the developing chicken heart. *Development* (*Cambridge, UK*) **120,** 871–883.

Zhang, H., and Bradley, A. (1996). Mice deficient for BMP2 are nonviable and have defects in amnion/chorion and cardiac development. *Development* (*Cambridge, UK*) **122,** 2977–2986.

Zou, Y., Evans, S., Chen, J., Kuo, H., Harvey, R., and Chien, K. (1997). CARP, a cardiac ankyrin repeat protein, is downstream in the Nkx2-5 homeobox gene pathway. *Development* (*Cambridge, UK*) **124,** 793–804.

11

Contribution of Neural Crest to Heart and Vessel Morphology

Margaret L. Kirby
Developmental Biology Program, Institute of Molecular Medicine and Genetics, Medical College of Georgia, Augusta, Georgia 30912

I. Introduction

Neural crest is composed of a pleuripotent cell population that originates from most of the craniocaudal length of the neural tube and migrates to various locations during early embryonic development (Le Douarin, 1982; Horstadius, 1950; Bronner-Fraser, 1994; Bronner-Fraser and Fraser, 1989). The cell lineage determinants within this population are partly acquired at their original axial location, and additional information is encountered in their migratory pathway and at terminal sites (Fraser and Bronner-Fraser, 1991; Stern *et al.*, 1991). While neural crest extends cranially to caudally in the neural axis, the neural crest cranial to somite 5 has a greater potential for generating mesenchymal cells than neural crest caudal to somite 5 and has been called cranial neural crest (Le Douarin, 1982; Horstadius, 1950). Neurons, glia (Schwann and support cells), and melanocytes are generated from the entire length of the neural crest (Le Douarin, 1982; Horstadius, 1950).

Numerous *in vivo* and *in vitro* studies have assessed whether certain neural crest cell lineages are determined before migration (Le Douarin, 1990; Fraser and Bronner-Fraser, 1991; Ito and Sieber-Blum, 1991; Artinger and Bronner-Fraser, 1992; Erickson and Perris, 1993; Horstadius, 1950; Nakamura and Ayer-Le Lièvre, 1982; Wachtler, 1984; Weston, 1982; Steller, 1995; Riccardi, 1991; Noden, 1978, 1980). In general, certain individual cells are determined for their lineage, i.e., melanocyte, neural, or mesenchymal before migration, whereas others are established during migration or after they reach their terminal destination. While this can be said for individual cells, it appears that neither entire lineages nor cell populations are fully committed prior to migration, such that a significant proportion of cells leaving the neural folds must be influenced by factors in the migration pathway before establishing their identity. It is not known whether a certain number of cells leaving the neural folds are determined to follow a neuronal or smooth muscle lineage. Certain cellular characteristics within a particular lineage, i.e., neurotransmitter identity, are not decided until neuronal precursors reach their terminal destination.

In the trunk, the mesoderm is segmented and is thought to provide craniocaudal positional information to the neural tube (Lumsden and Krumlauf, 1996). In the head, neural crest-derived mesenchyme is thought to carry information for craniocaudal identity of the pharyngeal arch derivatives (Lumsden and Krumlauf, 1996). Some studies indicate that genes involved in positional identity are somewhat autonomous from those that establish cell lineage. If neural crest destined for one arch is transplanted to another location it develops many of the structures that it should have expressed in its original location rather than being converted to produce structures appropriate for the new location (Noden, 1988; Kirby, 1989).

Much of the information regarding cranial neural crest is derived from chick embryos using quail-to-chick chimeras, *in situ* marking, and ablation of the neural folds (d'Amico-Martel and Noden, 1983; Noden, 1983; Le Douarin *et al.,* 1993; Couly *et al.,* 1992; Le Douarin and Jotereau, 1975; Miyagawa-Tomita *et al.,* 1991; Phillips *et al.,* 1987; Lumsden *et al.,* 1991; Nelson *et al.,* 1995; Waldo *et al.,* 1996; Hollway *et al.,* 1997). Marking techniques recently have been used in rodents and information from the chick appears to be in general agreement with the cultured mouse and rat models (Serbedzija *et al.,* 1990, 1991, 1992; Fukiishi and Morriss-Kay, 1992; Smits-van Prooije *et al.,* 1986; Jaenisch, 1985; Löfberg, 1985; Tan and Morriss-Kay, 1986). The neural crest cells that originate from the posterior rhombencephalon (rhombomeres 6–8) migrate to pharyngeal arches 3, 4, and 6 (Fig. 1) and from there to the heart, where they participate in outflow septation and form the cardiac ganglia of the heart. This crest also populates the esophagus en route to forming the parasympathetic enteric plexus of the the midgut and hindgut (Kirby and Waldo, 1995; Peters-Van der Sanden *et al.,* 1993). Cells remaining in arches 3, 4, and 6 support development of the pharyngeal (aortic) arch arteries (Fig. 2) and the glands derived from this region: the thymus and parathyroids (Fig. 3). Neural crest-derived cells in the caudal arches, in contrast to those in the rostral arches (1 and 2), have few skeletal derivatives but instead support vascular development of the arteries that will become the major arterial conduits of the definitive, adult cardiovascular system (aorta, carotid, and subclavian arteries, as well as the ductus arteriosus, another derivative that closes shortly after birth; Fig. 2). Comparable pharyngeal arch arteries develop in the cranial pharyngeal arches but regress. It is possible that regression occurs because the mesenchyme of these arches is largely diverted to the formation of craniofacial skeleton and accessory structures. Possibly, patterning genes expressed by the neural crest cells migrating into pharyngeal arches 1 and 2 do not support blood vessel maintenance, whereas this is a primary function of the cells populating arches 3, 4, and 6 (Qiu *et al.,* 1997; Kirby *et al.,* 1997).

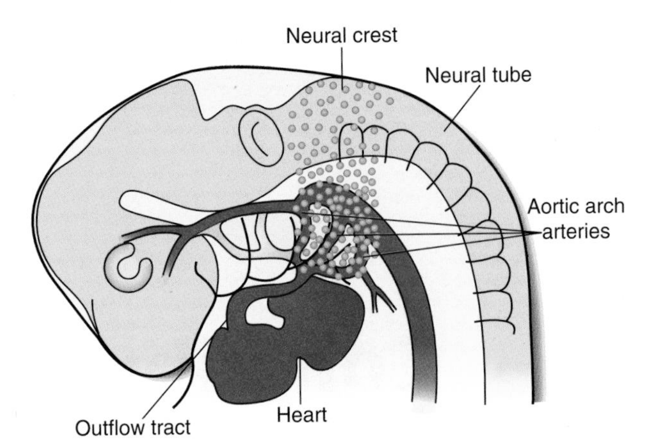

Figure 1 Diagram illustrating the migration of cardiac neural crest from rhombomeres 6–8 into pharyngeal arches 3, 4, and 6. Aortic arch arteries 3, 4, and 6 traverse these arches and connect the aortic sac and cardiac outflow tract ventrally with the dorsal aorta dorsally.

The caudal pharyngeal arches form the thymus and parathyroid glands via the interaction of endoderm–ectoderm and neural crest-derived mesenchyme (Le Douarin and Jotereau, 1975). While the thyroid anlage is initially formed in the cranial pharynx, it migrates into the caudal pharynx and is dependent on mesenchyme derived from the caudal arches for normal development. Removal of the premigratory cardiac neural crest results in poor development of these glands (Bockman and Kirby, 1984; Kuratani and Bockman, 1990).

II. The Neural Crest Ablation Model in Chick Embryos

The morphological and functional phenotype of the neural crest ablation model has been characterized extensively since the first ablation studies published in 1983 showed that the development of the outflow septum of the heart and patterning of the great arteries was affected (Fig. 4). It was also determined that the glandular derivatives of the pharynx (pharyngeal arches/

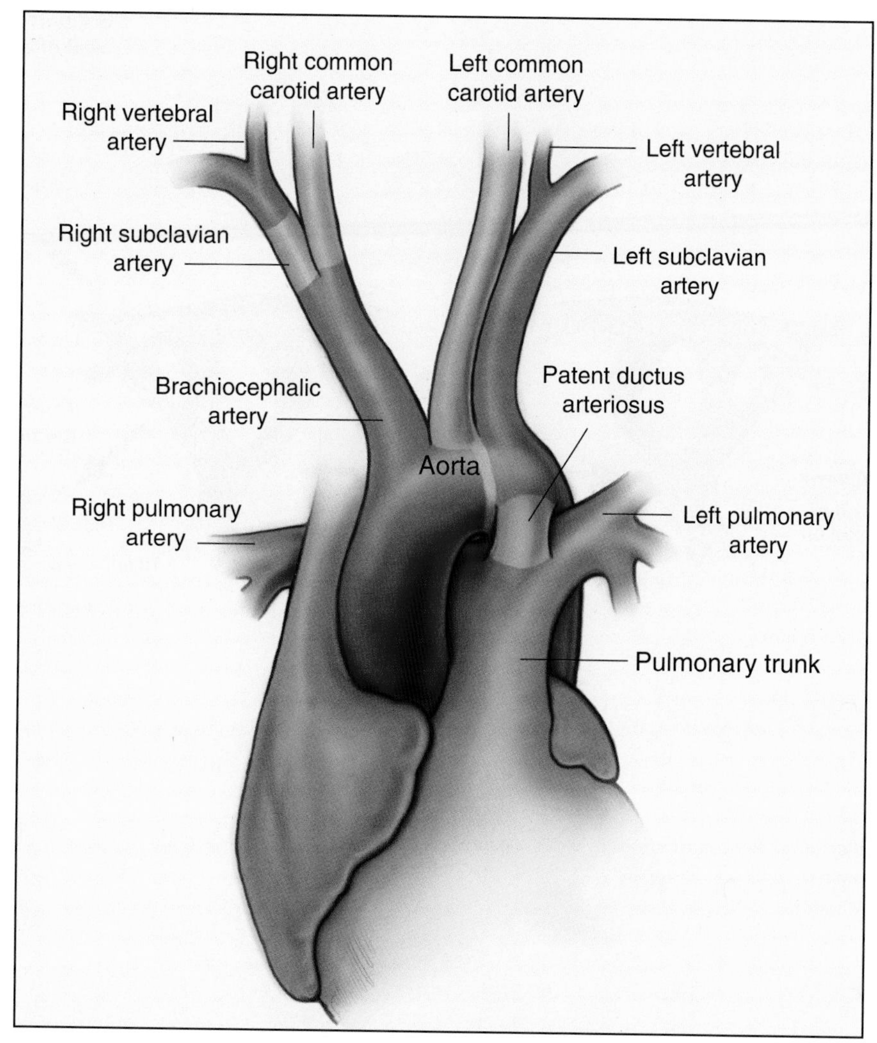

Figure 2 The cardiac outflow tract and great arteries depicted for a human fetus. The ductus arteriosus closes after birth. The carotid arteries are derived from aortic arch artery No. 3. The arch of the aorta is partially from the left fourth aortic arch artery, whereas the right fourth aortic arch artery contributes to the proximal right subclavian artery. The ductus arteriosus is from the left sixth aortic arch artery.

pouches 3 and 4), i.e., the thyroid, thymus, and parathyroid glands, were variably present or hypoplastic (Fig. 4; Nishibatake *et al.,* 1987; Kirby and Bockman, 1984; Kirby, 1987, 1988a; Kirby and Waldo, 1990, 1995; Kirby and Creazzo, 1995). Ablation of the cardiac neural crest (midotic placode to somite 3) has been variously performed using vibrating needle, tungsten needle, and laser, and the phenotypes produced are identical. Reconstitution of the preotic crest to the midbrain has been shown to occur; however, postotic neural crest is not reconstituted by any cell population within the neural tube (Kirby *et al.,* 1985a, 1993; Nishibatake *et al.,* 1987; Besson *et al.,* 1986; Suzuki and Kirby, 1997; Scherson *et al.,* 1993; Sechrist *et al.,* 1995). The absence of reconstitution has also been shown directly by labeling ventral neural tube cells or the neural crest cranial or caudal to the lesion site after the ablation, in which case no labeled cells can be seen leaving the ventral neural tube. Furthermore, neural crest cells do not migrate into the region of the gap from intact neural tube adjacent to the damaged area (Suzuki and Kirby, 1997). In contrast, extensive reconstitution of the neural crest by the remaining ventral neural tube occurs following ablation of neural crest in the preotic rhombencephalon or caudal mesencephalon (Sechrist *et al.,* 1995; Scherson *et al.,* 1993). This explains why a distinct phenotype is present after cardiac neural crest ablation but not necessarily after ablation of the neural crest migrating into the craniofacial region (McKee and Ferguson, 1984).

Interestingly, although there is no reconstitution of the postotic neural crest, the neurons of the cardiac ganglia are reconstituted from another source, the epipha-

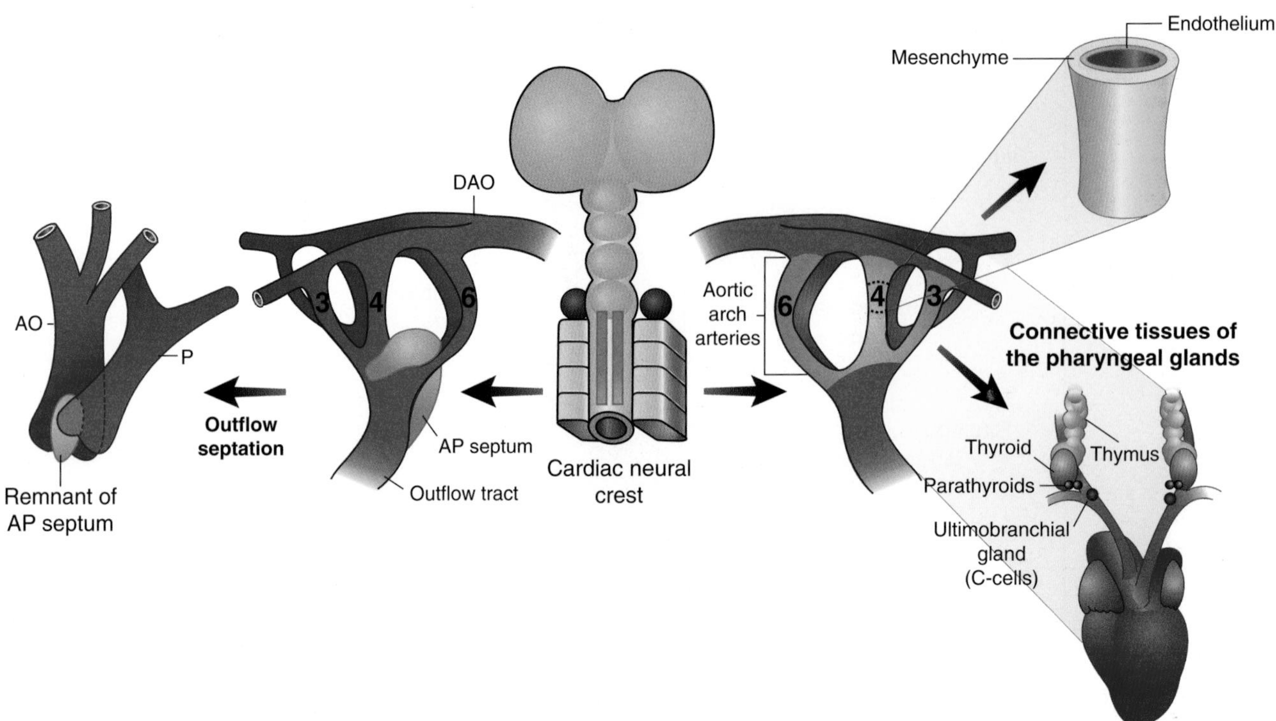

Figure 3 Diagram illustrating the contributions of cardiac neural crest to the chick outflow septation (left) and pharyngeal derivatives (right). AO, aorta; P, pulmonary trunk; DAO, dorsal aorta; AP, aorticopulmonary.

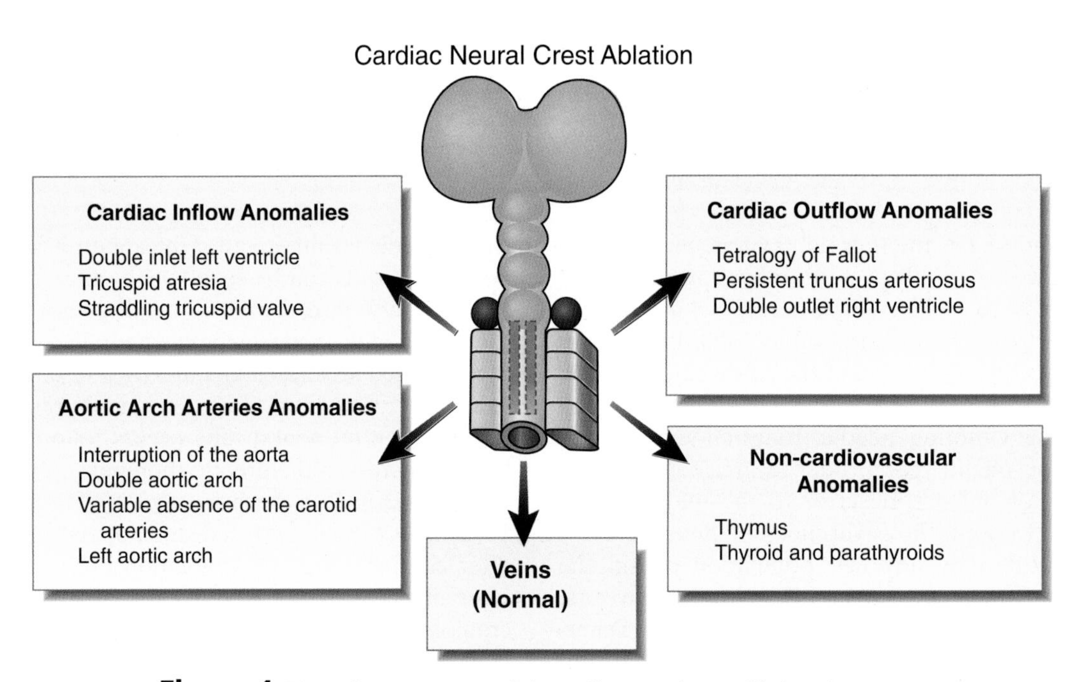

Figure 4 The major components of the cardiac neural crest ablation phenotype.

ryngeal placodes (nodose and glossopharyngeal; Kirby, 1988b). Thus, in the ablation model the heart has an undivided outflow vessel but cardiac parasympathetic innervation is approximately normal morphologically and functionally (Kirby *et al.,* 1985b, 1989; Kirby, 1988b).

Heart defects are present in about 90% of embryos surviving to Days 8-11 after neural crest ablation (Nishibatake *et al.,* 1987). The most prevalent defects involve the cardiac outflow tract and include double-outlet right ventricle (DORV), tetralogy of Fallot, persistent truncus arteriosus, and ventricular septal defect. Cardiac inflow anomalies occur occasionally. These include double-inlet left ventricle, straddling tricuspid valve, and tricuspid atresia (Nishibatake *et al.,* 1987). The phenotype seen in virtually every embryo after neural crest ablation is abnormal patterning of the great arteries derived from the aortic arch arteries (Tomita *et al.,* 1991; Männer *et al.,* 1996). Without the presence of neural crest-derived mesenchyme in the pharyngeal arches to support the development of the aortic arch arteries, the persistence of these vessels is unpredictable and variable, with the patterning in each embryo being unique. Most of the patterns include regression of an artery that should persist, although in some, abnormal persistence of arch 2 arteries appears to have substituted for one of the more caudal arteries that is missing (Bockman *et al.,* 1987, 1989; Rosenquist *et al.,* 1989). Since every embryo is unique, it is not possible to provide quantitative information regarding the patterns other than that patterning anomalies are present in 100% of embryos following cardiac neural crest ablation, whereas persistent truncus arteriosus (PTA) is present in only 90%. In this regard, aortic arch anomalies are more characteristic of neural crest ablation than any other portion of the phenotype.

It is hypothesized that a threshold amount of neural crest cells with mesenchymal potential must reach the caudal arches for normal induction of the pharyngeal endoderm for gland development in the case of the thymus and parathyroid glands and for support of stromal development in the case of the thyroid gland (Kuratani and Bockman, 1990; Bockman and Kirby, 1984).

III. Phenocopies of Neural Crest Ablation in Mammals

The neural crest ablation phenotype has a number of genetically based mimics in mice that have been useful in beginning to understand the molecular nature of the various disturbances that can lead to this phenotype.

A. Induced Mutations in Mice

1. Splotch

The Splotch mutant was first described in 1954 in a heterozygous mouse with a white belly spot (Auerbach, 1954). The Splotch allele is associated with a deletion or mutation in the *Pax-3* gene (Chalepakis *et al.,* 1994; Conway *et al.,* 1997). *Pax-3* is a member of the paired family of transcription factors that contain two DNA binding sites referred to as the paired box and the homeobox. In its heterozygous state the gene has been associated with Waardenburg's syndrome in humans, which has a mild phenotype consisting of deafness and pigmentary deficiencies (Chalepakis *et al.,* 1994). The absence of one allele of the *Pax-3* gene is not associated with cardiovascular defects in either humans or mice (Chalepakis *et al.,* 1994; Tassabehji *et al.,* 1992, 1994). However, in the homozygous state in mice, this mutation results in embryonic lethality at around Day 14 of gestation (Conway *et al.,* 1997). Homozygotes have a complete cardiac neural crest ablation phenotype: persistent truncus arteriosus, aortic arch anomalies, and hypoplasia or aplasia of the thymus and parathyroid and thyroid glands (Franz, 1989; Epstein, 1996; Conway *et al.,* 1997). The embryos have more global neural tube defects in addition to the cardiac neural crest ablation phenotype which makes this mutant an imperfect mimic of cardiac neural crest ablation. The cardiovascular phenotype has recently been associated with a deficiency of neural crest-derived cells traversing the pharyngeal arches and migrating into the cardiac outflow tract (Conway *et al.,* 1997).

2. Patch

The patch mutant mouse has an even more global phenotype than that of Splotch but in some respects mimics the neural crest ablation model. The patch mutation has not been completely characterized but appears to be a deletion that encompasses part of the *PDGF-α* receptor gene as well as the locus control region for the c-*kit* gene, encoding the receptor for Steel factor (Orr-Urtreger *et al.,* 1992; Wehrle-Haller *et al.,* 1996). Because both *PDGF-α* receptor and c-*kit* are expressed by neural crest cells, the patch phenotype cannot be ascribed to a single gene. As in the Splotch mouse, the patch heterozygote has a white belly spot. The homozygous condition is embryonic lethal with two periods of lethality: one prior to 11 days postcoitum and the other late during gestation. The mice that survive beyond Day 11 die prior to birth of multiple defects (Orr-Urtreger *et al.,* 1992). Steel factor is a known survival and proliferative factor for melanocytes (Reid *et al.,* 1996). Its receptor, c-*kit,* a member of the PDGF receptor family, is expressed in neural crest cells from the onset of migration, with expression continuing during migration (Besmer *et al.,* 1993). In the patch mutant, c-*kit* is ectopically expressed in the somites and lateral plate mesoderm. It is thought that competition for limited amounts of steel factor/kit ligand on the lateral neural crest migration pathway alters melanocyte dispersal and survival (Wehrle-Haller *et al.,* 1996). Soluble steel factor is required for lateral dispersal of the melanocyte precursors in the dermis and for survival in the initial staging area for migration (Wehrle-Haller and Weston, 1995), whereas membrane-bound steel factor promotes melanocyte survival in the dermis (Wehrle-Haller and Weston, 1995). In *Xenopus, PDGF-α* receptor is also expressed by cranial neural crest cells prior to migration and during migration into the pharyngeal arches. PDGF is expressed by the otic vesicle and both neural and pharyngeal ectoderm. The patch mutant has a number of cardiovascular anomalies that include persistent truncus arteriosus and abnormal patterning of the persisting aortic arch arteries, which suggests that PDGF plays a role in production of the neural crest ablation phenotype. Targeted mutation of the *PDGF-α* receptor has recently been reported and there appear to be similar patterns of prenatal lethality to the patch mutant. However, while the cardiovascular phenotype has not been investigated thoroughly, heart and aortic arch anomalies are not an obvious part of the phenotype (Soriano, 1997).

B. Null Mutations and Overexpression in Transgenic Mice

1. Endothelin

The endothelin family consists of three known ligands and two receptors. The endothelin ligands 1–3 (ET-1, ET-2, and ET-3) effect cellular responses via the endothelin-A or endothelin-B receptors (ETR-A and ETR-B). ET-1 has highest affinity for the ETR-A whereas ET-3 response is mediated by ETR-B. ETR-B is expressed by neural crest cells before and during migration at all levels of the neural axis (Nataf *et al.,* 1996). In culture, ET-3 stimulation results in proliferation of pleuripotent neural crest cells and melanocyte progenitors. The latter is a synergistic function with steel factor to promote both survival and proliferation. ET-3 also induces differentiation of melanoblasts into mature melanocytes (Reid *et al.,* 1996). Mutations in either *ET-3* or *ETR-B* result in pigment defects. In humans, a missense mutation in ETR-B has been identified in a family with Hirschsprung's disease (Puffenberger *et al.,* 1994). The mutated gene maps to human chromosome 13q22 and corresponds to the lethal spotting locus in the mouse mutant, which is a model of Hirschsprung's disease (Puffenberger *et al.,* 1994; Pavan *et al.,* 1995). Additionally, targeted disruptions of *ET-3* or *ETR-B* are associated with congenital megacolon. As noted previously, *ETR-B* is expressed by all levels of neural crest before and during migration and ET3 enhances proliferation of neural crest cells in the pleuripotent stage (Nataf *et al.,* 1996). While the ET-3/ETR-B system appears to be a reasonable candidate for supporting neural crest cells in cardiovascular development, the phenotype appears to be limited to neural crest cells that form the distal-most enteric plexus and melanocytes.

ET-1 is a 21-amino acid peptide that induces vasoconstriction and cell proliferation as well as increases cardiac contractility. It is expressed in the endothelium of the arch arteries and endocardial cushions as well as the pharyngeal arch epithelium. ET-1 is processed from a 38-amino acid inactive propeptide called "big endothelin-1" by endothelin-converting enzyme-1 (ECE-1) (Xu *et al.,* 1994; Torres *et al.,* 1997). ECE-1 is a novel membrane-bound metalloprotease (Xu *et al.,* 1994). When the *ET-1* gene is disrupted, the homozygous animals are smaller than normal and have hypoplastic thymus and thyroid glands that are not fused in the midline (Kurihara *et al.,* 1995a). In addition, the thymus does not descend to its appropriate position (Kurihara *et al.,* 1995b). A rather small proportion of homozygotes have aortic arch anomalies, whereas a larger percentage (50%) have ventricular septal and outflow tract defects. The percentage of homozygotes with cardiovascular anomalies can be enhanced by administering a blocking antibody or selective ETR-A antagonists during embryogenesis. Thus, it is thought that circulating maternal ET-1 or other endothelin isoforms may provide functional redundancy in the endothelin system (Kurihara *et al.,* 1995b).

This assumption is supported by the fact that disruption of the gene for endothelin-converting enzyme-1 (*ECE-1*) causes a much more severe phenotype in a larger number of mutant embryos. Homozygotes lack enteric ganglia and have heart and craniofacial defects (Pasini *et al.,* 1996). ECE-1 is a membrane-bound metalloprotease that activates both ET-1 and -3 (Pasini *et al.,* 1996). In addition to its effects on crest, ET-1 is the only known substance produced locally that upregulates cardiac contractility by increasing isometric force and decreasing actomyosin ATPase activity (Winegrad, 1997). Only *ETR-A* mRNA is found in cardiac myocytes with the receptor protein expressed at approximately 53,000 sites per cell (Hilal-Dandan *et al.,* 1997). The receptor is coupled to phosphoinositide hydrolysis and adenylyl cyclase, and ET-1 has been shown to elicit a three-fold increase in MAPK activity.

2. Hoxa-3

Null mutation of the *hoxa-3* gene results in death during the first few hours after birth. These mice have hypoplastic and/or aplastic thymus, parathyroid, and thyroid development and reduced submaxillary glandular tissue even though normal numbers of neural crest cells arrive in the pharyngeal arches (Manley and Capecchi, 1995). Hoxa-3 is expressed in the pharyngeal endoderm as well as in the neural crest, although it is not known if expression is necessary in one or both cell types for glandular development to occur. The *hoxa-3* mutant is interesting in that gland abnormalities partially mimic the neural crest ablation phenotype but no heart malformations have been found. The mice also have a wide range of tracheal abnormalities such that inspired air is routed to the stomach rather than the lungs after birth. Thus, these animals are unable to inflate their lungs, which results in persistence of the fetal pulmonary-to-systemic vascular shunts. The patency of the ductus arteriosus is dependent on prostaglandins, which are degraded in the lungs after initiation of respiration which is also associated with a dramatic decrease in pulmonary vascular pressure. Without respiration, the ductus arteriosus remains as a large patent channel connecting the left pulmonary artery to the thoracic aorta (Ruano and Kidd, 1991).

It is not known if the third aortic arch arteries develop normally in the *hoxa-3* mutant. However, treatment of chick premigratory cardiac neural crest with antisense oligonucleotides to the *hox-3* paralogous group followed by reimplantation leads to regression of the third arch artery. In this experimental paradigm, neither the treatment nor the patterning abnormality is associated with cardiac outflow defects. This and the lack of predictable arch artery patterning in the ablation phenotype suggest that neural crest cells carry instructive information regarding aortic arch patterning, but that the same patterning instructions are not necessare for cardiac outflow septation (Kirby *et al.*, 1997).

3. Retinoic Acid Receptor and Retinoid X Receptor

The retinoids have been associated with defective development that is reminiscent of neural crest ablation. Retinoic acid as a teratogen is discussed later, so this discussion will be limited to the receptor knockouts in mice, which produce phenotypes very similar to those seen after exposure of embryos to retinoic acid excess or deficiency (see Chapter 13).

Retinoid signals are transmitted via two families of receptors composed of nuclear proteins that act as transcriptional regulators when complexed with a retinoic acid ligand (J.-Y. Chen *et al.*, 1996). The retinoic acid receptor (RAR) family is activated by all-*trans* RA and 9-*cis* RA, whereas the retinoid X receptor (RXR) family is activated by 9-*cis* RA as well as a variety of other nonretinoid ligands, including vitamin D and thyroid hormone. There are three isoforms in each family, designated α, β, and γ. Further variants (called subtypes) are produced within these groups of the RAR receptors by differential splicing, and there is a specific spatiotemporal distribution of each isoform during embryogenesis. It has been shown in cultured cells that RA-responsive transcription is controlled by RAR–RXR heterodimers (J.-Y. Chen *et al.*, 1996; Lu *et al.*, 1997). Null mutations that affect all the subtypes of a single isoform (RARα or RARγ) show postnatal growth deficiency and mortality but otherwise have no apparent morphological defects. However, compound mutations in RAR isoforms mimic all the abnormalities seen in vitamin A deficiency in addition to several that are not seen.

Phenotype synergy is observed when the *RXRα* mutation is introduced into the *RARβ* or *RARγ* mutant background. *RXRα/RARβ* double mutants also have several malformations not seen in single mutants. This and other evidence from null mutations suggests that RXR–RAR heterodimers mediate retinoid signaling (Kastner *et al.*, 1994; Krezel *et al.*, 1996). RXRs serve as markers for undifferentiated neural derivatives of the trunk neural crest (Rowe *et al.*, 1991), and overexpression of RXRs and RARs in *Xenopus* results in formation of ectopic primary neurons. In the chick, RXRγ receptor transcripts are a good marker for migrating neural crest cells. Transcripts are gradually restricted to the differentiating neural derivatives, whereas expression is lost in the ectomesenchymal derivatives by stage 15 (Rowe and Brickell, 1995). In mice, RXRα null mutations result in ocular and cardiac malformations and *in utero* death (Kastner *et al.*, 1994). When *RARα* or *RARγ* mutations are combined with the *RXRα* mutant background, more severe ocular defects occur, and PTA and aortic arch defects appear (Kastner *et al.*, 1994). *RARα/RARβ* and *RARα/RXRα* double mutants have PTA, DORV, and aortic arch anomalies (Sucov *et al.*, 1995). In addition, *RXRα* null embryos have thin myocardium perhaps caused by premature differentiation of the ventricular myocardium, indicating that the RXR receptors are also important in development of the myocardium (Gruber *et al.*, 1996).

C. Human Syndromes

1. DiGeorge and Velocardiofacial Syndromes

Van Mierop and Kusche (1986) and Bockman and Kirby (1984) first suggested the association of DiGeorge Syndrome with neural crest mesenchyme in pharyngeal arches 3 and 4. Both DiGeorge syndrome and Velocardiofacial syndrome (VCFS) provide phenocopies of cardiac neural crest ablation. The most distinct features of these syndromes are interrupted aortic arch, outflow tract malformation (PTA or some degree of overriding aorta), hypoplastic thymus with some degree of immunocompromise, and hypoparathyroidism. Additional features that are seen are mild craniofacial irregularities, hypothyroidism, and, in the case of VCFS, facial clefting, i.e., cleft lip and palate, and mental retardation or in some instances episodes of psychosis (Karayiorgou *et al.,* 1995; Papolos *et al.,* 1996). These syndromes have been linked to a microdeletion in chromosome 22q11 (see Chapter 26). Other forms of these syndromes that have been reported and are associated with similar microdeletions are conotruncal anomaly face syndrome (Matsuoka *et al.,* 1994) and right-sided aortic arch (Strong, 1968). The name CATCH-22 has been proposed for syndromes with these manifestations that have the 22q11 microdeletion (Wilson *et al.,* 1993).

The search for the gene or genes underlying these physical traits has been intense and several candidates have emerged. The first to be described that had an embryonic expression pattern consistent with the spatial and temporal genesis of the defects was *Hira* (originally called *Tuple1*). *Tuple1* was renamed *Hira* because of its similarity with *Hir1* and *-2,* yeast genes that encode repressors of histone gene transcription (Lamour *et al.,* 1995; Scamps *et al.,* 1996). *Hira* is highly conserved in mouse, chick, and human. In both the chick and mouse embryos it is expressed in the neural plate, neural tube, neural crest, and mesenchyme of the head and pharyngeal arches (Roberts *et al.,* 1997; Halford *et al.,* 1993).

Clathrin heavy chain and a human homolog of the *Drosophila* segment polarity gene *dishevelled* (called *DVL*) are other genes located in the 22q11 microdeleted region (Wadey *et al.,* 1995; Pizzuti *et al.,* 1996). Human DVL is expressed in the fetal thymus and heart but its embryonic expression pattern is still not known (Pizzuti *et al.,* 1996). A very interesting gene located within the deleted region codes for an *armadillo*-like message called *ARV* (*armadillo repeat deleted in VCFS*) (Sirotkin *et al.,* 1997), a member of the β-catenin family whose closest relative is murine p120^{CAS}. The conceptually translated protein has 10 armadillo tandem repeats and a coiled coil domain and it is thought to play a role in protein–protein interactions at adherins junctions. The message is expressed ubiquitously in all fetal and adult tissues which would probably eliminate it as a serious candidate except for the fact that it shares 3′ untranslated sequence with the gene for *catechol-O-methyl transferase* which is located just proximal to *ARV* and is encoded on the opposite strand. Since these syndromes can have a psychotic component, it is interesting that one candidate for producing the structural defects might vie for transcription with a gene coding for a synthetic enzyme in a neurotransmitter pathway known to be active in cells derived from neural crest.

While the advances in understanding this neural crest ablation-type phenotype linked to a microdeletion of chromosome 22q11 are very exciting, it is important to keep in mind that phenocopies of these syndromes are linked to microdeletions on two other chromosomes. In a recent report from prenatal diagnosis of a fetus with a chromosome 17p13 deletion, the fetus was found to have multiple anomalies that included several characteristics of the DiGeorge syndrome, including thymic hypoplasia and DORV. In addition, the fetus showed polyhydramnios and intrauterine growth retardation (Greenberg *et al.,* 1988). Terminal deletions of chromosome 10 resulting in the loss of p13;p14 are associated with hypoparathyroidism and other manifestations of DiGeorge syndrome and VCFS (Daw *et al.,* 1996).

2. CHARGE Syndrome

The CHARGE phenotype includes *c*oloboma, *h*eart disease, *a*tresia of choanae, *r*etardation of physical and mental development, *g*enital hypoplasia, and *e*ar anomalies and/or deafness (Siebert *et al.,* 1985). This constellation of defects appears to have an identifiable DiGeorge/VCFS basis with multiple other anomalies. Thyroid and parathyroid glands are frequently absent and accompany outflow anomalies and aortic arch artery malformations. Malformation of the foregut, reproductive organs, kidneys, limbs and digits, and brain, including pituitary gland, with lung abscesses and focal hepatic necrosis suggests an etiology from an earlier developmental stage or involving a gene that has a broader expression pattern than those involved in DiGeorge/VCFS. To date, there is no animal model of this syndrome and no gene linkages have been reported.

3. Fetal Alcohol Syndrome

Alcohol exposure during the time when neural crest cells are populating the frontonasal process and caudal pharyngeal arches can cause a phenotype similar to that seen in DiGeorge syndrome and has been proposed as a causative agent in DiGeorge syndrome (Sulik *et al.,* 1986). Ethanol can exert a teratogenic effect by disrupting microtubules and microfilaments that would interfere with migration (Hassler and Moran, 1986), decreasing mitochondrial respiration (Nyquist-Battie and Freter, 1988), or excessive cell death in selected cell populations, including craniofacial neural crest, which might be caused by heightened membrane fluidity (S. Y. Chen *et al.,* 1996).

4. Retinoic Acid Embryopathy

Interestingly, the RA system is one of the few in which either absent or excess signaling is associated with similar phenotypes in humans and other mammals. While the defects that occur in retinoid embryopathy are not confined to the cardiovascular system (see

Chapter 28), heart and arch development are widely studied in animal models that employ RA. Cardiac and aortic arch anomalies in offspring of rats with RA deficiency or excess are directly correlated with the anomalies in humans (Wilson and Warkany, 1950; Rothman *et al.*, 1995). The lower jaw, palate, limbs, vertebrae, and tail are consistently abnormal after RA treatment. In cultured mouse embryos, RA causes a reduction in the size of arches 1 and 2 (Goulding and Pratt, 1986). Retinoic acid exposure of mice causes complete transposition of the great arteries which appears to be associated with hypoplasia of the conal but not the atrioventricular cushions (Yasui *et al.*, 1995; Nakajima *et al.*, 1996). In the chick, RA exposure causes a range of cardiac outflow defects from overriding aorta to DORV (Broekhuizen *et al.*, 1992; Bouman *et al.*, 1995). However, transposition is not seen in the chick.

While the mechanism of RA teratogenicity is not known, several clues have been identified recently. The head and hindbrain are especially sensitive to retinoid exposure and some members of the *hox* gene family (i.e. *Hoxa-1* and *-b-1* genes) contain retinoid-responsive elements (Langston *et al.*, 1997). Thus, RA treatment appears to alter the segmental expression of the *hox* gene code which in turn causes transformation of pharyngeal segmental identity, the hindbrain and otocyst are shifted anteriorly relative to foregut, and the preotic neural crest is retarded in migration (Marshall *et al.*, 1992). RA also retards neural crest cell migration *in vitro* but does not affect viability or DNA synthesis (Maxwell *et al.*, 1982). It is likely that RA alters the regional identity of cranial crest cells (Lee *et al.*, 1995). Unresolved data show that the effect may be on neural crest and cells committed to somitic mesoderm (Yasuda *et al.*, 1986), or it may be selective to cells undergoing migration rather than affecting a particular cell lineage (Thorogood *et al.*, 1982). There is a large accumulation of labeled RA in neural crest derivatives in the pharyngeal arches (Dencker *et al.*, 1990), which also express cellular RA-binding protein CRABP (Vaessen *et al.*, 1990; Maden *et al.*, 1990). HNK-1 expression, which is characteristic of migrating neural crest cells in avians, disappears as CRABP expression appears in the same cell population (Maden *et al.*, 1991). Even so, the regions affected in retinoid embryopathy are not correlated with CRABP-I or -II, so other factors must be responsible for the teratogenic effect (Horton and Maden, 1995).

IV. DORV and Red Herrings

Many mutations and experimental manipulations result in DORV; thus, it really should not be classified as a "neural crest ablation phenotype." The most characteristic signs of the cardiovascular portion of the neural crest ablation phenotype are aortic arch anomalies and persistent truncus arteriosus. Currently, teratogens or mutations known to result in persistent truncus arteriosus cause death or abnormal migration of the neural crest cells such that there are not enough to support normal development of the aortic arch arteries and outflow septation. However, many other experimental manipulations besides neural crest ablation, in addition to teratogens and single or multiple gene mutations as discussed previously, can produce DORV. Thus, while this anomaly is sometimes seen after neural crest ablation, it should not be viewed as unique to the neural crest ablation model or necessarily produced by malfunction of the neural crest.

A. Teratogens and Experimental Manipulations

In the most rigid definition of DORV neither of the semilunar valves is in fibrous continuity with either atrioventricular valve, both arterial trunks arise from the morphologically right ventricle, and a ventricular septal defect is present (Hagler *et al.*, 1968). A less rigid definition requires that the pulmonary trunk and at least half of the aorta emerge from the right ventricle (Lev *et al.*, 1972; Anderson *et al.*, 1974). If less than half of the aorta arises from the right ventricle, then it can be classified as overriding aorta. DORV is less easily explained than persistent truncus arteriosus, which is caused by an abnormally small number of neural crest cells reaching the outflow tract resulting in the absence of an outflow septum. DORV can be produced experimentally by venous or arterial ligature (Rychter, 1962), RA exposure (Bouman *et al.*, 1995), phenobarbital treatment (Nishikawa *et al.*, 1986), electric shock (Chon *et al.*, 1980), and a number of single gene mutations (Olson and Srivastava, 1996). All the experimental interventions or application of teratogens occur on or before Day 3 in the chick and do not appear to interfere with septation of the outflow tract because in DORV, the majority of the cardiac outflow septum is formed as can be seen by the presence of individual aortic and pulmonary trunks. However, the aorta and aortic semilunar valve override the ventricular septum such that the aortic root receives blood from the right ventricle rather than the left ventricle as it should. Thus, this defect can be viewed as a problem of alignment of the outflow tract rather than a problem of septation. Alignment of the inflow (atrioventricular canal) and outflow (conotruncal) portions of the tubular heart is associated with the processes of looping, convergence, and wedging. Looping allows expansion of the region of the heart tube that will become the left and right ventricles, whereas convergence brings the inflow and outflow portions of the tube into approximately the same craniocaudal plane. Wedging brings the aortic side of the conotruncus to nestle between the mitral and tricuspid valves. Wedging ensures that the mitral and aortic semilunar valves are in fibrous continuity. The outflow septum forms after convergence and before and during the process of wedging. Thus, the etiology of DORV may be temporally as well as spatially distinct from that of persistent truncus arteriosus.

B. Mutations

In order to ascribe a gene mutation that causes DORV to neural crest function the gene expression should be confined to neural crest cells. DORV is a prominent component of the phenotypes of four null mutations. These include null mutations in genes encoding the RAR/RXRs, neurofibromatosis-1 (NF-1), neurotrophin-3 (NT-3), and nonmuscle myosin heavy chain-B (NMHC-B).

1. RAR/RXRs

The RAR/RXR mutants have already been discussed in the context of the neural crest ablation phenotype. However, since RARs are expressed in cardiac myocytes as well as neural crest cells, the portions of the phenotypes contributed by the two regions cannot be separated without spatial and/or temporal control over the expression of the individual receptors.

2. NF-1

The neurofibromatosis gene *NF-1* codes for neurofibromin protein, a GTPase-activating protein that acts as a regulator of the *ras* signal transduction pathway. A targeted disruption of the *NF-1* gene leads to multiple developmental abnormalities: hyperplasia of the sympathetic ganglia and DORV in the homozygous mutant (Brannan *et al.,* 1994) while heterozygotes have an accelerated onset of tumor formation. Neural crest and placode cells isolated from *NF-1* mutants survive in culture in the absence of neurotrophins, whereas their wild-type counterparts die without NGF or BDNF (Vogel *et al.,* 1995). NF-1 functions as a tumor suppressor (von Deimling *et al.,* 1995; Vogel *et al.,* 1995). The myocardium has disoriented, poorly developed myofibrils which compromise heart function and result in pericardial and pleural effusion—signs of heart failure (Brannan *et al.,* 1994). Neurofibromin acts as a negative mediator of neurotrophin-mediated signaling for survival of neurons derived from neural crest or placodes (Vogel *et al.,* 1995). Neurofibromin is not known to be expressed in developing myocardium and since it is specifically expressed by neural crest derivatives, it may be a candidate for a neural crest-induced DORV. The neural crest cell population destined for the heart forms both cardiac ganglia and the outflow septum. If too many cells are allocated to the neuronal cell lineage, the septal cell lineage might be decreased. On the other hand, hyperplasia of autonomic innervation might be related to an altered level of circulating or local neurotransmitters which could affect cardiac contractility. Agents that cause changes in cardiac contractility, i.e., caffeine and isoproterenol, most frequently result in DORV.

3. NT-3

NT-3 receptors are expressed in the heart and by neural crest cells. *NT3*$^{-/-}$ animals die perinatally with cardiovascular defects (Donovan *et al.,* 1996). Abnormalities include defective chamber size, right ventricular dilation, incomplete outflow septation, ventricular and atrial septal defects, thinning of the atrial wall with consequent atrial dilation, premature closure of the ductus arteriosus, "defects" in the sinus venosus, aneurysmal dilation and attenuation of the tunica media of the pulmonary veins, dilation of the pulmonary artery with subpulmonic stenosis, overriding aorta, thickened semilunar valves, and dilation of the atrioventricular annuli. Pulmonary edema is present with congestion and intraalveolar hemorrhage. At 9.5 days postcoitum (pc) there is already a delay in myofibril organization of the truncus arteriosus and hypoplasia of the sinus venosus with atrial enlargement, and the tyrosine kinase receptor C (TrkC) level is decreased in the ventricles. Most of these changes represent much more global cardiac damage. Certainly, the earliest changes are in the myocardium or tubular heart and not in the neural crest. Thus, the DORV produced is more likely to be attributed to a myocardial origin rather than neural crest.

4. Nonmuscle Myosin Heavy Chain-B

Nonmuscle myosin heavy chain II has two isoforms, A and B, that have some overlapping expression domains as well as independent ones (Sellers and Goodson, 1995; Goodson and Spudich, 1993). The isoforms differ in the rate at which they hydrolyze ATP and propel actin filaments in motility assays (Kelley *et al.,* 1996). NMHC-B is localized near the plasma membrane of some cells, whereas NMHC-A is in stress fibers. Myocytes from newborn mice and primary cultures of embryonic chick cardiac myocytes contain only NMHC-B (Rhee *et al.,* 1994; Conrad *et al.,* 1995). Ablation of the gene for NMHC-B results in dextroposed aorta with the aorta overriding the right ventricle with a ventricular septal defect and also infundibular hypertrophy, which causes muscular obstruction of the pulmonary outflow. The cardiac myocytes are moderately disorganized, hypertrophied in both left and right ventricles as early as 12.5 days pc and contain disorganized myofibrils. Recent studies indicate that NMHC-B is coexpressed with smooth muscle α-actin in neural crest-derived cells that form the outflow septation complex (W. H. Lamers, personal communiction). Thus, disruption of this gene could lead to defective septation via neural crest or myocardial development.

V. Conclusions and Future Perspectives

The neural crest ablation model provides a valuable framework for beginning to understand the actions of various teratogens and gene mutations in cardiac and great vessel development. The primary signs of neural crest malfunction are PTA and aortic arch anomalies with or without anomalous development of the glands derived from the caudal pharynx. The best phenocopy of cardiac neural crest ablation is in human DiGeorge syndrome, and the gene or genes that underlie these anomalies will be very informative about the function of cardiac neural crest cells.

On the other hand, DORV is most likely associated with myocardial dysfunction during the critical period of looping, convergence, and wedging. The myocardium is affected through myocardial receptors, circulating factors, or hemodynamic alterations. Much confusion has been generated recently by referring to DORV as a neural crest-related defect without sufficient evidence that this is indeed the case. It is essential in analyzing new cardiovascular phenotypes in mutant and transgenically altered animals to make the distinction between neural crest and myocardially generated alterations in outflow tract development.

References

Anderson, R. H., Wilkinson, J. L., Arnold, R., Becker, A. E., and Lubkiewicz, K. (1974). Morphogenesis of bulboventricular malformations. II. Observations on malformed hearts. *Br. Heart J.* **36,** 948.

Artinger, K. B., and Bronner-Fraser, M. (1992). Partial restriction in the developmental potential of late emigrating avian neural crest cells. *Dev. Biol.* **149,** 149–157.

Auerbach, R. (1954). Analysis of the developmental effects of a lethal mutation in the house mouse. *J. Exp. Zool.* **127,** 305–329.

Besmer, P., Manova, K., Duttlinger, R., Huang, R. J., Packer, A., Gyssler, C., and Bachvarova, R. F. (1993). The *kit*-ligand (steel factor) and its receptor c-*kit/W:* Pleiotropic roles in gametogenesis and melanogenesis. *Development (Cambridge), Suppl.,* pp. 125–137.

Besson, W. T. I., Kirby, M. L., Van Mierop, L. H. S., and Teabeaut, J. R. I. (1986). Effects of cardiac neural crest lesion size at various embryonic ages on incidence and type of cardiac defects. *Circulation* **73,** 360–364.

Bockman, D. E., and Kirby, M. L. (1984). Dependence of thymus development on derivatives of the neural crest. *Science* **223,** 498–500.

Bockman, D. E., Redmond, M. E., Waldo, K., Davis, H., and Kirby, M. L. (1987). Effect of neural crest ablation on development of the heart and arch arteries in the chick. *Am. J. Anat.* **180,** 332–341.

Bockman, D. E., Redmond, M. E., and Kirby, M. L. (1989). Alteration of early vascular development after ablation of cranial neural crest. *Anat. Rec.* **225,** 209–217.

Bouman, H. G. A., Broekhuizen, M. L. A., Mieke, A., Baasten, J., Gittenberger-de Groot, A. C., and Wenink, A. C. G. (1995). Spectrum of looping disturbances in stage 34 chicken hearts after retinoic acid treatment. *Anat. Rec.* **243,** 101–108.

Brannan, C. I., Perkins, A. S., Vogel, K. S., Ratner, N., Nordlund, M. L., Reid, S. W., Buchberg, A. M., Jenkins, N. A., Parada, L. F,. and Copeland, N. G. (1994). Targeted disruption of the neurofibromatosis type-1 gene leads to developmental abnormalities in heart and various neural crest-derived tissues. *Genes Dev.* **8,** 1019–1029.

Broekhuizen, M. L. A., Wladimiroff, J. W., Tibboel, D., Poelmann, R. E., Wenink, A. C. G. and Gittenberger-de Groot, A. C. (1992). Induction of cardiac anomalies with all trans retinoic acid in the chick embryo. *Cardiol. Young* **2,** 311–317.

Bronner-Fraser, M. (1994). Neural crest cell formation and migration in the developing embryo. *FASEB J.* **8,** 699–706.

Bronner-Fraser, M., and Fraser, S. (1989). Developmental potential of avian trunk neural crest cells in situ. *Neuron* **3,** 755–766.

Chalepakis, G., Goulding, M., Read, A., Strachan, T,. and Gruss, P. (1994). Molecular basis of splotch and Waardenburg *Pax-3* mutations. *Proc. Natl. Acad. Sci. U.S.A.* **91,** 3685–3689.

Chen, J.-Y., Clifford, J., Zusi, C., Starrett, J., Tortolani, D., Ostrowski, J., Reczek, P. R., Chambon, P,. and Gronemeyer, H. (1996). Two distinct actions of retinoid-receptor ligands. *Nature (London)* **382,** 819–822.

Chen, S. Y., Yang, B., Jacobson, K., and Sulik, K. K. (1996). The membrane disordering effect of ethanol on neural crest cells in vitro and the protective role of GM1 ganglioside. *Alcohol* **13,** 589–595.

Chon, Y., Ando, M., and Takao, A. (1980). Spectrum of hypoplastic right ventricle in chick experimentally produced by electrical shock. *In* (R. Van Praagh, and A. Takao, eds.), "Etiology and Morphogenesis of Congenital Heart Disease" pp. 249–264. Futura Publ. Co., Mt. Kisco, NY.

Conrad, A. H., Jaffredo, T., and Conrad, G. W. (1995). Differential localization of cytoplasmic myosin II isoforms A and B in avian interphase and dividing embryonic and immortalized cardiomyocytes and other cell types *in vitro. Cell Motil. Cytoskeleton* **31,** 93–112.

Conway, S. J., Henderson, D. J., and Copp, A. J. (1997). *Pax3* is required for cardiac neural crest migration in the mouse: Evidence from the *splotch* (Sp^{2H}) mutant. *Development (Cambridge, UK)* **124,** 505–514.

Couly, G. F., Coltey, P. M., and Le Douarin, N. M. (1992). The developmental fate of the cephalic mesoderm in quail-chick chimeras. *Development (Cambridge, UK)* **114,** 1–15.

d'Amico-Martel, A., and Noden, D. M. (1983). Contributions of placodal and neural crest cells to avian cranial peripheral ganglia. *Am. J. Anat.* **166,** 445–468.

Daw, S. C. M., Taylor, C., Kraman, M., Call, K., Mao, J. I., Schuffenhauer, S., Meitinger, T., Lipson, T., Goodship, J., and Scambler, P. (1996). A common region of 10p deleted in DiGeorge and velocardiofacial syndromes. *Nat. Genet.* **13,** 458–460.

Dencker, L., Annerwall, E., Busch, C., and Eriksson, U. (1990). Localization of specific retinoid-binding sites and expression of cellular retinoic-acid-binding protein (CRABP) in the early mouse embryo. *Development (Cambridge, UK)* **110,** 343–352.

Donovan, M. J., Hahn, R., Tessarollo, L., and Hempstead, B. L. (1996). Identification of an essential nonneuronal function of neurotrophin 3 in mammalian cardiac development. *Nat. Genet.* **14,** 210–213.

Epstein, J. A. (1996). *Pax3,* neural crest and cardiovascular development. *Trends Cardiovasc. Med.* **6,** 255–261.

Erickson, C. A., and Perris, R. (1993). The role of cell-cell and cell-matrix interactions in the morphogenesis of the neural crest. *Dev. Biol.* **159,** 60–74.

Franz, T. (1989). Persistent truncus arteriosus in the Splotch mutant mouse. *Anat. Embryol.* **180,** 457–464.

Fraser, S. E., and Bronner-Fraser, M. (1991). Migrating neural crest cells in the trunk of the avian embryo are multipotent. *Development (Cambridge, UK)* **112,** 913–920.

Fukiishi, Y., and Morriss-Kay, G. M. (1992). Migration of cranial neural crest cells to the pharyngeal arches and heart in rat embryos. *Cell Tissue Res.* **268,** 1–8.

Goodson, H. V., and Spudich, J. A. (1993). Molecular evolution of the myosin family: relationships derived from comparisons of amino acid sequences. *Proc. Natl. Acad. Sci. U.S.A.* **90,** 659–663.

Goulding, E. H., and Pratt, R. M. (1986). Isotretinoin teratogenicity in mouse whole embryo culture. *J. Craniofacial Genet. Dev. Biol.* **6,** 99–112.

Greenberg, F., Courtney, K. B., Wessels, R. A., Huhta, J., Carpenter, R. J., Rich, D. C., and Ledbetter, D. H. (1988). Prenatal diagnosis of deletion 17p13 associated with DiGeorge anomaly. *Am. J. Med. Genet.* **31,** 1–4.

Gruber, P., Kubalak, S. W., Pexieder, T., Sucov, H. M., Evans, R. M., and Chien, K. R. (1996). RXRa deficiency confers genetic susceptibility to aortic sac, conotruncal, atrioventricular cushion, and ventricular muscle defects. *J. Clin. Invest.* **98,** 1332–1343.

Hagler, D. J., Ritter, D. G., and Puga, F. J. (1968). Double-outlet right ventricle. *In* "Moss' Heart Disease in Infants, Children, and Adolescents" (F. H. Adams and G. C. Emmanouilides, eds.), 3rd ed., pp. 351–369. Williams & Wilkins, Baltimore, MD.

Halford, S., Wilson, D. I., Daw, S. C., Roberts, C., Wadey, R., Kamath, S., Wickremasinghe, A., Burn, J., Goodship, J., and Mattei, M. G. (1993). Isolation of a gene expressed during early embryogenesis from the region of 22q11 commonly deleted in DiGeorge syndrome. *Hum. Mol. Genet.* **2,** 1577–1582.

Hassler, J. A., and Moran, D. J. (1986). Effectives of ethanol on the cytoskeleton of migrating and differentiating neural crest cells: Possible role in teratogenesis. *J. Craniofacial Genet. Dev. Biol. Suppl.* **2,** 129–136.

Hilal-Dandan, R., Ramirez, M. T., Villegas, S., Gonzalez, A., Endo-Mochizuki, Y., Brown, J. H., and Brunton, L. L. (1997). Endothelin ETA receptor regulates signaling and ANF gene expression via multiple G protein-linked pathways. *Am. J. Physiol.: Heart Circ. Physiol.* **272,** H130–H137.

Hollway, G. E., Suthers, G. K., Haan, E. A., Thompson, E., David, D. J., Gecz, J., and Mulley, J. C. (1997). Mutation detection in *FGFR2* craniosynostosis syndromes. *Hum. Genet.* **99,** 251–255.

Horstadius, S. (1950). "The Neural Crest. Its Properties and Derivatives in the Light of Experimental Research." Oxford University Press, London.

Horton, C., and Maden, M. (1995). Endogenous distribution of retinoids during normal development and teratogenesis in the mouse embryo. *Dev. Dyn.* **202,** 312–323.

Ito, K., and Sieber-Blum, M. (1991). *In vitro* clonal analysis of quail cardiac neural crest development. *Dev. Biol.* **148,** 95–106.

Jaenisch, R. (1985). Mammalian neural crest cells participate in normal embryonic development on microinjection into post-implantation mouse embryos. *Nature (London)* **318,** 181–183.

Karayiorgou, M., Morris, M. A., Morrow, B., Shprintzen, R. J., Goldberg, R., Borrow, J., Gos, A., Nestadt, G., Wolyniec, P. S., Lasseter, V. K., Eisen, H., Childs, B., Kazazian, H. H., Kucherlapati, R., Antonarakis, S. E., Pulver, A. E., and Housman, D. E. (1995). Schizophrenia susceptibility associated with interstitial deletions of chromosome 22q11. *Proc. Natl. Acad. Sci. U.S.A.* **92,** 7612–7616.

Kastner, P., Grondona, J. M., Mark, M., Gansmuller, A., LeMeur, M., Decimo, D., Vondsch, J.-L., Dollé, P., and Chambon, P. (1994). Genetic analysis of RXRa developmental function: Convergence of RXR and RAR signaling pathways in heart and eye morphogenesis. *Cell (Cambridge, Mass.)* **78,** 987–1003.

Kelley, C. A., Sellers, J. R., Gard, D. L. Bui, D., Adelstein, R. S., and Baines, I. C. (1996). Xenopus nonmuscle myosin heavy-chain isoforms have different subcellular localizations and enzymatic activities. *J. Cell Biol.* **134,** 675–687.

Kelly, R., and Buckingham, M. (1997). Manipulating myosin light chain 2 isoforms *in vivo*—A transgenic approach to understanding contractile protein diversity. *Circ. Res.* **80,** 751–753.

Kirby, M. (1989). Plasticity and predetermination of the mesencephalic and trunk neural crest transplanted into the region of cardiac neural crest. *Dev. Biol.* **134,** 402–412.

Kirby, M. L. (1987). Cardiac morphogenesis: Recent research advances. *Pediatr. Res.* **21,** 219–224.

Kirby, M. L. (1988a). Role of extracardiac factors in heart development. *Experientia* **44,** 944–950.

Kirby, M. L. (1988b). Nodose placode contributes autonomic neurons to the heart in the absence of cardiac neural crest. *J. Neurosc.* **8,** 1089–1095.

Kirby, M. L., and Bockman, D. E. (1984). Neural crest and normal development: A new perspective. *Anat. Rec.* **209,** 1–6.

Kirby, M. L., and Creazzo, T. L. (1995). Cardiovascular development. Neural crest and new perspectives. *Cardiol. Rev.* **3,** 226–235.

Kirby, M. L., and Waldo, K. L. (1990). Role of the neural crest in congenital heart disease. *Circulation* **82,** 332–340.

Kirby, M. L., and Waldo, K. L. (1995). Neural crest and cardiovascular patterning. *Circ. Res.* **77,** 211–215.

Kirby, M. L., Turnage, K. L., and Hays, B. M. (1985a). Characterization of conotruncal malformations following ablation of "cardiac" neural crest. *Anat. Rec.* **213,** 87–93.

Kirby, M. L., Aronstam, R. S., and Buccafusco, J. J. (1985b). Changes in cholinergic parameters associated with conotruncal malformations in embryonic chick hearts. *Circ.Res.* **56,** 392–401.

Kirby, M. L., Creazzo, T. L., and Christiansen, J. L. (1989). Chronotropic responses of chick hearts to field stimulation following various neural crest ablations. *Circ.Res.* **65,** 1547–1554.

Kirby, M. L., Kumiski, D. H., Myers, T., Cerjan, C., and Mishima, N. (1993). Backtransplantation of chick cardiac neural crest cells cultured in LIF rescues heart development. *Dev. Dyn.* **198,** 296–311.

Kirby, M. L., Hunt, P., Wallis, K. T., and Thorogood, P. (1997). Normal development of the cardiac outflow tract is not dependent on normal patterning of the aortic arch arteries. *Dev. Dyn.* **208,** 34–47.

Krezel, W., Dupé, V., Mark, M., Dierich, A., Kastner, P., and Chambon, P. (1996). RXRg null mice are apparently normal and compound RXRa$^{+/-}$/RXRb$^{-/-}$/RXRg$^{-/-}$ mutant mice are viable. *Proc. Natl. Acad. Sci. U.S.A.* **93,** 9010–9014.

Kuratani, S., and Bockman, D. E. (1990). Impaired development of the thy-mic primordium after neural crest ablation. *Anat. Rec.* **228,** 185–190.

Kurihara, Y., Kurihara, H., Maemura, K., Kuwaki, T., Kumada, M., and Yazaki, Y. (1995a). Impaired development of the thyroid and thymus in endothelin-1 knockout mice. *J. Cardiovasc. Pharmacol.* **26** (Suppl. 3), S13-S16.

Kurihara, Y., Kurihara, H., Oda, H., Maemura, K., Nagai, R., Ishikawa, T., and Yazaki, Y. (1995b). Aortic arch malformations and ventricular septal defect in mice deficient in endothelin-1. *J. Clin. Invest.* **96,** 293–300.

Lamour, V., Lecluse, Y., Desmaze, C., Spector, M., Bodescot, M., Aurias, A., Osley, M. A., and Lipinski, M. (1995). A human homolog of the S. cerevisiae HIR1 and HIR2 transcriptional repressors cloned from the DiGeorge syndrome critical region. *Hum. Mol. Genet.* **4,** 791–799.

Langston, A. W., Thompson, J. R., and Gudas, L. J. (1997). Retinoic acid-responsive enhancers located 3′ of the Hox A and Hox B homeobox gene clusters—Functional analysis. *J. Biol. Chem.* **272,** 2167–2175.

Le Douarin, N. M. (1982). "The Neural Crest." Cambridge University Press, Cambridge, UK.
Le Douarin, N. M. (1990). Cell lineage segregation during neural crest ontogeny. *Ann. N.Y. Acad. Sci.* **599,** 131–140.
Le Douarin, N. M., and Jotereau, F. V. (1975). Tracing of cells of the avian thymus through embryonic life in interspecific chimeras. *J. Exp. Med.* **142,** 17–40.
Le Douarin, N. M., Ziller, C., and Couly, G. F. (1993). Patterning of neural crest derivatives in the avian embryo: *In vivo* and *in vitro* studies. *Dev. Biol.* **159,** 24–49.
Lee, Y. M., Osumi-Yamashita, N., Ninomiya, Y., Moon, C. K., Eriksson, U., and Eto, K. (1995). Retinoic acid stage-dependently alters the migration pattern and identity of hindbrain neural crest cells. *Development (Cambridge, UK)* **121,** 825–837.
Lev, M., Bharati, S., Meng, L., Liberthson, R. R., Paul, M. H., and Idriss, F. A. (1972). A concept of double-outlet right ventricle. *J. Thorac. Cardiovasc. Surg.* **64,** 271.
Löfberg, J. (1985). The axolotl embryo as a model for studies of neural crest cell migration. *Axolotl Newsl.* **14,** 10–18.
Lu, H.-C., Eichele, G., and Thaller, C. (1997). Ligand-bound RXR can mediate retinoid signal transduction during embryogenesis. *Development (Cambridge, UK)* **124,** 195–203.
Lumsden, A., and Krumlauf, R. (1996). Patterning the vertebrate neuraxis. *Science* **274,** 1109–1115.
Lumsden, A., Sprawson, N., and Graham, A. (1991). Segmental origin and migration of neural crest cells in the hindbrain region of the chick embryo. *Development (Cambridge, UK)* **113,** 1281–1291.
Maden, M., Ong, D. E., and Chytil, F. (1990). Retinoid-binding protein distribution in the developing mammalian nervous system. *Development (Cambridge, UK)* **109,** 75–80.
Maden, M., Hunt, P., Eriksson, U., Kuroiwa, A., Krumlauf, R., and Summerbell, D. (1991). Retinoic acid-binding protein, rhombomeres and the neural crest. *Development (Cambridge, UK)* **111,** 35–44.
Manley, N. R., and Capecchi, M. R. (1995). The role of *Hoxa-3* in mouse thymus and thyroid development. *Development (Cambridge, UK)* **121,** 1989–2003.
Männer, J., Seidl, W., and Steding, G. (1996). Experimental study on the significance of abnormal cardiac looping for the development of cardiovascular anomalies in neural crest-ablated chick embryos. *Anat. Embryol.* **194,** 289–300.
Marshall, H., Nonchev, S., Sham, M. H., Muchamore, I., Lumsden, A., and Krumlauf, R. (1992). Retinoic acid alters hindbrain Hox code and induces transformation of rhombomeres 2/3 into a 4/5 identity. *Nature (London)* **360,** 737–741.
Matsuoka, R., Takao, A., Kimura, M., Imamura, S., Kondo, C., Joh-o, K., Ikeda, K., Nishibatake, M., Ando, M., and Momma, K. (1994). Confirmation that the conotruncal anomaly face syndrome is associated with a deletion within 22q11.2. *Am. J. Med. Genet.* **53,** 285–289.
Maxwell, G. D., Sietz, P. D., and Rafford, C. E. (1982). Synthesis and accumulation of putative neurotransmitters by cultured neural crest cells. *J. Neurosci.* **2,** 879–888.
McKee, G. J., and Ferguson, M. W. J. (1984). The effects of mesencephalic neural crest cell extirpation on the development of chicken embryos. *J. Anat.* **3,** 491–512.
Miyagawa-Tomita, S., Waldo, K., Tomita, H., and Kirby, M. L. (1991). Temporospatial study of the migration and distribution of cardiac neural crest in quail-chick chimeras. *Am. J. Anat.* **192,** 79–88.
Nakajima, Y., Morishima, M., Nakazawa, M., and Momma, K. (1996). Inhibition of outflow cushion mesenchyme formation in retinoic acid-induced complete transposition of the great arteries. *Cardiovasc. Res.* **31,** E77–E85.
Nakamura, H., and Ayer-Le Lièvre, C. S. (1982). Mesectodermal capabilities of the trunk neural crest of birds. *J. Embryol. Exp. Morphol.* **70,** 1–18.
Nataf, V., Lecoin, L., Eichmann, A., and Le Douarin, N. M. (1996). Endothelin-B receptor is expressed by neural crest cells in the avian embryo. *Proc. Natl. Acad. Sci. U.S.A.* **93,** 9645–9650.
Nelson, R. M., Venot, A., Bevilacqua, M. P., Linhardt, R. J., and Stamenkovic, I. (1995). Carbohydrate-protein interactions in vascular biology. *Annu. Rev. Cell Biol.* **11,** 601–631.
Nishibatake, M., Kirby, M. L., and Van Mierop, L. H. S. (1987). Pathogenesis of persistent truncus arteriosus and dextroposed aorta in the chick embryo after neural crest ablation. *Circulation* **75,** 255–264.
Nishikawa, T., Bruyere, J., H. J., Takagi, Y., and Gilbert, E. F. (1986). The teratogenic effect of phenobarbital on the embryonic chick heart. *J. Appl. Toxicol.* **6,** 91–94.
Noden, D. M. (1978). The control of avian cephalic neural crest cytodifferentiation. I. Skeletal and connective tissues. *Dev. Biol.* **67,** 296–312.
Noden, D. M. (1980). The migration and cytodifferentiation of cranial neural crest cells. *In* "Current Research Trends in Prenatal Craniofacial Development" (R. Pratt and R. L. Christiansen, eds.), pp. 3–25. Elsevier/North-Holland, Amsterdam.
Noden, D. M. (1983). The role of neural crest in patterning of avian cranial skeletal, connective, and muscle tissues. *Dev. Biol.* **96,** 144–165.
Noden, D. M. (1988). Interactions and fates of avian craniofacial mesenchyme. *Development (Cambridge, UK)* **103,** Suppl., 121–140.
Nyquist-Battie, C., and Freter, M. (1988). Cardiac mitochondrial abnormalities in a mouse model of the Fetal Alcohol Syndrome. *Alcoholism: Clin. Exp. Res.* **12,** 264–267.
Olson, E. N., and Srivastava, D. (1996). Molecular pathways controlling heart development. *Science* **272,** 671–676.
Orr-Urtreger, A., Bedford, M. T., Do, M.-S., Eisenbach, L., and Lonai, P. (1992). Developmental expression of the a receptor for platelet-derived growth factor, which is deleted in the embryonic lethal *Patch* mutation. *Development (Cambridge, UK)* **115,** 289–303.
Papolos, D. F., Faedda, G. L., Veit, S., Goldberg, R., Morrow, B., Kucherlapati, R., and Shprintzen, R. J. (1996). Bipolar spectrum disorders in patients diagnosed with velo-cardio-facial syndrome: Does a hemizygous deletion of chromosome 22q11 result in bipolar affective disorder? *Am. J. Psychiatry* **153,** 1541–1547.
Pasini, B., Ceccherini, I., and Romeo, G. (1996). *RET* mutations in human disease. *Trends Genet.* **12,** 138–144.
Pavan, W. J., Liddell, R. A., Wright, A., Thibaudeau, G., Matteson, P. G., McHugh, K. M., and Siracusa, L. D. (1995). A high-resolution linkage map of the lethal spotting locus: A mouse model for Hirschsprung disease. *Mamm. Genome* **6,** 1–7.
Peters-Van der Sanden, M. J. H., Kirby, M. L., Gittenberger-de Groot, A. C., Tibboel, D., Mulder, M. P., and Meijers, C. (1993). Ablation of various regions within the avian vagal neural crest has differential effects on ganglion formation in the fore-, mid- and hindgut. *Dev. Dyn.* **196,** 183–194.
Phillips, M. T., Kirby, M. L., and Forbes, G. (1987). Analysis of cranial neural crest distribution in the developing heart using quail-chick chimeras. *Circ. Res.* **60,** 27–30.
Pizzuti, A., Novelli, G., Mari, A., Ratti, A., Colosimo, A., Amati, F., Penso, D., Sangiuolo, F., Calabrese, G., Palka, G., Silani, V., Gennarelli, M., Mingarelli, R., Scarlato, G., Scambler, P., and Dallapiccola, B. (1996). Human homologue sequences to the *Drosophila dishevelled* segment-polarity gene are deleted in the DiGeorge syndrome. *Am. J. Hum. Genet.* **58,** 722–729.
Puffenberger, E. G., Hosoda, K., Washington, S. S., Nakao, K., DeWit, D., Yanagisawa, M., and Chakravarti, A. (1994). A missense mutation of the endothelin-B receptor gene in multigenic Hirschsprung's disease. *Cell (Cambridge, Mass.)* **79,** 1257–1266.
Qiu, M. S., Bulfone, A., Ghattas, I., Meneses, J. J., Christensen, L., Sharpe, P. T., Presley, R., Pedersen, R. A., and Rubenstein, J. L. R. (1997). Role of the Dlx homeobox genes in proximodistal patterning of the branchial arches: Mutations of Dlx-1, Dlx-2, and Dlx-1 and -2 alter morphogenesis of proximal skeletal and soft tissue

structures derived from the first and second arches. *Dev. Biol.* **185,** 165–184.

Reid, K., Turnley, A. M., Maxwell, G. D., Kurihara, Y., Kurihara, H., Bartlett, P. F., and Murphy, M. (1996). Multiple roles for endothelin in melanocyte development: Regulation of progenitor number and stimulation of differentiation. *Development (Cambridge, UK)* **122,** 3911–3919.

Rhee, D., Sanger, J. M., and Sanger, J. W. (1994). The premyofibril: Evidence for its role in myofibrillogenesis. *Cell Motil. Cytoskeleton* **28,** 1–24.

Riccardi, V. M. (1991). Neurofibromatosis: Past, present and future. *N. Engl. J. Med.* **324,** 1283–1285.

Roberts, C., Daw, S. C. M., Halford, S., and Scambler, P. J. (1997). Cloning and developmental expression analysis of chick *Hira* (*Chira*), a candidate gene for DiGeorge syndrome. *Hum. Mol. Genet.* **6,** 237-245.

Rosenquist, T. H., Kirby, M. L., and Van Mierop, L. H. S. (1989). Solitary aortic arch artery. A result of simultantous ablation of cardiac neural crest and nodose placode in the avian embryo. *Circulation* **80,** 1469–1475.

Rothman, K. J., Moore, L. L., Singer, M. R., Nguyen, U.-S. D. T., Mannino, S., and Milunsky, A. (1995). Teratogenicity of high vitamin A intake. *N. Engl. J. Med.* **333,** 1369–1373.

Rowe, A., and Brickell, P. M. (1995). Expression of the chicken ret-inoid X receptor-gamma gene in migrating cranial neural crest cells. *Anat. Embryol.* **192,** 1–8.

Rowe, A., Eager, N. S. C., and Brickell, P. M. (1991). A member of the RXR nuclear receptor family is expressed in neural-crest-derived cells of the developing chick peripheral nervous system. *Development (Cambridge, UK)* **111,** 771–778.

Ruano, G., and Kidd, K. K. (1991). Coupled amplification and sequencing of genomic DNA. *Proc. Natl. Acad. Sci. USA* **88,** 2815–2819.

Rychter, Z. (1962). Experimental morphology of the aortic arches and the heart loop in chick embryos. *Adv. Morphog.* **2,** 333–371.

Scamps, C., Lorain, S., Lamour, V., and Lipinski, M. (1996). The HIR protein family: Isolation and characterization of a complete murine cDNA. *Biochim. Biophys. Acta* **1306,** 5–8.

Scherson, T., Serbedzija, G., Fraser, S., and Bronner-Fraser, M. (1993). Regulative capacity of the cranial neural tube to form neural crest. *Development (Cambridge, UK)* **118,** 1049–1062.

Sechrist, J., Nieto, M. A., Zamanian, R. T., and Bronner-Fraser, M. (1995). Regulative response of the cranial neural tube after neural fold ablation: Spatiotemporal nature of neural crest regeneration and up-regulation of *Slug*. *Development (Cambridge, UK)* **121,** 4103–4115.

Sellers, J. R., and Goodson, H. V. (1995). Motor proteins 2: Myosin. *Protein Profile* **2,** 1323–1423.

Serbedzija, G. N., Fraser, S. E., and Bronner-Fraser, M. (1990). Pathways of trunk neural crest cell migration in the mouse embryo as revealed by vital dye labelling. *Development (Cambridge, UK)* **108,** 605–612.

Serbedzija, G. N., Burgan, S., Fraser, S. E., and Bronner-Fraser, M. (1991). Vital dye labelling demonstrates a sacral neural crest contribution to the enteric nervous system of chick and mouse embryos. *Development (Cambridge, UK)* **111,** 857–866.

Serbedzija, G. N., Bronner-Fraser, M., and Fraser, S. E. (1992). Vital dye analysis of cranial neural crest cell migration in the mouse embryo. *Development (Cambridge, UK)* **116,** 297–307.

Siebert, J. R., Graham, J. M., and MacDonald, C. (1985). Pathologic features of the CHARGE association: Support for involvement of the neural crest. *Teratology* **31,** 331–336.

Sirotkin, H., O'Donnell, H., DasGupta, R., Halford, S., St. Jore, B., Puech, A., Parimoo, S., Morrow, B., Skoultchi, A., Weissman, S. M., Scambler, P., and Kucherlapati, R. (1997). Identification of a new human catenin gene family member from the region deleted in velo-cardio-facial syndrome. *Genomics* **41,** 75–83.

Smits-van Prooije, A. E., Poelmann, R. E., Dubbeldam, J. A., Mentink, M. M. T., and Vermeij-Keers, C. (1986). Wheat germ agglutinin-gold as a novel marker for mesectoderm formation in mouse embryos cultured *in vitro*. *Stain Technol.* **61,** 97–106.

Soriano, P. (1997). The PDGFα receptor is required for neural crest cell development and for normal patterning of the somites. *Development (Cambridge, UK)* **124,** 2691–2700.

Steller, H. (1995). Mechanisms and genes of cellular suicide. *Science* **267,** 1445–1449.

Stern, C. D., Artinger, K. B., and Bronner-Fraser, M. (1991). Tissue interactions affecting the migration and differentiation of neural crest cells in the chick embryo. *Development (Cambridge, UK)* **113,** 207–216.

Strong, W. B. (1968). Familial syndrome of right-sided aortic arch, mental deficiency, and facial dysmorphism. *J. Pediatr.* **73,** 882–888.

Sucov, H. M., Luo, J.-M., Evans, R. M., and Giguère, V. (1995). Outflow tract and aortic arch malformations in retinoic acid receptor double-mutant embryos implicate a defect in the differentiation of the cardiac neural crest. *Circulation 92* (Suppl. I), I-F(abstr).

Sulik, K. K., Johnston, M. C., Daft, P. A., Russell, W. E., and Dehart, D. B. (1986). Fetal alcohol syndrome and DiGeorge anomaly: Critical ethanol exposure periods for craniofacial malformations as illustrated in an animal model. *Am. J. Med. Genet., Suppl.* **2,** 97–112.

Suzuki, H. R., and Kirby, M. L. (1997). Absence of neural crest regeneration from the postotic neural tube. *Dev. Biol.* **184,** 222–233.

Tan, S .S., and Morriss-Kay, G. M. (1986). Analysis of cranial neural crest cell migration and early fates in postimplantation rat chimeras. *J. Embryol. Exp. Morphol.* **98,** 21–58.

Tassabehji, M., Read, A. P., Newton, V. E., Harris, R., Balling, R., Gruss, P., and Strachan, T. (1992). Waardenburg's syndrome patients have mutations in the human homologue of the *Pax-3* paired box gene. *Nature (London)* **355,** 635–636.

Tassabehji, M., Newton, V. E., Leverton, K., Turnbull, K., Seemanova, E., Kunze, J., Sperling, K., Strachan, T., and Read, A. P. (1994). PAX3 gene structure and mutations: Close analogies between Waardenburg syndrome and the *Splotch* mouse. *Hum. Mol. Genet.* **3,** 1069–1074.

Thorogood, P., Smith, L., Nicol, A., McGinty, R., and Garrod, D. (1982). Effects of vitamin A on the behavior of migratory neural crest cells *in vitro*. *J. Cell Sci.* **57,** 331–350.

Tomita, H., Connuck, D. M., Leatherbury, L., and Kirby, M. L. (1991). Relation of early hemodynamic changes to final cardiac phenotype and survival after neural crest ablation in chick embryos. *Circulation* **84,** 1289–1295.

Torres, M., Stoykova, A., Huber, O., Chowdhury, K., Bonaldo, P., Mansouri, A., Butz, S., Kemler, R., and Gruss, P. (1997). An a-*E-catenin* gene trap mutation defines its function in preimplantation development. *Proc. Natl. Acad. Sci. U.S.A.* **94,** 901–906.

Vaessen, M.-J., Meijers, J. H. C., Bootsma, D., and van Kessel, A. D. (1990). The cellular retinoic-acid-binding protein is expressed in tissues associated with retinoic-acid-induced malformations. *Development (Cambridge, UK)* **110,** 371–378.

Van Mierop, L. H. S.,; and Kutsche, L. M. (1986).Cardiovascular anomalies in DiGeorge syndrome and importance of neural crest as a possible pathogenetic factor. *Am. J. Cardiol.* **58,** 133–137.

Vogel, K. S., Brannan, C. I., Jenkins, N. A., Copeland, N. G., and Parada, L. F. (1995). Loss of neurofibromin results in neurotrophin-independent survival of embryonic sensory and sympathetic neurons. *Cell (Cambridge, Mass.)* **82,** 733–742.

von Deimling, A., Krone, W., and Menon, A. G. (1995). Neurofibromatosis type 1: Pathology, clinical features and molecular genetics. *Brain Pathol.* **5,** 153–162.

Wachtler, F. (1984). On the differentiation and migration of some non-neuronal neural crest derived cell types. *Anat. Embryol.* **170,** 161–168.

Wadey, R., Daw, S., Taylor, C., Atif, U., Kamath, S., Halford, S., O'Donnell, H., Wilson, D., Goodship, J., Burn, J., and Scambler, P. (1995). Isolation of a gene encoding an integral membrane protein from the vicinity of a balanced translocation breakpoint associated with DiGeorge syndrome. *Hum. Mol. Genet.* **4,** 1027–1033.

Waldo, J. A. (1992). Neural crest cell develpment. Prog. Clin. Biol. Res. 85, 359–379. New York: Alan R. Liss, Inc.

Wehrle-Haller, B., and Weston, J. A. (1995). Soluble and cell-bound forms of steel factor activity play distinct roles in melanocyte precursor dispersal and survival on the lateral neural crest migration pathway. *Development (Cambridge, UK)* **121,** 731–742.

Wehrle-Haller, B., Morrison-Graham, K., and Weston, J. A. (1996). Ectopic c-kit expression affects the fate of melanocyte precursors in *patch* mutant embryos. *Dev. Biol.* **177,** 463–474.

Weston, J. A. (1982). Neural crest cell development. *Prog. Clin. Biol. Res.* **85,** 359–379. New York: Alan R. Liss, Inc.

Wilson, D. I., Burn, J., Scambler, P., and Goodship, J. (1993). DiGeorge syndrome: Part of CATCH 22. *J. Med. Genet.* **30,** 852–856.

Wilson, J. G., and Warkany, J. (1950). Cardiac and aortic arch anomalies in offspring of vitamin A deficient rats correlated with similar human anomalies. *Pediatrics* **5,** 708–725.

Winegrad, S. (1997). Endothelial cell regulation of contractility of the heart. *Annu. Rev. Physiol.* **59,** 505–525.

Xu, D., Emoto, N., Giaid, A., Slaughter, C., Kaw, S., DeWit, D., and Yanagisawa, M. (1994). ECE-1: A membrane-bound metalloprotease that catalyzes the proteolytic activation of big endothelin-1. *Cell (Cambridge, Mass.)* **78,** 473–785.

Yasuda, Y., Okamoto, M., Konishi, H., Matsuo, T., Kihara, T., and Tanimura, T. (1986). Developmental anomalies induced by all-trans retinoic acid in fetal mice: I. Macroscopic findings. *Teratology* **34,** 37–49.

Yasui, H., Nakazawa, M., Morishima, M., Miyagawa-Tomita, S., and Momma, K. (1995). Morphological observations on the pathogenetic process of transposition of the great arteries induced by retinoic acid in mice. *Circulation* **91,** 2478–2486.

12

Development of the Conduction System of the Vertebrate Heart

Antoon F. M. Moorman and Wouter H. Lamers

Department of Anatomy and Embryology, The Cardiovascular Research Institute Amsterdam, Academic Medical Center, University of Amsterdam, 1105 AZ Amsterdam, The Netherlands

I. Introduction

To understand the development of the conduction system, we must first consider the anatomical arrangement of structures that are responsible for the coordinated contraction of the formed heart from apex to base (Fig. 1) (Davies *et al.*, 1983). The sinus node, or cardiac pacemaker, is located subepicardially at the entrance of the superior caval vein to the right atrium. Its electrical impulse spreads via the ordinary atrial myocardium to reach the atrioventricular junctions, where further propagation is interrupted by fibrous insulation, except at that point where a specialized atrioventricular bundle penetrates the atrioventricular junction through the central fibrous body. The atrioventricular node, located on the atrial side of the atrioventricular junction, can be considered as the end of the atrial conduction axis. As is essential for proper heart function, a delay in the propagation of the impulse into the ventricles is produced by the atrioventricular node, allowing ventricular chambers to be filled while the atria contract. The impulse is then conveyed via the ventricular conduction system, comprising atrioventricular bundle and bundle branches, toward the peripheral Purkinje fibers that activate the ventricular myocardium from apex to base. To this end, the atrioventricular bundle and bundle branches are insulated from the ordinary ventricular myocardium by fibrous tissue.

This electrical configuration follows an ancient pattern in vertebrate evolution and is already realized in

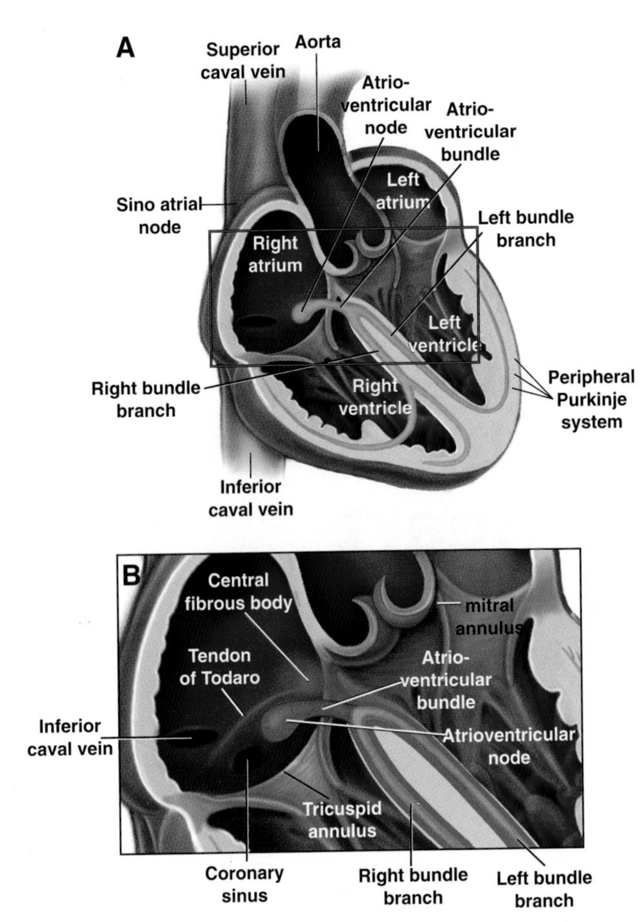

Figure 1 Disposition of the cardiac conduction system in the cardiac environment. Note the fibrous insulation (blue) of atrioventricular bundle and left and right bundle branch. (A) Overview; (B) detail.

the fish heart, which contains a single circuit in which the pacemaker is localized at the intake side of the heart. The delay in impulse propagation is realized at the atrioventricular junction and the ventricle is activated from apex to base (Randall, 1970). In lower vertebrates, the participating components are histologically ill-defined, as is the case in developing mammalian hearts in which atria and ventricles have not yet become insulated by fibrous tissue and a distinct conduction system is not present. Nevertheless, an electrocardiogram can already be derived from the tubular chicken heart at a stage when the atrial and ventricular chambers have just started to develop (van Mierop, 1967). This highlights the primary developmental imperative of developing a sequential pattern of activation of atria and ventricles rather than development of a distinct conduction system. This, in turn, implies that the coordinated contraction of the heart, as reflected in the electrocardiogram, is established by the arrangement of functionally different myocardial compartments. This mode of development also suggests that the conduction system develops from existing myocardium rather than from an extracardiac source (Moorman *et al.,* 1998).

The origin of the conduction system has always been a highly contentious topic. At the base of the controversies lies the conventional strict distinction made between working myocardium and conduction system in the formed heart. We believe that it has been insufficiently appreciated that all myocytes display, to variable degrees, contractile, conductive, and pacemaking properties, and that the correct topological disposition of functionally distinct myocyte populations realizes the coordinated contraction of the developing heart. Due to the expression in conduction system of proteins normally associated with neural structures (Gorza *et al.,* 1994), it was suggested that neural crest might give rise to the conduction system (Gorza *et al.,* 1988). However, neural crest cells arrive at the heart far later than the point at which an electrocardiogram can be recorded (Paff *et al.,* 1966; Kirby *et al.,* 1993). Thus, the discussion of the origin of the conduction system departs from the more fundamental question of how specification and arrangement of functionally different myocardial compartments leads to the coordinated contraction of the heart.

II. Cardiac Polarity and Development of the Sinus Node

The primary heart tube originates from the folding of the cardiogenic mesoderm around the developing endocardial tubes, as indicated in Fig. 2. While being formed, the primary heart tube displays polarity along the anteroposterior axis. The leading pacemaker—the sinus node equivalent—is always localized at the intake of the heart (Satin *et al.,* 1988; van Mierop, 1967), and through it an efficient unidirectional wave of myocardial contractions is ensured. The atrial phenotype becomes prevalent at the posterior (upstream) side of the heart tube and the ventricular phenotype at the anterior (downstream) side (Moorman *et al.,* 1995; Moorman and Lamers, 1994; Lyons, 1994). It may be that opposing gradients of gene expression cause functional differences between the upstream and downstream parts of the heart tube, where the upstream, atrial side starts to develop into the drainage pool of the embryo and the downstream, ventricular side into the muscular pump that generates systemic pressure. In the primitive chordate heart of Tunicates (Urochordates), anteroposterior polarity has not evolved and a fixed position for the leading pacemaker is not evident. Thus, blood is pumped alternately in both directions (Kuhl and van Hasselt, 1822). The molecular signals that impose polarity upon

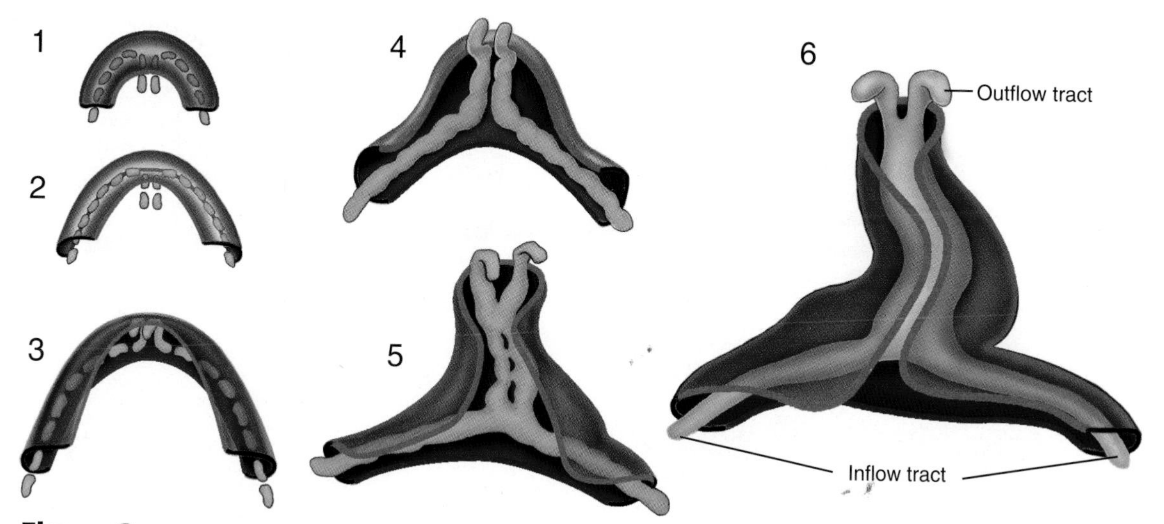

Figure 2 The folding of the horseshoe-shaped cardiogenic plate into the primary heart tube in six successive stages (1–6). Note that owing to the folding process, the anterior margin (blue) becomes positioned caudally by which the primary myocardium folds around the developing endocardial tubes (yellow) to form the myocardial heart tube. The originally posteriorly located myocardial margin is indicated in red.

the vertebrate cardiac tube are discussed in other chapters of this book. Remarkably, in the chicken, the first contractions are not observed at the intake of the heart but rather in that part of the tube where the ventricles will develop. This contradictory observation does not imply that the pacemaker jumps from the ventricular to the intake part of the heart tube, but rather coupled excitation and contraction is realized first in the ventricular part of the heart tube, before the inflow part has this capacity (van Mierop, 1967). Slightly later, the electrical phenotype of cells from the future sinus region differs from that of the cardiac tube, or future ventricles. The future sinus cells display prepotentials resembling those of the adult pacemaker, whereas the prepotentials of the future ventricular cells resemble those of the adult ventricles (Kamino, 1991; Meda and Ferroni, 1959).

Chicken embryos have been the model of choice for studies on the development of electrical activity because they are experimentally well accessible (Lamers *et al.,* 1991; Kamino, 1991). In a few hours of chicken development (7–10 somites, equivalent to a human embryonic age of ~20 days), a single pacemaking area becomes established at the inflow tract of the heart, as monitored by the use of voltage-sensitive dyes (Hirota *et al.,* 1979). Pacemaker dominance along the axis of the heart tube increases in an anteroposterior direction (Satin *et al.,* 1988; van Mierop, 1967), as does the frequency of the intrinsic beat rate (Satin *et al.,* 1988; Kamino *et al.,* 1981; Barry, 1942). In both birds and mammals, as soon as the sinus venosus has formed (~25 days in man) the leading pacemaker area is found at the right side, but initially it is most frequently detected at the left side (Goss, 1942; Sakai *et al.,* 1983; Kamino *et al.,* 1981; de Haan, 1959). One should appreciate that both right and left inflow tracts will eventually become incorporated into the right atrium, that the entire inflow area of the embryonic heart is relatively small in comparison to the size of the adult sinus node, and that the exact site of the leading pacemaker is not entirely fixed (Bouman *et al.,* 1968, 1978). Thus, the entire inflow area represents a more or less homogeneous pacemaking area, which is consistent with the observation of nodal-like cells in the myocardium surrounding the distal portion of the pulmonary veins in adult rat (Brunton and Fayrer, 1874; Cheung, 1980; Masani, 1986).

In man, as in other mammals, the sinus node becomes apparent (Carnegie stage 15; ~5 weeks of development) in myocardium surrounding the future superior caval vein (de Groot *et al.,* 1988; Virágh and Challice, 1980). Why and how the leading pacemaker area becomes transformed into a nodal structure is not clear. In chicken and lower vertebrates, the leading pacemaker remains indistinct as a loosely arranged conglomerate of venous sinus myocytes (Szabó *et al.,* 1986; de Groot *et al.,* 1985; Canale *et al.,* 1986; Davies, 1930).

III. Chamber Formation and Development of the Atrioventricular Node

With the development of distinct atrial and ventricular chambers (Fig. 3), their functional differences be-

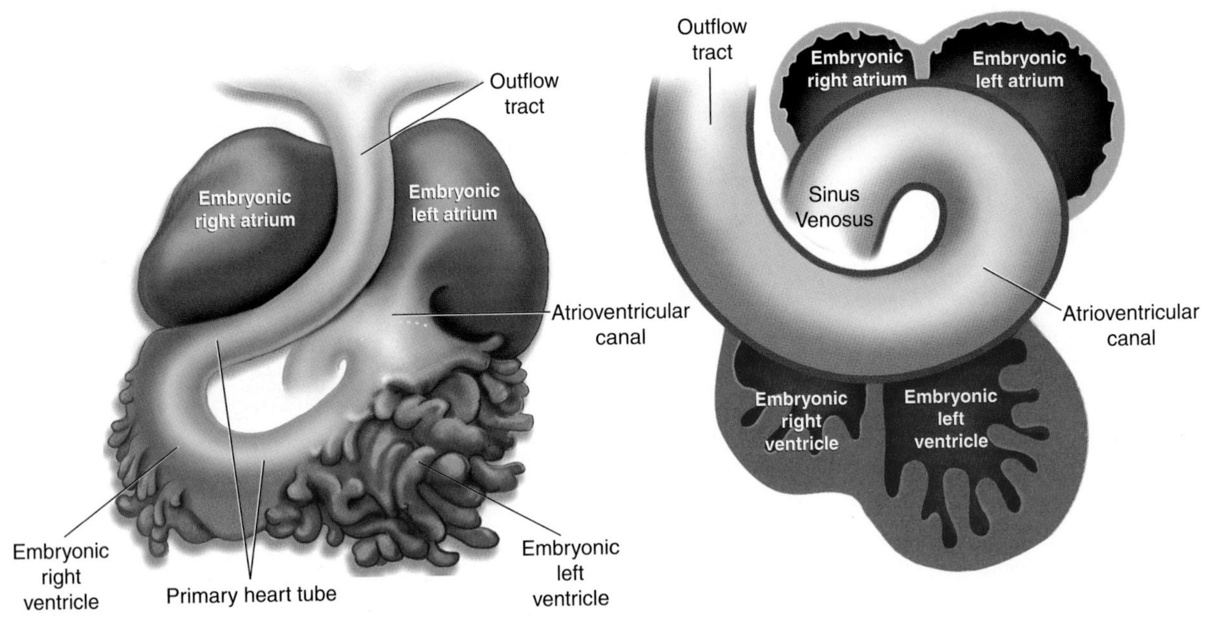

Figure 3 (Left) Cast of the lumen of a human embryonic heart of about 4 weeks of development after Streeter (1987). Note the smooth-walled myocardium at the inner curvature of the heart, representing in our view the "original primary heart tube," from which the atrial and ventricular chambers balloon out. (Right) Heart of the same stage after de Jong *et al.* (1997), indicating the distinct segments.

come unequivocally manifest. Thus, the primary heart tube, which displays peristaltoid contraction waves (Patten and Kramer, 1933; Patten, 1949) owing to the slow propagation of the cardiac impulse (de Jong *et al.*, 1992; Hirota *et al.*, 1983; Arguello *et al.*, 1986), is converted into the more efficient chamber pump in which atrial and ventricular chambers contract rapidly, because of the fast conduction of the cardiac impulse, but still sequential, because of the delay of the cardiac impulse in the atrioventricular canal (de Jong *et al.*, 1992; Arguello *et al.*, 1986; Lieberman and Paes de Carvalho, 1965a,b). We have labeled the myocardium of the early heart tube the "primary myocardium" to distinguish it from the latter (atrial and ventricular) working myocardium (Moorman and Lamers, 1994). The fast-conducting atrial and ventricular chambers remain flanked by the slow-conducting primary myocardium of the original heart tube, resulting in a structure with alternating slow- and fast-conducting compartments comprising inflow tract (slow), in which the sinus node will develop; atria (fast); atrioventricular canal (slow), where the atrioventricular node will develop; ventricles (fast); and outflow tract (slow). This configuration of the embryonic heart ensures the sequential activation of the atrial and ventricular chambers, as is indicated by the presence of an electrocardiogram, even though the histologically well-defined components of the conduction system present in the formed mammalian heart are absent (van Mierop, 1967; Paff *et al.*, 1968). The slow-conducting flanking segments of primary myocardium in the embryonic heart also function as sphincters that substitute for the valves that are yet to develop.

In many studies the atrioventricular canal has been identified as a zone of slow conduction, functioning as the equivalent of the atrioventricular node in hearts in which the atrial and ventricular chambers have not yet been insulated by fibrous tissue (de Jong *et al.*, 1992; Arguello *et al.*, 1986, 1988; Paff *et al.*, 1968; Lieberman and Paes de Carvalho, 1965a,b; van Mierop, 1967). In mammals, a distinct atrioventricular node can be distinguished around Carnegie stage 15 (~5 weeks of human development)(Virágh and Challice, 1977; Virágh and Porte, 1973). In chicken, however, the atrioventricular node remains diffuse and the whole atrioventricular junction has been supposed to fulfill the role of the atrioventricular node (Szabó *et al.*, 1986). Also, in mammals, this area may maintain characteristics of the embryonic atrioventricular canal, consistent with observations in pig and dog hearts in which the entire lower left and right atrial rim just above the fibrous annulus displays nodal-like action potentials and low abundance of connexin 43 (van Kempen *et al.*, 1991, 1996; McGuire *et al.*, 1996).

Segments of slow conduction (remaining primary myocardium) also persist at both extremities of the embryonic heart. This is particularly clear in the outflow tract due to its length and can be recognized as a "C wave" in an electrocardiogram (de Jong *et al.*, 1992; Paff and Boucek, 1975). In the formed dog heart, the musculature surrounding the right ventricular outflow tract retains a sphincter-like function during ventricular relaxation to support the pulmonary semilunar valves (Brock, 1964). Most interestingly, this part of the heart maintains a unique transcriptional potential in the formed heart (Franco *et al.,* 1997).

IV. Primary Myocardium and the Nodal Phenotype

Cardiomyocytes of the primary heart tube are small and display a loosely arranged myofibrillar apparatus and sarcoplasmic reticulum, indicating an immature contractile function. These early cells have strong autorhythmicity (Satin *et al.,* 1988; van Mierop, 1967; Canale *et al.,* 1986), indicating that they are poorly coupled, an essential feature of pacemaker function (Joyner and van Capelle, 1986). Therefore, it is not surprising that nodal myocytes resemble in many aspects myocytes of the primary myocardium. Their presence in the myocardium of the inflow tract and atrioventricular canal first becomes apparent when surrounding myocytes differentiate into atrial working myocardium. With development, both inflow tract and atrioventricular canal become incorporated into the atrium and provide the precursors of the sinus and atrioventricular nodes, respectively. This leaves the intriguing but as yet unresolved question of how a number of these cells are prevented from differentiating in atrial direction and become committed to a nodal phenotype. Although our understanding of the nodal phenotype is far from complete, three important nodal characteristics can be distinguished: (i) slow intercellular conduction, (ii) poorly developed contractile apparatus, and (iii) unique cytoskeleton.

First, in both primary myocardium and the sinus and atrioventricular nodes of many species, the number and size of gap junctions is small (van Kempen *et al.,* 1991; Arguello *et al.,* 1988; Virágh and Challice, 1980; Fromaget *et al.,* 1992; Gros *et al.,* 1978), and the major cardiac gap-junctional proteins connexin 43 and connexin 40, and their encoding mRNAs, are rare or undetectable (Moorman *et al.,* 1998; Shirinsky *et al.,* 1992; Gros *et al.,* 1994; Gourdie *et al.,* 1992; van Kempen *et al.,* 1991). This feature clearly distinguishes nodal myocytes from the surrounding atrial working myocardium.

Second, in both primary myocardium and nodes, coexpression of α- and β-myosin heavy chains has been observed in several species (e.g., Komuro *et al.,* 1987; Gorza *et al.,* 1986; for a review see Moorman *et al.,* 1998). Because nodal cells display heterogeneity in expression of the β-myosin isoform, expression of the myosin isoforms cannot be used as a reliable marker for nodal myocytes. Also, expression of the embryonic or slow skeletal troponin I isoform persists in atrioventricular nodal myocytes (Schiaffino *et al.,* 1993; Gorza *et al.,* 1993). In agreement with this, a 4200-nucleotide DNA region upstream of the human *slow skeletal troponin I* gene is able to confer expression of a reporter gene in adult mouse atrioventricular node (Zhu *et al.,* 1995).

Third, in most species, the cytoskeletal protein desmin is expressed at higher levels in nodes and the ventricular conduction system compared to the working myocardium (Eriksson *et al.,* 1979; Thornell and Eriksson, 1981). High levels of desmin are correlated with the morphologically well-differentiated Purkinje fibers of hoofed animals, whereas low levels are found in the morphologically poorly differentiated ventricular conduction system of the rat. Desmin is expressed in the early mouse and rat myocardium (Jing *et al.,* 1997; Baldwin *et al.,* 1991), making it a poor marker for the conduction system in these animals. However, 1 kb of the human *desmin* promoter will drive expression of a *LacZ* transgenic reporter gene in the cardiac conduction system of mice (Li *et al.,* 1993), as judged by *in toto* enzymatic staining, an important result which merits histological verification.

The rabbit heart is unique in that it expresses a distinct conduction system-specific marker. Both the nodes and the ventricular conduction system are characterized by the expression of neurofilament protein, which colocalizes with desmin (Vitadello *et al.,* 1990, 1996; Gorza and Vitadello, 1989). It is not expressed in the myocardium of the primary heart tube. The extensive expression of neurofilament protein in atrial myocardium relates to a long-standing debate on the presence of internodal tracts (Liebman, 1985; Davies *et al.,* 1983; Janse and Anderson, 1974). Although there is now a consensus that specialized tracts insulated from surrounding myocardium by fibrous tissue do not exist, the extensive expression of neurofilament protein in the atrium is suggestive of some form of distinctive structure, either extensions of the nodes themselves or specialized fast-conducting tissue. Consistent with these possibilities, Leu-7 (HNK-1) is expressed in internodal tracts forming a network across the roof of the right and left atrium in the developing human and rat heart (Aoyama *et al.,* 1995; Nakagawa *et al.,* 1993; Ikeda *et al.,* 1990). Also, in the chicken, an atrial Purkinje network has been reported (de Groot *et al.,* 1985, 1987; Virágh *et al.,* 1989). The central question is whether these cells represent preferential pathways of conduction. The molecular markers are now available to settle this much debated issue.

V. The Trabecular Ventricular Component and Development of the Ventricular Part of the Conduction System

The description of cardiac development given to date has resulted in the following scenario. The tubular heart becomes divided into five functionally distinct segments owing to the development of fast-conducting atrial and ventricular segments of working myocardium, flanked by slow-conducting inflow tract, atrioventricular canal, and outflow tract myocardium. These segments most likely arise as a result of some form of molecular polarity over the anteroposterior axis of the heart. In the context of cardiac looping separate left and right atrial and ventricular chambers are formed. Thus, the total number of segments may be seven. Consistent with this description are the recent analyses of *LacZ* transgene expression under the control of various truncated promoters that distinguish the distinct component parts (Franco *et al.,* 1998) and evolutionary considerations derived from measurements of the conduction velocities in junctional areas of the amphibian heart, which are remarkably similar to those measured in the flanking segments of the embryonic chicken heart (de Jong *et al.,* 1992; Alanis *et al.,* 1973). These junctional areas of the amphibian heart have a sphincter function (Canale *et al.,* 1986; Benninghoff, 1923), similar to the situation in embryonic chicken or mammalian hearts. Historically, the junctional region of the amphibian heart has been dubbed "specialized," although in view of its development origin and histology, this terminology could be considered as a misnomer.

The previously mentioned building plan of the heart, constructed from seven distinct building blocks, accounts for the (early) electrocardiogram but not for the activation of the ventricles from apex to base. In the formed heart, the atrioventricular bundle, bundle branches, and the peripheral Purkinje network are responsible for this mode of excitation. Although these structures are not present in lower vertebrate hearts, such as those of fish, their single ventricle is also activated from apex to base, indicating that the substrate for preferential conduction toward the apex is already present in primitive vertebrate hearts (Canale *et al.,* 1986). A clue as to how this ventricular electrical architecture is realized may come from the observation that the spongy trabecular myocardium of the fish ventricle has remarkably similar properties to the trabecular ventricular compartment of the embryonic mammalian and chicken heart, in which preferential conduction has been demonstrated (Chuck *et al.,* 1997; de Jong *et al.,* 1992). Based on recent studies on the patterns of gene expression in the embryonic ventricles and on genetic studies, we suggest that the ventricular conduction system originates from the trabecular ventricular component, including the interventricular myocardium (Moorman *et al.,* 1998).

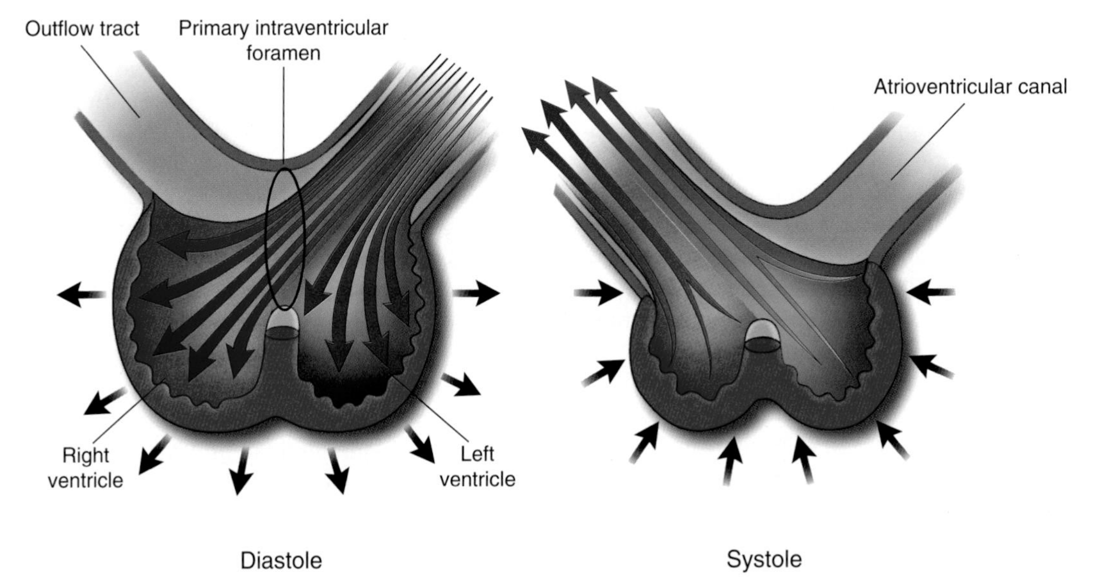

Figure 4 The blood flow through the embryonic heart during the heart cycle. With relaxation of the ventricles, the primary interventricular foramen is used as the entrance of the right ventricle, which in the formed heart has become the right atrioventricular orifice. With contraction the same foramen is used as the outflow of the left ventricle, which in the formed heart has become the left ventricular outlet. The remodeling of the primary interventricular foramen is depicted in Fig. 5.

During formation of the four-chambered heart, the serially arranged cardiac segments become arranged in parallel. Thus, the right atrium becomes directly connected with the right ventricle and the left ventricle with the outflow tract. Atrial and ventricular working myocardium differentiate from primary heart tube myocardium and balloon out at the outer curvature of the looped heart to form the atrial and ventricular chambers (Fig. 3). The ventricular septum is formed by apposition of myocardial cells at the outer side, leaving a primary foramen at the inner curvature in between the right and left ventricular parts of the heart tube. It is the remodeling of this primary foramen and of the inner curvature that plays a pivotal role in the septational process (Goor *et al.,* 1972; Lamers *et al.,* 1992).

One should appreciate that the primary ventricular foramen constitutes the entrance to the right ventricle and the outlet of the left ventricle, a topographical position that will essentially not change with further development (Fig. 4). It is equally important to appreciate that at the region of the inner curvature, interventricular myocardium is also atrioventricular canal myocardium and constitutes the atrioventricular conduction axis, as demonstrated by the expression of GlN2 in human (Wessels *et al.,* 1992), *Msx2* in chicken (Chan-Thomas *et al.,* 1993), and Leu-7 (HNK-1) in rat (Aoyama *et al.,* 1995; Nakagawa *et al.,* 1993) and human (Ikeda *et al.,* 1990). GlN2 and Leu-7 (HNK-1) recognize a complex carbohydrate moiety present on a series of molecules involved in cell adhesion, whereas *Msx2* is a *Drosophila* muscle-related homeobox gene. These studies show that the entire ventricular conduction system is derived from one source, namely, the ring of myocardium encircling the primary interventricular foramen and the trabecular ventricular component (Fig. 5). Expression is never observed in the compact ventricular component. Rightward expansion of the atrioventricular canal brings this ring of myocardium to a position where it can make contact with the atrial septum, through which the atrioventricular conduction axis becomes established. The leftward expansion of the outflow tract connects the outlet with the left ventricle. Remarkably, in the adult chicken heart, the entire system is still present (Davies, 1930). In humans, however, only the atrioventricular bundle (and bundle branches) remain. Based on its position in the chicken heart, the assumed location of the primary interventricular myocardium, had it not regressed, can be positioned in the adult human heart (Fig. 6). The right atrioventricular ring bundle is recognizable in the neonatal but not in the formed heart; it forms the lower rim of the right atrium just above the atrioventricular annulus, but that part encircling the left ventricular outlet regresses with fetal development (Wessels *et al.,* 1992). Thus, the atrioventricular bundle, bundle branches, and peripheral network take origin from the interventricular myocardium and the trabecular ventricular components. They form a drape upon and astride the developing ventricular septum extending into the ventricular trabecules (Virágh and Challice, 1977; Vassall-Adams, 1982). Evidence is accumulating that the trabecular ventricular compartment constitutes a distinct transcriptional domain separate from the compact ventricular myocardium (Franco *et al.,* 1996). After extensive remodeling, this trabecular ventricular component will form the connection between the atrial and compact ventricular components, allowing electrical input at the atrial side and output at the peripheral ventricular side. Insulating fibrous tissue will otherwise separate the atrial from the ventricular myocardium as well as the atrioventricular bundle and bundle branches from the compact ventricular myocardium. Because the trabecular ventricular conduction component (atrioventricular bundle and bundle branches) becomes insulated from working, this component represents another distinct tissue compartment within the developing heart.

The molecular phenotypes of the trabecular ventricular component and of the ventricular part of the conduction system display a similar, although ambiguous character. On the one hand, advanced intercellular conduction has been developed in contrast to the condition in the myocardium of the primary heart tube and of the nodes (Chuck *et al.,* 1997; de Jong *et al.,* 1992); on the other hand, the contractile machinery is less well developed and is more similar to primary myocardium (Canale *et al.,* 1986). Many so-called "atrial-specific" genes are expressed in the trabecular ventricular component and also in the ventricular conduction system (Franco *et al.,* 1996). A striking example is the expression of atrial natriuretic factor, which is highly abundant in the trabecular component of the developing ventricles but is absent from the compact myocardium (Zeller *et al.,* 1987; Toshimori *et al.,* 1987; Thompson *et al.,* 1986). Consistent with the idea of a trabecular origin of the ventricular conduction system, atrial natriuretic factor is also present in the fetal and adult ventricular conduction system but not in the nodes (Skepper, 1989; Pucci *et al.,* 1992; Anand-Srivastava *et al.,* 1989; Toshimori *et al.,* 1988; Hansson and Forsgren, 1993; Wharton *et al.,* 1988). As previously mentioned, the expression of cytoskeletal proteins, such as desmin and neurofilament, is distinct in the ventricular conduction system compared to the surrounding compact myocardium.

In rodents, both connexin 43 (Yancey *et al.,* 1992; Dahl *et al.,* 1995; Fromaget *et al.,* 1992; Gourdie *et al.,* 1992; van Kempen *et al.,* 1991) and connexin 40 (Delorme *et al.,* 1995, 1997; van Kempen *et al.,* 1995; Gourdie *et al.,* 1993; Bastide *et al.,* 1993; Gros *et al.,* 1994) are abundant in the ventricular conduction system, although in small mammals, such as the rat, connexin 43 is absent from

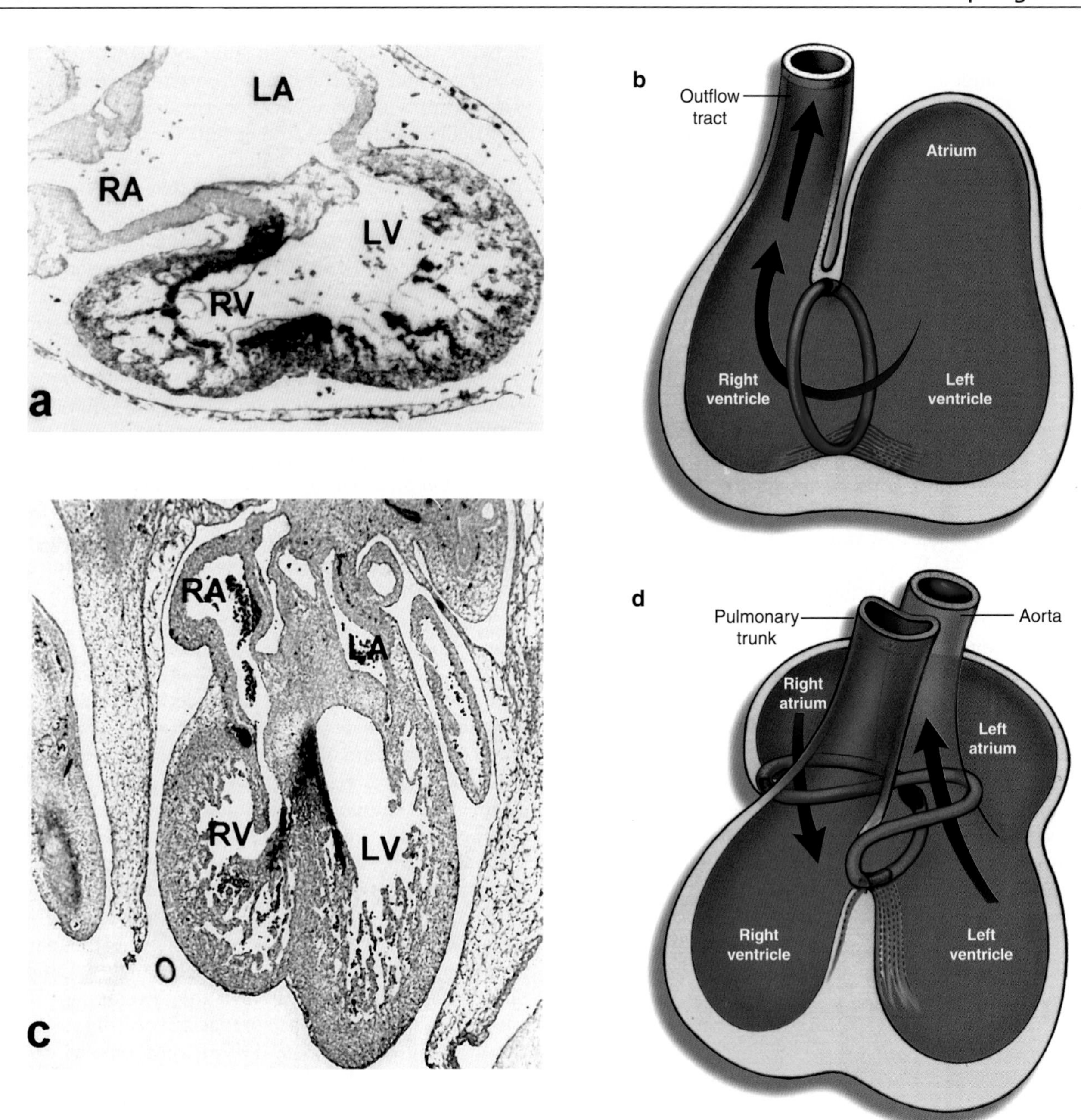

Figure 5 Formation of the ventricular conduction system. (a) Section of a human embryonic heart of 5 weeks of development stained for G1N2, a marker for the developing human conduction system (Wessels *et al.,* 1992). Staining is observed in the myocardium surrounding the primary interventricular foramen and in the ventricular trabecular component. The atrioventricular bundle will develop from the dorsal part of the ring, whereas bundle branches and the peripheral conduction system will develop from the trabecular component. (b) The G1N2 staining after reconstruction of all the sections. (c) Section of a human embryonic heart of 7 weeks of development stained for G1N2. The atrioventricular canal has expanded to the right, by which the right atrium has become connected with the right ventricle, and the outflow tract has expanded to the left, by which the aorta has become connected with the left ventricle. These morphogenetic movements could be followed due to the marker of the myocardium surrounding the primary interventricular foramen G1N2. The G1N2 staining after reconstruction of all the sections. With further development this staining disappears and in the human heart the parts of the conduction system surrounding the right atrioventricular junction and the left ventricular outlet will disappear. Most interestingly, they still exist in the adult chicken heart (Davies, 1930).

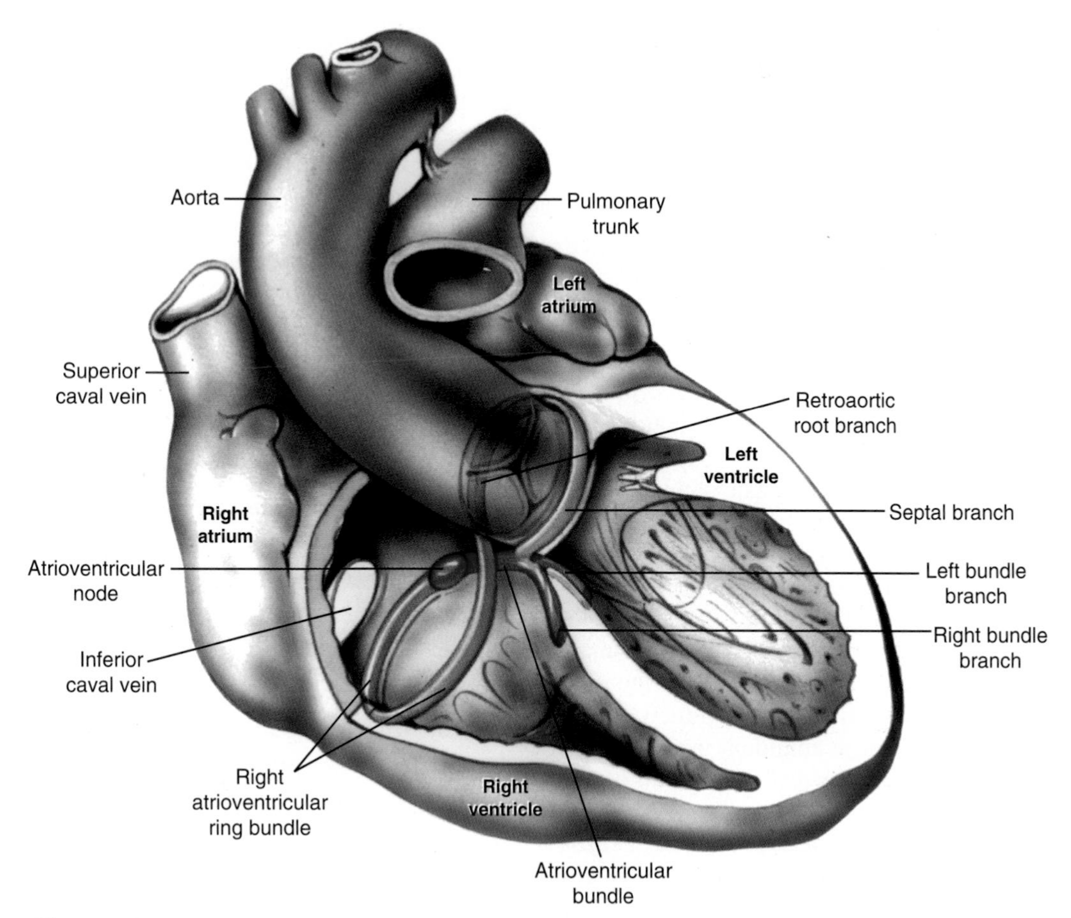

Figure 6 An adult human heart indicating the position of the "former primary interventricular myocardium," parts of which give rise to the ventricular conduction system, whereas other parts are no longer present in the adult human heart but still exist in the adult chicken heart (see text for details).

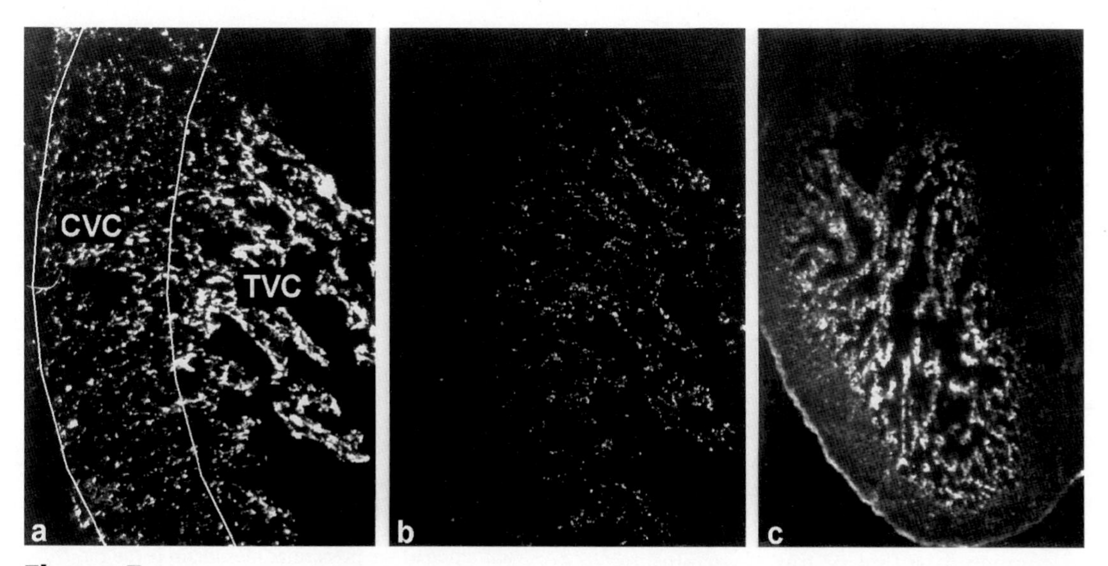

Figure 7 Protein expression pattern of desmin (a), connexin 43 (b), and connexin 40 (c) in the embryonic mouse heart of 12 days of development. Connexin 40 and 43 are clearly more highly expressed in the trabecular ventricular compartment (TVC) supporting the notion that the propagation of the impulse is preferentially achieved by the trabecules. CVC, compact ventricular compartment (photographs kindly provided by Dr. Daniel Gros, Marseille, France).

the proximal part of the system. Both proteins are expressed at higher levels in the trabecular ventricular component than in the compact myocardial component (Fig. 7), underscoring the notion of a trabecular origin of the ventricular part of the conduction system.

Coexpression of the α- and β-myosin heavy chain isoforms in the ventricular conduction system has been reported for a variety of species, from chicken to man (Kuro-o *et al.,* 1986; Komuro *et al.,* 1987; Sartore *et al.,* 1978, 1981; Dechesne *et al.,* 1987). Because coexpression of these isoforms is a characteristic of the myocardium of the primary heart tube, this feature can be dubbed "embryonic." In chicken, the peripheral conduction system is uniquely characterized by the expression of the slow tonic myosin isoform (González-Sánchez and Bader, 1985), although expression starts relatively late in development.

The cytoskeletal protein desmin identifies the ventricular conduction system. This is most clear in the well-developed conduction system of hoofed animals (Virtanen *et al.,* 1990; Forsgren *et al.,* 1980). In line with a trabecular origin of the ventricular conduction system, desmin expression is more abundant in the trabecular than in the compact myocardial component of the embryonic ventricles in mouse (Fromaget *et al.,* 1992). In the developing rabbit heart, the expression of neurofilament protein is an unambiguous marker for the developing ventricular conduction system (Gorza *et al.,* 1988; Vitadello *et al.,* 1990; Gorza and Vitadello, 1989). Here, neurofilament protein colocalizes with desmin and HNK-1, and in the embryonic ventricle it is abundant in the interiorly localized ventricular trabecular component.

VI. Perspectives

The development of the conduction system has always been a quarrelsome topic. This controversy originated in the indiscriminate extrapolation of adult concepts of the mammalian conduction system onto the embryonic situation, perhaps representing a modern example of the preformationist's mode of thought. In the formed mammalian heart, the conduction system and working myocardium are generally considered as two separate entities. As described in this chapter, this is difficult to reconcile with the embryonic heart, in which no distinctive conduction system is present and yet coordinated contraction of atria and ventricles is achieved. Moreover, the mammalian condition has been overemphasized compared to that of lower vertebrates, in which a well-defined cardiac conduction system is also lacking, similar to the mammalian embryonic heart, and yet proper cardiac contraction is realized. Our primary message is that the formation of the vertebrate cardiac building plan encompasses of necessity the formation of the "cardiac conduction system." It involves the proper formation and arrangement of myocardial building blocks within the tubular heart that display posteroanterior polarity of pacemaker activity. The myocardial building blocks represent disparate cardiomyocyte populations with distinct functional and transcriptional qualities. They allow for the sequential activation of atria and ventricles and for the activation of the ventricles from apex to base, irrespective of whether a histologically distinct conduction system is present (mammals) or not (lower vertebrates). How these building blocks are specified and remodeled, what the nature is of the interactions that occur between surrounding myocardial and nonmyocardial cells, and how some primary myocardial cells differentiate into nodal tissue instead of working myocardium is largely unknown. In addition, much remains to be learned about the factors that specify the formation and distinct transcriptional potential of the separate building blocks of the early heart tube. Molecular genetics and morphology, in combination with lineage analysis, will eventually uncover the key pathways for the fashioning of the vertebrate heart.

Acknowledgments

We are grateful to Mr. Frits de Jong for stimulating discussions and to Dr. Daniel Gros for providing Fig. 7. Research in the authors' lab is supported by the Netherlands Heart Foundation (NHS).

References

Alanis, J., Benitez, D., Lopez, E., and Martinez-Palomo, A. (1973). Impulse propagation through the cardiac junctional regions of the axolotl and the turtle. *Jpn. J. Physiol.* **23,** 149–164.

Anand-Srivastava, M. B., Thibault, G., Sola, C., Fon, E., Ballak, M., Charbonneau C., Haile-Meskel, H., Garcia, R., Genest, J., and Cantin, M. (1989). Atrial natriuretic factor in Purkinje fibers of rabbit heart. *Hypertension* **13,** 789–798.

Aoyama, N., Tamaki, H., Kikawada, R., and Yamashina, S. (1995). Development of the conduction system in the rat heart as determined by Leu-7 (HNK-1) immunohistochemistry and computer graphics reconstruction. *Lab. Invest.* **72,** 355–366.

Arguello, C., Alanis, J., Pantoja, O., and Valenzuela, B. (1986). Electrophysiological and ultrastructural study of the atrioventricular canal during the development of the chick embryo. *J. Mol. Cell. Cardiol.* **18,** 499–510.

Arguello, C., Alanis, J., and Valenzuela, B. (1988). The early development of the atrioventricular node and bundle of His in the embryonic chick heart. An electrophysiological and morphological study. *Development* **102,** 623–637.

Baldwin, H. S., Jensen, K. L., and Solursh, M. (1991). Myogenic differentiation of the precardiac mesoderm in the rat. *Differentiation* **47,** 163–172.

Barry, A. (1942). Intrinsic pulsation rates of fragments of embryonic chick heart. *J. Exp. Zool.* **91,** 119–130.

Bastide, B., Neyses, L., Ganten, D., Paul, M., Willecke, K., and Traub,

O. (1993). Gap junction protein connexin 40 is preferentially expressed in vascular endothelium and conductive bundles of rat myocardium and is increased under hypertensive conditions. *Circ. Res.* **73,** 1138–1149.

Benninghoff, A. (1923). Über die Beziehungen des Reitzles tungssystems under papillar muskeln zu den Konturfasern des Herzschlanches. *Verh. Anat. Gesellsch.* **57,** 125–208.

Bouman, L. N., Gerlings, E. D., Biersteker, P. A., and Barke, F. I. M. (1968). Pacemaker shift in the sinoatrial node during vagal stimulation. *Pflügers Arch.-Eur. J. Physiol.* **302,** 255–267.

Bouman, L. N., Mackaay, A. J. C., Bleeker, W. K., and Becker, A. E. (1978). Pacemaker shifts in the sinus node: Effects of vagal stimulation, temperature and reduction of extracellular calcium. In *The Sinus Node* (F. I. M. Barke, Ed.), p. 245. Nijhoff, Den Haag, The Netherlands.

Brock R. C. (1955). Control mechanisms in the outflow tract of the right ventricle in health and disease. *Guys Hosp. Reports* **104,** 356–388.

Brock, R. C. (1964). In *Gray's Anatomy.* Longmans, London.

Brunton, T. S., and Fayrer, J. (1874). Note on independent pulsation of the pulmonary veins and vena cava. *Proc. R. Soc. of London B* **25,** 174–176.

Canale, E. D., Campbell, G. R., Smolich, J. J., and Campbell J. H. (1986). *Cardiac muscle.* Springer–Verlag, Berlin.

Chan-Thomas, P. S., Thompson, R. P., Robert, B., Yacoub, M. H., and Barton, P. J. R. (1993). Expression of homeobox genes Msx-1 (Hox-7) and Msx-2 (Hox-8) during cardiac development in the chick. *Dev. Dyn.* **197,** 203–216.

Cheung, D. W. (1980). Electrical activity of the pulmonary vein and its interaction with the right atrium in the guinea pig. *J. Physiol.* **314,** 445–456.

Chuck, E. T., Freeman, D. M., Watanabe, M., and Rosenbaum, D. S. (1997). Changing activation sequence in the embryonic chick heart: Implications for the development of the His-Purkinje system. *Circ. Res.* **81,** 470–476.

Dahl, E., Winterhager, E., Traub, O., and Willecke, K. (1995). Expression of gap junction genes, connexin40 and connexin43, during fetal mouse development. *Anat. Embryol.* **191,** 267–278.

Davies, F. (1930). The conducting system of the bird's heart. *J. Anat.* **64,** 129–146.

Davies, M. J., Anderson, R. H., and Becker, A. E. (1983). The Conduction System of the Heart. Butterworths, London.

Dechesne, C. A., Leger, J. O. C., and Leger, J. J. (1987). Distribution of α- and β-myosin heavy chains in the ventricular fibers of the postnatal developing rat. *Dev. Biol.* **123,** 169–178.

de Groot, I. J. M., Hardy, G. P. M. A., Sanders, E., Los, J. A., and Moorman, A. F. M. (1985). The conducting tissue in the adult chicken atria: A histological and immunohistochemical analysis. *Anat. Embryol.* **172,** 239–245.

de Groot, I. J. M., Sanders, E., Visser, S. D., Lamers, W. H., de Jong, F., Los, J. A., and Moorman, A. F. M. (1987). Isomyosin expression in developing chicken atria: A marker for the development of conductive tissue? *Anat. Embryol.* **176,** 515–523.

de Groot, I. J. M., Wessels, A., Virágh, S., Lamers, W. H., and Moorman, A. F. M. (1988). The relation between isomyosin heavy chain expression pattern and the architecture of sinoatrial nodes in chicken, rat and human embryos. *In "Sarcomeric* and *nonsarcomeric* Muscles: Basic and Applied Research Prospects for the 90s" (U. Carraro Ed.), p. 305. Unipress, Padova.

de Haan, R. L. (1959). Cardia bifida and the development of pacemaker function in the early chick heart. *Dev. Biol.* **1,** 586–602.

de Jong, F., Opthof, T., Wilde, A. A. M., Janse, M. J., Charles, R., Lamers, W. H., and Moorman, A. F. M. (1992). Persisting zones of slow impulse conduction in developing chicken hearts. *Circ. Res.* **71,** 240–250.

De Jong, F., Virágh, S., and Moorman, A. F. M. (1997). Cardiac development: A morphologically integrated molecular approach. *Cardiol. Young* **7,** 131–146.

Delorme, B., Dahl, E., Jarry-Guichard, T., Marics, I., Briand, J. P., Willecke, K., Gros, D., and Théveniau-Ruissy, M. (1995). Developmental regulation of connexin40 gene expression in mouse heart correlates with the differentiation of the conduction system. *Dev. Dyn.* **204,** 358–371.

Delorme, B., Dahl, E., Jarry-Guichard, T., Briand, J. P., Willecke, K., Gros, D., and Théveniau-Ruissy, M. (1997). Expression pattern of connexin gene products at the early developmental stages of the mouse cardiovascular system. *Circ. Res.* **81,** 423–437.

Eriksson, A., Thornell, L. E., and Stigbrand, T. (1979). Skeletin immunoreactivity in heart Purkinje fibers from several species. *J. Histochem. Cytochem.* **27,** 1604–1609.

Forsgren, S., Thornell, L. E., and Eriksson, A. (1980). The development of the Purkinje fibre system in the bovine fetal heart. *Anat. Embryol.* **159,** 125–135.

Franco, D., Jing, Y., Wagenaar, G. T. M., Lamers, W. H., and Moorman, A. F. M. (1996). The trabecular component of the embryonic ventricle. In *The Developing Heart* (B. Ost'ádal, M. Nagano, N. Takeda, and N. S. Dhalla, Eds.), p. 51. Lippincott-Raven, New York.

Franco, D., Kelly, R., Moorman, A. F. M., and Buckingham, M. (1997). Regionalised transcriptional domains of myosin light chain 3F transgenes in the embryonic mouse heart: Morphogenetic implications. *Dev. Biol.* **188,** 17–33.

Franco, D., Lamers, W. H., and Moorman, A. F. M. (1998). Patterns of gene expression in the developing myocardium: Towards a morphologically integrated transcriptional model. *Cardiovasc. Res.* **38,** 25–53.

Fromaget, C., El Aoumari, A., and Gros, D. (1992). Distribution pattern of connexin 43, a gap junctional protein, during the differentiation of mouse heart myocytes. *Differentiation* **51,** 9–20.

González-Sánchez, A., and Bader, D. (1985). Characterization of a myosin heavy chain in the conductive system of the adult and developing chicken heart. *J. Cell Biol.* **100,** 270–275.

Goor, D. A., Dische, R., and Lillehei, C. W. (1972). The conotruncus. I. Its normal inversion and conus absorption. *Circulation* **46,** 375–384.

Gorza, L., and Vitadello, M. (1989). Distribution of conduction system fibres in the developing and adult rabbit heart, revealed by an antineurofilament antibody. *Circ. Res.* **65,** 360–369.

Gorza, L., Sartore, S., Thornell, L. E., and Schiaffino, S. (1986). Myosin types and fiber types in cardiac muscle. III. Nodal conduction tissue. *J. Cell Biol.* **102,** 1758–1766.

Gorza, L., Schiaffino, S., and Vitadello, M. (1988). Heart conduction system: A neural crest derivative? *Brain Res.* **457,** 360–366.

Gorza, L., Ausoni, S., Mercial, N., Hastings, K. E. M., and Schiaffino, S. (1993). Regional differences in troponin I isoform switching during rat heart development. *Dev. Biol.* **156,** 253–264.

Gorza, L., Vettore, S., and Vitadello, M. (1994). Molecular and cellular diversity of heart system myocytes. *Trends Cardiovasc. Med.* **4,** 153–159.

Goss, C. M. (1942). The physiology of the embryonic mammalian heart before circulation. *Am. J. Physiol.* **137,** 146–152.

Gourdie, R. G., Green, R. C., Severs, N. J., and Thompson, R. P. (1992). Immunolabeling patterns of gap junction connexins in the developing and mature rat heart. *Anat. Embryol.* **185,** 363–378.

Gourdie, R. G., Severs, N. J., Green, C. R., Rothery, S., Germroth, P., and Thompson, R. P. (1993). The spatial distribution and relative abundance of gap-junctional connexin40 and connexin43 correlate to functional properties of components of the cardiac atrioventricular conduction system. *J. Cell Sci.* **105,** 985–991.

Gros, D., Mocquard, J. P., Challice, C. E., and Schrével, J. (1978). Formation and growth of gap junctions in mouse mycoardium during ontogenesis. *J. Cell Sci.* **30,** 45–61.

Gros, D., Jarry-Guichard, T., Ten Velde, I., De Mazière, A., van Kempen, M. J. A., Davoust, J., Briand, J. P., Moorman, A. F. M., and Jongsma, H. J. (1994). Restricted distribution of connexin 40, a gap junctional protein, in mammalian heart. *Circ. Res.* **74,** 839–851.

Hansson, M., and Forsgren, S. (1993). Presence of immunoreactive atrial natriuretic peptide in nerve fibres and conduction system of the bovine heart. *Anat. Embryol.* **188,** 331–337.

Hirota, A., Fujii, S., and Kamino, K. (1979). Optical monitoring of spontaneous electrical activity of 8-somite embryonic chick heart. *Jpn. J. Physiol.* **29,** 635–639.

Hirota, H., Sakai, T. M., Fujii, S., and Kamino, K. (1983). Initial development of conduction patterns of spontaneous action potential in early embryonic precontractile chick heart. *Dev. Biol.* **99,** 517–523.

Ikeda, T., Iwasaki, K., Shimokawa, I., Sakai, H., Ito, H., and Matsuo, T. (1990). Leu-7 immunoreactivity in human and rat embryonic hearts, with special reference to the development of the conduction tissue. *Anat. Embryol.* **182,** 553–562.

Janse, M. J., and Anderson, R. H. (1974). Specialized internodal atrial pathways—Fact or fiction? *Eur. J. Cardiol.* **2,** 117–136.

Jing, Y., Markman, M. W. M., Wagenaar, G. T. M., Blommaart, P. J. E., Moorman, A. F. M., and Lamers, W. H. (1997). Expression of the smooth-muscle proteins alpha smooth-muscle actin and calponin, and of the intermediate filament protein desmin are parameters of cardiomyocyte maturation in the prenatal rat heart. *Anat. Rec.* **249,** 495–505.

Joyner, R. W., and van Capelle, F. J. L. (1986). Propagation through electrically coupled cells. How a small SA node drives a large atrium. *Biophys. J.* **50,** 1157–1164.

Kamino, K. (1991). Optical approaches to ontogeny of electrical activity and related functional organization during early heart development. *Physiol. Rev.* **71,** 53–91.

Kamino, K., Hirota, A., and Fujii, S. (1981). Localization of pacemaking activity in early embryonic heart monitored using voltage-sensitive dye. *Nature* **290,** 595–597.

Kirby, M. L., Kumiski, D. H., Myers, T., Cerjan, C., and Mishima, N. (1993). Backtransplantation of cardiac neural crest cells cultured in LIF rescues heart development. *Dev. Dyn.* **198,** 296–311.

Komuro, I., Nomoto, K., Sugiyama, T., Kurabayashi, M., Takaku, F., and Yazaki, Y. (1987). Isolation and characterization of myosin heavy chain isozymes of the bovine conduction system. *Circ. Res.* **61,** 859–865.

Kuhl, H., and van Hasselt, J. C. (1822). Uittreksels uit brieven van de Heeren Kuhl en van Hasselt, aan de Heeren CJ Timminck, Th van Swinderen en W de Haan. *Algemeene Konst en Letter-Bode* **1,** 115–117.

Kuro-o, M., Tsuchimochi, H., Ueda, S., Takaku, F., and Yazaki, Y. (1986). Distribution of cardiac myosin isozymes in human conduction system. *J. Clin. Invest.* **77,** 340–347.

Lamers, W. H., de Jong, F., de Groot, I. J. M., and Moorman, A. F. M., (1991). The development of the avian conduction system. *Eur. J. Morphol.* **29,** 233–253.

Lamers, W. H., Wessels, A., Verbeek, F. J., Moorman, A. F. M., Virágh, S., Wenink, A. C. G., Gittenberger-de Groot, A. C., and Anderson, R. H. (1992). New findings concerning ventricular septation in the human heart—their implications for maldevelopment. *Circulation* **86,** 1194–1205.

Li, Z., Marchand, P., Humbert, J., Babinet, C., and Paulin, D. (1993). Desmin sequence elements regulating skeletal muscle-specific expression in transgenic mice. *Development* **117,** 947–959.

Lieberman, M., and Paes de Carvalho, A. (1965b). The electrophysiological organization of the embryonic chick heart. *J. Gen. Physiol.* **49,** 351–363.

Lieberman, M., and Paes de Carvalho, A. (1965). The spread of excitation in the embryonic chick heart. *J. Gen. Physiol.* **49,** 365–379.

Liebman, J. (1985). Are there internodal tracts? Yes. *Int. J. Cardiol.* **7,** 174–185.

Lyons, G. E. (1994). In situ analysis of the cardiac muscle gene program during embryogenesis. *Trends Cardiovasc. Med.* **4,** 70–77.

Masani, F. (1986). Node-like cells in the myocardial layer of the pulmonary vein of rats: An ultrastructural study. *J. Anat.* **145,** 133–142.

McGuire, M. A., De Bakker, J. M. T., Vermeulen, J. T., Moorman, A. F. M., Loh, P., Thibault, B., Vermeulen, J. L. M., Becker, A. E., and Janse, M. J. (1996). Atrioventricular junctional tissue. Discrepancy between histological and electrophysiological characteristics. *Circulation* **94,** 571–577.

Meda, E., and Ferroni, A. (1959). Early functional differentiation of heart muscle cells. *Experientia* **15,** 427–428.

Moorman, A. F. M., de Jong, F., Denyn, M. M. F. J., and Lamers, W. H. (1998). Development of the cardiac conduction system. *Circ. Res.* **82,** 629–644.

Moorman, A. F. M., and Lamers, W. H. (1994). Molecular anatomy of the developing heart. *Trends Cardiovasc. Med.* **4,** 257–264.

Moorman, A. F. M., Vermeulen, J. L. M., Koban, M. U., Schwartz, K., Lamers, W. H., and Boheler, K. R. (1995). Patterns of expression of sarcoplasmic reticulum Ca^{2+} ATPase and phospholamban mRNAs during rat heart development. *Circ. Res.* **76,** 616–625.

Nakagawa, M., Thompson, R. P., Terracio, L., and Borg, T. K. (1993). Developmental anatomy of HNK-1 immunoreactivity in the embryonic rat heart: Co-distribution with early conduction tissue. *Anat. Embryol.* **187,** 445–460.

Paff, G. H., and Boucek, R. J. (1975). Conal contributions to the electrocardiogram of chick embryo hearts. *Anat. Rec.* **182,** 169–174.

Paff, G. H., Boucek, R. J., and Glander, T. P. (1966). Acetylcholinesterase-acetylcholine, an enzyme system essential to rhythmicity in the preneural embryonic chick heart. *Anat. Rec.* **154,** 675–684.

Paff, G. H., Boucek, R. J., and Harrell, T. C. (1968). Observations on the development of the electrocardiogram. *Anat. Rec.* **160,** 575–582.

Patten, B. M. (1949). Initiation and early changes in the character of the heartbeat in vertebrate embryos. *Physiol. Rev.* **29,** 31–47.

Patten, B. M., and Kramer, T. C. (1933). The initiation of contraction in the embryonic chicken heart. *Am. J. Anat.* **53,** 349–375.

Pucci, A., Wharton, J., Arbustini, E., Grasso, M., Diegoli, M., Needleman, P., Viganò, M., Moscoso, G., and Polak, J. M. (1992). Localization of brain and atrial natriuretic peptide in human and porcine heart. *Int. J. Cardiol.* **34,** 237–247.

Randall, D. J. (1970). The circulatory system. In *Fish Physiology* (W. S. Hoar and D. J. Randall, Eds.), Academic Press, New York.

Sakai, T., Hirota, A., Fujii, S., and Kamino, K. (1983). Flexibility of regional pacemaking priority in early embryonic heart monitored by simultaneous optical recording of action potentials from multiple sites. *Jpn. J. Physiol.* **33,** 337–350.

Sartore, S., Pierobon-Bormioli, S., and Schiaffino, S. (1978). Immunohistochemical evidence for myosin polymorphism in the chicken heart. *Nature* **274,** 82–83.

Sartore, S., Gorza, L., Pierobon-Bormioli, S., Dalla Libera, L., and Schiaffino, S. (1981). Myosin types and fiber types in cardiac muscle; 1. Ventricular myocardium. *J. Cell Biol.* **88,** 226–233.

Satin, J., Fujii, S., and de Haan, R. L. (1988). Development of cardiac heartbeat in early chick embryos is regulated by regional cues. *Dev. Biol.* **129,** 103–113.

Schiaffino, S., Gorza, L., and Ausoni, S. (1993). Troponin isoform switching in the developing heart and its functional consequences. *Trends Cardiovasc. Med.* **3,** 12–17.

Shirinsky, V. P., Biryukov, K. G., Hettasch, J. M., and Sellers, J. R. (1992). Inhibition of the relative movement of actin and myosin by caldesmon and calponin. *J. Biol. Chem.* **267,** 15886–15892.

Skepper, J. N. (1989). An immunocytochemical study of the sinuatrial node and atrioventricular conducting system of the rat for atrial natriuretic peptide distribution. *Histochem. J.* **21,** 72–78.

Streeter, G. L. (1987). *In* "Developmental Stages in Human Embryos" (R. O'Rahilly and F. Müller, Eds.), Carnegie Institute of Washington, Washington, DC.

Szabó, E., Virágh, S., and Challice, C. E. (1986). The structure of the atrioventricular conducting system in the avian heart. *Anat. Rec.* **215,** 1–9.

Thompson, R. P., Simson, J. A. V., and Currie, M. G. (1986). Atriopeptin distribution in the developing rat heart. *Anat. Embryol.* **175,** 227–233.

Thornell, L. E., and Eriksson, A. (1981). Filament systems in the Purkinje fibers of the heart. *Am. J. Physiol.* **241,** H291–H305.

Toshimori, H., Toshimori, K., Oura, C., and Matsuo, H. (1987). Immunohistochemical study of atrial natriuretic polypeptides in the embryonic, fetal and neonatal rat heart. *Cell Tissue Res.* **248,** 627–633.

Toshimori, H., Toshimori, K., Oura, C., Matsuo, H., and Matsukara, S. (1988). Immunohistochemical identification of Purkinje fibers and transitional cells in a terminal portion of the impulse-conducting system of porcine heart. *Cell Tissue Res.* **253,** 47–53.

van Kempen, M. J. A., Fromaget, C., Gros D., Moorman, A. F. M., and Lamers, W. H. (1991). Spatial distribution of connexin-43, the major cardiac gap junction protein, in the developing and adult rat heart. *Circ. Res.* **68,** 1638–1651.

van Kempen, M. J. A., Ten Velde, I., Wessels, A., Oosthoek, P. W., Gros, D., Jongsma, H. J., Moorman, A. F. M., and Lamers, W. H. (1995). Differential connexin distribution accommodates cardiac function in different species. *Microsc. Res. Techn.* **31,** 420–436.

van Kempen, M. J. A., Vermeulen, J. L. M., Moorman, A. F. M., Gros, D. B., Paul, D. L., and Lamers, W. H. (1996). Developmental changes of connexin40 and connexin43 mRNA-distribution patterns in the rat heart. *Cardiovasc. Res.* **32,** 886–900.

van Mierop, L. H. S. (1967). Localization of pacemaker in chick embryo heart at the time of initiation of heartbeat. *Am. J. Physiol.* **212,** 407–415.

Vassall-Adams, P. R. (1982). The development of the atrioventricular bundle and its branches in the avian heart. *J. Anat.* **134,** 169–183.

Virágh, S., and Challice, C. E. (1977). The development of the conduction system in the mouse embryo heart. II. Histogenesis of the atrioventricular node and bundle. *Dev. Biol.* **56,** 397–411.

Virágh, S., and Challice, C. E. (1980). The development of the conduction system in the mouse embryo heart. III. The development of sinus muscle and sinoatrial node. *Dev. Biol.* **80,** 28–45.

Virágh, S., and Porte, A. (1973). The fine structure of the conducting system of the monkey heart (macaca mulatta). I. The sino-atrial node and internodal connections. *Zeitschr. Zellforsch.* **145,** 191–211.

Virágh, S., Szabó, E., and Challice, C. E. (1989). Formation of primitive myo- and endocardial tubes in the chicken embryo. *J. Mol. Cell. Cardiol.* **21,** 123–137.

Virtanen, I., Närvänen, O., and Thornell, L. E. (1990). Monoclonal antibody to desmin purified from cow Purkinje fibers reveals a cell-type specific determinant. *FEBS Lett.* **267,** 176–178.

Vitadello, M., Matteoli, M., and Gorza, L. (1990). Neurofilament proteins are co-expressed with desmin in heart conduction system myocytes. *J. Cell Sci.* **97,** 11–21.

Vitadello, M., Vettore, S., Lamar, E., Chien, K. R., and Gorza, L. (1996). Neurofilament M mRNA is expressed in conduction system myocytes of the developing and adult rabbit heart. *J. Mol. Cell. Cardiol.* **28,** 1833–1844.

Wessels, A., Vermeulen, J. L. M., Verbeek, F. J., Virágh, S., Kálmán, F., Lamers, W. H., and Moorman, A. F. M. (1992). Spatial distribution of "tissue-specific" antigens in the developing human heart and skeletal muscle: III. An immunohistochemical analysis of the distribution of the neural tissue antigen G1N2 in the embryonic heart; implications for the development of the atrioventricular conduction system. *Anat. Rec.* **232,** 97–111.

Wharton, J., Anderson, R. H., Springall, D., Power, R. F., Rose, M., Smith, A., Espejo, R., Khagani, A., Wallwork, J., Yacoub, M. H., and Polak, J. M. (1988). Localization of atrial natriuretic peptide immunoreactivity in the ventricular myocardium and conduction system of the human fetal and adult heart. *Br. Heart J.* **60,** 267–274.

Yancey, S. B., Biswal, S., and Revel, J. P. (1992). Spatial and temporal patterns of distribution of the gap junction protein connexin43 during mouse gastrulation and organogenesis. *Development* **114,** 203–212.

Zeller, R., Bloch, K. D., Williams, B. S., Arceci, R. J., and Seidman, C. E. (1987). Localized expression of the atrial natriuretic factor gene during cardiac embryogenesis. *Genes Dev.* **1,** 693–698.

Zhu, L., Lyons, G. E., Juhasz, O., Joya, J. E., Hardeman, E. C., and Wade, R. (1995). Developmental regulation of troponin I isoform genes in striated muscles of transgenic mice. *Dev. Biol.* **169,** 487–503.

13

Retinoids in Heart Development

Steven W. Kubalak* and Henry M. Sucov†

*Cardiovascular Developmental Biology Center, Department of Cell Biology and Anatomy, Medical University of South Carolina, Charleston, South Carolina 29425

†Department of Cell and Neurobiology, Institute for Genetic Medicine, University of Southern California School of Medicine, Los Angeles, California 90033

I. Introduction

The deleterious consequences of nutritional deficiency on the adult visual system have been recognized since antiquity, when Hippocrates wisely suggested eating beef liver as a cure for night blindness (Mandel and Cohn, 1985). In the early parts of this century, defined animal diets lacking what is now recognized as vitamin A were formulated which resulted in these same visual deficiencies, leading to the biochemical isolation of plant carotenes and retinyl esters derived from animal sources as potent corrective agents. The term vitamin A refers to a large class of compounds which have bioactive properties, of which carotenoids (β-carotene and similar compounds) and retinoids (retinol and its derivatives) are the two primary subclasses of natural occurring compounds.

Clues to the function of retinoids in heart development came initially from studies of the effects of vitamin A-deficient diets on the pups of pregnant rats described by Warkany and colleagues almost 50 years ago (Wilson and Warkany, 1949; Wilson *et al.*, 1953). Since that time, numerous laboratories have published studies addressing specific questions regarding the effects of retinoids on the developing embryonic avian and mammalian heart. Two general experimental strategies have been employed in these efforts. Retinoid-deficient diets (or loss-of-function genetic manipulation of retinoid receptors; described later) cause heart defects because of failures in endogenous retinoid-dependent developmental processes. Alternatively, ectopic administration of an excess of retinoids results in defects which reflect the competence of the developing heart to respond to retinoids, although this response might not necessarily occur in normally developing embryos or might not occur to as great an extent. The data suggest that retinoids influence multiple events throughout the process of cardiogenesis, and that the effects observed are dependent on when during development the embryo is experimentally manipulated.

II. Vitamin A Metabolism

Dietary vitamin A is converted to retinol (vitamin A alcohol) and stored as retinyl esters. Retinol itself serves no known function except as a metabolic precursor. Successive oxidative reactions convert retinol to retinaldehyde and then to retinoic acid (RA). The form of vitamin A used in the visual system is retinaldehyde, which serves as a substrate for forming light-sensitive rhodopsin. For a long time, it was assumed that essentially all other biological properties of vitamin A are represented by retinoic acid, or downstream metabolites, based on the observation that retinoic acid could restore viability, but not visual function, to vitamin A-deficient animals (Thompson *et al.,* 1964). In recent years, several additional bioactive retinoids have been discovered. 9-*cis* RA (Heyman *et al.,* 1992) and 3,4-didehydro-RA (Thaller and Eichele, 1990) have been demonstrated to be bioactive forms of vitamin A rather than inactive by-products of all-*trans* RA metabolism. A number of retinol metabolites oxidized at the C-4 position have been shown (Achkar *et al.,* 1996; Blumberg *et al.,* 1996) to be bioactive molecules in their own right rather than by-products of retinol metabolism. Finally, a novel class of "retroretinoids" has been discovered which regulate certain aspects of immune function (Buck *et al.,* 1991, 1993).

As described later, the major pathway in which vitamin A has biological consequences is through interaction with receptors of the nuclear receptor family of ligand-activated transcription factors. All-*trans* RA, 9-*cis* RA, 3,4-didehydro-RA, 4-oxo-retinaldehyde, and 4-oxo-retinol have all been demonstrated to act as ligands for the retinoid receptors. All-*trans* RA is generally believed to be the most commonly utilized bioactive retinoid; the biological significance of some of these additional compounds remains to be demonstrated. Perhaps the most insightful evidence comes from the concentration of these compounds in tissues. For example, 9-*cis* RA is present only in trace amounts in most tissues (postnatal and embryonic), suggesting that its biological role may be restricted to very specific tissues or times. In early *Xenopus* embryos, the major retinoid is 4-oxo-retinaldehyde, with very little if any all-*trans* or 9-*cis* RA present (Blumberg *et al.,* 1996) suggesting that 4-oxo-retinaldehyde is the major bioactive retinoid in this system at this time. There are no data available which describe the distribution of retinoids in the developing heart.

Surprisingly little is known of the enzymes which synthesize bioactive retinoids from retinol. A number of dehydrogenases and oxoreductases have been identified with retinoid selectivity or specificity (Duester, 1996; Boerman and Napoli, 1996; Niederreither *et al.,* 1997). There is disagreement as to the role for these enzymes in general retinoid metabolism, and the extent to which any of these is involved in heart development is currently under investigation (see Chapter 28).

III. Vitamin A Receptors

All known biological requirements for vitamin A, with the exception of retinaldehyde in the visual system and retroretinoids in the immune system, appear to be mediated by the RA receptors. These receptors are members of the nuclear receptor family of ligand-dependent transcription factors (Evans, 1988), which act by recognizing specific DNA sequences in the promoters of responsive genes and directly induce transcriptional activation.

The retinoid receptor family is divided into two subfamilies, the RARs and the RXRs, each with three members—α, β, and γ. Both subfamilies have ligand affinities in the nanomolar range, as is generally the case for nuclear receptor–ligand interactions. All the known natural ligands are recognized by the RARs, whereas it appears that a 9-*cis* conformation is required for the RXRs to recognize natural ligands.

The RXR subgroup of the retinoid receptors has an additional, and critical, biological function. Many of the nonsteroid nuclear receptors (including the RARs, the thyroid hormone and vitamin D receptors, and others) bind DNA as a heterodimeric complex with RXR. The RXRs, in other words, serve as a common heterodimeric partner in mediating several different hormonal signaling pathways, independent of their role as retinoid receptors (Yu *et al.,* 1991). It is generally assumed that retinoid signaling (through either RAR–RXR heterodimers or RXR homodimers) can be modulated by the concentration of other ligands and receptors via competition for RXRs.

Each of the six retinoid receptor genes is expressed as multiple transcripts, differing in the 5′ untranslated region and amino-terminal coding sequence, and fused to common exons encoding the bulk of the respective receptor protein (Leid *et al.*, 1992). These isoforms are numbered 1, 2, etc. (i.e., RARα1, RARα2, etc.). These transcripts originate from alternative independent promoters, resulting in an overall gene expression pattern which represents the sum of two or more isoform transcripts.

Several *in situ* hybridization studies have characterized the expression patterns of the various receptor genes or gene isoforms in the mouse and chick embryos (Ruberte *et al.*, 1991; Mangelsdorf *et al.*, 1992; Dolle *et al.*, 1994). The retinoid receptors are widely expressed in the embryo, consistent with the multiple roles of vitamin A in development. The RARα and RXRβ genes are ubiquitously expressed in the embryo and in the adult, and the RXRα gene is ubiquitously expressed (although at a very low level) in the early to midgestation embryo [i.e., prior to embryonic day (E) 16.5, a time when most aspects of morphogenesis are complete]. The remaining three receptor genes show a more restricted expression pattern. In the developing heart, RARγ has the most distinctive expression pattern, being highly expressed in the endocardial cushions (Ruberte *et al.*, 1991).

A variety of mutations in the vitamin A receptor genes have been stably introduced into the mouse germline by targeted gene disruption. Mutations are available which eliminate all transcripts from a gene or which eliminate individual gene isoforms. Mutations are now available in all six genes, as summarized in Table I. These genetic studies lead to several conclusions which are relevant to the interpretation of receptor gene expression patterns. First, many if not all of the receptor genes are expressed in the developing heart, as evidenced by cardiac phenotypes in their absence, even if the level of expression is below detection by *in situ* hybridization. This is true for RARβ, RARγ, and possibly for RXRγ; RARα, RXRα, and RXRβ are detectable a priori. Second, the spatial and quantitative patterns of receptor gene expression (as detected by *in situ* hybridization) do not obviously correlate to the genetically defined function of these genes. Thus, RXRα and RXRβ are comparably expressed at a similar very low level in the early ventricles, even though RXRα alone is required for ventricular maturation. Similarly, by genetic criteria both RARα and RARγ are comparably involved in ventricular chamber development, although RARα mRNA is much more abundantly expressed. As currently understood, the pattern of receptor gene expression is not of primary significance in understanding the involvement of vitamin A in heart development.

A parameter more important than receptor expression, and yet one which is barely understood, is likely to be the presentation of ligand for these receptors. Local production and diffusion from a source is believed to be the primary manner in which retinoids are used to signal to target tissues (see Chapter 28), although retinoids can also be distributed via embryonic circulation in a classical endocrine manner. In addition, in mammalian embryos, transplacental transfer of RA from maternal circulation may also be a component of normal fetal retinoid biology. Certainly, in mammalian teratogenic studies in which RA is administered to pregnant females, embryonic defects arise by placental transfer and consequent inappropriate activation of embryonic retinoid-responsive pathways.

Table I Mouse Germline Mutations in the Retinoic Acid Receptor Genes

Mutations	Reference
Complete gene mutations	
RARα	Lufkin *et al.* (1993)
RARβ	Luo *et al.* (1995)
RARγ	Lohnes *et al.* (1993); Iulianella and Lohnes (1997)
RXRα	Sucov *et al.* (1994); Kastner *et al.* (1994); Dyson *et al.* (1995); Gruber *et al.* (1996)
RXRβ	Kastner *et al.* (1996)
RXRγ	Krezel *et al.* (1996)
Isoform-specific mutations	
RARα1	Li *et al.* (1993); Lufkin *et al.* (1993)
RARβ2	Mendelsohn *et al.* (1994b)
RARγ2	Lohnes *et al.* (1993)

Note: Various combinations of receptor mutations with cardiovascular phenotypes are described in Kastner *et al.* (1994); Mendelsohn *et al.* (1994a); Luo *et al.* (1996); and Lee *et al.* (1997).

IV. Retinoids in Precardiac Fields

The bilateral precardiac fields are positioned on either side of Henson's node in the primitive streak stage embryo. These fields coalesce to form the primitive heart tube which immediately begins to form the normally right-directed looped heart. Factors that influence the heart-forming fields during development are of great interest since it is at this time when the disposition of the left–right asymmetry is established (see Chapters 21 and 22). An endogenous local source of RA and other peptide growth factors is present in the primitive streak stage embryo, i.e., Henson's node in the chick, and as such any of these factors may influence the patterning of the heart tube. Indeed, RA exposure causes cardiac malformations in the early chick embryo (Osmond *et al.*, 1991; Chen and Solursh, 1992; Dickman and Smith, 1996). Placing RA-soaked beads to the right of Henson's node randomizes looping and results in a significant percentage of hearts looping to the left instead of the right (Dickman and Smith, 1996). Other malformations observed after RA treatment include situs inversus and cardia bifida. It is noteworthy that similar malformations are observed in the vitamin A-deficient chick and quail embryos (Thompson, 1969; Thompson *et al.*, 1969; Heine *et al.*, 1985; Dersch and Zile, 1993) and together indicates that vitamin A and its metabolites can have a profound influence on early stages of heart morphogenesis.

In the chick, the asymmetry of the precardiac fields is reflected in the expression of several proteins, including two matrix proteins [JB3, which may be a fibrillin isotype, or fibrillin-associated protein, and the heart-specific lectin-associated matrix protein hLAMP-1 (Smith *et al.*, 1997; see Chapter 10)], the morphogen sonic hedgehog (Shh), and the activin receptor IIa (Act-RIIa) (Levin *et al.*, 1995). Both JB3 and hLAMP-1 are expressed in the endocardial cushion tissues and have been implicated in endothelial-to-mesenchyme transformation (Wunsch *et al.*, 1994; Sinning and Hewitt, 1996). Normal expression of JB3 at Hamburger and Hamilton (HH) stages 3 or 4 is symmetric within the heart fields of the embryo (Wunsch *et al.*, 1994) as is the expression of the homeobox transcription factor cNkx2.5 at stages 7 or 8 (Schultheiss *et al.*, 1995; see Chapters 4 and 7). By HH stage 5, JB3 expression is greater in the right cardiac field, with hLAMP-1 expression predominating in the left (Smith *et al.*, 1997). The expression of the chick homeobox gene cNkx2.5 remains symmetric at this time of development (Schultheiss *et al.*, 1995). Smith *et al.* (1977) showed that the left and right precardiac fields in the chick differ in their sensitivity to RA, as assessed by the expression of JB3 and hLAMP-1. When RA treatment alters looping, the levels of these matrix proteins lose their corresponding asymmetry. RA treatment does not alter the expression patterns of nodal and sonic hedgehog in the chick, and null mutant mice for both Shh (Chiang *et al.*, 1996) and Act-RIIa Matzuk *et al.*, 1995) display normal asymmetry of cardiac looping. These studies suggest that RA can influence cardiac morphogenesis prior to the formation of the linear heart tube, and that RA is involved in normal processes of laterality and looping. JB3 and hLAMP-1 may play an important role in these processes.

Other gene products have also been implicated in the initial orientation of the heart tube, as evidenced by looping abnormalities at later stages of development. These include the situs inversus mutation (Yokoyama *et al.*, 1993), the iv mutation (VanKeuren *et al.*, 1991), and an extracellular matrix-associated malformation (Yost, 1992; see Chapters 21 and 22). In these three examples, the corresponding genes have not yet been identified. The fact that retinoids can influence the direction of looping (Osmond *et al.*, 1991; Smith *et al.*, 1997) implicates retinoic acid or its metabolites as significant players in setting the initial symmetry of the heart tube and suggests that these genes may interact in some manner with RA signaling processes.

V. Retinoids and Axial Specification

The coalescence of the heart-forming fields is also the time during development when segmental identity of the various regions of the heart tube is established. It is well-known that axial information during vertebrate development can be altered by RA treatment, as assessed by homeobox gene and protein expression in the neural tube (Kessel and Gruss, 1991; Conlon, 1995). Therefore, it would not be surprising if RA could also affect anteroposterior patterning in the early heart tube. Anterior cardiac structures are represented by the outflow tract and common ventricle, whereas posterior structures are composed of the common atria and inflow tract. Consequently, molecules that effect lineage diversification in the early heart tube can be viewed as effecting anterioposterior patterning. Indeed, there is evidence that RA may be involved in initial cardiac lineage diversification of atrial and ventricular myocytes. The earliest known chamber-restricted markers identifying atrial and ventricular cell lineages are the myofibrillar proteins atrial myosin heavy chain-1 (AMHC1, chick) and ventricular myosin light chain-2 (MLC-2v, mouse) (Sweeney *et al.*, 1987; O'Brien *et al.*, 1993). In the chick, treatment with RA shifts expression of AMHC1 into the more anterior progenitor cells (Yutzey *et al.*, 1994). Equivalent studies in the mouse have not been performed; however, in the RXRα null mouse, MLC-2v expression remains restricted to the ventricular chambers, whereas the atrial isoform (MLC-2a) is aberrantly expressed in the ventricles (Dyson *et al.*, 1995; see Chapter 15). In the chick, once the diversified lineages have been established, subsequent exposure to RA no longer alters the atrial gene expression pattern (Yutzey *et al.*, 1994). Recently, two helix–loop–helix transcription factors, dHAND and eHAND, have been identified which delineate the right and left ventricular chambers, respectively (Srivastava *et al.*, 1997; see Chapter 9). Whether these transcription factors are influenced by retinoic acid has not been reported.

VI. Retinoids in Endocardial Cushion Development

Endocardial cushion tissue initially forms as regionally restricted thickenings of cardiac jelly between the endocardial and myocardial cell layers in the primitive heart tube. Early in development, these expansions are filled with extracellular matrix components, the origin of which is thought to be primarily from the surrounding myocardium. Shortly thereafter, about E 8.5–9.0 in the mouse and HH stages 4–7 in the chick, endothelial cells of the endocardium transform into mesenchyme and migrate into the cushions (Markwald *et al.*, 1977; see Chapter 10). The superior and inferior atrioventricular (AV) cushions of the AV canal develop between the common atrial chamber and the primitive left ventricle. AV cushion formation is analogous to processes in the maturation of the conotruncal ridges in that there is an initial deposition of matrix molecules (adherons) in the cardiac jelly (Mjaatvedt and Markwald, 1989) followed by the transformation of endothelial cells into mesenchyme and migration into the cushion-forming tissues (Eisenberg and Markwald, 1995). Further expansion of the AV cushions results in the differentiation of mesenchyme into fibrous connective tissue of the mature valves (Lamers *et al.*, 1995). Myocardial cells may contribute to the mature phenotype of the inlet valves (particularly the tricuspid valve) by invading cushion tissue (Lamers *et al.*, 1995). Maturation of these structures divides the AV canal into right (tricuspid) and left (mitral) orifices.

Based on studies in both avian and murine models, these early stages of formation and transformation of cushion tissue can be influenced by local levels of RA. Quail embryos cultured in the absence of RA have, besides cardia bifida, hearts with an underdeveloped endocardium (Heine *et al.*, 1985). Further evidence that RA is required for the initial formation of myocardial and endocardial lineages comes from studies of the quail QCE-6 mesodermal tissue culture cell line (Eisenberg and Bader, 1995). These cells differentiate into both myocardial and endocardial lineages upon treatment with a combination of RA, basic fibroblast growth factor, transforming growth factor (TGF)-β2, and TGF-β3. Mouse embryos treated with RA at E9.5 and then cultured for 24 hr possess AV cushions that are reduced in size compared to sham-treated embryos (Davis and Sadler, 1981). By coincubating embryos with BrdU, the authors demonstrated that the mitotic activity of mesenchymal cells was reduced in embryos exposed to retinoic acid. In cushion explants from E9.5 mouse embryos, endoththelial-to-mesenchymal transformation was reduced after RA treatment (Nakajima *et al.*, 1996). These studies indicate the importance of maintaining appropriate RA levels during the initial formation of the heart tube.

RXRα-deficient mice display several defects of both the conotruncus and the AV canal. In the conotruncus, these malformations ranged from hypoplastic to the complete absence of conotruncal ridges to double-outlet right ventricle (Gruber *et al.*, 1996). The AV canal also demonstrated a range of defects; however, there was never a complete absence of cushion tissue. Defects in this region included hypoplastic cushions, incomplete or absent fusion of the superior and inferior AV cushions, cleft mitral valve, and cleft tricuspid valve. Additionally, heterozygous embryos had an intermediate phenotype, indicating a gene dosage effect of RXRα. Thus, this receptor may confer a genetic susceptibility for phenotypes analogous to congenital heart disease. Other combinations of retinoid receptor mutations also result in malformations of both the outflow tract and the AV canal.

Members of the TGF-β family have been strongly implicated in the epithelial-to-mesenchyme transformation process and may serve as candidate genes that mediate the downstream effects of RA in the early heart tube. TGF-β1, -2, and -3 are all expressed in the early embryonic chick heart (Choy *et al.*, 1991; Potts *et al.*, 1992). TGF-β1 protein is expressed in the heart throughout cardiogenesis, it is initially found on the epicardial and endocardial surfaces and cardiac jelly, then later in development in the endocardial cushions, and, finally, in the heart valve leaflets (Choy *et al.*, 1991). TGF-β2 and -3 are expressed in AV canal tissue (Potts *et al.*, 1992). *In vitro* assays of isolated endocardial cushion tissue indicate that TGF-β1 can stimulate endothelial–mesenchymal transformation (Rongish and Little, 1995). Furthermore, polyclonal antibodies prepared against extracellular proteins present in AV and outflow tract cardiac jelly inhibit endothelial transformation to mesenchyme and TGF-β1 expression in these assays (Nakajima *et al.*, 1996). TGF-β1 or -2, in combination with an explant of ventricular myocardium, will produce transformation of cultured AV canal endothelial cells *in vitro*, where epidermal growth factor and basic fibroblast growth factor do not have the same effect (Potts and Runyan, 1989). Antisense oligonucleotides targeted to TGF-β3 inhibit *in vitro* mesenchymal cell formation (Runyan *et al.*, 1992. In mouse cardiogenesis, TGF-β1 is expressed in cells overlying endocardial cushion tissue (Akhurst *et al.*, 1990). TGF-β2 is expressed at high levels around the outflow tract and AV canal of the heart at E8.5 in all cells which eventually differentiate into cardiomyocytes, and TGF-β2 protein expression persists throughout valve development (Dickson *et al.*, 1993). TGF-β3 RNA was not found in the early embryonic mouse heart (Runyan *et al.*, 1992). Mutation of mouse TGF-β2 results in conotruncal defects which are consistent with an effect on cushion mesenchyme (Sanford *et al.*, 1997). There are only limited data regarding the intersection of RA and TGF-β pathways in the heart: RA deficiency in mice results in a downregulation of TGF-β1 (Bavik *et al.,* 1996), whereas RA treatment results in an increased expression of TGF-β1 in endocardial and mesechymal cells and a decrease in TGF-β2 expression (Mahmood *et al.,* 1992). Thus, the TGF-β class of growth factors represents candidate downstream effectors of RA signaling in these processes.

Other molecules may also be downstream of RA signaling during the early stages of formation of the heart tube. As mentioned previously, cushion explant experiments indicate that the endothelial–mesenchymal transformation is reduced after RA treatment (Nakajima *et al.,* 1996). These authors showed that the expression of fibronectin and type I collagen was reduced in RA-treated cushion explants. Vitamin A-deficient quail embryo demonstrate an elevated level and wider distribution of expression of the transcription factor Msx-1 during early stages of development (Chen *et al.,* 1995). This gene is expressed in the chick heart in AV cushion tissue but not in the outflow tract cushions (Chan-Thomas *et al.,* 1993). GATA-4 transcription factor is found in the endocardium, endocardial cushion tissue, and the myocardium of the mouse E9.5 heart tube (Heikinheimo *et al.,* 1994; see Chapter 17) and in mouse F9 cells is upregulated (although not directly) by RA (Arceci *et al.,* 1993).

VII. Retinoids in Looping and Wedging

The heart tube assumes its D-loop configuration soon after the bilateral heart fields fuse. It is then important that looping proceeds in a predictable fashion since abnormal looping results in septal defects of both the conotruncus and the AV canal. A critical aspect in the formation of the mature four-chambered heart, including the correct orientation of the inflow and outflow tracts, is the process referred to as "wedging" (Kirby and Waldo, 1995). This process, during the later stages of looping (and simultaneously with septation), brings the aorta over the left ventricle between the mitral and tricuspid valves. Moreover, it results in the convergence of the outflow septum, the ventricular septum, and AV cushion tissue, culminating in the formation of a normally septated four-chambered heart (Kirby and Waldo, 1995). Inappropriate looping and wedging events can result in a variety of cardiac defects, including ventricular septal defects, double-inlet left ventricle, and double-outlet right ventricle (Kirby and Waldo, 1995).

Retinoic acid treatment of stage 15 chick embryos leads to cardiac malformations, including ventricular septal defects and double-outlet right ventricle (Bouman *et al.,* 1995), which may indicate an alteration in the looping process. Mouse embryos treated with RA at E8.5 display these defects and transposition, which have been proposed to arise from an initial disturbance in looping (Pexieder *et al.,* 1995). Several combinations of mutations of retinoid receptor genes in mice result in conotruncal malformations that might be explained by looping and/or wedging disturbances at earlier stages of development.

There are few molecular insights to account for looping defects caused by RA manipulation. Mice carrying a null mutation for the homeobox gene Nkx2.5 lack a normally looped heart (Lyons *et al.,* 1995; see Chapter 7), although whether RA is involved in pathways which include Nkx2.5 is not yet known.

VIII. Retinoids in Ventricular Maturation

At the time of completion of looping of the early heart, the ventricular chambers are thin-walled, have already been specified to left and right fates, and are separated by the bulboventricular groove which demarks the axis of the future interventricular septum. Maturation of the ventricles involves several processes: development of the trabecular component of the myocardium, thickening of the outer chamber wall through assembly of the compact zone of laterally oriented myocytes, and formation of the muscular portion of the ventricular septum.

One of the prominent malformations recognized as arising from maternal nutritional vitamin A deprivation (Wilson and Warkany, 1949) was an underdevelopment of the compact zone of the ventricular chamber wall. Defects in the muscular portion of the ventricular septum were also observed, which is most likely a related phenomenon in that this portion of the septum is derived from proliferation and involution of ventricular cardiomyocytes from the chamber wall. In these vitamin A-deficient embryos, the trabecular layer was not noted to be affected.

Mutation of the RXRα gene results in a virtually identical ventricular hypoplastic phenotype. In wild-type mouse embryos at E11.5, the ventricular chamber wall is two or three cell diameters thin, and there is no obvious anatomical difference between wild-type and RXRα-deficient embryos. By E12.5, thickening of the compact zone has begun in wild-type embryos and progresses steadily as the embryo ages, although mutants remain thin-walled. RXRα mutant embryos die in midgestation, mostly between E14.5 and 15.5, and lethality can be correlated with and almost certainly results from grossly decreased cardiac performance

(Dyson *et al.*, 1995). The trabecular layer is affected in RXRα$^{-/-}$ embryos, although not nearly as dramatically as is the compact zone, and it is unclear to what extent this contributes to the lethal phenotype.

Several models have been proposed to explain these phenotypes. Ventricular hypoplasia resulting from nutritional deficiency was originally explained as a retardation of the growth and differentiation of the cardiac muscle (Wilson and Warkany, 1949). Retardation or a complete arrest in the developmental process which causes compact zone formation is consistent with the anatomical appearance of RXRα$^{-/-}$ and nutritionally deficient hearts in that the ventricles of these experimental models resemble the thin-walled ventricles of the earlier embryo. In addition, ventricular expression of the myosin light chain 2a (MLC-2a) gene is abnormally persistent in RXRα$^{-/-}$ embryos through E14.5 (Dyson *et al.*, 1995; see Chapter 15), which is consistent with an arrest in development at around E11.5 when this gene is normally expressed in the ventricle.

A completely different model for explaining the ventricular hypoplasia phenotype is that the ventricular chamber is partially misspecified to an atrial fate (Dyson *et al.*, 1995). The atrial developmental program is to remain thin-walled and to express an atrial-specific gene expression program, including MLC-2a. However, the ventricles of RXRα$^{-/-}$ and nutritional vitamin A-deficient embryos are clearly ventricular in fate in that they are both trabeculated, and other chamber-specific markers are normally expressed in RXRα$^{-/-}$ mutant hearts. If misspecification to an atrial fate is a component of the RXRα$^{-/-}$ phenotype, it is only a partial and very selective aspect of specification which is affected. Still another model to explain the mutant phenotype is that ventricular cardiomyocytes prematurely differentiate in the absence of retinoid signaling. Support for this model comes from the observation (Kastner *et al.*, 1994) that the ultrastructural organization of the contractile apparatus in compact zone myocytes of RXRα$^{-/-}$ embryos is more highly organized than that in wild-type embryos and thus resembles what is seen in trabecular myocytes and in older and more differentiated compact zone myocytes. Because differentiated cardiomyocytes are postmitotic, this model also provides an explanation for the thin ventricular chamber wall. This model would seem to be in direct opposition to one in which development is arrested at an earlier stage, but these may be the same if developmental arrest causes premature differentiation.

Surprisingly, a large number of mouse mutations have a similar if not identical ventricular phenotype as is seen in RXRα$^{-/-}$ embryos. These genes include the cell cycle regulator N-myc (Moens *et al.*, 1993), the transcription factors TEF-1 (Chen *et al.*, 1994) and WT-1 (Kreidberg *et al.*, 1993), the G protein-coupled receptor kinase βARK-1 (Jaber *et al.*, 1996), and the cell surface receptor gp130 (Yoshida *et al.*, 1996). Whether there is any convergence between these several signaling pathways and RA signaling remains unknown.

In contrast to studies in which the prevention of an endogenous retinoid-dependent program by either nutritional deficiency of ligand or genetic deficiency of receptor dramatically alters ventricular morphology, the effects of RA excess on the ventricles are much less pronounced. An excess of RA does not obviously affect mouse ventricular maturation, although in RA-treated chick embryos, Dickman and Smith (1996) noted a myocardial thickening.

IX. Retinoids in Neural Crest and Aorticopulmonary Septation

Cells derived from the neural crest are responsible for initiating formation of the aorticopulmonary (A/P) septum and ultimately form the smooth muscle layer around the mature derivatives of the aortic arch arteries (LeDouarin, 1982; Kirby *et al.*, 1983). The cardiac neural crest domain has been mapped to the caudal hindbrain and rostral spinal cord region of the neural tube. Deficiencies in this neural crest population result in persistent truncus arteriosus (PTA; representative of a failure in formation of the A/P septum) and in various aortic arch artery defects as well as defects in the thymus, thyroid, and parathyroid organs.

Retinoids are clearly involved in the morphogenesis of the aortic arch arteries and the A/P septum. The nutritional vitamin A-deficiency studies (Wilson and Warkany, 1949) documented an impressive array of aortic arch artery defects, generally involving inappropriate persistence or regression of the paired early aortic arch artery network. Particularly prominent were several common human pediatric entities, such as retroesophageal right subclavian artery and right-sided aortic arch. Mutations of various retinoic acid receptor genes also result in numerous aortic arch defects. Persistent truncus arteriosus was a rare phenotype in the nutritional vitamin A deficiency study (probably rare because of experimental limitations) but can be recovered at up to 100% penetrance in different retinoid receptor mutation backgrounds.

RA excess treatment leads to a spectrum of defects which overlap the deficiency phenotypes (Shenefelt, 1972; Taylor *et al.*, 1980; Pexieder *et al.*, 1995). Thus, aortic arch anomalies are prominently seen after RA excess exposure, and while PTA is not a vitamin A-excess phenotype, conotruncal malformations such as double-outlet right ventricle are recovered. These conotruncal defects may arise because of defects in looping (described previously) or through perturbation of the neural crest population. Deficiencies in the thymus and parathyroid organs are also seen following RA excess exposure, consistent with an effect on the domain of the cardiac neural crest.

However, a strong arguement can be made that the consequences of vitamin A deficiency and excess are accounted for through different mechanisms. In the mouse, RA excess is given at E8.5–9.5 (when axial identity of the neural tube is established) to induce cardiovascular defects (Pexieder *et al.,* 1995), whereas the requirement for endogenous vitamin A signaling appears to be between E9.5 and 10.5 (when neural crest cells have migrated away from the neural tube and are resident in the outflow tract) (Wilson *et al.,* 1953). RA excess induces thymic and parathyroid hypoplasia (Lammer *et al.,* 1985), whereas nutritional (Wilson *et al.,* 1953) or genetic deficiency (Lee *et al.,* 1997) does not result in hypoplasia of these organs. Finally, RA excess at E8.5 alters the axial identity of cells in the hindbrain and spinal cord (Kessel and Gruss, 1991; Conlon, 1995), concurrent with the establishment of a new pattern of Hox gene expression and prior to the onset of crest cell migration away from the dorsal neural tube. In contrast, genetic deficiency does not cause homeotic transformations in embryos with PTA (Luo *et al.,* 1996), and embryos lacking RARγ have homeotic transformations but do not have PTA (Lohnes *et al.,* 1993). Consequently, while the effects of RA excess are likely to result from misspecification of crest cell identity in the neural tube, the effects of deficiency are more consistent with a failure in a later stage of crest cell migration or differentiation.

A number of other genes have been implicated in cardiac neural crest cell differentiation. The Pax-3 (Franz, 1989) and the endothelin-1 (Kurihara *et al.,* 1995) genes have been shown in mice to be required for normal outflow tract and aortic arch artery morphogenesis, although Pax-3 is expressed in the neural crest population (Conway *et al.,* 1997) and ET-1 is expressed in the endothelium of the aortic arches (Kurihara *et al.,* 1995). The dHAND gene is also expressed in neural crest, and mutation of this gene results in a failure of the aortic arch arteries to form (Srivastava *et al.,* 1997). The human pediatric condition known as DeGeorge syndrome results in conotruncal defects as well as thymic and parathyroid deficiencies, very similar to the phenotype seen after retinoic acid excess exposure (Greenberg, 1993), although the gene or genes responsible for DeGeorge syndrome have not been definitively identified (Gong *et al.,* 1996). Whether retinoic acid regulates the expression of any of these genes has not yet been explored.

X. Concluding Remarks

It is clear from the previous discussion that retinoids have a broad role in directing normal cardiovascular morphogenesis, and that the developing heart is extremely sensitive to perturbation (in either direction) of retinoid signaling. Previously, a popular model held that RA worked by establishing positional identity within a "developmental field," such as the limb bud, or in principle the heart. Recent analyses (as described previously) are far more supportive of a conceptualization of RA acting in multiple independent events during the course of heart development, even though these events are closely linked in space and time. What is lacking is a molecular explanation to account for the nature of the defects which are induced by either vitamin A deficiency or excess paradigms, although numerous candidate genes are available for testing. A critical goal for the future is the identification of upstream and downstream processes which dictate the activation of and response to retinoid signaling. The large number of heart morphogenic processes which are sensitive to RA, and the similarity of the experimental phenotypes to what is seen in clincal cases of congenital heart defects, suggest that this will be a fruitful and worthwhile area of pursuit.

References

Achkar, C. C., Derguini, F., Blumberg, B., Langston, A., Levin, A. A., Speck, J., Evans, R. M., Bolado, J. J., Nakanishi, K., and Buck, J. (1996). 4-Oxoretinol, a new natural ligand and transactivator of the retinoic acid receptors. *Proc. Natl. Acad. Sci. U.S.A.* **93,** 4879–4884.

Akhurst, R. J., Lehnert, S. A., Faissner, A., and Duffie, E. (1990). TGF beta in murine morphogenetic processes: The early embryo and cardiogenesis. *Development (Cambridge, UK)* **108,** 645–656.

Arceci, R. J., King, A. A., Simon, M. C., Orkin, S. H., and Wilson, D. B. (1993). Mouse GATA-4: A retinoic acid-inducible GATA-binding transcription factor expressed in endodermally derived tissues and heart. *Mol. Cell. Biol.* **13,** 2235–2246.

Bavik, C., Ward, S. J., and Chambon, P. (1996). Developmental abnormalities in cultured mouse embryos deprived of retinoic by inhibition of yolk-sac retinol binding protein synthesis. *Proc. Natl. Acad. Sci. U.S.A.* **93,** 3110–3114.

Blumberg, B., Bolado, J. J., Derguini, F., Craig, A. G., Moreno, T. A., Chakravarti, D., Heyman, R. A., Buck, J., and Evans, R. M. (1996). Novel retinoic acid receptor ligands in Xenopus embryos. *Proc. Natl. Acad. Sci. U.S.A.* **93,** 4873–4878.

Boerman, M. H., and Napoli, J. L. (1996). Cellular retinol-binding protein-supported retinoic acid synthesis. Relative roles of microsomes and cytosol. *J. Biol. Chem.* **271,** 5610–5616.

Bouman, H. G., Broekhuizen, M. L., Baasten, A. M., Gittenberger-de Groot, A.C., and Wenink, A. C. (1995). Spectrum of looping disturbances in stage 34 chicken hearts after retinoic acid treatment. *Anat. Rec.* **243,** 101–108.

Buck, J., Derguini, F., Levi, E., Nakanishi, K., and Hammerling, U. (1991). Intracellular signaling by 14-hydroxy-4,14-retro-retinol. *Science* **254,** 1654–1656.

Buck, J., Grun, F., Derguini, F., Chen, Y., Kimura, S., Noy, N., and Hammerling, U. (1993). Anhydroretinol: A naturally occurring inhibitor of lymphocyte physiology. *J. Exp. Med.* **178,** 675–680.

Chan-Thomas, P. S., Thompson, R. P., Robert, B., Yacoub, M. H., and Barton, P. J. (1993). Expression of homeobox genes Msx-1 (Hox-7) and Msx-2 (Hox-8) during cardiac development in the chick. *Dev. Dyn.* **197,** 203–216.

Chen, Y., and Solursh, M. (1992). Comparison of Hensen's node and retinoic acid in secondary axis induction in the early chick embryo. *Dev. Dyn.* **195,** 142–151.

Chen, Y., Kostetskii, I., Zile, M. H., and Solursh, M. (1995). Comparative study of Msx-1 expression in early normal and vitamin A-deficient avian embryos. *J. Exp. Med.* **272,** 299–310.

Chen, Z., Friedrich, G. A., and Soriano, P. (1994). Transcriptional enhancer factor 1 disruption by a retroviral gene trap leads to heart defects and embryonic lethality in mice. *Genes Dev.* **8,** 2293–2301.

Chiang, C., Litingtung, Y., Lee, E., Young, K. E., Corden, J. L., Westphal, H., and Beachy, P. A. (1996). Cyclopia and defective axial patterning in mice lacking Sonic hedgehog gene function. *Nature (London)* **383,** 407–413.

Choy, M., Armstrong, M. T., and Armstrong, P. B. (1991). Transforming growth factor-beta 1 localized within the heart of the chick embryo. *Anat. Embryol.* **183,** 345–352.

Conlon, R. A. (1995). Retinoic acid and pattern formation in vertebrates. *Trends Genet.* **11,** 314–319.

Conway, S. J., Henderson, D. J., and Copp, A. J. (1997). Pax3 is required for cardiac neural crest migration in the mouse: Evidence from the splotch (Sp2H) mutant. *Development (Cambridge, UK)* **124,** 505–514.

Davis, L. A., and Sadler, T. W. (1981). Effects of vitamin A on endocardial cushion development in the mouse heart. *Teratology* **24,** 139–148.

Dersch, H., and Zile, M. H. (1993). Induction of normal cardiovascular development in the vitamin A-deprived quail embryo by natural retinoids. *Dev. Biol.* **160,** 424–433.

Dickman, E. D., and Smith, S. M. (1996). Selective regulation of cardiomyocyte gene expression and cardiac morphogenesis by retinoic acid. *Dev. Dyn.* **206,** 39–48.

Dickson, M. C., Slager, H. G., Duffie, E., Mummery, C. L., and Akhurst, R. J. (1993). RNA and protein localizations of TGF beta 2 in the early mouse embryo suggest an involvement in cardiac development. *Development (Cambridge, UK)* **117,** 625–639.

Dolle, P., Fraulob, V., Kastner, P., and Chambon, P. (1994). Developmental expression of murine retinoid X receptor (RXR) genes. *Mech. Dev.* **45,** 91–104.

Duester, G. (1996). Involvement of alcohol dehydrogenase, short-chain dehydrogenase/reductase, aldehyde dehydrogenase, and cytochrome P450 in the control of retinoid signaling by activation of retinoic acid synthesis. *Biochemistry* **35,** 12221–12227.

Dyson, E., Sucov, H. M., Kubalak, S. W., Schmid-Schonbein, G. W., DeLano, F. A., Evans, R. M., Ross, J., and Chien, K. R. (1995). Atrial-like phenotype is associated with embryonic ventricular failure in RXRα-/- mice. *Proc. Natl. Acad. Sci. U.S.A.* **92,** 7386–7390.

Eisenberg, C. A., and Bader, D. (1995). QCE-6: A clonal cell line with cardiac myogenic and endothelial cell potentials. *Dev. Biol.* **167,** 469–481.

Eisenberg, L. M., and Markwald, R. R. (1995). Molecular regulation of atrioventricular valvuloseptal morphogenesis. *Circ. Res.* **77,** 1–6.

Evans, R. M. (1988). The steroid and thyroid hormone receptor superfamily. *Science* **240,** 889–895.

Franz, T. (1989). Persistent truncus arteriosus in the Splotch mutant mouse. *Anat. Embryol.* **180,** 457–464.

Gong, W., Emanuel, B. S., Collins, J., Kim, D. H., Wang, Z., Chen, F., Zhang, G., Roe, B., and Budarf, M. L. (1996). A transcription map of the DiGeorge and velo-cardio-facial syndrome minimal critical region on 22q11. *Hum. Mol. Genet.* **5,** 789–800.

Greenberg, F. (1993). DiGeorge syndrome: An historical review of clinical and cytogenetic features. *J. Med. Genet.* **30,** 803–806.

Gruber, P. J., Kubalak, S. W., Pexieder, T., Sucov, H. M., Evans, R. M., and Chien, K. R. (1996). RXR alpha deficiency confers genetic susceptibility for aortic sac, conotruncal, atrioventricular cushion, and ventricular muscle defects in mice. *J. Clin. Invest.* **98,** 1332–1343.

Heikinheimo, M., Scandrett, J. M., and Wilson, D. B. (1994). Localization of transcription factor GATA-4 to regions of the mouse embryo involved in cardiac development. *Dev. Biol.* **164,** 361–373.

Heine, U. I., Roberts, A. B., Munoz, E. F., Roche, N. S., and Sporn, M. B. (1985). Effects of retinoid deficiency on the development of the heart and vascular system of the quail embryo. *Virchows Arch. B* **50,** 135–152.

Heyman, R. A., Mangelsdorf, D. J., Dyck, J. A., Stein, R. B., Eichele, G., Evans, R. M., and Thaller, C. (1992). 9-cis retinoic acid is a high affinity ligand for the retinoid X receptor. *Cell (Cambridge, Mass.)* **68,** 397–406.

Iulianella, A., and Lohnes, D. (1997). Contribution of retinoic acid receptor gamma to retinoid-induced craniofacial and axial defects. *Dev. Dyn.* **209,** 92–104.

Jaber, M., Koch, W. J., Rockman, H., Smith, B., Bond, R. A., Sulik, K. K., Ross, J. J., Lefkowitz, R. J., Caron, M. G., and Giros, B. (1996). Essential role of beta-adrenergic receptor kinase 1 in cardiac development and function. *Proc. Natl. Acad. Sci. U.S.A.* **93,** 12974–12979.

Kastner, P., Grondona, J. M., Mark, M., Gansmuller, A., LeMeur, M., Decimo, D., Vonesch, J. L., Dolle, P., and Chambon, P. (1994). Genetic analysis of RXR alpha developmental function: Convergence of RXR and RAR signaling pathways in heart and eye morphogenesis. *Cell (Cambridge, Mass.)* **78,** 987–1003.

Kastner, P., Mark, M., Leid, M., Gansmuller, A., Chin, W., Grondona, J. M., Decimo, D., Krezel, W., Dierich, A., and Chambon, P. (1996). Abnormal spermatogenesis in RXR beta mutant mice. *Genes Dev.* **10,** 80–92.

Kessel, M., and Gruss, P. (1991). Homeotic transformations of murine vertebrae and concomitant alteration of Hox codes induced by retinoic acid. *Cell (Cambridge, Mass.)* **67,** 89–104.

Kirby, M. L., and Waldo, K. L. (1995). Neural crest and cardiovascular patterning. *Circ. Res.* **77,** 211–215.

Kirby, M. L., Gale, T. F., and Stewart, D. E. (1983). Neural crest cells contribute to normal aorticopulmonary septation. *Science* **220,** 1059–1061.

Kreidberg, J. A., Sariola, H., Loring, J. M., Maeda, M., Pelletier, J., Housman, D., and Jaenisch, R. (1993). WT-1 is required for early kidney development. *Cell (Cambridge, Mass.)* **74,** 679–691.

Krezel, W., Dupe, V., Mark, M., Dierich, A., Kastner, P., and Chambon, P. (1996). RXR gamma null mice are apparently normal and compound RXR alpha $^{+/-}$/RXR beta $^{-/-}$/RXR gamma $^{-/-}$ mutant mice are viable. *Proc. Natl. Acad. Sci. U.S.A.* **93,** 9010–9014.

Kurihara, Y., Kurihara, H., Oda, H., Maemura, K., Nagai, R., Ishikawa, T., and Yazaki, Y. (1995). Aortic arch malformations and ventricular septal defect in mice deficient in endothelin-1. *J. Clin. Invest.* **96,** 293–300.

Lamers, W. H., Viragh, S., Wessels, A., Moorman, A. F., and Anderson, R. H. (1995). Formation of the tricuspid valve in the human heart. *Circulation* **91,** 111–121.

Lammer, E. J., Chen, D. T., Hoar, R. M., Agnish, N. D., Benke, P. J., Braun, J. T., Curry, C. J., Fernhoff, P. M., Grix, A. J., Lott, I. T., Richard, J. M., and Sun, S. C. (1985). Retinoic acid embryopathy. *N. Engl. J. Med.* **313,** 837–841.

LeDouarin, N. M. (1982). "The Neural Crest." Cambridge University Press. Cambridge, UK.

Lee, R. Y., Luo, J., Evans, R. M., Giguere, V., and Sucov, H. M. (1997). Compartment-selective sensitivity of cardiovascular morphogenesis to combinations of retinoic acid receptor gene mutations. *Circ. Res.* **80,** 757–764.

Leid, M., Kastner, P., and Chambon, P. (1992). Multiplicity generates diversity in the retinoic acid signalling pathways. *Trends Biochem. Sci.* **17,** 427–433.

Levin, M., Johnson, R. L., Stern, C. D., Kuehn, M., and Tabin, C. (1995). A molecular pathway determining left-right asymmetry in chick embryogenesis. *Cell (Cambridge, Mass.)* **82,** 803–814.

Li, E., Sucov, H. M., Lee, K. F., Evans, R. M., and Jaenisch, R. (1993). Normal development and growth of mice carrying a targeted disruption of the alpha 1 retinoic acid receptor gene. *Proc. Natl. Acad. Sci. U.S.A.* **90,** 1590–1594.

Lohnes, D., Kastner, P., Dierich, A., Mark, M., LeMeur, M., and Chambon, P. (1993). Function of retinoic acid receptor gamma in the mouse. *Cell (Cambridge, Mass.)* **73,** 643–658.

Lufkin, T., Lohnes, D., Mark, M., Dierich, A., Gorry, P., Gaub, M. P., LeMeur, M., and Chambon, P. (1993). High postnatal lethality and testis degeneration in retinoic acid receptor alpha mutant mice. *Proc. Natl. Acad. Sci. U.S.A.* **90,** 7225–7229.

Luo, J., Pasceri, P., Conlon, R. A., Rossant, J., and Giguère, V. (1995). Mice lacking all isoforms of retinoic acid receptor β develop normally and are susceptible to the teratogenic effects of retinoic acid. *Mech. Dev.* **53,** 61–71.

Luo, J., Sucov, H. M., Bader, J.-A., Evans, R. M., and Giguere, V. (1996). Compound mutants for retinoic acid receptor (RAR) β and RARα1 reveal developmental functions for multiple RARβ isoforms. *Mech. Dev.* **55,** 33–44.

Lyons, I., Parsons, L. M., Hartley, L., Li, R., Andrews, J. E., Robb, L., and Harvey, R. P. (1995). Myogenic and morphogenetic defects in the heart tubes of murine embryos lacking the homeo box gene Nkx2-5. *Genes Dev.* **9,** 1654–1666.

Mahmood, R., Flanders, K. C., and Morriss-Kay, G. (1992). Interactions between retinoids and TGF beta in mouse morphogenesis. *Development (Cambridge, UK)* **115,** 67–74.

Mandel, H. G., and Cohn, V. H. (1985). Fat-soluble vitamins. *In* "The Pharmacological Basis of Therapeutics." (A. G. Gilman, L. S. Goodman, T. W. Rall, and F. Murad, eds.), 7th, ed., p. 1573. Macmillan, New York.

Mangelsdorf, D. J., Borgmeyer, U., Heyman, R. A., Zhou, J. Y., Ong, E. S., Oro, A. E., Kakizuka, A., and Evans, R. M. (1992). Characterization of three RXR genes that mediate the action of 9-cis retinoic acid. *Genes Dev.* **6,** 329–344.

Markwald, R. R., Fitzharris, T. P., and Manasek, F. J. (1977). Structural development of endocardial cushions. *Am. J. Anat.* **148,** 85–120.

Matzuk, M. M., Kumar, T. R., and Bradley, A. (1995). Different phenotypes for mice deficient in either activins or activin receptor type II. *Nature (London)* **374,** 356–360.

Mendelsohn, C., Lohnes, D., Decimo, D., Lufkin, T., LeMeur, M., Chambon, P., and Mark, M. (1994a). Function of the retinoic acid receptors (RARs) during development. (II) Multiple abnormalities at various stages of organogenesis in RAR double mutants. *Development (Cambridge, UK)* **120,** 2749–2771.

Mendelsohn, C., Mark, M., Dolle, P., Dierich, A., Gaub, M. P., Krust, A., Lampron, C., and Chambon, P. (1994b). Retinoic acid receptor beta 2 (RAR beta 2) null mutant mice appear normal. *Dev. Biol.* **166,** 246–258.

Mjaatvedt, C. H., and Markwald, R. R. (1989). Induction of an epithelial-mesenchymal transition by an in vivo adheron-like complex. *Dev. Biol.* **136,** 118–118.

Moens, C. B., Stanton, B. R., Parada, L. F., and Rossant, J. (1993). Defects in heart and lung development in compound heterozygotes for two different targeted mutations at the N-myc locus. *Development (Cambridge, UK)* **119,** 485–499.

Nakajima, Y., Morishima, M., Nakazawa, M., and Momma, K. (1996). Inhibition of outflow cushion mesenchyme formation in retinoic acid-induced complete transposition of the great arteries. *Cardiovasc. Res.* **31,** E77–E85.

Niederreither, K., McCaffery, P., Drager, U. C., Chambon, P., and Dolle, P. (1997). Restricted expression and retinoic acid-induced downregulation of the retinaldehyde dehydrogenase type 2 (RALDH-2) gene during mouse development. *Mech. Dev.* **62,** 67–78.

O'Brien, T. X., Lee, K. J., and Chien, K. R. (1993). Positional specification of ventricular myosin light chain 2 expression in the primitive murine heart tube. *Proc. Natl. Acad. Sc.i U.S.A.* **90,** 5157–5161.

Osmond, M. K., Butler, A. J., Voon, F. C., and Bellairs, R. (1991). The effects of retinoic acid on heart formation in the early chick embryo. *Development (Cambridge, UK)* **113,** 1405–1417.

Pexieder, T., Blanc, O., Pelouch, V., Ostadalova, I., Milerova, M., and Ostadal, B. (1995). Late fetal development of retinoic acid induced transposition of the great arteries. *In* "Developmental Mechanisms of Heart Disease." (M. R. Clark and A. Takas, eds.), pp. 297–307. Futura Publ. Co., Armonk, NY.

Potts, J. D., and Runyan, R. B. (1989). Epithelial-mesenchymal cell transformation in the embryonic heart can be mediated, in part, by transforming growth factor beta. *Dev. Biol.* **134,** 392–401.

Potts, J. D., Vincent, E. B., Runyan, R. B., and Weeks, D. L. (1992). Sense and antisense TGF beta 3 mRNA levels correlate with cardiac valve induction. *Dev. Dyn.* **193,** 340–345.

Rongish, B. J., and Little, C. D. (1995). The extracellular matrix during heart development. *Experientia* **51,** 873–882.

Ruberte, E., Dolle, P., Chambon, P., and Morriss-Kay, G. (1991). Retinoic acid receptors and cellular retinoid binding proteins. II. Their differential pattern of transcription during early morphogenesis in mouse embryos. *Development (Cambridge, UK)* **111,** 45–60.

Runyan, R. B., Potts, J. D., and Weeks, D. L. (1992). TGF-beta 3-mediated tissue interaction during embryonic heart development. *Mol. Reprod. Dev.* **32,** 152–159.

Sanford, L. P., Ormsby, I., Gittenberger-de Groot, A. C., Sariola, H., Friedman, R., Boivin, G. P., Cardell, E. L., and Doetschman, T. (1997). TGFβ2 knockout mice have multiple developmental defects that are nonoverlapping with other TGFβ knockout phenotypes. *Development (Cambridge, UK)* **124,** 2659–2670.

Schultheiss, T. M., Xydas, S., and Lassar, A. B. (1995). Induction of avian cardiac myogenesis by anterior endoderm. *Development (Cambridge, UK)* **121,** 4203–4214.

Shenefelt, R. E. (1972). Morphogenesis of malformations in hamsters caused by retinoic acid: relation to dose and stage at treatment. *Teratology* **5,** 103–118.

Sinning, A. R., and Hewitt, C. C. (1996). Identification of a 283-kDa protein component of the particulate matrix associated with cardiac mesenchyme formation. *Acta. Anat.* **155,** 219–230.

Smith, S. M., Dickman, E. D., Thompson, R. P., Sinning, A. R., Wunsch, A. M., and Markwald, R. R. (1997). Retinoic acid directs cardiac laterality and the expression of early markers of precardiac asymmetry. *Dev. Biol.* **182,** 162–171.

Srivastava, D., Thomas, T., Lin, Q., Kirby, M. L., Brown, D., and Olson, E. N. (1997). Regulation of cardiac mesodermal and neural crest development by the bHLH transcription factor, dHAND. *Nat. Genet.* **16,** 154–160.

Sucov, H. M., Dyson, E., Gumeringer, C. L., Price, J., Chien, K. R., and Evans, R. M. (1994). RXR alpha mutant mice establish a genetic basis for vitamin A signaling in heart morphogenesis. *Genes Dev.* **8,** 1007–1018.

Sweeney, L., Zak, R., and Manasek, F. J. (1987). Transitions in cardiac isomyosin expression during differentiation of embryonic chick heart. *Circ. Res.* **61,** 287–295.

Taylor, I. M., Wiley, M. J., and Agur, A. (1980). Retinoic acid-induced heart malformations in the hamster. *Teratology* **21,** 193–197.

Thaller, C., and Eichele, G. (1990). Isolation of 3,4-didehydroretinoic acid, a novel morphogenetic signal in the chick wing bud. *Nature (London)* **345,** 815–819.

Thompson, J. N. (1969). The role of vitamin A in reproduction. *In* "The Fat Soluble Vitamins," pp. 267–281. University of Wisconson Press, Madison.

Thompson, J. N., Howell, J., and Pitt, G. A. J. (1964). Vitamin A and reproduction in rats. *Proc. R. Soc. London* **159,** 510–535.

Thompson, J. N., Howell, J. M. C., Pitt, G. A. J., and McLaughlin, C. I. (1969). The biological activity of retinoic acid in the domestic fowl and the effects of vitamin A deficiency on the chick embryo. *Br. J. Nutr.* **23,** 471–490.

VanKeuren, M. L., Layton, W. M., Iacob, R. A., and Kurnit, D. M. (1991). Situs inversus in the developing mouse: Proteins affected by the iv mutation (genocopy) and the teratogen retinoic acid (phenocopy). *Mol. Reprod. Dev.* **29,** 136–144.

Wilson, J. G., and Warkany, J. (1949). Aortic arch and cardiac anomalies in the offspring of vitamin A deficient rats. *Am. J. Anat.* **85,** 113–155.

Wilson, J. G., Roth, C. B., and Warkany, J. (1953). An analysis of the syndrome of malformations induced by maternal vitamin A deficiency. Effects of restoration of vitamin A at various times during gestation. *Am. J. Anat.* **92,** 189–217.

Wunsch, A. M., Little, C. D., and Markwald, R. R. (1994). Cardiac endothelial heterogeneity defines valvular development as demonstrated by the diverse expression of JB3, an antigen of the endocardial cushion tissue. *Dev. Biol.* **165,** 585–601.

Yokoyama, T., Copeland, N. G., Jenkins, N. A., Montgomery, C. A., Elder, F. F., and Overbeek, P. A. (1993). Reversal of left-right asymmetry: A situs inversus mutation. *Science* **260,** 679–682.

Yoshida, K., Taga, T., Saito, M., Suematsu, S., Kumanogoh, A., Tanaka, T., Fujiwara, H., Hirata, M., Yamagami, T., Nakahata, T., Hirabayashi, T., Yoneda, Y., Tanaka, K., Wang, W. Z., Mori, C., Shiota, K., Yoshida, N., and Kishimoto, T. (1996). Targeted disruption of gp130, a common signal transducer for the interleukin 6 family of cytokines, leads to myocardial and hematological disorders. *Proc. Natl. Acad. Sci. U.S.A.* **93,** 407–411.

Yost, H. J. (1992). Regulation of vertebrate left-right asymmetries by extracellular matrix. *Nature (London)* **357,** 158–161.

Yu, V. C., Delsert, C., Andersen, B., Holloway, J. M., Devary, O. V., Naear, A. M., Kim, S. Y., Boutin, J. M., Glass, C. K., and Rosenfeld, M. G. (1991). RXR beta: A coregulator that enhances binding of retinoic acid, thyroid hormone, and vitamin D receptors to their cognate response elements. *Cell (Cambridge, Mass.)* **67,** 1251–1266.

Yutzey, K. E., Rhee, J. T., and Bader, D. (1994). Expression of the atrial-specific myosin heavy chain AMHC1 and the establishment of anteroposterior polarity in the developing chicken heart. *Development (Cambridge, UK)* **120,** 871–883.

14

Molecular Mechanisms of Vascular Development

Ondine Cleaver and Paul A. Krieg

Institute for Cellular and Molecular Biology and Department of Zoology, University of Texas at Austin, Austin, Texas 78712

I. Introduction

The general pattern of the embryonic vascular system is highly conserved between vertebrates (Fig. 1), most likely reflecting the vascular patterns of an ancestral species. Considering the conservation of the basic vascular pattern in different vertebrates, it is also likely that the cellular and molecular mechanisms underlying the patterning of the vascular architecture are also conserved in different species. The defining cell type of the vascular system is the endothelial cell, which comprises the lining of the entire circulatory system, including the heart and all veins and arteries. The morphogenesis of the embryonic vasculature begins with the appearance of angioblasts in most mesodermal tissues (Fig. 2A). Angioblasts are defined as endothelial precursor cells which have not yet incorporated into the endothelial tissue of blood vessels, and throughout this chapter we will use the terms angioblast and endothelial precursor cell interchangeably. Sometime after their specification in the mesoderm, the angioblasts associate into vascular cords (Fig. 2B), which subsequently mature into the blood vessels that form the primary vascular system of the embryo. Angioblast assembly into a vessel occurs via a process termed vasculogenesis, or the coalescence of free angioblasts into vascular primordia (Fig. 2C). After the initial vasculature is established, it is elaborated and extended throughout the embryo as a result of a process termed angiogenesis (Fig. 2D). Still later, the embryonic vascular system is extensively modified by endothelial remodeling, which involves both the enlargement and the splitting of existing vessels and extension of new vessels into avascular tissue (Fig. 2E). Remodeling can also include the regression or complete

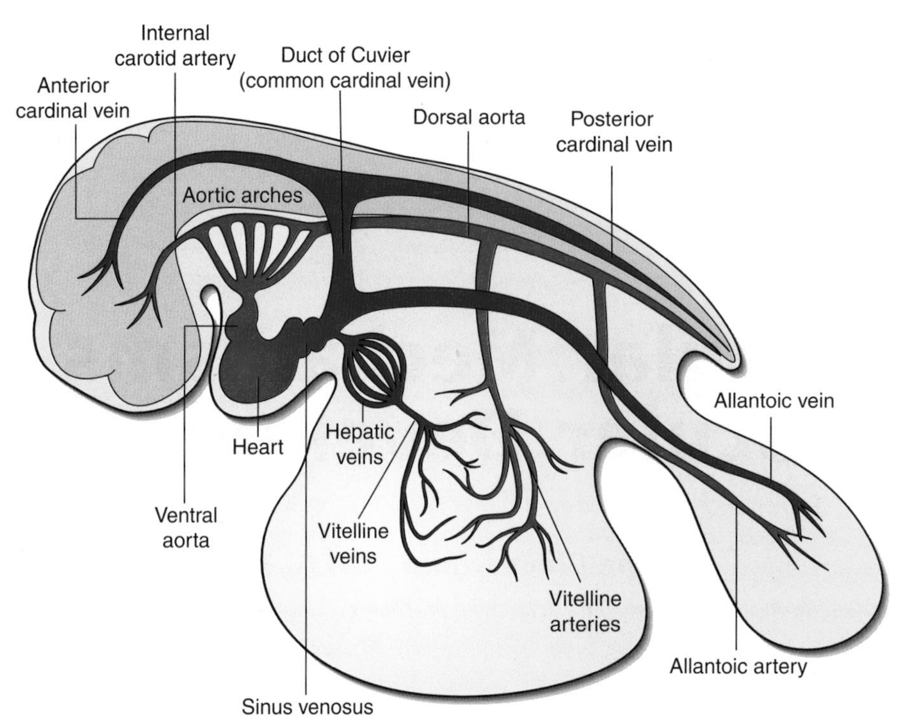

Figure 1 Major features of the vertebrate cardiovascular system. Endothelial cells create the inner lining of all vessels in the vertebrate embryo, including the major vessels (represented here in a simplified manner) and the minor vessels, such as small arterioles, venules, and capillaries (not shown). Major vessels include the endocardium, the dorsal aortae, the cardinal veins (anterior and posterior), the carotid arteries, the vitelline arteries and veins, the ducts of Cuvier, and the allantoic vein and artery. Generally, the blood flows from the heart, up the ventral aortae, through the aortic arches, and to the rest of the embryo. Blood then returns to the sinus venosus of the heart via the vitelline veins and the ducts of Cuvier.

disappearance of existing vessels. The final stage of vascular development is maturation, which involves a dramatic reduction in endothelial cell proliferation, morphological changes to endothelial cell shape and organization, and the recruitment of vascular wall components.

Driven in part by the study of blood vessel development during tumor angiogenesis, a great deal of recent research has focused on the function of growth factors and their receptors in regulating the proliferation and development of vascular tissue. Despite rapid progress, the molecular mechanisms underlying many aspects of embryonic vascular development remain unclear. For example, very little is known concerning the precise origins of angioblast precursor cells in the embryo or of the nature of the signals responsible for patterning the embryonic vasculature. Ultimately, knowledge of the function and regulation of the molecules involved in vascular growth and development will not only lead to an understanding of this key embryonic process but will also facilitate the development of diagnostic and therapeutic applications relevant to a wide range of vascular pathologies. This chapter will focus on the molecules involved in regulation of embryonic vascular development. It is important to state from the outset that, although many of the molecules involved in vascular development have been identified, our understanding of the precise roles of these molecules is still in its infancy.

II. Origin of Endothelial Cells

A. The Extraembryonic and Intraembryonic Angioblasts

Until recently, our knowledge of the origins and development of the vascular system was based almost exclusively on studies employing classic embryological techniques. Early studies, primarily using the avian embryo, suggested that the embryonic vasculature might originate following invasion of the embryo by vessels from extraembryonic tissues (His, 1868, 1900). This hypothesis was in part based on the fact that the first overt

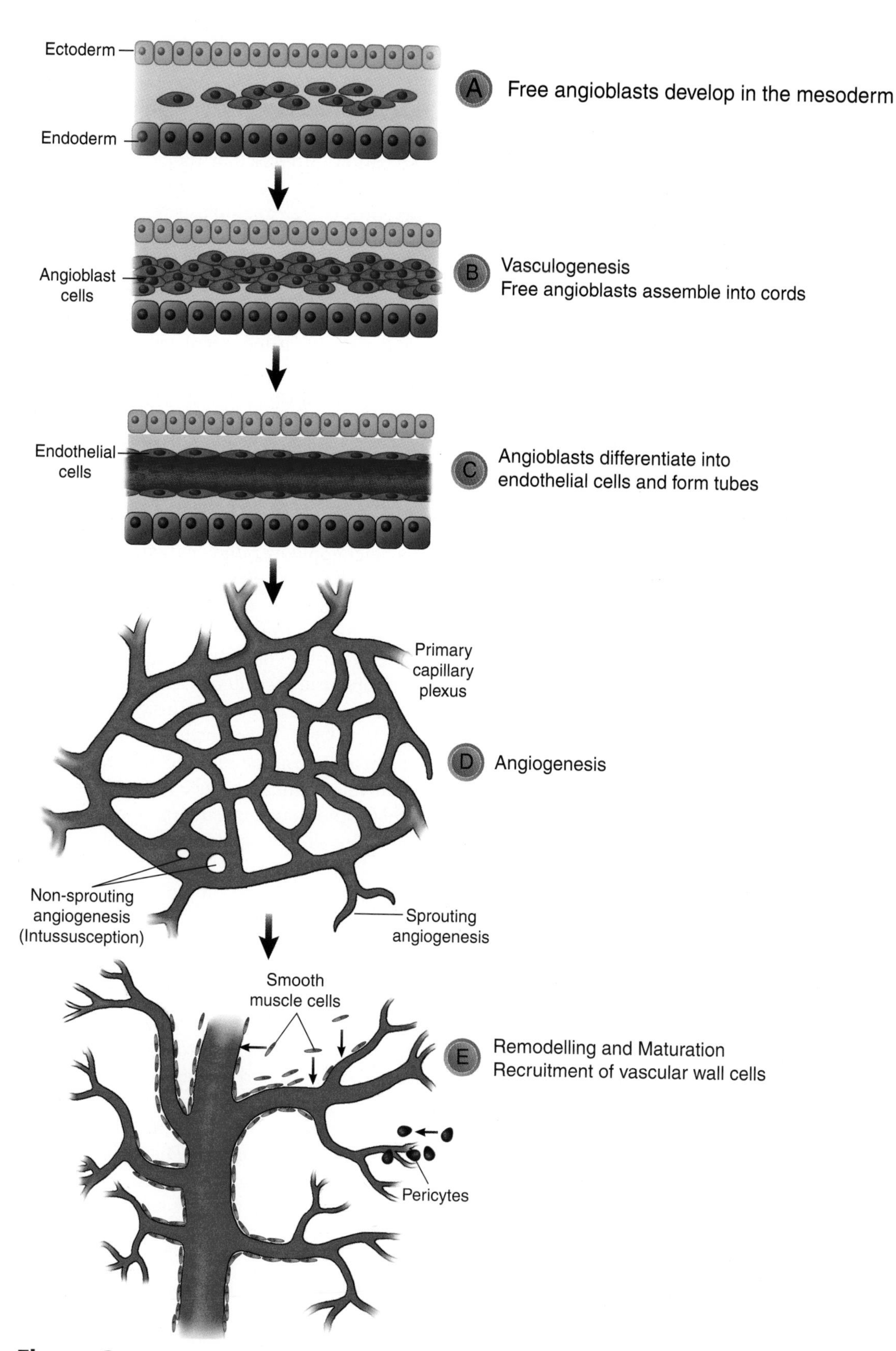

Figure 2 Schematic representation of the major processes involved in vascular development. Initially, (A) angioblasts differentiate from the mesoderm and then form cords either at the location where they emerge or at a distant location, following migration (B). (C) The endothelial cells in the cords now differentiate and form tubes (i.e., acquire patency). (D) The primary vascular plexus is then extended and elaborated by angiogenesis. (E) Vascular remodeling occurs, resulting in the formation of large and small vessels. Finally, the endothelium matures and mesenchymal cells are recruited to become components of the vascular wall.

signs of blood vessel development are evident on the yolk sac, outside the embryo proper, and that directed branching of vessels occurs toward the embryo. However, other researchers observed isolated endothelial vesicles within the embryonic splanchnic mesoderm and challenged this view (Evans, 1909; Sabin, 1917). Subsequent studies have demonstrated that the embryonic vasculature does indeed develop from intraembryonic precursors (Hahn, 1909; Stockard, 1915; Reagan, 1915; McClure, 1921; Dieterlen-Lièvre, 1975; 1984; Pardanaud *et al.,* 1987; Noden, 1989; Wilms *et al.,* 1991). For example, in experiments using the Japanese quail, Reagan (1915) separated the area pellucida (embryo proper) from the area opaca (yolk sac) prior to the formation of vessels and observed isolated endothelial differentiation within the embryo. This conclusively demonstrated that the embryonic vasculature arose *in situ* intraembryonically rather than by invasion of vascular precursor cells from extraembryonic tissues. Other researchers refined this proposition by analysis of the formation of major blood vessels, particularly in the head and heart region. Once again, these vessels were shown to form by the assembly of endothelial cells arising within the embryo rather than by colonization from an extraembryonic source (Jolley, 1940; Rosenquist, 1970; Johnston *et al.,* 1979).

B. Endothelial Origins in the Mesoderm

The precise origin of endothelial cells in the embryo has long remained elusive, although it is now clear that endothelial cell differentiation occurs exclusively in the mesoderm (Coffin and Poole, 1988; Noden, 1989). Furthermore, it appears that secondary embryonic inductions are not required since determination and differentiation of endothelial cells is independent of gastrulation (Azar and Eyal-Giladi, 1979; Zagris, 1980; Mitrani and Shimoni, 1990; Christ *et al.,* 1991). The early stages of endothelial development in the embryo were first revealed in studies using scanning electron microscopy (SEM) to directly examine developing blood vessels. Initial development of vessels, such as the dorsal aorta, was observed at the mesoderm–endoderm interface with the assembly of clusters and vesicles of endothelial precursor cells immediately preceding actual vessel formation (Meier, 1980; Hirakow and Hiruma, 1981, 1983). However, the SEM studies were fundamentally limited by the fact that endothelial cells could not be identified until they associated into groups. Therefore, the characterization of monoclonal and polyclonal antibodies specific to endothelial cells in the quail, such as MB1 (Peault *et al.,* 1983; LaBastie *et al.,* 1986) and QH-1 (Pardanaud *et al.,* 1987), provided a major breakthrough in the study of vasculogenesis. These antibodies facilitate the identification of endothelial cells soon after they arise in the mesoderm and long before their association into vascular tissue. Subsequent studies have revealed that the major embryonic blood vessels arise as cords of free angioblasts which coalesce, *in situ,* at the location where they will mature (Pardanaud *et al.,* 1987; Coffin and Poole, 1988; Peault *et al.,* 1988; Poole and Coffin, 1989).

Following the identification of QH-1, transplantation experiments using quail–chick chimeras refined our knowledge of the tissues containing endothelial precursor cells. Transplants of various quail tissues into chick embryos showed that all intraembryonic mesodermal tissues, except the prechordal plate, contain migrating endothelial precursor cells (Noden, 1989). Since this study, other groups have identified later embryonic tissues that contain angioblasts, including the splanchnopleure, the paraxial mesoderm, Hensen's node, the primitive streak, and, in rare cases, the somatopleuric mesoderm (Wilms *et al.,* 1991; Feinberg and Noden, 1991; Wilting *et al.,* 1992; Pardanaud and Dieterlen-Lièvre, 1993; Brand-Saberi *et al.,* 1994). Somatopleuric mesoderm, however, is thought to be initially colonized by extrinsic precursors (Pardanaud *et al.,* 1989). Recent studies suggest that angioblasts from different embryonic tissues may not represent equivalent lineages since angioblasts derived from the splanchnopleuric mesoderm may have the option of developing into hematopoietic cells, whereas those derived from the somatopleuric mesoderm appear to be restricted to the development of endothelial cells (Pardanaud *et al.,* 1996).

As originally suggested by Wilt (1965) and subsequently supported by numerous researchers, including Pardanaud *et al.,* (1989) and Palis *et al.* (1995), *in situ* differentiation of endothelial cells occurs primarily in mesoderm which is in contact with underlying endoderm. Organs and tissues devoid of endoderm, such as kidney and the brain, must be vascularized by angiogenesis because no resident angioblasts exist in these tissues (Pardanaud and Dieterlen-Lièvre, 1993). In addition to inducing angioblasts in the adjacent mesoderm, the endoderm is thought to pattern the endothelium later during development, creating regional differences in the properties of the overlying endothelial cell layer. For example, this patterning property of the endoderm occurs during endocardial development when the endoderm or endodermally secreted molecules are thought to modulate the competence of endocardial cells to respond to a myocardial-derived signal. This signal induces a regional epithelial–mesenchymal transformation of atrioventricular endothelium during valve formation (Bolender and Markwald, 1991).

III. Molecules Involved in Early Mesodermal Differentiation of Endothelial Cells

A. Fibroblast Growth Factor

Members of the fibroblast growth factor (FGF) family (especially bFGF) play a critical role in the induction of the mesodermal germ layer during the earliest stages of embryogenesis. In *Xenopus laevis,* they are potent inducers of ventral mesoderm, which will go on to form the blood islands and some muscle tissues (Slack *et al.,* 1987; Godsave *et al.,* 1988; Knöchel *et al.,* 1989; Isaacs *et al.,* 1992; Tannahill *et al.,* 1992). However, FGFs are also required for the development of certain dorsal types of mesoderm, suggesting a synergism between FGF and the transforming growth factor-β (TGF-β) family of growth factors, which are required for dorsal mesoderm formation (Cornell and Kimelman, 1994; LaBonne and Whitman, 1994). Experiments using an *in vitro* system of avian epiblast cell culture have shown that FGF induces expression of the receptor tyrosine kinase gene *flk-1* (Flamme *et al.,* 1995b) (known as *flk-1* in the mouse, *VEGFR2* in the chick, and *KDR* in humans), which is a very early marker of the endothelial cell lineage (Matthews *et al.,* 1991; Quinn *et al.,* 1993; Eichmann *et al.,* 1993; Millauer *et al.,* 1993; Yamaguchi *et al.,* 1993). The inducibility, however, is limited to the first 24 hr of culture. This led the authors to suggest that FGFs are important very early in the establishment of the endothelial cell lineage. In support of this suggestion, the vasculogenic mesoderm and endothelial cells fail to develop in *Xenopus* embryos lacking the FGF-receptor-1 activity (Amaya *et al.,* 1991). Despite the crucial function of FGFs in mesoderm induction and possibly in the induction of *flk-1* expression, these molecules are not believed to function during subsequent morphogenesis of the vasculature (as will be discussed later).

B. Flk-1/VEGF

Angioblast differentiation in the mesoderm requires the activity of vascular endothelial growth factor (VEGF) and its receptor, Flk-1. VEGF and Flk-1 act in a paracrine signaling system, with VEGF expression generally restricted to the endoderm and ectoderm and Flk-1 expression in the mesodermal endothelial cells (Breier *et al.,* 1995; Flamme *et al.,* 1995a; Cleaver *et al.,* 1997). Deficiency of either molecule results in significant vascular abnormalities. In the case of *VEGF,* loss of function of even a single copy of the gene is lethal. Heterozygous mice carrying a single functional copy of the *VEGF* gene die on Embryonic (E) Day 10.5 because of severe perturbation of vessel development, including the disruption of dorsal aorta formation (Carmeliet *et al.,* 1996a; Ferrara *et al.,* 1996). Differentiation of endothelial cells, growth of existing vessels, lumen formation, and spatial organization of vessels are also significantly impaired. This dramatic heterozygous phenotype suggests that threshold levels of VEGF are critical for most steps of vascular development. Homozygous mutant mice die at the same developmental stage as the heterozygous mutant mice; however, they show much more severe vascular abnormalities and tissue necrosis (Carmeliet *et al.,* 1996a). At E8.5, mutant mice lack the dorsal aorta over its entire length. These embryos also show reduced (but significant) expression of endothelial markers (*flk-1, flt-1, tie-2,* and *PECAM/CD31*), suggesting that endothelial cell development is delayed but not completely eliminated.

Ablation of Flk-1 function in mice causes severe disruption of angioblast development, leading to a total absence of blood vessel formation (Shalaby *et al.,* 1995). In addition, these embryos lack the hemangioblastic cell lineage and do not develop blood. These experiments demonstrate the central role of the VEGF/Flk-1 signaling pathway for vascular development. The difference between the VEGF and Flk-1 mutant phenotypes has led to speculation that another Flk-1 ligand (such as VEGF-B or VEGF-C) may be active during early mesoderm induction and could partially rescue the VEGF knockout (Breier and Risau, 1996). It seems extremely likely, therefore, that Flk-1 and VEGF are required for angioblast differentiation, but that the amounts and the activity of the different VEGF ligands determine angioblast survival (Risau, 1997).

IV. The Blood Islands

The earliest discernible vascular structures in the avian embryo are the blood islands, which arise in the yolk sac outside the embryo (Romanoff, 1960). Blood islands have also been observed in the mesodermal layer of the murine yolk sac and in the ventral-most mesoderm of the *Xenopus* embryo (Haar and Ackerman, 1971; Bertwistle *et al.,* 1996). Teleosts such as zebrafish do not have yolk sac blood islands and embryonic erythropoiesis occurs in the intermediate cell mass, which forms during gastrulation (Al-Adhami and Kunz, 1977; Colle-Vandevelde, 1963). The blood island anlagen give rise to hemangioblastic focal aggregations, in which the peripheral cells differentiate into endothelial cells and the inner cells become blood cells (Wilt, 1974; Pardanaud *et al.,* 1987; Peault *et al.,* 1988). The distinct developmental potential of these cell types has been shown in experiments in which the inner cells are removed and blood formation is precluded without affecting the development of vascular structures (Goss, 1928). Later in development, after the blood islands have formed in the splanchnopleura, they anastomose (make connections) to form a continuous primary vascular network (Jolley, 1940; Houser *et al.*, 1961; Haar and Ackerman, 1971). In the avian embryo, endothelial cells in intraembryonic tissues differentiate shortly after the appearance of the blood islands in extraembryonic tissue. The differentiation of these intraembryonic angioblasts is not usually associated with a concomitant differentiation of hematopoietic cells. The only exception is a small region of the aorta in which paraaortic clusters and intraaortic clusters are candidate sites for the production of intraembryonic hematopoietic cells (Miller and McWhorter, 1914; Cormier *et al.*, 1986; Olah *et al.*, 1988; Garora-Porrero *et al.*, 1995; Godin *et al.*, 1995; Tavian *et al.*, 1996; Dieterlen-Lièvre *et al.*, 1997).

In *Xenopus*, a single blood island is found in the ventral-most mesoderm, where a patch of hematopoietic precursors which will give rise to the embryonic blood is surrounded at the edges by a wide ring of endothelial cell precursors expressing the endothelial marker *flk-1* (Kau and Turpen, 1983; Bertwistle *et al.*, 1996; Cleaver *et al.*, 1997). It is not clear whether these lineages share a common precursor. Although the amphibian blood island is functionally equivalent to the blood islands of the chick and mouse, it differs in three-dimensional structure and is also intraembryonic. This is not surprising since the morphology of blood islands varies greatly between different classes of vertebrates and even between different species within a class (Goss, 1928). Later during development, the dorsal lateral plate mesoderm of *Xenopus* gives rise to hematopoietic precursors; however, these exclusively contribute to adult blood (Kau and Turpen, 1983; Maeno *et al.*, 1985). This region of mesoderm expresses *flk-1* as well as members of the *GATA* gene family which mark blood precursors (Bertwistle *et al.*, 1996; Cleaver *et al.*, 1997). Again, it is unclear if these blood and vascular lineages share a common precursor.

V. The Hemangioblast

The intimate temporal and spatial association of hematopoietic and endothelial cell development has led to the hypothesis that both lineages arise from a common precursor. This putative precursor cell has been called the hemangioblast (His, 1900; Sabin, 1920; Murray, 1932; Wagner, 1980). Although the hemangioblast theory is nearly a century old, the existence of a cell with both endothelial and hematopoietic potential has not yet been confirmed. However, a number of lines of evidence suggest that the hemangioblast exists. Studies of the receptor tyrosine kinase gene *flk-1* demonstrate that it is expressed in the extraembryonic yolk sac blood islands that contain both hematopoietic and endothelial lineages. This expression, however, is only maintained in the endothelial precursors (Matthews *et al.*, 1991; Eichmann *et al.*, 1993; Yamaguchi *et al.*, 1993; Millauer *et al.*, 1993; Palis *et al.*, 1995; Dumont *et al.*, 1995). Gene ablation studies subsequently demonstrated that mice mutant in the *flk-1* gene develop neither blood nor vascular tissue (Shalaby *et al.*, 1995), suggesting that a single cell type may be affected early during development. These expression and functional analyses imply that Flk-1 is present and necessary in both the hematopoietic and endothelial cell types, but the signals that determine the commitment to the angioblastic or hematopoietic cell lineages remain unknown.

In addition to *flk-1*, the hematopoietic and endothelial cell lineages express other genes in common during early embryogenesis. These include the Tie and Tek (Tie-2) receptor tyrosine kinases (Dumont *et al.*, 1992; Partanen *et al.*, 1992; Iwama *et al.*, 1993; Sato *et al.*, 1993; Schnurch and Risau, 1993; Korhonen *et al.*, 1994), the QH1 and MB1 antigens (Peault *et al.*, 1983; LaBastie *et al.*, 1986; Pardanaud *et al.*, 1987), TGF-β1 (Akhurst *et al.*, 1990), the transcription factor c-ets-1 (Pardanaud and Dieterlen-Lièvre, 1993), the cell adhesion molecules PECAM-1 (Newman *et al.*, 1990; Baldwin *et al.*, 1994) and CD34 (Fina *et al.*, 1990), the angiotensin-converting enzyme (ACE) (Caldwell *et al.*, 1976), the von Willebrand factor (Jaffe *et al.*, 1973; Hormia *et al.*, 1984), the cell adhesion glycoproteins P-selectin and E-selectin (Gotsch *et al.*, 1994), and the transcription factor SCL/TAL-1 (Begley *et al.*, 1989; Kallianpur, 1994). In many cases, expression of the genes that encode these molecules is maintained in only one lineage. For instance, during early development of the mouse, SCL/TAL-1 is expressed in both the hematopoietic and the endothelial cell lineages. However, SCL/TAL-1 appears to be required only in the hematopoietic cells (Porcher *et al.*, 1996). Homozygous mutant mice do not develop any of the hematopoietic cell lineages, such as red cells, myeloid cells, mast cells, or T and B cells. It is possible that the early expression of SCL/TAL-1 marks hemangioblastic cells but that the endothelial and blood cell lineages diverge very early and have different requirements for continuing SCL/TAL-1 expression.

Another line of evidence supporting the hemangioblast hypothesis is the spontaneous formation of hematopoietic and angioblastic cell lineages in avian epiblast cultured *in vitro*, which suggests that committed hemangioblast precursors may already be present in the epiblast prior to gastrulation (Murray, 1932; Azar and Eyal-Giladi, 1979; Zagris, 1980; Mitrani *et al.*, 1990). However, when avian epiblast cells are dissociated, cultured *in vitro,* and allowed to form embryoid bodies, no blood island differentiation is observed (Yablonka-Reuveni, 1989; Flamme and Risau, 1992). Addition of FGF to dissociated avian epiblast cells restores the capacity to form blood islands, suggesting that a single growth factor is necessary for the induction of both cell lineages or possibly for the induction of a common hemangioblast precursor (Flamme and Risau, 1992; Krah *et al.,* 1994). This is in contrast to mouse embryonic stem cell-derived embryoid bodies in which blood islands develop without the addition of FGF (Risau *et al.,* 1988). The difference in the appearance of blood islands in the avian and mouse systems has not been explained, but it may involve different endogenous levels of FGF. Thus, it is possible that FGF is necessary for the induction of the hemangioblast cell lineage in both the avian epiblast and the mouse embryoid bodies but that FGF activity is lost from the epiblast upon dissociation.

VI. Endothelial Proliferation

Once endothelial cells differentiate in the embryo, they proliferate and migrate before assembling into blood vessels. It is not until the vascular network has matured in the adult that endothelial cells become quiescent. In the mature vasculature, their turnover is extremely slow. The factors regulating endothelial cell proliferation have been extensively investigated. Both FGF and VEGF are mitogens of capillary endothelial cells in culture (Folkman and Shing, 1992), but FGF is not endothelial cell specific since it also stimulates the growth of a number of other cell types, including smooth muscle cells, fibroblasts, and certain epithelial cells (Shing *et al.,* 1984; Esch *et al.,* 1985). On the other hand, VEGF is a potent and specific mitogen for endothelial cells (Ferrara and Henzel, 1989; Gospodarowicz *et al.,* 1989; Levy *et al.,* 1989; Keyt *et al.,* 1996; for review, see Ferrara *et al.,* 1992). Addition of exogenous VEGF to developing embryos causes dramatic alterations to vascular structures. For example, ectopic VEGF leads to the formation of hyperfused vessels and ectopic vascular development in quail embryos (Drake and Little, 1995), increases in vascular density in the chick limb bud (Flamme *et al.,* 1995b), increased endothelial proliferation in the chick chrorioallantoic membrane (Wilting *et al.,* 1996), and the formation of ectopic vascular structures in the frog embryo (Cleaver *et al.,* 1997). Another growth factor, platelet-derived growth factor (PDGF), is also implicated in endothelial proliferation since it acts as an inducer of angiogenesis *in vivo* and is chemotactic for endothelial cells (Miyazono *et al.,* 1987; Ishikawa *et al.,* 1989; Battegay *et al.,* 1994). The endothelial cells of capillaries express both PDGF-B and its receptor, PDGF-β, suggesting an autocrine stimulatory system (Hermansson *et al.,* 1988; Holmgren *et al.,* 1991). This suggestion is reinforced by *in vitro* experiments which indicate that PDGF influences the angiogenic proliferation of endothelial cells in an autocrine fashion (Ishikawa *et al.,* 1989; Battegay *et al.,* 1994).

Certain factors have been demonstrated to inhibit the angiogenic proliferation of endothelial cells (Klagsbrun and D'Amore, 1991), including thrombospondin (Good *et al.,* 1990), platelet factor IV (Taylor and Folkman, 1982), γ-interferon (Friesel *et al.,* 1987), protamine (Taylor and Folkman, 1982), angiostatin (O'Reilley *et al.,* 1994), and TNF-α (Folkman and Shing, 1992). TGF-β is also a negative regulator of endothelial cell growth, having been shown to inhibit both endothelial cell proliferation (Heimark *et al.,* 1986; Muller *et al.,* 1987; Antonelli-Orlidge *et al.,* 1989) and migration (Sato and Rifkin, 1989). Hyaluronic acid (HA) also downregulates endothelial cell proliferation. In the avian limb bud, ectoderm-derived HA is thought to be responsible for maintaining a subectodermal avascular zone, and ectopic HA has been shown to induce avascular zones (Feinberg and Beebe, 1983). Interestingly, degradation products of HA seem to promote angiogenic proliferation in healing wounds (West *et al.,* 1985; Thompson *et al.,* 1992).

VII. Assembly of Blood Vessels

The formation of the mature vascular system is achieved by a coordination of vasculogenesis and angiogenesis (Risau *et al.,* 1988; Pardanaud *et al.,* 1989). Vasculogenesis is almost exclusively limited to the establishment of the primary vascular plexus in the embryo, whereas angiogenesis extends and remodels the primitive embryonic vasculature and is required for the normal growth of embryonic tissues. Vasculogenesis and angiogenesis are two distinctly different cellular mechanisms and it is likely that they are regulated by different molecular mechanisms.

A. Vasculogenesis

The earliest step in the development of the vascular system is the specification of mesodermal cells to become endothelial cells. These cells soon organize into a primitive vascular plexus via the process of vasculogenesis (Fig. 3A). Vasculogenesis is defined as the coalescence of free angioblasts into loose cords or the fusion of blood islands (Sabin, 1917; Poole and Coffin, 1989). Some definitions also state that this assembly of angioblasts must occur *in situ,* i.e., in the absence of signif-

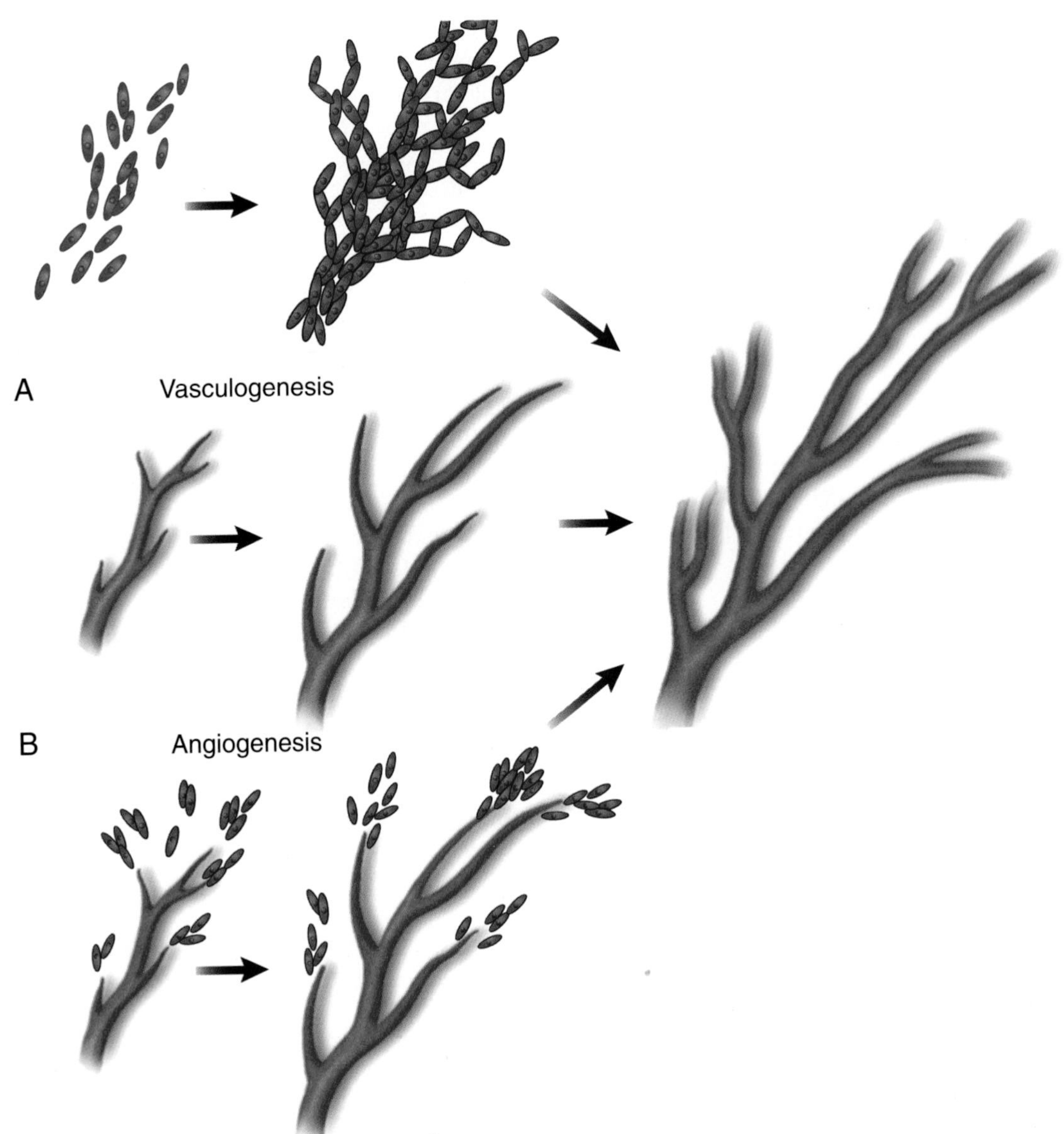

Figure 3 Schematic representation of the basic mechanisms of vascular development. (A) Vasculogenesis is the aggregation of angioblasts in the mesoderm to form blood vessels. Angioblasts either coalesce at the location where they emerge from the mesoderm or they migrate through tissues and form blood vessels at a distant site. (B) Angiogenesis involves the formation of new vessels from preexisting vessels. One form of angiogenesis, sprouting angiogenesis, is responsible for the growth of blood vessels into most developing organs. It involves both the proliferation and the migration of endothelial cells at the tips of the angiogenic sprouts. (C) In some cases, for example, the lung, both vasculogenesis and angiogenesis occur concurrently.

icant cell migration, but as will be explained later, this is not always the case. The vascular cords later mature into embryonic blood vessels at the same position that they originally assembled. Vasculogenesis is therefore responsible for the formation of the primordia of the major blood vessels as well as a rather homogeneous capillary network, which will eventually be modeled into the mature vascular network.

Formation of the blood islands, the dorsal aortae, the endocardium, and the cardinal and vitelline veins is accomplished by vasculogenesis (Nelsen, 1953; Coffin and Poole, 1988; Poole and Coffin, 1989; Pardanaud *et al.*, 1987; 1989; Kadokawa *et al.*, 1990; Coffin *et al.*, 1991; Risau and Flamme, 1995). Although, as a general rule, establishment of the vasculature of most organs occurs by angiogenesis (see below), the vascular network of certain endodermal organs, including liver, lung, pancreas, stomach/intestine, and spleen, occurs by vasculogenesis (Sherer, 1991; Pardanaud *et al.,* 1989). Overall, vasculogenesis involves a closely coordinated and sequential series of steps, including differentiation, migration, adhesion, and maturation, that results in the coalescence of individual migratory angioblasts into a continuous tubular endothelium (Coffin and Poole, 1988).

The development of the endocardium by vasculogenic mechanisms has been most completely described in the quail and mouse embryo. Experiments in quail, using the QH-1 antibody, show that the endocardium arises by the coalescence of migrating angioblasts that form vascular cords which then mature into a vessel (Coffin and Poole, 1988). These angioblasts originate around the headfold, begin to form cords at the periphery of the anterior intestinal portal (AIP), and then grow craniomediad to fuse and form the single straight tube of the endothelium of the heart. The ventral edge of the AIP thus provides a path for the migration of the endocardial precursor cells. Interestingly, blockage experiments show that alternative sources of angioblasts can contribute to the endocardium when the migratory path of the normal cardiac endothelial cells is obstructed. Thus, signals in the local environment, rather than an inherent developmental program, appear to be important in the recruitment of angioblasts into endocardial morphogenesis. In the mouse embryo, a slightly different series of events leads to endocardial development and evidence suggests that the single endocardium does not form by fusion of two chords of endothelial precursor cells. In this case, it appears that an extensive vascular plexus, lying adjacent to the promyocardial layer, undergoes remodeling to form a single endothelial tube (De Ruiter *et al.,* 1992).

The fusion of blood islands into a capillary plexus via vasculogenesis seems to require additional vasculogenic factors present in the embryo. The formation of a capillary plexus will not occur in embryoid bodies derived from mouse embryonic stem cells unless they are first implanted in the peritoneum of host mice or on a chick chorioallantoic membrane, which suggests that factors are required for vasculogenesis which are not present in the embryoid bodies (Risau *et al.,* 1988; Risau, 1991). This hypothesis is supported by other experimental observations. For example, retinoic acid-deficient quail embryos show normal blood island development in the yolk sac, but these never anastomose to create the vessels which would normally fuse with the intraembryonic vasculature (Heine *et al.,* 1985). Also, in chimeric mice expressing the polyoma *middle-T* oncogene, blood islands develop into large hemangiomas that never undergo fusion by vasculogenesis (Williams *et al.,* 1988). Although the roles of retinoic acid and the *middle-T* oncogene on blood island development are not known, these observations suggest that some signal normally present in the early embryo is required for plexus formation via vasculogenesis.

Based on experiments in quail, Poole and Coffin (1991) distinguished two types of vasculogenesis. In vasculogenesis type I, the angioblasts associate to form a mature vessel *in situ* at the location where they differentiate in the mesoderm. In vasculogenesis type I, there is no significant migration of angioblasts. In vasculogenesis type II, angioblasts may migrate significant distances from their original location and then associate into a vessel at a distant location. In the quail embryo, the dorsal aorta arises via vasculogenesis type I and the endocardium, and posterior cardinal veins arise via vasculogenesis type II (Poole and Coffin, 1991). In addition, the perineural vascular plexus in the avian embryo arises via vasculogenesis type II (Noden, 1989). In contrast to the situation in birds, studies of the frog *Xenopus* show that the cardinal veins develop via vasculogenesis type I. In this case, groups of angioblasts originate in the lateral plate mesoderm under the somites and coalesce into vessels at this site on each side of the embryo. Very soon after the appearance of the angioblasts in the lateral plate, a subset of the cells migrate toward the dorsal midline, under the somites, and associate to form a single dorsal aorta at a new location under the notochord. Therefore, in *Xenopus* the dorsal aorta is formed via vasculogenesis type II (Cleaver and Krieg, 1998).

B. Angiogenesis

After the establishment of the primitive vascular plexus, vascular structures are extended and propagated into avascular tissues via a process called sprouting angiogenesis. In addition, the structure of the primitive vascular plexus is modified by the splitting (or fusion) of established vessels via a process called nonsprouting angiogenesis, or intussusception (Folkman and Klagsburn, 1987; Klagsbrun and D'Amore, 1991; Patan *et al.,* 1996). These two types of angiogenesis involve distinct mechanisms for achieving blood vessel modifications (Risau, 1997). The features distinguishing each type of angiogenesis are described.

The first type of angiogenesis involves true sprouting of capillaries from preexisting blood vessels of the established primary vascular plexus (Fig. 3B). In this case, proteolytic degradation of the extracellular matrix is coupled with proliferation of the sprouting endothelial cells to allow the formation of a coherent extension from the primary vessel. Addition of new cells to the tip of an angiogenic sprout occurs by mitotic proliferation of the existing endothelial cells (Wagner, 1980), and these endothelial cells exhibit extensive migratory ability (Clark and Clark, 1939). In angiogenic extensions in the brain, the endothelial cells at the tip of the sprout exhibit a number of filiform processes that may serve to seek out and fuse with other such processes (Klosovskii, 1963). It is not known how individual sprouts find each other to establish necessary connections, but the filamentous or bulbous tips that characterize different angiogenic sprouts may represent different pathfinding mechanisms (Abell, 1946; Wagner, 1980; Wilting and Christ, 1997). As the new vessel extends and takes shape, endothelial cells begin to differentiate and the basement membrane forms along the newly sprouting structure (Ausprunk and Folkman, 1977). This endothelial differentiation involves the formation of a lumen and functional maturation of the endothelial cells.

The vascularization of many extraembryonic and intraembryonic tissues occurs by sprouting angiogenesis. Tissues include the yolk sac, embryonic kidney, thymus, brain, limb bud, and choroid plexus (Ekblom *et al.,* 1982; Le Lièvre and Le Douarin, 1975; Bär, 1980; Joterau and Le Douarin, 1978; Stewart and Wiley, 1981; Wilson, 1983). In addition, the intersomitic veins and arteries are formed by sprouting angiogenesis (Coffin and Poole, 1988). In some instances, angiogenesis can occur simultaneously with vasculogenesis (Fig. 3C), for example, during development of the vasculature of the lung (Baldwin, 1996). Factors that induce angiogenesis *in vitro* have been isolated from many of these tissues or organs (Risau and Ekblom, 1986; Risau, 1986). Sprouting angiogenesis is also the predominant mechanism of vessel formation in the later embryo and in the adult. This includes the neovascularization that accompanies the normal processes of somatic growth, corpus luteum formation, placental formation, and tissue regeneration (Sariola *et al.,* 1983; Folkman and Klagsburn, 1987; Kadokawa *et al.,* 1990; Klagsbrun and D'Amore, 1991; Augustin *et al.,* 1995). In adults, sprouting angiogenesis is also linked to pathological processes, such as tumor growth, inflammatory reactions, wound healing, and diabetic retinopathies (Sholley *et al.,* 1984; Folkman and Shing, 1992; Ferrara, 1995; Folkman, 1995; Hanahan and Folkman, 1996).

The second mechanism of angiogenesis is called intussusception, or nonsprouting angiogenesis, and it involves the splitting of preexisting vessels and the consequent expansion of capillary networks (Short, 1950; Caduff *et al.,* 1986; Burri and Tarek, 1990; Patan *et al.,* 1992, 1993, 1996). Intussusception occurs by proliferation of endothelial cells within a vessel. This proliferation results in the formation of a large lumen that is then split by the insertion of tissues columns, called tissue pillars or posts (Fig. 4A). These pillars are formed when a disk-like zone of contact is established between opposite walls of a vessel, in a manner reminiscent of a

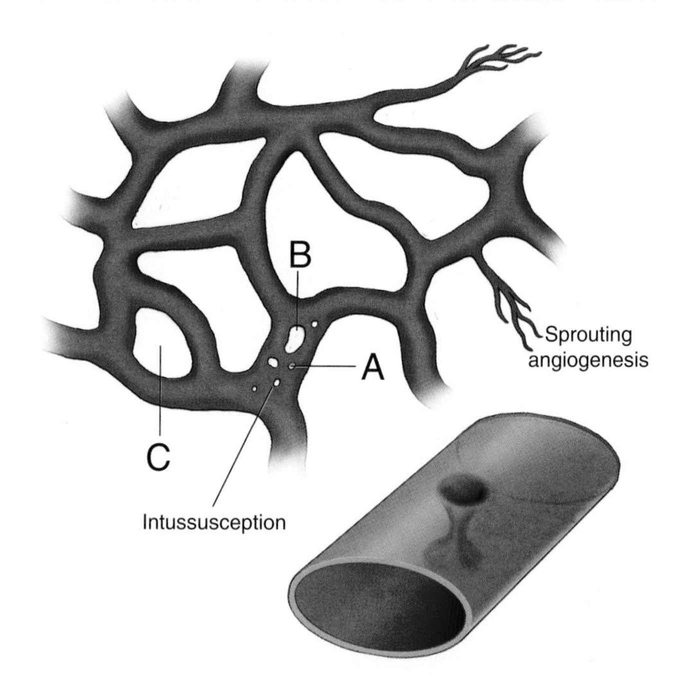

Figure 4 Intussusception, or nonsprouting angiogenesis. In capillary networks, sprouting angiogenesis and nonsprouting angiogenesis, or intussusception, can occur simultaneously. Intussusception involves the formation of transcapillary pillars or posts which split a capillary blood vessel into two. Initially, the pillar creates a small intervascular space (A), but the space subsequently enlarges (B and C) and forms a much larger intervascular region. After they have formed, the resulting intervascular spaces cannot be distinguished from those created by sprouting angiogenesis.

stalactite extending from the roof of a cave toward the floor. In the zone of contact between the opposite endothelial walls, intercellular junctions are formed between the endothelial cells. The contact zone then becomes perforated centrally, forming a canal within the pillar. This canal becomes invaded by cytoplasmic processes of myofibroblasts and pericytes and is eventually stabilized by the deposition of connective tissue fibers such as collagen. It has been demonstrated that the pillar can create a perilous obstacle for circulating erythrocytes, which tend to become damaged as they collide with it (Patan *et al.,* 1993). As a result of this process, the pillar splits a single vessel into two vessels at this location. The pillar can now enlarge along the length of the vessel, fully splitting a single vessel into two (Fig. 4B–4C). Intussusception can occur concurrently with sprouting angiogenesis (Patan *et al.,* 1996; Risau, 1997). In the developing lung, both these processes occur; however, intussusception predominates. This is in distinct contrast to the brain, in which sprouting angiogenesis predominates. Intussusception also occurs in the avian yolk sac, the myocardium, and the chorioallantoic membrane (Flamme *et al.,* 1992;

Short, 1950; Burri and Tarek, 1990; Van Groningen *et al.*, 1991; Patan *et al.*, 1993, 1996).

A second type of nonsprouting angiogenesis has been described as "the intercalated growth of blood vessels" (Folkman, 1987). This type of vascular growth results from the mitosis of endothelial cells within a vessel and leads to the increase in vessel diameter or length. This process is important during healing of endothelial wounds (Reidy and Schwartz, 1981) and in the development of the coronary arteries (Bogers *et al.*, 1989). Sprouting angiogenesis, intussusception, and intercalated growth have been described by Wilting and Christ (1997) as distinct mechanisms of angiotropic growth (development of endothelial cells from preexisting ones), whereas the development of endothelial cells from angioblasts has been defined as angioblastic development.

C. Molecules Involved in Vasculogenesis and Angiogenesis

A large number of molecules are capable of modulating vasculogenic and angiogenic activity (D'Amore and Klagsbrun, 1989; Folkman and Klagsbrun, 1987). These molecules can be soluble, associated with the cell membrane, or associated with the extracellular matrix. Examples of soluble modulators of vascular development include a diverse group of low-molecular-weight factors such as prostaglandins (Form and Auerbach, 1983), nicotinamide (Kull *et al.*, 1987), 1-butyryl-glycerol (Dobson *et al.*, 1990), and okadaic acid (Oikawa *et al.*, 1992). However, the actual role of these molecules in angiogenesis has been difficult to ascertain and we limit discussion to other molecules that have been more fully characterized.

Molecules involved in vasculogenesis or angiogenesis whose biochemical and biological properties have been well described include many polypeptide angiogenic factors. Examples include angiogenin (Fett *et al.*, 1985) and angiotropin (Hockel *et al.*, 1988), as well as polypeptide growth factors such as FGFs (Friesel and Maciag, 1995; Montesano *et al.*, 1986), PDGFs (Holmgren *et al.*, 1991), and TGF-α (Heimark *et al.*, 1986). Some growth factors are specific for endothelial cells and are responsible for specific aspects of endothelial cell behavior; consequently, their receptors are endothelial cell-specific markers. We will discuss four receptor tyrosine kinases, Flk-1, Flt-1, Tie-1, and Tie-2, which are abundantly expressed in the endothelium during both vasculogenesis and angiogenesis (Mustonen and Alitalo, 1995). In addition to growth factors and their receptors, extracellular matrix components have been shown to influence specific aspects of endothelial cell behavior. Fibronectin and laminin, for example, have been implicated in endothelial cell proliferation and subsequent maturation, respectively. Finally, we will discuss cell adhesion molecules, such as certain integrins and cadherins, which also play an important role in vascular development.

1. Growth Factors and Their Receptors

a. Flk-1. The receptor tyrosine kinase Flk-1, a high-affinity receptor for VEGF, is critical for both vasculogenesis and angiogenesis. Flk-1 is initially present in precursors to both blood and endothelium, but it soon becomes restricted to endothelial precursor cells. Expression of flk-1 is particularly high during embryonic neovascularization and during tumor angiogenesis (Millauer et al., 1993; Yamaguchi *et al.*, 1993; Plate *et al.*, 1993; Dumont *et al., 1995;* Flamme *et al.*, 1995a; Liao *et al.*, 1997; Fouquet *et al.*, 1997; Sumoy *et al.*, 1997; Cleaver *et al.*, 1997). The importance of Flk-1 for vascular development is dramatically revealed in gene targeting experiments. Mice which lack the function of Flk-1 die between E 8.5 and 9.5 as a result of defects in the development of both endothelial and hematopoietic cell lineages (Shalaby *et al.*, 1995). More specifically, endothelial precursor cells do not coalesce into blood vessels by vasculogenic aggregation. In separate experiments, a Flk-1 dominant-negative construction has been used to eliminate Flk-1 function in glioblastoma cells subcutaneously implanted into nude mice (Millauer *et al.*, 1994). The angiogenic growth of vascular tissue in these tumors is significantly inhibited, demonstrating the importance of Flk-1 in angiogenesis in general and solid tumor growth in particular.

b. Flt-1. The receptor tyrosine kinase Flt-1 (also called VEGFR1) is closely related to Flk-1 and shows similarities to Flk-1 in both overall structure and expression distribution (Shibuya *et al.*, 1990; de Vries *et al.*, 1992). Like Flk-1, it is a high-affinity receptor for VEGF and also for the VEGF-related growth factor, placental growth factor (de Vries *et al.*, 1992; Waltenberger *et al.*, 1994). Flt-1 expression is associated with vascular development in mouse embryos and with neovascularization in wound healing (Peters *et al.*, 1993). Unlike Flk-1, however, Flt-1 is maintained in the differentiated endothelium of adult vascular tissues, suggesting that it has a function in the quiescent endothelia of mature vessels. Targeted mutation of the *flt-1* gene results in embryos which develop endothelial cells in both intra- and extraembryonic tissues, but these endothelial cells do not properly assemble and organize into vessels (Fong *et al.*, 1995). In these mutant embryos, all examined vascular structures, including the major embryonic vessels, extraembryonic vessels, endocardium, and capillary networks, are disrupted. These results suggest that it is vasculogenesis, rather than endothelial cell specification, which is impaired since endothelial cell formation and proliferation are unaffected by the disruption of the *flt-1* gene. In fact, an increase in endothelial cell number is reported in the yolk sac and the endocardium that may be the result of a failure of contact inhibition. It has thus been suggested that the Flt-1 signaling pathway may be involved in regulating the adhesion of endothelial cells to each other or to the extracellular matrix (Fong *et al.*, 1995).

c. Tie-2, Angiopoietin-1, and Angiopoietin-2. Another receptor tyrosine kinase important in vasculogenesis is the receptor Tie-2 (or Tek). In the mouse, *tie-2* is expressed in endothelial precursors shortly after the onset of *flk-1* expression (Dumont *et al.*, 1992, 1995). Mice lacking Tie-2 function die at E 10.5, with defects in the integrity of the endothelium and defects in cardiac development (Dumont *et al.*, 1994). These mutant embryos show multiple signs of vascular hemorrhage, possibly due to the failure of either endothelial proliferation or survival. Mutant embryos also exhibit distended yolk sac vessels and a ruptured and disorganized dorsal aorta. The relative number of endothelial cells decreases as development proceeds, with almost 75% less endothelial cells present in the embryo by E 9.0 relative to heterozygous controls. These experiments suggest that Tie-2, although not absolutely required for the differentiation of the endothelial cell lineage, is necessary for the expansion and maintenance of the lineage as vessels form by vasculogenesis and the embryo grows in size.

Independent experiments have demonstrated that Tie-2 is also necessary for sprouting angiogenesis (Sato *et al.*, 1995). In addition to the early defects in endothelial cell survival, targeted mutation in the mouse leads to a failure of proper formation of the endothelial lining of the heart and to an absence of capillary angiogenesis in the neurectoderm (Sato *et al.*, 1995). The mutant mice display uniformly dilated vessels in the perineural plexus, abnormal and dilated vascular network formation in the yolk sac, and a failure of extensive branching of vessels in the myocardium. Large and small vessels are not distinguishable, suggesting a defect in vascular remodeling. Because of these abnormalities in lumen diameter, it has been proposed that Tie-2 either modulates the activity of VEGF, which in turn regulates both intussusceptive and sprouting angiogenesis, or is involved in recruitment of the vascular cell wall components, which play a key role in endothelial integrity. Tie-2 may therefore be involved in the regulation of endothelial cell angiogenesis, in addition to proliferation, maturation, and survival.

Two ligands for Tie-2 have been identified and are called the angiopoietins (Davis *et al.*, 1996; Suri *et al.*, 1996). Angiopoietin-1 is expressed in close proximity to developing blood vessels in the mouse embryo, including the mesenchyme surrounding most vessels and the myocardium of the heart (Suri *et al.*, 1996). It does not, however, directly promote the proliferation of endothelial cells or tube formation *in vitro* (Davis *et al.*, 1996). Targeted mutation of the *angiopoietin-1* gene results in an overall phenotype very similar to that observed in *tie-2* mutant mice—basically a vascular network lacking complexity of branching and heterogeneity of vessel size. In addition, these mice exhibit a severe failure in the recruitment of vascular cell wall components. This observation suggests that this receptor–ligand pair is important for both the angiogenic processes that occur after initial embryonic vasculogenesis and subsequent events of vascular maturation (Suri *et al.*, 1996). Angiopoietin-2 is a second ligand for Tie-2, and in this case it appears to act as an antagonist to Tie-2 function (Maisonpierre *et al.*, 1997). While angiopoietin-2 binds Tie-2 with an affinity equivalent to angiopoietin-1, it does not activate Tie-2, and so its overall effect is to inhibit angiopoietin-1 signaling. Angiopoietin-2 expression in the embryo is detected in the smooth muscle layer underlying the endothelium and also in the dorsal aorta and the major aortic arches. The expression occurs in a punctate pattern that possibly represents either endothelial cells or endothelium-associated mesenchymal cells. In the adult, angiopoietin-2 is associated with sites of vascular remodeling. In summary, Tie-2 and its ligands appear to regulate vascular development after endothelial specification and after establishment of the primitive vascular network.

d. FGF. The fibroblast growth factor family is composed of mitogens for a broad range of cell types and mediators of many developmental and pathophysiological processes. Of the nine members of the family, FGF-1 (aFGF) and FGF-2 (bFGF) are both modifiers of angiogenesis (Fernig and Gallagher, 1994; Friesel and Maciag, 1995). aFGF secreted by a rat bladder carcinoma cell line promotes angiogenesis both *in vitro* and *in vivo* (Jouanneau *et al.*, 1995). In experiments in which capillary or aortic endothelial cells are stimulated with aFGF, the capillary endothelial cells respond by increased proliferation (D'Amore and Smith, 1993). These same populations of small and large vessel endothelial cells are stimulated by bFGF. However, despite the proliferative activity of aFGF and bFGF *in vitro,* it appears unlikely that these molecules play a significant role in the development of the vascular system in the embryo after the initial specification of endothelial cells. This interpretation is based on the lack of detectable levels of *FGF–receptor* transcripts in endothelial cells of embryonic tissues (Heuer *et al.,* 1990; Wanaka *et al.,* 1991; Peters *et al.,* 1992). Furthermore, careful analysis of blood vessel growth during brain angiogenesis shows no correlation with the temporal and spatial expression of the FGFs (Emoto *et al.,* 1989). While it is true that the endothelium of certain large vessels does express FGF–receptors and responds to FGFs *in vivo* (Lindner *et al.,* 1990; Peters *et al.,* 1992; Liaw and Schwartz, 1993), this might reflect a function in regenerative mechanisms rather than in embryonic vascular morphogenesis.

e. VEGF. VEGF is known to be mitogenic for endothelial cells and it has been suggested that it may also be chemotactic for endothelial precursors (Breier and Risau, 1996). VEGF is expressed in regions of the embryo that are undergoing both vasculogenesis and angiogenesis (Breier *et al.,* 1992). In general, the *VEGF* gene is expressed in endodermal or ectodermal tissues, whereas *flk-1* is expressed in the adjacent mesoderm (Dumont *et al.,* 1995; Flamme *et al.,* 1995a; Cleaver *et al.,* 1997). Together, VEGF and Flk-1 are thought to be responsible for both primary vessel formation by vasculogenesis and angiogenic invasion of developing organs. In gene ablation studies, mice which lack a single *VEGF* allele die on about E 10.5. These mutant mice show gross abnormalities in vascular development, including defects in the *in situ* differentiation of endothelial cells, sprouting angiogenesis, lumen formation, the formation of large vessels, and the spatial organization of the vasculature (Carmeliet *et al.,* 1996a; Ferrara *et al.,* 1996). The extensive defects evident in this heterozygous lethal phenotype are unprecedented for a targeted gene disruption, and this implies that tight regulation of VEGF levels is essential for correct vascular morphogenesis.

The role of VEGF in angiogenesis has been extensively analyzed. In the early quail embryo, VEGF is strongly expressed in the yolk sac and is likely to be responsible for promoting extensive sprouting of blood vessels (Flamme *et al.,* 1995a). In mouse, VEGF is also found in organs, such as the kidney, which are not juxtaposed to Flk-1-expressing endothelial cells and which are known to be vascularized by angiogenesis (Dumont *et al.,* 1995). It is likely, therefore, that VEGF causes blood vessels to grow into the developing kidney from adjacent vascular structures. Analysis of VEGF in the brain reveals that the spatial and temporal expression correlates well with the ingrowth of blood vessel sprouts into the ventricular neurectoderm layer (Breier *et al.,* 1992; Millauer *et al.,* 1993; Breier and Risau, 1996). VEGF has also been shown to be important in hypoxia-induced angiogenesis (Shweiki *et al.,* 1992; Plate *et al.,* 1992). In experiments using the feline and murine retina, hypoxia upregulates VEGF levels in astrocytes that migrate into the retina, presumably guiding capillary growth (Stone *et al.,* 1995). Angiogenic sprouting from nearby vessels then extends into the retinal tissue. Once the tissue is vascularized and the oxygen need is met, the production of VEGF declines. This upregulation is probably due to increased transcription of the *VEGF* gene mediated by hypoxia-inducible factor-1 (Liu *et al.,* 1995); however, *VEGF* expression is also modulated by a stabilization of *VEGF* mRNA (Ikeda *et al.,* 1995; Semenza, 1996).

f. PDGF. Although the precise role of PDGF and its receptors in vascular development has not been determined, PDGF appears to be involved in endothelial proliferation and angiogenesis. PDGF occurs as a homodimer or heterodimer of two isoforms, PDGF-A and PDGF-B (Beck and D'Amore, 1997). While both PDGF-B and PDGF-A have been implicated in vascular maturation and vascular wall development, PDGF-B is likely to be involved in the autocrine stimulation of endothelial cells and angiogenesis. In the human placenta, transcripts for *PDGF-B* and its high-affinity receptor, *PDGF-β,* are both present in the capillary endothelial cells, implying an autocrine signaling system (Holmgren *et al.,* 1991). The endothelium of larger vessels, however, maintains PDGF-B expression but does not express the PDGF-β receptor, suggesting that a switch from autocrine to paracrine signaling occurs as the endothelium begins to recruit mesenchymal cells into the developing vascular wall. Several other experiments show that capillary endothelial cells are competent to respond to PDGF, presumably due to the presence of the PDGF-β receptor. When PDGF-BB is added to dermal wounds or to the chick chorioallantoic membrane, an increase in capillary density can be observed (Pierce *et al.,* 1992; Risau *et al.,* 1992). In experiments using bovine aortic endothelial cells, which spontaneously undergo angiogenesis *in vitro,* PDGF-BB was shown to be necessary for cord and tube formation of these cells (Battegay *et al.,* 1994). Receptors for PDGF-BB were found exclusively on the extending sprouts and forming endothelial tubes but not on the surrounding endothelial cells of the cultured monolayer. Expression increased as the cells further organized into tube-like structures. Antibodies which blocked the activity of PDGF-BB significantly reduced angiogenic activity, whereas antibodies against PDGF-AA had no effect. Other studies, however, suggest that the role of PDGF on endothelial cells may not always be direct and that PDGF may influence angiogenesis indirectly through the stimulation of other cells. For example, when myofibroblasts and endothelial cells are cultured together *in vitro,* PDGF stimulates the myofibroblasts to secrete an unspecified endothelial growth factor which then causes vasculogenic aggregation of the endothelial cells into cords (Sato *et al.,* 1993).

2. Extracellular Matrix and Cell Adhesion Molecules

In vitro experiments show that extracellular matrix (ECM) can modulate growth and differentiation of endothelial cells and also influence their migration (Risau and Lemmon, 1988; Ausprunk *et al.,* 1991). All these processes are essential for both vasculogenesis and angiogenesis. Extracellular matrix components, such as the adhesive glycoproteins fibronectin, laminin, and vitronectin, the collagens type I, II, IV, and V, the proteoglycans, and the glycosaminoglycans, comprise the environment in which angioblasts migrate and organize into the cords which will form the primary vascular plexus. A number of studies have analyzed the distribution of extracellular matrix molecules in order to determine the correlation with vascular development (Risau and Lemmon, 1988; Ausprunk *et al.,* 1991; Little *et al.,* 1989; Drake *et al.,* 1990). Other *in vitro* studies have directly assayed the ability of these molecules to stimulate endothelial cell proliferation, migration, differentiation, or vascular wall cell recruitment. These studies have included two-dimensional assays, three-dimensional collagen gel assays, and serum-free explant cultures of rat aorta (Bischoff, 1995; Grant *et al.,* 1990). Just as the dynamic changes in the composition of the extracellular matrix are important for endothelial behavior, so are the adhesive receptors that regulate the interactions of endothelial cells with their environment. A summary of observations concerning selected extracellular matrix components and cell adhesion molecules and their influence on vascular development is presented later.

a. Fibronectin. Vasculogenesis, the assembly of vessels from free angioblasts, takes place in a fibronectin-rich extracellular matrix (Mayer *et al.,* 1981; Risau and Lemmon, 1988; for review, see Hynes, 1990). In the chick yolk sac, as neighboring blood islands fuse into the primary vascular plexus, they appear to approach each other using fibronectin-rich extensions (Mayer *et al.,* 1981). As soon as the basic vascular network is established, fibronectin decreases in the vicinity of developing blood vessels and endothelial cells begin to produce laminin and collagen IV in increasing amounts. This dynamic pattern of expression of extracellular matrix components occurs during avian blood vessel development in general (Risau and Lemmon, 1988) and more specifically during the development of the endocardium (Drake *et al.,* 1990) and the chick chorioallantoic membrane (Ausprunk *et al.,* 1991). Mice which lack a functional *fibronectin* gene exhibit severe defects in blood vessel and heart development, including a failure of the fusion of the two cardiac primordia and in some cases a complete absence of the endocardium and the dorsal aorta (George *et al.,* 1993). In addition, the extraembryonic vasculature does not develop and blood island development is disrupted, with blood cells accumulating in the exocoelomic cavity. These results emphasize the critical role of fibronectin in the proliferative and migratory events in early vasculogenesis and angiogenesis.

b. Collagens. Members of the collagen family of extracellular matrix components appear to possess different regulatory activities during vascular development. Endothelial tube formation *in vitro* is associated with the deposition of various collagens, including collagens type I and III–V (Iruela-Arispe *et al.,* 1991). When capillary endothelial cells are cultured on interstitial collagens, such as collagens type I and III, rapid proliferation occurs in all directions (Madri and Williams, 1983). However, when basement membrane collagens such as collagen type IV are used as the culture substrate, endothelial cells aggregate and form highly organized tube-like structures. In similar experiments, endothelial cells grown in a three-dimensional collagen type I matrix organized into networks of branching and anastomosing tubes (Montesano *et al.,* 1983). Inhibition of collagen deposition, collagen triple-helix formation, or collagen cross-linking has been shown to prevent angiogenesis (Ingber, 1991). Loss of *collagen type I α-chain* gene function results in the rupture of blood vessels in the developing embryonic vasculature (Löhler *et al.,* 1984). It seems likely that other collagen components of the extracellular matrix will play similar roles during embryonic vascular development and during pathological neovascularization.

c. Integrins ($\alpha_5\beta_1$ *and* $\alpha_v\beta_3$). Integrins are perhaps the best characterized of the cell adhesion molecules involved in vascular development (Luscinckas and Lawler, 1994; Strömblad and Cheresh, 1996). Integrins generally mediate cell–ECM interactions and occasionally cell–cell adhesion. They are heterodimers which consist of an α subunit and a noncovalently associated β subunit, both of which are integral membrane proteins. Many different α and β subunits exist (15 and 8, respectively), and many of these can associate to form different functional receptors (Baldwin, 1996). Large vessel endothelial cells express $\alpha_2\beta_1$, $\alpha_3\beta_1$, $\alpha_5\beta_1$, and $\alpha_v\beta_3$, whereas microvascular endothelial cells express $\alpha_1\beta_1$, $\alpha_6\beta_1$, $\alpha_6\beta_4$, and $\alpha_v\beta_5$ (Luscinckas and Lawler, 1994). These vascular integrins serve as receptors for collagen, laminin, fibronectin, and thrombospondin.

Integrin $\alpha_5\beta_1$ is the receptor for fibronectin, and blocking the function of either subunit results in major defects in early vasculogenesis. Mouse embryos in which integrin α_5 function has been ablated are defective in blood vessel and blood island formation (Yang *et al.*, 1993). This is a phenotype similar to that observed in fibronectin loss of function experiments. Homozygous mutant embryos exhibit severely disrupted blood vessel formation and blood cells can be found accumulating in the exocoelemic space. Embryos then die on E 10 or 11 due to numerous morphological defects. The importance of integrin $\alpha_5\beta_1$ is also demonstrated by experiments in quail embryos, in which blocking the binding of β_1 to its ligands with an anti-integrin antibody (CSAT) results in vasculogenic defects, including failure of lumen formation in the dorsal aorta (Drake *et al.*, 1992). Overall, it appears that loss of integrin $\alpha_5\beta_1$ function causes vasculogenesis to be arrested after the stage when angioblasts form cords but before they have organized into patent tubes.

The β_3 family of integrins is essential for normal angiogenesis and vascular cell survival. For example, integrin $\alpha_v\beta_3$, which interacts with a wide variety of extracellular matrix components, is one of several members of the integrin family that is expressed in endothelial cells and which may be involved in angiogenesis and the maintenance of newly formed capillaries. This receptor complex is observed at the tips of newly formed sprouting blood vessels in human wound granulation tissue but is absent from normal skin (Brooks *et al.*, 1994a; Clark *et al.*, 1996). As the vessels mature, $\alpha_v\beta_3$ expression declines to an undetectable level. During angiogenesis in the chick chorioallantoic membrane, $\alpha_v\beta_3$ expression levels show a fourfold increase. When antibodies are used to block the activity of $\alpha_v\beta_3$ in this assay system, neovascularization is inhibited, whereas preexisting vessels are unaffected (Brooks *et al.*, 1994a). In control experiments, antibodies against the related integrin $\alpha_v\beta_5$ had no effect. Subsequent experiments using integrin $\alpha_v\beta_3$ antagonists showed that apoptosis of proliferative angiogenic endothelial cells occurs when the interaction of $\alpha_v\beta_3$ with its natural substrates (such as vitronectin, fibrin, and fibronectin) is disrupted (Brooks *et al.*, 1994b).

d. Vascular Endothelial Cadherin. Vascular endothelial cadherin (VE-cadherin or cadherin-5) mediates calcium-dependent homophilic binding at adherens junctions between cells and is associated with catenins and the actin cytoskeleton (Dejana *et al.*, 1995; Dejana, 1996). VE-cadherin is expressed specifically in endothelial cells and appears to play a role in the formation of interendothelial junctions (Suzuki *et al.*, 1991; Lampugnani *et al.*, 1992; Breier *et al.*, 1996). Agents that increase monolayer permeability (such as thrombin and elastase) cause a significant decrease in VE-cadherin at cell boundaries, suggesting a specific role in the control of endothelium permeability (Lampugnani *et al.*, 1992). Expression analysis in mouse embryos reveals *VE-cadherin* gene expression in the endothelial precursor cells of the earliest vascular structures, the blood islands, and later in the vasculature of all organs examined, including the endocardium, the dorsal aorta, the intersomitic vessels, and the brain capillaries (Breier *et al.*, 1996). Chinese hamster ovary cells and L292 cells that are transfected with the *VE-cadherin* gene *in vitro* are inhibited from proliferating (Caveda *et al.*, 1996). Gene targeting experiments in which *VE-cadherin* is disrupted in mouse ES-derived embryoid bodies reveal that endothelial cells remain dispersed and fail to organize into vascular structures (Vittet *et al.*, 1997). Together, the functional studies and the embryonic expression pattern suggest that VE-cadherin plays a role in the early events of vasculogenesis.

D. Endothelial Cell Migration

Endothelial cell migration is required during both vasculogenesis and angiogenesis. Many studies have confirmed the migratory ability of angioblasts, especially during vasculogenesis type II, which occurs preceding the formation of some structures of the primitive vascular plexus (Dieterlen-Lièvre, 1984; Noden, 1988, 1990; Poole and Coffin, 1989; Christ *et al.*, 1990; Wilting *et al.*, 1995). From experiments using chick–quail chimeras, it is clear that angioblasts originating in transplanted tissue are highly invasive and may migrate quickly over long distances (Noden, 1989, 1990). These cells invade the surrounding mesenchyme and contribute to the formation of veins, arteries, and capillaries surrounding the site of implantation. Individual angioblasts have the potential to migrate significant distances from their site of origin, and migratory distances up to 400 μm have been observed (Klessinger and Christ, 1996). Regardless of the source of the tissue comprising the transplant, the transplanted endothelial cells become normally integrated into newly forming vessels in adjacent tissues in the host, indicating that regulatory signals responsible for patterning the vasculature reside within individual embryonic tissues and not in the endothelial cells themselves. Despite the highly invasive character of angioblasts, many studies have demonstrated that they never cross the midline of the embryo (Hahn, 1908; Wilting *et al.*, 1995; Wilting and Christ, 1996). Using transplantation experiments similar to those of Noden (1989, 1990), Klessinger and Christ (1996) showed that the notochord appears to be the source of signals which create this barrier.

The migration of angioblasts is required for normal vasculogenesis and immediately precedes the formation of certain vascular structures in the avian embryo, such as the endocardium, the ventral aortae, and the cardinal and intersomitic veins (Poole and Coffin, 1991). Using the QH1 antibody, angioblasts can be observed to migrate individually and in groups toward the sites where they will subsequently assemble into a vessel (Poole and Coffin, 1991). As mentioned previously, endocardial angioblasts migrate from the periphery of the embryo toward the midline, where the heart will form. Blockage experiments which interrupt this path demonstrate that angioblast migration is critical for endocardial formation. Angioblast migration also precedes cardinal vein formation, when clusters of cells migrate mediad, from lateral regions, into the somatopleure, where they will aggregate and organize into vessels along the Wolffian duct (Poole and Coffin, 1991).

Endothelial cell migration is also an important part of angiogenesis. Angiogenic sprouts are generated by mitoses in preexisting endothelial tissue, such as the wall of a blood vessel, and this proliferation is accompanied by extensive migratory activity (Clark and Clark, 1939). Sprouts advance into the interstitium by ameboid migratory activity, and the distal tip of the sprout may invade surrounding tissue or fuse with the endothelium of an adjacent vessel (Wagner, 1980). For instance, in the chick, endothelial cells of the perineural plexus form capillary sprouts that degrade the perineural basement membrane, and then they migrate into and invade the neurectoderm (Bär, 1980).

Recent studies in the frog, *X. laevis,* have also demonstrated the importance of endothelial cell migration in the formation of the primary embryonic vasculature (Cleaver and Krieg, 1998). Using expression of *flk-1* to mark vascular precursor cells, angioblasts are first detected in the late neurula stage embryo in distinct portions of the cephalic region, the lateral plate, and the ventral-most mesoderm. Endothelial cells that will form the cardinal veins lie immediately ventral to the somites on each side of the embryo, but no endothelial cells are present at the position of the future dorsal aorta. At the early tail bud stage, a subset of the lateral endothelial cells actively mi-grates under the somites toward the midline, where they form the dorsal aorta, ventral to the hypochord. The hypochord is a transient structure found in amphibian and fish embryos that lies immediately ventral to the notochord (Gibson, 1910; Lofberg and Collazo, 1997). Once the angioblasts have taken up their position under the hypochord, expression of *flk-1* declines dramatically in both the dorsal aorta and the cardinal veins and these vessels begin to mature. A similar observation of mediad cell migration has been made in zebrafish (Stockard, 1915; Al-Adhami and Kunz, 1977), and, recently, the signals directing this migration have been shown to originate in the vicinity of the notochord (Fouquet *et al.,* 1997). It is likely that the dorsal aorta in fish forms by a comparable mechanism involving endothelial cell migration. In contrast to the situation observed in quail and mouse (Poole and Coffin, 1991; Coffin *et al.,* 1991), the dorsal aorta in the frog and zebrafish arises by vasculogenesis type II (migration of endothelial cells followed by assembly), whereas the posterior cardinal veins in the frog arise by vasculogenesis type I (*in situ* assembly without migration).

E. Molecules Involved in Endothelial Cell Migration

1. Flk-1/VEGF

In frog embryos, the hypochord is a concentrated source of *VEGF* expression (Cleaver *et al.,* 1997). It seems possible that the directed migration of *flk-1*-expressing endothelial cells from the lateral plate mesoderm toward the midline of the frog embryo occurs in response to a VEGF signal gradient. This possibility is supported by the observation that the hypochord primarily expresses the diffusible form of VEGF (O. Cleaver and P. A. Krieg, unpublished results) and that exogenous VEGF can cause aberrant migration and proliferation of endothelial cells in the frog embryo (Cleaver *et al.,* 1997). Furthermore, work in mouse suggests that Flk-1 has a function in the migration of endothelial cells. In homozygous *flk-1* mutant mice, in which the first *flk-1* exons are replaced by a promoterless *β-galactosidase* gene, no mature endothelial or hematopoietic cells are present (Shalaby *et al.,* 1995). However, expression of *β-galactosidase* driven by the *flk-1* promoter is detected at high levels in the region of the connecting stalk and in an aortic arch. In these mutant embryos, it is possible that angioblasts cannot migrate from these sites to the locations where the elements of the primary vascular structure would normally differentiate. In support of this hypothesis, experiments using mouse chimeras show that expression of *flk-1* is required for mesodermal precursor cells to exit the posterior primitive streak and reach their correct location in the blood islands, where they will develop into endothelial and hematopoietic progenitors (Shalaby *et al.,* 1997). Overall, Flk-1 and VEGF signaling appears to be required for endothelial cell migration during early establishment of the vascular system.

2. Fibronectin

Fibronectin is implicated in endothelial cell motility during vascular development (Noden, 1991). First, *in vitro* experiments have demonstrated that fibronectin can stimulate the migration of vascular endothelial cells (Bowersox and Sorgente, 1982; Ungari *et al.*, 1985; Krug *et al.*, 1987). Second, when the distribution of fibronectin is analyzed in the chick embryo or the chick chorioallantoic membrane, it is associated with both migrating endothelial cells, prior to their coalescence into vessels, and with the early steps of angiogenesis, when capillaries are extending and invading avascular tissue (Risau and Lemmon, 1988; Ausprunk *et al.*, 1991). Third, application of a pentapeptide (GRGDS), which blocks the fibronectin receptor on endothelial cells, results in significant inhibition of endothelial cell migration both *in vivo* and *in vitro* (Christ *et al.*, 1990; Nicosia and Bonanno, 1991). One specific example involves the mediad migration of precardiac mesoderm along a gradient of fibronectin (Linask and Lash, 1986). When fibronectin blocking reagents are applied, or the path of migration is rotated by 180°, the growth of precardiac mesoderm is completely blocked (Linask and Lash, 1988a,b).

3. Integrin $\alpha_v\beta_3$

In addition to a role in maintaining, or stabilizing, early vascular structures (discussed previously), integrin $\alpha_v\beta_3$ is also implicated in endothelial cell migration and in proteolytic modification of the extracellular matrix. Invasive angiogenic cells must degrade the underlying basement membrane to extend new sprouts into adjacent tissues. $\alpha_v\beta_3$ has been shown to colocalize with the active form of the matrix-degrading enzyme, matrix metalloproteinase-2, in growing blood vessels and the two have been shown to bind specifically to each other *in vitro* (Brooks *et al.*, 1996). In addition, vitronectin has binding sites for both the $\alpha_v\beta_3$ integrin and the plasminogen activator inhibitor-1 (PAI-1). Since these binding sites overlap (Stefansson and Lawrence, 1996), binding of PAI-1 to vitronectin interferes with binding to its receptor, the $\alpha_v\beta_3$ integrin. It has been suggested that plasminogen activator may bind to PAI-1, displacing it from vitronectin and therefore inducing cell migration by allowing the receptor–ligand interaction. Additional experiments demonstrate that VEGF can upregulate the expression of $\alpha_v\beta_3$ integrin and plasminogen activator, and that both plasminogen activator and PAI-1 are upregulated in migrating endothelial cells (Pepper and Montesano, 1990; Pepper *et al.*, 1991). These interactions provide a molecular basis for the coordination of cell migration and matrix degradation.

VIII. Vascular Pruning and Remodeling

Once the primary capillary plexus is established in the embryo, it is rapidly remodeled and modified by angiogenesis and matures into the familiar continuum of larger and smaller blood vessels. One of the processes by which this characteristic architecture is acquired has been called pruning, by analogy to the process of trimming a tree (Risau, 1997). Pruning was first described in the embryonic retina and involves the removal of excess endothelial cells which form redundant channels (Ashton, 1966). Blood flow generally ceases in these excess capillaries, the lumens are obliterated, and the endothelial cells retract toward adjacent capillaries, leaving behind thin extensions called intercapillary bridges. These endothelial cells do not appear to die by apoptosis (Augustin *et al.*, 1995), but their precise fate is unclear. It has been suggested that they may reassemble into additional vessels or simply dedifferentiate and contribute to alternative tissues (Risau, 1997). One study shows that they may dedifferentiate to become either muscular or supportive components of the vascular cell wall (Ashton, 1966).

In addition to the trimming of excess endothelial cells, the embryonic vasculature undergoes dynamic changes in morphology, called remodeling (Beck and D'Amore, 1997; Risau, 1997). Once the vascular system is mature, it is relatively stable and undergoes angiogenic remodeling only in female reproductive tissues, during wound healing, or during pathological processes such as tumor growth. Remodeling is not well understood, but it is known to involve the growth of new vessels and the regression of others as well as changes in the diameter of vessel lumens and vascular wall thickness. Blood flow is known to be a key regulator of vessel maintenance since unperfused capillaries regress preferentially. Endothelial rearrangements also seem to occur in response to local tissue demands. For example, if a tissue is highly vascularized, and therefore prone to be hyperoxygenized, vessel regression will usually follow (Ashton, 1966; Risau, 1997). However, if a tissue is oxygen deprived, it will usually stimulate angiogenic invasion. Overall, it is likely that only a small number of embryonic blood vessels persist into adulthood (Risau, 1995), with most capillaries of the embryonic plexus regressing at some time in development to allow the differentiation of other tissues. Examples include the regression of capillaries in prechondrogenic regions to allow the differentiation of cartilage (Hallmann *et al.*, 1987) and the regression of the hyaloid vasculature to allow the development of the vitreous body in the eye, which is necessary for proper vision (Latker and Kuwubara, 1981).

IX. Remodeling, Patterning, and Maturation

A variety of factors are involved in the maturation of blood vessels and in the dramatic changes that occur after the circulation of blood cells has been established. In many cases the larger vessels, such as arteries or veins, develop from the fusion of capillaries after the formation of the primary vascular plexus. Early vessels are generally characterized by the presence of thick and plump endothelial cells with tenuous adherence and incomplete basement membrane formation, but this changes as blood flow increases and endothelial cells mature (Wagner, 1980). Anastomoses disappear, capillaries may split by intussusception, the direction of blood flow may reverse many times within a given vessel, and adherence between endothelial cells increases dramatically (Sabin, 1920; Clark and Clark, 1939; Wagner, 1980). As vessels mature, a basement membrane forms, gradually thickens, and becomes less heterogeneous (Wolff and Bär, 1972). Vessels also become shaped by the mechanical forces generated by the circulation, molding the development of large vessels, such as the vitelline veins and arteries (Murphy and Carlson, 1978; Franke *et al.,* 1984; reviewed in Resnick and Gimbrone, 1995). It is also known that hemodynamic forces can cause changes in the expression of a number of endothelial genes, including *PDGF, FGF, TGFβ,* and *tissue factor,* and that these forces can also modify endothelial cell adherence (Griendling and Alexander, 1996; Resnick and Gimbrone, 1995). It is important to note, however, that the determination of growth and pattern within the vasculature does not depend solely on blood pressure since growth and elaboration of blood vessels does proceed in the absence of a heart (Chapman, 1918). Overall, it is likely that final maturation of the vasculature requires interaction of endothelial cells with each other, with the extracellular matrix environment, and with adjacent mesenchymal support cells, such as pericytes and smooth muscle cells.

As the vascular system develops, the initial plexus becomes remodeled into a complex and heterogeneous array of blood vessels, including larger vessels, such as veins and arteries, and smaller vessels, such as venules, arterioles, and capillaries. The endothelia that line these different blood vessels have very different properties (Kumar *et al.,* 1987). For example, the endothelium of large vessels has an important role in controlling vasoconstriction and vasodilation and in the regulation of blood pressure. On the other hand, the endothelium of the small vessels plays a critical role in the exchange of gases and nutrients with the tissues (Risau, 1995). Within the small vessels, capillary endothelium is subdivided into three different phenotypes—continuous, discontinuous, and fenestrated (Bennet *et al.,* 1959; Risau, 1995)—and these morphological differences reflect different permeability requirements of these vessels in different tissues. Continuous capillaries are composed of endothelial cells which are perforated by the vessel lumen (intraendothelial canalization) and have therefore been called "seamless endothelia" (Wolff and Bär, 1972). The lumen formation has been postulated to result from extensive vacuolization and subsequent fusion of vacuoles (Sabin, 1920; Wagner, 1980). Continuous capillaries are found in the central nervous system, the lymph nodes, and muscle. Discontinuous capillaries have clustered pores, each of a diameter of approximately 80–200 μm, located at each end of the endothelial cell. This type of capillary is found in the liver, bone marrow, and spleen. Lastly, fenestrated capillaries exhibit large pores containing a highly permeable diaphragm and are found in the kidney glomeruli, the choroid plexus, the endocrine glands, and the gastrointestinal tract. Fenestrated capillaries are more permeable to low-molecular-weight hydrophilic molecules and this is consistent with their presence in tissues involved in secretion, filtration, and absorption (Levick and Smaje, 1987). VEGF, which is also known as vascular permeability factor, has been shown to induce increased permeability and endothelial fenestration in vascular endothelium that is not normally fenestrated (Roberts and Palade, 1995). While the molecular mechanism remains unknown, local signaling is believed to be responsible for imposing these different vascular phenotypes (Risau, 1991).

As the vascular endothelium begins to mature, endothelial cells synthesize multiple extracellular matrix proteins which form a basement membrane. This membrane lies along the basal surface of endothelial cells throughout the vascular system, forming a sleeve around most blood vessels. Basement membranes are composed primarily of fibronectin, laminin, entactin/nidogen, collagen, and a heparin sulfate proteoglycan (Grant *et al.,* 1990), and this important extracellular matrix structure is believed to maintain cell polarity and to regulate endothelial cell behaviors such as proliferation, adhesion, and differentiation (Grant *et al.,* 1990). During embryological blood vessel development, it seems likely that extracellular matrix deposition helps to establish the basic patterning of the primary vascular plexus and the formation of the basement membrane is an early indication of blood vessel maturation. In the adult, basement membranes provide stability to blood vessels, which are only modified in the event of injury, cyclical changes in the reproductive system of females, or in response to pathological conditions. For example, when tumor cells metastasize, one of the first steps is the degradation of the endothelial basement membrane and the subsequent invasion of adjacent tissues.

Following the morphological changes associated with pruning and remodeling of the vascular plexus, mesenchymal support cells are recruited to lend mechanical and physiological support to the endothelium. Pericytes are recruited to the capillaries, and the vascular wall forms around larger vessels by the addition of smooth muscle cells and adventitial fibroblasts (Le Lièvre and Le Douarin, 1975; Schwartz and Liaw, 1993). Pericytes are cells which exist in close association with the endothelium of the capillaries, but which cover only a fraction of their surface (Fig. 5C). Their specific function is unclear, but they may modulate the behavior of endothelial cells, probably by regulating their permeability (De Oliveira, 1966) and their proliferation (Crocker *et al.,* 1970) and by helping to maintain their integrity (Rhodin, 1968). Pericytes and endothelial cells are the only cell types included in the mature capillaries and postcapillary venules (Orlidge and D'Amore, 1987). The absence of pericytes has been correlated with growth and proliferation of endothelial cells during neovascularization (De Oliveira, 1966), whereas a higher density of pericytes is observed on quiescent capillaries (Tilton *et al.,* 1985). *In vitro* experiments have shown that pericytes can inhibit capillary endothelial cell growth (Orlidge and D'Amore, 1987) and that this inhibition is mediated by TGF-β (Antonelli-Orlidge *et al.,* 1989).

Upon maturation, larger vessels begin to be surrounded by cellular and extracellular matrix components which comprise the vascular wall. As the endothelium matures, it is thought to signal nearby mesenchymal cells to join the growing vascular wall, and then contact with the endothelium causes these cells to differentiate. Much like the capillaries, which recruit pericytes, the larger vessels recruit a different type of vascular supportive cell, the smooth muscle cell (SMC), which is essential for the physiological properties of these vessels. Early SMCs express smooth muscle α-actin (Gabbiani *et al.,* 1981; Owens and Thompson, 1986), and later they express additional differentiation genes, such as *SM22* and *calponin* (Duband *et al.,* 1993).

The final result of the growth, differentiation, and maturation of the vascular endothelium is a network of highly specialized blood vessels. The major vessels, arteries, and veins are surrounded by a sheath of supportive cells, called the vascular wall, that maintains the integrity of the endothelium and contains the high blood pressure (Fig. 5). Three main layers have been identified in the major blood vessels (Rhodin, 1980). The tunica intima is the innermost layer and is composed of the endothelium (which lines the lumen of the blood vessel) the basement membrane, and the internal elastic tissue. The tunica media surrounds the tunica intima. This layer is often very thick and is composed mostly of SMCs with some elastic tissue. The tunica media is already rather well formed in the fetus and does not significantly change into adulthood. Lastly, the tunica adventitia surrounds the inner layers with fibrous connective tissue, elastic tissue, and mesenchymal cells. The composition of the vascular wall is specific for different types of blood vessels. Arteries, which experience very high pressure, are surrounded by a very thick smooth muscle cell layer (Fig. 5B). Veins, in contrast, which conduct blood under much lower pressure, have much less smooth muscle and elastic tissue in their vascular walls; therefore, they can stretch considerably, becoming a temporary reservoir of blood (Fig. 5A).

A. Molecules Involved in Vessel Maturation and Patterning

1. PDGF

In addition to playing a role in angiogenesis, PDGF is important for the recruitment of vascular wall components (Beck and D'Amore, 1997). A number of studies of mouse and human tissues show that PDGF-B is expressed in the endothelial cells of large vessels, whereas its receptor, PDGF-β, is found in adjacent mesenchyme (Holmgren *et al.,* 1991; Shinbrot *et al.,* 1994). This observation strongly implies that the endothelium secretes PDGF to recruit and stimulate the proliferation of mesenchymal cells in the immediate vicinity. In smaller vessels, both PDGF and its receptor are coexpressed in capillary endothelial cells, suggesting that an autocrine stimulatory pathway may promote cell proliferation. Experiments involving the targeted mutation of the *PDGF-B* and *PDGF-β* genes generally support a role for this signaling system in vascular wall cell recruitment (Leveen *et al.,* 1994; Soriano, 1994). Mice mutant for either gene display a range of anatomical and histological abnormalities, including an absence of capillary tufts in the renal glomerulus and dilation of the heart and blood vessels. Mutant mice die at about the time of birth from fatal hemorrhages. The timing of the hemorrhages approximately correlates with the increase in embryonic blood pressure. The kidney defect is attributed to a lack of mesangial cells, which bear many structural and functional similarities to capillary pericytes, and the hemorrhages and vessel dilation are attributed to a lack of pericytes throughout the capillary network (Beck and D'Amore, 1997). Additional work suggests that PDGF-BB secreted from endothelial cells does indeed chemotactically recruit and stimulate proliferation of SMCs (Beck and D'Amore, 1997).

While PDGF-B has been implicated in the maturation of the microvasculature, PDGF-A is likely to be involved in the maturation of larger vessels. In support of this possibility, *in vitro* experiments have shown that

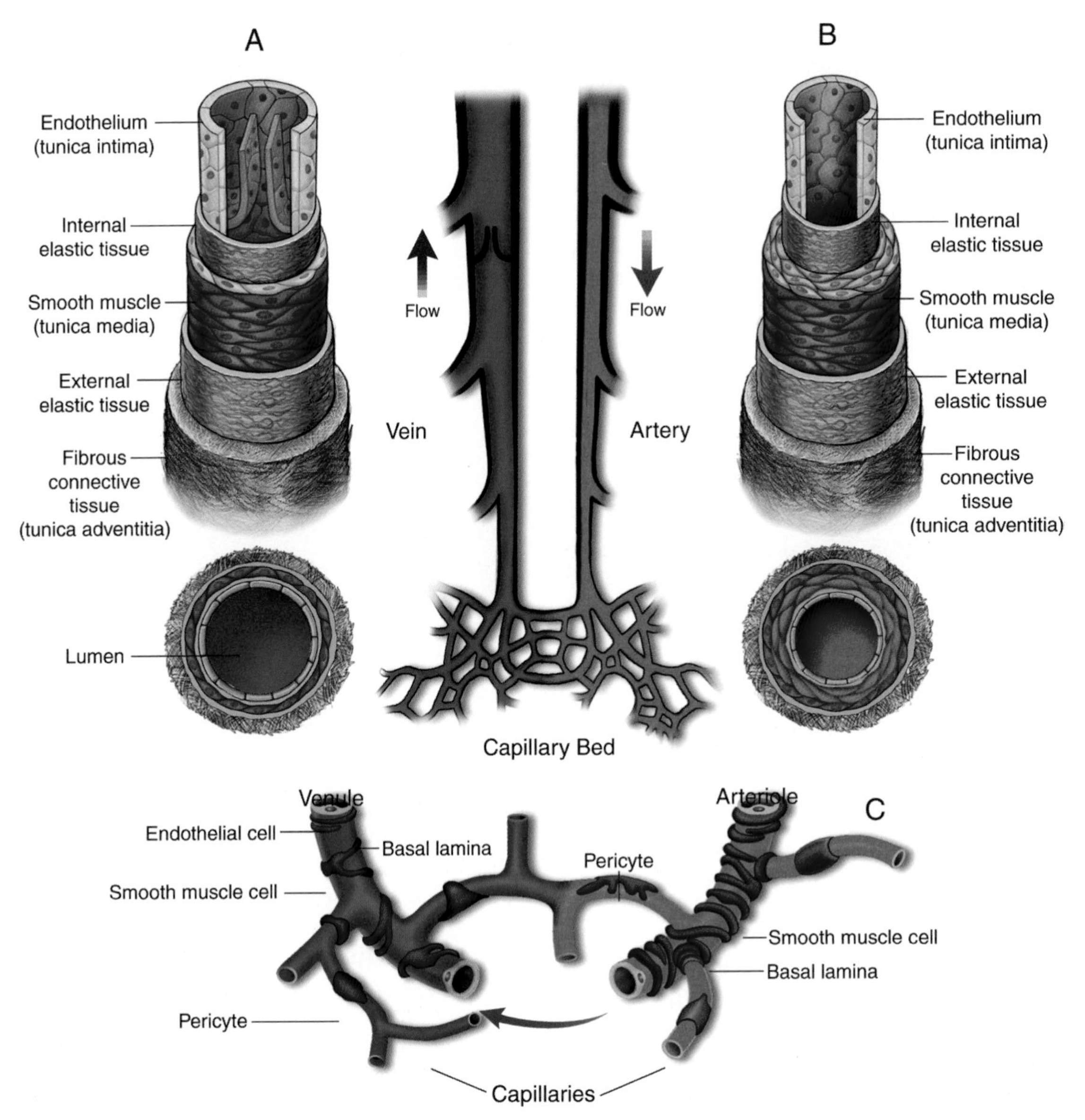

Figure 5 Schematic drawing of the mesenchymal derivatives associated with the major structures of the circulatory system. (A) Vascular wall components of the veins, including the endothelial tunica intima, the smooth muscle tunica media, and the extracellular matrix tunica adventitia. (B) Vascular wall components of the arteries, including the same vascular wall layers as those of the veins. Arteries, however, have a much thicker tunica media. (C) The capillary network consists almost exclusively of an endothelial layer. Pericytes associate with the capillaries covering only a small fraction of their surface area.

PDGF-AA is a better mitogen for smooth muscle cells than for pericytes (D'Amore and Smith, 1993). In addition, mutation of the PDGF-α receptor in the *Patch* mouse results in abnormalities in cardiovascular development, involving a significant reduction of layers of smooth muscle cells surrounding the endothelium (Schatteman *et al.*, 1995).

2. TGF-β

Contact between endothelial cells and SMCs or pericytes leads to the activation of *TGF-β* expression (Antonelli-Orlidge *et al.*, 1989). TGF-β in turn causes a number of effects, including the inhibition of proliferation and migration of endothelial cells (Orlidge and D'Amore, 1987; Sato and Rifkin, 1989), the induction of SMC and pericyte differentiation (Rohovsky *et al.*, 1996), and the stimulation of extracellular matrix deposition (Basson *et al.*, 1992). Overall, these events signal the differentiation and maturation of the developing blood vessels. When the function of TGF-β is disrupted in mice, 50% of homozygous mutant mice and 25% of heterozygous mice show defects in both vasculogenesis and hematopoiesis (Dickson *et al.,* 1995). Endothelial proliferation, however, is not affected in these mutant mice, suggesting that the primary vascular defect lies in the terminal differentiation of the endothelium rather than in the initial differentiation. Similar defects are observed in mice lacking the TGF-β receptor type II (Oshima *et al.,* 1996), offering further evidence for a role of the TGF-β signaling pathway during maturation of the primary vascular structure into the fully developed blood vessel.

3. Tie-2, Angiopoietin-1, and Angiopoietin-2

As described previously, Tie-2 is a receptor tyrosine kinase expressed in the vascular endothelium. In addition to being important in the early events of vasculogenesis and angiogenesis (Dumont *et al.,* 1994), it is also required for vascular remodeling (Sato *et al.,* 1995). Targeted mutation of the *tie-2* gene results in a disorganized vasculature and the absence of angiogenic sprouting into the neurectoderm. Furthermore, there is little distinction between the large and small blood vessels in both the head region and the yolk sac (Sato *et al.,* 1995). The ligands for Tie-2 are angiopoietin-1 (Davis *et al.,* 1996) and angiopoietin-2 (Maisonpierre *et al.,* 1997). Mice lacking angiopoietin-1 function have defects very similar to those of mice lacking the function of the Tie-2 receptor (Suri *et al.,* 1996). Endothelial cells in mutant embryos are also poorly associated with smooth muscle cells or pericytes, which are present in reduced numbers. These endothelial cells are abnormally rounded and collagen-like fibers are disorganized, indicating that these endothelial cells have not acquired polarity and are not differentiated. Therefore, Tie-2 and angiopoietin-1 appear to act as key regulators of the formation of the vascular wall and of endothelial–matrix interactions in addition to playing a role in remodeling. Angiopoietin-2 is an antagonist to angiopoietin-1 and Tie-2 (Maisonpierre *et al.,* 1997). Analysis of *angiopoietin-2* expression indicates that it is present only at sites of vascular remodeling, such as the dorsal aorta and the aortic branches, and its expression pattern is quite distinct from that of *angiopoietin-1.* In the adult, expression of *angiopoietin-2* is high in the ovary, placenta, and uterus, tissues which normally undergo remodeling in the adult. When *angiopoietin-2* is overexpressed in the blood vessels of transgenic mice, the resulting defects in the vasculature closely resembled those of Tie-2 or angiopoietin-1-deficient embryos.

Recently, a missense mutation has been mapped to the kinase domain of the *tie-2* gene in two unrelated human families in which affected members suffer from inherited venous malformations (Vikkula *et al.,* 1996). These affected individuals develop vein-like structures that are deficient in components of the vascular wall, especially SMCs. The identified mutation is most likely an activating mutation because overexpression studies using insect cells have demonstrated that this mutant form of the receptor has increased activity, showing up to 10 times higher autophosphorylation activity.

A model has been developed which proposes a role for a number of these molecules in the maturation of blood vessels (Folkman and D'Amore, 1996). In this model (Fig. 6), mesenchymal cells produce angiopoietin-1, which activates the Tie-2 receptor on nearby endothelial cells. In response to the Tie-2 activation, the endothelial cells release a PDGF signal (Fig. 6A), which acts to recruit nearby mesenchymal cells (Fig. 6B). In the case of pericytes, this signal is PDGF-BB, and in the case of SMCs, the signal is PDGF-AA or heparin-binding epidermal growth factor. Once the mesenchymal cells have contacted the endothelium, TGF-β is activated. The presence of TGF-β serves to reduce the proliferation of both endothelial and vascular wall cells, to induce their differentiation, and to stimulate extracellular matrix deposition (Fig. 6C).

4. Tie-1

Tie-1 is a receptor tyrosine kinase that is closely related to Tie-2. Like Tie-2, Tie-1 is important for both the survival and the integrity of endothelial cells following their differentiation (Puri *et al.,* 1995; Sato *et al.,* 1995). Expression of *tie-1* in the embryo is specific to endothelial cells and their precursors and is detected in both proliferative and quiescent endothelial cells (Korhonen *et al.,* 1994; Partanen *et al.,* 1992; Iwama *et al.,* 1993; Hatva *et al.,* 1995). Unlike Flk-1, but like other endothelial tyrosine kinase receptors, Tie-1 can be detected in most nonproliferating adult endothelial cells. Mouse embryos homozygous for a disrupted *tie-1* gene are phenotypically indistinguishable from their heterozygous littermates until about E 13. 0, when the mutant mice begin to die as a result of multiple vascular defects (Puri *et al.,* 1995; Sato *et al.,* 1995). Mutant embryos show edema and localized hemorrhaging and presumably die due to loss of the integrity of the microvasculature. Thus, Tie-1 is not necessary for the early steps of endothelial cell differentiation or subsequent vasculogenesis, but it appears to be required for later aspects of endothelial cell survival, maintenance, or proliferation.

5. Extracellular Matrix Molecules

In addition to a role in vasculogenesis, angiogenesis, and endothelial cell migration, *in vitro* studies suggest that the extracellular matrix also influences the differentiation of endothelial cells (Kubota *et al.,* 1988). For

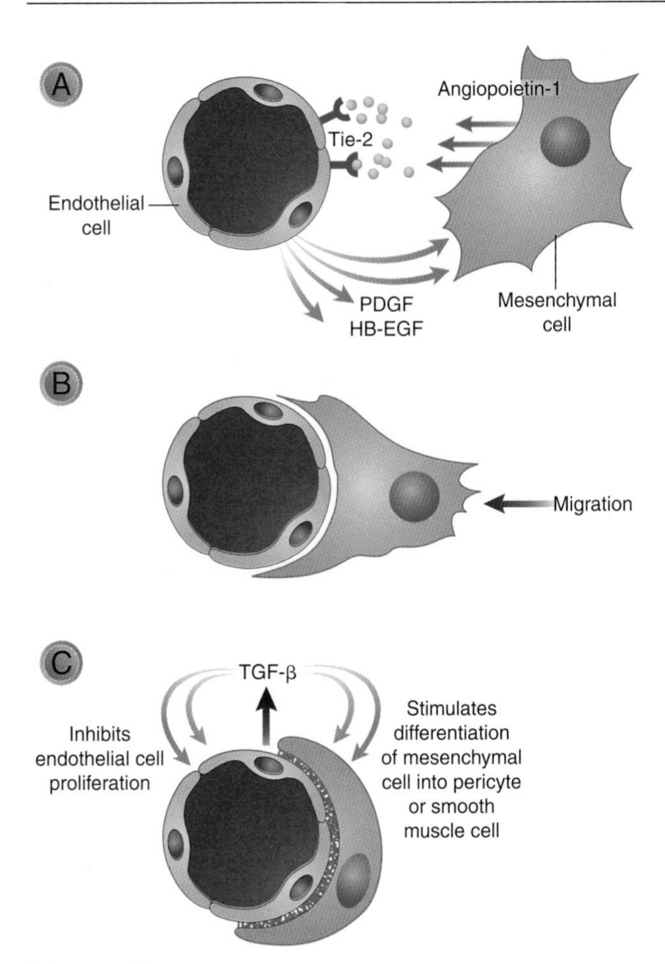

Figure 6 Model for the recruitment of the cellular vascular wall components. Angiopoietin-1 is secreted by mesenchymal cells and binds to the Tie-2 receptor located on the endothelial cells. This receptor activation triggers the release of factors from the endothelium that cause a chemotactic attraction of mesenchymal cells. These factors may include PDGF-AA or HB-EGF for the recruitment of large vessel vascular wall components, such as smooth muscle cells, or PDGF-BB for the recruitment of pericytes to the capillaries. When these mesenchymal cells contact the endothelium, TGF-β is activated and causes vessel maturation (adapted from Folkman and D'Amore, 1996).

example, a clear correlation between dynamic changes in extracellular matrix composition and endothelial cell maturation has been established. As described previously, the accumulation of fibronectin around endothelial cells is associated with their proliferation and migration. However, in both the chick embryo and the chick chorioallantoic membrane, as endothelial cells begin to mature levels of fibronectin gradually decrease, whereas there is a corresponding increase in the levels of surrounding laminin and type IV collagen (Risau and Lemmon, 1988; Ausprunk *et al.,* 1991). Fibronectin thus appears to be associated with the early steps of endothelial development, whereas laminin may be an early marker for vascular maturation (Risau, 1991). This is supported by observations of the extracellular matrix of the chick chorioallantoic membrane, where laminin appears in the walls of blood vessels 2 days later than fibronectin and then gradually increases in concentration with time. Unlike fibronectin, which is more widely distributed, laminin is confined to the basal lamina of the chorioallantoic membrane which is characteristic of differentiated vessels (Ausprunk *et al.,* 1991). Laminin has also been shown to promote the differentiation of capillary and venous endothelial cells *in vitro,* to inhibit endothelial cell proliferation (Kubota *et al.,* 1988), and to maintain the contractile activity of cultured SMCs (Hedin *et al.,* 1988). In separate studies, however, laminin has also been associated with endothelial proliferation and migration (Form *et al.,* 1986) and so its precise role during vascular maturation remains uncertain. An extracellular component that appears even later in the chorioallantoic membrane basal lamina is collagen IV (Ausprunk *et al.,* 1991). Cultured aortic endothelial cells are more adhesive to a substrate composed of collagen IV than one composed of laminin, and it has been suggested that it might stabilize vessel walls during vessel maturation (Herbst *et al.,* 1988; Ausprunk *et al.,* 1991). Appearance of collagen IV is correlated with lumen formation and differentiation of endothelial cells (Form *et al.,* 1986).

6. Coagulation Factors

a. Tissue Factor. Tissue factor (TF) is a cytokine receptor with high affinity for the plasma factor VII/VIIa. It is also the primary initiator of blood coagulation and plays a key role in hemostasis (Mackman, 1995). Immunohistochemical localization of the TF protein reveals its presence in the tunica adventitia of major blood vessels and in the epidermal and mucosal cells of the body surface. Inactivation of the *tissue factor* gene in mice results in embryonic lethality between E 9.5 and 10.5 due to defective circulation from the yolk sac to the embryo, leading to wasting and necrosis (Carmeliet *et al.,* 1996b; Bugge *et al.,* 1996). The most obvious abnormality in these mutant embryos is the presence of an irregular capillary plexus which replaces the large vitelline vessels, suggesting a defect in vasculogenic patterning. However, these capillaries are enlarged and fused, implying an additional defect in remodeling. This disorganized plexus may result from a lack of proper contacts between the endodermal and mesodermal tissue layers and consequently, in the inductive signals between the two tissue layers. In addition, examination of these vessels reveals the absence of smooth muscle α-actin-expressing vascular wall cells in proximity to the endothelium. These mesenchymal cells are thought to be the precursors of pericytes and SMCs which support and maintain integrity of the endothelium. This observation may explain why the onset of compromised vascular development occurs at the time vessels acquire a smooth muscle lining and undergo the stress of blood pressure. Thus, in addition to the defects in early vascular development, these embryos also show a defect in the recruitment of vascular wall cells and in proper maturation.

b. Factor V. Coagulation factor V, which is an essential cofactor for the formation of blood clots, is also involved in vascular development. Targeted mutation of the *factor V* gene results in the death of approximately half of the homozygous embryos at E 9 or 10 due to defects in the development of the yolk sac vasculature (Cui *et al.,* 1996). The yolk sacs of these embryos are strikingly abnormal and contain a large vascular plexus filled with blood. Littermates which survive until birth soon die as the result of massive hemorrhages. These phenotypes are in contrast to the milder effects observed in human factor V deficiencies (Tracy and Mann, 1987), in afibrinogenemic humans (Bi *et al.,* 1995), and in mice mutant in the *fibrinogen* gene (Suh *et al.,* 1995). However, the phenotype of the factor V deficiency does resemble that of mice lacking the *thrombin receptor* gene (Connolly *et al.,* 1996), in which half of the homozygous mutant mice die at E 9.0 or 10 due to similar vascular abnormalities. The authors suggest that factor V may be required for the generation of the thrombin receptor and that some signal transduced through the thrombin receptor is necessary for early vascular development (Cui *et al.,* 1996).

B. Additional Genes Implicated in Vascular Development

Many of the genes reviewed so far have proven to be excellent markers for vascular endothelium and their roles in vascular development has been examined and characterized. However, there remain a large number of genes whose expression patterns imply a role in endothelial cell behavior, but about which we have very little functional data. For example, ACE and von Willebrand factor were among the first markers used to study endothelial development, but their roles during early vascular formation are not clear (Ryan *et al.,* 1976; Ewenstein *et al.,* 1987). In addition, PECAM and VCAM, which are cell adhesion molecules, have also been used as vascular markers. PECAM is first expressed by angioblasts immediately following their differentiation from the mesoderm and continues to be abundant in both capillary and large vessel endothelium (Newman *et al.,* 1990; Baldwin *et al.,* 1994; Vecchi *et al.,* 1994). VCAM induces angiogenesis *in vivo* and migration of human endothelial cells *in vitro* (Koch *et al.,* 1995), and it is also involved in normal vascular development since targeted mutation results in failure of placental development and a failure of the subepicardial vascular development (Kwee *et al.,* 1995). The G protein-coupled receptor, X-msr, is a marker of vascular endothelial precursor cells in the frog embryo (Devic *et al.,* 1996). X-msr is related to angiotensin receptor II and it is expressed in most large and small vessels during embryonic development.

C. Transcription Factors Expressed in the Endothelium

Little is known about the transcription factors that control the development of the embryonic endothelium. However, homeobox genes are known to be critical for the development of many specific organs, and Newman *et al.* (1997) have shown that the *Xenopus* homeobox gene *XHex* is expressed in vascular endothelial cells long before their organization into the primary vascular plexus. Expression of *XHex* begins shortly after the onset of *flk-1* expression, raising the possibility that it lies downstream of the Flk-1/VEGF signaling pathway. XHex may play a role in regulating proliferation of angioblasts since overexpression of XHex results in abnormal vascular structures due to an increase in the number of vascular endothelial cells (Newman *et al.,* 1997). Another transcription factor which is expressed in endothelium is c-ets-1. Originally described as an oncogene, studies in the avian embryo have subsequently shown that it is expressed at high levels in the angioblasts of the blood islands, in the splanchnopleuric mesoderm during vasculogenesis, and later in the intraembryonic endothelial cells (Pardanaud and Dieterlen-Lièvre, 1993). c-ets-1 may also be involved in extracellular matrix degradation because consensus binding motifs are present in the promoter regions of certain metalloproteinases (Pepper and Montesano, 1990). Finally, the transcription factor SCL/TAL-1, mentioned previously, is a very early marker of the endothelial cell lineage in the embryo (Drake *et al.,* 1997) and is also present in hematopoietic precursor cells. The role of SCL/TAL-1 in vascular development is unclear, however, since mouse gene ablation studies have shown no evidence for vascular defects in homozygous mutant animals (Porcher *et al.,* 1996).

X. Conclusion

The characterization of a diverse range of molecules that play a role in the regulation of endothelial cell specification, proliferation, migration, and differentiation has significantly improved our understanding of vascular development. However, many questions remain to be answered at the molecular level. How and when do mesodermal cells become fated to develop into angioblasts? What are the inductive interactions between mesoderm and endoderm which specify angioblast commitment? Are there different populations of angioblasts destined to become different parts of the vascular system or is angioblast fate plastic, depending on environmental cues? If so, what are these environmental cues? What subtle orchestration of extracellular matrix, cell–cell adhesion, and growth factor signals is necessary for endothelial proliferation and maturation *in vivo?* What controls the remodeling of embryonic blood vessels to form the complex array of arteries, veins, and capillaries in the mature organism? Perhaps most important, what are the fundamental molecular mechanisms regulating vasculogenesis and angiogenesis? Clearly, our knowledge of the mechanisms underlying vascular development is still in its early stages. However, the application of molecular techniques is likely to result in rapid progress in understanding this critical embryological process. Significantly, the lessons learned from the study of embryonic vascular development in model organisms are likely to be directly applicable to understanding the vascular development associated with a wide range of human pathologies.

Acknowledgments

This work was supported by NIH Grant HD25746 to PAK. We thank Antone Jacobson for critical reading of the manuscript and for many helpful suggestions.

References

Abell, R. G. (1946). The permeability of blood capillary sprouts and newly formed capillaries as compared to that of older capillaries. *Am J Physiol* **147,** 237–241.

Akhurst, R. J., Lehnert, S. A., Faissner, A. J., and Duffie, E. (1990). TGF-beta in murine morphogenetic processes: The early embryo and cardiogenesis. *Development* (*Cambridge, UK*) **108,** 645–656.

Al-Adhami, M. A., and Kunz, Y. W. (1977). Ontogenesis of haematopoietic sites in *Brachydanio rerio* (Hamilton-Buchanan). *Dev. Growth Differ.* **19,** 171–179.

Amaya, E., Musci, T. J., and Kirshner, M. W. (1991). Expression of a dominant negative mutant of the FGF-receptor disrupts mesoderm formation in *Xenopus* embryos. *Cell* (*Cambridge, Mass.*) **66,** 257–270.

Antonelli-Orlidge, A., Saunders, K. B., Smith, S. R., and D'Amore, P. A. (1989). An activated form of TGF-β is produced by cocultures of endothelial cells and pericytes. *Proc. Natl. Acad. Sci. U.S.A.* **86,** 4544–4548.

Ashton, N. (1966). Oxygen and the growth and development of retinal vessels. *Am. J. Ophthalmol.* **62,** 412–435.

Augustin, H. G., Braun, K., Telemenakis, I., Modlich, U., and Kuhn, W. (1995). Ovarian angiogenesis: Phenotypic characterization of endothelial cells in a physiological model of blood vessel growth and regression. *Am. J. Pathol.* **147,** 339–351.

Ausprunk, D. H., and Folkman, J. (1977). Migration and proliferation of endothelial cells in preformed and newly formed blood vessels during tumor angiogenesis. *Microvasc. Res.* **14,** 53–65.

Ausprunk, D. H., Dethlefsen, S. M., and Higgins, E. R. (1991). Distribution of fibronectin, laminin and type IV collagen during development of blood vessels in the chick chorioallantoic membrane. *Issues Biomed.* **14,** 93–108.

Azar, Y., and Eyal-Giladi, H. (1979). Marginal zone cells-the primitive streak-inducing component of the primary hypoblast in the chick. *J. Embryol. Exp. Morphol.* **52,** 79–88.

Baldwin, H. S. (1996). Early embryonic vascular development. *Cardiovasc. Res.* **31,** E34–E45.

Baldwin, H. S., Hong, M. S., Hong-Chin, Y., DeLisser, H. M., Chung, A., Mickanin, C., Trask, T., Kirschbaum, N. E., Newman, P. J., Albelda, S. M., and Buck, C. A. (1994). Platelet endothelial cell adhesion molecule-1 (PECAM/CD31): Alternatively spliced, functionally distinct isoforms expressed during mammalian cardiovascular development. *Development* (*Cambridge, UK*) **120,** 2539–2553.

Bär, T. (1980). The vascular system of the cerebral cortex. *Adv. Anat. Embryol. Cell Biol.* **59,** 1–62.

Basson, C. T., Kocher, O., Basson, M. D., Asis, A., and Madri, J. A. (1992). Differential modulation of vascular cell integrin and extracellular matrix expression in vitro by TGF-β1 correlates with reciprocal effects on cell migration. *J. Cell Physiol.* **153,** 118–128.

Battegay, E. J., Rupp, J., Iruela-Arispe, L., Sage, E. H., and Pech, M. (1994). PDGF-BB modulates endothelial proliferation and angiogenesis in vitro via PDGF β-receptors. *J. Cell Biol.* **125,** 917–928.

Beck, L., Jr., and D'Amore, P. A. (1997). Vascular development: Cellular and molecular regulation. *FASEB J.* **11,** 365–373.

Begley, C. G., Aplan, P. D., Denning, S. M., Haynes, B. F., Waldmann, T. A., and Kirsch, I. R. (1989). The gene SCL is expressed during early hematopoiesis and encodes a differentiation-related DNA-binding motif. *Proc. Natl. Acad. Sci. U.S.A.* **86,** 10128–10132.

Bennet, H. S., Luft, J. H., and Hampton, J. C. (1959). Morphological classifications of vertebrate blood capillaries. *Am. J. Physiol.* **196,** 381–390.

Bertwistle, D., Walsley, M. E., Read, E. M., Pizzey, J. A., and Patient, R. K. (1996). GATA factors and the origins of adult and embryonic blood in *Xenopus:* Responses to retinoic acid. *Mech. Dev.* **57,** 199–214.

Bi, L., Lawler, A. M., Antonarakis, S. E., High, K. A., Gearhart, J. D., and Kazazian, H. H., Jr. (1995). Targeted disruption of the mouse factor VIII gene produces a model of haemophilia A. *Nat. Genet.* **10,** 119–121.

Bischoff, J. (1995). Approaches to studying cell adhesion molecules in angiogenesis. *Trends Cell Biol.* **5,** 69–74.

Bogers, A. J. J. C., Gittenberger-de Groot, A. C., Poelmann, R. E., Peault, B. M., and Huysmans, H. A. (1989). Development of the origin of the coronary arteries, a matter of ingrowth or outgrowth? *Anat. Embryol.* **180,** 437–441.

Bolender, D. L., and Markwald, R. R. (1991). Endothelial formation and transformation in early avian heart development: Induction by proteins organized into adherons. *Issues Biomed.* **14,** 109–124.

Bowersox, J. C., and Sorgente, N. (1982). Chemotaxis of aortic endothelial cells in response to fibronectin. *Cancer Res* **42,** 2547–2551.

Brand-Saberi, B., Köntges, G., Wilting, J., and Christ, B. (1994). Die angiogene Potenz von Somiten und Somatopleura bei Vogelembryonen. *Verh. Anat. Ges.* **89,** 128–129.

Breier, G., and Risau, W. (1996). The role of vascular endothelial growth factor in blood vessel formation. *Trends Cell Biol.* **6,** 454–456.

Breier, G., Albrecht, U., Sterrer, S., and Risau, W. (1992). Expression of vascular endothelial growth-factor during embryonic angiogenesis and endothelial-cell differentiation. *Development (Cambridge, UK)* **114,** 521–532.

Breier, G., Clauss, M., and Risau, W. (1995). Coordinate expression of VEGF-receptor 1 (flt-1) and its ligand suggests a paracrine regulation of murine development. *Dev. Dyn.* **204,** 228–239.

Breier, G., Breviario, F., Caveda, L., Berthier, R., Schnürch, H., Gotsch, U., Vestweber, D., Risau, W., and Dejana, E. (1996). Molecular cloning and expression of murine vascular endothelial-cadherin in early stage development of cardiovascular system. *Blood* **87,** 630–641.

Brooks, P. C., Clark, R. A. F., and Cheresh, D. A. (1994a). Requirement of vascular integrin $\alpha_v\beta_3$ for angiogenesis. *Science* **264,** 569–571.

Brooks, P. C., Montgomery, A. M. P., Rosenfeld, M., Reisfeld, R. A., Hu, T., Klier, G., and Cheresh, D. A. (1994b). Integrin $\alpha_v\beta_3$ antagonists promote tumor regression by inducing apoptosis of angiogenic blood vessels. *Cell (Cambridge, Mass.)* **79,** 1157–1164.

Brooks, P. C., Stromblad, S., Sanders, L. C., von Schalscha, T. L., Aimes, R. T., Stetler-Stevenson, W. G., Quigley, J. P., and Cheresh, D. A. (1996). Localization of matrix metalloproteinase MMP-2 to the surface of invasive cells by interaction with integrin alpha v beta 3. *Cell (Cambridge, Mass.)* **85,** 683–693.

Bugge, T. H., Xiao, Q., Kombrinck, K. W., Flick, M. J., Holmback, K., Danton, M. J., Colbert, M. C., Witte, D. P., Fujikawa, K., Davie, E. W., and Degen, J. L. (1996). Fatal embryonic bleeding events in mice lacking tissue factor, the cell-associated initiator of blood coagulation. *Proc. Natl. Acad. Sci. U.S.A.* **93,** 6258–6263.

Burri, P. H., and Tarek, M. R. (1990). A novel mechanism of capillary growth in the rat pulmonary microcirculation. *Anat Rec.* **228,** 35–45.

Caduff, J. H., Fischer, L. C., and Burri, P. H. (1986). Scanning electron microscope study of the developing microvasculature in the postnatal rat lung. *Anat. Rec.* **216,** 154–164.

Caldwell, P. R., Seegal, B. C., Hsu, K. C., Das, M., and Soffer, R. L. (1976). Angiotensin-converting enzyme: Vascular endothelial localization. *Science* **191,** 1050–1051.

Carmeliet, P., Ferreira, V., Breier, G., Pollefeyt, S., Kieckens, L., Gertsenstein, M., Fahrig, M., Vandenhoeck, A., Harpal, K., Eberhardt, C., Declercq, C., Pauling, J., Moons, L., Collen, D., Risau, W., and Nagy, A. (1996a). Abnormal blood vessel development and lethality in embryos lacking a single VEGF allele. *Nature (London)* **380,** 435–439.

Carmeliet, P., Mackman, N., Moons, L., Luther, T., Gressens, P., Van Vlaenderen, I., Demunck, H., Kasper, M., Breier, G., Evrard, P., Müller, M., Risau, W., Edgington, T., and Collen, D. (1996b). Role of tissue factor in embryonic blood vessel development. *Nature (London)* **383,** 73–75.

Caveda, L., Martin-Padura, I., Navarro., P., Breviario, F., Corada, M., Gulino, D., Lampugnani, M. G., and Dejana, E. (1996). Inhibition of cultured cell growth by vascular endothelial cadherin (cadherin-5/VE-cadherin). *J. Clin. Invest.* **98,** 886–893.

Chapman, W. B. (1918). The effect of heart-beat upon the development of the vascular system of the chick. *Am. J. Anat.* **23,** 175–203.

Christ, B., Poelmann, R. E., Mentink, M. M. T., and Gittenberger-de Groot, A. C. (1990). Vascular endothelial cells migrate centripetally within embryonic arteries. *Anat. Embryol.* **181,** 333–339.

Christ, B., Grim, M., Wilting, J., von K. K., and Wachtler, F. (1991). Differentiation of endothelial cells in avian embryos does not depend on gastrulation. *Acta Histochem.* **91,** 193–199.

Clark, E. R., and Clark, E. L. (1939). Microscopic observations on the growth of blood capillaries in the living mammal. *Am. J. Anat.* **64,** 251–301.

Clark, R. A. F., Tonnesen, M. G., Gailit, J., and Cheresh, D. A. (1996). Transient functional expression of $\alpha_v\beta_3$ on vascular cells during wound repair. *Am. J. Pathol.* **148,** 1407–1421.

Cleaver, O., and Krieg, P. A. (1998). VEGF mediates angioblasts migration during development of the dorsal aorta in *Xenopus. Development*, in press.

Cleaver, O., Tonissen, K. F., Saha, M. S., and Krieg, P. A. (1997). Neovascularization of the *Xenopus* embryo. *Dev. Dyn.* **210,** 66–77.

Coffin, J. D., and Poole, T. (1988). Embryonic vascular development: Immunohistochemical identification of the origin and subsequent morphogenesis of the major vessel primordia in quail embryos. *Development (Cambridge, UK)* **102,** 735–748.

Coffin, J. D., Harrison, J., Schwartz, S., and Heimark, R. (1991). Angioblast differentiation and morphogenesis of the vascular endothelium in the mouse embryo. *Dev. Biol.* **148,** 51–62.

Colle-Vandevelde, A. (1963). Blood anlage in teleostei. *Nature (London)* **198,** 1223.

Connolly, A. J., Ishihara, H., Kahn, M. L., Farese, R. V., Jr., and Coughlin, S. R. (1996). Role of the thrombin receptor in development and evidence for a second receptor. *Nature (London)* **381,** 516–519.

Cormier, F., de Paz, P., and Dieterlen-Lièvre, F. (1986). In vitro detection of cells with monocytic potentiality in the wall of the chick embryo aorta. *Dev. Biol.* **118,** 167–175.

Cornell, R. A., and Kimelman, D. (1994). Activin-mediated mesoderm induction requires FGF. *Development (Cambridge, UK)* **120,** 453–462.

Crocker, D. J., Murad, T. M., and Greer, J. C. (1970). Role of the pericyte in wound healing. An ultrastructural study. *Exp. Mol. Pathol.* **13,** 51–65.

Cui, J., O'Shea, K. S., Purkayastha, A., Sauder, T. L., and Ginsburg, D. (1996). Fatal haemorrhage and incomplete block to embryogenesis in mice lacking coagulation factor V. *Nature (London)* **384,** 66–68.

D'Amore, P. A., and Klagsbrun, M. (1989). Angiogenesis: Factors and mechanisms. *In* "The Pathobiology of Neoplasia" (A. E. Sirica, ed.). Plenum, New York.

D'Amore, P. A., and Smith, S. R. (1993). Growth factor effects on cells of the vascular wall: A survey. *Growth Factors* **8,** 61–75.

Davis, S., Aldrich, T. H., Jones, P. F., Acheson, A., Compton, D. L., Jain, V., Ryan, T. E., Bruno, J., Radziejewski, C., Maisonpierre, P. C., and Yancopoulos, G. D. (1996). Isolation of angiopoietin-1, a ligand for the TIE2 receptor, by secretion-trap expression cloning. *Cell (Cambridge, Mass.)* **87,** 1161–1169.

Dejana, E. (1996). Endothelial adherens junctions: Implications in the control of vascular permeability and angiogenesis. *J. Clin. Invest.* **98,** 1949–1953.

Dejana, E., Corada, M., and Lampugnani, M. G. (1995). Endothelial cell-to-cell junctions. *FASEB J.* **9,** 910–918.

De Oliveira, F. (1966). Pericytes in diabetic retinopathy. *Br. J. Ophthalmol.* **50,** 134–143.

De Ruiter, M. C., Poelmann, R. E., Van der Plas-de Vries, I., Mentink, M. M. T., and Gittenberger-de Groot, A. C. (1992). The development of the myocardium and endocardium in mouse embryos: Fusion of two heart tubes? *Anat. Embryol.* **185,** 461–473.

Devic, E., Paquereau, L., Vernier, P., Knibiehler, B., and Audigier, Y. (1996). Expression of a new G protein-coupled receptor X-msr is associated with an endothelial lineage in *Xenopus laevis. Mech. Dev.* **59,** 129–140.

de Vries, C., Escobedo, J. A., Ueno, H., Houck, K., Ferrara, N., and Williams, L. T. (1992). The fms-like tyrosine kinase, a receptor for vascular endothelial growth factor. *Science* **255,** 989–991.

Dickson, M. C., Martin, J. S., Cousins, F. M., Kulkarni, A. B., Karlsson, S., and Akhurst, R. J. (1995). Defective haematopoiesis and vasculogenesis in transforming growth factor β-1 knock out mice. *Development (Cambridge, UK)* **121,** 1845–1854.

Dieterlen-Lièvre, F. (1975). On the origin of hemopoietic stem cells in the avian embryo. *J. Embryol. Exp. Morphol.* **33,** 607–619.

Dieterlen-Lièvre, F. (1984). Emergence of intra-embryonic blood stem cells in avian chimeras by means of monoclonal antibodies. *Dev. Comp. Immunol.* **3,** 75–80.

Dieterlen-Lièvre, F., Godin, I., and Pardanaud, L. (1997). Where do hematopoietic stem cells come from? *Int. Arch. Allergy Immunol.* **112,** 3–8.

Dobson, D. E., Kambe, A., Block, E., Dion, T., Lu, H., Castellot, J. J., Jr., and Spiegelman, B. M. (1990). 1-Butyryl-glycerol: A novel angiogenesis factor secreted by differentiating adipocytes. *Cell (Cambridge, Mass)* **61,** 223–230.

Drake, C. J., and Little, C. D. (1995). Exogenous vascular endothelial growth factor induces malformed and hyperfused vessels during embryonic neovascularization. *Proc. Natl. Acad. Sci. U.S.A.* **92,** 7657–7661.

Drake, C. J., Davis, L. A., Walters, L., and Little, C. D. (1990). Avian vasculogenesis and the distribution of collagens I, IV, laminin, and fibronectin in the heart primordia. *J. Exp. Zool.* **255,** 309–322.

Drake, C. J., Davis, L. A., and Little, C. D. (1992). Antibodies to β1-integrins cause alterations of aortic vasculogenesis, in vivo. *Dev. Dyn.* **193,** 83–91.

Drake, C. J., Brandt, S. J., Trusk, T. C., and Little, C. D. (1997). TAL1/SCL is expressed in endothelial precursor cells/angioblasts and defines a dorsal-to-ventral gradient of vasculogenesis. *Dev. Biol.* **192,** 17–30.

Duband, J.-L., Gimona, M., Scatena, M., Sartore, S., and Small, J. V. (1993). Calponin and SM22 as differentiation markers of smooth muscle: Spatiotemporal distribution during avian embryonic development. *Differentiation (Berlin)* **55,** 1–11.

Dumont, D. J., Yamaguchi, T. P., Conlon, R. A., Rossant, J., and Breitman, M. L. (1992). Tek, a novel tyrosine kinase gene located on mouse chromosome 4, is expressed in endothelial cells and their presumptive precursors. *Oncogene* **7,** 1471–1480.

Dumont, D. J., Gradwohl, G., Fong, G. H., Puri, M. C., Gertsenstein, M., Auerbach, A, and Breitman, M. L. (1994). Dominant-negative and targeted null mutations in the endothelial receptor tyrosine kinase, tek, reveal a critical role in vasculogenesis of the embryo. *Genes Dev.* **8,** 1897–1909.

Dumont, D. J., Fong, G.-H., Puri, M. C., Gradwohl, G., Alitalo, K., and Breitman, M. L. (1995). Vascularization of the mouse embryo: A study of flk-1, tek, tie, and vascular endothelial growth factor expression during development. *Dev. Dyn.* **203,** 80–92.

Eichmann, A., Marcelle, C., Bréant, C., and Le Douarin, N. M. (1993). Two molecules related to the VEGF receptor are expressed in the early endothelial cells during avian embryonic development. *Mech. Dev.* **42,** 33–48.

Ekblom, P., Sariola, H., Karkinen, M., and Saxen, L. (1982). The origin of the glomerular endothelium. *Cell Differ.* **11,** 35–39.

Emoto, N., Gonzalez, A. M., Walicke, P. A., Wada, E., Simmons, D. M., Shimasaki, S., and Baird, A. (1989). Basic fibroblast growth factor (FGF) in the central nervous system: Identification of specific loci of basic FGF expression in the rat brain. *Growth Factors* **2,** 21–29.

Esch, F., Baird, A., Ling, N., Ueno, N., Hill, F., Denoroy, L., Klepper, R., Gospodarowicz, D., Bohlen, P., and Guillemin. R. (1985). Primary structure of bovine pituitary basic fibroblast growth factor (FGF) and comparison with the amino-terminal sequence of bovine brain acidic FGF. *Proc. Natl. Acad. Sci. U.S.A.* **82,** 6507–6511.

Evans, H. M. (1909). On the development of the aorta, cardinal and umbilical veins, and other blood vessels of embryos from capillaries. *Anat. Rec.* **3,** 498–518.

Ewenstein, B. M., Warhol, M. J., Handin, R. L., and Pober, J. S. (1987). Composition of the von Willebrand factor storage organelle (Weibel-Palade body) isolated from cultured human umbilical vein endothelial cells. *J. Cell Biol.* **104,** 1423–1433.

Feinberg, R. N., and Beebe, D. C. (1983). Hyaluronate in vasculogenesis. *Science* **220,** 1177–1179.

Feinberg, R. N., and Noden, D. M. (1991). Experimental analysis of blood vessel development in the avian wing bud. *Anat. Rec.* **231,** 136–144.

Fernig, D. G., and Gallagher, J. T. (1994). Fibroblast growth factors and their receptors: An information network controlling tissue growth, morphogenesis and repair. *Prog. Growth Factor Res.* **5,** 353–377.

Ferrara, N. (1995). The role of vascular endothelial growth factor in pathological angiogenesis. *Breast Cancer Res. Treat.* **36,** 127–137.

Ferrara, N., and Henzel, W. J. (1989). Pituitary follicular cells secrete a novel heparin-binding growth factor specific for vascular endothelial cells. *Biochem. Biophys. Res. Commun.* **161,** 851–858.

Ferrara, N., Houck, K., Jakeman, L., and Leung, D. W. (1992). Molecular and biological properties of the vascular endothelial growth factor family of proteins. *Endocr. Rev.* **13,** 18–32.

Ferrara, N., Carver-Moore, K., Chen, H., Dowd, M., Lu, L., O'Shea, K. S., Powell-Braxton, L., Hillan, K. J., and Moore, M. W. (1996). Heterozygous embryonic lethality induced by targeted inactivation of the VEGF gene. *Nature (London)* **380,** 439–442.

Fett, J. W., Strydom, D. J., Lobb, R. R., Alderman, E. M., Bethune, J. L., Riordan, J. F., and Vallee, B. L. (1985). Isolation and characterization of angiogenin, an angiogenic protein from human carcinoma cells. *Biochemistry* **24,** 5480–5486.

Fina, L., Molgaard, H. V., Robertson, D., Bradley, N. J., Monaghan, P., Delia, D., Sutherland, D. R., Baker, M. A., and Greaves, M. F. (1990). Expression of the CD34 gene in vascular endothelial cells. *Blood* **75,** 2417–2426.

Flamme, I., and Risau, W. (1992). Induction of vasculogenesis and hematopoiesis in vitro. *Development (Cambridge, UK)* **116,** 435–439.

Flamme, I., Messerli, M., Risau, W., Jacob, M., and Jacob, H. J. (1992). "Vascular Growth in the Extraembryonic Mesoderm of Avian Embryos. Formation and Differentiation of Early Embryonic Mesoderm," pp. 323–335. Plenum, New York.

Flamme, I., Breier, G., and Risau, W. (1995a). Expression of vascular endothelial growth factor (VEGF) and VEGF-receptor 2 (flk-1) during induction of hemangioblastic precursors and vascular differentiation in the quail embryo. *Dev. Biol.* **169,** 699–712.

Flamme, I., von Reutern, M., Drexler, H. C. A., Syed-Ali, S., and Risau, W. (1995b). Overexpression of vascular endothelial growth factor in the avian embryo induces hypervascularization and increased vascular permeability without alterations of embryonic pattern formation. *Dev. Biol.* **171,** 399–414.

Folkman, J. (1995). Angiogenesis in cancer, vascular, rheumatoid and other disease. *Nat. Med.* **1,** 27–31.

Folkman, J. (1987). Thrombosis and Haemostasis (M. Verstraete, J. Vermylen, R. Lijnene, J. Arnout, Eds.) Leuven: Leuven Univ. Press.

Folkman, J., and D'Amore, P. A. (1996). Blood vessel formation: What is the molecular basis? *Cell (Cambridge, Mass.)* **87,** 1151–1155.

Folkman, J., and Klagsbrun, M. (1987). Angiogenic factors. *Science* **235,** 442–447.

Folkman, J., and Shing, Y. (1992). Angiogenesis. *J. Biol. Chem.* **267,** 10931–10934.

Fong, G. H., Rossant, J., Gertsenstein, M., and Breitman, M. L. (1995). Role of the Flt-1 receptor tyrosine kinase in regulating the assembly of vascular endothelium. *Nature (London)* **376,** 66–70.

Form, D. M., and Auerbach, R. (1983). PGE2 and angiogenesis. *Proc. Soc. Exp. Biol. Med.* **172,** 214–218.

Form, D. M., Pratt, B. M., and Madri, J. A. (1986). Endothelial cell proliferation during angiogenesis: In vitro modulation by basement membrane components. *Lab. Invest.* **55,** 521–530.

Fouquet, B., Weinstein, B. M., Serluca, F. C., and Fishman, M. C. (1997). Vessel patterning in the embryo of the zebrafish: Guidance by notochord. *Dev. Biol.* **183,** 37–48.

Franke, R.-P., Gräfe, M., Schnittler, H., Seiffge, D., and Mittermayer,

C. (1984). Induction of human vascular endothelial stress fibres by fluid sheer stress. *Nature (London)* **307,** 648–649.

Friwawl, R., and Maciag, T. (1995). Molecular mechanisms of angiogenesis: Fibroblast growth factor signal transduction. *FASEB J.* **9,** 919–925.

Friesel, R., Komoriya, A., and Maciag, T. (1987). Inhibition of endothelial cell proliferation by gamma-interferon. *J. Cell Biol.* **104,** 689–696.

Gabbiani, G., Schmid, E., Winter, S., Chaponnier, C., De Chastonay, C., Vandekerckhove, J., Weber, K., and Franke, W. W. (1981). Vascular smooth muscle cells differ from other smooth muscle cells: Predominance of vimentin filaments and a specific a-type actin. *Proc. Natl. Acad. Sci. U.S.A.* **78,** 298–302.

Garcia-Porrero, J. A., Godin, I. E., and Dieterlen-Lièvre, F. (1995). Potential intraembryonic hemogenic sites at pre-liver stages in the mouse. *Anat. Embryol.* **192,** 425–435.

George, E. L., Georges-Labouesse, E. N., Patel-King, R. S., Rayburn, H., and Hynes, R. O. (1993). Defects in mesoderm, neural tube and vascular development in mouse embryos lacking fibronectin. *Development (Cambridge, UK)* **119,** 1079–1091.

Gibson, W. T. (1910). The development of the hypochord in Raia batis; with a note upon the occurrence of the epibranchial groove in amniote embryos. *Anat. Anz.* **35,** 407–428.

Godin, I., Dieterlen-Lièvre, F., and Cumano, A. (1995). Emergence of multipotent hemopoietic cells in the yolk sac and paraaortic splanchnopleura in mouse embryos, beginning at 8.5 days postcoitus. *Proc. Natl. Acad. Sci. U.S.A.* **92,** 773–777.

Godsave, S. F., Isaacs, H. V., and Slack, J. M. W. (1988). Mesoderm-inducing factors: A small class of molecules. *Development (Cambridge, UK)* **102,** 555–566.

Good, D. J., Polverini, P. J., Rastinejad, F., LeBeau, M. M., Lemons, R., Frazier, W. A., and Bouck, N. P. (1990). A tumor suppressor-dependent inhibitor of angiogenesis is immunologically and functionally indistinguishable from a fragment of thrombospondin. *Proc. Natl. Acad. Sci. U.S.A.* **87,** 6624–6628.

Gospodarowicz, D., Abraham, J. A., and Schilling, J. (1989). Isolation and characterization of a vascular endothelial cell mitogen produced by pituitary-derived folliculo stellate cells. *Proc. Natl. Acad. Sci. U.S.A.* **86,** 7311–7315.

Goss, C. M. (1928). Experimental removal of the blood island of Amblystoma punctatum embryos. *J. Exp. Zool.* **52,** 45–61.

Gotsch, U., Jager, U., Dominis, M., and Vestweber, D. (1994). Expression of P-selectin on endothelial cells is upregulated by LPS and TNF-alpha in vivo. *Cell Adhes. Commun.* **2,** 7–14.

Grant, D. S., Kleinman, H. K., and Martin, G. R. (1990). The role of basement membranes in vascular development. *Ann. N.Y. Acad. Sci.* **588,** 61–72.

Griendling, K. K., and Alexander, R. W. (1996). Endothelial control of the cardiovascular system: Recent advances. *FASEB J.* **10,** 283–292.

Haar, J. L., and Ackerman, G. A. (1971). A phase and electron microscopic study of vasculogenesis and erythropoiesis in the yolk sac of the mouse. *Anat. Rec.* **170,** 199–224.

Hahn, H. (1908). Experimentelle Studien über die Entstehung des Blutes and der ersten Geässe beim Hühnchen. *Anat. Rec.* **33,** 153–170.

Hahn, H. (1909). Exoerimentelle Studien uber die Entstehung des Blutes und der erstenGefasse beim Huhnchen. *Wilhelm Roux' Arch. Entwicklungsmech. Org.* **27,** 337–433.

Hallmann, R., Feinberg, R. N., Latker, C. H., Sasse, J., and Risau, W. (1987). Regression of blood vessels precedes cartilage differentiation during chick limb development. *Differentiation (Berlin)* **34,** 98–105.

Hanahan, D., and Folkman, J. (1996). Patterns and emerging mechanisms of the angiogenic switch during tumorigenesis. *Cell (Cambridge, Mass.)* **86,** 353–364.

Hatva, E., Kaipainen, A., Mentula, P., Jääskeläinen, J., Paetau, A., Haltia, M., and Alitalo, K. (1995). Expression of endothelial cell-specific receptor tyrosine kinases and growth factors in human brain tumours. *Am. J. Pathol.* **146,** 368–378.

Hedin, U., Bottger, B. A., Forsberg, E., Johansson, S., and Thyberg, J. (1988). Diverse effects of fibronectin and laminin on phenotypic properties of cultured arterial smooth muscle cells. *J. Cell Biol.* **107,** 307–315.

Heimark, R. L., Twardzik, D. R., and Schwartz, S. M. (1986). Inhibition of endothelial regeneration by type-beta transforming growth factor from platelets. *Science* **233,** 1078–1080.

Heine, U. I., Roberts, A. B., Muno, E. F., Roche, N. S., and Sporn, M. B. (1985). Effects of retinoid deficiency on the development of the heart and vascular system of the quail embryo. *Virchows Arch. B* **50,** 135–152.

Herbst, T. J., McCarthy, J. B., Tsilibary, E. C., and Furcht, L. T. (1988). Differential effects of laminin, intact type IV collagen and specific domains of type IV collagen on endothelial cell adhesion and migration. *J. Cell Biol.* **106,** 1365–1373.

Hermansson, M., Nister, M., Betsholtz, C., Heldin, C.-H., Westermark, B., and Funa, K. (1988). Endothelial cell hyperplasia in human glioblastoma: Coexpression of mRNA for patelet-derived growth factor (PDGF) B chain and PDGF receptor suggests autocrine growth stimulation. *Proc. Natl. Acad. Sci. U.S.A.* **85,** 7748–7752.

Heuer, J. G., von Bartheld, C. S., Kinoshita, Y., Evers, P. C., and Bothwell M. (1990). Alternating phases of FGF receptor and NGF receptor expression in the developing chicken nervous system. *Neuron* **5,** 283–296.

Hirakow, R., and Hiruma, T. (1981). Scanning electron microscopy study on the development of primitive blood vessels in chick embryos at the early somite stage. *Anat. Rec.* **163,** 299–306.

Hirakow, R., and Hiruma, T. (1983). TEM-studies on development and canalization of the dorsal aorta in the chick embryo. *Anat. Embryol.* **166,** 307–315.

His, W. (1868). "Untersuchungen über die erste Anlage des Wirbeltierleibes." Vogel, Lepzig.

His, W. (1900). Lecithoblast und Angioblast der Wirbeltiere. *Abh. Math-Phys. Kl. Saechs. Ges.* **26,** 171–328.

Hockel, M., Jung, W., Vaupel, P., Rabes, H., Khaledpour, C., and Wissler, J. H. (1988). Purified monocyte-derived angiogenic substance (angiotropin) induces controlled angiogenesis associated with regulated tissue proliferation in rabbit skin. *J. Clin. Invest.* **82,** 1075–1090.

Holmgren, L., Glaser, A., Pfeifer-Ohlsson, S., and Ohlsson, R. (1991). Angiogenesis during human extraembryonic development involves the spatiotemporal control of PDGF ligand and receptor gene expression. *Development (Cambridge, UK)* **113,** 749–754.

Hormia, M., Lehto, M.-P., and Virtanen, I. (1984). Intracellular localization of factor VIII-related antigen and fibronectin in cultures human endothelial cells: Evidence for divergent routes of intracellular translocation. *Eur. J. Cell Biol.* **33,** 217–228.

Houser, J. W., Ackerman, G. A., and Knouff, R. (1961). Vasculogenesis and erythropoiesis in the living yolk sac of the chick embryo: A phase microscopic study. *Anat. Rec.* **140,** 29–44.

Hynes, R. O. (1990). "Fibronectins." Springer-Verlag, New York.

Ikeda, E., Achen, M. G., Breier, G., and Risau, W. (1995). Hypoxia-induced transcriptional activation and increased mRNA stability of vascular endothelial growth factor in C6 glioma cells. *J. Biol. Chem.* **270,** 19761–19766.

Ingber, D. (1991). Extracellular matrix and cell shape: Potential control points for inhibition of angiogenesis. *J. Cell Biochem.* **47,** 236–241.

Iruela-Arispe, M. L., Hasselaar, P., and Sage, H. (1991). Differential expression of extracellular proteins is correlated with angiogenesis in vitro. *Lab. Invest.* **64,** 174–186.

Isaacs, H. V., Tannahill, D., and Slack, J. M. (1992). Expression of a novel FGF in the Xenopus embryo. A new candidate-inducing factor for mesoderm formation and anteroposterior specification. *Development (Cambridge, UK)* **114,** 711–720.

Ishikawa, F., Miyazono, K., Hellman, U., Drexler, H., Wernstedt, C., Hagiwara, K., Usuki, K., Takaku, F., Risau, W., and Heldin, C.-H. (1989). Identification of angiogenic activity and the cloning and expression of platelet-derived endothelial cell growth factor. *Nature (London)* **338,** 557–562.

Iwama, A., Hamaguchi, I., Hashiyama, M., Murayama, Y., Yasunaga, K., and Suda, T. (1993). Molecular cloning and characterization of mouse TIE and TEK receptor tyrosine kinase genes and their expression in hematopoietic stem cells. *Biochem. Biophys. Res. Commun.* **195,** 301–309.

Jaffe, E. A., Hoyer, L. W., and Nachman, R. L. (1973). Synthesis of anti-hemophilic factor antigen by cultured human endothelial cells. *J. Clin. Invest.* **52,** 2757–2764.

Johnston, M. C., Noden, D. M., Hazelton, R. D., Coulombre, J. L., and Coulombre A. J. (1979). Origins of avian ocular and periocular tissues. *Exp. Eye Res.* **29,** 27–43.

Jolley, J. (1940). Recherches sur la formation du système vascular de l'embryon. *Arch. Anat. Microsc.* **35,** 295–361.

Jotereau, F., and Le Douarin, N. (1978). The developmental relationship between osteocytes and osteoclasts: A study using the quail-chick nuclear marker in endochondral ossification. *Dev. Biol.* **63,** 253–265.

Jouanneau, J., Moens, G., Montesano, R., and Thiery, J. P. (1995). FGF-1 but not FGF-4 secreted by carcinoma cells promotes in vitro and in vivo angiogenesis and rapid tumor proliferation. *Growth Factors* **12,** 37–47.

Kadokawa, Y., Suemori, H., and Nakatsuji, N. (1990). Cell lineage analyses of epithelia and blood vessels in chimeric mouse embryos by use of an embryonic stem cell line expressing the beta-galactosidase gene. *Cell Differ. Dev.* **29,** 187–194.

Kallianpur, A. R., Jordan, J. E., and Brandt, S. J. (1994). The SCL/TAL-1 gene is expressed in progenitors of both the hematopoietic and vascular systems during embryogenesis. *Blood* **83,** 1200–1208.

Kau, C., and Turpen, J. B. (1983). Dual contribution of embryonic ventral blood island and dorsal lateral plate mesoderm during ontogeny of hemopoietic cells in *Xenopus laevis. J. Immunol.* **131,** 2262–2266.

Keyt, B. A., Berleau, L. T., Nguyen, H. V., Chen, H., Heinsohn, H., Vandlen, R., and Ferrara, N. (1996). The carboxyl-terminal domain (111-165) of vascular endothelial growth factor is critical for its mitogenic potency. *J. Biol. Chem.* **271,** 7788–7795.

Klagsbrun, M., and D'Amore, P. A. (1991). Regulators of angiogenesis. *Annu. Rev. Physiol.* **53,** 217–239.

Klessinger, S., and Christ, B. (1996). Axial structures control laterality in the distribution pattern of endothelial cells. *Anat. Embryol.* **193,** 319–330.

Klosovskii, B. N. (1963). "The Development of the Brain." Pergamon, Elmsford, NY.

Knöchel, W., Grunz, H., Loppnow-Blinde, B., and Tiedemann H. (1989). Mesoderm induction and blood formation by angiogenic growth factors and embryonic inducing factors. *Blut* **59,** 207–213.

Koch, A. E., Halloran, M. M., Haskell, C. J., Shah, M. R., and Polverini, P. J. (1995). Angiogenesis mediated by soluble forms of E-selectin and vascular cell adhesion molecule-1. *Nature (London)* **376,** 517–519.

Korhonen, J., Polvi, A., Partanen, J., and Alitalo, K. (1994). The mouse tie receptor tyrosine kinase in endothelial cells during neovascularization. *Blood* **80,** 2548–2555.

Krah, K., Mironov, V., Risau, W., and Flamme, I. (1994). Induction of vasculogenesis in quail blastodisc-derived embryoid bodies. *Dev. Biol.* **164,** 123–132.

Krug, E. L., Mjaatvedt, C. H., and Markwald, R. R. (1987). Extracellular matrix from embryonic myocardium elicits an early morphogenetic event in cardiac endothelial differentiation. *Dev Biol Apr* **120,** 348–355.

Kubota, Y., Kleinman, H. K., Martin, G. R., and Lawlley, T. J. (1988). Role of laminin and basement membrane in the morphological differentiation of human endothelial cells into capillary-like structures. *J. Cell Biol.* **107,** 1589–1598.

Kull, F. C., Jr., Brent, D. A., Parikh, I., and Cuatrecasas, P. (1987). Chemical identification of a tumor-derived angiogenic factor. *Science* **236,** 843–845.

Kumar, I., West, D. C., and Ager, A. (1987). Heterogeneity in endothelial cells from large vessels and microvessels. *Differentiation (Berlin)* **36,** 57–70.

Kwee, L., Baldwin, H. S., Shen, H. M., Stewart, C. L., Buck, C., Buck, C. A., and Labow, M. A. (1995). Defective development of the embryonic and extraembryonic circulatory systems in vascular cell adhesion molecule (VCAM-1) deficient mice. *Development (Cambridge, UK)* **121,** 489–503.

LaBastie, M. C., Poole, T. J., Peault, B. M., and Le Douarin, N. M. (1986). MB1, a quail leukocyte-endothelium antigen: Partial characterization of the cell surface and secrested forms in cultured endothelial cells. *Proc. Natl. Acad. Sci. U.S.A.* **83,** 9016–9020.

LaBonne, C., and Whitman, M. (1994). Mesoderm induction by activin requires FGF-mediated intracellular signals. *Development (Cambridge, UK)* **120,** 463–472.

Lampugnani, M. G., Resnati, M., Raiteri, M., Pigott, R., Pisacane, A., Houen, G., Ruco, L. P., and Dejana, E. (1992). A novel endothelial-specific membrane protein is a marker of cell-cell contacts. *J. Cell Biol.* **118,** 1511–1522.

Latker, C. H., and Kuwabara, T. (1981). Regression of the tunica vasculosa lentis in the postnatal rat. *Invest. Ophthalmol. Visual Sci.* **21,** 689–699.

Le Lièvre, C. S., and Le Douarin, N. M. (1975). Mesenchymal derivatives of the neural crest: Analysis of chimeric quail and chick embryos. *J. Embryol. Exp. Morphol.* **34,** 125–154.

Leveen, P., Pekny, M., Gebre-Medhin, S., Swolin, B., Larsson, E., and Betsholtz, C. (1994). Mice deficient for PDGF B show renal, cardiovascular and hematological abnormalities. *Genes Dev.* **8,** 1875–1887.

Levick, J. R., and Smaje, L. H. (1987). An analysis of the permeability of a fenestra. *Microvasc. Res.* **33,** 233–256.

Levy, A., Tamargo, R., Brem, H., and Nathans, D. (1989). An endothelial cell growth factor from the mouse neuroblastoma cell line NB41. *Growth Factors* **2,** 9–19.

Liao, W., Bisgrove, B. W., Sawyer, H., Hug, B., Bell, B., Peters, K., Grunwald, D. J., and Stainier, D. Y. (1997). The zebrafish gene cloche acts upstream of a flk-1 homologue to regulate endothelial cell differentiation. *Development (Cambridge, UK)* **124,** 381–389.

Liaw, L., and Schwartz, S. M. (1993). Comparison of gene expression in bovine aortic endothelium in vivo versus in vitro-differences in growth regulatory molecules. *Arterioscler. Thromb.* **13,** 985–993.

Linask, K., and Lash, J. W. (1986). Precardiac cell migration: Fibronectin localization at mesoderm-endoderm interface during directional movement. *Dev. Biol.* **114,** 87–101.

Linask, K., and Lash, J. W. (1988a). A role for fibronectin in the migration of avian precardiac cells. I. Dose-dependent effects of fibronectin antibody. *Dev. Biol.* **129,** 315–323.

Linask, K., and Lash, J. W. (1988b). A role for fibronectin in the migration of avian precardiac cells. II. Rotation of the heart-forming region during different stages and its effects. *Dev. Biol.* **129,** 324–329.

Lindner, V., Majack, R. A., and Reidy, M. A. (1990). Basic fibroblast growth factor stimulates endothelial regrowth and proliferation in denuded arteries. *J. Clin. Invest.* **85,** 2004–2008.

Little, C. D., Piquet, D. M., Davis, L. A., Walters, L., and Drake, C. J. (1989). Distribution of laminin, collagen type IV, collagen type I, and fibronectin in chicken cardiac jelly/basement membrane. *Anat. Rec.* **224,** 417–425.

Liu, Y. X., Cox, S. R., Morita, T., and Kourembanas, S. (1995). Hypoxia regulates vascular endothelial growth factor gene expression in endothelial cells. Identification of a 5′ enhancer. *Circ. Res.* **77,** 638–643.

Lofberg, J., and Collazo, A. (1997). Hypochord, an enigmatic embryonic structure: Study of the Axolotl embryo. *J. Morphol.* **232,** 57–66.

Löhler, J., Timpl, R., and Jaenisch, R. (1984). Embryonic lethal mutation in mouse collagen I gene causes rupture of blood vessels and is associated with erythropoietic and mesenchymal cell death. *Cell (Cambridge, Mass.)* **38,** 597–607.

Luscinskas, F. W., and Lawler, J. (1994). Integrins as dynamic regulators of vascular function. *FASEB J.* **8,** 929–938.

Mackman, N. (1995). Regulation of the tissue factor gene. *FASEB J.* **9,** 883–889.

Madri, J. A., and Williams, S. K. (1983). Capillary endothelial cell cultures: Phenotypic modulation by matrix components. *J. Cell Biol.* **97,** 153–165.

Maeno, M., Todate, A., and Katagiri, C. (1985). The localization of precursor cells for larval and adult hemopoietic cells in *Xenopus laevis* in two regions of embryos. *Dev. Growth Differ.* **27,** 137–148.

Maisonpierre, P. C., Suri, C., Jones, P. F., Bartunkova, S., Wiegand, S. J., Radziejewski, C., Compton, D., McClain, J., Aldrich, T. H., Papadopoulos, N., Daly, T. J., Davis, S., Sato, T. N., and Yancopoulos, G. D. (1997). Angiopoietin-2, a natural antagonist for Tie2 that disrupts in vivo angiogenesis. *Science* **277,** 55–60.

Matthews, W., Jordan, C. T., Gavin, M., Jenkins, N. A., Copeland, N. G., and Lemischka, I. R. (1991). A receptor tyrosine kinase cDNA isolated from a population of enriched primitive hematopoietic cells and exhibiting close genetic linkage to c-kit. *Proc. Natl. Acad. Sci. U.S.A.* **88,** 9026–9030.

Mayer, B. W., Hay, E. D., and Hynes, R. O. (1981). Immunocytochemical localization of fibronectin in embryonic chick trunk and area vasculosa. *Dev. Biol.* **82,** 267–286.

McClure, C. F. W. (1921). The endothelial problem. *Anat. Rec.* **22,** 219–237.

Meier, S. (1980). Development of the chick embryo mesoblast: Pronephros, lateral plate and early vasculature. *J. Embryol. Exp. Morphol.* **55,** 291–306.

Millauer, B., Wizigmann-Voos, S., Schnürch, H., Martinez, R., Møller, N. P. H., Risau, W., and Ullrich, A. (1993). High affinity VEGF binding and developmental expression suggest flk-1 as a major regulator of vasculogenesis and angiogenesis. *Cell (Cambridge, Mass.)* **72,** 835–846.

Millauer, B., Shawver, L. K., Plate, K. H., Risau, W., and Ullrich, A. (1994). Glioblastoma growth inhibited in vivo by a dominant-negative Flk-1 mutant. *Nature (London)* **367,** 576–579.

Miller, A. M., and McWhorter, J. E. (1914). Experiments on the development of blood vessels in the area pellucida and embryonic body of the chick. *Anat. Rec.* **8,** 203–227.

Mitrani, E., and Shimoni, Y. (1990). Induction by soluble factors of organized axial structures in chick epiblasts. *Science* **247,** 1092–1094.

Mitrani, E., Gruenbaum, Y., Shohat, H., and Ziv, T. (1990). Fibroblast growth factor during mesoderm induction in the early chick embryo. *Development (Cambridge, UK)* **109,** 387–393.

Miyazono, K., Okabe, T., Urabe, A., Takaku, F., and Heldin, C.-H. (1987). Purification and properties of an endothelial cell growth factor from human platelets. *J. Biol. Chem.* **262,** 4098–4103.

Montesano, R., Orci, L., and Vassalli, P. (1983). In vitro rapid organization of endothelial cells into capillary-like networks is promoted by collagen matrices. *J. Cell Biol.* **97,** 1648–1652.

Montesano, R., Vassalli, J. D., Baird, A., Guillemin, R., and Orci, L. (1986). Basic fibroblast growth factor induces angiogenesis in vitro. *Proc. Natl. Acad. Sci. U.S.A.* **83,** 7297–7301.

Muller, G., Behrens, J., Nussbaumer, U., Bohlen, P., and Birchmeier, W. (1987). Inbitory action of transforming growth factor beta on endothelial cells. *Proc. Natl. Acad. Sci. U.S.A.* **84,** 5600–5604.

Murphy, M. E., and Carlson, E. C. (1978). An ultrastructural study of developing extra cellular matrix in vitelline blood vessels of the early chick embryo. *Am. J. Anat.* **151,** 345–376.

Murray, P. D. F. (1932). The development in vitro of the blood of the early chick embryo. *Proc. R. Soc. B London, Ser. B* **111,** 497–521.

Mustonen, T., and Alitalo, K. (1995). Endothelial receptor tyrosine kinases involved in angiogenesis. *J. Cell Biol.* **129,** 895–898.

Nelsen, O. E. (1953). "Comparative Embryology of the Vertebrates." Blakiston, New York.

Newman, C. S., Chia, F., and Krieg, P. A. (1997). The XHex homeobox gene is expressed during development of the vascular endothelium: Overexpression leads to an increase in vascular endothelial cell number. *Mech. Dev.* **66,** 83–93.

Newman, P. J., Berndt, M. C., Gorski, J., White, G. C., Lyman, S., Paddock, C., and Muller, W. A. (1990). PECAM-1 (CD31) cloning and relation to adhesion molecules of the immunoglobulin gene superfamily. *Science* **247,** 1219–1222.

Nicosia, R. F., and Bonanno, E. (1991). Inhibition of angiogenesis in vitro by Arg-Gly-Asp-containing synthetic peptide. *Am. J. Pathol.* **138,** 829–833.

Noden, D. M. (1988). Interactions and fates of avian craniofacial mesenchyme. *Development (Cambridge, UK)* **103,** S121–S140.

Noden, D. M. (1989). Embryonic origins and assembly of blood vessels. *Am. Rev. Respir. Dis.* **140,** 1097–1103.

Noden, D. M. (1990). Origins and assembly of avian embryonic blood vessels. *Ann. N.Y. Acad. Sci.* **588,** 236–249.

Noden, D. M. (1991). Development of Craniofacial Blood Vessels. The Development of the Vascular System. (Feinberg, R. N., Sherer, G. K., Auerbach, R., Eds.) *Issues Biomed.* Basel, Karger, **14,** 1–24.

Oikawa, T., Suganuma, M., Ashino-Fuse, H., and Shimamura, M. (1992). Okadaic acid is a potent angiogenesis inducer. *Jpn. J. Cancer Res.* **83,** 6–9.

Olah, I., Medgyes, J., and Glick, B. (1988). Origin of aortic cell clusters in the chicken embryo. *Anat. Rec.* **222,** 60–68.

O'Reilley, M. S., Holmgren, L., Shing, Y., Chen, C., Rosenthal, R. A., Moses, M., Lane, W. S., Cao, Y., Sage, E. H., and Folkman, J. (1994). Angiostatin: A novel angiogenesis inhibitor that mediates the suppression of metastases by a Lewis lung carcinoma. *Cell (Cambridge, Mass.)* **79,** 315–328.

Orlidge, A., and D'Amore, P. A. (1987). Inhibition of capillary endothelial cell growth by pericytes and smooth muscle cells. *J. Cell Biol.* **105,** 1455–1462.

Oshima, M., Oshima, H., and Taketo, M. M. (1996). TGF-b receptor type II deficiency results in defects of yolk sac hematopoiesis and vasculogenesis. *Dev. Biol.* **179,** 297–302.

Owens, G. K., and Thompson, M. M. (1986). Developmental changes in isoactin expression in rat aortic smooth muscle cells in vivo. *J. Biol. Chem.* **261,** 13373–13380.

Palis, J., McGrath, K. E., and Kingsley, P. D. (1995). Initiation of haematopoiesis and vasculogenesis in murine yolk sac explants. *Blood* **86,** 156–163.

Pardanaud, L., and Dieterlen-Lièvre, F. (1993). Emergence of endothelial and hematopoietic cells in the avian embryo. *Anat. Embryol.* **187,** 107–114.

Pardanaud, L., Altman, C., Kitos, P., Dieterlen-Lièvre, F., and Buck, C. A. (1987). Vasculogenesis in the early quail blastodisc as studied with a monoclonal antibody recognizing endothelial cells. *Development (Cambridge, UK)* **100,** 339–349.

Paradanaud, L., Yassine, F., and Dieterlen-Lièvre, F. (1989). Relationship between vasculogenesis, angiogenesis and hematopoiesis during avian ontogeny. *Development* (*Cambridge, UK*) **105,** 473–485.

Pardanaud, L., Luton, D., Prigent, M., Bourcheix, L. M., Catala, M., and Dieterlen-Lièvre, F. (1996). Two distinct endothelial lineages in ontogeny, one of them related to hemopoiesis. *Development* (*Cambridge, UK*) **122,** 1363–1371.

Partanen, J., Armstrong, E., Mäkelä, T., Korhonen, J., Sandberg, M., Renkonen, R., Knuutila, S., Huebner, K., and Alitalo, K. (1992). A novel endothelial cell surface receptor tyrosine kinase with extracellular epidermal growth factor homology domains. *Mol. Cell. Biol.* **12,** 1698–1707.

Patan, S., Alvarez, M. J., Schittny, J. C., and Burri, P. H. (1992). Intussusceptive microvascular growth: A common alternative to capillary sprouting. *Arch. Histol. Cytol., Suppl.* **55,** 65–75.

Patan, S., Haenni, B., and Burri, P. H. (1993). Evidence for intussusceptive capillary growth in the chicken chorio-allantoic membrane (CAM). *Anat. Embryol.* **187,** 121–130.

Patan, S., Haenni, B., and Burri, P. H. (1996). Implementation of intussusceptive microvascular growth in the chicken chorioallantoic membrane (CAM): 1. Pillar formation by folding of the capillary wall. *Microvasc. Res.* **51,** 80–98.

Peault, B. M., Thiery, J.-P., and Le Douarin, N. M. (1983). Surface marker for hemopoietic and endothelial cell lineages in quail that is defined by a monoclonal antibody. *Proc. Natl. Acad. Sci. U.S.A.* **80,** 2976–2980.

Peault, B. M., Cotley, M., and Le Douarin, N. M. (1988). Ontogenic emergence of a quail leukocyte/endothelium cell surface antigen. *Cell Differ.* **23,** 165–174.

Pepper, M. S., and Montesano, R. (1990). Proteolytic balance and capillary morphogenesis. *Cell Differ. Dev.* **32,** 319–328.

Pepper, M. S., Ferrara, N., Orci, L., and Montesano, R. (1991). Vascular endothelial growth factor (VEGF) induces plasminogen activators and plasminogen activator inhibitor-1 in microvascular endothelial cells. *Biochem. Biophys. Res. Commun.* **181,** 902–906.

Peters, K. G., Werner, S., Chen, G., and Williams, L. T. (1992). Two FGF receptor genes are differentially expressed in epithelial and mesenchymal tissues during limb formation and organogenesis in the mouse. *Development* (*Cambridge, UK*) **114,** 233–243.

Peters, K. G., de Vries, C., and Williams, L. T. (1993). Vascular endothelial growth factor receptor expression during embryogenesis and tissue repair suggests a role in endothelial differentiation and blood vessel growth. *Proc. Natl. Acad. Sci. U.S.A.* **90,** 8915–8919.

Pierce, G. F., Tarpley, J. E., Yanagihara, D., Mustoe, T. A., Fox, G. M., and Thomason, A. (1992). Platelet-derived growth factor (BB homodimer), transforming growth factor-beta 1, and basic fibroblast growth factor in dermal wound healing. Neovessel and matrix formation and cessation of repair. *Am. J. Pathol.* **140,** 1375–1388.

Plate, K. H., Breier, G., Weich, H. A., and Risau, W. (1992). Vascular endothelial growth factor is a potential tumour angiogenesis factor in human gliomas in vivo. *Nature* (*London*) **359,** 843–845.

Plate, K. H., Breier, G., Millauer, B., Ullrich, A., and Risau, W. (1993). Up-regulation of vascular endothelial growth factor and its cognate receptors in a rat glioma model of tumor angiogenesis. *Cancer Res.* **53,** 5822–5827.

Poole, T. J., and Coffin, J. D. (1988). Developmental angiogenesis: Quail embryonic vasculature. *Scanning Microsc.* **2,** 443–448.

Poole, T. J., and Coffin, J. D. (1989). Vasculogenesis and angiogenesis: Two distinct morphogenetic mechanisms establish embryonic vascular pattern. *J. Exp. Zool.* **251,** 224–231.

Poole, T. J., and Coffin, J. D. (1991). Morphogenetic mechanisms in Avian vascular development. *Issues Biomed.* **14,** 25–37.

Porcher, C., Swat, W., Rockwell, K., Fujiwara, Y., Alt, F. W., and Orkin, S. H. (1996). The T cell leukemia oncoprotein SCL/tal-1 is essential for development of all hematopoietic lineages. *Cell* (*Cambridge, Mass.*) **86,** 47–57.

Puri, M. C., Rossant, J., Alitalo, K., Bernstein, A., and Partanen, J. (1995). The receptor tyrosine kinase TIE is required for integrity and survival of vascular endothelial cells. *EMBO J.* **14,** 5884–5891.

Quinn, T. P., Peters, K. G., de Vries, C., Ferrara, N., and Williams, L. T. (1993). Fetal liver kinase 1 is a receptor for vascular endothelial growth factor and is selectively expressed in vascular endothelium. *Proc. Natl. Acad. Sci. U.S.A.* **90,** 7533–7537.

Reagan, F. P. (1915). Vascularization phenomena in fragments of embryonic bodies completely isolated from yol-sac entoderm. *Anat. Rec.* **9,** 329–341.

Reidy, M. A., and Schwartz, S. M. (1981). Endothelial regeneration. III. Time course of intimal changes after small defined injury to rat aortic endothelium. *Lab. Invest.* **44,** 301–308.

Resnick, N., and Gimbrone, M. A. (1995). Hemodynamic forces are complex regulators of endothelial gene-expression. *FASEB J.* **9,** 874–882.

Rhodin, J. A. (1968). Ultrastructure of mammalian venous capillaries, venules and small collecting veins. *J. Ultrastruct. Res.* **25,** 452–500.

Rhodin, J. A. G. (1980). Architecture of the vessel wall. *In* "Handbook of Physiology" S. R. Geiger, (ed.), Sect. 2, Vol. II. Am. Physiol. Soc., Bethesda, MD.

Risau, W. (1986). Developing brain produces an angiogenesis factor. *Dev. Biol.* **83,** 3855–3859.

Risau, W. (1991). Vasculogenesis, angiogenesis and endothelial cell differentiation during embryonic development. *Issues Biomed.* **14,** 58–68.

Risau, W. (1995). Differentiation of the endothelium. *FASEB J.* **9,** 926–933.

Risau, W. (1997). Mechanisms of angiogenesis. *Nature* (*London*) **386,** 671–674.

Risau, W., and Ekblom, P. (1986). Production of a heparin-binding angiogenesis factor by the embryonic kidney. *J. Cell Biol.* **103,** 1101–1107.

Risau, W., and Flamme, I. (1995). Vasculogenesis. *Annu. Rev. Cell Dev. Biol.* **11,** 73–91.

Risau, W., and Lemmon, V. (1988). Changes in the vascular extracellular matrix during embryonic vasculogenesis and angiogenesis. *Dev. Biol.* **125,** 441–450.

Risau, W., Sariola, H., Zerwes, H.-G., Sasse, J., Ekblom, P., Kemler, R., and Doetschman, T. (1988). Vasculogenesis and angiogenesis in embryonic stem cell-derived embryoid bodies. *Development* (*Cambridge, UK*) **102,** 471–478.

Risau, W., Drexler, H., Mironov, V., Smits, A., Siegbahn, A., Funa, K., and Heldin, C. H. (1992). Platelet-derived growth factor is angiogenic in vivo. *Growth Factors* **7,** 261–266.

Roberts, W. G., and Palade, G. E. (1995). Increased microvascular permeability and endothelial fenestration induced by vascular endothelial growth factor. *J. Cell Sci.* **108,** 2369–2379.

Rohovsky, S. A., Hirschi, K. K., and D'Amore, P. A. (1996). Growth factor effects on a model of vessel formation. *Surg. Forum.* **47,** 390–391.

Romanoff, A. L. (1960). The hematopoietic, vascular and lymphatic systems. *In* "The Avian Embryo: Structural and Functional Development" (A. L. Romanoff, ed.), pp. 571–663. Macmillan, New York.

Rosenquist, G. C. (1970). Aortic arches in the chick embryo: origin of the cells as determined by radioautographic mapping. *Anat. Rec.* **168,** 351–359.

Ryan, U. S., Ryan, J. W., and Chiu, A. (1976). Kininase II (angiotensin converting enzyme) and endothelial cells in culture. *Adv. Exp. Med. Biol.* **70,** 217–227.

Sabin, F. R. (1917). Origin and development of the primitive vessels of the chick and the pig. *Carnegie Contrib. Embryol.* **6,** 61–124.

Sabin, F. R. (1920). Studies on the origin of the blood vessels and of red blood corpuscles as seen in the living blastoderm of chick during the second day of incubation. *Carnegie Contrib. Embryol.* **9,** 215–262.

Sariola, H., Ekblom, P., Lehtonen, E., and Sazen, L. (1983). Differentiation and vascularization of the metanephric kidney grafted on the chorioallantoic membrane. *Dev. Biol.* **96,** 427–435.

Sato, N., Beitz, J. G., Kato, J., Yamamoto, M., Clark, J. W., Calabresi, P., Raymond, A., and Frackelton, A. R., Jr. (1993). Platelet-derived growth factor indirectly stimulates angiogenesis in vitro. *Am. J. Pathol.* **142,** 1119–1130.

Sato, T. N., Qin, Y., Kozak, C. A., and Audus, K. (1993). tie-1 and tie-2 define another class of putative receptor tyrosine kinase genes expressed in early embryonic vascular system. *Proc. Natl. Acad. Sci. U.S.A.* **90,** 9355–9358.

Sato, T. N., Tozawa, Y., Deutsch, U., Wolburg-Buchholz, K., Fujiwara, Y., Gendron-Maguire, M., Gridley, T., Wolburg, H., Risau, W., and Qin, Y. (1995). Distinct roles of the receptor tyrosine kinases Tie-1 and Tie-2 in blood vessel formation. *Nature (London)* **376,** 70–74.

Sato, Y., and Rifkin, D. B. (1989). Inhibition of endothelial cell movement by pericytes and smooth muscle cells: Activation of a latent transforming growth factor-beta 1-like molecule by plasmin during co-culture. *J. Cell Biol.* **109,** 309–315.

Schatteman, G. C., Motley, S. T., Effmann, E. L., and Bowen-Pope, D. F. (1995). Platelet-derived growth factor receptor alpha subunit deleted patch mouse exhibits severe cardiovascular dysmorphogenesis. *Teratology* **51,** 351–366.

Schnurch, H., and Risau, W. (1993). Expression of tie-2, a member of a novel family of receptor tyrosine kinases, in the endothelial cell lineage. *Development (Cambridge, UK)* **119,** 957–968.

Schwartz, S. M., and Liaw, L. (1993). Growth control and morphogenesis in the development and pathology of arteries. *J. Cardiovasc. Pharmacol.* **21,** S31–S49.

Semenza, G. L. (1996). Transcriptional regulation by hypoxia-inducible factor 1—Molecular mechanisms of oxygen homeostasis. *Trends Cardiovasc. Med.* **6,** 151–157.

Shalaby, F., Rossant, J., Yamaguchi, T. P., Gertsenstein, M., Wu, X.-F., Breitman, M. L., and Schuh, A. (1995). Failure of blood-island formation and vasculogenesis in flk-1 deficient mice. *Nature (London)* **376,** 62–66.

Shalaby, F., Ho, J., Stanford, W. L., Fisher, K.-D., Schuh, A. C., Schwartz, L., Bernstein, A., and Rossant, J. (1997). A requirement for Flk-1 in primitive and definitive hematopoiesis and vasculogenesis. *Cell (Cambridge, Mass.)* **89,** 981–990.

Sherer, G. K. (1991). Vasculogenic Mechanisms and Epithelio-Mesenchymal Specificity in Endodermal Organs. The Development of the Vascular System. (Feinberg, R. N., Sherer, G. K., Auerbach, R., Eds.) *Issues Biomed.* Basel, Karger, **14,** 37–57.

Shibuya, M., Yamaguchi, S., Yamane, A., Ikeda, T., Tojo, A., Matsushime, H., and Sato, M. (1990). Nucleotide sequence and expression of a novel human receptor-type tyrosine kinase gene (flt) closely related to the fms family. *Oncogene* **5,** 519–524.

Shinbrot, D., Peters, K. G., and Williams, L. T. (1994). Expression of the platelet-derived growth factor β receptor during organogenesis and tissue differentiation in the mouse embryo. *Dev. Dyn.* **199,** 169–175.

Shing, Y., Folkman, J., Sullivan, R., Butterfield, C., Murray, J., and Klagsbrun, M. (1984). Heparin affinity: Purification of a tumor-derived capillary endothelial cell growth factor. *Science* **223,** 1296–1298.

Sholley, M. M., Ferguson, G. P., Seibel, H. R., Montour, J. L., and Wilson, J. D. (1984). Mechanisms of neovascularization. *J. Lab. Invest.* **51,** 624–634.

Short, R. H. D. (1950). Alveolar epithelium in relation to growth of the lung. *Philos. Trans. R. Soc. London, Ser. B* **235,** 35–87.

Shweiki, D., Itin, A., Soffer, D., and Keshet, E. (1992). Vascular endothelial growth factor induced by hypoxia may mediate hypoxia-initiated angiogenesis. *Nature (London)* **359,** 843–845.

Slack, J. M., Darlington, B. G., Heath, J. K., and Godsave, S. F. (1987). Mesoderm induction in early Xenopus embryos by heparin-binding growth factors. *Nature (London)* **326,** 197–200.

Soriano, P. (1994). Abnormal kidney development and hematological disorders in PDGF β receptor mutant mice. *Genes Dev.* **8,** 1888–1896.

Stephansson, S., and Lawrence, D. A. (1996). The serpin PAI-1 inhibits cell migration by blocking integrin α_{v3} binding to vitronectin. *Nature* **383,** 441–443.

Stewart, P. A., and Wiley, M. J. (1981). Developing nervous tissue induces formation of blood-brain barrier characteristics in invading endothelial cells: A study using quail-chick transplantation chimeras. *Dev. Biol.* **84,** 183–192.

Stockard, C. R. (1915). The origin of blood and vascular endothelium in embryos without a circulation of the blood and in the normal embryo. *Am. J. Anat.* **18,** 227–327.

Stone, J., Itin, A., Alon, T., Pe'er, J., Gnessin, H., Chan-Ling, T., and Keshet, E. (1995). Development of retinal vasculature is mediated by hypoxia-induced vascular endothelial growth factor (VEGF) expression by neuroglia. *J. Neurosci.* **15,** 4738–4747.

Strömblad, S., and Cheresh, D. A. (1996). Cell adhesion and angiogenesis. *Trends Cell Biol.* **6,** 462–467.

Suh, T. T., Holmback, K., Jensen, N. J., Daugherty, C. C., Small, K., Simon, D. I., Potter, S., and Degen, J. L. (1995). Resolution of spontaneous bleeding events but failure of pregnancy in fibrinogen-deficient mice. *Genes Dev.* **9,** 2020–2033.

Sumoy, L., Keasey, J. B., Dittman, T. D., and Kimelman, D. (1997). A role for notochord in axial vascular development revealed by analysis of phenotype and the expression of VEGR-2 in zebrafish flh and ntl mutant embryos. *Mech. Dev.* **63,** 15–27.

Suri, C., Jones, P. F., Patan, S., Bartunkova, S., Maisonpierre, P. C., Davis, S., Sato, T. N., and Yancopoulos, G. D. (1996). Requisite role of angiopoietin-1, a ligand for the TIE2 receptor, during embryonic angiogenesis. *Cell (Cambridge, Mass.)* **87,** 1171–1180.

Suzuki, S., Sano, K., and Tanihara, H. (1991). Diversity of the cadherin family: Evidence for eight new cadherins in nervous tissue. *Cell. Regul.* **2,** 261–270.

Tannahill, D., Isaacs, H. V., Close, M. J., Peters, G., and Slack, J. M. (1992). Developmental expression of the Xenopus int-2 (FGF-3) gene: Activation by mesodermal and neural induction. *Development (Cambridge, UK)* **115,** 695–702.

Tavian, M., Coulombel, L., Luton, D., Clemente, H. S., Dieterlen-Lièvre, F., and Peault, B. (1996). Aorta-associated CD34+ hematopoietic cells in the early human embryo. *Blood* **87,** 67–72.

Taylor, S., and Folkman, J. (1982). Protamine is an inhibitor of angiogenesis. *Nature (London)* **297,** 307–312.

Thompson, W. D., Smith, E. B., Stirk, C. M., Marshall, F. I., Stout, A. J., and Kocchar, A. (1992). Angiogenic activity of fibrin degradation products is located in fibrin fragment E. *Am. J. Pathol.* **168,** 47–53.

Tilton, R. G., Miller, E. J., Kilo, C., and Williamson, J. R. (1985). Pericyte form and distribution in rat retinal and uveal capillaries. *Invest. Ophthalmol. Visual Sci.* **26,** 68–73.

Tracy, P. B., and Mann, K. G. (1987). Abnormal formation of the prothrombinase complex: Factor V deficiency and related disorders. *Hum. Pathol.* **18,** 162–169.

Ungari, S., Katari, R. S., Alessandri, G., and Gullino, P. M. (1985). Cooperation between fibronectin and heparin in the mobilization of capillary endothelium. *Invasion Metastasis* **5,** 193–205.

Van Groningen, J. P., Wenink, A. C. G., and Testers, L. H. M. (1991). Myocardial capillaries: Increase in number by splitting of existing vessels. *Anat. Embryol.* **184,** 65–70.

Vecchi, A., Garlanda, C., Lampugnani, M. G., Resnati, M., Matteucci,

C., Stoppacciaro, A., Schnurch, H., Risau, W., Ruco, L., Mantovani, A., and Dejana, E. (1994). Monoclonal antibodies specific for endothelial cells of mouse blood vessels. Their application in the identification of adult and embryonic endothelium. *Eur. J. Cell Biol.* **63,** 247–254.

Vikkula, M., Boon, L. M., Carraway, K. L., III, Calvert, J. T., Diamonti, A. J., Goumnerov, B., Pasyk, K. A., Marchuk, D. A., Warman, M. L., Cantley, L. C., Mulliken, J. B., and Olsen, B. R. (1996). Vascular dysmorphogenesis caused by an activating mutation in the receptor tyrosine kinase TIE2. *Cell (Cambridge, Mass.)* **87,** 1181–1190.

Vittet, D., Buchou, T., Schweitzer, A., Dejana, E., and Huber, P. (1997). Targeted null-mutation in the vascular endothelial-cadherin gene impairs the organization of vascular-like structures in embryoid bodies. *Proc. Natl. Acad. Sci. U.S.A.* **94,** 6273–6278.

Wagner, R. C. (1980). Endothelial cell embryology and growth. *Adv. Microcirc.* **9,** 45–75.

Waltenberger, J., Claesson-Welsh, L., Siegbahn, A., Shibuya, M., and Heldin, C.-H. (1994). Different signal transduction properties of KDR and Flt1, two receptors for vascular endothelial growth factor. *J. Biol. Chem.* **269,** 26988–26995.

Wanaka, A., Milbrandt, J., and Johnson, E. M., Jr. (1991). Expression of FGF receptor gene in rat development. *Development (Cambridge, UK)* **111,** 455–468.

West, D. C., Hampson, I. N., Arnold, F., and Kumar, S. (1985). Angiogenesis induced by degradation products of hyaluronic acid. *Science* **228,** 1324–1326.

Williams, R. L., Courtneidge, S. A., and Wagner, E. F. (1988). Embryonic lethalities and endothelial tumors in chimeric mice expressing polyoma virus middle T oncogene. *Cell (Cambridge, Mass.)* **52,** 121–131.

Wilms, P., Christ, B., Wilting, J., and Wachtler, F. (1991). Distribution and migration of angiogenic cells from grafted avascular intraembryonic mesoderm. *Anat. Embryol.* **183,** 371–377.

Wilson, D. (1983). The origin of the endothelium in the developing marginal vein of the chick wing-bud. *Cell Differ.* **13,** 63–67.

Wilt, F. H. (1965). Erythropoiesis in the chick embryo: The role of the endoderm. *Science* **147,** 1588–1590.

Wilt, F. H. (1974). The beginning of erythropoiesis in the yolk sac of the chick embryo. *Ann. N.Y. Acad. Sci.* **241,** 99–112.

Wilting, J., and Christ, B. (1996). Embryonic angiogenesis: A review. *Naturwissenschaften* **83,** 153–164.

Wilting, J., Christ, B., Grim, M., and Wilms, P. (1992). "Angiogenic Capacity of Early Avian Mesoderm. Formation and Differentiation of Early Embryonic Mesoderm." Plenum, New York.

Wilting, J., Brand-Saberi, B., Huang, R., Zhi, Q., Köntges, G., Ordahl, C. P., and Christ, B. (1995). Angiogenic potential of the avian somite. *Dev. Dyn.* **202,** 165–171.

Wilting, J., Birkenhäger, R., Eichmann, A., Kurz, H., Martiny-Baron, G., Marme, D., McCarthy, J. E. G., Christ, B., and Weich, H. A. (1996). $VEGF_{121}$ induces proliferation of vascular endothelial cells and expression of flk-1 without affecting lymphatic vessels of the chorioallantoic membrane. *Dev. Biol.* **176,** 76–85.

Wilting, J., and Christ, B. (1997). Embryonic Angiogenesis: A Review. *Natur Wissenschaftenaufsätze* **83,** 153–164.

Wolff, J. R., and Bär, T. (1972). 'Seamless' endothelia in brain capillaries during development of the rat's cerebral cortex. *Brain Res.* **41,** 17–24.

Yablonka-Reuveni, Z. (1989). The emergence of the endothelial cell lineage in the chick embryo can be detected by uptake of actylated low density lipoprotein and the presence of von Willebrand-like Factor. *Dev. Biol.* **132,** 230–240.

Yamaguchi, T. P., Dumont, D. J., Conlon, R. A., Breitman, M. L., and Rossant, J. (1993). Flk-1, a flt-related receptor tyrosine kinase is an early marker for endothelial cell precursors. *Development (Cambridge, UK)* **118,** 489–498.

Yang, J. T., Rayburn, H., and Hynes, R. O. (1993). Embryonic mesodermal defects in α_5 integrin-deficient mice. *Development (Cambridge, UK)* **119,** 1093–1105.

Zagris, N. (1980). Erythroid cell differentiation in unincubated chick blastoderm in culture. *J. Embryol. Exp. Morphol.* **58,** 209–216.

V

Genetic Control of Muscle Gene Expression

15

The MLC-2 Paradigm for Ventricular Heart Chamber Specification, Maturation, and Morphogenesis

Vân Thi Bich Nguyêñ-Trân, Ju Chen, Pilar Ruiz-Lozano, and Kenneth Randall Chien

Department of Medicine, Center for Molecular Genetics, and American Heart Association–Bugher Foundation Center for Molecular Biology, University of California, San Diego, La Jolla, California 92093

I. Background

The formation of distinct atrial and ventricular chambers is a critical step in vertebrate cardiogenesis. Following the fusion of the cardiac primordia and subsequent development of a linear heart tube, vertebrate cardiogenesis proceeds through a series of looping events and involutions that ultimately result in formation of four cardiac chambers and acquisition of regional-specific properties of atrial, ventricular, and specialized conduction system cells (Fishman *et al.*, 1998; see Chapters 10 and 12). Although it has been the focus of several decades of embryological analysis, relatively little is known about the specific molecular and positional cues that dictate the discrete steps in cardiac chamber specification, maturation, and morphogenesis. Our approach to this problem has centered on the atrial and ventricular chamber-specific isoforms of the myosin light chain-2 (MLC-2) protein which are encoded by two separate genes, the atrial *MLC-2a* and ventricular

MLC-2v genes. To identify the molecular mechanisms that regulate the expression of MLC-2a and MLC-2v, which are genetic markers closely linked to the onset of distinct stages of chamber morphogenesis, we have utilized a combinatorial approach including *in vitro* ventricular muscle cell assay systems (Zhu *et al.*, 1991; Navankasattusas *et al.*, 1992) and genetically manipulated mouse models (Lee *et al.*, 1992; Arber *et al.*, 1997; Chen *et al.*, 1998a,b).

This chapter focuses on the molecular paradigms that illustrate how the expression of two isoforms of the regulatory or "phosphorylable" MLC-2 proteins, namely, the cardiac ventricular/slow twitch isoform (MLC-2v) and the cardiac atrial isoform (MLC-2a), is controlled. In addition, the unique role of *MLC-2v* in cardiac contractile function will be discussed in the context of the other myosin light chain isoforms.

A. MLC-2 as a Contractile Protein

Despite having distinct embryological origins and transcriptional regulatory pathways, both cardiac and skeletal muscle have considerable overlap in the contractile protein isoforms they express (Schiaffino and Reggiani, 1996). The functional diversity of these distinct muscle cells is the result of the differential expression of interchangeable sarcomeric components (Kelly and Buckingham, 1997). Sarcomeric myosin, the major protein of the thick filament in muscle, provides the molecular motor for force production during myocardial contraction. Each myosin hexamer is composed of one pair of heavy chains (MHCs), which make up the "tail" and the globular "head," whereas while the two pairs of light chains (MLCs) are associated with the two "hinge" regions of the myosin molecule (Fig. 1). The three-dimensional structure of myosin indicates that the "alkali" or essential light chains (MLC-1 and MLC-3) are located at the terminal half of the neck, whereas the regulatory or phosphorylable light chain (MLC-2) is located at the neck–tail junction (Rayment *et al.*, 1993a,b; Spudich, 1994).

There are three MLC-2 isoforms in mammalian striated muscle: a fast skeletal muscle isoform (MLC-2f; Nudel *et al.*, 1984), a cardiac ventricular and slow twitch skeletal muscle isoform (MLC-2v; Henderson *et al.*, 1989), and a cardiac atrial isoform (MLC-2a; Hailstones *et al.*, 1992; Kubalak *et al.*, 1994). There is no direct evidence that the highly conserved structure of these three individual isoforms reflects a unique functional requirement in distinct muscle cell types. In skeletal myosin the most striking mechanical effect of MLC-2 is on the physiological speed of shortening of the actomyosin complex because removal of MLC-2 significantly decreases the velocity of actin movement (Lowey *et al.*, 1993). However, until recently the precise role of MLC-2 in vertebrate cardiac muscle was much less well-known. Alterations in MLC-2 expression have been correlated with the onset of cardiac morphogenetic defects during embryogenesis (Sucov *et al.*, 1994; Dyson *et al.*, 1995; Lyons *et al.*, 1995). In addition, point mutations in *MLC-2v* have been shown to be associated with a genetic form of human cardiomyopathy (Poetter *et al.*, 1996).

Figure 1 Structure of the myosin molecule. Myosin hexamers are composed of one pair of heavy chains (MHCs) which form the tail and globular head. Two pairs of light chains are associated with the two hinge regions of the myosin: The alkali or essential light chains (ELC: MLC-1 and MLC-3) are located at the terminal half of the neck, whereas the regulatory or phosphorylatable light chain (RLC: MLC-2) is located at the neck–tail junction [adapted from Lodish, *Molecular Cell Biology*, (Berk, Zipursky, Matsudaira, and Darnell, Eds.), Baltimore].

II. Temporal and Regional Expression of *MLC-2v* and *MLC-2a* during Cardiac Morphogenesis

A. *MLC-2v* Expression Marks Ventricular Specification

Even though a number of cardiac markers are expressed uniformly throughout the developing heart tube, it has become increasingly clear that there are specific genes which display a more restricted pattern of expression in the individual segments destined to form the outflow and inflow tracts as well as the atrial and ventricular chambers (Olson and Srivastava, 1996; see Chapters 7 and 9). In studies using an *in vitro* model of cardiogenesis in embryonic stem (ES) cells, *MLC-2v* expression is found to occur independently of heart tube formation, suggesting that ventricular specification occurs relatively early in mammalian cardiogenesis (Miller-Hance *et al.,* 1993). The *MLC-2v* gene is the earliest known ventricular-specific marker in the vertebrate heart (O'Brien *et al.,* 1993) and displays bilateral symmetry in a restricted zone of the cardiogenic plate. As early as Embryonic Day 8.0 postcoitum (pc), *MLC-2v* expression is detected in the ventricular portion of the heart tube, without detectable expression in the atrial or sinus venosus regions (Fig. 2). The proximal outflow tract of the heart tube also expresses a minimal level of *MLC-2v* at this time, but by 9 or 10 days pc, expression becomes restricted to the ventricles. By Day 11 pc, prior to the completion of septation, *MLC-2v* expression is restricted to the ventricular region at and below the level of the atrioventricular cushion.

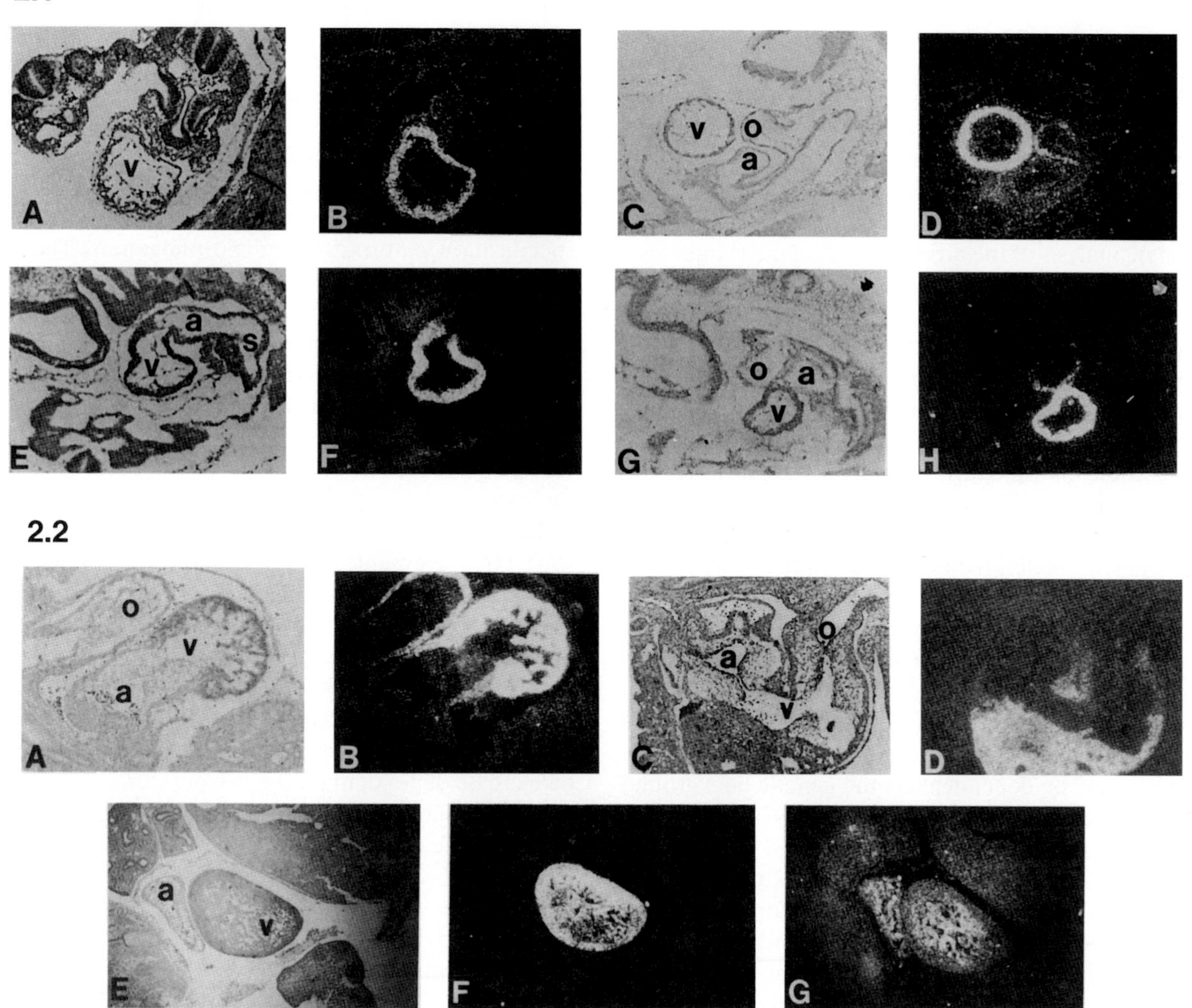

Figure 2 Localization of the *MLC-2v* messenger RNA in the mouse embryo. The distribution of the mRNA for MLC-2v was determined by *in situ* hybridization on sagittal sections of mouse embryos. Micrographs of paraffin sections of sequential stages of mouse development are shown in bright field to illustrate the integrity of the tissues and dark field to detect hybridization. (2.1) MLC-2v expression in early mouse embryos. Day 8 pc sagittal, bright field (A) and dark field (B); Day 8.5 midsagittal, bright (C) and dark field (D); Day 9 pc parasagittal, bright (E) and dark field (F); Day 9 pc sagittal, bright (G) and dark field (H). (2.2) MLC-2v expression in later stages of mouse development. Day 10 pc midsagittal, bright (A) and dark field (B); Day 11 pc parasagittal, bright (C) and dark field (D); Day 13 pc sagittal, bright (E) and dark field (F). a, atria; v, ventricle; o, outflow tract; s, sinus venosus. [From O'Brien *et al.,* 1993, with permission. Copyright (1993) National Academy of Sciences, U.S.A.]

B. *MLC-2a* Downregulation Marks Ventricular Maturation

During the course of cardiogenesis, the atrial and ventricular cell lineages display distinct electrophysiological, biochemical, endocrine, and contractile properties, which are dependent on the expression of a subset of chamber-specific genes. Numerous chamber-restricted genes acquire their regional pattern relatively late during cardiogenesis, usually after the completion of septation or during the neonatal period (de Groot *et al.,* 1989; Sassoon *et al.,* 1988; Lyons *et al.,* 1990). This suggests that physiologic or hemodynamic changes in the various cardiac chambers may act as switches to regulate the cardiac gene program, similar to the positive and negative regulation in adult myocardium following exposure to mechanical or adrenergic stimuli (Komuro and Yazaki, 1993; Chien *et al.,* 1993).

As noted previously, analysis of the expression of the *MLC-2* genes has shown that *MLC-2v* is restricted to the ventricular segment of the linear heart tube, prior to the septation and the formation of distinct cardiac chambers (O'Brien *et al.,* 1993). The cloning and expression analysis of the atrial *MLC-2a* gene (Hailstones *et al.,* 1992; Kubalak *et al.,* 1994) provided further evidence that chamber-specific expression may occur prior to completion of septation. Northern blot, reverse transcriptase-linked polymer*ase* chain reaction (PCR), RNase protection, and Western blot analysis revealed that *MLC-2a* is expressed in an atrial-restricted pattern in the adult mouse heart, with a very low level of expression in the aorta and no detectable levels in the ventricle, skeletal muscle, uterus, or liver. During mouse development, *in situ* hybridization studies showed that *MLC-2a* is expressed in both the atrial and the ventricular segments at early stages of development. Downregulation of *MLC-2a* in the ventricular chamber takes place between Day 9 pc (prior to the completion of septation) and Day 14 pc (Fig. 3). However, the exact time frame of *MLC-2a* downregulation in the ventricular chamber may be dependent on the genetic background (Gruber *et al.,* 1996; Kastner *et al.,* 1997a). Recently, *MLC-2a* expression has been described in different mouse models, i.e., in gene targeted studies of the *retinoid X receptor α* (*RXRα*) gene (Dyson *et al.,* 1995) and the *βARK-1* gene (P. Ruiz-Lozano and K. R. Chien, unpublished observations). In both strains, *MLC-2a* expression fails to be downregulated in the ventricular chamber. Whether this failure represents a maturational arrest or a premature differentiation in these mutants is currently unknown.

Comparison of *MLC-2a* and *MLC-2v* transcripts during cardiogenesis in differentiating ES cells revealed that *MLC-2a* is expressed in embryoid bodies as early as Day 6 compared to Day 9 for *MLC-2v* (Miller-Hance *et al.,* 1993). Although the process of ventricular specification occurs between Days 8 and 10 of *in vitro* cardiogenesis in the ES cell model system, the expression of the *MLC-2a* gene occurs at a much earlier stage (Fig. 4). Detectable *MLC-2a* expression at Day 6 of *in vitro* differentiation corresponds to the activation of the earliest known markers of the cardiac muscle gene program (Lints *et al.,* 1993), including a cardiac member of the homeobox gene family (Komuro and Izumo, 1993). Thus, *MLC-2a* and *MLC-2v* serve as genetic markers for atrial and ventricular chamber specification in both *in vitro* and *in vivo* contexts (see Chapter 19).

III. MLC-2v as a Paradigm for Gene Regulation

A. Combinatorial Pathways

During the past several years, the *MLC-2v* gene has been used as a model system to identify the molecular pathways that may play a role in ventricular chamber specification, maturation, and morphogenesis. The regulatory elements within the *MLC-2v* promoter which confer ventricular specificity have been identified, as have several of the transcription factors which occupy these sites and form a combinatorial pathway for regulation of the ventricular muscle gene program (Fig. 5). These studies have encompassed transient assays in cultured cardiac muscle cells as well as transgenic and gene targeted mice.

In transient assays, a single copy of the 28-base pair (bp) HF-1 element, which has both HF-1a and HF-1b/MEF-2 sites, is sufficient to confer cardiac cell-specific expression in primary cultures of ventricular muscle cells (Zhu *et al.,* 1991). In independent lines of transgenic mice, the 250-bp *MLC-2v* promoter can confer ventricular specificity to a luciferase reporter gene (Lee *et al.,* 1992). The maintenance of this specificity appears to be dependent on the two adjacent, positive regulatory elements (HF-1a and HF-1b/MEF-2) (Zhu *et al.,* 1993; Lee *et al.,* 1994).

The *MLC-2v* promoter is not downregulated by cotransfection with an *Id* expression vector, which is known to negatively regulate the activity of E box-dependent promoters in cardiac and skeletal muscle (Evans *et al.,* 1993). In addition, the ventricular muscle gene program, as assessed by the expression of *MLC-2v,* is not dominantly activated in cardiac myocyte Fibroblast heterokaryons (Evans *et al.,* 1994). In fact, the expression of an MLC-2–luciferase fusion gene is extinguished in these heterokaryons, indicating that the ventricular muscle phenotype is recessive, and that negative

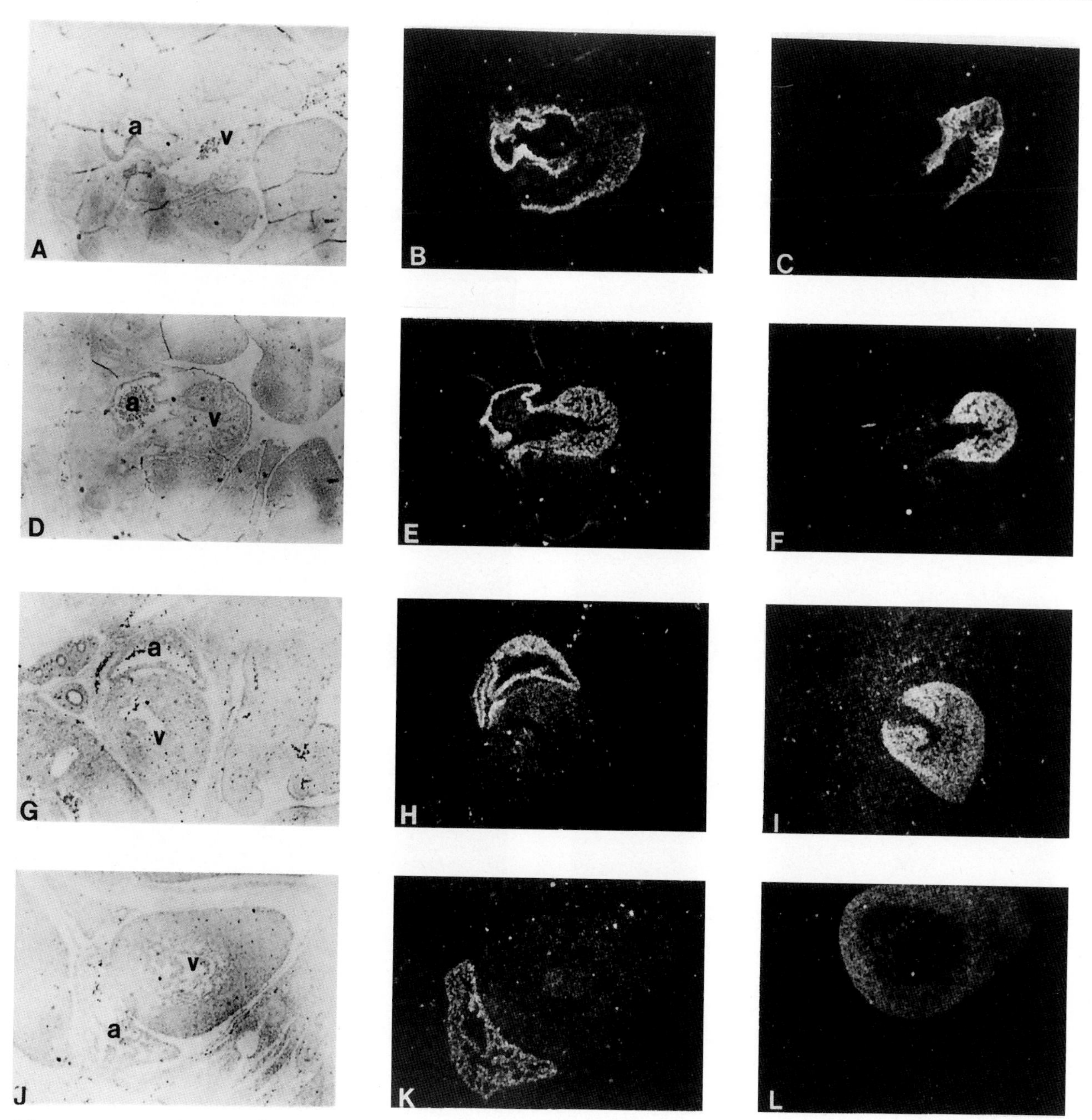

Figure 3 Comparison of MLC-2v and MLC-2a expression during mouse development. Riboprobes for the MLC-2a and MLC-2v were used to determine the expression of both genes during murine heart development. Day 11 pc sagittal, bright (A) and dark field (B, E) hybridized with the MLC-2a probe or with the MLC-2v probe, respectively (C, F). Day 12 pc sagittal (G) hybridized with MLC-2a probe (H) or MLC-2v (I). Day 14 pc sagittal (J) hybridized with MLC-2a (K) or MLC-2v (L). Note the progressive restriction of the MLC-2a signal to the atrial chamber and compare to the early ventricular specification of the MLC-2v message. (From Kubalak *et al.*, 1994, with permission.)

regulators of the ventricular muscle gene program may exist. Thus, it appears that cardiac muscle specificity for the *MLC-2v* promoter most likely requires combinatorial interactions between a variety of regulatory elements located within the 250-bp *MLC-2v* promoter fragment that are distinct from those previously described for the skeletal muscle gene program.

B. Regulation of Ventricular Specification and the Establishment of an Anterior–Posterior Gradient of Expression of *MLC-2v*

To delineate the precise combinatorial pathways which confer chamber specificity, a systematic analysis

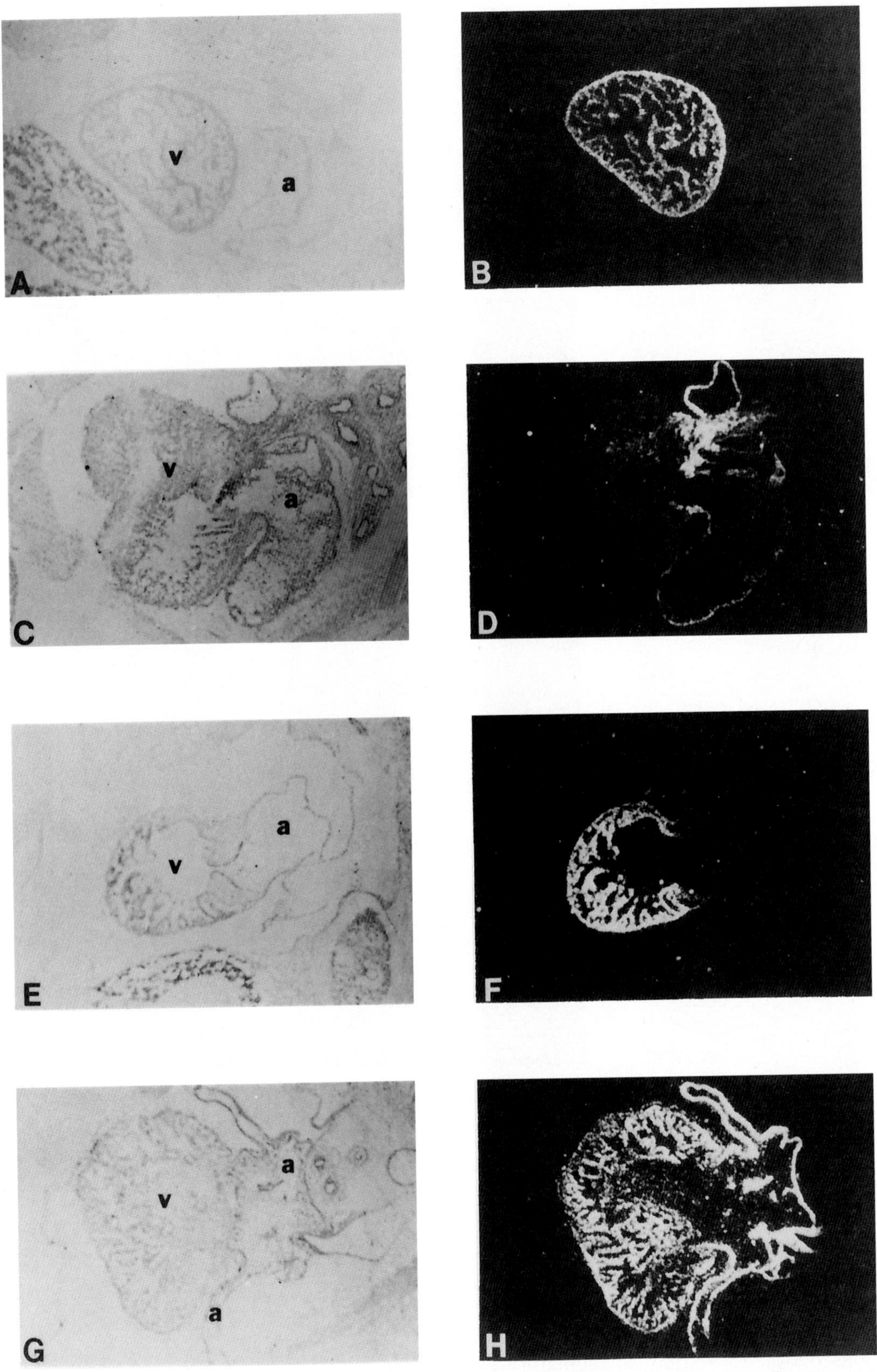

Figure 4 MLC-2a and MLC-2v expression in the wild-type and RXRα mutant mouse embryos. (A–D) Day 13.5 wild-type mouse embryos hybridized with MLC-2v (B) or MLC-2a (D). (E–H) Day 13.5 RXRα mutant mouse embryos hybridized with MLC-2v (F) or MLV2a (H). Note the persistent expression of MLC-2a in the ventricles of the RXRα mutant mouse. [From Dyson *et al.*, 1995, with permission. Copyright (1995) National Academy of Sciences, U.S.A.]

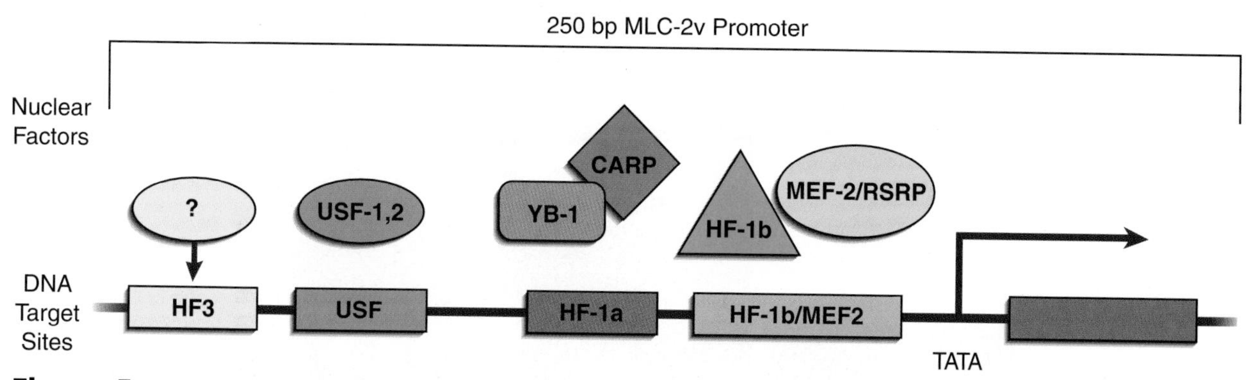

Figure 5 Organization of the MLC-2v promoter. A 250-bp fragment of the MLC-2v promoter is sufficient to confer ventricular muscle-specific gene expression. This 250-bp fragment contains three evolutionary conserved elements (HF-1a, -1b, and -3) and an E box site for the USF transcription factors (Navakasattusas *et al.,* 1994). Transcription factors binding to the conserved elements have been isolated: The HF-1b/MEF-2 site binds to the zinc finger protein HF-1b (Zhu *et al.,* 1993). HF-1a is a binding site for YB-1, where the CARP protein may act as a cofactor (Zou and Chien, 1995). The HF-3 interacting factors that regulate the transcription of MLC-2v are currently unknown.

of multiple independent transgenic mice harboring the *MLC-2v* promoter/β-galactosidase reporter transgenes was performed (Ross *et al.,* 1996). Both the 250-bp *MLC-2v* promoter fragment and a dimerized 28-bp subelement (HF-1), which contains binding sites for HF-1a and HF-1b/MEF-2 factors, can direct ventricular-specific reporter expression as early as the endogenous gene, i.e., at Day 7.5–8.0 pc (Fig. 6). The transgene expression was restricted to the heart and appeared to display an anterior–posterior gradient. The duplex HF-1 oligonucleotide completely recapitulated the pattern of expression seen with the intact 250-bp *MLC-2v* promoter fragment (Fig. 6). While the endogenous gene is expressed uniformly throughout both ventricles, the transgenes were expressed in a right ventricular–conotruncal dominant fashion, suggesting that they contain only a subset of the elements that respond to positional information in the developing heart tube. To determine whether the ventricular-specific expression of the transgene was dependent on regulatory genes already shown to be required for correct ventricular differentiation, the 250-bp MLC-2v/lacZ transgenic animals were bred into both *RXRα* and *Nkx2-5* null backgrounds. Despite the almost total absence of endogenous *MLC-2v* transcripts in *Nkx2-5* null embryos, transgene expression occurred normally in both null backgrounds (Ross *et al.,* 1996).

Based on these results, a minimal combinatorial element, composed of known *trans*-acting factor binding sites, has been shown to confer ventricle-specific transgene expression during murine cardiogenesis. Two copies of the 28-bp *cis* regulatory element (HF-1) can alone confer ventricular specificity. HF-1 contains adjacent regulatory elements that bind two distinct factors, termed HF-1a and HF-1b/MEF-2. Both these elements have been shown to be critical for the maintenance of the 250-bp *MLC-2v* promoter activity in transient assays and in transgenic mice (Zhu *et al.,* 1991; Lee *et al.,* 1992). The *trans*-acting factors that bind to the HF-1 element are neither ventricular nor cardiac chamber specific and will be discussed in subsequent sections (Fig. 7).

C. *Trans*-Acting Factors

1. YB-1

Studies in transgenic mice and transient assays in cultured ventricular muscle cells have identified a critical role for the HF-1a regulatory element, which acts synergistically with the HF-1b/MEF-2 element to maintain ventricular chamber-specific expression of the *MLC-2v* gene. When a rat neonatal heart cDNA expression library was screened with the HF-1a binding site, EFI_A, the rat homolog of human YB-1, was isolated (Zou and Chien, 1995). Purified recombinant EFI_A/YB-1 protein binds to the HF-1a site in a sequence-specific manner. Detailed studies of the specificity of binding suggest that YB-1, in conjunction with a cardiac restricted cofactor, occupies a subset of the HF-1a contact points made by the endogenous cardiac nuclear factor(s). An-

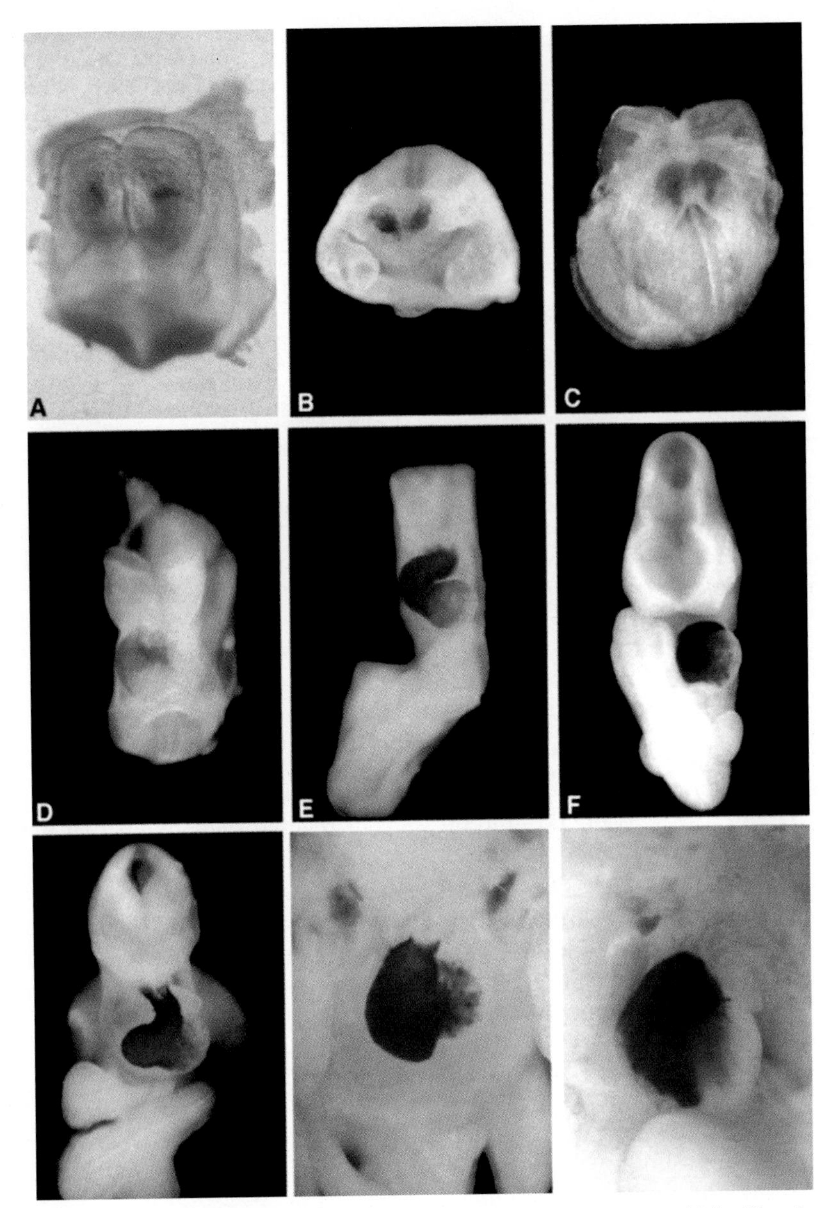

Figure 6 Combinatorial pathways confer cardiac ventricular specificity. Timed transgenic MLC-2v/lacZ embryos stained for β-galactosidase activity. (A) Day 7–7.5 pc, early headfold; (B) Day 7.5–8; (C) Day 8–8.5; (D) Day 9–9.5 (E) Day 9–9.5; (F) Day 9.5–10; (G) Day 11–11.5; (H) Day 12–12.5; (I) Day 16–16.5. Note that β-galactosidase staining is initially evident in a bilaterally symmetrical manner at the earliest time points in cardiogenesis (A–C) in the presumptive cardiac precursor cells flanking the midline. By Day 8.0 pc, as the linear heart tube forms (D), asymmetric β-gal staining is evident in the conotruncal and bulboventricular sections of the forming heart. This pattern is persistent through the remainder of cardiogenesis (E–I) and is defined in the septated heart by Day 12 pc (H).

tiserum against *Xenopus* YB-3, which is 100% identical in the DNA-binding domain and 89% identical in the overall amino acid sequence to rat EFI_A, can specifically abolish a component of the endogenous HF-1a complex in rat cardiac myocyte nuclear extracts. In cotransfection assays, EFI_A/YB-1 can increase the 250-bp *MLC-2v* promoter activity in a cardiac cell context in a manner that is dependent on the HF-1a site. EFI_A/YB-1 complexes with an unknown protein in cardiac myocyte nuclear extracts to form the endogenous HF-1a binding complex. Taken together, these data suggest that EFI_A/YB-1 and its associated protein partners bind

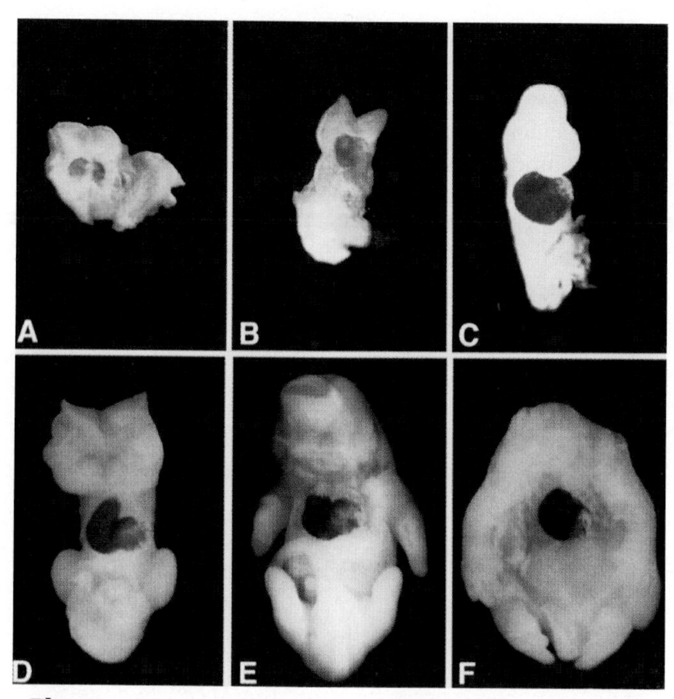

Figure 7 A 28-bp HF-1 element confers ventricular-specific expression in transgenic mice. Timed transgenic HF-1/lacZ embryos stained for β-galactosidase activity. (A) Day 7–7.5 pc, late headfold stage/early foregut pocket; (B) Day 8–8.5 pc; (C) Day 9–95 pc; (D) Day 10–10.5; (E) Day 12–12.5; (F) Day 14–14.5. Note that the transgene expression is restricted to the heart and with an anteroposterior gradient of expression similar to the MLC-2v/lacZ shown in Fig. 6.

to the HF-1a site and, in conjunction with HF-1b, mediate ventricular chamber-specific expression of the *MLC-2v* gene.

2. USF-1

To determine the potential regulatory role of other *cis*-acting elements located within the 250-bp *MLC-2v* promoter, *in vivo* footprinting was used to identify sites outside of the HF-1a and HF-1b sequence (Navankasattusas *et al.,* 1994). A novel regulatory element, known as MLE-1, which contains a conserved core sequence (CACGTG), was identified in the *MLC-2v* promoter. This CACGTG motif, which conforms to the consensus E box site (CANNTG), is identical to the core sequence of the major late transcriptional factor (USF) binding site in the major late promoter of adenovirus (Carthew *et al.,* 1985; Sawadogo and Roeder, 1985). Transient assays of the luciferase reporter genes that have point mutations in the MLE-1 sites demonstrate the importance of this *cis* regulatory element in the transcriptional activation of the *MLC-2v* gene in cardiac ventricular muscle cells. Furthermore, USF, a member of the basic helix–loop–helix (HLH) leucine zipper family, is shown to bind to MLE-1, HF-1a, and to PRE B sites. This suggests that it is a component of protein complexes that coordinately control the expression of *MLC-2v* and α-myosin heavy chain genes. Systematic targeting of the *USF-1* and *USF-2* genes should be valuable in determining how these HLH zipper proteins participate in the regulation of the ventricular muscle gene program during mammalian cardiogenesis.

3. CARP

As described previously, our laboratory has shown that the ubiquitous transcriptional factor, YB-1, binds to the HF-1a site in conjunction with a cofactor (Zou and Chien, 1995). To identify potential interacting cofactors, YB-1 was used as a bait in a yeast two-hybrid screening of a rat neonatal heart cDNA library. As a result, the gene encoding for cardiac ankyrin repeat protein (*CARP*) was isolated and characterized (Zou *et al.,* 1997). Coimmunoprecipitation studies revealed that CARP forms a physical complex with YB-1 in cardiac myocytes. Transfection assays in neonatal rat ventricular myocytes show that CARP can negatively regulate a HF-1-TK minimal promoter in a HF-1 sequence-dependent manner. In addition, when fused to a GAL4 DNA-binding domain, CARP displays a transcriptional inhibitory activity in both cardiac and noncardiac cell contexts. During murine embryogenesis, endogenous *CARP* expression is detected at Day 8.5 pc specifically in the heart. The expression of *CARP* appears to be temporally and spatially regulated in the myocardium. *CARP* expression in *Nkx2-5* mutant (−/−) embryos is found to be selectively and significantly reduced, suggesting that *CARP* is downstream of the homeobox gene *Nkx2-5* in the cardiac regulatory program (see Chapter 7).

CARP represents a cardiac-restricted nuclear coregulatory of the cardiac muscle gene program and apparently acts as a negative transcriptional regulatory cofactor. The recent generation of CARP-deficient mice should be informative as to the precise role of CARP and related family members in the control of cardiac growth and development.

D. Hierarchical Regulation of *MLC-2v*

The *tinman* gene is a member of the homeobox gene family that is required for formation of the *Drosophila* dorsal vessel, the equivalent of a heart-like organ. Recent studies have identified a family of related genes in the mammalian heart. *Nkx2-5,* a divergent homolog of the *Drosophila tinman* gene, is a marker of the early stages of the vertebrate heart field (Harvey, 1996). *Nkx2-5* is expressed in the lateral plate mesoderm in the cardiogenic region of mouse (Komuro and Izumo, 1993; Lints *et al.,* 1993), frog (Tonissen *et al.,* 1994), fish (Chen and Fishman, 1996), and avian (Schultheiss *et al.,* 1995) embryos. Targeted disruption of *Nkx2-5* in mouse blocks heart development at the looping stage but does not prevent the formation of the heart tube (Lyons *et al.,* 1995; see Chapter 7). This implies that if *Nkx2-5* is critical for the initiation of cardiac cell fate decisions in the mammalian heart, there must be redundant pathways, perhaps *via* other members of this family. Experiments performed in zebrafish suggest that the propensity to become heart is controlled, in part, by achieving a threshold level *Nkx2-5* (Fishman and Chien, 1997). However, *Nkx2-5* by itself seems to be insufficient to specify myocardial cell fate determination. The presence of additional signals has also been suggested by studies examining the ectopic expression of *Nkx2-5* by transplantation or by transfection into fibroblasts because these cells fail to acquire a cardiac phenotype (Chen and Fishman, 1996). *MLC-2v* is a putative downstream target gene for *Nkx2-5* since the expression of this cardiac myosin light chain isoform gene is selectively downregulated in *Nkx2-5* mutant (-/-) mouse embryos (Lyons *et al.,* 1995). This suggests a position for *Nkx2.5* early in the hierarchial pathway that regulates the ventricular muscle cell gene program (Lyons *et al.,* 1995).

E. MEF-2 Pathways

As discussed previously, recent studies have documented that a 28-bp *cis* regulatory element in the *MLC-2v* promoter, which is composed to two discrete regulatory elements (HF-1a and HF-1b/MEF-2), can confer ventricular chamber specificity during murine cardiogenesis (Ross *et al.,* 1996). The HF-1b/MEF-2 element is of particular interest because a number of cardiac and skeletal muscle promoters require an intact MEF-2 site for the maintenance of muscle specificity and because contransfection studies have documented the ability of MEF-2 factors to *trans*-active muscle promoters (Ianello *et al.,* 1991; Zhu *et al.,* 1991; Nakatsuji *et al.,* 1992; Navankasattusas *et al.,* 1992).

Four *MEF-2* genes, referred to as *MEF-2A–D,* have been identified in vertebrate species (Olson *et al.,* 1995; see Chapter 8). The *MEF-2* genes are members of the related-serum response factor (RSRF) protein family. In mice, *MEF-2B* and *MEF-2C* are expressed in the precardiac mesoderm at Embryonic Day 7.75 pc, followed by expression of *MEF-2A* and *MEF-2D* (Edmonson *et al.,* 1994; Molkentin *et al.,* 1996). The MEF-2A and MEF-2C proteins play a critical role in the regulation of both the cardiac and skeletal muscle gene programs. In *Drosophila,* the loss of its single *dMEF-2* gene prevents formation of cardiac, visceral, and skeletal muscle (Lilly *et al.,* 1994). In mouse, mutation of the *MEF-2C* gene interrupts expression of a small subset of cardiac genes and alters heart development at the looping stage (Lin *et al.,* 1997). In contrast, mutant mice deficient for the *MEF-2B* gene have normal cardiac development. Interestingly, although *MLC-2v* requires the MEF-2 site in the 28-bp HF-1 regulatory element to maintain ventricular specificity, during *in vivo* cardiogenesis in the *MEF-2C*$^{-/-}$ embryos cardiac and chamber-specific expression of *MLC-2v* is maintained, indicating that the AV chambers were specified correctly (Lin *et al.,* 1997). Thus, the possibility exists that other members of the MEF-2 family have redundant functions with regard to *MLC-2v* expression. Alternatively, other distinct cardiac transcription factors may occupy this site (Olson *et al.,* 1995).

IV. The Role of *MLC-2v* in Embryonic Heart Function

Recent studies in gene targeted mice have led to the suggestion that the MLC-2v isoform may have a unique function in the maintenance of cardiac contractility and ventricular chamber morphogenesis during mammalian cardiogenesis. To define the role of MLC-2v in mammalian cardiac muscle function, the *MLC-2v* gene was disrupted by homologous recombination in mice (Chen *et al.,* 1998b). To examine the physiological function of *MLC-2v, in vivo* cardiac function was monitored in living *MLC-2v* mutant embryos using miniaturized physiological techniques (Kubalak *et al.,* 1996; Dyson *et al.,* 1995; Tanaka *et al.,* 1997).

The results of from these experiments demonstrate that *MLC-2v* deficiency results in embryonic lethality at Day 12.5 pc, and the hearts of these mutant embryos display massive cardiac enlargement, wall thinning, chamber dilation, and pleural effusions (Chen *et al.,* 1998b). Analysis of global cardiac function in *MLC-2v* homozygotes (−/−) reveals that the left ventricular ejection fraction is significantly reduced when compared with that of wild-type and heterozygous littermates, whereas the left ventricular end diastolic volumes are significantly increased by Day 12.5 pc. These results suggest that cardiac dysfunction may result in progressive embryonic heart failure in mice lacking *MLC-2v.*

Ultrastructural analysis of the left ventricular free walls of wild-type and *MLC-2v* mutant mice shows abnormalities in the myofibrillar organization in *MLC-2v*$^{-/-}$ embryos. During the early stages of cardiac chamber development, MLC-2a protein levels are elevated in the mutant ventricles, suggesting that *MLC-2a* may partially replace and compensate for the absence of *MLC-2v* in the developing ventricular myocyte. Immunohistochemical analysis and confocal microscopy of *MLC-2a* expression in *MLC-2v*$^{-/-}$ embryos document an increase in MLC-2a protein level and *MLC-2a* incorporation into the nascent myofibrils of these mutant mice (Fig. 8). Despite this substitution of *MLC-2a,* the deficiency of *MLC-2v* is still not fully compensatory, as

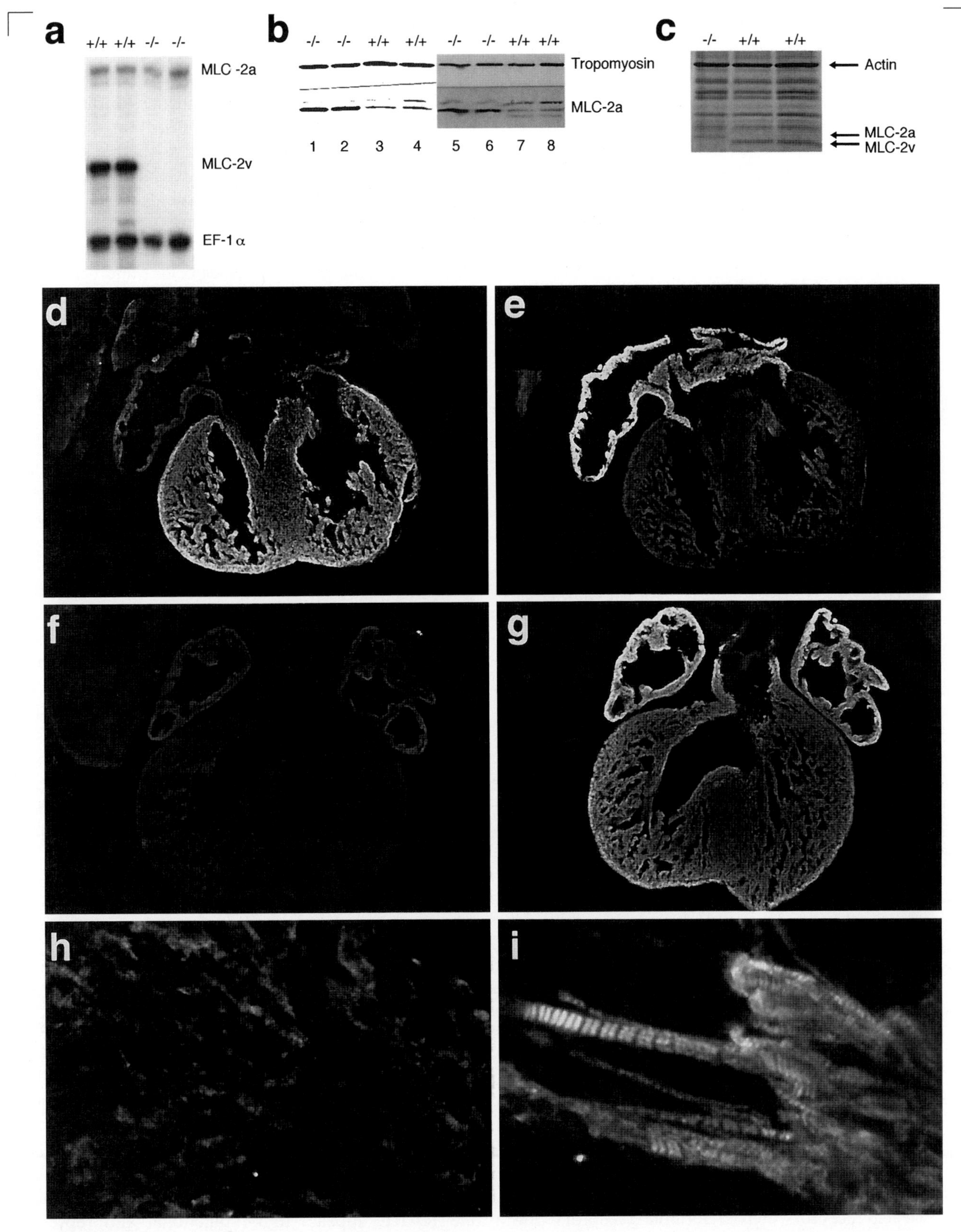

Figure 8 Effect of MLC-2v mutation on MLC-2a expression. Measurement of MLC-2v and MLC-2a mRNAs and proteins in the context of the MLC-$2v^{-/-}$ embryos. (a) RNase protection analysis on isolated ventricles on Day 12 embryos using riboprobes for MLC-2a, MLC-2v, and a control probe EF-1α. (b) Western blot analysis of total protein (1–14) and purified myofilament proteins (5–8) from E 12 embryonic ventricles using a polyclonal antibody for MLC-2a and monoclonal for tropomyosin. (C) Myofilamental proteins from E 12 ventricle, analyzed by SDS–polyacrilamide and Coomassie brilliant blue staining. (d–i) Immunohistochemical analysis of MLC-2v (d and f) and MLC-2a (e, g–i) in the wild-type (d, e, h) and MLC-2v mutant embryos (f, g, i), using confocal microscopy. (From Chen *et al.,* 1998b, with permission.)

shown by the lack of normal sarcomeric structure and by a form of embryonic dilated cardiomyopathy in *MLC-2v* mutant mice. Thus, it appears that the *MLC-2v* isoform may have an essential function in the maintenance of cardiac contractility. Also, a chamber-specific combinatorial code for sarcomeric assembly may exist which ultimately requires MLC-2v in ventricular muscle cells. Further studies on embryos that harbor combined *MLC-2a* and *MLC-2v* deficiencies and rescue studies with various MLC-2 isoforms should provide a rigorous analysis of this concept.

V. *MLC-2a* as a Negatively Regulated Ventricular Muscle Gene

A. Assessment of Ventricular Specification and Maturation in RXRa$^{-/-}$ Embryos

Several studies have established the use of the atrial and ventricular chamber-specific *MLC-2* genes as molecular markers for the process of chamber maturation and specification (O'Brien *et al.*, 1993; Kubalak *et al.*, 1994). Within the embryonic and adult heart, the ventricular isoform *MLC-2v* is expressed exclusively in the ventricular chamber, with negligible expression in the atrial compartment. This regional specification of *MLC-2v* expression occurs early during early cardiogenesis because specification of *MLC-2v* expression occurs early during early cardiogenesis because specification of *MLC-2v* expression to the ventricular segment is found as early as the linear heart tube (Day 8.5 pc; O'Brien *et al.*, 1993). The expression of the atrial isoform, *MLC-2a*, appears to be uniform throughout the linear heart tube (Kubalak *et al.*, 1994; Ruiz-Lozano *et al.*, 1997a) and is then selectively downregulated in the ventricular chamber during expansion of the compact zone and onset of trabeculation (Kubalak *et al.*, 1994). In this manner, loss of *MLC-2a* expression is an indicator of the process of ventricular maturation, which occurs during Days 10.5–13.5 pc.

B. Retinoids and Downregulation of *MLC-2a* Expression

Among heterodimeric nuclear receptors, the RXR plays a central role as an obligate heterodimerization partner. Heterodimeric partners for RXR include retinoic acid receptors, thyroid hormone receptors, vitamin D receptors, and peroxisomal proliferator-activated receptors (Mangelsdorf *et al.*, 1995). There are three known mammalian isoforms of RXR (α, β, and γ) which are encoded by different genes that display distinct patterns of temporal and spatial expression (Dollé *et al.*, 1994). There is no *in vitro* evidence for differences in DNA binding or heterodimerization properties among the three RXRs. However, at the physiological level, RXRα seems to be the main RXR implicated in the developmental function of retinoids (Kastner *et al.*, 1997b; Lee *et al.*, 1997; see Chapter 13). RXRα null mutant mice die of embryonic heart failure at Embryonic Day 13.5 pc and display obvious ocular and cardiovascular malformations, with thinning of the ventricular wall, decreased trabeculation, persistent ventricular expression of the atrial marker *MLC-2a* (Sucov *et al.*, 1994; Kastner *et al.*, 1994; Dyson *et al.*, 1995), and modifications in the expression of a panel of genes (Ruiz-Lozano *et al.*, 1998).

To determine whether muscle cells from *RXRα*$^{-/-}$ hearts were defective in either ventricular specification or maturation, *in situ* hybridization analyses were performed with *MLC-2a* and *MLC-2v* RNA probes. The experiments revealed that restriction of *MLC-2v* expression to the ventricular compartment of the heart in the *RXRα*$^{-/-}$ knockout embryos was appropriately maintained (Fig. 4). However, in adjacent sections from the same embryos, we found that there was aberrant, persistent expression of *MLC-2a* in the relatively thin-walled ventricular chambers. In contrast, wild-type littermates had already downregulated *MLC-2a* in the ventricular chamber and displayed background levels of expression. These studies provide further evidence for a qualitative defect in the ventricular muscle cells found in the *RXRα*$^{-/-}$ mice, in which there appears to be persistent expression of an atrial marker, a thin-walled atrial-like phenotype of the ventricular muscle cells.

The atrial-like phenotype in the hearts of *RXRα* null mutant mice may imply a change in the transcriptional program of their ventricular cells. The question arises as to whether this reflects a direct effect of RXRα homodimer or heterodimer pathways on the expression of atrial or ventricular genes. To address this question, analysis of the regulatory sequences of the *MLC-2a* promoter was performed (Doevendans, 1997). In the 5′ regulatory region of the *MLC-2a* gene is a highly regulated TATA-less promoter that contains consensus sequences for the binding of ubiquitous and cardiac-specific transcription factors, such as MEF-2, GATA-4, and SRF. In particular, the *MLC-2a* promoter contains four retinoid response elements (RAREs) (Fig. 9). At least one RARE appears to be functional and represses an *MLC-2a*–luciferase reporter gene in ventricular myocytes in a dose-dependent manner. The retinoid-mediated repression of the MLC-2a promoter maps to a direct repeat or DR1 element (Ruiz-Lozano *et al.*,

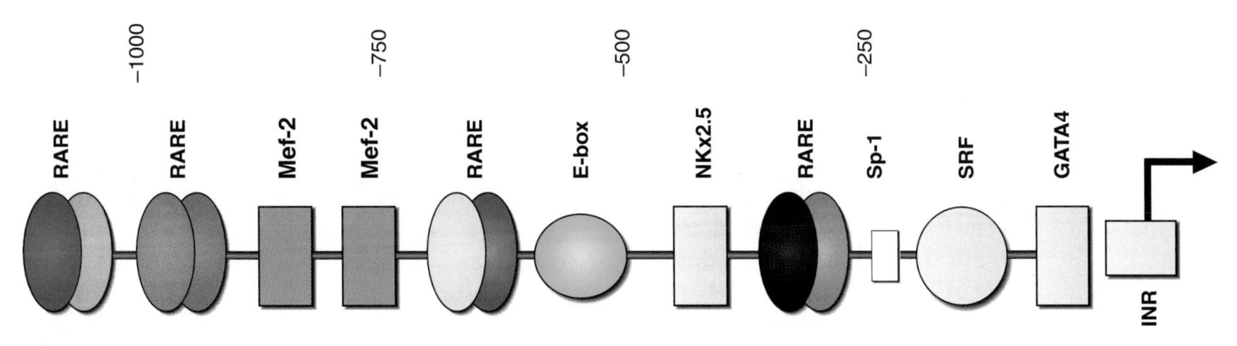

Figure 9 Localization of retinoid response elements in the MLC-2a promoter. Sequence analysis of the 1-kb proximal sequence of the *MLC-2a* gene showed a TATA-less promoter containing consensus sequences for a variety of transcription factors, including MEF-2, Nkx2.5, and Sp-1. It also contains four pairs of hexameric sequences of putative binding sites for the retinoid family of nuclear receptors, retinoic acid response element (RARE).

1997b) that can be a site for binding of homo- or heterodimeric nuclear receptors. The ratio between homo- and heterodimers would determine the switch from transcriptional inhibition to transcriptional activation (P. Ruiz-Lozano and K. R. Chien, unpublished observations).

VI. Induction of MLC-2v and MLC-2a during Hypertrophy

In the face of increased workload, the heart adapts via a variety of hormonal and mechanical stimuli with a hypertrophic response in the ventricular myocardium. The hypertrophic response is characterized by an increase in the size of individual cardiac cells, an increase in the assembly of individual contractile proteins into the myofilaments, a selective activation of gene expression of contractile proteins within individual myocardial cells (MLC-2a and MLC-2v), and the reexpression of embryonic markers, such as the atrial natriuretic factor (ANF). Classically, the *ANF* gene is induced by most of the known stimuli, whereas the *MLC-2* genes are thought to have more selective responses (Iwaki *et al.,* 1990; Knowlton *et al.,* 1991, 1995).

The selectivity of the hypertrophic response to a particular subset of contractile proteins can be explained, in part, by an adaptational response of certain transcription factors to these physiological stimuli. For example, ventricular tissues of the spontaneously hypertensive rat suffer changes in the content of two DNA-binding transcription factors, myocyte enhancer factor (MEF-2) and CArG-binding factor (SRF), and these changes parallel the evolution of cardiac hypertrophy (Doud *et al.,* 1995). The transcriptional regulation of *MLC-2v* in response to hypertrophic stimuli has been finely dissected in transient expression assays. The region sufficient for both cardiac-specific and α-adrenergic-inducible expression was originally determined to be the 250-bp promoter fragment (Zhu *et al.,* 1991) and was subsequently mapped to the conserved 28-bp element (HF-1) in the rat cardiac *MLC-2* gene.

The hypertrophic response is characterized not only by changes in gene expression of the two atrial genes (*ANF* and *MLC-2v*) but also by the reorganization of contractile proteins into additional sarcomeric units. Evidence exists that these two events are controlled by different regulatory pathways. For example, activation of an estradiol-regulated Raf-1 protein kinase has been shown to activate the mitogen-activated protein (MAP) kinase and the activity of the *ANF* and *MLC-2v* promoters (Thorburn *et al.,* 1994). However, activation of Raf-1 is not sufficient to induce the organization of actin into sarcomeric units, suggesting that a distinct response to the hypertrophic stimulus regulates actin organization. In contrast, transcriptional changes associated with mutations in *ras* displayed an enhanced ventricular hypertrophic phenotype, i.e., increased wall thickness, selective induction of natriuretic peptide genes, and myofibrillar disarray (Hunter *et al.,* 1995).

Similarly, obstructive hypertrophy was observed in an echo-selected substrain of mice transgenic for an *MLC-2v-ras* fusion gene (Gottshall *et al.*, 1997). These *ras* mutant mice also display a reactivation of *MLC-2a* expression during the course of left ventricular hypertrophy (Fig. 10). This observation raises the question as to whether there are shared or divergent signaling pathways that mediate the induction of atrial genes during hypertrophy (*ANF* and *MLC-2a*). This question is of interest because the possibility exists that a single molecular switch can reactivate these two genes in parallel. Whether the increase in MLC-2a expression represents an additional hypertrophic response and whether it correlates with myofibrillar disarray, is currently under determination.

VII. *MLC-2v* Promoter as a Tool for Lineage Tracing and Cardiac-Restricted Gene Targeting via Cre/LoxP Strategies

Gene knockout technology, involving gene targeting in ES cells, provides a powerful tool for the functional identification of the gene product and has had a fundamental impact in biology and basic medicine. The mutant mouse generated by the conventional knockout technique involves germline transmission of the null mutation. Therefore, all cell types in this mutant mouse carry the mutant gene. Because many genes have multiple roles in different tissues and different developmental stages, the inactivation of a gene in all cell types makes it difficult to attribute an abnormal phenotype to a particular type of cell or tissue, especially if there is embryonic lethality. However, inactivation of a gene in a certain lineage or at a certain stage of development would allow a more detailed dissection of gene function (Rossant and Nagy, 1995). The Cre/loxP system has been used successfully to achieve this in the T cell lineage (Gu *et al.*, 1994) and in specific subregions and specific cell types of the brain (Tsien *et al.*, 1996).

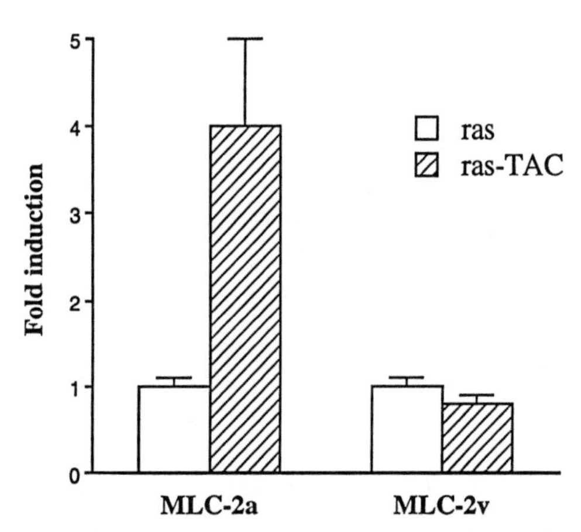

Figure 10 Reactivation of MLC-2a expression during ras-induced ventricular hypertrophy. Quantitation of Northern blot experiments showing the expression of MLC-2a and MLC-2v in the control ras-transgenic mice (ras) and the transverse aortic constriction-induced hypertrophy (TAC). Note the specific response of MLC-2a to TAC stimuli.

Finding an appropriate promoter is the key to success for developing the technology of ventricular myocyte-restricted gene targeting. The ventricular myocyte-specific promoter, *MLC-2v,* has been very well characterized by our laboratory. Any transgene driven by this promoter, such as the endogenous *MLC-2v* itself, will initially be expressed at about Embryonic Day 8 pc and be restricted to the ventricular chambers throughout embryonic life and into adulthood. However, transgene expression driven by the *MLC-2v* promoter was found to be patchy and more restricted to the right ventricle. Thus, we decided to introduce the Cre cDNA into the endogenous locus of *MLC-2v* (Fig. 11). The mouse carrying Cre cDNA under endogenous *MLC-2v* control (henceforth referred to as MLC-$2v^{CreK1}$) was generated as described by Chen *et al.* (1998b). Mice homozygous for MLC-$2v^{CreK1}$ are embryonic lethal because of the disruption of *MLC-2v,* which is essential for maintenance of embryonic heart function. However, the heterozygous mice are indistinguishable from their wild-type littermates.

To determine the specificity and efficiency of Cre/loxP recombination, the MLC-$2v^{creK1}$ heterozygous mice were crossed into homozygous $RXR\alpha^{flox}$ mice, which carry a normal *RXRα* gene flanked by loxP sites, which are the targets for Cre recombination (Chen and Chien, 1997). Intercrosses between "floxed RXRα" and the MLC-2v Cre mouse lines confirm cardiac-restricted gene targeting by PCR and Southern analysis. The deleted *RXRα* allele was detectable only in ventricular cardiomyocytes and in slow skeletal muscle, recapitulating the endogenous MLC-2v expression pattern. Embryonic lethality of the RXRα deficiency was not observed in the progeny of these intercrosses, indicating that there is no critical requirement for RXRα in the muscle per se. Additional studies are being conducted to determine if aortic sac, cushion defects, or other postnatal phenotypes are evident in these animals. The MLC-2v Cre/LoxP system should allow for cardiac-restricted targeting of any floxed alleles of interest.

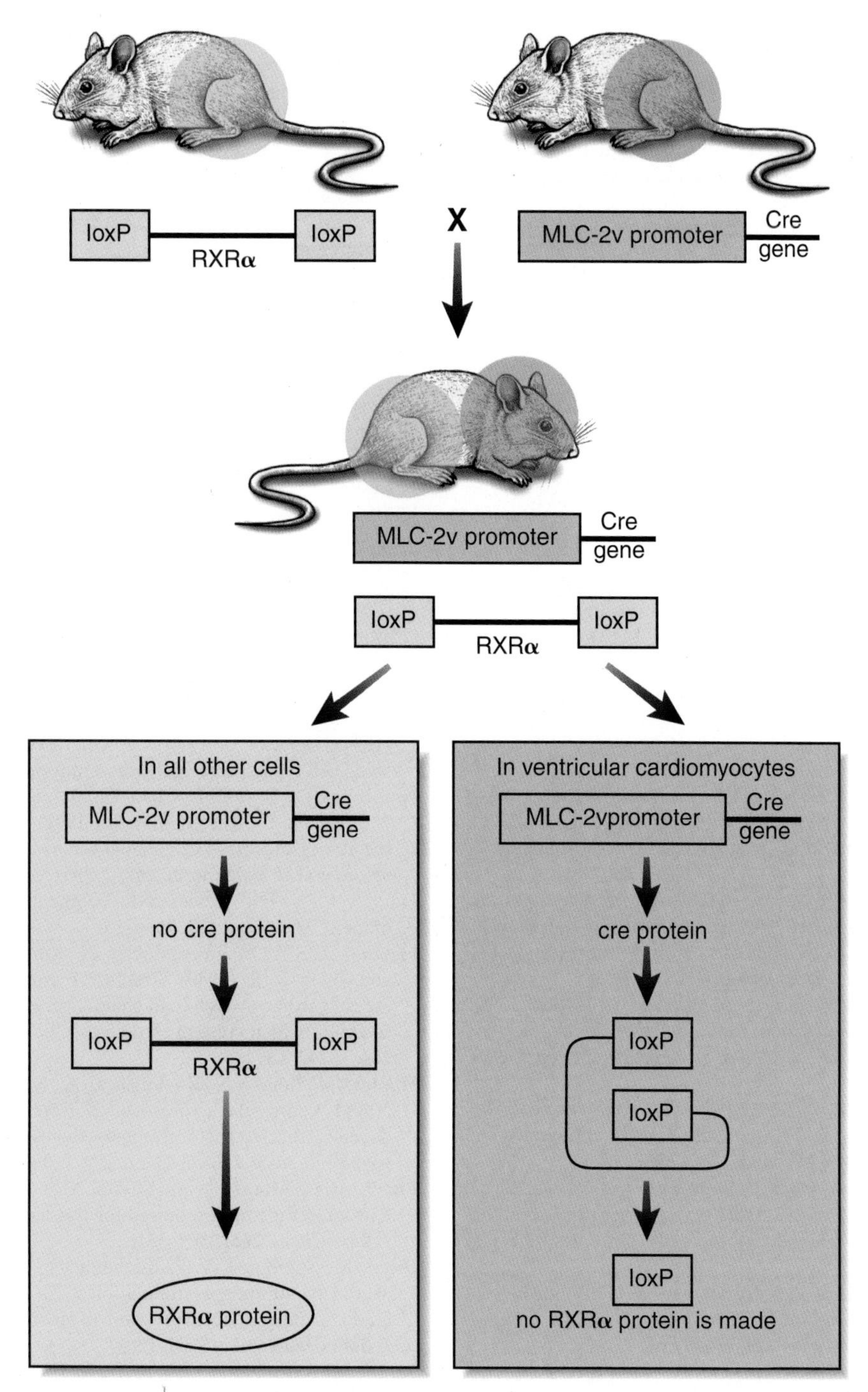

Figure 11 Strategy to generate a ventricular-specific RXRα mutant mouse by the use of the MLC-2v promoter expressing Cre recombinase. The cDNA encoding the Cre recombinase of the phage P1 was cloned into the *MLC-2v* locus so that the Cre recombinase expression would recapitulate the expression of the endogenous *MLC-2v* mRNA. In a separate mouse, intronic sequences of the *RXRα* gene flanking exon 4 were modified by addition of Cre recombinase recognition sites. The offspring of this intercross would result in mice which express Cre recombinase specifically in ventricular myocytes and, therefore, the *RXRα* gene would be specifically mutated in ventricular myocytes. The rest of the cells (that do not contain transcribed Cre) would express a wild-type *RXRα* gene.

VIII. Conclusions and Future Prospectives

The identification of the molecular pathways that guide chamber morphogenesis during cardiac development will provide a basic understanding of how chambers are formed and of how disruptions of these pathways can lead to congenital morphogenic defects and other diseases in the postnatal heart. New insights into cardiac developmental biology will be obtained when the genetic analysis of complex cardiovascular phenotypes in different genetically engineered animals becomes possible.

References

Arber, S., Hunter, J. J., Ross, J., Jr., Hongo, M., Sansig, G., Borg, J., Perriard, J.-C., Chien, K. R., and Caroni, P. (1997). MLP-deficient mice exhibit a disruption of cardiac cytoarchitectural organization, dilated cardiomyopathy, and heart failure. *Cell (Cambridge, Mass.)* **88,** 393–403.

Carthew, R. W., Chodosh, L. A., and Sharp, P. A. (1985). An RNA polymerase II transcription factor binds to an upstream element in the adenovirus major late promoter. *Cell (Cambridge, Mass.)* **43,** 439–448.

Chen, J., and Chien, K. R. (1997). Cardiac restricted gene targeting of the RXRα gene reveals an essential, non-cell autonomous requirement of retinoid signals in ventricular chamber morphogenesis. *Circulation* **276,** I-501.

Chen, J., Kubalak, S. W., and Chien, K. R. (1998a). Ventricular muscle-restricted targeting of the RXRα gene reveals a non-cell autonomous requirement in cardiac chamber morphogenesis. Development *(Cambridge UK)* **125,** 1943–1949.

Chen, J., Kubalak, S. W., Minamisawa, S., Price, R. L., Becker, K. D., Hickey, R., Ross, J., Jr., and Chien, K. R. (1998b). Selective requirement of myosin light chain-2v in embryonic heart function. *J. Biol. Chem.* **273,** 1252–1256.

Chen, J.-N., and Fishman, M. C. (1996). Zebrafish *tinman* homolog demarcates the heart field and initiates myocardial differentiation. *Development (Cambridge, UK)* **122,** 3809–3816.

Chien, K. R., Zhu, H., Knowlton, K. U., Miller-Hance, W., van Bilsen, M., O'Brien, T. X., and Evans, S. M. (1993). Transcriptional regulation during cardiac growth and development. *Annu. Rev. Physiol.* **55,** 77–95.

de Groot, I. J., Lamers, W. H., and Moorman, A. F. (1989). Isomyosin expression patterns during rat heart morphogenesis: an immunohistochemical study. *Anat. Rec.* **224,** 365–373.

Doevendans, P. A. F. M. (1997). Cardiac specific gene expression of the regulatory myosin light chains. Doctoral Dissertation, University of Maastricht.

Dollé, P., Fraulob, V., Kastner, P., and Chambon, P. (1994). Developmental expression of murine retinoid X receptor (RXR) genes. *Mech. Dev.* **45**(2), 91–104.

Doud, S. K., Pan, L. X., Carleton, S., Marmorstein, S., and Siddiqui, M. A. (1995). Adaptational response in transcription factors during development of myocardial hypertrophy. *J. Mol. Cell. Cardiol.* **27,** 2359–2372.

Dyson, E., Sucov, H. M., Kubalak, S. W., Schmid-Schönbein, G. W., DeLano, F. A., Evans, R. M., Ross, J. Jr., and Chien, K. R. (1995). Atrial-like phenotype is associated with embryonic ventricular failure in retinoid X receptor alpha −/− mice. *Proc. Natl. Acad. Sci. U.S.A.* **92,** 7386–7390.

Edmonson, D. G., Lyons, G. E., Martin, J. F., and Olson, E. N. (1994). Mef2 gene expression marks the cardiac and skeletal muscle lineages during mouse embryogenesis. *Development (Cambridge, UK)* **120,** 1251–1263.

Evans, S. M., Walsh, B. A., Newton, C. B., Thorburn, J. S., Gardner, P. D., and van Bilsen, M. (1993). Potential role of helix-loop-helix proteins in cardiac gene expression. *Circ. Res.* **73,** 569–578.

Evans, S. M., Tai, L. J., Tan, V. P., Newton, C. B., and Chien, K. R. (1994). Heterokaryons of cardiac myocytes and fibroblasts reveal the lack of dominance of the cardiac muscle phenotype. *Mol. Cell. Biol.* **14,** 4269–4279.

Fishman, M. C., and Chien, K. R. (1997). Fashioning the vertebrate heart: Earliest embryonic decisions. *Development (Cambridge, UK)* **124,** 2099–2117.

Fishman, M. C., Olson, E. N., and Chien, K. R. (1998). Molecular advances in cardiovascular development. *In* "Molecular Basis of Cardiovascular Disease" (K. R. Chien, ed.). Saunders, Philadelphia, (in press).

Gottshall, K. R., Hunter, J. J., Tanaka, N., Becker, K. D., Ross, J., Jr., and Chien, K. R. (1997). *Ras* dependent pathways induce obstructive hypertrophy in echo-selected transgenic mice. *Proc. Natl. Acad. Sci. U.S.A.* **94,** 4710–4715.

Gruber, P. J., Kubalak, S. W., Pexieder, T., Sucov, H. M., Evans, R. M., and Chien, K. R. (1996). RXR alpha deficiency confers genetic susceptibility for aortic sac, conotruncal, atrioventricular cushion, and ventricular muscle defects in mice. *J. Clin. Invest.* **98,** 1332–1343.

Gu, H., Marth, J. D., Orban, P. C., Mossman, H., and Rajewsky, C. (1994). Deletion of a DNA polymerase β gene segment in T cells using cell type-specific gene targeting. *Science* **265,** 103–106.

Hailstones, D., Barton, P., Chan-Thomas, P., Sasse, S., Sutherland, C., Harderman, E., and Gunning, P. (1992). Differential regulation of the atrial isoforms of myosin light chains during striated muscle development. *J. Biol. Chem.* **267,** 23295–23300.

Harvey, R. P. (1996). Review: NK-2 homeobox genes and heart development. *Dev. Biol.* **178,** 203–216.

Henderson, S. A., Spencer, M., Sen, A., Kumar, C., Siddiqui, M. A. Q., and Chien, K. R. (1989). Structure, organization, and expression of the rat cardiac myosin light chain-2 gene: Identification of a 250 bp fragment which confers cardiac specific expression. *J. Biol. Chem.* **264,** 18142–18148.

Hunter, J. J., Tanaka, N., Rockman, H. A., Ross, J., Jr., and Chien, K. R. (1995). Ventricular expression of a *MLC-2v-Ras* fusion gene induces cardiac hypertrophy and selective diastolic dysfunction in transgenic mice. *J. Biol. Chem.* **270,** 23173–23178.

Ianello, R. C., Mar, J. H., and Ordahl, C. P. (1991). Characterization of a promoter element required for transcription in myocardial cells. *J. Biol. Chem.* **266,** 3309–3316.

Iwaki, K., Sukhatme, V. P., Shubeita, H., and Chien, K. R. (1990). α- and β- adrenergic stimulation induces distinct patterns of immediate early gene expression in neonatal rat myocardial cells. *J. Biol. Chem.* **265,** 13809–13817.

Kastner, P., Grondona, J. M., Mark, M., Gansmuller, A., LeMeur, M., Decimo, D., Vonesch, J. L., Dollé, P., and Chambon, P. (1994). Genetic analysis of RXR alpha developmental function: Convergence of RXR and RAR signaling pathways in heart and eye morphogenesis. *Cell (Cambridge, Mass.)* **78,** 987–1003.

Kastner, P., Messaddeq, N., Mark, M., Wendling, O., Grondona, J. M., Ward, S., Ghyselinck, N., and Chambon, P. (1997a). Vitamin A deficiency and mutatins of RXRa, RXR b and RXR g lead to early differentiation of embryonic ventricular cardiomyocytes. *Development (Cambridge, UK)* **124,** 4749–4758.

Kastner, P., Mark, M., Ghyselinck, N., Krezel, W., Dupé, V., Grondona, J. M., and Chambon, P. (1997b). Genetic evidence that the retinoid signal is transduced by heterodimeric RXR/RAR functional units during mouse development. *Development (Cambridge, UK)* **124,** 313–326.

Kelly, R., and Buckingham, M. (1997). Manipulating myosin light chain 2 isoforms in vivo. A transgenic approach to understanding contractile protein diversity. *Circ. Res.* **80,** 751–753.

Knowlton, K. U., Baracchini, E., Ross, R. S., Harris, A. N., Henderson, S. A., Evans, S. M., Glembotski, C. C., and Chien, K. R. (1991). Co-regulation of the Atrila Netriuretic Factor and cardiac Myosin Light Chain genes during a α-adrenergic stimulation of neonatal rat ventricular cells. *J. Biol. Chem.* **266,** 7759–7768.

Knowlton, K. U., Rockman, H. A., Itani, M., Vovan, A., Seidman, C. E., and Chien, K. R. (1995). Divergent pathways mediate the induction of ANF transgenes in neonatal and hypertrophyc ventricular myocardium. *J. Clin. Invest.* **96,** 1311–1318.

Komuro, I., and Izumo, S. (1993). Csx: A murine homeobox-containing gene specifically expressed in the developing heart. *Proc. Natl. Acad. Sci. U.S.A.* **90,** 8145–8149.

Komuro, I., and Yazaki, Y. (1993). Control of cardiac gene expression by mechanical stress. *Annu. Rev. Physiol.* **55,** 55–75.

Kubalak, S. W., Miller-Hance, W. C., O'Brien, T. X., Dyson, E., and Chien, K. R., (1994). Chamber specification of atrial myosin light chain-2 expression precedes septation during murine cardiogenesis. *J. Biol. Chem.* **269,** 16961–16970.

Kubalak, S. W., Doevendans, P. A., Rockman, H. A., Hunter, J. J., Tanaka, N., Ross, J., Jr., and Chien, K. R. (1996). Molecular analysis of cardiac muscle diseases via mouse genetics. *Methods Mol. Gene.* **8,** 470–487.

Lee, K. J., Ross, R. S., Rockman, H. A., Harris, A. N., O'Brien, T. X., van Bilsen, M., Shubeita, H. E., Kandolf, R., Brem, G., Price, J., and Chien, K. R. (1992). Myosin light chain-2 luciferase transgenic mice reveal distinct regulatory programs for cardiac and skeletal muscle-specific expression of a single contractile protein gene. *J. Biol. Chem.* **267,** 15875–15885.

Lee, K. J., Hickey, R., Zhu, H., and Chien, K. R. (1994). Positive regulatory elements (HF-1a and HF-1b) and a novel, negative regulatory element (HF-3) mediate ventricular muscle-specific expression of myosin light-chain 2-luciferase fusion genes in transgenic mice. *Mol. Cell. Biol.* **14,** 1220–1229.

Lee, R. Y., Luo, J., Evans, R. M., Giguère, V., and Sucov, H. M. (1997). Compartment-selective sensitivity of cardiovascular morphogenesis to combinations of retinoic acid receptor gene mutations. *Circ. Res.* **80,** 757–764.

Lilly, B., Galewsky, S., Firulli, A. B., Schulz, R. A., and Olson, E. N. (1994). D-MEF2: An MADS box transcription factor expressed in differentiating mesoderm and muscle cell lineages during Drosophila embryogenesis. *Proc. Natl. Acad. Sci. U.S.A.* **91,** 5662–5666.

Lin, Q., Schwarz, J., Bucana, C., and Olson, E. N. (1997). Control of mouse cardiac morphogenesis and myogenesis by transcriptional factor MEF2C. *Science* **276,** 1404–1407.

Lints, T. J., Parsons, L. M., Hartley, L., Lyons, I., and Harvey, R. P. (1993). *Nkx2.5*: A novel murine homeobox gene expressed in early heart progenitor cells and their myogenic descendants. *Development (Cambridge, UK)* **119,** 419–431.

Lowey, S., Waller, G. S., and Trybus, K. M. (1993). Skeletal muscle myosin light chains are essential for physiological speeds of shortening. *Nature (London)* **365,** 454–456.

Lyons, G. E., Schiaffino, S., Sassoon, D., Barton, P., and Buckingham, M. (1990). Developmental regulation of myosin gene expression in mouse cardiac muscle. *J. Cell Biol.* **111,** 2427–2436.

Lyons, I., Parsons, L. M., Harley, L., Li, R., Andrews, J. E., Robb, L., and Harvey, R. P. (1995). Myogenic and morphogenetic defects in the heart tubes of murine embryos lacking the homeo box gene Nkx2-5. *Genes Dev.* **9,** 1654–1666.

Mangelsdorf, D. J., Thummel, C., Beato, M., Herrlich, P., Schultz, G., Umesono, K., Blumberg, B., Kastner, P., Mark, M., Chambon, P., and Evans, R. M. (1995). The nuclear receptor superfamily: The second decade. *Cell (Cambridge, Mass.)* **83,** 835–839.

Miller-Hance, W. C., LaCorbiere, M., Fuller, S. J., Evans, S. M., Lyons, G., Schmidt, C., Robbins, J., and Chien, K. R. (1993). In vitro chamber specification during embryonic stem cell cardiogenesis: Expression of the ventricular myosin light chain-2 gene is independent of heart tube formation. *J. Biol. Chem.* **268,** 25244–25252.

Molkentin, J. D., Firulli, A., Black, B., Lyons, G. E., Edmonson, D., Hustad, C. M., Copeland, N., Jenkins, N., and Olson, E. N. (1996). MEF2B is a potent transactivator expressed in early myogenic lineages. *Mol. Cell. Biol.* **16,** 3814–3824.

Nakatsuji, Y., Hidaka, K., Tsujino, S., Yamamoto, Y., Mukai, T., Yanagihara, T., Kishimoto, T., and Sakoda, S. (1992). A single MEF-2 site is a major positive regulatory element required for transcription of the muscle-specific subunit of the human phosphoglycerate mutase gene in skeletal and cardiac muscle cells. *Mol. Cell. Biol.* **12,** 4384–4390.

Navankasattusas, S., Zhu, H., Garcia, A., Evans, S. M., and Chien, K. R. (1992). A ubiquitous factor (HF-1a) and a distinct muscle factor (HF-1b/MEF-2) form an E-box independent pathway for cardiac muscle gene expression. *Mol. Cell. Biol.* **12,** 1469–1479.

Navankasattusas, S., Sawadogo, M., van Bilsen, M., Dang, C. V., and Chien, K. R. (1994). The basic helix-loop-helix protein upstream stimulating factor regulates the cardiac ventricular myosin light-chain 2 gene via independent cis regulatory elements. *Mol. Cell. Biol.* **14** (11), 7331–7339.

Nudel, U., Calvo, J. M., Shanni, M., and Levy, Z. (1984). The nucleotide sequence of a rat myosin light chain 2 gene. *Nucleic Acids Res.* **12,** 7175–7186.

O'Brien, T. X., Lee, K. J., and Chien, K. R. (1993). Positional specification of ventricular myosin light chain 2 expression in the primitive murine heart tube. *Proc. Natl. Acad. Sci. U.S.A.* **90,** 5157–5161.

Olson, E. N., and Srivastava, D. (1996). Molecular pathways controlling heart development. *Science* **272,** 671–675.

Olson, E. N., Perry, M., and Schulz, R. A. (1995). Regulation of muscle differentiation by the MEF2 family of MADS box transcription factors. *Dev. Biol.* **172,** 2–14.

Poetter, K., Jiang, H., Hassanzadeh, S., Master, S. R., Chang, A., Dalakas, M. C., Rayment, I., Sellers, J. R., Fananapazir, L., and Epstein, N. D. (1996). Mutations in either the essential or regulatory light chains of myosin are associated with a rare myopathy in human heart and skeletal muscle. *Nat. Genet.* **13,** 63–69.

Rayment, I., Rypniewski, W. R., Schmidt-Base, K., Smith, R., Tomchick, D. R., Benning, M. M., Winkelmann, D. A., Wesenberg, G., and Holden, H. M. (1993a). Three-dimensional structure of myosin subfragment-1: A molecular motor. *Science* **261,** 50–58.

Rayment, I., Holden, H. M., Whittaker, M., Yohn, C. B., Lorenz, M., Holmes, K. C., Milligan, R. A. (1993b). Structure of the actin-myosin complex and its implications for muscle contraction. *Science* **261,** 58–65.

Ross, R. S., Navansakasattusas, S., Harvey, R. P., and Chien, K. R. (1996). An HF-1a/HF-1b/MEF-2 combinatorial element confers cardiac ventricular specificity and establishes an anterior-posterior gradient of expression. *Development (Cambridge, UK)* **122,** 1799–1809.

Rossant, J., and Nagy, A. (1995). Genome engineering: the new mouse genetics. *Nat. Med.* **1** (16), 592–594.

Ruiz-Lozano, P., Doevendans, P., Brown, A., Gruber, P. J., and Chien, K. R. (1997a). Developmental and tissue restricted expression of a murine spliceosome associated protein gene mSAP49 during cardiogenesis. *Dev. Dyn.* **208,** 482–490.

Ruiz-Lozano, P., Merki, E. E., Doevendans, P. A., Zhou, M. D., and Chien, K. R. (1997b). Rxr Homodimer and Rxr/ppar heterodimer pathways negatively regulate an atrial marker, MLC-2a, at the transcriptional level in ventricular muscle cells. *Circulation* **96,** I-241.

Ruiz-Lozano, P., Smith, S. M., Perkins, G., Kubalak, S. W., Boss, G. R., Sucov, H. M., Evans, R. M., and Chien, K. R. (1998). Energy deprivation and a deficiency in downstream metabolic target genes during the onset of embryonic heart failure in RXRα −/− embryos. *Development (Cambridge, UK)* **125,** 533–544.

Sassoon, D. A., Garner, I., and Buckingham, M. (1988). Transcripts of alpha-cardiac and alpha-skeletal actins are early markers for myogenesis in the mouse embryo. *Development (Cambridge, UK)* **104,** 155–164.

Sawadogo, M., and Roeder, R. G. (1985). Interaction of a gene-specific transcriptional factor with the adenovirus major late promoter upstream of the TATA box region. *Cell (Cambridge, UK)* **43,** 165–175.

Schiaffino, S., and Reggiani, C., (1996). Molecular diversity of myofibrillar proteins: Gene regulation and functional significance. *Physiol. Rev.* **76,** 371–423.

Schultheiss, T. M., Xydas, S., and Lassar, A. B. (1995). Induction of avian cardiac myogenesis by anterior endoderm. *Development (Cambridge, UK)* **120,** 4203–4214.

Spudich, J. A. (1994). How molecular motors work. *Nature (London)* **372,** 515–517.

Sucov, H. M., Dyson, E., Gumeringer, C. L., Price, J., Chien, K. R., and Evans, R. M. (1994). RxRa mutant mice establish a genetic basis for vitamin A signaling in heart morphogenesis. *Genes Dev.* **8,** 1007–1018.

Tanaka, N., Mao, L., DeLano, F. A., Sentianin, E. M., Chien, K. R., Schmid-Schönbein, G. W., and Ross, J. Jr. (1997). Left ventricular volumes and function in the embryonic mouse heart. *Am. J. Physiol.* **273,** H1368–H1376.

Thorburn, J., McMahon, M., and Thorburn, A. (1994). Raf-1 kinase activity is necessary and sufficient for gene expression changes but not sufficient for cellular morphology changes associated with cardiac myocyte hypertrophy. *J. Biol. Chem.* **269,** 30580–30586.

Tonissen, K. F., Drysdale, T. A., Lints, T. J., Harvey, R. P., and Krieg, P. A. (1994). *XNkx-2.3,* a *xenopus* gene related to *Nkx-2.5* and *tinman:* Evidence for a conserved role in cardiac development. *Dev. Biol.* **162,** 325–328.

Tsien, J. Z., Chen, D. F., Gerber, D., Tom, C., Mercer, E. H., Anderson, D. J., Mayford, M., Kandel, E. R., and Tonegawa, S. (1996). Subregion- and cell type-restricted gene knockout in mouse brain. *Cell (Cambridge, Mass.)* **87** (7), 1317–1326.

Zhu, H., Garcia, S., Ross, R. S., Evans, S. M., and Chien, K. R. (1991). A conserved 28 bp element (HF-1) within the rat cardiac myosin light chain-2 gene confers cardiac specific and α-adrenergic inducible expression in cultured neonatal rat myocardial cells. *Mol. Cell. Biol.* **11,** 2273–2281.

Zhu, H., Nguyêñ, V. T. B., Brown, A., Pourhosseini, A., Garcia, A. V., van Bilsen, M., and Chien, K. R. (1993). A novel, tissue restricted zinc finger protein (HF1-b) binds to the cardiac regulatory element (HF1-b/MEF-2) in the rat myosin light chain-2 gene. *Mol. Cell. Biol.* **13,** 4432–4444.

Zou, Y., and Chien, K. R. (1995). EFIA/YB-1 is a component of cardiac HF-1A binding activity and positively regulates transcription of the myosin light chain-2 v gene. *Mol. Cell. Biol.* **15,** 2972–2982.

16

Serum Response Factor–*NK* Homeodomain Factor Interactions, Role in Cardiac Development

James M. Reecy, Narasimhaswamy S. Belaguli, and Robert J. Schwartz

Department of Cell Biology, Baylor College of Medicine, Houston, Texas 77030

I. Introduction

Defining the molecular basis underlying the establishment and maintenance of cardiac muscle differentiation presents a fundamental challenge in the study of developmental biology and molecular genetics. Important progress in the field of muscle developmental biology has been made with the identification of myogenic factors which program mesodermal cells to a myogenic fate and the identification of other regulatory elements and factors that impart regulated gene expression. These myogenic regulatory factors, the MyoD family of basic helix–loop–helix factors, are expressed exclusively in skeletal muscle and are not thought to be involved in the transcriptional control of ventricular or atrial muscle cells. Therefore, what other factors could regulate cardiac gene expression? This review will focus on the recently identified *tinman* gene family members, vertebrate *Nkx2-5* and the closely related *Nkx-2.8,* and their roles in specification of the cardiac lineage and patterning of the embryonic heart. We will also examine the biological role of serum response factor (SRF), a purported ubiquitous transcription factor which is required for the expression of contractile protein genes, such as the α-actins. Serum response factor was recently demonstrated to be highly expressed in embryonic muscle lineages (Croissant *et al.,* 1996). In addition, this chapter will explore SRF's interaction with Nkx2-5. Nkx2-5 and SRF are capable of activating cardiogenic gene activity in fibroblasts (C. Y. Chen *et al.,* 1996; Chen and Schwartz, 1996). Finally, we attempt to explain how Nkx2-5 and SRF function in a phylogenetically conserved pathway that does not utilize the skeletal myogenic MyoD family and is instrumental for cardiac cell differentiation.

II. *NK* Homeodomain Factors

Homeobox genes have been studied extensively in *Drosophila,* in which they are involved in the commitment of undifferentiated cells to a specific developmental pathway and play an important role in pattern formation (Hunt and Krumlauf, 1992; Sekelsky *et al.*, 1995). Recently, the *NK* homeobox gene family (*NK1/S59, NK2/vnd, NK3/bagpipe, NK4/msh-2/tinman,* and *H6*) was identified (Kim and Nirenberg, 1989; Stadler *et al.*, 1995; see Chapter 5). These genes do not map to clusters, although *NK3/bagpipe* and *NK4/tinman* are paired, and have temporal and spatial patterns of expression during development which are distinct from the *Hox* genes. *Drosophila msh-2/NK4/tinman* is expressed in the primitive mesoderm; however, later in development expression becomes restricted to the developing dorsal vessel, which is the insect equivalent of the vertebrate heart, and visceral mesoderm (Bodmer *et al.*, 1990; Azpiazu and Frasch, 1993; Bodmer, 1993). Inactivation of *tinman* does not affect mesoderm invagination or dorsal spreading but results in the loss of dorsal vessel formation in the *Drosophila* embryo. In addition, *tinman* is known to regulate *NK3/bagpipe* expression in the visceral mesoderm (Azpiazu and Frasch, 1993) and the expression of the *Drosophila* MEF-2 factor, dMEF-2 (Gajewski *et al.*, 1997). These observations suggest that *tinman* is a likely marker for cardiac mesoderm induction and may be involved in cardiac mesoderm patterning.

A. Developmental Regulation of Vertebrate *Nkx2-5*

The murine *NK2*-like homeobox gene, *Nkx2-5/Csx,* is expressed in early cardiac progenitor cells prior to cardiogenic differentiation and continues to be expressed throughout adulthood (Komuro and Izumo, 1993; Lints *et al.*, 1993; see Chapter 7). Superimposed upon the appearance of *Nkx2-5* in cardiac progenitor cells is the sequential expression of the cell type-restricted *cardiac α-actin* and *myosin heavy chain* genes (Lints *et al.*, 1993). *Nkx2-5* cDNA clones have been identified in other vertebrates, such as zebrafish (Chen and Fishman, 1996; Lee *et al.*, 1996; see Chapter 6), *Xenopus* (Tonissen *et al.*, 1994; see Chapter 3), and chickens (Schultheiss *et al.*, 1995; see Chapter 4), and these genes are highly related in sequence and expression patterns to mouse *Nkx2-5*. For example, in *Xenopus, Nkx2.5* expression was first observed in a diffuse pattern in the anterior portion of the embryo. As development proceeded, the region of expression remained centered on the presumptive heart but became increasingly smaller, until it eventually corresponded to differentiated heart tissue. Similarly, in zebrafish embryos, *nkx2.5* expression demarcates the heart field and initiates myocardial differentiation (Chen and Fishman, 1996).

Knockout experiments with *Nkx2-1* (Kimura *et al.*, 1996) and *Nkx2-5* (Lyons *et al.*, 1995) have demonstrated that *Nkx* genes are required for organogenesis which suggests that other *Nkx2* family members should have similar roles in development. Embryos homozygous for the disrupted *Nkx2-5* allele displayed heart morphogenetic defects at Embryonic Day 8.5 (see Chapter 7). A beating linear heart tube developed, but looping morphogenesis, which is a critical determinant of heart formation, was not initiated in *Nkx2-5* null mice. Although these mice die at 9 days of embryonic development, the expression of several cardiogenic-restricted genes which mark the differentiation of early cardiac myocytes did not appear to be affected. Potential redundancy with another *Nkx2* gene which is coexpressed during early heart development, such as the recently identified chicken *Nkx-2.8* gene (Reecy *et al.*, 1997), *Xenopus Nkx2.3* (Tonnisen *et al.*, 1994), or zebrafish *nkx2.7* (Lee *et al.*, 1996), may explain this phenomena. Thus, there may be several vertebrate *tinman* homologs.

B. Role for Bone Morphogenetic Proteins and Wnts in *Nkx2-5* Gene Expression

Decapentaplegic (dpp) and wingless (wg) are two secreted factors which are primarily expressed in the ectoderm and serve as inductive signals for mesoderm patterning, in particular the cardiac mesoderm. Decapentaplegic, which is closely related to the vertebrate bone morphogenetic proteins (BMP) and related members of the transforming growth factor-β superfamily of secreted polypeptide signaling molecules, is an important *Drosophila* dorsalizing morphogen and a mediator of inductive interaction between mesoderm and endoderm (Sekelsky *et al.*, 1995; Graff, 1997). Decapentaplegic is required for the maintenance of *tinman* expression in the dorsal mesoderm (Frasch, 1995). *Decapentaplegic* binds to membrane-bound receptors with serine and threonine kinase activity, which in turn activate Mads (mothers against decapentaplegic) factors (Sekelsky *et al.*, 1995; Thomsen, 1996). In vertebrates, smads, which are Mads-like factors, have been shown to enter the nucleus (Niehrs, 1996) and interact with winged helix/forkhead transcription factors (Y. Chen *et al.*, 1996). Recently, MADR1, a MAD-related protein, was shown to function in BMP-2 signaling pathway (Hoodless *et al.*, 1996), whereas Thomsen (1996) showed that Mads is an embryonic ventralizing agent that acts downstream of the BMP-2/4 receptor.

Similarly, *wg* is also directly involved in heart formation. Elimination of *wg* function shortly after gastrulation, at a time when *tinman* becomes restricted to the dorsal mesoderm, results in the selective loss of heart progenitor cells with little effect on other mesodermal derivatives (Park *et al.*, 1996). Jagla *et al.* (1997) identified two homeobox genes which are required for dorsal vessel formation in *Drosophila.* Expression of *ladybird early* and *ladybird late* is regulated by both *wg* and *tinman.* In addition, *ladybird* is required for maintenance of epidermal *wg* expression. Furthermore, Bodmer *et al.* (1990) demonstrated that the helix–loop–helix protein, twist, regulates the appearance of *tinman.* Thus, in *Drosophila* signaling through both dpp and wg signaling pathways appears to be required for *tinman* expression and the appearance of cardiac mesodermal precursor cells.

Bone morphogenic proteins appear to have an early role in signaling the formation of mesoderm in vertebrate embryos (Hogan, 1996). Schultheiss *et al.* (1997) demonstrated that ectopic application of BMP-2, -4, and -7 to regions of chick embryos medial to the normal *cNkx-2.5* expression pattern that are not usually fated to become heart tissue allowed for the induction of *cNkx-2.5* expression (see Chapter 4). Lough *et al.* (1996) found that BMP-2 or fibroblast growth factor-4 (FGF-4) treatment of noncardiac mesoderm alone was not sufficient to elicit beating heart cells, but treatment with both BMP-2 and FGF-4 could support cardiogenesis. BMPs may facilitate cardiogenesis by both paracrine and autocrine mechanisms. BMP-2 is expressed in the myocardium of the common atrium, whereas BMP-4 is expressed in the body wall which surrounds the pericardial cavity. BMP-5 is expressed in the ventricular myocardium, and BMP-7 is expressed in the myocardium of both atrial and ventricular chambers (Dudley and Robertson, 1997). Little or no mesoderm differentiation occurs in BMP-4 knockout mice. In addition, the mesoderm marker, *Brachyury,* is not expressed (Winnier *et al.,* 1995). Furthermore, some mice deficient for BMP-2 lacked *Nkx2-5* expression and failed to enter early phases of heart development (Zhang and Bradley, 1996). Thus, BMPs appear to influence cardiogenesis and *Nkx2-5* expression early in embryogenesis.

C. Embryonic Expression of Avian *Nkx2-8*

Low-stringency cDNA library screening, genomic library screening, and 5′ RACE strategies were used to identify a novel member of the *Nkx-2* family in chickens. Comparison of the homeodomain amino acid sequence with other NK family members revealed a 92% amino acid sequence identity to *Xenopus,* chicken, and zebrafish Nkx-2.5 homeobox sequence (Fig. 1). Chicken *Nkx-2.8* transcripts are first observed by *in situ* hybridization at Hamburger and Hamilton (HH) stage 7 in the splanchnopleure (see Fig. 2 for a comparison of *cNkx-2.5* and *Nkx-2.8* expression patterns). At stage 10+, the *Nkx-2.8* gene is expressed in the linear heart tube and the dorsal half of the vitelline vein, similar to *cNkx-2.5.* However, after looping (HH stage 13), *Nkx-2.8* is no longer expressed in the heart tube but is expressed in the ventral portion of the pharynx. *cNkx-2.5* is still expressed throughout the looped heart tube at this time. At stage 15, *Nkx-2.8* is expressed in the epithelium of the developing branchial arches and the truncus arteriosus. By HH stage 17, *Nkx-2.8* expression is detectable in lateral endoderm of the second and third pharyngeal pouches, the posterior portion of the aortic sac, and the sinus venosus. Interestingly, *Nkx-2.8* is not expressed in the myocardium after the heart tube has begun to loop. In *Nkx2-5* knockout mice, the heart tube fails to loop. Thus, in *Nkx2-5* null mice, it is possible that *Nkx-2.8* or some other *NK2*-like gene may compensate for the loss of *Nkx2-5* expression and allow for the development of the heart to the linear tube stage. In the chicken, during early heart formation, *cNkx-2.5* and *-2.8* are expressed in partially overlapping domains in the lateral plate mesoderm and linear heart tube (Figs. 3B and 3C). However, after the linear heart tube has undergone the looping process, *cNkx-2.3, -2.5,* and *-2.8* are expressed in distinct and partially overlapping domains.

These data are consistent with a model in which *Nkx-2.8* performs a unique temporally and spatially restricted function in the developing embryonic heart and pharyngeal region. Moreover, *Nkx-2.8* may have a redundant role with *cNkx-2.5* in the coalescing chicken heart tube. In addition, *Nkx-2.8* may play an important role in the transcriptional program(s) that underlies thymus formation. It is thus possible that the position and identity of these organ rudiments are determined by the combinatorial expression of *Nkx* genes (Fig. 3A). Whether *Nkx* gene expression, individually or in combination, is sufficient to specify a cardiac cell type or define a specific pattern of cardiac gene expression remains to be determined.

D. *Nkx* Code

The partially overlapping expression pattern of *Hox* genes in embryos has lead to the concept of a "*Hox* code" (Kessel and Gruss, 1991). The term *Hox* code means that a particular combination of *Hox* genes is functionally active in a region and thereby specifies the developmental fate of this region. The existence of eight *Nkx-2* family members, their overlapping DNA-binding specificity, and most important, their partially overlapping patterns of expression raises the possibility of an "*Nkx* code" (Reecy *et al.,* 1997). Figure 3 is a diagram of known *Nkx2* gene expression patterns in the developing pharyngeal primordium and heart (Lazzaro *et al.,* 1991; Lints *et al.,* 1993; Tonissen *et al.,* 1994; Evans *et al.,* 1995; Schultheiss *et al.,* 1995; Buchberger *et al.,* 1996; Lee *et al.,* 1996; Reecy *et al.,* 1997). In the pharyngeal region,

Organism	Gene name	1 10 20 30 40 50 60	Percent homology
Chicken	cNkx-2.8	RRKPRVLFSQTQVLELERRFKQQKYLSALEREHLANVLQLTSTQVKIWFQNRRYKCKRQR	100
Frog	XNkx-2.5	----------A--Y----------------P--D------K---------------------	92
Chicken	cNkx-2.5	----------A--Y----------------P--D------K---------------------	92
Fish	nkx2.5	----------A--Y----------------P--D------K---------------------	92
Mouse	Nkx2-3	----------A--F---------R----P-------S-K---------------------	90
Chicken	cNkx-2.3	----------A--F---------R----P-------S-K---------------------	90
Frog	XNkx-2.3	----------A--F---------R----P-------S-K---------------------	90
Fish	nkx2.3	----------A--F---------R----P------ST-K---------------------	88
Fish	nkx2.7	-------------F---------R----P--D---LA-K---------------------	88
Human	CSX	----------A--Y---------R----P--DQ--S--K---------------------	87
Mouse	Nkx2-5	----------A--Y---------R----P--DQ--S--K---------------------	87
Mouse	Nkx2-6	Q--S------A---A--------R--T-P------SA------------------S-S--	82
Canine	TTF-1	---R------A--Y--------------P------SMIH--P-------------M---A	82
Human	TTF-1	---R------A--Y--------------P------SMIH--P-------------M---A	82
Rat	TTF-1	---R------A--Y--------------P------SMIH--P-------------M---A	82
Flatworm	Dth-2	---R-I----A-IY--------------P-------LIN--P---------H------SQ	78
Mouse	Nkx-2.2	K--R-----KA-TY------R--R----P------SLIR--P---------H---M--A-	72
Leech	Lox-10	---R-I----A-IY------R-------P------TFIG--P---------H---T-KSK	72
Fish	nk2.2	K--R-----KA-TY------R--R----P------SILR--P---------H---M--A-	72
Flatworm	Dth-1	K--R-----KK-I-----H-R-K-----P-------LIG-SP---------H---M--AH	70
Fruitfly	NK2/vnd	K--R----TKA-TY------R--R----P------SLIR--P---------H---T--AQ	68
Fruitfly	NK4/*tin*	K---------A------C--RLK---TGA---II-QK-N-SA-------------S--GD	68
Frog	XeNK2	K--R----SKA-TY------R--R----P------SLIR--P---------H---T--AQ	68
Flatworm	ceh-22	K--R----TKA-TY------RS-R----P---A--MQIR--P---------H---T-KSH	65
Fruitfly	bagpipe	KKRS-AA--HA--F-----FA--R---GP--SEM-KS-R--E-------------T--KQ	60
Mouse	Nkx3-1	QKRS-AA--HT--I----KFSH------P--A---KN-K--E-------------T--KQ	63
Mouse	Bapx1	KKRS-AA--HA--I------NH-R---GP--AD--AS-K--K-------------T--RQ	62
		Helix I Helix II Helix III	

Figure 1 Comparison of the homeodomain region of chicken Nkx-2.8 with those of other homeodomain proteins showing the most similarity. The homeodomain amino acid sequence is represented along with the gene name and species of origin. All sequences are compared to chicken Nkx-2.8 and identical amino acids are indicated by a dash. The leucine at position 29 (boldface) is unique to Nkx-2.8. The predicted homeodomain helixes are indicated at the bottom. The percentage amino acid identity of each homeodomain to Nkx-2.8 is indicated on the right. References for individual genes are as follows: *XNkx2.5,* Tonissen *et al.* (1994); *cNkx-2.5,* Schultheiss *et al.* (1995); *nkx2.3/nkx2.5/nkx2.7,* Lee *et al.* (1996); *Nkx2-1/Nkx2-2/Nkx2-3/Nkx2-4,* Price *et al.* (1992); *cNkx-2.3,* Buchberger *et al.* (1996); *XNkx2.3,* Evans *et al.* (1995); *CSX,* Komuro and Izumo (1993); *Nkx2-3/Nkx2-5/Nkx2-6,* Lints *et al.* (1993); canine, human, and rat *TTF-1,* Mizuno *et al.* (1991), Saiardi *et al.* (1995), and Van Renterghem *et al.* (1995); *Lox-10,* Nardelli-Haefliger and Shankland (1993); *nk2.2,* Barth and Wilson (1995); *vnd,* Jimenez *et al.* (1995); *msh-2,* Bodmer *et al.* (1990); *bagpipe,* Azpiazu and Frasch (1993); *ceh-22,* Okkema and Fire (1994); *XeNK2,* Saha *et al.* (1993); *Dth-1/Dth-2,* Garcia-Fernandez *et al.* (1991); *Nkx3-1,* Sciavolino *et al.* (1997); *Bapx1,* Tribioli *et al.* (1997). Adapted from Reecy *et al.* (1997).

Figure 2 *In situ* hybridization comparison of *cNkx-2.5* and chicken *Nkx-2.8.* The HH stage is indicated on the left. *In situ* probes are indicated in the upper right corners. (A) Stage 8, ventral view; *Nkx-2.8* expression was detected in the forming heart tubes. (B) Stage 8+, ventral view; *cNkx-2.5* expression was detected in the forming heart tube and the pharyngeal region. (C) Stage 7, transverse section; weak *Nkx-2.8* expression was detected in the splanchnopleura. (D) Stage 7, transverse section; *cNkx-2.5* expression is seen in the lateral plate mesoderm as well as the adjacent ectoderm and endoderm. (E) Stage 10, 11 somite, ventral view; *Nkx-2.8* is expressed in the linear heart tube, the vitelline veins, with a low level of expression in the telencephalon. (F) Stage 10, ventral view; the predominant signal is found in the tubular heart. (G) *Nkx-2.8* expression was detected in the myocardium and ventral pharyngeal endoderm lateral to the dorsal mesocardium. (H) Stage 10, transverse section; *cNkx-2.5* expression is detected in the myocardium but not the endocardium and is also found in the floor of the pharynx and pharyngeal ectoderm. (I) Stage 17, side view; *Nkx-2.8* expression was detected in the second and third pharyngeal pouches but not in the heart tube. (J) Stage 14, side view; *cNkx-2.5* expression is detected only in the heart. CT, cardiac tube; Ec, endocardium; Ect, ectoderm; End, endoderm; H, heart; Mes, mesoderm; Mc, myocardium; NG, neural groove, NT, neural tube; pC, pericardiac coelom; PP, pharyngeal pouch; So, somite. [Parts A, C, E, G, and I are from Reecy *et al.,* 1997, with permission; Parts B, D, F, H, and J are from Schultheiss *et al.,* 1995, with permission.]

cNkx-2.3 and *-2.5* are expressed in broad, partially overlapping domains primarily within ectoderm and endoderm. By contrast, *Nkx2-1* and *-2.8* expression is spatially more restricted; transcripts are found at the site at which the anlagen for thyroid, tonsils, thymus, and lung reside. Thus, the position and identity of these organ rudiments may be determined by the combinatorial expression of *Nkx* genes. Whether the genes depicted in Fig. 3 are sufficient to define a unique *Nkx* code, or whether other (e.g., zebrafish *nkx-2.7*; Lee *et al.,* 1996) and/or yet to be discovered family members are required, remains an open question. In addition, the ex-

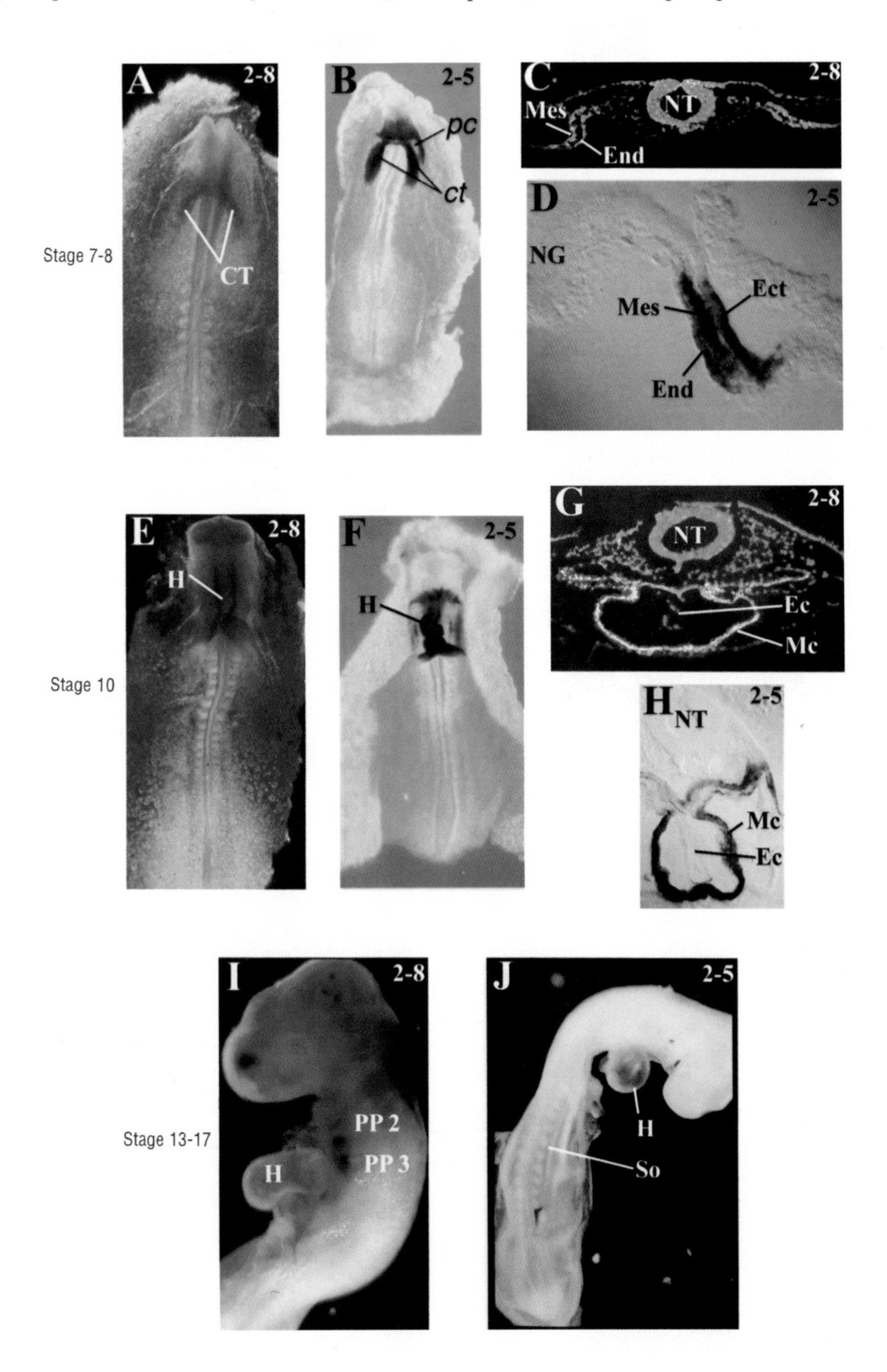

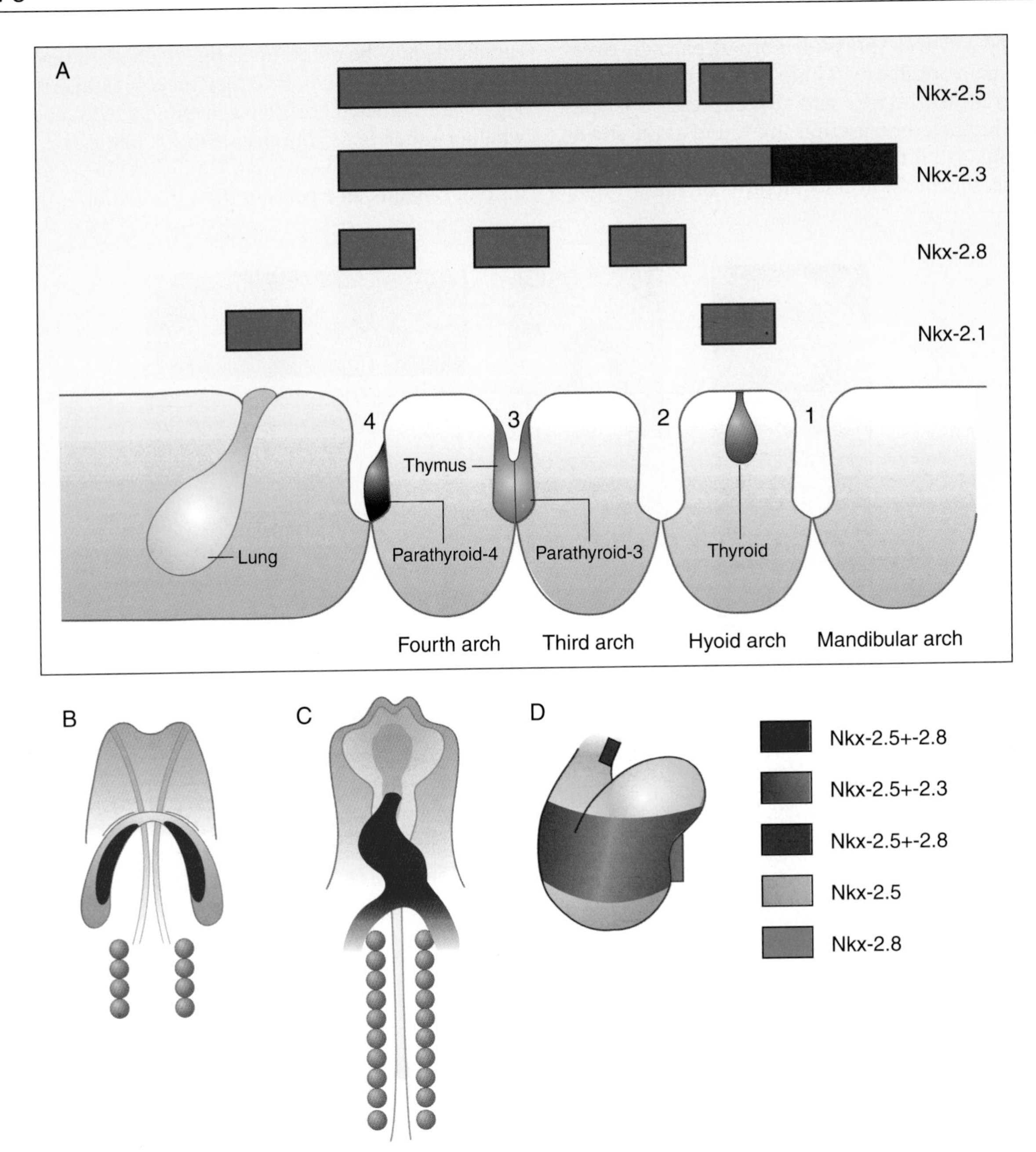

Figure 3 *Nkx* code. A summary of *Nkx* expression patterns in the pharyngeal region and developing chicken heart. (A) Expression pattern of *Nkx* genes in the pharyngeal region. *Nkx2-3* and *-2-5* are expressed in broad patterns, whereas *Nkx2-1* and *-2.8* are expressed in a more localized manner. In contrast to chicken *Nkx-2.3,* zebrafish *nkx2.3* expression extended more anteriorly, which is indicated by the purple box. Only one side of the pharyngeal region is depicted with the pharyngeal pouches numbered 1–4. The pharyngeal arches are indicated with dotted lines and the organ rudiments arising from the pharyngeal endoderm. (B–D) Expression pattern of chicken *Nkx2* genes in the developing heart: B, HH stage 7; C, HH stage 10; D, late stage. The genes that correspond to a given color are indicated. References for expression domains are as follows: mouse *Nkx2-1* Lazzaro *et al.* (1991); *Xenopus,* chicken, and zebrafish *Nkx-2.3,* Evans *et al.* (1995), Buchberger *et al.* (1996), and Lee *et al.* (1996); mouse, chicken, *Xenopus,* and zebrafish *Nkx2-5,* Lints *et al.* (1993), Tonnisen *et al.* (1994), Schultheiss *et al.* (1995), and Lee *et al.* (1996); zebrafish *Nkx2-7,* Lee *et al.* (1996). Reproduced from Reecy *et al.* (1997).

pression patterns of vertebrate *Nkx2* genes are not always conserved [compare *cNkx2-3* expression in chickens (Buchberger *et al.,* 1996), *Xenopus* (Evans *et al.,* 1995), and zebrafish (Lee *et al.,* 1996)]. Of note, numerous *Hox* genes are also expressed in migrating neural crest cells within the branchial arch region (Hunt and Krumlauf, 1992) and such *Hox* genes may pattern this part of the embryo, perhaps in conjunction with the *Nkx2* genes. Several *Hox* genes are expressed in the thyroid as well as in the thymus primordium (Gaunt, 1988), which suggests a direct involvement in the specification of these tissues.

E. Characterization of the NK-2 Homeobox Family

The homeodomain amino acid sequence of Nkx-2.8 and its closest relatives are shown in Fig. 1. Amino acid groupings within the NK-related homeodomains not only support the close evolutionary relationship between the *NK2/vnd, NK3/bagpipe,* and *NK4/tinman* genes but also support functional divergence within this group. There are at least six mouse *Nkx2* genes which have significant homology to *Drosophila NK2* (Harvey, 1996; Fig. 1). Within the homeodomain sequences, Nkx2-5 was shown to be most closely related to the newly identified chick Nkx-2.8 (Fig. 1; Reecy *et al.,* 1997). The tyrosine at position 54, which may influence DNA binding site specificity (Damante *et al.,* 1994), is unique among the *Drosophila* vnd, bagpipe and tinman proteins and their closest relatives.

Another distinguishing feature of NK-2 relatives is the conservation of a 17-amino acid (aa) motif carboxy terminal to the homeodomain (Fig. 4). This NK-2 specific domain contains a central cluster of invariant hydrophobic amino acids (VPVLV) which may have a role in protein–protein interaction. Mutagenesis of this domain in Nkx2-5 influences its ability to bind to DNA *in vitro* (J. Sepulveda and R. Schwartz, unpublished observation). Analogous to the repression domains identified in

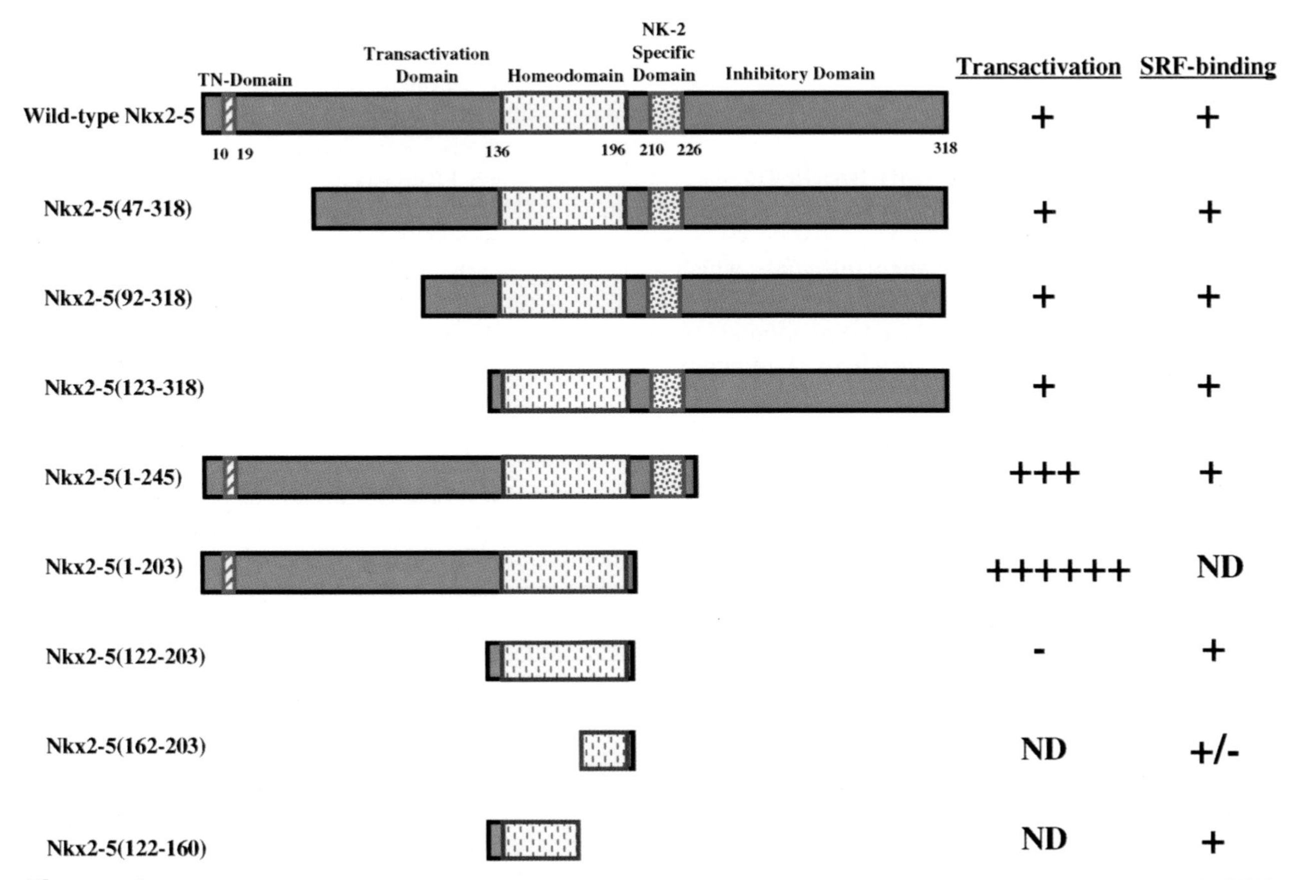

Figure 4 Schematic diagram of wild-type Nkx2-5 and mutant proteins. This diagram depicts the positions of the homeodomain (HD), TN domain, and NK-2 specific domain in the Nkx2-5 protein. The NK-2-specific domain (dense stripes) and 5′ decapetide (TN domain) are conserved in all Nkx2 homologs. Wild-type Nkx2-5 and deletion mutants are used in transactivation assays and mapping Nkx2-5/SRF interactions. Numbers indicate the amino acid position. ND, not determined; ± transcriptional activity and ability to functionally interact with SRF (adapted from Chen and Schwartz, 1995).

the *Drosophila* repressor kruppel (Licht *et al.,* 1990), an Nkx2-5 inhibitory domain (Fig. 4; Chen and Schwartz, 1995) is also rich in alanine and proline residues. The Nkx2-5 segment from amino acids 266 to 282 (AAYPAAPPAAQPPAASA) is reminiscent of the short runs of AAAA and PAA peppered throughout the 84-aa repression domain of kruppel (Licht *et al.,* 1990). These short stretches of hydrophobic amino acids are relatively unstructured. Like the sequences which comprise the inhibitory domain, the Nkx2-5 activation domain is composed of alanine- and proline-rich sequences, but it is also enriched in acidic aspartate and glutamate residues. We have demonstrated that Nkx2-5 can serve as a modest transcription activator in transfection assays when analyzed with reporter genes which have multimerized NKE binding sites (Figs. 4 and 7). However, the *trans*-activation activity of Nkx2-5 was stimulated at least 50-fold or more when its C-terminal inhibitory domain was deleted. However, the inhibitory domain has not been precisely mapped. These findings suggest that potential hydrophobic interactions between the inhibitory and activation domains might block access of transcription initiation factors to this highly charged moiety. Thus, it is likely that conformational changes in Nkx2-5 protein structure might transduce Nkx2-5 into an operative transcriptional activator.

F. Nkx2 Homeodomain Binds to Novel Target Sequences

Nkx2 factors are DNA-binding proteins which are capable of activating transcription. The three-dimensional structure of the 60-aa homeodomain is composed of three α-helices, in which helix II and helix III form a helix–turn–helix motif (Scott *et al.,* 1989). DNA-binding assays (Chen and Schwartz, 1995) indicated that Nkx2-5 bound as a monomer to various DNA targets. Sequence comparison and *in vitro* DNA-binding assays have revealed that Nkx2-1, also called thyroid transcription factor-1 (TTF-1; Civitareale *et al.,* 1989; Javaux *et al.,* 1992), and Nkx2-5 bind to a specific Nkx2 response elements (NKE) sequences containing 5′-TNAAGTG-3′ (Guazzi *et al.,* 1990; Damante and Di Lauro, 1991; Bohinski *et al.,* 1994; Chen and Schwartz, 1995), which is different from the motif 5′-TAAT-3′ recognized by the Hox protein family (Damante and Di Lauro, 1991).

NKEs are present in the Myo-2 enhancer, which binds CEH-22, an *Nkx2* homolog which regulates pharyngeal myosin heavy chain expression in *Caenorhabditis elegans* (Okkema and Fire, 1994), and mammalian lung specific surfactant gene promoters (Bohinski *et al.,* 1994; Table I). Ray *et al.* (1996) demonstrated that Nkx2-1 and Nkx2-5 bound the same NKEs and were capable of transactivating the murine *clara cell-specific 10-kDa protein (mCC10)* promoter. Certain serum response elements (SRE) within the *cardiac α-actin* promoter can serve as both high-affinity (SRE_2 and SRE_3) and intermediate strength binding targets (SRE_1 and SRE_4) for bacterially expressed Nkx2-5 (Chen and Schwartz, 1996). Durocher *et al.* (1996) reported that two NKE sites within the proximal region of the cardiac *atrial natriuretic factor* promoter directed high levels of cardiac-specific promoter activity. In addition, Gajewski *et al.* (1997) showed that two NKE sites direct *dmef-2* expression in response to *tinman.* Thus, NK-2 homologs bind to similar DNA sequences, but how do NK-2 homologs regulate gene transcription in a cell-specific manner?

III. Serum Response Factor

A. Serum Response Factor Serves as a Positive Transfactor in Myogenesis

SRF, a 67-Kda DNA-binding protein, was first cloned from a HeLa cDNA library (Norman *et al.,* 1988). SRF was generally presumed to be a ubiquitous and constitutive transcription factor which bound to SREs as a homodimer. SRF binds as a dimer and symmetrically contacts various SRE elements with a consensus sequence $CC(A/T)_6GG$ (Fig. 5A). SRF is member of an ancient DNA-binding protein family which shares a highly conserved DNA-binding/dimerization domain of 90 aa termed the MADS box (Sommer *et al.,* 1990; Fig. 5B). SRF, yeast transcription factors MCM1 and ARG80, and several plant proteins, such as Deficiens, all have a related MADS box and similar DNA sequence binding specificity (Norman *et al.,* 1988). In addition, SRF-related proteins (RSRF/MEF-2) constitute a subfamily of the MADS box family of transcription factors. MEF-2 factors contain the MADS box and an adjacent MEF-2 box. MEF-2 proteins bind to MEF-2 sites, $CTA(A/T)_4TAG$, which can be found in the regulatory regions of both nonmuscle and muscle-specific genes (Pollock and Treisman, 1991; for review, see Olson *et al.,* 1995, and Chapter 8).

Despite their similarities, MADS box proteins have evolved to perform diverse functions, such as specification of mating type in yeast, homeotic activities in plants, pulmonary development in *Drosophila,* and elaboration of mesodermal structures in vertebrates. Interestingly, the overall structural divergence of SRF proteins among evolutionarily distant species of animals appear to be related to differences in their spatial expression pattern. For example, pruned, the SRF homolog in *Drosophila* (Guillemin *et al.,* 1996), is the most divergent member with sequence conservation limited

Table I Conservation of NK-2 Homeodomain DNA-Binding Targets

Homeobox gene	Target gene	Sequence	Reference
Nkx2-1	Rat thyroglobin protein	−160 TACTCAAGTA	
		−133 GACTCAAGTA	Civitareale *et al.* (1989)
		−69 CAGTCAAGTG	
	Mouse clara cell-specific protein (mCC10)	−258 CCAGAGAGG	
		−282 CTGGAGTGCT	Ray *et al.* (1996)
		−344 CCTGAAGGGT	
	Murine surfactant protein C	−186 TAGGCCAAGGGCCTTGGGG	Kelly *et al.* (1996)
	Consensus	GNNCACTCAAG	Guazzi *et al.* (1990)
Nkx2-5	Avian cardiac α-actin SRE2	−95 CATTCATGGG	
	Avian cardiac α-actin SRE3	−164 CATCTAAGGC	C. Y. Chen *et al.* (1996)
	Atrial naturetic factor		
	Box A		
	Rat	CCGCAAGTG	
	Human	CTGCAAGTG	
	Mouse	CTGCAAGTG	
			Durocher *et al.* (1996)
	Box B		
	Rat	CAGAATGG	
	Human	CAGAATGG	
	Mouse	CAGAATGG	
	Consensus	TNAAGTG	Chen and Schwartz (1995)
		CA(A/T)TAATTN	
ceh-22	*C. elegans* Myo-2 enhancer	CGCTAAAGTG	Okkema and Fire (1994)
tinman	*D-mef2 IIA237*	+21 CTCAAGTGG	
		+93 CCACTTGAG	Gajewski *et al.* (1997)

A

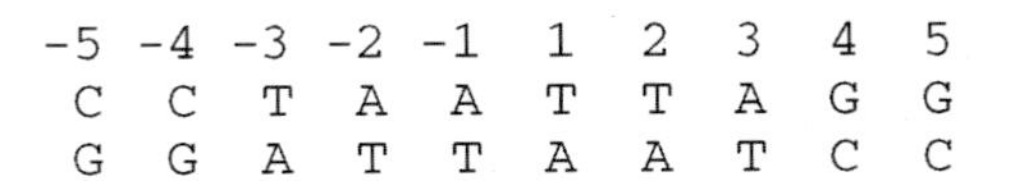

B

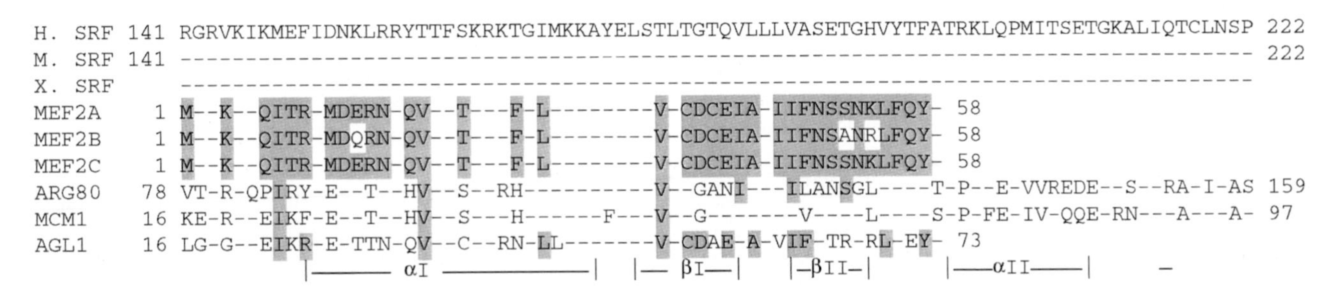

Figure 5 A typical serum response element and aligned sequences of MADS domains. (A) The protypical serum response element composed of two half sites encompassing 10 nucleotides. (B) Protein sequences of different MADS domains of human (H), mouse (M), and *Xenopus* (X) SRF, MEF2A, -B, and -C, yeast ARG80, MCM1, and plant AGL1 that were aligned to show high degree of sequence conservation across broad evolutionary time. Standard single amino acid letter codes were used in alignments. The human SRF α and β secondary structure assignments are indicated: αI, aa. 153–179; βI, aa 182–188; βII, aa 194–198; αII, aa 209–219 (adapted from Pellegrini *et al.*, 1995).

only to the MADS box domain. *Drosophila* SRF expression is localized to the tracheal system (Affolter *et al.,* 1994), which is different from that of *Xenopus,* avian, and murine SRF. These *SRF* genes are more unified in structure and tissue expression pattern.

The *SRF* mRNA expression pattern in the avian embryo is similar to the expression pattern of the related *MEF-2* genes. Like *MEF-2, SRF* is preferentially expressed in cardiac, skeletal, and smooth muscle tissues (Yu *et al.,* 1992; Croissant *et al.,* 1996). We have observed differential *SRF* gene expression during avian embryogenesis (Fig. 6) (Croissant *et al.,* 1996). SRF mRNA, protein, and DNA-binding activity are enriched in striated and smooth muscle tissues. Lower but significant levels of SRF were detected in neuronal tissues, whereas, almost undetectable levels of SRF were observed in liver tissue of late-stage avian embryos. In addition, high levels of *SRF* mRNA were present in precardiac splanchnic mesoderm and dorsal somitic mesoderm. SRF protein was also shown to be selectively expressed in the myocardium during heart morphogenesis and in the myotomal segment of the anterior somites in avian embryos. Thus, high-level *SRF* expression appears to coincide with the expression of cardiac, skeletal, and smooth muscle α-actins, which are early markers for terminal differentiation of striated and smooth muscle (Ruzicka and Schwartz, 1988).

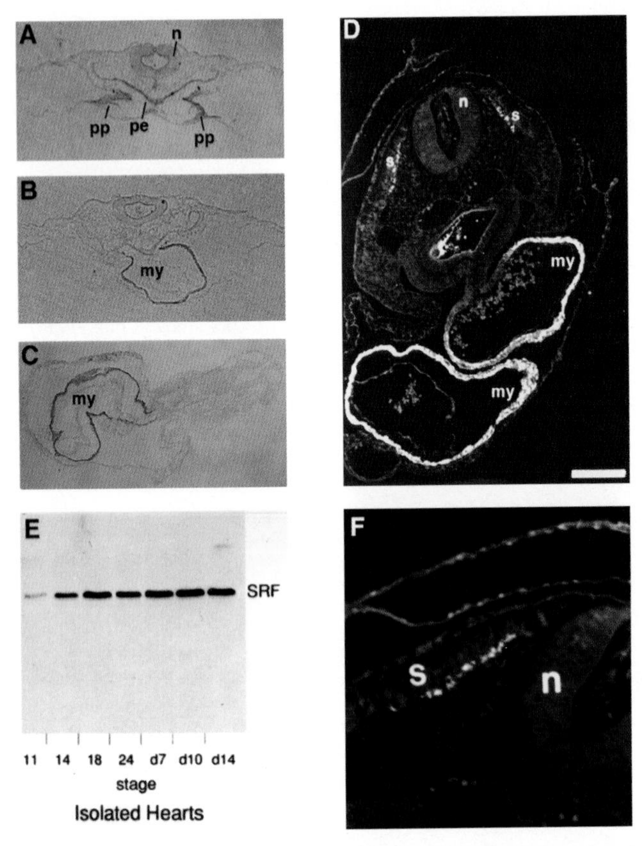

Figure 6 Serum response factor is expressed primarily in striated and smooth muscle tissues which coincides with the accumulation of cardiac α-actin transcripts during the formation of the embryonic heart and somites. (A–D) Transverse sections of HH stage 8 (A), HH stage 10 (B), HH stage 12 (C), and HH stage 15 (D). Avian embryos were probed with a rabbit SRF antibody produced against bacterially expressed human SRF to detect SRF localization during myocardial development. SRF protein was localized to the fusing paired primordium of the developing heart (A) and throughout the myocardium in later developmental stages (B–D). (E) Western analysis showing the upregulation of SRF expression during cardiogenesis. Protein, isolated from chick hearts collected at various stages of development, was probed with the same antibody used in immunohistochemisty. (F) Higher magnification of the stage 15 embryo (D) shows SRF localization to the myotome in the somites. A–C, immunoperoxidase; D,F, immunofluorescence; E, ECL detection; my, myocardium; n, neural tube; pe, pharyngeal endoderm; pp, promyocardium; s, somite (adapted from Croissant *et al.,* 1996).

Nuclear proteins have been shown to interact with α-actin CArG/SRE sequences (Walsh and Schimmel, 1987; Gustafson *et al.,* 1988; Walsh and Schimmel, 1988; Boxer *et al.,* 1989a; Sommer *et al.,* 1990). A CArG box-binding factor reported to regulate the transcription of α-actin genes has been shown to be indistinguishable from the serum response factor that binds to the c-*fos* SRE (Treisman, 1986; Gustafson *et al.,* 1988; Boxer *et al.,* 1989b). Croissant *et al.* (1996) detected a significant increase in *SRF* transcripts and protein mass during primary myogenesis in culture which correlated with the induction of skeletal and cardiac α-actins. *SRF* expression was repressed in replicating myoblasts; however, SRF mRNA and protein levels increased approximately 40-fold during the progression from primary myoblast cultures to postdifferentiated myotubes. SRF protein levels were highest in postreplicative myotubes and were localized to the nucleus. These events preceded the induction of *skeletal α-actin* expression during myogenic differentiation (Lee *et al.,* 1992; Croissant *et al.,* 1996). Concordant with its role as a positive regulator of *α-actin* genes, Vandromme *et al.* (1992) demonstrated that the microinjection of SRF antibodies prevented the myoblast–myotube progression and the expression of the myogenic factors myogenin and MyoD. SRF can also transactivate *α-actin* gene transcription under conditions that block myogenic differentiation (Lee *et al.,* 1992). In contrast, an SRF mutant (SRFpm1) which inhibited DNA binding, but not dimerization of monomeric SRF subunits, blocked transcriptional activation of the *skeletal α-actin* gene (Croissant *et al.,* 1996). We have recently observed that stable transfections of Sol 8 and C_2C_{12} myogenic cell lines with SRFpm1 inhibited myoblast fusion and postreplicative myogenic differentiation. In addition, SRFpm1 transfected cells were unable to express *myogenein* and *skeletal α-actin* gene activity (data not shown). Dominant-negative SRF activity could be complemented by the formation of nonfunctional SRF heterodimers or

SRFpm1 homodimers which compete with endogenous SRF for the binding of relevant accessory factors. These complexes could even provide nonfunctional targets for phosphorylation events. Soulez *et al.* (1996) elegantly showed that the induced expression of antisense *SRF* blocked both the growth and the differentiation of C2 myoblasts. Antisense *SRF* effectively inhibited the appearance of MyoD, myogenin, and myosin heavy chain protein. These studies add support to the notion that SRF plays an integral role during myogenic differentiation. Therefore, selective enrichment of *SRF* gene activity in embryonic myocardial, skeletal, and smooth muscle-derived cell types should dispel any notion that SRF is a ubiquitous factor with a limited range of expression. Taken together, these observations clearly support a tissue-restricted pattern of *SRF* gene activity and an obligatory role for SRF during striated muscle differentiation.

B. SRF–Accessory Protein Interactions

Since SREs play a primary role in the regulation of early response genes such as c-*fos* (Treisman, 1986) and the myogenic-restricted *α-actin* genes (Minty and Kedes, 1986; Boxer *et al.,* 1989b; Mohun *et al.,* 1989, 1991; Chow *et al.,* 1991; Lee *et al.,* 1991), how does SRF convey its specificity? Recent studies indicate the versatile nature of SRF, which can act as a platform to interact with other regulatory proteins and thus change its DNA-binding activity and ultimately the regulation of specific gene programs. Studies regarding the regulation of the c-*fos* gene by SRF have led to the identification of several SRF accessory factors, including SAP-1, Elk-1, and Phox-1 (Schroter *et al.,* 1990; Pollock and Treisman, 1991; Dalton and Treisman, 1992). All these SRF accessory factors appear to potentiate SRF's transcriptional activity on the c-*fos* SRE, although the mechanisms are somewhat different. Similarly, the function of yeast protein MCM1, which like SRF is also a MADS family member, is influenced by the recruitment of an array of accessory factors which either activate or repress genes in a cell type-specific and temporal pattern (Herskowitz, 1989). MCM1 can interact with MATa2, a homeodomain protein, and bind to specific operator sequences (Herskowitz, 1989). Grueneberg *et al.* (1992) demonstrated that human SRF interacts with a novel human homeodomain protein, Phox, which was identified by expression cloning in yeast. Phox is identical to MHox and is expressed in mesodermally derived cells (Cserjesi *et al.,* 1992). Phox interacts with SRF to enhance the exchange of SRF with its binding site in the c-*fos* promoter and does not require specific homeodomain DNA-binding activity. However, Phox/Mhox was unable to activate the *cardiac α-actin* promoter in the presence of SRF. Thus, MADS box family members are capable of interaction with homeobox proteins, which ultimately alters the function of the MADS box protein.

C. Synergistic Transactivation of the *cardiac α-actin* Gene by Nkx2-5 and SRF

We demonstrated that the expression of either Nkx2-5 or SRF stimulated *cardiac α-actin* promoter activity about 3- to 5-fold. However, the combined transfections of Nkx2-5 and SRF expression vectors led to robust reporter gene activity, which was elevated by 15- to 30-fold over control levels (Fig. 7; Chen and Schwartz, 1995, 1996; C. Y. Chen *et al.,* 1996). In contrast, these transcription factors did not stimulate the SRE-deficient promoters of the herpes simplex thymidine kinase gene and the SV40 early gene. MHox, a murine homolog of human Phox-1 (Cserjesi *et al.,* 1992; Grueneberg *et al.,* 1992) could not substitute for Nkx2-5 in cotransfection assays and inhibited SRF-stimulated promoter activity. Thus, the failure of MHox, a paired-liked homeodomain factor, to stimulate the *cardiac α-actin* promoter demonstrates that homeodomain factors are not functionally equivalent. These results suggest that Nkx2-5 and SRF cotransactivation was restricted to the *cardiac α-actin* promoter and that the transcriptional synergy with SRF on the *cardiac α-actin* promoter may be limited to a specific class of homeodomain factors.

We questioned how this mechanism was mediated. Even though Nkx2-5 can bind weakly to some SREs, we found that activation of a minimal promoter consisting of a single SRF binding site was dependent on SRF. When Nkx2-5-binding activity was blocked by a point mutation in the third helix of the homeodomain, SRF was still capable of recruiting mutated Nkx2-5 to the *cardiac α-actin* promoter. Investigation of protein–protein interactions demonstrated that Nkx2-5 could bind to SRF in the absence of DNA as soluble protein complexes isolated from cardiac myocyte nuclei. In addition, Nkx2-5 and SRF could be detected coassociated in binding complexes on the proximal cardiac α-actin SRE. Recruitment of Nkx2-5 to an SRF was dependent on SRF DNA-binding activity and could be blocked by the dominant-negative SRF_{pm1} mutant, which can dimerize with wild-type SRF monomers but cannot itself bind to DNA. The interactive regions between Nkx2-5 and SRF were mapped to N-terminus/helix I and helix II/helix III regions of the Nkx2-5 homeodomain and to an N-terminal extension of the MADS box (Chen and Schwartz, 1996; Figs. 4 and 8).

The recent elucidation of the X-ray crystal structure of SRF bound to DNA (Pellegrini *et al.,* 1995) provides an explanation for the mutually inclusive binding

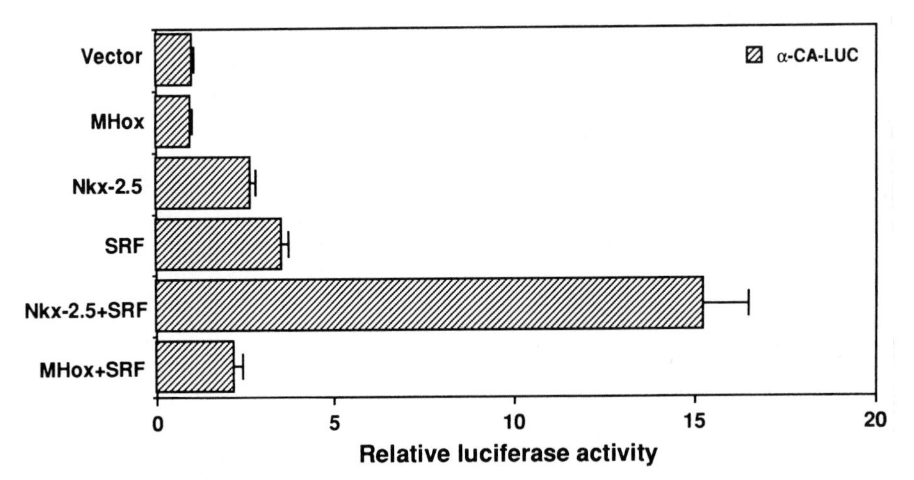

Figure 7 SRF with Nkx2-5 but not MHox coactivates the cardiac actin promoter. The *cardiac α-actin* promoter linked to a luciferase reporter (α-CA-LUC) was cotransfected with an Nkx2-5 or an MHox expression vector alone or in combination with SRF into mouse C3H10T1/2 cells. MHox, a paired-liked homeodomain factor, was unable to transactive the *cardiac α-actin* promoter. In contrast, Nkx2-5 and SRF cotransfection increased *cardiac α-actin* promoter activity (15-fold) under identical conditions. Results were averages of two independent duplicate transfection experiments normalized to β-gal activity, an internal standard. The relative value of each promoter obtained with an empty expression vector was set at 1 (reproduced from Chen and Schwartz, 1996, with permission).

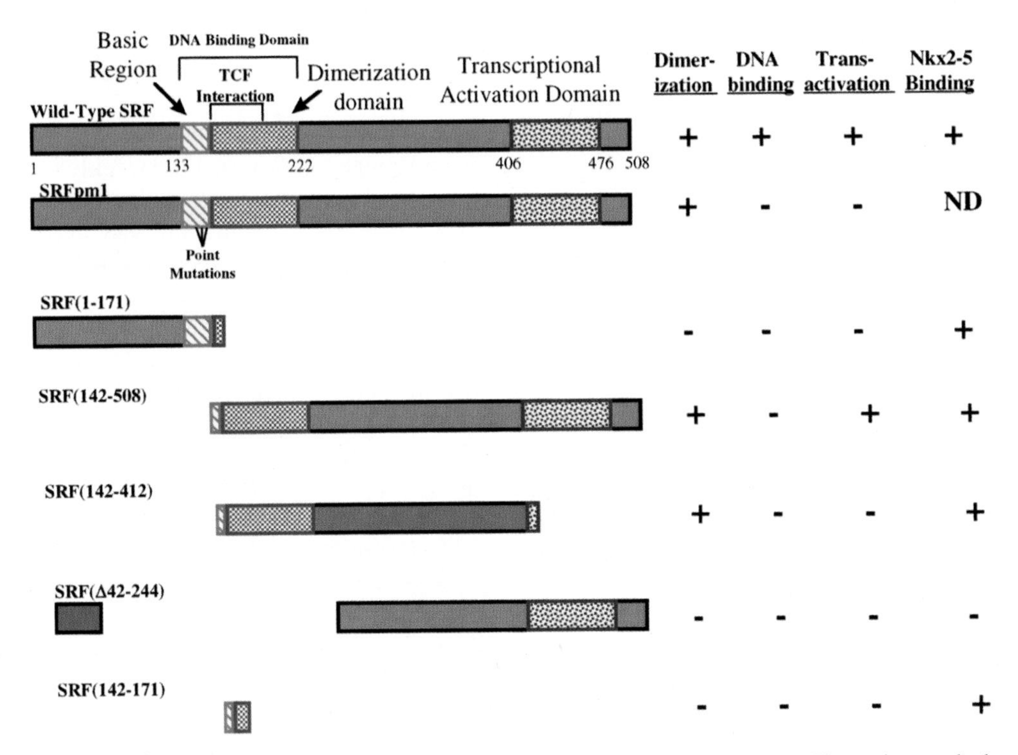

Figure 8 Schematic diagram of wild-type SRF, SRFpm1, and deletion mutants. The region marked DNA-binding domain is sufficient for both DNA binding and ternary complex formation recruitment. The triple-point mutant SRFpm1 converted Arg143, Lys145, and Leu146 to Ile143, Ala145, and Gly146, respectively. Numbers indicate amino acid position. ND, not determined; ± ability to bind DNA, transcriptional activity, and ability to functionally interact with Nkx2-5 (adapted from Chen and Schwartz, 1996, with permission).

of Nkx2-5 and SRF to a single SRE. A novel DNA-binding motif, a coiled coil formed by the MADS box α1 helix (aa 153–179), lies parallel and on top of a narrow DNA major groove and makes contact with the phosphate backbone on an SRE half site, whereas the unstructured N-terminal extension from the αI helix (aa 132–152) makes critical base contacts in the minor groove. Dimerization of the MADS box occurs above the αI helix by a structure composed of two β-sheets in the monomer that interact with the same unit in its partner. A second αII helix in the C-terminal portion of the MADS box, stacked above these β-sheets, completes this stratified structure and is involved with Elk-1 binding. Bending of the DNA toward the protein allows the N-terminal part of each α-helix to make groove contacts at the edge of the SRE, thus leaving an unobstructed major groove in the center of the SRE, which could allow for the simultaneous interaction with homeodomain proteins such as Phox-1 and Nkx2-5. The fact that Phox-1/SRF and Nkx2-5/SRF interactions both required the amino-terminal arm/helix 1/helix 2 region of the Nkx2-5 homeobox leads us to propose a model whereby the amino terminus and helix 3 are responsible for homeodomain–DNA interactions, whereas helixes 1 and 2 are probably mediating protein–protein interactions of homeobox proteins among themselves and with other protein factors, which might be an essential requirement to regulate their ability to act as activators, coactivators, or repressors (Chen and Schwartz, 1996).

Under appropriate extracellular signals, the formation of the SRF/Phox-1/Elk1 tertiary complex on the c-*fos* SRE can transactivate the c-*fos* promoter, In contrast, Phox-1 and Elk-1 cannot activate the *cardiac α-actin* promoter (data not shown). Possibly, during early cardiogenesis, paired-like homeodomain genes such as *Phox-1, MHox,* or *S8* (Leussink *et al.,* 1995; Opstelten *et al.,* 1991) form nonproductive inhibitory complexes with SRF on the *α-actin* promoter SREs. The high level of S8 expression in the endocardial cushions, in fact, correlates well with regions of the heart which do not express *Nkx2-5, α-actin* genes, and other contractile genes. Thus, SRF/Mhox complexes may serve a repressive role to block contractile activity in regions of the heart that will form septations and valves. Nkx2-5 might actually compete, through its SRF-interactive subdomains, with Phox-1 for SRF MADS box binding. The outcome of Nkx2-5 and SRF interactions may simply preclude Phox-1 and Elk-1 from the complex which allows for the activation of the *cardiac α-actin* promoter in cardiac myocytes. Consistent with this idea, Nkx2-5 blocked the serum-inducible expression of a minimal c-*fos* SRE-containing promoter in transient transfection assays (N. Belaguli and R. J. Schwartz, unpublished data).

Therefore, in cardiac myocytes precursors, *cardiac α-actin* gene activation may require increased levels of SRF in combination with the coappearance of Nkx 2-5 to foster cooperative transfactor complex formation. We recently showed that the increase in nuclear localized SRF/Nkx2-5 complexes competed off negative-acting factors such as YY1, which allows for the saturation of the multiple SREs with positive-acting SRF complexes which activated the *cardiac α-actin* promoter (Chen and Schwartz, 1997). Thus, we believe that, in cell culture transactivation assays it is necessary to express SRF to increase the cellular levels of SRF in 10T1/2 fibroblasts so that SRF/Nkx2-5 may compete off inhibitory complexes. Conversely, the Nkx2-5/SRF complexes that are activated in nonreplicating cardiac myocytes might serve to repress the c-*fos* promoter through the formation of nonproductive complexes on its SRE. Therefore, SRF might be able to mediate accessory factor interactions with certain homeodomain factors that either activate or repress transcription.

In addition to SRF, MEF-2 protein, another member of MADS family, has been shown to interact with other regulatory factors (Molkentin *et al.,* 1995). The recent demonstration that MEF2 and myogenic basic helix–loop–helix proteins can associate on DNA raises the possibility that these two classes of myogenic transcriptional factors collaborate to induce muscle-specific transcription of MEF-2-dependent genes. In addition, certain MEF-2 sites can be bound by the mesodermal homeodomain protein MHox, which suggests that there may be cross talk between MEF-2 proteins and other regulators. Therefore, a common property of MADS proteins could be their ability to cooperate with other transcriptional regulators to control gene expression. The emerging view is that MADS box proteins provide specific regulatory activity via protein–protein interactions.

D. Cardiogenesis, an Nkx-2-Dependent Paradigm

An attractive hypothesis from the analysis of NK-2 homologs is that these homeodomain factors function in phylogenetically conserved pathways in muscle cell types that do not utilize the MyoD family (Fig. 9). Ectopic expression of Nkx2-5 in 10T1/2 fibroblasts demonstrated that downstream targets (i.e., *cardiac α-actin*) were not directly activated by Nkx2-5 alone but required the collaboration of additional factors, such as SRF (Chen and Schwartz, 1996). Whether the vertebrate *Nkx2-5* or other *Nkx2*-related genes in combination with *SRF* are sufficient to play the primary role in heart specification remains to be determined.

Inhibition of *GATA-4* expression by antisense transcripts blocks development of beating cardiac muscle

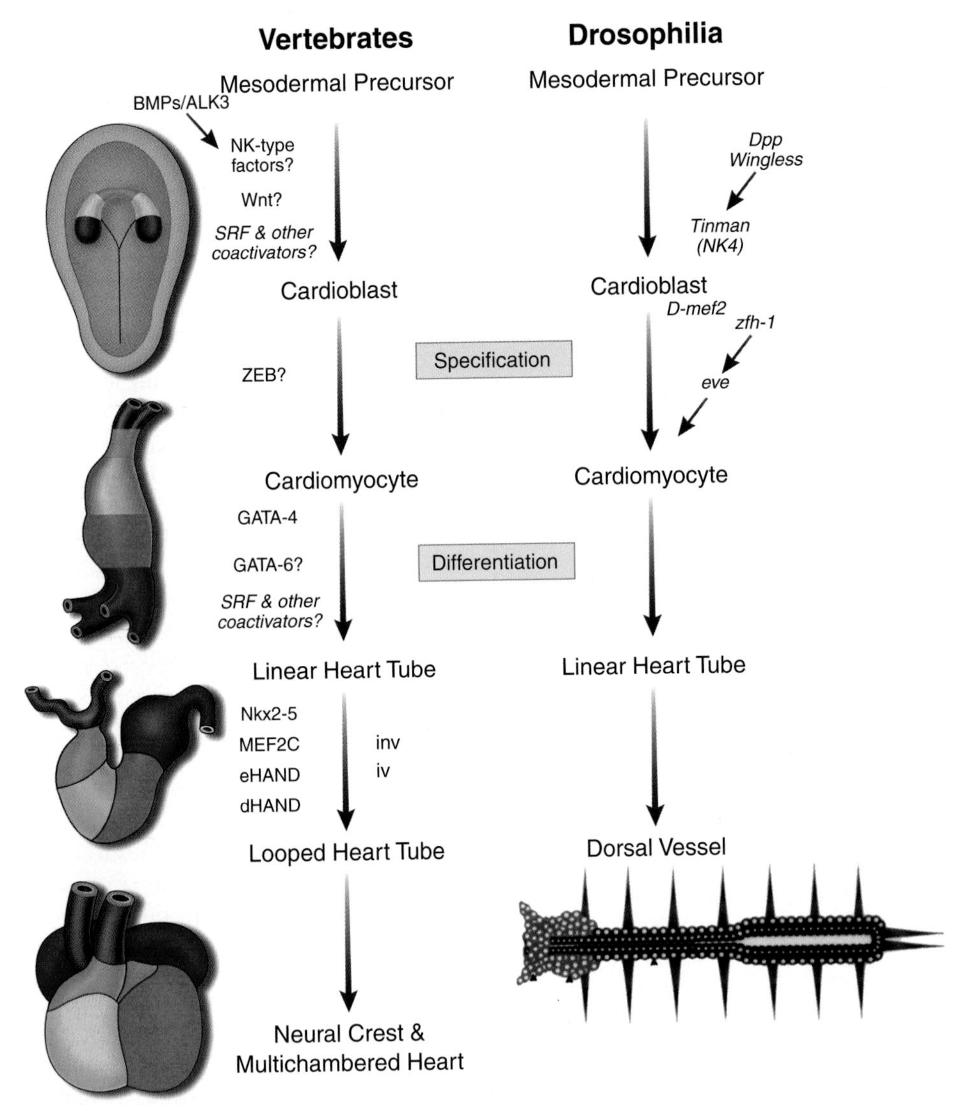

Figure 9 A phylogenetically conserved pathway for heart development. The developmental points at which *Drosophila* and vertebrate transcription factors and secretory proteins influence cardiac development have been indicated. Assignment of position was based on mutant phenotypes. Nuclear transcription factors and secretory signaling proteins followed by a question mark indicate potential regulatory factors. *BMP-2,* Zhang and Bradley (1996); *BMP-4,* Winnier *et al.* (1995); *dHAND* and *eHAND,* Srivastava *et al.* (1995); *GATA-4,* Molkentin *et al.* (1997) and Kuo *et al.* (1997); *MEF2C,* Lin *et al.* (1997); *Nkx2-5,* Lyons *et al.* (1995); del 22q11, Goldmuntz and Emanuel (1997); *iv,* Layton (1976); *inv,* Yokoyama *et al.* (1993); *D-mef2,* Bour *et al.* (1995), Lilly *et al.* (1995), and Ranganayakulu *et al.* (1995); *Dpp,* Frasch (1995); *wg,* Wu *et al.* (1995) and Park *et al.* (1996); *tinman,* Bodmer, (1993) and Azpiazu and Frasch (1993); *zfh-1,* Lai *et al.* (1993).

cells and interferes with expression of cardiac muscle markers in the pluripotent P19 embryonic carcinoma cell line (Grepin *et al.,* 1995), which can be differentiated into beating cardiac muscle cells (McBurney *et al.,* 1982). In contrast, cardiac myocytes are present in *GATA-4* knockout mice. However, the linear heart tube failed to form (Kuo *et al.,* 1997; Molkentin *et al.,* 1997; see Chapter 17). In addition, the right ventricle of the heart failed to form in *MEF-2C* knockout mice (Lin *et al.,* 1997). Thus, we believe that the zinc finger-containing GATA-4 factor (Arceci *et al.,* 1993; Grepin *et al.,* 1995) and MADS box family members related to *Drosophila* dMEF-2 (Lilly *et al.,* 1994) and vertebrate *MEF-2C* (Martin *et al.,* 1993) genes are also well suited to be high in the hierarchical order of regulatory factors that, in combination with Nkx2-5 and SRF, may specify the cardiac cell lineage. In fact, recent studies (Durocher *et al.,* 1997; Sepulveda *et al.,* 1998) indicate that Nkx2-5 and

GATA-4 form complexes that coactivate cardiac-specific genes, such as *atrial natruiretic factor* and *cardiac α-actin.*

E. Gene Therapy for Heart Disease

Differentiated cardiomyocytes are incapable of undergoing cell replication (McGill and Brooks, 1995). Thus, diseased or damaged cardiomyocytes cannot be replaced in the adult heart. In contrast, cardiac fibroblast are capable of cell replication and are present in large quantities. Thus, if cardiac fibroblast could be converted to cardiomyocytes through genetic manipulation, damaged heart tissue could be repaired as well as spare fibroblast wound repair or scaring. However, to date the expression of no single transcription factor has been able to convert a differentiated cell into a cardiac myocyte in a similar manner to the MyoD family of transcription factors in skeletal muscle (Davis *et al.*, 1987). Based on our Nkx2-5 and SRF cotransfection experiments, cardiac fibroblasts may be able to be converted into cardiomyocytes. However, we have not observed spontaneous contraction in cotransfected cells. Thus, it appears that additional factors must be expressed in order to obtain a complete cardiac myocyte phenotype. Alternatively, the master regulatory gene that controls cardiac myocyte determination remains to be identified.

Acknowledgments

This work was supported by Grants NIH R01 HL50422 and P01 HL49953. The authors thank Dr. Thomas Schultheiss for the chicken Nkx-2.5 *in situ* figures.

References

Affolter, M., Montagne, J., Walldorf, U., Groppe, J., Kloter, U. X., LaRosa, M., and Gehring, W. J. (1994). The Drosophila SRF homologue is expressed in a subset of tracheal cells and maps within a genomic region required for tracheal development. *Development* (*Cambridge, UK*) **120,** 743–753.

Arceci, R. J., King, A. A., Simon, M. C., Orkin, S. H., and Wilson, D. B. (1993). Mouse GATA-4: A retinoic acid-inducible GATA-binding transcription factor expressed in endodermally derived tissues and heart. *Mol. Cell. Biol.* **13,** 2235–2246.

Azpiazu, N., and Frasch, M. (1993). Tinman and bagpipe: Two homeo box genes that determine cell fates in the dorsal mesoderm of Drosophila. *Genes Dev.* **7,** 1325–1340.

Barth, K. A., and Wilson, S. W. (1995). Expression of zebrafish nk2.2 is influenced by sonic hedgehog/vertebrate hedgehog-1 and demarcates a zone of neuronal differentiation in the embryonic forebrain. *Development* (*Cambridge, UK*) **121,** 1755–1768.

Bodmer, R. (1993). The gene tinman is required for specification of the heart and visceral muscles in Drosophila. *Development* (*Cambridge, UK*) **118,** 719–729.

Bodmer, R., Jan, L. Y., and Jan, Y. N. (1990). A new homeobox-containing gene, msh-2, is transiently expressed early during mesoderm formation of Drosophila. *Development* (*Cambridge, UK*) **110,** 661–669.

Bohinski, R. J., Di Lauro, R., and Whitsett, J. A. (1994). The lung-specific surfactant protein B gene promoter is a target for thyroid transcription factor 1 and hepatocyte nuclear factor 3, indicating common factors for organ-specific gene expression along the foregut axis. *Mol. Cell. Biol.* **14,** 5671–5681.

Bour, B. A., O'Brien, M. A., Lockwood, W. L., Goldstein, E. S., Bodmer, R., Taghert, P. H., Abmayr, S. M., and Nguyen, H. T. (1995). *Drosophila MEF2,* a transcription factor that is essential for myogenesis. *Genes Dev.* **9,** 730–741.

Boxer, L. M., Miwa, T., Gustafson, T. A., and Kedes, L. (1989a). Identification and characterization of a factor that binds to two human sarcomeric actin promoters. *J. Biol. Chem.* **264,** 1284–1292.

Boxer, L. M., Prywes, R., Roeder, R. G., and Kedes, L. (1989b). The sarcomeric actin CArG-binding factor is indistinguishable from the c-fos serum response factor. *Mol. Cell. Biol.* **9,** 515–522.

Buchberger, A., Pabst, O., Brand, T., Seidl, K., and Arnold, H. H. (1996). Chick NKx-2.3 represents a novel family member of vertebrate homologues to the Drosophila homeobox gene tinman: Differential expression of cNKx-2.3 and cNKx-2.5 during heart and gut development. *Mech. Dev.* **56,** 151–163.

Chen, C. Y., and Schwartz, R. J. (1995). Identification of novel DNA binding targets and regulatory domains of a murine tinman homeodomain factor, nkx-2.5. *J. Biol. Chem.* **270,** 15628–15633.

Chen, C. Y., and Schwartz, R. J. (1996). Recruitment of the tinman homolog Nkx-2.5 by serum response factor activates cardiac alpha-actin gene transcription. *Mol. Cell. Biol.* **16,** 6372–6384.

Chen, C. Y., and Schwartz, R. J. (1997). Competition between negative acting YY1 versus positive acting serum response factor and tinman homologue NKx-2.5 regulates cardiac α-actin promoter activity. *Mol. Endo.* **11,** 812–822.

Chen, C. Y., Croissant, J., Majesky, M., Topouzis, S., McQuinn, T., Frankovsky, M. J., and Schwartz, R. J. (1996). Activation of the cardiac alpha-actin promoter depends upon serum response factor, Tinman homologue, Nkx-2.5, and intact serum response elements. *Dev. Genet.* **19,** 119–130.

Chen, J. N., and Fishman, M. C. (1996). Zebrafish tinman homolog demarcates the heart field and initiates myocardial differentiation. *Development* (*Cambridge, UK*) **122,** 3809–3816.

Chen, Y., Lebrun, J. J., and Vale, W. (1996). Regulation of transforming growth factor beta- and activin-induced transcription by mammalian Mad proteins. *Proc. Nat. Acad. Sci. U.S.A.* **93,** 12992–12997.

Chow, K. L., Hogan, M. E., and Schwartz, R. J. (1991). Phased cis-acting promoter elements interact at short distances to direct avian skeletal alpha-actin gene transcription. *Proc. Nat. Acad. Sci. U.S.A.* **88,** 1301–1305.

Civitareale, D., Lonigro, R., Sinclair, A. J., and Di Lauro, R. (1989). A thyroid-specific nuclear protein essential for tissue-specific expression of the thyroglobulin promoter. *EMBO J.* **8,** 2537–2542.

Croissant, J. D., Kim, J. H., Eichele, G., Goering, L., Lough, J., Prywes, R., and Schwartz, R. J. (1996). Avian serum response factor expression restricted primarily to muscle cell lineages is required for alpha-actin gene transcription. *Dev. Biol.* **177,** 250–264.

Cserjesi, P., Lilly, B., Bryson, L., Wang, Y., Sassoon, D. A., and Olson, E. N. (1992). MHox: A mesodermally restricted homeodomain protein that binds an essential site in the muscle creatine kinase enhancer. *Development* (*Cambridge, UK*) **115,** 1087–1101.

Dalton, S., and Treisman, R (1992). Characterization of SAP-1, a protein recruited by serum response factor to the c-fos serum response element. *Cell* (*Cambridge, Mass.*) **68,** 597–612.

Damante, G., and Di Lauro, R. (1991). Several regions of Antennapedia and thyroid transcription factor 1 homeodomains contribute to DNA binding specificity. *Proc. Nat. Acad. Sci. U.S.A.* **88,** 5388–5392.

Damante, G., Fabbro, D., Pellizzari, L., Civitareale, D., Guazzi, S., Polycarpou-Schwartz, M., Cauci, S., Quadrifoglio, F., Formisano, S., and Di Lauro, R. (1994). Sequence-specific DNA recognition by the thyroid transcription factor-1 homeodomain. *Nucleic Acids Res.* **22,** 3075–3083.

Davis, R. L., Weintraub, H., and Lassar, A. B. (1987). Expression of a single transfected cDNA converts fibroblasts to myoblasts. *Cell (Cambridge, Mass.)* **51,** 987–1000.

Dudley, A. T., and Robertson, E. J. (1997). Overlapping expression domains of bone morphogenetic protein family members potentially account for limited tissue defects in BMP7 deficient embryos. *Dev. Dyn.* **208,** 349–362.

Durocher, D., Chen, C. Y., Ardati, A., Schwartz, R. J., and Nemer, M. (1996). The atrial natriuretic factor promoter is a downstream target for Nkx-2.5 in the myocardium. *Mol. Cell. Biol.* **16,** 4648–4655.

Durocher, D., Charron, F., Warren, R., Schwartz, R. J., and Nemer, M. (1997). The cardiac transcription factors Nkx2-5 and GATA-4 are mutual cofactors. *EMBO J.* **16,** 5687–5696.

Evans, S. M., Yan, W., Murillo, M. P., Ponce, J., and Papalopulu, N. (1995). Tinman, a Drosophila homeobox gene required for heart and visceral mesoderm specification, may be represented by a family of genes in vertebrates: XNkx-2.3, a second vertebrate homologue of tinman. *Development (Cambridge, UK)* **121,** 3889–3899.

Frasch, M. (1995). Induction of visceral and cardiac mesoderm by ectodermal *Dpp* in the early *Drosophila* embryo. *Nature (London)* **374,** 464–467.

Gajewski, K., Kim, Y., Lee, Y. M., Olson, E. N., and Schulz, R. A. (1997). D-mef2 is a target for Tinman activation during Drosophila heart development. *EMBO J.* **16,** 515–522.

Garcia-Fernandez, J., Baguna, J., and Salo, E. (1991). Planarian homeobox genes: Cloning, sequence analysis, and expression. *Proc. Natl. Acad. Sci. U.S.A.* **88,** 7338–7342.

Gaunt, S. J. (1988). Mouse homeobox gene transcripts occupy different but overlapping domains in embryonic germ layers and organs: A comparison of Hox-3.1 and Hox-1.5. *Development (Cambridge, UK)* **103,** 135–144.

Goldmuntz, E., and Emanuel, B. S. (1997). Genetics disorders of cardiac morphogenesis: The DiGeorge and velocardiaofacial syndromes. *Circ. Res.* **80,** 437–443.

Graff, J. M. (1997). Embryonic patterning: To BMP or not to BMP, that is the question. *Cell (Cambridge, Mass.)* **89,** 171–174.

Grepin, C., Robitaille, L., Antakly, T., and Nemer, M. (1995). Inhibition of transcription factor GATA-4 expression blocks in vitro cardiac muscle differentiation. *Mol. Cell. Biol.* **15,** 4095–4102.

Grueneberg, D. A., Natesan, S., Alexandre, C., and Gilman, M. Z. (1992). Human and Drosophila homeodomain proteins that enhance the DNA-binding activity of serum response factor. *Science* **257,** 1089–1095.

Guazzi, S., Price, M., De Felice, M., Damante, G., Mattei, M. G., and Di Lauro, R. (1990). Thyroid nuclear factor 1 (TTF-1) contains a homeodomain and displays a novel DNA binding specificity. *EMBO J.* **9,** 3631–3639.

Guillemin, K., Groppe, J., Ducker, K., Treisman, R., Hafen, E., Affolter, M., and Krasnow, M. A. (1996). The pruned gene encodes the Drosophila serum response factor and regulates cytoplasmic outgrowth during terminal branching of the tracheal system. *Development (Cambridge, UK)* **122,** 1353–1362.

Gustafson, T. A., Miwa, T., Boxer, L. M., and Kedes, L. (1988). Interaction of nuclear proteins with muscle-specific regulatory sequences of the human cardiac alpha-actin promoter. *Mol. Cell. Biol.* **8,** 4110–4119.

Harvey, R. P. (1996). NK-2 homeobox genes and heart development. *Dev. Biol.* **178,** 203–216.

Herskowitz, I. (1989). A regulatory hierarchy for cell specialization in yeast. *Nature (London)* **342,** 749–757.

Hogan, B. L. (1996). Bone morphogenetic proteins in development. *Curr. Opin. Genet. Dev.* **6,** 432–438.

Hoodless, P. A., Haerry, T., Abdollah, S., Stapleton, M., O'Connor, M. B., Attisano, L, and Wrana, J. L. (1996). MADR1, a MAD-related protein that functions in BMP2 signaling pathways. *Cell (Cambridge, Mass.)* **85,** 489–500.

Hunt, P., and Krumlauf, R. (1992). Hox codes and positional specification in vertebrate embryonic axes. *Annu. Rev. Cell Biol.* **8,** 227–256.

Jagla, K., Frasch, M., Jagla, T., Dretzen, G., Bellard, F., and Bellard, M. (1997). *Ladybird,* a new component of the cardiogenic pathway in *Drosophila* required for diversification of heart precursors. *Development (Cambridge, UK)* **124,** 3471–3479.

Javaux, F., Bertaux, F., Donda, A., Francis-Lang, H., Vassart, G., Di-Lauro, R., and Christophe, D. (1992). Functional role of TTF-1 binding sites in bovine thyroglobulin promoter. *FEBS Lett.* **300,** 222–226.

Jimenez, F., Martin-Morris, L., Velasco, L., Chu, H., Sierra, J., Rosen, D., and White, K. (1995). *Vnd,* a gene required for early neurogenesis of *Drosophila,* encodes a homeodomain protein. *EMBO J.* **14,** 3487–3495.

Kelly, S., Bachurske, C., Burhans, M., and Glasser, S. (1996). Transcription of the lung-specific surfactant protein C gene is mediated by thyroid transcription Factor 1. *J. Biol. Chem.* **271,** 6881–6888.

Kessel, M., and Gruss, P. (1991). Homeotic transformations of murine vertebrae and concomitant alteration of Hox codes induced by retinoic acid. *Cell (Cambridge, Mass.)* **67,** 89–104.

Kim, Y., and Nirenberg, M. (1989). Drosophila NK-homeobox genes. *Proc. Natl. Acad. Sci. U.S.A.* **86,** 7716–7720.

Kimura, S., Hara, Y., Pineau, T., Fernandez-Salguero, P., Fox, C. H., Ward, J. M., and Gonzalez, F. J. (1996). The T/ebp null mouse: Thyroid-specific enhancer-binding protein is essential for the organogenesis of the thyroid, lung, ventral forebrain, and pituitary. *Genes Dev.* **10,** 60–69.

Komuro, I., and Izumo, S. (1993). Csx: A murine homeobox-containing gene specifically expressed in the developing heart. *Proc. Natl. Acad. Sci. U.S.A.* **90,** 8145–8149.

Kuo, C. T., Morrisey, E. E., Anandappa, R., Sigrist, K., Lu, M. M. X., Parmacek, M. S., Soudais, C., and Leiden, J. M. (1997). GATA4 transcription factor is required for ventral morphogenesis and heart tube formation. *Genes Dev.* **11,** 1048–1060.

Lai, Z., Rushton, E., Bate, M., and Rubin, G. M. (1993). Loss of function of the *Drosophila zfh-1* gene results in abnormal development of mesodermally derived tissues. *Proc. Natl. Acad. Sci. U.S.A.* **90,** 4122–4126.

Layton, W. M. (1976). Random determination of a developmental process: Reversal of normal visceral asymmetry in the mouse. *J. Hered.* **67,** 336–338.

Lazzaro, D., Price, M., De Felice, M., and Di Lauro, R. (1991). The transcription factor TTF-1 is expressed at the onset of thyroid and lung morphogenesis and in restricted regions of the foetal brain. *Development (Cambridge, UK)* **113,** 1093–1104.

Lee, K. H., Xu, Q., and Breitbart, R. E. (1996). A new tinman-related gene, nkx2.7, anticipates the expression of nkx2.5 and nkx2.3 in zebrafish heart and pharyngeal endoderm. *Dev. Biol.* **180,** 722–731.

Lee, T. C., Chow, K. L., Fang, P., and Schwartz, R. J. (1991). Activation of skeletal alpha-actin gene transcription: The cooperative formation of serum response factor-binding complexes over positive cis-acting promoter serum response elements displaces a negative-acting nuclear factor enriched in replicating myoblasts and nonmyogenic cells. *Mol. Cell. Biol.* **11,** 5090–5100.

Lee, T. C., Shi, Y., and Schwartz, R. J. (1992). Displacement of BrdUrd-induced YY1 by serum response factor activates skeletal alpha-acting transcription in embryonic myoblasts. *Proc. Natl. Acad. Sci. U.S.A.* **89,** 9814–9818.

Leussink, B., Brouwer, A., Khattabi, M., Poelmann, R., Gittenberger-de Groot, A., and Meijlink, F. (1995). Expression potterns of the paired-related homeobox genes *Mhox/Prxl* and *S8/Prx2* suggest roles in development of the heart and the forebrain. *Mech. Dev.* **52,** 51–64.

Licht, J. D., Grossel, M. J., Figge, J., and Hansen, U. M. (1990). Drosophila Kruppel protein is a transcriptional repressor. *Nature (London)* **346,** 76–79.

Lilly, B., Galewsky, S., Firulli, A. B., Schulz, R. A., and Olson, E. N. (1994). D-MEF2: A MADS box transcription factor expressed in differentiating mesoderm and muscle cell lineages during Drosophila embryogenesis. *Proc. Nat. Acad. Sci. U.S.A.* **91,** 5662–5666.

Lilly, B., Zhao, B., Rangenayakula, G., Patterson, B. M., Schulz, R. A., and Olson, E. N. (1995). Requirement of MADS domain transcription factor D-MEF2 for muscle formation in *Drosophila. Science* **276,** 688–693.

Lin, Q., Schwarz, J., Bucana, C., and Olson, E. N. (1997). Control of mouse cardiac morphogenesis and myogenesis by transcription factor MEF2C. *Science* **276,** 1404–1407.

Lints, T. J., Parsons, L. M., Hartley, L., Lyons, I., and Harvey, R. P. (1993). Nkx-2.5: A novel murine homeobox gene expressed in early heart progenitor cells and their myogenic descendants. *Development (Cambridge, UK)* **119,** 419–431.

Lough, J., Barron, M., Brogley, M., Sugi, Y., Bolender, D. L., and Zhu, X. (1996). Combined BMP-2 and FGF-4, but neither factor alone, induces cardiogenesis in non-precardiac embryonic mesoderm. *Dev. Biol.* **178,** 198–202.

Lyons, I., Parsons, L. M., Hartley, L., Li, R., Andrews, J. E., Robb, L., and Harvey, R. P. (1995). Myogenic and morphogenetic defects in the heart tubes of murine embryos lacking the homeo box gene Nkx2-5. *Genes Dev.* **9,** 1654–1666.

Martin, J. F., Schwarz, J. J., and Olson, E. N. (1993). Myocyte enhancer factor (MEF) 2C: A tissue-restricted member of the MEF-2 family of transcription factors. *Proc. Natl. Acad. Sci. U.S.A.* **90,** 5282–5286.

McBurney, M. W., Jones-Villeneuve, E. M., Edwards, M. K., and Anderson, P. J. (1982). Control os muscle and neuronal differentiation in a cultured embryonal carcinoma cell line. *Nature (London)* **99,** 165–167.

McGill, C. J., and Brooks, G. (1995). Cell cycle control mechanisms and their role in cardiac growth. *Cardiovasc. Res.* **30,** 557–569.

Minty, A., and Kedes, L. (1986). Upstream regions of the human cardiac actin gene that modulate its transcription in muscle cells: Presence of an evolutionarily conserved repeated motif. *Mol. Cell. Biol.* **6,** 2125–2136.

Mizuno, K., Gonzalez, F. J., and Kimura, S. (1991). Thyroid-specific enhancer-binding protein (T/EBP): cDNA cloning, functional characterization, and structural identity with thyroid transcription factor TTF-1. *Mol. Cell. Biol.* **11,** 4927–4933.

Mohun, T. J., Taylor, M. V., Garrett, N., and Gurdon, J. B. (1989). The CArG promoter sequence is necessary for muscle-specific transcription of the cardiac actin gene in Xenopus embryos. *EMBO J.* **8,** 1153–1161.

Mohun, T. J., Chambers, A. E., Towers, N., and Taylor, M. V. (1991). Expression of genes encoding the transcription factor SRF during early development of Xenopus laevis: Identification of a CArG box-binding activity as SRF. *EMBO J.* **10,** 933–940.

Molkentin, J. D., Black, B. L., Martin, J. F., and Olson, E. N. (1995). Cooperative activation of muscle gene expression by MEF2 and myogenic bHLH proteins. *Cell* **83,** 1125–1136.

Molkentin, J. D., Lin, Q., Duncan, S. A., and Olson, E. N. (1997). Requirement of the transcription factor GATA4 for heart tube formation and ventral morphogenesis. *Genes Dev.* **11,** 1061–1072.

Nardelli-Haefliger, D., and Shankland, M. (1993). Lox10, a member of the NK-2 homeobox gene class, is expressed in a segmental pattern in the endoderm and in the cephalic nervous system of the leech Helobdella. *Development (Cambridge, UK)* **118,** 877–892.

Niehrs, C. (1996). Growth factors. Mad connection to the nucleus. *Nature (London)* **381,** 561–562.

Norman, C., Runswick, M., Pollock, R., and Treisman, R. (1988). Isolation and properties of cDNA clones encoding SRF, a transcription factor that binds to the c-fos serum response element. *Cell (Cambridge, Mass.)* **55,** 989–1003.

Okkema, P. G., and Fire, A. (1994). The Caenorhabditis elegans NK-2 class homeoprotein CEH-22 is involved in combinatorial activation of gene expression in pharyngeal muscle. *Development (Cambridge, UK)* **120,** 2175–2186.

Olson, E. N., Perry, M., and Schulz, R. A. (1995). Regulation of muscle differentiation by the MEF2 family of MADS box transcription factors. *Dev. Biol.* **172,** 2–14.

Opstelten, D.-J. E., Volgels, R., Robert, B., Kalkhoven, E., Zwartkruis, F., DeLaff, L., Destree, O. H., Deschamps, J., Lawson, K. A., and Meijlink, F. (1991). The mouse homeobox gene, S8, is expressed during embryogenesis predominantly in mesenchyme. *Mech. Dev.* **34,** 29–42.

Park, M., Wu, X., Golden, K., Axelrod, J. D., and Bodmer, R. (1996). The wingless signaling pathway is directly involved in Drosophila heart development. *Dev. Biol.* **177,** 104–116.

Pellegrini, L., Tan, S., and Richmond, J. S. (1995). Structure of serum response factor core bound to DNA. *Nature (London)* **376,** 490–498.

Pollock, R., and Treisman, R. (1991). Human SRF-related proteins: DNA-binding properties and potential regulatory targets. *Genes Dev.* **5,** 2327–2341.

Price, M., Lazzaro, D., Pohl, T., Mattei, M. G., Ruther, U., Olivo, J. C., Duboule, D., and Di Lauro, R. (1992). Regional expression of the homeobox gene Nkx-2.2 in the developing mammalian forebrain. *Neuron* **8,** 241–255.

Ranganayakulu, G., Schulz, R. A., and Olson, E. N. (1995). A series of mutation in the D-MEF2 transcription forctor reveals multiple functions in larval and adult myogenesis in *drosophila. Dev. Biol.* **171,** 169–181.

Ray, M. K., Chen, C. Y., Schwartz, R. J., and DeMayo, F. J. (1996). Transcriptional regulation of a mouse Clara cell-specific protein (mCC10) gene by the NKx transcription factor family members thyroid transcription factor 1 and cardiac muscle-specific homeobox protein (CSX). *Mol. Cell. Biol.* **16,** 2056–2064.

Reecy, J. M., Yamada, M., Cummings, K., Sosic, D., Chen, C.-Y., Eichele, G., Olson, E. N., and Schwartz, R. J. (1997). Chicken Nkx-2.8: A novel homeobox gene expressed in early heart progenitor cells and pharyngeal pouch-2 and -3 endoderm. *Dev. Biol.* **188,** 295–311.

Ruzicka, D. L., and Schwartz, R. J. (1988). Sequential activation of alpha-actin genes during avian cardiogenesis: Vascular smooth muscle alpha-actin gene transcripts mark the onset of cardiomyocyte differentiation. *J. Cell Biol.* **107,** 2575–2586.

Saha, M. S., Michel, R. B., Gulding, K. M., and Grainger, R. M. (1993). A Xenopus homeobox gene defines dorsal-ventral domains in the developing brain. *Development (Cambridge, UK)* **118,** 193–202.

Saiardi, A., Tassi, V., De Fillipis, V., and Civitareale, D. (1995). Cloning and sequence analysis of human thyroid transcription factor 1. *Biochim. Biophys. Acta* **1261,** 307–310.

Sciavolino, P., Abrams, E., Yang, L., Austenberg, L., Shen, M., and Abate-Shen, C. (1997). Tissue-specific expression of murine *Nkx3.1* in the male urogenital system. *Dev. Dyn.* **209,** 127–138.

Schroter, H., Mueller, C. G., Meese, K., and Nordheim, A. (1990). Synergism in ternary complex formation between the dimeric glycoprotein p67SRF, polypeptide p62TCF and the c-fos serum response element. *EMBO J.* **9,** 1123–1130.

Schultheiss, T. M., Xydas, S., and Lassar, A. B. (1995). Induction of avian cardiac myogenesis by anterior endoderm. *Development (Cambridge, UK)* **121,** 4203–4214.

Schultheiss, T. M., Burch, J. B., and Lassar, A. B. (1997). A role for bone morphogenetic proteins in the induction of cardiac myogenesis. *Genes Dev.* **11,** 451–462.

Scott, M. P., Tamkun, J. W., and Hartzell, G. W., 3rd (1989). The structure and function of the homeodomain. *Biochim. Biophys. Acta* **989,** 25–48.

Sekelsky, J. J., Newfeld, S. J., Raftery, L. A., and Chartoff, E. H. X., and Gelbart, W. M. (1995). Genetic characterization and cloning of mothers against dpp, a gene required for decapentaplegic function in Drosophila melanogaster. *Genetics* **139,** 1347–1358.

Sepulveda J. L., Belaguli, N., Nigam, V., Chen, C. Y., Nemer, M., and Schwartz, R. J. (1998). GATA-4 and NKx-2.5 coactivate NKx-2 DNA binding targets: Role for regulating early cardiac gene expression. *Mol. Cell. Biol.* **18,** 3405–3415.

Sommer, H., Beltran, J. P., Huijser, P., Pape, H., Lonnig, W. E. X., Saedler, H., and Schwarz-Sommer, Z. (1990). Deficiens, a homeotic gene involved in the control of flower morphogenesis in Antirrhinum majus: The protein shows homology to transcription factors. *EMBO J.* **9,** 605–613.

Soulez, M., Tuil, D., Kahn, A., and Gilgenkrantz, H. (1996). The serum response factor (SRF) is needed for muscle-specific activation of CArG boxes. *Biochem. Biophys. Res. Commun.* **219,** 418–422.

Srivastava, D., Cserjesi, P., and Olson, E. N. (1995). A subclass of bHLH protein required for cardiac morphogenesis. *Science* **270,** 1995–1999.

Stadler, H. S., Murray, J. C., Leysens, N. J., Goodfellow, P. J., and Solursh, M. (1995). Phylogenetic conservation and physical mapping of members of the H6 homeobox gene family. *Mamm. Genome* **6,** 383–388.

Thomsen, G. H. (1996). Xenopus mothers against decapentaplegic is an embryonic ventralizing agent that acts downstream of the BMP-2/4 receptor. *Development (Cambridge, UK)* **122,** 2359–2366.

Tonissen, K. F., Drysdale, T. A., Lints, T. J., Harvey, R. P., and Krieg, P. A. (1994). XNkx-2.5, a Xenopus gene related to Nkx-2.5 and tinman: Evidence for a conserved role in cardiac development. *Dev. Biol.* **162,** 325–328.

Treisman, R. (1986). Identification of a protein-binding site that mediates transcriptional response of the c-fos gene to serum factors. *Cell (Cambridge, Mass.)* **46,** 567–574.

Tribioli, C., Frasch, M., and Lufkin, T. (1997). *Bapx1:* An evolutionary conserved homologue of the *Drosophila bagpipe* homeobox gene is expressed in splanchnic mesoderm and the ambryonic skeleton. *Mech. Dev.* **65,** 145–162.

Vandromme, M., Gauthier-Rouviere, C., Carnac, G., and Lamb, N. X., and Fernandez, A. (1992). Serum response factor p67SRF is expressed and required during myogenic differentiation of both mouse C2 and rat L6 muscle cell lines. *J. Cell Biol.* **118,** 1489–1500.

Van Renterghem, P., Dremier, S., Vassart, G., and Christophe, D. (1995). Study of TTF-1 gene expression in dog thyrocytes in primary culture. *Mol. Cell. Endocrinol.* **112,** 83–93.

Walsh, K., and Schimmel, P. (1987). Two nuclear factors compete for the skeletal muscle actin promoter. *J. Biol. Chem.* **262,** 9429–9432.

Walsh, K., and Schimmel, P. (1988). DNA-binding site for two skeletal actin promoter factors is important for expression in muscle cells. *Mol. Cell. Biol.* **8,** 1800–1802.

Winnier, G., Blessing, M., Labosky, P. A., and Hogan, B. L. (1995). Bone morphogenetic protein-4 is required for mesoderm formation and patterning in the mouse. *Genes Dev.* **9,** 2105–2116.

Wu, X., Golden, K., and Bodmer, R. (1995). Heart development in *Drosophila* requires the segment polarity gene *wingless. Dev. Biol.* **169,** 619–628.

Yokoyama, T., Copeland, N. G., Jenkins, N. A., Montgomery, C. A., Elder, F. F., and Overbeek, P. A. (1993). Reversal of left-right asymmetry: A *situs inversus* mutation. *Science* **260,** 679–682.

Yu, Y. T., Breitbart, R. E., Smoot, L. B., Lee, Y., Mahdavi, V., and Nadal-Ginard, B. (1992). Human myocyte-specific enhancer factor 2 comprises a group of tissue-restricted MADS box transcription factors. *Genes Dev.* **6,** 1783–1798.

Zhang, H., and Bradley, A. (1996). Mice deficient for BMP2 are nonviable and have defects in amnion/chorion and cardiac development. *Development (Cambridge, UK)* **122,** 2977–2986.

17

GATA Transcription Factors and Cardiac Development

Michael S. Parmacek and Jeffrey M. Leiden

Departments of Medicine and Pathology, University of Chicago, Chicago, Illinois 60637

I. Introduction

The development of the mammalian heart is a complex process that involves the specification of multiple cell lineages, including endocardial cells, cardiomyocytes, and cells of the coronary vasculature, and their subsequent precisely orchestrated assembly into the mature four-chambered cardiac structure. Despite its relative complexity, the major steps of cardiac development take place over a remarkably short period of early embryonic development, beginning at approximately Embryonic Day (E) 7.5 and culminating by E11 in the mouse (Fishman and Chien, 1997; Kuo *et al.,* 1997; Olson and Srivastava, 1996). The correct assembly of a beating heart and the concomitant establishment of the fetal circulation is required for the viability of the early mouse embryo. Defects in cardiovascular morphogenesis result in embryonic lethality between E8.5 and 12 or, if less severe, in congenital heart disease.

For descriptive purposes, cardiac development can be divided into three temporally and spatially distinct stages (Kuo *et al.,* 1997). The first stage, which begins at approximately E7.0 in the mouse, involves the specification of the cardiomyocyte lineage(s) from cells of the splanchnic mesoderm (see Chapter 1). At E7.0, the formation of the intraembryonic coelom splits the embryonic mesoderm into splanchnic and somatic components (Fig. 1, top). In response to signals from the underlying endoderm, a crescent-shaped portion of the splanchnic mesoderm located in the anterior and dorsolateral region of the embryo then differentiates into cuboidal precardiac mesoderm or procardiomyocytes. These cells express contractile proteins but fail to assemble these proteins into myofibrils and therefore do not contract. The second stage of cardiac development begins at approximately E8.0 and involves the migration of the specified procardiomyocytes from the dorsal and anterior regions of the embryo to the ventral midline to form a linear heart tube (Fig. 1, middle). This procardiomyocyte migration is intimately linked to a complex series of morphogenic events that together form many of the ventral structures of the developing embryo, including the anterior intestinal portal and foregut, the pericardial cavity and heart tube, and the ventral closure of the yolk sac (Fig. 1, bottom). Conceptually, this process can be viewed as involving two folding events: (i) the rostral-to-caudal folding of head and adjacent precardiac mesoderm which positions the precardiac mesoderm caudal to the head fold, and (ii) the lateral-to-ventral folding and subsequent midline fusion of the precardiac splanchnic mesoderm which positions the heart tube in the ventral midline of the developing thoracic cavity (Kuo *et al.,* 1997). By E8.5, the linear heart tube becomes lined by endocardial cells and attached to the dorsal wall of the pericardial cavity by the dorsal mesocardium. The third stage of cardiac development occurs between E8.5 and 11 as the linear heart tube undergoes looping and septation to generate the mature four-chambered cardiac structure.

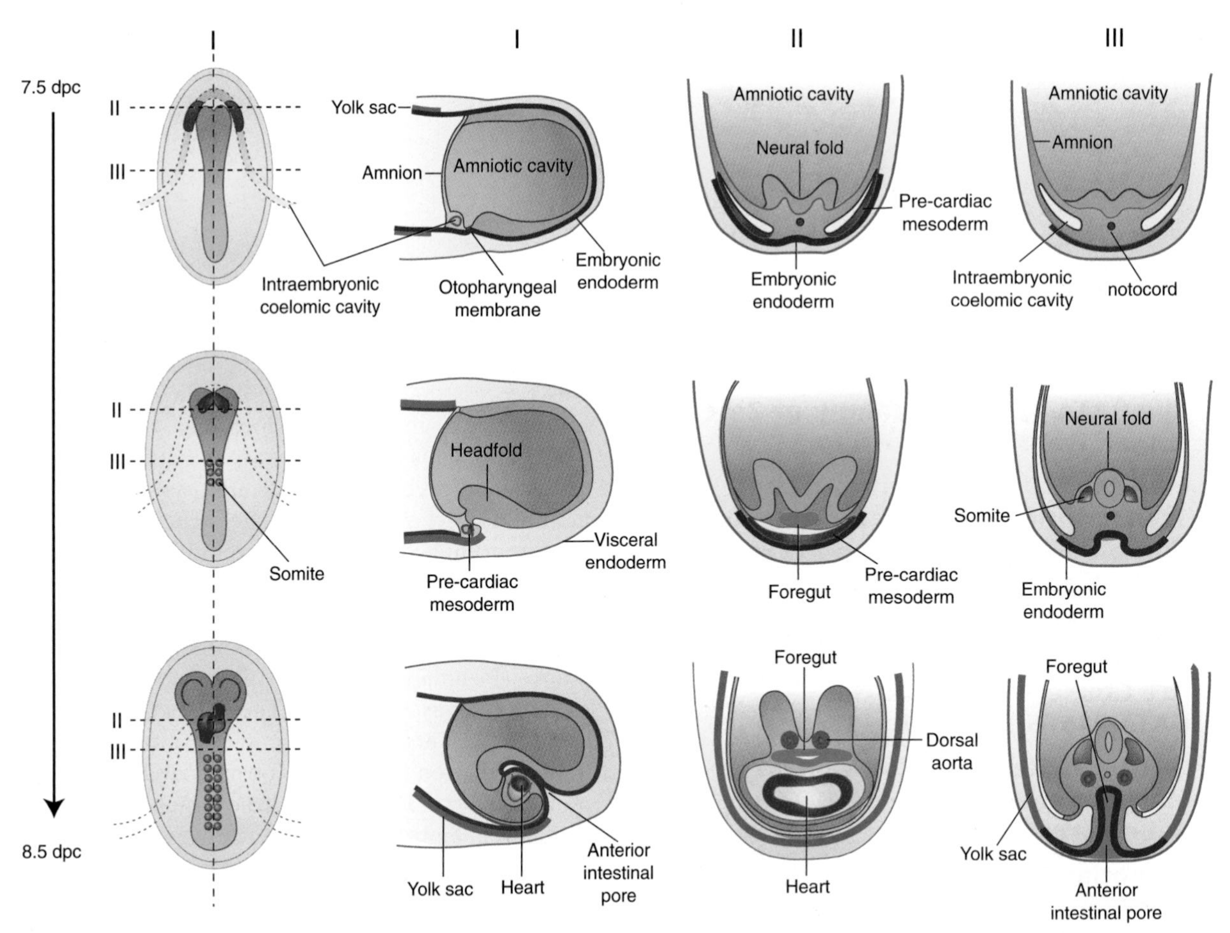

Figure 1 Schematic representation of heart tube formation during murine embryogenesis. (Left) Dorsal view of the developing mouse embryo between 7.5 (top) and 8.5 (bottom) days pc. I, midline sagittal sections of the same embryos; II, transverse sections through the same embryos at the level of the developing heart; III, transverse sections through the same embryos at the level of the developing foregut. The precardiac mesoderm and linear heart tube are shown in red and the somites are shown in solid black. The neural tube is shown by the vertical blue shading. The arrows indicate the direction of embryo folding.

By analogy with other developmental processes, it was reasonable to assume that specific sets of nuclear transcription factors regulate each stage of cardiac development. During the past 5 years, several different approaches have been used to identify such transcription factors. First, a number of groups isolated and characterized a series of cardiac-specific transcriptional regulatory elements and used these elements to identify cardiac-restricted DNA-binding proteins (Amacher *et al.*, 1993; Argentin *et al.*, 1994; Biben *et al.*, 1994, 1996; Cserjesi *et al.*, 1994; Donoviel *et al.*, 1996; Grepin *et al.*, 1994; Iannello *et al.*, 1991; Ip *et al.*, 1994; Knotts *et al.*, 1994; Larkin *et al.*, 1996; Lee *et al.*, 1994; Molkentin *et al.*, 1994, 1996; Navankasattusas *et al.*, 1992; Parmacek *et al.*, 1992; Sartorelli *et al.*, 1990; Seidman *et al.*, 1988; Subramaniam *et al.*, 1991; Wanker *et al.*, 1996; Zhu *et al.*, 1991, 1993). In a complementary approach, novel cardiac transcription factors were isolated on the basis of their sequence similarities to proteins that were known to play important roles in muscle development in mammals, reptiles, and insects (Arceci *et al.*, 1993; Bodmer, 1993; Kelley *et al.*, 1993; Komuro and Izumo, 1993; Laverriere *et al.*, 1994; Lints *et al.*, 1993; Morrisey *et al.*, 1996; Srivastava *et al.*, 1997; Tonissen *et al.*, 1994). Together, these studies have identified several families of transcription factors as potential regulators of cardiac development, including the GATA zinc finger proteins (Arceci *et al.*, 1993; Kelley *et al.*, 1993; Laverriere *et al.*, 1994; Morrisey *et al.*, 1996), the MEF2 and SRF family of MADS box proteins (Cserjesi *et al.*, 1994; Gossett *et al.*, 1989; Lee *et al.*, 1994; Pollock and Treisman, 1991; Zhu *et al.*, 1991; 1993), the bHLH proteins, dHAND and eHAND (Srivastava *et al.*, 1997), homeodomain proteins of the Nkx family (Bodmer, 1993; Komuro and Izumo, 1993; Lints *et al.*, 1993; Tonissen *et al.*, 1994), and TEA domain proteins such as TEF-1 (Azakie *et al.*, 1996; Farrance and Ordahl, 1996; Mar and Ordahl, 1990; Stewart *et al.*, 1994; Xiao *et al.*, 1991). Each of these proteins has been shown to bind to one or more cardiac-specific transcriptional elements and to be expressed in the developing mammalian heart.

An understanding of the precise role of each of these transcription factors in regulating cardiac development has only recently started to emerge from the results of gene targeting studies in flies and mice. Thus, for example, we now know that proper looping and septation of the linear heart tube to form a four-chambered heart requires Nkx2.5, dHAND, and eHAND (Biben and Harvey, 1997; see Chapters 7 and 9). However, these proteins are not required for either cardiac myocyte determination or the assembly of the linear heart tube because both processes occur normally in mice containing targeted mutations of the *Nkx2.5, eHAND,* and *dHAND* genes. Similarly, *MEF2C* appears to be required for looping morphogenesis of the embryonic heart and/or formation of the right ventricle (Lin *et al.*, 1997; Ross *et al.*, 1996; see Chapter 8).

In this chapter, we summarize the current understanding of the role of three GATA family transcription factors, GATA-4, -5, and -6, in cardiovascular development. Although each of these proteins is expressed in the developing cardiovascular system, definitive evidence concerning their roles in cardiac development, as determined from gene targeting studies, is available only for GATA-4 and GATA-5. Moreover, the question of potential redundancies between the proteins has not yet been definitively addressed by producing mice lacking two or more GATA family members. Thus, much remains to be learned about the role of these transcription factors in cardiovascular development and this review should, accordingly, be viewed as a work in progress rather than as a completed story. Many investigators have contributed to our understanding of the role of GATA proteins in mammalian development. In some cases it has not been possible to include all the relevant citations. We apologize for any inadvertent omissions of specific references.

II. The GATA Family of Zinc Finger Transcription Factors

Members of the GATA family of zinc finger transcription factors play key roles in transducing nuclear events that modulate cell lineage differentiation during vertebrate development (Orkin, 1992; Simon, 1995; Weiss and Orkin, 1995). To date, six related vertebrate GATA proteins have been identified, each of which is expressed in a developmentally regulated lineage-restricted fashion (Arceci *et al.*, 1993; Dorfman *et al.*, 1992; Evans and Felsenfeld, 1988; Ho *et al.*, 1991; Kelley *et al.*, 1993; Ko *et al.*, 1991; Laverriere *et al.*, 1994; Morrisey *et al.*, 1996; 1997a; Tsai *et al.*, 1989; Wilson *et al.*, 1990). All GATA proteins contain a conserved Cys-X_2-Cys-X_{17}-Cys-X_2-Cys type IV zinc finger DNA-binding domain that recognizes and binds to the consensus motif (A/T-G-A-T-A-A/G) (Ko and Engel, 1993; Mericka and Orkin, 1993; Omichinski *et al.*, 1993; Yang *et al.*, 1994). Previous studies have demonstrated that the C-terminal zinc finger and adjacent basic domain of GATA-1 are required for sequence-specific DNA-binding activity, whereas the N-terminal finger increases the affinity of GATA-1 for its cognate binding motif (Martin *et al.*, 1990; Trainor *et al.*, 1996). The zinc fingers of GATA-1 also mediate homo- and heterodimerization with other transcriptional activators, including Sp1,

EKLF, and RBTN2 (Crossley *et al.,* 1995; Merika and Orkin, 1995; Osada *et al.,* 1995; Yang and Evans, 1995).

GATA-1, -2, and -3 are expressed in overlapping subsets of hematopoietic cells and their precursors as well as in several nonhematopoietic tissues (Table I). GATA-1 is expressed in mature erythrocytes, multipotent hematopoietic progenitor cells, megakaryocytes, mast cells, and the Sertoli cells of the testes (Martin *et al.,* 1990; Romeo *et al.,* 1990; Yamamoto *et al.,* 1990). GATA-2 is expressed in hematopoietic stem cells and progenitors, immature erythroid cells, mast cells, megakaryocytes, endothelial cells, and the developing brain (Dorfman *et al.,* 1992; Zon *et al.,* 1993). GATA-3 is expressed in T lymphocytes and in specific regions of the central and peripheral nervous systems (Ho *et al.,* 1991). Functionally important GATA sites have been identified in transcriptional regulatory elements that control the expression of erythroid, lymphoid, myeloid, and endothelial cell-specific genes (Weiss and Orkin, 1995).

Much has been learned about the function of GATA-1, -2, and -3, respectively, in controlling differentiation of the hematopoietic cell lineages from pluripotent stem cells through gene targeting experiments (Simon, 1995; Weiss and Orkin, 1995). GATA-1 is required for normal erythroid development (Pevny *et al.,* 1991; Simon *et al.,* 1992). Primitive erythroid precursors are not produced in *GATA-1*$^{-/-}$ mice (Pevny *et al.,* 1991) and *GATA-1*$^{-/-}$ embryonic stem (ES) cells do not contribute to the mature erythroid compartment of chimeric mice (Weiss *et al.,* 1994). Disruption of the *GATA-2* locus results in a global defect in hematopoiesis with severe quantitative defects in both primitive and definitive erythropoiesis as well as defects in myelopoiesis and lymphopoiesis, suggesting that GATA-2 plays a critical role in differentiation of an early hematopoietic progenitor or stem cell (Tsai *et al.,* 1994). GATA-3 is required for the development of the T cell lineage and also appears to control *Th2* cytokine gene expression (Ting *et al.,* 1996; Zheng and Flavell, 1997). In addition, *GATA-3*$^{-/-}$ embryos exhibit severe deformities of the spinal cord and brain (Pandolfi *et al.,* 1995). Taken together, these studies demonstrated that although GATA-1, -2, and -3 are developmentally coexpressed in overlapping subsets of hematopoietic cells, each protein subserves a unique function in the developing embryo. These functions include (i) activation of lineage-specific target genes, (ii) distinct steps in restricting the developmental potential of multipotent hematopoietic stem cells, and (iii) direct or indirect regulation of programmed cell death during hematopoietic development. Of note, these studies also suggest that GATA-1, -2, and -3 may be functionally redundant with respect to the activation of lineage-specific genes in cells that coexpress more than one family member.

Table I Cellular Distribution of GATA Family Members

Family member	Tissue distribution
GATA-1	Erythroid, mast, megakaryocytic lineage Hematopoietic progenitor cells Testes
GATA-2	Mast, megakaryocytic lineages Early erythroid cells, endothelial cells Hematopoietic progenitor cells Embryonic brain
GATA-3	T lymphocytes, endothelial cells Embryonic brain and adult central and peripheral nervous systems Placenta, kidney, adrenal gland
GATA-4	Embryonic and adult heart Gut epithelium, embryonic liver Testes, ovaries
GATA-5	Embryonic heart Gut epithelium Embryonic lung (mesenchyme and bronchial SMCs) Embryonic urogenital ridge and bladder SMCs
GATA-6	Embryonic and adult heart Gut epithelium Vascular SMCs Embryonic lung (bronchial epithelium) Embryonic urogenital ridge and bladder SMCs

III. The GATA-4/5/6 Subfamily of Zinc Finger Transcription Factors

Following the isolation and characterization of GATA-1, -2, and -3, cDNA cross-hybridization studies identified three previously undescribed vertebrate GATA factors that were named GATA-4, -5, and -6 (Arceci *et al.,* 1993; Heikinheimo *et al.,* 1994; Kelley *et al.,* 1993; Laverriere *et al.,* 1994; Morrisey *et al.,* 1996, 1997a; Tamura *et al.,* 1993). Each of these proteins contains two type IV zinc fingers that are related closely to those of GATA-1, -2, and -3. Moreover, GATA-4, -5, and -6, but not GATA-1, -2, and -3, share low-level amino acid sequence identity across regions located within their N termini (Morrisey *et al.,* 1997b). Preliminary characterization revealed that the genes encoding GATA-4, -5, and -6 are expressed in an overlapping pattern within the precardiac mesoderm, heart, and gut epithelium (Arceci *et al.,* 1993; Kelley *et al.,* 1993; Laverriere *et al.,* 1994) (Table I). Based on their amino acid sequence identities, and their overlapping patterns of expression, the *GATA-4, -5,* and *-6* genes have been subclassified as a separate subfamily of GATA factors (Laverriere *et al.,* 1994).

Detailed analyses of the temporal and spatial patterns of expression of *GATA-4, -5,* and *-6* gene expression in staged murine embryos revealed that each gene displays a unique pattern of expression during vertebrate development (Heikinheimo *et al.,* 1994; Morrisey *et al.,* 1996, 1997a). In the early mouse embryo, *GATA-6* transcripts are first detectable within the visceral endoderm at the advanced egg cylinder stage (E6.5) (E. Morrisey and M. Parmacek, unpublished observation). At E7.5, the *GATA-4* and *-6* genes are both expressed in the primitive streak mesoderm, the mesoderm subjacent to the headfold region (which gives rise to the cardiogenic plate), and the extraembryonic visceral and parietal endoderm. In contrast, between E7.0 and 8.0, the *GATA-5* gene is expressed in a more spatially restricted fashion within the precardiac mesoderm that is limited to the cardiogenic plate (Heikinheimo *et al.,* 1994; Morrisey *et al.,* 1996, 1997a).

At E9.5, concomitant with the onset of regular beating of the primitive heart and connection of the embryonic and yolk sac vasculatures, the *GATA-4, -5,* and *-6* genes are expressed in both endocardium and myocardium of the primitive atria, ventricle, and truncus arteriosus, or cardiac outflow tract (Figs. 2A–2C). In addition, all three genes are expressed in the underlying septum transversum, which gives rise to cardiac myocytes and cells of the embryonic liver (Figs. 2A–2C). At midgestation (E12.5), the *GATA-4* and −*6* genes continue to be expressed throughout the embryonic heart (Figs. 2D and 2F). In contrast, *GATA-5* mRNA is ob-

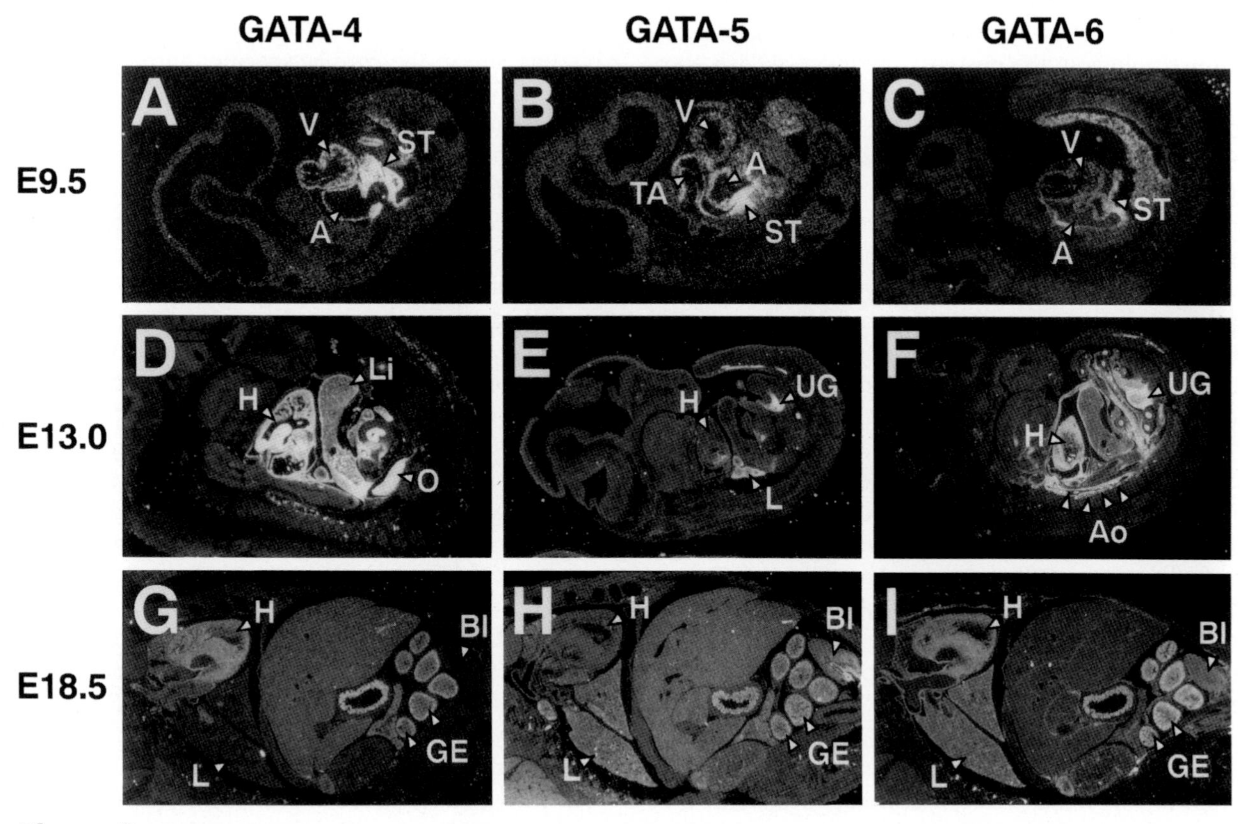

Figure 2 The temporal and spatial patterns of *GATA-4, -5,* and *-6* gene expression during murine embryonic development. *In situ* hybridization analyses were performed using the control GATA-4 antisense riboprobe (A, D, G), the GATA-5 antisense riboprobe (B, E, H), or the GATA-6 antisense riboprobe (C, F, I) on staged E9.5 (A–C), E13.0 (D–F), and E18.5 (G and H) embryos. (A–C) In E9.5 embryos, the GATA-4 (A), -5 (B), and -6 (C) riboprobes hybridized (white staining) to the atria (A) and ventricle (V), and truncus arteriousus (TA) of the two-chambered heart, the septum transversum (ST), and the cells lining the midgut and hindgut regions. Magnification = ×12.5. (D–F) In E13.0 embryos, both the GATA-4 (D) and GATA-6 (F) riboprobes hybridized to the heart (H). In contrast, the GATA-5 riboprobe hybridized predominantly to the atrial endocardium (E). All three riboprobes hybridized to the gut epithelium, whereas the GATA-4 riboprobe hybridized to the embryonic liver (Li) and ovary (O), and the GATA-5 and -6 riboprobes hybridized to the urogenital ridge (UG) and lung bud (L) (D). Only the GATA-6 riboprobe hybridized to the dorsal aorta (Ao; F). Magnification = ×3.1–3.5. (G–I) In E18.5 embryos, the GATA-4 (G) and -6 (I) riboprobe hybridized to the heart (H) and stomach and intestinal epithelium (GE). In contrast, *GATA-5* mRNA was not detected within the late fetal heart (H). Only the GATA-6 riboprobe hybridized to the aorta, small arteries, and veins (I and data not shown), whereas the GATA-5 and -6 riboprobes hybridized to the bladder (Bl) (H and I). Magnification = ×3.1.

served exclusively within the atria and becomes restricted primarily to endocardial cells lining the atria and endocardial cushions (Fig. 2E). Shortly thereafter (E16.5), *GATA-5* gene expression within the heart is extinguished (Fig. 2H and data not shown). Thus, the *GATA-4* and *-6* genes are expressed throughout the precardiac mesoderm and embryonic heart, and these genes continue to be expressed throughout pre and postnatal development. In contrast, the *GATA-5* gene is expressed transiently within the primitive heart, becomes restricted to the atrial endocardium by E12.5, and is extinguished by E16.5.

Each member of the GATA-4/5/6 subfamily is also coexpressed in the epithelial cells lining the primitive and postnatal gastrointestinal tract (Figs. 2G–2I; Morrisey *et al.,* 1996, 1997a). However, outside of the gastrointestinal tract, each gene has a unique temporally and spatially restricted pattern of expression. Coincident with the onset of vasculogenesis (Fig. 2F), and continuing throughout the postnatal period, *GATA-6* is expressed in arterial and venous smooth muscle (Fig. 2E; Morrisey *et al.,* 1996; Narita *et al.,* 1998; Suzuki *et al.,* 1996). *GATA-6* is also expressed transiently within the embryonic bronchial epithelium as well as in the urogenital ridge and the smooth muscle cells (SMCs) of the bladder wall (Figs. 2F and 2I). In contrast, at midgestation (E13.0), the *GATA-5* gene is expressed most abundantly within the primitive lung bud and urogenital ridge (Fig. 2E). Within the embryonic lung, the *GATA-5* gene is expressed initially in the pulmonary mesenchyme (at E13.0) and subsequently within bronchiolar SMCs (at E18.5). In addition, *GATA-5* is coexpressed with *GATA-6* in SMCs of the bladder wall (Figs. 2H and 2I). Thus, *GATA-5* and *-6* are expressed in overlapping, but distinct, tissue-restricted subsets of SMCs. This finding suggests that previously unrecognized transcriptional programs may distinguish SMC sublineages and that GATA-5 and -6 may play important roles in controlling the specification and/or development of these different SMC lineage(s).

In summary, *GATA-4, -5,* and *-6* are three of the earliest markers of the cardiac myocyte lineage(s). Each protein is expressed in the precardiac mesoderm and primitive heart tube at least 6–12 hr prior to the expression of genes encoding cardiac-specific contractile protein isoforms. Moreover, *GATA-4, -5,* and *-6* are expressed at least as early as the cardiac-specific homeobox gene *Nkx2-5* (Biben and Harvey, 1997; Harvey, 1996; Komuro and Izumo, 1993; Lints *et al.,* 1993) and prior to the bHLH proteins dHAND and eHAND (Srivastava *et al.,* 1997). Interestingly, *GATA-4* and *-6* are also expressed in the visceral endoderm, a tissue that plays an inductive role in cardiac myocyte specification within the subjacent lateral plate mesoderm. *GATA-4* and *-6* continue to be expressed throughout the myocardium and endocardium throughout the life of the organism, whereas *GATA-5* gene expression within the heart is extinguished by E16.5. Finally, in extracardiac tissues, *GATA-4, -5,* and *-6* are each expressed in unique temporal and spatial patterns in higher vertebrates.

IV. Multiple Cardiac-Specific Transcriptional Regulatory Elements Contain Functionally Important GATA-Binding Sites

The first evidence that GATA factors might play an important role in cardiac myocyte differentiation was the finding that a functionally important nuclear protein binding site in the cardiac-specific murine *cardiac troponin C* (*cTnC*) promoter, designated CEF-1, contained an embedded GATA motif (WGATAR), and that oligonucleotides corresponding to CEF-1 bound at least one cardiac lineage-restricted nuclear protein complex (Parmacek *et al.,* 1992). Molecular characterization of CEF-1-binding activities revealed that the cardiac muscle cell lineage-restricted nuclear protein complex did in fact contain the zinc finger transcription factor GATA-4 (Ip *et al.,* 1994). Functional analyses of the *cTnC* promoter revealed that a mutation that abolished GATA-4 binding resulted in a 90% reduction in transcriptional activity (Ip *et al.,* 1994). Most important, as shown in Fig. 3, forced expression of GATA-4 in NIH 3T3 cells could transactive the 124-base pair *cTnC* promoter by 40- to 50-fold and this transactivation was dependent on a functional GATA binding site in CEF-1.

Functionally important GATA sites have also been identified in other cardiac-specific transcriptional regulatory elements including the cardiac α- and β-*myosin heavy chain* promoters and the *ANP* and *B-type natriuretic peptide* (*BNP*) promoters. Each of these cardiac-specific transcriptional regulatory elements can also be transactivated by forced expression of GATA-4 in noncardiac muscle cells (Grepin *et al.,* 1994; Huang *et al.,* 1995; Molkentin *et al.,* 1994). Although not functionally tested, consensus GATA binding sites (WGATAR) are also present in regulatory elements controlling the expression of other cardiac-specific genes, including the mouse *cardiac troponin I* promoter, the chicken *cardiac troponin T* promoter, and the rat *phospholamban* promoter. Recent studies demonstrated that forced expression of GATA-5 or GATA-6 also transactivates the cardiac-specific *cTnC* promoter–enhancer in noncardiac

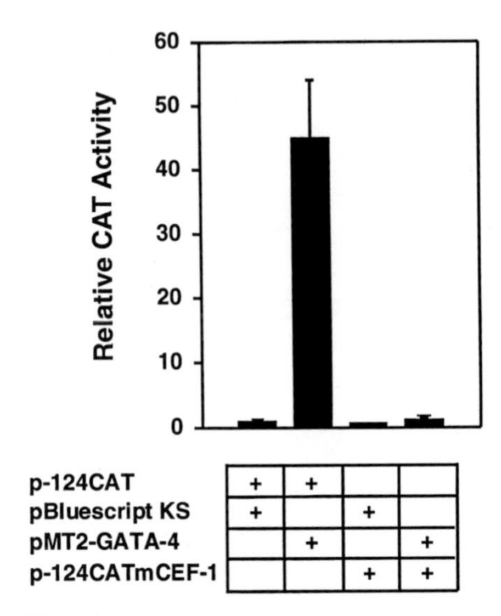

p-124CAT	+	+		
pBluescript KS	+		+	
pMT2-GATA-4		+		+
p-124CATmCEF-1			+	+

Figure 3 GATA-4-modulated transactivation of the cTnC promoter–enhancer in NIH 3T3 cells. NIH 3T3 cells were transfected with 24 μg of the p-124CAT reporter plasmid containing the wild-type cTnC promoter–enhancer (p-124CAT+) or the p-124CATmCEF-1 reporter plasmid containing a five-nucleotide substitution in the CEF-1, GATA-4 binding site (p-124CATmCEF-1 +), and either 6 μg of the control plasmid pBluescript KS (pBluescript +) or 6 μg of the GATA-4 expression plasmid pMT2-GATA-4 (p-124CATmCEF-1 +). All transfections also contained 5 μg of the pMSVβgal reference plasmid. Forty-eight hours after transfection, CAT and β-galactosidase activities were determined. CAT activities, corrected for differences in transfection efficiencies, were normalized to the CAT activity obtained following transfection of the p-124CAT plasmid with the pBluescript KS control plasmid, which produced 0.1% acetylation. The data are presented as relative CAT activities ± SEM.

muscle cells (Morrisey *et al.,* 1996; 1997a), suggesting that at least *in vitro* the activation of cardiac-specific target genes by GATA-4, -5, and -6 is redundant. Taken together, these data strongly suggested that members of the GATA-4/5/6 subfamily of transcription factors play important roles in controlling the expression of multiple cardiac-specific genes.

This model has gained additional support from both *in vitro* loss-of-function and *in vivo* gain-of-function analyses. For example, expression of antisense *GATA-4* transcripts in pluripotent P19 embryonal carcinoma cells blocked retinoic acid-inducible expression of genes encoding cardiac-specific contractile proteins (Grepin *et al.,* 1995). In addition, forced expression of GATA-4 in P19 cells accelerated cardiogenesis and increased the number of cardiac myocytes during *in vitro* differentiation of P19 cells (Grepin *et al.,* 1997). Moreover, injection of *GATA-4* mRNA into *Xenopus* oocytes resulted in the premature expression of genes encoding *cardiac α-actin* and *α-myosin heavy chain* (Jiang and Evans, 1996). Together, these data strongly suggested that GATA-4 and/or other GATA factors function (directly or indirectly) to regulate tissue-specific gene expression during vertebrate cardiac development.

V. GATA-4 Activates Transcription via Two Novel Domains That Are Conserved within the GATA-4/5/6 Subfamily

Although genetic studies have identified important roles for the GATA proteins in vertebrate development, relatively little is understood about the molecular mechanisms that control the functional activity of each GATA factor. Consistent with previous analyses of the GATA-1, -2, and -3 proteins, the conserved C-terminal zinc finger and adjacent basic domain [amino acids (aa) 251–324] of the murine GATA-4 protein are necessary and sufficient for sequence-specific DNA-binding activity (Morrisey *et al.,* 1997b). In addition, this domain contains a nuclear localization signal (Morrisey *et al.,* 1997b). Taken together, these data demonstrate that the GATA-4 C-terminal zinc finger and basic domain are bifunctional, modulating both DNA-binding and nuclear targeting activities. Both these domains are conserved in yeast single-finger GATA proteins, suggesting that this important bifunctional domain has been conserved through ancient evolution (Cunningham and Cooper, 1991; Fu and Marzluf, 1990).

Structure–function analyses of the murine GATA-4 protein led to the identification of two independent transcriptional activation domains within the N terminus of the protein (Morrisey *et al.,* 1997b). Activation domain I (ADI) (aa 1–74) is a neutral proline-rich motif (Fig. 4A), whereas ADII (aa 130–177) has a p*I* of 6.20 and contains three proline, five serine, and four tyrosine residues (Fig. 4B). Both ADI and ADII are conserved from *Xenopus laevis* to humans. Interestingly, highly related activation domains are also present in the *Xenopus* and chicken GATA-5 and -6 proteins (Figs. 4A and 4B). Functional analyses confirmed that these regions of GATA-5 and -6 function as *bona fide* transcriptional activators *in vivo* (Morrisey *et al.,* 1997b). In contrast, amino acid sequence identity was not observed between ADI or ADII and the previously identified transcriptional activation domains in the GATA-1 and -3 proteins (Martin and Orkin, 1990; Yang and Evans, 1992; Yang *et al.,* 1994). Interestingly, both ADI and ADII contain conserved tyrosine and serine residues, suggesting that posttranslational modifications of GATA-4, -5, and -6 may play an important role in regulating their functional activity Figs. 4A and 4B (arrowheads).

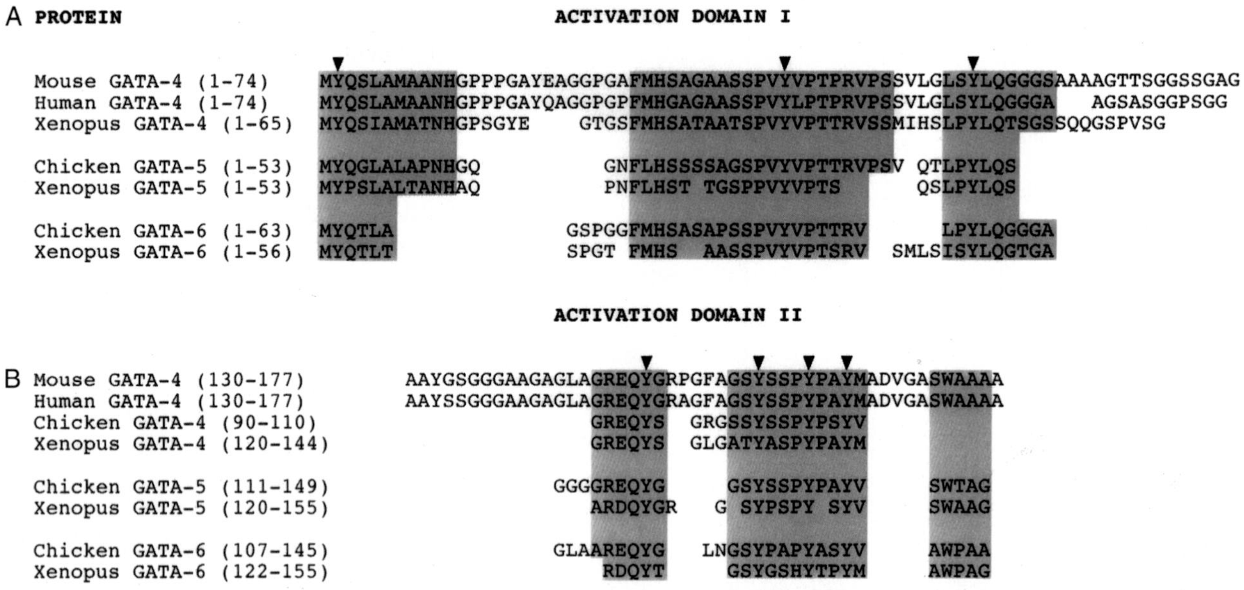

```
A PROTEIN                                     ACTIVATION DOMAIN I

Mouse GATA-4 (1-74)      MYQSLAMAANHGPPPGAYEAGGPGAFMHSAGAASSPVYVPTPRVPSSVLGLSYLQGGGSAAAAGTTSGGSSGAG
Human GATA-4 (1-74)      MYQSLAMAANHGPPPGAYQAGGPGPFMHGAGAASSPVYLPTPRVPSSVLGLSYLQGGGA   AGSASGGPSGG
Xenopus GATA-4 (1-65)    MYQSIAMATNHGPSGYE    GTGSFMHSATAATSPVYVPTTRVSSMIHSLPYLQTSGSSQQGSPVSG

Chicken GATA-5 (1-53)    MYQGLALAPNHGQ          GNFLHSSSSAGSPVYVPTTRVPSV QTLPYLQS
Xenopus GATA-5 (1-53)    MYPSLALTANHAQ          PNFLHST TGSPPVYVPTS      QSLPYLQS

Chicken GATA-6 (1-63)    MYQTLA               GSPGGFMHSASAPSSPVYVPTTRV       LPYLQGGGA
Xenopus GATA-6 (1-56)    MYQTLT               SPGT FMHS  AASSPVYVPTSRV  SMLSISYLQGTGA

                                              ACTIVATION DOMAIN II

B Mouse GATA-4 (130-177)      AAYGSGGGAAGAGLAGREQYGRPGFAGSYSSPYPAYMADVGASWAAAA
  Human GATA-4 (130-177)      AAYSSGGGAAGAGLAGREQYGRAGFAGSYSSPYPAYMADVGASWAAAA
  Chicken GATA-4 (90-110)                    GREQYS  GRGSSYSSPYPSYV
  Xenopus GATA-4 (120-144)                   GREQYS  GLGATYASPYPAYM

  Chicken GATA-5 (111-149)                GGGGREQYG     GSYSSPYPAYV     SWTAG
  Xenopus GATA-5 (120-155)                   ARDQYGR   G SYPSPY SYV     SWAAG

  Chicken GATA-6 (107-145)                GLAAREQYG   LNGSYPAPYASYV     AWPAA
  Xenopus GATA-6 (122-155)                    RDQYT     GSYGSHYTPYM     AWPAG
```

Figure 4 Amino acid sequence alignment of activation domains I (top) and II (bottom) in members of the GATA-4/5/6 subfamily. The amino acid sequences of the deduced human, murine, *Xenopus,* and chicken GATA-4, -5, and -6 proteins were aligned using the multiple sequence alignment protocol computer algorithm. Only the conserved regions of the protein are shown in this alignment. Subdomains that are conserved across each family member and across species are boxed in gray. Tyrosine residues that are conserved within each activation domain are indicated with arrowheads.

VI. GATA-4 Is Required for Ventral Morphogenesis and Heart Tube Formation

As described previously *in vitro* studies had suggested an important role for one or more GATA proteins in regulating the transcription of multiple cardiac genes and by inference the processes of cardiomyocyte specification and/or cardiac morphogenesis. To more precisely determine the role of the GATA proteins in cardiovascular development *in vivo,* targeted mutations of the *GATA-4, -5,* and *-6* genes were created in ES cells and these ES cells were used to produce mice harboring null mutations in each of the GATA genes (Kuo *et al.,* 1997; Molkentin *et al.,* 1997). The phenotypes of these mice are summarized below.

GATA-4-deficient mice produced using two independent gene targeting strategies displayed identical phenotypes (Kuo *et al.,* 1997; Molkentin *et al.,* 1997). Heterozygous *GATA-4*$^{+/-}$ mice were viable, fertile, and displayed normal cardiovascular development and function. In contrast, homozygous deficient (*GATA-4*$^{-/-}$) embryos died between E8.5 and 10.5 and displayed severe defects in both rostral-to-caudal and lateral-to-ventral folding which were reflected in a generalized disruption of the ventral body pattern (schematically depicted in Kuo *et al.,* 1997). Perhaps most strikingly, these embryos displayed markedly aberrant cardiac morphogenesis. The initial stage of cardiac development, specification of cardiac myocytes from splanchnic mesoderm, was normal in the *GATA-4*$^{-/-}$ embryos. Like wild-type littermates, the *GATA-4*$^{-/-}$ embryos developed primitive cardiac myocytes in the dorsolateral region of the embryos that expressed the normal array of genes encoding contractile proteins including *myosin heavy chain, myosin light chains 2A* and *2V, cardiac troponin C, cardiac troponin I,* and *atrial natriuretic factor* (Fig. 5 and data not shown). In contrast, the GATA-4-deficient embryos displayed severe abnormalities of the second stage of cardiac development. Specified procardiomyocytes failed to migrate from the anterior and dorsal region of the embryo to the ventral midline to form the linear heart tube. Instead, those *GATA-4*$^{-/-}$ embryos that survived until E10.5 formed one or two aberrant cardiac structures in the dorsolateral and anterior regions of the embryo, anterior to the head fold (Fig. 6 and data not shown). These aberrant cardiac structures expressed the full array of contractile proteins but failed to form a functional circulatory system resulting in embryonic lethality between E 8.0 and 10.5. Taken together, these results demonstrated that GATA-4 is not required for the specification of the cardiac myocyte lineage, but instead appears to regulate the second stage of cardiac morphogenesis—the rostral-to-caudal and lateral-to-ventral migration of procardiomyocytes to form the ventral heart tube.

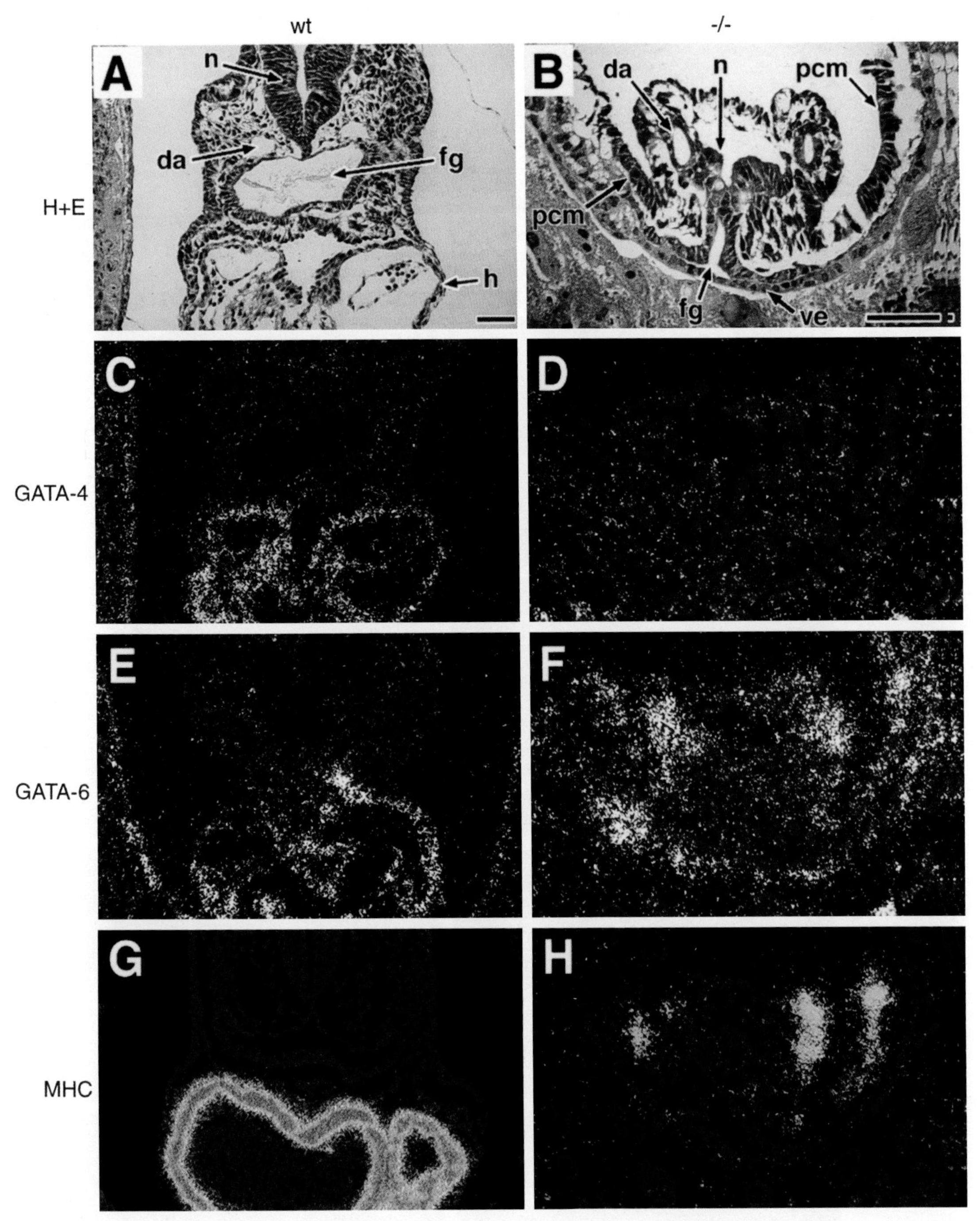

Figure 5 Cardiac defects in E8.5 *GATA-4*$^{-/-}$ embryos. Histological (A, B) and *in situ* hybridization (C–H) analyses of E8.5 *wild-type* (wt) and *GATA-4*$^{-/-}$ (−/−) embryos. (A) Hematoxylin and eosin (H & E) stained transverse section of an E8.5 *wild-type* embryo. Note the dorsally located neural tube (n), dorsal aortas (da), foregut (fg), and the ventrally located heart (h). (B) H & E stained transverse section of an E8.5 *GATA-4*$^{-/-}$ embryo. Note the dorsally located neural tube (n), dorsal aortas (da), ventrally located invaginating foregut (fg), visceral endoderm (ve), and laterally located precardiac mesoderm (pcm). (C, D) *In situ* hybridizations using a GATA-4 antisense cRNA probe. Note the *GATA-4* expression in the developing heart tube of an E8.5 *wild-type* embryo and the lack of *GATA-4* expression in the *GATA-4*$^{-/-}$ embryo. (E, F) *In situ* hybridizations using a GATA-6 antisense cRNA probe. *GATA-6* is expressed in an overlapping pattern with *GATA-4* in the developing heart tube and in the visceral endoderm of the *wild-type* embryo. In the *GATA-4*$^{-/-}$ embryo, *GATA-6* is expressed in visceral endoderm, precardiac mesoderm, and in regions surrounding the dorsal aortas lateral to the neural fold. (G, H) *In situ* hybridizations using an MHC antisense cRNA probe. *MHC* is expressed in the heart tube of the *wild-type* embryo and in regions lateral to the neural fold as well as in the precardiac mesoderm of the *GATA-4*$^{-/-}$ embryo. Different magnifications and photographic exposures of the wt and *GATA-4*$^{-/-}$ embryos were used to ensure adequate visualization of each of the probes in both embryos. Scale bars = 33 μM.

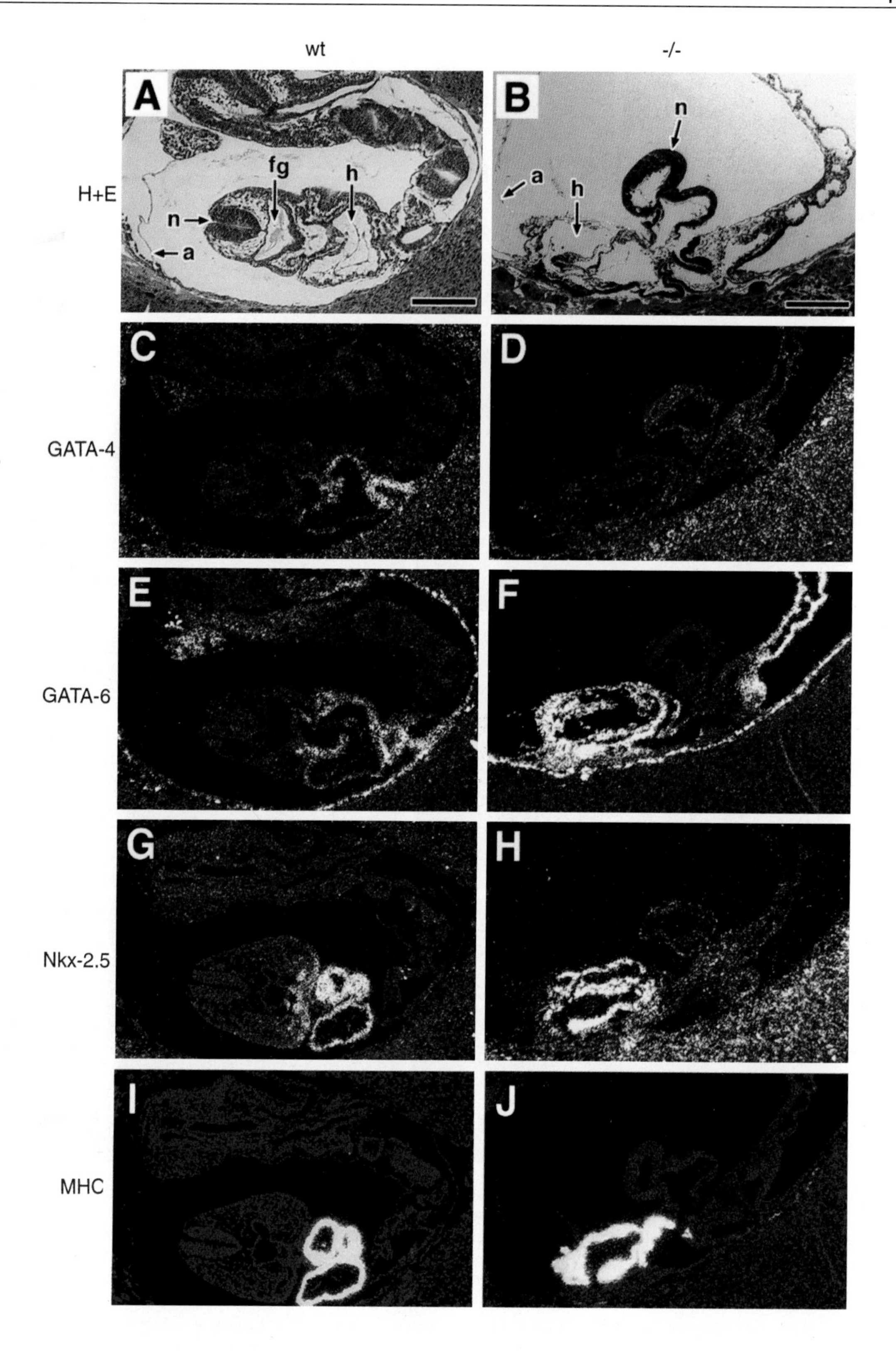
wt
-/-
A
B
H+E
fg
h
n
a
C
D
GATA-4
E
F
GATA-6
G
H
Nkx-2.5
I
J
MHC

Consistent with the hypothesis that the cardiac abnormalities observed in the $GATA\text{-}4^{-/-}$ embryos reflected a more generalized defect in ventral morphogenesis, these embryos also displayed severe abnormalities of other ventral structures (Kuo *et al.*, 1997; Molkentin *et al.*, 1997). For example, they lacked a ventral pericardial cavity and displayed defective formation of the anterior intestinal portal and foregut which appeared as a thin fragmented slit in the $GATA\text{-}4^{-/-}$ embryos compared to the normal tube-like structure seen in wild-type E9–10.5 embryos. Despite these defects, the $GATA\text{-}4^{-/-}$ embryos contained differentiated foregut epithelium as measured by expression of *HNF3α* and *HNF3β*. Although *in vitro* evidence suggested that GATA-4 is required for visceral endoderm and yolk sac formation (Bielinska *et al.*, 1996; Soudais *et al.*, 1995), the GATA-4-deficient embryos contained intact visceral endoderm that expressed *GATA-6* and yolk sacs that expressed high levels of both *AFP* and *HNF4* (Kuo *et al.*, 1997). However, both the yolk sac and the amnion failed to close around the ventral portion of the $GATA\text{-}4^{-/-}$ embryos (Kuo *et al.*, 1997; Molkentin *et al.*, 1997). Ventral closure of these membranes is tightly linked to the same lateral-to-ventral folding process that is responsible for heart tube and foregut formation, suggesting that this defect also reflected abnormal ventral morphogenesis.

One important question raised by the cardiac developmental phenotype of the $GATA\text{-}4^{-/-}$ embryos was whether this phenotype reflected an inability of the GATA-4-deficient cardiomyocytes to respond to ventral morphogenic signals or, alternatively, defects in the morphogenic signals themselves. To address this question directly, β-galactosidase-expressing $GATA\text{-}4^{-/-}$ ES cells were injected into wild-type C57BL/6 blastocysts to produce chimeric mice. Histological analyses of such chimeric embryos demonstrated contribution of the $GATA\text{-}4^{-/-}$ ES cells to the pericardium, endocardium, and myocardium (Fig. 7). This finding suggested that the defect in cardiac morphogenesis seen in the absence of GATA-4 represented a defective morphogenic signal rather than an intrinsic failure of cardiomyocyte responsiveness. The identity of the aberrant morphogenic signal in the $GATA\text{-}4^{-/-}$ embryos is unknown. However, possibilities include a soluble morphogen, an extracellular matrix molecule, or a protease that is required for normal cardiomyocyte migration. Such a molecule might be produced by procardiomyocytes themselves or by the visceral endoderm or splanchnic mesoderm, the three tissues that are known to express GATA-4 during early embryogenesis. These possibilities are currently under investigation.

In contrast to the results observed in the $GATA\text{-}4^{-/-}$ animals, GATA-5-deficient mice do not display a cardiovascular phenotype (J. D. Molkentin and E. N. Olson, unpublished observation). Thus, despite the fact that *GATA-5* is expressed in the early developing heart, and that it can bind to and transactivate cardiac specific promoters, it is not required for cardiovascular development. As discussed later, however, this does not imply that GATA-5 is not involved in cardiac development because it is possible that GATA-4 and/or GATA-6 may serve redundant functions that can rescue the cardiovascular phenotype of the GATA-5-deficient animals. This possibility is currently being investigated by crossing the $GATA\text{-}5^{+/-}$ mice with $GATA\text{-}4^{+/-}$ and $GATA\text{-}6^{+/-}$ mice to obtain $GATA\text{-}5^{-/-}$–$GATA\text{-}4^{+/-}$, $GATA\text{-}5^{-/-}$–$GATA\text{-}4^{-/-}$, $GATA\text{-}5^{-/-}$–$GATA\text{-}6^{+/-}$, and $GATA\text{-}5^{-/-}$–$GATA\text{-}6^{-/-}$ mice. Preliminary analyses of mice carrying null mutations of the *GATA-6* gene demonstrated that homozygous $GATA\text{-}6^{-/-}$ embryos die before E7.5 (E. Morrisey and M. Parmacek, unpublished observation). The cause of their lethality and their cardiovascular phenotype is currently under investigation.

VII. Potential Functional Redundancy and Cross Talk between GATA-4 and GATA-6

GATA-4, -5, and -6 are each expressed in the early developing heart and can each bind to and transactivate multiple cardiac-specific transcriptional regulatory ele-

Figure 6 Cardiac defects in E10.5 $GATA\text{-}4^{-/-}$ embryos. Histological (A, B) and *in situ* hybridization (C–J) analyses of E9.0 *wild-type* (wt) and E10.5 $GATA\text{-}4^{-/-}$ (−/−) embryos. (A, B) Hematoxylin and eosin (H & E)-stained sagittal sections. a, amnion; n, neural fold; fg, foregut; h, heart. The $GATA\text{-}4^{-/-}$ embryo is partially outside of the amniotic cavity and the $GATA\text{-}4^{-/-}$ heart is located anterior to the head fold. (C, D) *In situ* hybridizations using a GATA-4 antisense cRNA probe. *GATA-4* is expressed in the heart tube and outflow track of 9.0-day pc wild-type embryo but not in the $GATA\text{-}4^{-/-}$ embryo. (E, F) *In situ* hybridizations using a GATA-6 antisense cRNA probe. *GATA-6* is expressed in developing heart tube and outflow track, the visceral endoderm of the *wild-type* embryo, and in the visceral endoderm, mutant heart, and regions posterior to the neural fold in the $GATA\text{-}4^{-/-}$ embryo. Different photographic exposures of the *wt* and $GATA\text{-}4^{-/-}$ embryos were used to ensure adequate visualization of GATA-6 expression in both embryos. (G, H) *In situ* hybridizations using a Nkx-2.5 antisense cRNA probe. *Nkx-2.5* is expressed in the heart tube of the *wild-type* embryo and in the mutant heart of the $GATA\text{-}4^{-/-}$ embryo. (I, J) *In situ* hybridizations using a MHC antisense cRNA probe. *MHC* is expressed in the heart tube of the *wild-type* embryo and in the mutant heart of the $GATA\text{-}4^{-/-}$ embryo. Scale bars = 100 μ*M*.

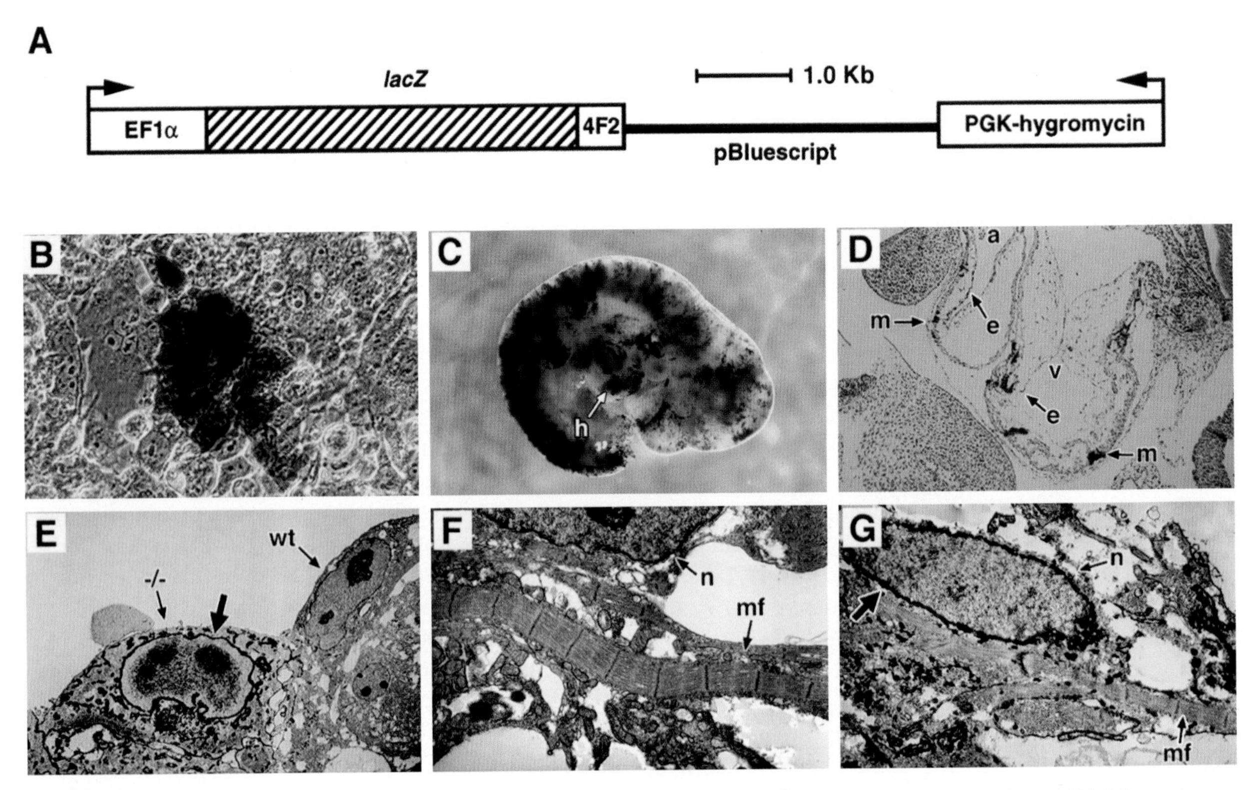

Figure 7 Contribution of β-gal$^+$-*GATA-4*$^{-/-}$ ES cells to the hearts of *GATA-4*$^{-/-}$*–C57BL/6* chimeric embryos. (A) Schematic representation of the lacZ expression vector which contains the *EF1α* promoter (*EF1α*) and *4F2 heavy chain* first intron enhancer (*4F2*) driving transcription of lacZ as well as a PGK-hygromycin cassette. (B) X-gal staining of *GATA-4*$^{-/-}$ ES cells stably transfected with the lacZ expression vector. Note the blue ES cell colony growing on a monolayer of nontransfected fibroblasts. (C) X-gal staining of an E10.5 β-*gal*$^+$–*GATA-4*$^{-/-}$ chimeric embryo showing blue cells in all areas of the embryo including the developing heart (h). (D) Histological analysis of β-*gal*$^+$–*GATA-4*$^{-/-}$ chimeric embryo. β-*gal*$^+$–*GATA*-4$^{-/-}$ cardiac myocytes (m) and endocardial cells (e) are present in both the atrium (a) and the ventricle (v). (E–G) Electron microscopic analysis of myocardium in a β-*gal*$^+$–*GATA-4*$^{-/-}$ chimeric embryo. (E) Note the wild-type cardiac myocyte (wt) next to a β-*gal*$^+$–*GATA*-4$^{-/-}$ cardiac myocyte (−/−). (F, G) Note the cardiac contractile myofibers (mf) in both wild-type (F) and β-*gal*$^+$–*GATA-4*$^{-/-}$ (G) cardiac myocytes. The large arrows point to the electron-dense crystalloid X-gal reaction products surrounding the nuclei (n) in the β-*gal*$^+$–*GATA*-4$^{-/-}$ cardiac myocytes.

ments. Thus, it was of interest to study the patterns of *GATA-5* and *-6* expression in the *GATA-4*$^{-/-}$ embryos. By *in situ* hybridization, the expression of *GATA-5* appeared to be both qualitatively and quantitatively normal in the GATA-4-deficient mice. In contrast, the E 8.0–10.5 *GATA-4*$^{-/-}$ embryos expressed significantly higher levels of *GATA-6* in both cardiac myocytes and extraembryonic membranes compared to wild-type age-matched embryos. Moreover, the *GATA-4*$^{-/-}$ embryos displayed high levels of *GATA-6* expression in the body wall posterior to the head fold, a region of the embryo which does not normally express detectable *GATA-6*. Finally, *GATA-6* expression was also significantly elevated in embryoid bodies differentiated *in vitro* from *GATA-4*$^{-/-}$ ES cells. Taken together, these findings suggested that GATA-4 and GATA-6 may belong to a common developmental pathway in which GATA-4 normally downregulates the expression of GATA-6. Interestingly, similar relationships have recently been described between GATA-2 and GATA-1, which regulate the development of the hematopoietic stem cell and erythroid lineages, respectively, and between the myogenic bHLH proteins that regulate skeletal myogenesis. These findings also raised the possibility that GATA-6 overexpression may partially rescue the phenotype of the GATA-4-deficient mice in those tissues that coexpress both proteins (i.e., heart and gut) and/or that overexpression of GATA-6 was itself responsible for at least some of the developmental defects observed in the GATA-4-deficient mice. In this regard, it will be of interest to study *GATA-4* expression in the *GATA-6*$^{-/-}$ embryos and to produce mice with targeted mutations in both the *GATA-4* and *GATA-6* genes. Such studies are currently in progress.

VIII. Summary and Future Directions

During the past 5 years, much has been learned about the role of GATA transcription factors in cardiovascular development. We now know that there are three closely related GATA family members (GATA-4, -5, and -6) expressed in temporally distinct patterns in the developing mammalian heart. In addition, *GATA-6* is expressed in both arterial and venous SMCs, suggesting a potential role for this protein in the development of the vasculature. GATA-4, -5, and -6 can each bind to the transcriptional regulatory regions of multiple cardiac promoters and enhancers and transactivate these transcriptional regulatory elements in nonmuscle cells. Each of these factors share two unique and independent transcriptional activation domains. Gene targeting experiments have demonstrated a necessary role for GATA-4 in regulating the second stage of cardiac development—the migration of specified procardiomyocytes from the dorsolateral region of the embryo to form the ventral linear heart tube. Interestingly, this defect appears to reflect a generalized requirement for GATA-4 in ventral morphogenesis rather than a cardiac-restricted phenotype. Chimera experiments suggest that GATA-4 is required to initiate and/or maintain the ventral morphogenic signal rather than to act as a regulator of cardiomyocyte responsiveness to this signal. GATA-6 is also required for early embryonic viability and its precise role in cardiovascular development is currently under investigation. In contrast, GATA-5 does not appear to be required for the development of the heart and vasculature. Finally, preliminary evidence suggests that GATA-4 and GATA-6 may belong to a common developmental pathway in which GATA-4 normally downregulates the expression of GATA-6 in the heart and extraembryonic membranes.

As is often the case, recent progress in understanding the role of GATA proteins in cardiac development has raised as many (or more) questions than it has answered. It is likely that there are significant functional redundancies between GATA-4, -5, and -6 which can only be fully understood by producing animals lacking two or more GATA proteins. The regulation of GATA-4, -5, and -6 expression during early heart development, including potential cross talk between the different proteins, is of great interest. However, to date it has been difficult to identify and characterize the promoter regions of the *GATA-4, -5,* and *-6* genes. This difficulty may reflect the existence of distant and complex transcriptional regulatory elements in these genes. Although the GATA proteins can each bind to many cardiac promoters and enhancers, we currently know relatively little about the *bona fide* transcriptional targets for each of the proteins *in vivo.* Subtractive hybridization techniques using embryos and embryoid bodies deficient in each of the proteins may be useful in this regard. GATA-4, -5, and -6 each contain conserved transcriptional activation domains, suggesting that they may interact with common transcriptional regulatory proteins. The identities of these proteins remain unknown. However, recent studies have suggested that GATA-4 can interact with MEF-2 and Nkx-2.5. Other interacting partners may be identified using yeast two-hybrid approaches. Given the importance of GATA proteins in mammalian development and the recent significant progress in this area, the next several years should lead to a much more detailed molecular understanding of the role of these interesting proteins in cardiovascular development.

References

Amacher, S. L., Buskin, J. N., and Hauschka, S. D. (1993). Multiple regulatory elements contribute differentially to *muscle creatine kinase* enhancer activity in skeletal and cardiac muscle. *Mol. Cell. Biol.* **13,** 2753–2764.

Arceci, R. J., King, A. A., Simon, M. C., Orkin, S. H., and Wilson, D. B. (1993). Mouse GATA-4: A retinoic acid-inducible GATA-binding transcription factor expressed in endodermally derived tissues and heart. *Mol. Cell. Biol.* **13,** 2235–2246.

Argentin, S., Ardati, A., Tremblay, S., Lihrmann, I., Robitaille, L., Drouin, J., and Nemer, M. (1994). Developmental stage-specific regulation of *atrial natriuretic factor* gene transcription in cardiac cells. *Mol. Cell. Biol.* **14,** 777–790.

Azakie, A., Larkin, S. B., Farrance, I. K., Grenningloh, G., and Ordahl, C. P. (1996). DTEF-1, a novel member of the transcription enhancer factor-1 (TEF-1) multigene family. *J. Biol. Chem.* **271,** 8260–8265.

Biben, C., and Harvey, R. P. (1997). Homeodomain factor Nkx2-5 controls left/right asymmetric expression of bHLH gene *eHAND* during murine heart development. *Genes Dev.* **11,** 1357–1369.

Biben, C., Kirschbaum, B. J., Garner, J., and Buckingham, M. (1994). Novel muscle-specific enhancer sequences upstream of the *cardiac actin* gene. *Mol. Cell. Biol.* **14,** 3504–3513.

Biben, C., Hadchouel, J., Tajbakhsh, S., and Buckingham, M. (1996). Developmental and tissue-specific regulation of the murine *cardiac actin* gene *in vivo* depends on distinct skeletal and cardiac muscle-specific enhancer elements in addition to the proximal promoter. *Dev. Biol.* **173,** 200–212.

Bielinska, M., Narita, N., Heikinheimo, M., Porter, S. B., and Wilson, D. B. (1996). Erythropoiesis and vasculogenesis in embryoid bodies lacking visceral yolk sac endoderm. *Blood* **88,** 3720–3730.

Bodmer, R. (1993). The gene *tinman* is required for specification of the heart and visceral muscles in *Drosophila. Development (Cambridge, UK)* **118,** 719–729.

Crossley, M., Merika, M., and Orkin, S. H. (1995). Self-association of the erythroid transcription factor GATA-1 mediated by its zinc finger domains. *Mol. Cell. Biol.* **15,** 2448–2456.

Cserjesi, P., Lilly, B., Hinkley, C., Perry, M., and Olson, E. N. (1994). Homeodomain protein MHox and MADS protein myocyte enhancer-binding factor-2 converge on a common element in the *muscle creatine kinase* enhancer. *J. Biol. Chem.* **269,** 16740–16745.

Cunningham, T. S., and Cooper, T. G. (1991). Expression of the DAL80 gene, whose product is homologous to the TATA factors and is a negative regulator of multiple nitrogen catabolic genes in *Saccharaomyces cerevisiae,* is sensitive to nitrogen catabolite repression. *Mol. Cell. Biol.* **11,** 6205–6215.

Donoviel, D., Shield, M. A., Buskin, J. N., Haugen, H. S., Clegg, C. H., and Hauschka, S. D. (1996). Analysis of *muscle creatine kinase* gene regulatory elements in skeletal and cardiac muscles of transgenic mice. *Mol. Cell. Biol.* **16,** 1649–1658.

Dorfman, D. M., Wilson, D. B., Bruns, G. A., and Orkin, S. H. (1992). Human transcription factor GATA-2. Evidence for regulation of *preproendothelin-1* gene expression in endothelial cells. *J. Biol. Chem.* **267,** 1279–1285.

Evans, T., M., R., and Felsenfeld, G. (1988). An erythrocyte-specific DNA-binding factor recognizes a regulatory sequence common to all chicken globin genes. *Proc. Natl. Acad. Sci. U.S.A.* **85,** 5976–5980.

Farrance, I. K., and Ordahl, C. P. (1996). The role of transcription enhancer factor-1 (TEF-1) related proteins in the formation of M-CAT binding complexes in muscle and non-muscle tissues. *J. Biol. Chem.* **271,** 8266–8274.

Fishman, M. C., and Chien, K. R. (1997). Fashioning the vertebrate heart: Earliest embryonic decisions. *Development (Cambridge, UK)* **124,** 2099–2117.

Fu, Y.-H., and Marzluf, G. A. (1990). *nit-2,* the major nitrogen regulatory gene of *Neurospora crassa,* encodes a protein with a putative zinc finger DNA-binding domain. *Mol. Cell. Biol.* **10,** 1056–1065.

Gossett, L. A., Kelvin, D. J., Sternberg, E. A., and Olson, E. N. (1989). A new myocyte-specific enhancer-binding factor that recognizes a conserved element associated with multiple muscle-specific genes. *Mol. Cell. Biol.* **9,** 5022–5033.

Grepin, C., Dagnino, L., Robitaille, L., Haberstroh, L., Antakly, T., and Nemer, M. (1994). A hormone-encoding gene identifies a pathway for cardiac but not skeletal muscle gene transcription. *Mol. Cell. Biol.* **14,** 3115–3129.

Grepin, C., Robitaille, L., Antakly, T., and Nemer, M. (1995). Inhibition of transcription factor GATA-4 expression blocks in vitro cardiac muscle differentiation. *Mol. Cell. Biol.* **15,** 4095–4102.

Grepin, C., Nemer, G., and Nemer, M. (1997). Enhanced cardiogenesis in embryonic stem cells overexpressing the GATA-4 transcription factor. *Development (Cambridge, UK)* **124,** 2387–2395.

Harvey, R. P. (1996). NK-2 homeobox genes and heart development. *Dev. Biol.* **178,** 203–216.

Heikinheimo, M., Scandrett, J. M., and Wilson, D. B. (1994). Localization of transcription factor GATA-4 to regions of the mouse embryo involved in cardiac development. *Dev. Biol.* **164,** 361–373.

Ho, I. C., Vorhees, P., Marin, N., Oakley, B. K., Tsai, S. F., Orkin, S. H., and Leiden, J. M. (1991). Human GATA-3: A lineage-restricted transcription factor that regulates the expression of the T cell receptor alpha gene. *EMBO J.* **10,** 1187–1192.

Huang, W. Y., Cukerman, E., and Liew, C. C. (1995). Identification of a GATA motif in the *cardiac alpha-myosin heavy-chain*-encoding gene and isolation of a human GATA-4 cDNA. *Gene* **155,** 219–223.

Iannello, R. C., Mar, J. H., and Ordahl, C. P. (1991). Characterization of a promoter element required for transcription in myocardial cells. *J. Biol. Chem.* **266,** 3309–3316.

Ip, H. S., Wilson, D. B., Heikinheimo, M., Tang, Z., Ting, C. N., Simon, M. C., Leiden, J. M., and Parmacek, M. S. (1994). The GATA-4 transcription factor transactivates the *cardiac muscle-specific troponin* C promoter-enhancer in nonmuscle cells. *Mol. Cell. Biol.* **14,** 7517–7526.

Jiang, Y., and Evans, T. (1996). The *Xenopus GATA-4/5/6* genes are associated with cardiac specification and can regulate cardiac-specific transcription during embryogenesis. *Dev. Biol.* **174,** 258–270.

Kelley, C., Blumberg, H., Zon, L. I., and Evans, T. (1993). GATA-4 is a novel transcription factor expressed in endocardium of the developing heart. *Development (Cambridge, UK)* **118,** 817–827.

Knotts, S., Rindt, H., Neumann, J., and Robbins, J. (1994). *In vivo* regulation of the mouse *beta myosin heavy chain* gene. *J. Biol. Chem.* **269,** 31275–31282.

Ko, L. J., and Engel, J. D. (1993). DNA-binding specificities of the GATA transcription factor family. *Mol. Cell. Biol.* **13,** 4011–4022.

Ko, L. J., Yamamoto, M., Leonard, M. W., George, K. M., Ting, P., and Engel, J. D. (1991). Murine and human T-lymphocyte GATA-3 factors mediate transcription through a cis-regulatory element within the human *T-cell receptor delta* gene enhancer. *Mol. Cell. Biol.* **11,** 2778–2784.

Komuro, I., and Izumo, S. (1993). Csx: A murine homeobox-containing gene specifically expressed in the developing heart. *Proc. Natl. Acad. Sci. U.S.A.* **90,** 8145–8149.

Kuo, C. T., Morrisey, E. E., Anandappa, R., Sigrist, K., Lu, M. M., Parmacek, M. S., Soudais, C., and Leiden, J. M. (1997). GATA-4 transcription factor is required for ventral morphogenesis and heart tube formation. *Genes Dev.* **11,** 1048–1060.

Larkin, S. B., Farrance, I. K., and Ordahl, C. P. (1996). Flanking sequences modulate the cell specificity of M-CAT elements. *Mol. Cell. Biol.* **16,** 3742–3755.

Laverriere, A. C., MacNeill, C., Mueller, C., Poelmann, R. E., Burch, J. B., and Evans, T. (1994). GATA-4/5/6, a subfamily of three transcription factors transcribed in developing heart and gut. *J. Biol. Chem.* **269,** 23177–23184.

Lee, K. J., Hickey, R., Zhu, H., and Chien, K. (1994). Positive regulatory elements (HF-1a and HF-1b) and a novel negative regulatory element (HF-3) mediate ventricular muscle-specific expression of *myosin light-chain* 2-luciferase fusion genes in transgenic mice. *Mol. Cell. Biol.* **14,** 1220–1229.

Lin, Q., Schwarz, J., Bucana, C., and Olson, E. N. (1997). Control of mouse cardiac morphogenesis and myogenesis by transcription factor MEF2C. *Science* **276,** 1404–1407.

Lints, T. J., Parsons, L. M., Hartley, L., Lyons, I., and Harvey, R. P. (1993). *Nkx-2.5:* A novel murine homeobox gene expressed in early heart progenitor cells and their myogenic descendants. *Development (Cambridge, UK)* **119,** 419–431.

Mar, J. H., and Ordahl, C. P. (1990). M-CAT binding factor, a novel trans-acting factor governing muscle-specific transcription. *Mol. Cell. Biol.* **10,** 4271–4283.

Martin, D. L., and Orkin, S. H. (1990). Transcriptional activation and DNA binding by the erythroid factor GF-1/NF-E1/Eryf 1. *Genes Dev.* **4,** 1886–1898.

Martin, D. I., Zin, L. I., Mutter, G., and Orkin, S. H. (1990). Expression of an erythroid transcription factor in megakaryocytic and mast cell lineages. *Nature (London)* **344,** 444–446.

Merika, M., and Orkin, S. H. (1993). DNA-binding specificity of GATA family transcription factors. *Mol. Cell. Biol.* **13,** 3999–4010.

Merika, M., and Orkin, S. H. (1995). Functional synergy and physical interactions of the erythroid transcription factor GATA-1 with the Kruppel family proteins Sp1 and EKLF. *Mol. Cell. Biol.* **15,** 2437–2447.

Molkentin, J. D., Kalvakolanu, D. V., and Markham, B. E. (1994). Transcription factor GATA-4 regulates cardiac muscle-specific expression of the *α-myosin heavy chain* gene. *Mol. Cell. Biol.* **14,** 4947–4957.

Molkentin, J. D., Jobe, S. M., and Markham, B. E. (1996). *α-myosin heavy chain* gene regulation: Delineation and characterization of the cardiac muscle-specific enhancer and muscle-specific promoter. *J. Mol. Cell. Cardiol.* **28,** 1211–1225.

Molkentin, J. D., Lin, Q., Duncan, S. A., and Olson, E. N. (1997). Requirement of the transcription factor GATA4 for heart tube formation and ventral morphogenesis. *Genes Dev.* **11,** 1061–1072.

Morrisey, E. E., Ip, H. S., Lu, M. M., and Parmacek, M. S. (1996). GATA-6: A zinc finger transcription factor that is expressed in multiple cell lineages derived from lateral mesoderm. *Dev. Biol.* **177,** 309–322.

Morrisey, E. E., Ip, H. S., Tang, Z., Lu, M. M., and Parmacek, M. S. (1997a). GATA-5: A transcriptional activitor expressed in a novel temporally and spatially-restricted pattern during embryonic development. *Dev. Biol.* **183,** 21–36.

Morrisey, E. E., Ip, H. S., Tang, Z., and Parmacek, M. S. (1997b). GATA-4 activates transcription via two novel domains that are conserved within the GATA-4/5/6 subfamily. *J. Biol. Chem.* **272,** 8515–8524.

Narita, N., Heikinheimo, M., Bielinska, M., White, R. A., and Wilson, D. B. (1998). The gene for transcription factor GATA-6 resides on mouse chromosome 18 and is expressed in myocardium and vascular smooth muscle. *Genomics* **36,** 345–348.

Navankasattusas, S., Zhu, H., Garcia, A. V., Evans, S. M., and Chien, K. R. (1992). A ubiquitous factor (HF-1a) and a distinct muscle factor (HF-1b/MEF-2) form an E-box-independent pathway for cardiac muscle gene expression. *Mol. Cell. Biol.* **12,** 1469–1479.

Olson, E. N., and Srivastava, D. (1996). Molecular pathways controlling heart development. *Science* **272,** 671–675.

Omichiaski, J. G., Clore, G. M., Schaad, O., Felsenfeld, G., Trainor, C., Apella, E., Stahl, S. J., and Gronenborn, A. M. (1993). NMR structure of a specific DNA complex of Zn-containing DNA-binding domain of GATA-1. *Science* **261,** 438–446.

Orkin, S. H. (1992). GATA-binding transcription factors in hematopoietic cells. *Blood* **80,** 575–581.

Osada, H., Grutz, G., Axelson, H., Forster, A., and Rabbitts, T. H. (1995). Association of erythroid transcription factors: Complexes involving the LIM protein RBTN2 and the zinc-finger protein GATA1. *Proc. Natl. Acad. Sci. U.S.A.* **92,** 9585–9589.

Pandolfi, P. P., Roth, M. E., Karis, A., Leonard, M. W., Dzierzak, E., Grosveld, F. G., Engel, J. D., and Lindenbaum, M. H. (1995). Targeted disruption of the *GATA3* gene causes severe abnormalities in the nervous system and in fetal liver haematopoiesis. *Nat. Genet.* **11,** 40–44.

Parmacek, M. S., Vora, A. J., Shen, T., Barr, E., Jung, F., and Leiden, J. M. (1992). Identification and characterization of a cardiac-specific transcriptional regulatory element in the *slow/cardiac troponin C* gene. *Mol. Cell. Biol.* **12,** 1967–1976.

Pevny, L., Simon, M. C., Robertson, E., Klein, W. H., Tsai, S. F., D'Agati, V., Orkin, S. H., and Costantini, F. (1991). Erythroid differentiation in chimaeric mice blocked by a targeted mutation in the gene for transcription factor GATA-1. *Nature (London)* **349,** 257–260.

Pollock, R., and Treisman, R. (1991). Human SRF-related proteins: DNA-binding properties and potential regulatory targets. *Genes Dev.* **5,** 2327–2341.

Romeo, P. H., Prandini, M. H., Joulin, V., Mignotte, V., Prenant, M., Vainchenker, W., Marguerie, G., and Uzan, G. (1990). Megakaryocytic and erthrocytic lineages share specific transcription factors. *Nature (London)* **344,** 447–449.

Ross, R. S., Navankasattusas, S., Harvey, R. P., and Chien, K. R. (1996). An HF-1a/HF-1b/MEF-2 combinatorial element confers cardiac ventricular specificity and established an anterior-posterior gradient of expression. *Development (Cambridge, UK)* **122,** 1799–809.

Sartorelli, V., Webster, K. A., and Kedes, L. (1990). Muscle-specific expression of the *cardiac α-actin* gene requires MyoD1, CArG-box binding factor, and Sp1. *Genes Dev.* **4,** 1811–1822.

Seidman, C. E., Wong, D. W., Jarcho, J. A., Bloch, K. D., and Seidman, J. G. (1988). Cis-acting sequences that modulate artrial natriuretic factor gene expression. *Proc. Natl. Acad. Sci. U.S.A.* **85,** 4104–4108.

Simon, M. C. (1995). Gotta have GATA. *Nat. Genet.* **11,** 9–11.

Simon, M. C., Pevny, L., Wiles, M. V., Keller, G., Costantini, F., and Orkin, S. H. (1992). Rescue of erythroid development in gene targeted *GATA-1* mouse embryonic stem cells. *Nat. Genet.* **1,** 92–98.

Soudais, C., Bielinska, M., Heikinheimo, M., MacArthur, C. A., Narita, N., Saffitz, J. E., Simon, M. C., Leiden, J. M., and Wilson, D. B. (1995). Targeted mutagenesis of the transcription factor *GATA-4* gene in mouse embryonic stem cells disrupts visceral endoderm differentiation. *Development (Cambridge, UK)* **121,** 3877–3888.

Srivastava, D., Thomas, T., Lin, Q., Kirby, M. L., Brown, D., and Olson, E. N. (1997). Regulation of cardiac mesodermal and neural crest development by the bHLH transcription factor, dHAND. *Nat. Genet.* **16,** 154–160.

Stewart, A. F., Larkin, S. B., Farrance, I. K., Mar, J. H., Hall, D. E., and Ordahl, C. P. (1994). Muscle-enriched TEF-1 isoforms bind M-CAT elements from muscle-specific promoters and differentially activate transcription. *J. Biol. Chem.* **269,** 3147–3150.

Subramaniam, A., Jones, W. K., Gulick, J., Wert, S., Neumann, J., and Robbins, J. (1991). Tissue-specific regulation of the *α-myosin heavy chain* gene promoter in transgenic mice. *J. Biol. Chem.* **266,** 24613–24620.

Suzuki, E., Evans, T., Lowry, J., Truong, L., Bell, D. W., Testa, J. R., and Walsh, K. (1996). The human *GATA-6* gene: Structure, chromosomal location, and regulation of expression by tissue-specific and mitogen-responsive signals. *Genomics* **38,** 283–290.

Tamora, S., Wang, X.-H., Maeda, M., Futai, M., (1993). Gastric DNA-binding proteins recognize upstream sequence motifs of parietal cell-specific genes. *Proc. Natl. Acad. Sci. U.S.A.* **90,** 10876–10880.

Ting, C.-N., Olson, M. C., and Leiden, J. M. (1996). The GATA-3 transcription factor is required for the differentiation of the T cell lineage in mice. *Nature (London)* **384,** 474–478.

Tonissen, K. F., Drysdale, T. A., Lints, T. J., Harvey, R. P., and Krieg, P. A. (1994). *XNkx-2.5,* a Xenopus gene related to *Nkx-2.5* and *tinman:* Evidence for a conserved role in cardiac development. *Dev. Biol.* **162,** 325–328.

Trainor, C. D., Omichinski, J. G., Vandergon, T. L., Gronenborn, A. M., Clore, G. M., and Felsenfeld, G. (1996). A palindromic regulatory site within vertebrate *GATA-1* promoters requires both zinc fingers of the GATA-1 DNA-binding domain for high affinity interaction. *Mol. Cell. Biol.* **16,** 2238–2247.

Tsai, F. Y., Keller, G., Kuo, F. C., Weiss, M., Chen, J., Rosenblatt, M., Alt, F. W., and Orkin, S. H. (1994). An early haematopoietic defect in mice lacking the transcription factor GATA-2. *Nature (London)* **371,** 221–226.

Tsai, S. F., Martin, D. I. K., Zon, L. I., D'Andrea, A. D., Wong, G. G., and Orkin, S. H. (1989). Cloning of cDNA for the major DNA-binding protein of the erythroid lineage through expression in mammalian cells. *Nature (London)* **339,** 446–451.

Wanker, M., Boheler, K. R., Fiszman, M. Y., and Schwartz, K. (1996). Molecular cloning and analysis of the human *cardiac sarco (endo)plasmic reticulum Ca-ATPase (SERCA2)* gene promoter. *J. Mol. Cell. Cardiol.* **28,** 2139–2150.

Weiss, M. J., and Orkin, S. H. (1995). GATA transcription factors: Key regulators of hematopoiesis. *Exp. Hematol.* **23,** 99–107.

Weiss, M. J., Keller, G., and Orkin, S. H. (1994). Novel insights into erythroid development revealed through *in vitro* differentiation of GATA-1 embryonic stem cells. *Genes Dev.* **8,** 1184–1197.

Wilson, D. B., Dorfman, D. M., and Orkin, S. H. (1990). A nonerythroid GATA-binding protein is required for function of the human *preproendothelin-1* promoter in endothelial cells. *Mol. Cell. Biol.* **10,** 4854–4862.

Xiao, J. H., Davidson, I., Matthes, H., Garnier, J. M., and Chambon, P. (1991). Cloning, expression, and transcriptional properties of the human enhancer factor TEF-1. *Cell (Cambridge, Mass.)* **65,** 551–568.

Yamamoto, M., Ko, L. J., Leonard, M. W., Beug, H., Orkin, S. H., and Engel, J. D. (1990). Activity and tissue-specific expression of the transcription factor NF-E1 multigene family. *Genes Dev.* **4,** 1650–1662.

Yang, H.-Y., and Evans, T. (1992). Distinct roles for the two cGATA-1 finger domains. *Mol. Cell. Biol.* **12,** 4562–4570.

Yang, H. Y., and Evans, T. (1995). Homotypic interactions of chicken GATA-1 can mediate transcriptional activation. *Mol. Cell. Biol.* **15,** 1353–1363.

Yang, Z., Gu, L., Romeo, P.-H., Bories, D., Motohashi, H., Yamamoto, M., and Engel, J. D. (1994). Human GATA-3 trans-activation, DNA-binding, and nuclear localization activities are organized into distinct structural domains. *Mol. Cell. Biol.* **14,** 2201–2212.

Zheng, W.-P., and Flavell, R. A. (1997). The transcription factor GATA-3 is necessary and sufficient for *Th2* cytokine gene expression in CD4 T cells. *Cell (Cambridge, Mass.)* **89,** 587–596.

Zhu, H., Garcia, A., Ross, R., Evans, S., and Chien, K. (1991). A conserved 28-bp element (HF-1) in the rat *cardiac myosin light-chain-2* gene confirms cardiac-specific and alpha-adrenergic-inducible expression in cultured neotatal rat myocardial cells. *Mol. Cell. Biol.* **11,** 2273–2281.

Zhu, H., Nguyen, V. T. B., Brown, A. B., Pourhosseini, A., Garcia, A., van Bilsen, M., and Chien, K. (1993). A novel, tissue-restricted zinc finger protein (HF-1b) binds to the cardiac regulatory element (HF-1b/MEF-2) in the rat *myosin light-chain 2* gene. *Mol. Cell. Biol.* **13,** 4432–4444.

Zon, L. I., Yamaguchi, Y., Yee, K., Albee, E. A., Kimura, A., Bennett, J. C., Orkin, S. H., and Ackerman, S. J. (1993). Expression of mRNA for the GATA-binding proteins in human eosinophils and basophils: Potential role in gene transcription. *Blood* **81,** 3234–3241.

18

Multiple Layers of Control in Transcriptional Regulation by MCAT Elements and the TEF-1 Protein Family

Sarah B. Larkin[1] and Charles P. Ordahl

Department of Anatomy and Cardiovascular Research Institute, University of California, San Francisco, California 94143

I. Introduction

The transcriptional regulation of vertebrate gene expression involves controls that reflect different levels of biological organization. On the one hand, a gene whose product is required for fundamental cellular structure–function, such as *β-actin,* must be accessible to the transcriptional machinery of all cell types. On the other hand, tissue-specific genes, such as the *skeletal α-actin* (*αSkA*) gene, are inactive in all but a few specific cell types in which their transcription is activated, often to very high levels. How does the transcriptional regulatory apparatus of each cell discriminate between such genes? This can be explained in part by the requirement for binding between the DNA regulatory elements of some cell-specific genes and transcription factors that are themselves cell specific, such as the skeletal muscle-specific MyoD family of transcriptional regulatory proteins (Emerson, 1993; Funk *et al.,* 1991; Lassar and Munsterberg, 1994; Olson, 1990, 1993; Weintraub, *et al.,* 1991). In other cases, however, the DNA sequence elements governing cell-specific expression bind nuclear proteins that are present in many, if not all, cell types. For example, skeletal muscle-specific activity of the *αSkA* gene promoter appears to be governed by serum

[1]Present address: Amylin Pharmaceuticals, Inc., San Diego, CA 92121.

response factors which are ubiquitous (Carson *et al.*, 1996; Lee *et al.*, 1992; MacLellan *et al.*, 1994).

This latter case may also be true for cardiac-specific gene transcription because many cardiac gene promoters are governed by MCAT elements, a distinct class of DNA sequence transcriptional regulatory elements that are involved in at least two distinct transcriptional regulatory systems. One system is cell/tissue-specific because MCAT elements govern the expression of a wide array of cardiac genes (Table I) in cell-specific, hormone-responsive, and growth/hypertrophy-related fashions. The second regulatory system, however, is cell nonspecific and is best exemplified by gene regulation in the SV40 and related viruses in which MCAT-dependent transcription is linked to systems controlling cell growth and division.

This review will focus on the interactions between MCAT elements and the TEF-1 family of transcriptional regulatory proteins, with particular emphasis on the roles of these components in cardiac-specific transcription. We will summarize recent evidence that some MCAT elements function by binding not only transcription enhancer-1 (TEF-1) family members but also a second, as yet undefined, class of binding factors that carry significance in the cobinding interactions that control different modes of MCAT-dependent gene regulation.

A. Muscle-Specific Genes Controlled by MCAT Elements

The list of muscle-specific genes whose promoters are partially or completely regulated by MCAT elements includes sarcomeric proteins and other cell structural proteins, enzymes, and membrane channels (Table I). The majority of these genes are expressed in the myocardium, such as the *cardiac troponin T* and *C* (*cTnT* and *cTnC*) genes and the α and β *myosin heavy chain genes* (αMHC, and βMHC). Others are expressed in skeletal muscle and vascular smooth muscle, such as the β isoform of the *acetylcholine receptor* ($AChR\beta$), the c-*mos* genes, and the *vascular smooth muscle* α *actin* ($VSM\alpha A$) genes (see Table I for a complete listing and references).

MCAT elements are found in complex promoters frequently containing SP1 sites, A/T-rich elements (CarG boxes or MEF-2 binding sites) and E boxes (Fig. 1). The majority of these MCAT-containing promoters contain a single copy of the MCAT element; however, the *cTnT*, αMHC and rat βMHC and c-*mos* promoters contain 2 copies (Mar *et al.*, 1988; Mar and Ordahl, 1988; Shimizu *et al.*, 1992b; Gupta *et al.*, 1994; Molkentin and Markham, 1994), and the *nitric oxide synthase gene* (*NOS*) promoter contains 10: 1 perfect copy and 9 additional imperfect copies, though promoter activity has not been established for any of these *NOS* MCAT elements (Hall *et al.*, 1994). It has been shown that multiple MCAT copies confer more than additive activity of single copies in muscle cells (Flink *et al.*, 1992) and spacing changes between the two MCATs in the *cTnT* promoter result in changes in promoter strength (Mar and Ordahl, 1990), indicating that synergistic interactions occur between MCAT elements, at least in certain contexts. Similarly, MCAT elements in nonmuscle promoters synergize and this synergy is also disrupted by spacing introduced between MCAT pairs (Fromental, *et al.*, 1988; Ondek *et al.*, 1988). In the nonmuscle context MCAT elements have been shown to be capable of synergy with another unrelated promoter site (Fromental, *et al.*, 1988; Ondek *et al.*, 1988). The synergy in nonmuscle cells appears to result at least in part from cooperative binding to the correctly spaced MCATs; however, binding to multimerized MCATs from the *cTnT* promoter is not cooperative (Davidson *et al.*, 1988; Mar and Ordahl, 1990; and S. B. Larkin, unpublished observation).

All MCAT elements are not equivalent: The properties of the two MCAT elements of the *cTnT* promoter have been examined individually in artificial promoters and shown to differ, such that the distal-most is a much stronger stimulator of muscle-specific gene activation (Larkin *et al.*, 1996). Moreover, not all MCAT elements can independently confer muscle specificity (Larkin *et al.*, 1996).

A subset of MCAT elements are also inducible promoter elements. These MCAT sites are the targets of second messenger cascades arising from the action of several serum mitogens interacting with their receptors at the cell membrane and of hypertrophic signals that are likely to be mediated by such mitogens and/or α_1-adrenergic receptor stimulation. MCAT elements in the βMHC and αSkA gene promoters are the target of α-adrenergic upregulation probably via activation of a protein kinase C (PKC) cascade (Kariya *et al.*, 1993, 1994; Karns *et al.*, 1995). An MCAT element in the αSkA gene promoter, in conjunction with its serum response elements (CarG boxes), mediates responses in the myocardium to the serum mitogen transforming growth factor (TGF-β) (MacLellan *et al.*, 1994). This MCAT element also mediates αSkA gene upregulation on stretch-induced hypertrophy in skeletal muscle (Carson *et al.*, 1996). The hypertrophic upregulation of the *brain natriuretic peptide* (*BNP*) gene promoter is also mediated via its MCAT element (Thuerauf and Glembotski, 1997).

The $VSM\alpha A$ gene is expressed *de novo* in myofibroblasts in response to stimuli involved in wound healing and certain neoplasias (Skalli and Gabbiani, 1988).

Table I Muscle-Specific MCAT Sites, Non-Muscle-Specific MCAT Sites Bound by TEF-1, and MCAT-like Sites Not Bound by TEF-1*

Promoter/position	MCAT and flanking sequence	Tissue†	Reference
	Muscle-Specific MCAT Sites		
Cardiac troponin T, chick			Mar *et al.* (1988); Mar and Ordahl (1988)
-66S	GCGCCGGGCA**CATTCCT**GCTGCTCTGC	Sk, C	
-89S	ACAAGTGTTG**CATTCCT**CTCTGGGCGC		
β myosin heavy chain			
Human (-284AS)	CCAGGCCTCA**CATTCCA**CAGCTGGCAG	C	Flink *et al.* (1992)
Rabbit (-264AS)	TATCTCCTCG**CATTCCA**CTGCCTGTGG	Sk	Shimizu *et al.* (1992a)
Rat			
-267AS	ATATCCCTTA**CATTCCA**CAGCTCAC	C	Thompson *et al.* (1991)
-196S	CTGAACATGC**CATACCA**CAACAATGAC	↑ PKC	Kariya *et al.* (1994)
α myosin heavy chain, rat			
-236AS	TCCTTGGGCA**CATTCCT**CCTCCCCAAA	Sk, C	Molkentin and Markham (1994); Gupta *et al.* (1994)
-42AS	CTTTATAGCT**CATTCCA**CGTGCCTGCT	↑ cAMP	
myosin light chain 2, chick (-35AS)	AAATACACCC**CATTCCA**GGCTAAAAAT	C[a]	Qasba *et al.* (1992)
s/cTnC, mouse (-50 AS)	CCTCCTGCTA**CATTCCC**AGCCCAGCCC	C[a]	Parmacek and Leiden (1989)
VSM α actin, mouse (-175AS)	TCTTCCACTG**CATTCCT**CTGTTCTGCT	VSM, ↑ serum	Cogan *et al.* (1995); Sun *et al.* (1995)
Sk α actin			
Chick (-60S)	AGCTTGCCG**CATTCCT**GGGGGCCGGG	C, ↑ TGF-β, ↑ PKC	MacLellan *et al.* (1994)
Mouse (-63S)	AGGGCAGCAA**CATTCTT**TCGGGGCGGT		Karns *et al.* (1995)
AChRβ, rat (-43S)	ACAGGTGCA**CATTCCT**GGGCGCCTCG	Sk	Berberich *et al.* (1993)
PGAM-M, human (-72AS)	TGCCAATCAG**CATTCCA**GGCGGTGGCA	Sk, C[a]	Nakatsuji *et al.* (1992)
c-*mos,* rat			
-1561S	AGGCTTTATC**CATTTCT**GAGATAAAGA	Sk	Lenormand and Leibovitch (1995)
-1541S	ATAAAGATTT**CATTTCT**AATCTCAGTA		
AE3 CL^-/HCO_3^- channel, rodent/human (-190S)	GATCTCGGCA**CATTCCT**TCCATCTTAT	C[a]	Linn *et al.* (1995)
brain natriuretic peptide			
Rat (-101AS)	TTATCAGACA**CATTCCT**GCCTGCTGAG	C, ↑ PKC	Thuerauf and Glembotski (1997)
Human (-117S)	CGGAGGGGCT**CATTCCC**GGGCCCTGAT	C[a]	LaPointe *et al.* (1996)
nitric oxide synthase, human (-256AS)	GTTGGGATAG**CATTCCT**TAATGGGACA*	Sk[a]	Hall *et al.* (1994)
	Non-Muscle-Specific MCAT Sites Bound by TEF-1		
SV40 *GTIIC,* viral (263AS)	AACTGACACA**CATTCCA**CAGCTGGTTC	HeLa Cells	Zenke *et al.* (1986)
SV40 *SphI/SphII,* viral (201/210AS)	TGAGATG**CATGCTT**TG**CATACTT**CTGCCTG	HeLa Cells	Zenke *et al.* (1986)
papillomavirus, viral			
7688S	CAACGCCTTA**CATACCG**CTGTTAGGCA	Keratinocytes	Ishiji *et al.* (1992)
7705S	CTGTTAGGCA**CATATTT**TTGGCTTGTT*		
involucrin, human			
-557	CACCAAAAGA**CAGAGTT**TATTTCCACA	Keratinocytes	Takahashi *et al.* (1995)
-542	TTTATTTCCA**CATAGGA**TGGAGTTAAA		
-437	TATCTTCAAA**CATATAA**CCCAGCCTGG		
polyomavirus mutant, viral (5238AS)	GGGTGGAAA**CATTCCA**GGCCTGGGTG	ES, EC, 2-cell embryos	Melin *et al.* (1993) DasGupta *et al.* (1993)
BOXF1, mouse (-328AS)	CTTATCCACG**CATTCCA**TTGTTGTCAA	EC cells	Kihara-Negishi *et al.* (1993)
	MCAT-Like Sites, Not Bound by TEF-1		
MCK, mouse (-1279S)	GGACACCCGA**GATGCCT**GGTTATAATT	Sk	Fabre-Suver and Hauschka (1996)
hCS	CTGGACCA**CATTCCA**GTCTAAT	Placenta	Jiang *et al.* (1994)
proliferin, mouse (-207/-218AS)	ATT**CATCTCA**TGTT**CATGCTC**TGACTAA	Placenta, fibroblasts, ↑ serum	Groskopf and Linzer (1994)
tRNA$^{(Ser)Sec}$, X. Laevis (-205S)	GAAGTACCAG**CATGCCT**CGCGCGCGTG	Oocytes	Myslinski *et al.* (1992)

* MCAT motifs are shown in bold face text. Ten nucleotides of sequence flanking the MCAT on each side are shown in every case where this information is published. Nominal E-box sequences are underlined.

† Sk, C, or VSM indicate the tissue type in which the MCAT-containing promoter was tested; skeletal myocytes or derived cell lines, cardiomyocytes, or vascular smooth muscle cells, respectively. Other notations refer to MCATs that have been demonstrated to be the target response elements for the stimuli noted: cyclic AMP (cAMP); β protein kinase C (PKC); or transforming growth factor β (TGFβ). [a] In top panel denotes an MCAT element whose function has not yet been shown.

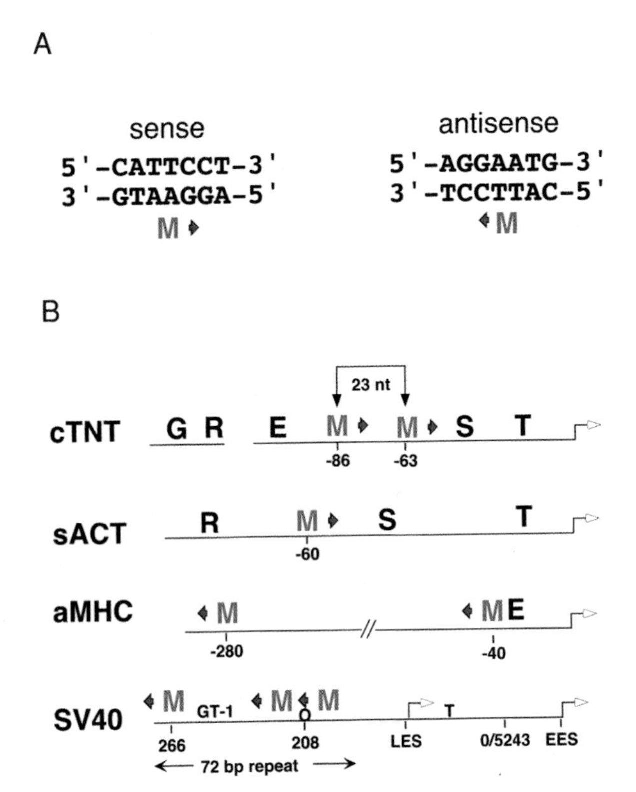

Figure 1 MCAT elements and complex MCAT-dependent promoters. (a) Nucleotide sequence of the core MCAT motif shown in both orientations (see Table I). Arrows designate sense or antisense orientation as indicated in B. (b) Diagram of MCAT-dependent promoters and neighboring regulatory sites. G, GATA binding sites; R, RSRF binding sites; E, E box site; S, SP1 binding site; T, TATA box; O, octamer binding site; LES, late early start of transcription; EES, early early start of transcription; M, MCAT motif.

The *VSMαA* inducibility can be reproduced in part *in vitro* by serum stimulation of fibroblasts, and this was shown to be mediated by MCAT and CarG elements of the VSMaA promoter (Cogan *et al.*, 1995).

Not all MCATs function as mitogen response elements, and those that do are not equivalent in their responsiveness: Hypertrophic induction of the *αMHC* gene seems to be mediated through the more distal of its two MCAT elements (Molkentin and Markham, 1994). However, serum stimulation or cAMP induction of the *αMHC* promoter is directed through its other, proximal MCAT site (Gupta *et al.*, 1994). Moreover, this same serum stimulation that induces the *αMHC* gene results in downregulation of the *βMHC* gene, whose fetal expression and adult hypertrophic induction is also regulated by MCAT elements (Gupta *et al.*, 1994).

Thus, different muscle-specific MCAT elements have distinct and separable properties. These properties may be enhanced and compensated for by the promoter settings in which they occur, with reference to other promoter elements, or by additional MCATs.

B. Non-Muscle-Specific Promoters Controlled by MCAT Elements

The MCAT consensus sequence derived from muscle promoters (Table I) is CATTCCT, and this nucleotide sequence was shown to be optimal in the context of the *cTnT* promoter (Table II; Farrance *et al.*, 1992). Replacements are tolerated at several positions including sites that render the MCAT core motif identical to the GTIIC, SphI, and SphII motifs of the Simian virus (SV) 40 enhancer (Table II). The latter variants are bona fide MCAT sites since both these viral sites and the muscle MCAT sites bind members of the *TEF-1* gene family (Farrance *et al.*, 1992; Farrance and Ordahl, 1996; Gupta *et al.*, 1994; Ishiji *et al.*, 1992; Kariya *et al.*, 1993; Molkentin and Markham, 1994; Shimizu *et al.*, 1993; Xiao *et al.*, 1991; Yockey *et al.*, 1996). The SV40 enhancer is active in a variety of nonmuscle cell types, including HeLa cells and P19 and F9 teratocarcinoma cells (Davidson *et al.*, 1986; Fromental, *et al.*, 1988; Ondek *et al.*, 1987). MCAT variant sites also govern nonmuscle-specific expression in the human *papillomavirus-16 E6* and *E7* oncogene that is expressed in skin and keratinized epithelium cells (Ishiji *et al.*, 1992).

Several diverged MCAT elements appear in the promoter of the *involucrin* gene that is expressed in terminally differentiated keratincytes. These sites have

Table II Activity of MCAT Point Mutation Variants

Base no.	Base in MCAT	Mutation	CAT activity (% of wild-type)	Found in
1	C	G	46 ± 12	
		A	61 ± 8	
		T	11 ± 3	
2	A	G	19 ± 4	
		T	10 ± 6	
		C	7 ± 2	
3	T	G	14 ± 3	
		A	51 ± 10	
		C	9 ± 3	
4	T	G	42 ± 10	SphI
		A	26 ± 2	SphII
		C	6 ± 1	
5	C	G	58 ± 7	
		A	11 ± 4	
		T	17 ± 4	
6	C	G	7 ± 3	
		A	25 ± 4	
		T	32 ± 8	SphI/II
7	T	G	49 ± 9	
		A	63 ± 9	GTIIC
		C	33 ± 8	

been shown to bind TEF-1, however, the contribution of the sites to the activity of the promoter has not been demonstrated (Takahashi *et al.,* 1995). MCAT variant sites are major regulatory elements of the *human chorionic somatomammotropin* (*hCS*) enhancer that activates transcription in a placental-specific fashion (Jacquemin *et al.,* 1994; Jiang *et al.,* 1994). A *NOS* implicated in NO production in multiple tissues, including central and peripheral neuronal tissue, kidney, adrenal medulla, pancreatic β cells, skeletal muscle, and the male sexual organ, was recently cloned and characterized (Hall *et al.,* 1994). Among other transcription factor binding sites in its promoter region (AP-2, Ets, NF-1, NRF-1, CREB, and NF-κB) is one perfect MCAT element (CATTCCT) and, strikingly, multiple imperfect copies that may control its expression in both the muscle and the nonmuscle tissues in which this protein is active (Hall *et al.,* 1994). Thus, the MCAT site is capable of directing gene expression in defined ways that differ from context to context: muscle-specific, both striated and smooth; non-tissue-specific; and placental-specific.

II. The TEF-1 Family of MCAT Binding Factors

Most MCAT elements characterized to date have been shown to bind the TEF-1 family of transcription factors (Xiao *et al.,* 1991; Farrance *et al.,* 1992; Farrance and Ordahl, 1996; Gupta *et al.,* 1994; Ishiji *et al.,* 1992; Kariya *et al.,* 1993; Molkentin and Markham, 1994; Shimizu *et al.,* 1993; Yockey *et al.,* 1996). The first cloned member of the family was TEF-1, identified as the protein binding to the GTIIC and Sph motifs of the SV40 enhancer in HeLa cell extracts (Davidson *et al.,* 1988; Xiao *et al.,* 1991). TEF-1 is a prototypic member of the *TEA* DNA-binding domain family of transcription factors that are involved in developmental events in animals and plants (Andrianopoulos and Timberlake, 1991; Bürglin, 1991; Fig. 2A). Recently, the first cloning of three distinct *TEF-1* cDNAs from a single organism, chick (Azakie *et al.,* 1996), allowed a number of TEF-1 cDNAs reported in other organisms to be catagorized with reference to the chick *TEF-1* multigene family (Table III lists the published vs the proposed systemic nomenclature; Xiao *et al.,* 1991; Blatt and DePamphilis, 1993; Shimizu *et al.,* 1993; Jacquemin *et al.,* 1994, 1996; Stewart *et al.,* 1994, 1996; Yasunami *et al.,* 1995; Azakie *et al.,* 1996; Hsu *et al.,* 1996; Yockey *et al.,* 1996). The *TEA* DNA-binding domain is extremely well conserved throughout all members of the TEF-1 family in multiple organisms, with a few notable exceptions (Fig. 7). The recognition that TEF-1 was actually a family of proteins prompted us to suggest a coordinate naming scheme, including renaming the original member of the group NTEF-1 (*Nominal* TEF-1, that for which the family is named) and this is the nomenclature used in this review.

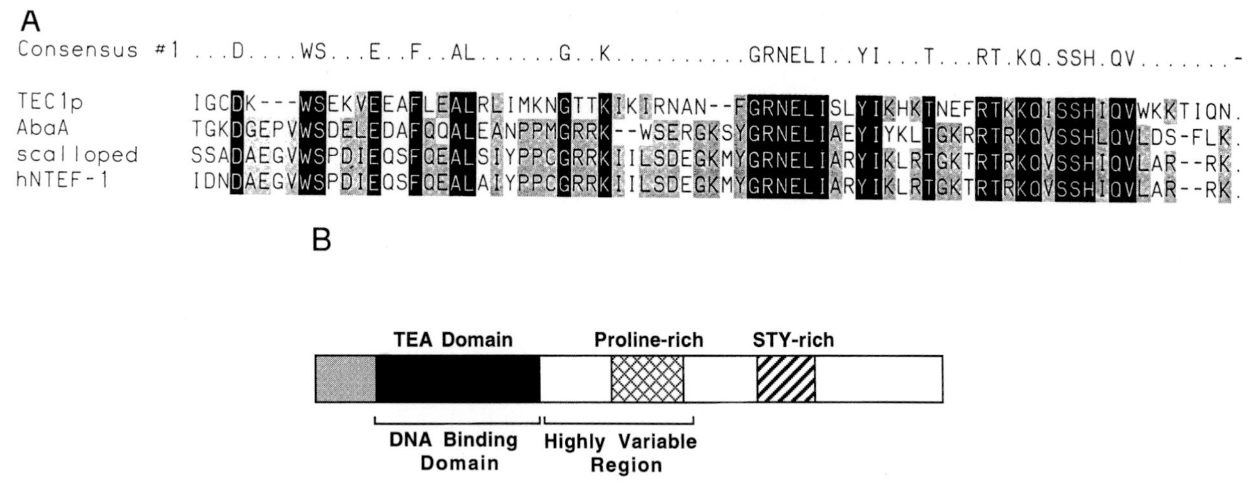

Figure 2 Structural features of the TEF-1 family. (a) Alignment of TEA domains from the prototypic members of the class: yeast TEC1, *Aspergillus nidulans* AbaA, human NTEF-1, and *drosophila* Scalloped (Campbell *et al.,* 1992; Laloux *et al.,* 1990; Mirabito *et al.,* 1989; Xiao *et al.,* 1991). Residues that are shared by all are shown in black and gray shading indicates where the chemical nature of residues is conserved in three or more proteins. A consensus sequence is shown above the alignment, where the consensus residue is that found in all proteins. (b) TEF-1 structural domains. Diagramatic representation showing the overall structure of the TEF-1 family of transcription factors (not drawn to scale). The N terminus of the protein (gray) is highly variable and may contain part of a multipartite transcriptional activating function. The TEA domain is a highly conserved domain that contains the DNA-binding function of the protein (black). This domain is followed by a highly variable domain, including a proline-rich domain (cross-hatched) that may be another part of the multipartite activating domain. The C terminus of the protein is also well conserved and contains a serine-, threonine-, and tyrosine-rich domain (striped) that may be an additional part of the multipartite activating domain.

Table III Systematic Nomenclature of Vertebrate TEF-1 Family Members

General name	NTEF-1	RTEF-1	DTEF-1	ETEF-1
Chicken	cNTEF-1[a]	cRTEF-1[a]	cDTEF-1[a]	—
Human	hTEF-1[b]	hTEF-3[c,d]	hTEF-5[c]	hTEF-4[c]
Mouse	MTEF-1[e]	mTEF-3[c] FR-19[f] TEFR-1[h]	mTEF-5[c]	mTEF-5[c] ETF[g]

[a]Azakie *et al.* (1996).
[b]Xiao *et al.* (1991).
[c]Jacquemin *et al.* (1996).
[d]Stewart *et al.* (1996).
[e]Blatt and DePamphilis (1993) and Shimizu *et al.* (1993).
[f]Hsu *et al.* (1996).
[g]Yasunami *et al.* (1995).
[h]Yockey *et al.* (1996).

Four members of the TEF-1 family have been described to date, though it appears likely that there are more. They are all well conserved at the amino acid level throughout most of the length of the protein and in particular in the DNA-binding domain, the TEA domain. Individual members of the family are also well conserved through evolution (Figs 3–6). All members of the family have distinct spatiotemporal expression patterns and none so far is ubiquitiously expressed. Two family members, NTEF-1 and RTEF-1, are skeletal muscle-enriched and absent from several other tissues, DTEF-1 is enriched in cardiac muscle and likewise absent from certain tissues, and ETEF-1 has a clearly defined spatiotemporal pattern of expression mostly confined to embryonic neural tissues (Table IV; Azakie *et al.*, 1996; Yasunami *et al.*, 1995).

A. NTEF-1

The first cloned member of the family is now called NTEF-1. Originally cloned from HeLa cells (Xiao *et al.*, 1991), NTEF-1 was subsequently cloned from mouse, chick, and rat (Shimizu *et al.*, 1993; Blatt and DePamphilis, 1993; Chen *et al.*, 1994; Azakie *et al.*, 1996; P. Simpson, personal communication; Fig. 3). Avian NTEF-1 is 96.5% identical to human NTEF-1, and mouse and human NTEF-1 are 99% identical at the amino acid level (Xiao *et al.*, 1991; Blatt and DePamphilis, 1993; Azakie *et al.*, 1996; plus exon VTSM: Shimizu *et al.*, 1993). Such a high level of conservation may indicate a conservation of important function. Northern blots of *NTEF-1* transcripts indicate that NTEF-1 is expressed in HeLa and F9 embryonic teratoma cell lines but absent in lymphoid-derived cell lines (Xiao *et al.*, 1991). In tissue blots human NTEF-1 is found to be enriched in skeletal muscle, found at lower levels in pancreas, placenta, and heart and at trace levels in brain, and is undetectable in liver, lung, and kidney (Stewart, *et al.*, 1996). Two reports of Northern blots for mouse NTEF-1 differ substantially with respect to the level found in heart, lung, and brain, but enrichment in skeletal muscle and kidney is reported in both (Blatt and DePamphilis, 1993; Shimizu *et al.*, 1993). *In situ* hybridization in Embryonic (E) Day 8.5–9.5 mouse embryos shows widespread expression of NTEF-1 but, by E 10.5, preferential expression in discrete tissues, including the mitotic layer of the neuroepithelium, and in the developing myocardium. By E 15.5 NTEF-1 expression persists in these tissues but is also found in skeletal muscle anlagen and subsequently in a range of other tissues (Jacquemin *et al.*, 1996). This pattern, as well as the finding that MCAT-binding activity is detected as early as the two-cell embryo stage (Melin *et al.*, 1993), may suggest multiphasic functions for NTEF-1, as well as for the other TEF-1 family members, in embryogenesis and organogenesis.

Several alternatively spliced cDNA isoforms of NTEF-1 have been described in human and mouse (Shimizu *et al.*, 1993; Xiao *et al.*, 1991); however, the function of these alternative spliced regions is not yet known. Many alternatively spliced forms of NTEF-1 have been found in rat tissues, with widely differing transactivational potentials and differential tissue distribution (P. Simpson, personal communication). It is not known if these isoforms will be found in other organisms and they have not been described in the other rodent model (mouse), in which evolutionary relationship suggests they should exist.

To date, NTEF-1 is the only member of the *TEF-1* gene family that has been inactivated in the genome.

Table IV Tissue Distributions of TEF-1 Family Members

Name	TEF-1 iso-mRNAs and tissues						
	Cardiac muscle	Skeletal muscle	Smooth muscle	Brain	Lung	Liver	Kidney
RTEF-1	+++	+++	+	−	+	−	−
DTEF-1	+++	+	++	−	++	−	+
NTEF-1	+	++	+	+	+	+	+

Figure 3 Alignment of known NTEF-1 proteins. NTEF-1 proteins are shown from human, mouse, and chick (Blatt and DePamphilis, 1993; Shimizu *et al.*, 1993; Stewart *et al.*, 1996; Xiao *et al.*, 1991). Residues that are shared by two or more proteins are shown in black. A consensus sequence for NTEF-1 is shown above the alignment, where the consensus residue is shared by at least 75% (three or more) of the proteins; otherwise a period is shown. The alignment was produced using the MegAlign program of the DNASTAR package.

The NTEF-1 knockout was identifed in a screen for developmentally regulated genes using an insertional mutagenesis approach called promoter trapping (Friedrich, 1991). A promoterless βGal/Neo construct, inserted into the genome of embryonic stem (ES) cells at random, may be activated by an upstream active promoter and therefore selected by neomycin resistance. Thus, a locus active in early develpment is marked but inactived by the insertion. Expression patterns of the endogenous gene promoter can be tracked by βGal staining and the function of the gene inferred, at least in part, by the phenotype of the homozygous null mice.

NTEF-1 insertional mutant heterozygotes express LacZ in early preimplantation cleavage-stage embryos (consistent with reports that TEF-1 is active in two-cell mouse embryos; Melin *et al.*, 1993) and at all stages of development tested through to adulthood. Staining heterozygotes at E 12 is ubiquitous. Adult heterozygotes express LacZ at relatively high levels in kidney, lung, uterus, heart, and skeletal muscle and at lower levels in other tissues, including brain, liver, thymus, and spleen. Neither lymphocytes nor whole blood expressed the marker gene, consistent with the inactivity of TEF-1-dependent enhancers in lymphoid cells (Davidson *et al.*, 1986; Fromental, *et al.*, 1988; Ondek *et al.*, 1987, 1988; Zenke *et al.*, 1986).

Despite the expression of NTEF-1 in preimplantation embryos, NTEF-1 homozygous null mice survive early development and are overtly normal until E 9.5. However, homozygous null mice die between E 10.5 and 11.5 apparently from cardiac insufficiency. The primitive cardiac tubes form and fuse normally but late cardiac maturation appears to be impaired, leading to a thin ventricular wall and reduced trabeculation, a slowed and weakened heartbeat, edema, and eventually necrosis and resorbtion. Since many cardiac-specific sarcomeric structural genes are regulated by TEF-1 binding sites, it was hypothesized that this impairment was directly due to a deficiency in the expression of these *TEF-1* target genes. However, the protein products of the TEF-1 target genes, *cTnT,* and *cTnI,* and cardiac myosins, as well as *cTnC* mRNA, were examined and found to be present at normal levels. Thus, NTEF-1 is not absolutely required for early development or for further activation of early cardiac structural genes. NTEF-1, however, may be required for late events in cardiogenesis. This observation begged the question of redundancy with other members of the TEF-1 family.

Transcripts for other members of the family are to be found in heart, and still others may be described that are also expressed very early in development and that are responsible for early TEF-1 functions in the whole organism and in early heart and skeletal muscle organogenesis.

B. RTEF-1

RTEF-1, the second member of the TEF-1 family to be identified, was originally cloned from avian muscle in the search for an MCAT binding factor that is responsible for the activity of the *cTnT* promoter (Farrance *et al.*, 1992; Mar and Ordahl, 1990; Stewart *et al.*, 1994) and has now been cloned from multiple organisms (Stewart *et al.*, 1994, 1996; Frigerio *et al.*, 1995; Hsu *et al.*, 1996; Yockey *et al.*, 1996; Fig. 4). There is 89% conservation between RTEF-1 cloned from avian cells and that later isolated from human cells (Stewart *et al.*, 1994, 1996). RTEF-1 proteins in both chick and mouse appear to be responsible for muscle-specific gel-shifted species generated on MCAT elements, although in neither species is the protein strictly muscle-specific (Farrance and Ordahl, 1996; Yockey *et al.*, 1996).

Two cDNA clones named *TEFR-1a* and *-b* (most likely splice isoforms), which were cloned as the MCAT-binding factor responsible for muscle specificity of the *βMHC* promoter, are thought to be murine RTEF-1 homologs (Yockey *et al.*, 1996). Identical cDNAs named *FR-19A* and *-B* were almost simultaneously cloned as a delayed early response gene for fibroblast growth factor-1 (FGF-1; acidic FGF) in murine fibroblasts (Hsu *et al.*, 1996). These murine family members (*TEFR-1a*, and *-b* and *FR-19A* and *-B*) and a later mouse clone derived by PCR cloning (*mTEF-3;* Jacquemin *et al.*, 1996) are more similar to chick RTEF-1 than any other cloned TEF-1 family member and group with them in phylogenetic trees derived by parsimonious methods (Stewart, *et al.*, 1996). Strangely, however, all these clones have five changes in the TEA domain relative to NTEF-1 (Hsu *et al.*, 1996; Jacquemin *et al.*, 1996; Yockey *et al.*, 1996). Neither human nor avian RTEF-1 have changes in the TEA domain relative to NTEF-1 (Stewart *et al.*, 1994, 1996). It seems surprising that rodent RTEF-1 should differ so markedly from human RTEF-1 when avian is relatively better conserved in the TEA domain, and it is possible that a more closely related RTEF-1 form remains to be found in rodents, perhaps a splice variant of those cloned to date.

The changes in the TEA domain may have implications for binding site selectivity. The TEA domain in the *drosophila scalloped* TEA domain protein has only a single substitution in its TEA domain relative to human NTEF-1 (Fig. 7B) and this substitution has important consequences for sequence-specific DNA binding (Hwang *et al.*, 1993). Both RTEF-1 variants noted previously have been shown to be able to bind MCAT sites (Farrance and Ordahl, 1996; Stewart *et al.*, 1994; Yockey *et al.*, 1996); but so far there has been no systematic attempt to compare their affinity for a range of MCAT sites of differing core and flanking sequence. Interestingly, unlike NTEF-1, RTEF-1 with or without the five-amino acid changes does not exhibit cooperative binding to tandemly repeated GTIIC elements, which may in part explain the divergence between the cooperative binding observed on multimerized *cTnT*-derived MCAT elements vs GTIIC elements using muscle and HeLa nuclear extracts, respectively (Davidson *et al.*, 1988; Fromental, *et al.*, 1988; Jacquemin *et al.*, 1996; S. B. Larkin, unpublished observation).

Chick *RTEF-1* transcripts in adult and late embryonic tissues are enriched in both skeletal and cardiac muscle and found at successively lower levels in gizzard (a smooth muscle-enriched tissue), liver, and brain (Stewart *et al.*, 1994). This differential mRNA abundance corresponds to the levels of MCAT-binding proteins detected by gel shift and southwestern blot (Farrance *et al.*, 1992). *RTEF-1* transcripts in adult human tissues are enriched in skeletal muscle and pancreas and at lower levels in heart, placenta, and kidney, and these transcripts are undetectable in brain, liver, and lung (Stewart, *et al.*, 1996). *RTEF-1* transcripts in adult mouse tissues show enrichment in lung followed by skeletal muscle and heart, kidney and skin (Hsu *et al.*, 1996; Yockey *et al.*, 1996).

RTEF-1, like NTEF-1, is found throughout the early mouse embryo (E 7.5–8.5) but at later stages (E 9.5–12.5) is found only in skeletal muscle precursors (Jacquemin *et al.*, 1996). The transcripts are found from E 9.5 in the somitic myotome in the craniocaudal progression that marks the determined skeletal muscle lineage (Jacquemin *et al.*, 1996; Yockey *et al.*, 1996).

The RTEF-1 clones from mouse and chick both exist as multiple alternatively spliced isoforms, including the curious splicing seen by Yockey *et al.* in which both the 5′ and 3′ ends of the TEFR-1a and -1b clones differ markedly (Hsu *et al.*, 1996; Stewart *et al.*, 1994; Yockey *et al.*, 1996). Splicing variants in the coding region of chick and mouse RTEF-1 modify the region just downstream of the TEA domain (immediately following the amino acid residues KLK), however, the exons found in these positions differ. This splicing of RTEF-1 appears to have implications for the transactivational potential of the chick and mouse isoforms (Stewart *et al.*, 1994; Yockey *et al.*, 1996). In the chick *RTEF-1* cDNAs, the additional exon may confer an activation function, whereas in mouse the effect of the additional exon is dependent on the protein context, resulting in opposite effects in different deletion mutants.

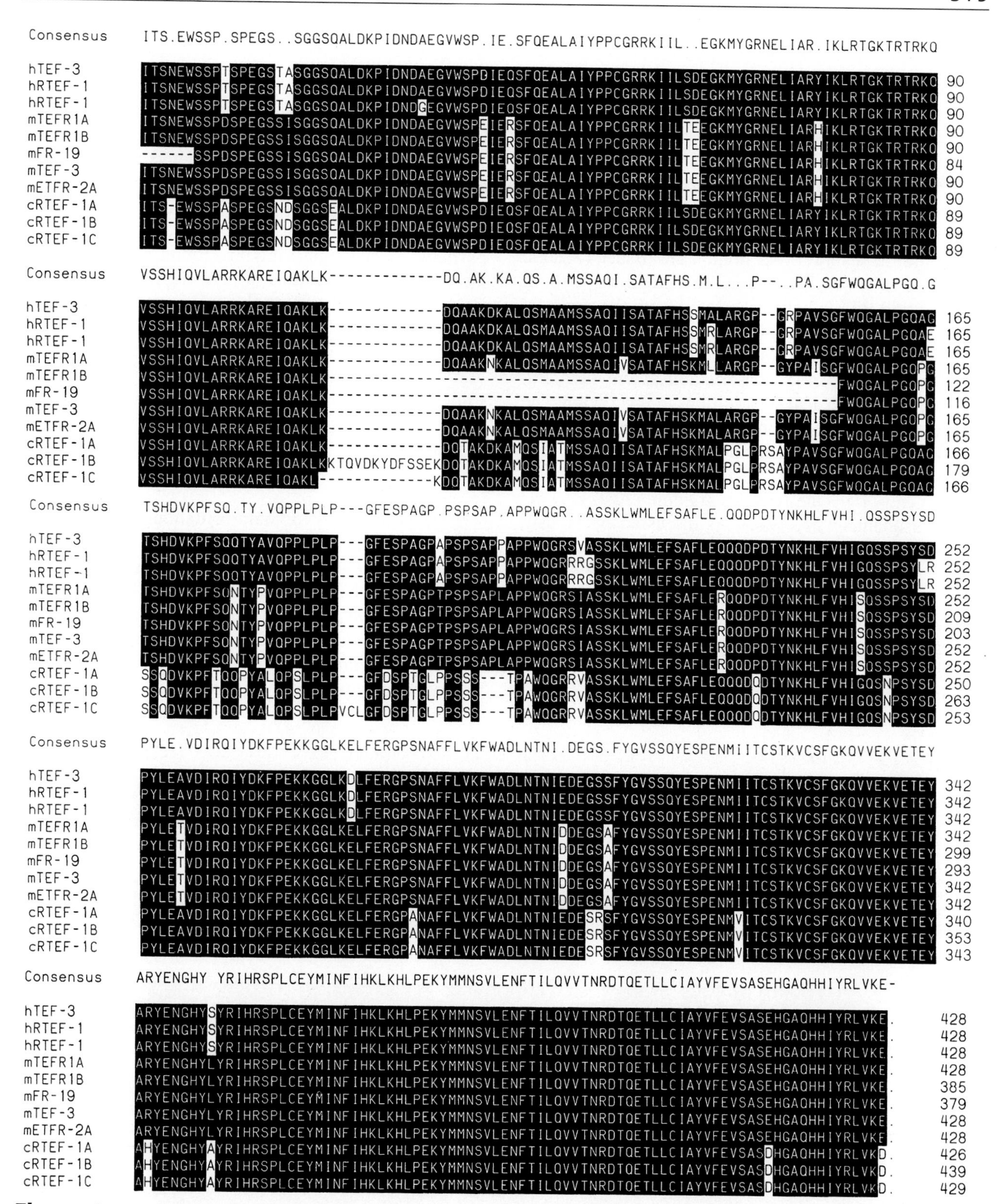

Figure 4 Alignment of all known RTEF-1 proteins. RTEF-1 proteins, and isoforms thereof, are shown from human, mouse and chick (Hsu *et al.*, 1996; Jacquemin *et al.*, 1996; Stewart *et al.*, 1994, 1996; Yockey *et al.*, 1996). Residues that are shared by the majority (five or more) of the proteins are shown in black. A consensus sequence for RTEF-1 is shown above the alignment, where the consensus residue is that shared by 75% (eight or more) of the proteins; otherwise, a period is shown.

C. DTEF-1

DTEF-1 was also cloned first from chicken, and recently a homolog has been reported in mouse (Azakie *et al.,* 1996; Yasunami *et al.,* 1996; Fig. 5). The chick cDNA was isolated in a search for TEF-1-related family members that might contribute to TEF-1 activity in heart, and in fact in chick DTEF-1 shows a high level of expression in the heart and not in skeletal muscle. It is also found in several other nonmuscle tissues (Azakie *et al.,* 1996). Chick DTEF-1 is 72% identical to both NTEF-1 and RTEF-1 clones isolated from the same organism and was thus classified as a novel family member. Curiously, splicing isoforms of chick DTEF-1 exist that have either perfect conservation of the TEA domain or two amino acid substitutions in this region. The mouse *DTEF-1* cDNA contains a single substitution in the TEA domain at the same position as one of the substitutions in the chick *DTEF-1A* cDNA, although a lysine instead of glycine replaces arginine (Azakie *et al.,* 1996; Yasunami *et al.,* 1996). As discussed previously, these substitutions may have implications for the binding specificity of these isoforms. That this member of the TEF-1 family is also subject to alternative splicing that results in two different versions of the TEA domain strengthens the notion that the differences seen between murine RTEF-1 and chick and human RTEF-1 might also result from splicing events.

DTEF-1 transcripts are enriched in cardiac muscle, with lower levels found in lung, and gizzard (a smooth muscle-enriched tissue) and very low levels found in kidney and skeletal muscle, and *DTEF-1* transcripts are undetectable in liver and brain tissues (Azakie *et al.,* 1996). Mouse *DTEF-1* mRNA is detectable by Northern blot in embryos from E8 and has a broad tissue distribution consistent with expression in vascular smooth muscle (Yasunami *et al.,* 1996). In the chick, presumed partially spliced transcripts are found at much higher levels in skeletal muscle which might indicate post-transcriptional control of the tissue representation of DTEF-1. Since alternative splicing is widespread in the TEF-1 family and most members initiate translation at nonstandard initiation codons, there is ample reason to believe that this family is subject to extensive post-transcriptional regulation.

D. ETEF-1

Embryonic TEA domain-containing factor (ETF) was first cloned in the search for cDNAs preferentially expressed in neural precursor cells and later by a PCR screen for cDNAs containing related TEA domains (Yasunami *et al.,* 1995; Jacquemin *et al.,* 1996; Fig. 6). We proposed renaming this clone ETEF-1 to reflect both its membership of the *TEF-1* gene family and its predominant embryonic expression pattern. Sequence

Figure 5 Alignment of all known DTEF-1 proteins. DTEF-1 proteins and their isoforms are shown from mouse and chick (Azakie *et al.,* 1996; Yasunami *et al.,* 1996). Residues that are shared by 67% (two or more) of the proteins are shown in black. A consensus sequence for DTEF-1 is shown above the alignment, where the consensus residue is that shared by 67% (two or more) of the proteins; otherwise, a period is shown.

Figure 6 Alignment of all known ETEF-1 proteins. ETEF-1 proteins are shown from human and mouse (Jacquemin *et al.*, 1996; Yasunami *et al.*, 1995). Residues that are shared by 67% (two or more) of the proteins are shown in black. A consensus sequence for ETEF-1 is shown above the alignment, where the consensus residue is that shared by 67% (two or more) of the proteins; otherwise, a period is shown.

comparison with existing family members indicated that this murine cDNA encodes a protein that is only 66% identical to murine NTEF-1 and 65% identical to chick RTEF-1, thus making this, overall, the most diverged member of the TEF-1 family. However, unlike the murine *RTEF-1* cDNAs and chick *DTEF-1A*, within the TEA domain *ETEF-1* is identical to all other known TEF-1 family members.

Unlike all other members of the family, *ETEF-1* transcripts do not appear to be enriched in muscle. In one account *ETEF-1* transcripts peak at E 10 of embryonic development, with high expression in the embryonic hindbrain from which the adult cerebellum is derived and weaker expression in the distal tips of limb buds and tail (Yasunami *et al.*, 1995). In another report, this cDNA, independently cloned, is also found by *in situ* hybridization in the developing neuroepithelium of the brain and spinal cord through multiple days of embryonic development but is also found in other developing organ systems, including the lungs, bladder, and facial and gut mesenchyme. *ETEF-1* transcripts appear to be absent from both developing heart and skeletal muscle (Jacquemin *et al.*, 1996). In adult tissues very low mRNA levels are found in brain by Northern blot, but mRNA levels are found in no other tissue tested (Yasunami *et al.*, 1995).

E. Non-TEF-1 MCAT Binding Factors

Some MCAT sites may bind a protein(s) distinct from TEF-1 as distinguished by immunological criteria (Table I). The *hCS* enhancer contains multiple MCAT-related sites that apparently bind a factor (CSEF-1) that is judged to be a non-TEF-1-related protein by virtue of its size (~33kDa compared to TEF-1 family members in the ~50kDa size range), thermal stability, and nonimmunoreactivity with a polyclonal antibody raised against chicken RTEF-1 that also recognizes human and rat NTEF-1 (Jiang and Eberhardt, 1995, 1996). CSEF-1 is found in BeWo choriocarcinoma cells, COS-1 cells, and at low levels in HeLa cells. However, this factor has not been cloned, so it may yet prove to be a diverged member of the TEF-1 family or a member of the larger TEA domain family of proteins.

The *muscle creatine kinase* (*MCK*) enhancer contains a sequence, Trex (CATCTCG), that is distantly related to the MCAT heptamer and that was shown to be important for the activity of the enhancer in skeletal myoblasts and myocytes but not cardiomyocytes. This sequence was shown to be a target for binding of a factor unrelated to TEF-1, judging by sequence specificity and immunoreactivity (Fabre-Suver and Hauschka, 1996). Likewise, the *proliferin* and the *Xenopus laevis* *tRNA*$^{(Ser)Sec}$ gene promoters have SphI/II-like sequences that appear to be important in their activity; however, in both cases the binding species has been shown by sequence specificity and in the former case also by immunoreactivity to be distinct from TEF-1 (Groskopf and Linzer, 1994; Myslinski *et al.,* 1992). In the cases of the *MCK* enhancer and tRNA promoter this may simply be due to the very diverged nature of the MCAT sites (CATCTCG, and CATACACN$_9$CATGCTG, respectively).

III. TEF-1 Structure–Function

A. The TEA Class of DNA Binding Domains

NTEF-1 contains a DNA-binding domain toward its N terminus called the TEA domain (Xiao *et al.,* 1991; Hwang *et al.,* 1993; Fig. 2B). The *TEF-1* DNA-binding domain is a prototype of the TEA domain class of DNA-binding domains (Andrianopoulos and Timberlake, 1991; Bürglin, 1991; Fig. 2A). The other prototypic members of this class are: the yeast transcription factor TEC-1, which regulates *Ty1* enhancer activity (Laloux *et al.,* 1990); the transcriptional activator of *Aspergillus nidulans* sporulation-specific genes, AbaA (Mirabito *et al.,* 1989); and the *drosophila* transcription factor Scalloped (sd), which is required for sensory organ differentiation (Campbell *et al.,* 1992). Recently, human NTEF-1 was shown to be capable of functionally replacing the *sd* gene in *drosophila* mutants (Deshpande *et al.,* 1997). Thus, the TEA domain is a conserved protein domain found to be involved in developmental events throughout evolution.

Predictive analysis of the TEA domain indicated three α-helices (Bürglin, 1991) or one α-helix and two β sheets (Andrianopoulos and Timberlake, 1991). The helices/sheets were mutagenized in NTEF-1 by introduction of structure-breaking prolines: disruption of the first or third helix destroyed or reduced, respectively, sequence-specific site binding (Hwang *et al.,* 1993). However, mutagenesis of the second helix had no apparent effect. The deletion of the C-terminus domain or the serine, threonine, and tyrosine (S,T,Y)-rich domain also interfered with protein binding, although a fragment of protein containing only the TEA domain is necessary and sufficient for efficient binding, indicating modulating interactions between several domains of the intact protein. Finally, some degree of sequence specificity is conferred by residue 48 of helix 1 that is an alanine in all TEF-1 family members but a serine in the *drosophila* TEA domain protein Scalloped (Hwang *et al.,* 1993). The TEA domain of Scalloped is otherwise perfectly conserved and interestingly does not bind GTIIC. Substitution of this serine with alanine in Scalloped only partially restores the ability to bind GTIIC unless the carboxy terminus is deleted, at which point the mutant regains the ability to bind the GTIIC element. However, surprisingly, substitution of the Ala48 with a serine in NTEF-1 does not affect its binding (Hwang *et al.,* 1993). Thus, TEF-1 family members with amino acid substitutions in the TEA domain may have distinct binding site specificies, but until their relative affinities for a range of MCAT variants are directly tested, no prediction can be made.

B. TEF-1 Contains a Multipartite Activation Domain

Study of the activation potential of NTEF-1 proved difficult since the protein was found to be entirely inactive upon transfection into lymphoid-derived cells lines in which the protein is normally not expressed. This has been suggested to be due to the action of a negatively acting TATA-associated factor (Chaudhary *et al.,* 1994) or the lack of positively acting factors required for TEF-1 transactivation (Xiao *et al.,* 1991). Moreover, attempts to superactivate with transfected NTEF-1 or RTEF-1 in cell types in which endogenous TEF-1 is active produced profound repression of reporters bearing TEF-1 binding sites but not those based on unrelated sites (Hwang *et al.,* 1993; Ishiji *et al.,* 1992; Shimizu *et al.,* 1993; Stewart *et al.,* 1994; Xiao *et al.,* 1991). This failure to demonstrate transactivation experimentally has been proposed to result from the limiting presence of a bridging factor(s) that allows TEF-1 to interact with the general transcriptional machinery at the TATA box. Such factors have been variously called coactivators or transcriptional intermediary factors. Proteins that seem to function in this way have been cloned for the Oct factors, nuclear receptors, and CREB (Arany *et al.,* 1994; Chiba *et al.,* 1994; Guarente, 1995; Luo and Roeder, 1995; Meier, 1996). These are apparently distinct from the TATA-binding protein (TBP)-associated factors that form a tight complex with the TBP in higher eukaryotes (Tansey and Herr, 1997). Repression that is proposed to result from the competition of coactivators away from the promoter by excess concentrations of transfected transactivators which would otherwise functionally interact with them is termed "squelching." Progress to date in the characterization and isolation of TEF-1-specific coactivators is detailed later.

The conclusion that NTEF-1 is indeed an activator of transcription and not simply a repressor is based on studies done in which NTEF-1 and later RTEF-1, or portions thereof, were linked to the heterologous DNA-binding domain (DBD) of the yeast transcription factor GAL4 (Hwang *et al.,* 1993; Stewart *et al.,* 1994; Xiao *et al.,* 1991; Yockey *et al.,* 1996). UAS_G (GAL4 binding site)-based reporters are used to monitor chimeric activity. These chimeras retain the ability to repress TEF-1 binding site-based reporters but do not as effectively self-squelch. The reasons for this are not clear but might reflect a reduced requirement for coactivators resulting from a different interaction of the GAL-NTEF-1 chimera with its UAS_G binding site. In any case, using this system certain NTEF-1 constructs were shown to activate UAS_G-based reporters by as much as 60-fold (Hwang *et al.,* 1993).

Further dissection of NTEF-1 activation domains utilized variously deleted NTEF-1 fused to the GAL4 DBD (Hwang *et al.,* 1993). These studies revealed a complex activation function located in multiple interdependent domains in the protein. The additional presence of the natural TEF-1 DBD (the TEA domain) was found to depress the activation function of the chimera, and its inactivation by deletion resulted in a more active chimera (paradoxically, this did not result in a more potent squelching ability of the TEA domain-deleted mutant while otherwise transactivation strength is well correlated with the degree of activation efficiency of the various mutants). The transcriptional activity of NTEF-1 appeared to be resident in at least three subdomains, none of which alone are active. Certain combinations of two such subdomains could produce a chimera of reduced activity but for full activity any deletion other than the TEA domain resulted in some diminution of activity (Hwang *et al.,* 1993). Thus, NTEF-1 may require an intact three-dimensional structure to function effectively, both for binding and for maximal activity. In this sense, it may prove an exception to the accepted notion of modular and separable domains in transcription factors.

The three NTEF-1 activation subdomains have respectively an acidic character (the N-terminus domain) or are rich in prolines or in S,T,Y residues (Fig. 2B). Such characteristics have been noted in activation domains of other transcription factors, such as VP16, CTF/NF-1, Oct-2, GHF-1, and myogenin (Treizenberg, 1995; Theill *et al.,* 1989; Tanaka and Herr, 1990; Schwarz *et al.,* 1992). These domains are conserved in their overall characteristics in other members of the family (Fig. 7). Curiously, both the acidic and the proline-rich domains are the most poorly conserved domains of the TEF family in actual amino acid sequence (Fig. 7). The S,T,Y- rich region, by contrast, is well conserved. A deletion study has been undertaken of one other member of the TEF-1 family (mouse TEFR-1; Yockey *et al.,* 1996) in which a deletion of the N terminus, including the entire acidic domain, increases the activation capacity of GAL4-mTEFR1 chimeras. Moreover, deletion of the proline-rich region does not abolish mTEFR1 activity. Thus, the function of the NTEF-1-defined activation subdomains may not be conserved through other members of the family.

The extreme C terminus of NTEF-1 may also contribute to activation function. Deletion of this domain disables binding of the resultant GAL-NTEF-1 chimeras, despite the presence of the heterologous GAL4 DBD, and thus its function in activation is deduced indirectly by the inability of C terminus deleted mutants to squelch endogenous TEF-1 activity (Hwang *et al.,* 1993). Similarly, deletion of the extreme C-terminus of an otherwise highly active GAL-mTEFR-1 deletion abolishes its activity; however, the ability of this chimera to bind DNA was not tested (Yockey *et al.,* 1996).

C. Repression by TEF-1

In the context of the *hCS* enhancer, a TEF-1 protein appears to function as a represser of basal *hCS* transcription by competition with a distinct factor, CSEF-1, for their common binding site (Jiang and Eberhardt, 1995). Multiple copies of an SV40-derived GTIIC element can activate transcription from the *hCS* promoter in COS cells (which contain CSEF-1 binding activity but little TEF-1) but not in HeLa cells (which contain TEF-1 but very little CSEF-1 binding activity). BeWo cells are a placental cell line in which the *hCS* enhancer normally functions and that contain both TEF-1 and CSEF-1 binding activities. In these cells, the GTIIC-based artificial enhancer–promoter reporters are moderately active relative to their activity in COS cells. Introduction into BeWo cells of antisense oligonucleotides targeted to TEF-1 coding sequences produces a threefold increase in activity of wild-type or GTIIC multimeric enhancer constructs, consistent with the idea that TEF-1 represses the activity of these enhancers via competition for its binding site (Jiang and Eberhardt, 1995). However, basal transcription of several non-MCAT-containing promoters is also upregulated in BeWo cells by the reduction of TEF-1 (Jiang and Eberhardt, 1996). Direct interaction of human NTEF-1 with recombinant TBP via at least the NTEF-1 proline-rich domain was found (Jiang and Eberhardt, 1996). Thus, interference by TEF-1 with TBP binding was proposed to account for the inhibition of basal transcription in these cells. Direct binding of TEF-1 to TBP remains controversial, however.

The *involucrin* gene, which encodes a precursor protein for the cornified envelope in terminally differentiated keratinatocytes, has three highly divergent MCAT sites in its promoter (CAGAGTT, CATAGGA, and CATATAA) that nevertheless appear to bind TEF-1 (Takahashi *et al.,* 1995). TEF-1 had previously been im-

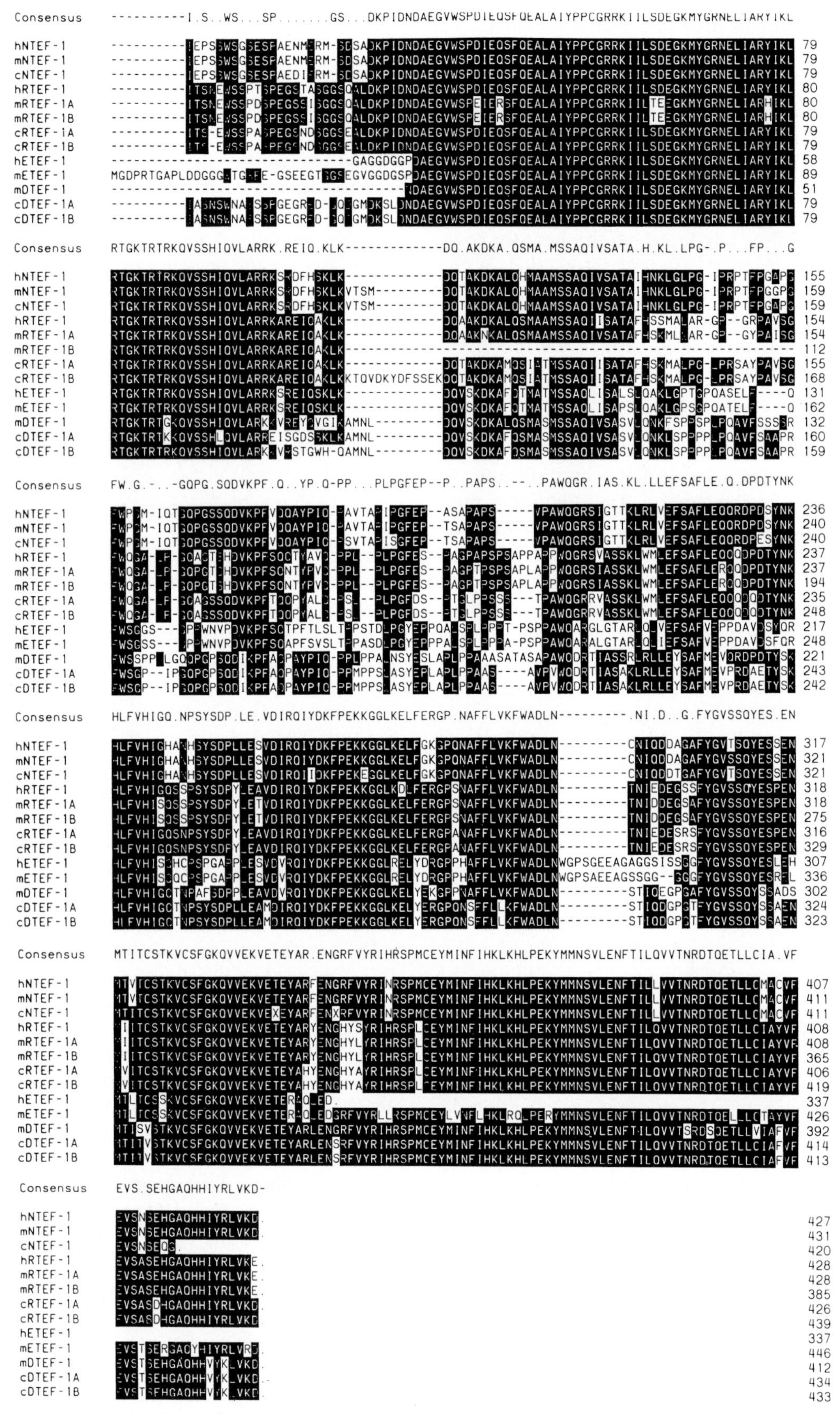

Figure 7 Alignment of the TEF-1 family. Representative members of NTEF-1, RTEF-1, DTEF-1, and ETEF-1, as well as splicing isoforms of each, are shown from each species where they have been cloned (for references, see Figs. 3–6). The most frequently occurring residues are shown in black. A consensus sequence for the TEF-1 family is shown above the alignment, where the consensus residue is that shared by 58% (7 or more) of the proteins; otherwise, a period is shown.

plicated in the activity of the human *papillomavirus-16 E 6* and *E 7* oncogene enhancer in keratinocytes (Ishiji *et al.,* 1992). Following cotransfection with TEF-1 or GAL4-TEF-1, the *involucrin* promoter was found to be repressed and the authors thus conclude that endogenous TEF-1 normally functions to repress this promoter (Takahashi *et al.,* 1995). However, it should be noted that the concentrations of both TEF-1 and GAL4-TEF-1, used in this study have been seen to result in repression of other TEF-1-responsive promoters (Hwang *et al.,* 1993; Ishiji *et al.,* 1992; Xiao *et al.,* 1991). This repression by the GAL4-TEF-1 chimera in particular, which lacks the capacity to bind the MCAT-containing promoter, has been taken as evidence of squelching of endogenous TEF-1 activity by competition for the proposed intermediary factors or coactivators rather than active repression, although experimental evidence directly supporting this hypothesis is still absent. Therefore, the conclusion that TEF-1 normally represses this promoter must be treated with caution.

IV. Regulation of TEF-1 Expression

A. Transcription: the NTEF-1 Promoter

A promoter has been isolated and partially characterized for human *NTEF-1* (Boam *et al.,* 1995). The promoter lacks a TATA box but contains an initiator element, SP1 sites, and ATF-1 binding sites. The *NTEF-1* promoter is active in HeLa cells but inactive in lymphoid cell lines, as would be predicted by the representation of *NTEF-1* mRNA in these cell types. *In vitro,* however, the promoter is active in both lymphoid and HeLa cell extracts (Boam *et al.,* 1995). SP1 and ATF-1 were shown to bind their respective sites in both cell types; thus the question of how the tissue specificity of this promoter is achieved *in vivo* remains. The authors speculate that either posttranscriptional destabilization of the *NTEF-1* transcripts or inactivation of the promoter by CpG island methylation occurs in the lymphoid cells. The promoter was shown to rely on the most proximal SP1 site for localization of the transcriptional start at the initiator element. This function of the SP1 could not be functionally replaced by substitution of a TATA box. This is similar to the function assigned to SP1 in early zygotic gene activation (Nothias, *et al.,* 1995) when a limited number of transcription factors are available for the activation of transcription. Further research will be required to determine what other transcriptional regulatory elements or posttranscriptional mechanisms are responsible for the muscle enrichment seen with NTEF-1, or isoforms thereof, in a variety of organisms (Blatt and DePamphilis, 1993; Shimizu *et al.,* 1993; Stewart et al, 1996).

B. Translation of TEF-1 Family Members

ETEF-1 is the sole member of the TEF-1 family to date that appears to initiate translation at a methionine residue (Figs. 6 and 7; Jacquemin *et al.,* 1996; Yasunami *et al.,* 1995). Non-Met codons are rare and most frequently found in potent regulatory factors (Hann, 1994). This has been proposed to represent a means of subjecting these potent proteins to more stringent regulation within the cell and thereby to maintain them at low relative concentrations. Site-directed mutagenesis of human NTEF-1 identified an isoleucine (AUU) in a reasonably favorable Kozak consensus as the most likely initiation codon (Xiao *et al.,* 1991). However, start codons ending in a nucleotide other than G were not found for any other eukaryotic RNA polymerase II gene in a database of translational signals (Dalphin *et al.,* 1996). Moreover, similar mutagenesis experiments in chick RTEF-1 did not support the usage of an isoleucine in an equivalent position as its initiation codon: thus, an upstream leucine in a favorable Kozak consensus was tentatively identified as its translation start site (I. K. Farrance and S. B. Larkin, unpublished observation). In all other cases the initiation start site has not been directly tested, but start sites have been assigned by analogy with human NTEF-1 at isoleucines.

C. TEF-1 Induction and Modification

Many MCAT sites are the target of extracellular signals (Gupta *et al.,* 1994; Kariya *et al.,* 1994; MacLellan *et al.,* 1994; Cogan *et al.,* 1995; Karns *et al.,* 1995; Sun *et al.,* 1995). Since most MCAT sites have been shown to be bound by TEF-1 family members, these proteins are candidate downstream targets for the second messenger cascades initiated by the extracellular signals (Farrance *et al.,* 1992; Farrance and Ordahl, 1996; Gupta *et al.,* 1994; Ishiji *et al.,* 1992; Kariya *et al.,* 1993; Molkentin and Markham, 1994; Shimizu *et al.,* 1993; Xiao *et al.,* 1991; Yockey *et al.,* 1996). Indeed, the MCAT elements that mediate the TGF-β inducibility of the *αSkA* gene, the cAMP inducibility of the cardiac *αMHC* gene and the serum stimulation of the *VSMαA* gene have all been shown to bind TEF-1 proteins (Cogan *et al.,* 1995; Gupta *et al.,* 1994; MacLellan *et al.,* 1994).

There is evidence that in several systems TEF-1-related factors are upregulated and/or posttranslationally modified following certain stimuli. The pathway is perhaps best worked out for upregulation of the *βMHC* gene in the mammalian heart following hypertrophy. The cardiac hypertrophic response, including the fetal gene switch with upregulation of the *βMHC* gene, can be reproduced by *in vitro* by α1-adrenergic stimulation of cardiac myocytes (Long *et al.,* 1991; Simpson *et al.,*

1989; Waspe *et al.,* 1990). *βMHC* gene upregulation is mimicked by transfection of a consitutively active mutant of β PKC and both the former extracellular and the later presumed intracellular intermediary stimuli mediate their effects through an MCAT site that was shown to bind a TEF-1-related factor (Kariya *et al.,* 1993, 1994). Phosphorylated TEF-1 protein levels may be increased by hypertrophic stimuli (P. Simpson, personal communication). The hypertrophic response of the cardiac *αMHC* gene was also shown to be associated with a fivefold increase in its MCAT binding factor that is antigenically related to chick RTEF-1 (Molkentin and Markham, 1994). Furthermore, there is a transitory increase in TEF-1-related proteins that bind to the *αSkA* gene promoter upon slow twitch muscle hypertrophy (Carson *et al.,* 1996). In fibroblasts that are stimulated by the serum mitogen acidic fibroblast growth factor *RTEF-1* transcription is activated *de novo* with the kinetics of a delayed early response gene (Hsu *et al.,* 1996). Thus, the extracellular signals result in an increase in TEF-1 protein levels, a probable posttranslational modification that may impact its transactivation capacity, or both.

TEF-1 proteins may also be subject to regulation that confers an added level of tissue specificity. RTEF-1 protein (Farrance and Ordahl, 1996) or transcripts (Yockey *et al.,* 1996) appear in multiple tissues. However, RTEF-1 proteins in both chick and mouse appear to be responsible for muscle-specific gel-shifted species generated on MCAT elements (Farrance and Ordahl, 1996; Shimizu *et al.,* 1993; Yockey *et al.,* 1996). The MCAT elements of the *cTnT* gene promoter have been shown to bind both NTEF-1 and RTEF-1 in chick muscle and nonmuscle cells; however, in muscle cells the RTEF-1/MCAT complex migrates in a unique position under nondenaturing electrophoretic conditions (gel shift conditions), whereas the NTEF-1/MCAT complexes are unaffected (Farrance and Ordahl, 1996). The RTEF-1 protein and one of the two NTEF-1 isoforms that bind these *cTnT* MCATs are phosphorylated; however, dephosphorylation does not alter the position of the muscle-specific RTEF-1/MCAT complex. Furthermore, no muscle-specific splicing has been shown for chick *RTEF-1* transcripts, unlike the splicing seen with *MEF-2* RNAs (Stewart *et al.,* 1994; Yu *et al.,* 1992). Thus, the muscle-specific shift in this model may be due to some other posttranslational modification, an as yet undescribed splicing event, or an interaction with an additional factor specifically in muscle cells.

Likewise, the distal MCAT element of the *βMHC* gene in mouse differentiated muscle cell lines is bound by both NTEF-1 and RTEF-1 proteins (Shimizu *et al.,* 1993; Yockey *et al.,* 1996). This distal MCAT element of the *βMHC* gene was earlier shown to form both ubiquitous (A1) and muscle-specific (A2) gel shift complexes (Shimizu *et al.,* 1993). It appears that NTEF-1 and possibly one isoform of RTEF-1 (*TEFR-1a*) form the ubiquitous complex (Shimizu *et al.,* 1993; Yockey *et al.,* 1996) whereas another isoform of RTEF-1 (*TEFR-1b*) is likely to be responsible for the muscle-specific MCAT gel shift complex (Yockey *et al.,* 1996). These *RTEF-1* transcripts are expressed only in differentiated muscle cell lines, but expression is not as tightly regulated *in vivo* since the transcripts are found in several nonmuscle tissues. Therefore, the *βMHC* muscle-specific complex may result from regulation at the level of the promoter or RNA stability, but some posttranscriptional or posttranslational modulation may also occur *in vivo.*

D. TEF-1 Activity in Cleavage-Stage Embryonic Development

Mutations were generated in the polyomavirus enhancer and screened for their activity in undifferentiated ES cells and in one- and two-cell stage mouse embryos (Melin *et al.,* 1993). The enhancer configuration that was identified as "embryo responsive" by this approach is a tandem duplication of a region containing a mutation resulting in an MCAT site. Thus, some member of the TEF-1 family is active at an early stage of development, prior to formation of somites or precardiac mesoderm which contain skeletal and cardiac muscle precursor cells. TEF-1 activity in early development may, therefore, serve a more generalized function. This may involve novel members of the family (e.g., ETEF-1, which is expressed early in development in nonmuscle tissues) or the absence at this stage of a specific repressor of TEF-1 activity in nonmuscle cell types. Such activation does not require NTEF-1 since knockout of this member of the TEF-1 family results in embryonic death relatively late in development, with early events apparently unaffected.

Interestingly, SP1 sites are also active in ES cells and cleavage-stage mouse embryos. It appears that the SP1 sites rather than the TATA box of a model promoter are required for mediation of enhancer stimulation of gene expression at these early stages (Nothias, 1995). Interestingly, SP1 sites are found associated with MCATs in the majority of MCAT-dependent promoters and enhancers and have been implicated in the activity of many of these (Berberich *et al.,* 1993; Karns *et al.,* 1995; MacLellan *et al.,* 1994; Mar and Ordahl, 1990; Shimizu *et al.,* 1992b; Thompson *et al.,* 1991; Zenke *et al.,* 1986).

V. Potential Modalities of MCAT-Dependent Gene Regulation

A. The Complexity of the TEF-1 Family

As outlined previously, MCAT elements govern a wide variety of transcriptional responses; cell specific, cell nonspecific, hormone inducible, etc. Such regulatory complexity can be explained, at least in part, by the various levels of complexity of the TEF-1 family of MCAT binding factors. The tissue-restricted pattern of *TEF-1* gene expression (Table IV) provides a partial explanation. Noteworthy, in particular, is the partially overlapping checkerboard pattern of high level of *RTEF-1* and *DTEF-1* mRNA expression in different tissues and the relatively ubiquitous expression of *NTEF-1*.

A second level of TEF-1 diversity arises from pre-mRNA alternative splicing both in and near the conserved, multifunctional TEA domain, but the consequences of this have not been extensively explored. The TEF-1 isoproteins shown in Table III appear to initiate translation at multiple and/or atypical start codons that may further contribute to diversity. It would appear likely, therefore, that TEF-1 isoprotein diversity plays a major role in governing the multiple types of transcriptional regulation noted for MCAT-dependent promoters. For instance, southwestern analysis indicated that RTEF-1 binds one MCAT element with higher affinity than either of two NTEF-1 isoforms (Farrance and Ordahl, 1996), suggesting that different isoforms recognize different promoters based on the nucleotide sequences of their MCAT binding sites.

Finally, the fact that TEF-1 family members can be induced and/or modified by phosphorylation or addition of other functional groups in response to extracellular signals or in a tissue-specific fashion carries further implications for both transactivation potency and binding specificity and affinity. Notable in this regard is the difference in MCAT–RTEF-1A interaction in skeletal muscle versus nonmuscle (Farrance and Ordahl, 1996).

B. Evidence for the Existence of TEF-1-Specific Coactivators

Coactivators and transcriptional intermediary factors are non-DNA binding factors that bridge between a DNA-bound transcription factor and some component of the basal transcriptional machinery to promote increased productive preinitiation complex formation and/or RNA polymerase recruitment to the start site. Such proteins may be required by TEF-1 proteins or a subset thereof, at least in some instances. This putative interaction would introduce a further level of complexity and possible heterogeneity into TEF-1 protein-mediated transcriptional activity.

The GTIIC and Sph motif binding protein was implicated in the activity of the SV40 enhancer in several cell lines but notably not in cells of lymphoid origin, in which the octamer motif that overlaps the Sph motifs is selectively active and replaces their function (Davidson *et al.,* 1988; Fromental *et al.,* 1988; Ondek *et al.,* 1987, 1988). Subsequently it was shown that *NTEF-1* mRNA was undetectable in lymphoid-derived cell lines and functionally inactive even when artificially introduced into them (Xiao *et al.,* 1991). The lack of a positively acting coactivator or the presence of a negatively acting TATA-associated factor in lymphoid cells have been suggested as the cause (Chaudhary *et al.,* 1994; Xiao *et al.,* 1991). Evidence for the existence of TEF-1-specific coactivators, in several cell types in which endogenous TEF-1 is apparently normally active, is indicated by the ability of overexpressed NTEF-1 to effectively squelch reporter activity based on TEF-1 binding sites but not those based on unrelated sites (Hwang *et al.,* 1993; Ishiji *et al.,* 1992; Xiao *et al.,* 1991).

It has been reported that a GAL-NTEF-1 chimera interacts with two chromatographically distinct TFIID complexes. Chimeras containing either the strong activating domain in the viral protein VP16 or the C-terminus activating domain of the estrogen receptor interact with only one of these TFIID complexes. The second is apparently specific to GAL-NTEF-1 (Brou *et al.,* 1993). Factors present in this TFIID fraction are required for activation by GAL-NTEF-1 in an *in vitro* reconstituted transcription system.

Chromatographic purification and reconstitution of activated transcription *in vitro* was used to search for candidate coactivators for TEF-1 in HeLa TFIID fractions and to attempt to understand the basis of the inactivity of TEF-1 in lymphoid cells (Chaudhary *et al.,* 1994). Interestingly, the lymphoid cell-derived TFIID fractions were found to be capable of mediating TEF-1 activation once separated by immunopurification from an apparently cell-specific activity called negative factor-1 (NEF-1). Addition of NEF-1-containing fractions blocks activation by GAL-NTEF-1 but not by GAL-VP16 (Chaudhary *et al.,* 1994). Thus, the inactivity of NTEF-1 in lymphoid cells appears to be due to the presence of an inhibitory factor associated with a particular TFIID fraction that selectively blocks its activity rather than to the lack of a particular coactivator.

A second activity is found in lymphoid cells and surprisingly also in HeLa cells (in which GAL-NTEF-1 is active) that selectively inhibits GAL-NTEF-1 activity in reconstituted transcription reactions (NEF-2; Chaudhary *et al.,* 1995). NEF-2 is chromotographically distinct from NEF-1 but blocks transcriptional activation by GAL-NTEF-1 but not by GAL-VP16 in a similar fashion. Inhibition of TEF-1 activation by NEF-2-containing fractions can be relieved by addition of a partially purified TFIID fraction but not by recombinant TBP. Moreover, NEF-2 can not inhibit GAL-NTEF-1-mediated activation if added after preinitiation complexes have formed (Chaudhary *et al.*, 1995). This suggests that NEF-2 blocks enhancement of preinitiation complex formation that is brought about by interaction of GAL-NTEF-1 with elements of the TFIID complex by competing for the element of the TFIID complex normally bound by NTEF-1. However, since GAL-NTEF-1 is in fact active in HeLa cells the function of the NEF-2 activity in these cells is uncertain. It cannot be ruled out that NEF-1 is another transcription factor that mediates its transactivation by interaction with the same TFIID component as TEF-1 and its inhibition of GAL-NTEF-1 in this enriched context is largely artefactual. However, the interesting possibility arises that it may be involved in mediating the selective inhibition of TEF-1 on certain muscle-specific type MCAT elements in nonmuscle cells, such as HeLa cells, (Larkin *et al.*, 1996).

In parallel to indications that TEF-1, interacts with the basal transcriptional machinery via some component of the TFIID complex are other reports that TEF-1 can bind directly to TBP (Gruda *et al.*, 1993; Jiang and Eberhardt, 1996), whereas one study does not support direct binding of TEF-1 to TBP (Berger *et al.*, 1996).

C. TEF-1 Interacts with the Promiscuous Viral Activators T Antigen and IE86

The large tumor antigen (TAg) of SV40 activates many cellular and viral promoters and has transforming activity that is probably mediated by its ability to bind to and sequester the retinoblastoma (Rb) protein, its family members, and p53 (Fanning and Knippers, 1992; Levine, 1993; Moran, 1993) as well as promoting viral genome replication. TAg activates the SV40 late promoter; this activation does not require direct DNA binding by TAg but instead has been shown to require the TEF-1 binding sites. Column studies using immobilized GST-TAg show that TAg can directly bind TEF-1 and TBP as well as TFIIB and SP1 (Gruda *et al.*, 1993; Johnston *et al.*, 1996). TEF-1/TAg interaction requires the TEA domain of TEF-1 and a domain of TAg overlapping but distinct from that mediating its binding to DNA and the Rb protein (Berger *et al.*, 1996; Gruda *et al.*, 1993). This interaction appears to disrupt the binding of TEF-1 to DNA coincident with activation of the promoter. Mutation of the TEF-1 binding sites also appeared to activate the late promoter in this study, implying that TEF-1 is acting as a repressor in this context (Berger *et al.*, 1996). Unlike the early promoter, which is activated by TEF-1 sites, the late promoter contains an initiator element in place of a TATA box. However, other studies have indicated that intact TEF-1 sites are necessary and sufficient to mediate the activation of the late promoter by TAg (Casaz *et al.*, 1991; Gilinger and Alwine, 1993; Rice and Cole, 1993). This inconsistency remains to be resolved.

TEF-1 is apparently able to serve a similar function for the cytomegalovirus (CMV) immediate early (IE) protein IE86 as that described for TAg (Lukac *et al.*, 1994). IE86 promiscuously activates a variety of viral and cellular promoters and has been shown to be capable of activating via simple artificial promoters containing only a TATA box and any one of several upstream elements, such as a TEF-1 element. Column studies of GST fusions of these proteins show that they directly bind both the TBP protein and TEF-1 as well as SP1 (Lukac *et al.*, 1994).

TAg is able to replace the function of the TATA-associated factor $TEF_{II}250$ in a temperature-sensitive mutant (Damania and Alwine, 1996). Thus, one may hypothesize that TAg, and by extension the CMV IE86 protein, may function as a coactivator, bridging TEF-1 to the general transcriptional machinery. Whether the interaction with TAg and IE86 mimics a necessary interaction by TEF-1 with a cellular coactivator protein is unknown. However, two studies indicate that TEF-1 directly binds TBP (Gruda *et al.*, 1993; Jiang and Eberhardt, 1996). The binding of TEF-1 to TBP may indicate that activation by TEF-1 does not necessarily require a bridging factor.

The ability of TEF-1 to directly interact with multiple different transcriptionally active proteins, including NEF-1 and NEF-2, TAg, some component of the TFIID complex, and possibly TBP directly, reflects its multiple functions on different promoters in different cellular contexts.

D. TEF-1-Binding Partners: An Additional Level of Regulatory Complexity

An additional level of regulatory complexity is suggested by recent evidence indicating that binding of a second factor to MCAT flanking sequences further modulates the activity of the TEF-1 protein bound to the heptameric core MCAT motif (Fig. 8). The flanking sequence of the *cTnT* MCAT elements has been shown to direct the binding of a second factor that apparently renders these sites highly active in muscle cells but inactive in nonmuscle cells (Larkin *et al.*, 1996).

Mutation of the flanking sequences or their substitution by sequences found flanking the MCAT sites in the

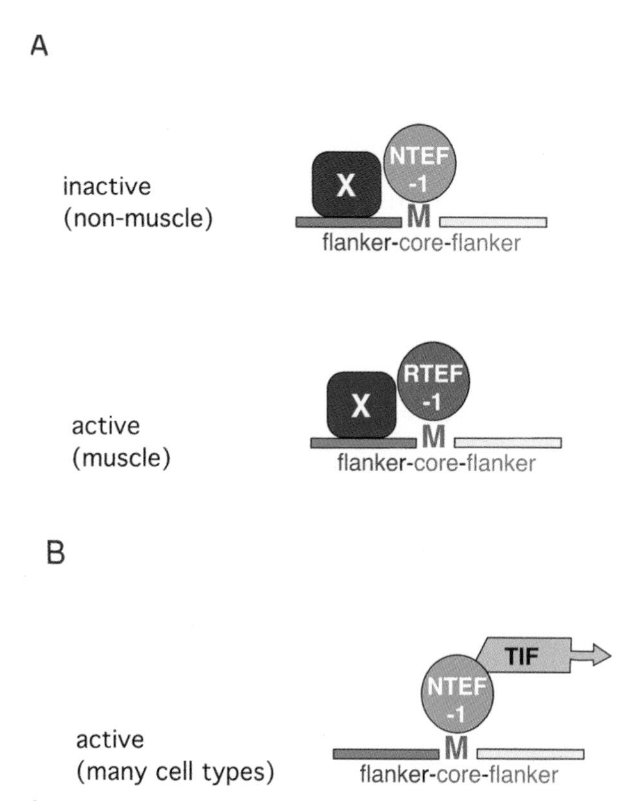

Figure 8 TEF-1 association with other nuclear proteins. (a) Cooperative binding between TEF-1 and protein(s) binding to MCAT flanking sequences both activates and represses transcription in a cell- and tissue- specific fashion. In the simplest scenario illustrated, the positive or negative activity of the promoter is determined by the bound TEF-1 family member and/or by X. X is a protein that has been shown to exist but whose identity is unknown. (b) TEF-1 activity in the SV40 promoter. The absence of a requirement for flanker binding suggests that any TEF-1 family member is able to activate this promoter. However, a transcriptional intermediary factor(s) (TIF) may also be required.

SV40 enhancer generates MCAT elements that are active in both muscle and nonmuscle cell types. The formation of a ternary complex comprising an MCAT element, TEF-1, and a factor that specifically recognizes the MCAT flanking sequence has been demonstrated in gel shift experiments, and footprinting shows protein binding extending over this flanking region. Although the identity of this second protein(s) remains to be determined, we hypothesize that the muscle specificity of the two MCAT elements from the *cTnT* promoter is dependent on the cooperative interaction of this second factor with TEF-1 and nucleotides flanking the MCAT core motif.

How general is the role of TEF-1 cobinding partners in the regulatory complexity of MCAT elements? The cardiac *αMHC* MCAT element is also a hybrid site that binds a TEF-1 protein and an additional protein(s) in the size range 25–28 kDa (Gupta *et al.,* 1994). The possible function of the additional binding protein(s) in the cardiac *αMHC* hybrid MCAT site has not been directly addressed, but since this site mediates the inducibility of the gene by cAMP it is possible that this second protein(s), rather than the TEF-1 component, is the target of this modulation. Another MCAT-dependent promoter, the *βMHC* gene promoter, is down regulated by cAMP (Gupta *et al.,* 1994). This differential effect of cAMP might occur via a different secondary component or by a direct modulation of TEF-1 in a MCAT complex that contains a different protein or lacks a second protein.

VI. Summary and Conclusions

MCAT elements are responsible for diverse transcriptional regulatory responses of both muscle-specific and non-muscle-specific promoters. Two classes of MCAT transcriptional regulatory elements can be distinguished. One class of MCAT elements governs gene expression with a high level of cell and tissue specificity as exemplified by the *cTnT* gene promoter (Fig. 8A). This class of MCAT-dependent genes includes many cardiac genes that are cardiac specific and, in some cases, hormone responsive. A second class of MCAT-dependent genes, exemplified by the SV40 promoter (Fig. 8B), exhibits a low level of cell and tissue specificity.

MCAT elements bind members of the TEF-1 family of transcriptional activators through the core seven-nucleotide motif, CATTCCT. The *TEF-1* genes are the vertebrate members of the highly conserved TEA domain family of regulatory proteins that govern gene activity in a wide variety of plants and animals. There are at least four vertebrate *TEF-1* genes, some of which are expressed in a tissue-restricted fashion, with alternative splicing adding to the complexity of expressed TEF-1 proteins. These proteins are likely to bind differentially to different MCAT elements leading to distinct spatiotemporal patterns of transactivation, dependent on the cognate TEF-1 isoprotein. For instance, results from the author's laboratory indicate that RTEF-1 is the likely binding factor governing the muscle-specific expression of the *cTnT* gene, whereas NTEF-1 is almost certainly the principal regulatory protein governing the SV40 promoter (Fig. 8).

Some evidence points to the existence of coactivators that, in some instances, may affect the ability of TEF-1 proteins to interact productively with the basal transcriptional machinery. To date, coactivators have not been successfully isolated but are proposed to exhibit tissue specificity and may also consist of multiple members with distinct characteristics. If TEF-1 coactivators exist, their pattern of expression and specificity for the different TEF-1 isoproteins and MCAT variants would add another layer of regulatory complexity to the TEF-1 system.

Recent work also shows that additional DNA-binding proteins are required for TEF-1 family members to act on transcription. In the case of the *cardiac troponin T* gene promoter, a second, unknown protein (X in Fig. 8) appears to bind cooperatively to flanking sequences and to the adjacent TEF-1 moiety bound to the core motif (Fig. 8A). Since NTEF-1 predominates in nonmuscle tissue, we have hypothesized that the X–NTEF-1 complex is repressive. By contrast, in muscle tissue, in which RTEF-1 is abundant, the X–RTEF-1 complex stimulates transcription. The diversity of flanking sequences present in the MCAT elements of muscle promoters (Table I) suggests that many different proteins may be responsible for modulating MCAT-dependent promoter activity in skeletal and cardiac muscle cells. In the case of the ubiquitously expressed SV40 promoter, the binding of a second protein to flanking sequences appears not to be required and activation through core motif binding may be accomplished by either NTEF-1 or RTEF-1. In the case of the SV40 promoter, however, the activity of a coactivator may be required for transcriptional activation (Fig. 8B). Coactivators by definition do not bind directly to DNA and thus are distinct from the factor(s) interacting with MCAT flanking sequence. Future work should elucidate the modes of activity of MCAT elements via analysis of the diversity and structure of TEF-1 proteins as well as other protein factors with which they cooperate to stimulate and repress transcription.

Acknowledgments

The authors thank their colleagues for sharing research materials and unpublished observations during the work leading up to the writing of the manuscript. We are grateful to Paul Simpson, David Gardner, and Christopher Glembotski for sharing their observations prior to publication. We also thank the members of the Ordahl lab, in particular Alison Butler, for generously sharing their ideas and critical comments during the preparation of the manuscript. Work in the author's laboratories was funded by grant GM32018 from the National Institutes of Health.

References

Andrianopoulos, A., and Timberlake, W. E. (1991). ATTS, a new and conserved DNA binding domain. *Plant Cell* **3,** 747–748.

Arany, Z., Sellers, W. R., Livingston, D. M., and Eckner, R. (1994). E1A-associated p300 and CREB-associated CBP belong to a conserved family of coactivators [letter]. *Cell (Cambridge, Mass.)* **77,** 799–800.

Azakie, A., Larkin, S. B., Farrance, I. K., Grenningloh, G., and Ordahl, C. P. (1996). DTEF-1, a novel member of the transcription enhancer factor-1 (TEF-1) multigene family. *J. Biol. Chem.* **271,** 8260–8265.

Berberich, C., Durr, I., Koenen, M., and Witzemann, V. (1993). Two adjacent E box elements and a MCAT box are involved in the muscle specific regulation of the rat acetylcholine receptor beta subunit gene. *Eur. J. Biochem.* **216,** 395–404.

Berger, L. C., Smith, D. B., Davidson, I., Hwang, J. J., Fanning, E., and Wildeman, A. G. (1996). Interaction between T antigen and TEA domain of the factor TEF-1 derepresses simian virus 40 late promoter in vitro: Identification of T-antigen domains important for transcription control. *J. Virol.* **70,** 1203–1212.

Blatt, C., and DePamphilis, M. L. (1993). Striking homology between mouse and human transcription enhancer factor-1 (TEF-1). *Nucleic Acids Res.* **21,** 747–748.

Boam, D. S., Davidson, I., and Chambon, P. (1995). A TATA-less promoter containing binding sites for ubiquitous transcription factors mediates cell type specific regulation of the gene for transcription enhancer factor-1 (TEF-1). *J. Biol. Chem.* **270,** 19487–19494.

Brou, C., Chaudhary, S., Davidson, I., Lutz, Y., Wu, J., Egly, J. M., Tora, L., and Chambon, P. (1993). Distinct TFIID complexes mediate the effect of different transcriptional activators. *Embo. J.* **12,** 489–499.

Bürglin, T. R. (1991). The TEA domain: A novel, highly conserved DNA-binding motif. *Cell (Cambridge, Mass.)* **66,** 11–12.

Campbell, S., Inamdar, M., Rodrigues, V., Raghavan, K. V., Palazzolo, M., and Chovnick, A. (1992). The *scalloped* gene encodes a novel, evolutionarily conserved transcription factor required for sensory organ differentiation in Drosophila. *Genes Dev.* **6,** 367–379.

Carson, J. A., Schwartz, R. J., and Booth, F. W. (1996). SRF and TEF-1 control of chicken skeletal alpha-actin gene during slow-muscle hypertrophy. *Am. J. Physiol.* **270,** C1624–C1633.

Casaz, P., Sundseth, R., and Hansen, U. (1991). Trans activation of the simian virus 40 late promoter by large T antigen requires binding sites for the cellular transcription factor TEF-1. *J. Virol.* **65,** 6535–6543.

Chaudhary, S., Brou, C., Valentin, M. E., Burton, N., Tora, L., Chambon, P., and Davidson, I. (1994). A cell specific factor represses stimulation of transcription in vitro by transcriptional enhancer factor 1. *Mol. Cell. Biol.* **14,** 5290–5299.

Chaudhary, S., Tora, L., and Davidson, I. (1995). Characterization of a HeLa cell factor which negatively regulates transcriptional activation in vitro by transcriptional enhancer factor-1 (TEF-1). *J. Biol. Chem.* **270,** 3631–3637.

Chen, Z., Friedrich, G. A., and Soriano, P. (1994). Transcriptional enhancer factor 1 disruption by a retroviral gene trap leads to heart defects and embryonic lethality in mice. *Genes. Dev.* **8,** 2293–2301.

Chiba, H., Muramatsu, M., Nomoto, A., and Kato, H. (1994). Two human homologues of Saccharomyces cerevisiae SWI2/SNF2 and Drosophila brahma are transcriptional coactivators cooperating with the estrogen receptor and the retinoic acid receptor. *Nucleic Acids Res* **22,** 1815–1820.

Cogan, J. G., Sun, S., Stoflet, E. S., Schmidt, L. J., Getz, M. J., and Strauch, A. R. (1995). Plasticity of vascular smooth muscle alpha-actin gene transcription. Characterization of multiple, single-, and double-strand specific DNA-binding proteins in myoblasts and fibroblasts. *J. Biol. Chem.* **270,** 11310–11321.

Dalphin, M. E., Brown, C. M., Stockwell, P. A., and Tate, W. P. (1996). Transderm: A database of translational signals. *Nucleic Acids Res.* **24,** 216–218.

Damania, B., and Alwine, J. C. (1996). TAF-like function of SV40 large T antigen. *Genes Dev.* **10,** 1369–1381.

DasGupta, S., Shivakumar, C. V., and Das, G. C. (1993). Identification of proteins binding to the F441 locus of polyomavirus B enhancer that are required for its activity in embryonic carcinoma cells. *J. Gen. Virol.* **74,** 597–605.

Davidson, I., Fromental, C., Augereau, P., Wildeman, A., Zenke, M., and Chambon, P. (1986). Cell-type specific protein binding to the enhancer of simian virus 40 in nuclear extracts. *Nature (London)* **323,** 544–548.

Davidson, I., Xiao, J. H., Rosales, R., Staub, A., and Chambon, P. (1988). The HeLa cell protein TEF-1 binds specifically and cooperatively to two SV40 enhancer motifs of unrelated sequence. *Cell (Cambridge, Mass.)* **54,** 931–942.

Deshpande, N., Chopra, A., Rangarajan, A., Shashidhara, L. S., Rodrigues, V., and Krishna, S. (1997). The human transcription factor, TEF-1, can substitute for Drosophila *scalloped* during wingblade development. *J. Biol. Chem.* **272,** 10664–10668

Emerson, C. P., Jr. (1993). Skeletal myogenesis: genetics and embryology to the fore. *Curr. Opin. Genet. Dev.* **3,** 265–274.

Fabre-Suver, C., and Hauschka, S. D. (1996). A novel site in the muscle creatine kinase enhancer is required for expression in skeletal but not cardiac muscle. *J. Biol. Chem.* **271,** 4646–4652.

Fanning, E., and Knippers, R. (1992). Structure and function of simian virus 40 large T antigen. *Annu. Rev. Biochem.* **61,** 55–85.

Farrance, I. K., and Ordahl, C. P. (1996). The role of transcription enhancer factor-1 (TEF-1) related proteins in the formation of MCAT binding complexes in muscle and non-muscle tissues. *J. Biol. Chem.* **271,** 8266–8274.

Farrance, I. K., Mar, J. H., and Ordahl, C. P. (1992). MCAT binding factor is related to the SV40 enhancer binding factor, TEF-1. *J. Biol. Chem.* **267,** 17234–17240.

Flink, I. L., Edwards, J. G., Bahl, J. J., Liew, C. C., Sole, M., and Morkin, E. (1992). Characterization of a strong positive cis-acting element of the human beta-myosin heavy chain gene in fetal rat heart cells. *J. Biol. Chem.* **267,** 9917–9924.

Friedrich, G. a. S., P. (1991). Promoter traps in embryonic stem cells: A genetic screen to identify and mutate developmental genes in mice. *Genes Dev.* **5,** 1513–1523.

Frigerio, J. M., Berthezene, P., Garrido, P., Ortiz, E., Barthellemy, S., Vasseur, S., Sastre, B., Seleznieff, I., Dagorn, J. C., and Iovanna, J. L. (1995). Analysis of 2166 clones from a human colorectal cancer cDNA library by partial sequencing. *Hum. Mol. Genet.* **4,** 37–43.

Fromental, C., Kanno, M., Nomiyama, H., and Chambon, P. (1988). Cooperativity and hierarchical levels of functional organization in the SV40 enhancer. *Cell (Cambridge, Mass.)* **54,** 943–953.

Funk, W. D., Ouellette, M., and Wright, W. E. (1991). Molecular biology of myogenic regulatory factors. *Mol. Biol. Med.* **8,** 185–195.

Gilinger, G., and Alwine, J. C. (1993). Transcriptional activation by simian virus 40 large T antigen: Requirements for simple promoter structures containing either TATA or initiator elements with variable upstream factor binding sites. *J. Virol.* **67,** 6682–6688.

Groskopf, J. C., and Linzer, D. I. (1994). Characterization of a delayed early serum response region. *Mol. Cell. Biol.* **14,** 6013–6020.

Gruda, M. C., Zabolotny, J. M., Xiao, J. H., Davidson, I., and Alwine, J. C. (1993). Transcriptional activation by simian virus 40 large T antigen: Interactions with multiple components of the transcription complex. *Mol. Cell. Biol.* **13,** 961–969.

Guarente, L. (1995). Transcriptional coactivators in yeast and beyond. *Trends Biochem. Sci.* **20,** 517–521.

Gupta, M. P., Gupta, M., and Zak, R. (1994). An E-box/MCAT hybrid motif and cognate binding protein(s) regulate the basal muscle specific and cAMP-inducible expression of the rat cardiac alpha-myosin heavy chain gene. *J. Biol. Chem.* **269,** 29677–29687.

Hall, A. V., Antoniou, H., Wang, Y., Cheung, A. H., Arbus, A. M., Olson, S. L., Lu, W. C., Kau, C. L., and Marsden, P. A. (1994). Structural organization of the human neuronal nitric oxide synthase gene (NOS1). *J. Biol. Chem.* **269,** 33082–33090.

Hann, S. R. (1994). Regulation and function of non-AUG-initiated proto-oncogenes. *Biochimie* **76,** 880–886.

Hsu, D. K. W., Guo, Y., Alberts, G. F., Copeland, N. G., Gilbert, D. J., Jenkins, N. A., Peifley, K. A., and Winkles, J. A. (1996). Identification of a murine TEF-1-related gene expressed after mitogenic stimulation of quiescent fibroblasts and during myogenic differentiation. *J. Biol. Chem.* **271,** 13786–13795.

Hwang, J. J., Chambon, P., and Davidson, I. (1993). Characterization of the transcription activation function and the DNA binding domain of transcriptional enhancer factor-1. *Embo. J.* **12,** 2337–2348.

Ishiji, T., Lace, M. J., Parkkinen, S., Anderson, R. D., Haugen, T. H., Cripe, T. P., Xiao, J. H., Davidson, I., Chambon, P., and Turek, L. P. (1992). Transcriptional enhancer factor (TEF)-1 and its cell specific co-activator activate human papillomavirus-16 E6 and E7 oncogene transcription in keratinocytes and cervical carcinoma cells. *Embo. J.* **11,** 2271–2281.

Jacquemin, P., Oury, C., Belayew, A., and Martial, J. A. (1994). A TEF-1 binding motif that interacts with a placental protein is important for the transcriptional activity of the hCS-B enhancer. *DNA Cell Biol.* **13,** 1037–1045.

Jacquemin, P., Hwang, J. J., Martial, J. A., Dolle, P., and Davidson, I. (1996). A novel family of developmentally regulated mammalian transcription factors containing the TEA/ATTS DNA binding domain. *J. Biol. Chem.* **271,** 21775–21785.

Jiang, S. W., and Eberhardt, N. L. (1995). Involvement of a protein distinct from transcription enhancer factor-1 (TEF-1) in mediating human chorionic somatomammotropin gene enhancer function through the GT-IIC enhanson in choriocarcinoma and COS cells. *J. Biol. Chem.* **270,** 13906–13915.

Jiang, S.W., and Eberhardt, N. L. (1996). TEF-1 transrepression in BeWo cells is mediated through interactions with the TATA-binding protein, TBP. *J. Biol. Chem.* **271,** 9510–9518.

Jiang, S. W., Eberhardt, N. L., Cripe, T. P., Xiao, J. H., Davidson, I., Chambon, P., and Turek, L. P. (1994). The human chorionic somatomammotropin gene enhancer is composed of multiple DNA elements that are homologous to several SV40 enhansons. *J. Biol. Chem.* **269,** 10384–10392.

Johnston, S. D., Yu, X. M., and Mertz, J. E. (1996). The major transcriptional transactivation domain of simian virus 40 large T antigen associates nonconcurrently with multiple components of the transcriptional preinitiation complex. *J. Virol.* **70,** 1191–1202.

Kariya, K., Farrance, I. K., and Simpson, P. C. (1993). Transcriptional enhancer factor-1 in cardiac myocytes interacts with an alpha 1-adrenergic- and beta-protein kinase C-inducible element in the rat beta-myosin heavy chain promoter. *J. Biol. Chem.* **268,** 26658–26662.

Kariya, K., Karns, L. R., and Simpson, P. C. (1994). An enhancer core element mediates stimulation of the rat beta-myosin heavy chain promoter by an alpha 1-adrenergic agonist and activated beta-protein kinase C in hypertrophy of cardiac myocytes. *J. Biol. Chem.* **269,** 3775–3782.

Karns, L. R., Kariya, K., and Simpson, P. C. (1995). MCAT, CArG, and SP1 elements are required for alpha 1-adrenergic induction of the skeletal alpha-actin promoter during cardiac myocyte hypertrophy. Transcriptional enhancer factor-1 and protein kinase C as conserved transducers of the fetal program in cardiac growth. *J. Biol. Chem.* **270,** 410–417.

Kihara-Negishi, F., Tsujita, R., Negishi, Y., and Ariga, H. (1993). BOX DNA: A novel regulatory element related to embryonal carcinoma differentiation. *Mol. Cell. Biol.* **13,** 7747–7756.

Laloux, I., Dubois, E., Dewerchin, M., and Jacobs, E. (1990). TEC1, a gene involved in the activation of Ty1 and Ty1-mediated gene expression in Saccharomyces cerevisiae: Cloning and molecular analysis. *Mol. Cell. Biol.* **10,** 3541–3550.

LaPointe, M. C., Wu, G., Garami, M., Yang, X. P., Gardner, D. G. (1996). Tissue-specific expression of the human brain natriuretic peptide gene in cardiac myocytes. *Hypertension* **27,** 715–722.

Larkin, S. B., Farrance, I. K., and Ordahl, C. P. (1996). Flanking sequences modulate the cell specificity of MCAT elements. *Mol. Cell. Biol.* **16,** 3742–3755.

Lassar, A., and Munsterberg, A. (1994). Wiring diagrams: Regulatory circuits and the control of skeletal myogenesis. *Curr. Opin. Cell. Biol.* **6,** 432–442.

Lee, T. C., Shi, Y., and Schwartz, R. J. (1992). Displacement of BrdU-induced YY1 bt serum response factor activates skeletal alpha actin transcription in embryonic myoblasts. *Proc. Natl. Acad. Sci. U.S.A.* **89,** 9814–9818.

Lenormand, J. L., and Leibovitch, S. A. (1995). Identification of a novel regulatory element in the c-mos locus that activates transcription in somatic cells. *Biochem. Biophys. Res. Commun.* **210,** 181–188.

Levine, A. J. (1993). The tumour suppressor genes. *Annu. Rev. Biochem.* **62,** 623–651.

Linn, S. C., Askew, G. R., Menon, A. G., and Shull, G. E. (1995). Conservation of an AE3 Cl-/HCO3- exchanger cardiac specific exon and promoter region and AE3 mRNA expression patterns in murine and human hearts. *Circ. Res.* **76,** 584–591.

Long, C. S., Kariya, K., Karns, L., and Simpson, P. C. (1991). Sympathetic activity: Modulator of myocardial hypertrophy. *J. Cardiovasc. Pharmacol.* **17,** Suppl. 2; S20–S24.

Lukac, D. M., Manuppello, J. R., and Alwine, J. C. (1994). Transcriptional activation by the human cytomegalovirus immediate-early proteins: Requirements for simple promoter structures and interactions with multiple components of the transcription complex. *J. Virol.* **68,** 5184–5193.

Luo, Y,. and Roeder, R. G. (1995). Cloning, functional characterization, and mechanism of action of the B-cell specific transcriptional coactivator OCA-B. *Mol. Cell. Biol.* **15,** 4115–4124.

MacLellan, W. R., Lee, T. C., Schwartz, R. J., and Schneider, M. D. (1994). Transforming growth factor-beta response elements of the skeletal alpha-actin gene. Combinatorial action of serum response factor, YY1, and the SV40 enhancer-binding protein, TEF-1. *J. Biol. Chem.* **269,** 16754–16760.

Mar, J. H., and Ordahl, C. P. (1988). A conserved CATTCCT motif is required for skeletal muscle specific activity of the cardiac troponin T gene promoter. *Proc. Natl. Acad. Sci. U.S.A.* **85,** 6404–6408.

Mar, J. H., and Ordahl, C. P. (1990). MCAT binding factor, a novel *trans*-acting factor governing muscle specific transcription. *Mol. Cell. Biol.* **10,** 4271–4283.

Mar, J. H., Antin, P. B., Cooper, T. A., and Ordahl, C. P. (1988). Analysis of the upstream regions governing expression of the chicken cardiac troponin T gene in embryonic cardiac and skeletal muscle cells. *J. Cell. Biol.* **107,** 573–585.

Meier, C. A. (1996). Co-activators and co-repressors: Mediators of gene activation by nuclear hormone receptors. *Eur. J. Endocrinol.* **134,** 158–159.

Melin, F., Miranda, M., Montreau, N., DePamphilis, M. L., and Blangy, D. (1993). Transcription enhancer factor-1 (TEF-1) DNA binding sites can specifically enhance gene expression at the beginning of mouse development. *Embo. J.* **12,** 4657–4666.

Mirabito, P. M., Adams, T. H., and Timberlake, W. E. (1989). Interactions of three sequentially expressed genes control temporal and spatial specificity in Aspergillus development. *Cell* (*Cambridge, Mass.*) **57,** 859–868.

Molkentin, J. D., and Markham, B. E. (1994). An MCAT binding factor and an RSRF-related A-rich binding factor positively regulate expression of the alpha-cardiac myosin heavy-chain gene in vivo. *Mol. Cell. Biol.* **14,** 5056–5065.

Moran, E. (1993). DNA tumor virus transforming proteins and the cell cycle. *Curr. Opin. Genet. Dev.* **3,** 63–70.

Myslinski, E., Krol, A., and Carbon, P. (1992) Optimal tRNA((Ser)Sec) gene activity requires an upstream SPH motif. *Nucleic Acids Res.* **20,** 203–209.

Nakatsuji, Y., Hidaka, K., Tsujino, S., Yamamoto, Y., Mukai, T., Yanagihara, T., Kishimoto, T., and Sakoda, S. (1992). A single MEF-2 site is a major positive regulatory element required for transcription of the muscle specific subunit of the human phosphoglycerate mutase gene in skeletal and cardiac muscle cells. *Mol. Cell. Biol.* **12,** 4384–4390.

Nothias, J.-Y., Majumder, S., Kaneko, K. J., and dePamphilis, M. L. (1995). Regulation of gene expression at the beginning of mammalian development. *J. Biol. Chem.* **270,** 22077–22080.

Olson, E. N. (1990). MyoD family: A paradigm for development? *Genes Dev.* **4,** 1454–1461.

Olson, E. N. (1993). Regulation of muscle transcription by the MyoD family. The heart of the matter. *Circ. Res.* **72,** 1–6.

Ondek, B., Shepard, A., and Herr, W. (1987). Discrete elements within the SV40 enhancer region display different cell specific enhancer activities. *Embo. J.* **6,** 1017–1025.

Ondek, B., Gloss, L., and Herr, W. (1988). The SV40 enhancer contains two distinct levels of organization. *Nature* (*London*) **333,** 40–45.

Parmacek, M. S., and Leiden, J. M. (1989). Structure and expression of the murine slow/cardiac troponin C gene. *J. Biol. Chem.* **264,** 13217–13225.

Qasba, P., Lin, E., Zhou, M. D., Kumar, A., and Siddiqui, M. A. (1992). A single transcription factor binds to two divergent sequence elements with a common function in cardiac myosin light chain-2 promoter. *Mol. Cell. Biol.* **12,** 1107–1116.

Rice, P. W., and Cole, C. N. (1993). Efficient transcriptional activation of many simple modular promoters by simian virus 40 large T antigen. *J. Virol.* **67,** 6689–6697.

Schwarz, J. J., Chakraborty, T., Martin, J., Zhou, J., and Olson, E. N. (1992). The basic region of myogenin cooperates with two transcription activation domains to induce muscle specific transcription. *Mol. Cell. Biol.* **12,** 266–275.

Shimizu, N., Dizon, E., and Zak, R. (1992a). Both muscle specific and ubiquitous nuclear factors are required for muscle specific expression of the myosin heavy-chain b gene in cultured cells. *Mol. Cell. Biol.* **12,** 619–630.

Shimizu, N., Prior, G., Umeda, P. K., and Zak, R. (1992b). *Cis*-acting elements responsible for muscle specific expression of the myosin heavy chain β gene. *Nucleic Acids Res.* **20,** 1793–1799.

Shimizu, N., Smith, G., and Izumo, S. (1993). Both a ubiquitous factor mTEF-1 and a distinct muscle specific factor bind to the MCAT motif of the myosin heavy chain beta gene. *Nucleic Acids Res.* **21,** 4103–4110.

Simpson, P. C., Long, C. S., Waspe, L. E., Henrich, C. J., and Ordahl, C. P. (1989). Transcription of early developmental isogenes in cardiac myocyte hypertrophy. *J. Mol. Cell. Cardiol.* **21,** Suppl 5, 79–89.

Skalli, O., and Gabbiani, G. (1988). The biology of the myofibroblast. Relationship to wound contraction and fibrocontractive diseases. *In* "The Molecular and Cellular Biology of Wound Repair" (Clark, R. A. F. and Henson, P. M., eds.), pp. 373–402. Plenum, New York and London.

Stewart, A. F., Larkin, S. B., Farrance, I. K., Mar, J. H., Hall, D. E., and Ordahl, C. P. (1994). Muscle-enriched TEF-1 isoforms bind MCAT elements from muscle specific promoters and differentially activate transcription. *J. Biol. Chem.* **269,** 3147–3150.

Stewart, A. F., Richard, C. W., 3rd, Suzow, J., Stephan, D., Weremowicz, S., Morton, C. C., and Adra, C. N. (1996). Cloning of human RTEF-1, a transcription enhancer factor-1-related gene preferentially expressed in skeletal muscle: Evidence for an ancient multigene family. *Genomics* **37,** 68–76.

Sun, S., Stoflet, E. S., Cogan, J. G., Strauch, A. R., and Getz, M. J. (1995). Negative regulation of the vascular smooth muscle alpha-actin gene in fibroblasts and myoblasts: Disruption of enhancer function by sequence specific single-stranded-DNA-binding proteins. *Mol. Cell. Biol.* **15,** 2429–2436.

Takahashi, H., Kobayashi, H., Matsuo, S., and Iizuka, H. (1995). Repression of involucrin gene expression by transcriptional enhancer factor 1 (TEF-1). *Arch. Dermatol. Res.* **287,** 740–746.

Tanaka, M., and Herr, W. (1990). Differential transcriptional activation by Oct-1 and Oct-2: Interdependent activation domains induce Oct-2 phosphorylation. *Cell (Cambridge, Mass.)* **60,** 375–386.

Tansey, W. P., and Herr, W. (1997). TAFs: Guilt by association? *Cell (Cambridge, Mass.)* **88,** 729–732.

Theill, L. E., Castrillo, J.-L., Wu, D., and Karin, M. (1989). Dissection of the functional domains of the pituitary-specific transcription factor GHF-1. *Nature (London)* **342,** 945–948.

Thompson, W. R., Nadal-Ginard, B., and Mahdavi, V. (1991). A MyoD1-independent muscle specific enhancer controls the expression of the β-myosin heavy chain gene in skeletal and cardiac muscle cells. *J. Biol. Chem.* **266,** 22678–22688.

Thuerauf, D. J., and Glembotski, C. C. (1997). Differential effects of protein kinase C, Ras, and Raf-1 kinase on the induction of the cardiac B-type natriuretic peptide gene through a critical promoter-proximal MCAT element. *J. Biol. Chem.* **272,** 7464–7472.

Treizenberg, S. J. (1995). Structure and function of transcriptional activation domains. *Curr. Opin. Genet. Dev.* **5,** 190–196.

Waspe, L. E., Ordahl, C. P., and Simpson, P. C. (1990). The cardiac beta-myosin heavy chain isogene is induced selectively in alpha 1-adrenergic receptor-stimulated hypertrophy of cultured rat heart myocytes. *J. Clin. Invest.* **85,** 1206–14.

Weintraub, H., Davis, R., Tapscott, S., Thayer, M., Krause, M., Benezra, R., Blackwell, T. K., Turner, D., Rupp. R., Hollenberg, S., Zhuang, Y., and Lassar, A. (1991). The *myoD* gene family: Nodal point during specification of the muscle cell lineage. *Science* **251,** 761–766.

Xiao, J. H., Davidson, I., Matthes, H., Garnier, J.-M., and Chambon, P. (1991). Cloning, expression, and transcriptional properties of the human enhancer factor TEF-1. *Cell (Cambridge, Mass.)* **65,** 551–568.

Yasunami, M., Suzuki, K., Houtani, T., Sugimoto, T., and Ohkubo, H. (1995). Molecular characterization of cDNA encoding a novel protein related to transcriptional enhancer factor-1 from neural precursor cells. *J. Biol. Chem.* **270,** 18649–18654.

Yasunami, M., Suzuki, K., and Ohkubo, H. (1996). A novel family of TEA domain-containing transcription factors with distinct spatiotemporal expression patterns. *Biochem. Biophys. Res. Commun.* **228,** 365–370.

Yockey, C. E., Smith, G., Izumo, S., and Shimizu, N. (1996). cDNA cloning and characterization of murine transcriptional enhancer factor-1-related protein 1, a transcription factor that binds to the MCAT motif. *J. Biol. Chem.* **271,** 3727–3736.

Yu, Y.-T., Breitbart, R. E., Smoot, L. B., Lee, Y., Mahdavi, V., and Nadal-Ginard, B. (1992). Human myocyte specific enhancer factor 2 comprises a group of tissue-restricted MADS box transcription factors. *Genes Dev.* **6,** 1783–1798.

Zenke, M., Grundstrom, T., Matthes, H., Wintzerith, M., Schatz, C., Wildeman, A., and Chambon, P. (1986). Multiple sequence motifs are involved in SV40 enhancer function. *EMBO J.* **5,** 387–397.

VI

Heart Patterning: The Anterior–Posterior Axis

19

Regionalization of Transcriptional Potential in the Myocardium

Robert G. Kelly,[*] Diego Franco,[†] Antoon F. M. Moorman,[†] and Margaret Buckingham[*]

*[*]CNRS URA 1947, Department of Molecular Biology, Pasteur Institute, 75724 Paris Cedex 15, France
[†]Department of Anatomy and Embryology, Academic Medical Center, University of Amsterdam, 1105 AZ Amsterdam, The Netherlands*

I. Introduction

The fully formed vertebrate heart consists of a series of functionally distinct compartments which are the products of complex morphogenetic processes during embryonic development. In order to discuss gene expression in the developing heart, it is necessary to summarize the major events of cardiac morphogenesis. Figure 1 presents a schema for cardiac development in higher vertebrates (Larsen, 1993). Cardiomyocytes first differentiate within the splanchnic lateral mesoderm on either side of the midline in the cranial region of the embryo at Embryonic (E) Day 7 in the mouse (see Chapter 1). As a result of cephalic and lateral folding of the embryo, left and right precardiac regions fuse at the anterior edge of the cardiac crescent to form a primitive heart tube. The cardiac tube is transiently organized along the anterior–posterior (A–P) axis with a rostral arterial pole and a caudal venous pole, and subsequently loops to the right at E8 in the mouse, with the originally caudal inflow region moving in an anterior and dorsal direction to generate an S-shaped heart tube. Primitive venous inflow tract (IFT), atrial, atrioventricular canal (AVC), ventricular, and outflow tract (OFT) regions are apparent at this stage. The looped heart rapidly develops right and left atrial appendages, and

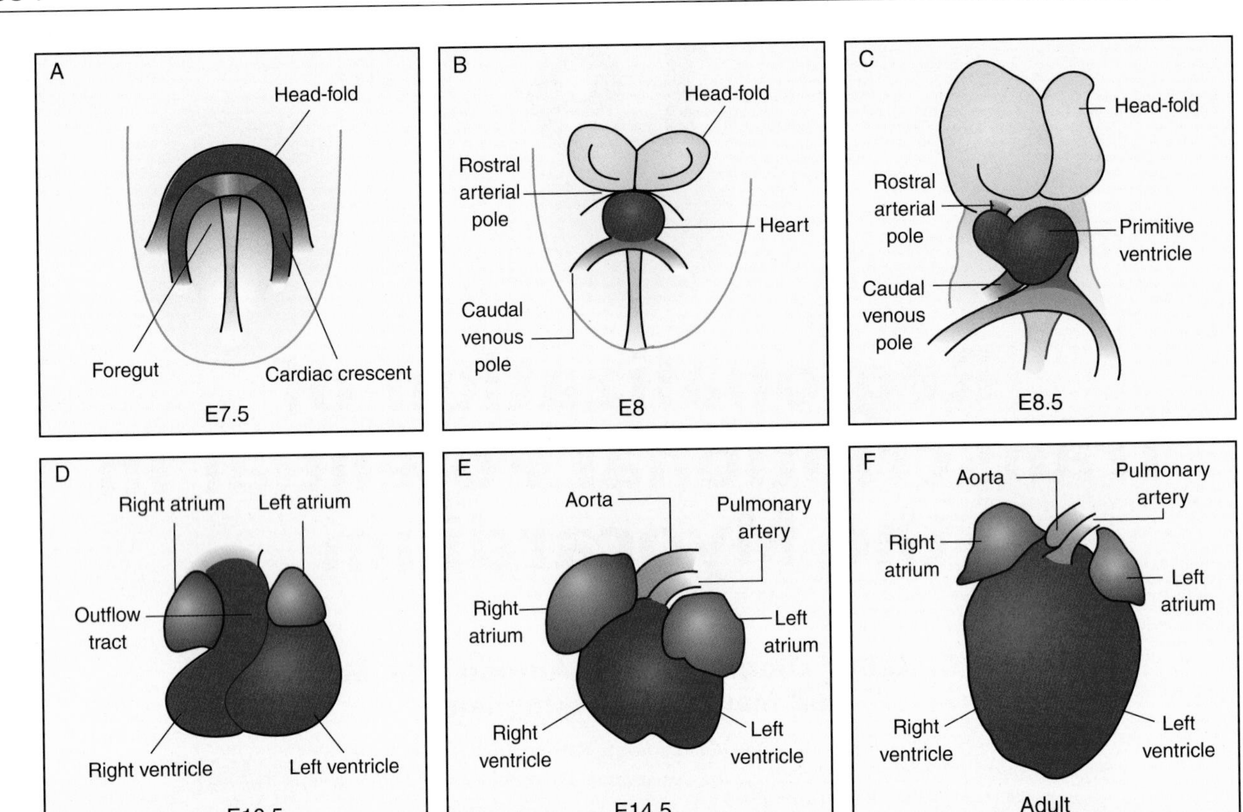

Figure 1 Schema of heart development in the mouse. (A) Myocardiocytes first differentiate in the cardiac crescent from precursors in the sphlancnic lateral mesoderm at E7.5. (B) By E8 a tubular heart forms at the anterior of the crescent with rostral arterial and caudal venous poles (outflow and inflow, respectively). (C) The heart subsequently loops to the right, with the venous pole moving anteriorally and dorsally; at this stage a primitive ventricle with arterial outflow and venous inflow regions is evident. (D) In the embryonic heart at E10.5, primitive right and left atria and right and left ventricles are apparent; at this stage the heart has an extended outflow tract. (E) During late embryonic and fetal development the primitive cardiac chambers grow and septate, and the outflow tract divides into aortic and pulmonary trunks. (F) External view of an adult mouse heart showing the relative positions of the principal cardiac chambers and the positions of the aorta and pulmonary artery. Not to scale.

the embryonic right ventricle expands distally to the initially larger embryonic left ventricle; at the same time the heart tube continues to extend as myocardium is added to both rostral and caudal ends, generating the distal OFT and venous IFT. Subsequently, ventricular, atrial, and OFT septa converge in the AV region such that by birth four distinct chambers direct separate pulmonary and systemic blood flow (see Chapter 12).

II. The Development of the Heart and the Onset of Atrial and Ventricular Transcriptional Regionalization

How and when is regionalization developed in the myocardium? In this section, we will discuss the emergence of atrial and ventricular transcriptional diversity along the A–P axis of the early heart. Differentiation within the heart tube follows a craniocaudal gradient (Icardo, 1996). Myocardial genes tend to be coexpressed throughout the embryonic heart at the earliest stages of heart tube formation; in birds, however, anterior to posterior fusion of bilateral cardiac precursors results in a linear heart tube within which regional differences in expression of a subset of cardiac markers are already detectable (De Jong *et al.,* 1987, 1990, Yutzey *et al.,* 1994). Atrial myosin heavy chain (MHC) transcripts are expressed predominantly in the posterior heart tube as the cardiac primordia fuse, consistent with early restriction of atrial MHC protein and suggesting that atrial and ventricular lineages are diversified when they first differentiate (Sweeney *et al.,* 1987; De Jong *et al.,* 1990; Yutzey *et al.,* 1994; Yutzey and Bader, 1995). In the mouse, significant regionalization of gene expression within the developing heart emerges from the time of cardiac looping (Lyons *et al.,* 1990;

Lyons, 1994). The temporal profile of chamber-specific restriction of cardiac genes is highly variable: Some markers are restricted early in development, for example, the regulatory myosin light chain 2V (*MLC2V*) gene which is predominantly expressed in the primitive ventricle from E8.5 (O'Brien *et al.*, 1993); other myosin genes encoding alkali MLCs continue to be expressed throughout the myocardium after looping (Fig. 2; Table I; see Chapter 15). A subset of markers appears to be initially coexpressed in a graded manner; the MHC genes, for example are expressed in inverse gradients in the embryonic heart, with α*MHC* being expressed at a higher level at the arterial pole and β*MHC* at the venous pole of the tubular heart (De Jong *et al.*, 1987; De Groot *et al.*, 1989; Lyons *et al.*, 1990). The β*MHC* gene shows relatively early restriction of transcription, becoming confined predominantly to the ventricular compartment by E10.5 in the mouse (Fig. 2; Lyons *et al.*, 1990). *MLC1V* transcripts, in contrast, continue to be detectable in atrial cells until E15 and *MLC1A* transcripts remain detectable in the ventricles even after birth (Lyons *et al.*, 1990). Changes in myocardial gene expression do not necessarily occur in a synchronized fashion even within a cardiac compartment. For example, the decrease in levels of *MLC1A* in fetal ventricles takes place in a trans-mural gradient from epicardium to endocardium, reflecting transcriptional differences between the trabeculated and compact myocardial components of the ventricle (Lyons *et al.*, 1990; Franco *et al.*, 1997b). Further developmental changes in expression pattern take place during fetal and early postnatal development; for example, β*MHC* is downregulated and α*MHC* upregulated in the ventricles of the mouse heart at birth.

Transcriptional differences between atria and ventricles are consistent with the different contractile, electrophysiological, and pharmacological characteristics of atrial and ventricular cardiomyocytes. For example, different MHC isoforms have different biochemical properties: β*MHC* has a lower ATPase activity consistent with the lower shortening velocity and rate of tension development in ventricular compared to atrial myocardium. In the embryonic chicken heart the expression of MHC protein isoforms correlates with local differences in the contractile pattern of the heart (De Jong *et al.*, 1987). α*MHC* and β*MHC* continue to be coexpressed in particular regions of the embryonic heart marked by peristaltic contraction and slow conduction velocity, such as the AVC and OFT (De Jong *et al.*, 1987; Moorman and Lamers, 1994). Other cardiac genes, such as phospholamban and sarcoplasmic reticulum calcium ATPase, implicated in modulating calcium ion concentration, are expressed in inverse A–P gradients in the embryonic heart (Moorman *et al.*, 1995). While these patterns of gene expression are likely to be restricted in response to underlying transcription factor gradients

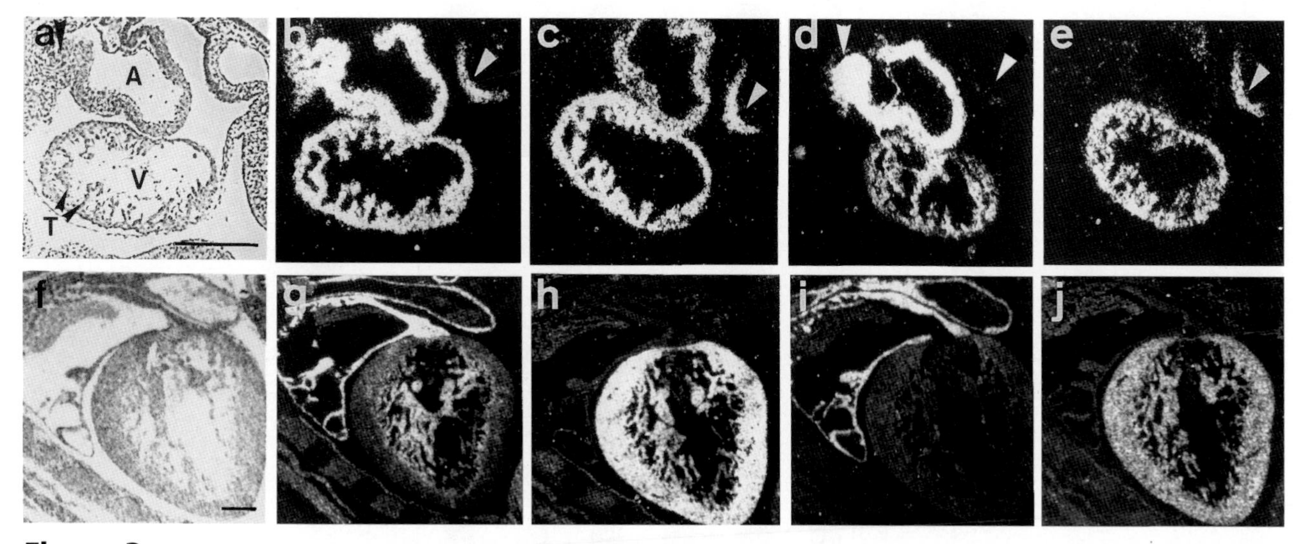

Figure 2 *In situ* localization of myosin transcripts in developing mouse hearts at E10.5 (a–e) and E15.5 (f–j). (a and f) Bright field images. A, atria; V, ventricle, T, trabeculae. (b and g) *MLC1A* transcripts are expressed throughout the embryonic heart and are detectable in the atria and trabeculated portion of the ventricle at fetal stages. (c and h) *MLC1V* transcripts are expressed in both cardiac compartments at E10.5 and become restricted to the ventricle by E16.5. (d and i) α*MHC* transcripts are expressed in the embryonic atria and at a low level in the embryonic ventricle and become restricted to the atria by E15.5 (e and j) β*MHC* transcripts are restricted to the ventricular compartment by E10.5; note that β*MHC* messages continue to be found in the OFT region (arrowheads). Scale bars = 200 μm. [Reproduced from Lyons *et al.*, *The Journal of Cell Biology* (1990) **111**, 2427–2436, by copyright permission of The Rockefeller University Press.]

Table I Myosin Gene Expression in the Developing Mouse Myocardium

	Tubular heart (E8.5)		Embryonic (E10.5)					Foetal (E15.5)		
Gene	Arterial	Venous	OFT	V	AVC	A	IFT	V	A	Reference
MHCα										Lyons *et al.* (1990)
MHCβ										Lyons *et al.* (1990)
MLC1A										Lyons *et al.* (1990)
MLC1V										Lyons *et al.* (1990)
MLC2A										Kubalak *et al.* (1994) Franco *et al.* (1997b)
MLC2V										O'Brien *et al.* (1993) Franco *et al.* (1998a)

Note: In situ hybridization data for myosin heavy and light chain transcript accumulation are illustrated at three stages of development—the tubular heart (E8.5), the embryonic heart (E10.5), and the fetal heart (E15.5). A, arterial; V, venous; OFT, outflow tract; AVC, atrioventricular canal; IFT, in-flow tract.

Relative expression levels are graded as follows: ■ > ■ > ■ > ■ > □ .

within the embryonic heart, changes in isoform expression in adult hearts subjected to pressure overload point to the role that hemodynamic factors play in the regionalization of cardiac gene expression. The functional significance of switches in MLC and other sarcomeric components is less well established, although ongoing isoform replacement experiments in myocytes *in vitro* and in the adult heart *in vivo* are addressing the role that different isoforms play in sarcomeric function (Gulick *et al.*, 1997).

III. Left and Right Cardiac Compartments: Regionalized Transgene Expression

Regionalization of transcriptional potential within the myocardium has been recently demonstrated to be more extensive than predicted from the analysis of atrial- and ventricular-specific gene expression. Transcriptional differences between the left and right sides of the embryonic heart have been detected which persist throughout cardiac development and in adult hearts, and which provide support for a segmental model of heart tube formation (see Chapter 9). Such left–right (L–R) differences were first detected with β-galactosidase reporter genes in transgenic mice containing cardiac regulatory elements from different heart-specific genes and have since been extended to a number of endogenous genes. These findings confirm that left and right cardiac compartments differ in their transcriptional potential, at least from the time of looping and several days before septation.

Transgenes showing regionalization of reporter gene expression in the embryonic myocardium are documented in Table II. Regulatory sequences from the promoter and enhancers of the cardiac α-actin gene direct preferential expression of a reporter gene encoding nuclear localizing β-galactosidase (*nlacZ*) in the primitive left ventricle and right more than left atrium of the heart (Biben *et al.*, 1996). This pattern is also seen with the fast alkali myosin light chain 3F (*MLC3F*) promoter and 3′ enhancer, which is transcribed in cardiac as well as skeletal muscle, although the MLC3F protein does not accumulate in the former (Kelly *et al.*, 1995). A complementary pattern is seen with other transgenes controlled by upstream sequences from the regulatory myosin light chain *MLC2V* (Ross *et al.*, 1996), desmin (Kuisk *et al.*, 1996), dystrophin (Kimura *et al.*, 1997), or *SM22α* genes (Moessler *et al.*, 1996; Li *et al.*, 1996; Kim *et al.*, 1997), and by sequences flanking the *hdf* transgene insertion site (Yamamura *et al.*, 1997). These transgenes are expressed predominantly in the primitive right ventricle and adjacent OFT. In Fig. 3 expression of an *MLC3F–nlacZ-2E* transgene (Kelly *et al.*, 1995) in the left ventricle and right atrium of the embryonic heart is compared with the right ventricular and OFT expression of a desmin transgene (Kuisk *et al.*, 1996). Regionalized β-galactosidase expression reflects regionalized distribution of *nlacZ* transcripts, confirming that this regulation is at the level of transcription (Fig. 4a; Kelly *et al.*, 1995). Furthermore, since the regionalized expression profiles are reproduced in different lines containing the same constructs, the *cis*-acting regulatory sequences included in the transgenes are likely to be responsible for the regionalization of reporter gene expression.

Right–left asymmetry of transgene expression is not an absolute phenomenon in that some labeled cells can be detected in the neighboring negative cardiac compartments, although for some transgenes the boundary between expressing and nonexpressing domains is sharper than for others. Another important distinction between different regionalized cardiac transgenes is the temporal progression of regionalization: In some cases expression is extinguished during fetal development (e.g., the *SM22α* and *hdf* transgenes; Moessler *et al.*, 1996; Yamamura *et al.*, 1997), whereas in other cases the profile remains essentially the same from embryonic to adult stages. For example, in adult hearts of *MLC3F–nlacZ-2E* transgenic mice myocardial cells of the left ventricle and septum are β-galactosidase-positive, and only a few positive cells are occasionally observed in the free wall of the right ventricle; expression in the right atrium is significantly more pronounced than that in the left atrium (Fig. 5; Kelly *et al.*, 1995). This is the same overall pattern of transgene expression as that seen at E10.5 (Fig. 3). The precise extent of this labeling varies between individual mice within one transgenic line, suggesting that the exact position of the myocardial fiber-containing cells which have activated the transgene differs from mouse to mouse. Interestingly, in the embryonic heart of *MLC3F–nlacZ-2E* mice the myocardium in the AVC region is also positive, demonstrating continuity of transgene expression between the primitive left ventricle and right atria (Fig. 3d). This continuum of expression may define a "primitive myocardium" predating the acquisition of a pulmonary circulation.

At what stage is the regionalization of transgenes first detectable in the embryonic mouse heart? In most reported cases this would appear to be from the stage when cardiac looping is first initiated. In the case of the *MLC3F–nlacZ-2E* transgene, marked regionalization emerges after looping (at E9), although the transgene is expressed from E7.5 (Franco *et al.*, 1997). Regional differences in transgene expression which emerge at the time of looping may reflect an underlying positional

Table II Transgenes Regionalized in the Embryonic Myocardium

Transgene	Outflow tract	Ventricles		AVC	Atria		Inflow tract	Cis acting sequences	Reference
		Right	Left		Right	Left			
MLC3F-nlacZ-2E								−2 to +2kb + 3' enhancer	Kelly *et al.* (1995)
MLC3F-nlacZ-9								−9 to +2kb	Franco *et al.* (1997)
α-cardiac actin-nlacZ								−9 to +0.8kb	Biben *et al.* (1996)
MLC2V-lacZ								−250 bp (requires MLE1/USF site) or 2xHF-1 seq (HF1a/EF1a, HF1b/MEF2)	Ross *et al.* (1996)
Desmin-nlacZ								−976 to +200bp (requires MEF2 site)	Kuisk *et al.* (1996)
Dystrophin-lacZ								−900 to +22bp	Kimura *et al.* (1997)
SM22α-lacZ								−445, −441 or −280 to +41 (requires SRE/SME3 site)	Moessler *et al.* (1996) Li *et al.* (1996) Kim *et al.* (1997)
hdf-lacZ								−1.6kb Hoxa-1/insertion site	Yamamura *et al.* (1997)
TnIs-CAT								−4.2kb to +12bp	Zhu *et al.* (1995)
GATA-6-lacZ								−9.2 to +0.8kb	He and Burch (1997)
GATA-6-lacZ								−1.5 to +0.8kb	He and Burch (1997)

Note: Reporter gene expression data for transgenes regionalized in the embryonic mouse myocardium (approximately E10.5). Regionalized expression levels are graded (in the case of α-cardiac actin and *MLC2V* transgenes, lower expression represents the extension of transgene expression into adjacent domains in a subset of transgenic lines).

Relative expression levels are graded as follows: ▬ > ▬ > ▬ > ▭ .

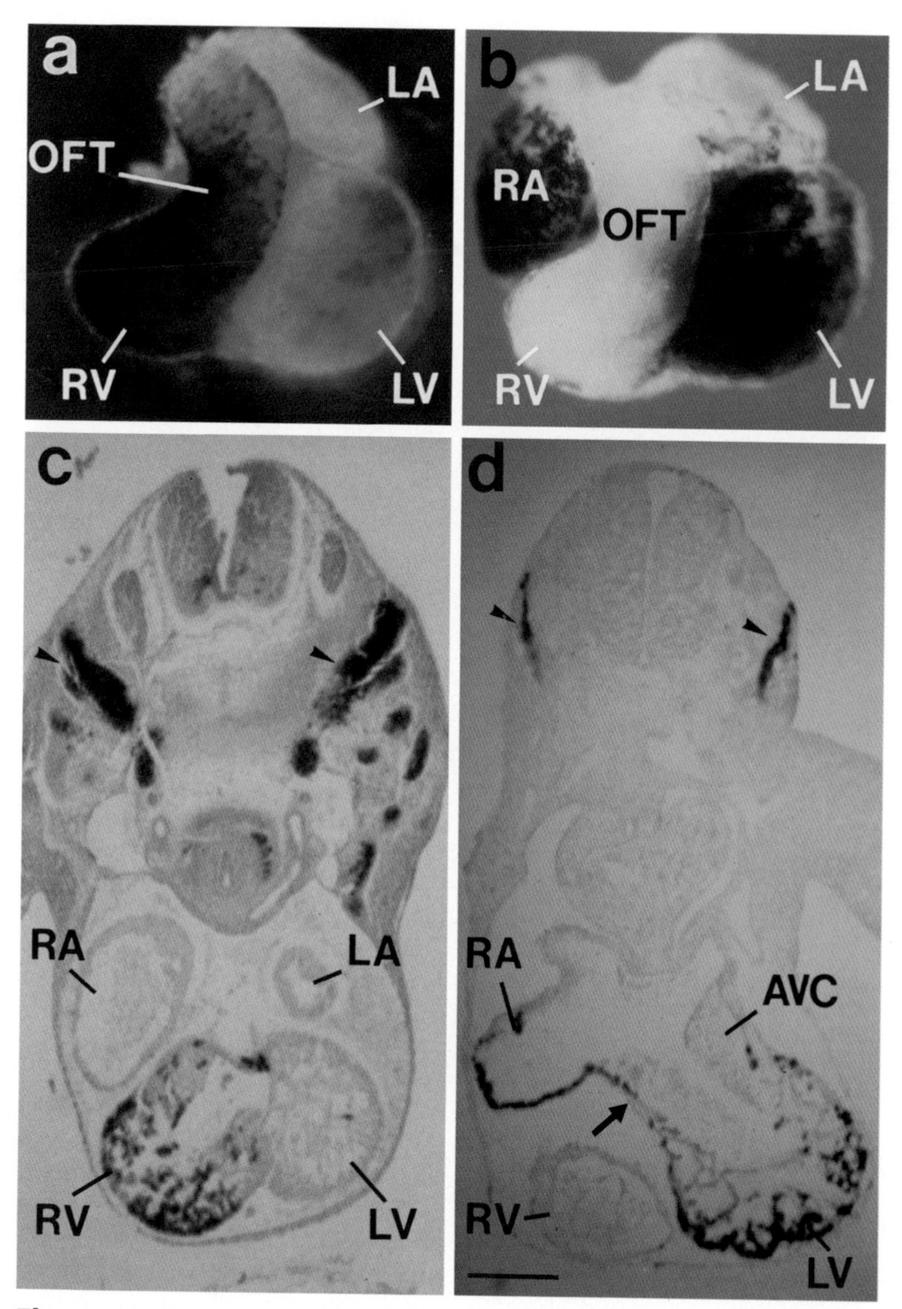

Figure 3 Two transgenes showing complementary regionalized expression patterns in the embryonic myocardium. (a and c) A *desmin–lacZ* transgene is expressed in the embryonic right ventricle (RV) and outflow tract (OFT) as visualized by whole mount X-gal staining of an isolated heart at E10.5 and X-gal staining of a cryostat section at E11.5 (Kuisk *et al.*, 1996). The embryonic left ventricle (LV) and left atrium (LA) are β-galactosidase negative. (b and d) The *MLC3F–nlacZ-2E* transgene is expressed in the right atrium (RA) and left ventricle (E10.5 whole mount and E11.5 section); there are a few β-galactosidase-positive cells in the left atrium; the embryonic right ventricle and OFT are β-galactosidase negative (Kelly *et al.*, 1995). Note continuity of *MLC3F–nlacZ* expression along the atrioventricular canal (AVC) between the right atria and left ventricle (arrow). Both transgenes are expressed in skeletal muscle at this stage (arrowheads). Scale bar = 200 μm. [Panels a and c reproduced from *Developmental Biology* (1996) **174,** 1–13 by permission of Academic Press.]

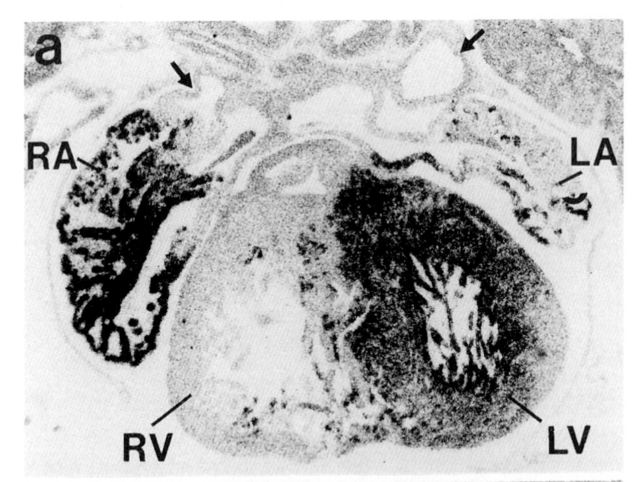

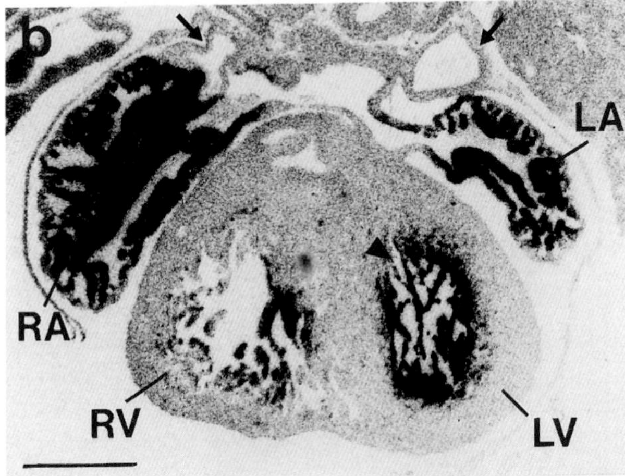

Figure 4 *In situ* localization of *nlacZ* and atrial natriuretic factor (*ANF*) transcripts in *MLC3F–nlacZ-2E* transgenic hearts. (a) At E14.5 *nlacZ* transcripts are confined to the right atrium (RA) and left ventricle (LV), with low-level expression in the left atrium (LA), confirming that regionalized β-glactosidase expression reflects regionalized transcription. Virtually no *nlacZ* transcripts are observed in the right ventricle (RV). (b) *ANF* transcripts are expressed predominantly in the atria but are also detected in the ventricles, where they are present at a higher level in the left than right chamber; at E14.5 *ANF* transcripts accumulate exclusively in the trabeculated component of the ventricles (arrowhead). Note that *nlacZ* and *ANF* transcripts are absent from the caval veins (arrows). Scale bar = 400 μm.

prepattern rather than arising *de novo;* lack of early regionalization is consistent with the observation that most early cardiac markers are broadly expressed in the embryonic mouse myocardium. In Fig. 6, E7.75 hearts from desmin (right ventricular dominant, Kuisk *et al.,* 1996) and *MLC3F–nlacZ-2E* (left ventricular dominant expression) transgenes are compared. The anterior region of the as yet unlooped heart tube appears to express the *desmin* transgene at a higher level than the *MLC3F–nlacZ-2E* transgene, suggesting that regionalization may initiate at this early stage, at least for some markers (including the *MLC2V* transgene; Ross *et al.,* 1996). Detailed direct comparison of different transgenic lines is required to address this issue.

A. Left and Right Cardiac Compartments: Regionalized Endogenous Gene Expression

Observations with transgenes which are regionally expressed in the heart confirm and refine the finding emerging from other transgenic studies that the expression of a gene in diverse cell types or at different times during development is under the control of separable *cis*-acting elements. Thus, expression of a gene transcribed throughout the heart may be the product of several different regulatory elements, each active in a specific myocardial subdomain. The endogenous *MLC2V, desmin,* and *SM22α* genes are known to be expressed in the entire ventricular compartment (and atria, for *desmin* and *SM22α*); elements conferring left ventricular expression of the endogenous *MLC2V* gene, for example, must therefore be absent from the 250-base pair (bp) promoter included in the transgene. Speculatively, these transcriptional subdomains may reflect the increasing complexity of heart anatomy during vertebrate evolution, such that recently evolved chambers or heart structures invoke transcriptional pathways which differ from those controlling gene expression in more "primitive" regions of the heart. There are, however, a number of endogenous genes which have been shown to be expressed, at least transiently, in a differential manner between left and right sides of the heart. These include the atrial natriuretic factor (*ANF*) gene, which is expressed in embryonic ventricles and atria and which becomes restricted to the atria during fetal development. During embryonic development *ANF* transcripts accumulate in the left ventricle to a higher level than in the right ventricle (Fig. 4b; Zeller *et al.,* 1987); *ANF* transcripts are further regionalized since they are expressed at a high level in the ventricular trabeculae and at a low level in the compact myocardium, supporting the idea that these regions have distinct transcriptional properties (Fig. 4b). A low level of ventricular *ANF* is detectable in the adult heart (1% of atrial levels), and this is detected predominantly in the left ventricle (Gardner *et al.,* 1986). The onset of transcription of the M-isoform of creatine phosphokinase (*MCK*) at E12.5 in the mouse embryo is more pronounced in the right ventricle than in the left, with expression subsequently extending to the entire myocardium (Lyons *et al.,* 1991; Lyons, 1994); MCK protein also accumulates in the right ventricle before the left ventricle in the developing rat heart (Has-

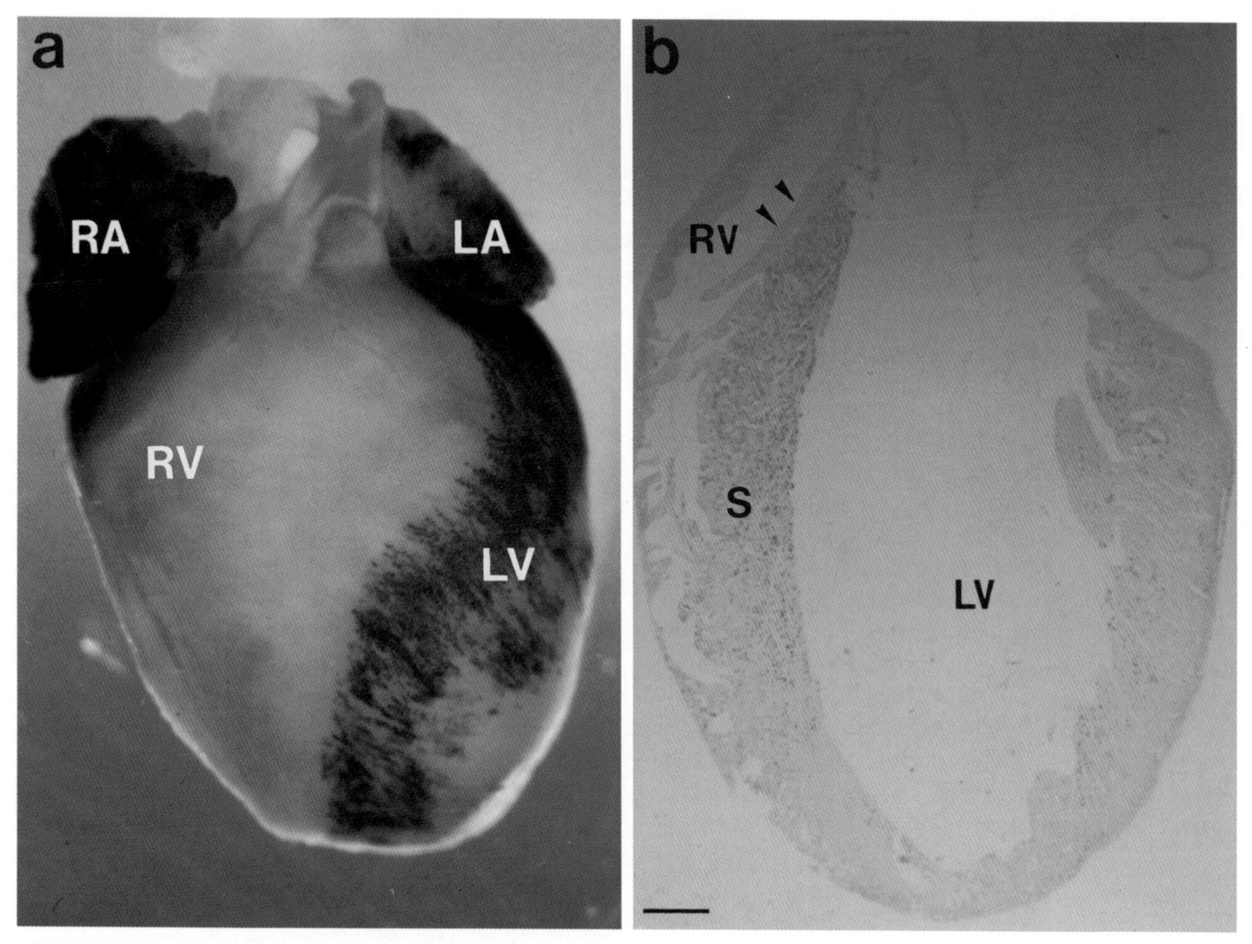

Figure 5 *MLC3F–nlacZ-2E* transgene expression in the adult heart. (a) External view of X-gal-stained heart: β-Galactosidase is expressed predominantly in the left ventricle (LV) and right atrium (RA), with low-level expression in the left atrium (LA). β-Galactosidase expression is excluded from the right ventricle (RV). (b) An X-gal stained cryostat section illustrates that the interventricular septum (S) is composed of transgene expressing and nonexpressing (arrowheads) regions on the left and right ventricular faces, respectively; *nlacZ*-positive nuclei are observed in the left but not in the right free ventricular wall. Scale bar = 500 μm. Panel a reproduced from Kelly *et al., The Journal of Cell Biology* (1995) **129,** 383–396, by copyright permission of the Rockefeller University Press.

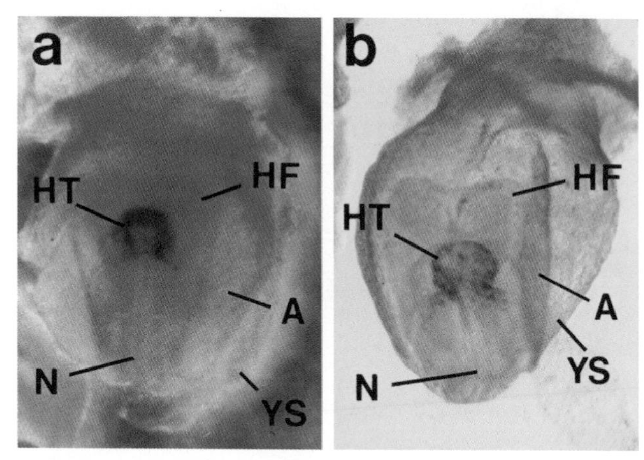

Figure 6 Comparison of *desmin–lacZ* (a; Kuisk *et al.,* 1996) and *MLC3F–nlacZ-2E* (b; Franco *et al.,* 1997) transgene expression in X-gal-stained E7.75 embryos, showing β-galactosidase-expressing cardiomyocytes in the unlooped heart (HT). N, notochord; HF, headfold; A, amnion; YS, yolk sac. [Reproduced from *Developmental Biology* (1995) **174,** 1–13 (panel a) and (1997) **188,** 17–33 (panel b) by permission of Academic Press.]

selbaink *et al.,* 1990). The development of whole mount *in situ* hybridization facilitates visualization of L–R differences which might be missed using *in situ* hybridization on tissue sections. An example is the *MLC2A* gene (see Chapter 15), which exhibits stronger expression in the right ventricle than the left prior to restriction to the atrial compartment and OFT during embryonic development (D. Franco, R. G. Kelly, and P. Zammit, unpublished observations). Other endogenous mouse genes expressed in a L–R regionalized manner at the looping stage include the transcription factors *e-hand* and *d-hand* (see Chapter 9). Most endogenous genes which have been shown to be regionalized are differentially expressed only transiently, and the fraction of endogenous genes which display L–R differences in the mature mouse heart remains to be determined. Immunofluorescent studies of MHC isoform distribution within the rat and rabbit heart have revealed differences in the frequency of αMHC-labeled fibers between right and left ventricles; cellular heterogeneity within the ventricular myocardium may therefore be a general feature of the mammalian heart (Gorza *et al.,* 1981; Sartore *et al.,* 1981; Bougnavet *et al.,* 1984; Litten *et al.,* 1985).

B. Further Regionalization: The OFT and Other Myocardial Transcriptional Compartments

In addition to revealing L–R transcriptional differences in the embryonic heart, regionalized transgene and endogenous gene expression patterns provide evidence for further levels of subcompartmentalization within the myocardium. The outflow tract, for example, appears to be a distinct transcriptional compartment of the embryonic heart: a second *MLC3F* transgene including the MLC3F promoter and an intronic enhancer element (*MLC3F–nlacZ-9*) expresses *nlacZ* in the entire embryonic myocardium except for the OFT and IFT (Fig. 7a and 7b; Franco *et al.,* 1997). In the early heart tube, prior to E9, this transgene is expressed throughout the myocardium with a similar distribution to the *MLC3F–nlacZ-2E* transgene (Fig. 8; Franco *et al.,* 1997). As suggested previously, looping is therefore followed by restriction of regional transcription along the A–P axis of the heart, possibly by cranial addition of non-transgene-expressing anterior myocardium (Virágh and Challice, 1973; Argüello *et al.,* 1975). Further evidence for the OFT being a distinct transcriptional domain comes from analysis of a slow troponin *I*(*TnIs*) *CAT* transgene which is expressed in the embryonic heart, although at a low level in the ventricles, and not in the OFT; the endogenous gene, in contrast, is expressed throughout the embryonic heart, including the OFT (Zhu *et al.,* 1995). Specific regulatory sequences required for *TnIs* transcription in the OFT are presumably absent from this transgene construct. A subset of endogenous cardiac genes are differentially expressed across the OFT–right ventricle (RV), boundary, including *MLC2A* (expressed in the atria and OFT; Fig. 7c) and *MLC2V* [expressed in both ventricles and at low levels in the OFT (Fig. 7d); Franco *et al.,* 1998a]. In the chick, smooth muscle α-actin, which is initially expressed throughout the heart tube, becomes restricted to the OFT after looping (Ruzicka and Schwartz, 1988).

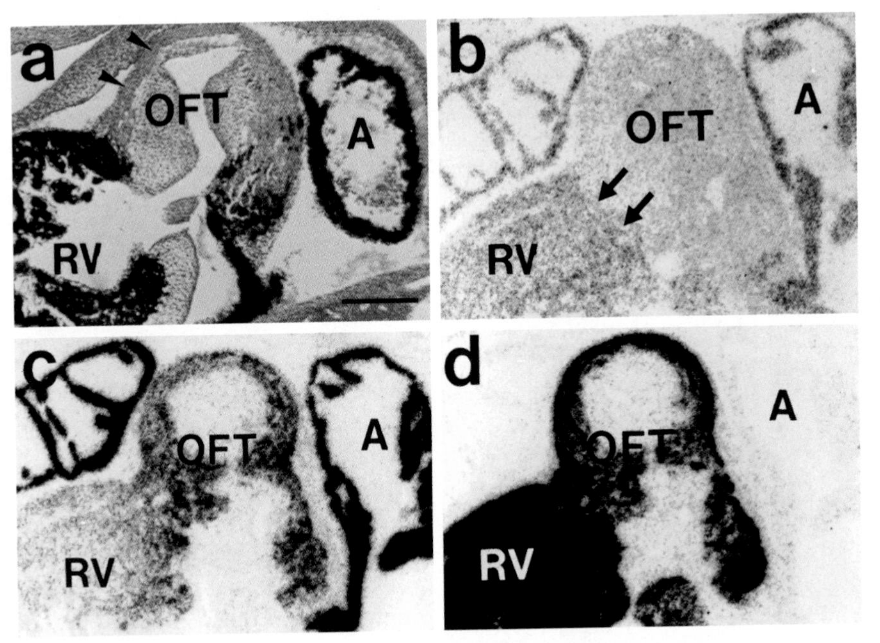

Figure 7 The outflow tract (OFT) is a distinct transcriptional compartment. (a) X-gal-stained cryostat section showing β-galactosidase activity in *MLC3F–nlacZ-9* transgenic embryos in the atria (A) and ventricles (RV) but not in the OFT at E14.5; arrowheads denote β-galactosidase negative myocardial cells of the OFT. (b) β-Galactosidase regionalization reflects the distribution of *nlacZ* transcripts detected by *in situ* hybridization. There is a sharp boundary between transgene expressing and nonexpressing cardiomyocytes (arrows). (c) *MLC2A* transcripts are localized in the atria and OFT but not in the ventricular compartment. (d) *MLC2V* transcripts, in contrast, are expressed at a high level in the ventricles and lower level in the OFT. Scale bar = 200 μm.

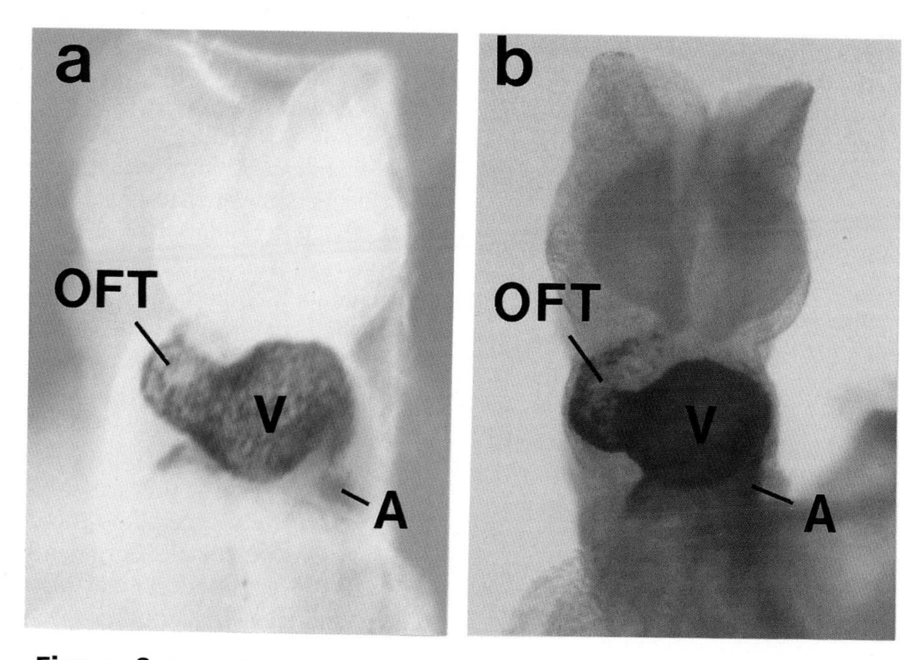

Figure 8 Comparison of *MLC3F–nlacZ-2E* (a) and *MLC3F–nlacZ-9* (b) transgenes in the looped heart at E8.5. β-Galactosidase expression is similar in these two X-gal-stained embryos, despite later divergence in expression patterns. The *MLC3F–nlacZ-9* transgene, unlike the *MLC3F–nlacZ-2E* transgene, is expressed in the embryonic right ventricle at later stages of cardiac development (see Figs. 3 and 7). OFT, forming outflow tract; V, primitive ventricle; A, future atrial region.

Other genes show more subtle regionalization of expression within the myocardium. Examples include the transient restriction of carbonic anhydrase II expression to an anterior domain of the left ventricular wall (Vuillemin and Pexeider, 1997) and the expression of a subset of genes in the myocardium overlying the atrioventricular region of the embryonic heart, including *Bmp4* and *desert hedgehog* (Jones *et al.,* 1991; Bitgood and McMahon, 1995). Another example of regionalization of transgene expression is provided by a chick *GATA6–lacZ* construct which is expressed exclusively in the AVC of embryonic transgenic mouse hearts (He and Burch, 1997). Interestingly, a larger *GATA6* transgene is also expressed in the RV (He and Burch, 1997), and the endogenous gene is expressed throughout the myocardium (Morrisey *et al.,* 1997). This provides further evidence for a modular basis of transcription for a gene expressed throughout the myocardium. The cardiac inflow region is also composed of distinct gene expression domains: Analysis of different *MLC3F–nlacZ* transgene constructs reveals at least four transcriptional domains at the venous pole of the developing heart (Franco *et al.,* 1998b). *ANF* transcripts accumulate in atrial myocytes but not in the myocardium of the caval or pulmonary veins (Fig. 4), whereas other cardiac markers, including *MLC1A* and α*MHC,* are expressed in the caval and pulmonary myocardium from fetal stages (Lyons *et al.,* 1990; Jones *et al.,* 1994). What is the functional significance of this extensive transcriptional subcompartmentalization within the heart? In specialized regions of the embryonic myocardium, such as the AVC or OFT, transcriptional differences probably reflect the different contractile and conductive properties of these regions or the role of these regions in endocardial cushion development (Moorman and Lamers, 1994; Eisenberg and Markwald, 1995). It remains to be seen to what extent further regionalization of gene expression is functionally important for the physiological requirements of the developing myocardium. The functional requirements of left and right ventricles are clearly different in postnatal hearts, and the regional expression of genes encoding sarcomeric proteins or muscle-specific enzymes may reflect such differences. Furthermore, each ventricular and atrial compartment is characterized by different anatomical regions in the mature heart and has specific inlet and outlet connections; such a degree of differentiation is likely to require multiple subdomains of gene expression which may be prefigured in the fetal and embryonic heart. It appears likely that continued analysis of new transgenic lines will uncover further levels of subcompartmentalization within the developing heart.

IV. Myocardial Regionalization: Anterior–Posterior and Left–Right Patterning

The above profiles of transcriptional regionalization in the developing heart are underscored by substantially earlier patterning events. Evidence for this comes from embryological analysis of the early events of heart tube formation and investigations into the time of specification and determination of anterior and posterior regions of the early heart tube, largely in avian systems. There are two temporally distinct inputs into regionalization of the early heart: patterning of precardiac cells along the A–P axis and the subsequent interpretation of L–R positional information which leads to looping of the heart tube.

A. A Rostrocaudal Prepattern

There is substantial evidence that the embryonic heart is patterned along the A–P axis. The heart tube forms predominantly in a rostrocaudal sequence by addition of precardiac material at the posterior end (Patten, 1922; DeHaan, 1963; Rosenquist and DeHaan, 1966), although there is also a significant extension of the myocardium at the arterial pole of the early heart tube (Virágh and Challice, 1973; Argüello *et al.*, 1975). Myocardial differentiation within the heart tube also occurs in a craniocaudal gradient (Litvin *et al.*, 1992; Han *et al.*, 1992), and precardiac cells have been shown to progress from a caudal BrdU- and TPA-sensitive to rostral insensitive state, indicative of ongoing differentiation along the A–P axis of the early heart (Gonzalez-Sanchez and Bader, 1990; Montgomery *et al.*, 1994). The precardiac splanchnic mesoderm develops as a coherent mesothelium in both chick and mouse embryos (Rosenquist and DeHaan, 1966; Manasek, 1968; Kaufman and Navaratnam, 1981; De Ruiter *et al.*, 1992). The A–P position of cells within this sheet is likely to be maintained until differentiation, suggesting that defined regions in the heart tube are prefigured by boundaries in the precardiac mesoderm (Stalsberg and DeHaan, 1969). Indeed, different regions of the tubular heart (OFT, ventricles, atria, and IFT) have been shown to originate from a series of subdivisions along the A–P axis of the precardiac mesoderm (Stalsberg and DeHaan, 1969). Rostrocaudal position may in fact be specified at the time of gastrulation since the order in which prospective cardiogenic cells migrate through the primitive streak correlates with their position along the A–P axis of the heart tube (Garcia-Martinez and Schoenwolf, 1993; see Chapter 1). These cells, however, are not irreversibly committed to their respective rostrocaudal fates until differentiation (Inagaki *et al.*, 1993). The earliest physiological manifestation of A–P differences in the heart is an intrinsic gradient in beat rate at the onset of contractility (DeHaan, 1963; Van Mierop, 1966). Explants from caudal (prospective sinus venosus) precardiac cells beat faster than more rostral explants (in addition, the left caudal pacemaker region beats faster than the right; DeHaan, 1963; Satin *et al.*, 1988). Elegant transplantation experiments in chick embryos have shown that beat rate is determined by regional cues from tissue surrounding the precardiac mesoderm, and therefore that prior to differentiation the mesoderm is not stably coded for future beat rate (Satin *et al.*, 1988), consistent with the observation that the rostrocaudal fate of postgastrulation precardiac cells is labile (Inagaki *et al.*, 1993). The source of positional information outside the premyocardiac cells remains to be determined, but the adjacent endoderm may play a role in A–P patterning, and signals may be propagated by planar and vertical means (see Inagaki *et al.*, 1993). The precise correlation between A–P position and the emergence of different cardiac lineages (atrial and ventricular cells) remains to be established. Mouse embryonic stem cells, which are derived from pregastrulation embryos, will differentiate into embryoid body aggregates containing a diversity of cell types including cardiomyocytes, among which both atrial and ventricular phenotypes can be identified (Maltsev *et al.*, 1993). No positional information is available for cells within embryoid bodies, and, at least in this case, cell–cell interactions are likely to play an important role in lineage diversification.

In summary, the rostrocaudal patterning of precardiac cells may be specified at the time of gastrulation but not irreversibly determined until differentiation. The acquisition of subregional positional information along the A–P axis of the early heart, resulting in the restricted expression domains later revealed by transgenes, may arise at the same stage of development as the acquisition of anterior (atrial) or posterior (ventricular) identity. Isolation of subregional markers in the chick heart will address the issues of the developmental stage at which future left and right precardiac populations become specified and whether these populations represent different cardiac lineages.

B. Anterior–Posterior Patterning: The Role of Retinoic Acid

The importance of positional information along the A–P axis of the heart is further suggested by experiments with retinoic acid (RA), which is known to perturb patterning along the A–P axis of the embryo (Durston *et al.*, 1989; Kessel and Gruss, 1991). Normal cardiac development in rats exhibits a stage-dependent requirement for vitamin A (Wilson and Warkany, 1949), and RA treatment of early embryos causes severe cardiac malformations in diverse species, including deletion or ablation of cardiac tissue or perturbed precardiac cell migration leading to cardiac bifida (Osmond *et al.*, 1991; Stainier and Fishman, 1992; Drysdale *et al.*, 1994; see Chapter 13). In the zebrafish, RA exposure perturbs the A–P axis of the heart at doses which do not affect the rest of the body axis (Stainier and Fishman, 1992). These authors noted a continuous gradient of RA sensitivity along the A–P axis of the zebrafish heart such that increased exposure to RA caused the sequential and progressive deletion of cardiac compartments, affecting anterior-most compartments first (OFT at low doses and then the ventricle, atria, and sinus venosus; Stainier and Fishman, 1992). Partially deleted chambers were also observed, suggesting the A–P polarity within the early heart tube is to some extent independent of atrial or ventricular assignment (Stainier and Fishman, 1992). Application of RA to chick embryos has a posteriorizing effect on the cardiac tube, increasing the expression domain of the atrial myosin heavy chain AMHC1 and subsequently leading to a range of cardiac malformations (Yutzey *et al.*, 1994; Osmond *et al.*, 1991; Dickman and Smith, 1996). These experiments suggest, consistent with the findings in zebrafish embryos, that sensitivity to RA is greater at the arterial than the venous pole. Yutzey *et al.* (1994) proposed that the increase in atrial myogenic cells reflects a conversion of myocytes from one myocardiac lineage (anterior and ventricular) to another (posterior and atrial). This is supported by the finding that ventricular to atrial conversion by RA can occur in explanted cardiac tissue up to the time of myocardial differentiation (Yutzey *et al.*, 1995). RA activity is mediated by multiple receptors, and simple and compound RA receptor mutant mice have been shown to exhibit cardiac malformations, including ventricular, OFT, and aortic arch defects (Mendelsohn *et al.*, 1994; Sucov *et al.*, 1994; Kastner *et al.*, 1997). Many *RAR* and *RXR* mutants exhibit neural crest defects which are likely to repercuss on the heart at the level of formation of the great arteries and outflow septum. Combinations of *RAR* and *RXR* mutations, however, result in selective defects in particular cardiovascular compartments, suggestive of differential expression or function of RA receptors in different domains of the developing heart (Lee *et al.*, 1997). The effect of RA on patterning of the embryonic A–P axis is associated with perturbation of homeobox gene expression (McGinnis and Krumlauf, 1992). Homeobox genes provide positional identity within the embryo and are therefore promising candidate genes which might mediate A–P patterning within the heart (Kern *et al.*, 1995; Thomas and Barton, 1997). No detailed studies of clustered homeobox gene expression in the myocardium have been reported. A subset of these genes, however, including members of the *Hoxa* cluster, are known to be expressed in the embryonic heart (Gaunt, 1988; Patel *et al.*, 1992). Whereas most single *Hox* mutations have no cardiac defects, *Hoxa-5* mutant mice do exhibit cardiac malformations; however, these malformations may be secondary to pharyngeal arch perturbations (Chisaka and Capecchi, 1991).

C. Anterior–Posterior Patterning: A Segmented Embryonic Heart

Consistent with the transient organization of the developing heart along the A–P axis and the effect of RA on heart development, a number of embryological experiments in the chick have led to a segmental model of heart tube formation (De la Cruz *et al.*, 1977, 1989; Stalsberg and DeHaan, 1969). According to these *in vivo* labeling experiments using iron oxide particle, indian ink, and radioactive markers, the embryonic heart develops sequentially along the A–P axis, with primitive cardiac segments, or regions, being added successively, predominantly in a caudal direction but also anteriorly. The acquisition of regional identity is likely to be influenced by the time at which particular cardiac segments are added to the growing heart. The first region of the heart tube to be formed is the apical region of the right ventricle, followed by the apical region of the left ventricle (De la Cruz *et al.*, 1989). Future left and right ventricles therefore lie in series rather than in parallel in the early heart tube (De Vries and Saunders, 1962). The heart extends by addition of myocardium contributing to adjacent heart regions: the OFT cranially and the AVC and atria caudally (Virágh and Challice, 1973; Argüello *et al.*, 1975). The expression patterns uncovered by regionalized transgenes are consistent with this segmental model of heart development, by which embryological units of the early heart tube contribute not to entire cardiac chambers but to specific anatomical subregions of the definitive heart (De la Cruz *et al.*, 1989).

Support for a segmental model of heart development comes from the analysis of mutant zebrafish and mice with cardiac patterning defects. Chamber-specific defects in zebrafish hearts have emerged in large-scale mutation screens, and phenotypes include deletion of the ventricular chamber in *lonely atrium* and *pandora* mutants and development of the ventricle within the atria in the axial patterning *heart and soul* mutation (Chen *et al.*, 1996; Stainier *et al.*, 1996); certain zebrafish contractility mutants also show regionalized phenotypes specific to atrial or ventricular chambers. The mouse *hdf* insertional mutation is characterized by failure of the right ventricle and outflow tract to develop correctly from the looping stage, consistent with abnormal development of the most anterior segment of the primary heart tube (Yamamura *et al.,* 1997). Overexpression of genes on mouse chromosome 13, where the *hdf* mutation is located, can result in double-outlet right ventricle, a defect associated with incorrect development of the first segment (Vuillemin *et al.,* 1991). The *hdf* and zebrafish mutations support a segmental model of cardiac development by which the embryonic heart tube is generated by the progressive addition of distinct segments under separate genetic control.

D. Left–Right Patterning

The second input into regionalization within the embryonic heart, namely L–R asymmetry (laterality), is superimposed on the A–P patterned heart tube prior to looping. It is important to distinguish embryonic L–R asymmetry from later L–R chamber identity. Stalsberg (1969) showed that the contribution of the right and left precardiac fields to the chick heart tube varies along the rostrocaudal axis such that the right side dominates rostrally and the left side caudally, possibly due to an asymmetry in cell number between the two cardiac primordia. Transcriptional regionalization emerging at the time of looping is therefore likely to have inputs from both rostrocaudal and L–R patterning processes. The time at which chick precardiac mesoderm cells acquire L–R positional information is subsequent to the acquisition of rostrocaudal information (Hoyle *et al.,* 1992). Looping takes place with an initial leftward displacement of the heart tube axis due to asymmetric development of the caudal region of the heart, which precedes overt rightward looping (Biben and Harvey, 1997). The heart also rotates along the A–P axis at looping such that the ventral wall is derived predominantly from the left precardiac field.

The first manifestation of L–R asymmetry in the mouse heart may be mediated by genes encoding signaling molecules such as *nodal* and *lefty,* which show left-handed expression in lateral mesoderm, extending cranially up to the level of developing progenitors of the caudal heart tube (Meno *et al.,* 1996; Collignon *et al.,* 1996). In mouse mutants (*iv* and *inv*) with situs inversus phenotypes the expression of these genes is either randomized (*iv*) or right-sided only (*inv*), resulting in aberrant heart looping and inverted cardiac laterality (King and Brown, 1997; Levin, 1997; see Chapters 21 and 22). The notochord is likely to play a critical role in the initial setting up and maintenance of L–R asymmetry in addition to a role in patterning the A–P and dorsoventral axes (Danos and Yost, 1996; Lohr *et al.,* 1997). In the zebrafish, BMP4 expression is asymmetric at the time of initiation of fusion of the heart tube and is located on the left-hand side of the sinoatrial region of the heart tube prior to looping; BMP4 expression, and the subsequent direction of looping, is perturbed in zebrafish notochord mutants (Fishman and Chien, 1997). A number of extracellular matrix proteins are asymmetrically expressed in the chick precardiac mesoderm and may contribute to the process of asymmetric heart looping. These include hLAMP1 and flectin in the left precardiac mesoderm and a fibrillin-related protein JB3 and QHI in endocardial endothelial cells on the right side (Tsuda *et al.,* 1996; Sugi and Markwald, 1996; Smith *et al.,* 1997). Interestingly, RA beads implanted adjacent to the right precardiac field randomize the direction of looping and perturb hLAMP1 and JB3 expression; left-side RA application induces laterality defects only at high concentrations (Smith *et al.,* 1997). Thus, the regionalized expression profiles of a subset of genes in the precardiac mesoderm and the early heart tube are mediated by laterality signals. Laterality is also likely to influence the restriction of initially A–P-oriented expression domains, in particular in refining expression across the inner and outer curvature of the looped heart.

Regionalized patterns of gene expression within the myocardium thus result from differential read out of a previously established positional code, which is the product of an overlap between regionalization along the rostrocaudal axis and regionalization as a result of laterality signals. The time at which transcriptional regionalization emerges varies considerably among different myocardial markers and between species. For most myocardial markers regionalization occurs after the onset of looping and can occur substantially after looping, as in the case of the transient right ventricular expression of *MCK* transcripts at E12.5 (Lyons, 1994). The relative importance of A–P and L–R patterning differs in different regions of the embryonic heart. Left and right ventricular identity is clearly defined by initial position along the A–P axis. In contrast, the role of laterality appears to be important in determining left and right atrial identity since a significant proportion of *iv/iv* embryos exhibit atrial isomerism (two right or two left atria), based both on morphological evidence (Seo *et al.,* 1992) and *MCL3F–nlacZ-2E* transgene expression pattern (D. Franco, R. G. Kelly, M. Buckingham, and N. Brown, unpublished observations; see Chapter 25). An interesting possibility is that the differential input of A–P and L–R axes into ventricular and atrial sidedness reflects the stage at which these different cardiac compartments are incorporated into the growing heart tube.

V. The Molecular Basis of Regionalization

Regionalization along the A–P axis of the heart appears to be predominantly achieved at the transcriptional level, although posttranscriptional mechanisms are known to contribute to the control of cardiac gene expression (Gorza *et al.,* 1993). The distribution of β-galactosidase in *MLC3F–nlacZ-2E* hearts mirrors the regionalization of *nlacZ* transcripts confirming that these subdomains of β-galactosidase activity reflect differences in transcriptional potential (Fig. 4a). The following is therefore a key question: What are the transcriptional activators and repressors which mediate regional gene expression within the myocardium? Two major approaches to identify these transcription factors are being followed: the definition of *cis*-acting elements (and *trans*-acting factors) in the regulatory regions of asymmetrically expressed transgenes, and the mapping of expression patterns and functional analysis of genes important for transcription in cardiomyocytes or which are known to mediate positional information (such as homeobox genes). Recently, research has focused on four families of transcription factors which have been implicated in combinatorial regulation of cardiac gene expression: homeodomain-containing proteins of the Nkx family, basic helix–loop–helix (bHLH) E box-binding factors e- and d-hand, MADS proteins MEF2 and SRF, and zinc finger-containing proteins GATA-4, -5, and -6 (Olson and Srivastava, 1996; Lyons, 1996).

A. Cis-Acting Elements from Regionalized Transgenes

Of the transgenes which show regionalization in the heart, analysis of transcription factor target sites has been carried out *in vivo* in three cases. Whereas a 250-bp mouse *MLC2V* promoter confers right ventricular expression on a *lacZ* reporter gene *in vivo,* none of the *trans*-acting factors which interact with this sequence has been reported to show regionalization within the myocardium. An E box motif, which is bound by the transcription factor *USF,* has been shown to be required for promoter activity in ventricular muscle of transgenic mice (Ross *et al.,* 1996). A 28-bp region of the *MLC2V* promoter, containing HF1a and HF1b/MEF2 sites (but not the *USF*-binding E box motif), confers ventricular specificity on a *lacZ* reporter gene driven by a minimal promoter sequence, maintaining preferential transgene expression in the right ventricle (Ross *et al.,* 1996). Members of the MEF2 family and a novel cofactor of HF-1a, cardiac ankyrin repeat protein (CARP), are, however, all expressed throughout the myocardium (Zou *et al.,* 1997). MEF2 has also been implicated in the right ventricular regionalization of *desmin* transgene expression since cardiac muscle expression requires a MEF2 site (see Chapter 8). An adjacent E box, however, which is necessary for skeletal muscle expression, is dispensable in the heart, with cardiac transgene expression remaining regionalized (Kuisk *et al.,* 1996). Thirdly, a serum response element (SRE), a target of serum response factor (see Chapter 16), has been shown to be essential for expression of an *SM22α* transgene in cardiac and smooth muscle (Kim *et al.,* 1997; Li *et al.,* 1997). Since mutation of the E box in the *MLC2V* promoter, the MEF2 site in the *desmin* promoter, and the SRE in the *SM22α* promoter all abolish cardiac transgene expression, it is not possible to assess the role of these elements in regionalization per se, as opposed to their role in heart-specific expression. The sequences specifically important for regionalization, such as a target site for a left or right ventricular repressor, could lie elsewhere in these constructs. The HF-1a and HF1b/MEF2 element, in contrast, must itself confer regionalization and is thus a promising target of further research.

B. Cardiac Transcription Factors: Expression Patterns

The expression profiles of members of transcription factor families which play a role in myocardial transcription have been extensively analyzed and, to date, at the transcript level, only three have been documented to show regionalization within the myocardium. These are *e-* and *d-hand* in the mouse heart and *Nkx2.8* in the chick (Biben and Harvey, 1997; G. Lyons, personal communication; Srivastava *et al.,* 1997; Boettger *et al.,* 1997; Brand *et al.,* 1997; Reecy *et al.,* 1997; see Chapters 7, 9, and 16). Although their role as transcription factors remains unproven, *e-* and *d-hand* have a bHLH domain which would permit binding to E box motifs present in the regulatory regions of many cardiac genes. In the developing mouse heart*e-hand* transcripts are present in the left-hand caudal myocardium from the onset of looping; Fig. 9 illustrates the differential expression of *e-hand* in the embryonic ventricle at E10.5. This pattern is superimposed on an earlier symmetrical expression of the gene (Biben and Harvey, 1997). Left handed *e-hand* expression is downstream of laterality signals since it is reversed in mirror-image (*inv*) mouse hearts, and *e-hand* is therefore a candidate molecule for interpreting laterality signals which drive heart looping to the right (Biben and Harvey, 1997). In addition to a role in looping, *e-hand* may also, as a transcription factor, be responsible for left-handed regionalization of transgene

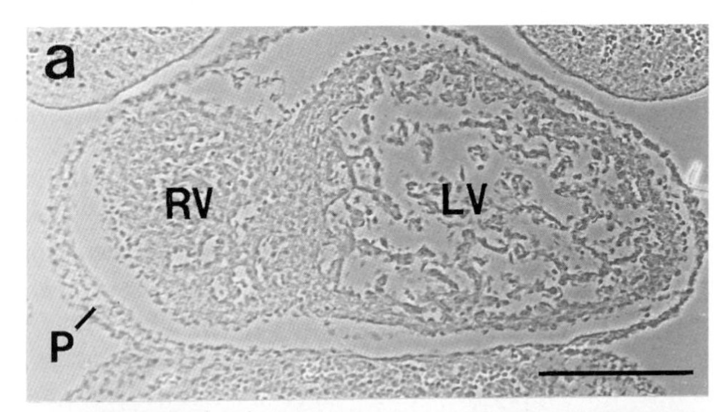

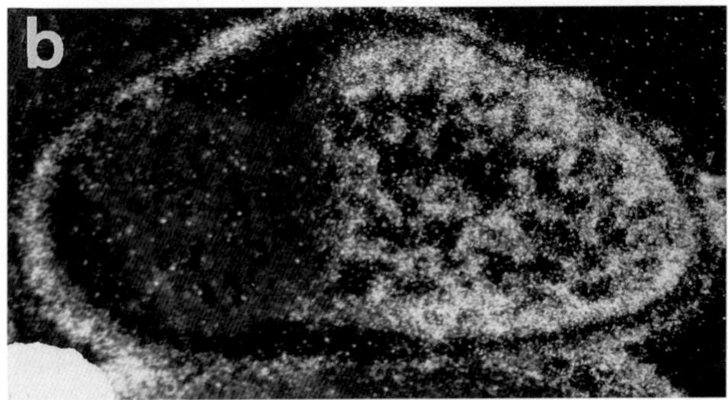

Figure 9 *In situ* localization of *e-hand* transcripts in the embryonic heart at E10.5. Bright field (a) and dark field (b) images showing *e-hand* expression in the embryonic left ventricle (LV) but not the embryonic right ventricle (RV). *e-hand* is also expressed in the pericardium (P). Scale bar = 200 μm.

expression, presumably acting directly or indirectly through regulatory sites in the transgenes. *d-hand* is also regionalized in the mouse heart, being more concentrated on the right-hand side and the OFT region of the looping heart (Biben and Harvey, 1997; Srivastava *et al.*, 1997). Like *e-hand,* the profile of *d-hand* expression is dynamic, and both genes are expressed in the septating OFT myocardium at later stages of development. Interestingly, in the chick heart, *e-* and *d-hand* expression profiles overlap more extensively and are not asymmetric (Srivastava *et al.,* 1995). Although the homeodomain transcription factor *Nkx2.5* is expressed throughout the embryonic heart (Lyons *et al.,* 1995), transcripts of the related gene, *Nkx2.8,* are regionalized in the chick heart from the heart tube stage, becoming restricted to OFT and atria and subsequently to extracardiac regions in the branchial arches (Boettger *et al.,* 1997; Brand *et al.,* 1997; Reecy *et al.,* 1997). In addition, Brand *et al.* (1997) report that *Nkx2.8* expression shows a lateral bias in the tubular heart, with transcripts being more abundant on the right side. *Nkx2.5* and *-2.8* may therefore act combinatorially to confer different transcriptional properties on different regions of the myocardium. Positional identity may also be conferred by other homeobox genes and it remains to be seen whether nested expression domains of clustered *Hox* genes are present in the heart.

C. Cardiac Transcription Factors: Functional Analysis

The analysis of targeted mutations in cardiac transcription factor genes has also provided insights into the molecular basis of regionalization in the myocardium. *Nkx2.5* mutant mouse embryos have a cardiac looping defect and die at E10.5 (Lyons *et al.,* 1995; see Chapter 7). In support of a possible role for Nkx proteins in regionalization within the heart, the expression of a subset of markers is perturbed in the early myocardium of *Nkx2.5* null embryos; for example, CARP expression, normally detected throughout the heart tube, is modified in a graded manner along the embryonic heart. *CARP* transcripts are absent from the OFT and downregulated in more caudal cardiac compartments (Zou *et al.,* 1997). This result was recapitulated with a 2.5-kb *CARP* promoter in transgenic mice and suggests that despite the equivalent expression of *Nkx2.5* along the heart tube, transcription in different cardiac regions is more or less *Nkx2.5* dependent. Interestingly, despite this change in *CARP* expression, the right ventricular-dominant expression of *MLC2V* transgenes is unperturbed in *Nkx2.5* mutant mice, showing that regional expression of the *MLC2V–lacZ* transgene does not require complete looping and suggesting that transgene regionalization is the product of restriction along the A–P axis of the heart (Ross *et al.,* 1996). *SM22α* expression is also regionally perturbed in *Nkx2.5* mutant hearts and is lost in the OFT and ventricle and maintained in the IFT and perhaps the atrial region (C. Biben and R. Harvey, personal communication). *Nkx2.8* expression does not compensate for *Nkx2.5* in patterning *CARP* and *SM22α* expression. Furthermore, a subset of downstream cardiac markers are absent in these mutant mice, including the endogenous *MLC2V* gene and two markers expressed in the left ventricle (*ANF* and *e-hand*), suggesting that *Nkx2.5* is essential for activation (and possibly regionalization) of these genes (Biben and Harvey, 1997).

Mouse embryos mutant for one of the hand proteins, d-hand, have also been generated (see Chapter 9). The phenotype of d-hand null embryos is embryonic lethal due to a severe defect in cardiac morphogenesis—the absence of a right ventricle (Srivastava *et al.,* 1997). This is the first direct evidence for a cardiac transcription factor with a segment-specific role and suggests that *d-hand* is essential for development of the right ventricular compartment. The remaining ventricle in mutant hearts expresses *e-hand,* which is not normally expressed in the right ventricle and cannot therefore compensate for the *d-hand* mutation, suggesting that these two factors act together to define right and left ventricular identity. In the chick, when expression of both hand genes is abolished in antisense experiments, heart morphogenesis is blocked at the looping stage (Srivastava *et al.*, 1995).

Despite the apparently even distribution of *MEF2C* transcripts within the heart, MEF2C mutant mice exhibit a phenotype similar to the *d-hand* mutation, namely, loss of the right ventricle, in addition to the complete absence of an extracardiac vascular system (Lin *et al.*, 1997). A subset of downstream markers are absent in MEF2C mutant hearts, including *ANF* transcripts (which are normally L–R regionalized), whereas others, including *MLC2V*, are expressed at normal levels. Other gene products are upregulated in mutant hearts, such as MEF2B (which may lead to partial compensation). Evidence that heart chambers may be misspecified in MEF2C mutant mice comes from analysis of *SM22α* transgene expression on the MEF2C mutant background. Whereas this transgene is normally expressed in the OFT and RV, it is now expressed in the atria, suggesting that MEF2C contributes to regionalization within the heart (E. Olson, personal communication; see Chapter 8). As in the case of the *d-hand* mutation, it remains to be seen whether the loss of the right ventricle in MEF2C null mice results from the deletion of myocardiocytes fated to contribute to the RV or the conversion of right ventricular precursors to another cell fate. This question also arises for other mouse and zebrafish mutants in which cardiac compartments are lost and for RA–mediated truncations of the early heart. In the case of the latter, explant experiments strongly suggest that conversion of cell fate is occurring (Yutzey *et al.*, 1995). In characterized mouse mutants which have perturbed cardiac compartmentalization, the phenotype becomes apparent at the time of looping (Lyons *et al.*, 1995; Srivastava *et al.*, 1997; Yamamura *et al.*, 1997); this is the stage at which significant regionalization of myocardial gene expression is first detected.

Continued analysis of these and other cardiac transcription factors will lead to a clearer definition of the interactions occurring between different transcriptional activators and repressors which contribute to defining regionalization of the myocardium. A key question is how these transcription factors establish, interpret, and integrate the A–P and L–R myocardial prepatterns which anticipate the regionalization of myocardial markers. The basis of transcription factor regionalization and the mechanism by which transcription factors expressed throughout the heart (although in most cases this has been demonstrated only at the transcript level) mediate regionalized gene expression remain unknown.

VI. Morphogenetic Implications of Regionalized Transcriptional Potential in the Heart

The identification of extensive transcriptional subcompartmentalization within the myocardium is of major significance for our understanding of cardiogenesis. In addition to reflecting transcriptional diversity, which is likely to drive cardiac function and morphogenesis during development, regionalized markers allow the movements of subpopulations of myocardial cells to be examined throughout organogenesis. Although there are obvious caveats in using transcriptional markers to follow the fate of cell populations, where the transgene expression profile appears to be nondynamic, these markers can complement other approaches to identify the contribution of different regions of the embryonic heart to the structure of the fetal and adult heart. Analysis of several transgenic mouse lines listed in Table II has already provided examples of the potential importance of such markers. We will discuss several examples of the use of these markers in the analysis of normal and abnormal mouse heart development.

A. Regionalized Markers to Follow Normal Cardiac Development

MLC3F–nlacZ transgenes have been used to follow the contribution of transgene-expressing and nonexpressing transcriptional domains to the mature heart during cardiac morphogenesis (Franco *et al.*, 1997). As illustrated in Fig. 3d, the *MLC3F–nlacZ-2E* transgene is expressed in the left ventricle, right atrium, and AVC of the embryonic heart at a stage when the atrial and ventricular myocardium are contiguous. Subsequently, the atrial and ventricular myocardium become insulated, and β-galactosidase-positive cells are found at the base of the atria; this is particularly clear in the case of the largely nonexpressing left atrium. This region has distinct properties from the rest of the atrial myocardium, such as slow conduction rates and low connexin 43 levels (Van Kempen *et al.*, 1996; McGuire *et al.*, 1996). The pattern of β-galactosidase expression in *MLC3F–nlacZ-2E* hearts supports a model in which the AVC becomes incorporated into this region of the atria (Wessels *et al.*, 1996). A second example of the use of regionalized transgene markers to follow transgene-expressing and nonexpressing cell populations at different developmental stages is provided by *MLC3F–nlacZ-9* mice, which express *nlacZ* in both ventricles but not in the OFT or IFT of the embryonic heart. The "fate" of the outflow tract has been the subject of extensive debate, and it is not clear to which regions of the adult heart it gives rise, despite detailed embryological and anatomical examination (Pexeider, 1995). Franco *et al.* (1997) have shown that the outlet region of the adult RV is negative for *MLC3F–nlacZ-9* transgene expression (Fig. 10). By monitoring transgene expression through fetal development, these authors observed a β-galactosidase-negative population of myocytes in the developing RV, suggesting that the β-galactosidase-negative region in the embryonic heart becomes incorporated into the ventricle as the morphological OFT disappears, and contributes to the outlet region of the definitive RV (illustrated in Fig. 11); a portion of the embryonic OFT also contributes to the subaortic region of the LV (Franco *et al.,* 1997). Support for these conclusions comes from the study of transgenes showing a complementary expression pattern (embryonic OFT-positive), such as a *MLCIV–nlacZ* transgene (R. G. Kelly and M. Buckingham, unpublished observations).

B. Regionalized Markers to Follow Abnormal Cardiac Development

In addition to the analysis of normal heart development, regionalized transgene expression can be used to follow abnormal heart development by genetic crosses between transgenic mice and particular mouse mutants. Many common cardiac malformations in man arise from defects affecting one or two segments of the embryonic heart; it is therefore important to assess the contribution of different primitive segments to malformed heart structures in mouse models. The right ventricular and OFT regionalized *MLC2V–nlacZ* transgene has been studied in RXRα null mutant mice (Ross *et al.,* 1996). Retinoids have been shown to play an important role in cardiogenesis, and the *RXRα* mutation is characterized by multiple heart defects, including abnormal ventricular morphogenesis and misexpression of chamber-specific markers (Sucov *et al.,* 1994; Dyson *et al.,* 1995; see Chapter 13). In the RXRα null background *MLC2V* transgene expression is essentially normal and shows graded ventricular specificity. Analysis of crosses between L–R ventricular or OFT regionalized transgenes and mice with OFT septation defects such as *splotch* or *Trisomy 16* should also be informative and enable the contribution of left and right myocardiocytes to the defective regions of the heart to be assessed. Atrial isomerism is a feature of human visceral heterotaxy syndrome (Ho *et al.,* 1991) and occurs at high frequency in mice with laterality defects such as *iv/iv*

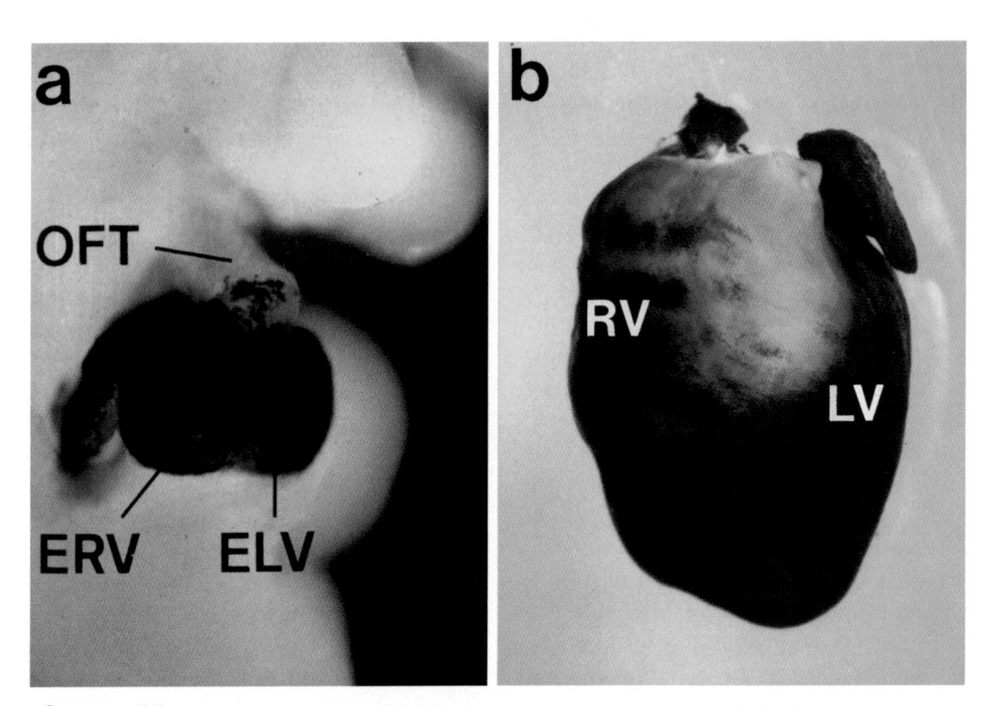

Figure 10 Expression pattern of the *MLC3F–nlacZ-9* transgene in embryonic and adult mouse hearts. (a) At E10.5 both atria and the embryonic left (ELV) and right (ERV) ventricles are β-galactosidase-positive, whereas the outflow tract (OFT) is nonexpressing. (b) In the adult heart, both atria, the left ventricle (LV), and part of the right ventricle (RV) are β-galactosidase-positive. The outlet region of the right ventricle is β-galactosidase-negative and may be embryologically derived from the embryonic OFT. This supposition is supported by analysis of fetal transgenic hearts (see Fig. 7 and 11). [Reproduced with permission from "Genetic Control of Heart Development" (1997). (E. N. Olson, R. P. Harvey, R. A. Schulz and J. S. Altman, (Eds.) HFSP, Strasbourg.]

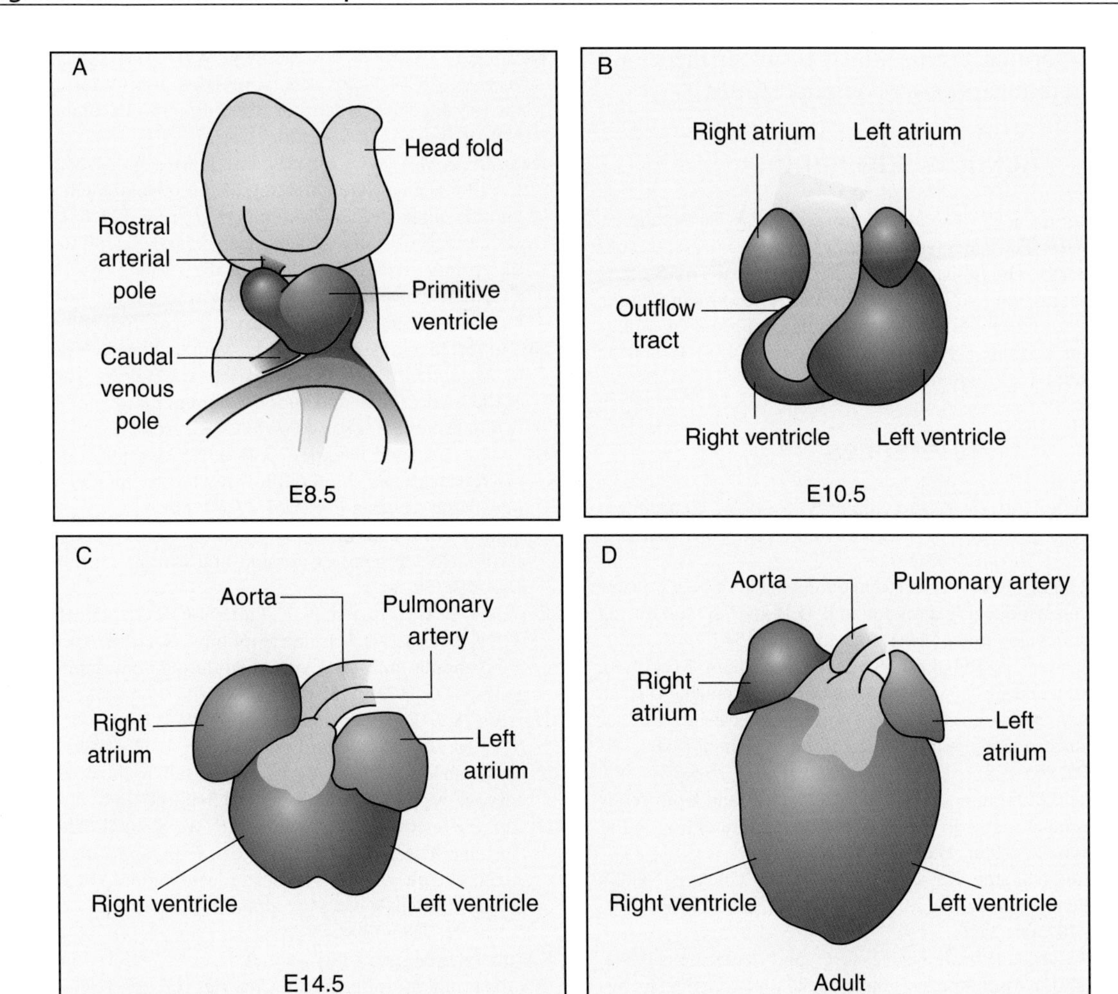

Figure 11 *MLC3F–nlacZ-9* transgene expression during embryonic and fetal development and in the adult heart illustrating the absorption of a population of β-galactosidase myocardiocytes into the right ventricle from the embryonic outflow tract. (a) *MLC3F–nlacZ-9* expression is similar to *MLC3F–nlacZ-2E* expression in the looped heart at E8.5. (b) Expression is subsequently observed in both left and right atrial and ventricular compartments of the embryonic heart, whereas the outflow tract is β-galactosidase-negative. (c) During fetal development a population of β-galactosidase-negative myocardiocytes is incorporated into the developing right ventricle. (d) A population of β-galactosidase-negative myocardiocytes persists in the infundibulum of the right ventricle and in the subaortic region of the left ventricle in the adult heart.

where it has previously been detected on a morphological basis (Seo *et al.*, 1992). *MLC3F–nlacZ-2E* transgenic mice which express *nlacZ* predominantly in the right atrium (and LV) from early stages of embryogenesis have been crossed with *iv/iv* mice. In addition to providing insights into the process of determination of cardiac chamber identity, this cross facilitates the identification of embryos exhibiting atrial isomerism prior to their identification on morphological grounds (D. Franco, R. Kelly, M. Buckingham, and N. Brown, unpublished observations; see Chapter 25).

These examples indicate the potential provided by regional markers for certain types of morphogenetic analysis during normal and abnormal cardiac development. We have termed transgenic animals which define subcompartments of the myocardium "cardiosensor" mice in order to emphasize their importance in the fine analysis and consequent understanding of cardiac morphogenesis. The complicated nature of this process makes it particularly desirable to have markers such as those provided by *lacZ* transgenes which can be easily visualized, by whole mount detection, in three dimensions. The use of such regionalized markers to monitor populations of transgene-expressing or nonexpressing cells throughout heart development complements other approaches to heart morphogenesis such as *in vivo* labeling or classical cell lineage analyses. Cardiosensor mice can therefore contribute significantly to our understanding of the complex processes of cardiac morphogenesis and may also provide valuable insight into the early patterning events which result in the emergence of different cardiomyocyte populations.

Acknowledgments

We are grateful to Dr. Yassemi Capetanaki for Figs. 3a, 3c, and 6a, Dr. Gary Lyons for Fig. 9, and colleagues who communicated results and manuscripts prior to publication. We also thank Professor Stefano Schiaffino for discussion and Dr. Peter Zammit for comments on the manuscript. Research in M. B.'s laboratory is supported by grants from the Pasteur Institute, CNRS, AFM, and Grant PL 950228, from the EC Biotechnology Programme.

References

Argüello, C., De la Cruz, M. V., and Gomez, C. S. (1975). Experimental study of the formation of the heart tube in the chick embryo. *J. Embryol. Exp. Morphol.* **33,** 1–11.

Biben, C., and Harvey, R. P. (1997). Homeodomain factor Nkx2-5 controls left/right asymmetric expression of bHLH gene eHand during murine heart development. *Genes Dev.* **11,** 1357–1369.

Biben, C., Hadchouel, J., Tajbakhsh, S., and Buckingham, M. (1996). Developmental and tissue-specific regulation of the murine cardiac actin gene in vivo depends on distinct skeletal and cardiac muscle-specific enhancer elements in addition to the proximal promoter. *Dev. Biol.* **173,** 200–212.

Bitgood, M. J., and McMahon, A. P. (1995). Hedgehog and Bmp genes are coexpressed at many diverse sites of cell-cell interaction in the mouse embryo. *Dev. Biol.* **172,** 126–138.

Boettger, T., Stein, S., and Kessel, M. (1997). The chicken Nkx2.8 homeobox gene: A novel member of the Nk-2 gene family. *Dev. Genes Evol.* **207,** 65–70.

Bougnavet, P., Leger, J., Pons, F., Dechesne, C., and Leger, J. L. (1984). Fiber types and myosin types in human atrial and ventricular myocardium. *Circ. Res.* **55,** 794–804.

Brand, T., Andree, B., Schneider, A., Buchberger, A., and Arnold, H.-H. (1997). Chicken Nkx2.8, a novel homeobox gene expressed during early heart and foregut formation. *Mech. Dev.* **64,** 53–59.

Chen, J.-N., Haffter, P., Odenthal, J., Vogelsang, E., Brand, M., Van Eeden, F. J. K., Furutani-Seiki, M., Granao, M., Hammerschmidt, M., Heisenberg, C.-P., Jiang, Y.-J., Kane, D. A., Kelsh, R. N., Mullins, M. C., and Nüsslein-Volhard, C. (1996). Mutations affecting the cardiovascular system and other internal organs in zebrafish. *Development (Cambridge, UK)* **123,** 293–302.

Chisaka, O., and Capecchi, M. R. (1991). Regionally restricted developmental defects resulting from targeted disruption of the mouse homeobox gene hox-1.5. *Nature (London)* **350,** 473–479.

Collignon, J., Varlet, I., and Robertson, E. J. (1996). Relationship between asymmetric nodal expression and the direction of embryonic turning. *Nature (London)* **381,** 155–158.

Danos, M. C., and Yost, H. J. (1996). Role of notochord in specification of cardiac left-right orientation in zebrafish and Xenopus. *Dev. Biol.* **177,** 96–103.

De Groot, I. J. M., Lamers, W. H., and Moorman, A. F. M. (1989). Isomyosin expression pattern during heart morphogenesis: An immunohistochemical study. *Anat. Rec.* **224,** 365–373.

DeHaan, R. L. (1963). Regional organisation of prepacemaker cells in the cardiac primordia of the early chick embryo. *J. Embryol. Exp. Morphol.* **11,** 65–76.

De Jong, F., Geerts, W. J. C., Lamers, W. H., Los J. A., and Moorman, A. F. M. (1987). Isomyosin expression patterns in tubular stages of chicken heart development: A 3-D immunohistochemical analysis. *Anat. Embryol.* **177,** 81–90.

De Jong, F., Geerts, W. J. C., Lamers, W. H., Los, J. A., and Moorman, A. F. M. (1990). Isomyosin expression patterns during formation of the tubular chicken heart: A three-dimensional immunohistochemical analysis. *Anat. Rec.* **226,** 213–227.

De la Cruz, M. V., Sanchez Gomez, C., Arteaga, M. M., and Argüello, C. (1977). Experimental study of the development of the truncus and the conus in the chick embryo. *J. Anat.* **123,** 661–686.

De la Cruz, M. V., Sanchez-Gomez, C., and Palomino, M. A. (1989). The primitive cardiac regions in the straight tube heart (stage 9) and their anatomical expression in the mature heart: An experimental study in the chick embryo. *J. Anat.* **165,** 121–131.

De Ruiter, M. C., Poelmann, R. E., Vander Plas-de Vries, I., Mentink, M. M., and Gittenberger-de Groot, A. C. (1992). The development of the myocardium and endocardium in mouse embryos. Fusion of two heart tubes? *Anat. Embryol.* **185,** 461–473.

De Vries, P. A., and Saunders, J. B. de C. M. (1962). Development of the ventricles and spiral outflow tract in the human heart. *Carnegie Inst. Wash. Contrib. Embryol.* **37,** 87–114.

Dickman, E. D., and Smith, S. M. (1996). Selective regionalisation of cardiomyocyte gene expression and cardiac morphogenesis. *Dev. Dyn.* **206,** 39–48.

Drysdale, T. A., Tonissen, K. F., Patterson, K. D., Crawford, M. J., and Krieg, P. A. (1994). Cardiac troponin I is a heart-specific marker in the Xenopus embryo: Expression during abnormal heart morphogenesis. *Dev. Biol.* **165,** 432–441.

Durston, A. J., Timmermans, J. P., Hage, W. J., Hendriks, H. F., de Vries, N. J., Heideveld, M., and Nieuwkoop, P. D. (1989). Retinoic acid causes an anteroposterior transformation in the developing central nervous system. *Nature (London)* **340,** 140–144.

Dyson, E., Sucov, H. M., Kubalak, S. W., Schmid-Schonbein, G. W., De Lano, F. A., Evans, R. M., Ross, J., Jr., and Chien, K. R. (1995). Atrial-like phenotype is associated with embryonic ventricular failure in retinoid X receptor alpha −/− mice. *Proc. Natl. Acad. Sci. U.S.A.* **92,** 7386–7390.

Eisenberg, L. M., and Markwald, R. R. (1995). Molecular regulation of valvoseptal morphogenesis. *Circ. Res.* **1,** 1–16.

Fishman, M. C., and Chien, K. R. (1997). Fashioning the vertebrate heart: Earliest embryonic decisions. *Development (Cambridge, UK)* **124,** 2099–2117.

Franco, D., Kelly, R., Lamers, W., Buckingham, M., and Moorman, A. F. M. (1997a). Regionalized transcriptional domains of myosin light chain 3F transgenes in the embryonic mouse heart: Morphogenetic implications. *Dev. Biol.* **188,** 17–33.

Franco, D., Lamers, W. H., and Moorman, A. F. M. (1998a). Patterns of gene expression in the developing heart: Towards a morphologically integrated transcriptional model. *Cardiovasc. Res.* **38,** 25–53.

Franco, D., Kelly, R., Zammit, P., Buckingham, M., and Moorman, A. F. M. (1998b). Regionalisation of transcriptional potential in the atrial myocardium: morphogenetic implications. In preparation.

Garcia-Martinez, V., and Schoenwolf, G. C. (1993). Primitive-streak origin of the cardiovascular system in avian embryos. *Dev. Biol.* **159,** 706–719.

Gardner, D. G., Deschepper, C. F., Ganong, W. F., Hane, S., Fiddes, J., Baxter, J. D., and Lewicki, J. (1986). Extra-atrial expression of the gene for atrial natriuretic factor. *Proc. Natl. Acad. Sci. U.S.A.* **83,** 6697–6701.

Gaunt, S. J. (1988). Mouse homeobox gene transcripts occupy different but overlapping domains in embryonic germ layers and organs: A comparison of Hox-3.1 and Hox-1.5. *Development (Cambridge, UK)* **103,** 135–144.

Gonzalez-Sanchez, A., and Bader, D. (1990). In vitro analysis of cardiac progenitor cell differentiation. *Dev. Biol.* **139,** 197–209.

Gorza, L., Pauletto, P., Pessina, A. C., Sartore, S., and Schiaffino, S. (1981). Isomyosin distribution in normal and pressure-overloaded rat ventricular myocardium. *Circ. Res.* **49,** 1003–1009.

Gorza, L., Ausoni, S., Merciai, N., Hastings, K. E., and Schiaffino, S. (1993). Regional differences in troponin I isoform switching during rat heart development. *Dev. Biol.* **156,** 253–264.

Gulick, J., Hewett, T. E., Klevitsky, R., Buck, S. H., Moss, R. L., and Robbins, J. (1997). Transgenic remodelling of the regulatory myosin light chains in the mammalian heart. *Circ. Res.* **80,** 655–664.

Han, Y., Dennis, J. E., Cohen-Gould, L., Bader, D. M., and Fischman, D. A. (1992). Expression of sarcomeric myosin in the presumptive myocardium of chicken embryos occurs within six hours of myocyte commitment. *Dev. Dyn.* **193,** 257–265.

Hasselbaink, H. D. J., Labruyere, W. T., Moorman, A. F. M., and Lamers, W. H. (1990). Creatine kinase isozyme expression in prenatal rat heart. *Anat. Embryol.* **182,** 195–203.

He, S., and Burch, J. B. (1997). The chicken GATA-6 locus contains multiple control regions that confer distinct patterns of heart-region-specific expression in transgenic mouse embryos. *J. Biol. Chem.* **272,** 28550–28556.

Ho, S. Y., Cook, A., Anderson, R. H., Allan, L. D., and Fagg, N. (1991). Isomerism of the atrial appendages in the fetus. *Pediatr. Pathol.* **11,** 589–608.

Hoyle, C., Brown, N. A., and Wolpert, L. (1992). Development of left/right handedness in the chick heart. *Development (Cambridge, UK)* **115,** 1071–1078.

Icardo, J. M. (1996). Developmental biology of the vertebrate heart. *J. Exp. Zool.* **275,** 144–161.

Inagaki, T., Garcia-Martinez, V., and Schoenwolf, G. C. (1993). Regulative ability of the prospective cardiogenic and vasculogenic areas of the primitive streak during avian gastrulation. *Dev. Dyn.* **197,** 57–68.

Jones, C. M., Lyons, K. M., and Hogan, B. L. (1991). Involvement of Bone Morphogenetic Protein-4 (BMP-4) and Vgr-1 in morphogenesis and neurogenesis in the mouse. *Development (Cambridge, UK)* **111,** 531–542.

Jones, W. K., Sanchez, A., and Robbins, J. (1994). Murine pulmonary myocardium: Developmental analysis of cardiac gene expression. *Dev. Dyn.* **200,** 117–128.

Kastner, P., Mark, M., Ghyselinck, N., Krezel, W., Dupe, V., Grondona, J. M., and Chambon, P. (1997). Genetic evidence that the retinoid signal is transduced by heterodimeric RXR/RAR functional units during mouse development. *Development (Cambridge, UK)* **124,** 313–326.

Kaufman, M. H., and Navaratnam, V. (1981). Early differentiation of the heart in mouse embryos. *J. Anat.* **133,** 235–246.

Kelly, R., Alonso, S., Tajbakhsh, S., Cossu, G., and Buckingham, M. (1995). Myosin light chain 3F regulatory sequences confer regionalized cardiac and skeletal muscle expression in transgenic mice. *J. Cell Biol.* **129,** 383–396.

Kern, M. J., Argao, E. A., and Potter, S. S. (1995). Homeobox genes and heart development. *Trends Cardiovasc. Med.* **5,** 47–54.

Kessel, M., and Gruss, P. (1991). Homeotic transformations of murine vertebrae and concomitant alteration of Hox codes induced by retinoic acid. *Cell (Cambridge, Mass.)* **67,** 89–104.

Kim, S., Ip, H. S., Lu, M. M., Clendenin, C., and Parmacek, M. S. (1997). A serum response factor-dependent transcriptional regulatory program identifies distinct smooth muscle sublineages. *Mol. Cell. Biol.* **17,** 2266–2278.

Kimura, S., Abe, K., Susuki, M., Ogawa, M., Yoshioka, K., Kaname, T., Miike, T., and Yamamura, K. (1997). A 900bp genomic region from the mouse dystrophin promoter directs lacZ reporter expression only to the right heart of transgenic mice. *Dev. Growth Differ.* **39,** 257–265.

King, T., and Brown, N. A. (1997). Embryonic asymmetry: Left TGFβ at the right time? *Curr. Biol.* **7,** R212–R215.

Kubalak, S. W., Miller-Hance, W. C., O'Brien, T. X., Dyson, E., and Chien, K. R. (1994). Chamber specification of atrial myosin light chain-2 expression precedes septation during murine cardiogenesis. *J. Biol. Chem.* **269,** 16961–16970.

Kuisk, I. R., Li, H., Tran, D., and Capetanaki, Y. (1996). A single MEF2 site governs desmin transcription in both heart and skeletal muscle during mouse embryogenesis. *Dev. Biol.* **174,** 1–13.

Larsen, W. J. (1993). "*Human Embryology*" Chapter 7, p. 131–165. Churchill-Livingstone, London.

Lee, R. Y., Luo, J., Evans, R. M., Giguerre, V., and Sucov, H. M. (1997). Compartment-selective sensitivity of cardiovascular morphogenesis to combinations of retinoic acid receptor gene mutations. *Circ. Res.* **80,** 757–764.

Levin, M. (1997). Left-right asymmetry in vertebrate embryogenesis. *BioEssays* **19,** 287–296.

Li, L., Miano, J. M., Mercer, B., and Olson, E. N. (1996). Expression of the SM22alpha promoter in transgenic mice provides evidence for distinct transcriptional regulatory programs in vascular and visceral smooth muscle cells. *J. Cell Biol.* **132,** 849–859.

Li, L., Liu, Z., Mercer, B., Overbeek, P., and Olson, E. N. (1997). Evidence for serum response factor-mediated regulatory networks governing SM22α transcription in smooth, skeletal and cardiac muscle cells. *Dev. Biol.* **187,** 311–321.

Lin, Q., Schwarz, J., Bucana, C., and Olson, E. N. (1997). Control of mouse cardiac morphogenesis and myogenesis by transcription factor MEF2C. *Science* **276,** 1404–1407.

Litten, R. Z., Martin, B. J., Buchtal, R. H., Low, R. B., and Alpert, N. R. (1985). Heterogeneity of myosin isozyme content of rabbit heart. *Circ. Res.* **57,** 406–414.

Litvin, J., Montgomery, M., Gonzalez-Sanchez, Bisaha, J. G., and Bader, D. (1992). Commitment and differentiation of cardiac myocytes. *Trends Cardiovasc. Med.* **2,** 27–32.

Lohr, J. L., Danos, M. C., and Yost, H. J. (1997). Left-right asymmetry of a nodal-related gene is regulated by dorsoanterior midline structures during Xenopus development. *Development (Cambridge, UK)* **124,** 1465–1472.

Lyons, G. E. (1994). In situ analysis of the cardiac muscle gene program during embryogenesis. *Trends Cardiovasc. Med.* **4,** 70–77.

Lyons, G. E. (1996). Vertebrate heart development. *Curr. Opin. Genet. Dev.* **6,** 454–460.

Lyons, G. E., Schiaffino, S., Sassoon, D., Barton, P., and Buckingham, M. (1990). Developmental regulation of myosin gene expression in mouse cardiac muscle. *J. Cell Biol.* **111,** 2427–2436.

Lyons, G. E., Muhlebach, S., Moser, A., Masood, R., Paterson, B. M., Buckingham, M. E., and Perriard, J. C. (1991). Developmental regulation of creatine kinase gene expression by myogenic factors in embryonic mouse and chick skeletal muscle. *Development (Cambridge, UK)* **113,** 1017–1029.

Lyons, I., Parsons, L. M., Hartley, L., Li, R., Andrews, J. E., Robb, L., and Harvey, R. P. (1995). Myogenic and morphogenetic defects in the heart tubes of murine embryos lacking the homeo box gene Nkx2-5. *Genes Dev.* **9,** 1654–1666.

Maltsev, V. A., Rohwedel, J., Hescheler, J., and Wobus, A. M. (1993). Embryonic stem cells differentiate in vitro into cardiomyocytes representing sinusnodal, atrial and ventricular cell types. *Mech. Dev.* **44,** 41–50.

Manasek, F. J. (1968). Embryonic development of the heart: I. A light and electron microscopic study of myocardial development in the early chick embryo. *J. Morphol.* **125,** 329–366.

McGinnis, W., and Krumlauf, R. (1992). Homeobox genes and axial patterning. *Cell (Cambridge, Mass.)* **68,** 283–302.

McGuire, M. A., de Bakker, J. M. T., Vermeulen, J. T., Moorman, A. F. M., Loh, P., Thibault, B., Vermeulen, J. L. M., Becker A. E., and Janse, M. J. (1996). Atrioventricular Junctional Tissue. Discrepancy between histological and electrophysiological characteristics. *Circulation* **94,** 571–577.

Mendelsohn, C., Lohnes, D., Decimo, D., Lufkin, T., Le Meur, M., Chambon, P., and Mark, M. (1994). Function of the retinoic acid receptors (RARs) during development (II). Multiple abnormalities at various stages of organogenesis in RAR double mutants. *Development (Cambridge, UK)* **120,** 2749–2771.

Meno, C., Saijoh, Y., Fujii, H., Ikeda, M., Yokoyama, T., Yokoyama, M., Toyoda, Y., and Hamada, H. (1996). Left-right asymmetric expression of the TGF beta-family member lefty in mouse embryos. *Nature (London)* **381,** 151–155.

Moessler, H., Mericskay, M., Li, Z., Nagl, S., Paulin, D., and Small, J. V. (1996). The SM 22 promoter directs tissue-specific expression in arterial but not in venous or visceral smooth muscle cells in transgenic mice. *Development (Cambridge, UK)* **122,** 2415–2425.

Montgomery, M. O., Litvin, J., Gonzalez-Sanchez, A., and Bader, D. (1994). Staging of commitment and differentiation of avian cardiac myocytes. *Dev. Biol.* **164,** 63–71.

Moorman, A. F. M., and Lamers, W. H. (1994). Molecular anatomy of the developing heart. *Trends Cardiovasc. Med.* **4,** 257–264.

Moorman, A. F. M., Vermeulen, J. L. M., Schwartz, K., Lamers, W. H., and Boheler, K. R. (1995). Patterns of expression of sarcoplasmic reticulum Ca++-ATPase and phospholamban mRNA during rat heart development. *Circ. Res.* **76,** 616–625.

Morrisey, E. E., Ip, H. S., Tang, Z., Lu, M. M., and Parmacek, M. S. (1997). GATA-5: A transcriptional activator expressed in a novel temporally and spatially-restricted pattern during embryonic development. *Dev. Biol.* **183,** 21–36.

O'Brien, T. X., Lee, K. J., and Chien, K. R. (1993). Positional specification of ventricular myosin light chain 2 expression in the primitive murine heart tube. *Proc. Natl. Acad. Sci. U.S.A.* **90,** 5157–5161.

Olson, E. N., and Srivastava, D. (1996). Molecular pathways controlling heart development. *Science* **272,** 671–676.

Osmond, M. K., Butler, A. J., Voon, F. C., and Bellairs, R. (1991). The effects of retinoic acid on heart formation in the early chick embryo. *Development (Cambridge, UK)* **113,** 1405–1417.

Patel, C. V., Gorski, D. H., Le Page, D. F., Lincecum, J., and Walsh, K. (1992). Molecular cloning of a homeobox transcription factor from adult aortic smooth muscle. *J. Biol. Chem.* **267,** 26085–26090.

Patten, B. M. (1992). Formation of the cardiac loop in the chick. *Am. J. Anat.* **30,** 373–397.

Pexeider, T. (1995). Conotruncus and its septation at the advent of the molecular biology era. *In* "Developmental Mechanisms of Heart Disease" (E. B. Clark, R. R. Markwald, and A. Takao, eds.), pp. 227–247. Futura Publ. Co., Armonk, NY.

Reecy, J. M., Yamada, M., Cummings, K., Sosic, D., Chen, C.-Y., Eichele, G., Olson, E. N., and Schwartz, R. J. (1997). Chicken Nkx2-8: a novel homeobox gene expressed in early heart progenitor cells and branchial cleft -2 and -3 endoderm. *Dev. Biol.* **188,** 295–311.

Rosenquist, G. C., and DeHaan, R. L. (1996). Migration of precardiac cells in the chick embryo: A radioautographic study. *Carnegie Inst. Wash. Contrib. Embryol.* **38,** 111–121.

Ross, R. S., Navankasattusas, S., Harvey, R. P., and Chien, K. R. (1996). An HF-1a/HF-1b/MEF-2 combinatorial element confers cardiac ventricular specificity and established an anterior-posterior gradient of expression. *Development (Cambridge, UK)* **122,** 1799–1809.

Ruzicka, D. L., and Schwartz, R. J. (1988). Sequential activation of alpha-actin genes during avian cardiogenesis: Vascular smooth muscle alpha-actin gene transcripts mark the onset of cardiomyocyte differentiation. *J. Cell Biol.* **107,** 2575–2586.

Sartore, S., Gorza, L., Pierbon-Bormioli, S., Dalla-Libera, L., and Schiaffino, S. (1981). Myosin types and fibre types in cardiac muscle. I. The ventricular myocardium. *J. Cell Biol.* **88,** 226–233.

Satin, J., Fujii, S., and DeHaan, R. L. (1988). Development of cardiac beat rate in early chick embryos is regulated by regional cues. *Dev. Biol.* **129,** 103–113.

Seo, J. W., Brown, N. A., Ho, S. Y., and Anderson, R. H. (1992). Abnormal laterality and congenital cardiac anomalies. Relations of visceral and cardiac morphologies in the iv/iv mouse. *Circulation* **86,** 642–650.

Smith, S. M., Dickman, E. D., Thompson, R. P., Sinning, A. R., Wunsch, A. M., and Markwald, R. R. (1997). Retinoic acid directs cardiac laterality and the expression of early markers of precardiac asymmetry. *Dev. Biol.* **182,** 162–171.

Srivastava, D., Cserjesi, P., and Olson, E. N. (1995). A subclass of bHLH proteins required for cardiac morphogenesis. *Science* **270,** 1995–1999.

Srivastava, D., Thomas, T., Lin, Q., Kirby, M. L., Brown, D., and Olson, E. N. (1997). Regulation of cardiac mesodermal and neural crest development by the bHLH transcription factor dHAND. *Nat. Genet.* **16,** 154–160.

Stainier, D. Y. R., and Fishman, M. C. (1992). Patterning the zebrafish heart tube: acquisition of anteroposterior polarity. *Dev. Biol.* **153,** 91–101.

Stainier, D. Y. R., Fouquet, B., Chen, J.-N., Warren, K. S., Weinstein, B. M., Meiler, S. E., Mohideen, M.-A. P. K., Neuhauss, S. C. F., Solnica-Krezel, L., Schier, A. F., Zwartkruis, F., Stemple, D. L., Malicki, J., Driever, W., and Fishman, M. C. (1996). Mutations affecting the formation and function of the cardiovascular system in the zebrafish embryo. *Development (Cambridge, UK)* **123,** 285–292.

Stalsberg, H. (1969). The origin of heart asymmetry: Right and left contributions to the early chick embryo heart. *Dev. Biol.* **19,** 109–127.

Stalsberg, H., and DeHaan, R. L. (1969). The precardiac areas and formation of the tubular heart in the chick embryo. *Dev. Biol.* **19,** 128–159.

Sucov, H. M., Dyson, E., Gumeringer, C. L., Price, J., Chien, K. R., and Evans, R. M. (1994). RXR alpha mutant mice establish a genetic basis for vitamin A signaling in heart morphogenesis. *Genes Dev.* **8,** 1007–1018.

Sugi, Y., and Markwald, R. R. (1996). Formation and early morphogenesis of endocardial endothelial precursor cells and the role of endoderm. *Dev. Biol.* **175,** 66–83.

Sweeney, L. J., Zak, R., and Manasek, F. J. (1987). Transitions in cardiac isomyosin expression during differentiation of the embryonic chick heart. *Circ. Res.* **61,** 287–295.

Thomas, P., and Barton, P. R. J. (1997). Homeobox genes in cardiac development. *Annu. Cardiac Surg.* pp. 21–29.

Tsuda, T., Philp, N., Zile, M. H., and Linask, K. K. (1996). Left-right asymmetric localization of flectin in the extracellular matrix during heart looping. *Dev. Biol.* **173,** 39–50.

Van Kempen, M. J., Vermeulen, J. L. M., Moorman, A. F. M., Gros, D., Paul, D. L., and Lamers, W. H. (1996). Developmental changes of connexin40 and connexin43 mRNA distribution patterns in the rat heart. *Cardiovasc. Res.* **32,** 886–900.

Van Mierop, L. H. S. (1996). Location of pacemaker in chick embryo heart at the time of initiation of heartbeat. *Am. J. Physiol.* **212,** 407–415.

Virágh, S., and Challice, C. E. (1973). Origin and differentiation of cardiac muscle cells in the mouse. *J. Ultrastruct. Res.* **42,** 1–24.

Vuillemin, M., and Pexeider, T. (1997). Carbonic anhydrase II expression pattern in mouse embryonic and fetal heart. *Anat. Embryol.* **195,** 267–277.

Vuillemin, M., Pexieder, T., and Winking, H. (1991). Pathogenesis of various forms of double outlet right ventricle in mouse fetal trisomy 13. *Int. J. Cardiol.* **33,** 281–304.

Wessels, A., Markman, M. W., Vermeulen, J. L. M., Anderson, R. H., Moorman, A. F. M., and Lamers, W. H. (1996). The development of the atrioventricular junction in the human heart. *Circ. Res.* **78,** 110–117.

Wilson, J. G., and Warkany, J. (1949). Aortic-arch and cardiac anomalies in the offspring of vitamin A deficient rats. *Am. J. Anat.* **92,** 113–155.

Yamamura, H., Zhang, M., Markwald, R. R., and Mjaatvedt, C. H. (1997). A heart segmental defect in the anterior-posterior axis of a transgenic mutant mouse. *Dev. Biol.* **186,** 58–72.

Yutzey, K. E., and Bader, D. (1995). Diversification of cardiomyogenic cell lineages during early heart development. *Circ. Res.* **77,** 216–219.

Yutzey, K. E., Rhee, J. T., and Bader, D. (1994). Expression of the atrial-specific myosin heavy chain AMHC1 and the establishment of anteroposterior polarity in the developing chicken heart. *Development* (*Cambridge, UK*) **120,** 871–883.

Yutzey, K. E., Gannon, M., and Bader, D. (1995). Diversification of cardiomyogenic cell lineages in vitro. *Dev. Biol.* **170,** 531–541.

Zeller, R., Bloch, K. D., Williams, B. S., Arceci, R. J., and Seidman, C. E. (1987). Localized expression of the atrial natriuretic factor gene during cardiac embryogenesis. *Genes Dev.* **1,** 693–698.

Zhu, L., Lyons, G. E., Juhasz, O., Joya, J. E., Hardeman, E. C., and Wade, R. (1995). Developmental regulation of troponin I isoform genes in striated muscles of transgenic mice. *Dev. Biol.* **169,** 487–503.

Zou, Y., Evans, S., Chen, J., Kuo, H.-C., Harvey, R. P., and Chien, K. (1997). CARP, a cardiac ankyrin repeat protein, is downstream in the Nkx2-5 homeobox gene pathway. *Development* (*Cambridge, UK*) **124,** 793–804.

20

Chamber-Specific Gene Expression and Regulation during Heart Development

Gang Feng Wang and Frank E. Stockdale

Department of Medicine, Stanford University School of Medicine, Stanford, California 94305

I. Heart Formation and Early Atrial and Ventricular Regionalization

A vertebrate heart consists of two general compartments—the atria and ventricles—which differ in morphology and electrophysiology and in the repertoire of muscle contractile protein genes that each expresses (DeHaan, 1965; Icardo and Manasek, 1992; Lyons, 1994). Fate-mapping studies of the early chicken embryo have shown that the precursors to these compartments are present in the rostral half of the primitive streak at stage 3 just behind Henson's node (Yatskievych *et al.,* 1997; for review, see Yutzey and Bader, 1995; see Chapter 3). By stage 4, two separated regions of cardiogenic mesoderm reside on each side of the primitive streak. Within this mesoderm a pattern is established such that the atria will form from the cells in the caudal cardiogenic mesoderm, whereas the ventricles will form from the rostral cardiogenic mesoderm. Between stages 7 and 10, the bilateral regions of precardiac mesoderm fuse at the midline to form a tubular heart. Looping and a complicated process of cardiac morphogenesis affecting the tubular heart reverse the positions of the rostral and caudal regions such that the atria take up their final rostral position. Septation of the tubular heart, a process that begins independently in the atrioventricular canal, primitive atrium, and primitive ventricle, leads to the formation of the definitive four-chamber heart at about stage 30. Thus, by Embryonic (E) Day 6 the primitive atrium and ventricle of the chicken embryo have been divided into right and left chambers and a four-chamber heart emerges (Icardo and Manasek, 1992).

Separate atrial and ventricular cell lineages can be identified soon after gastrulation in the chicken embryo (Yutzey *et al.,* 1995), whereas in the zebrafish embryo diversification of the atrial and ventricular lineages begins at the midblastula stage (Stainier and Fishman, 1992). *In vivo,* the earliest regional difference between conus, ventricular, and atrial precursors can be experimentally detected by a gradient in the beat rate by stage 5 in chicken embryos (Satin *et al.,* 1988). At this time, the conus, ventricular, and atrial precursor cells are arranged in a rostral–caudal sequence (Fig. 1). When these regions are divided from one another by microsurgery, each region differentiates into a spontaneous beating vesicle with a characteristic beat rate (Fig. 1). The caudal-most vesicle, formed from the region of atrial precursors, has the fastest beating rate followed by the prospective ventricular precursors, whereas the rostral vesicle formed from the region of prospective conus has the slowest rate. These regional differences in beat rate appear to be determined by the positional cues within the embryo between stages 5 and 7 (Satin *et al.,* 1988). That the prospective atria and ventricles are under separate genetic and developmental control is suggested by the *lonely atrium* and *pandora* mutations in the zebrafish, in which the ventricular but not the atrial chamber is deleted (J. N. Chen *et al.,* 1996; Stainier *et al.,* 1996; see Chapter 6). Mechanisms for atrial or ventricular lineage diversification are not known.

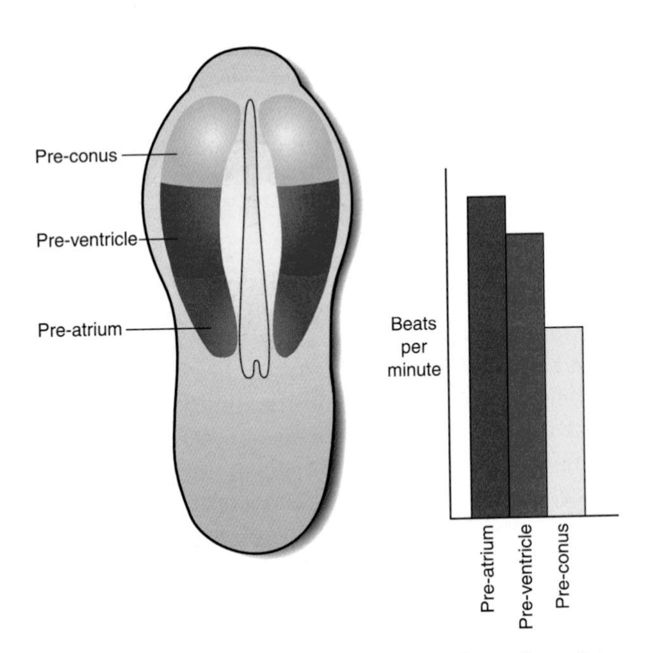

Figure 1 Detection of differences in precardiac regions of stages 5–7 chicken embryos (Satin *et al.,* 1988). The precardiac mesoderm was divided into three segments: preatrium, preventricle, and preconus. Each segment forms a separate beating vesicle 24 hr after microsurgical separation from the other regions. There is a rostral–caudal beating rate gradient with the preatrial region being the most rapid and the preconus region beating the slowest.

II. Developmental Expression Patterns of Atrial and Ventricular Chamber-Specific Genes

The regional differences in endogenous gene transcription between the prospective atria and ventricles begin to emerge after tubular heart formation (Lyons, 1994). Several genes have been identified which are expressed in an atrial or ventricular chamber-specific fashion at some time during vertebrate embryo development (Lyons, 1994). In the adult, a number of atrial-restricted genes have been identified; however, with the exception of *AMHC1* (Yutzey *et al.,* 1994), all the atrial chamber-specific genes are initially expressed throughout the tubular heart. The timing to reach atrial chamber-specific expression of these genes during development is highly variable. *AMHC1,* encoding an atrial-specific myosin heavy chain (MyHC), is the earliest atrial-specific marker. *AMHC1* is first expressed in the posterior region of the fusing chicken heart, the future atrial compartment, by stage 9 before the bilateral regions of cardiac mesoderm have completed fusing into the single primitive tubular heart and prior to the commencement of beating (Yutzey *et al.,* 1994). AMHC1 shares high sequence homology to both α- and β-MyHC sequences (Yutzey *et al.,* 1994), as does the quail homolog of AMHC1, slow MyHC 3 (Nikovits *et al.,* 1996). In contrast to AMHC1 expression in the chicken, *slow MyHC 3* in the quail is initially expressed throughout the tubular heart (Wang *et al.,* 1996). As the heart chamberizes, expression of the *slow MyHC 3* gene in the ventricles is downregulated, whereas expression in the atria is maintained. Based on RNA *in situ* hybridization and Northern analysis, *slow MyHC 3* transcripts are confined to the atria by E 10 (Nikovits *et al.,* 1996). Another MyHC, recognized by monoclonal antibody F18 (Evans *et al.,* 1988), has atrial chamber-specific expression in the E 3 quail embryo (Fig. 2a), a time when *slow MyHC 3* is still expressed in the prospective ventricles and outflow track (Fig. 2b). In zebrafish, a myosin heavy chain has been shown to be atrial chamber restricted in its expression at the 26-somite stage (Stainier and Fishman, 1992).

In mammals, the earliest gene to demonstrate atrial restriction is *myosin light chain 2a* (*MLC-2a*) (Kubalak *et al.,* 1994; see Chapter 15). Similarly to *slow MyHC 3, MLC-2a* is initially expressed throughout the tubular heart at E 8 in the mouse embryo and is downregulated in the ventricular segment during chamber formation. The downregulation of *MLC-2a* in the ventricular chamber is initiated by E 9 and is completed by E 12. In contrast, the *MLC-1a* gene shows relatively late chamber restriction during mouse development. Downregu-

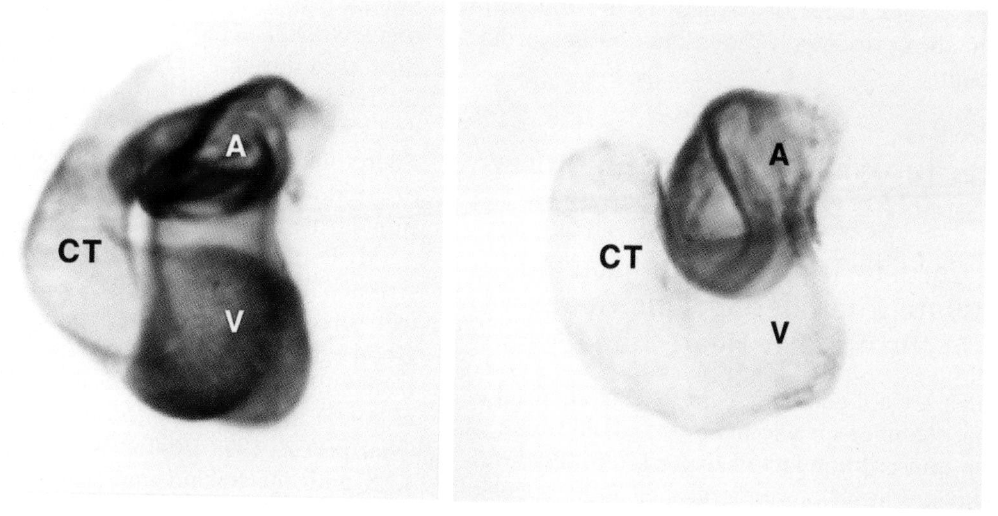

Figure 2 Expression of two atrial chamber-specific MyHC markers in the hearts of the quail embryo. (Left) Slow MyHC 3 is expressed in the primitive right and left atria (A) and at lower levels in the primitive right and left ventricles (V) and conus truncus (CT) as seen by monoclonal antibody NA8 staining (Chen *et al.*, 1997) of the tubular heart of the E 3.5 quail embryo. (Right) A MyHC, recognized by the monoclonal antibody F 18, is expressed in the primitive atrium (A) but not the primitive ventricle (V) or conus truncus (CT) of the tubular heart of an E 3 quail embryo.

lation of *MLC-1a* in the ventricles begins during fetal development but detectable levels are observed in the ventricles even after birth (Lyons, 1994).

Although their expression is not strictly atrial chamber specific, *α-MyHC* and *atrial natriuretic factor* (*ANF*) have been used extensively to study heart development. *α-MyHC* transcripts are detected at E 8 in the mouse embryo throughout the tubular heart when *MLC-1a* and *MLC-2a* are also first expressed (Lyons, 1994). Levels of *α-MyHC* expression in the atria remain high throughout development into adulthood. Initially coexpressed with *β-MyHC* in the ventricles, beginning at about E 10.5 *α-MyHC* expression decreases and reaches its lowest level by E 16.5. Subsequently, *α-MyHC* expression begins to increase again in the ventricle and ultimately replace β-MyHC in all postnatal ventricular cardiomyocytes (Lyons *et al.*, 1990). Agents and mechanical stresses that induce cardiac hypertrophy can lead to reexpression of *β-MyHC* and upregulation of *ANF* expression in the ventricles.

ANF mRNA is first detected in a subpopulation of cardiomyocytes at E 8 of mouse development (Zeller *et al.*, 1987). Throughout embryonic and fetal development, *ANF* is expressed in both atrial and ventricular cardiomyocytes. ANF expression declines rapidly in ventricular cardiomyocytes soon after birth (Argentin *et al.*, 1994). However, low levels of *ANF* transcripts are still detectable in the adult ventricles, amounting to approximately 1% of adult atrial levels, in which *ANF* transcripts account for 1–3% of all mRNA species (Seidman *et al.*, 1991).

At specific times during development, expression of *MLC-1v, MLC-2v,* and *β-MyHC* becomes ventricular chamber specific (Lyons, 1994). All these ventricular markers are activated in both the prospective atria and the ventricles of the E 8 mouse embryo. *MLC-2v* shows the earliest restriction at E 8.5 and becomes ventricular chamber specific by E 10.5 (O'Brien *et al.*, 1993). *β-MyHC* is quickly downregulated in the atria at E 9.5 and is ventricular chamber specific by E 10.5 (Lyon, 1994). As noted previously, β-MyHC is an embryonic form of MyHC and in the ventricles is replaced by α-MyHC soon after birth. Expression of *MLC-1v* becomes ventricular chamber specific by E 13.5 (Lyons, 1994).

A feature common to atrial and ventricular chamber-specific markers is that they are expressed in both the prospective atria and ventricles at the tubular heart stage. Initially activated by cardiac transcription factors that are expressed throughout the tubular heart, chamber-specific expression of these markers is achieved by downregulation of the gene in the other chamber. If chamber-specific gene expression is achieved primarily through negative regulation, this raises two important points. First, because one group of genes becomes restricted to the atrial chambers (e.g., *ANF, AMHC1,* and *MLC-2a*) while a second group becomes restricted to the ventricular chambers (e.g., *β-MyHC* and *MLC-2v*), changes in the transcriptional machinery of atrial and ventricular cells must be transduced differently by atrial- and ventricular-expressed genes. Second, downregulation of chamber-specific gene expression occurs at different times in development for each gene, suggesting a complex series of changes in the transcriptional regulators present at various times during cardiac development.

III. Mechanisms of Atrial Chamber-Specific Gene Expression

A. Cardiac Transcription Factors Activate Chamber-Specific Genes at the Tubular Heart Stage

Although several genes are expressed in an atrial or ventricular chamber-specific manner, initially none of the known transcription factors expressed in the developing heart are atrial or ventricular chamber specific, i.e., all are initially expressed in both prospective atria and ventricles. However, two basic helix–loop–helix transcriptional regulators, dHAND and eHAND, in the mouse (Srivastava *et al.,* 1997) and the homeobox gene *Nkx2-8* in the chick (Brand *et al.,* 1997; Reecy *et al.,* 1997) do show subsequent restriction to specific chambers. Early in mammalian cardiogenesis, dHAND and eHAND become restricted to the ventricular segment of the heart tube and are expressed in a complementary fashion in regions fated to form the right and left ventricles, respectively (Srivastava *et al.,* 1997; Biben and Harvey, 1997). Mice lacking *dHAND* die *in utero* exhibiting a phenotype in which a right ventricle fails to form, suggesting that dHAND mediates differentiation of the right ventricle (Srivastava *et al.,* 1997; see Chapter 9). On the other hand, *eHAND* is expressed in a left-dominant pattern in the developing myocardium. Its expression is likely to be controlled by Nkx2-5 because eHAND expression is absent in *Nkx2-5* null mice in which cardiac looping does not occur (Biben and Harvey, 1997; see Chapter 7).

Although their structure and nuclear localization suggests that the HAND proteins are transcriptional regulators, there is no evidence that expression of atrial and ventricular chamber-specific markers is dependent on the expression of these two factors. In the case of the *dHAND* null mouse, normal levels of expression of the atrial markers *ANF* and *MLC-2a* and the ventricular marker *MLC-2v* are observed in the prospective atria and left ventricle (Srivastava *et al.,* 1997). It is not clear if dHAND regulates chamber-specific markers in the prospective right ventricle in which it normally is expressed because the right ventricle is absent in the *dHAND* null mouse. Unlike the related homeobox gene *Nkx2-5, Nkx2-8* (Brand *et al.,* 1997; Reecy *et al.,* 1997) becomes restricted to the prospective atria of the late tubular heart after its initial expression throughout the tubular heart. Downstream targets of Nkx2-8 are not known. The chamber-restricted transcription factors, dHAND, eHAND, and Nkx2-8, are all transiently expressed during early cardiogenesis and are absent in the fetal and adult heart (Srivastava *et al.,* 1997; Biben and Harvey, 1997; Brand *et al.,* 1997; Reecy *et al.,* 1997). Therefore, they must not be responsible for maintaining of atrial and ventricular chamber-specific gene expression during fetal development or in the adult.

At the tubular heart stage, then, what are the transcription factors responsible for activation of genes which will subsequently display chamber-specific restriction? Good candidates are Nkx2-5 and Nkx2-8 (Komuro and Izumo, 1993; Lints *et al.,* 1993; Brand *et al.,* 1997; Reecy *et al.,* 1997), myocyte enhancer binding factor-2c (MEF2C) (Martin *et al.,* 1993; Edmondson *et al.,* 1994), and GATA-4, -5, and -6 (Heikinheimo *et al.,* 1994; Morrisey *et al.,* 1997; Laverriere *et al.,* 1994; Jiang and Evans, 1996).

Nkx2-5 (Lints *et al.,* 1993) and *MEF2C* (Martin *et al.,* 1993) are homologs of *Drosophila tinman* and *D-MEF2,* respectively, which are required for heart and mesoderm formation in the *Drosophila* embryo (Bodmer, 1993; Lilly *et al.,* 1995; Lin *et al.,* 1997; see Chapter 5). *Nkx2-5* is first expressed early in the precardiac mesoderm of the E 7.5 mouse embryo and it appears to be the earliest marker of the embryonic heart field of zebrafish (Chen and Fishman, 1996). *Nkx2-5* null mice die early in development exhibiting a thin ventricular wall, the cardiomyocytes of which do not express *MLC-2v* (Lyons *et al.,* 1995), suggesting that Nkx2-5 is essential for the activation of this ventricular chamber-specific marker. Additionally, *Nkx2-5* null mice do not express the eHAND (Biben and Harvey, 1997) and have markedly reduced amounts of CARP (Zou *et al.,* 1997). Nkx2-5 can synergize with other cardiac transcription factors, such as serum response factor and GATA-4, to activate cardiac gene expression (Chen and Schwartz, 1996; Durocher *et al.,* 1997). Unlike *Nkx2-5,* transcripts of a related homeodomain transcription factor, *Nkx2-8,* become restricted to the outflow track and atria (Brand *et al.,* 1997; Reecy *et al.,* 1997; see Chapter 16), making Nkx2-8 a good candidate as a factor that might be involved in activation of genes which will subsequently show atrial chamber-specific restriction.

Members of the MEF2 family of MADS (MCM1, agamous, deficiens, and serum response factor) box transcription factors are expressed in the skeletal, cardiac, and smooth muscle lineages of vertebrate and *Drosophila* embryos. In *Drosophila* embryos in which the single *D-MEF2* gene is mutated, somatic, cardiac, and visceral muscle cells do not differentiate, but cardiac and skeletal myoblasts are found to be normally specified and positioned (Lilly *et al.,* 1995). One murine MEF2 family member, MEF2C, has a key role in the development of the right ventricle and in the regulated expression of several cardiac genes as revealed by genetic knockout. In the *MEF2C* null mouse the heart tube does not undergo looping morphogenesis, there is no expression of dHAND after looping, and consequently the right ventricle fails to form (Lin *et al.*, 1997; see Chapter 8). This phenotype is similar to that observed in *dHAND* knockout mice suggesting MEF2C controls expression of *dHAND*. The remaining ventricular cardiomyocytes expressed MLC-2v, demonstrating that expression of this ventricular-restricted marker is not dependent on MEF2C or dHAND expression. In contrast, cardiomyocytes of *MEF2C* null mice did not express α-MyHC or ANF, supporting the notion that MEF2C may be required for the initial activation of these atrial chamber-restricted markers (Lin *et al.*, 1997). However, as further evidence of complex cardiac regulatory pathways, expression of the atrial-restricted *MLC-2a* gene was unaffected in the *MEF2C* null mice.

Another family of genes important in early cardiac development encodes the GATA family of transcription factors (see Chapter 17). Three GATA factors, GATA-4, GATA-5, and GATA-6, are found in the mouse, chicken, and *Xenopus* heart (Arceci *et al.*, 1993; Heikinheimo *et al.*, 1994; Morrisey *et al.*, 1997; Laverriere *et al.*, 1994; Jiang and Evans, 1996). The GATA factors contain a highly conserved DNA-binding domain consisting of two zinc fingers which directly bind to the DNA sequence element (A/T)GATA(A/G). Studies *in vitro* have implicated GATA-4 in the regulation of several cardiac genes, including *α-MyHC, cardiac troponin-c* (*cTNC*), *ANF*, and *brain natriuretic peptide* (*BNP*) (Grepin *et al.*, 1994; Ip *et al.*, 1994; Molkentin *et al.*, 1994; Thuerauf *et al.*, 1994). *In vivo*, injection of *GATA-4, GATA-5*, or *GATA-6* RNA into *Xenopus* embryos can activate expression of the *α-cardiac actin* and *α-MyHC* genes (Jiang and Evans, 1996), suggesting that genes of the *GATA* family encode cardiac transcription activators. Support for this idea comes from experiments in which antisense *GATA-4* oligonucleotides were introduced into P 19 embryonic carcinoma cells. *GATA-4* is expressed in P 19 cells during differentiation into cardiomyocytes, but the addition of antisense *GATA-4* oligonucleotides blocks this differentiation (Grepin *et al.*, 1995). Knockout experiments show that *GATA-4* is required for ventral fusion of the precardiac mesoderm to form the tubular heart (Kuo *et al.*, 1997; Molkentin *et al.*, 1997). However, nearly all cardiac genes are expressed in *GATA-4* null mice. Another *GATA* family member, *GATA-6*, is upregulated in the *GATA-4* null mouse, in which it has been postulated to replace *GATA-4* function in the activation of cardiac genes (Kuo *et al.*, 1997; Molkentin *et al.*, 1997). These observations suggest that cardiac GATA transcription factors are involved in cardiac differentiation and regulation of cardiac gene expression but do not demonstrate a role for GATA family members in restricting expression to atrial or ventricular chambers.

Nkx2-5, MEF2C, and *GATA-4* are expressed throughout the tubular heart as well as in both atrial and ventricular chambers of the adult heart (Lints *et al.*, 1993; Martin *et al.*, 1993; Arceci *et al.*, 1993), suggesting that they may be involved not only in activation but also in maintenance of cardiac gene expression. Studies have identified binding sites for these transcriptional regulators in the promoters of cardiac genes and demonstrated the importance of these factors for cardiac-specific expression. However, the role of these transcriptional regulators in restricting expression to atrial or ventricular chambers is unclear. For example, Nkx2-5 upregulates rat ANF expression in cultured rat atrial cardiomyocytes (Durocher *et al.*, 1996), but since Nkx2-5 is present in the adult ventricle, why does *ANF* become restricted to the atria? In contrast, GATA-4 appears to be a positive regulator of *ANF* (Grepin *et al.*, 1994) and slow MyHC 3 in cultured ventricular cardiomyocytes (Wang *et al.*, 1996). A constellation of observations suggests that the GATA and MEF2 factors are functional activators in the ventricular chamber. First, mutation of either the MEF2 or the GATA motif in the MLC3 promoter attenuates promoter activity in the mouse heart *in vivo* (McGrew *et al.*, 1996). Second, a MEF2 site confers ventricular-specific expression of a *MLC-2v* transgene in transgenic mice (Ross *et al.*, 1996). Third, GATA-4 regulates *α-MyHC, cTNC, ANF*, and *BNP* in cultured ventricular cardiomyocytes (Grepin *et al.*, 1994; Ip *et al.*, 1994; Molkentin *et al.*, 1994; Thuerauf *et al.*, 1994).

Therefore, how do atrial chamber-specific or -restricted markers, such as *slow MyHC3, ANF*, and *α-MyHC*, achieve chamber-restricted expression during development if the transcriptional activators of these genes are still functional in the ventricles? We propose that there are ventricular chamber-specific inhibitors which downregulate the expression of the atrial chamber-specific or -restricted markers within the ventricle. Evidence to support this hypothesis comes from studies on the atrial chamber-specific expression of *slow MyHC 3* gene and perhaps other genes.

B. Do Nuclear Hormone Receptors Specify Atrial Chamber-Specific Gene Expression?

The nuclear hormone receptor superfamily of transcriptional regulators has been implicated in many critical aspects of vertebrate development (Chambon, 1994; Mangelsdorf and Evans, 1995). Members of this family include the retinoic acid receptors (RXRs and

RARs), vitamin D receptors (VDRs), and the thyroid hormone receptors (TR). Ligands which bind these receptors have well-documented effects on cardiogenesis and cardiac gene expression. For example, thyroid hormone (T_3) has been postulated to be a primary physiological regulator of cardiac *MyHC* gene expression in mammals. Around the time of birth the *β-MyHC* is downregulated in the ventricles concomitant with the upregulation of the *α-MyHC* gene. This change in gene expression is correlated with a surge in circulating levels of thyroid hormone. *In vitro,* thyroid hormone stimulates the expression of the *α-MyHC* gene (Gustafson *et al.,* 1987) while it inhibits *β-MyHC* expression in primary cultures of rat cardiomyocytes (Edwards *et al.,* 1994; Gosteli-Peter *et al.,* 1996). Similar results were obtained in adult hearts when T_3 levels were experimentally altered *in utero* (Morkin, 1993).

An especially important role in early aspects of heart formation has been ascribed to retinoic acid (RA)(see Chapter 13). In zebrafish embryos, increasing durations of exposure to RA causes progressive sequential deletion first of the bulbous arteriosus and then (in order) the ventricle, the atrium, and finally the sinus venosus, indicating that the ventricle, or the anterior tissue domain, is more sensitive to the effects of RA than the atrium (Stainier and Fishman, 1992). In birds, RA treatment produces an expansion of the posterior (atrial) domain of the heart at the expense of the anterior domain, again suggesting that the presumptive ventricle is particularly sensitive to the RA (Yutzey *et al.,* 1994). These experiments suggest a role for RA in the commitment of cardiac precursors to different cell lineages—atrial or ventricular. Additional observations implicate RA in the process of cardiomyocyte differentiation and regulation of cardiac gene expression. In an *in vitro* model system of cardiac muscle cell hypertrophy, RA at physiological concentrations suppresses the increase in cell size and upregulation of ANF, markers for cardiac hypertrophy (Zhou *et al.,* 1995). RA can inhibit both endothelin-1-induced and α-adrenergic-induced signal transduction pathways leading to cardiomyocyte hypertrophy (Zhou *et al.,* 1995). Recently, RA was shown to block differentiation of the myocardium after heart specification in *Xenopus* (Drysdale *et al.,* 1997), suggesting that RA may suppress the function of cardiac transcription factors which activate differentiation.

The thyroid hormone, retinoid, and vitamin D_3 family of ligands produce their effects by binding to cognate TR, RAR, or VDR receptors. Binding studies suggest that these receptors bind to regulatory regions of genes in the form of heterodimers with a common partner, the RXRs, although there are responses in experimental settings to RXR homodimers. There are two families of RARs consisting of six genes, *RARα* (*α1* and *α2*), *RARβ* (*β1–β4*), *RARγ* (*γ1* and *γ2*) and the *RXRα*, *RXRβ*, and *RXRγ* receptors (Mangelsdorf and Evans, 1995). The RAR family responds to all-*trans* RA but not to 9-*cis* RA, whereas the RXR family responds to 9-*cis* RA and high concentrations of all-*trans* RA. Genetic evidence suggests that the retinoic signal is transduced by heterodimeric RXR/RAR receptors (Kastner *et al.,* 1997). The nuclear hormone receptors act through direct binding to *h*ormone *r*esponding *e*lements (HREs)—specific DNA sequences consisting of two copies of the hexad consensus sequence, AGGTCA. These direct repeats are separated by a specific numbers of nucleotides and the number of nucleotides between repeats is important in defining the HRE for different receptors (Umesono *et al.,* 1991). For example, the *cis* elements to which VDR, TR, and RAR bind are composed of direct repeats of AGGTCA separated by 3, 4, or 5 nucleotides, respectively.

The mechanism by which the thyroid hormone and the RARs regulate gene expression illustrates that the action of nuclear hormone receptors as transcriptional regulators can activate or suppress target genes by interaction with other nuclear proteins. For example, in CV-1 monkey kidney cells, in the absence of its ligand, TR or RAR act as transcriptional repressors by interacting with a corepressor called SMRT (J. D. Chen *et al.,* 1996). In the presence of their ligands, however, these two receptors act as potent activators of transcription. This differs from the action of thyroid hormone on β-MyHC expression in the heart, in which in the presence of the ligand presumably the TR inhibits gene expression (Edwards *et al.,* 1994).

Knockout mice have been useful for demonstrating the importance of the RAR family in cardiogenesis. Mice lacking the *RARα, RARβ, RARγ, RXRβ, or RXRγ* genes appear to have normal hearts (Kastner *et al.,* 1995). However, double mutants, such as *RARα/RARγ,* have a thinner "spongy" myocardium. These results not only reveal an importance for RARs in heart development but also imply a functional redundancy among some of the family members (Mendelsohn *et al.,* 1994). Unlike the other single RAR gene knockouts, the *RXRα* knockout does show a cardiac phenotype. This phenotype is similar to mice deficient in vitamin A, a metabolic precursor of RA. *RXRα* null mutants die between E 10.5 and 17.5 of development from cardiac failure (Kastner *et al.,* 1994; Sucov *et al.,* 1994). Interestingly, in the *RXRα* null mice, *MLC-2a* is expressed in the ventricles at a time when wild-type mice express MLC-2a in an atrial-restricted fashion (Sucov *et al.,* 1994; see Chapters 13 and 15), suggesting that RXRα not only has a role in heart morphogenesis but also may act directly or indirectly on the inhibition of *MLC-2a* expression in the ventricles. Detailed analysis of *RXRα* null embryos shows ventricular septal, atrioventricular cushion, and conotruncal ridge defects, with a double-outlet right ventricle, a patent aorticopulmonary window, and persistent truncus arteriosus (Gruber *et al.*, 1996). Heterozygous *RXRα* embryos display a predisposition for trabecular and papillary muscle, ventricular septal, conotruncal ridge and atrioventricular cushion defects, and pulmonic stenosis (Gruber *et al.*, 1996).

As previously stated, thyroid hormone plays a very important role in cardiac gene expression. There are two classes of thyroid hormone receptors, α (with α1 and α2 isoforms) and β. Only α1 and β mediate the functions of thyroid hormones. Homozygous inactivation of the *TR alpha* gene abrogates the production of both TR α1 and TR α2 isoforms (α2 does not bind T_3) and leads to death in mice within 5 weeks after birth (Fraichard *et al.*, 1997). By 2 weeks of life, the *TR alpha* null mice exhibit growth arrest and delayed maturation of the small intestine and bone, mimicking the symptoms of progressive hypothyroidism. Production of thyroid hormone, substantially upregulated at the time of weaning, is dependent on *TR alpha* gene products. Patterns of *α-* and *β-MyHC* expression in the *TR alpha* null mice have not been reported, and thus the importance of the TR α1 and β receptors in cardiac gene expression remains to be elucidated.

The VDR is of particular interest in cardiac development because this receptor appears to be important in the expression of the atrial chamber-specific marker, slow MyHC 3 (Wang *et al.*, 1996). A principal function of VDR and its ligand, 1,25-dihydroxyvitamin D_3 (VD3), is the regulation of calcium and phosphorus homeostasis in higher vertebrates. High-affinity receptors for VD3 have been described in the myocardium by Waters and colleagues (1986), and others find cardiac hypertrophy and alterations in cardiac contractility associated with VD3 deficiency (Weishaar and Simpson, 1987). Like other nuclear hormone receptors, this receptor can activate gene expression through heterodimer formation with RXRs (Mangelsdorf and Evans, 1995). While the cardiovascular system does not represent a classical target for the ligands of the VDR, Gardner and colleagues (Wu *et al.*, 1996) have shown that both cardiomyocytes and vascular smooth muscle cells respond to agents that bind the VDR receptor. They find that high concentrations of VD3 or RA can reduced endothelin-stimulated hypertrophy of ventricular cardiomyocytes as well as inhibit ANF secretion in a dose-dependent fashion. When both ligands (VD3 and RA) are combined the effect is additive. These investigators conclude that the ligand-bound vitamin D and retinoid receptors are capable of suppressing the activation of a number of genes associated with hypertrophy of neonatal rat cardiomyocytes. The effects on cell growth and gene expression are likely an indirect result of ligand binding because there do not appear to be typical VDR-like elements in the promoter region of the human *ANF* gene (Wu *et al.*, 1996), nor has binding of radiolabeled VDR to its regulatory region been observed. It should be noted that to date the RXR or VDR has not been shown to act directly on cardiac-expressed genes as either a positive or negative transcriptional regulator.

Recently, the *VDR* gene has been knocked out (Yoshizawa *et al.*, 1997). Null mice exhibited impaired bone formation, uterine hypoplasia, and growth retardation after weaning and died within 15 weeks after birth. There is no report of cardiac anomalies in the *VDR* null mice nor any description of the cardiac isoforms expressed. In light of recent observations documenting the importance of VDR-binding sequences in cardiac gene expression, it will be interesting to examine the repertoire of cardiac genes expressed in the *VDR* null mice. Some of the observations regarding the effects of VD3 and the VDR on the heart may be secondary to the action of this hormone on calcium metabolism; however, these results also support the general position that the VDR plays a role in the regulation of cardiac gene-specific expression independent of the role usually attributed to it in physiology.

Published reports suggest that nuclear hormone receptors are ubiquitously distributed in the tissues of the developing embryo. In contrast, the effect of ligands show regional differences. Even where receptors show no obvious endogenous differences in expression, as in the developing heart, experimental manipulations suggest that nuclear hormone receptor ligands can have different effects in the ventricles and atria. The nuclear hormone receptors and their ligands are likely to have important roles not only in cardiac morphogenesis but also in the control of expression of genes associated with cardiac muscle contraction.

C. Cardiac *cis* Elements Regulate Chamber-Specific Gene Expression

Cardiac-specific genes encode proteins that are crucial to the functioning of the primitive tubular heart as well as the chamberized heart (Jones *et al.*, 1996). Understanding the *cis* elements required for initiation and maintenance of cardiac gene expression is essential to an understanding of cardiac function because these proteins determine the rate and force of contraction. Furthermore, since chamber-restricted expression of cardiac genes is an early event in the process of atria and ventricle formation, an understanding of the mechanisms regulating cardiac gene expression may shed light on the process whereby the atria and ventricles become demarcated. The regulation of cardiac gene expression, especially genes intregal to cardiac compartment function, has been fruitfully studied in early mammalian cardiac development. The genes that have been studied include *myosin light chains* (Kelly *et al.,* 1995; O'Brien *et al.,* 1993; McGrew *et al.,* 1996), *α-MyHC* (Gustafson *et al.,* 1987; Morkin, 1993; Subramaniam *et al.,* 1991, 1993), *β-MyHC* (Cribbs *et al.,* 1989; Rindt *et al.,* 1993; Thompson *et al.,* 1991), *cTnT* (Mar *et al.,* 1988; Christensen *et al.,* 1993; Iannello *et al.,* 1991), *MCK* (Amacher *et al.,* 1993), *ANF* (Argentin *et al.,* 1994; Seidman *et al.,* 1991), and *cardiac α-actin* (Moss *et al.,* 1994; Sartorelli *et al.,* 1992). While not exclusively so, much of this work on the *cis* elements in cardiac gene expression has focused on delineation of those elements that interact with transcription factors to control gene expression in the ventricle.

The most thoroughly analyzed ventricular compartment-specific gene is *MLC-2v,* which becomes restricted to the ventricle between E 8.5 and 10.5 of mouse development (O'Brien *et al.,* 1993). Work by Chien and colleagues (Ross *et al.,* 1996; Zou *et al.,* 1997) on the mouse *MLC-2v* gene serves as a model in the identification of *cis-* and *trans-*acting factors that restrict the expression of genes to the ventricular chamber (see Chapter 15). These investigators have demonstrated that a 28-base pair (bp) portion of the promoter containing two protein binding sites, HF-1a and MEF2, is sufficient to confer ventricular-specific expression on a reporter gene in transgenic mice (Ross *et al.,* 1996; Zou *et al.,* 1997). Binding of YB-1, a CCAAT box-binding transcription factor to the HF-1a site, functions as a positive regulator of *MLC-2v* expression in cardiac muscle. Recently, this group has reported that interactions of the CARP gene product with YB-1 to form a complex may act as a negative regulator of the HF-1a-dependent pathway of cardiac gene expression (Zou *et al.,* 1997). Although these results indicate that CARP is an attractive candidate for the downregulation of *MLC-2v* in the atria, in fact neither CARP nor any of the known HF-1a and MEF2 site binding proteins show a chamber-restricted expression pattern (Zou *et al.,* 1997).

Less is known about the *cis* elements which regulate cardiac genes in the atria. Most atrially restricted genes are expressed in the ventricle at some time during development, and where it has been investigated, *cis* elements important for cardiac gene expression in the ventricles also regulate expression in the atria. Studies on the regulation of *α-cardiac MyHC, MLC3,* and *ANF* in the heart serve as examples. Regulation of the *α-cardiac MyHC* gene has been extensively studied *in vivo* and *in vitro* (Gustafson *et al.,* 1987; Subramaniam *et al.,* 1993; Molkentin and Markham, 1993; Molkentin *et al.,* 1994). A 2-kilobase (kb) upstream region of the *α-cardiac MyHC* gene directs cardiac tissue-specific expression of a transgene throughout the heart, and a *t*hyroid hormone-*r*esponsive *e*lement, TRE2, is important for reporter activity in both the atrium and the ventricle (Subramaniam *et al.,* 1991, 1993). In cardiac cell culture (primarily ventricular cardiomyocytes) TRE, MEF2, and GATA-4 binding sites have documented regulatory functions for *α-cardiac MyHC* expression (Gustafson *et al.,* 1987; Molkentin and Markham, 1993; Molkentin *et al.,* 1994). In the case of the *MLC3* gene, Buckingham and colleagues (Kelly *et al.,* 1995) have reported that a *MLC3–LacZ* transgene, containing a 2-kb promoter region and a 260-bp enhancer sequence of the *MLC3* gene, is expressed in a spatially restricted manner within the atria and left ventricular compartments in adult mice (see Chapter 19). This work identifies transcriptional differences between cardiomyocytes in the left and right ventricles of the heart as well as between atrial cardiomyocytes and at least right ventricular cardiomyocytes. The *cis* elements responsible for *MLC3* transgene expression in the atria and the left ventricle have not been identified. However, mutation of either a MEF2 or GATA motif, in the promoter of the endogenous *MLC3* gene, attenuates its activity in the heart (McGrew *et al.,* 1996). Finally, a 2.4-kb regulatory region of the *ANF* gene directs reporter expression equally well in both atrial and ventricular cardiomyocytes of transgenic mice (Seidman *et al.,* 1991). Transfection of cardiomyocytes demonstrated that GATA and Nkx2-5 binding sites located within the promoter are important in regulating *ANF* expression (Durocher *et al.,* 1996). To date, work with neither cultured cardiomyocytes nor transgenic mice has identified genetic elements which impart atrial chamber-specific expression on the *α-cardiac MyHC, MLC3,* or *ANF* genes.

A new myosin heavy chain gene, *slow MyHC3,* expressed in the quail heart as well as in embryonic slow skeletal muscle, has been identified and characterized (Nikovits *et al.,* 1996; Wang *et al.,* 1996). A comparison of 3′ untranslated sequences suggests that *slow MyHC 3* encodes the quail homolog of *AMHC1* (Yutzey *et al.,* 1994). Initially, the *slow MyHC 3* gene is expressed throughout the tubular heart. As the heart chamberizes, expression of *slow MyHC 3* in the ventricles is downregulated, whereas expression in the atria is maintained (Wang *et al.,* 1996; G. Wang *et al.,* unpublished result). Although slow MyHC 3/AMHC1 shows greatest sequence similarity to the α- and β-cardiac MyHC isoforms, the expression pattern of *slow MyHC 3* during heart development is different from that of its mammalian counterparts. The early downregulation of *slow MyHC 3* in the ventricles is more similar to that observed for the mammalian *MLC-2a* gene (Kubalak *et al.,* 1994) than to the pattern of postnatal downregulation described for *ANF* expression (Argentin *et al.,* 1994).

By delineating the *cis* elements required for atrial chamber-specific expression of the *slow MyHC 3* gene, clues to the mechanism(s) regulating its expression pattern may become apparent. Four regions of the *slow MyHC 3* promoter that contain either positive and negative *cis* elements have been identified in experiments performed both *in vitro* and *in vivo* (Wang *et al.,* 1996). One of these regions, 160 bp of the 5′ flanking sequence between -840 and -680 bp relative to the transcriptional initiation site—designated as atrial regulatory domain 1 (ARD1)—functions as an atrial-specific enhancer in primary cardiomyocyte cultures as well as in the embryo (Fig. 3A). The function of the ARD1 *in vivo* has been investigated by using an avian retroviral vector, RCAN/PCAT/F, to deliver an ARD1/SV40 promoter/CAT construct into embryos. In infected avian embryos, inclusion of the ARD1 element increases expression of the CAT reporter 12.3-fold in the atria compared to noncardiac tissues and 4.6-fold when compared to the ventricles (Wang *et al.,* 1996).

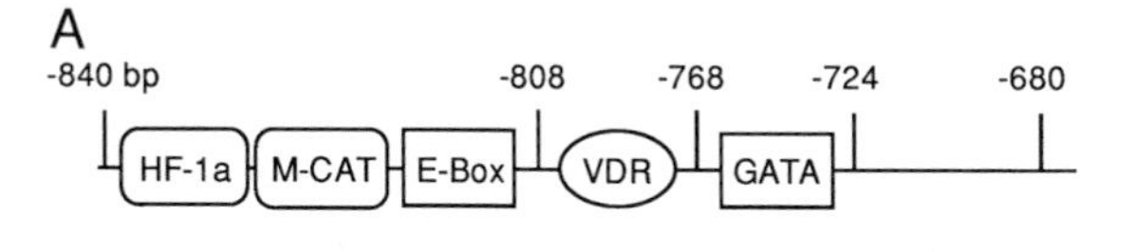

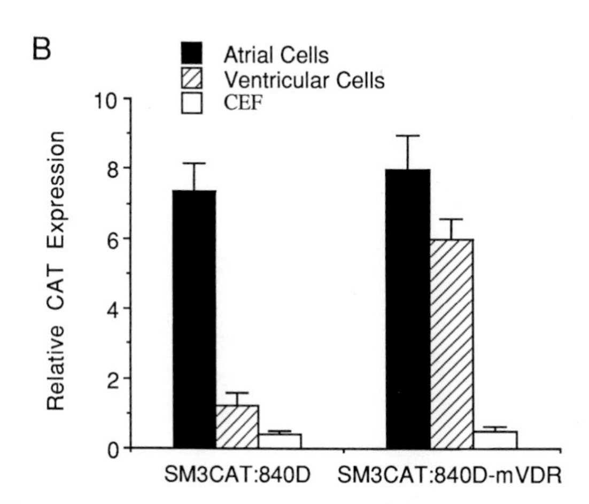

Figure 3 The vitamin D receptor (VDR)-like element specifies atrial chamber-specific expression of *slow MyHC 3* gene by inhibition of *slow MyHC 3* expression in the ventricles. (A) The regulatory motifs found in the ARD1 enhancer. (B) Expression of SM3CAT:840D, which contains 840 bp of sequence upstream from the *slow MyHC 3* gene transcription start site, is atrial specific because it is actively inhibited in the ventricular cardiomyocytes by the VDR-like element. Mutation of the VDR-like element between -801 and -796, in the context of SM3CAT:840D to generate SM3CAT:840D-mVDR, leads to upregulation in the ventricular cardiomyocytes but not in the atrial cardiomyocytes or chicken embryonic fibroblasts (CEF).

Within the ARD1 are several sequence motifs with homology to functionally defined regulatory elements identified in other cardiac genes. These include, from 5′ to 3′, HF-1a (Ross *et al.,* 1996), M-CAT (Mar and Ordahl, 1988), E box (Sartorelli *et al.,* 1992), VDR or RAR-like element (Mangelsdorf and Evans, 1995), and GATA elements (Grepin *et al.,* 1994; Ip *et al.,* 1994; Molkentin *et al.,* 1994; Thuerauf *et al.,* 1994). To define the *cis* elements that regulate atrial-specific expression, deletions were made which removed portions of ARD1. Deletion of the HF-1a, M-CAT, and E box motifs did not change atrial-specific reporter expression following transfection into atrial cardiomyocytes, ventricular cardiomyocytes, or chicken embryonic fibroblasts (CEFs). However, further deletion which removes the VDR-like element as well as HF-1a, M-CAT, and E box elements results in an increase of reporter expression in ventricular cardiomyocytes to a level equal to that observed in atrial cardiomyocytes (Wang *et al.,* 1996). Furthermore, when compared to a construct with an intact VDR-like motif, mutations within the VDR-like motif of the ARD1 of the endogenous promoter for *slow MyHC 3* result in upregulation of the reporter expression in ventricular cardiomyocytes (Fig. 3B). These results suggest that the HF-1a, M-CAT, and E box motifs within ARD1 are not essential for atrial-specific expression of *slow MyHC 3.* Instead, atrial-specific expression of *slow MyHC 3* is directly related to the function of a 40-bp sequence, which includes a VDR-like element, within the *slow MyHC 3* promoter. The VDR-like motif acts to negatively regulate *slow MyHC 3* expression specifically in ventricular cardiomyocytes but has no apparent role in positively regulating expression in atrial cardiomyocytes. Additional experiments have shown that the GATA element within the *slow MyHC 3* enhancer confers heart specificity to expression of this gene but has no role in downregulating *slow MyHC 3* expression in the ventricles (G. Wang and F. Stockdale, unpublished data). In the transition from the tubular heart to the chamberized quail heart, there is increased inhibitory action of the VDR-like element in ventricular cardiomyocytes (G. Wang and F. Stockdale, unpublished data). The VDR-like element is much less effective at inhibiting reporter expression in ventricular cardiomyocytes isolated from E3 hearts than in ventricular cardiomyocytes isolated from E6 hearts. This increased inhibitory action coincides with developmental downregulation of *slow MyHC 3* in the ventricles.

An inhibitory factor(s) was also found to regulate *ANF* expression in ventricular cardiomyocytes. A Nkx2-5 response element, termed NKE, is responsible for *ANF* promoter construct expression in atrial cardiomyocytes (Durocher *et al.,* 1996). Interestingly, a deletion removing a small region of the *ANF* promoter, containing the NKE, leads to upregulation of a reporter in cultured ventricular cardiomyocytes but not in cultured atrial cardiomyocytes. This suggests that the NKE, or an adjacent site, binds an inhibitor restricted to ventricular cardiomyocytes (Durocher *et al.,* 1996). These results also focus attention on members of the Nkx family as potential regulators of chamber-specific expression.

As the *cis* elements regulating expression of cardiac genes become more defined, it will be possible to determine to what extent divergent and overlapping pathways regulate expression of the atrial-specific *slow MyHC 3* gene and other atrial chamber-specific or restricted genes. The finding that expression of the *slow MyHC 3* gene becomes confined to the atrium concurrent with embryonic heart chamberization makes this gene unique and provides a model to investigate the mechanism(s) establishing regional differences in the heart.

D. Models of Atrial Chamber-Specific Expression of Cardiac Genes

Two models can depict regulation of chamber-specific expression of cardiac genes during embryonic development: an activation model and an inhibition model (Fig. 4). In the activation model of chamber-specific expression, chamber-specific genes are activated and maintained by chamber-specific transcription factors. Few transcription factors have been identified which show a significant difference in expression between atria and ventricles. With the possible exception of *AMHC1,* chamber-specific markers are initially expressed throughout the tubular heart. Chamber specificity is initially achieved by downregulation of genes in atria or ventricles. In principle, chamber-specific expression could be achieved if a transcription factor, such as Nkx2-5, GATA-4, or MEF2C, that initiates or maintains expression is selectively downregulated in one of the two chambers. Currently, there is no reported evidence to support this as a general mechanism. We favor the inhibition model on the basis of studies of the regulation of *slow MyHC 3* gene and on studies of *ANF* regulation. In this model a positive factor(s), acting through a positive *cis* element, is required during development for activation of heart-specific expression. In the absence of a negative factor(s), cardiac genes can be expressed throughout the heart. The appearance of a negative factor(s) which functions through a negative *cis* element is responsible for subsequent downregulation. In early stages of heart development, positive factors emerging in the absence of a negative factor(s) induce cardiac-specific but not chamber-specific expression. Subsequently, differential expression of a negative factor(s) at specific developmental stages leads to downregulation of atrial chamber-specific genes in the ventricles or of ventricular chamber-specific genes in the atria. In the case of atrial-specific gene expression, the negative factor(s) is ventricular chamber specific. This model more accurately describes atrial chamber-specific expression of *slow MyHC 3* and the *ANF* genes. Members of the nuclear hormone receptor family appear to have an important role in chamber-specific gene expression through inhibition. For example, thyroid hormone, presumably acting through its receptor, inhibits *β-MyHC* expression in ventricular cardiomyocytes (Edwards *et al.,* 1994). Expression of the *ANF* gene appears to be subject to vitamin D_3-dependent inhibition (Li and Gardner, 1994). Additionally, deletion or mutation of a

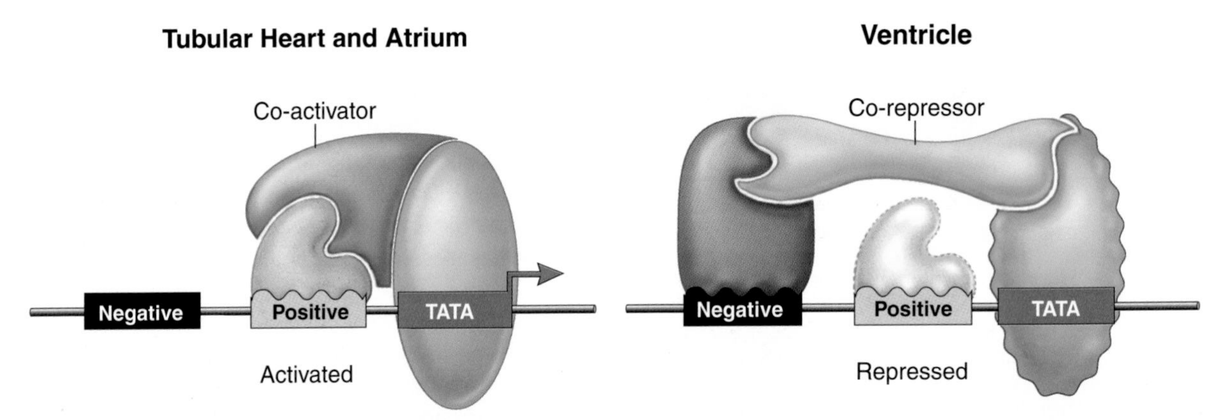

Figure 4 Inhibition model of atrial chamber-specific expression. A positive factor(s) is required to activate atrial chamber-specific expression throughout the tubular heart and atrium by binding specifically to the positive element. We postulate that expression in the atria is maintained by the positive factor(s). Expression in the ventricles is inhibited by a negative factor(s), although the positive factor(s) is still present in the ventricles and may or may not continue to occupy the positive element. We postulate that activation and repression of transcription are mediated through the actions of a coactivator and corepressor, respectively.

VDR-like motif in the *slow MyHC 3* promoter suggests that a vitamin D or RAR is involved in the inhibition of this gene in ventricular cardiomyocytes (Wang *et al.*, 1996). Thus, it appears that nuclear hormone receptor family members may play a role as negative regulators in ventricular cardiomyocytes. Whether members of the nuclear hormone receptor family generally serve as inhibitors of chamber-specific expression during heart development needs further study.

Acknowledgment

We thank Dr. William Nikovits, Jr., for very helpful discussions.

References

Amacher, S. L., Buskin, J. N., and Hauschka, S. D. (1993). Multiple regulatory elements contribute differentially to muscle creatine kinase enhancer activity in skeletal and cardiac muscle. *Mol. Cell. Biol.* **13,** 2753–2764.

Arceci, R. J., King, A. A., Simon, M. C., Orkin, S. H., and Wilson, D. B. (1993). Mouse GATA-4: A retinoic acid-inducible GATA-binding transcription factor expressed in endodermally derived tissues and heart. *Mol. Cell. Biol.* **13,** 2235–2246.

Argentin, S., Ardati, A., Tremblay, S., Lihrmann, I., Robitaille, L., Drouin, J., and Nemer, M. (1994). Developmental stage-specific regulation of atrial natriuretic factor gene transcription in cardiac cells. *Mol. Cell. Biol.* **14,** 777–790.

Biben, C., and Harvey, R. P. (1997). Homeodomain factor Nkx2-5 controls left/right asymmetric expression of bHLH gene *eHand* during murine heart development. *Genes Dev.* **11,** 1357–1369.

Bodmer, R. (1993). The gene *tinman* is required for specification of the heart and visceral muscles in *Drosophila. Development (Cambridge, UK)* **118,** 719–729; published erratum: **119,**(3), 969.

Brand, T., Andree, B., Schneider, A., Buchberger, A., and Arnold, H. H. (1997). Chicken *NKx2-8,* a novel homeobox gene expressed during early heart and foregut development. *Mech. Dev.* **64,** 53–59.

Chambon, P. (1994). The retinoid signaling pathway: Molecular and genetic analyses. *Semin. Cell Biol.* **5,** 115–125.

Chen, C. Y., and Schwartz, R. J. (1996). Recruitment of the *tinman* homolog Nkx-2.5 by serum response factor activates cardiac alpha-actin gene transcription. *Mol. Cell Biol.* **16,** 6372–6384.

Chen, J. D., Umesono, K., and Evans, R. M. (1996). SMRT isoforms mediate repression and anti-repression of nuclear receptor heterodimers. *Proc. Natl. Acad. Sci. U.S.A.* **93,** 7567–7571.

Chen, J. N., and Fishman, M. C. (1996). Zebrafish *tinman* homolog demarcates the heart field and initiates myocardial differentiation. *Development (Cambridge, UK)* **122,** 3809–3816.

Chen, J. N., Haffter, P., Odenthal, J., Vogelsang, E., Brand, M., van Eeden, F. J., Furutani-Seiki, M., Granato, M., Hammerschmidt, M., Heisenberg, C. P., Jiang, Y. J., Kane, D. A., Kelsh, R. N., Mullins, M. C., and Nüsslein-Volhard, C. (1996). Mutations affecting the cardiovascular system and other internal organs in zebrafish. *Development (Cambridge, UK)* **123,** 293–302.

Chen, Q., Moore, L. A., Wick, M., and Bandman, E. (1997). Identification of a genomic locus containing three slow myosin heavy chain genes in the chicken. *Biochim. Biophys. Acta* **1353,** 48–56.

Christensen, T. H., Prentice, H., Gahlmann, R., and Kedes, L. (1993). Regulation of the human *cardiac/slow-twitch troponin* C gene by multiple, cooperative, cell-type-specific, and MyoD-responsive elements. *Mol. Cell. Biol.* **13,** 6752–6765.

Cribbs, L. L., Shimizu, N., Yockey, C. E., Levin, J. E., Jakovcic, S., Zak, R., and Umeda, P. K. (1989). Muscle-specific regulation of a transfected rabbit myosin heavy chain beta gene promoter. *J. Biol. Chem.* **264,** 10672–10678.

DeHaan, R. L. (1965). Morphogenesis of the vertebrate heart. In "Organogenesis" (R. L. DeHaan and H. Ursprung, (eds.), pp. 377–419. Holt, Rinehart & Winston, New York.

Drysdale, T. A., Patterson, K. D., Saha, M., and Krieg, P. A. (1997). Retinoic acid can block differentiation of the myocardium after heart specification. *Dev. Biol.* **188,** 205–215.

Durocher, D., Chen, C. Y., Ardati, A., Schwartz, R. J., and Nemer, M. (1996). The atrial natriuretic factor promoter is a downstream target for Nkx-2.5 in the myocardium. *Mol. Cell. Biol.* **16,** 4648–4655.

Durocher, D., Charron, F., Warren, R., Schwartz, R. J., and Nemer, M. (1997). The cardiac transcription factors Nkx2-5 and GATA-4 are mutual cofactors [in process citation]. *EMBO J.* **16,** 5687–5696.

Edmondson, D. G., Lyons, G. E., Martin, J. F., and Olson, E. N. (1994). *Mef2* gene expression marks the cardiac and skeletal muscle lineages during mouse embryogenesis. *Development (Cambridge, UK)* **120,** 1251–1263.

Edwards, J. G., Bahl, J. J., Flink, I. L., Cheng, S. Y., and Morkin, E. (1994). Thyroid hormone influences beta myosin heavy chain (beta MHC) expression. *Biochem. Biophys. Res. Commun.* **199,** 1482–1488.

Evans, D., Miller, J. B., and Stockdale, F. E. (1988). Developmental patterns of expression and coexpression of myosin heavy chains in atria and ventricles of the avian heart. *Dev. Biol.* **127,** 376–383.

Fraichard, A., Chassande, O., Plateroti, M., Roux, J. P., Trouillas, J. Dehay, C., Legrand, C., Gauthier, K., Kedinger, M., Malaval, L., Rousset, B., and Samarut, J. (1997). The *T3R* alpha gene encoding a thyroid hormone receptor is essential for post-natal development and thyroid hormone production. *EMBO J.* **16,** 4412–4420.

Gosteli-Peter, M. A., Harder, B. A., Eppenberger, H. M., Zapf, J., and Schaub, M. C. (1996). Triiodothyronine induces over-expression of alpha-smooth muscle actin, restricts myofibrillar expansion and is permissive for the action of basic fibroblast growth factor and insulin-like growth factor I in adult rat cardiomyocytes. *J. Clin. Invest.* **98,** 1737–1744.

Grepin, C., Dagnino, L., Robitaille, L., Haberstroh, L., Antakly, T., and Nemer, M. (1994). A hormone-encoding gene identifies a pathway for cardiac but not skeletal muscle gene transcription. *Mol. Cell. Biol.* **14,** 3115–3129.

Grepin, C., Robitaille, L., Antakly, T., and Nemer, M. (1995). Inhibition of transcription factor GATA-4 expression blocks in vitro cardiac muscle differentiation. *Mol. Cell. Biol.* **15,** 4095–4102.

Gruber, P. J., Kubalak, S. W., Pexieder, T., Sucov, H. M., Evans, R. M., and Chien, K. R. (1996). RXR alpha deficiency confers genetic susceptibility for aortic sac, conotruncal, atrioventricular cushion, and ventricular muscle defects in mice. *J. Clin. Invest.* **98,** 1332–1343.

Gustafson, T. A., Markham, B. E., Bahl, J. J., and Morkin, E. (1987). Thyroid hormone regulates expression of a transfected alpha-myosin heavy-chain fusion gene in fetal heart cells. *Proc. Natl. Acad. Sci. U.S.A.* **84,** 3122–3126.

Heikinheimo, M., Scandrett, J. M., and Wilson, D. B. (1994). Localization of transcription factor GATA-4 to regions of the mouse embryo involved in cardiac development. *Dev. Biol.* **164,** 361–373.

Iannello, R. C., Mar, J. H., and Ordahl, C. P. (1991). Characterization of a promoter element required for transcription in myocardial cells. *J. Biol. Chem.* **266,** 3309–3316.

Icardo, J. M., and Manasek, F. J. (1992). Cardiogenesis: Development mechanisms and embryology. *In* "The Heart and Cardiovascular System" (H. A. Fozzard, E. Haber, R. B. Jennings, A. M. Katz, and H. E. Morgan, eds.), pp. 1563–1586. Raven Press, New York.

Ip, H. S., Wilson, D. B., Heikinheimo, M., Tang, Z., Ting, C. N., Simon, M. C., Leiden, J. M., and Parmacek, M. S. (1994). The GATA-4 transcription factor transactivates the cardiac muscle-specific troponin C promoter-enhancer in nonmuscle cells. *Mol. Cell. Biol.* **14,** 7517–7526.

Jiang, Y., and Evans, T. (1996). The *Xenopus GATA-4/5/6* genes are associated with cardiac specification and can regulate cardiac-specific transcription during embryogenesis. *Dev. Biol.* **174,** 258–270.

Jones, W. K., Grupp, I. L., Doetschman, T., Grupp, G., Osinska, H., Hewett, T. E., Boivin, G., Gulick, J., Ng, W. A., and Robbins, J. (1996). Ablation of the murine *alpha myosin heavy chain* gene leads to dosage effects and functional deficits in the heart. *J. Clin. Invest.* **98,** 1906–1917.

Kastner, P., Grondona, J. M., Mark, M., Gansmuller, A., LeMeur, M., Decimo, D., Vonesch, J. L., Dolle, P., and Chambon, P. (1994). Genetic analysis of RXR alpha developmental function: Convergence of RXR and RAR signaling pathways in heart and eye morphogenesis. *Cell (Cambridge, Mass.)* **78,** 987–1003.

Kastner, P., Mark, M., and Chambon, P. (1995). Nonsteroid nuclear receptors: What are genetic studies telling us about their role in real life? *Cell (Cambridge, Mass.)* **83,** 859–869.

Kastner, P., Mark, M., Ghyselinck, N., Krezel, W., Dupe, V., Grondona, J. M., and Chambon, P. (1997). Genetic evidence that the retinoid signal is transduced by heterodimeric RXR/RAR functional units during mouse development. *Development (Cambridge, UK)* **124,** 313–326.

Kelly, R., Alonso, S., Tajbakhsh, S., Cossu, G., and Buckingham, M. (1995). Myosin light chain 3F regulatory sequences confer regionalized cardiac and skeletal muscle expression in transgenic mice. *J. Cell Biol.* **129,** 383–396.

Komuro, I., and Izumo, S. (1993). *Csx:* A murine homeobox-containing gene specifically expressed in the developing heart. *Proc. Natl. Acad. Sci. U.S.A.* **90,** 8145–8149.

Kubalak, S. W., Miller-Hance, W. C., O'Brien, T. X., Dyson, E., and Chien, K. R. (1994). Chamber specification of atrial myosin light chain-2 expression precedes septation during murine cardiogenesis. *J. Biol. Chem.* **269,** 16961–16970.

Kuo, C. T., Morrisey, E. E., Anandappa, R., Sigrist, K., Lu, M. M., Parmacek, M. S., Soudais, C., and Leiden, J. M. (1997). GATA4 transcription factor is required for ventral morphogenesis and heart tube formation. *Genes Dev.* **11,** 1048–1060.

Laverriere, A. C., MacNeill, C., Mueller, C., Poelmann, R. E., Burch, J. B., and Evans, T. (1994). GATA-4/5/6, a subfamily of three transcription factors transcribed in developing heart and gut. *J. Biol. Chem.* **269,** 23177–23184.

Li, Q., and Gardner, D. G. (1994). Negative regulation of the human *atrial natriuretic peptide* gene by 1,25-dihydroxyvitamin D_3. *J. Biol. Chem.* **269,** 4934–4939.

Lilly, B., Zhao, B., Ranganayakulu, G., Paterson, B. M., Schulz, R. A., and Olson, E. N. (1995). Requirement of MADS domain transcription factor D-MEF2 for muscle formation in Drosophila. *Science* **267,** 688–693.

Lin, Q., Schwarz, J., Bucana, C., and Olson, E. N. (1997). Control of mouse cardiac morphogenesis and myogenesis by transcription factor MEF2C. *Science* **276,** 1404–1407.

Lints, T. J., Parsons, L. M., Hartley, L., Lyons, I., and Harvey, R. P. (1993). *Nkx-2.5:* A novel murine homeobox gene expressed in early heart progenitor cells and their myogenic descendants. *Development (Cambridge, UK)* **119,** 969.

Lyons, G. E. (1994). In situ analysis of the cardiac muscle gene program during embryogenesis. *Trends Cardiovasc. Med.* **4,** 70–77.

Lyons, G. E., Schiaffino, S., Sassoon, D., Barton, P., and Buckingham, M. (1990). Developmental regulation of myosin gene expression in mouse cardiac muscle. *J. Cell. Biol.* **111,** 2427–2436.

Lyons, I., Parsons, L. M., Hartley, L., Li, R., Andrews, J. E., Robb, L., and Harvey, R. P. (1995). Myogenic and morphogenetic defects in the heart tubes of murine embryos lacking the homeo box gene Nkx2-5. *Genes Dev.* **9,** 1654–1666.

Mangelsdorf, D. J., and Evans, R. M. (1995). The RXR heterodimers and orphan receptors. *Cell (Cambridge, Mass.)* **83,** 841–850.

Mar, J. H., and Ordahl, C. P. (1988). A conserved CATTCCT motif is required for skeletal muscle-specific activity of the cardiac troponin T gene promoter. *Proc. Natl. Acad. Sci. U.S.A.* **85,** 6404–6408.

Mar, J. H., Antin, P. B., Cooper, T. A., and Ordahl, C. P. (1988). Analysis of the upstream regions governing expression of the chicken *cardiac troponin T* gene in embryonic cardiac and skeletal muscle cells. *J. Cell Biol.* **107,** 573–585.

Martin, J. F., Schwarz, J. J., and Olson, E. N. (1993). Myocyte enhancer factor (MEF) 2C: A tissue-restricted member of the MEF-2 family of transcription factors. *Proc. Natl. Acad. Sci. U.S.A.* **90,** 5282–5286.

McGrew, M. J., Bogdanova, N., Hasegawa, K., Hughes, S. H., Kitsis, R. N., and Rosenthal, N. (1996). Distinct gene expression patterns in skeletal and cardiac muscle are dependent on common regulatory sequences in the MLC1/3 locus. *Mol. Cell. Biol.* **16,** 4524–4534.

Mendelsohn, C., Lohnes, D., Decimo, D., Lufkin, T., LeMeur, M., Chambon, P., and Mark, M. (1994). Function of the retinoic acid receptors (RARs) during development (II). Multiple abnormalities at various stages of organogenesis in RAR double mutants. *Development (Cambridge, UK)* **120,** 2749–2771.

Molkentin, J. D., and Markham, B. E. (1993). Myocyte-specific enhancer-binding factor (MEF-2) regulates alpha-cardiac myosin heavy chain gene expression in vitro and in vivo. *J. Biol. Chem.* **268,** 19512–19520.

Molkentin, J. D., Kalvakolanu, D. V., and Markham, B. E. (1994). Transcription factor GATA-4 regulates cardiac muscle-specific expression of the alpha-myosin heavy-chain gene. *Mol. Cell. Biol.* **14,** 4947–4957.

Molkentin, J. D., Lin, Q., Duncan, S. A., and Olson, E. N. (1997). Requirement of the transcription factor GATA4 for heart tube formation and ventral morphogenesis. *Genes Dev.* **11,** 1061–1072.

Morkin, E. (1993). Regulation of myosin heavy chain genes in the heart. *Circulation* **87,** 1451–1460.

Morrisey, E. E., Ip, H. S., Tang, Z., and Parmacek, M. S. (1997). GATA-4 activates transcription via two novel domains that are conserved within the GATA-4/5/6 subfamily. *J. Biol. Chem.* **272,** 8515–8524.

Moss, J. B., McQuinn, T. C., and Schwartz, R. J. (1994). The avian cardiac alpha-actin promoter is regulated through a pair of complex elements composed of E boxes and serum response elements that bind both positive- and negative-acting factors. *J. Biol. Chem.* **269,** 12731–12740.

Nikovits, W., Jr., Wang, G. F., Feldman, J. L., Miller, J. B., Wade, R., Nelson, L., and Stockdale, F. E. (1996). Isolation and characterization of an avian slow myosin heavy chain gene expressed during embryonic skeletal muscle fiber formation. *J. Biol. Chem.* **271,** 17047–17056.

O'Brien, T. X., Lee, K. J., and Chien, K. R. (1993). Positional specification of ventricular myosin light chain 2 expression in the primitive murine heart tube. *Proc. Natl. Acad. Sci. U.S.A.* **90,** 5157–5161.

Reecy, J. M., Yamada, M., Cummings, K., Sosic, D., Chen, C. Y., Eichele, G., Olson, E. N., and Schwartz, R. J. (1997). Chicken *Nkx-2.8:* A novel homeobox gene expressed in early heart progenitor cells and pharyngeal pouch-2 and -3 endoderm. *Dev. Biol.* **188,** 295–311.

Rindt, H., Gulick, J., Knotts, S., Neumann, J., and Robbins, J. (1993). *In vivo* analysis of the murine *beta-myosin heavy chain* gene promoter. *J. Biol. Chem.* **268,** 5332–5338.

Ross, R. S., Navankasattusas, S., Harvey, R. P., and Chien, K. R. (1996). An HF-1a/HF-1b/MEF-2 combinatorial element confers cardiac ventricular specificity and established an anterior-posterior gradient of expression. *Development (Cambridge, UK)* **122,** 1799–1809.

Sartorelli, V., Hong, N. A., Bishopric, N. H., and Kedes, L. (1992). Myocardial activation of the human *cardiac alpha-actin* promoter by helix-loop-helix proteins. *Proc. Natl. Acad. Sci. U.S.A.* **89,** 4047–4051.

Satin, J., Fujii, S., and DeHaan, R. L. (1988). Development of cardiac beat rate in early chick embryos is regulated by regional cues. *Dev. Biol.* **129,** 103–113.

Seidman, C. E., Schmidt, E. V., and Seidman, J. G. (1991). cis-dominance of rat *atrial natriuretic factor* gene regulatory sequences in transgenic mice. *Can. J. Physiol. Pharmacol.* **69,** 1486–1492.

Srivastava, D., Thomas, T., Lin, Q., Kirby, M. L., Brown, D., and Olson, E. N. (1997). Regulation of cardiac mesodermal and neural crest development by the bHLH transcription factor, dHAND [see comments]. *Nat. Genet.* **16,** 154–160.

Stainier, D. Y., and Fishman, M. C. (1992). Patterning the zebrafish heart tube: Acquisition of anteroposterior polarity. *Dev. Biol.* **153,** 91–101.

Stainier, D. Y., Fouquet, B., Chen, J. N., Warren, K. S., Weinstein, B. M., Meiler, S. E., Mohideen, M. A., Neuhauss, S. C., Solnica-Krezel, L., Schier, A. F., Zwartkruis, F., Stemple, D. L., Malicki, J., Driever, W., and Fishman, M. C. (1996). Mutations affecting the formation and function of the cardiovascular system in the zebrafish embryo. *Development (Cambridge, UK)* **123,** 285–292.

Subramaniam, A., Jones, W. K., Gulick, J., Wert, S., Neumann, J., and Robbins, J. (1991). Tissue-specific regulation of the *alpha-myosin heavy chain* gene promoter in transgenic mice. *J. Biol. Chem.* **266,** 24613–24620.

Subramaniam, A., Gulick, J., Neumann, J., Knotts, S., and Robbins, J. (1993). Transgenic analysis of the thyroid-responsive elements in the *alpha-cardiac myosin heavy chain* gene promoter. *J. Biol. Chem.* **268,** 4331–4336.

Sucov, H. M., Dyson, E., Gumeringer, C. L., Price, J., Chien, K. R., and Evans, R. M. (1994). RXR *alpha* mutant mice establish a genetic basis for vitamin A signaling in heart morphogenesis. *Genes Dev.* **8,** 1007–1018.

Thompson, W. R., Nadal-Ginard, B., and Mahdavi, V. (1991). A MyoD1-independent muscle-specific enhancer controls the expression of the beta-myosin heavy chain gene in skeletal and cardiac muscle cells. *J. Biol. Chem.* **266,** 22678–22688.

Thuerauf, D. J., Hanford, D. S., and Glembotski, C. C. (1994). Regulation of rat brain natriuretic peptide transcription. A potential role for GATA-related transcription factors in myocardial cell gene expression. *J. Biol. Chem.* **269,** 17772–17775.

Umesono, K., Murakami, K. K., Thompson, C. C., and Evans, R. M. (1991). Direct repeats as selective response elements for the thyroid hormone, retinoic acid, and vitamin D_3 receptors. *Cell (Cambridge, Mass.)* **65,** 1255–1266.

Walters, M. R., Wicker, D. C., and Riggle, P. C. (1986). 1,25-Dihydroxyvitamin D_3 receptors identified in the rat heart. *J. Mol. Cell. Cardiol.* **18,** 67–72.

Wang, G. F., Nikovits, W., Schleinitz, M., and Stockdale, F. E. (1996). Atrial chamber-specific expression of the slow myosin heavy chain 3 gene in the embryonic heart. *J. Biol. Chem.* **271,** 19836–19845.

Weishaar, R. E., and Simpson, R. U. (1987). Vitamin D_3 and cardiovascular function in rats. *J. Clin. Invest.* **79,** 1706–1712.

Wu, J., Garami, M., Cheng, T., and Gardner, D. G. (1996). $1,25(OH)_2$ vitamin D_3, and retinoic acid antagonize endothelin-stimulated hypertrophy of neonatal rat cardiac myocytes. *J. Clin. Invest.* **97,** 1577–1588.

Yatskievych, T., Ladd, A., and Antin, P. (1997). Induction of cardiac myogenesis in avian pregastrula epiblast: The role of the hypoblast and activin. *Development (Cambridge, UK)* **124,** 2561–2570.

Yoshizawa, T., Handa, Y., Uematsu, Y., Takeda, S., Sekine, K., Yoshihara, Y., Kawakami, T., Arioka, K., Sato, H., Uchiyama, Y., Masushige, S., Fukamizu, A., Matsumoto, T., and Kato, S. (1997). Mice lacking the vitamin D receptor exhibit impaired bone formation, uterine hypoplasia and growth retardation after weaning. *Nat. Genet.* **16,** 391–396.

Yutzey, K. E., and Bader, D. (1995). Diversification of cardiomyogenic cell lineages during early heart development. *Circ. Res.* **77,** 216–219.

Yutzey, K. E., Rhee, J. T., and Bader, D. (1994). Expression of the atrial-specific myosin heavy chain AMHC1 and the establishment of anteroposterior polarity in the developing chicken heart. *Development (Cambridge, UK)* **120,** 871–883.

Yutzey, K., Gannon, M., and Bader, D. (1995). Diversification of cardiomyogenic cell lineages in vitro. *Dev. Biol.* **170,** 531–541.

Zeller, R., Bloch, K. D., Williams, B. S., Arceci, R. J., and Seidman, C. E. (1987). Localized expression of the atrial natriuretic factor gene during cardiac embryogenesis. *Genes Dev.* **1,** 693–698.

Zhou, M. D., Sucov, H. M., Evans, R. M., and Chien, K. R. (1995). Retinoid-dependent pathways suppress myocardial cell hypertrophy. *Proc. Natl. Acad. Sci. U.S.A.* **92,** 7391–7395.

Zou, Y., Evans, S., Chen, J., Kuo, H. C., Harvey, R. P., and Chien, K. R. (1997). CARP, a cardiac ankyrin repeat protein, is downstream in the Nkx2-5 homeobox gene pathway. *Development (Cambridge, UK)* **124,** 793–804.

VII

Heart Patterning: Left–Right Asymmetry

21

Establishing Cardiac Left–Right Asymmetry

H. Joseph Yost

Huntsman Cancer Institute Center for Children, Departments of Oncological Sciences and Pediatrics, University of Utah, Salt Lake City, Utah 84112

I. Overview of Cardiac Development Preceding Looping

All vertebrates share analogous developmental steps preceding the formation and looping of a cardiac tube (Icardo, 1989; Brown and Wolpert, 1990; Yost, 1991, 1995b). Differences in the geometry of the embryos, for example, spherical *Xenopus* embryos compared to discoid avian embryos, result in distinct spatial positioning of the cardiac primordia with respect to other primordia in the embryo. However, the general steps in cardiac morphogenesis appear to be similar. The developmental timing of these steps in humans, mice, and frogs is compared in Fig. 1.

The presumptive cardiac cells arise as a pair of lateral mesoderm primordia at the beginning of gastrulation. Explant and tissue recombination experiments in *Xenopus* indicate that signals derived from the dorsal midline mesoderm and endoderm cells induce dorsolateral mesoderm cells to become precardiac mesoderm cells (Sater and Jacobson, 1989, 1990; Muslin and Williams, 1991; Nascone and Mercola, 1995; see Chapter 3). The molecular identities of these cell–cell signals are not known. However, experiments in chick implicate *BMP*s and *activin* (Schultheiss *et al.,* 1997; Yatskievych *et al.,* 1997) and cardiac mesoderm can be induced in *Xenopus* animal cap ectoderm explants by high levels of activin (Logan and Mohun, 1993). There has been controversy regarding the roles of endoderm in cardiac mesoderm induction. However, results from *GATA-4* chimeras substantiate a role for endoderm in the development of cardiac cells (Kuo *et al.,* 1997; Molkentin *et al.,* 1997; Narita *et al.,* 1997; see Chapter 17).

During gastrulation, the precardiac mesoderm involutes with other axial mesoderm so that at the end of gastrulation the cardiac primordia are in a dorsoante-

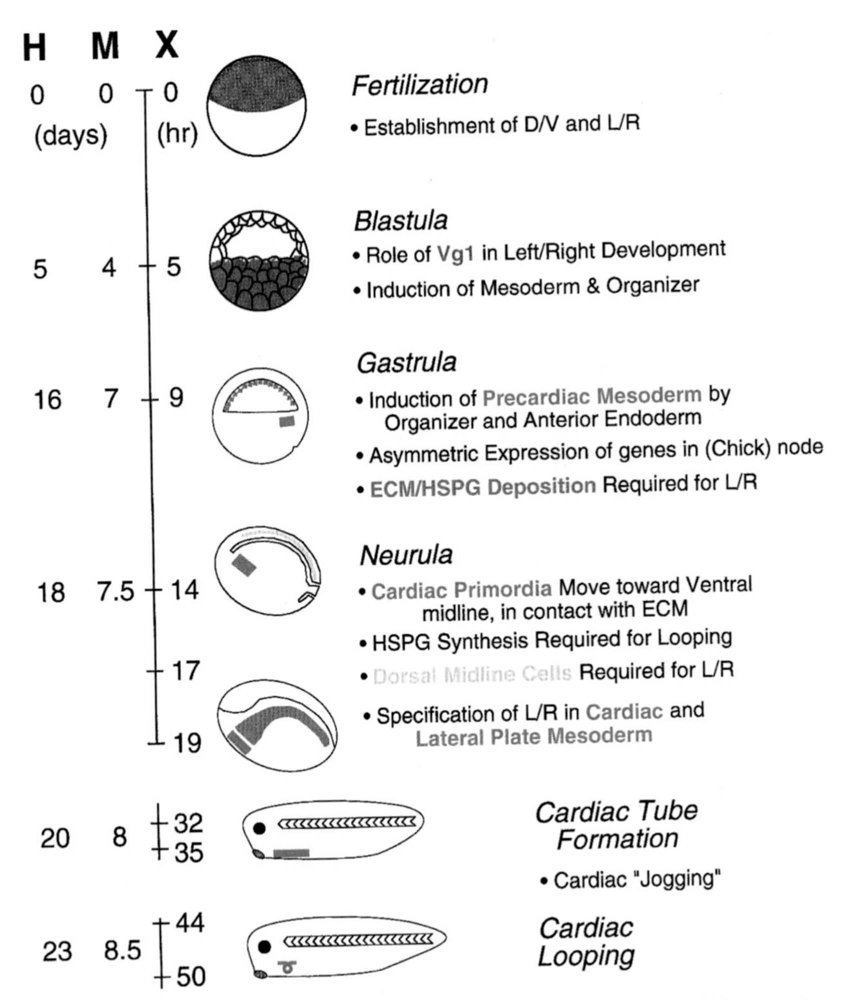

Figure 1 Timeline of cardiac L–R development in human, mouse, and *Xenopus* development. Days postfertilization in human (H) and mouse (M) embryos and hours postfertilization in *Xenopus* (X) embryos are listed on the left. Schematic lateral views of *Xenopus* embryos and landmarks in development are shown on the right. The animal hemisphere in *Xenopus* eggs (top) is indicated by darker pigmentation than that of the vegetal hemisphere. *Vg1* RNA and protein are localized in the vegetal hemisphere (violet). At the blastula stage, the blastocoel is formed as an internal space within the animal hemisphere. The precardiac mesoderm primordia (indicated by red bar) are just lateral of the dorsal midline and ingress during gastrulation. At the beginning of gastrulation, HSPG-containing matrix (indicated by green x's) is deposited on the basal surface of animal hemisphere cells (the roof of the blastocoel). At the end of gastrulation, cardiac primordia are in a dorsoanterior position, and midline (notochord) cells differentiate (orange). During neurula stages, the cardiac primordia move ventrally in contact with the matrix deposited during gastrulation and fuse at the ventral midline. *Nodal* is expressed in the left lateral plate mesoderm (blue). Subsequently, the cardiac mesoderm forms a cardiac tube that jogs (in mice and zebrafish) and loops along the embryonic L–R axis (adapted and updated from Yost, 1991).

rior position. In the chick, the node is asymmetric in morphology (Wetzel, 1929; Hoyle *et al.*, 1992; Cooke, 1995) and in gene expression patterns (Levin *et al.*, 1995). Previous fate mapping experiments suggest that there are asymmetric contributions of cells to regions of the cardiac mesoderm primordia (Zwirner and Kuhlo, 1964; Stalsberg, 1969), but more refined techniques are required to assess whether differential cell contributions are important for left–right (L–R) development. In *Xenopus*, extracellular matrix and cell surface heparan sulfate proteoglycans are deposited on the animal hemisphere ectoderm basolateral surface during gastrulation and have been implicated in L–R development (Yost, 1992).

During the neurula stages, the paired heart primordia move from a dorsolateral position to the ventral midline. In some vertebrates, the paired cardiac primordia form primitive tubes before they fuse at the ventral midline. Fusion occurs in an anterior to posterior progression. In *Xenopus,* the cardiac primordia move as paired sheets of cells across the animal hemisphere ectoderm and matrix to reach the ventral midline where they fuse into a continuous sheet (Keller, 1975, 1976; Yost, 1990, 1992). Transcription factors of the homeobox-containing *tinman* (*Nkx* or *Csx*) family and the *GATA-4/5/6* family are expressed in cardiogenic mesoderm, and to various extents in adjacent tissues, beginning during this period (Kelley *et al.,* 1993; Komuro and Izumo, 1993; Logan and Mohun, 1993; Drysdale *et al.,* 1994; Tonissen *et al.,* 1994; Evans *et al.,* 1995; Cleaver *et al.,* 1996; Jiang and Evans, 1996; Gove *et al.,* 1997; see Chapter 3). Explants of presumptive cardiac mesoderm (and overlying ectoderm) from gastrula-stage embryos are capable of forming beating cardiac tissue in culture (Sater and Jacobson, 1989). Both the expression of early cardiac molecular markers and the ability to form cardiac mesoderm in explants indicate that specification of mesoderm to form cardiomyocytes has occurred by the end of gastrulation.

Mesoderm cells acquire cardiac mesoderm identity during gastrulation. Left–right patterning within the cardiac field occurs during the open neural plate stages. It is quite striking that cells in the heart primordia are specified for left or right well before L–R morphogenesis occurs. In *Xenopus,* this occurs when the primordia is a simple sheet of mesoderm, shortly after fusion of the paired cardiac primordia at the ventral midline (Danos and Yost, 1995). Explanted cardiac primordia form cardiac tubes that loop *in vitro.* Explants made at the end of gastrulation have randomized orientation of cardiac looping. Explants made a few hours later, as the neural tube closes, form a cardiac tube that loops in the normal L–R orientation *in vitro.* Thus, the transmission of L–R signals from other parts of the embryo to the cardiac primordium is complete by the time the neural tube closes.

The cardiac tube is relatively symmetric along the ventral midline, contains distinct endocardium and myocardium, and extends anteriorly to the prospective aortic arches and posteriorly to the liver diverticulum. This apparent bilateral symmetry is broken by consistent looping of the tube to the right side of the embryo, occurring at 44–50 hr postfertilization in *Xenopus,* approximately 8.5 days in mice, and 23 days in humans. Recently, it has been observed that the cardiac tube undergoes an earlier morphological expression of L–R asymmetry. In zebrafish and mice, the cardiac tube appears to rapidly and transiently shift to the left (Biben and Harvey, 1997; Chen *et al.,* 1997). In zebrafish, the cardiac tube shifts back to a central position before looping to the right. This transient shift has been termed cardiac "jogging" (Chen *et al.,* 1997) and appears to be highly predictive of the subsequent orientation of cardiac looping, at least in zebrafish (Chen *et al.,* 1997; see Chapter 6). Subsequent cardiac tube looping places the prospective atrium, derived from the posterior portion of the cardiac tube, dorsal and anterior to the prospective ventricle. By this stage the heart is circulating blood, and valve and septation formation is under way.

II. Left–Right Nomenclature

It is useful to make a nomenclatural distinction between "left–right axis formation" and "left–right development." Use of the term axis formation reflects the widely held but poorly enunciated view that there is an early, embryowide (global) mechanism for establishing a difference between left body side and right body side and that this underlying mechanism directs the subsequent L–R development and morphogenesis of every asymmetric organ system in a coordinated fashion. The result of normal L–R axis formation is the consistent L–R orientation of internal organs with respect to the orthogonal body axes, which appears to be highly conserved in vertebrates. It is useful to limit the term L–R axis formation to the initial mechanisms that establish embryowide (global) L–R asymmetries across an otherwise bilaterally symmetric body plan. In contrast, L–R development is the reception and playing out of this global axial information by individual primordia, such as groups of cells that form the heart, segments of the gut, or brain. In general, there are four outcomes of embryological or genetic manipulations that alter L–R development, each of which is consistent with having altered either L–R axis formation or the response of individual primordia to L–R axis information.

A. Inversion

In rare situations, L–R axis information can be "inverted," such that genes normally expressed on the left are now expressed only on the right, and the orientation of all the organs is a mirror image of normal. This biological enantiomer of a normal embryo occurs in *inv/inv* mice (Yokoyama *et al.,* 1993) and in *situs inversus totalis* in humans (Bowers *et al.,* 1996; see Chapters 22 and 25). The molecular identities of the mutated genes have not been reported. Experimentally, complete inversion has only been produced by injection of *BVg1* in specific cell lineages in *Xenopus* (Hyatt and Yost, 1998).

B. Global Randomization

Elimination of normal L–R axis information throughout the embryo leads to "global randomization" of the orientation of each individual organ with respect to the orthogonal axes and with respect to each other. In the absence of L–R axial information, many organ primordia appear to retain an intrinsic ability to generate L–R asymmetry, but their orientation is randomized with respect to the orthogonal axes (dorsal–ventral and anterior–posterior) and with respect to what is occurring in other asymmetric organs. For experimental manipulations that appear to eliminate L–R axis information, individual embryos can have a heart in normal, reversed, or indeterminate orientation and intestinal coiling in normal, reversed, or indeterminate orientation, with no correlation between the orientation of the heart and intestine (Yost, 1992; Hyatt *et al.,* 1996). Similarly, the orientation of the heart is not statistically concordant with the orientation of the viscera in *iv/iv* mice (Brown and Wolpert, 1990; Seo *et al.,* 1992) and in human *heterotaxia* (Bowers *et al.,* 1996).

C. Isolated Randomization

A third outcome is randomization of the L–R orientation of an individual organ without alteration of the normal orientation of surrounding organs and perhaps without the alteration of normal asymmetric expression patterns of upstream genes. This is seen as a result of experimental manipulation of small patches of extracellular matrix across which individual primordia migrate in *Xenopus* embryos (Yost, 1992) and in human cases of isolated laterality defects, such as isolated cardiac reversals and isolated asplenia/polysplenia (Bowers *et al.,* 1996). In these cases, it is likely that L–R axis formation and the production of global L–R signals are normal but that the affected organ primordium has lost its ability either to receive the global L–R signals or to use those signals during organ morphogenesis.

D. Failure of L–R Morphogenesis

A fourth outcome is failure to generate L–R asymmetry. The most simple example is a cessation of cardiac tube looping. When *Xenopus* embryos are treated with β-xyloside to block proteoglycan synthesis during the neurula stages, the cardiac tube fails to generate L–R asymmetry later in development (Yost, 1990). Treatment of chick embryos with antisense oligonucleotides against two transcription factors that are normally expressed in the heart primordium (and elsewhere), *dHAND* and *eHAND* (Srivastava *et al.,* 1995), or knockout of homeobox-containing *Nkx2-5* in mice (Lyons *et al.,* 1995), results in cessation of cardiac tube looping. Due to the developmental stages during which these manipulations have an effect, it is likely that these treatments block the pathways that lead to organ L–R morphogenesis and not the earlier steps of L–R axis formation.

III. Discovery of Molecular Signals That Precede Cardiac L–R Specification

One of the most significant recent findings in cardiac L–R development is that the earliest molecular events that determine cardiac orientation occur by cell–cell signaling pathways that do not involve the cardiac primordia. Genes that are not expressed within perspective heart cells have a profound effect on the formation of the heart. This was seen in the observation that perturbations of extracellular matrix in cells that do not contribute to the heart lead to loss of normal L–R development in the cardiac (and other) organ primordia in *Xenopus* (Yost, 1992). However, the breakthrough in describing the molecular identities of genes that regulate cardiac orientation occurred with the description of several genes that are expressed asymmetrically near the time of gastrulation in the chick. Most of these genes are not expressed in cardiac precursor cells, but experimental alterations of the expression patterns of some of these genes can alter cardiac orientation (Levin *et al.,* 1995). Subsequently, genes in various vertebrates have been described that are either asymmetrically expressed along the L–R axis or perturb L–R development when either ectopically expressed or mutated. An attempt to synthesize the pathway of genes that lead to L–R morphogenesis, according to their temporal and spatial expression patterns in various vertebrates, is shown in Fig. 2. Experiments with some of these genes are described in the following sections.

A. Studies in Chick and the Role of the Node

Hensen's node (HN) of chick embryos is analogous to the Organizer in frog embryos. This group of cells is located at the midline and drives both gastrulation and the progressive formation of the anterior–posterior axis. The cells within the center of the node, during and after gastrulation, give rise to the notochord and other

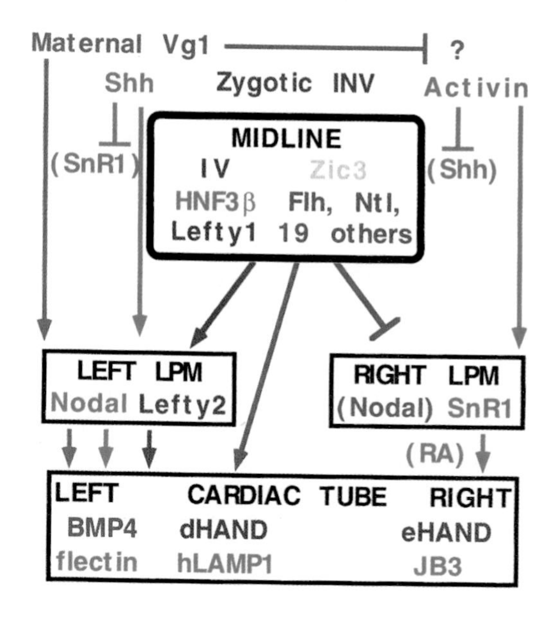

Figure 2 Genes implicated in L–R development. Genes from *Xenopus* (blue), chick (green), zebrafish (red), mouse (violet), and human (orange) are shown with the earliest genes in the pathway at the top. Positions of genes are assigned according to their relative expression periods in development and known interactions. Proposed positive and negative interactions are indicated by arrows and bars, respectively. A gene in parentheses indicates suppression of that gene by upstream signals. Genes important for both midline development and cardiac L–R development are in the upper central box (labeled midline). The pathway of genes within the midline has not been elucidated for L–R development. Genes expressed in left or right lateral plate mesoderm (LPM) are indicated in the side boxes. Genes expressed asymmetrically in the cardiac tube, which are probably effectors of asymmetric organogenesis, are in the lower box. Note that the color assigned to each gene indicates that organism in which the gene was first discovered or whether the gene has been implicated in L–R development in other organisms (see text for details and references).

axial structures. The node displays a transient but consistent morphological L–R asymmetry (Cooke, 1995). By RNA *in situ* hybridization analysis, at least six RNAs have been shown to be asymmetrically distributed along the L–R axis in chick embryos during gastrulation and neurula stages, predominantly either in or near HN. It is striking that most of these asymmetrically expressed genes encode cell–cell signaling proteins that have been implicated in other embryonic patterning events. Cell signaling genes *activin βB* (Levin *et al.*, 1997), *Sonic hedgehog* (*Shh*), and a *Nodal*-related gene (*cNR1*) (Levin *et al.*, 1995) are asymmetrically expressed, as are a transmembrane receptor for activin (*cAct-RIIa*) and transcription factors *HFN3β* (Levin *et al.*, 1995) and *Snr1* (Isaac *et al.*, 1997). Other genes, including the cell signaling gene *cWnt-8c* and the receptor patched (*PTC*), have asymmetric expression patterns in HN (Levin, 1997), but altered expression experiments with these genes have not been reported.

Based on the developmental timing of asymmetric expression patterns, a cascade of signals has been proposed from studies of chick embryos (Levin *et al.*, 1995) (Fig. 2, green). *Activin βB,* a member of the transforming growth factor-β (TGF-β) superfamily, is expressed on the right side of HN from stage 3 to stage 5^+ (Levin *et al.*, 1997). This precedes and probably induces the expression of the activin receptor *cAct-RIIa* on the right side of primitive streak and then in the ectoderm on the right side of HN (Levin *et al.*, 1995), and downregulates the expression of *shh* (a member of the *hedgehog* family of cell signaling peptides) on the right side. Left-only expression of *shh* in the ectoderm of HN from stages 4^+ through 7 induces the reappearance of chick *cNR1* at stage 7 just on the left side, lateral and anterior to the *Shh*-expressing region in HN. The expression of *cNR1* then expands in the left, but not the right, lateral plate mesoderm. Slightly later and more posterior, the transcription factor *snail-related zinc finger gene* (*cSnr1*) is expressed in right lateral plate mesoderm and the right cardiac primordium before fusion with the left primordium (Isaac *et al.*, 1997). *cSnr1* is also expressed symmetrically in the segmenting somites and pharyngeal endoderm. The mutually exclusive asymmetric expressions of *cNR1* and *cSnr1* then signal the heart and visceral primordia to provide orientation during subsequent asymmetric organogenesis.

This proposed cascade has been partially tested by ectopic expression experiments (Levin *et al.*, 1995, 1997) in which candidate proteins were released from beads soaked in purified protein or from pellets of cells transformed to express a candidate protein. Beads or cells were implanted on the right or left side of HN in the embryo, and subsequent gene expression patterns and organ orientation were assessed. Ectopic expression of *activin* on the left side induces *cAct-RIIa* and suppresses *shh* on the left side. This results in bilateral expression of *cAct-RIIa,* the absence of *shh* on either side, the absence of *cNR1* on either side, and randomized heart orientation (Levin *et al.*, 1995). However, the observation that ectopic *activin* expression results in a variety of symmetric and asymmetric expression patterns of *cNR1* and *cSnr1* suggests that ectopic *activin* destabilizes the L–R pathway and leads to independent reinitiation of the pathway on each side of the embryo (Isaac *et al.*, 1997). Ectopic expression of *follistatin,* which antagonizes the activity of activin (and some other TGF-β members) resulted in bilateral expression of *shh* in 5 of 20 cases (Levin *et al.*, 1997). The effects of follistatin on heart orientation have not been reported.

Ectopic expression of *shh* on the right side induces *cNR1* expression in the right lateral plate (giving bilateral expression), eliminates right-sided expression of *cSnr1,* and randomizes cardiac L–R orientation (Levin *et al.,* 1995; Isaac *et al.,* 1997). Expression of *cNR1* on the right side (giving bilateral *nodal* expression) results in a higher frequency of bilaterally symmetric or inverted hearts (Levin *et al.,* 1997). In each of these experiments, control implants of either untreated beads or nontransfected cells had no significant effect, and implants on contralateral sides had no significant effect. Treatment with *cSnr1* antisense oligonucleotides before heart tube formation did not alter *cNR1* asymmetric expression but resulted in a 30–50% incident of heart reversals, indicating that *cSnr1* is downstream of *cNR1* or in a parallel pathway. The results from ectopic expression experiments and antisense oligonucleotide treatments indicate that in chick embryos there is a cascade of cell–cell signaling factors, from *activin* to *shh* to *cNR1* and *cSnr1,* that establish L–R orientation prior to heart morphogenesis.

B. Conserved Expression Pattern of Nodal in Lateral Plate Mesoderm

It is not known whether the pathway elucidated in chick embryos is conserved in other vertebrates. Some of the genes shown to have asymmetric expression patterns in chick embryos have been shown to have symmetric expression patterns in other embryos. This could be due to the technical limitations of RNA *in situ* hybridization, transient asymmetric expression patterns that could be missed, gene expression regulation at the posttranscriptional level, or a nonconserved pathway in chick embryos. Currently, *nodal* has the earliest known molecular expression pattern that is conserved in all vertebrates examined; *cNR1* in chick (Levin *et al.,* 1995), *Xnr-1* in *Xenopus* (Hyatt *et al.,* 1996; Lowe *et al.,* 1996; Lohr *et al.,* 1997), and *nodal* in mouse (Collignon *et al.,* 1996; Lowe *et al.,* 1996; Meno *et al.,* 1996). In all three organisms, altered expression of *nodal* is correlated with altered cardiac L–R orientation. Furthermore, in *Xenopus* and chick embryos, ectopic expression of *nodal* alters cardiac orientation, indicating that *nodal* is directly in the pathway of cardiac L–R development (Levin *et al.,* 1997; Sampath *et al.,* 1997). Right-sided injection of a plasmid that expresses *Xnr-1* after the midblastula stages results in perturbed L–R development in the heart and viscera. Left-sided injections or ectopic expression of *nodal*-related genes *Xnr-2* and *Xnr-3* does not affect L–R development (Sampath *et al.,* 1997). Thus, it appears that asymmetric expression of *nodal* in the left lateral plate mesoderm is upstream of normal cardiac L–R morphogenesis in vertebrates. *Nodal* expression appears to be proximal to the specification of cardiac L–R asymmetry and near the end of the L–R signaling pathway in the embryo (Fig. 2). While it is possible that the signaling cascades upstream of *nodal* asymmetric expression are distinct in different vertebrates, it seems unlikely that the highly conserved cardiac orientation and asymmetric expression of *nodal* are products of the evolutionary convergence of distinct signaling pathways that each produce asymmetric *nodal* expression.

Studies of the hedgehog family of cell signaling molecules provide an example of the difficulties in comparative analysis of L–R development among vertebrate embryos. As described previously, *shh* is asymmetrically expressed in chick embryos and ectopic expression of *shh* alters subsequent asymmetric gene expression and cardiac orientation (Levin *et al.,* 1995). However, laterality defects have not been reported in mice embryos that are mutant for *shh* (Chiang *et al.,* 1996) nor in the zebrafish mutant *sonic-you* (Brand *et al.,* 1996; Chen *et al.,* 1997). Although *hedgehog* RNAs have not been found to be asymmetrically expressed in *Xenopus,* zebrafish (Ekker *et al.,* 1995), or mice (Collignon *et al.,* 1996), there is evidence that perturbation of a *hedgehog*-related signaling pathway in *Xenopus* and zebrafish can alter L–R development (Chen *et al.,* 1997; Sampath *et al.,* 1997). *Banded hedgehog* (*Xbhh*) is a member of the hedgehog family of signaling proteins that is expressed in the gastrula stage, and the *Xbhh-N* construct encodes the active N-terminal cleavage product, ensuring its strong expression (Ekker *et al.,* 1995). Right-sided injections of *Xbhh-N* RNA at the four-cell stage, but not left-sided injections, perturb *Xnr-1* expression (Sampath *et al.,* 1997). Similarly, injection of *sonic hedgehog* RNA into zebrafish embryos at the one- or two-cell stage alters asymmetric expression of *BMP4* in the heart and L–R cardiac morphogenesis (Chen *et al.,* 1997). Thus, ectopic expression results in chick, *Xenopus,* and zebrafish suggest that a *hedgehog*-like signaling pathway is involved in cardiac L–R development, but mutant analysis eliminates some members of the *hedgehog* family from direct roles in L–R development.

IV. Initiation of the L–R Axis before Node Formation

A. Early Signals in *Xenopus* and the "Left–Right Coordinator"

While considerable progress has been made in describing later events in L–R development (Fig. 2), comparably little is known about the mechanism that initiates L–R axis formation. It is not known how the initial asymmetry that establishes the L–R axis is generated, but it is likely that this occurs before the asymmetric cell–cell signaling cascade develops in the node. To explore earlier cell signaling events, I shift attention from the chick to the frog (*Xenopus laevis*) embryo, which can be manipulated immediately after fertilization. A distinct experimental advantage of amphibian embryos is that specific cell lineages can be targeted for ectopic expression by microinjection of large cells up to the 64-cell stage. RNAs encoding molecules that are candidates for early L–R signaling have been injected into individual cells at the 16-cell stage. The results suggest that cell lineages established at the 16-cell stage are distinct in their response to ectopic gene expression (Hyatt *et al.*, 1996; Hyatt and Yost, 1998). The most striking results indicate that *Vg1*, a member of the TGF-β family of signaling proteins, plays a role in initiating the L–R axis.

Vg1 RNA was the first RNA found to be asymmetrically distributed in vertebrate oocytes or embryos, in this case along the animal–vegetal axis (Weeks and Melton, 1987), and has been implicated in both mesoderm formation (Dale *et al.*, 1993; Thomsen and Melton, 1993) and L–R development (Hyatt *et al.*, 1995). During oogenesis, *Vg1* RNA is synthesized and localized to the vegetal hemisphere so that it is inherited in vegetal cells in blastula-stage embryos. The vegetal cells give rise to the embryonic endoderm. Like other members of the TGF-β family, Vg1 is synthesized as a pro-protein and is processed to an active mature form. If it occurs, processing of pro-Vg1 is in a restricted region of the *Xenopus* embryo. Vg1 pro-protein is readily detected in left and right cells of blastula-stage embryos (H. J. Yost, data unpublished), but mature protein is difficult to detect in either normal embryos or embryos injected with *Vg1* RNA to synthesize more pro-protein (Dale *et al.*, 1993; Thomsen and Melton, 1993). Mature Vg1 protein can be made in detectable amounts by injection of chimeric RNAs that provide different processing sites, derived from other members of the TGF-β family, fused to the coding region of the mature Vg1 peptide (Dale *et al.*, 1993; Thomsen and Melton, 1993; Kessler and Melton, 1995). For example, BVg1 is a chimera that provides mature Vg1 peptide via the BMP protein processing pathway and AVg is a chimera utilizing the activin pathway.

Injection of a modified *Vg1* construct (*BVg1* or *AVg*) that ensures high levels of *Vg1* expression has significant effects on L–R development. Injection of 16-cell embryos in the dorsal vegetal cell on the right side, but not the left side, of the dorsal midline randomizes heart and gut orientation and induces expression of *Xenopus nodal* (*Xnr-1*) in the right lateral plate (Hyatt *et al.*, 1996; Hyatt and Yost, 1998). The effects of *Vg1* expression are more dramatic when *BVg1* or *AVg* are injected into right lateral cells, 90° from the midline, in the 16-cell embryo: The L–R axis of the embryo is inverted, *Xnr-1* is repressed in the left lateral plate mesoderm and induced in the right lateral plate mesoderm, and the L–R orientation of the heart and gut is inverted (Hyatt and Yost, 1998). This is the only reported experimental manipulation that fully inverts the L–R axis.

The current working model is that the processing of Vg1 pro-protein into mature Vg1 is spatially regulated along the L–R axis, and that the processing of Vg1 initiates the L–R signaling cascade (Fig. 3). In this model, endogenous Vg1 pro-protein is processed on the left side but not the right side of the early embryo. Correspondingly, *Vg1* RNA injection into either left or right cells has no effect on L–R development. This is expected if the overexpressed Vg1 pro-protein cannot be processed on the right side. By using the BMP processing pathway (*BVg1*) or the activin processing pathway (*AVg*), the hypothesized constraint in Vg1 protein processing can be bypassed and mature Vg1 protein can be produced on the right side.

The observation that the L–R axis can be inverted by ectopic Vg1 expression, and the ability to alter L–R development by other perturbations of the *Vg1* signaling pathway (Hyatt *et al.*, 1996; Hyatt and Yost, 1998), led to the suggestion that there is a group of cells on the left lateral side that is capable of organizing the L–R axis in the embryo via the *Vg1* signaling pathway and interacting with the Spemann Organizer to establish the highly conserved relationships of the embryonic axes. This group of cells has been termed the Left–Right Coordinator (L–R Coordinator) (Hyatt and Yost, 1998) in analogy with the Spemann Organizer (Spemann and Mangold, 1924) that regulates dorsoanterior development (see Section V,D).

These results suggest that asymmetric protein processing in the embryo can initiate L–R axis formation (Fig. 3). How is this asymmetry established? In *Xenopus*, dorsal–ventral axis formation is initiated by a cytoplasmic/cortical rotation in the single cell during the first cell cycle (Gerhart *et al.*, 1989). The rotation is normally dependent on a transient array of microtubules in the vegetal hemisphere of the fertilized egg during the first cell cycle (Gerhart *et al.*, 1989). Disruption of the vegetal microtubule array results in no rotation and ventralized embryos. Dorsal–ventral axis formation can be "rescued" in embryos in the absence of the vegetal microtubule arrays by tilting the eggs during the first cell cycle. Tilting leads to gravity-driven cytoplasmic rotation which rescues dorsal–ventral axis formation. These results indicate that it is the rotation per se, not the microtubules or sperm entry position, that establishes the dorsal–ventral axis. Strikingly, in embryos in which cytoplasmic rotation and dorsal–ventral axis for-

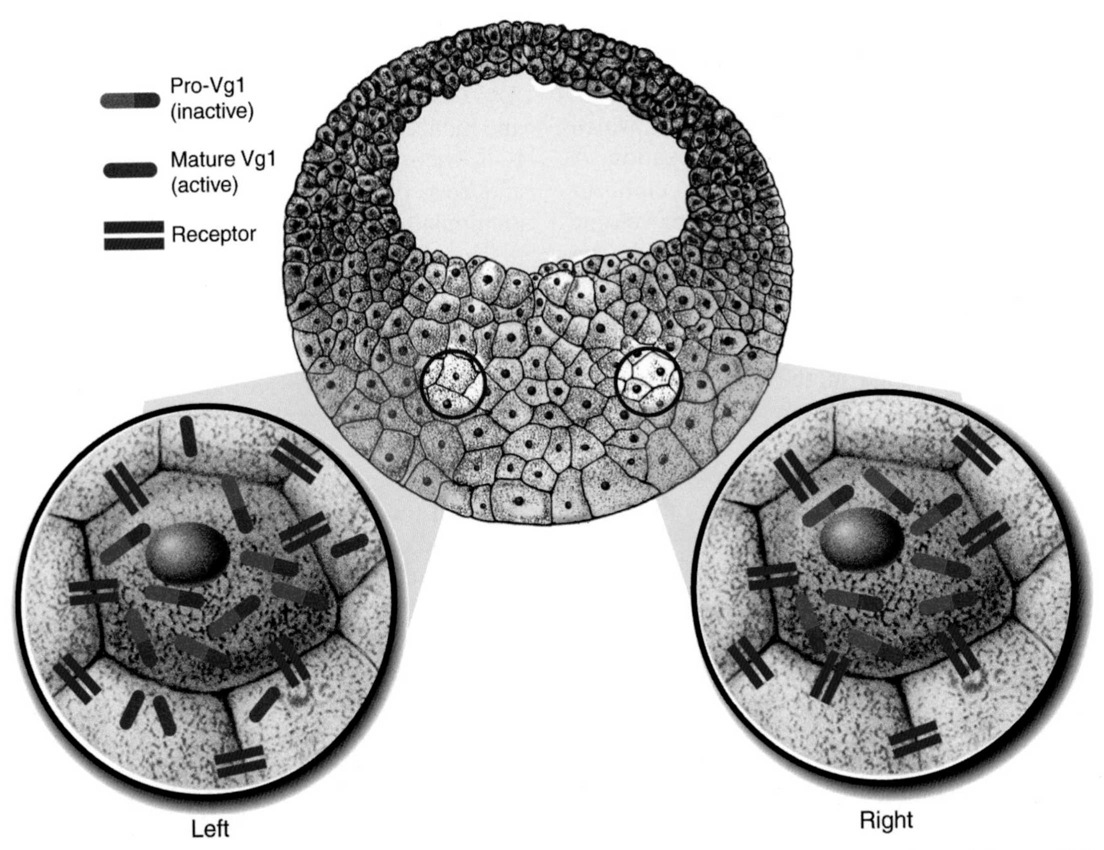

Figure 3 Model of asymmetric pro-Vg1 processing. *Vg1* RNA is stored in the vegetal hemisphere of the egg. *Vg1* RNA and pro-Vg1 protein (red/blue bars) and the unidentified Vg1 receptor (green bars) are present in left and right cells (Weeks and Melton, 1987; Dale *et al.*, 1993; Thomsen and Melton, 1993). It is hypothesized that processing of Vg1 into mature, active ligand (blue bars) occurs mostly on the left side. Expression of mature Vg1 activates the L–R coordinator (Hyatt *et al.*, 1996; Hyatt and Yost, 1998).

mation are "rescued" by tilting, normal L–R axis formation is not rescued (Yost, 1991). This suggests that the initiation of L–R axis formation occurs concurrently with the initiation of the dorsal–ventral axis, and that initiation of L–R axis formation is dependent on the transient microtubule array during the first cell cycle.

B. Early Signals in Chick and Mice

Two recent observations in chick embryos concur with the suggestion that L–R asymmetries originate in lateral cells (analogous to the L–R Coordinator in *Xenopus*) and are imposed upon the Organizer (or HN in chick). First, a node with normal L–R expression of *Shh* can be regenerated in an embryo from which the endogenous node was deleted (Psychoyos and Stern, 1996). Second, the asymmetric expression pattern of *Shh* in a node that was surgically inverted early in embryogenesis is oriented with respect to the lateral tissues and not with respect to the initial node orientation (Pagan and Tabin, 1998).

Insertional mutation of the mouse *INV* gene presents a surprising phenotype; *inv/inv* embryos have fully inverted L–R orientation of the heart, gut, and body axis (Yokoyama *et al.*, 1993; see Chapter 22). Although the molecular identity of *INV* has not been reported, and thus the timing of expression is unknown, *INV* is upstream of *nodal* (Collignon *et al.*, 1996; Lowe *et al.*, 1996). In Fig. 2, *INV* is placed in proximity with maternal *Vg1* because both *inv/inv* mice and *Xenopus* embryos in which *Vg1* is ectopically expressed in specific cell lineages (Hyatt and Yost, 1998) have the same phenotype; full inversion of cardiac and visceral L–R orientation and inversion of *nodal* expression pattern.

V. Role of the Midline

The L–R orientation of cardiac looping in vertebrates is conserved with respect to the anterior–posterior (cranial–caudal) and dorsal–ventral (back–front) axes. To an external observer, the L–R axis can be defined once the other axes are identified. This can lead to a mistaken assumption that establishing the other two axes is sufficient to establish the L–R axis. However, to a cell within an embryo, the cartesian coordinates defined by these axes are insufficient to provide information indicating that it is on the right or left side of the embryo unless the coordinates also provide vectoral information (Yost, 1995b). For example, a cell that has axial information indicating that it is at a position 30% along the anterior–posterior axis and position 80% on the dorsal–ventral axis can be on the left or right side at those defined coordinates.

The observation that reduced dorsoanterior development, produced by diverse treatments, is correlated with loss of normal L–R development suggests that a developmental mechanism coordinates the alignment of the three axes (Danos and Yost, 1995). A growing body of evidence implicates cells in the dorsal midline of the embryo, specifically notochord cells or cells induced by the notochord, in the coordination of the three geometric axes. It should be noted that cells at the midline of the embryo, for example, notochord and neural floorplate cells, are derived from analogous central structures during gastrulation: the Organizer (in frog), HN (in chick), the node (in mice), and the dorsal shield (in zebrafish).

A. Studies in *Xenopus* and Zebrafish

In *Xenopus,* direct extirpation of prospective notochord and floorplate cells during the open neural plate stages (stage 15–20) results in randomization of cardiac orientation (Danos and Yost, 1996) and bilateral expression of *Xnr-1* in lateral plate mesoderm (Lohr *et al.,* 1997). In control experiments, mock dissections or extirpations of midline tissues after the neural tube has closed but before cardiac tube formation do not alter cardiac orientation or *nodal* asymmetric expression. This provides embryological evidence that midline cells are required for L–R development during a period limited to the open neural plate stages but does not identify genes within the midline cells that are essential for this pro-cess. These stages concur with the period during which L–R orientation is specified in the cardiac primordia (Danos and Yost, 1995).

These embryological observations led to the hypothesis that zebrafish embryos with notochord defects would have randomized cardiac L–R development. Initially, two zebrafish mutants, *no tail* (*ntl*) and *floating head* (*flh*), with well-characterized midline defects that include the absence of notochord (Halpern *et al.,* 1993; Talbot *et al.,* 1995) were shown to have randomized cardiac orientation (Danos and Yost, 1996). *Ntl* encodes the zebrafish homolog of the mouse *brachyury* gene, a transcription regulatory factor expressed in dorsal midline mesoderm (Schulte-Merker *et al.,* 1994). *Flh* encodes a homeobox-containing transcription regulatory factor, a homolog of *Xenopus Xnot* (von Dassow *et al.,* 1993) and chick *Cnot* (Stein and Kessel, 1995), that is expressed in the organizer (or node) and the notochord (Talbot *et al.,* 1995). The observation that mutation of genes expressed in the dorsal cells results in randomization of L–R orientation of cardiac looping provides genetic evidence for the role of midline cells in L–R development (Danos and Yost, 1996).

Mutational analysis in zebrafish will provide more insight into the pathways that regulate cardiac development (Fishman and Chien, 1997; see Chapter 6). One of the most exciting efforts in developmental genetics has been the generation and characterization of a large collection of zebrafish mutants that are embryonic lethals and that have specific visible defects in various aspects of organogenesis (for initial reports of mutant screens in Tubingen and Boston, see *Development* **123**). In order to identify interactions between cardiac L–R development and other embryonic events, many zebrafish mutants were rescreened for alterations in cardiac L–R development (Chen *et al.,* 1997). This was aided by observations of two events that precede cardiac L–R looping. *BMP4,* a member of the TGF-β family of cell signaling molecules, is uniformly expressed in cardiac primordia. After cardiac primordia fusion, *BMP4* RNA becomes asymmetrically expressed, more abundantly on the left side of the heart tube. Subsequently, the prospective atrial end of the newly fused cardiac tube "jogs" to the left and then returns to the midline, followed by looping of the cardiac tube (predominantly the prospective ventricle) to the right. Experiments in which *BMP4* distribution was perturbed indicate that *BMP4* asymmetric expression is necessary for normal cardiac jogging and looping orientation (Chen *et al.,* 1997).

Among the mutants screened, only some dorsal–ventral mutants or midline mutants have laterality defects. Others have normal L–R development (Chen *et al.,* 1997). From the observation that some mutations affecting midline development have associated laterality defects and other midline mutants do not, one can speculate the early genes share a common pathway for midline development and L–R development. Manipulations that alter this common anteriodorsal and L–R pathway should yield coordinated failures in the development of both axes. At some point, before cardiac L–R specification during neural tube closure (Danos and Yost, 1996; Lohr *et al.,* 1997), the midline development pathway and the L–R development pathway diverge. Further genetic analysis of zebrafish mutants and the molecular identification of the genes should elucidate the mechanisms by which midline development influences L–R development.

B. Studies in Mice and Humans

The midline expression patterns of the genes *IV*, *lefty*, *HNF3β*, and *ZIC3*, underscore the importance of midline cells in cardiac L–R development. The *iv/iv* homozygote has randomized cardiac visceral L–R orientation, a phenotype that has greatly influenced modeling of L–R development (Brown and Wolpert, 1990; Seo *et al.*, 1992). Recently, the mouse *IV* gene has been shown to encode a dynein, termed L–R dynein (LRD), that is expressed early in embryogenesis and in the node at the midline (Supp *et al.*, 1997). Asymmetric expression of LRD has not been reported and it is unknown when the *IV* gene acts in the L–R pathway. However, genetic evidence indicates that *IV* is upstream of cardiac L–R orientation (Brueckner *et al.*, 1989) and both *nodal* and *lefty* asymmetric expression (Lowe *et al.*, 1996; Meno *et al.*, 1997; Fig. 2). Dyneins interact with microtubules for a variety of intracellular functions and for cilia motility. Patients with Kartegener's syndrome (Afzelius, 1976) have laterality defects and dynein defects in their cilia, predisposing them to respiratory problems and male infertility.

Two closely related genes, *lefty-1* and *lefty-2*, are asymmetrically expressed in mice at approximately the same stages as *nodal* (Meno *et al.*, 1996, 1997). *Lefty-1* is expressed in midline tissue, on the left side of the floorplate of the developing neural tube. *Lefty-2* is expressed in the left lateral plate mesoderm. Both of these genes are downstream of *IV* and *INV* (Meno *et al.*, 1996, 1997); phenotypes in *lefty* knockouts have not been reported. Mouse embryos that are double heterozygous for mutations in *HNF3β* and *nodal* display laterality defects (Collignon *et al.*, 1996). *HNF3β* is expressed in the midline in mice (Collignon *et al.*, 1996) and is transiently asymmetric in chick (Levin *et al.*, 1995).

Molecular genetic evidence from humans with familial situs ambiguous has identified an X-linked gene, *ZIC3* (Gebbia *et al.*, 1997; see Chapter 27). *ZIC3* is a member of the *ZIC* (zinc finger protein of the cerebellum) transcription factor family first described for specific expression in the adult mouse cerebellum and related to the *Drosophila* pair-rule gene *odd-paired* (Aruga *et al.*, 1996). Although asymmetric expression patterns have not been reported, *ZIC3* is expressed in the primitive streak and early neural tissues in mouse and *Xenopus* embryos (Aruga *et al.*, 1996; Nagai *et al.*, 1997; Nakata *et al.*, 1997), perhaps placing it in the midline pathway for L–R development (Fig. 2).

There is strong genetic evidence from zebrafish, mice, and humans that genes expressed in the midline are important for cardiac L–R development, yet asymmetric RNA expression patterns have not been detected for most of these genes. To date, evidence indicates that the genes in the midline pathway are upstream of asymmetric gene expression in the lateral plate mesoderm, and the asymmetric expression of lateral plate mesoderm genes is upstream of cardiac L–R specification. Several mechanisms have been postulated by which midline genes regulate asymmetric gene expression in the lateral plate mesoderm, including mechanical elongation of the embryo by the midline (notochord) (Yost, 1995a), which then might align molecules in an extracellular matrix known to be essential for both heart and gut L–R orientation (Yost, 1992). Alternatively, midline cells may act either as a source for signaling molecules that are transmitted to the lateral plate or as a barrier or sink (antagonist) for L–R signaling molecules. One interpretation of results in conjoined twins (see Section V,C) suggests that cell–cell signals on one side of the embryo cannot cross over the embryonic midline barrier but can influence adjacent tissues, including a conjoined embryo. If the midline emits either positive-inducing signals or antagonistic signals, it is not clear whether signals are asymmetrically emitted from the midline or whether symmetric midline signals interact with asymmetric signals or an asymmetric receptor distribution that is generated elsewhere in the embryo to give left-sided *nodal* expression and to orient heart and gut development.

A mutation in the gap junction gene *connexin 43* has been implicated in some human cardiac laterality defects (Britz-Cunningham *et al.*, 1995) but has not been detected in large screens of other individuals (Casey and Ballabio, 1995; Splitt *et al.*, 1995; Gebbia *et al.*, 1996; Penman Splitt *et al.*, 1997). However, it is interesting to note that a connexin gene in zebrafish is specifically expressed in the notochord and is downstream from the *ntl* gene (Essner *et al.*, 1996), which is necessary for both midline development and cardiac L–R development (Danos and Yost, 1996). The specific roles of gap junctions in L–R development await work on model systems.

The correlation of midline defects in frogs or fish with cardiac laterality defects suggests that complex syndromes involving both neural development defects and cardiac defects in humans might have a common developmental mechanism and genetic pathway. Individuals with laterality defects should be examined for subtle defects in neural tube development, and individuals with neural tube defects should be examined for laterality defects. As with some zebrafish midline mutants and extirpations of neural tissue after neural tube closure in *Xenopus*, some cases should have isolated neural tube defects without cardiac defects. These cases reflect the observation that only specific stages of early neural development and a subset of the genetic pathway for midline development are necessary for cardiac L–R development. This also suggests that prenatal treatments, such as diet supplements, that reduce the frequency of early neural tube defects (Butterworth and Bendich, 1996) also might be palliative for an underlying propensity for cardiac defects.

C. Left–Right Development in Conjoined Twins

Conjoined twins in humans are rare and laterality defects are just one of many circumstances that require medical intervention in these individuals (Spencer, 1996). The general rule for conjoined twins is that the twin on the left has normal cardiac L–R development (Fig. 4A) and the twin on the right has randomized cardiac orientation (Fig. 4B). This is true in human twins joined at the trunk (Cunniff *et al.,* 1988; Layton, 1989; Burn, 1991; Spencer, 1992; Levin *et al.,* 1996), in spontaneous chick twins joined at the trunk (Levin *et al.,* 1996), in amphibian twins formed by dividing a blastula-stage embryo in half with a hair loop (Spemann and Falkenberg, 1919), and in *Xenopus* twins formed by organizer transplantation (Nascone and Mercola, 1997). Conjoined twins are readily generated by injection of a wide variety of signaling molecules that induce the formation of an ectopic organizer (Hyatt *et al.,* 1996; Nascone and Mercola, 1997). Results from injections of most signaling molecules follow the general rule of "left twin normal, right twin random" (Figs. 4A and 4B). In most cases, L–R orientation in induced (secondary) axes is not ascertained (Figs. 4A and 4B) due to concerns of the effect of incomplete midlines on L–R development (Danos and Yost, 1995, 1996; Lohr *et al.,* 1997).

There are five known exceptions to the left twin normal, right twin random rule (Fig. 4, indicated in blue) that help to assess the role of the L–R Coordinator and might be indicative of molecules involved in the L–R pathway. Three of these involve *activin* or *BVg1,* molecules implicated in L–R Coordinator function. The only known example of randomization in the left-sided twin occurs when *activin* RNA injections are used to induce the secondary axis on the right side (Fig. 4A; Hyatt *et al.,* 1996). When secondary axes are formed on the right side by either *activin* or *BVg1* injections, L–R development in the right-sided twin is normal (Fig. 4B; Hyatt *et al.,* 1996). The ability of *BVg1* and *activin* to alter L–R development in twins indicates the involvement of these molecules in L–R Coordinator function. Left–right development in the secondary axes induced by *BVg1* or *activin* has not been assessed.

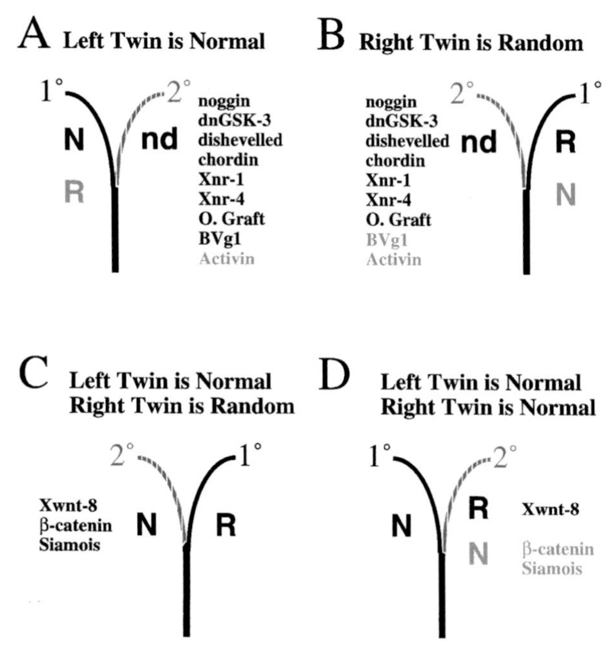

Figure 4 Summary of conjoined twin results in *Xenopus.* Schematic dorsal view of twins, with anterior on top, left twin on left, and right twin on right. Primary embryonic axis (1°, indicated by black line) and secondary induced axis (2°, indicated by red stripped line) are show in their respective positions with resulting cardiac L–R development indicated; N, normal orientation; nd, not determined; R, randomized. (A, B) In twins induced by most molecules (in black letters), the left twin has normal cardiac orientation (A) and the right twin has randomized cardiac orientation (B). Exceptions are indicated in blue. In most cases, L–R orientation in the induced (2°) axis is not determined due to variable anteriodorsal development (see text). (C, D) For *Xwnt-8, β-catenin,* and *Siamois,* L–R development can be scored in the fully induced (2°) axis. For *Xwnt-8,* as with most other injected molecules, the twin on the left is normal and the twin on the right is randomized, regardless of whether the secondary axis makes the left (C) or right (D) twin. The exceptional cases (indicated in blue) are secondary axes that make a twin on the right with normal L–R development (summary of results from Hyatt *et al.,* 1996, and Nascone and Mercola, 1997).

The other two exceptions to the left twin normal, right twin random were found in right-sided secondary axes induced by *β-catenin* or *Siamois* (Nascone and Mercola, 1997). Injection of members of the *Wnt* signaling pathway (Moon *et al.,* 1997), *Xwnt-8, β-catenin,* and *Siamois* give complete secondary axes that have been assessed for L–R development (Nascone and Mercola, 1997). In twins formed by *Xwnt-8* injection, both primary and secondary axes follow the general rule (Figs. 4C and 4D). In twins formed by *β-catenin* and *Siamois,* the left-sided twin is normal, whether it was derived from the primary axis (Fig. 4D) or the secondary axis (Fig. 4C). However, when a secondary axis induced by *β-catenin* or *Siamois* forms a right-sided twin, it has normal L–R development (Fig. 4D). It is puzzling that right-sided secondary axes induced by downstream members of the *Wnt* pathway have normal L–R development, whereas right-sided secondary axes induced by Xwnt-8, which is upstream of *β-catenin* and *Siamois* in the *Wnt* pathway (Moon *et al.,* 1997), are randomized.

There are at least three interpretations of these results, depending on whether L–R signaling originates within the organizer or is extraneous to the organizer (e.g., in the L–R Coordinator): First, perhaps the organizer has intrinsic L–R asymmetry (i.e., is not dependent on a L–R Coordinator), and only *β-catenin* and *Siamois* form an organizer that contains intrinsic L–R information, whereas an upstream inducer of these molecules, *Xwnt-8,* cannot (Nascone and Mercola, 1997; Fig. 4D). This does not explain why in all cases, except for *BVg1* and *activin* injections, the twin on the right is randomized, even when the right-sided twin is derived from the endogenous organizer (i.e., the primary axis) (Figs. 4B and 4C). If the organizer has independent and intrinsic L–R information, what mechanism randomizes these primary axes? It should be noted that injection of *Xwnt-8* or *Siamois* RNA does not perturb L–R development in the primary axis when injected into vegetal cells that are proximal to the primary axis (Hyatt *et al.,* 1996), suggesting that it is the formation of a secondary twin on the right, not proximity to cells containing these molecules, that perturbs L–R development in the twin on the left. Second, it is possible that the twin on the right is perturbed by signals emanating from the twin on the left (Levin *et al.,* 1996). In this case, the secondary axes formed by *β-catenin* or *Siamois* on the right are impervious to interference from the twin on the left side. Finally, the L–R Coordinator is initiated by *Vg1* signaling on the left of the organizer or midline (Hyatt and Yost, 1998) and perhaps provides L–R information to the organizer (see Section IV,A). If the L–R Coordinator is not duplicated in conjoined twins, then only the twin on the left would receive information from the L–R Coordinator. The twin on the right would be separated from L–R Coordinator signals by the intervening midline (see section V) and not acquire L–R axis information. In most cases, the axis on the right cannot regenerate a new L–R Coordinator and is therefore randomized. Perhaps axes induced by *β-catenin* or *Siamois* are capable of forming a new L–R Coordinator. A model for the interactions between the midline and the L–R Coordinator is proposed in the following section.

D. A Model for Interaction between the Midline and L–R Coordinator

In conjoined twins, it is possible that the Organizer or midline cells on the left interfere with L–R signaling in the twin on the right either by separating the twin on the right from the L–R Coordinator (Hyatt *et al.,* 1996: Hyatt and Yost, 1998) or by juxtaposing signals from the adjacent Organizer or midline (Levin *et al.,* 1996, Levin *et al.,* 1997). These possibilities concur with the proposal that signals from the midline repress *Xnr-1* expression on the right of the midline (see Section IV,a; Danos and Yost, 1995, 1996; Lohr *et al.,* 1997) and thereby regulate cardiac and visceral L–R development. On the left side in normal embryos, the L–R Coordinator is dominant and *Xnr-1* is induced. On the right side of normal embryos, interaction with the midline supersedes the L–R Coordinator activity and *Xnr-1* is not induced (Fig. 5A). In embryos in which midline is removed before *Xnr-1* specification, *Xnr-1* is induced on both sides due to the absence of antagonistic signals from the midline (Fig. 5B). Ectopic expression of Vg1 in dorsovegetal cells on the right side induces *Xnr-1* on the right side, resulting in embryos with randomized cardiac orientation (Hyatt *et al.,* 1996; Fig. 5C). Expression of Vg1 in lateral vegetal cells on the right side inverts the L–R axis (Hyatt *et al.,* 1996; Fig. 5C). It is predicted that elimination of Vg1 in the embryo would result in no expression of *Xnr-1* in lateral plate mesoderm and randomization of cardiac orientation (Fig. 5D). In conjoined twins, if the L–R Coordinator is not duplicated the embryo on the left would have normal L–R development, due to proximity with the L–R Coordinator. *Xnr-1* expression in the right twin is suppressed (Nascone and Mercola, 1997; Hyatt and Yost, 1998), perhaps due to the antagonistic action of the midlines (Fig. 5E).

A test of the model of antagonistic interactions between the midline and the L–R Coordinator has recently been performed (Hyatt and Yost, 1998). *Xwnt-8*-induced secondary axis formation on the left results in randomization of right-sided twin (Fig. 4C; Nascone and Mercola, 1997). The cardiac L–R orientation in right-sided twins (the primary axis) can be "rescued" to normal by placing an L–R Coordinator (via *BVg1* injection) on the left side of the primary axis and can be fully inverted by injection of *BVg1* on the right side of the primary axis. These results suggest that the placement of a new L–R Coordinator, via *Vg1* expression, can overcome the randomizing effects of having an axis on the left and can regulate L–R axis formation in the twin on the right (Hyatt and Yost, 1998).

VI. Looping Morphogenesis

A. Biomechanics of Looping

The asymmetric expression of *nodal* in the lateral plate mesoderm is temporally and spatially proximal to the specification of cardiac asymmetry. The challenge in understanding the proximal events of the specifica-

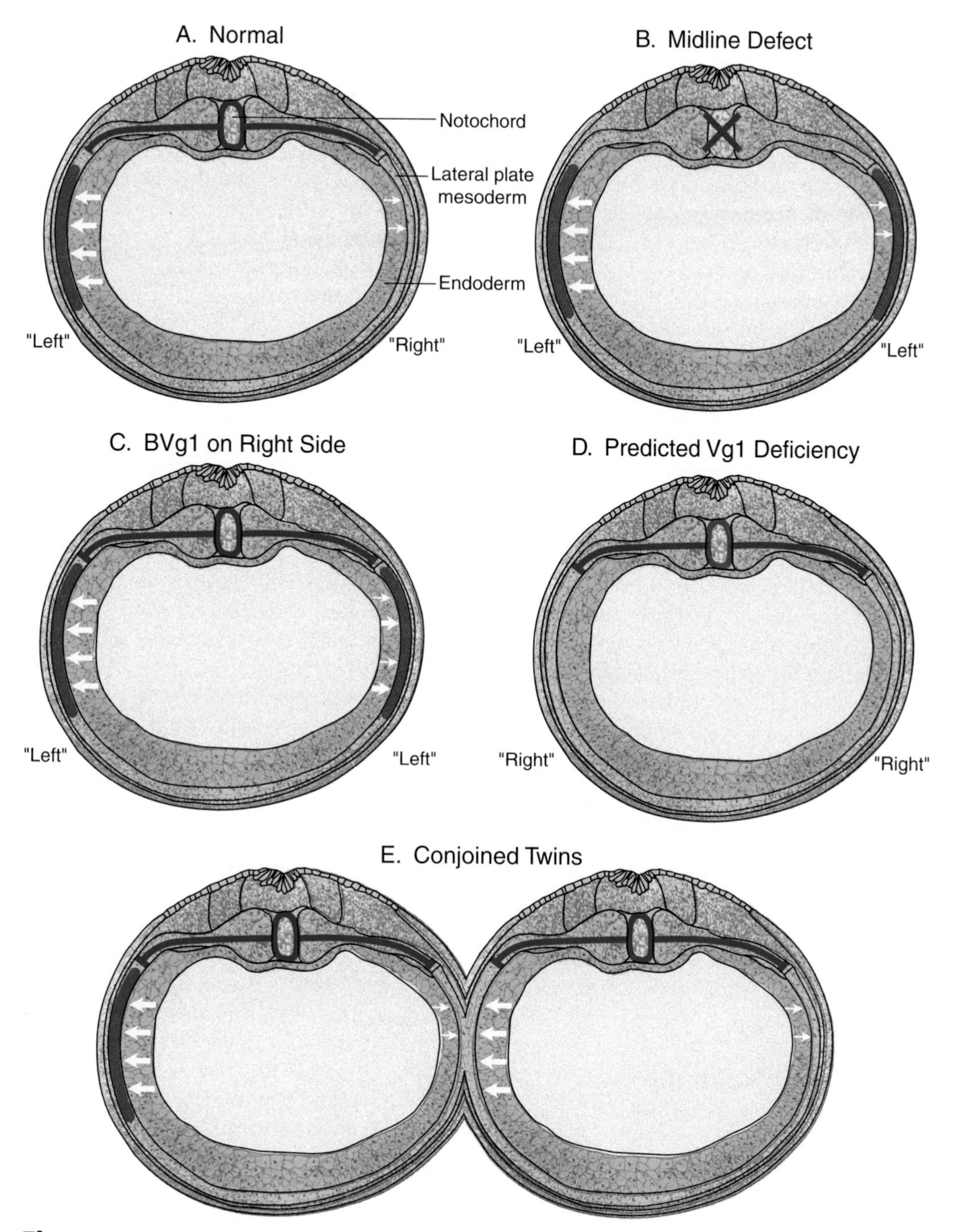

Figure 5 Model of interactions between midline and L–R Coordinator. Schematic cross sections of *Xenopus* embryos with dorsal at top and embryo's left to the left. (A) In normal embryos, the L–R Coordinator (strong *Vg1*) on the left side (represented by blue arrows in endoderm cells) supersedes the antagonistic signals (represented by green bars) from the midline cells, and *nodal* (*Xnr-1,* represented by red) is induced in the left lateral plate mesoderm. On the right side, the antagonistic signals from the midline supersede any remnant L–R signal (weak *Vg1*), and *nodal* is not induced. (B) In embryos from which midline cells are removed, the absence of antagonistic signal allows *Xnr-1* to be induced on both sides of the embryo. (C) Expression of Vg1 ligand (via *BVg1* or *AVg* RNA injection) in dorsovegetal cells induces Xnr-1 in right lateral plate mesoderm. (D) It is predicted that deficiency in Vg1 or pro-Vg1 processing would result in loss of Xnr-1 expression in lateral plate mesoderm and randomization of cardiac L–R orientation. (E) In conjoined twins, interactions between the two midlines antagonize any residual L–R Coordinator activity in the junction between the conjoined twins. This prevents expression of nodal on either side in the right-sided twin. In the absence of asymmetric nodal expression, cardiac L–R orientation is randomized in the twin on the right side.

tion of cardiac L–R orientation will be to discover how the asymmetric expression of cell–cell signaling molecules can influence the subsequent biomechanics of cardiac tube looping such that the cardiac tube eventually loops away from the side of the embryo that expresses *nodal* and toward the side of the embryo that expresses *Snr1* (Fig. 2). However, the biomechanics of cardiac tube looping are poorly understood (Taber *et al.,* 1995).

The recent description of cardiac jogging morphogenesis in zebrafish (Chen *et al.,* 1997) indicates that cardiac looping is a very dynamic process and that a detailed analysis of cardiac tube behavior will be necessary before morphogenesis can be understood. One could imagine several morphogenetic mechanisms by which a relatively symmetric tube of cells breaks symmetry. First, more cells could accumulate on one side of the tube by differential cell division rates, differential cell size, or different cell contributions from the two sides of the embryo. Asymmetric accumulation of cells would cause the tube to bend toward one side. In amphibians and chick, fate map studies suggest that there is a differential contribution of cells from the left and right sides along the length of the cardiac primordium (Zwirner and Kuhlo, 1964; Stalsberg, 1969), but detailed fate mapping with respect to the signaling pathways that regulate L–R morphogenesis awaits further study. Cardiac tube looping could be driven by cell shape changes, perhaps due to differential contraction of circumapical actin cytoskeleton (Taber *et al.,* 1995), leading to flattened cells at the outer curvature of the loop and constricted cells in the inner curvature. Treatment of chick embryos with cytochalasin B, an inhibitor of actin polymerization, blocks cardiac looping (Itasaki *et al.,* 1991), implicating the cytoskeleton in cardiac loop morphogenesis.

B. Genes Expressed during Cardiac Morphogenesis

Given the increasing number of genes in the L–R signaling pathway that are expressed in noncardiac cells and that precede the specification of cardiac L–R asymmetry (Fig. 2), it is striking that very few genes have been identified that show L–R asymmetric expression in the cardiac primordium. Expression of *BMP4,* a member of the TGF-β superfamily of cell signaling molecules, is initially symmetric in the cardiac tube in zebrafish and then becomes stronger on the left (Chen *et al.,* 1997). This expression pattern is correlated with the direction of cardiac jogging and looping in wild-type and laterality mutants (Chen *et al.,* 1997) and is therefore directly responsive to upstream signals proximal to cardiac L–R specification.

Three extracellular matrix proteins are expressed asymmetrically in the cardiac jelly of the looping heart; flectin (Tsuda *et al.,* 1996) and hLAMP1 (Smith *et al.,* 1997) on the left and JB3 on the right (Smith *et al.,* 1997). The expression of these molecules is responsive to retinoid-sensitive pathway, but the retinoid-sensitive step appears to be proximal to specification of cardiac L–R development, downstream from *nodal* (Fig. 2). Retinoic acid (RA) treatment on the right results in randomized cardiac orientation (Chen and Solursh, 1992; Smith *et al.,* 1997) and altered hLAMP and JB3 expression (Smith *et al.,* 1997), whereas treatment on the left only does so at high concentrations. In quail embryos that are deficient for vitamin A, a precursor for retinoids, cardiac looping is perturbed and flectin expression in the heart is disorganized (Dersch and Zile, 1993; Tsuda *et al.,* 1996). RA treatment does not significantly alter asymmetric expression of *Shh* in the node and *cNR1* (*nodal*) in the lateral plate mesoderm (Smith *et al.,* 1997), and vitamin A-deficient embryos have normal *Shh* asymmetric expression (Chen *et al.,* 1996).

It is unclear whether the basic helix–loop–helix transcription factors *dHAND* and *eHAND* play a role in L–R development. *dHAND* and *eHAND* expression patterns overlap significantly in chick heart (Srivastava *et al.,* 1995; see Chapter 9). Application of antisense nucleotides that decrease the expression of both *eHAND* and *dHAND* arrest cardiac tube looping (Srivastava *et al.,* 1995). In mice, *eHAND* and *dHAND* are expressed in the left and right chambers, respectively, in complementary patterns (Srivastava *et al.,* 1997). Knockout of the *dHAND* gene results in elimination of the right ventricle (Srivastava *et al.,* 1997). Knockout of the homeobox gene *Nkx2-5* results in arrested cardiac looping (Lyons *et al.,* 1995) and down regulation of left-sided *eHAND* expression, suggesting that these transcription factors are part of the cardiac L–R morphogenetic pathway (Biben and Harvey, 1997; see Chapter 7). The roles of transcription factors and their target genes in cardiac looping is an exciting area of investigation.

VII. Conclusions and Future Perspectives

Significant advances in our understanding of the molecular basis for cardiac L–R development have been made in the past 3 years; several genes and embryonic tissue interactions during embryogenesis have been identified. The intermediate steps in a cell–cell signaling pathway are being elucidated in several species, but it is not yet clear to what extent a genetic pathway for L–R development is conserved in vertebrates. The candidate genes elucidated in embryos of zebrafish, chick, frogs, and mice will be used for studies of human families with laterality defects. Conversely, genes identified by mapping in human families will be further characterized in other vertebrate model systems. Results from studies of the earliest stages of embryogenesis reveal that a L–R difference is in place; we currently do not understand the origin of the L–R axis. Research in this field continues to push the following fundamental question in biology back in developmental time, perhaps to events in the single-cell embryo: What is the mechanism that transforms symmetry into asymmetry and initiates the L–R axis?

References

Afzelius, B. A. (1976). A human syndrome caused by immotile cilia. *Science* **193,** 317–319.

Aruga, J., Nagai, T., Tokuyama, T., Hayashizaki, Y., Okazaki, Y., Chapman, V. M., and Mikoshiba, K. (1996). The mouse *zic* gene family. Homologues of the *Drosophila* pair-rule gene odd-paired. *J. Biol. Chem.* **271,** 1043–1047.

Biben, C., and Harvey, R. P. (1997). Homeodomain factor *Nkx2-5* controls left/right asymmetric expression of bHLH gene *eHand* during murine heart development. *Genes Dev.* **11,** 1357–1369.

Bowers, P. M., Brueckner, M., and Yost, H. J. (1996). Laterality disturbances. *Prog. Pediatr. Cardiol.* **6,** 53–62.

Brand, M., Heisenberg, C. P., Warga, R. M., Pelegri, F., Karlstrom, R. O., Beuchle, D., Picker, A., Jiang, Y. J., Furutani-Seiki, M., van Eeden, F. J., Granato, M., Haffter, P., Hammerschmidt, M., Kane, D. A., Kelsh, R. N., Mullins, M. C., Odenthal, J., and Nusslein-Volhard, C. (1996). Mutations affecting development of the midline and general body shape during zebrafish embryogenesis. *Development* **123,** 129–142.

Britz-Cunningham, S. H., Shah, M. M., Zuppen, C. W., and Fletcher, W. H. (1995). Mutations of the connexin43 gap-junction gene in patients with heart malformations and defects of laterality. *N. Engl. J. Med.* **332,** 1323–1329.

Brown, N. A., and Wolpert, L. (1990). The development of handedness in left/right asymmetry. *Development* **109,** 1–9.

Brueckner, M., D'Eustachio, P., and Horwich, A. L. (1989). Linkage mapping of a mouse gene, *iv,* that controls left–right asymmetry of the heart and viscera. *Proc. Natl. Acad. Sci. USA* **86,** 5035–5038.

Burn, J. (1991). Disturbance of morphological laterality in humans. *In* "Ciba Foundation Symposium" (G. R. Bock and J. Marsh, Eds.), Vol. 162, pp. 282–296. Wiley, Chichester, UK.

Butterworth, C. E., Jr., and Bendich, A. (1996). Folic acid and the prevention of birth defects. *Annu. Rev. Nutr.* **16,** 73–97.

Casey, B., and Ballabio, A. (1995). Connexin43 mutations in sporadic and familial defects of laterality. *N. Engl. J. Med.* **333,** 941. [See Discussion, pp. 941–942]

Chen, J.-N., van Eeden, F. J. M., Warren, K. S., Chin, A., Nusslein-Volhard, C., Haffter, P., and Fishman, M. C. (1997). Left–right pattern of cardiac *BMP4* may drive asymmetry of the heart in zebrafish. *Development* **124,** 4373–4382.

Chen, Y., and Solursh, M. (1992). Comparison of Hensen's node and retinoic acid in secondary axis induction in the early chick embryo. *Dev. Dyn.* **195,** 142–151.

Chen, Y. P., Dong, D., Kostetskii, I., and Zile, M. H. (1996). Hensen's node from vitamin A-deficient quail embryo induces chick limb bud duplication and retains its normal asymmetric expression of *sonic hedgehog* (*shh*). *Dev. Biol.* **173,** 256–264.

Chiang, C., Litingtung, Y., Lee, E., Young, K. E., Corden, J. L., Westphal, H., and Beachy, P. A. (1996). Cyclopia and defective axial patterning in mice lacking *Sonic hedgehog* gene function. *Nature* **383,** 407–413.

Cleaver, O. B., Patterson, K. D., and Krieg, P. A. (1996). Overexpression of the tinman-related genes *XNkx-2.5* and *XNkx-2.3* in *Xenopus* embryos results in myocardial hyperplasia. *Development* **122,** 3549–3556.

Collignon, J., Varlet, I., and Robertson, E. J. (1996). Relationship between asymmetric *nodal* expression and the direction of embryonic turning. *Nature* **381,** 155–158.

Cooke, J. (1995). Vertebrate embryo handedness, *Nature* **374,** 681.

Cunniff, C., Jones, K. L., Jones, M. C., Saunder, B., Shepard, T., and Benirschke, K. (1998). Laterality defects in conjoined twins: Implications for normal asymmetry in human embryogenesis. *Am. J. Med. Genet.* **31,** 669–677.

Dale, L., Matthews, G., and Colman, A. (1993). Secretion and mesoderm-inducing activity of the *TGFβ*-related domain of Xenopus *Vg1. EMBO J.* **12,** 4471–4480.

Danos, M. C., and Yost, H. J. (1995). Linkage of cardiac left–right asymmetry and dorsal–anterior development in *Xenopus. Development* **121,** 1467–1474.

Danos, M. C., and Yost, H. J. (1996). Role of notochord in specification of cardiac left–right orientation in zebrafish and Xenopus. *Dev. Biol.* **177,** 96–103.

Dersch, H., and Zile, M. H. (1993). Induction of normal cardiovascular development in the vitamin A-deprived quail embryo by natural retinoids. *Dev. Biol.* **160,** 424–433.

Drysdale, T. A., Tonissen, K. F., Patterson, K. D., Crawford, M. J., and Krieg, P. A. (1994). Cardiac troponin I is a heart-specific marker in the *Xenopus* embryo: Expression during abnormal heart morphogenesis. *Dev. Biol.* **165,** 432–441.

Eisenberg, L. M., and Markwald, R. R. (1995). Molecular regulation of atrioventricular valvuloseptal morphogenesis. *Circ. Res.* **77,** 1–6.

Ekker, S. C., McGrew, L. L., Lai, C. J., Lee, J. J., von Kessler, D. P., Moon, R. T., and Beachy, P. A. (1995).Distinct expression and shared activities of members of the *hedgehog* gene family of *Xenopus laevis. Development* **121,** 2337–2347.

Essner, J. J., Laing, J. G., Beyer, E. C., Johnson, R. G., and Hackett, P. B., Jr. (1996). Expression of zebrafish connexin43.4 in the notochord and tail bud of wild-type and mutant *no tail* embryos. *Dev. Biol.* **177,** 449–462.

Evans, S. M., Yan, W., Murillo, M. P., Ponce, J., and Papalopulu, N. (1995). tinman, a *Drosophila* homeobox gene required for heart and visceral mesoderm specification, may be represented by a family of genes in vertebrates: *XNkx-2.3,* a second vertebrate homologue of *tinman. Development* **121,** 3889–3899.

Fishman, M., and Chien, K. (1997). Fashioning the vertebrate heart: Earliest embryonic decisions. *Dev. Suppl.* **124,** 2099–2117.

Gebbia, M., Towbin, J. A., and Casey, B. (1996). Failure to detect connexin43 mutations in 38 cases of sporadic and familial *heterotaxy. Circulation* **94,** 1909–1912.

Gebbia, M., Ferrero, G. B., Pilia, G., Bassi, M. T., Aylsworth, A., Penman-Splitt, M., Bird, L. M., Bamforth, J. S., Burn, J., Schlessinger, D., Nelson, D. L., and Casey, B. (1997). X-linked *situs* abnormalities result from mutations in *ZIC3. Nature Genet.* **17,** 305–308.

Gerhart, J., Danilchik, M., Doniach, T., Roberts, S., Rowning, B., and Stewart, R. (1989). Cortical rotation of the Xenopus egg: Consequences for the anteroposterior pattern of embryonic dorsal development. *Development* **107,** 37–51.

Gove, C., Walmsley, M., Nijjar, S., Bertwistle, D., Guille, M., Partington, G., Bomford, A., and Patient, R. (1997). Overexpression of *GATA-6* in *Xenopus* embryos blocks differentiation of heart precursors. *EMBO J.* **16,** 355–368. [published erratum appears in *EMBO J.* 1997 **16**(7), 1806–1807]

Halpern, M. E., Ho, R. K., Walker, C., and Kimmel, C. B. (1993). Induction of muscle pioneers and floor plate is distinguished by the zebrafish *no tail* mutation. *Cell* **75,** 99–111.

Hoyle, C., Brown, N. A., and Wolpert, L. (1992). Development of left/right handedness in the chick heart. *Development* **115,** 1071–1078.

Hyatt, B. A., and Yost, H. J. (1998). The left–right coordinator: The role of *Vg1* in organizing left–right axis formation. *Cell* **93,** 37–46.

Hyatt, B. A., Lohr, J. L., and Yost, H. J. (1996). Initiation of vertebrate left–right axis formation by material *Vgl. Nature* **384,** 62–65.

Icardo, J. M. (1989). Heart anatomy and developmental biology. *Experientia* **44,** 910–919.

Isaac, A., Sargent, M. G., and Cooke, J. (1997). Control of vertebrate left–right asymmetry by a *snail*-related zinc finger gene. *Science* **275,** 1301–1304.

Itasaki, N., Nakamura, H., Sumida, H., and Yasuda, M. (1991). Actin bundles on the right side in the caudal part of the heart tube play a role in dextro-looping in the embryonic chick heart *Anat. Embryol.* **183,** 29–39.

Jiang, Y., and Evans, T. (1996). The *Xenopus GATA-4/5/6* genes are associated with cardiac specification and can regulate cardiac-specific transcription during embryogenesis. *Dev. Biol.* **174,** 258–270.

Keller, R. E. (1975). Vital dye mapping of the gastrula and neurula of *Xenopus laevis.* I. Prospective areas and morphogenetic movements of the superficial layer. *Dev. Biol.* **42,** 222–241.

Keller, R. E. (1976). Vital dye mapping of the gastrula and neurula of *Xenopus laevis.* II. Prospective areas and morphogenetic movements of the deep layer. *Dev. Biol.* **51,** 118–137.

Kelley, C., Blumberg, H., Zon, L. I., and Evans, T. (1993). *GATA-4* is a novel transcription factor expressed in endocardium of the developing heart. *Development* **118,** 817–827.

Kessler, D. S., and Melton, D. A. (1995). Induction of dorsal mesoderm by soluble, mature Vg1 protein. *Development* **121,** 2155–2164.

Komuro, I., and Izumo, S. (1993). *Csx:* A murine homeobox-containing gene specifically expressed in the developing heart. *Proc. Natl. Acad. Sci. USA* **90,** 8145–8149.

Kuo, C. T., Morrisey, E. E., Anandappa, R., Sigrist, K., Lu, M. M., Parmacek, M. S., Soudais, C., and Leiden, J. M. (1997). GATA4 transcription factor is required for ventral morphogenesis and heart tube formation. *Genes Dev.* **11,** 1048–1060.

Layton, W. M. (1989). *Situs inversus* in conjoined twins. *Am. J. Med. Genet.* **34,** 297.

Levin, M. (1997). Left–right asymmetry in vertebrate embryogenesis. *Bioessays* **19,** 287–296.

Levin, M., Johnson, R. L., Stern, C. D., Kuehn, M., and Tabin, C. (1995). A molecular pathway determining left–right asymmetry in chick embryogenesis. *Cell* **82,** 803–814.

Levin, M., Roberts, D. J., Holmes, L. B., and Tabin, C. (1996). Laterality defects in conjoined twins. *Nature* **384,** 321.

Levin, M., Pagan, S., Roberts, D. J., Cooke, J., Kuehn, M. R., and Tabin, C. J. (1997). Left/right patterning signals and the independent regulation of different aspects of *situs* in the chick embryo. *Dev. Biol.* **189,** 57–67.

Logan, M., and Mohun, T. (1993). Induction of cardiac muscle differentiation in isolated animal pole explants of *Xenopus* laevis embryos. *Development* **118,** 865–875.

Lohr, J. L., Danos, M. C., and Yost, H. J. (1997). Left–right asymmetry of a nodal-related gene is regulated by dorsoanterior midline structures during *Xenopus* development. *Development* **125,** 1465–1472.

Lowe, L. A., Supp, D. M., Sampath, K., Yokoyama, T., Wright, C. V. E., Potter, S. S., Overbeek, P., and Kuehn, M. R. (1996). Conserved left–right asymmetry of *nodal* expression and alterations in murine *situs inversus. Nature* **381,** 158–161.

Lyons, I., Parsons, L. M., Hartley, L., Li, R., Andrews, J. E., Robb, L., and Harvey, R. P. (1995). Myogenic and morphogenetic defects in the heart tubes of murine embryos lacking the homeobox gene *Nkx2-5. Genes Dev.* **9,** 1654–1666.

Meno, C., Saijoh, Y., Fujii, H., Ikeda, M., Yokoyama, T., Yokoyama, M., Toyoda, Y., and Hamada, H. (1996). Left–right asymmetric expression of the *TGFβ*-family member lefty in mouse embryos. *Nature* **381,** 151–155.

Meno, C., Ito, Y., Saijoh, Y., Matsuda, Y., Tashiro, K., Kuhara, S., and Hamada, H. (1997). Two closely-related left–right asymmetrically expressed genes, *lefty-1* and *lefty-2:* Their distinct expression domains chromosomal linkage and direct neuralizing activity in *Xenopus* embryos. *Genes Cells* **2,** 513–524.

Molkentin, J. D., Lin, Q., Duncan, S. A., and Olson, E. N. (1997). Requirement of the transcription factor *GATA4* for heart tube formation and ventral morphogenesis. *Genes Dev.* **11,** 1061–1072.

Moon, R. T., Brown, J. D., and Torres, M. (1997). *WNTs* modulate cell fate and behavior during vertebrate development. *Trends Genet.* **13,** 157–162.

Muslin, A. J., and Williams, L. T. (1991). Well-defined growth factors promote cardiac development in axolotl mesodermal explants. *Development* **112,** 1095–1101.

Nagai, T., Aruga, J., Takada, S., Gunther, T., Sporle, R., Schughart, K., and Mikoshiba, K. (1997). The expression of the mouse *Zic1, Zic2,* and *Zic3* gene suggests an essential role for *Zic* genes in body pattern formation. *Dev. Biol.* **182,** 299–313.

Nakata, K., Nagai, T., Aruga, J., and Mikoshiba, K. (1997). *Xenopus Zic3,* a primary regulator both in neural and neural crest development. *Proc. Natl. Acad. Sci. USA* **94,** 11980–11985.

Narita, N., Bielinska, M., and Wilson, D. B. (1997). Wild-type endoderm abrogates the ventral developmental defects associated with *GATA-4* deficiency in the mouse. *Dev. Biol.* **189,** 270–274.

Nascone, N., and Mercola, M. (1995). An inductive role for the endoderm in *Xenopus* cardiogenesis. *Development* **121,** 515–523.

Nascone, N., and Mercola, M. (1997). Organizer induction determines left–right asymmetry in *Xenopus. Dev. Biol.* **189,** 68–78.

Pagan-Westphal, S. M., and Tabin, C. J. (1998). The transfer of left–right positional information during chick embryogenesis. *Cell* **93,** 25–35.

Penman Splitt, M., Tsai, M. Y., Burn, J., and Goodship, J. A. (1997). Absence of mutations in the regulatory domain of the gap junction protein connexin 43 in patients with visceroatrial *heterotaxy. Heart* **77,** 369–370.

Psychoyos, D., and Stern, C. D. (1996). Restoration of the organizer after radical ablation of Hensen's node and the anterior primitive streak in the chick embryo. *Development* **122,** 3263–3273.

Sampath, K., Cheng, A. M., Frisch, A., and Wright, C. V. (1997). Functional differences among *Xenopus nodal*-related genes in left–right axis determination. *Development* **124,** 3293–3302.

Sater, A. K., and Jacobson, A. G. (1990). The specification of heart mesoderm occurs during gastrulation in *Xenopus laevis. Development* **105,** 821–830.

Sater, A. K., and Jacobson, A. G. (1990). The role of the dorsal lip in the induction of heart mesoderm in *Xenopus laevis. Development* **108,** 461–470.

Schulte-Merker, S., van Eeden, F. J. M., Halpern, M. E., Kimmel, C. B., and Nusslein-Volhard, C. (1994). *no tail* (*ntl*) is the zebrafish homologue of the mouse *T* (*Brachyury*) gene. *Development* **120,** 1009–1015.

Schultheiss, T. M., Burch, J. B., and Lassar, A. B. (1997). A role for bone

morphogenetic proteins in the induction of cardiac myogenesis. *Genes Dev.* **11,** 451–462.
Seo, J. W., Brown, N. A., Ho, S. Y., and Anderson, R. H. (1992). Abnormal laterality and congenital cardiac anomalies. Relations of visceral and cardiac morphologies in the *iv/iv* mouse. *Circulation* **86,** 642–650.
Smith, S. M., Dickman, E. D., Thompson, R. P., Sinning, A. R., Wunsch, A. M., and Markwald, R. R. (1997). Retinoic acid directs cardiac laterality and the expression of early markers of precardiac asymmetry. *Dev. Biol.* **182,** 162–171.
Spemann, H., and Falkenberg, H. (1919). Uber asymmetriche Entwicklung und *situs inversus* viscerum bei Zwillingen und Doppelbildungen. *Arch. Entwicklungsmechanik Organismem.* **45,** 371–422.
Spemann, H., and Mangold, H. (1924). Uber induction von embryonalanlagen durch implantation artfremder organisatoren. *Roux. Arch.* **100,** 599–638.
Spencer, R. (1992). Conjoined twins: Theoretical embryologic basis. *Teratology* **45,** 591–602.
Spencer, R. (1996). Anatomic description of conjoined twins: A plea for standardized terminology. *J. Pediatr. Surg.* **31,** 941–944.
Splitt, M. P., Burn, J., and Goodship. J. (1995). *Connexin43* mutations in sporadic and familial defects of laterality. *N. Engl. J. Med.* **333,** 941. [see Discussion, pp. 941–942]
Srivastava, D., Cserjesi, P., and Olson, E. N. (1995). A subclass of bHLH proteins required for cardiac morphogenesis. *Science* **270,** 1995–1999.
Srivastava, D., Thomas, T., Lin, Q., Kirby, M. L., Brown, D., and Olson, E. N. (1997). Regulation of cardiac mesodermal and neural crest development by the bHLH transcription factor, *dHAND. Nature Genet.* **16,** 154–160.
Stalsberg, H. (1969). The origin of heart asymmetry: Right and left contributions to the early chick embryo heart. *Dev. Biol.* **19,** 109–127.
Stein, S., and Kessel, M. (1995). A homeobox gene involved in node, notochord and neural plate formation of chick embryos. *Mech. Dev.* **49,** 37–48.
Supp, D. M., Witte, D. P., Potter, S. S., and Brueckner, M. (1997). Mutation of an axonemal dynein affects left–right asymmetry in *inversus viscerum* mice. *Nature* **389,** 963–966.
Taber, L. A., Lin, I. E., and Clark, E. B. (1995). Mechanics of cardiac looping. *Dev. Dyn.* **203,** 42–50.
Talbot, W. S., Trevarrow, B., Halpern, M. E., Melby, A. E., Farr, G., Postlethwait, J. H., Jowett, T., Kimmel, C. B., and Kimelman, D. (1995). A homeobox gene essential for zebrafish notochord development. *Nature* **378,** 150–157.
Thomsen, G. H., and Melton, D. A. (1993). Processed Vg1 protein is an axial mesoderm inducer in *Xenopus. Cell* **74,** 433–441.
Tonissen, K. F., Drysdale, T. A., Lints, T. J., Harvey, R. P., and Krieg, P. A. (1994). *XNkx-2.5,* a *Xenopus* gene related to *Nkx-2.5* and *tinman:* Evidence for a conserved role in cardiac development. *Dev. Biol.* **162,** 325–328.
Tsuda, T., Philp, N., Zile, M. H., and Linask, K. K. (1996). Left–right asymmetric localization of flectin in the extracellular matrix during heart looping. *Dev. Biol.* **173,** 39–50.
von Dassow, G., Schmidt, J. E., and Kimelman, D. (1993). Induction of the Xenopus organizer: Expression and regulation of *Xnot,* a novel FGF and activin-regulated homeobox gene. *Genes Dev.* **7,** 355–366.
Weeks, D. L., and Melton, D. A. (1987). A maternal mRNA localized to the vegetal hemisphere in Xenopus eggs codes for a growth factor related to TGFβ. *Cell* **51,** 861–867.
Wetzel, R. (1929). Untersuchungen am Hunchen. Die Entwicklung des Keims wharend der ersten beiden Bruttage. *Arch. Entwichklungsmech* **119,** 188–321.
Yatskievych, T. A., Ladd, A. N., and Antin, P B. (1997). Induction of cardiac myogenesis in avian pregastrula epiblast: The role of the hypoblast and activin. *Development* **124,** 2561–2570.
Yokoyama, T., Copeland, N. G., Jenkins, N. A., Montgomery, C. A., Elder, F. F., and Overbeek, P. A. (1993). Reversal of left–right asymmetry: A *situs inversus* mutation. *Science* **260,** 679–682.
Yost, H. J. (1990). Inhibition of proteoglycan synthesis eliminates left–right asymmetry in *Xenopus laevis* cardiac looping. *Development* **110,** 865–874.
Yost, H. J. (1991). Development of the left–right axis in amphibians. *In* "Biological Asymmetry and Handedness" (G. R. Bock and J. Marsh, Eds.), Vol. 162, pp. 165–181. Wiley, Chichester, UK.
Yost, H. J. (1992). Regulation of vertebrate left–right asymmetries by extracellular matrix. *Nature* **357,** 158–161.
Yost, H. J. (1995a). Breaking symmetry: Left–right cardiac development in *Xenopus laevis. In* "Fourth International Symposium on Etiology & Morphogenesis of Congenital Heart Disease—Developmental Mechanisms" (R. M. Markwald, E. B. Clark, and A. Takao, Eds.), pp. 505–511. Futura, New York.
Yost, H. J. (1995b). Vertebrate left–right development. *Cell* **82,** 689–692.
Zwirner, R., and Kuhlo, B. (1964). *Roux' Arch. Entwicklungsmechanik Organismem* **155,** 511–524.

22

Left–Right Asymmetry and Cardiac Looping

Kumud Majumder* **and Paul A. Overbeek**

Department of Cell Biology, Baylor College of Medicine, Houston, Texas 77030

I. Introduction

This review focuses on the astounding progress that has been made in the past 3 or 4 years in the identification of genes that may play roles in the specification of left–right (L–R) asymmetry and cardiac looping in early vertebrate embryos. We summarize the morphological changes that occur during early heart development, genes that have been identified to be asymmetrically expressed before and during heart looping, embryonic manipulations that can affect looping, and mutations that randomize or reverse the direction of looping. We also propose a novel molecular model for the initial determination of L–R asymmetry. Since heart looping represents the first morphological manifestation of L–R asymmetry in vertebrates, an intriguing goal for research on cardiac embryogenesis is to elucidate the specific signals that are used to translate molecular L–R asymmetry into architectural L–R asymmetry.

II. Early Cardiac Morphogenesis

Heart morphogenesis is a multistep process that starts with the organization of cardiac mesoderm into a cardiac crescent (Fig. 1A). The lateral wings of the cardiac crescent migrate to the ventral midline and fuse in a craniocaudal direction to form the linear heart tube (Fig. 1B). As the heart tube continues to grow within the confines of the pericardiac cavity, the tube must loop to fit into the available space. In all vertebrates, the heart tube normally loops to the right side so that the anterior portion of the outflow tract is shifted to the right and subsequently becomes the right ventricle, whereas the

*Present address: CuraGen Corporation, Alachua, FL 32615.

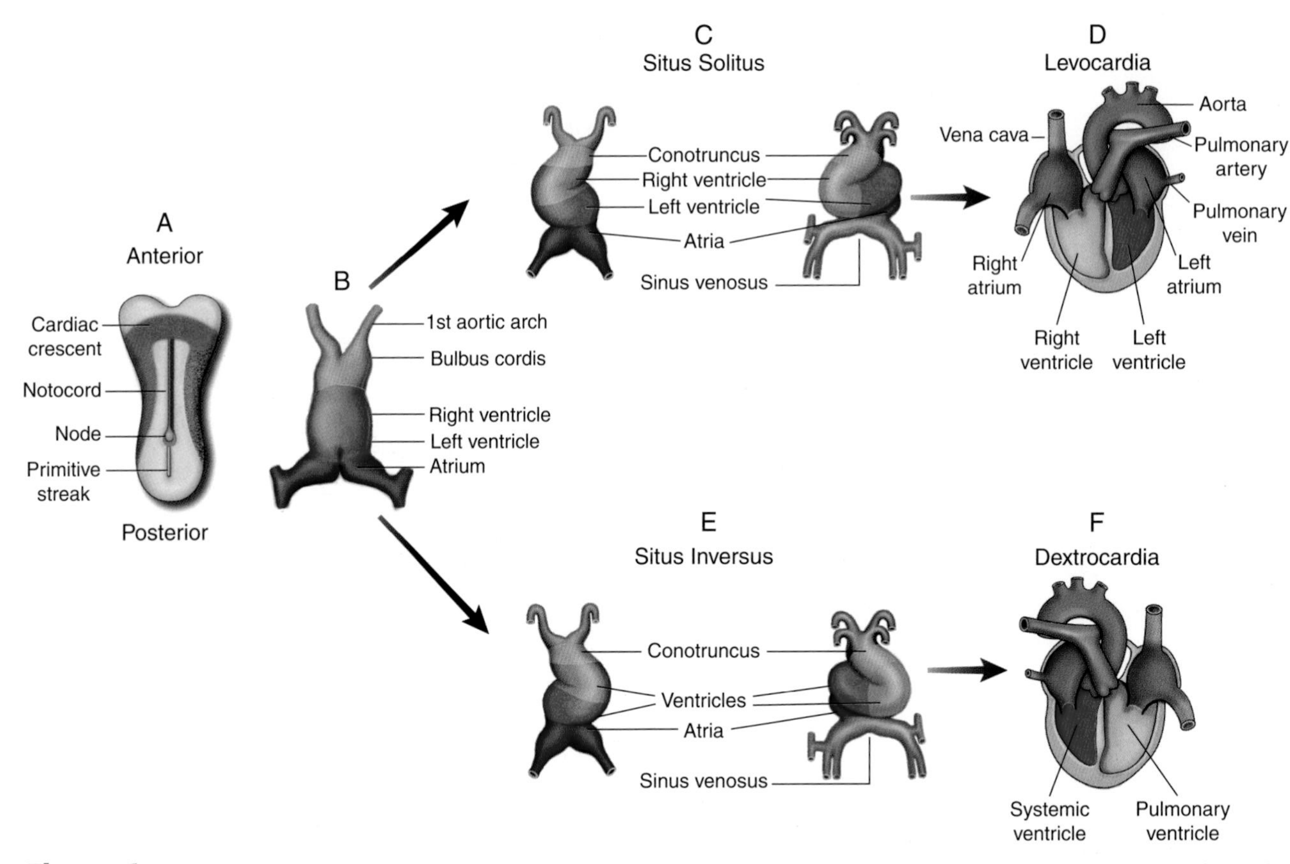

Figure 1 Early cardiac morphogenesis in vertebrate embryos. (A) Schematic drawing of a ventral view of a mouse embryo prior to the formation of the definitive heart. The cardiac crescent, which will form the heart tube, is shown in red. The flanking left and right lateral plate mesoderm which appear to play an important role in L–R axis specification are indicated in blue and brown, respectively (see also Fig. 2). Migration of the lateral wings of the cardiac crescent to the ventral middle produces a bilaterally symmetric linear heart tube (B) which functions as a primitive heart. Blood flows from the inflow tracts at the bottom (sinus venosus) to the outflow tracts at the top (bulbus cordis and aortic arches). (C) Normal looping of the heart tube to the right. The atria (purple) and inflow tracts (magenta) move dorsally and anteriorly behind the ventricular regions. (D) Mature normal four-chambered heart following septation. The vena cavae are linked to the right atrium, the pulmonary veins feed into the left atrium, the right ventricle pumps blood to the pulmonary artery, and the left ventricle is connected to the aorta. (E, F) Although the heart tube normally loops to the right, looping to the left can result in the formation of a heart that has a mirror-image reversal of L–R polarity. As a result the pulmonary ventricle becomes located on the left and the systemic ventricle is positioned on the right (adapted from Overbeek, 1997).

inferior segment of the outflow tract moves to the left and subsequently becomes the left ventricle (Figs. 1C and 1D). At the same time, there is a dorsal and anterior shift of the inflow tract within the pericardial cavity (Fig. 1C). The left branch of the inflow tract becomes the left atrium, whereas the right branch becomes the right atrium (Figs. 1C and 1D). Interestingly, the atrial and ventricular regions are specified even before looping takes place (Bisaha and Bader, 1991; Mikawa *et al.*, 1992; Yutzey and Bader, 1995; Olson and Srivastava, 1996; Srivastava *et al.*, 1997). The determination of these lineages appears to be based on the anteroposterior polarity of the heart progenitors in very early embryos (Yutzey and Bader, 1995).

Looping of the heart tube to the left instead of to the right can initiate a pathway of cardiac morphogenesis that produces a heart that is a mirror image of normal (Figs. 1E and 1F), a condition referred to as dextrocardia or situs inversus. The situs inversus heart can function like a normal heart. However, alterations in cardiac L–R axis specification are often associated with cardiovascular defects (Olson, 1997).

III. Asymmetric Gene Expression Prior to Cardiac Looping

Although vertebrate organisms have a consistent, genetically specified, L–R laterality, until recently no genes had been shown to be expressed with a L–R asymmetry during early vertebrate development. However, Tabin and coworkers (Levin *et al.*, 1995) discovered a set of genes that are asymmetrically transcribed near Hensen's node during gastrulation and neurulation in chicken embryos. The L–R asymmetrically expressed genes included *activin receptor IIa* (*cAct-RIIa*), *sonic hedgehog* (*Shh*), the transcription factor *HNF-3β*, and the chicken nodal gene (*cNR-1*). Subsequent studies have identified additional genes that are asymmetrically transcribed in chicken embryos, including *activin βB, cWnt-8c* (a Wnt family member), *patched* (a receptor for *Shh*), follistatin (an activin-binding protein), and cSnR-1 (a snail-related transcription factor) (Isaac *et al.,* 1997; Levin, 1997). Other genes, including *cNot, FGF4, goosecoid, Msx1,* and *Hoxb-8,* were found to be expressed symmetrically suggesting that the asymmetric distribution of certain mRNAs was not due to morphological asymmetry of the node region itself (Levin, 1997). The asymmetric patterns of gene expression are often transient and may be preceded by symmetrical gene expression (Table I). The asymmetric expression of these genes occurs well before the embryos exhibit morphological L–R asymmetries (Levin *et al.,* 1995).

The initial studies by Levin *et al.* (1995) suggested that *cAct-RIIa* was expressed preferentially on the right side of Hensen's node, whereas *Shh* was subsequently localized primarily to the left side of the node. The *cNR-1* gene was expressed on the same side as *Shh* initially near the node, and subsequently throughout the left lateral plate mesoderm. In order to test whether asymmetric expression of activin and/or *Shh* was important for L–R axis specification, Levin *et al.* (1995) implanted beads carrying activin or *Shh* into early chick embryos. Left-sided implantation of activin beads was found to downregulate *Shh* and *cNR-1* expression on the left side, whereas right-sided implantation of *Shh*

Table I Molecular Markers for Heart Looping in Normal Chicken Embryo

Marker	Developmental stage[a]	Signal location		Comments on expression pattern
		Left	Right	
Activin βB	3		+	Right side of HN
cAct-RIIa	4		+	More strongly on the primitive ridge on right side of primitive streak
	4^+		+	Only on right side of HN and only in ectoderm
	5	+	+	Coincident with areas having heart forming potential
cAct-RIIb	4^-	+	+	In primitive streak, but excluded from HN
	4^+	+	+	Strongly in HN and both sides of the HN
HNF-3β	4^-	+		Transient, asymmetric in small part of primitive ridge, posterior to HN
	5	+	+	Symmetric in HN, but has an asymmetric left-sided tail
Shh	3	+	+	Throughout HN
	4^+–6	+		In ectoderm only of HN
	7	+		In larger domain in LPM and notochord cells but not in regressing HN
cSnR	5	+	++	In lateral presumptive head territories opposite HN
	6–7	+	+++	Predominantly in cardiogenic right LPM; symmetrical in left and right somites
	10	+	+++	Predominantly in right LPM; symmetrical in right and left somites
cNR-1	4^-	+	+	Symmetrical in and lateral to midline two-third of primitive streak, but not in HN
	4^+	–	–	Signal disappears
	6^+–7	+		Signal superimposes with that of Shh
	7–11	+		On the left side, just lateral and anterior to HN followed by a larger patch in LPM; at stage 9 a smaller medial region of expression appears eventually in the right side as well
Flectin	7^+–8^-	++	+	Mainly in left pericardiac mesoderm and heart tube
	9	++	+	Predominantly in matrix of left myocardial basal lamina
	12–14	++	+	Predominantly in outer convex side of looping heart, ECM of myocardium, myocardial basal lamina and cardiac jelly
hLAMP1	4–5^-	+		Only in HN precardiac region
	6–8	++	+	In precardiac and ventral foregut region (signal intensity, left : right = 1.5 : 1)
JB3	3–4	+	+	In precardiac population emerging from primitive streak
	5–7	+	++	Predominantly in right precardiac region (signal intensity, left : right = 1 : 1.5); symmetric in presomites

[a]Stages are according to Hamburger and Hamilton (1951). Abbreviations used: HN, Hensen's node; LPM, lateral plate mesoderm; ECM, extracellular matrix. Table is based on Levin *et al.* (1995), Levin (1997), Isaac *et al.* (1997), Smith *et al.* (1997), and Tsuda *et al.* (1996).

beads induced *cNR-1* expression in the right lateral plate mesoderm. The anterior edge of *nodal* expression in the left lateral plate mesoderm is adjacent to the cells of the cardiac crescent which are fated to become cardiac tissue and which express the marker gene *Nkx2.5* (Harvey, 1996; Levin, 1997). Levin *et al.* (1995) proposed the following model for L–R axis specification. Normally, right-sided expression of activin (or a related protein) stimulates cAct-RIIa on the right side, which in turn suppresses *Shh* expression on the right, but *Shh* continues to be expressed on the left. *Shh* then induces left-sided cNR-1 expression, which initiates a signaling cascade leading to morphological asymmetry and rightward looping of the cardiac tube.

In mice the activin genes, the activin receptors IIa (*Act-RIIa*) and IIb (*Act-RIIb*), and the activin-binding protein follistatin do not show L–R asymmetries of expression. The genes have all been inactivated by targeted mutagenesis (Matzuk *et al.,* 1995a,b,c; Oh and Li, 1997). Except for the *Act-RIIb* mutants, mice with mutations in these genes do not show any obvious cardiac abnormalities. The *Act-RIIb*-deficient mice show complicated cardiac defects including right isomerism, malposition of great arteries, and ventricular and septal defects (Oh and Li, 1997). These knockout studies suggest that mice and chickens may use activins and/or activin receptors differently during the embryonic specification of L–R polarity.

In contrast to the activins, *nodal* genes in different vertebrate organisms are consistently expressed in the left but not the right lateral plate mesoderm (Levin *et al.,* 1995; Lowe *et al.,* 1996). A schematic drawing of *nodal* expression in the mouse embryo is shown in Fig. 2. Asymmetric expression of *nodal* in mice has been observed by whole mount *in situ* hybridization and by monitoring the expression of a *lacZ* reporter targeted into the *nodal* gene (Collignon *et al.,* 1996).

In mouse embryos an additional transforming growth factor-β (TGF-β) homolog, termed *lefty,* has been found to be asymmetrically expressed at the 2 to 4-somite stage of development (Meno *et al.,* 1996). *Lefty* is asymmetrically expressed in both the floorplate of the neuroectoderm and the lateral plate mesoderm (Meno *et al.,* 1996). Expression in the lateral plate mesoderm overlaps with *nodal* expression and extends from the primitive streak to the caudal region of the developing heart tube (Fig. 2). King and Brown (1997) hypothesized that although cardiac looping mainly involves the rostral heart (outflow tract and ventricles), looping is preceded by or even perhaps driven by morphological and proliferative asymmetry in the inflow tract region. Hence, the localization of *nodal* and *lefty* transcripts in areas adjacent to the future inflow tracts (Fig. 2) may be significant. Recent studies have revealed

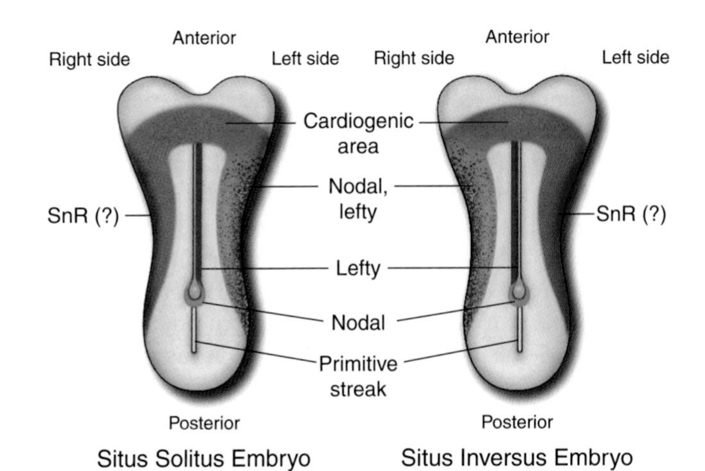

Figure 2 Asymmetric patterns of expression of *nodal* and *lefty* (blue) in the lateral plate and neural floorplate of 4-somite stage situs solitus and situs inversus embryos are shown. The expression of *SnR* (brown) in the opposite lateral plate is indicated, but it is based on the pattern of *cSnR* expression in chick embryos (Isaac *et al.,* 1997). The mouse equivalent of *cSnR* is not known.

that the mouse genome actually encodes two closely linked *lefty* genes (*lefty-1* and *lefty-2*) which are expressed preferentially in the mesoderm (*lefty-1*) or neural tube (*lefty-2*) (Meno *et al.,* 1997). The *nodal* and *lefty* genes are both expressed transiently in the lateral plate mesoderm and are turned off before cardiac looping becomes apparent. The mechanisms that regulate appearance and disappearance of the *lefty* and *nodal* transcripts are not known. In addition, the molecular events that link asymmetric expression of *Shh* near the node to asymmetric *nodal* and *lefty* expression in the lateral plate mesoderm remain unresolved. The receptors for *nodal* and *lefty* also remain unidentified, although the phenotype of the *Act-RIIβ* knockout mice (Oh and Li, 1997) is consistent with the possibility that this gene might encode the *nodal* or *lefty* receptor. The *Act-RIIβ* null mutant mice show right atrial isomerism, suggesting that they have problems recognizing the signals that normally specify development of the left side of the heart.

The chicken gene encoding the zinc finger transcription factor *cSnR* has been shown to have a pattern of expression that becomes restricted to the right lateral plate mesoderm, nearly complementary to the pattern of *nodal* expression (Isaac *et al.,* 1997; Fig. 2). In chickens, expression of *cSnR* may be linked to the activin pathway of L–R polarity determination since implantation of activin beads on the left side of the embryo randomizes *cSnR* expression (Isaac *et al.,* 1997). Suppression of *cSnR* mRNA using antisense oligonucleotides did not lead to randomization of *nodal* expression (Isaac *et al.,* 1997). However, both of the previous treatments randomized the direction of heart looping. Thus, *cSnR* is considered to act downstream of or parallel to the *nodal* gene (Isaac *et al.*, 1997).

In *Xenopus*, *Vg1* (another *TGF-β* family member) has been implicated in L–R patterning. Hyatt *et al.* (1996) injected RNA encoding *BVg1* (which contains the BMP-2 amino-terminal peptide domain fused to mature Vg1) into blastomeres on the right or left side of 16-cell stage embryos. They also injected RNA encoding a dominant-negative truncated activin receptor that is known to block Vg1 signaling (Kessler and Melton, 1995). Expression of *BVg1* on the right or disruption of receptor activity on the left was found to randomize cardiac and visceral L–R orientation and to alter the expression pattern of *Xnr-1* (*Xenopus nodal related-1*) (Hyatt *et al.*, 1996), suggesting that Vg1 signaling on the left side of the embryo is essential for normal L–R asymmetry. Injection of mRNAs encoding activin, noggin, Wnt family members, or the transcription factor *siamois* had no effect on the direction of heart looping (Hyatt *et al.*, 1996). Interestingly, injection of mRNA encoding Vg1 with its own propeptide domain, rather than the BMP-2 prodomain, also did not induce changes in L–R specification. These results suggest that proteolytic activation of Vg1 may occur specifically on the left side of the midline during normal embryonic development.

Although the bone morphogenetic proteins BMP-2 and BMP-4 appear to be important for cardiac differentiation (Zhang and Bradley, 1996; Winnier *et al.*, 1995; Schultheiss *et al.*, 1997), the BMP proteins are not known to influence L–R patterning in mouse embryos.

IV. Gene Expression within the Embryonic Heart

Several transcription factors are known to play critical roles in heart development. For example, mice with a targeted mutation in the *MEF-2C* gene die at about 9.0 days postcoitus (dpc) and show cardiac defects including a failure to undergo normal rightward looping (Lin *et al.*, 1997). The right ventricle does not form, and the mutant hearts show downregulation of expression of the *dHAND* gene, implying that MEF-2C is an essential regulator of cardiac myogenesis and right ventricular development (Lin *et al.*, 1997).

In *Drosophila*, the *tinman* gene is essential for development of the dorsal vessel, a primitive heart-like structure. A *tinman*-related homeobox gene termed *Nkx2.5* or *Csx* has been identified in vertebrates (Komuro and Izumo, 1993; Lints *et al.*, 1993). *Nkx2.5* is one of the earliest markers for the vertebrate cardiac system. Null mutants of *Nkx2.5* have been generated (Lyons *et al.*, 1995; Biben and Harvey, 1997). Although heart tube formation occurs in the mutant embryos, looping of the cardiac tube fails to proceed normally, and there is a lack of expression of the myogenic genes *eHAND*, *MLC-2v*, and *CARP* (cardiac ankyrin repeat protein) in the myocardial region.

Two members of the bHLH family of transcription factors, termed *dHAND* (*Hxt2*) and *eHAND* (*Hxt1*), are expressed in the heart progenitor cells, the looping heart tube, and cardiac neural crest-derived cells (Srivastava *et al.*, 1995; Cserjesi *et al.*, 1995; Hollenberg *et al.*, 1995; Cross *et al.*, 1995). The function of these HAND proteins in chick embryos was studied using antisense oligonucleotides. Simultaneous treatment with antisense oligos for both genes led to arrested development at the looping heart tube stage. *In situ* hybridizations showed homogenous expression of both factors in the looping heart. In contrast, in mouse embryos asymmetric expression patterns have been observed (Srivastava *et al.*, 1997; Biben and Harvey, 1997; see Chapters 7 and 9). At 7.75 dpc both *dHAND* and *eHAND* are expressed throughout the precardiogenic mesoderm and in other regions of the embryo. After the formation of the linear heart tube, *eHAND* gene expression is turned off in the region of the tube that will become the right ventricle (Srivastava *et al.*, 1997). During heart looping, the future right ventricle shows *dHAND* expression, whereas the future left ventricle shows *eHAND* expression. Therefore, beginning at the earliest stages of cardiac looping, regions of the heart tube that are destined to form different compartments appear to have distinctive molecular addresses. This suggests that the HAND proteins may provide transcriptional control of genes involved in cardiac looping and chamber-specific myogenesis. Targeted deletion of the *dHAND* gene in mice results in embryonic hearts which begin to loop in the correct direction but fail to form a right ventricle and have only a single left-sided ventricular chamber (Srivastava *et al.*, 1997). It will be interesting to determine whether expression of *eHAND* under the control of the *dHAND* promoter will cure these cardiac defects.

The zinc finger transcription factors GATA-4, GATA-5, and GATA-6 also play important roles in cardiogenesis (Laverriere *et al.*, 1994; Grepin *et al.*, 1995; Jiang and Evans, 1996; Morrisey *et al.*, 1997; Kuo *et al.*, 1997; Molkentin *et al.*, 1997; see Chapter 17).

Despite recent progress in understanding the developmental roles of these cardiac transcription factors, the molecular pathways that are used to direct or determine cardiac looping remain unidentified. It is unclear how the initiation of heart looping is linked to the chamber specification process.

Several extra-cellular matrix (ECM) proteins are asymmetrically localized during early cardiac development. These include flectin (Tsuda *et al.,* 1996), hLAMP1 (heart-specific lectin-associated matrix protein-1), and JB3 (Smith *et al.,* 1997). In normal chick embryos, flectin is expressed predominantly in the left pericardiac mesoderm on the basal side of the myocardium and within the cardiac jelly (Tsuda *et al.,* 1996). The hLAMP1 and JB3 proteins are distributed asymmetrically within the pericardiac fields at the head process stage, with hLAMP1 distribution being left biased and JB3 expression right biased (Smith *et al.,* 1997).

Retinoic acid (RA) may play an important role in cardiac looping (see Chapter 13). Vitamin A-deficient quail embryos do not show proper heart looping (Dersch and Zile, 1993; Twal *et al.,* 1995). Ectopic application of RA has been found to alter hLAMP1 and JB3 distribution (Smith *et al.,* 1997) and to change the direction of heart looping, thereby indicating that both these ECM components can act as useful molecular markers for identifying the direction of heart looping. The defects in heart looping in vitamin A-deficient embryos or in mice bearing null alleles of RA receptors or retinoid X-receptors do not appear to arise from changes in L–R axis specification (Mark, 1997).

V. Mutations in Genes That Affect L–R Asymmetry

Defects in dynein arms have been found in humans with immotile cilia syndrome, a condition in which 50% of the patients develop with situs inversus (Afzelius, 1995). The *iv* (inversus viscerum) mutation in mice also results in randomization of the direction of heart looping (Hummel and Chapman, 1959; McGarth *et al.,* 1992; Seo *et al.,* 1992; Schreiner *et al.,* 1993). Recently, the *iv* gene has been identified and found to encode an axonemal dynein (Supp *et al.,* 1997). The *iv* mutation is thought to impair aspects of intracellular transport that are essential for the initial determination of embryonic L–R asymmetry.

The protein encoded by the *iv* locus gene has been named left–right dynein (lrd). The *iv* mutation is a point mutation that maps to a conserved amino acid within the dynein protein. Transcripts of the *lrd* gene are present at the blastocyst stage of development. At embryonic day 7.5, *lrd* is preferentially expressed in the node of mouse embryos (Supp *et al.,* 1997), prior to the onset of asymmetric expression of *nodal* or *lefty* (Meno *et al.,* 1996; Collignon *et al.,* 1996). Interestingly, other genes that have been implicated in L–R asymmetry, for example, *HNF-3β,* and *Shh,* are also expressed in the mouse node. The exact role of *lrd* in cardiac looping is yet to be determined. However, the random heart looping seen in *iv/iv* mice may be a direct consequence of random L–R asymmetry determination (Lowe *et al.,* 1996).

Among the known L–R laterality mutations, the *inv* (*inversion of embryonic turning*) mutation is interesting and unique in that it causes a consistent reversal of L–R polarity (Yokoyama *et al.,* 1993). Mutants homozygous for the *inv* mutation show a consistent reversal of the L–R polarity of heart looping and embryonic turning (Yokoyama *et al.,* 1993, 1995). The *inv* mutation was generated by random insertional mutagenesis in a family of transgenic mice (Yokoyama *et al.,* 1993).

Recently, a 500-kb yeast artificial chromosome (YAC) from the *inv* region of mouse chromosome 4 was used to generate transgenic mice by microinjection into mouse embryos. Matings were then used to transfer the YAC transgenes into the *inv* genetic background. For two different integration sites, the transgenic YAC rescued the *situs inversus* phenotype (K. Majumder and P. Overbeek, personal communication). This genetic complementation indicates that coding and control elements for the *inv* gene are contained within the 500-kb YAC. A candidate *inv* cDNA has recently been identified. The candidate gene is transcribed in wild-type and YAC-rescued mice but is defective in homozygous *inv* mutants (P. Overbeek and W. Dai, personal communication). A novel model for L–R axis specification involving the *inv* gene is proposed in Section VIII.

The expression patterns of *lefty* (Meno *et al.,* 1996) and *nodal* (Collignon *et al.,* 1996; Lowe *et al.,* 1996) in normal, *iv/iv,* and *inv/inv* mice show that there is a consistent correlation between the side on which each gene is expressed and the direction of cardiac looping. For example, *inv/inv* embryos express *nodal* and *lefty* on the right side rather than the left side (Fig. 2). The patterns of *nodal* expression in *iv/iv* mice are particularly fascinating. Four different patterns of expression in the lateral plate mesoderm are seen: left-side only, right-side only, both sides, and neither side (Lowe *et al.,* 1996). Presumably, a bilaterally symmetric *nodal* expression pattern implies that the normal system for L–R axis specification has been disrupted, and that embryonic tissues will not be provided with consistent L-R polarity information. Interestingly, *Shh* misexpression in chick embryos also results in symmetric *nodal* expression, random heart looping, and a heterotaxic phenotype (Levin, 1997).

Other mutations which show randomization of L–R asymmetry include the *Ft* (*Fused toes*) (van der Hoeven *et al.,* 1994) and *Dh* (*Dominant hemimelia*) (Biddle *et al.,* 1991) mutations in mice and *Hyd* (*Hydrocephalus*) mutation in rats (Torikata *et al.,* 1991). The *Hyd* rats have respiratory ciliary defects, with dynein arms missing, somewhat analogous to human immotile cilia syndromes. Randomization of heart polarity and L–R asymmetry has also been observed in mice homozygous for a null mutation of *Mgat-1* (Metzler *et al.,* 1994). *Mgat-1* encodes an enzyme that is involved in the biosynthesis of N-linked oligosaccharides. The mutant phenotype suggests that complex oligosaccharides may play a role in L–R axis specification.

In humans, mutations that affect cardiac L–R polarity are present in families with immotile cilia syndrome (Moreno and Murphy, 1991) and in families with X-linked heterotaxia (*HTX*) (Casey *et al.,* 1993). The *HTX* gene has recently been cloned and encodes *ZIC3,* a zinc finger transcription factor (Gebbia *et al.,* 1997). The mechanism of action of the *HTX* gene is not known.

Human twins who are conjoined along the midline axis show an elevated incidence of situs inversus in the right twin (Cunniff *et al.,* 1988; Fig. 3). Levin *et al.* (1996), based on their studies in chicken embryos, proposed a model for "cross talk" between twinned embryos (Fig. 3). The twin on the left side would be predicted to synthesize an inhibitor of Shh (e.g., activin) on its right side. If the inhibitor diffuses to the adjacent twin, it will inhibit Shh expression on the left side of that twin (Fig. 3). As a result, Shh expression in the right twin would be inhibited on both sides resulting in loss of *nodal* induction and random situs specification (Fig. 3). This is consistent with the observations in dicephalic (joined laterally at the chest) human twins (Levin *et al.,* 1996).

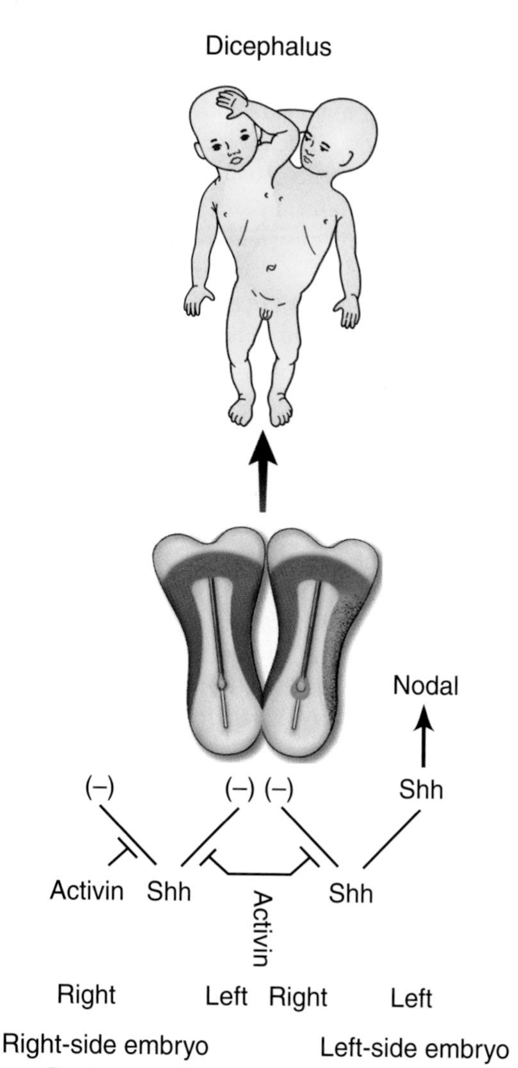

Figure 3 Proposed molecular pathway leading to laterality defects in human conjoined twins. The predicted patterns of activin and *Shh* expression are shown at the bottom. Activin expressed on the right side of the left embryo is proposed to inhibit Shh expression on the left side of the right embryo resulting in the loss of *nodal* (blue) expression in the embryo on the right and a subsequent increased likelihood of L–R axis reversal (adapted from Levin *et al.,* 1996).

VI. Embryonic Alterations of Cardiac Looping in Vertebrates

In *Xenopus* embryos it has been demonstrated that disruptions of the notochord and floorplate lead to bilateral *Xnr-1* expression (Lohr *et al.,* 1997) and randomization of cardiac looping (Danos and Yost, 1996). Zebrafish mutants having notochord and midline defects have been identified, and in two of the mutants, floating head (*flh*) and no tail (*ntl*), cardiac inversions have been observed (Halpern *et al.,* 1993; Talbot *et al.,* 1995). The function(s) of the notochord and floorplate may in part be mediated by the fibronectin-rich extracellular matrix (Yost, 1992; see Chapter 21). One hypothesis is that the notochord and floorplate provide a barrier to diffusion of L–R morphogens such as Vg-1, *nodal,* and/or *lefty* (Lohr *et al.,* 1997). Using cells stained with lysine–rhodamine–dextran dye, Khaner (1996) has shown that cells do not cross the dorsal midline in chicken embryos, indicating the presence of a midline barrier.

The effects of adrenergic receptors on cardiac looping have been studied in a culture system using presomite rat embryos (Flynn *et al.,* 1993). Fifty percent of the rat embryos show situs inversus when cultured in serum-free medium. The incidence of situs inversus increases to 73% when the α-adrenergic antagonist l-phenylephrine is added to the medium, indicating that adrenergic receptors may influence the events that determine normal L–R asymmetry in vertebrate embryos (Flynn *et al.,* 1993).

VII. Signal Transduction

During the past few years considerable progress had been made in the identification of genes that play roles in L–R specification. The consistent pattern of *nodal* expression in the left lateral plate mesoderm in different species indicates that molecular pathways of L–R determination are conserved across species. Nonetheless, not much is known about the receptors and signal transduction pathways that are used upstream and downstream from *nodal* in L–R axis specification. Presumably, *patched* receptors are used to recognize the Shh signals (Levin, 1997), and receptors of the TGF-β family are stimulated by activin, *Vg1*, *nodal*, *lefty-1*, and *lefty-2* proteins. However, the mechanisms of signal transduction and the roles of downstream transcription factors, such as the *Gli* and *Smad* proteins, remain undefined. The regulation of *cSnR-1* also remains to be clarified. In terms of understanding cardiac morphogenesis, it is hoped that future experiments will provide information about how activin or *nodal* signaling is linked to heart looping, *eHAND* and *dHAND* expression, and organization of the cardiac ECM.

VIII. Initiation of L–R Asymmetry: An Axis Conversion Model

Two fundamental questions remain unanswered: How is L–R asymmetry first established and how does the embryo use L–R axis information to determine morphological asymmetries? Although embryos are likely to use their anterioposterior (A–P) and dorsoventral (D–V) axes to help specify L–R asymmetry (Brown and Wolpert, 1990; Wood, 1997), the events and signals that connect all the axes remain unidentified.

The development of handed asymmetry in vertebrates is likely to be a multistep process (Brown and Wolpert, 1990). Based on the recent identification of a candidate cDNA for *inv,* we have devised a simplified model that provides an explicit proposal for a molecular pathway that may be used to initiate L–R asymmetry (Fig. 4). Our model assumes that the D–V and A–P axes are specified at the embryonic midline prior to the initiation of L–R asymmetry. Once these axes are established, epithelial cells along the bilateral midline will have a unique concordance between their apical–basal polarity and the D–V axis of the embryo (Fig. 4). Although the molecules that signal A–P asymmetry along the midline have not been defined, it is possible that A–P axis information is encoded in a consistent molecular orientation for extracellular matrix proteins (Fig. 4). We propose that the midline cells express a transmembrane protein that can bind to the oriented extracellular proteins and that can be connected to the intracellular cytoskeletal architecture through the function of the *inv* protein, thereby providing a molecular system that can transduce extracellular polarity along one axis into intracellular polarity along a different axis. The essential components of the model are a transmembrane "axis sensor" complex and an intracellular "axis conversion" complex which would include the *inv* protein (Fig. 4).

Currently, the model is purely hypothetical, but it has certain features that make it unique and potentially relevant. First, and contrary to all previous models, it proposes a genetically specified pathway for initiating L–R asymmetry that is linked molecularly to the prior establishment of D–V and A–P axes but does not presuppose any type of preceding L–R asymmetry. The consistent specification of L–R asymmetry is determined by the molecular structures of the axis sensor and the axis conversion proteins. Second, this model provides an updated prespective on the "F" protein proposed by Brown and Wolpert (1990). In the current model, the system that establishes L–R polarity inside the cell is oriented by sterospecific binding of the transmembrane axis sensor, thereby establishing a multiprotein membrane-embedded chiral complex that is stably asymmetric along all axes of the embryo. Finally, the model makes the distinctive prediction that positional information can be provided not only by gradients of diffusible morphogens but also by stable polarization of the components of the ECM. Directional organization within the ECM could be used in general to convey positional information to polarized epithelial and neuronal cells.

The L–R polarization of the midline epithelial cells would represent only the first step in the developmental cascade that specifies morphological L–R asymmetry. Polarization of the cytoskeleton, and particularly the microtubules, in the midline epithelial cells could result in directional intracellular transport perhaps mediated specifically by *lrd* (Supp *et al.,* 1997). Asymmetric cell division, asymmetric cellular growth, or asymmetric communication between adjacent cells might be used to establish a localized asymmetry, and then subsequently to direct the asymmetric expression of diffusible extracellular signals (such as *nodal* or *lefty*) that would coordinate later asymmetric morphogenesis of the visceral organs.

To illustrate this model, assume that cell division resulted in the preferential localization of a repressor of activin expression in cells to the left of the midline. Cells to the right would then produce more activin than cells on the left and a L–R morphogen gradient would be initiated analogous to the original model of Brown and

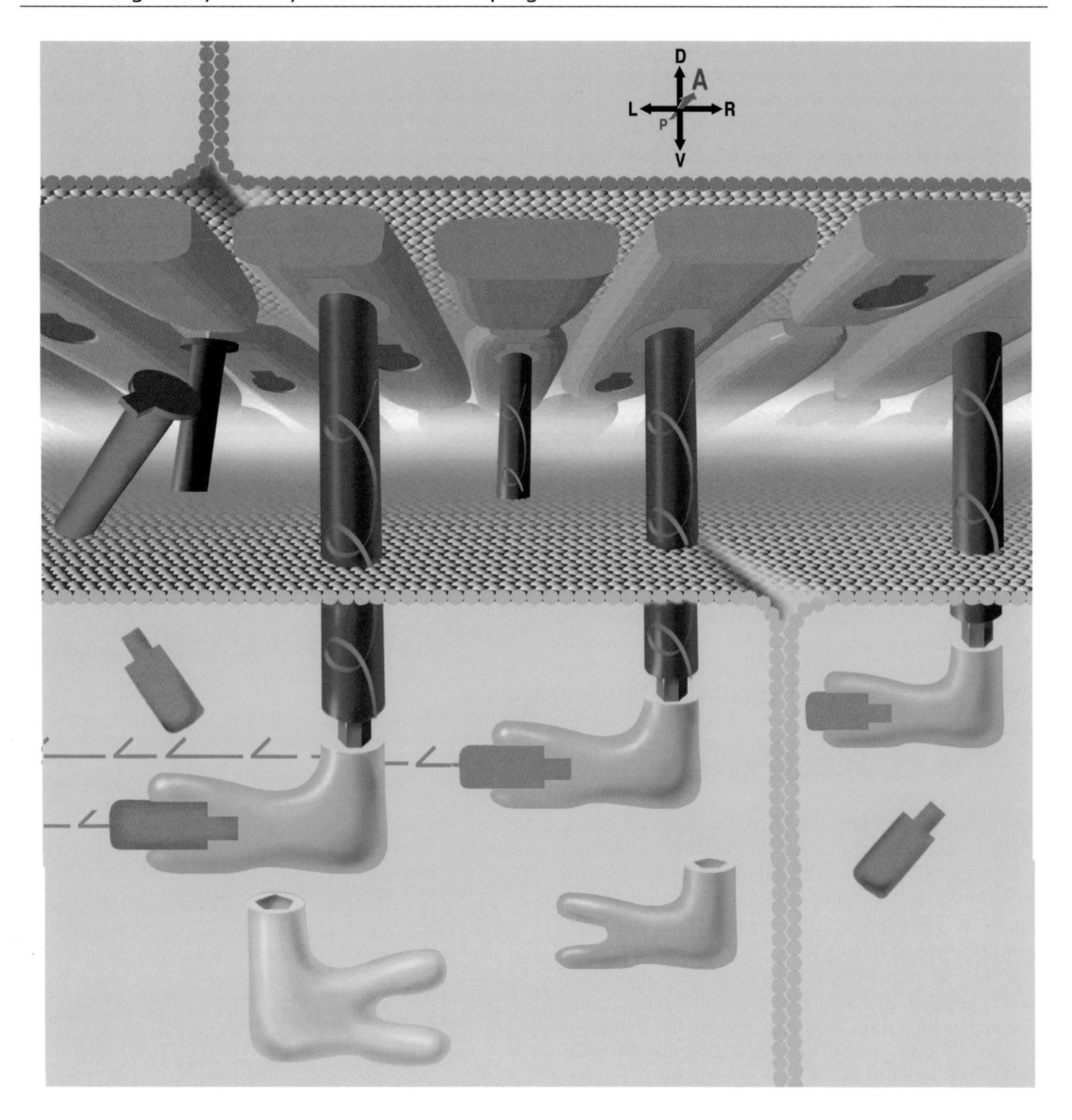

Figure 4 Schematic representation of a molecular model for converting D–V and A–P asymmetry into L–R asymmetry. The diagram depicts an interface between two layers of epithelial cells (aqua and pink) that are located along the midline of a vertebrate embryo. The view is looking down the A–P axis of the embryo. The D–V axis of the embryo matches the apical-basal polarity of the epithelial cells. Once the A–P axis of the embryo is established, the epithelial cells are hypothesized to synthesize/organize an extracellular matrix (ECM) (indicated by the red bars) that includes (one or more) proteins that are oriented in a consistent pattern along the A–P axis of the embryo. Alignment along the A–P axis establishes a consistent orientation for binding sites (depicted as indentations) within the ECM. The putative "axis sensor" proteins are depicted in blue. These are transmembrane proteins (analogous to integrins) that have both an ECM-binding domain (docking is indicated by the green cap) and also an intracellular domain that can serve as a docking site for an intracellular "axis conversion" system (lighter blue) that is predicted to include the inv protein. The axis conversion system is postulated to serve as an organizing center (magenta) for polarized cytoskeletal proteins such as microtubules (indicated by arrows). An inverted cytoskeletal orientation could be provided by a postulated default organizer (yellow). This model proposes that sterospecific immobilization of a transmembrane (axis sensor) protein can be used to convert extracellular polarity along one specific axis into intracellular asymmetry along a different axis. The model is unique in that it describes a molecularly feasible system to specify consistent L–R asymmetry based only on previously specified D–V and A–P asymmetries.

Wolpert (1990). This initial establishment of L–R asymmetry could be followed by downstream events that specify asymmetric distribution of molecules such as *nodal, lefty,* cSnR, and flectin. In homozygous *iv* mutants the dynein defects would lead to randomization of the distribution of the repressor and therefore random distribution of the morphogen, resulting in a 50% chance that either side will produce sufficient morphogen to trigger the downstream pathways of axis determination. These predictions agree well with the observed patterns of *nodal* expression in *iv* mutant mice (Lowe *et al.,* 1996). Experimental manipulations that result in loss of ECM polarity are also predicted to result in a randomization of the process for initiating L–R asymmetry. One important feature of this model is the prediction that cells along the midline of normal vertebrate embryos will show a consistent L–R cellular polarity.

The axis sensor–axis converter model does not directly explain the consistent L–R axis reversal seen in the *inv/inv* mutants. One possibility is that the mutated *inv* locus encodes a novel protein that helps to specify a reversed cellular polarity in the absence of wild-type *inv.* A more likely possibility is that the *inv* mutation is a null mutant, implying the existence of other proteins that can consistently specify a reversed cellular and/or embryonic polarity when *inv* is missing (Fig. 4).

IX. Concluding Remarks

Looping of the heart is the first morpological indication of L–R asymmetry in vertebrates. Several proteins, including TGF-β-related signaling molecules, bHLH transcription factors, zinc finger transcription factors, and ECM components, are asymmetrically expressed both prior to and during looping of the heart. It is not clear how asymmetric expression of these proteins is used to regulate or coordinate cell proliferation, adhesion, and/or migration to induce asymmetric looping of the heart. Other critical questions also remain unanswered. When and how is L–R asymmetry first specified? How is molecular L–R asymmetry converted into architectural asymmetry? Why did evolution decide that vertebrates would benefit from cardiac looping to the right?

In summary, there are still mysteries to unravel and models to build before the secrets to asymmetry of the heart will finally yield.

Acknowledgments

Research on the *inv* mutation was supported by Grant HL49953 from the NIH. The authors thank John Ellsworth for his patient secretarial assistance.

References

Afzelius, B. A. (1985). The immotile-cilia syndrome: A microtubule-associated defect. *CRC Crit. Rev. Biochem.* **19,** 63–87.

Biben, C., and Harvey, R. P. (1997). Homeodomain factor Nkx2.5 controls left/right asymmetric expression of bHLH gene eHAND during murine heart development. *Genes Dev.* **11,** 1357–1369.

Biddle, F. G., Jung, J. D., and Eales, B. A. (1991). Genetically determined variation in the azygos vein in the mouse. *Teratology* **44,** 675–683.

Bisaha, J. G., and Bader, D. (1991). Identification and characterization of a ventricular specific avian myosin heavy chain, VMHC1: expression in differentiating cardiac and skeletal muscle. *Dev. Biol.* **148,** 355–364.

Brown, N. A., and Wolpert, L. (1990). The development of handedness in left/right asymmetry. *Development* **109,** 1–9.

Casey, B., Devoto, M., Jones, K. L., and Ballabio, A. (1993). Mapping a gene for familial situs abnormalities to human chromosome Xq24–q27.1. *Nature Genet.* **5,** 403–407.

Collignon, J., Varlet, I., and Robertson, E. J. (1996). Relationship between asymmetric nodal expression and the direction of embryonic turning. *Nature* **381,** 155–158.

Cross, J. C., Flannery, M. L., Blanar, M. A., Steingrimsson, E., Jenkins, N. A., Copeland, N. G., Rutter, W. J., and Werb, Z. (1995). Hxt encodes a basic helix–loop–helix transcription factor that regulates trophoblast cell development. *Development* **121,** 2513–2523.

Cserjesi, P., Brown, D., Lyons, G. E., and Olson, E. N. (1995). Expression of the novel basic helix–loop–helix gene eHAND in neural crest derivatives and extraembryonic membranes during mouse development. *Dev. Biol.* **170,** 664–678.

Cunniff, C., Jones, K. L., Jones, M. C., Saunders, B., Shepard, T., and Benirschke, K. (1988). Laterality defects in conjoined twins: Implications for normal asymmetry in human embryogenesis. *Am. J. Med. Genet.* **31,** 669–677.

Danos, M. C., and Yost, H. J. (1996). Role of notochord in specification of cardiac left–right orientation in zebrafish and *Xenopus. Dev Biol.* **177,** 96–103.

Dersch, H., and Zile, M. H. (1993). Induction of normal cardiovascular development in the vitamin A-deprived quail embryo by natural retinoids. *Dev. Biol.* **160,** 424–433.

Flynn, T. J., Gibson, R. R., and Johannessen, J. N. (1993). Evidence that spontaneous *situs inversus* in cultured neural plate staged rat embryos is additive with and not mediated through adrenergic mechanisms. *Teratology* **48,** 161–168.

Gebbia, M., Gerrero, G. B., Pilia, G., Bassi, M. T., Aylsworth, A. S., Penman-Splitt, M., Bird, L. M., Bamforth, J. S., Burn, J., Schlessinger, D., Nelson, D. L., and Casey, B. (1997). X-linked *situs* abnormalities result from mutations in ZIC3. *Nature Genet.* **17,** 305–309.

Grepin, C., Robitaille, L., Antakly, T., and Nemer, M. (1995). Inhibition of transcription factor GATA-4 expression blocks in vitro cardiac muscle differentiation. *Mol. Cell Biol.* **15,** 4095–4102.

Halpern, M. E., Ho, R. K., Walker, C., and Kimmel, C. B. (1993). Induction of muscle pioneers and floor plate is distinguished by the zebrafish *no tail* mutation. *Cell* **75,** 99–111.

Hamburger, V., and Hamilton, H. L. (1951). A series of normal stages in the development of the chick embryo. *J. Morphol.* **88,** 49–67.

Harvey, R. P. (1996). NK-2 homeobox genes and heart development. *Dev. Biol.* **178,** 203–216.

Hollenberg, S. M., Sternglanz, R., Cheng, P. F., and Weintraub, H. (1995). Identification of a new family of tissue-specific basic helix–loop–helix proteins with a two-hybrid system. *Mol. Cell Biol.* **15,** 3813–3822.

Hummel, K. P., and Chapman, D. B. (1959). Visceral inversion and associated anomalies in the mouse. *J. Heredity* **50,** 10–13.

Hyatt, B. A., Lohr, J. L., and Yost, H. J. (1996). Initiation of vertebrate left–right axis formation by maternal Vg1. *Nature* **384,** 62–66.

Isaac, A., Sargent, M., and Cooke, J. (1997). Control of vertebrate left–right asymmetry by a snail related zinc finger gene. *Science* **275,** 1301–1304.

Itasaki, N., Nakamura, H., Sumida, H., and Yasuda, M. (1991). Actin bundles on the right side in the caudal part of the heart tube play a role in dextro-looping in the embryonic chick heart. *Anat. Embryol. (Berlin)* **183,** 29–39.

Jiang, Y., and Evans, T. (1996). The Xenopus GATA-4/5/6 genes are associated with cardiac specification and can regulate cardiac-specific transcription during embryogenesis. *Dev. Biol.* **174,** 258–270.

Kessler, D. S., and Melton, D. A. (1995). Induction of dorsal mesoderm by soluble, mature Vg1 protein. *Development* **121,** 2155–2164.

Khaner, O. (1996). Axis formation in half blastoderm of the chick: Stage at separation and the relative postions of fused halves influence axis development. *Roux's Arch. Dev. Biol.* **205,** 364–370.

King, T., and Brown, N. A. (1997). Embryonic asymmetry: Left TGFβ at the right time? *Curr. Biol.* **7,** R212–R215.

Komuro, I., and Izumo, S. (1993). *Csx:* A murine homeobox-containing gene specifically expressed in the developing heart. *Proc. Natl. Acad. Sci. USA* **90,** 8145–8149.

Kuo, C. T., Morrisey, E. E., Anandappa, R., Sigrist, K., Lu, M. M., Parmacek, M. S., Soudais, C., and Leiden, J. M. (1997). GATA4 transcription factor is required for ventral morphogenesis and heart tube formation. *Genes Dev.* **11,** 1048–1060.

Laverriere, A. C., MacNeill, C., Mueller, C., Poelmann, R. E., Burch, J. B., and Evans, T. (1994). GATA-4/5/6, a subfamily of three transcription factors transcribed in developing heart and gut. *J. Biol. Chem.* **269,** 23177–23184.

Levin, M. (1997). Left–right asymmetry in vertebrate embryogenesis. *BioEssays* **19,** 287–296.

Levin, M., Johnson, R. L., Stern, C. D., Kuehn, M., and Tabin, C. (1995). A molecular pathway determining left–right asymmetry in chick embryogenesis. *Cell* **82,** 803–814.

Levin, M., Roberts, D. J., Holmes, L. B., and Tabin, C. (1996). Laterality defects in conjoined twins. *Nature* **384,** 321.

Lin, Q., Schwarz, J., Bucana, C., and Olson, E. N. (1997). Control of mouse cardiac morphogenesis and myogenesis by transcription factor MEF2C. *Science* **276,** 1404–1407.

Lints, T. J., Parsons, L. M., Hartley, L., Lyons, I., and Harvey, R. P. (1993). *Nkx-2.5:* A novel murine homeobox gene expressed in early heart progenitor cells and their myogenic descendants. *Development* **119,** 419–431.

Lohr, J. L., Danos, M. C., and Yost, H. J. (1997). Left–right asymmetry of a *nodal* related gene is regulated by dorsoanterior midline structures during *Xenopus* development. *Development* **124,** 1465–1472.

Lowe, L. A., Supp, D. M., Sampath, K., Yokoyama, T., Wright, C. V., Potter, S. S., Overbeek, P., and Kuehn, M. R. (1996). Conserved left–right asymmetry of *nodal* expression and alterations in murine *situs inversus. Nature* **381,** 158–161.

Lyons, I., Parsons, L. M., Hartley, L., Li, R., Andrews, J. E., Robb, L., and Harvey, R. P. (1995). Myogenic and morphogenetic defects in the heart tubes of murine embryos lacking the homeobox gene *Nkx2.5. Genes Dev.* **9,** 1654–1666.

Mark, M. (1997). Retinoic acid receptors in cardiac development. In *Genetic Control of Heart Development* (E. N. Olson, R. P. Harvey, R. A. Schultz, and J. S. Altman, Eds.), pp. 154–160. HFSP, Strasbourg.

Matzuk, M. M., Kumar, T. R., Vassalli, A., Bickenbach, J. R., Roop, D. R., Jaenisch, R., and Bradley, A. (1995a). Functional analysis of activins during mammalian development. *Nature* **374,** 354–356.

Matzuk, M. M., Kumar, T. R., and Bradley, A. (1995b). Different phenotypes for mice deficient in either activins or activin receptor type II. *Nature* **374,** 356–360.

Matzuk, M. M., Lu, N., Vogel, H., Sellheyer, K., Roop, D. R., and Bradley, A. (1995c). Multiple defects and perinatal death in mice deficient in follistatin. *Nature* **374,** 360–363.

McGarth, J., Horwich, A. L., and Brueckner, M. (1992). Duplication/deficiency mapping of situs inversus viscerum (iv), a gene that determines left–right asymmetry in the mouse. *Genomics* **14,** 643–648.

Meno, C., Saijoh, Y., Fujii, H., Ikeda, M., Yokoyama, T., Yokoyama, M., Toyoda, Y., and Hamada, H. (1996). Left–right asymmetric expression of the TGF beta family member *lefty* in mouse embryos. *Nature* **381,** 151–155.

Meno, C., Ito, Y., Saijoh, Y., Matsuda, Y., Tashiro, K., Kuhara, S., and Hamada, H. (1997). Two closely-related left–right asymmetrically expressed genes, *lefty-1* and *lefty-2:* Their distinct expression domains, chromosomal linkage and direct neuralizing activity in *Xenopus* embryos. *Genes Cells* **2,** 513–524.

Metzler, M., Gertz, A., Sarkar, M., Schachter, H., Schrader, J. W., and Marth, J. D. (1994). Complex asparagine-linked oligosaccharides are required for morphogenic events during post-implantation development. *EMBO J.* **13,** 2056–2065.

Mikawa, T., Borisov, A., Brown, A. M. C., and Fischman, D. A. (1992). Clonal analysis of cardiac morphogenesis in the chicken embryo using a replication-defective retrovirus: I. Formation of the ventricular myocardium. *Dev. Dyn.* **193,** 11–23.

Molkentin, J. D., Lin, Q., Duncan, S. A., and Olson, E. N. (1997). Requirement of the transcription factor GATA4 for heart tube formation and ventral morphogenesis. *Genes Dev.* **11,** 1061–1072.

Moreno, A. and Murphy, E. A. (1991). Inheritence of Kartagener's syndrome. *Am. J. Med. Genet.* **8,** 305–313.

Morrisey, E. E., Ip, H. S., Lu, M. M., and Parmacek, M. S. (1996). GATA-6: A zinc finger transcription factor that is expressed in multiple cell lineages derived from lateral mesoderm. *Dev. Biol.* **177,** 309–322.

Morrisey, E. E., Ip, H. S., Tang, Z., Lu, M. M., and Parmacek, M. S. (1997). GATA-5: A transcriptional activator expressed in a novel temporally and spatially-restricted pattern during embryonic development. *Dev. Biol.* **183,** 21–36.

Oh, S. P., and Li, E. (1997). The signaling pathway mediated by the type IIB activin receptor controls axial patterning and lateral asymmetry in the mouse. *Genes Dev.* **11,** 1812–1826.

Olson, E. N. (1997). Things are developing in cardiology. *Circ. Res.* **80,** 604–606.

Olson, E. N., and Srivastava, D. (1996). Molecular pathway controlling heart development. *Science* **272,** 671–676.

Overbeek, P. A. (1997). Right and left go dHAND and eHAND. *Nature Genet.* **16,** 119–121.

Price, R. L., Chintanowonges, C., Shiraishi, I., Borg, T. K., and Terracio, L. (1996). Local and regional variations in myofibrillar patterns in looping rat hearts. *Anat. Rec.* **245,** 83–93.

Schreiner, C. M., Scott, W. J. Jr., Supp, D. M., and Potter, S. S. (1993). Correlation of forelimb malformation asymmetries with visceral organ situs in the transgenic mouse insertional mutation *legless. Dev. Biol.* **158,** 560–562.

Schultheiss, T. M., Burch, J. B., and Lassar, A. B. (1997). A role of bone morphogenic proteins in the induction of cardiac myogenesis. *Genes Dev.* **11,** 451–462.

Seo, J. W., Brown, N. A., Ho, S. Y., and Anderson, R. H. (1992). Abnormal laterality and congenital cardiac anomalies. Relations of visceral and cardiac morphologies in the iv/iv mouse. *Circulation* **86,** 642–650.

Smith, S. M., Dickman, E. D., Thompson, R. P., Sinning, A. R., Wunsch, A. M., and Markwald, R. R. (1997). Retinoic acid directs cardiac laterality and the expression of early markers of precardiac asymmetry. *Dev. Biol.* **182,** 162–171.

Srivastava, D., Cserjesi, P., and Olson, E. N (1995). A subclass of bHLH proteins required for cardiac morphogenesis. *Science* **270,** 1995–1999.

Srivastava, D., Thomas, T., Lin, Q., Kirby, M. L., Brown, D., and Olson, E. N. (1997) Regulation of cardiac mesoderm and neural crest development by the bHLH transcription factor, dHAND. *Nature Genet.* **16,** 154–160.

Supp, D. M., Witte, D. P., Potter, S. S., and Brueckner, M. (1997). Mutation of an axonemal dynein affects left–right asymmetry in *inversus viscerum* mice. *Nature* **389,** 963–966.

Taber, I. A., Lin, I. E., and Clark, E. B. (1995). Mechanics of cardiac looping. *Dev. Dyn.* **203,** 42–50.

Talbot, W. S., Trevarrow, B., Halpern, M. E., Melby, A. E., Farr, G., Postlethwait, J. H., Jowett, T., Kimmel, C. B., and Kimelman, D. (1995). A homeobox gene essential for zebrafish notochord development. *Nature* **378,** 150–157.

Thomas, T., Yamagishi, H., Overbeek, P. A., Olson, E. N., and Srivastava, D. (1998). The bHLH factors, dHAND and eHAND, specify pulmonary and systemic cardiac ventricles independent of left–right sidedness. *Dev. Biol.* **196,** 228–236.

Torikata, C., Kijimoto, C., and Koto, M. (1991). Ultrastructure of respiratory cilia of WIC-Hyd male rats. An animal model for human immotile cilia syndrome. *Am. J. Pathol.* **138,** 341–347.

Tsuda, T., Philp, N., Zile, M. H., and Linask, K. K. (1996). Left–right asymmetric localization of flectin in the extracellular matrix during heart looping. *Dev. Biol.* **173,** 39–50.

Twal, W., Roze, L., and Zile, M. H. (1995). Anti-retinoic acid monoclonal antibody localizes all-trans-retinoic acid in target cells and blocks normal development in early quail embryo. *Dev. Biol.* **168,** 225–234.

van der Hoeven, F., Schimmang, T., Volkmann, A., Mattei, M-G., Kyewski, B., and Ruther, U. (1994). Programmed cell death is affected in the novel mouse mutant Fused toes (Ft). *Development* **120,** 2601–2607.

Winnier, G., Blessing, M., Labosky, P. A., and Hogan, B. L. (1995). Bone morphogenetic protein-4 is required for mesoderm formation and patterning in the mouse. *Genes Dev.* **9,** 2105–2116.

Wood, W. B. (1997). Left–right asymmetry in animal development. *Annu. Rev. Cell Dev. Biol.* **13,** 53–82.

Yokoyama, T., Copeland, N. G., Jenkins, N. A., Montgomery, C. A., Elder, F. F., and Overbeek, P. A. (1993). Reversal of left–right asymmetry: A situs inversus mutation. *Science* **260,** 679–682.

Yokoyama, T., Harrison, W. R., Elder, F. F. B., and Overbeek, P. A. (1995). Molecular analysis of the *inv* insertional mutation. In Developmental Mechanisms of Heart Disease (E. B. Clark, R. R. Markwald, and A. Takao, (Eds.), pp. 513–520. Futura, Armonk, NY.

Yost, H. J. (1992) Regulation of vertebrate left–right asymmetries by extracellular matrix. *Nature* **357,** 158–161.

Yutzey, K. E., and Bader, D. (1995). Diversification of cardiomyogenic cell lineages during early heart development. *Circ. Res.* **77,** 216–219.

Zhang, H., and Bradley, A. (1996). Mice deficient for BMP2 are nonviable and have defects in amnion/chorion and cardiac development. *Development* **122,** 2977–2986.

Zou, Y. M., Evans, S., Chen, J., Kuo, H. C., Harvey, R. P., and Chien, K. R. (1997). CARP, a cardiac ankyrin repeat protein, is downstream in the NKx2.5 homeobox gene pathway. *Development* **124,** 793–804.

VIII

Cell Proliferation in Cardiovascular Development and Disease

23

The Cardiac Cell Cycle

W. Robb MacLellan* **and Michael D. Schneider†,‡**

*Molecular Cardiology Unit, Departments of Medicine, †Cell Biology, and ‡Molecular Physiology & Biophysics, and *Houston Veterans Affairs Medical Center, Baylor College of Medicine, Houston, Texas 77030*

I. Introduction

Historically, initial insights into the regulatory mechanisms underlying cell cycle progression arose from studies in unicellular protozoa and, subsequently, in *Xenopus* oocytes (Pardee *et al.*, 1978). These nominally simple systems established a foundation and structural framework with which to understand more complex, multicellular ones and have led progressively to a general model for the mammalian cell cycle (Hunter and Pines, 1994; Sherr, 1994; Weinberg, 1995; Stillman, 1996). In response to mitogenic signals, a series of interdependent, cyclical events is orchestrated, beginning with the coordinated expression and activation of cyclin–dependent protein kinases (Cdks), whose enzymatic activity in turn is governed by regulated expression of periodically expressed proteins, known as cyclins (Fig. 1). Thus, cyclins are synthesized and activated in a cell cycle-dependent manner and function as the regulatory subunit of cyclin–Cdk complexes, which then phosphorylate a series of molecules necessary for cell cycle progression. In addition to these two series of positive-acting regulators, several tumor suppressor proteins and at least two classes of Cdk inhibitors have been identified that provide negative regulation to the cell cycle. This chapter will focus on recent progress in the field of

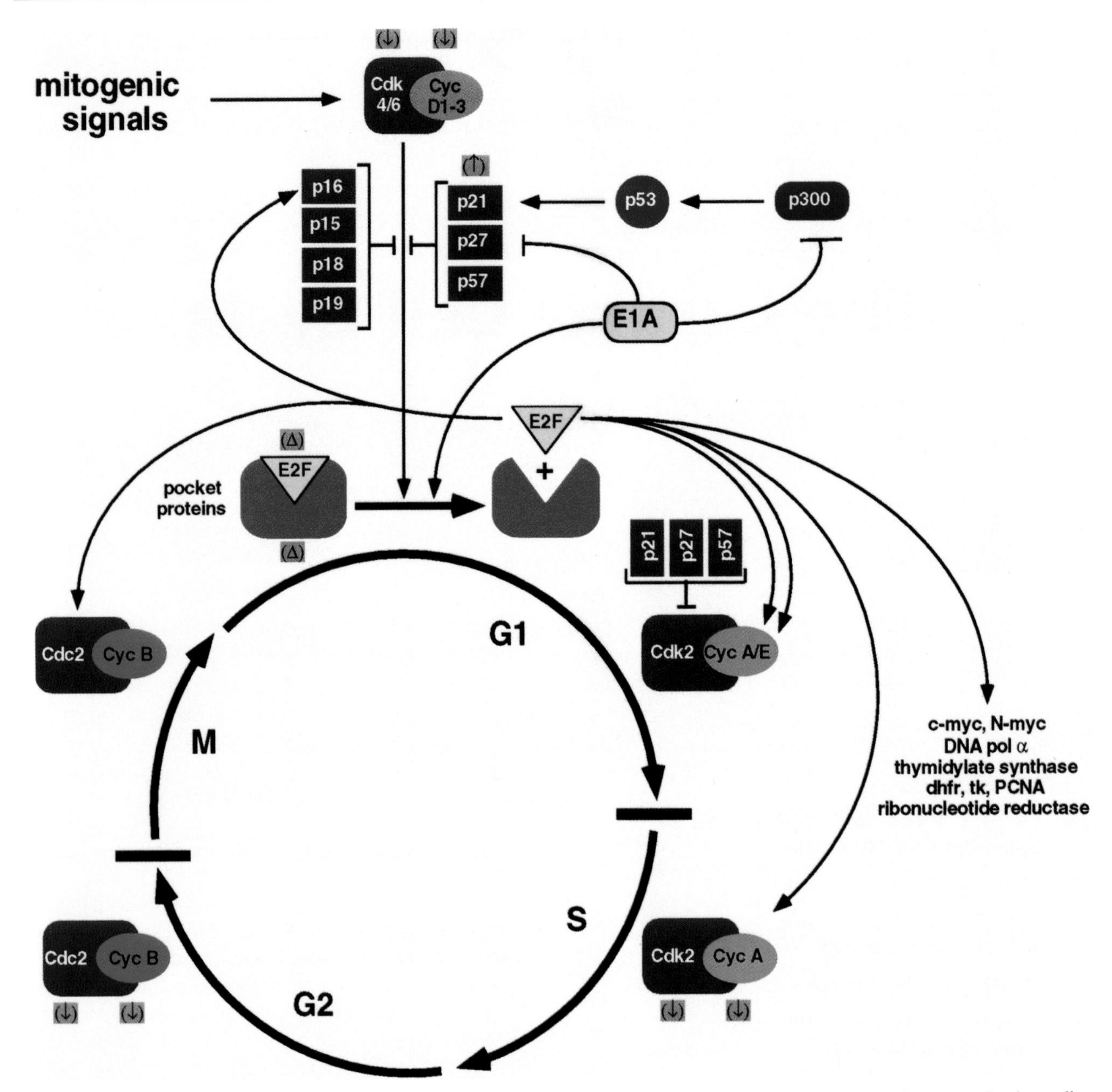

Figure 1 Schematic representation of the principal classes of proteins that have been implicated in cell cycle progression in cardiac myocytes. The figure emphasizes the postulated central role of pocket proteins in the processes of cell cycle control and terminal differentiation and reactivation of the cardiac cell cycle by E1A or E2F-1. Demonstrated changes in the expression of cell cycle regulators in adult ventricular myocytes include (↓) downregulation of cyclins D1–3, cyclin A, cyclin B, Cdk4, Cdk2, and Cdc2; (↑) upregulation of the Cdk inhibitor, p21; and (Δ) changes in E2F and pocket protein composition (see text for details). Potential mechanisms for the impact of adenoviral E1A protein on the cardiac cell cycle include interference with pocket proteins, p300, p27, and other, uncharacterized targets.

eukaryotic cell cycle regulation as it pertains to cardiac muscle and on the specific implications of the cardiac cell cycle for understanding cardiac development and disease.

II. Cardiac Myocyte Proliferation *in Utero* and *in Ovo*

Cell proliferation is a prerequisite for the generation and structural modeling of all tissues. Cardiac development likewise involves the robust proliferation of cardiac myocytes during embryonic life, culminating in the completion of cycling soon after birth, virtually abrogating the potential for further hyperplastic growth. Whether postnatal maturation of ventricular myocytes represents an absolute, irreversible exit from the cell cycle, as is true for normal skeletal muscle myotubes (Gu *et al.,* 1993), or whether adult cardiac myocytes retain a limited capacity for cell cycle reentry is controversial, as will be discussed in detail later. Recent analyses of cardiac myocyte growth during murine development indicate that cardiac myocyte DNA synthesis occurs in two distinct phases (Soonpaa and Field, 1997). The first phase occurs during fetal life and is maximal at the earliest time points measured, with labeling indexes of 33% in embryonic day (E) 12 ventricular myocardium. At this time, karyokinesis and cytokinesis were matched, resulting in an expansion of the cardiac myocyte population. The second phase occurs early in neonatal life and peaks 4–6 days after birth. By contrast, during this phase, karyokinesis occurs in the absence of cytokinesis, resulting in binucleation of ventricular myocytes, as is also seen during postnatal growth of the rat heart (Clubb and Bishop, 1984; F. Q. Li *et al.,* 1996). Mechanisms responsible for the characteristic uncoupling of nuclear division from cell division in neonatal cardiac myocytes are unknown; however, expression of several cell cycle regulators (cyclins D1–3, cyclin A, Cdk2, Cdk4, and Cdc2) persists during this period (Yoshizumi *et al.,* 1995; Soonpaa *et al.,* 1997).

III. Growth Factors, Cytokines, and Receptors

Although numerous potential mitogens (as well as antimitogens) are expressed in the developing ventricle, often an essential role for these molecules in cardiac myocyte proliferation has not been proven. Members of the fibroblast growth factor (FGF) family are an exception in this regard. FGF receptors (FGFR) are expressed in the embryonic mouse heart and are downregulated during development as cardiac myocytes lose their capacity to proliferate (Jin *et al.,* 1994). FGF-1 and FGF-2 are expressed in developing rat myocardium (Spirito *et al.,* 1991) and are produced by cardiac myocytes themselves (Weiner and Swain, 1989). FGF-1 can induce cell cycle reentry by cultured neonatal rat myocytes (Parker *et al.,* 1990), although the results are more variable for FGF-2 (Parker *et al.,* 1990; Pasumarthi *et al.,* 1996). In avians, FGF-1, FGF-2, and FGF-4 each can trigger the proliferation of precardiac mesoderm and are expressed in the early heart, appearing first in endoderm, followed by myocardium (Zhu *et al.,* 1996). The remarkable complexity of the FGF family is highlighted by differential patterns of expression for FGF-12 (in the embryonic atrium) versus FGF-13 (in the embryonic atrium and ventricle) (Hartung *et al.,* 1997). Other polypeptide growth factors have also been implicated as mitogens for precardiac mesoderm [activin-A, insulin, insulin-like growth factor (IGF)-I, and IGF-II] or as antimitogens [transforming growth factor-β (TGF-β)] (Antin *et al.,* 1996), at least *in vitro.*

To test more conclusively the dependence of cardiac myocytes' proliferation on FGF *in ovo,* a dominant-inhibitory (kinase-defective) FGFR1 was introduced into the embryonic chick ventricle by inoculation of a recombinant retrovirus (Mima *et al.,* 1995). This truncated receptor forms heterodimers with all known FGFRs, prevents autophosphorylation in *trans,* and effectively blocks FGF-initiated signaling. The truncated receptor decreased clonal expansion by the infected myocytes, relative to a control virus, indicating that endogenous FGF indeed is a necessary mitogen for proliferating ventricular muscle cells. (An ambiguity inherent to such studies, however, is the potential counterhypothesis that FGF acts, in part or instead, as a survival factor in this context.)

The gradual accumulation of transgenic mice with abnormalities of cardiac growth allows insight into at least a subset of regulators for this process, illustrated by the loss of trabeculae associated with neural crest defects seen in mice lacking the epidermal growth factor (EGF) homolog, neuregulin (Meyer and Birchmeier, 1995), or its receptors, Erb2 and Erb4 (Gassmann *et al.,* 1995; K. F. Lee *et al.,* 1995). However, the factors that regulate embryonic cardiac myocyte proliferation in mammals are largely speculative, because loss-of-function mutations, which often provide the best way to gain insight into protein function *in vivo,* can be confounded by multiple factors in addition to functional redundancy, including premature lethality, systemic defects, and maternal rescue. For example, FGFR1-deficient embryos produced by homologous recombination successfully generate mesoderm but display severe early growth defects, making a direct assessment of this receptor's role in normal cardiac development impossible by conventional, global knockout techniques (Yamaguchi *et al.,* 1994). Another limitation, peculiar to secreted proteins, is the potential for maternal rescue of a "null" phenotype by transplacental delivery of a growth factor, as shown for TGF-β1 (Letterio *et al.,* 1994). One further difficulty to interpreting the results of certain transgenic models is the mutually dependent interaction of the heart with the fetal circulation and extracardiac tissues: cardiac growth is sensitive not only to defects primarily affecting the myocardium but also to any extrinsic change that affects its work load, such as mutations altering vasculogenesis or body mass.

In part for these reasons, determining that a particular ligand is necessary and sufficient to modulate cardiac growth can be problematic. Illustrative examples of this principle include the complexity of transgenic systems necessary to gain results for two receptor–ligand systems, IGF-I and interleukin (IL)-6. IGF-I-deficient mice display a generalized growth deficiency without a specific cardiac phenotype and die perinatally of respiratory failure (Baker *et al.,* 1993; Liu *et al.,* 1993). To circumvent the postnatal lethality that results from a complete absence of IGF-I, mice were created that express reduced levels of IGF-I (~30% of normal): Mice with a partial deficiency in IGF-I do survive to adulthood and have an intermediate size when compared to wild-type or IGF-I-deficient animals (Lembo *et al.,* 1996). The mice display normal left ventricular (LV) mass when corrected for body weight, suggesting that embryonic cardiac myocyte proliferation was appropriate for the animals' size, and exhibit a normal hypertrophic response to load produced by banding the ascending aorta (Lembo *et al.,* 1996). Conversely, cardiomegaly, resulting from ventricular myocyte hyperplasia, was found to result in transgenic mice with IGF-I targeted to the heart via the α myosin heavy chain (α-MHC) promoter (Reiss *et al.,* 1996a), and in mice lacking the IGF-II receptor, which normally binds and inactivates IGF-I (Lau *et al.,* 1994). These two observations concur with *in vitro* evidence suggesting that IGF-I modulates myocyte proliferation rather than hypertrophy (Kajstura *et al.,* 1994). In summary, while it is clear that IGF-I is capable of inducing cardiac myocyte proliferation both *in vitro* and *in vivo,* it has not been proven that IGF-I is required for normal cardiac development or adaptive growth.

Gain- and loss-of-function mutations have been applied, similarly, to the cardiac role of gp130, a signaling partner that is shared in common by a variety of heteroduplex receptors, including those for interleukin IL-6, leukemia inhibitory factor, cardiotropin-1, ciliary neurotrophic factor, and oncostatin M. Mice deficient for gp130 die *in utero,* between E12.5 and birth (Yoshida *et al.,* 1996), with ventricular hypoplasia as the most striking abnormality and, presumably, the cause of embryonic demise. Since IL-6 is produced by cultures of fetal cardiac myocytes (Metcalf *et al.,* 1995) and is upregulated in adult myocardium in response to pathophysiological stimuli (Kukielka *et al.,* 1995), transgenic mice were created that overexpress IL-6 or the IL-6 receptor in the heart to explore the function of this cytokine as a potential mediator of hypertrophy (Hirota *et al.,* 1995). Surprisingly, whereas neither manipulation alone was sufficient for cardiac hypertrophy, dramatic myocardial enlargement occurred when the two lines were crossed. This suggests that neither active IL-6 nor its receptor is present in normal circumstances (at least not at sufficient concentrations), but that over expression of both in tandem can constitute the fully functional ligand–receptor system, acting in concert with endogenous gp130. While these studies do not detract from a more generic role of gp130-mediated signaling in cardiac development or disease, the direct physiological relevance of IL-6 itself as a cardiac growth factor is questionable since neither the ligand nor its specific receptor is expressed in the developing ventricle. The full spectrum of physiologically relevant ligands for gp130 remains to be established for cardiac muscle, but an especially intriguing member of this family is cardiotrophin-1, a novel cytokine which was cloned on the basis of its ability to induce cardiac hypertrophy in cultured cells, signals through the gp130 pathway, is expressed in the developing heart tube (followed by a more widespread distribution), and promotes both myocyte survival and proliferation (Pennica *et al.,* 1995; Sheng *et al.,* 1996).

Various transgenic animals that express growth-promoting factors under the control of cardiac-restricted promoters have resulted in a phenotype of increased ventricular size and myocyte hypertrophy. Overexpression of the angiotensin II type I receptor (AT1) by the α-MHC promoter results in massive atrial enlargement secondary to hyperplasia and early mortality presumably as a result of disturbances in heart rate and atrioventricular conduction (Hein *et al.,* 1997). Since angiotensin II induces a hypertrophic response in cultured neonatal ventricular myocytes (Sadoshima *et al.,* 1993) and appears to mediate neonatal cardiac growth in adaptation to load (Beinlich *et al.,* 1991), it is surprising that morphology of the ventricles and of ventricular myocytes in transgenic animals was normal despite elevated levels of AT1 receptor. This phenotype obviously contrasts with the ventricular hypertrophy produced in adult transgenic mice by forced expression of a constitutively active α-1B adrenergic receptor using similar α-MHC promoter sequences (Milano *et al.,* 1994): α-MHC is expressed in the early heart tube (Lyons *et al.,*

1990), is chiefly expressed in the atria during embryogenesis, and then markedly upregulated in the ventricle after birth, in part via the postnatal rise in thyroid hormone concentrations (Subramaniam *et al.,* 1993). Because these two G protein-coupled receptors share similar downstream effectors, a deficiency of downstream signaling intermediaries may be less likely to explain the divergent phenotypes than are other explanations. In principle, the AT1 receptor phenotype could reflect relative levels of the active ligand in the atria versus the ventricle of newborn mice or differential expression of AT1 receptor-specific cofactors that determine the growth response.

Notably, despite the attractiveness of ascribing a role to AT1 and α-1B in cardiac myocyte growth, it must be remembered that genetic manipulations that enhance cardiac growth need not imply a requirement for the proteins during either myocardial hyperplasia or hypertrophy in normal circumstances. This disparity is highlighted by the reportedly normal cardiac phenotype in angiotensinogen-deficient mice (Niimura *et al.,* 1995) as well as AT1a (Ito *et al.,* 1995) and AT-2 (Hein *et al.,* 1995; Ichiki *et al.,* 1995) receptor knockouts. Again, such results suggest that members of the renin–angiotensin system may be sufficient to provoke cardiac growth, but are not necessarily required. A contrasting, essential role was shown for β-adrenergic receptor kinase-1, one of several G protein-coupled receptor kinases (GRKs), which mediate agonist-dependent phosphorylation and desensitization of G protein-coupled receptors (Jaber *et al.,* 1996): mice lacking βARK1 die prior to E15.5, with a hypoplastic myocardium similar to that discussed for the transcription factors N-myc, TEF-1, RXRα, and WT-1 (see Chapters 13, 15, and 18).

IV. Signal Transducers

While little is known of the intracellular pathways coupling membrane events to the nucleus for cardiac muscle cell proliferation, the small GTP-binding protein, Ras, is a widely utilized and pivotal component of mitogenic signaling pathways which also mediates signaling for other categories of trophic signals. Ras is necessary for serum-stimulated DNA synthesis (Mulcahy *et al.,* 1985), acting at more than one point during the G_1 phase of the cell cycle (Dobrowolski *et al.,* 1994). Recently, the mechanism by which Ras regulates cell cycle progression has been partially clarified by the finding that cyclin D, the retinoblastoma protein, Rb, and E2F all are downstream targets of the Ras-effector pathway (Peeper *et al.,* 1997). The members and operation of this cyclin–Rb–E2F axis are detailed later.

Interestingly, although Ras-dependent pathways also have been shown to be involved in the activation of hypertrophy—both in cultured neonatal cardiac myocytes (Sadoshima and Izumo, 1993, 1996; Thorburn *et al.,* 1993; Abdellatif *et al.,* 1994; Abdellatif and Schneider, 1997) and in transgenic mice (Hunter *et al.,* 1995)—hyperplasia was not seen in either setting. The lack of hyperplasia is surprising given the central role of Ras in mediating proliferative signals for other systems as well as the known ability of neonatal cardiac myocytes to exit G_1 in response to other genetic manipulations (Kirshenbaum and Schneider, 1995; Kirshenbaum *et al.,* 1996), and to mitogenic stimuli (Ueno *et al.,* 1988; Kirshenbaum *et al.,* 1996) at least in some preparations; cf. Simpson *et al.,* 1982. A mechanistic explanation for this discrepancy in the impact of Ras is not obvious, although lineage-specific differences are known to occur in Ras-dependent growth effects, typified by Ras-dependent growth arrest and neuronal differentiation in PC12 pheochromocytoma cells, in which the protein mediates the action of nerve growth factor (H. Z. Li *et al.,* 1996). While differences in Ras-dependent autocrine factors are one potential basis for such differences, cardiac-specific effects of the Ras GTPase-activating protein, Ras-GAP, have been reported, indicating that differences exist between cell types in the use of particular Ras effector proteins (Abdellatif and Schneider, 1997). Similar disparities—hyperplasia of sympathetic ganglia but hypoplastic myocardium—were seen in mice lacking NF-1, a related protein encoded by the neurofibromatosis type-1 gene (Brannan *et al.,* 1994). However, it is unproven whether the cardiac phenotype in this case indicates a requirement for NF-1 in ventricular myocytes versus alterations in neural crest derived cells.

V. Transcription Factors

One of the earliest reported targeted mutations affecting cardiac development was a disruption of the N-myc gene, causing embryonic lethality around E11, characterized in part by hypoplastic ventricular myocardium and ventricular septal defects (Charron *et al.,* 1992; Moens *et al.,* 1993; Sawai *et al.,* 1993). N-myc belongs to the family of sequence-specific DNA-binding proteins encoded by Myc protooncogenes, which like other immediate- early genes are postulated to act as "third messengers" for mitogen- and other ligand-dependent signals (Kretzner *et al.,* 1992; Eisenman and Cooper, 1995). Forced expression of c-myc in cultures of skeletal myocytes can inhibit or delay the expression of skeletal muscle-specific genes (Schneider *et al.,* 1987; Miner and Wold, 1991). To circumvent the early embryonic lethality of this complete N-myc deficiency, compound heterozygotes carrying one inactivated and one hypomorphic N-myc gene were created that expressed N-myc at ~15% of normal levels (Moens *et al.,* 1993). These mice survived to E14 and displayed a selective defect of the compact (subepicardial) layer of ventricular myocardium, sparing the inner trabecular layer, which correlates with the observed expression of N-myc in the compact layer alone. This spatial distribution is the antithesis of the failure of trabeculation, seen with the deletion of neuregulin or its receptors, and bears strong similarity to the human congenital abnormality, noncompaction of the ventricle (Chin *et al.,* 1990). However, these result do not by themselves differentiate between two equally plausible roles for N-myc—namely, direct involvement in myocyte cell cycle progression versus delaying the differentiation of subepicardial myocytes, thus prolonging the period of proliferative competency. Conversely, transgenic mice with forced expression of c-myc in myocardium develop ventricular myocyte hyperplasia (Jackson *et al.,* 1990). However, the Myc transgene was not sufficient for sustained proliferation after birth.

Although the role of retinoids and retinoic acid receptors in cardiac development is detailed in Chapter 13, it is germane to the current discussion of the cardiac cell cycle that mice lacking RXRα die in utero between day E13.5 and 16.5, with their principal cardiac defect appearing to be the failure of normal proliferation in the compact ventricular layer between E14.5 and 16.5 (Kastner *et al.,* 1994; Sucov *et al.,* 1994): Trabecular myocardium also was diminished to a lesser extent. Chamber hypoplasia was commonly associated with muscular ventricular septal defects in the homozygous null mice. However, this reduction in cell number was not accompanied by corresponding defects in mitotic index or the prevalence of DNA synthesis in cardiac myocytes of wild-type versus RXRα-deficient embryos. The caveat raised by this negative result is worth noting, given the number of additional embryonic-lethal mutations that likewise result in hypoplastic myocardium as the conspicuous cardiac defect. Relevant examples include the neurofibromatosis gene (Brannan *et al.,* 1994; Jacks *et al.,* 1994), the TEA domain transcription factor TEF-1 (Chen *et al.,* 1994), the Wilm's tumor protein WT-1 (Kreidberg *et al.,* 1993), in addition to several growth factors and receptors discussed earlier. Hence, appropriate caution should be exercised before concluding a direct, critical role for these proteins in cardiac myocyte proliferation. A "hypoplastic" phenotype could conceivably result from decreased myocyte proliferation but could equally arise from a failure of cardiac myocyte recruitment, increased cell death, precocious terminal differentiation, or deficient recruitment of key complimentary lineages.

VI. G_1 Cyclins and Cyclin-Dependent Kinases

The cell cycle comprises a sequence of highly organized, ordered events that occur repeatedly in the development of complex organisms. Therefore, it is not surprising that a series of checkpoints has arisen to ensure that each round of DNA replication and chromosome segregation is completed with the correct timing and fidelity. In recent years, the identification of specific proteins that positively or negatively regulate cell cycle progression has begun to resolve the mechanisms underscoring the growth arrest common to such diverse processes as DNA damage, terminal differentiation, and cellular senescence.

Integration of external growth signals with passage through G_1 into S phase in mammals is sequentially regulated by the G_1 cyclins (D1–3, E, and A), particularly D-type cyclins. These factors and their catalytic partners, Cdk2, Cdk4, and Cdk6, are induced and activated by mitogenic signals (Sherr, 1994). Cyclin D-dependent kinase activity is critical for early G_1 phase progression but dispensable in late G_1 for S phase entry in many cultured cell lines (Baldin *et al.,* 1993; Quelle *et al.,* 1993), and it is believed to act chiefly by hyperphosphorylation and inactivation of "pocket" proteins, for which the archetype is Rb. Cyclin D1 and D3 have been shown to directly interact with Rb through a classic LXCXE motif shared by many Rb-binding proteins and viral proteins capable of inactivating pocket proteins (Dowdy *et al.,* 1993). The notion that cyclin D-dependent phosphorylation of Rb is necessary for G_1 exit is supported by observation that these kinases are dispensable for the G_1/S transition in cells lacking functional Rb (Medema *et al.,* 1995).

Rb and other pocket proteins, in turn, are assumed to exert their effects on the cell cycle through the binding and inhibition of the E2F family of transcription factors (Qin *et al.,* 1995). Quiescent cells express an active, hypophosphorylated form of Rb, whereas cells in S and/or M phases contain hyperphosphorylated/inactive Rb. Therefore, during G_1, phosphorylation of Rb liberates free E2F, resulting in upregulation of E2F-dependent genes; including DNA polymerase α, dihydrofolate reductase, thymidylate synthetase, cyclin A and Cdc2, leading to cell cycle progression (Sherr, 1994; Weinberg, 1995). Dissociation of E2F from the Rb pocket alleviates transcriptional repression by Rb tethered to the E2F binding sites (Weintraub *et al.,* 1995). Cyclin E associates with Cdk2 in late G_1, where its kinase activity is also critical for the G_1/S transition and early S phase progression (Koff *et al.,* 1992; Ohtsubo *et al.,* 1995). Blocking cyclin E activity inhibits S phase entry and, in contrast to cyclin D1, cyclin E was required for the G_1/S transition even in cells lacking functional Rb. With the onset of S phase, cyclin A combines with Cdk2, localizing to sites of DNA replication, and supports DNA synthesis, as shown for cell cycle reentry triggered by SV40 large T antigen in postmitotic myotubes as well as for normally cycling cells (Cardoso *et al.,* 1993). While substrates for Cdks exist in addition to pocket proteins, in many cases their functional significance is still unproven; conversely, multiple cyclin–Cdk complexes can phosphorylate Rb (D1–Cdk4, E–Cdk2, and A–Cdk2), albeit at differing residues (Kitagawa *et al.,* 1996; Zarkowska and Mittnacht, 1997). Whereas either phosphorylation of T821 by cyclin A-Cdk2 or phosphorylation of T826 by cyclin D1-Cdk4 can prevent the binding of LXCXE proteins that interact with the Rb "pocket," only cyclin A-Cdk2 causes the dissociation of preexisting LXCXE-Rb protein complexes (Zarkowska and Mittnacht, 1997); similarly, only cyclin D1-Cdk4, not cyclin E-Cdk2, phosphorylated S780, which prevented binding to E2F-1 (Kitagawa *et al.,* 1996). Phosphorylation by Cdks also controls the intrinsic transcriptional repressor domain shared by all three pocket proteins (Chow *et al.,* 1996).

Members of the D-cyclin family are expressed in distinct tissue-restricted patterns, suggesting specialized roles for these proteins in certain tissues. This is supported by the observation that cyclin D1- and D2-deficient mice, although displaying no cardiac phenotype, have tissue-specific defects primarily affecting retina and breast versus ovaries and testes, respectively (Sicinski *et al.,* 1995, 1996). All three D-type cyclin mRNAs are readily detected in embryonic and neonatal ventricles but are downregulated in the adult heart (Soonpaa *et al.,* 1997). This pattern of expression is paralleled by the proteins' abundance, at least for cyclin D1 and D3, which are detected in embryonic myocardium but are downregulated during the neonatal period, correlating well with cardiac myocytes' exit from the cell cycle (Soonpaa *et al.,* 1997; Kang *et al.,* 1997). Similarly, Cdk4, mediating the early G_1 transition, and Cdk2, active later in G_1, both are expressed in the embryonic ventricle when cardiac myocytes are proliferating but not in adult hearts. Thus, G_1 cyclins each are downregulated in adult myocardium, as might have been predicted on the basis of cell cycle arrest; however, this entails downregulation of the three D cyclin genes (not merely cycle-dependent protein degradation) and is accompanied by marked downregulation of the cyclins' protein targets, G_1 Cdks.

Functional proof for the importance of diminished cyclin expression was established using transgenic mice which overexpress cyclin D1, under the control of the α-MHC promoter. Targeted expression of cyclin D1 in myocardium caused a 40% increase in heart weight that was ascribed to an approximately twofold increase in cardiac myocyte number at 14 days (Soonpaa *et al.,* 1997). Persistent DNA synthesis was confirmed in adult ventricular muscle; however, while the relative labeling index of 0.05% is well above rates observed in control animals (<0.0003%), the absolute magnitude for this effect is small. There is no evidence of the possibility that cyclin D1 might suffice to enhance both DNA synthesis and cytokinesis since increased DNA content per myocyte was observed, nuclear enlargement was conspicuous, and mitoses were not reported. Hence, this process is reminiscent of endoreduplication (S phases without intervening mitoses), which is common to many invertebrates, such as in the transition to polyteny in *Drosophila* (Smith and Orr-Weaver, 1991), which is associated with the loss of the mitotic cyclins A and B but continued periodic expression of cyclin E. Whether a similar mechanism is responsible in transgenic cardiac myocytes is not yet known. In contrast to skeletal muscle, where forced expression of cyclin D1, A, or E inhibited myogenic differentiation (Rao *et al.,* 1994; Skapek *et al.,* 1995), several markers of cardiac differentiation were normal in the D1 transgenic hearts. Interestingly, cyclin D1 caused MyoD phosphorylation and could block skeletal myocyte differentiation even in the presence of a nonphosphorylatable form of Rb (two properties unlike those shown by cyclins A and E), indicating the existence of both Rb-dependent and Rb-independent pathways (Skapek *et al.,* 1995, 1996). It is ambiguous whether the normal extent of differentiation in cyclin D1 transgenic hearts, rather than the block produced by cyclin D1 in skeletal muscle models, results from a lineage-specific difference in fundamental mechanisms for differentiation (i.e., the absence of a mutually exclusive relationship between differentiation and proliferative growth) or alternatively is secondary to technical differences such as the respective levels of cyclin expression in these very different experimental systems.

By contrast with the D cyclins, A-type cyclins are implicated at both the G_1/S and G_2/M transitions of the cell cycle and associate with Cdk2 and Cdc2 at these phases, respectively. Preventing cyclin A function or expression at the appropriate phase can arrest cells at either transition selectively, suggesting its essential role in both events (Pagano *et al.,* 1992). Two A-type cyclins have been identified: A1 (limited to germ cells) and A2 (expressed widely) (Howe *et al.,* 1995). Cyclin A is found in proliferating cardiac myocytes but is downregulated in the adult heart (Yoshizumi *et al.,* 1995; Kang *et al.,* 1997). Like cyclin A, cyclin B and Cdc2 (the mitotic Cdk) are markedly downregulated in adult hearts (Kang *et al.,* 1997; Soonpaa *et al.,* 1997). To address the *in vivo* function of cyclin A, mice were created with a targeted deletion of cyclin A2 (Murphy *et al.*, 1997): The homozygous deficiency of A2 was embryonic lethal at E5.5, demonstrating the necessity for cyclin A2 in normal proliferation and development. The ability of cyclin A null embryos to develop even to the postimplantation stage may be related to a maternal pool of cyclin A2 protein or an unexpected role for cyclin A1 during early embryogenesis. This phenotype differs markedly from results for cyclin D1- and D2-deficient mice, in which the essential role of these cyclins appeared to be cell type specific. It is conjectural whether this difference can be explained by the greater potential for redundancy among the three D-type cyclins.

VII. Cdk Inhibitors

The cloning of $p21^{WAF1/Cip1}$ and the subsequent identification of a much broader array of Cdk inhibitors (p15, p16, p18, p19, p27, and p57) elucidated a novel mechanism by which to regulate cellular proliferation in mammalian cells beyond merely the abundance of the cyclins and their target kinases. The set of Cdk inhibitors, in turn, has now been shown to comprise two functionally disparate groups. One class, the Ink4 family, includes four members—$p15^{Ink4b}$, $p16^{Ink4a}$, $p18^{Ink4c}$, and $p19^{Ink4d}$—which have selective inhibitory activity against Cdk4 and Cdk6 (Sherr and Roberts, 1995). These proteins competitively bind G_1 Cdks, preventing their interaction with cyclin D and effectively inhibiting their catalytic activity. In keeping with this mechanism of action, their ability to inhibit cell cycle progression is Rb dependent (Guan *et al.*, 1994; Lukas *et al.*, 1995). p15 was originally cloned as a TGF-β-responsive gene, and appears to be ubiquitously expressed in normal tissues including heart (Quelle *et al.*, 1995). p18 message was likewise expressed in all tissues examined, although levels varied greatly, with the greatest abundance in cardiac and skeletal muscle (Guan *et al.*, 1994); p18 induction also was especially prominent during skeletal myocyte differentiation *in vitro* (Franklin and Xiong, 1996). In contrast, p16 is expressed only in spleen, lung, and, to a lesser extent, liver (Quelle *et al.*, 1995), whereas p19 is also expressed selectively, primarily in the bone marrow and spleen, with no detectable transcripts in the heart (Hirai *et al.*, 1995). While markedly differing expression patterns suggest that the four INK4 Cdk inhibitors may have developed to performed specialized roles, no differences in their inhibitory action have yet been identified. p16 null mice have abnormal hematopoiesis and are predisposed to the development of sarcomas and B cell lymphomas but apparently have a normal cardiac phenotype, consistent with the gene's distribution of expression (Serrano *et al.*, 1996). In short, the role of INK4 factors in cardiac development and growth control is unknown and may require combinatorial inactivation since there is overlap in the expression and, at least, the *in vitro* functions of p15 and p18.

The Cip/Kip family includes three members, $p21^{WAF1/Cip1}$, $p27^{Kip1}$, and $p57^{Kip2}$ (Harper *et al.*, 1993; Polyak *et al.*, 1994b; Toyoshima and Hunter, 1994; Matsuoka *et al.*, 1995). These proteins show significant homology within their amino terminus, a domain required for their Cdk inhibitory action (Polyak *et al.*, 1994b), but diverge considerably in their carboxy-terminal ends. These proteins inhibit cyclin–Cdk complexes by binding to the cyclin subunit and blocking the active site of the associated Cdk (Russo *et al.*, 1996). The prototypical member, p21, was cloned via three divergent strategies —its physical association with Cdk2 (Harper *et al.*, 1993), upregulation by p53 (el-Deiry *et al.*, 1994), and accumulation in senescent cells (Noda *et al.*, 1994). p21 can bind and inhibit a broad range of cyclin–Cdk complexes (Xiong *et al.*, 1993; Sherr and Roberts, 1995) as well as other proteins involved in cellular progression, such as proliferating cell nuclear antigen (PCNA) (Waga *et al.*, 1994). The ability of p21 to inhibit cyclin–Cdk activity contrasts with its presence in normal cycling cells (Zhang *et al.*, 1993) and in catalytically active cyclin–Cdk complexes (Zhang *et al.*, 1994). This apparent paradox has been recently resolved and the mechanism of p21 action clarified by the observation that p21, at low levels, can promote the physical association between a cyclin and Cdk partner, whereas at higher ratios it inhibits formation of the cyclin–Cdk complex and, hence, Cdk activity (Harper *et al.*, 1995; LaBaer *et al.*, 1997).

Expression studies show p21 is induced in the differentiation of a variety of postmitotic cell types including cardiac muscle cells (Parker *et al.*, 1995). A functional role for p21 *in vivo* was recently demonstrated using mice that lack p21 through targeted gene disruption. Grossly, these mice appear normal with no discernible developmental defects and no increased incidence of tumor growth. More subtly, fibroblasts from the p21-deficient mice demonstrate defective G_1 checkpoint control in response to radiation-induced DNA damage (Brugarolas *et al.*, 1995; Deng *et al.*, 1995), and keratinocytes from p21 null mice show increased tumorigenicity *in vivo* after transformation by Ras (Missero *et al.*, 1996). The lack of a dramatic phenotype in p21 null mice is presumably secondary to redundancies in the expression and action of Cdk inhibitors, although differences between p21- and p27-deficient cells have already been noted (Missero *et al.*, 1996).

$p27^{Kip1}$ was cloned through its ability to interact with cyclin D1 (Toyoshima and Hunter, 1994) and by its induction during growth arrest triggered by TGF-β or cell–cell contact (Polyak *et al.,* 1994a): These two approaches highlight the complementary strategies through which the Cdk inhibitors have been defined. Overexpression of p27 arrests cell in G_1, and, like p21, p27 binds more avidly to cyclin–Cdk complexes than to Cdks alone. p27 can inhibit cyclin D-, E-, A-, or B-dependent kinases; however, it is mainly complexed with cyclin E/Cdk2 in proliferating cells (Toyoshima and Hunter, 1994). p27 expression was found in all tissues examined, with abundant message in human heart (Polyak *et al.,* 1994a). In contrast to the minimal phenotype of p21-deficient mice, p27 null mice showed generalized hyperplasia with an ~20% increase in heart weight (Kiyokawa *et al.,* 1996).

Like other members of this family, p57 is upregulated in cells exiting the cell cycle. However, its tissue distribution is more restricted than that of p21 or p27: 1.5- and 1.7-kb transcripts were readily detected in placenta, with lesser but abundant message detected in skeletal and cardiac muscle (M. H. Lee *et al.,* 1995; Matsuoka *et al.,* 1995). During development, p57 mRNA was detected in brain, skeletal muscle, and the lens as early as E9.5–14.5, generally in parallel with the expression of p21 in postmitotic cells (Matsuoka *et al.,* 1995). Message was first detected in the heart at E10.5, then increased in expression, peaking at E17.5 (Yan *et al.,* 1997). p57-deficient mice predominantly died in the perinatal period, displaying multiple developmental defects including short limbs, cleft palate, and gastrointestinal anomalies. Although no frank cardiac defect was observed, p57 mutant mice displayed an increase in apoptosis in heart muscle, as in other organs (Yan *et al.,* 1997).

In skeletal muscle, p21 is induced by MyoD, parallels myogenin induction *in vitro,* and correlates closely with terminal differentiation of skeletal muscle *in vivo* (Halevy *et al.,* 1995; Missero *et al.,* 1995; Parker *et al.,* 1995; Wang and Walsh, 1996). Strong evidence supporting a functional role for Cdk inhibitors in skeletal muscle differentiation comes from the forced expression of p21 or p16, either of which can overcome the block to myogenic differentiation imposed by mitogenic serum (Skapek *et al.,* 1995). Similarly, p21 also blocks repression of MyoD activity by cyclin D (Skapek *et al.,* 1995). Direct evidence of a similar role for p21 in cardiac muscle is lacking or, alternatively, a requirement for other Cdk inhibitors has yet to be established.

VIII. Pocket Proteins

Rb, p107, and p130 comprise a family of pocket proteins sharing sequence homology, common structural organization, and similar but distinguishable associated protein partners. Pocket proteins have been proposed as playing a critical role in proliferation and differentiation in most tissues (Weinberg, 1995), including cardiac and skeletal muscle (Gu *et al.,* 1993; Schneider *et al.,* 1994). The exact expression pattern for pocket protein family members in cardiac muscle is controversial, because studies concerning the expression of Rb in myocardium have reported conflicting results. Some authors have suggested that Rb message and protein are expressed as early as E12 in the developing ventricle, with Rb message remaining constant throughout development but with protein levels subsequently downregulated (Kim *et al.,* 1994; Soonpaa *et al.,* 1996). Other reports have failed to confirm the presence of Rb in fetal heart (Szekely *et al.,* 1992; Zacksenhaus *et al.,* 1996). Most reports confirm the presence of Rb in neonatal and adult myocardium, albeit at low levels and with the caveat that the critical data for localization to cardiac myocytes themselves often are lacking (Bernards *et al.,* 1989; Chen *et al.,* 1996). A provisional conclusion that Rb, even if important for other aspects of growth control or differentiation in cardiac muscle, is dispensable at least for initial morphogenesis and chamber formation is supported by the seemingly normal cardiac structure in Rb null embryos at E14.5, despite widespread cell death and abnormalities in other tissues, including defective hematopoiesis and neurogenesis (Clarke *et al.,* 1992; Jacks *et al.,* 1992; Lee *et al.,* 1992). Details of the cardiac phenotype in Rb-deficient mice are not available, however.

The related pocket protein, p107, is expressed in cardiac myocytes from embryonic ventricle and is downregulated after birth (Kim *et al.,* 1995; Soonpaa *et al.,* 1996). Two alternatively spliced forms of p107 of 2.4 and 4.9 kb were observed in some tissues, including the heart. Although their comparative biological significance is unknown, the protein encoded by the shorter transcript is predicted to lack the spacer and B domain of the pocket. This omission is conservatively expected to alter function since the B domain is implicated in the physical interaction with most p107-binding proteins and contributes to the pocket protein repressor domain cited earlier. Although the possibility has been reported that mitogenic signals might regulate p107 abundance rather than Rb protein function, subsequent studies indicate that the cyclin D–Cdk4 complex results in hyperphosphorylation of p107, dissociation of E2F-4 from p107, and inactivation of the ability of p107 to arrest cells in G_1 (Beijersbergen *et al.,* 1995; Xiao *et al.,* 1996).

The least studied pocket protein, p130, is expressed at low levels in the myocardium as early as E12.5 and remains present in the adult heart (Chen *et al.,* 1996). One recent report suggests that p130 along with Rb are the major components of pocket protein complexes with E2F in adult heart (Flink *et al.,* 1997). In marked contrast with Rb-deficient mice, p107- and p130- nullizygous mice are viable and appear phenotypically normal (Cobrinik *et al.,* 1996; Lee *et al.,* 1996). Abnormalities in chondrocyte growth and bone development were uncovered only after simultaneous inactivation of both p107 and p130 (Cobrinik *et al.,* 1996). The notion that each pocket protein has distinct roles in both cell cycle and tissue-specific regulation is supported by the recent observation for specific dysregulation of distinct E2F-dependent genes in $Rb^{-/-}$ versus $p107^{-/-}/p130^{-/-}$ embryonic fibroblasts (Hurford *et al.,* 1997). Thus, if pocket proteins play a role in cardiac determination or differentiation, Rb by itself or p107/p130 can suffice. Whether each is equally capable of supporting terminal differentiation remains to be determined.

Current expectations regarding this question are influenced by the outcome of related studies in skeletal muscle. MyoD can induce both myogenic gene transcription and, independently, cell cycle exit (Crescenzi *et al.,* 1990). Rb is known to interact with diverse cellular proteins apart from E2F and to modify the transcriptional activity of tissue-specific nuclear proteins through protein protein interaction (Taya, 1997; Wang, 1997), and physical association with pocket proteins is reportedly required for the transcriptional activity of MyoD (Gu *et al.,* 1993). The ability of pocket proteins to act as transcriptional cofactors seems to be a general property of this family, at least in muscle, since p107 can suffice in $Rb^{-/-}$ cells, to support transactivation by MyoD (Schneider *et al.,* 1994). Of critical importance, however, functional disparities were seen with regard to the reversibility of differentiation and of cell cycle exit. Normal, wild-type myotubes are refractory to mitogenic serum, with Rb remaining hypophosphorylated. In $Rb^{-/-}$ myotubes, by contrast, serum stimulation causes downregulation of the homolog, p107; dedifferentiation and cell cycle reentry ensue. For this reason, Rb has been taken as the basis for irreversible exit from the cell cycle in postmitotic myotubes. Developmental studies indicating that p107 and (hyperphosphorylated) Rb are coexpressed in neonatal myocardium but that (hypophosphorylated) Rb predominates in adult ventricular muscle give credence to the extrapolation that Rb holds the key to terminal differentiation in myocardium as well. Although cooperative physical association among transcription factors has recently been implicated in trans-activation by Nkx2.5 and serum response factor (Chen and Schwartz, 1996), it remains unknown if these or other cardiac specific factors depend on a pocket protein for their transcriptional activity.

These pivotal studies of myogenesis in Rb-null cells recently have been extended using MyoD to induce skeletal muscle differentiation in both p107- and p130-deficient mouse embryo fibroblasts (Novitch *et al.,* 1996). While the expression of early differentiation markers, such as p21 or myogenin, including myosin heavy chain, were attenuated, was normal in Rb-deficient cells, markers, indicative of later differentiation, defects that were not observed in cells lacking either p107 or p130. Moreover, only Rb-null skeletal myocytes were able to reenter the cell cycle in response to serum provocation. Thus, these results confirm in $p107^{-/-}$ and $p130^{-/-}$ myocytes the earlier conclusion that Rb alone is the basis of permanent growth arrest in skeletal myotubes. Although comparable studies have not been performed with cells isolated from Rb-deficient hearts, suggestive information on the role of Rb in terminal cardiac differentiation is available from a surrogate *in vitro* model system that recapitulates cardiac differentiation. Cardiac myocytes derived from Rb-null embryonic stem (ES) cells were reported to differentiate normally but, unlike cardiac myocytes formed from wild-type ES cells, retained the ability to synthesize DNA in response to mitogens (Schneider *et al.,* 1995).

IX. Viral Oncoproteins as a Probe for Cell Cycle Regulators in Cardiac Muscle: Pocket Protein- and p300-Dependent Pathways

A complementary means to dissect the importance of pocket proteins in mediating cardiac cell cycle control has been the use of certain products of DNA tumor viruses that bind the pocket and inactivate pocket protein function. These viral oncoproteins—SV40 large T antigen, adenoviral E1A protein, and human papilloma virus protein E7—drive quiescent cells to reenter the cell cycle—in large part by displacing a cellular transcription factor, E2F, from the pocket (Nevins, 1992; Moran, 1994). Overexpression of these proteins in differentiated cells can also result in the downregulation of tissue-specific genes (Webster *et al.,* 1988; Braun *et al.,* 1992; Gu *et al.,* 1993; Tiainen *et al.,* 1996). One mechanism of dedifferentiation is postulated to be disruption of pocket protein interactions with tissue-specific transcription factors (Gu *et al.,* 1993) (as discussed previously) or, alternately, cell cycle-dependent inactivation of these factors. The first of these viral proteins to be exploited in cardiac muscle was SV40 large T antigen. Transgenic mice were created expressing T antigen driven by a number of cardiac muscle-restricted promoters. T antigen driven by the atrial natriuretic factor promoter resulted in atrial myocyte hyperplasia, atrial tumors, and arrhythmias (Field, 1988). Although cardiac myocytes derived from the initial atrial tumors themselves were not proliferative in cell culture, proliferating myocytes were successfully derived from the subsequently propagated transplantable atrial tumor lines (Steinhelper *et al.,* 1990) and have proven to be useful in delineating cardiac proteins that bind to T antigen (Daud *et al.,* 1993). Ventricular hyperplasia resulted when T antigen was overexpressed in the heart using the αMHC promoter (Katz *et al.,* 1992), whereas expression controlled by the β-MHC promoter, which is also transcribed in skeletal muscle, resulted in cardiac and skeletal muscle myopathy (DeLeon *et al.,* 1994).

E1A, a structural and functional analog of SV40 large T antigen, has been used extensively both to block skeletal muscle differentiation and to selectively downregulate tissue-restricted genes and force cell cycle reentry in postmitotic cells (Webster *et al.,* 1988; Braun *et al.,* 1992; Caruso *et al.,* 1993; Tiainen *et al.,* 1996). E1A's ability to bind specific host cell proteins has been pivotal in deciphering mechanisms for cell cycle control and differentiation. Like T antigen, E1A prevents activation of muscle-specific promoters in myoblasts, blocks transactivation despite forced expression of myogenic basic helix–loop-helix proteins, and reactivates DNA synthesis in myotubes (Webster *et al.,* 1988; Braun *et al.,* 1992; Caruso *et al.,* 1993; Tiainen *et al.,* 1996). Interference with myogenic differentiation has been ascribed not only to the pocket protein-binding domain of E1A but also to the N-terminal site for binding the bromodomain transcription factor p300 (discussed below; Caruso *et al.,* 1993) or to alternative N-terminal interactions via amino acids 55–60 (EPDNEE) within conserved region 1 (Sandmoller *et al.,* 1996).

As predicted from the outcome of these studies, adenoviral E1A protein also has proven useful in cardiac myocytes to determine the importance of pocket protein function. Notably, E1A is sufficient to trigger cell cycle reentry (accompanied by apoptosis) in quiescent cardiac myocytes from embryonic (Liu and Kitsis, 1996) or neonatal (Kirshenbaum and Schneider, 1995) rats. Mutations that disrupt pocket protein binding remained effective for DNA synthesis, indicating the existence of an alternative pathway for cell cycle reentry provoked by E1A at either age. By contrast, N-terminal mutations that disrupt p300 binding markedly decreased the DNA synthesis in embryonic cardiac myocytes (Liu and Kitsis, 1996), with less impairment in neonatal myocytes. It has been suggested that p300-dependent effects predominate in embryonic cardiac myocytes and that pocket proteins might be more important only at later stages (Liu and Kitsis, 1996), although potential technical explanations for differing effects of the E1A proteins cannot be disregarded. Interestingly, E1A was also found to repress cardiac-restricted transcription, even in the presence of E1B, a viral homolog of Bcl-2, to prevent E1A-dependent apoptosis (Kirshenbaum and Schneider, 1995). Like DNA synthesis, transcriptional repression of cardiac-restricted promoters occurred using mutations that reportedly disrupt either p300 binding (R2G) or pocket protein binding (Y47H, C124G), whereas a mutation affecting both domains (R2G, C124G) was innocuous for both effects (Kirshenbaum and Schneider, 1995).

p300 is understood to be a structural and functional homolog of the transcriptional coactivator CBP (Eckner *et al.,* 1994; Dorsman *et al.,* 1997); both bind the cyclic AMP response element-binding protein (CREB). The amino terminus of E1A binds both p300 and CBP, and E1A proteins competent for this interaction are predicted to block p300/CBP-dependent gene expression. One attractive mechanism for the role of p300 in growth arrest is the requirement for p300 during the induction of the Cdk inhibitors, p21 and p15 (Missero *et al.,* 1995; Datto *et al.,* 1997). In addition to CREB, p300 directly binds other transcription factors, including MEF2C and MyoD (Eckner *et al.,* 1996; Yuan *et al.,* 1996; Puri *et al.,* 1997; Sartorelli *et al.,* 1997), which may specifically account for the ability of E1A to counter tissue-restricted expression in skeletal muscle cells even in the absence of disrupting pocket protein function. Given the pivotal role of MEF2C in cardiac organogenesis (Lin *et al.,* 1997), a similar mechanism can be readily envisioned in cardiac myocytes as well, though further work is needed to define the specifically associated proteins that bind this region of E1A after gene delivery to cardiac myocytes. Among other targets, p300 also has been found to serve as a coactivator for hypoxia-induced factor-1 α, which mediates induction of certain glycolytic enzymes, erythropoietin, and vascular endothelial growth factor under conditions of hypoxic stress (Arany *et al.,* 1996).

As cited earlier for skeletal muscle, it has recently been suggested that E1A inhibition of cardiac-restricted genes might require interactions with the amino terminus distinct from p300 binding (Bishopric *et al.,* 1997). This portion of the E1A interacts with a number of lower molecular weight cellular proteins distinct from p300 (Sang and Giordano, 1997) whose identity and function are unknown, as are potential binding partners for this region in cardiac muscle itself. One plausible set of candidates would comprise lineage-restricted cardiac transcription factors or their coactivators, discussed previously. As one alternative, E1A is known to bind directly to the Cdk inhibitor, p27, and thus to increase G_1 Cdk activity indirectly (Mal *et al.,* 1996). No mutagenesis was performed to identify the E1A domain involved. However, given the known ability of G_1 cyclins to downregulate muscle-specific genes, this provides an additional mechanism for E1A effects both on cell cycle progression and on gene transcription. Thus, E1A can lead to G_1 exit and inactivation of the myogenic pathway through multiple mechanisms (Figs. 1 and 2).

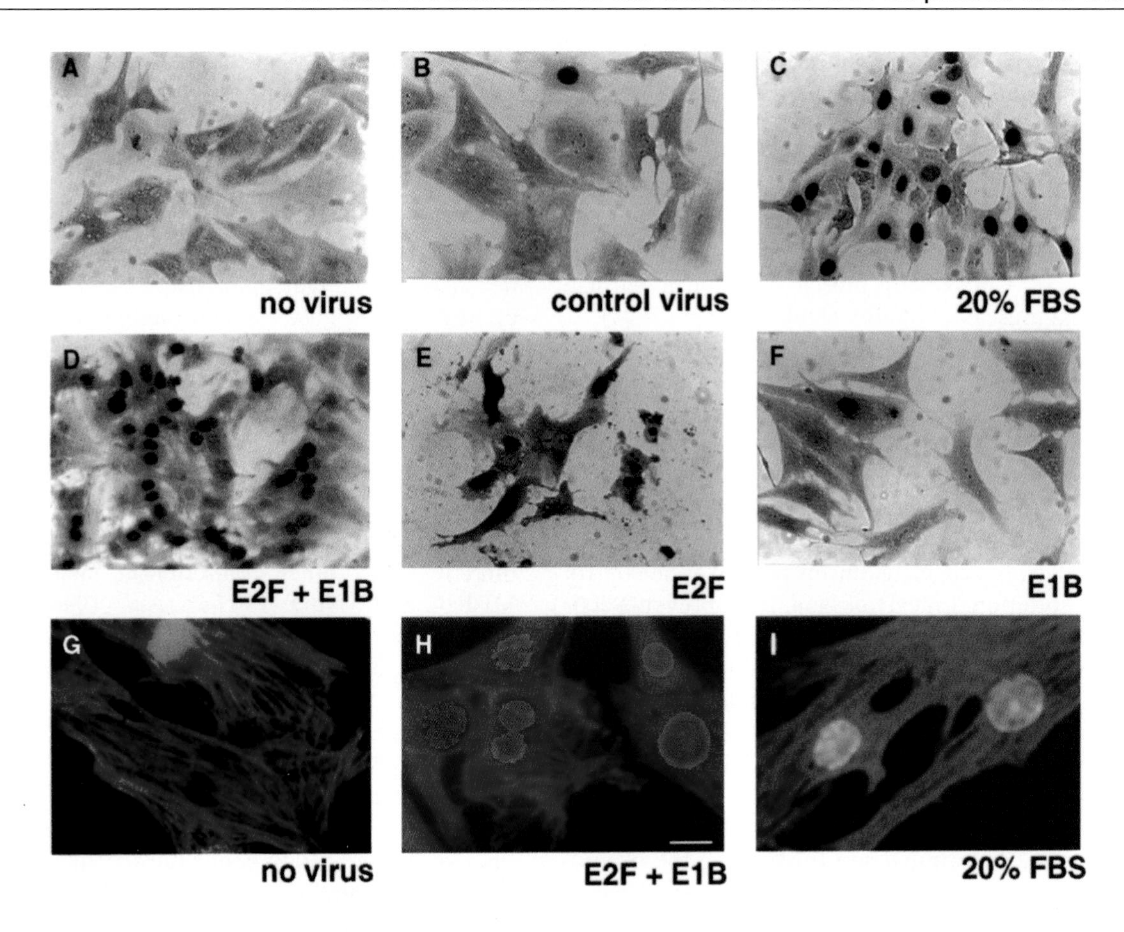

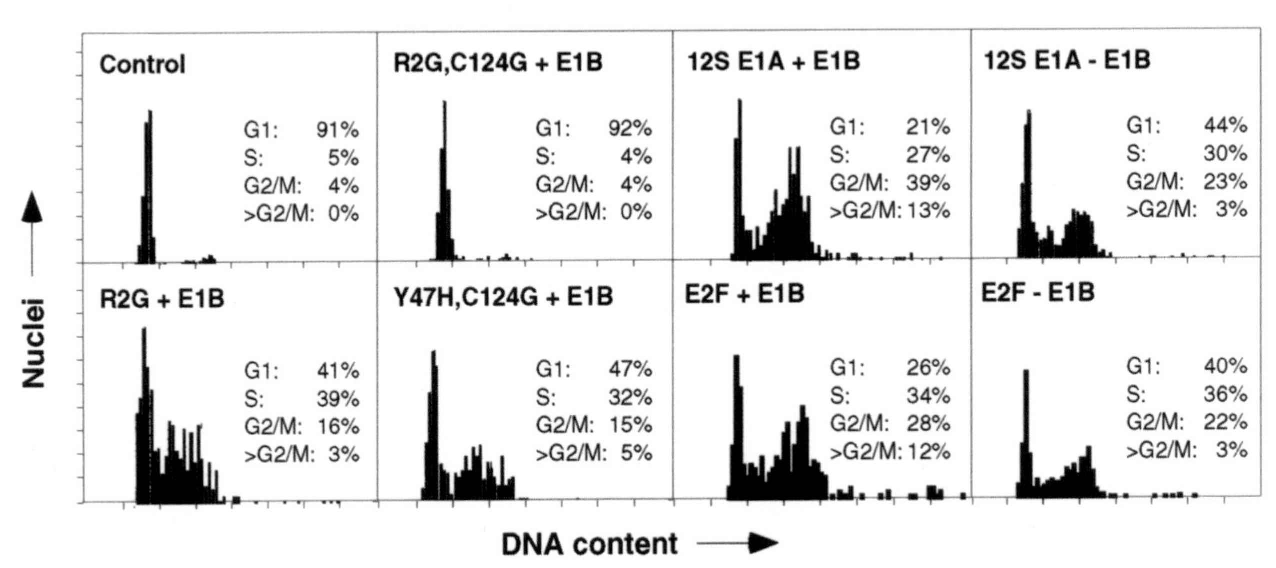

Figure 2 Adenoviral delivery of E1A or E2F-1 triggers G_1 exit in neonatal ventricular myocytes (Kirshenbaum *et al.*, 1996). (Top) DNA synthesis was demonstrated by immunoperoxidase (A–F) or immunofluorescence (G–I) staining for incorporation of BrdU. (Bottom) DNA content was measured by quantitative image analysis of Feulgen-stained nuclei. The indicated E1A mutations impair pocket protein binding (Y47H and C124G), p300 binding (R2G), or both domains (R2G and C124G).

X. E2F Transcription Factors

E2F denotes a family of heterodimeric transcription factors with five members, E2F-1 to -5, and the dimerization partners, DP-1 to -3. While less is known regarding the exact functional differences among these proteins, E2Fs-1 to -3 preferentially associate with Rb (Lees *et al.*, 1993), whereas E2F-4 has been found in complexes with p107 and p130 (Hijmans *et al.*, 1995; Moberg *et al.*, 1996) and E2F-5 binds p130 preferentially (Hijmans *et al.*, 1995). The developmental regulation of E2F and DP transcription factors has not been studied systematically during cardiac organogenesis; however, other temporal and spatial differences among E2F family members suggest, along with cited differences in affinity for specific pocket proteins, the likelihood that E2F proteins may serve distinguishable functions. Although relatively little direct functional evidence is available on this point, only E2F-1 triggers apoptosis, although E2F-2 (and, to a lesser extent, E2F-3) likewise can induce S phase (DeGregori *et al.*, 1997). The key importance of this finding is that cell cycle reentry and apoptosis are not linked obligatorily, as might have been surmised from E2F-1 alone. The ability of E2F-1 to drive quiescent cells through G_1 into S phase (Johnson *et al.*, 1993; Qin *et al.*, 1994) extends to neonatal cardiac myocytes *in vitro* and even to adult ventricular muscle both *in vitro* and *in vivo* (Kirshenbaum *et al.*, 1996; Agah *et al.*, 1997b). Thus, the block to cell cycle reentry in postmitotic myocardium might result from the inability to release E2F from pocket proteins at that stage, developmental regulation of the E2F isoforms (Itoh *et al.*, 1995), or both.

In cardiac myocytes, forced expression of E2F-1 also caused apoptosis unless accompanied by E1B or an equivalent antiapoptotic gene (Kirshenbaum *et al.*, 1996; Agah *et al.*, 1997b). E2F not only induced DNA synthesis and proliferating cell nuclear antigen but also inhibited cardiac-specific gene expression. Image analysis of Feulgen-stained nuclei confirmed an increase in tetraploid DNA content, proving the cells' accumulation in G_2/M and suggesting the presence of a second pivotal cell cycle checkpoint in the cardiac myocytes. Thus, E2F-1 was sufficient to provoke each of the effects seen earlier with the viral protein E1A (Fig. 2). As described previously, numerous studies have established the paradigm that E2F activity is necessary for G_1/S progression (Duronio *et al.*, 1995); however, this model may be simplistic given the challenge to this interpretation posed by the phenotype of mice nullizygous for E2F-1, which show widespread tissue hyperplasia and characteristic tumor formation, indicating, paradoxically, that E2F-1 also functions as a tumor suppressor (Field *et al.*, 1996; Yamasaki *et al.*, 1996). These discordant results from *in vitro* versus *in vivo* studies highlight the indispensable importance of genetic models to test predictions from *in vitro* results and, potentially, to gain novel insights into proteins' net function *in vivo.* Provisional interpretations of this unanticipated phenotype in E2F-1-deficient mice include the possibility that an optimal concentration of E2F-1 exists for quiescence and that, consequently, overexpression and deletion each might increase cell proliferation (Field *et al.*, 1996; Yamasaki *et al.*, 1996). Deregulated growth in the absence of E2F-1 could conceivably represent a deficiency in cell death (Weinberg, 1996), because E2F-1 induces apoptosis (Qin *et al.*, 1994; Kirshenbaum *et al.*, 1996; Agah *et al.*, 1997b). A third possibility worth noting might be the absence of an E2F-dependent negative feedback loop.

XI. The G_2/M Boundary

Even less is known in cardiac muscle of the regulators for entry into mitosis, even though conspicuous differences from other cell types (binucleation or polyploidy) make this issue inherently intriguing. Proteins required for chromatin condensation and breakdown of the nuclear lamina in other background include the G_2 cyclins A and B and Cdc2 (Swenson *et al.*, 1986; Pines and Hunter, 1989), with complex regulation of Cdc2 kinase activity in G_2/M via phosphorylation at multiple sites. Cdc2 remains inactivated until the end of G_2 through phosphorylation of Tyr15 by the mitotic inhibitor, Weel (Gould and Nurse, 1989; Krek and Nigg, 1991). Weel, in turn, is inactivated at the onset of mitosis by nim1 (Coleman *et al.*, 1993; Parker *et al.*, 1993). The cyclin B/Cdc2 complex phosphorylates and activates the Cdc25 phosphatase, which then dephosphorylates Cdc2 at the onset of mitosis. These linked events provide a negative feedback loop that limits the duration of active cyclin/Cdc2 complexes and is permissive for exit from M phase (Sebastian *et al.*, 1993). A Cdc2-independent pathway to mitosis can be provoked through forced expression of NIMA, a mitotic kinase in *Aspergillus nidulan,* which provokes premature M phase in vertebrate cells (Lu and Hunter, 1995). Indeed, a NIMA-like kinase activity recently was identified in mammalian cells (Lu and Hunter, 1995).

Previous work has established that the G_2/M boundary represents a critical checkpoint in cardiac cell cycle progression, which limits the efficacy of E1A and E2F-1 (Kirshenbaum and Schneider, 1995; Kirshenbaum *et al.*, 1996), strongly resembling the accumulation of Rb-deficient skeletal muscle cells in G_2 (Novitch *et al.*, 1996). The basis of this restriction is unknown and limited data are available regarding the expression of pertinent regulatory molecules. Cyclin A and Cdc2 are expressed in proliferating cardiac myocytes but are downregulated in the adult heart (Yoshizumi *et al.,* 1995; Soonpaa *et al.,* 1997); however, the expression of cyclin B and, equally important, the status of kinases controlling Cdc2 function are unknown. Similarly, neither expression nor activity of the M phase-specific, polo-like kinases have been systematically examined in relation to the postmitotic phenotype in myocardium (Glover *et al.,* 1996; Lane and Nigg, 1997).

XII. Reactivation of the Cell Cycle in Adult Ventricular Myocytes

While not the focus of this chapter, the capacity for DNA synthesis and mitosis of adult cardiac myocytes is relevant to the issue of their terminally differentiated state and remains controversial. Numerous studies have been published assessing the prevalence for DNA synthesis by adult cardiac myocytes under normal and pathological conditions, often providing discordant results. Estimates of DNA-labeling indexes range from 0.04% to <0.005% in the normal adult mouse ventricle (Rumyantsev, 1977; Soonpaa *et al.,* 1996, 1997; Nakagawa *et al.,* 1997; Soonpaa and Field, 1997). While there seems to be little spontaneous DNA labeling in the mouse, reports of a mitotic index as high as 2.5% in the normal rat heart (Anversa *et al.,* 1991) suggest that differences might exist between these species in the potential for activating cardiac myocyte DNA synthesis, mitosis, or both in pathophysiological settings (Oberpriller *et al.,* 1995). Several studies of labeling indexes in pathological conditions have challenged the dogma that cardiac myocytes are permanently withdrawn from the cell cycle and have no capacity for regeneration. Studies of rat cardiac myocytes after cardiac injury (Capasso *et al.,* 1992; Liu *et al.,* 1995) or with hypertrophy (Anversa *et al.,* 1990) reported the presence of mitotic myocytes, with an increased proportion of cells in the S and G_2/M phases. However, related studies in various rat and murine models of hypertrophy have failed to confirm this increase in DNA synthesis (Vliegen *et al.,* 1990; Kellerman *et al.,* 1992; Soonpaa and Field, 1997). Several reasons explain these apparently discordant results, beyond the obvious possibility of species-specific differences, including methodological differences, and variations in the nature or severity of trophic stimuli, perhaps with differing capacity to stimulate hyperplastic versus hypertrophic growth. Recent reports indeed suggest that cyclins E, A, and B, Cdk2, Cdc2, and PCNA are upregulated after infarction in the surviving myocytes (Reiss *et al.,* 1996b) along with Cdk2 and Cdc2 kinase activity. Similar upregulation was not observed in murine models of cardiac hypertrophy, in concurrence with the absence of cell cycle reentry discussed previously. Notwithstanding this controversy, even if a degree of hyperplasia were possible, the absolute capacity for adult cardiac myocytes to restore functional muscle mass through an increase in cell number is relatively meager, and genetic manipulations in efforts to augment this capacity merit further attention.

More generally, this discordance between models suggests the plausible hypothesis that the inability to reexpress (or activate) cell cycle regulators that are absent (or inactive) in the normal adult heart may provide a basis for the postmitotic state. Such a model has one advantage, compared to the notion that Rb alone provides the irreversible lock to the cell cycle, since the function of Rb protein should be wholly reversible upon hyperphosphorylation. The ability of E1A and E2F-1 to trigger G_1 exit in cardiac myocytes is consistent with this revisionist interpretation, since both act downstream from G_1 Cdks, Rb, and other pocket proteins (Kirshenbaum and Schneider, 1995; Kirshenbaum *et al.,* 1996; Agah *et al.,* 1997b). Indeed, in postmitotic skeletal muscle myotubes, the ability of SV40 large T antigen to override Rb and reactivate DNA synthesis is associated with—and plausibly contingent on—the reinduction of cyclins and Cdks (Wang and Nadal-Ginard, 1995). A recent, contrasting report contends that mitogenic serum cannot provoke DNA synthesis in neonatal cardiac myocytes despite an observed increase in phosphorylation of Rb, suggesting a block distal to G_1 Cdk activity, at least under some experimental conditions (Sadoshima *et al.,* 1997).

Although the systemic defects and early lethality associated with the absence of Rb preclude the use of the existing genomic knockout in testing the hypothesis that Rb provides the "lock" in postmitotic ventricular myocytes, recent advances in conditional knockout mutations make it possible to confine gene deletion to a particular lineage, to a predetermined time, and, potentially, to a delimited time and place using tissue-restricted or drug-inducible systems to direct the expression of Cre or Flp site-specific DNA recombinases (Gu *et al.,* 1994; Kühn *et al.,* 1995; see Chapter 15). The feasibility and overall utility of this approach are illustrated in Fig. 3 both for a cardiac-restricted Cre gene and for adenoviral delivery of Cre to postmitotic myocardium.

The clinical outcome of myocardial infarction is confounded both by the limited capacity to restore functional cardiac mass by adaptive hypertrophy and by deleterious effects associated with hypertrophy which further impair survival (Pfeffer, 1995). Current therapies for preventing the progression of hypertrophy to overt cardiac failure are aimed at ameliorating workload or improving myocytes' intrinsic contractility. It

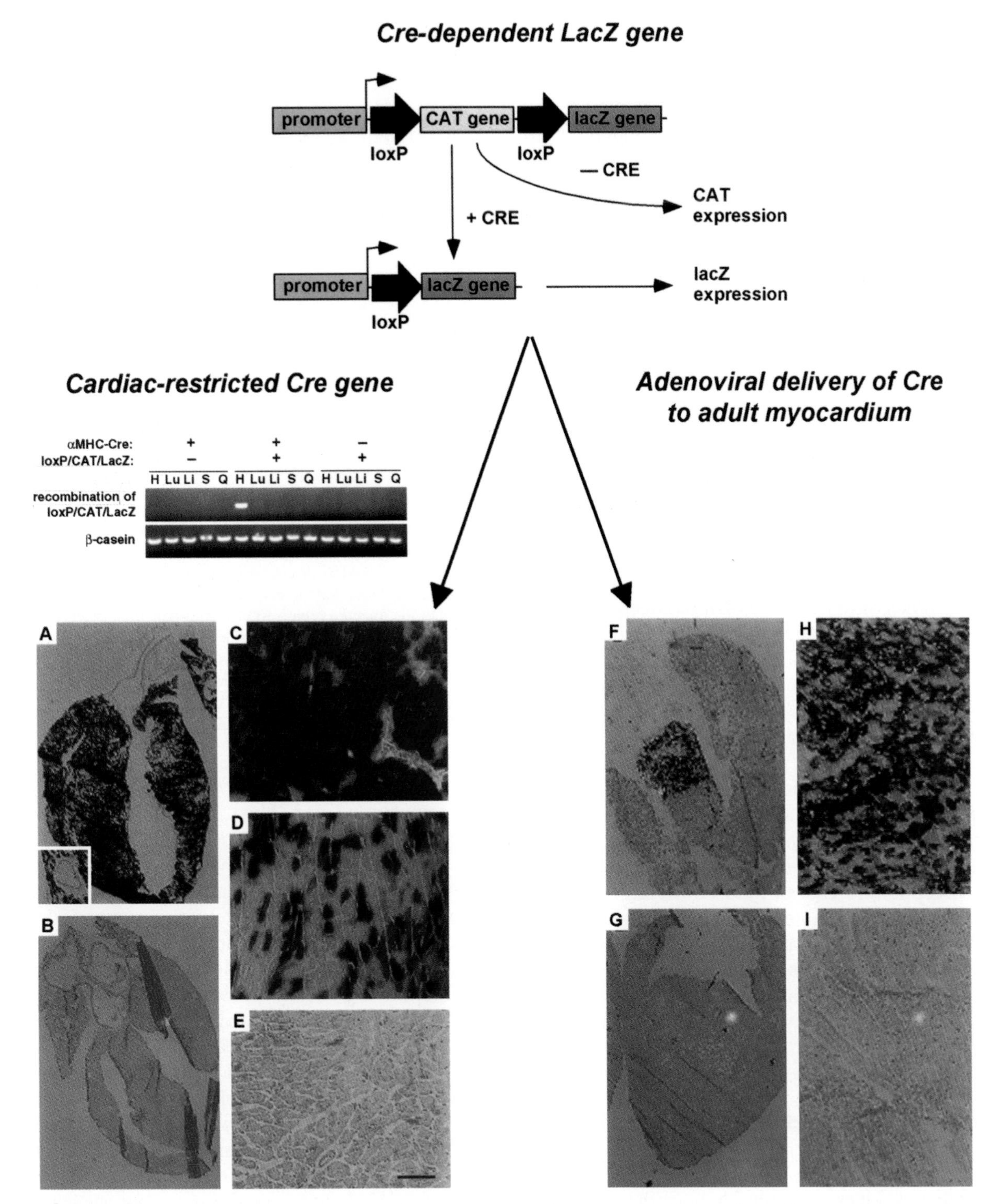

Figure 3 Cre-mediated gene recombination in cardiac muscle using cardiac-specific targeting of Cre recombinase or adenoviral delivery of Cre by direct injection of adult myocardium (Agah *et al.,* 1997a). The Cre-dependent LacZ reporter gene used for these validation studies is out of frame, downstream from the loxP-flanked CAT insert. Thus, LacZ expression is contingent on recombination between the loxP sites, excising CAT, and placing LacZ in frame. Recombination of the Cre-dependent reporter gene was strictly contingent on coinheritance or viral delivery of Cre. (Left) Cardiac-restricted recombination of the Cre-dependent reporter was shown using PCR primers outside the paired loxP sites and an elongation time too short for amplification of the intact gene. (A–E) The prevalence of recombination and specificity for cardiac myocytes were demonstrated by histochemical staining for LacZ. (A, C) αMHC-Cre$^+$/CAG-CATZ$^+$ (derived from a high-prevalence founder Cre line); (D) αMHC-Cre$^+$/CAG-CATZ$^+$ (derived from a lower prevalence founder Cre line); (B, E) αMHC-Cre$^-$/CAG-CATZ$^+$. (Right) The feasibility of *de novo* recombination in postmitotic cardiac muscle was determined by induction of the Cre-dependent LacZ transgene following intramyocardial injection of an adenoviral Cre vector (F, H) versus a control virus (G, I).

has been suggested that increasing myocardial mass through an increase in myocyte number could provide an alternate strategy to obviate the development of maladaptive growth. Initial attempts to accomplish this goal include cell implantation, such as grafting fetal cardiac myocytes (Soonpaa *et al.,* 1994; Klug *et al.,* 1996; Leor *et al.,* 1996), and converting nonmuscle cells to functional muscle in areas of scarring, using myogenic transcription factors (Murry *et al.,* 1996). One conceptual alternative would be direct manipulation of the cardiac myocyte cell cycle machinery to promote proliferation, although formidable impediments would need to be overcome.

With this caveat, the studies described previously for cultured cardiac myocytes demonstrating cell cycle reactivation both by viral proteins such as E1A and by mammalian E2F-1 have been promising (Kirshenbaum and Schneider, 1995; Kirshenbaum *et al.,* 1996; Liu and Kitsis, 1996; Bishopric *et al.,* 1997), and the findings in neonatal ventricular myocytes have already been extended to adult ventricular muscle in the case of E2F-1 (Agah *et al.,* 1997b). Importantly, the conclusion that E2F-1 is sufficient to trigger G_1 exit has been substantiated in "postmitotic" adult ventricular myocytes, not only in cell culture but also after direct injection into adult mouse myocardium *in vivo* (Fig. 4) (Agah *et al.,* 1997b). One noteworthy aspect of pursuing this approach in the mouse—viral gene delivery to genetically defined recipients—is the opportunity to explore functional gene interactions in combinatorial fashion. For example, E2F-dependent apoptosis (and other forms of apoptosis) are reportedly dependent on p53 in other settings; however, although p53-null mice were much more susceptible to G_1 exit induced by E2F-1, this increase was not attributable to relief of apoptosis and instead indicates a growth-inhibitory function of p53 in the adult myocardium. Further progress will require a more detailed understanding of the proteins whose absence, inactivity, or presence controls the G_2/M transition in postmitotic cells.

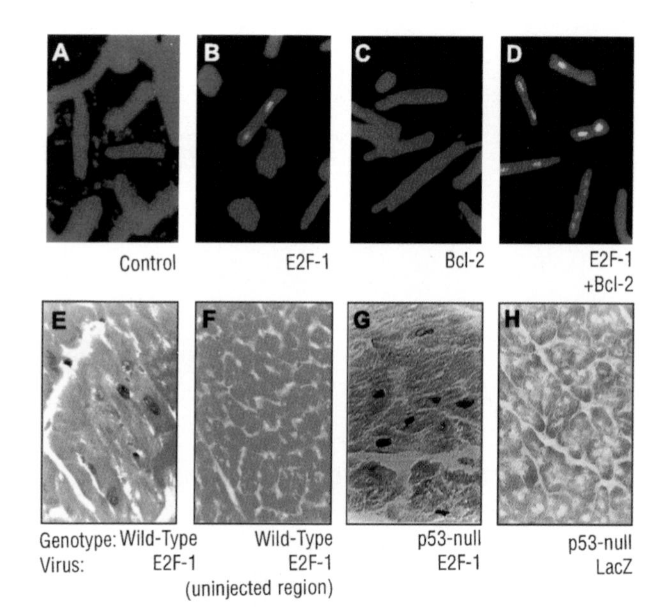

Figure 4 Cell cycle reentry in postmitotic cardiac muscle cells using adenoviral delivery of E2F-1 (Agah *et al.,* 1997b). (A–D) Adult rat ventricular myocytes in culture; (E–H) adult mouse myocardium *in vivo,* injected under direct visualization after thoracotomy. The greater prevalence for G_1 exit in p53-deficient mice (~30% of myocytes in the injected region 10% in wild-type mice) could not be ascribed to differences in virus uptake or expression or to a diminished frequency of E2F-induced apoptosis (40–45% for both genetic backgrounds). Together, these results demonstrate that E2F-1 suffices for cell cycle reentry in "postmitotic" cardiac myocytes and confirm a growth-inhibitory role of p53 in adult myocardium.

XIII. Conclusions and Perspectives

This chapter has attempted to summarize current understanding of cell cycle regulation in cardiac myocytes and the molecular connections among proliferation, differentiation, and apoptosis in the developing heart. Clinical syndromes in which these events are germane include, most obviously, hypoplastic heart syndromes and many common congenital heart defects but also the less familiar disorder, noncompaction. Moreover, it has been suggested, in other contexts, that hypertrophic growth may be a forme fruste of cell cycle reentry, with the imposition of a block at the G_1/S checkpoint through pocket protein action (Franch *et al.,* 1995); direct tests are needed, however, to assess the possible relevance of this model to agonist- and load-induced myocardial hypertrophy. Likewise, with increasing evidence for the prevalence of apoptosis in end-stage heart failure in humans, it becomes germane to determine whether any significant component of cell death in failing hearts can be ascribed to abortive cell cycle reentry. A link between cell cycle regulators and death signals that might not be considered in the context of cycling is shown by the degradation of Rb via an ICE-like protease during apoptosis (Janicke *et al.,* 1996) as well as by the ability of a cleavage-resistant Rb to protect cells from certain apoptotic triggers (Janicke *et al.,* 1996). Beyond this role in pathophysiology, apoptosis is also believed to contribute to normal cardiac morphogenesis, such as during the normal regression of the right ventricle upon unloading after birth (Kajstura *et al.,* 1995). It remains unknown whether differential rates of proliferation, apoptosis, or alternative events explain the mechanics of how looping morphogenesis proceeds (Biben and Harvey, 1997). Despite the many advances detailed here in both observational and mechanistic studies, critical deficiencies exist in the information now available. Can the postmitotic phenotype in adult ventricular muscle be entirely explained by a developmental increase in Cdk inhibitors, or do additional events impede cell cycle reentry in response to trophic signals? For seemingly redundant cell cycle regulators, does protein function differ *in vivo* in ways that might not be predicted from cell cycle studies alone? While E2F transcription factors are understood to induce multiple enzymes for DNA synthesis, what molecular events specifically couple the Cdk–Rb–E2F pathway to assembly of the DNA replication complex (Stillman, 1996)? Ultimately, the most generalized impact of the cardiac cell cycle in a clinical context is likely to be the postmitotic phenotype itself and the resulting deficiency in compensatory growth after injury—i.e., infarction as a "myocyte deficiency disease."

References

Abdellatif, M., and Schneider, M. D. (1997). An effector-like function of Ras GTPase-activating protein predominates in cardiac muscle cells. *J. Biol. Chem.* **272,** 525–533.

Abdellatif, M., MacLellan, W. R., and Schneider, M. D. (1994). p21 Ras as a governor of global gene expression. *J. Biol. Chem.* **269,** 15423–15426.

Agah, R., Frenkel, P. A., French, B. A., Michael, L. H., Overbeek, P. A., and Schneider, M. D. (1997a). Gene recombination in postmitotic cells: Targeted expression of Cre recombinase provokes cardiac-restricted, site-specific rearrangement in adult ventricular muscle in vivo. *J. Clin. Invest.* **100,** 169–179.

Agah, R., Kirshxenbaum, L. A., Truong, L. D., Chakraborty, S., Abdellatif, M., Michael, L. H., and Schneider, M. D. (1997b). Adenoviral delivery of E2F-1 directs cell cycle re-entry and p53-independent apoptosis in post-mitotic adult myocardium in vivo. *J. Clin. Invest.* **100,** 2722–2728.

Antin, P. B., Yatskievych, T., Dominguez, J. L., and Chieffi, P. (1996). Regulation of avian precardiac mesoderm development by insulin and insulin-like growth factors. *J. Cell. Physiol.* **168,** 42–50.

Anversa, P., Palackal, T., Sonnenblick, E. H., Olivetti, G., Meggs, L. G., and Capasso, J. M. (1990). Myocyte cell loss and myocyte cellular hyperplasia in the hypertrophied aging rat heart. *Circ. Res.* **67,** 871–885.

Anversa, P., Fitzpatrick, D., Argani, S., and Capasso, J. M. (1991). Myocyte mitotic division in the aging mammalian rat heart. *Circ. Res.* **69,** 1159–1164.

Arany, Z., Huang, L. E., Eckner, R., Bhattacharya, S., Jiang, C., Goldberg, M. A., Bunn, H. F., and Livingston, D. M. (1996). An essential role for p300/CBP in the cellular response to hypoxia. *Proc. Natl. Acad. Sci. U.S.A.* **93,** 12969–12973.

Baker, J., Liu, J. P., Robertson, E. J., and Efstratiadis, A. (1993). Role of insulin-like growth factors in embryonic and postnatal growth. *Cell (Cambridge, Mass.)* **75,** 73–82.

Baldin, V., Lukas, J., Marcote, M. J., Pagano, M., and Draetta, G. (1993). Cyclin D1 is a nuclear protein required for cell cycle progression in G1. *Genes Dev.* **7,** 812–821.

Beijersbergen, R. L., Carlee, L., Kerkhoven, R. M., and Bernards, R. (1995). Regulation of the retinoblastoma protein-related p107 by G(1) cyclin complexes. *Genes Dev.* **9,** 1340–1353.

Beinlich, C. J., Baker, K. M., White, G. J., and Morgan, H. E. (1991). Control of growth in the neonatal pig heart. *Am. J. Physiol.* **261**(Suppl. 4), 3–7.

Bernards, R., Schackleford, G. M., Gerber, M. R., Horowitz, J. M., Friend, S. H., Schartl, M., Bogenmann, E., Rapaport, J. M., McGee, T., and Dryja, T. P. (1989). Structure and expression of the murine retinoblastoma gene and characterization of its encoded protein. *Proc. Natl. Acad. Sci. U. S. A.* **86,** 6474–6478.

Biben, C., and Harvey, R. P. (1997). Homeodomain factor Nkx2-5 controls left/right asymmetric expression of bHLH gene eHand during murine heart development. *Genes Dev.* **11,** 1357–1369.

Bishopric, N. H., Zeng, G.-Q., Sato, B., and Webster, K. A. (1997). Adenovirus E1A inhibits cardiac myocyte-specific gene transcription via its amino terminus. *J. Biol. Chem.* **272,** 20584–20594.

Brannan, C. I., Perkins, A. S., Vogel, K. S., Ratner, N., Nordlund, M. L., Reid, S. W., Buchberg, A. M., Jenkins, N. A., Parada, L. F., and Copeland, N. G. (1994). Targeted disruption of the neurofibromatosis type-1 gene leads to developmental abnormalities in heart and various neural crest-derived tissues. *Genes Dev.* **8,** 1019–1029.

Braun, T., Bober, E., and Arnold, H. H. (1992). Inhibition of muscle differentiation by the adenovirus E1a protein: Repression of the transcriptional activating function of the HLH protein Myf-5. *Genes Dev.* **6,** 888–902.

Brugarolas, J., Chandrasekaran, C., Gordon, J. I., Beach, D., Jacks, T., and Hannon, G. J. (1995). Radiation-induced cell cycle arrest compromised by p21 deficiency. *Nature (London)* **377,** 552–557.

Capasso, J. M., Bruno, S., Cheng, W., Li, P., Rodgers, R., Darzynkiewicz, Z., and Anversa, P. (1992). Ventricular loading is coupled with DNA synthesis in adult cardiac myocytes after acute and chronic myocardial infarction in rats. *Circ. Res.* **71,** 1379–1389.

Cardoso, M. C., Leonhardt, H., and Nadal-Ginard, B. (1993). Reversal of terminal differentiation and control of DNA replication: Cyclin A and Cdk2 specifically localize at subnuclear sites of DNA replication. *Cell (Cambridge, Mass.)* **74,** 979–992.

Caruso, M., Martelli, F., Giordano, A., and Felsani, A. (1993). Regulation of MyoD gene transcription and protein function by the transforming domains of the adenovirus E1A oncoprotein. *Oncogene* **8,** 267–278.

Charron, J., Malynn, B. A., Fisher, P., Stewart, V., Jeannotte, L., Goff, S. P., Robertson, E. J., and Alt, F. W. (1992). Embryonic lethality in mice homozygous for a targeted disruption of the N-myc gene. *Genes Dev.* **6,** 2248–2257.

Chen, C. Y., and Schwartz, R. J. (1996). Recruitment of the tinman homolog nkx-2.5 by serum response factor activates cardiac alpha-actin gene transcription. *Mol. Cell. Biol.* **16,** 6372–6384.

Chen, G., Guy, C. T., Chen, H. W., Hu, N. P., Lee, E., and Lee, W. H. (1996). Molecular cloning and developmental expression of mouse p130, a member of the retinoblastoma gene family. *J. Biol. Chem.* **271,** 9567–9572.

Chen, Z., Friedrich, G. A., and Soriano, P. (1994). Transcriptional enhancer factor 1 disruption by a retroviral gene trap leads to heart defects and embryonic lethality in mice. *Genes Dev.* **8,** 2293–2301.

Chin, T. K., Perloff, J. K., Williams, R. G., Jue, K., and Mohrmann, R. (1990). Isolated noncompaction of left ventricular myocardium. A study of eight cases. *Circulation* **82,** 507–513.

Chow, K. N. B., Starostik, P., and Dean, D. C. (1996). The Rb family contains a conserved cyclin-dependent kinase-regulated transcriptional repressor motif. *Mol. Cell. Biol.* **16,** 7173–7181.

Clarke, A. R., Maandag, E. R., van Roon, M., van der Lugt, N. M., van der Valk, M., Hooper, M. L., Berns, A., and te Riele, H. (1992). Requirement for a functional Rb-1 gene in murine development. *Nature (London)* **359,** 328–330.

Clubb, F. J., Jr., and Bishop, S. P. (1984). Formation of binucleated myocardial cells in the neonatal rat: an index for growth hypertrophy. *Lab. Invest.* **50,** 571–577.

Cobrinik, D., Lee, M. H., Hannon, G., Mulligan, G., Bronson, R. T.,

Dyson, N., Harlow, E., Beach, D., Weinberg, R. A., and Jacks, T. (1996). Shared role of the pRB-related p130 and p107 proteins in limb development. *Genes Dev.* **10,** 1633–1644.

Coleman, T. R., Tang, Z., and Dunphy, W. G. (1993). Negative regulation of the wee1 protein kinase by direct action of the nim1/cdr1 mitotic inducer. *Cell (Cambridge, Mass.)* **72,** 919–929.

Crescenzi, M., Fleming, T. P., Lassar, A. B., Weintraub, H., and Aaronson, S. A. (1990). MyoD induces growth arrest independent of differentiation in normal and transformed cells. *Proc. Natl. Acad. Sci. U. S. A.* **87,** 8442–8446.

Datto, M. B., Hu, P. P. C., Kowalik, T. F., Yingling, J., and Wang, X. F. (1997). The viral oncoprotein E1A blocks transforming growth factor beta-mediated induction of p21/WAF1/Cip1 and p15/INK4B. *Mol. Cell. Biol.* **17,** 2030–2037.

Daud, A. I., Lanson, N. A., Claycomb, W. C., and Field, L. J. (1993). Identification of SV40 large T-antigen-associated proteins in cardiomyocytes from transgenic mice. *Am. J. Physiol.* **264,** H1693–H1700.

DeGregori, J., Leone, G., Miron, A., Jakoi, L., and Nevins, J. R. (1997). Distinct roles for E2F proteins in cell growth control and apoptosis. *Proc. Natl. Acad. Sci. U. S. A.* **94,** 7245–7250.

DeLeon, J. R., Federoff, H. J., Dickson, D. W., Vikstrom, K. L., and Fishman, G. I. (1994). Cardiac and skeletal myopathy in beta myosin heavy-chain simian virus 40 tsA58 transgenic mice. *Proc. Natl. Acad. Sci. U. S. A.* **91,** 519–523.

Deng, C. X., Zhang, P. M., Harper, J. W., Elledge, S. J., and Leder, P. (1995). Mice lacking p21(C/P1/WAF1) undergo normal development, but are defective in G1 checkpoint control. *Cell (Cambridge, Mass.)* **82,** 675–684.

Dobrowolski, S., Harter, M., and Stacey, D. W. (1994). Cellular ras activity is required for passage through multiple points of the G(0)/G(1) phase in BALB/c 3T3 cells. *Mol. Cell. Biol.* **14,** 5441–5449.

Dorsman, J. C., Teunisse, A. F. A. S., Zantema, A., and van der Eb, A. J. (1997). The adenovirus 12 E1A proteins can bind directly to proteins of the p300 transcription co-activator family, including the CREB-binding protein CBP and p300. *J. Gen. Virol.* **78,** 423–426.

Dowdy, S. F., Hinds, P. W., Louie, K., Reed, S. I., Arnold, A., and Weinberg, R. A. (1993). Physical interaction of the retinoblastoma protein with human D cyclins. *Cell (Cambridge, Mass.)* **73,** 499–511.

Duronio, R. J., O'Farrell, P. H., Xie, J. E., Brook, A., and Dyson, N. (1995). The transcription factor E2F is required for S phase during Drosophila embryogenesis. *Genes Dev.* **9,** 1445–1455.

Eckner, R., Ewen, M. E., Newsome, D., Gerdes, M., Decaprio, J. A., Lawrence, J. B., and Livingston, D. M. (1994). Molecular cloning and functional analysis of the adenovirus E1A-associated 300–kD protein (P300) reveals a protein with properties of a transcriptional adaptor. *Genes Dev.* **8,** 869–884.

Eckner, R., Yao, T. P., Oldread, E., and Livingston, D. M. (1996). Interaction and functional collaboration of p300/CBP and bHLH proteins in muscle and B-cell differentiation. *Genes Dev.* **10,** 2478–2490.

Eisenman, R. N., and Cooper, J. A. (1995). Signal transduction—beating a path to myc. *Nature (London)* **378,** 438–439.

el-Deiry, W. S., Harper, J. W., O'Connor, P. M., Velculescu, V. E., Canman, C. E., Jackman, J., Pietenpol, J. A., Burrell, M., Hill, D. E., Wiman, K. G., Mercer, W. E., Kastan, M. B., Kohn, K. W., Elledge, S. J., Kinzler, K. W., and Vogelstein, B. (1994). WAF1/CIP1 is induced in p53–mediated G1 arrest and apoptosis. *Cancer Res.* **54,** 1169–1174.

Field, L. J. (1988). Atrial natriuretic factor-SV40 T antigen transgenes produce tumors and cardiac arrhythmias in mice. *Science* **239,** 1029–1033.

Field, S. J., Tsai, F. Y., Kuo, F., Zubiaga, A. M., Kaelin, W. G., Livingston, D. M., Orkin, S. H., and Greenberg, M. E. (1996). E2F-1 functions in mice to promote apoptosis and suppress proliferation. *Cell (Cambridge, Mass.)* **85,** 549–561.

Flink, I., Oana, S., Maitra, N., Bahl, J. J., and Morkin, E. (1997). The cell cycle inhibitor, p21, is induced during terminal differentiation of cardiac myocytes. *Keystone Symp. Mol. Biol. Muscle Dev.,* Snowmass, CO, p. 27.

Franch, H. A., Shay, J. W., Alpern, R. J., and Preisig, P. A. (1995). Involvement of pRB family in TGF beta-dependent epithelial cell hypertrophy. *J. Cell Biol.* **129,** 245–254.

Franklin, D. S., and Xiong, Y. (1996). Induction of p18(INK4c) and its predominant association with CDK4 and CDK6 during myogenic differentiation. *Mol. Biol. Cell.* **7,** 1587–1599.

Gassmann, M., Casagranda, F., Orioli, D., Simon, H., Lai, C., Klein, R., and Lemke, G. (1995). Aberrant neural and cardiac development in mice lacking the ErbB4 neuregulin receptor. *Nature (London)* **378,** 390–394.

Glover, D. M., Ohkura, H., and Tavares, A. (1996). Polo kinase: The choreographer of the mitotic stage? *J. Cell Biol.* **135,** 1681–1684.

Gould, K. L., and Nurse, P. (1989). Tyrosine phosphorylation of the fission yeast cdc2+ protein kinase regulates entry into mitosis. *Nature (London)* **342,** 39–45.

Gu, H., Marth, J. D., Orban, P. C., Mossman, H., and Rajewsky, K. (1994). Deletion of a DNA polymerase beta gene segment in T cells using cell type-specific gene targeting. *Science* **265,** 103–106.

Gu, W., Schneider, J. W., Condorelli, G., Kaushal, S., Mahdavi, V., and Nadal-Ginard, B. (1993). Interaction of myogenic factors and the retinoblastoma protein mediates muscle cell commitment and differentiation. *Cell (Cambridge, Mass.)* **72,** 309–324.

Guan, K. L., Jenkins, C. W., Li, Y., Nichols, M. A., Wu, X., O'Keefe, C. L., Matera, A. G., and Xiong, Y. (1994). Growth suppression by p18, a p16INK4/MTS1– and p14INK4B/MTS2-related CDK6 inhibitor, correlates with wild-type pRb function. *Genes Dev.* **8,** 2939–2952.

Halevy, O., Novitch, B. G., Spicer, D. B., Skapek, S. X., Rhee, J., Hannon, G. J., Beach, D., and Lassar, A. B. (1995). Correlation of terminal cell cycle arrest of skeletal muscle with induction of p21 by MyoD. *Science* **267,** 1018–1021.

Harper, J. W., Adami, G. R., Wei, N., Keyomarsi, K., and Elledge, S. J. (1993). The p21 Cdk-interacting protein Cip1 is a potent inhibitor of G1 cyclin-dependent kinases. *Cell (Cambridge, Mass.)* l. **75,** 805–816.

Harper, J. W., Elledge, S. J., Keyomarsi, K., Dynlacht, B., Tsai, L. H., Zhang, P. M., Dobrowolski, S., Bai, C., Connell-Crowley, L., Swindell, E., Fox, M. P., and Wei, N. (1995). Inhibition of cyclin-dependent kinases by p21. *Mol. Biol. Cell.* **6,** 387–400.

Hartung, H., Feldman, B., Lovec, H., Coulier, F., Birnbaum, D., and Goldfarb, M. (1997). Murine FGF-12 and FGF-13: Expression in embryonic nervous system, connective tissue and heart. *Mech. Dev.* **64,** 31–39.

Hein, L., Barsh, G. S., Pratt, R. E., Dzau, V. J., and Kobilka, B. K. (1995). Behavioural and cardiovascular effects of disrupting the angiotensin II type-2 receptor gene in mice. *Nature (London)* **377,** 744–747.

Hein, L., Stevens, M. E., Barsh, G. S., Pratt, R. E., Kobilka, B. K., and Dzau, V. J. (1997). Overexpression of angiotensin AT(1) receptor transgene in the mouse myocardium produces a lethal phenotype associated with myocyte hyperplasia and heart block. *Proc. Natl. Acad. Sci. U.S.A.* **94,** 6391–6396.

Hijmans, E. M., Voorhoeve, P. M., Beijersbergen, R. L., Vantveer, L. J., and Bernards, R. (1995). E2F-5, a new E2F family member that interacts with p130 in vivo. *Mol. Cell. Biol.* **15,** 3082–3089.

Hirai, H., Roussel, M. F., Kato, J. Y., Ashmun, R. A., and Sherr, C. J. (1995). Novel INK4 proteins, p19 and p18, are specific inhibitors of the cyclin D-dependent kinases CDK4 and CDK6. *Mol. Cell. Biol.* **15,** 2672–2681.

Hirota, H., Yoshida, K., Kishimoto, T., and Taga, T. (1995). Continuous activation of gp130, a signal-transducing receptor component for interleukin 6–related cytokines, causes myocardial hypertrophy in mice. *Proc. Natl. Acad. Sci. U. S. A.* **92,** 4862–4866.

Howe, J. A., Howell, M., Hunt, T., and Newport, J. W. (1995). Identification of a developmental timer regulating the stability of embryonic cyclin A and a new somatic A-type cyclin at gastrulation. *Genes Dev.* **9,** 1164–1176.

Hunter, J. J., Tanaka, N., Rockman, H. A., Ross, J., and Chien, K. R. (1995). Ventricular expression of a MLC-2v-ras fusion gene induces cardiac hypertrophy and selective diastolic dysfunction in transgenic mice. *J. Biol. Chem.* **270,** 23173–23178.

Hunter, T., and Pines, J. (1994). Cyclins and cancer .2: Cyclin d and CDK inhibitors come of age. *Cell (Cambridge, Mass.)* **79,** 573–582.

Hurford, R. K., Cobrinik, D., Lee, M. H., and Dyson, N. (1997). pRB and p107/p130 are required for the regulated expression of different sets of E2F responsive genes. *Genes Dev.* **11,** 1447–1463.

Ichiki, T., Labosky, P. A., Shiota, C., Okuyama, S., Imagawa, Y., Fogo, A., Miimura, F., Ichikawa, I., Hogan, B. L. M., and Inagami, T. (1995). Effects on blood pressure and exploratory behavior of mice lacking angiotensin II type-2 receptor. *Nature (London)* **377,** 748–750.

Ito, M., Oliverio, M. I., Mannon, P. J., Best, C. F., Maeda, N., Smithies, O., and Coffman, T. M. (1995). Regulation of blood pressure by the type 1A angiotensin II receptor gene. *Proc. Natl. Acad. Sci. U. S. A.* **92,** 3521–3525.

Itoh, A., Levinson, S. F., Morita, T., Kourembanas, S., Brody, J. S., and Mitsialis, S. A. (1995). Structural characterization and specificity of expression of E2F-5: A new member of the E2F family of transcription factors. *Cell Mol. Biol. Res.* **41,** 147–154.

Jaber, M., Koch, W. J., Rockman, H., Smith, B., Bond, R. A., Sulik, K. K., Ross, J., Lefkowitz, R. J., Caron, M. G., and Giros, B. (1996). Essential role of beta-adrenergic receptor kinase 1 in cardiac development and function. *Proc. Natl. Acad. Sci. U. S. A.* **93,** 12974–12979.

Jacks, T., Fazeli, A., Schmitt, E. M., Bronson, R. T., Goodell, M. A., and Weinberg, R. A. (1992). Effects of an Rb mutation in the mouse. *Nature (London)* **359,** 295–300.

Jacks, T., Shih, T. S., Schmitt, E. M., Bronson, R. T., Bernards, A., and Weinberg, R. A. (1994). Tumour predisposition in mice heterozygous for a targeted mutation in Nf1. *Nat. Genet.* **7,** 353–361.

Jackson, T., Allard, M. F., Sreenan, C. M., Doss, L. K., Bishop, S. P., and Swain, J. L. (1990). The c-myc proto-oncogene regulates cardiac development in transgenic mice. *Mol. Cell. Biol.* **10,** 3709–3716.

Janicke, R. U., Walker, P. A., Lin, X. Y., and Porter, A. G. (1996). Specific cleavage of the retinoblastoma protein by an ICE-like protease in apoptosis. *EMBO J.* **15,** 6969–6978.

Jin, Y., Pasumarthi, K. B. S., Bock, M. E., Lytras, A., Kardami, E., and Cattini, P. A. (1994). Cloning and expression of fibroblast growth factor receptor-1 isoforms in the mouse heart: Evidence for isoform switching during heart development. *J. Mol. Cell. Cardiol.* **26,** 1449–1459.

Johnson, D. G., Schwarz, J. K., Cress, W. D., and Nevins, J. R. (1993). Expression of transcription factor E2F1 induces quiescent cells to enter S-phase. *Nature (London)* **365,** 349–352.

Kajstura, J., Cheng, W., Reiss, K., and Anversa, P. (1994). The IGF-1–IGF-1 receptor system modulates myocyte proliferation but not myocyte cellular hypertrophy in vitro. *Exp. Cell Res.* **215,** 273–283.

Kajstura, J., Mansukhani, M., Cheng, W., Reiss, K., Krajewski, S., Reed, J. C., Quaini, F., Sonneblick, E. H., and Anversa, P. (1995). Programmed cell death and expression of the protooncogene bcl-2 in myocytes during postnatal maturation of the heart. *Exp. Cell Res.* **219,** 110–121.

Kang, M. J., Kim, J. S., Chae, S. W., Koh, K. N., and Koh, G. Y. (1997). Cyclins and cyclin dependent kinases during cardiac development. *Mol. Cells* **7,** 360–366.

Kastner, P., Grondona, J. M., Mark, M., Gansmuller, A., LeMeur, M., Decimo, D., Vonesch, J. L., Dolle, P., and Chambon, P. (1994). Genetic analysis of RXR alpha developmental function: Convergence of RXR and RAR signaling pathways in heart and eye morphogenesis. *Cell (Cambridge, Mass.)* **78,** 987–1003.

Katz, E. B., Steinhelper, M. E., Delcarpio, J. B., Daud, A. I., Claycomb, W. C., and Field, L. J. (1992). Cardiomyocyte proliferation in mice expressing alpha-cardiac myosin heavy chain-SV40 T-antigen transgenes. *Am. J. Physiol.* **262,** H1867–H1876.

Kellerman, S., Moore, J. A., Zierhut, W., Zimmer, H. G., Campbell, J., and Gerdes, A. M. (1992). Nuclear DNA content and nucleation patterns in rat cardiac myocytes from different models of cardiac hypertrophy. *J. Mol. Cell. Cardiol.* **24,** 497–505.

Kim, K. K., Soonpaa, M. H., Daud, A. I., Koh, G. Y., Kim, J. S., and Field, L. J. (1994). Tumor suppressor gene expression during normal and pathologic myocardial growth. *J. Biol. Chem.* **269,** 22607–22613.

Kim, K. K., Soonpaa, M. H., Wang, H., and Field, L. J. (1995). Developmental expression of p107 mRNA and evidence for alternative splicing of the p107 (RBL1) gene product. *Genomics* **28,** 520–529.

Kirshenbaum, L. A., and Schneider, M. D. (1995). Adenovirus E1A represses cardiac gene transcription and reactivates DNA synthesis in ventricular myocytes, via alternative pocket protein- and p300–binding domains. *J. Biol. Chem.* **270,** 7791–7794.

Kirshenbaum, L. A., Abdellatif, M., Chakraborty, S., and Schneider, M. D. (1996). Human E2F-1 reactivates cell cycle progression in ventricular myocytes and represses cardiac gene transcription. *Dev. Biol.* **179,** 402–411.

Kitagawa, M., Higashi, H., Jung, H. K., Suzuki-Takahashi, I., Ikeda, M., Tamai, K., Kato, J., Segawa, K., Yoshida, E., Nishimura, S., and Taya, Y. (1996). The consensus motif for phosphorylation by cyclin D1–Cdk4 is different from that for phosphorylation by cyclin A/E-Cdk2. *EMBO J.* **15,** 7060–7069.

Kiyokawa, H., Kineman, R. D., Manovatodorova, K. O., Soares, V. C., Hoffman, E. S., Ono, M., Khanam, D., Hayday, A. C., Frohman, L. A., and Koff, A. (1996). Enhanced growth of mice lacking the cyclin-dependent kinase inhibitor function of p27(Kip1). *Cell (Cambridge, Mass.)* **85,** 721–732.

Klug, M. G., Soonpaa, M. H., Koh, G. Y., and Field, L. J. (1996). Genetically selected cardiomyocytes from differentiating embryonic stem cells form stable intracardiac grafts. *J. Clin. Invest.* **98,** 216–224.

Koff, A., Giordano, A., Desai, D., Yamashita, K., Harper, J. W., Elledge, S., Nishimoto, T., Morgan, D. O., Franza, B. R., and Roberts, J. M. (1992). Formation and activation of a cyclin E-cdk2 complex during the G1 phase of the human cell cycle. *Science* **257,** 1689–1694.

Kreidberg, J. A., Sariola, H., Loring, J. M., Maeda, M., Pelletier, J., Housman, D., and Jaenisch, R. (1993). WT-1 is required for early kidney development. *Cell (Cambridge, Mass.)* **74,** 679–91.

Krek, W., and Nigg, E. (1991). Mutations of p34cdc2 phosphorylation sites induce premature mitotic events in Hela cells: Evidence for a double block to p34cdc2 kinase activation in vertebrates. *EMBO J.* **10,** 3331–3341.

Kretzner, L., Blackwood, E. M., and Eisenman, R. N. (1992). Myc and max proteins possess distinct transcriptional activities. *Nature (London)* **359,** 426–429.

Kühn, R., Schwenk, F., Aguet, M., and Rajewsky, K. (1995). Inducible gene targeting in mice. *Science* **269,** 1427–1429.

Kukielka, G. L., Smith, C. W., Manning, A. M., Youker, K. A., Michael, L. H., and Entman, M. L. (1995). Induction of interleukin-6 synthesis in the myocardium. Potential role in postreperfusion inflammatory injury. *Circulation* **92,** 1866–1875.

LaBaer, J., Garrett, M. D., Stevenson, L. F., Slingerland, J. M., Sandhu, C., Chou, H. S., Fattaey, A., and Harlow, E. (1997). New functional activities for the p21 family of CDK inhibitors. *Genes Dev.* **11,** 847–862.

Lane, H. A., and Nigg, E. A. (1997). Cell-cycle control: POLO-like kinases join the outer circle. *Trends Cell Biol.* **7,** 63–68.

Lau, M. M. H., Stewart, C. E. H., Liu, Z., Bhatt, H., Rotwein, P., and Stewart, C. L. (1994). Loss of the imprinted IGF2/cation-independent mannose-6–phosphate receptor results in fetal overgrowth and perinatal lethality. *Genes Dev.* **8,** 2953–2963.

Lee, E. Y. H. P., Chang, C. Y., Hu, N. P., Wang, Y. C. J., Lai, C. C., Herrup, K., Lee, W. H., and Bradley, A. (1992). Mice deficient for Rb are nonviable and show defects in neurogenesis and haematopoiesis. *Nature (London)* **359,** 288–294.

Lee, K. F., Simon, H., Chen, H., Bates, B., Hung, M. C., and Hauser, C. (1995). Requirement for neuregulin receptor ErbB2 in neural and cardiac development. *Nature (London)* **378,** 394–398.

Lee, M. H., Reynisdottir, I., and Massague, J. (1995). Cloning of p57KIP2, a cyclin-dependent kinase inhibitor with unique domain structure and tissue distribution. *Genes Dev.* **9,** 639–649.

Lee, M. H., Williams, B. O., Mulligan, G., Mukai, S., Bronson, R. T., Dyson, N., Harlow, E., and Jacks, T. (1996). Targeted disruption of p107: Functional overlap between p107 and rb. *Genes Dev.* **10,** 1621–1632.

Lees, J. A., Saito, M., Vidal, M., Valentine, M., Look, T., Harlow, E., Dyson, N., and Helin, K. (1993). The retinoblastoma protein binds to a family of E2F transcription factors. *Mol. Cell. Biol.* **13,** 7813–7825.

Lembo, G., Rockman, H. A., Hunter, J. J., Steinmetz, H., Koch, W. J., Ma, L., Prinz, M. P., Ross, J., Chien, K. R., and Powell-Braxton, L. (1996). Elevated blood pressure and enhanced myocardial contractility in mice with severe IGF-1 deficiency. *J. Clin. Invest.* **98,** 2648–2655.

Leor, J., Patterson, M., Quinones, M. J., Kedes, L. H., and Kloner, R. A. (1996). Transplantation of fetal myocardial tissue into the infarcted myocardium of rat: A potential method for repair of infarcted myocardium? *Circulation* **94,** 332–336.

Letterio, J. J., Geiser, A. G., Kulkarni, A. B., Roche, N. S., Sporn, M. B., and Roberts, A. B. (1994). Maternal rescue of transforming growth factor-beta 1 null mice. *Science* **264,** 1936–1938.

Li, F. Q., Wang, X. J., Capasso, J. M., and Gerdes, A. M. (1996). Rapid transition of cardiac myocytes from hyperplasia to hypertrophy during postnatal development. *J. Mol. Cel.l Cardiol.* **28,** 1737–1746.

Li, H. Z., Kawasaki, H., Nishida, E., Hattori, S., and Nakamura, S. (1996). Ras-regulated hypophosphorylation of the retinoblastoma protein mediates neuronal differentiation in PC12 cells. *J. Neurochem.* **66,** 2287–2294.

Lin, Q., Schwarz, J., Bucana, C., and Olson, E. N. (1997). Control of mouse cardiac morphogenesis and myogenesis by transcription factor MEF2C. *Science* **276,** 1404–1407.

Liu, J. P., Baker, J., Perkins, A. S., Robertson, E. J., and Efstratiadis, A. (1993). Mice carrying null mutations of the genes encoding insulin-like growth factor-I (Igf-1) and type-1 IGF receptor (Igf1r). *Cell (Cambridge, Mass.)* **75,** 59–72.

Liu, Y., and Kitsis, R. N. (1996). Induction of DNA synthesis and apoptosis in cardiac myocytes by E1A oncoprotein. *J. Cell Biol.* **133,** 325–334.

Liu, Y., Cigola, E., Cheng, W., Kajstura, J., Olivetti, G., Hintze, T. H., and Anversa, P. (1995). Myocyte nuclear mitotic division and programmed myocyte cell death characterize the cardiac myopathy induced by rapid ventricular pacing in dogs. *Lab. Invest.* **73,** 771–787.

Lu, K. P., and Hunter, T. (1995). Evidence for a NIMA-like mitotic pathway in vertebrate cells. *Cell (Cambridge, Mass.)* **81,** 413–424.

Lukas, J., Parry, D., Aagaard, L., Mann, D. J., Bartkova, J., Strauss, M., Peters, G., and Bartek, J. (1995). Retinoblastoma-protein-dependent cell-cycle inhibition by the tumour suppressor p16. *Nature (London)* **375,** 503–506.

Lyons, G. E., Schiaffino, S., Sassoon, S., Barton, P., and Buckingham, M. (1990). Developmental regulation of myosin gene expression in normal mouse cardiac muscle. *J. Cell Biol.* **111,** 2427–2436.

Mal, A., Poon, R. Y. C., Howe, P. H., Toyoshima, H., Hunter, T., and Harter, M. L. (1996). Inactivation of p27(Kip1) by the viral E1A oncoprotein in TGF beta-treated cells. *Nature (London)* **380,** 262–265.

Matsuoka, S., Edwards, M. C., Bai, C., Parker, S., Zhang, P., Baldini, A., Harper, J. W., and Elledge, S. J. (1995). $p57^{KIP2}$, a structurally distinct member of the $p21^{CIP1}$ Cdk inhibitor family, is a candidate tumor suppressor gene. *Genes Dev.* **9,** 650–662.

Medema, R. H., Herrera, R. E., Lam, F., and Weinberg, R. A. (1995). Growth suppression by p16(ink4) requires functional retinoblastoma protein. *Proc. Natl. Acad. Sci. U. S. A.* **92,** 6289–6293.

Metcalf, D., Willson, T. A., Hilton, D. J., Di, R. L., and Mifsud, S. (1995). Production of hematopoietic regulatory factors in cultures of adult and fetal mouse organs: Measurement by specific bioassays. *Leukemia* **9,** 1556–1564.

Meyer, D., and Birchmeier, C. (1995). Multiple essential functions of neuregulin in development. *Nature (London)* **378,** 386–390.

Milano, C. A., Dolber, P. C., Rockman, H. A., Bond, R. A., Venable, M. E., Allen, L. F., and Lefkowitz, R. J. (1994). Myocardial expression of a constitutively active a1B- adrenergic receptor in transgenic mice induces cardiac hypertrophy. *Proc. Natl. Acad. Sci. U. S. A.* **91,** 10109–10113.

Mima, T., Ueno, H., Fischman, D. A., Williams, L. T., and Mikawa, T. (1995). Fibroblast growth factor receptor is required for in vivo cardiac myocyte proliferation at early embryonic stages of heart development. *Proc. Natl. Acad. Sci. U.S.A.* **92,** 467–471.

Miner, J. H., and Wold, B. J. (1991). c-myc inhibition of MyoD and myogenin-initiated myogenic differentiation. *Mol. Cell. Biol.* **11,** 2842–2851.

Missero, C., Calautti, E., Eckner, R., Chin, J., Tsai, L. H., Livingston, D. M., and Dotto, G. P. (1995). Involvement of the cell-cycle inhibitor Cip1/WAF1 and the E1A-associated p300 protein in terminal differentiation. *Proc. Natl. Acad. Sci. U.S.A.* **92,** 5451–5455.

Missero, C., Di, C. F., Kiyokawa, H., Koff, A., and Dotto, G. P. (1996). The absence of p21Cip1/WAF1 alters keratinocyte growth and differentiation and promotes ras-tumor progression. *Genes Dev.* **10,** 3065–3075.

Moberg, K., Starz, M. A., and Lees, J. A. (1996). E2F-4 switches from p130 to p107 and pRB in response to cell cycle reentry. *Mol. Cell. Biol.* **16,** 1436–1449.

Moens, C. B., Stanton, B. R., Parada, L. F., and Rossant, J. (1993). Defects in heart and lung development in compound heterozygotes for two different targeted mutations at the N-myc locus. *Development (Cambridge, Mass.)* **119,** 485–499.

Moran, E. (1994). Mammalian cell growth controls reflected through protein interactions with the adenovirus E1A gene products. *Semin. Virol.* **5,** 327–340.

Mulcahy, L. S., Smith, M. R., and Stacey, D. W. (1985. Requirement for ras proto-oncogene function during serum stimulated growth of NIH 3T3 cells. *Nature (London)* **313,** 241–243.

Murphy, M., Stinnakre, M. G., Senamaud-Beaufort, C., Winston, N. J., Sweeney, C., Kubelka, M., Carrington, M., Brechot, C., and Sobczak-Thepot, J. (1997). Delayed early embryonic lethality following disruption of the murine cyclin A2 gene. *Nat. Genet.* **15,** 83–86.

Murry, C. E., Kay, M. A., Bartosek, T., Hauschka, S. D., and Schwartz, S. M. (1996). Muscle differentiation during repair of myocardial necrosis in rats via gene transfer with MyoD. *J. Clin. Invest.* **98,** 2209–2217.

Nakagawa, M., Price, R. L., Chintanawonges, C., Simpson, D. G., Horacek, M. J., Borg, T. K., and Terracio, L. (1997). Analysis of heart development in cultured rat embryos. *J. Mol. Cell. Cardiol.* **29,** 369–379.

Nevins, J. R. (1992). E2F: A link between the Rb tumor suppressor protein and viral oncoproteins. *Science* **258,** 424–429.

Niimura, F., Labosky, P. A., Kakuchi, J., Okubo, S., Yoshida, H., Oikawa, T., Ichiki, T., Naftilan, A. J., Fogo, A., Inagami, T., Hogan, B. L. M., and Ichikawa, I. (1995). Gene targeting in mice reveals a requirement for angiotensin in the development and maintenance of kidney morphology and growth factor regulation. *J. Clin. Invest.* **96,** 2947–2954.

Noda, A., Ning, Y., Venable, S. F., Pereira-Smith, O. M., and Smith, J. R. (1994). Cloning of senescent cell-derived inhibitors of DNA synthesis using an expression screen. *Exp. Cell Res.* **211,** 90–98.

Novitch, B. G., Mulligan, G. J., Jacks, T., and Lassar, A. B. (1996). Skeletal muscle cells lacking the retinoblastoma protein display defects in muscle gene expression and accumulate in S and G(2) phases of the cell cycle. *J. Cell Biol.* **135,** 441–456.

Oberpriller, J. O., Oberpriller, J. C., Matz, D. G., and Soonpaa, M. H. (1995). Stimulation of proliferative events in the adult amphibian cardiac myocyte. *Ann. N. Y. Acad. Sci.* **752,** 30–46.

Ohtsubo, M., Theodoras, A. M., Schumacher, J., Roberts, J. M., and Pagano, M. (1995). Human cyclin E, a nuclear protein essential for the G(1)-to-S phase transition. *Mol. Cell. Biol.* **15,** 2612–2624.

Pagano, M., Pepperkok, R., Verde, F., Ansorge, W., and Draetta, G. (1992). Cyclin A is required at two points in the human cell cycle. *EMBO J.* **11,** 961–971.

Pardee, A. B., Dubrow, R., Hamlin, J. L., and Kletzien, R. F. (1978). Animal cell cycle. *Annu. Rev. Biochem.* **47,** 715–750.

Parker, L. L., Walter, S. A., Young, P. G., and Piwnica-Worms, H. (1993). Phosphorylation and inactivation of the mitotic inhibitor Wee1 by the nim1/cdr1 kinase. *Nature (London)* **363,** 736–738.

Parker, S. B., Eichele, G., Zhang, P. M., Rawls, A., Sands, A. T., Bradley, A., Olson, E. N., Harper, J. W., and Elledge, S. J. (1995). p53-independent expression of p21(Cip)1 in muscle and other terminally differentiating cells. *Science* **267,** 1024–1027.

Parker, T. G., Packer, S. E., and Schneider, M. D. (1990). Peptide growth factors can provoke "fetal" contractile protein gene expression in rat cardiac myocytes. *J. Clin. Invest.* **85,** 507–514.

Pasumarthi, K. B., Kardami, E., and Cattini, P. A. (1996). High and low molecular weight fibroblast growth factor-2 increase proliferation of neonatal rat cardiac myocytes but have differential effects on binucleation and nuclear morphology. Evidence for both paracrine and intracrine actions of fibroblast growth factor-2. *Circ. Res.* **78,** 126–36.

Peeper, D. S., Upton, T. M., Ladha, M. H., Neuman, E., Zalvide, J., Bernards, R., DeCaprio, J. A., and Ewen, M. E. (1997). Ras signalling linked to the cell-cycle machinery by the retinoblastoma protein. *Nature (London)* **386,** 177–181.

Pennica, D., King, K. L., Shaw, K. J., Luis, E., Rullamas, J., Luoh, S. M., Darbonne, W. C., Knutzon, D. S., Yen, R., Chien, K. R., Baker, J. B., and Wood, W. I. (1995). Expression cloning of cardiotrophin 1, a cytokine that induces cardiac myocyte hypertrophy. *Proc. Natl. Acad. Sci. U. S. A.* **92,** 1142–1146.

Pennica, D., Arce, V., Swanson, T. A., Vejsada, R., Pollock, R. A., Armanini, M., Dudley, K., Phillips, H. S., Rosenthal, A., Kato, A. C., and Henderson, C. E. (1996). Cardiotrophin-1, a cytokine present in embryonic muscle, supports long-term survival of spinal motoneurons. *Neuron* **17,** 63–74.

Pfeffer, M. A. (1995). Left ventricular remodeling after acute myocardial infarction. *Annu. Rev. Med.* **46,** 455–466.

Pines, J., and Hunter, T. (1989). Isolation of a human cyclin cDNA: Evidence for cyclin mRNA and protein regulation in the cell cycle and for interaction with p34cdc2. *Cell (Cambridge, Mass.)* **58,** 833–46.

Polyak, K., Kato, J. Y., Solomon, M. J., Sherr, C. J., Massagué, J., Roberts, J. M., and Koff, A. (1994a). p27(kip1), a cyclin-Cdk inhibitor, links transforming growth factor-beta and contact inhibition to cell cycle arrest. *Genes Dev.* **8,** 9–22.

Polyak, K., Lee, M. H., Erdjument-Bromage, H., Koff, A., Roberts, J. M., Tempst, P., and Massagúe, J. (1994b). Cloning of p27Kip1, a cyclin-dependent kinase inhibitor and a potential mediator of extracellular antimitogenic signals. *Cell (Cambridge, Mass.)* **78,** 59–66.

Puri, P. L., Avantaggiati, M. L., Balsano, C., Sang, N. L., Graessmann, A., Giordano, A., and Levrero, M. (1997). p300 is required for MyoD-dependent cell cycle arrest and muscle-specific gene transcription. *EMBO J.* **16,** 369–383.

Qin, X. Q., Livingston, D. M., Kaelin, W. G., and Adams, P. D. (1994). Deregulated transcription factor E2F-1 expression leads to S-phase entry and p53-mediated apoptosis. *Proc. Natl. Acad. Sci. U. S. A.* **91,** 10918–10922.

Qin, X. Q., Livingston, D. M., Ewen, M., Sellers, W. R., Arany, Z., and Kaelin, W. G. (1995). The transcription factor E2F-1 is a downstream target of Rb action. *Mol. Cell. Biol.* **15,** 742–755.

Quelle, D. E., Ashmun, R. A., Shurtleff, S. A., Kato, J. Y., Bar-Sagi, D., Roussel, M. F., and Sherr, C. J. (1993). Overexpression of mouse D-type cyclins accelerates G1 phase in rodent fibroblasts. *Genes Dev.* **7,** 1559–1571.

Quelle, D. E., Ashmun, R. A., Hannon, G. J., Rehberger, P. A., Trono, D., Richter, K. H., Walker, C., Beach, D., Sherr, C. J., and Serrano, M. (1995). Cloning and characterization of murine p16(INK4a) and p15(INK4b) genes. *Oncogene* **11,** 635–645.

Rao, S. S., Chu, C., and Kohtz, D. S. (1994). Ectopic expression of cyclin D1 prevents activation of gene transcription by myogenic basic helix-loop-helix regulators. *Mol. Cell. Biol.* **14,** 5259–5267.

Reiss, K., Cheng, W., Ferber, A., Kajstura, J., Li, P., Li, B. S., Olivetti, G., Homcy, C. J., Baserga, R., and Anversa, P. (1996a). Overexpression of insulin-like growth factor-1 in the heart is coupled with myocyte proliferation in transgenic mice. *Proc. Natl. Acad. Sci. U. S. A.* **93,** 8630–8635.

Reiss, K., Cheng, W., Giordano, A., Deluca, A., Li, B. S., Kajstura, J., and Anversa, P. (1996b). Myocardial infarction is coupled with the activation of cyclins and cyclin-dependent kinases in myocytes. *Exp. Cell Res.* **225,** 44–54.

Rumyantsev, P. P. (1977). Interrelations of the proliferation and differentiation process during cardiac myogenesis and regeneration. *Int. Rev. Cytol.* **51,** 187–273.

Russo, A. A., Jeffrey, P. D., Patten, A. K., Massague, J., and Pavletich, N. P. (1996). Crystal structure of the p27(Kip1) cyclin-dependent-kinase inhibitor bound to the cyclin A cdk2 complex. *Nature (London)* **382,** 325–331.

Sadoshima, J., and Izumo, S. (1993). Mechanical stretch rapidly activates multiple signal transduction pathways in cardiac myocytes: Potential involvement of an autocrine/paracrine mechanism. *EMBO J.* **12,** 1681–1692.

Sadoshima, J., and Izumo, S. (1996). The heterotrimeric G(q) protein-coupled angiotensin II receptor activates p21(ras) via the tyrosine kinase-Shc-Grb2–Sos pathway in cardiac myocytes. *EMBO J.* **15,** 775–787.

Sadoshima, J., Xu, Y. H., Slayter, H. S., and Izumo, S. (1993). Autocrine release of angiotensin-II mediates stretch-induced hypertrophy of cardiac myocytes in vitro. *Cell (Cambridge, Mass.)* **75,** 977–984.

Sadoshima, J., Aoki, H., and Izumo, S. (1997). Angiotensin II and serum differentially regulate expression of cyclins, activity of cyclin-dependent kinases, and phosphorylation of retinoblastoma gene product in neonatal cardiac myocytes. *Circ. Res.* **80,** 228–241.

Sandmoller, A., Meents, H., and Arnold, H. H. (1996). A novel E1A domain mediates skeletal-muscle-specific enhancer repression independently of pRB and p300 binding. *Mol. Cell. Biol.* **16,** 5846–5856.

Sang, N. L., and Giordano, A. (1997). Extreme N terminus of E1A oncoprotein specifically associates with a new set of cellular proteins. *J. Cell. Physiol.* **170,** 182–191.

Sartorelli, V., Huang, J., Hamamori, Y., and Kedes, L. (1997). Molecular mechanisms of myogenic coactivation by p300: Direct interaction with the activation domain of MyoD and with the MADS box of MEF2C. *Mol. Cell. Biol.* **17,** 1010–1026.

Sawai, S., Shimono, A., Wakamatsu, Y., Palmes, C., Hanaoka, K., and Kondoh, H. (1993). Defects of embryonic organogenesis resulting from targeted disruption of the N-myc gene in the mouse. *Development (Cambridge, Mass.)* **117,** 1445–55.

Schneider, J. W., Gu, W., Zhu, L., Mahdavi, V., and Nadal-Ginard, B. (1994). Reversal of terminal differentiation mediated by p107 in Rb(−/−) muscle cells. *Science* **264,** 1467–1471.

Schneider, J. W., Smith, T. W., and DeCaprio, J. A. (1995). Dual checkpoints ensure the developmental arrest of cell division in cardiac myocytes. *J. Invest. Med.* **43** (Suppl. 2), 267A (abstr.).

Schneider, M. D., Perryman, M. B., Payne, P. A., Spizz, G., Roberts, R., and Olson, E. N. (1987). Autonomous myc expression in transfected muscle cells does not prevent myogenic differentiation. *Mol. Cell. Biol.* **7,** 1973–1977.

Sebastian, B., Kakizuka, A., and Hunter, T. (1993). Cdc25M2 activation of cyclin-dependent kinases by dephosphorylation of threonine-14 and tyrosine-15. Proc. *Natl. Acad. Sci. U. S. A.* **90,** 3521–3524.

Serrano, M., Lee, H. W., Chin, L., Cordoncardo, C., Beach, D., and Depinho, R. A. (1996). Role of the INK4a locus in tumor suppression and cell mortality. *Cell (Cambridge, Mass.)* **85,** 27–37.

Sheng, Z. L., Pennica, D., Wood, W. I., and Chien, K. R. (1996). Cardiotrophin-1 displays early expression in the murine heart tube and promotes cardiac myocyte survival. *Development (Cambridge, Mass.)* **122,** 419–428.

Sherr, C. J. (1994). G1 phase progression: cycling on cue. *Cell (Cambridge, Mass.)* 79, 551–555.

Sherr, C. J., and Roberts, J. M. (1995). Inhibitors of mammalian G(1) cyclin-dependent kinases. *Genes Dev.* **9,** 1149–1163.

Sicinski, P., Donaher, J. L., Parker, S. B., Li, T. S., Gardner, H., Haslam, S. Z., Bronson, R. T., Elledge, S. J., and Weinberg, R. A. (1995). Cyclin D1 provides a link between development and oncogenesis in the retina and breast. *Cell (Cambridge, Mass.)* **82,** 621–630.

Sicinski, P., Donaher, J. L., Geng, Y., Parker, S. B., Gardner, H., Park, M. Y., Robker, R. L., Richards, J. S., McGinnis, L. K., Biggers, J. D., Eppig, J. J., Bronson, R. T., Elledge, S. J., and Weinberg, R. A. (1996). Cyclin D2 is an FSH-responsive gene involved in gonadal cell proliferation and oncogenesis. *Nature (London)* **384,** 470–474.

Simpson, P., McGrath, A., and Savion, S. (1982). Myocyte hypertrophy in neonatal rat heart cultures and its regulation by serum and by catecholamines. *Circ. Res.* **511,** 787–801.

Skapek, S. X., Rhee, J., Spicer, D. B., and Lassar, A. B. (1995). Inhibition of myogenic differentiation in proliferating myoblasts by cyclin D1–dependent kinase. *Science* **267,** 1022–1024.

Skapek, S. X., Rhee, J., Kim, P. S., Novitch, B. G., and Lassar, A. B. (1996). Cyclin-mediated inhibition of muscle gene expression via a mechanism that is independent of pRB hyperphosphorylation. *Mol. Cell. Biol.* **16,** 7043–7053.

Smith, A. V., and Orr-Weaver, T. L. (1991). The regulation of the cell cycle during Drosophila embryogenesis: The transition to polyteny. *Development (Cambridge, UK)* **112,** 997)–1008.

Soonpaa, M. H., and Field, L. J. (1997). Assessment of cardiomyocytes DNA synthesis in the normal and injured adult mouse heart. *Am. J. Physiol.* **272,** H220–H226.

Soonpaa, M. H., Koh, G. Y., Klug, M. G., and Field, L. J. (1994). Formation of nascent intercalated disks between grafted fetal cardiomyocytes and host myocardium. *Science* **264,** 98–101.

Soonpaa, M. H., Kim, K. K., Pajak, L., Franklin, M., and Field, L. J. (1996). Cardiomyocyte DNA synthesis and binucleation during murine development. *Am. J. Physiol.* **271,** H2183–H2189.

Soonpaa, M. H., Koh, G. Y., Pajak, L., Jing, S., Wang, H., Franklin, M. T., Kim, K. K., and Field, L. J. (1997). Cyclin D1 overexpression promotes cardiomyocyte DNA synthesis and multinucleation in transgenic mice. *J. Clin. Invest.* **99,** 2644–2654.

Spirito, P., Fu, Y.-M., Yu, Z.-X., Epstein, S. E., and Casscells, W. (1991). Immunhistochemical localization of basic and acidic fibroblast growth factors in the developing rat heart. *Circulation* **84,** 322–332.

Steinhelper, M. E., Lanson, N. A. J., Dresdner, K. P., Delcarpio, J. B., Wit, A. L., Claycomb, W. C., and Field, L. J. (1990). Proliferation in vivo and in culture of differentiated adult atrial cardiomyocytes from transgenic mice. *Am. J. Physiol.* **259,** H1826–H1834.

Stillman, B. (1996). Cell cycle control of DNA replication. *Science* **274,** 1659–64.

Subramaniam, A., Gulick, J., Neumann, J., Knotts, S., and Robbins, J. (1993). Transgenic analysis of the thyroid-responsive elements in the alpha-cardiac myosin heavy chain gene promoter. *J. Biol. Chem.* **268,** 4331–4336.

Sucov, H. M., Dyson, E., Gumeringer, C. L., Price, J., Chien, K. R., and Evans, R. M. (1994). RXR alpha mutant mice establish a genetic basis for vitamin A signaling in heart morphogenesis. *Genes Dev.* **8,** 1007–1018.

Swenson, K. I., Farrell, K. M., and Ruderman, J. V. (1986). The clam embryo protein cyclin A induces entry into M phase and the resumption of meiosis in Xenopus oocytes. *Cell (Cambridge, Mass.)* **47,** 861–870.

Szekely, L., Jiang, W. Q., Bulic-Jakus, F., Rosen, A., Ringertz, N., Klein, G., and Wiman, K. G. (1992). Cell type and differentiation dependent heterogeneity in retinoblastoma protein expression in SCID mouse fetuses. *Cell Growth Differ.* **3,** 149–156.

Taya, Y. (1997). RB kinases and RB-binding proteins: New points of view. *Trends Biochem. Sci.* **22,** 14–17.

Thorburn, A., Thorburn, J., Chen, S. Y., Powers, S., Shubeita, H. E., Feramisco, J. R., and Chien, K. R. (1993). HRas-dependent pathways can activate morphological and genetic markers of cardiac muscle hypertrophy. *J. Biol. Chem.* **268,** 2244–2249.

Tiainen, M., Spitkovsky, D., Jansendurr, P., Sacchi, A., and Crescenzi, M. (1996). Expression of E1A in terminally differentiated muscle cells reactivates the cell cycle and suppresses tissue-specific genes by separable mechanisms. *Mol. Cell. Biol.* **16,** 5302–5312.

Toyoshima, H., and Hunter, T. (1994). p27, a novel inhibitor of G1 cyclin-Cdk protein kinase activity, is related to p21. *Cell (Cambridge, Mass.)* **78,** 67–74.

Ueno, H., Perryman, M. B., Roberts, R., and Schneider, M. D. (1988). Differentiation of cardiac myocytes following mitogen withdrawal exhibits three sequential stages of the ventricular growth response. *J. Cell Biol.* **107,** 1911–1918.

Vliegen, H. W., Bruschke, A. V., and Van der Laarse, A. (1990). Different response of cellular DNA content to cardiac hypertrophy in human and rat heart myocytes. *Com. Biochem. Physiol. A.* **95A,** 109–114.

Waga, S., Hannon, G. J., Beach, D., and Stillman, B. (1994). The p21 inhibitor of cyclin-dependent kinases controls DNA replication by interaction with PCNA. *Nature (London)* **369,** 574–578.

Wang, J., and Nadal-Ginard, B. (1995). Regulation of cyclin and p34CDC2 expression during terminal differentiation of C2C12 myocytes. *Biochem. Biophys. Res. Commun.* **206,** 82–88.

Wang, J., and Walsh, K. (1996). Inhibition of retinoblastoma protein phosphorylation by myogenesis-induced changes in the subunit composition of the cyclin-dependent kinase 4 complex. *Cell Growth Differ.* **7,** 1471–1478.

Wang, J. Y. J. (1997). Retinoblastoma protein in growth suppression and death protection. *Curr. Opin. Genet. Dev.* **7,** 39–45.

Webster, K. A., Muscat, G. E., and Kedes, L. (1988). Adenovirus E1A products suppress myogenic differentiation and inhibit transcription from muscle-specific promoters. *Nature (London)* **332,** 553–557.

Weinberg, R. A. (1995). The retinoblastoma protein and cell cycle control. *Cell (Cambridge, Mass.)* **81,** 323–330.

Weinberg, R. A. (1996). E2F and cell proliferation: A world turned upside down. *Cell (Cambridge, Mass.)* **85,** 457–459.

Weiner, H. L., and Swain, J. L. (1989). Acidic fibroblast growth factor mRNA is expressed by cardiac myocytes in culture and the protein is localized to the extracellular matrix. *Proc. Natl. Acad. Sci. U.S.A.* **86,** 2683–2687.

Weintraub, S. J., Chow, K. N. B., Luo, R. X., Zhang, S. H., He, S., and Dean, D. C. (1995). Mechanism of active transcriptional repression by the retinoblastoma protein. *Nature (London)* **375,** 812–815.

Xiao, Z. X., Ginsberg, D., Ewen, M., and Livingston, D. M. (1996). Regulation of the retinoblastoma protein-related protein p107 by g(1) cyclin-associated kinases. *Proc. Natl. Acad. Sci. U. S. A.* **93,** 4633–4637.

Xiong, Y., Zhang, H., and Beach, D. (1993). Subunit rearrangement of the cyclin-dependent kinases is associated with cellular transformation. *Genes Dev.* **7,** 1572–1583.

Yamaguchi, T. P., Harpal, K., Henkmeyer, M., and Rossant, J. (1994). *fgfr-1* is required for embryonic growth and mesodermal patterning during mouse gastrulation. *Genes Dev.* **8,** 3032–3044.

Yamasaki, L., Jacks, T., Bronson, R., Goillot, E., Harlow, E., and Dyson, N. J. (1996). Tumor induction and tissue atrophy in mice lacking E2F-1. *Cell (Cambridge, Mass.)* **85,** 537–548.

Yan, Y. M., Lee, M. H., Massagué, J., and Barbacid, M. (1997). Ablation of the CDK inhibitor p57(Kip2) results in increased apoptosis and delayed differentiation during mouse development. *Genes Dev.* **11,** 973–983.

Yoshida, K., Taga, T., Saito, M., Suematsu, S., Kumanogoh, A., Tanaka, T., Fujiwara, H., Hirata, M., Yamagami, T., Nakahata, T., Hirabayashi, T., Yoneda, Y., Tanaka, K., Wang, W. Z., Mori, C., Shiota, K., Yoshida, N., and Kishimoto, T. (1996). Targeted disruption of gp130, a common signal transducer for the interleukin 6 family of cytokines, leads to myocardial and hematological disorders. *Proc. Natl. Acad. Sci. U. S. A.* **93,** 407–411.

Yoshizumi, M., Lee, W. S., Hsieh, C. M., Tsai, J. C., Li, J., Perrella, M. A., Patterson, C., Endege, W. O., Schlegel, R., and Lee, M. E. (1995). Disappearance of cyclin A correlates with permanent withdrawal of cardiomyocytes from the cell cycle in human and rat hearts. *J. Clin. Invest.* **95,** 2275–2280.

Yuan, W. C., Condorelli, G., Caruso, M., Felsani, A., and Giordano, A. (1996). Human p300 protein is a coactivator for the transcription factor MyoD. *J. Biol. Chem.* **271,** 9009–9013.

Zacksenhaus, E., Jiang, Z., Chung, D., Marth, J. D., Phillips, R. A., and Gallie, B. L. (1996). pRb controls proliferation, differentiation, and death of skeletal muscle cells and other lineages during embryogenesis. *Genes Dev.* **10,** 3051–3064.

Zarkowska, T., and Mittnacht, S. (1997). Differential phosphorylation of the retinoblastoma protein by G(1)/S cyclin-dependent kinases. *J. Biol. Chem.* **272,** 12738–12746.

Zhang, H., Xiong, Y., and Beach, D. (1993). Proliferating cell nuclear antigen and p21 are components of multiple cell cycle kinase complexes. *Mol. Biol. Cell.* **4,** 897–906.

Zhang, H., Hannon, G. J., and Beach, D. (1994). p21-containing cyclin kinases exist in both active and inactive states. *Genes Dev.* **8,** 1750–1758.

Zhu, X. L., Sasse, J., McAllister, D., and Lough, J. (1996). Evidence that fibroblast growth factors 1 and 4 participate in regulation of cardiogenesis. *Dev. Dyn.* **207,** 429–438.

24

Regulation of Vascular Smooth Muscle Differentiation and Cell Cycle

Kenneth Walsh, Harris R. Perlman, and Roy C. Smith

Division of Cardiovascular Research, St. Elizabeth's Medical Center, and Program in Cell, Molecular and Developmental Biology, Sackler School of Biomedical Studies, Tufts University School of Medicine, Boston, Massachusetts 02135

I. Developmental Origins of Vascular Smooth Muscle

The embryonic origin of the vascular system is best understood with respect to endothelial cell development (see Chapter 14). In general, endothelial precursors coalesce into tubes that form the primitive vascular tree (Coffin and Poole, 1988; Poole and Coffin, 1989). Subsequent to formation of the endothelial tubes, cells are recruited to the periphery of the endothelium and differentiate into vascular smooth muscle cells (VSMCs) (Manasek, 1971). The embryonic origin of the VSMCs is highly diverse, with the VSMCs of the great vessels being derived from the neural crest (Le Lièvre and Le Douarin, 1975), whereas those of the smaller vessels are derived from the lateral mesoderm-derived mesenchyme of the local organ (Hungerford *et al.,* 1996). The thoracic aorta is composed of ectodermal VSMCs, and the ablation of the cardiac neural crest leads to replacement with other neural ectoderm-derived cells (Kirby, 1988). However, ablation of all neural ectoderm precursors results in severely deformed aortic arches composed entirely of mesoderm-derived VSMCs (Rosenquist *et al.,* 1989). Hence, there is a preference for populating the thoracic aorta with VSMCs derived from the neural ectoderm, which appear to be a distinct subpopulation with essential developmental functions that cannot be provided by mesoderm-derived cells.

That multiple VSMC populations comprise the aorta

has been suggested by numerous studies demonstrating diverse growth and functional properties of cultured VSMCs. Whether these differences are related to the embryonic origin of these cultures has not received much attention. However, recently, ectoderm- and mesoderm-derived VSMCs isolated from the embryonic chick aorta were shown to have highly divergent growth and transcriptional responses to transforming growth factor-β (TGF-β) (Topouzis and Majesky, 1996). This property may be related to an observed difference in the extent of type II TGF-β receptor glycosylation. Therefore, different lineage-dependent responses of VSMCs to growth regulators may have an important role in vessel remodeling during development and pathogenesis.

II. VSMC Phenotypic Modulation

VSMCs are morphologically and functionally different at successive stages of embryonic development. Initially, smooth muscle cells resemble fibroblasts. However, as the organism matures VSMCs acquire contractile properties and modulate their growth characteristics. The VSMCs of embryonic vessels produce higher levels of growth factors and are more responsive to growth factor stimulation than VSMCs from mature vessels (Hultgårdh-Nilsson *et al.*, 1991). In addition, embryonic VSMCs tend to express nonmuscle isoforms of contractile proteins, whereas mature VSMCs express a higher proportion of smooth muscle-specific isoforms (Owens, 1995).

In addition to the complexity that derives from differences in embryonic origin and stage of development, mature VSMCs also exhibit plasticity with regard to phenotype (Campbell and Campbell, 1985; Chamley-Campbell *et al.*, 1979; Y. Chen *et al.*, 1997). Unlike mature cardiac and skeletal muscle cells, which are terminally differentiated and incapable of further cell cycle activity, mature VSMCs retain the ability to reenter the cell cycle in response to growth factor stimulation. An activated VSMC has functional and morphological properties similar to those of embryonic VSMCs, being capable of proliferation, migration, and extracellular matrix production, and in this manner VSMCs contribute to many vessel wall pathologies.

On the basis of cell culture models and *in vivo* observations, it has been proposed that VSMCs can exhibit a spectrum of phenotypes depending on environmental conditions (Campbell and Campbell, 1985; Chamley-Campbell *et al.*, 1979) (Fig. 1). This process is reversible and has been referred to as phenotypic modulation. At one end of the phenotypic spectrum are the highly differentiated contractile state cells, which represent the VSMCs of the normal adult vessel wall. At the other extreme, synthetic state cells are proliferative, fibroblast-like VSMCs associated with embryonic development or vessel wall lesions.

The conversion from the contractile to the synthetic state can also be observed within several days after VSMCs are dispersed for primary culture (Campbell *et al.*, 1988; Chamley-Campbell *et al.*, 1979). The cytoplasmic organization, as observed by transmission electron microscopy, is the key characteristic of the two states. The morphological characteristics of synthetic and contractile state cells are depicted in Fig. 1. The contractile state cell has a spindle-shaped cell body with a rigid appearance and a relatively small nucleus containing one or two nucleoli. The cytoplasm of contractile VSMCs, as depicted in Fig. 1, is largely occupied by thin and thick myofilaments with interspersed dense bodies and by small complement of synthetic organelles. In contrast to cardiac or skeletal muscle, the myofilaments of VSMCs are not arrayed in highly ordered sarcomeres and appear randomly oriented within the cytoplasm. The myofilaments and dense bodies are believed to be components of a contractile apparatus organized as shown in the schematic of dense body interaction in Fig. 1. The myofilament structures allow force transduction through the cytoplasm and across intercellular desmosome-like gap junctions as depicted in the cell junction schematic in Fig. 1.

In contrast, a synthetic or modulated smooth muscle cell has a larger cell body that is fibroblast-like in appearance and possesses an enlarged nucleus containing multiple nucleoli. The cytoplasmic organization of a synthetic or modulated smooth muscle cell is depicted in Fig. 1. The large nucleus is surrounded by synthetic and secretory organelles consistent with the increased secretory capacity of these cells. Also evident are lipid droplets resulting from increased cholesterol ester uptake. The considerable increase in synthetic organelles accompanies a decrease in myofilament content.

The concept of phenotypic modulation is well illustrated at the molecular level by changes in the regulation of the contractile protein, myosin heavy chain (Aikawa *et al.*, 1993; Owens, 1995). Several isoforms of the myosin heavy chain molecule are expressed by VSMCs. Non-muscle-specific forms, such as SMemb, are expressed during embryonic development of the aorta but decline in the neonate and adult. Two isoforms expressed in the aorta, SM1 and SM2, are smooth muscle specific and indicate a differentiated VSMC phenotype. Of these two, only the SM2 isoform is expressed postnatally, at times when VSMC proliferation has ceased. During atherogenesis, SM1 and SM2 levels decline while SMemb increases (Aikawa *et al.*, 1993). These changes in myosin heavy chain expression support the notion that VSMCs of diseased vessels regress to a primitive or embryonic phenotype.

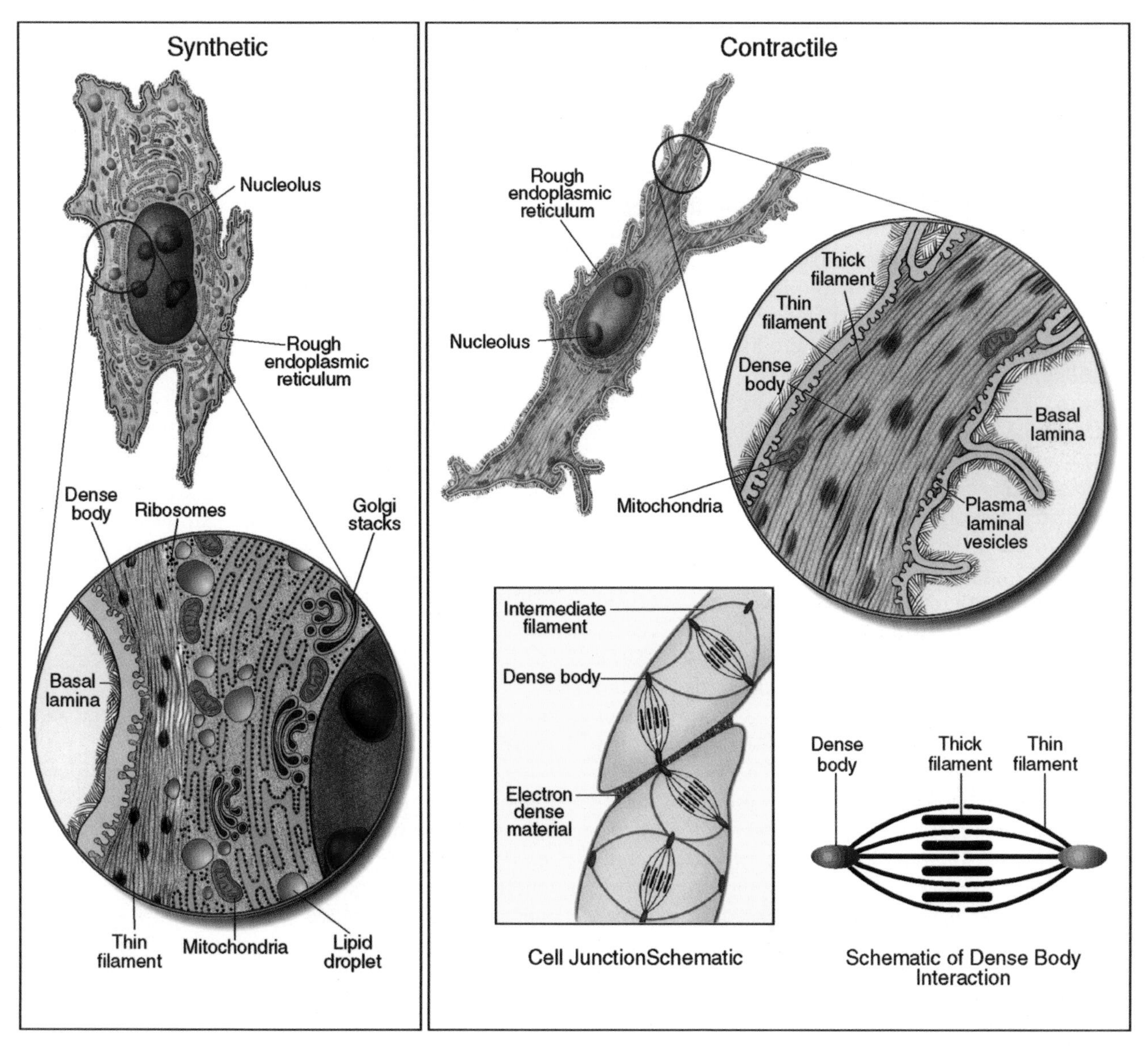

Figure 1 The phenotypic modulation of vascular smooth muscle cells. Depicted in this diagram are the alternate phenotypic states of vascular smooth muscle cells. The contractile state is typified by the cellular morphology observed in an enzyme-dispersed short-term culture of vascular smooth muscle cells. This phenotypic state is also attributed to the contractile smooth muscle cells found in the uninjured vessel wall. The molecular arrangement of the thin and thick filaments between dense bodies into a contractile apparatus is shown in dense body interaction schematic. The arrangement of these cells into a contractile network, continuous across the gap junctions between cells, is depicted in the cell junction schematic. The second phenotype is the synthetic state cell, typified by a vascular smooth muscle cell in culture (>8 days). This phenotypic state is also attributed to the proliferative smooth muscle cells found in a number of vascular lesions. An assignment to a phenotypic state is possible via electron microscopic analysis of the extent of the contractile apparatus and the quantity of synthetic organelles.

III. Mitogen-Regulated Transcription Factors

From the working hypothesis of phenotypic modulation emerges the concept that, in order to divide, a vascular myocyte may need to achieve a less differentiated state. Thus, transcription factors may function at nodal points to coordinate VSMC cell cycle activity and state of differentiation. Presumably, the expression of these transcription factors would be regulated by mitogens or cell cycle activity, and proliferation, differentiation, and/or migration of VSMCs should be altered when these factors are over- or underexpressed.

A. The *gax* Homeobox Gene

Homeobox genes represent a class of transcription factors that are important in modeling the body plan of a wide variety of organisms, including man, and at a cellular level they can control proliferation, differentiation, and migration. Some homeobox genes have been shown to induce a less differentiated phenotype and to promote uncontrolled cellular proliferation (Kawabe *et al.*, 1997; Song *et al.*, 1992), whereas others inhibit cell growth and promote the differentiated phenotype (Chawengsaksophak *et al.*, 1997; Suh *et al.*, 1994). Thus, homeobox genes can function as both positive and negative regulators of cell growth and may be anticipated to be important regulators, particularly in cells capable of modulating their state of differentiation in a reversible manner (Gorski *et al.*, 1993a).

Several homeobox genes have been isolated from adult vascular libraries (Gorski *et al.*, 1994; LePage *et al.*, 1994). One of these genes, *gax*, is expressed in both neuroectoderm- and mesoderm-derived tissues, including all the muscle lineages (Skopicki *et al.*, 1997). Expression of *gax* in smooth muscle is maximal in quiescent VSMCs and is rapidly downregulated when cells are stimulated with mitogens, including platelet-derived growth factor (PDGF) and angiotensin II (Gorski *et al.*, 1993b; Walsh and Perlman, 1996; Yamashita *et al.*, 1997). Similarly, expression of *gax* is rapidly downregulated in the neural crest-derived VSMCs of the carotid arteries following an injury that stimulates VSMC proliferation (Weir *et al.*, 1995). In cultured VSMCs, the extent of downregulation correlates with the ability of mitogens to stimulate DNA synthesis (Gorski *et al.*, 1993b). Moreover, *gax* expression is upregulated under conditions that lead to growth arrest, including serum deprivation (Gorski *et al.*, 1993b), and by exposure to either nitric oxide or C-type natriuretic peptide, which inhibit VSMC proliferation via the cGMP cascade (Yamashita *et al.*, 1997) (Fig. 2). Overall, the pattern of *gax* expression in VSMCs is similar to that of growth arrest-specific (*gas*) genes (Schneider *et al.*, 1988) and growth arrest and DNA damage-inducible (*gadd*) genes (Fornace *et al.*, 1992), some of which function as negative regulators of the cell cycle (Barone *et al.*, 1994; Del Sal *et al.*, 1992; Zhan, *et al.*, 1994).

The regulatory properties of the Gax homeoprotein have been examined at a molecular level *in vitro* (Smith *et al.*, 1997a). Growth inhibition is observed when bacterially produced Gax fusion protein is microinjected into VSMCs or fibroblasts. At early time points (<24 hr) *gax* overexpression blocks cell growth in the G_1 phase of the cell cycle. The growth inhibition by *gax* overexpression is mediated by the p53-independent upregulation of the cdk inhibitor p21, which occurs at the level of transcription, suggesting a transcriptional link. That p21 fulfills an essential function in *gax*-mediated growth arrest is supported by the finding that $p21^{-/-}$ MEFs are not susceptible to *gax*-induced growth inhibition. Hence, the p21 cdk inhibitor is an essential mediator of the cytostatic effects of *gax*.

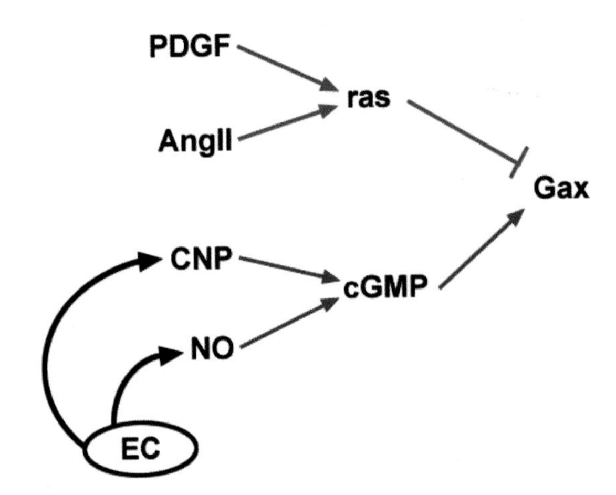

Figure 2 Regulation of *gax* expression in vascular smooth muscle cells. The pathways that regulate the expression of the homeobox gene *gax* are depicted. Substances that induce *gax* expression (red) include nitric oxide and c-type natriuretic peptide (CNP), endothelial cell (EC)-derived factors that inhibit VSMC proliferation. Factors that induce VSMC proliferation also suppress gax expression (green) including platelet-derived growth factor (PDGF) and angiotensin II (AngII).

gax overexpression also inhibits the proliferative response occurring upon vessel wall injury in the rat carotid artery model of balloon denudation (Smith *et al.*, 1997a). In this model, VSMCs of the media dedifferentiate, migrate, and proliferate to form a neointima that can partially occlude the artery (Clowes *et al.*, 1983) (Fig. 3). Two weeks following injury, saline or control virus-treated carotid arteries develop considerable neointimal thickening. In contrast, treatment with a *gax*-expressing adenovirus markedly reduces the proliferative response, with a decrease in intimal hyperplasia and a corresponding decrease in luminal narrowing. Furthermore, percutaneous adenovirus-mediated *gax* gene transfer to balloon-injured rabbit iliac arteries reveals that *gax*-treated arteries have larger lumen diameters at 1 month postinjury (Maillard *et al.*, 1997). These studies provide direct evidence that an endogenous homeobox gene is functionally relevant in controlling VSMC proliferation *in vivo*.

gax may have a similar regulatory role in the developing heart. *In situ* histochemical analyses in developing chick heart show Gax protein expression at relatively late times, when cardiomyocyte proliferation is declining (Fisher *et al.*, 1997). Similarly, *gax* is not detectable in blood vessels at early time points but is expressed at later times as VSMC proliferation declines (Skopicki *et al.*, 1997). Using a procedure developed to express transgenes from plasmid and adenoviral vectors in the

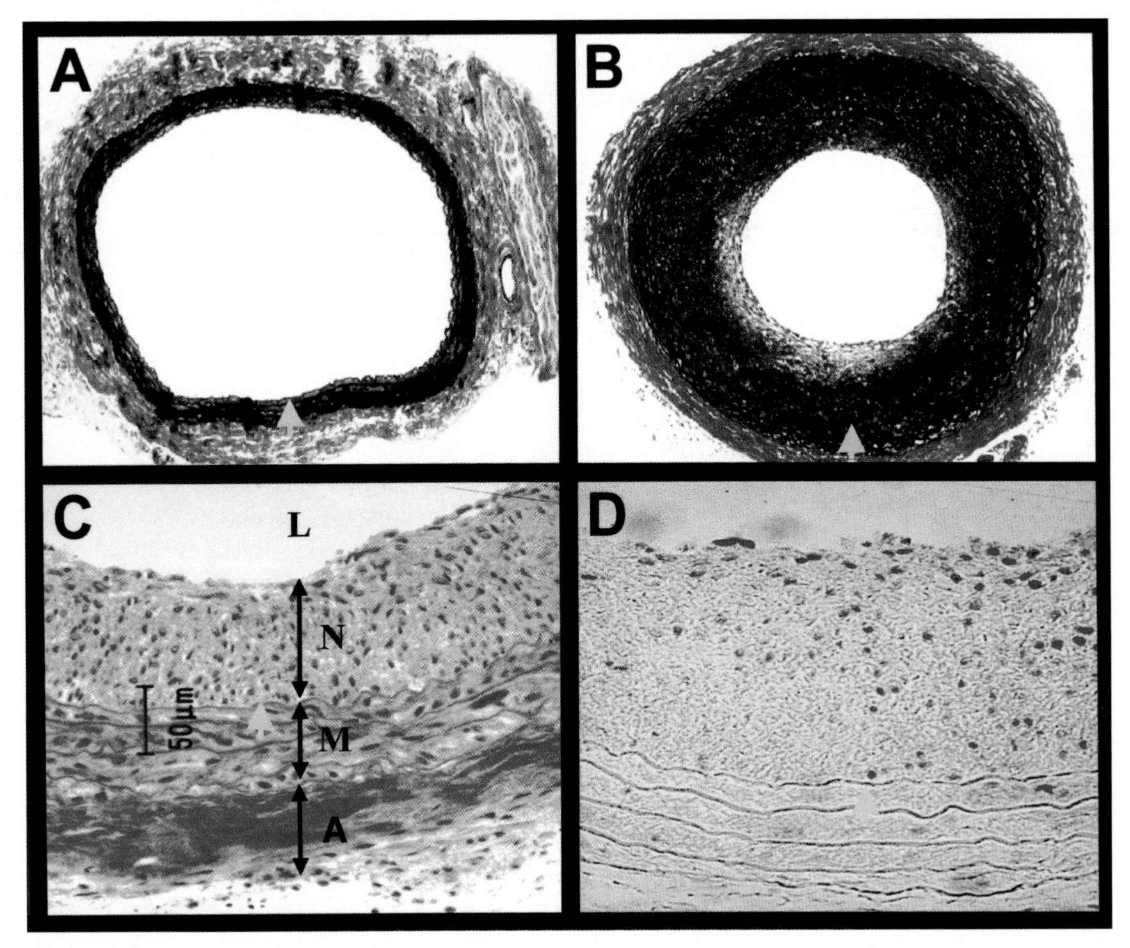

Figure 3 Neointimal formation in injured rat carotid arteries. Isolated segments of rat carotid arteries were subjected to a distending deendothelializing injury utilizing a balloon catheter. Following injury, blood flow was restored to the injured segment. At 2 weeks following injury the treated segments were harvested and fixed, and cross sections were treated with an elastic-trichrome stain to allow visualization of the elastic lamina. Arrows indicate the position of the internal elastic lamina, which lies beneath the endothelial layer. Shown are photomicrographs of (A) an uninjured artery which is free of neointima and (B) an injured segment which has developed an extensive neointimal layer of VSMCs. A high-power view of the vessel wall from an injured segment (C) stained with hematoxalin–eosin to allow visualization of cells and nuclei. The neointimal [N], medial [M], and adventitial [A] layers are demarcated and the lumen [L] of the vessel is indicated. The presence of proliferative VSMCs at the luminal edge of the neointima is demonstrated in (D) by proliferative cell nuclear antigen immunostaining of a lesion at 14 days following injury. Proliferative cells are distinguished by red nuclear staining.

developing chick heart tube (Fisher *et al.,* 1996; Fisher and Watanabe, 1996), Fisher *et al.* (1997) demonstrated that precocious expression of *gax* inhibits cardiomyocyte proliferation and thereby perturbs heart morphogenesis. In this study, forced *gax* expression inhibited cardiomyocyte proliferation as indicated by the decreased clonal expansion of infected cardiomyocytes and the decrease in percentage of cells positive for PCNA, a marker of cell cycle progression. Precocious *gax* expression results in alterations of heart morphology, including reduced ventricles with rounded apices, thinning of the compact zone, and coarse trabeculae. This result suggests that premature cardiomyocyte cell cycle exit, resulting from dysregulated *gax* expression, decreases the number of myocardial cells in the ventricle wall below levels required for proper formation of the trabeculae and the intravascular septum.

B. The MADS Box Transcription Factors SRF and MEF2

MADS (*M*CM1-*a*gamous-*d*eficiens-*S*RF) box family of proteins encodes for transcriptional regulators that have been implicated in the regulation of cell fate, cell growth, and tissue-specific gene expression (see Chapters 8 and 16). To date, the most widely studied MADS box protein is the *s*erum *r*esponse *f*actor (SRF). The SRF binding site consensus sequence is $CC(A/T)_6GG$, which has been dubbed a CArG motif (Minty and Kedes, 1986). Of the few smooth muscle-specific promoters examined to date, many contain CArG sequence motifs. These include SM α-actin (Blank *et al.*, 1992; Carroll *et al.*, 1988; Foster *et al.*, 1992; Simonson *et al.*, 1995), SM22α (Solway *et al.*, 1995), telokin (Herring and Smith, 1996), and smooth muscle myosin heavy chain genes (Katoh *et al.*, 1994). In the smooth muscle α-actin promoter, two CArG elements act synergistically to activate tissue-specific transcription in VSMCs as well as mesangial cells, contractile cells of the kidney with growth and differentiation properties similar to VSMCs (Simonson *et al.*, 1995; Shimizu *et al.*, 1995). Recently, evidence for the importance of dual CArG elements in SM22α gene transcription also has been provided (Kim *et al.*, 1997; Li *et al.*, 1997).

The binding sites for SRF occur in the promoters and enhancers of functionally different classes of genes. For example, CArG box elements control the expression of both differentiation-specific genes, such as skeletal α-actin (Taylor *et al.*, 1989; Walsh, 1989), and the immediate early genes required for cell cycle progression, such as the c-*fos* protooncogene (Treisman, 1992). It is paradoxical that a single class of DNA regulatory element can be involved in positive regulation of both cell cycle and differentiation-specific genes. This paradox in promoter regulation by CArG elements is also evident from studies on the SM α-actin gene. Surprisingly, in mesangial cells the CArG box elements of the SM-α-actin promoter are responsible for both tissue-specific expression in nonproliferating contractile cells and for markedly elevated expression when mesangial cells undergo phenotypic modulation (Simonson *et al.*, 1995) such as in glomerulosclerosis in humans (Johnson *et al.*, 1991).

Based on the considerations discussed previously, it is clear that CArG elements alone do not determine phenotype-specific expression in VSMCs and mesangial cells. Therefore, SRF regulation during phenotypic modulation may be relevant to this issue. While SRF is expressed in quiescent VSMCs, the DNA-binding activity of SRF is markedly induced by mitogen stimulation of quiescent VSMCs (Suzuki *et al.*, 1995), indicating that SRF levels may influence the phenotypic state. Other possibilities can also be envisioned; the transcriptional activity from CArG elements may be influenced by the interaction of SRF with ancillary proteins, including ets-related proteins (Dalton and Treisman, 1992; Hipskind *et al.*, 1991), homeodomain proteins (Grueneberg *et al.*, 1992), YY1 (Gualberto *et al.*, 1992), and others. Of particular note, expression of the ets-1 transcription factor is rapidly upregulated by mitogen activation of quiescent VSMCs in culture and by mechanical injury in rat carotid arteries (Hultgårdh-Nilsson *et al.*, 1996). Additionally, transcriptional activity may be modulated by distal DNA elements that are themselves subject to distinct regulatory controls in different cell types. This latter feature is also evident in the SM α-actin promoter in which distinct upstream sequences differentially repress CArG element activity in mesangial cells versus VSMCs (Nakano *et al.*, 1991; Simonson *et al.*, 1995).

The transcription factors dubbed *r*elated to *s*erum *r*esponse *f*actor (RSRF) or *m*yocyte *e*nhancer *f*actor (MEF2) have also been identified in VSMCs (Martin *et al.*, 1994; Suzuki *et al.*, 1995; Yu *et al.*, 1992), but their role in the regulation of VSMC proliferation and differentiation is not well understood. The MEF2/RSRF transcription factors are encoded by four genes: MEF2A–D (nomenclature as suggested by Breitbart *et al.*, 1993; see Chapter 8) that are alternatively spliced. The MEF2 proteins bind to a consensus A/T-rich sequence which is found in the control regions of both muscle-specific and growth factor-induced genes (Pollock and Treisman, 1991). A functional MEF2-binding site also occurs in the *gax* gene promoter (Andrés *et al.*, 1995). MEF2A DNA-binding activity in VSMCs is upregulated within 4 hr of mitogen stimulation (Suzuki *et al.*, 1995). This upregulation is sustained throughout all phases of the cell cycle and correlates with an increase in MEF2A protein but not mRNA. Pulse-chase experiments revealed no significant difference in MEF2A protein stability in quiescent and serum-induced cells, suggesting that mitogens activate MEF2A translation. Consistent with the *in vitro* observations, MEF2 transcription factors are also upregulated in the VSMCs of rat carotid arteries following injury (Firulli *et al.*, 1996). Collectively, these data indicate that MEF2, like SRF, may function to activate a genetic program associated with the phenotypic modulation of VSMCs.

IV. The GATA-6 Transcription Factor

GATA factors comprise a small family of transcriptional regulatory proteins that contain a highly conserved zinc finger DNA-binding domain (Laverriere *et al.*, 1994; see Chapter 17). Six vertebrate GATA factor genes have been identified thus far. The *GATA-1, -2,* and *-3* isoforms are required for development of hematopoietic cell lineages because the disruption of any of these genes results in deficiencies in hematopoiesis and embryonic lethality (Weiss and Orkin, 1995). The recently identified *GATA-4, -5,* and *-6* genes represent a subfamily of factors that are expressed in some cardiovascular tissues (Jiang and

Evans, 1996; Laverriere *et al.,* 1994; Suzuki *et al.,* 1996). At least one GATA factor is an important regulator of cardiovascular development since disruption of *GATA-4* in mice leads to embryonic lethality by abrogating foregut and centralized heart tube formation (Kuo *et al.,* 1997; Molkentin *et al.,* 1997). Furthermore, antisense oligonucleotides to *GATA-4* mRNA prevent differentiation of P19 embryonic carcinoma cells into cardiac myocytes (Soudais *et al.,* 1995). The *GATA-4* and *-6* factors exhibit overlapping expression patterns during embryogenesis of the heart (Morrisey *et al.,* 1996) though both are expressed throughout heart development only *GATA-6* is expressed in the vasculature. Following embryonic day 9.5 *GATA-6* is expressed continuously in vessels throughout and on into adulthood.

GATA-6 is the predominant GATA factor expressed in VSMCs cultured from adult vessels. In cultured VSMCs, *GATA-6* expression is rapidly and transiently downregulated following mitogen stimulation (Suzuki *et al.,* 1996). In rat carotid arteries, *GATA-6* expression is downregulated following an injury that stimulates VSMC cell cycle activity (H. R. Perlman and K. Walsh, unpublished observations). In a manner similar to *gax, GATA-6* overexpression induces p21 cdk inhibitor expression, in a p53-independent manner, and inhibits S-phase entry (Perlman *et al.,* 1998). Furthermore, the growth inhibitory activity of *GATA-6* is attenuated in fibroblasts deficient for p21, indicating that the upregulation of this cdk inhibitor is a functionally significant feature of growth arrest *in vitro.*

V. Cell Cycle Regulation in VSMCs

As described previously, mature VSMCs differ from cardiac and skeletal myocytes in that they are not terminally differentiated and they retain their ability to reenter the cell cycle in response to appropriate environmental stimuli. VSMC proliferation *in vivo* is best understood from studies using the rat carotid model of balloon injury. Uninjured vessels display a very low mitotic index. However, balloon injury results in the denudation of the endothelium and extensive death of smooth muscle cells in the underlying medial layer (Perlman *et al.,* 1997). Within hours, the remaining medial smooth muscle cells reenter the cell cycle in a relatively synchronous manner. Cell cycle regulatory protein expression in the VSMCs of the vessel wall is temporally and spatially regulated during injury-induced lesion formation. Analyses of the kinetics and activity of cell cycle regulatory factors have revealed a marked induction of the cdk2 cyclin-dependent kinase and its cyclin E regulatory subunit within 8 hr of injury (Wei *et al.,* 1997). This is followed by an induction of cyclin A at 1 or 2 days postinjury corresponding to a peak of proliferative activity in the media at 48 hr postinjury (Clowes *et al.,* 1983). The proliferating VSMCs repopulate the medial layers and can migrate across the inner elastic lamina layer to proliferate and form a neointimal layer (Fig. 3). The expression of cdk2 and cyclins A and E occurs initially in the medial VSMCs and then in the developing neointima. The importance of cell cycle regulatory molecules during intimal hyperplasia is underscored by the observation that injury-induced lesion formation is inhibited by molecules that can interfere with cell cycle progression. For example, intimal hyperplasia is inhibited by antisense cdk2 oligonucleotides (Abe *et al.,* 1994; Morishita *et al.,* 1994) and by overexpression of both the wild-type and mutant forms of the retinoblastoma protein (Chang *et al.,* 1995a; Smith *et al.,* 1997b).

The activity of cyclin-dependent kinases is modulated by a class of molecules referred to as cdk inhibitors, including $p27^{Kip1}$ and $p21^{Cip1/Waf1}$, which inhibit cdk kinase activity and are capable of preventing cell cycle progression and proliferation. Both p27 and p21 have been implicated in the regulation of VSMC proliferation both *in vitro* and *in vivo.* In VSMC cultures, p27 is found at low levels in proliferating cells but is induced when cultures are made quiescent by mitogen deprivation (D. Chen *et al.,* 1997). In the quiescent cultures, p27 is associated with cdk2 complexes and corresponds with a marked reduction in cdk2 kinase activity. *In vivo,* this inhibitor does not appear to be responsible for maintaining quiescence of VSMCs in uninjured arteries but may function by inducing VSMCs to return to a quiescent state following injury. When examined for p27 expression, uninjured arteries exhibit low levels of this inhibitor. In contrast, balloon-injured arteries exhibit a time-dependent increase in p27 expression that correlates with attenuated neointimal expansion. When balloon-injured rat carotid arteries are treated with an adenoviral expression vector encoding p27, neointima formation is significantly decreased. These results suggest that p27 may represent an endogenous mechanism for limiting VSMC proliferative response to vascular injury. A similar pattern of expression is observed for p21 in balloon-injured porcine femoral arteries. p21 expression is not observed in normal untreated arteries but is present in arteries shortly after injury and increases as the lesion expands and DNA synthesis declines (Yang *et al.,* 1996). In addition, treatment of injured arteries with a p21 adenovirus expression vector significantly decreases neointima formation in both the porcine femoral artery and rat carotid artery models (Chang *et al.,* 1995b). Thus, the induction of p21, like p27 induction during lesion formation, may limit neointimal growth.

In the media of the artery VSMCs are nonproliferative and the maintenance of this quiescent state may be facilitated by contact with matrix components through cell surface receptors of the integrin family (Hynes, 1992). An extracellular matrix composed of laminin and collagens I, III, and IV and other minor components (Thyberg *et al.,* 1990) surrounds the VSMCs of the artery. In normal arteries, the collagen exists in a native polymerized or fibrillar form, but at sites of inflammation or atherogenesis the extracellular matrix is perturbed and may provide a permissive environment for VSMC proliferation. Recently, the expression of the cdk inhibitors p21 and p27 in VSMCs has been shown to be regulated by the composition of the extracellular matrix (Koyama *et al.,* 1996). VSMCs cultured in mitogen-containing media on plates coated with the monomer form of type I collagen are proliferative and contain high levels of cdk2/cyclin E kinase activity, whereas cultures grown on the native polymerized form have greatly reduced kinase activity, cyclin A expression, and DNA synthesis. This downregulation of cdk2/cyclin E activity correlates with the association of the cyclin E complex with the cdk inhibitors p27 and p21. The inhibitory effects of polymerized collagen may be mediated by the α 2 integrin, which allows adhesion to both forms of collagen. Thus, extracellular matrix may play a key role in maintaining homeostasis of the vasculature, whereby integrins provide a sensing mechanism utilized by VSMCs to monitor tissue integrity within the artery.

Endothelial release of nitric oxide (NO) is believed to function as a physiological regulator of VSMC proliferation (Moncada *et al.,* 1991). Administration of the NO precursor, L-arginine, enhances vascular NO release and reduces the size of arterial lesions following injury (Schwarzacher *et al.,* 1997). Furthermore, restoration of endothelial cell NO synthase activity by gene transfer to rat carotid arteries denuded of endothelium reduces intimal hyperplasia (von der Leyen *et al.,* 1995). Recent studies have elucidated the effects of NO on VSMC cell cycle activity. One such effect is the upregulation of the general cdk inhibitor p21 (Ishida *et al.,* 1997), whereas another is the repression of cyclin A expression at the level of transcription (Guo *et al.,* 1998; see Chapter 23). Of note, parallels exist between the mechanisms of growth arrest induced by NO and the *gax* transcription factor (Smith *et al.,* 1997a) in that both upregulate p21 and repress cyclin A expression. Further, cGMP, which is believed to mediate NO-induced growth arrest, also upregulates *gax* expression (Yamashita *et al.,* 1997) (Figure 2).

VI. Regulation of VSMC Apoptosis

VSMC apoptosis occurs upon arterial remodeling during embryogenesis. In particular, the abdominal aorta undergoes extensive remodeling in response to a large decrease in blood flow when the placenta is lost at birth. Benedeck and Langille (1991) observed that the DNA content of the abdominal aorta of neonatal lambs was much lower than what would be predicted from the rates of VSMC mitosis. Based on these measurements, it was proposed that the VSMCs of this vessel undergo a high rate of apoptosis immediately after birth in response to the changes in blood flow. A high incidence of apoptosis has in fact been observed in these specimens after, but not before, birth (Cho *et al.,* 1995). Regionalized VSMC apoptosis has also been demonstrated during the remodeling of the human ductus arteriosus prior to birth (Slomp *et al.,* 1997). Of note, the VSMCs within these regions express fetal markers suggesting that dedifferentiation and apoptosis are associated processes.

VSMC apoptosis also appears to have a role in vessel wall remodeling during disease. VSMCs undergo a very low rate of turnover in the normal vessel wall. However, following acute vascular damage or in chronic vessel wall lesions, the frequency of apoptosis is more appreciable and approximately correlates with the proliferative capacity of the tissue. Specifically, VSMC apoptosis has been documented by TUNEL labeling and electron microscopic analysis in human atherectomy specimens (Bennett *et al.,* 1995; Geng and Libby, 1995; Isner *et al.,* 1995). In these lesions TUNEL positivity correlates with the expression of the caspase family members ICE (Geng and Libby, 1995) and CPP-32 (Mallat *et al.,* 1997).

Apoptotic VSMC death has also been documented in a number of animal models of vessel stenosis. In injured rat carotid arteries, the total accumulation of VSMCs in the neointima reaches a maximum at 2 weeks postinjury, but continuous cellular proliferation occurs for up to 12 weeks with no discernible increase in VSMC number. To reconcile these paradoxical observations, Clowes *et al.* (1983) proposed that the death of neointimal VSMCs could account for the lack of VSMC accumulation at later time points. Consistent with this hypothesis, it has recently been demonstrated that VSMC apoptosis occurs in the neointima from 7 to 30 days postinjury (Bochaton-Piallat *et al.,* 1995; Han *et al.,* 1995). Using TUNEL labeling as an indicator of apoptosis, a maximum of 40% apoptotic cell death within the two most luminal layers of the neointima was observed at 9 days postinjury (Han *et al.,* 1995). In another study, *in situ* end labeling (ISEL), a technique similar to TUNEL, revealed up to 14% apoptotic cell death in neointimal VSMCs at 20 days postinjury (Bochaton-Piallat *et al.,* 1995). In addition to ISEL positivity, the apoptotic phenotype was confirmed by identifying morphological structures that are characteristic of apoptosis by transmission electron microscopy. In these developed lesions little or no apoptosis is observed in the VSMCs of the medial layer and the apoptosis appears to be largely confined to the most luminal cell layers of the neointima (Fig. 4).

It has also been reported that balloon denudation of the vessel wall induces a rapid wave of medial VSMC apoptosis (Perlman *et al.,* 1997). Uninjured rat carotid vessels and vessels harvested at the time of injury (T = 0) do not display evidence of apoptosis. However, at 30 min postinjury as much as 70% of medial VSMCs appear apoptotic by both TUNEL staining and chromatin condensation. High frequencies of TUNEL-positive cells are observed at 0.5 and 1 hr following injury but not at 4 hr (Fig. 4). This time course of TUNEL positivity suggests that the medial VSMC population responds uniformly to the injury and cells are rapidly eliminated through phagocytosis or by extrusion into the lumen. Consistent with this hypothesis, there is a 65% loss of cellular density in the media at 4 hr postinjury, which is in good agreement with the percentage of TUNEL-positive cells detected at the earlier time points. Electron microscopic analyses of these sections revealed cells with morphological features characteristic of apoptosis, including chromatin condensation localized at the edges of the nuclear membrane and retention of organelle membrane integrity. Furthermore, the rapid on-

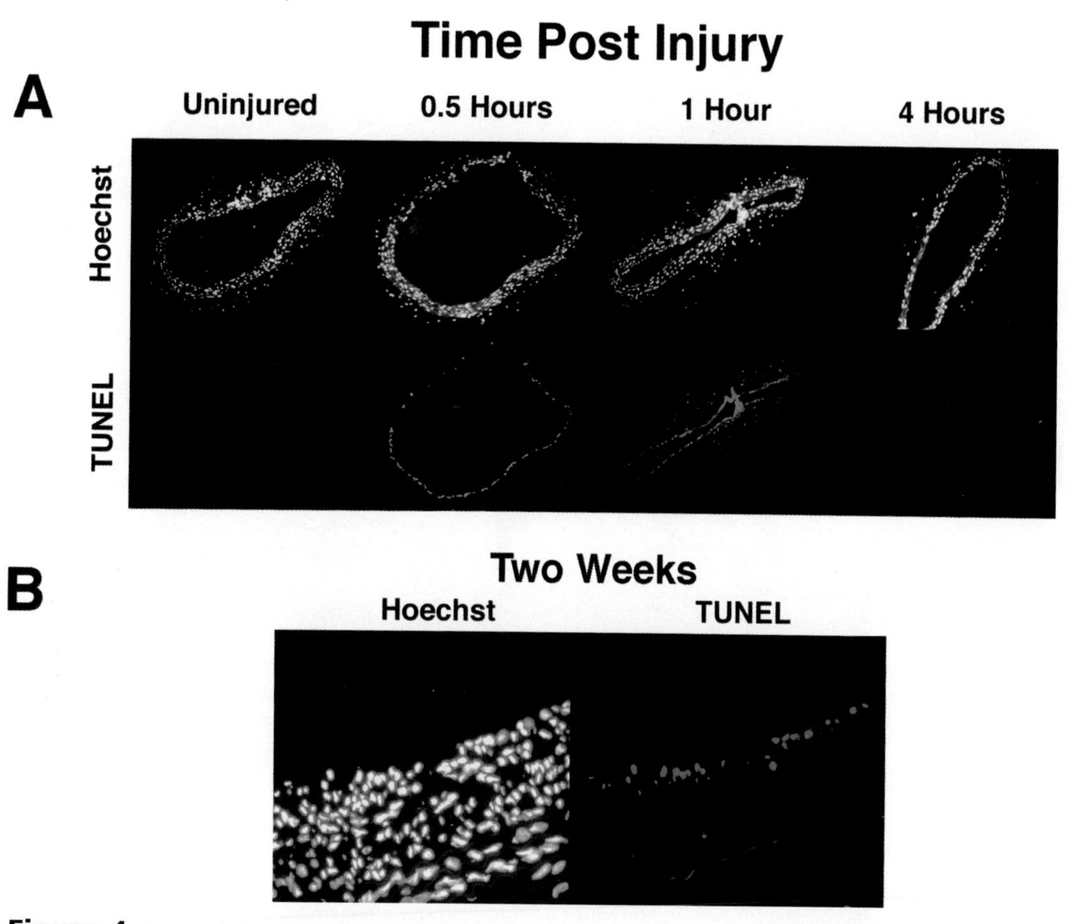

Figure 4 Apoptosis in balloon-injured rat carotid arteries. Rat carotid arteries were subjected to a distending deendothelializing injury utilizing a balloon angioplasty catheter. At the indicated times, injured segments were harvested and fixed. Cross sections were stained for immunofluorescent analysis with either Hoechst (blue) to visualize nuclear chromatin or with TUNEL (green) to visualize apoptotic nuclei undergoing chromatin degradation. (A) The rapid onset of medial VSMC apoptosis is evident by 0.5–1 hr following injury. By 4 hr postinjury the number of TUNEL-positive cells in the media has decreased dramatically. (B) Apoptosis of neointimal VSMCs is apparent at 4 weeks postinjury in photomicrographs of injured rat carotid arteries.

set of apoptosis correlated with a marked decrease in bcl-X expression at 1 hr postinjury in the most luminal layers of the injured media. These layers also exhibit the highest levels of both TUNEL-positive nuclei and cellular loss at later time points. The decrease in bcl-X expression suggests that modulations in the level of this protein may be a feature of the acute apoptotic response to vessel wall injury.

VII. Smooth Muscle in Human Atherosclerotic and Restenotic Lesions

Endothelial cells line the luminal surface of the vessel wall and function to prevent inflammation by serving as a nonadherent surface to circulating leukocytes. Upon stimulation by the inflammatory cytokine TNF-α, the endothelium expresses surface adhesion molecules that promote the tight adherence of leukocytes. In addition, TNF-α also downregulates Fas ligand expression on the endothelium, an event that appears necessary for the successful transendothelial migration of inflammatory cells (Sata and Walsh, 1998) (Fig. 5). Subendothelial T cells and macrophages secrete mitogens that promote the dedifferentiation and proliferation of VSMCs (Ross, 1993). As these lesions develop, synthetic smooth muscle cells produce prodigious quantities of the extracellular matrix that contributes to lesion mass (Riessen *et al.*, 1994). Phenotypically modified smooth muscle cells also express osteopontin and bone morphogenic protein 2α (Bostrom *et al.*, 1993; Giachelli *et al.*, 1993). Ultimately, the atherosclerotic

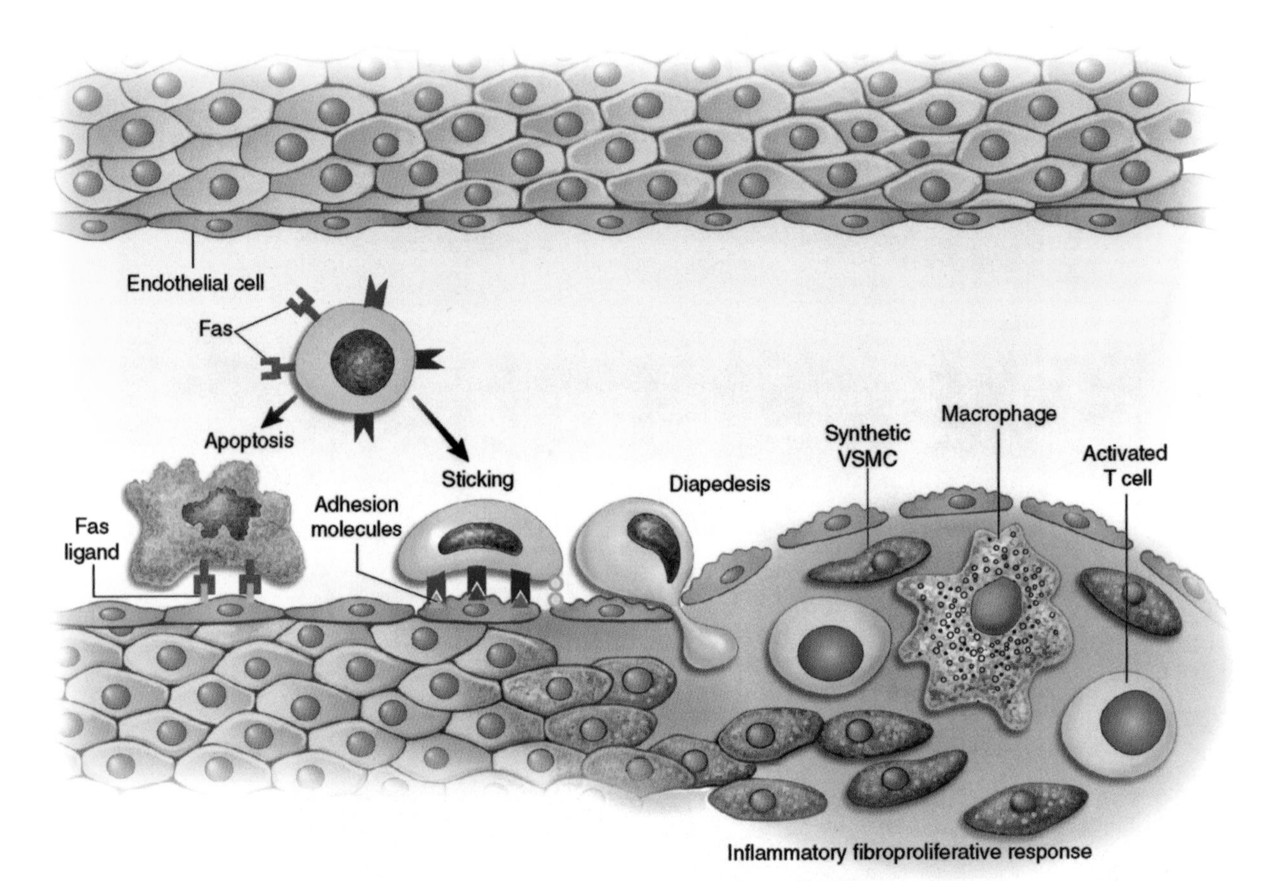

Figure 5 A putative mechanism for atherosclerotic lesion formation involving regulation of lymphocyte binding and extravasation. This drawing depicts a mechanism for FasL in preventing inflammatory–proliferative responses that may lead to VSMC hyperproliferation and vascular lesion formation. This longitudinal view of a vessel indicates the alternative fates a lymphocyte faces when interacting with the endothelial lining of the vessel wall. Lymphocytes express the Fas receptor on their cell surface and can interact with the Fas ligand present on nonactivated endothelial cells. Occupancy of the Fas receptor by its cognate ligand induces apoptosis of lymphocytes thus bound. An alternative fate can be envisioned when the endothelial cells of the vessel are activated by an inflammatory cytokine such as TNF-α. FasL expression is then downregulated in the endothelial cells and lymphocytes are able to bind and cross the endothelium via the process of diapedesis. The accumulation of lymphocytes, such as activated macrophages and T cells, in the media of the vessel wall can then stimulate the conversion of local VSMCs to a proliferative synthetic phenotype and thereby encourage the formation of stenotic lesions.

plaque becomes acellular and is composed primarily of matrix and calcified deposits, the remnants of pathological cellular activities that occurred at much earlier times.

Balloon angioplasty is commonly employed to restore blood flow through occluded atherosclerotic arteries; an estimated 400,000 procedures are being conducted each year in the United States. However, approximately 40% of these procedures fail within 6 months due to the development of a secondary, or restenotic, lesion. In contrast with atherosclerotic plaques, restenotic lesions are much more cellular in nature, containing phenotypically modulated VSMCs (Y. Chen *et al.,* 1997; Pickering *et al.,* 1993). Recently, the implantation of stents, prosthetic devices that mechanically prevent the constrictive remodeling of the vessel wall (Fig. 6), has become a popular clinical revascularization procedure because they reduce the incidence of complications during angioplasty and provide a near perfect angiographic result. An additional advantage of the stent is the greater gain in lumen diameter relative to conventional balloon angioplasty, which translates into a reduced incidence of restenosis. Thus, stent use has increased dramatically, with as many as 60% of all percutaneous revascularization procedures utilizing stent implantation. However, the complication of in-stent restenosis has only recently come to be appreciated. In contrast to balloon angioplasty-induced restenosis, which results from both constrictive remodeling and intimal hyperplasia, in-stent restenosis is exclusively a consequence of intimal hyperplasia. Thus, the secondary lesion that results from stent implantation in humans is remarkably cellular compared with atherosclerotic and restenotic lesions. The predominant cell type in atherectomy specimens from patients with in-stent restenosis is the phenotypically modulated smooth muscle cell (Kearney *et al.,* 1997). In addition, T cells and macrophages are also present along with large amounts of VSMC-derived matrix components. Analysis of these specimens reveals a high proportion of VSMCs positive for markers of proliferation: PCNA, cdk2, and cyclin E. As in the situation for the rat carotid artery model, the VSMCs within these lesions undergo a relatively high frequency of apoptotic cell death as indicated by TUNEL staining (Fig. 6). Therefore, the lesion that develops within a stent is dynamic with a high frequency of VSMC turnover.

In-stent restenosis, though it occurs less frequently than restenosis following ordinary balloon angioplasty, is proving very difficult to treat. Clinicians classify these lesions as malignant, massive, and diffuse. Typically, patients with this disorder undergo additional balloon angioplasty procedures at the site of in-stent restenosis. However, many of these patients go on to another cycle of restenosis. Thus, the unfortunate patient who develops this complication is trapped in a "revolving door" of repeat angioplasties with recurrent restenotic episodes. The problem of restenosis may ultimately be solved by a combination of pharmacological and mechanical approaches. In other words, the delivery of an antiproliferative agent, perhaps locally at the time of the angioplasty, would likely have a beneficial impact by preventing the formation of cellular lesions within stents and thereby reduce or eliminate restenosis.

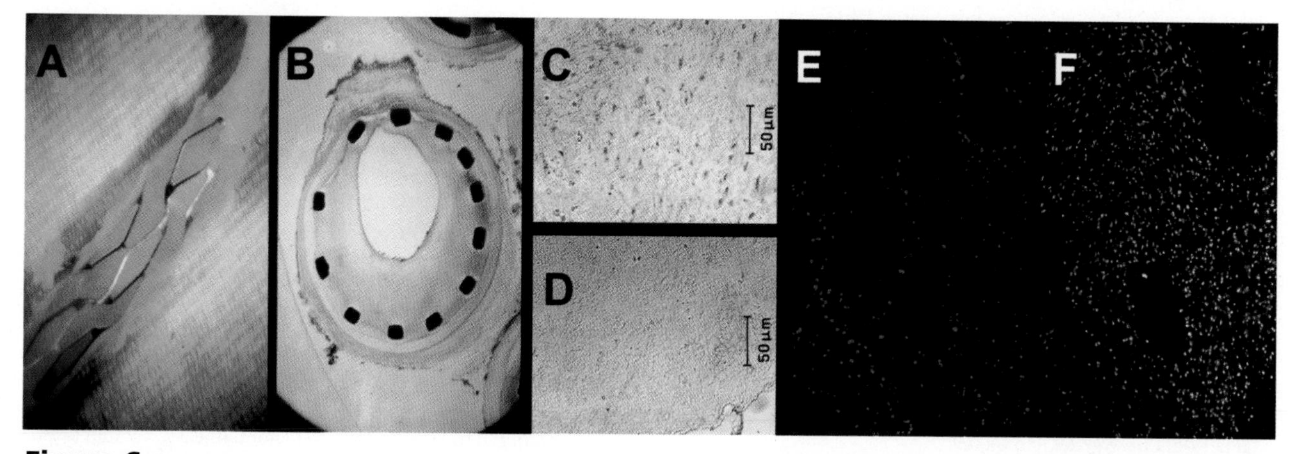

Figure 6 Demonstration of intimal hyperplasia and apoptosis in stented vessels. (A) An *ex vivo* view of the adventitial aspect of a rabbit iliac artery in which a stent has been implanted. (B) In-stent intimal hyperplasia is evident in this cross section of a stented rabbit iliac artery. The dark bars are cross sections of the stent that had been placed within the lumen of the vessel. (C) Human stenotic lesions obtained by directional atherectomy contain a large percentage of proliferative VSMCs. This clinical sample was analyzed for proliferative VSMCs by immunostaining for PCNA with proliferative cells being distinguished by red nuclear staining. A control for nonspecific binding was conducted utilizing a nonreactive IgG and is also shown (D). The stenotic lesions also have a high level of VSMC apoptosis as demonstrated by TUNEL analysis (green) for chromosome fragmentation in the same atherectomy sample (E). The cellularity of the sample is demonstrated by Hoechst dye which stains the chromatin blue (F).

Therefore, the development of therapeutic strategies to specifically target this lesion will likely benefit from a thorough understanding the regulatory pathways controlling VSMC differentiation, proliferation, and apoptosis.

VIII. Perspective

Compared with striated myocytes, smooth muscle cells have received relatively little attention from the scientific community. This is surprising in light of the important role that VSMCs play in the formation of blood vessel lesions, which in turn have a major impact on the function and viability of cardiomyocytes. A particularly intriguing property of mature VSMCs is their ability to modulate their state of differentiation, giving rise to proliferative cells that resembles their fibroblast-like progenitors. The VSMCs of the normal adult vessel wall are quiescent and display characteristics of a differentiated phenotype but can also dedifferentiate and reenter the cell cycle in the process of vascular regeneration or in response to injury. Therefore, the factors that regulate embryonic VSMC fate may also influence the genesis of atherosclerotic and restenotic lesions in adults. The analysis of the developmental regulatory networks that control VSMC fate will be of interest to both the academic and the clinical communities.

References

Abe, J., Zhou, W., Taguchi, J., Takuwa, N., Miki, K., Okazaki, H., Kurokawa, K., Kumada, M., and Takuwa, Y. (1994). Suppression of neointimal smooth muscle cell accumulation in vivo by antisense cdc2 and cdk2 oligonucleotides in rat carotid artery. *Biochem. Biophys. Res. Commun.* **198,** 16–24.

Aikawa, M., Palanisamy, N. S., Kuro-o, M., Kimura, K., Nakahara, K., Takewaki, S., Ueda, M., Yamaguchi, H., Yazaki, Y., Periasamy, M., and Nagai, R. (1993). Human smooth muscle myosin heavy chain isoforms as molecular markers for vascular development and atherosclerosis. *Circ. Res.* **73,** 1000–1012.

Andrés, V., Fisher, S., Wearsch, P., and Walsh, K. (1995). Regulation of *Gax* homeobox gene transcription by a combination of positive factors including MEF2. *Mol. Cell. Biol.* **15,** 4272–4281.

Barone, M. V., Crozat, A., Tabaee, A., Philipson, L., and Ron, D. (1994). CHOP (GADD153) and its oncogenic variant, TLS-CHOP, have opposing effects on the induction of G1/S arrest. *Genes Dev.* **8,** 453–464.

Benedeck, M. P., and Langille, B. L. (1991). Rapid accumulation of elastin and collagen in the aorta of sheep in the immediate perinatal period. *Circ. Res.* **69,** 1165–1169.

Bennett, M. R., Evan, G. I., and Schwartz, S. M. (1995). Apoptosis of human vascular smooth muscle cells derived from normal vessels and coronary atheroschlerotic plaques. *J. Clin. Invest.* **95,** 2266–2274.

Blank, R. S., McQuinn, T. C., Yin, K. C., Thompson, M. M., Takeyasu, K., Schwartz, R. J., and Owens, G. K. (1992). Elements of the smooth muscle α-actin promoter required in cis for transcriptional activation in smooth muscle. *J. Biol. Chem.* **267,** 984–989.

Bochaton-Piallat, M., Gabbiani, F., Redard, M., Desmouliere, A., and Gabbiani, G. (1995). Apoptosis participates in cellularity regulation during rat aortic intimal thickening. *Am. J. Pathol.* **146,** 1059–1064.

Bostrom, K., Watson, K. E., Horn, S., Wortham, C., Herman, I. M., and Demer, L. L. (1993). Bone morphogenetic protein expression in human arteroslerotic lesions. *J. Clin. Invest.* **91,** 1800–1809.

Breitbart, R. E., Liang, C., Smoot, L. B., Laheru, D. A., Mahdavi, V., and Nadal-Ginard, B. (1993). A fourth human MEF2 transcription factor, hMEF2D, is an early marker of myogenic lineage. *Development (Cambridge, UK)* **118,** 1095–1106.

Campbell, G. R., and Campbell, J. H. (1985). Smooth muscle phenotypic changes in arterial wall homeostasis: Implications for pathogenesis of atherosclerosis. *Exp. Mol. Path.* **42,** 139–162.

Campbell, G. R., Campbell, J. H., Manderson, J. A., Horrigan, S., and Rennick, R. E. (1988). Arterial smooth muscle: A multifunctional mesenchymal cell. *Arch. Pathol. Lab. Med.* **112,** 977–986.

Carroll, S. L., Bergsma, D. J., and Schwartz, R. J. (1988). A 29-nucleotide DNA segment containing an evolutionarily conserved motif is required in *cis* for cell-type-restricted repression of the chicken α-smooth muscle actin gene core promoter. *Mol. Cell. Biol.* **8,** 241–250.

Chamley-Campbell, J., Campbell, G. R., and Ross, R. (1979). The smooth muscle cell in culture. *Physiol. Rev.* **59,** 1–61.

Chang, M. W., Barr, E., Seltzer, J., Jiang, Y., Nabel, G. J., Nabel, E. G., Parmacek, M. S., and Leiden, J. M. (1995a). Cytostatic gene therapy for vascular proliferative disorders with a constitutively active form of the retinoblastoma gene product. *Science* **267,** 518–522.

Chang, M. W., Barr, E., Lu, M. M., Barton, K., and Leiden, J. M. (1995b). Adenovirus-mediated over-expression of the cyclin/cyclin-dependent kinase inhibitor, p21 inhibits vascular smooth muscle cell proliferation and neointima formation in the rat carotid artery model of balloon angioplasty. *J. Clin. Invest.* **96,** 2260–2268.

Chawengsaksophak, K., James, R., Hammond, V. E., Kontgen, F., and Beck, F. (1997). Homeostasis and intestinal tumors in Cdx2 mutant mice. *Nature (London)* **386,** 84–87.

Chen, D., Krasinski, K., Chen, D., Sylvester, A., Chen, J., Nisen, P. D., and Andrés, V. (1997). Down-regulation of cyclin-dependent kinase activity and cyclin A promoter activity in vascular smooth muscle cells by p27 (KIP-1), an inhibitor of neointima formation in the rat carotid artery. *J. Clin. Invest.* **99,** 2334–2341.

Chen, Y., Chen, Y., Lin, S., Chou, C., Mar, G., Chang, M., and Wang, S. (1997a). Electron Microscopic Studies of Phenotypic Modulation of Smooth Muscle Cells in Coronary Arteries of Patients with Unstable Angina Pectoris and Postangioplasty Restenosis. *Circulation* **95,** 1168–1175.

Cho, A., Courtman, D. W., and Langille, B. L. (1995). Apoptosis (programmed cell death) in arteries of the neonatal lamb. *Circ. Res.* **76,** 168–175.

Clowes, A. W., Reidy, M. A., and Clowes, M. M. (1983). Kinetics of cellular proliferation after arterial injury I: Smooth muscle cell growth in the absence of endothelium. *Lab. Invest.* **49,** 327–333.

Coffin, J. D., and Poole, T. J. (1988). Embryonic vascular development: Immunohistochemical identification of the origin and subsequent morphogenesis of the major vessel primordia in quail embryos. *Development (Cambridge, UK)* **102,** 735–748.

Dalton, S., and Treisman, R. (1992). Characterization of SAP-1, a protein recruited by serum response factor to the c-*fos* serum response element. *Cell (Cambridge, Mass.)* **68,** 597–612.

Del Sal, G., Ruaro, M. E., Philipson, L., and Schneider, C. (1992). The growth arrest-specific gene, *gas1,* is involved in growth suppression. *Cell (Cambridge, Mass.)* **70,** 595–607.

Firulli, A. B., Miano, J. M., Bi, W., Johnson, A. D., Casscells, W., Olson, E. N., and Schwarz, J. J. (1996). Myocyte Enhancer Binding Factor-2 expression and activity in vascular smooth muscle cells. *Circ. Res.* **78,** 196–204.

Fisher, S. A., and Watanabe, M. (1996). Expression of exogenous protein and analysis of morphogenesis in the developing chicken heart using an adenoviral vector. *Cardiovasc. Res.* **31,** E86–E95.

Fisher, S. A., Walsh, K., and Forehand, C. J. (1996). Characterization of cardiac gene cis-regulatory elements in the early stages of chicken heart morphogenesis. *J. Mol. Cell. Cardiol.* **28,** 113–122.

Fisher, S. A., Siwik, E., Branellec, D., Walsh, K., and Watanabe, M. (1997). Forced expression of the homeodomain protein Gax inhibits cardiomyocyte proliferation and perturbs heart morphogenesis. *Development (Cambridge, UK)* **124,** 4405–4413.

Fornace, A. J., Jackman, J., Hollander, M. C., Hoffman-Liebermann, B., and Liebermann, D. A. (1992). Genotoxic-stress-response genes and growth arrest genes: *gadd, MyD,* and other genes induced by treatments eliciting growth arrest. *Ann. N. Y. Acad. Sci.* **663,** 139–153.

Foster, D. N., Min, B., Foster, L. K., Stoflet, E. S., Sun, S., Getz, M. J., and Strauch, A. R. (1992). Positive and negative cis-acting regulatory elements mediate expression of the mouse vascular smooth muscle α-actin gene. *J. Biol. Chem.* **267,** 11995–12003.

Geng, J. Y., and Libby, P. (1995). Evidence for apoptosis in advanced human atheroma: colocalization with interleukin-1β converting enzyme. *Am. J. Pathol.* **147,** 251–266.

Giachelli, C. M., Bae, N., Almeida, M., Denhardt, D. T., Alpers, C. E., and Schwartz, S. M. (1993). Osteopontin is elevated during neointima formation in rat arteries and is a novel component of human artherosclerotic plaques. *J. Clin. Invest.* **92,** 1686–1696.

Gorski, D. H., Patel, C. V., and Walsh, K. (1993a). Homeobox transcription factor regulation in the cardiovascular system. *Trends Cardiovasc. Med.* **3,** 184–190.

Gorski, D. H., LePage, D. F., Patel, C. V., Copeland, N. G., Jenkins, N. A., and Walsh, K. (1993b). Molecular cloning of a diverged homeobox gene that is rapidly down-regulated during the G_0/G_1 transition in vascular smooth muscle cells. *Mol. Cell. Biol.* **13,** 3722–3733.

Gorski, D. H., LePage, D. F., and Walsh, K. (1994). Cloning and sequence analysis of homeobox transcription factor cDNAs with an inosine-containing probe. *BioTechniques* **16,** 856–865.

Grueneberg, D. A., Natesan, S., Alexandre, C., and Gilman, M. Z. (1992). Human and *Drosophila* homeodomain proteins that enhance the DNA-binding activity of the serum response factor. *Science* **257,** 1089–1095.

Gualberto, A., LePage, D. F., Pons, G., Mader, S. L., Park, K., Atchison, M. L., and Walsh, K. (1992). Functional antagonism between YY1 and the serum response factor. *Mol. Cell. Biol.* **12,** 4209–4214.

Guo, K., Andrés, V., and Walsh, K. (1998). Downregulation of cdk2 activity and cyclin A gene transcription by a nitric oxide-generating vasodilator in vascular smooth muscle cells. *Circulation* **97,** 2066–2072.

Han, D. K. M., Haudenschild, C. C., Hong, M. K., Tinkle, B. T., Leon, M. B., and Liau, G. (1995). Evidence for apoptosis in human atherogenesis and in a rat vascular injury model. *Am. J. Pathol.* **147,** 267–277.

Herring, B. P., and Smith, A. F. (1996). Telokin expression is mediated by a smooth muscle cell-specific promoter. *Am. J. Physiol.* **270,** C1656–C1665.

Hipskind, R. A., Rao, V. N., Mueller, C. G. F., Reddy, E. S. P., and Nordheim, A. (1991). Ets-related protein Elk-1 is homologous to the c-*fos* regulatory factor p62TCF. *Nature (London)* **354,** 531–534.

Hultgårdh-Nilsson, A., Krondahl, U., Querol-Ferrer, V., and Rigertz, N. R. (1991). Differences in growth factor response in smooth muscle cells isolated from adult and neonatal rats. *Differentiation (Berlin)* **47,** 99–105.

Hultgårdh-Nilsson, A., Cercek, B., Wang, J.-W., Naito, S., Lövdahl, C., Sharifi, B., Forrester, J. S., and Fagin, J. A. (1996). Regulated expression of the Ets-1 transcription factor in vascular smooth muscle cells *in vivo* and *in vitro*. *Circ. Res.* **78,** 589–595.

Hungerford, J. E., Owens, G. K., Argraves, W. S., and Little, C. D. (1996). Development of the aortic vessel wall as defined by vascular smooth muscle markers and extracellular matrix markers. *Dev. Biol. 178,* 375–392.

Hynes, R. O. (1992). Integrins: Versatility, modulation, and signalling in cell adhesion. *Cell (Cambridge, Mass.)* **69,** 11–24.

Ishida, A., Sasaguri, T., Kosaka, C., Nojima, H., and Ogata, J. (1997). Induction of the cyclin-dependent kinase inhibitor p21Sdi1/Cip1/-Waf1 by nitric oxide-generating vasodilator in vascular smooth muscle cells. *J. Biol. Chem.* **272,** 10050–10057.

Isner, J. M., Kearney, M., Bortman, S., and Passeri, J. (1995). Apoptosis in human atherosclerosis and restenosis. *Circulation.* **91,** 2703–2711.

Jiang, Y., and Evans, T. (1996). The Xenopus GATA-4/5/6 genes are associated with cardiac specification and can regulate cardiac-specific transcription during embryogenesis. *Dev. Biol.* **174,** 258–270.

Johnson, R. J., Iida, H., Alpers, C. E., Majesky, M. W., S.M., S., Pritzi, P., Gordon, K., and Gown, A. M. (1991). Expression of smooth muscle cell phenotype by rat mesangial cells in immune complex nephritis. Alpha-smooth muscle actin is a marker of mesangial cell proliferation. *J. Clin. Invest.* **87,** 847–858.

Katoh, Y., Loukianov, E., Kopras, E., Zilberman, A., and Periasamy, M. (1994). Identification of functional promoter elements in the rabbit smooth muscle myosin heavy chain gene. *J. Biol. Chem.* **269,** 30538–30545.

Kawabe, T., Muslin, A. J., and Korsmeyer, S. J. (1997). Hox11 interacts with protein phosphatases PP2A and PP1 and disrupts a G2/M cell-cycle checkpoint. *Nature (London)* **385,** 454–458.

Kearney, M., Pieczek, A., Haley, L., Losordo, D. W., Andrés, V., Schainfeld, R., Rosenfield, K., and Isner, J. M. (1997). Histopathology of in-stent restenosis in patients with peripheral artery disease. *Circulation* **95,** 1998–2002.

Kim, S., Ip, H. S., Lu, M. M., Clendenin, C., and Parmacek, M. S. (1997). A serum response factor-dependent transcriptional regulatory program identifies distinct smooth muscle cell sublineages. *Mol. Cell. Biol.* **17,** 2266–2278.

Kirby, M. L. (1988). Nodose placode provides ectomesenchyme to the developing chick heart in the absence of cardiac neural crest. *Cell Tissue Res.* **252,** 17–22.

Koyama, H., Raines, E. W., Bornfeldt, K. E., Roberts, J. M., and Ross, R. (1996). Fibrillar collagen inhibits arterial smooth muscle proliferation through regulation of cdk2 inhibitors. *Cell (Cambridge, Mass.)* **87,** 1069–1078.

Kuo, C. T., Morrisey, E. E., Anandappa, R., Sigrist, K., Lu, M. M., Parmacek, M. S., Soudais, C., and Leiden, J. M. (1997). GATA4 transcription factor is required for ventral morphogenesis and heart tube formation. *Genes Dev.* **11,** 1048–1060.

Laverriere, A. C., MacNeill, C., Mueller, C., Poelmann, R. E., Burch, J. B., and Evans, T. (1994). GATA-4/5/6, a subfamily of three transcription factors transcribed in developing heart and gut. *J. Biol. Chem.* **269,** 23177–23184.

Le Lièvre, C. S., and Le Douarin, N. M. (1975). Mesenchymal derivatives of the neural crest: Analysis of chimaeric quail and chich embryos. *J. Embryol. Exp. Morphol.* **34,** 125–154.

LePage, D. F., Altomare, D. A., Testa, J. R., and Walsh, K. (1994). Molecular cloning and localization of the human GAX gene to 7p21. *Genomics* **24,** 535–540.

Li, L., Liu, Z.-C., Mercer, B., Overbeek, P., and Olson, E. N. (1997). Evidence for serum response factor-mediated regulatory networks governing SM22a transcription in smooth, skeletal and cardiac muscle cells. *Dev. Biol.* **187,** 311–321.

Maillard, L., van Belle, E., Smith, R. C., Le Roux, A., Denèfle, P., Steg, G., Barry, J. J., Branellec, D., Isner, J. M., and Walsh, K. (1997). Percutaneous delivery of the gax gene inhibits vessel stenosis in a rabbit model of balloon angioplasty. *Cardiovasc. Res.* **35,** 536–546.

Mallat, Z., Ohan, J., Leseche, G., and Tedugi, A. (1997). Colocalization of CPP-32 with apoptotic cells in human atherosclerotic plaques. *Circulation* **96,** 424–428.

Manasek, F. J. (1971). The ultrastructure of embryonic myocardial blood vessels. *Dev. Biol.* **26,** 42–54.

Martin, J. F., Miano, J. M., Hustad, C. M., Copeland, N. G., Jenkins, N. A., and Olson, E. N. (1994). A MEF2 gene that generates a muscle-specific isoform via alternative splicing. *Mol. Cell. Biol.* **14,** 1647–1656.

Minty, A., and Kedes, L. (1986). Upstream regions of the human cardiac actin gene that modulate its transcription in muscle cells: Presence of an evolutionary conserved repeated motif. *Mol. Cell. Biol.* **6,** 2125–2136.

Molkentin, J. F., Lin, Q., Duncan, S. A., and Olson, E. N. (1997). Requirement of the transcription factor GATA4 for heart tube formation and ventral morphogenesis *Genes Dev.* **11,** 1061–1072.

Moncada, S., Palmer, R. M. J., and Higgs, E. A. (1991). NO: Physiology, pathophysiology and pharmacology. *Pharmacol. Rev.* **43,** 109–142.

Morishita, R., Gibbons, G. H., Ellison, K. E., Nakajima, M., von der Leynen, H., Zhang, L., Kaneda, Y., Ogihara, T., and Dzau, V. J. (1994). Intimal hyperplasia after vascular injury is inhibited by antisense cdk 2 kinase oligonucleotides. *J. Clin. Invest.* **93,** 1458–1464.

Morrisey, E. E., Ip, H. S., and Parmacek, M. S. (1996). GATA-6: A zinc finger transcription factor that is expressed in multiple cell lineages derived from lateral mesoderm. *Dev. Biol.* **177,** 309–323.

Nakano, Y., Nishihara, T., Sasayama, S., Miwa, T., Kamada, S., and Kakunaga, T. (1991). Transcriptional regulatory elements in the 5′ upstream and first intron regions of the human smooth muscle (aortic type) α-actin-encoding gene. *Gene* **99,** 285–289.

Owens, G. K. (1995). Regulation of differentiation of fascular smooth muscle cells. *Physiol. Rev.* **75,** 487–517.

Perlman, H., Maillard, L., Krasinski, K., and Walsh, K. (1997). Evidence for the rapid onset of apoptosis in medial smooth muscle cells following balloon injury. *Circulation* **95,** 981–987.

Perlman, H. R., Suzuki, E., Simonson, M., Smith, R. C., and Walsh, K. (1998). GATA-6 induces p21(CIP1) expression and G1 cell cycle arrest. *J. Biol. Chem.* **273,** 13713–13718.

Pickering, J. G., Weir, L., Jekanowski, J., Kearney, M. A., and Isner, J. M. (1993). Proliferative activity in peripheral and coronary atherosclerotic plaque among patients undergoing percutaneous revascularization. *J. Clin. Invest.* **91,** 1469–1480.

Pollock, R., and Treisman, R. (1991). Human SRF-related proteins: DNA-binding properties and potential regulatory targets. *Genes Dev.* **5,** 2327–2341.

Poole, T. J., and Coffin, J. D. (1989). Vasculogenesis and angiogenesis: Two distinct morphogenetic mechanisms establish embryonic vascular pattern. *J. Exp. Zool.* **251,** 224–231.

Riessen, R., Isner, J. M., Blessing, E., Loushin, C., Nikol, S., and Wight, T. N. (1994). Regional differences in the distribution of proteoglycans biglycan and decorin in the extracellular matrix of atherosclerotic and restenotic human coronary arteries. *Am. J. Pathol.* **144,** 962–974.

Rosenquist, T. H., Kirby, M. L., and Van Mierop, L. H. (1989). Solitary aortic arch artery. A result of surgical ablation of cardiac neural crest and nodose placode in the avian embryo. *Circulation* **80,** 1469–1475.

Ross, R. (1993). The pathogenesis of atherosclerosis: a perspective for the 1990s. *Nature (London)* **362,** 801–809.

Sata, M., and Walsh, K. (1998). TNFα regulation of Fas ligand expression on endothelium modulates leukocyte extravasation. *Nature Med. (London)* **4,** 415–420.

Schneider, C., King, R. M., and Philipson, L. (1988). Genes specifically expressed at growth arrest of mammalian cells. *Cell (Cambridge, Mass.)* **54,** 787–793.

Schwarzacher, S. P., Lim, T. T., Wang, B., Kernoff, R. S., Neibauer, J., Cooke, J. P., and Yeung, A. C. (1997). Local intramural delivery of L-arginine enhances nitric oxide generation and inhibits lesion formation after balloon angioplasty. *Circulation* **95,** 1863–1869.

Shimizu, R. T., Blank, R. S., Jervis, R., Lawrenz-Smith, S. C., and Owens, G. K. (1995). The smooth muscle alpha-actin gene promoter is differentially regulated in smooth muscle versus nonsmooth muscle cells. *J. Biol. Chem.* **270,** 7631–7643.

Simonson, M. S., Walsh, K., Kumar, C. C., Bushel, P., and Herman, W. H. (1995). Two proximal CArG elements regulate SM α-actin promoter, a genetic marker of activated phenotype of mesangial cells. *Am. J. Physiol. (Renal Fluid Electrolyte Physiol.)* **268**(37), F760–F769.

Skopicki, H. A., Lyons, G. E., Shatteman, G., Smith, R. C., Andrés, V., Schirm, S., Isner, J. M., and Walsh, K. (1997). Embryonic expression of the Gax homeodomain gene in cardiac, smooth and skeletal muscle. *Circ. Res.* **80,** 452–462.

Slomp, J., Gittenberger-de-Groot, A. C., Glukhova, M. A., van Musteren, C., Kockx, M. M., Schwarz, S. M., and Koteliansky, V. E. (1997). Differentiation, dedifferentiation, and apoptosis of smooth muscle cells during the development of the human ductus arteriosus. *Arterioscler. Thromb. Vasc. Biol.* **17,** 1003–1009.

Smith, R. C., Branellec, D., Gorski, D. H., Guo, K., Perlman, H., Dedieu, J.-F., Pastore, C., Mahfoudi, A., Denèfle, P., Isner, J. M., and Walsh, K. (1997a). p21^{CIP1}-mediated inhibition of cell proliferation by overexpression of the *gax* homeodomain gene. *Genes Dev.* **11,** 1674–1689.

Smith, R. C., Wills, K. N., Antelman, D., Perlman, H., Truong, L. N., Krasinski, K., and Walsh, K. (1997b). Adenoviral constructs encoding phosphorylation-competent full-length and truncated forms of the human retinoblastoma protein inhibit myocyte proliferation and neointima formation. *Circulation* **96,** 1899–1905.

Solway, J., Seltzer, J., Samaha, F. F., Kim, S., Alger, L. E., Niu, Q., Morrisey, E. E., Ip, H. S., and Parmacek, M. S. (1995). Structure and expression of a smooth muscle cell-specific gene, SM22a. *J. Biol. Chem.* **270,** 13460–13469.

Song, K., Wang, Y., and Sassoon, D. (1992). Expression of *Hox-7.1* in myoblasts inhibits terminal differentiation and induces cell transformation. *Nature (London)* **360,** 477–481.

Soudais, C., Bielinska, M., Heikinheimo, M., MacArthur, C. A., Narita, N., Saffitz, J. E., Simon, M. C., Leiden, J. M., and Wilson, D. B. (1995). Targeted mutagenesis of the transcription factor GATA-4 gene in mouse embryonic stem cells disrupts visceral endoderm differentiation in vitro. *Development (Cambridge, UK)* **121,** 3877–3888.

Suh, E., Chen, L., Taylor, J., and Traber, P. G. (1994). A homeodomain protein related to caudal regulates intestine-specific gene transcription. *Mol. Cell. Biol.* **14,** 7340–7351.

Suzuki, E., Guo, K., Kolman, M., Yu, Y.-T., and Walsh, K. (1995). Serum-induction of MEF2/RSRF expression in vascular myocytes is mediated at the level of translation. *Mol. Cell. Biol.* **15,** 3415–3423.

Suzuki, E., Evans, T., Lowry, J., Truong, L., Bell, D. W., Testa, J. R., and Walsh, K. (1996). The human GATA-6 gene: Structure, chromosomal location and regulation of expression by tissue-specific and mitogen-responsive signals. *Genomics* **38,** 283–290.

Taylor, M., Treisman, R., Garrett, N., and Mohun, T. (1989). Muscle-specific (CArG) and serum-responsive (SRE) promoter elements are functionally interchangeable in Xenopus embryos and mouse fibroblasts. *Development (Cambridge, UK)* **106,** 67–78.

Thyberg, J., Hedin, U., Sjölund, M., Palmberg, L., and Bottger, B. A. (1990). Regulation of differentiated properties and proliferation of arterial smooth muscle cells. *Arteriosclerosis* **10,** 966–990.

Topouzis, S., and Majesky, M. W. (1996). smooth muscle lineage diversity in the chick embryo. *Dev. Biol.* **178,** 430–445.

Treisman, R. (1992). Structure and function of serum response factor. *In* "Transcriptional Regulation" (S. L. McKnight and K. R.

Yamamoto, eds.), Cold Spring Harbor Lab. Press, Plainview, NY. pp. 881–905.

von der Leyen, H. E., Gibbons, G. H., Morishita, R., Lewis, N. P., Zhang, L., Nakajima, M., Kaneda, Y., Cooke, J. P., and Dzau, V. J. (1995). Gene therapy inhibiting neointimal vascular lesion: *In vivo* transfer of endothelial cell nitric oxide synthase gene. *Proc. Natl. Acad. Sci. U.S.A.* **92,** 1137–1141.

Walsh, K. (1989). Cross-binding of factors to functionally different promoter elements in c-fos and skeletal actin genes. *Mol. Cell. Biol.* **9,** 2191–2201.

Walsh, K., and Perlman, H. (1996). Molecular strategies to inhibit restenosis: Modulation of the vascular myocyte phenotype. *Semin. Intervent. Cardiol.* **1,** 173–179.

Wei, G. L., Krasinski, K., Kearney, M., Isner, J. M., Walsh, K., and Andrés, V. (1997). Temporally and spatially coordinated expression of cell cycle regulatory factors after angioplasty. *Circ. Res.* **80,** 418–426.

Weir, L., Chen, D., Pastore, C., Isner, J. M., and Walsh, K. (1995). Expression of GAX, a growth-arrest homeobox gene, is rapidly downregulated in the rat carotid artery during the proliferative response to balloon injury. *J. Biol. Chem.* **270,** 5457–5461.

Weiss, M., and Orkin, S. (1995). GATA transcription factors: Key regulators of hematopoiesis. *Exp. Hematol.* **23,** 99–107.

Yamashita, J., Itoh, H., Ogawa, Y., Tamura, N., Takaya, K., Igaki, T., Doi, K., Chun, T.-H., Inoue, M., Masatsugu, K., and Nakao, K. (1997). Opposite regulation of Gax homeobox expression by Angiotensin II and C-type natriuretic peptide. *Hypertension* **29,** 381–387.

Yang, Z., Simari, R., Perkins, N., Sang, H., Gordon, D., Nabel, G., and Nabel, E. (1996). Role of the p21cyclin-dependent kinase inhibitor in limiting intimal cell proliferation in response to arterial injury. *Proc. Natl. Acad. Sci. U.S.A.* **93,** 7905–7910.

Yu, Y.-T., Breitbart, R. E., Smoot, L. B., Lee, Y., Mahdavi, V., and Nadal-Ginard, B. (1992). Human myocyte-specific enhancer factor 2 comprises a group of tissue-restricted MADS box transcription factors. *Genes Dev.* **6,** 1783–1798.

Zhan, Q., Lord, K. A., Alamo, I. Jr., Hollander, M. C., Carrier, F., Ron, D., Kohn, K. W., Hoffman, B., Liebermann, D. A., Fornace, A. J., Jr. (1994). The gadd and MyD genes define a novel set of mammalian genes encoding acidic proteins that synergistically suppress cell growth. *Mol. Cell. Biol.* **14,** 2361–2371.

IX

Human Cardiac Development Defects

25

Symmetry and Laterality in the Human Heart: Developmental Implications

Nigel A. Brown* **and Robert H. Anderson**†

**Department of Anatomy and Developmental Biology, St. George's Hospital Medical School, London SW17 0RE, United Kingdom*

†Section of Pediatrics, Royal Brompton Campus, Imperial College School of Medicine, National Heart and Lung Institute, London, SW3 6LY, United Kingdom

I. Introduction

It has long been recognized that an abnormal arrangement of the internal organs is associated with characteristic congenital cardiac malformations (Martin, 1826; Abernethy, 1793; Van Mierop and Wigglesworth, 1962; Moller et al., 1967). This suggests that normal cardiac development is dependent on the pathways that define the layout of the body plan and in particular on the processes that distinguish left from right. In man, more than any other species, the anatomy of congenital cardiac malformation has been exhaustively recorded, and these deviations from normal are a rich source of information on development. The key to unraveling this information is a systematic understanding of morphology, both within the heart and in relationship to other organs. We will apply systematic morphology to identify the components of the heart that possess intrinsic left- or right-related structure and explore their implications for development.

A. The Concept of Symmetry versus Lateralization

Humans, like all deuterostomes and protostomes, are overtly bilaterally symmetrical. That is, they have a midline plane of symmetry, with their two sides being mirror images of one another. Concealed within this symmetrical frame is an internal body plan which is almost wholly lateralized. Lateralization is the same as asymmetry and simply means that the two sides differ. In theory, and in some reality, this lateralization can be random in a population, such that a structure is found as often on the left as it is on the right. Normally, however, the internal body is consistently handed so that a particular structure is always found on the same side. In principle, then, we can distinguish three different kinds of deviation from this norm: loss of handedness to give random lateralization; reversal of handedness which generates mirror imagery of the whole body plan, not to be confused with the relationship between individual sides in symmetry; and the loss of lateralization which produces symmetry.

There is no lack of concrete examples of these concepts in normal humans. The limbs perfectly illustrate symmetry; the right and left hands are mirror images of each other, and they can be termed an isomeric (strictly, enantiomeric) pair. Importantly, as for any isomeric pair, the relationships of their component parts (i.e., their topology) serves to distinguish between the right-sided and left-sided partner, irrespective of their position in space. For example, a right hand can still be identified as such, even if it were at the end of a left arm. Handed lateralization is most obvious in the unpaired organs: The liver is predominantly to the right, the stomach and spleen are exclusively on the left, and the gut has a characteristic pattern for the attachment of its mesentery to the body wall (Fig. 1A). The heart is positioned within the mediastinum so that two-thirds of its bulk is to the left, and its long axis points downwards and to the left. The aorta is left sided, whereas the inferior caval vein is to the right of the spine.

The lateralization of paired organs goes beyond their mere position. The right lung has three lobes compared to two in the left (Fig. 1A). The bronchus of the right lung is short and is crossed by the artery supplying the middle and lower lobes after its first branch, making this branch an eparterial bronchus. In contrast, the long left bronchus is crossed by the artery supplying the lower lobe before its first branch, which is, in consequence, hyparterial. The overall pattern of the totality of these normal-handed lateralizations is described as situs solitus, which simply means a usual arrangement (Fig. 1A).

It has been known for many years (Baker-Cohen, 1961) that about 1 in 8500 people show reversal of the handedness of this usual arrangement to give the mirror-imaged variant traditionally described as situs inversus (Fig. 1B). This reversal is not "upside-downness," so our preference is to describe the pattern as mirror-imaged arrangement. These individuals have, for example, three lung lobes supplied by a short eparterial bronchus on the left, the liver is left sided, and the stomach and spleen are found in the right upper quadrant of the abdomen (Fig. 1B). Those with this pure mirror-imaged arrangement will usually have mirror imaged, but otherwise normal, hearts.

That lateralization can be lost, as distinct from reversed, is a more recent enlightenment. In the middle part of this century, when pathologists and cardiologists started to take a particular interest in congenitally malformed hearts, it became clear that complex intracardiac lesions were found in patients who showed some of the features of mirror imagery. The hearts of such patients were frequently in the right half of the chest; the cardiac apex often pointed to the right, or directly downward; and the aortic arch was frequently right sided. These hearts tended to be lumped together with situs inversus (Campbell and Deucher, 1967), but even at this time, prescient observers such as Putschar and Manion (1956) noted a tendency to morphologic symmetry within the atriums, particularly in the structure of the atrial appendages. Conventional wisdom, nonetheless, dictated that these syndromes were distinguished from the usual and mirror-imaged arrangements on the basis of the supposed ambiguity of their abdominal organs, termed visceral heterotaxy or situs ambiguous (Van Mierop *et al.*, 1972). Two subsets were then identified within the group, depending on the state of the spleen (absent or multiple).

In reality, there is nothing ambiguous about the arrangement of the lungs and bronchial trees in these patients (Landing *et al.*, 1971). They are, in fact, isomeric. It is, however, difficult to be sure of their precise incidence (see Table I). In most of the patients lacking a spleen, both lungs are trilobed, and both bronchuses are short and eparterial (Fig. 1C). Similarly, in most with multiple spleens, the lungs each have two lobes and are supplied by bronchuses which are long and hyparterial (Fig. 1D). What of the structure of the heart? Is there evidence of atrial or ventricular isomerism? To answer these questions, objective criterions are needed to recognize the topology of the cardiac components.

II. Morphology of the Atriums and Ventricles

When determining the nature of any anatomic structure, it is good practice to use as the distinguishing feature the component of the structure which is least variable in its morphology, particularly when the structure itself is congenitally malformed (the "morphological method"; Lev, 1954; Van Praagh and Vlad, 1978). Applying this principle to the heart, the atrial chambers are

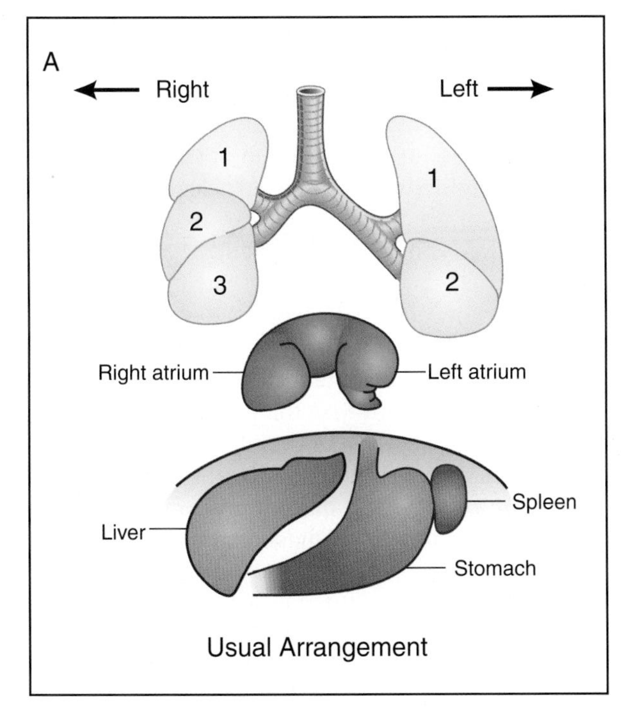

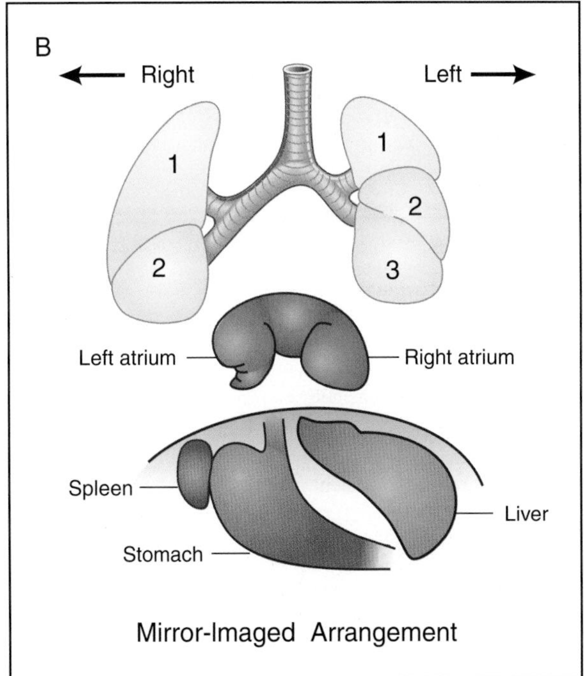

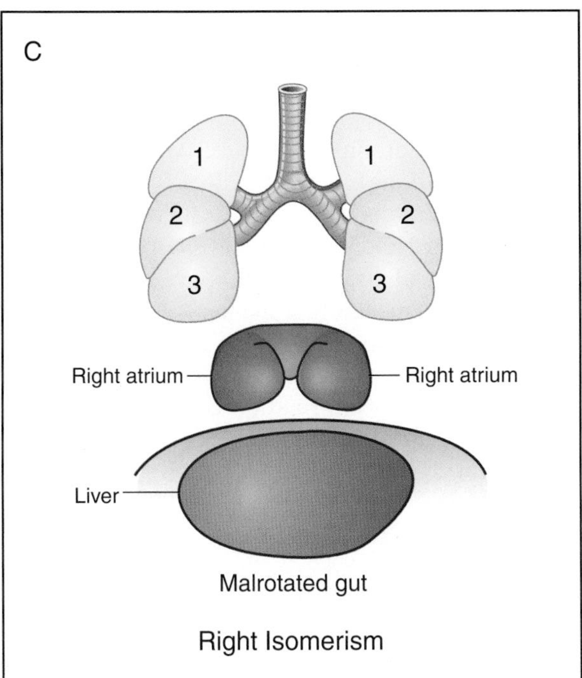

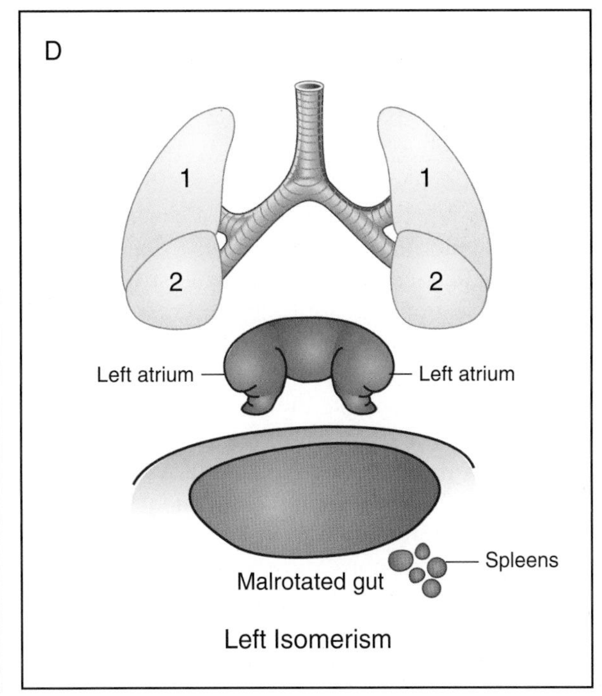

Figure 1 The arrangement (situs) of the main lateralized organs of the body: the lungs and bronchuses, the atrial appendages, liver, stomach, and spleen, shown in usual (solitus, A), mirror-imaged (inversus, B), right isomeric (C), and left isomeric (D) forms. 1–3, lobes of the lungs.

made up of venous components, vestibules, the septum, and the appendages. Of these, the venous components, vestibules, and the septum can all be markedly abnormal, or even totally lacking, when the heart itself is congenitally malformed. It is the appendages that are most constant when the atrial chambers are malformed, and the structure of the normal morphologically right appendage (Fig. 2A) is markedly different from that of the left appendage (Fig. 2B). The shape of the appendages can be distorted in malformed hearts but the extent of

Table I Incidence of Cardiac Laterality Defects in the Human[a]

• From a liveborn population encountered in a clinical setting 2,381 patients in the first year of life (Fyler *et al.*, 1980)	
Complete transposition	9.9%
"Heterotaxy"	4.0%
Corrected transposition	0.9%
• From an autopsied series, selecting patients in the first year of life 291 hearts (Hegerty *et al.*, 1985)	
Usual atrial arrangement	95.2%
Isomeric right appendages	3.8%
Isomeric left appendages	1.4%
Discordant VA connections	11.0%
Discordant AV connections	0.3%
• From an autopsied collection, assembled over an extensive period 1,842 hearts (Sharma *et al.*, 1988)	
Usual atrial arrangement	96.5%
Mirror-imagery	0.4%
Isomeric right appendages	1.9%
Isomeric left appendages	1.2%

[a]It is exceedingly difficult to provide precise figures for the incidence of isomerism and segmental mismatches in the general population. The figures quoted refer to incidences in selected populations of those with congenitally malformed hearts. In general, congenital cardiac malformations are found in about 8 of every 1,000 live births.

the pectinate muscles, relative to the vestibule of the atrioventricular junctions, remains as a distinguishing feature. In the normal morphologically right atrium (Fig. 3A), these pectinate muscles are extensive, running all around the vestibule to abut the septum at the crux in the region of the post-Eustachian sinus. In the normal morphologically left atrium, in contrast, the pectinate muscles are almost completely confined within the tubular appendage, and the smooth muscular vestibule of the mitral valve is directly continuous with the smooth-walled pulmonary venous component (Fig. 3B). A further feature of note is that the smooth-walled posterior vestibule of the morphologically left atrium contains a prominent transverse venous structure, the coronary sinus, which is absent on the right.

The extent of the pectinate muscles enables all appendages to be designated as morphologically right or morphologically left, including those in hearts with so-called situs ambiguous (Uemura *et al.*, 1995). Thus, all hearts can be assigned to one group with lateralized appendage arrangement (usual or mirror imagery) or another group with symmetry (right or left isomerism).

When the morphological method is applied to the ventricles, the significant component parts are the inlets, apical trabecular components, and outlets. Of these, it is the apical trabecular component which is not only most constant but also most reliable in distinguishing right and left ventricles; the right have coarse (Fig. 4A) and the left fine trabeculations (Fig. 4B). Thus, the ventricles can be recognized as being morphologically right or left, even when represented only by the trabecular parts and regardless of their position in the body. Obviously, the morphologically right ventricle is located on the right side of the body in usual arrangement and on the left in mirror-image arrangement. These arrangements can be determined in an isolated heart without any reference to the body by using the system of ventricular topology. The key is to first identify the morphologically right ventricle, then establish which hand, figuratively speaking, can be laid with the palm on its interventricular septal surface with the thumb in the inlet and the fingers in the outlet component. Normally, this is the right hand, but it is the left hand in mirror-imaged arrangement, thus defining right-hand and left-hand ventricular topology.

Unlike the atriums, it is exceedingly rare to find any evidence of symmetry within the ventricular mass. In all our experience of hearts with two ventricles, including those in which the atrial chambers connect to one dominant ventricle in the presence of a second incomplete and rudimentary ventricle, we have seen only one heart that contained two ventricles with apparently similar apical morphology. Hearts can, very rarely, have a solitary and indeterminate ventricle, which has coarse apical trabeculations typical of neither the right nor the left ventricle. Perhaps significantly, almost 80% of hearts with a solitary and indeterminate ventricle are found in the setting of isomeric morphologically right appendages. The position of the heart relative to the chest and the orientation of its apex convey no information regarding the internal makeup of the chambers nor do they provide any information concerning the arrangement of the atrial appendages.

III. Morphology of Transpositions

In complete transposition of the great arteries, and in congenitally corrected transposition, the atriums, ventricles, and outflow tracts are each lateralized, but not all with the same handedness, thus creating segmental mismatch. In congenitally corrected transposition, the cavities of the atrial and ventricular segments are joined together inappropriately across the atrioventricular junctions. In the most common variant, the atrial chambers are in the usual arrangement but are connected to ventricles which, topologically, are mirror imaged. The right-sided morphologically right atrium is joined to the right-sided morphologically left ventricle, and the left atrium is joined to the left-sided right ventricle. In the rarer variant, the atrial chambers are mirror imaged

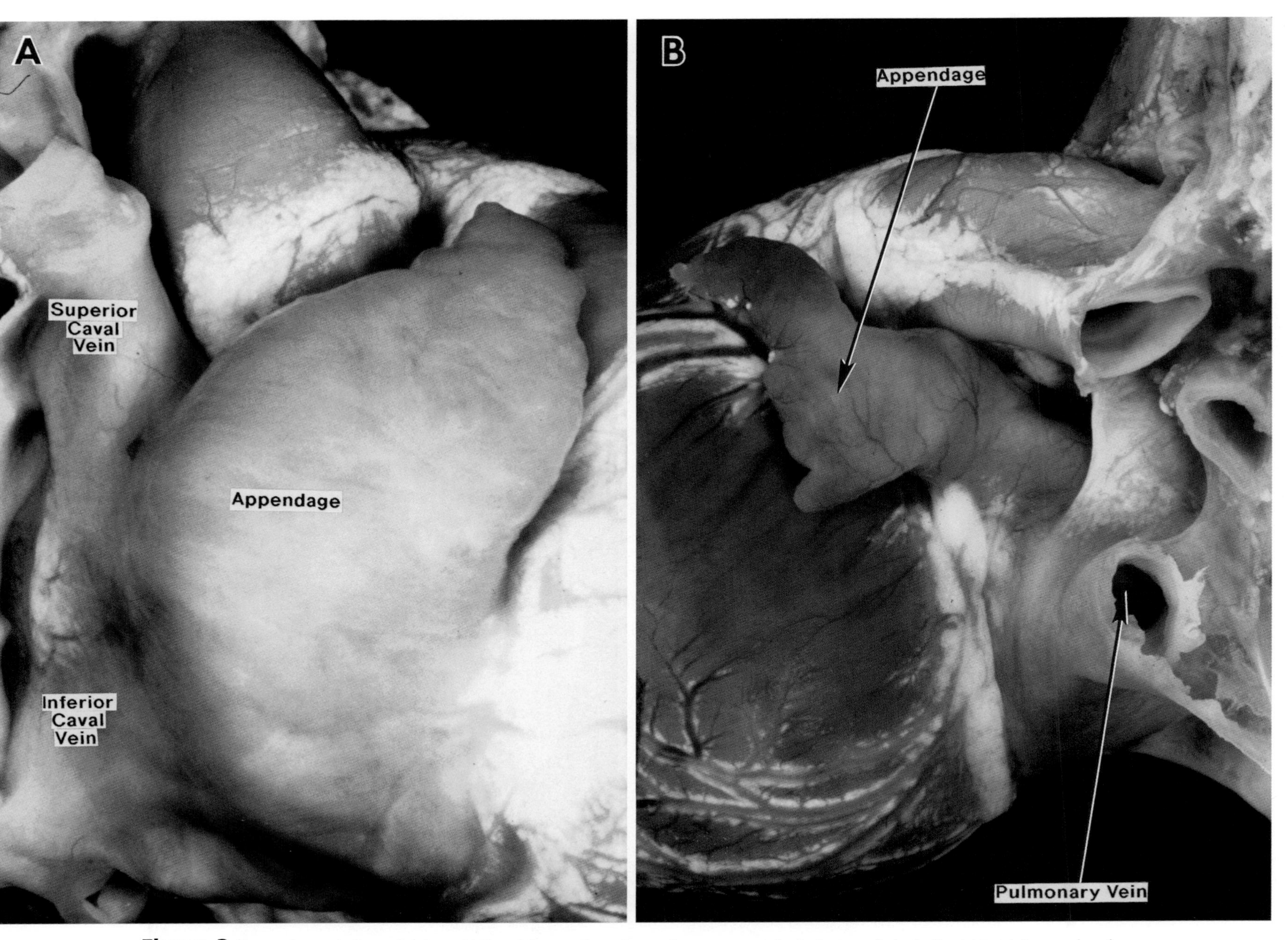

Figure 2 External morphology of right and left atrial appendages in usual arrangement. In A, the morphologically right atrial appendage is seen from its right side. It has a characteristic triangular shape and a broad junction with the systemic venous component across the terminal groove. In comparison, the morphologically left appendage (viewed from the left in B) is long and tubular, with a constricted junction with the pulmonary venous component.

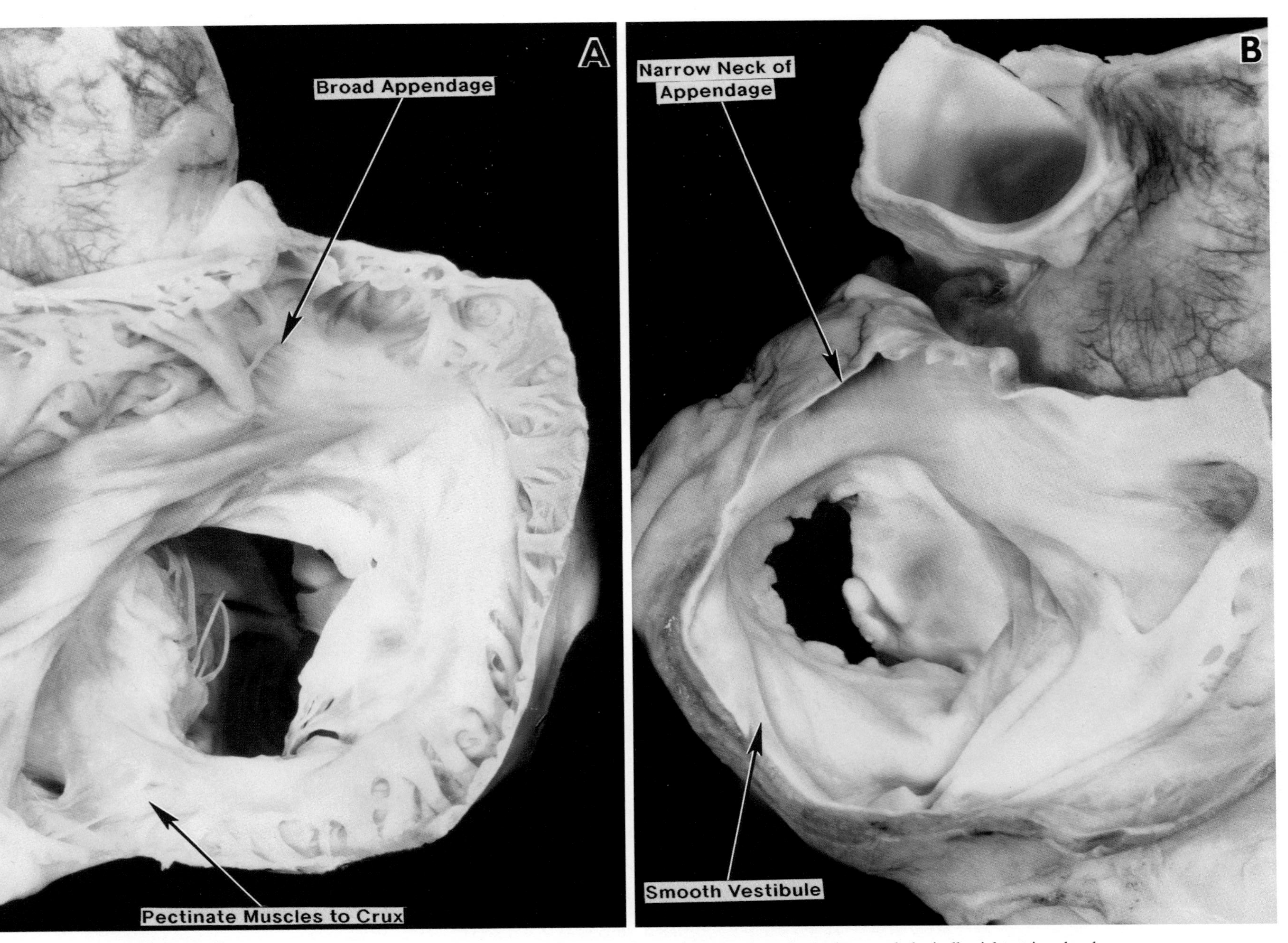

Figure 3 Internal morphology of right and left atrial appendages in usual arrangement. In A, the morphologically right atrium has been opened through the appendage to show the vestibule to the tricuspid valve. Note how the pectinate (comb-like) muscles extend all round the vestibule to the mouth of the coronary sinus. In B, the morphologically left atrium is viewed through the pulmonary venous component to show the vestibule to the mitral valve. Note that the appendage has a minimal relationship to the vestibule, which is continuous with the smooth venous component.

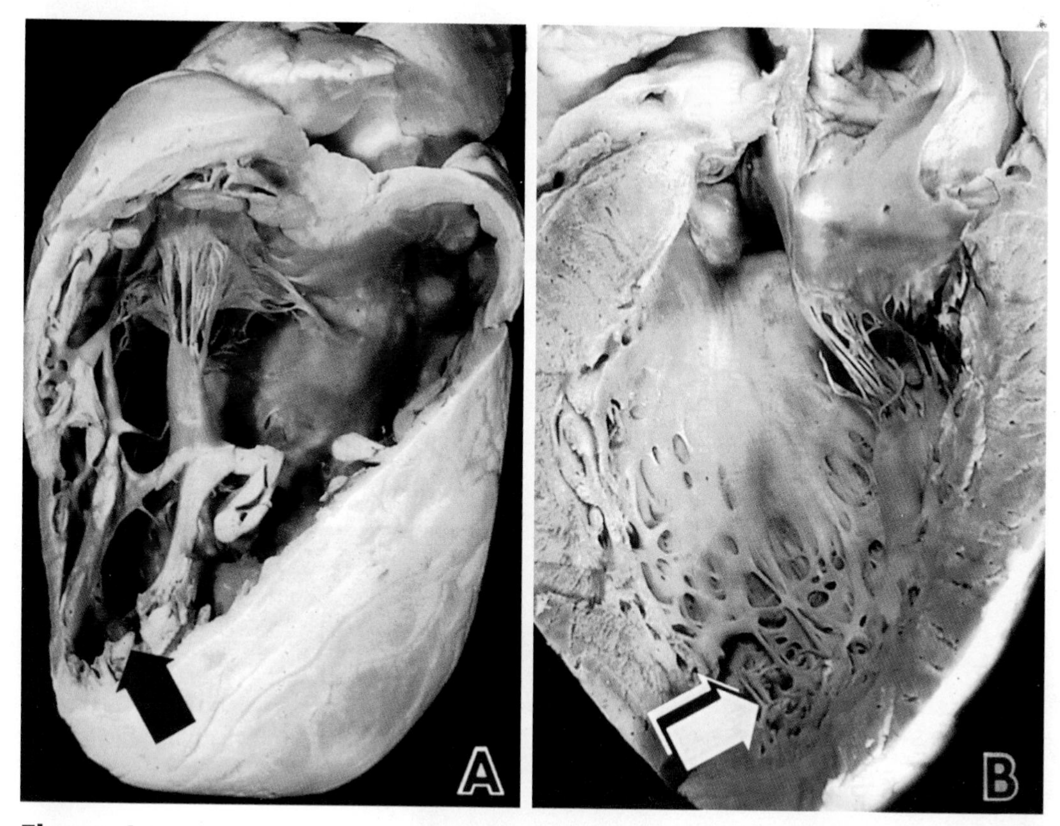

Figure 4 Internal morphology of right and left ventricles in usual arrangement (right-hand ventricular topology). The morphologically right ventricle (A) is viewed after removal of the parietal wall to show the inlet (tricuspid), coarse apical trabecular (arrow), and subpulmonary outlet components. The same view of the morphologically left ventricle (B) of the same heart shows inlet (mitral), fine apical trabecular (arrows), and aortic outlet components.

while the ventricles are normal. These abnormal connections are said to be discordant. In most instances, the ventriculoarterial connections are also discordant, hence the correction of the circulation. In hearts with any transposition of the great arteries, the pulmonary trunk arises from the morphologically left ventricle and the aorta from the morphologically right ventricle. Again, this can occur in usual and mirror-image arrangements and is called complete transposition.

IV. Morphology of Hearts with Isomeric Atrial Appendages

A. Right Isomerism

The pathognomonic feature of right isomerism is the pectinate muscles extending symmetrically around both atrioventricular junctions to meet at the crux (Fig. 5A). Most frequently, the atrial septum is also grossly deficient, often represented merely by a muscular strand spanning the common atrial cavity. There is a common atrioventricular valve in the majority of cases. Bilateral superior caval veins are found in most cases and, when present, each drains directly to the atrial roof, almost always being separated from the morphologically right appendage by bilateral terminal crests. In some instances, one of the bilateral superior caval veins may be atretic, or a unilateral superior caval vein may be either on the right or on the left. Irrespective of the number of superior caval veins present, the most characteristic feature of atrial anatomy is complete absence of the coronary sinus. As a result, the drainage of the coronary veins is grossly abnormal. In contrast to the absence of the coronary sinus, the inferior caval vein almost always drains directly to the atriums, sometimes bilaterally but more usually unilaterally, and can be connected to either the right- or the left-sided atrium. Interruption of the inferior caval vein is very rare. In comparison, the pulmonary veins are always connected in an anomalous fashion, even when they drain directly to the atriums, usually via a fibrous pulmonary venous sinus. In the majority of cases, drainage is extracardiac and can be supracardiac, infracardiac, or mixed. Self-evidently, drainage cannot occur through a coronary sinus.

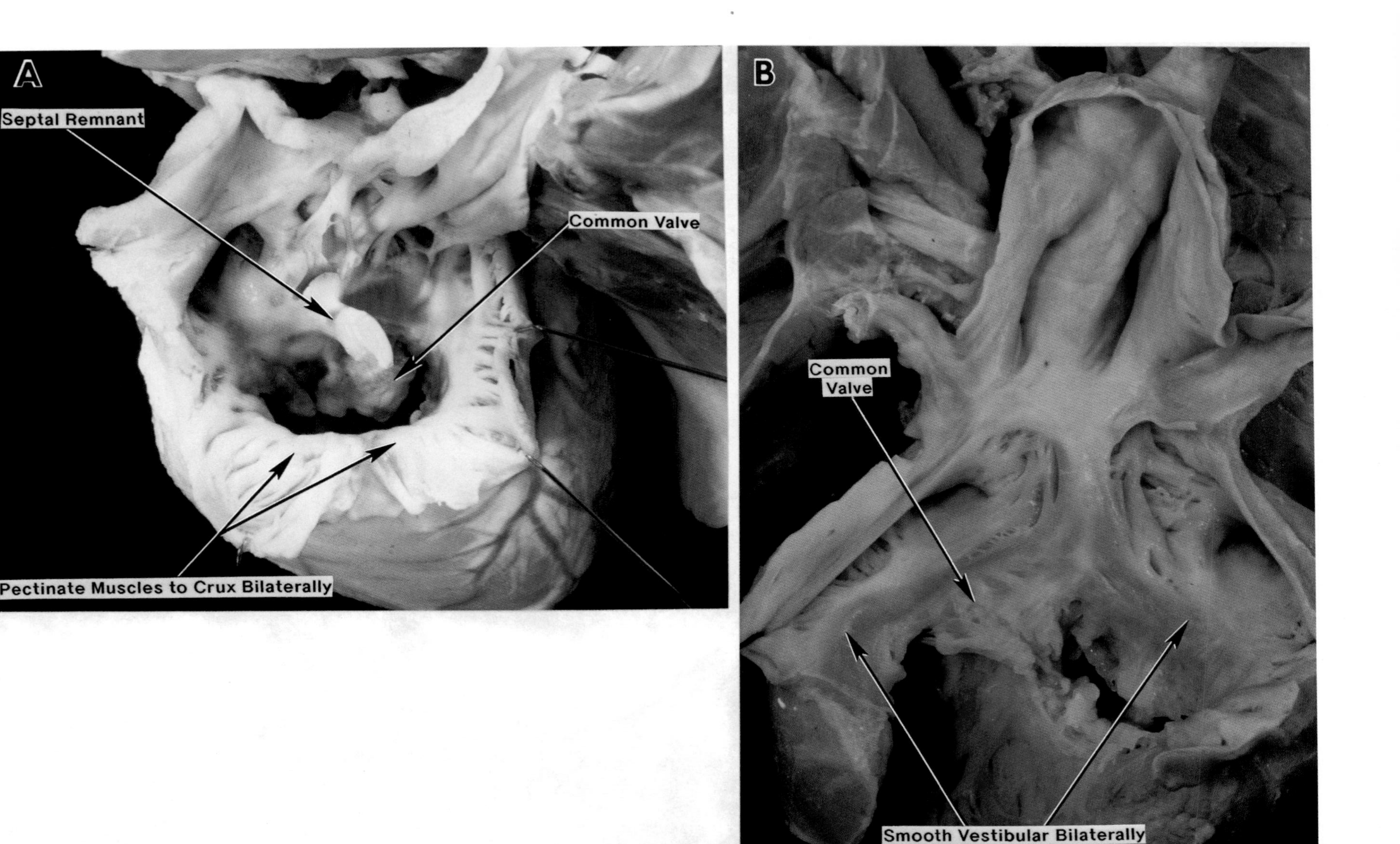

Figure 5 Internal morphology of the atriums in right and left isomerism, viewed by reflecting the atrial dome. Both hearts have a common atrioventricular valve, but it is the extent of the pectinate muscles which always serves to distinguish right and left isomerism. In right isomerism (A), pectinate muscles are present bilaterally. Note also the opening of the right superior caval vein relative to the terminal crest and the absence of the coronary sinus. In a heart with isomeric left atrial appendages (B) there are characteristic smooth posterior vestibules, confluent with the venous components, with superior caval veins on the right and left.

B. Left Isomerism

The atrial septum is usually much better formed in patients with isomeric left atrial appendages, although, on occasion, there can be a common chamber characterized by the bilateral smooth posterior vestibules (Fig. 5B). The pulmonary veins often connect bilaterally to the left-sided and right-sided extremities of the common atrial chamber but can drain to one or other atrium. The atrial septum tends to be intact, or else there is an atrioventricular septal defect present. Bilateral superior caval veins are found just as frequently as in right isomerism. Their connections to the atrial chambers are markedly different. Terminal crests are lacking, and the caval venous opening is usually sandwiched between the orifice of the appendage and the pulmonary venous connections, often marked by extensive folds. The coronary sinus may be present and can drain one superior caval vein as well as the venous return from the heart itself. The coronary sinus, nonetheless, is not universally present. In its absence, the termination of the coronary veins can be as abnormal as those in right isomerism (Uemura *et al.,* 1995). The major venous anomaly found in hearts with isomeric left appendages is interruption of the abdominal inferior caval vein, with continuation and drainage through the azygos system of veins.

C. Atrioventricular Connections in Isomerism

The majority of hearts with isomeric atrial appendages have a common atrioventricular junction. In right isomerism, this common junction is usually part of a double-inlet atrioventricular connection, to either a dominant left or a dominant right ventricle, but not infrequently to a solitary and indeterminate ventricle. Double-inlet ventricle can also occur through two separate atrioventricular valves, but this arrangement is much more rare, as is the absence of one of the atrioventricular connections. Right isomerism with a common atrioventricular junction can be part of a heart with biventricular connections. In this setting, because of the symmetry of the appendages, the morphological connections will be ambiguous irrespective of whether the right-sided atrium is connected to a morphologically right ventricle (right-hand topology) or to a morphologically left ventricle (left-hand topology). Thus, for full description of the atrioventricular connections it is necessary to specify both the type of isomerism present and the ventricular topology. Left isomerism is found more frequently with a common atrioventricular junction and separate right and left atrioventricular valves ("ostium primum" defect) or with separate right and left atrioventricular junctions guarded by tricuspid and mitral valves.

D. Ventriculoarterial Connections in Isomerism

Abnormalities of the ventricular outflow tracts are much more frequent in the setting of right isomerism but can also be found with left isomerism. It is rare to find concordant ventriculoarterial connections when there are right-sided appendages bilaterally. Usually, there is a double-outlet right ventricle, with bilateral infundibulums in most cases, or else the ventriculoarterial connections are discordant. In the majority of cases, these connections are then further complicated by pulmonary stenosis or atresia. When there is pulmonary atresia and right isomerism, almost always the pulmonary arteries are fed through the arterial duct, which may be present bilaterally. The ventricular outflow tracts are much more frequently normally arranged in the setting of isomeric left atrial appendages, although the concordant connections can be compromised by the presence of aortic stenosis, atresia, or coarctation.

The cardiac position, and the location of the cardiac apex and the aortic arch, can be abnormal in patients with isomerism of either the right or the left atrial appendages. The two subsets of isomerism, nonetheless, cannot be separated statistically on the basis of these features.

V. Implications of Morphology

We can draw a number of conclusions from the patterns of morphology observed in the human heart:

- The atriums are handed lateralized structures in which loss of lateralization leads to symmetry. This suggests that, during development, the atriums have pathways specifying "leftness" and "rightness."
- The right and left ventricles differ from each other but do not show true lateralization because in no circumstances is this difference lost to create symmetry. This suggests that the ventricles are patterned during development by some pathway not dependent on left–right differences.
- There are two topological arrangements of the ventricles, which can be independent of the lateralization of the atriums and can produce discordant atrioventricular connections. This suggests two separable developmental processes.
- In most cases of corrected transposition, the atriums are arranged appropriately, which suggests that atrial lateralization is more robust than the developmental control of ventricular topology.

- In most cases with discordant atrioventricular connections, the ventriculoarterial connections are also discordant, which can be interpreted as showing concordance between the atriums and outflow tracts. This may suggest that these two regions share some synteny in the lateralization process.
- In right isomerism, there is a common atrioventricular valve with a junction predominantly to one ventricle. The pulmonary venous portal is lost, the coronary sinus is not present, the atrial septum is poorly formed, and there is little extracardiac connection. These associations suggest that leftness is required for normal formation of the coronary sinus, atrial septum, pulmonary vein, and for expansion of the atrioventricular junction.

In the next section, we incorporate these conclusions with the current understanding of cardiac morphogenesis and gene expression in a model of early heart development.

VI. The Development of Heart Laterality

A. Morphology

It is striking how early morphological lateralization is established in the heart. By the stage of 25 somites [Embryonic (E) Day 10 in mouse, E12 in rat, and E25 in human], the left and right atriums and ventricles are already markedly different, the atrioventricular junction is placed to the left, and the outflow tract is on the right (Fig. 6). In contrast, the initial development of the heart, up to about 6 somites, is overtly symmetrical, resulting in the formation of a transiently straight tube made up of a caudal horseshoe (systemic venous tributaries and prospective atriums), central bulbar (ventricles), and rostral cylindrical (outflow tract) portions (Figs. 7A and 8). Dorsally, the heart is attached medially to the body in the horseshoe region, whereas the dorsal mesocardium more rostrally is breaking down. The first sign of lateralization is a slight swelling on the left side of the horseshoe, close to the junction with the central bulb, which we will call the caudal asymmetry (Fig. 7A; D. Bellomo and N. A. Brown, unpublished). We have followed the fate of this region after DiI injection and have shown that it contributes primarily to the atrioventricular junction (D. Bellomo and N. A. Brown, unpublished). Soon thereafter, this whole region on the left side appears to expand, whereas on the opposite side it invaginates to form a marked right-sided groove between the ventricular and atrial portions (Fig. 7B). As a result, the atrioventricular junction becomes left sided, the ventricular bulb rotates anticlockwise (viewed frontally) and the outflow is displaced to the right (Fig. 7C). Others have recently described these later changes as a leftward displacement of the cardiac axis in mouse (Biben and Harvey, 1997; see Chapter 7) or a leftward "jog" in zebrafish (see Chapter 6). The whole process described above has traditionally been called ventricular looping. This is misleading both because the initial

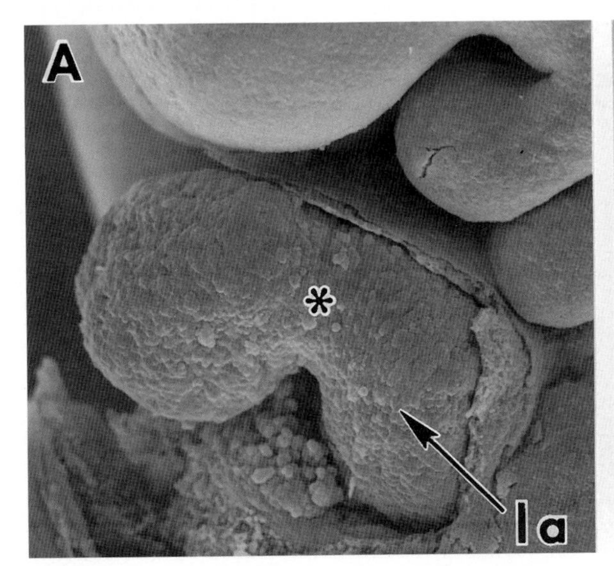

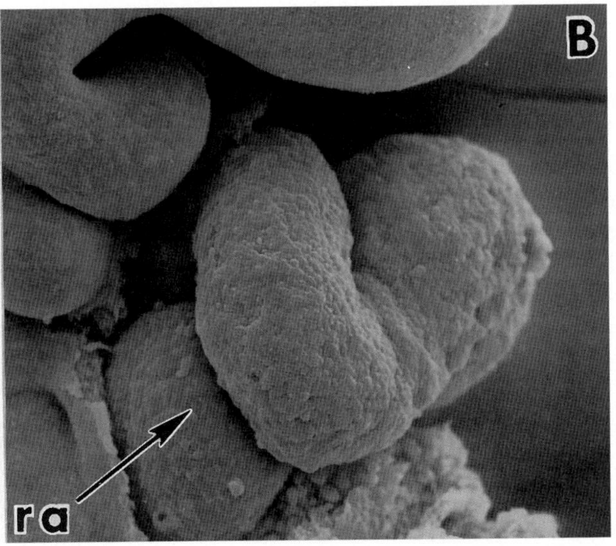

Figure 6 Scanning electron micrographs of the left (A) and right (B) sides of a 25-somite mouse embryo, with the pericardium removed. The atrial appendages are markedly different in the structure as soon as they are recognizable, with the left being flat-sided and directly related to the atrioventricular canal and the right being bulbous and lying beneath the outflow tract and outflow portion of the ventricular loop. la, left atrial appendage; ra, right atrial appendage; *, atrioventricular canal.

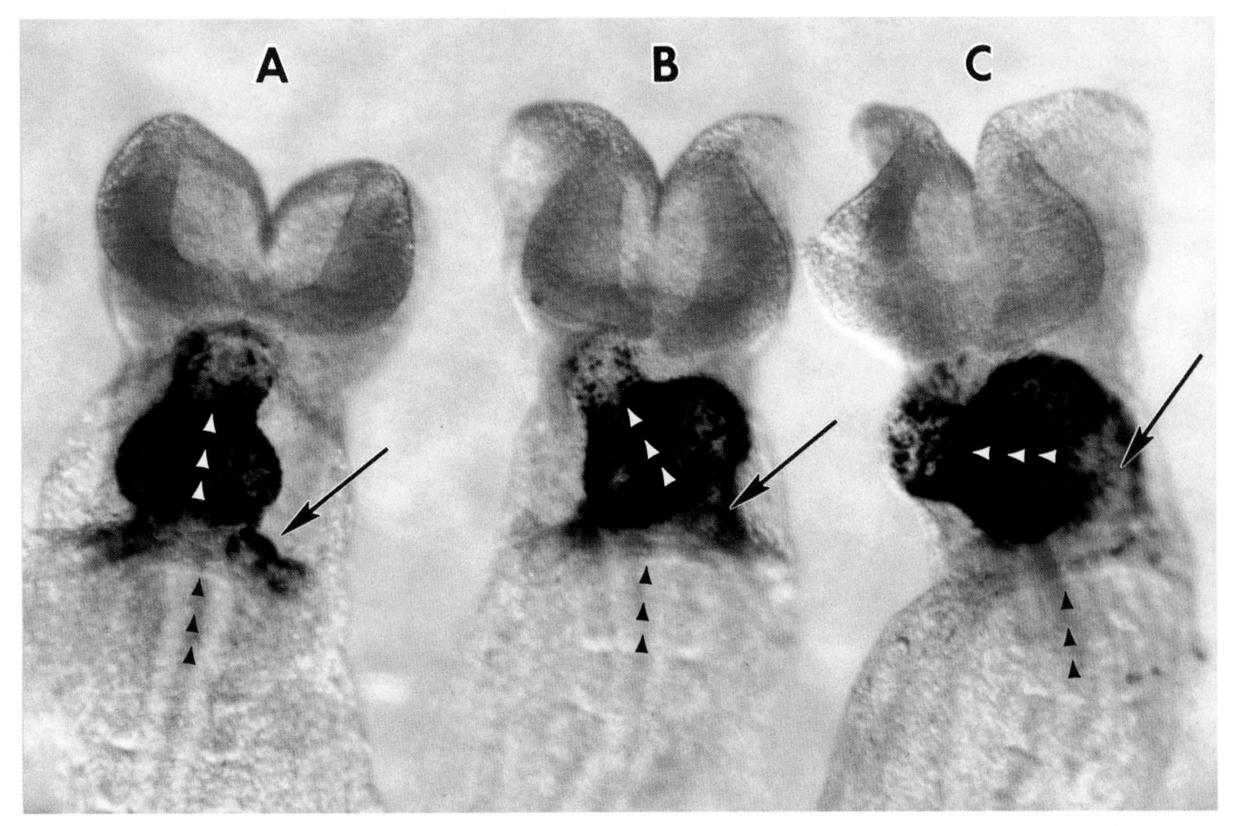

Figure 7 Frontal views of mouse embryos at 6- (A), 7- (B), and 8 (C)-somite stages. The mice carry an *nlacZ* reporter gene under the control of the 2E region of the regulatory sequences from the MLC3 promoter (see Chapter 19), used here just to visualize the extent of the myocardium. Note the initial caudal asymmetry at 6 somites (arrow in A), the first tilt of the cardiac midline at 7 somites (white arrowheads in B), and the left-sided expansion of the atrioventricular canal (arrows in B and C). The black arrowheads mark the embryonic midline (prospective floor plate). Note also the progressive closure of the cranial neural folds.

changes are not in the ventricles and the motion is not one of looping. Later in development the atrioventricular canal expands rightwards to connect the right atrium and ventricle.

On the basis of the developmental defects observed in human hearts, we suggest that the displacement of the putative atrioventricular canal, specification of atrial laterality, and rotation of the ventricular portion comprise a sequence of separable processes. Various combinations of defects in lateralization (reversal or randomization of handedness and loss of lateralization) in individual components can account for all the observed patterns of malformation (Fig. 9). However, it is not clear how these individual processes normally relate to one another. The formation of the caudal asymmetry is clearly the first observable event, and we suggest that this results in the leftward location of the atrioventricular canal, and that ventricular looping is a passive secondary change. It is not clear how the atrial appendages might acquire their handed identities. The right and left sides of the primary atrium are structurally very different. The left is flat and closely related to the atrioventricular canal while the right is outpouched (Fig. 6), apparently as a direct result of the left location of the junction with the ventricles. It is not known if this architectural disparity is necessary, or sufficient, to result in the different patterns of formation of the pectinate muscles. We have observed centrally located atrioventricular junctions in *iv/iv* embryos (D. Bellomo and N. A. Brown, unpublished data) with randomized ventricular topology and both left and right atrial isomerism (N. Brown and R. Anderson, unpublished data). In these circumstances, what determines whether the left or right atrial isomerism develops is unknown. Left isomerism is twice as common as right isomerism in late fetal *iv/iv* mice (Seo *et al.,* 1992), but right isomerism is much more common in NOD mice or following retinoic acid treatment, again for unknown reasons.

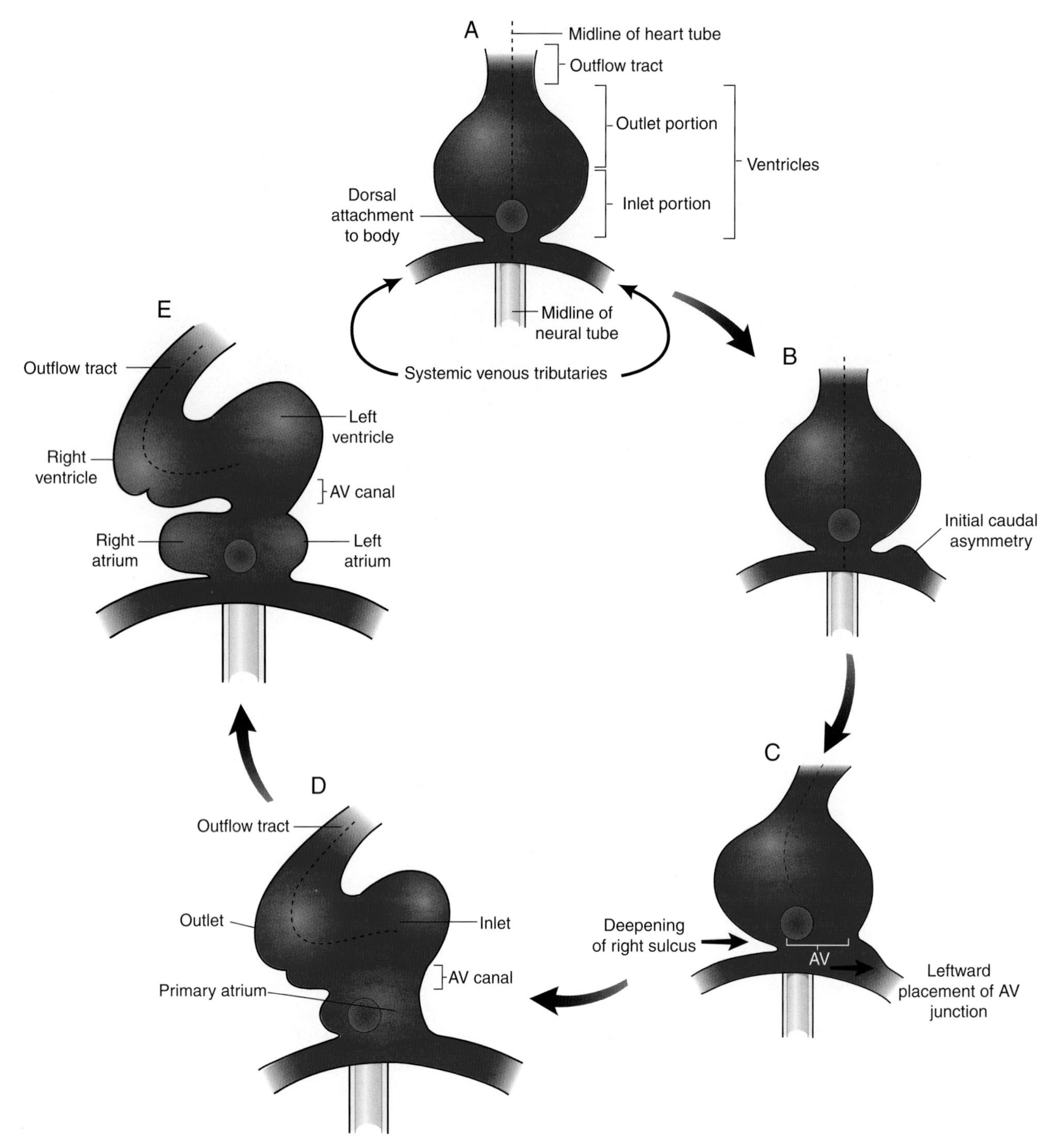

Figure 8 Diagrammatic representations of the establishment of morphological laterality in the mammalian heart over the period of about 5–12 somites. A, linear heart tube; B, initial caudal asymmetry; C, leftward placement of the atrioventricular junction, deepening of the right atrioventricular groove, and anticlockwise rotation of the ventricles; D, formation of a left-sided atrioventricular canal and development of the primary atrium; E, outpouching of the right atrial appendage. B–D are equivalent to the stages shown in Figs. 7A–7C.

B. Cell Properties

Although fate maps of the mammalian heart at this stage have yet to be made, it is likely that the left and right precardiac mesoderm areas make the left and right halves of the linear heart tube, at least approximately. We have injected DiI into the most rostral mesoderm, just lateral to the neural groove, at the mid- and late head-fold stages and cultured these embryos until after heart "looping." Figure 10 illustrates an embryo

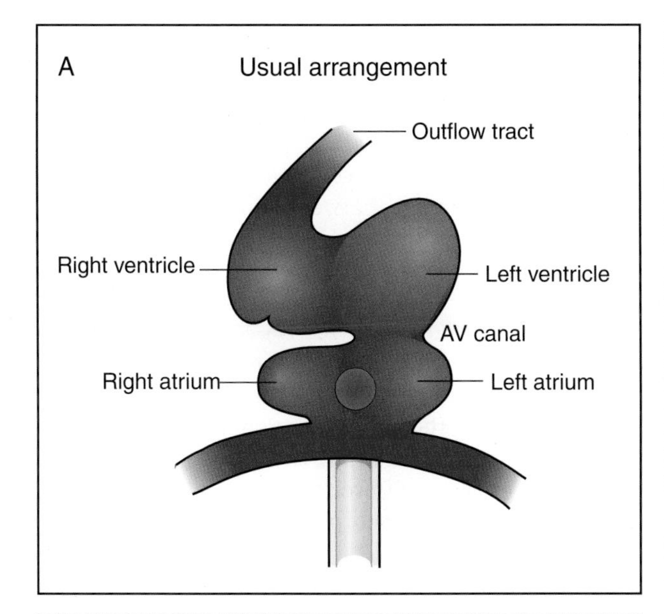

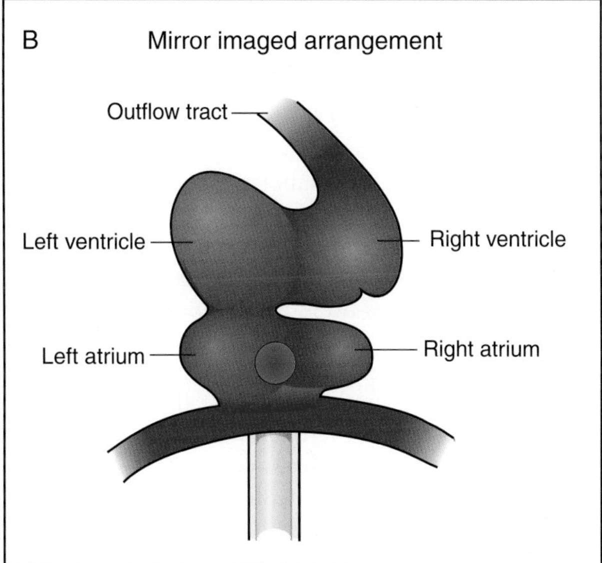

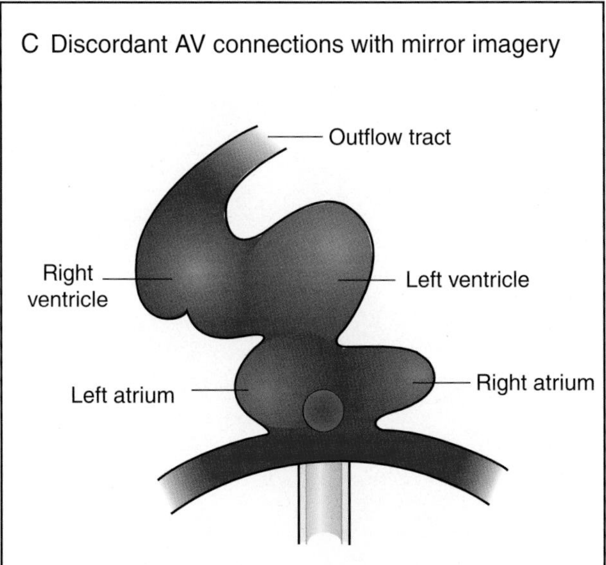

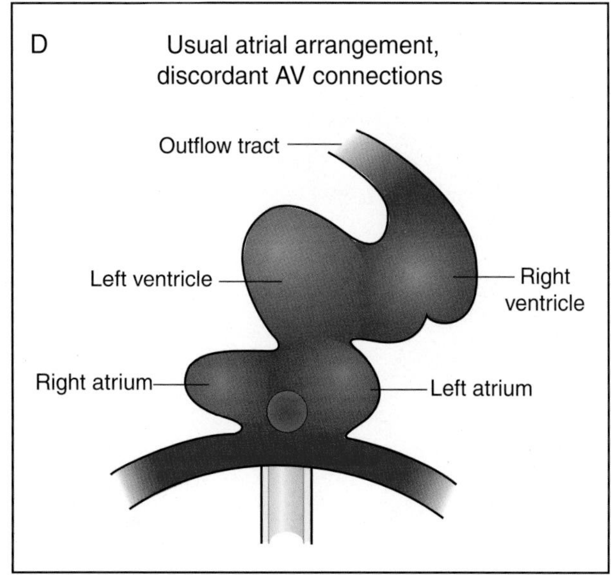

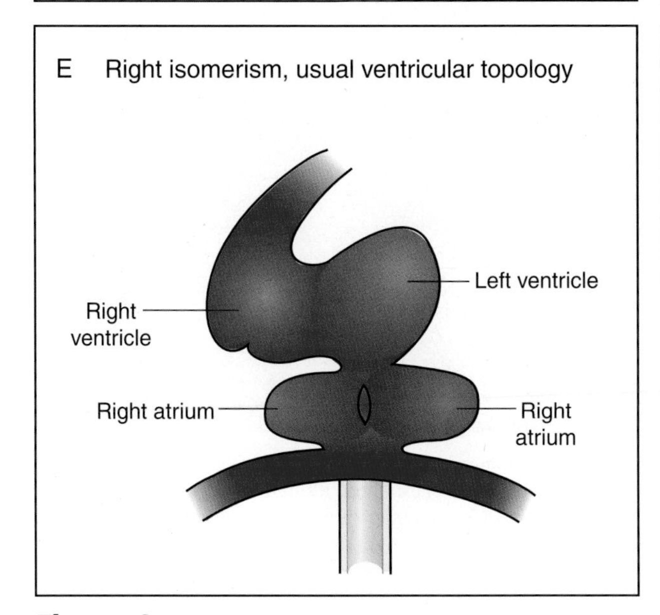

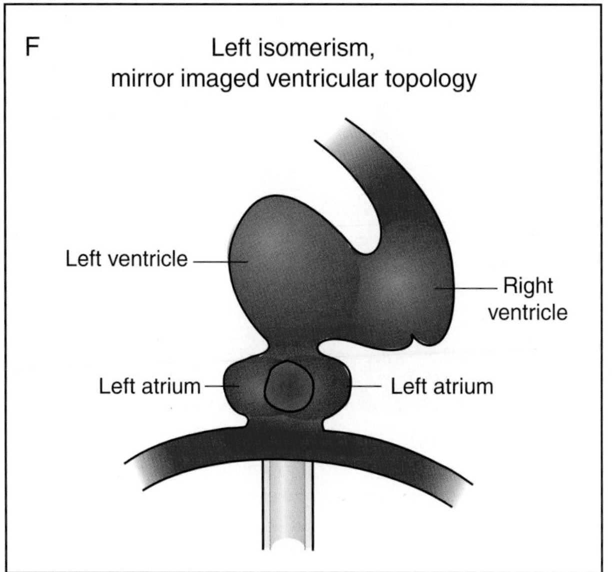

Figure 9 Diagrammatic representations of abnormalities in the establishment of morphological laterality in the mammalian heart. There are numerous different permutations of reversal of handedness and loss of lateralization in the atrial and ventricular portions. Not all are illustrated here, but rather a selection of those found and described here for human hearts are shown. Note that right and left isomerism can each be found with either right-hand or left-hand ventricular topology.

following labeling to the right mesoderm and shows labeled cells in the right side (outer curve) of the outflow tract and the outer portion of the ventricular loop. Injections of the left side label the inner curve (not shown).

It is obvious from this approximate cell fate, and from the sequence of morphological change, why the left and right ventricles are not lateralized. It is simply because they develop primarily from different rostrocaudal levels of the heart tube and not from the left and right sides. This does not suggest that the cell populations originating from the left and right sides contribute equally to the left and right ventricles or that these populations of cells are identical in all properties. Indeed, it is very likely that this is not the case. For example, the left and right sides of the tube contribute to the rostral and caudal aspects of the ventricular loop, respectively (Fig. 10). These regions will probably make different proportions of the two ventricles. As an example of a molecular difference between these two populations, *eHand* appears to provide an early molecular marker, with expression being enhanced on the left side of the ventricular bulb shortly after it looses its linear configuration (Biben and Harvey, 1997).

It is possible that the left and right horns of the horseshoe heart may not contribute equally to the atrioventricular or the atrial components. At a stage just prior to the formation of the caudal asymmetry, we have found a consistent difference in the pattern of cell proliferation (by BrdU incorporation) in the left and right horns (D. Bellomo and N. A. Brown, unpublished data). The significance is unknown, but the horns are close to

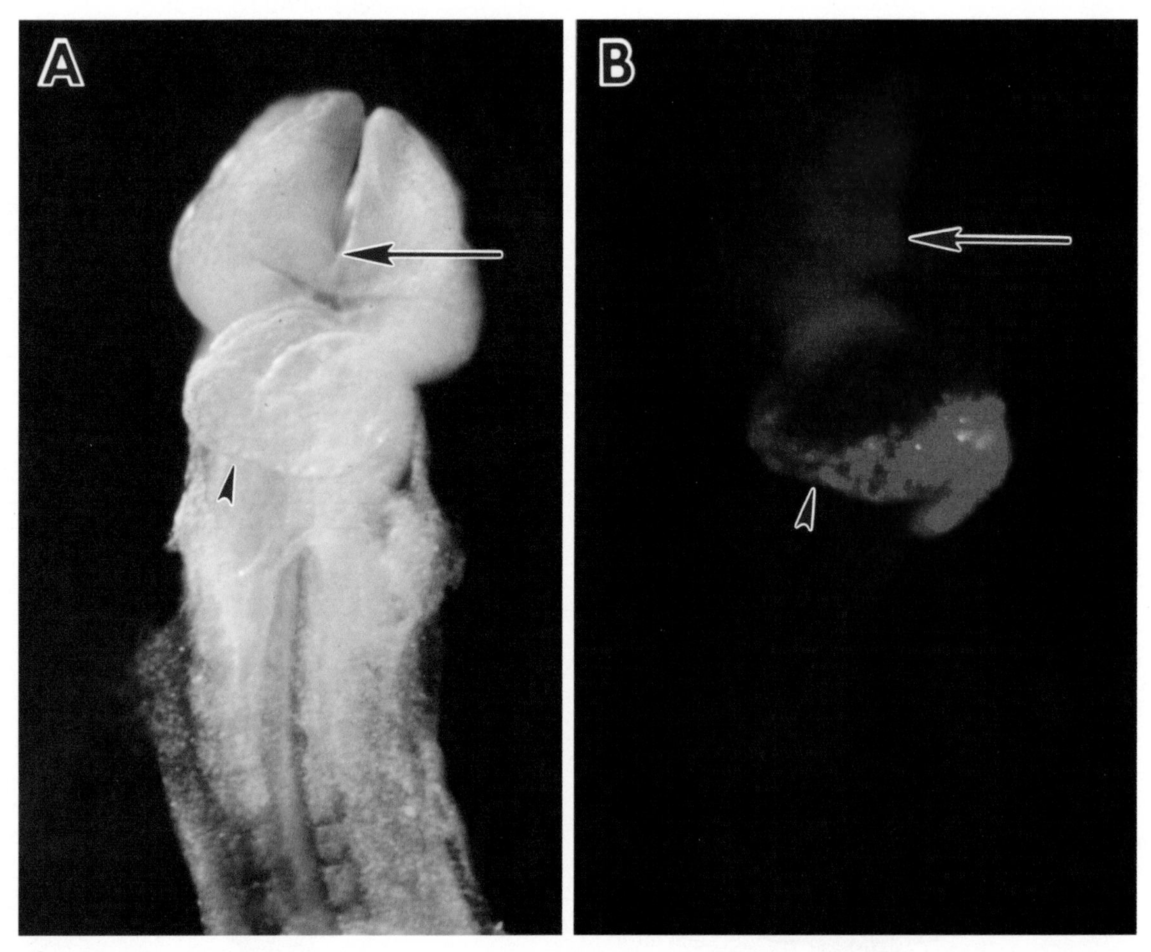

Figure 10 The fate of right-side precardiac mesoderm. This embryo was injected with a bolus of DiI, just lateral to the midline, at the most rostral extent of the mesoderm at the pre-somite mid-head-fold stage, then cultured for 24 hours to the 10 somite stage (compare the cranial folds with Fig. 7C). In the bright field view (A), "looping" is seen to be complete, and in the dark field view (B), fluorescent cells derived from the right mesoderm are seen in the right side (outer curve) of the outflow tract and in the caudal half (outer curve) of the ventricular loop. Very weak fluorescence is also visible in the right cranial fold mesoderm. For orientation, the interventricular groove (arrowhead) and rostral margin of the neuropore (arrow) are marked in both images.

regions with known molecular differences between left and right.

C. Molecular Signals

Recent advances in identifying the molecular components of left–right asymmetry are reviewed elsewhere (e.g. King and Brown, 1997). Of most relevance here is that transformation growth factor-β (TGF-β) family proteins (nodal and lefty-2) are expressed in the left lateral plate of mouse embryos with an anterior margin very close to, or within, splanchnic cardiac mesoderm at a stage just prior to formation of the caudal asymmetry. Also, in *iv/iv* and *inv/inv* mouse embryos, the lateralization of expression can be reversed or lost (giving either bilateral or no expression). However, the exact relationship between the expression of lefty and nodal and heart lateralization is not known because it has not been possible to correlate expression with cardiac morphology in individual embryos.

Regardless of the exact pathways, its seems very likely that a single early molecular asymmetrical signal is propagated to multiple signals (nodal, lefty, snail-related, etc.), with complex interrelationships, at later stages. Therefore, disruptions early in this pathway could lead to multiple effects later, and perhaps not necessarily to lateralization defects of the same type in different organs or even in different portions of the heart. For example, the mouse *iv* locus encodes a cytoplasmic dynein molecule (Supp *et al.,* 1997), which is likely to be an early component of the pathway. It is known to cause reversal and symmetry of both expression patterns and morphology. Further analysis of signals will be complex. Null mutants of nodal are lethal at early stages, and there seems to be an interaction with HNF3β in heterozygotes (Collignon *et al.,* 1996). There are multiple lefty genes, and a null mutation of *lefty-1* does not appear to affect ventricular topology but does result in abnormal expression of *lefty-2* and left isomerism of the lung (Meno *et al.,* 1998). This has been interpreted as showing *lefty-1* to be upstream of *lefty-2.*

The idea that a signal mediated by a TGF-β family protein is responsible for leftness is supported by the null mutation of the type IIB activin receptor (Oh and Li, 1997). Homozygous mutants have many of the features of right isomerism, including the arrangement of the atrial appendages, lung lobes, and bronchuses. Notably, none of the mice had mirror-image ventricular topology, supporting a separation of looping from atrial specification. Intriguingly, at least some of the hearts had an atrioventricular connection predominantly to the left side (see Fig. 3K in Oh and Li, 1997). This is comparable with right isomerism in humans and again suggests that expansion of the atrioventricular junction requires leftness.

References

Abernethy, J. (1793). Account of two instances of uncommon formations in the viscera of the human body. *Philos. Trans. R. Soc. London* **83,** 59–66.

Baker-Cohen, K. F. (1961). Visceral and vascular transpositions in fishes, and a comparison with similar anomalies in man. *Am. J. Anat.* **109,** 37–55.

Biben, C., and Harvey, R. (1997). Homeodomain factor Nkx2.5 controls left/right asymmetric expression of bHLH gene eHand during murine embryonic development. *Genes Dev.* **11,** 1357–1369.

Campbell, M., and Deuchar, D. C. (1967). Absent inferior vena cava, symmetrical liver, splenic agenesis and situs inversus and their embryology. *Br. Heart J.* **29,** 268–275.

Collignon, J., Varlet, I., and Robertson, E. J. (1996). Relationship between asymmetric nodal expression and the direction of embryonic turning. *Nature* (*London*) **381,** 155–158.

King, T., and Brown, N. A. (1997). Embryonic asymmetry: Left TGFβ at the right time? *Curr. Biol.* **7,** 212–215.

Landing, B. H., Lawrence, T. K., Payne, V. C., and Wells, T. R. (1971). Bronchial anatomy in syndromes with abnormal visceral situs, abnormal spleen and congenital heart disease. *Am. J. Cardiol.* **12,** 456–462.

Lev, M. (1954). Pathologic diagnosis of positional variations in cardiac chambers in congenital heart disease. *Lab. Invest.* **3,** 71–82.

Martin, G. (1826). Observation d'une déviation organique de l'estomac, d'une anomalie dans la situation et dans le configuration du coeur et des vaisseaux qui en partent ou qui s'y rendant. *Bull. Soc. Anat. Paris* **1,** 40–48.

Meno, C., Shimono, A., Saijoh, Y., Yashiro, K., Mochida, K., Ohishi, O., Noji, S., Kondoh, H., and Hamada, H. (1998). lefty-1 Is Required for Left-Right Determination as a Regulator of lefty-2 and nodal. *Cell* **94,** 287–297.

Moller, J. H., Nakib, A., Anderson, R. C., and Edwards, J. E. (1967). Congenital cardiac disease associated with polysplenia: A developmental complex of bilateral "left-sidedness." *Circulation* **36,** 789–799.

Oh, S. P., and Li, E. (1997). The signalling pathway mediated by the type IIB activin receptor controls axial patterning and lateral asymetry in the mouse. *Genes Dev.* **11,** 1812–1826.

Putschar, W. G. J., and Manion, W. C. (1956). Congenital absence of the spleen and associated anomalies. *Am. J. Clin. Pathol.* **26,** 429–470.

Seo, J. W., Brown, N. A., Ho, S. Y., and Anderson, R. H. (1992). Abnormal laterality and congenital cardiac anomalies: Relations of visceral and cardiac morphologies in the iv/iv mouse. *Circulation* **86,** 642–650.

Supp, D. M., Witte, D. P., Potter, S. S., and Brueckner, M. (1997). Mutation of an axonemal dynein in the left-right asymmetry in *inversus viscerum* mice. *Nature* (*London*) **389,** 963–966.

Uemura, H., Ho, S. Y., Devine, W. A., Kilpatrick, L. L., and Anderson, R. H. (1995). Atrial appendages and venoatrial connections in hearts from patients with visceral heterotaxy. *Ann. Thorac. Surg.* **60,** 561–569.

Van Mierop, L. H. S., and Wigglesworth, F. W. (1962). Isomerism of the cardiac atria in the asplenia syndrome. *Lab. Invest.* **11,** 1303–1315.

Van Mierop, L. H. S., Gessner, I. H., and Schiebler, G. L. (1972). Asplenia and polysplenia syndromes. *Birth Defects: Orig. Art. Ser.* 8(5), 36–44.

Van Praagh, R., and Vlad, P. (1978). Dextrocardia, mesocardia, and levocardia: The segmental approach to diagnosis in congenital heart disease. In "Heart Disease in Infancy and Childhood" (J. D. Keith, R. D. Rowe, and P. Vlad, eds.), 3rd ed., pp. 638–697. Macmillan, New York.

26

The Genetic Basis of Conotruncal Cardiac Defects: The Chromosome 22q11.2 Deletion

Beverly S. Emanuel,* Marcia L. Budarf,* and Peter J. Scambler†

*Division of Human Genetics and Molecular Biology, The Children's Hospital of Philadelphia, and the Department of Pediatrics, University of Pennsylvania School of Medicine, Philadelphia, Pennsylvania 19104

†Molecular Medicine Unit, Institute of Child Health, University College London, London WC1N 1EH, United Kingdom

I. History

A. DiGeorge Syndrome

Although cases linking congenital heart defect with hypoparathyroidism and thymic aplasia had been previously reported (Harington, 1829; Lobdell, 1959), DiGeorge (1965) was the first to propose that the concurrent absence of the thymus and parathyroids was not coincidental. He further suggested that loss of both organs might be due to a developmental perturbation of the third and fourth pharyngeal pouches (DiGeorge, 1965). While the initial descriptions of DiGeorge syndrome (DGS) focused on cell-mediated immune defects in association with hypoparathyroidism, it was soon realized that heart defects were present in the majority of cases (Freedom *et al.*, 1972; Conley *et al.*, 1979; Moerman *et al.*, 1980; Marmon *et al.*, 1984; Van Mierop and Kutsche, 1986). Thus, DGS was defined as consisting of the constellation of conotruncal cardiac anomalies with aplasia or hypoplasia of the thymus and parathyroid glands. Subsequent to these initial reports, the spectrum of clinical features associated with DGS has been expanded to include anomalies such as cleft palate, cleft lip, renal agenesis, neural tube defects, and hypospadias (Conley *et al.*, 1979). The main structures affected in DGS are indicated in Fig. 1.

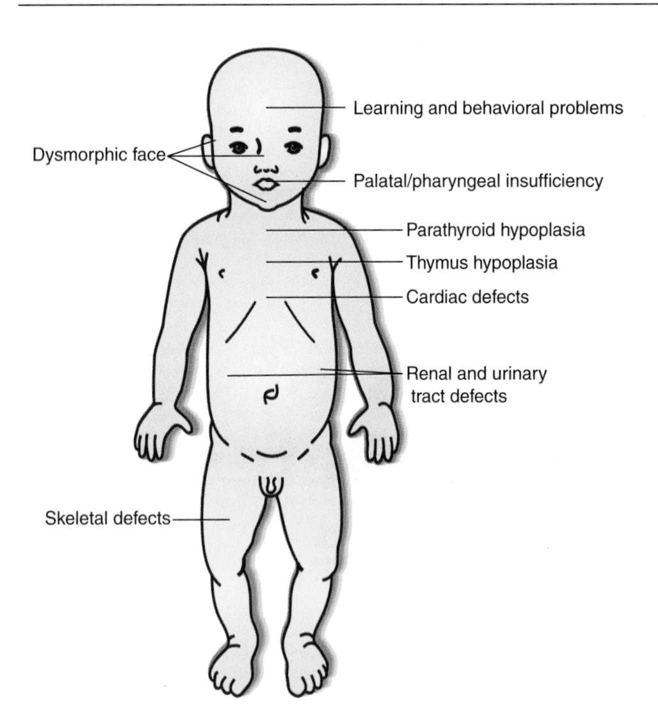

Figure 1 The main structures affected in DiGeorge, velocardiofacial, and conotruncal anomaly face syndromes.

The initial evidence linking DGS to 22q11 came from reports of DGS patients with unbalanced chromosomal translocations which resulted in the loss of the short arm and proximal long arm of chromosome 22 (22pter $\rightarrow$ q11) (de la Chapelle *et al.*, 1981; Kelley *et al.*, 1982; Greenberg *et al.*, 1984, 1988; Bowen *et al.*, 1986; Faed *et al.*, 1987). Further support for the etiologic involvement of chromosome 22 came from the identification of patients with cytogenetically visible interstitial deletions of 22q11 (Greenberg *et al.*, 1988; Mascarello *et al.*, 1989). Subsequently, molecular studies have demonstrated that approximately 90% of DGS patients with apparently normal chromosomes have microdeletions within 22q11 (Driscoll *et al.*, 1990, 1992a, 1993; Scambler *et al.*, 1991; Carey *et al.*, 1992). Historically, DGS had been thought to be a heterogeneous disorder with evidence for genetic, chromosomal, and teratogenic etiologies (Lammer and Opitz, 1986). However, molecular cytogenetic studies have demonstrated that deletions of chromosomal region 22q11 are the major cause of DGS. This has led to the suggestion that DGS is a disease caused by haploinsufficiency for a gene(s) which maps to the deleted region on 22q11.

B. Velocardiofacial and Conotruncal Anomaly Face Syndromes

The phenotypic description of velocardiofacial syndrome (VCFS) has been evolving over time. In 1968, Strong reported an association of right-sided aortic arch, mental deficiency, and facial dysmorphism segregating in two generations, suggesting autosomal dominant inheritance. Subsequently, Kinouchi reported an unusual facial appearance specifically seen in patients with conotruncal anomalies, and called it conotruncal anomaly face syndrome (CTAFS) (Kinouchi *et al.*, 1976). In 1978, Shprintzen *et al.*, independently reviewed 12 patients with overt or submucous clefts of the secondary palate, ventricular septal defects, typical craniofacial dysmorphia, and learning disabilities, and called the disorder VCFS. Subsequently, several investigators have emphasized the clinical variability of VCFS (Meinicke *et al.*, 1986; Lipson *et al.*, 1991). The presence of features commonly seen in DGS (neonatal hypocalcemia and decreased lymphoid tissue) in patients diagnosed with VCFS led to the proposal that these two disorders might share a common pathogenesis (Goldberg *et al.*, 1985; Stevens *et al.*, 1990).

The suggestion of a common etiology for DGS and VCFS prompted cytogenetic and molecular studies which demonstrated that VCFS is also associated with 22q11 microdeletions. Cytogenetic studies using high-resolution banding techniques detected interstitial deletions in 20% of VCFS patients (Driscoll *et al.*, 1992b). Furthermore, molecular studies with chromosome 22 probes, previously shown to be hemizygous in patients with DGS, have demonstrated that the majority of individuals with VCFS (>80%) have deletions of 22q11 (Driscoll *et al.*, 1992b, 1993; Kelly *et al.*, 1993). The deletion in VCFS encompasses the same region of 22q11 as in patients with DGS, lending additional support to the idea that these two disorders represent the same entity. As a result of advances in medical and surgical management of children with complex congenital cardiac disease, survival of children diagnosed with DGS has improved. These patients were previously subject to significant morbidity and mortality in the neonatal period. Thus, many of the features originally described in patients with VCFS have now been observed in the older children with a diagnosis of DGS.

CTAFS has striking similarities to DGS and VCFS and is characterized by the presence of conotruncal cardiac defects in association with a characteristic facial appearance. The facial features of patients with CTAFS include ocular hypertelorism, lateral displacement of the inner canthi, flat nasal bridge, small mouth, narrow palpebral fissures, bloated eyelids, and malformed ears. CTAFS has been well characterized in a population of Japanese patients (Kinouchi *et al.*, 1976; Kinouchi, 1980). The finding in CTAFS patients of deletions of the region of 22q11 deleted in DGS and VCFS confirmed an earlier suggestion that CTAFS and VCFS are the same entity (Shimizu *et al.*, 1984; Burn *et al.*, 1993; Matsuoka *et al.*, 1994).

C. Opitz G/BBB Syndrome

Opitz syndrome is characterized by hypospadias and hypertelorism (Opitz *et al.,* 1965). Patients may have other features, including cleft lip and palate, cardiac defects (coarctation of the aorta and atrial septal defect), laryngotracheoesophageal cleft, umbilical and inguinal hernias, cryptorchidism, imperforate anus, and facial dysmorphia such as telecanthus, prominent nasal bridge, or depressed nasal root with anteverted nares. Recent studies suggest that Opitz G/BBB syndrome is genetically heterogeneous with an autosomal-dominant form linked to chromosome 22 and an X-linked form at Xp22 caused by mutations in a RING finger gene *MID1* (Robin *et al.,* 1995a; Quaderi *et al.,* 1997). McDonald-McGinn *et al.* (1995) described four patients, including a father and son, with Opitz G/BBB syndrome and 22q11 deletions suggesting that in some cases deletions within 22q11 are causally related to Opitz syndrome. Additional studies are required to determine if the autosomal form of Opitz syndrome results from a mutation in a gene which maps within the deleted region or whether the Opitz G/BBB locus maps outside the DGS/VCFS chromosomal region.

D. Other Associations

Although the DiGeorge anomaly has been described in patients with other disorders, including Noonan syndrome and CHARGE association, the 22q11 deletion has not been found to be causally related (Robin *et al.,* 1995b; D. A. Driscoll, M. L. Budarf, and B. S. Emanuel, unpublished data). Wilson *et al.* (1993) described a single patient with both Noonan syndrome and DGS who had a 22q11 deletion. In a similar study, a patient with Noonan-like features was noted to have a 22q11 deletion (Robin *et al.,* 1995b). However, evaluation of an additional five patients failed to identify other deletions. Giannotti *et al.* (1994) suggested that Cayler cardiofacial syndrome should also be considered in the spectrum of disorders associated with the 22q11 deletion since a subset of these cardiac patients had deletions within 22q11. Retrospectively, examination of these patients demonstrated that they have features consistent with VCFS. Thus, the partial facial palsy which results in asymmetry of the lower lip should be considered one of the manifestations of the 22q11 deletion syndrome.

II. Incidence and Clinical Details

Based on clinical data, DGS had been estimated to occur in 1:20,000 live births when ascertained in a group of patients investigated for congenital heart disease (Muller *et al.,* 1988). Recently, using fluorescence *in situ* hybridization (FISH) for diagnosis, the incidence of the 22q11 deletion has been estimated to be approximately 1:4000 live births (Burn and Goodship, 1996). In addition, a French study of a regional birth defects registry reported figures ranging from 1/4500 in 1993 to 1/24000 in 1989 (Du Montcel *et al.,* 1996). However, since diagnosis of the deletion using FISH has only been available since 1993, it is likely that the figure cited by these authors for that year is more accurate. Nonetheless, all of the previous studies are likely to underestimate the true incidence because of ascertainment bias. A prospective study of consecutive, unselected live births will be required to determine the true population incidence of the deletion.

Between 10 and 25% of deletions are inherited, with the majority of deletions being inherited from the mother (Leana-Cox *et al.,* 1996; Ryan *et al.,* 1997; D. A. Driscoll, M. L. Budarf, and B. S. Emanuel, unpublished data). It is likely that this reflects a psychosocial rather than a biological phenomenon since the disorder occurs with equal frequency in males and females. Data from several studies differ with regard to the parent of origin of *de novo* deletions, with one study reporting that the majority are paternal in origin (Ryan *et al.,* 1997) and two others showing a maternal excess (Demczuk *et al.,* 1995b; Seaver *et al.,* 1994). When data for patients from these three studies is combined, there are 33 maternal and 30 paternal deletions. Data from a fourth cohort agrees with the combined data because they demonstrate no bias regarding parental origin (D. A. Driscoll, M. L. Budarf, and B. S. Emanuel, unpublished data). Furthermore, there is no evidence for major imprinting effects in DGS/VCFS. This is in agreement with the observation that both maternal and paternal uniparental disomy for chromosome 22 have been observed without obvious phenotypic effects (Schinzel *et al.,* 1994; Mimy *et al.,* 1995).

Within families in which the deletion is segregating, the disorder frequently is more severe in the younger generations. However, deletion size appears to be stable and the apparent anticipation is probably a consequence of ascertainment bias. Severely affected individuals will not reproduce and those with the milder phenotypes are more likely to go undiagnosed. Phenotypes can vary widely within families segregating a 22q11 deletion such that different, but often overlapping, abnormalities are observed (Wilson *et al.,* 1991, 1992; McLean *et al.,* 1993; Driscoll *et al.,* 1993; Leana-Cox *et al.,* 1996; Cuneo *et al.,* 1997). Indeed, there are reports of families in which a child with a normal heart has inherited a deletion from a parent with a cardiac defect (Holder *et al.,* 1993) and a pair of monozygotic twins discordant for phenotype (Goodship *et al.,* 1995). It is therefore likely that effects of genetic background, stochastic factors, and/or environmental influences have a role in determining the final phenotype.

A. Cardiovascular Defects

Approximately 75% of the cardiac defects in DGS consist of persistent truncus arteriosus (PTA), interrupted aortic arch type B (IAA), and tetralogy of Fallot (TOF) (Van Mierop and Kutsche, 1986). The most common types of CHD in VCFS include ventricular septal defects (VSD), TOF, and right aortic arch (Young *et al.*, 1980). The abnormalities seen in association with the 22q11 deletion can be classified based on failure of specific critical events during cardiac morphogenesis. For example, PTA is a reflection of failure of proper aorticopulmonary septation. Following convergence of the inflow and outflow tracts, if the outflow tract remains in an incorrect orientation with respect to the ventricles, defective "wedging" of the great vessels occurs which leads to a VSD (Kirby and Waldo, 1995). Such malalignment may also result in TOF, which is characterized by overriding aorta, VSD, pulmonary stenosis, and right ventricular hypertrophy. Defective growth and/or remodeling of the aortic arch artery system is likely to result in IAA, coarctation of the aorta, and aberrant origin of subclavian or pulmonary arteries. However, failure of neural crest cell migration or function appears to be a unifying feature for these anomalies as evidenced by the animal models.

Within the population of children with congenital heart disease many of these defects are relatively rare, whereas in patients with DGS they are quite common (Van Mierop and Kutsche, 1986). Thus, the detection of IAA, TOF, or PTA should alert the cardiologist to the possibility of a 22q11 deletion, particularly if any other relevant dysmorphic feature or anomaly is present. The high frequency of outflow tract abnormalities in DGS prompted several investigators to examine the possibility that nonsyndromic heart defects might also be caused by deletions of 22q11. Initially, the results were encouraging, with analyses of both familial (Wilson *et al.*, 1992) and sporadic (Goldmuntz *et al.*, 1993) cases suggesting a high incidence of 22q11 deletion associated with a range of isolated congenital heart defects. However, subsequent studies and further evaluation of the patients has led to the conclusion that no heart defect associated with a deletion of 22q11 is truly nonsyndromic, in the sense that these patients usually exhibit mild facial dysmorphism, sometimes with additional features (Amati *et al.*, 1995; McDonald-McGinn *et al.*, 1997a; Goldmuntz and Emanuel, 1997).

To determine the frequency of 22q11 deletions in a large sample of cardiac patients, 260 individuals with conotruncal defects were screened for the presence of the 22q11 deletion (Goldmuntz *et al.*, 1998). Deletions were found in 14/26 (53.6%) with IAA, 10/29 (34.5%) with PTA, and 20/131 (15.3%) with TOF. Of interest, deletions were observed in 2/5 (40%) patients with posterior malalignment VSD. Furthermore, it has been determined that the frequency of 22q11 deletions is higher in those patients with an aortic arch or vessel anomaly compared to patients with a normal left arch. Only 1/20 (5%) patients with double-outlet right ventricle (DORV) was found to have the deletion. The deletion-positive DORV patient had a subaortic VSD, right aortic arch, and isolated left pulmonary artery which is reminiscent of TOF with multiple arch anomalies. In a smaller study (Takahashi *et al.*, 1995), 0/8 patients with DORV had a deletion. Because there is significant anatomic variability within the category of DORV, it is possible that a small subgroup of these patients may have a deletion. Thus, a larger study of patients with DORV and detailed anatomic description seems to be warranted. In the study by Goldmuntz and colleagues (1998), none of 39 patients with d-transposition of the great arteries (TGA) were positive for the deletion. These results are similar to those of a smaller study that also found no deletions in 16 patients with d-TGA (Takahashi *et al.*, 1995). Other reports have noted d-TGA in association with unbalanced translocations (Kelley *et al.*, 1982) or interstitial 22q11 deletions [4/32 = 12.5%] (Melchionda *et al.*, 1995). These studies taken together suggest that the number of deletion-positive TGA patients is likely to be relatively small.

To better define the relative incidence and type of congenital malformations seen in association with this chromosomal abnormality, a European consortium collated information from 558 cases of 22q11 deletion (Ryan *et al.*, 1997). The information available regarding cardiac status for 545 patients is shown diagrammatically in Fig. 2. The data indicate a high frequency of cardiac defects in subjects with the 22q11 deletion, despite the fact that 25% of patients had no or an insignificant abnormality as determined by clinical examination and/or echocardiography. Forty-four deaths were recorded in this series of deleted patients, of which 43 were secondary to the complications of congenital heart disease. Death usually occurred within the first several months of life (55% at 1 month and 86% by 6 months). Thus, congenital heart defect is a major cause of morbidity and mortality in this group of patients. An additional 161 patients with the deletion were assessed at one center in the United States and 22% had normal cardiac examination (McDonald-McGinn *et al.*, 1997b).

B. Other Defects

Over 70 different congenital malformations have been reported in association with DGS/VCFS (Goldberg *et al.*, 1993). Hypocalcemia, present in 60% of cases, is usually present in the neonatal period (Ryan *et al.*, 1997),

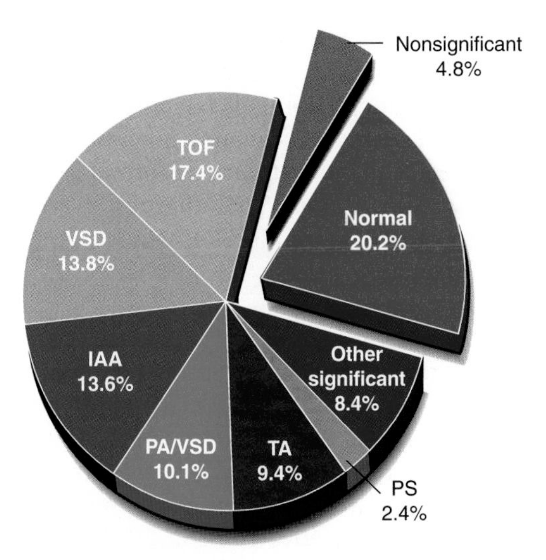

Figure 2 Relative incidence of congenital heart defect in the 22q11.2 deletion syndrome. TOF, tetralogy of Fallot; VSD, ventricular septal defect; IAA, interrupted aortic arch; PA, pulmonic atresia; TA, truncus arteriosus; PS, pulmonic stenosis.

although it can manifest as latent or frank hypocalcemic hypoparathyroidism in adolescence or young adulthood (Cuneo *et al.,* 1997). While many children are said to have recurrent infections, severe immunodeficiency with the 22q11 deletion is rare (Junker and Driscoll, 1995). Nonetheless, 80% of patients are immunocompromised to some extent (Sullivan *et al.,* 1998) and there is an increased prevalence of IgA deficiency (Smith *et al.,* 1998), confirming the presence of significant humoral deficits in this patient population. Furthermore, there is evidence that the prevalence of polyarthritis in patients with the deletion is markedly increased over the prevalence of polyarticular juvenile rheumatoid arthritis in the general population (Sullivan *et al.,* 1997). Genitourinary problems are more common than initially realized (Devriendt *et al.,* 1996), and abdominal ultrasound should be included in the assessment of 22q11 cases. One case of Potter sequence (intrauterine death due to bilateral renal insufficiency) with ovouterine agenesis and deletion 22q11 has been recorded (Devriendt *et al.,* 1997), and kidney defect or dysfunction was found in 36% of cases in the European study (Ryan *et al.,* 1997). Undescended testes were noted in 8% of males (Ryan *et al.,* 1997).

Facial dysmorphism is said to be typical, but it is clear that these features are not always apparent in infancy and become more pronounced as the child grows. The craniofacial features include a shallow midface, prominent nasal root, narrow alae nasi with prominent or dimpled nasal tip (Shprintzen *et al.,* 1978; Goldberg *et al.,* 1993; Gripp *et al.,* 1997), and hooded upper eyelids. In addition, protruberant ears and structural anomalies of the ear have been seen in many patients. However, it has been noted that there is a paucity of typical facial characteristics in African American patients with the deletion (McDonald-McGinn *et al.,* 1996). Other structural defects include tracheoesophageal fistulae, laryngeal clefts, and skeletal malformations (Cormier-Daire *et al.,* 1995; Ryan *et al.,* 1997; Ming *et al.,* 1997). Additional clinical problems result from velopharyngeal insufficiency (VPI), which may be accompanied by cleft palate. In a cohort of 96 deleted patients whose palatal status was evaluated, 11% had a normal palate, 30% had VPI, 25% were suspected of VPI but required follow-up, 14% had a submucosal cleft palate, 13% had overt cleft palate, 5% had a bifid uvula, and 2% had a cleft lip and palate (McDonald-McGinn *et al.,* 1997b). The palatal defects produce a nasal quality to the voice but lead to more practical problems when they result in feeding difficulties and speech delay, both of which require medical attention.

Learning and/or behavioral problems are common, although cognitive development is not as severely impaired as once thought. In the European study, 62% of patients were recorded as being within the normal range or having only mildly retarded learning (18% moderate or severe, and 20% no specific data) (Ryan *et al.,* 1997). Furthermore, psychoeducational evaluation of patients with the deletion revealed that they manifest a nonverbal learning disability, attaining higher verbal than performance IQ scores (Moss *et al.,* 1995, 1998; Swillen *et al.,* 1997). Behavioral problems and psychiatric manifestations have been increasingly recognized as part of the 22q11 deletion phenotype. This is an active area of research with schizophrenia (Shprintzen *et al.,* 1992; Karayiorgou *et al.,* 1995) and bipolar spectrum disorders (Papolos *et al.,* 1996; Lachman *et al.,* 1996), the main presenting psychoses. Brain malformations (arhinencephaly, absent corpus collosum, and cerebellar hypoplasia) have been reported, but they are rare and their significance is unclear (Mitnick *et al.,* 1994; Lynch *et al.,* 1995).

III. Development of Structures Affected in DGS/VCFS

DGS has occasionally been referred to as "III–IV pharyngeal pouch syndrome" (Lischner *et al.,* 1969; Robinson, 1975), to reflect the fact that the thymus is a derivative of the third pharyngeal pouch and the parathyroids are derivatives of the third and fourth pharyngeal pouches. However, this designation did not adequately take into account the facial and cardiovascular malformations seen so commonly in the 22q11 deletion syndrome. Results emerging from studies of early vertebrate development led to the realization that the main structures affected in DGS were reliant on a contribution from the rostral neural crest for their normal development. A detailed account of the role of the neural crest in cardiovascular development is given elsewhere in this book (see Chapter 11), so only a brief outline is given here in the context of DGS/VCFS.

Studies in the chick/quail chimera system and the mouse embryo have demonstrated that the neural crest cells migrate to the developing face and branchial arches and contribute to the formation of the thymus, parathyroids, thyroid, branchial arch artery system and the outflow tract of the heart (Fig. 3). The population of neural crest cells derived from hindbrain rhombomeres 6–8 (midotic placode to somite 3) is termed the cardiac crest as it continues its migration from the pharyngeal arches into the outflow tract. In chick embryos, ablation of this cardiac crest, or its replacement with presumptive crest from a different axial level, results in a series of cardiac defects reminiscent of those seen in, and relatively specific for, DGS: PTA, TOF, and IAA. In addition, abnormal looping and convergence in this system can produce DORV and double-inlet left ventricle (Kirby and Waldo, 1995) which, interestingly, are rarely seen in patients with the deletion (Goldmuntz *et al.,* 1998). These experimental manipulations also commonly result in noncardiac malformations typical of DGS, including thymic, parathyroid, and thyroid hypo/aplasia (Kirby *et al.,* 1983; Bockman and Kirby, 1984; Bockman *et al.,* 1987; Nishibatake *et al.,* 1987; Kirby, 1989).

Evaluation of mammalian models also indicates a likely role for the neural crest in the development of structures affected in the 22q11 deletion disorders. Expression studies suggest that *Hox* genes provide positional identity to cells within the hindbrain by the combinatorial expression of different *Hox* family members (Hunt and Krumlauf, 1991; Krumlauf, 1994). Disruption of *Hox* gene expression can cause defects overlapping with those of DGS/VCFS, most strikingly seen in the targeted disruption of *HoxA3* in the mouse (Chisaka and Capecchi, 1991; Manley and Capecchi, 1995). Mice homozygous for this mutation have parathyroid and thymic gland aplasia. They also have a variety of heart

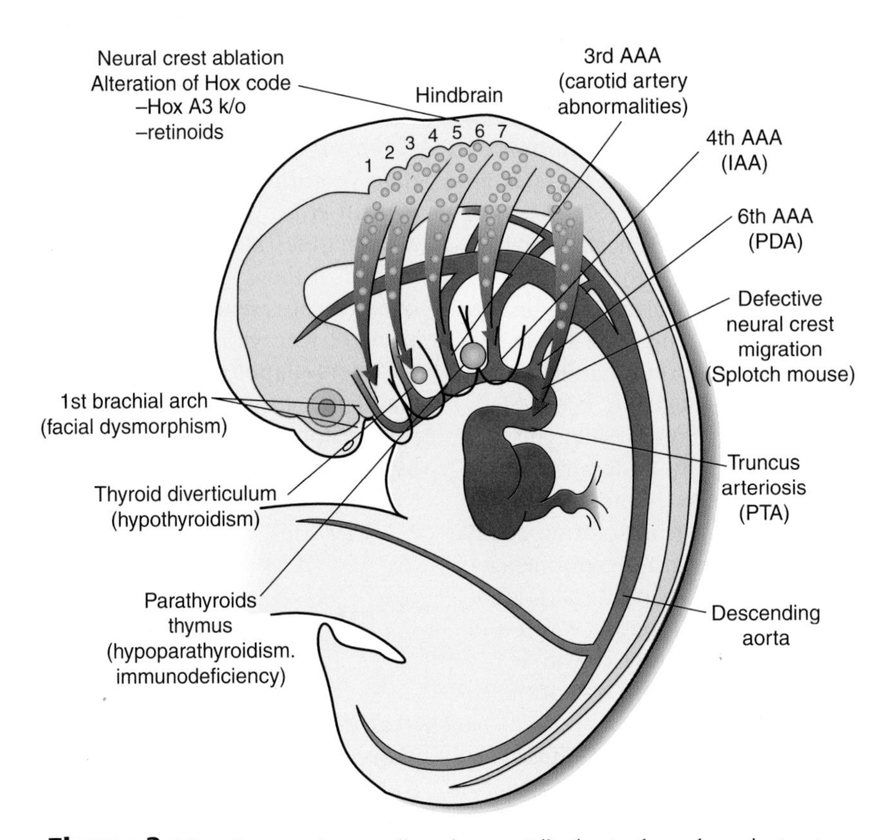

Figure 3 Migrating neural crest cells make a contribution to the embryonic structures affected in DiGeorge/velocardiofacial syndrome. The figure represents a human embryo at 4–6 weeks of gestation. The migration of neural crest cells to the branchial arch/pharyngeal pouch system and outflow tract is highlighted. Malformations associated with disruption of this process are indicated and, as explained in the text, these overlap substantially with the anomalies seen in the 22q11 deletion disorders.

defects, of which none are typical of DGS. However, since the *Hox* gene clusters do not map to chromosome 22, they may function upstream or downstream of the gene(s) in 22q11.2. Exposure of early human and mouse embryos to high concentrations of retinoids also produces a phenocopy of DGS (Lammer *et al.,* 1986). Although retinoids have many modes of action, one result of retinoid exposure is an alteration (posteriorization) of *Hox* gene expression within the hindbrain (Marshall *et al.,* 1992).

Another mouse mutant which demonstrates a phenotype similar to DGS/VCFS is the *Splotch* mouse. Several *Splotch* alleles have been described of which five have been characterized on a molecular level. The mutations involve the *Pax3* gene which maps to human chromosome 2 and is the gene mutated in Waardenburg syndrome (Strachan and Read, 1994). Embryos homozygous for the *Splotch*^{2H} allele die at around 13 or 14 days postcoitus with a constellation of defects, including PTA, VSD, and abnormal thymus, thyroid, and parathyroid glands, in addition to severe neural tube defects (Epstein, 1996). Using molecular markers, it has recently been shown that in *Splotch*^{2H} homozygotes the neural crest cells migrate normally from the occipital neural tube, but there is a deficiency or absence of migration into the outflow tract (Conway *et al.,* 1997) (Fig. 3). Taken together, these observations have led to the proposal that DGS/VCFS will also be a result of a defective neural crest contribution. While this hypothesis certainly has its attractions, particularly as far as the congenital heart defects are concerned, it cannot entirely explain the full spectrum of abnormalities seen (e.g., behavioral problems, genitourinary malformation, and skeletal malformation).

IV. Mapping the 22q11.2 Deletions

The aims of mapping the DGS/VCFS region of 22q11 have been threefold. First, it was necessary to determine whether there was any correlation between the phenotype and either the extent or position of the deletion. Second, by comparing deletions in different individuals it was hoped that the shortest region of deletion overlap (SRO) would establish the location of the gene (or genes) whose haploinsufficiency results in DGS/VCFS. Third, the isolation of mapping resources within this region would provide a platform for the eventual cloning and sequencing of the entire region, together with the identification of genes encoded within this sequence (see Section V). It was widely anticipated that these studies would eventually lead to the isolation of a gene carrying mutations in patients with DGS, VCFS, or CTAF who have no chromosomal rearrangement. Implicit was the assumption that the main features of DGS and related disorders are the consequence of haploinsufficiency of a single major gene.

Attempts to correlate the phenotype with the nature of the deletions have proved largely disappointing. The vast majority of patients have a large interstitial deletion of 22q11 encompassing 2 or 3Mb of sequence (Scambler *et al.,* 1991; Driscoll *et al.,* 1992a; Lindsay *et al.,* 1993; 1995; Morrow *et al.,* 1995; Driscoll and Emanuel, 1996; Gong *et al.,* 1996)—the typically deleted region (TDR). The proximal breakpoints of these deletions appear to be more clustered than the distal breakpoints which may be spread over approximately 1 Mb. The proximal breakpoint cluster (PBC) is shown in Fig. 4 as the proximal boundary of the TDR. However, individuals with the larger deletions seem no more severely affected than those with the smaller ones. Similarly, the phenotype of patients with loss of 22pter → q11 as a result of an unbalanced translocation (see Section I,A) is indistinguishable from that seen with deletions which encompass the TDR. These patients are missing additional chromosomal material proximal to the PBC.

Efforts have been directed toward determining the minimal DGS/VCFS critical region or the SRO in affected patients. Using breakpoint mapping data from a small number of balanced translocations, unbalanced translocations, and atypical deletions in patients with phenotypic features of DGS/VCFS, several different SROs have been reported. Figure 4 shows a schematic summarizing the results of the mapping and helps illustrate some of the major findings that have emerged from the efforts of several laboratories. However, the small patient numbers involved has meant that laboratories have analyzed various combinations of patients resulting in differences in the nomenclature. Furthermore, due to the variability of the DGS/VCFS phenotype, caution should be applied when interpreting data based on single patients. The summary presented here will attempt to resolve some of these issues.

The term DGS critical region (or DGCR) was initially used to describe the largest deletion interval (Kaplan and Emanuel, 1989) and was based on unbalanced translocations associated with DGS (see Section I,A). The first substantial refinement of this interval came with the mapping of the 22q11 break points of the unbalanced translocations of several cell lines which narrow the distal end of the DGCR (GM00980 and GM05878) (Halford *et al.,* 1993b). Further refinement of the position of the distal end of the DGCR was achieved with the mapping of another breakpoint from a patient with a 15;22 unbalanced translocation (Jaquez *et al.,* 1997). This interval has been referred to as the minimal DiGeorge critical region (MDGCR) (Budarf *et al.,* 1995a; Gong *et al.,* 1996; Gottlieb *et al.,* 1997). Re-

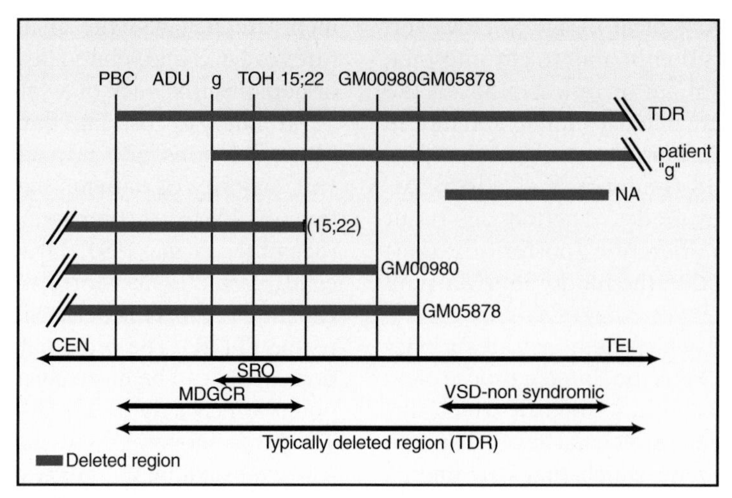

Figure 4 Deletions and rearrangements in DiGeorge/velocardiofacial syndrome. The blue bars indicate the relative extent of hemizygosity seen in different patients with deletions within 22q11. The vertical lines show the positions of important chromosome breakpoints discussed in the text. Patient/breakpoint names used in the text are given on the right of deletion bars and/or above breakpoint indicators. The exception is PBC, which indicates the proximal breakpoint cluster. The double-headed black arrow indicates the orientation of the region with respect to 22q. The red double-headed arrows delineate the various DiGeorge region intervals that have been described in the literature and demonstrate how they correlate with the chromosomal rearrangements. SRO, shortest region of deletion overlap; MDGCR, minimally deleted DiGeorge critical region; ADU-980, region between the ADU and GM00980 translocation break points; TDR, typically deleted region in the 22q11.2 deletion syndrome (see text for discussion).

finement of the proximal border was accomplished by identification of an atypical interstitial deletion in patient G. This patient's proximal breakpoint is located approximately 100 kb distal to the ADU t(2;22) breakpoint (Levy *et al.,* 1995). By combining the breakpoints from the t(15;22) and patient G, it is possible to establish a smaller SRO (Gottlieb *et al.,* 1997). However, this has not led to the identification of a gene that is altered in the nondeletion cases of DGS/VCFS.

The situation is further complicated by consideration of two balanced translocation patients, ADU and TOH. ADU has a DGS phenotype in association with a 2;22 translocation (Augusseau *et al.,* 1986), with the chromosome 22 breakpoint occurring within the proximal region of the DGCR (Budarf *et al.,* 1995a; Demczuk *et al.,* 1995a; Wadey *et al.,* 1995). Naturally, this breakpoint was the focus of intensive positional cloning efforts which have not, so far, identified a gene unequivocally responsible for DGS and the related conditions. The second patient, TOH, has features of VCFS in association with a balanced 21;22 translocation mapping to the SRO. TOH's translocation disrupts the *CLTCL* gene (Holmes *et al.,* 1997). There are two possible interpretations of these results. The first is that TOH's partial DGS/VCFS phenotype indicates that haploinsufficiency for more than one gene in the MDGCR may be etiologic for DGS/VCFS. Alternatively, these data could be explained by postulating that the ADU and TOH breakpoints exert position effects on a gene or genes within the TDR (Fig. 4). In other systems, position effects have been reported to exist over a considerable distance from a breakpoint.

Two additional atypical 22q11 interstitial deletions have recently been reported. The first was detected in a patient diagnosed with CTAF with TOF and pulmonary atresia. Although the deletion includes sequences hemizygous in other DGS/VCFS patients with the large deletion, this deletion is distal to and nonoverlapping with the previously described SROs (MDGCR, ADU-980, and SRO in Fig. 4; Kurahashi *et al.,* 1996). The second deletion (NA) was detected in a child born with a VSD that closed spontaneously and who otherwise has a normal history. NA is nondysmorphic and progressing well at school. NA's deletion is distal to and nonoverlapping with the other DGS/VCFS SROs (Fig. 4), but it does not extend as far distally as that of the patient described by Kurahashi (O'Donnell *et al.,* 1997). Given that a dysmorphic facies is now thought to be a universal finding in patients with deletions encompassing the DGCR, it is likely that the NA deletion does not directly disrupt a gene with major effect in DGS/VCFS. Furthermore, it is possible that the mild heart defect is due to factors other than the 22q11 deletion. The Kurahashi case is more problematic and, at face value, it would suggest that haploinsufficiency at a second locus within 22q11.2 is capable of producing DGS/VCFS.

Thus, the SRO mapping has indicated that the genetics of DGS/VCFS are not as simple as first envisaged. Despite searches in several laboratories worldwide, mutations affecting a single gene have yet to be discovered. The possibility remains that the 22q11.2 deletion syndrome is a consequence of haploinsufficiency for more than one gene and that techniques other than positional cloning will be required to understand the pathogenesis of the wide spectrum of abnormalities. To provide the resources necessary for both mutation screening and functional analysis an exhaustive (and exhausting) screen for genes within the deleted region has been undertaken and is described in the following section.

V. Gene Identification in the 22q11 Deleted Region

Numerous genes have been identified within the region most commonly deleted in the 22q11.2 deletion disorders. Although most of these genes have been suggested as potential candidates for the disorder, it has not been possible to unequivocally identify any single gene responsible for the observed phenotype or for any of the individual features. The following is a brief description of the transcripts that have been characterized. For simplicity, genes will be described by map position, proceeding toward the telomere. *DGCR6* is the gene mapping furthest proximal within the common interstitial deletion region. It is estimated to be 115 kb centromeric to the balanced t(2;22) of ADU (Demczuk *et al.,* 1996). Sequence comparison demonstrates that *DGCR6* has homology to the *Drosophila* gonadal protein and the laminin γ-1 chain. Although little is known about the function of the *Drosophila* gonadal protein, laminin is a well-characterized extracellular matrix protein. It has been suggested that *DGCR6* could play a role in neural crest cell migration (Demczuk *et al.,* 1996). Several transcripts have been identified in the immediate vicinity of the ADU t(2;22), but none have a high coding potential and the function of these sequences is not clear (Budarf *et al.,* 1995a; Gong *et al.,* 1996; Sutherland *et al.,* 1996). *LAN/DGCR2/IDD* maps immediately distal to the ADU breakpoint and has been shown to be widely expressed (Budarf *et al.,* 1995a; Demczuk *et al.,* 1995a; Wadey *et al.,* 1995; Gong *et al.,* 1996). This gene is predicted to encode a transmembrane protein with similarity to the LDL-receptor and C-type lectins (Demczuk *et al.,* 1995a;Wadey *et al.,* 1995; Gong *et al.,* 1996). A proposed role for LAN/DGCR2/IDD has been in cell–cell or cell–matrix interactions. A serine/threonine kinase gene was identified 8 kb distal to *LAN/DGCR2/IDD.* Northern blot analysis of adult tissues indicates that expression is limited to the testis (Gong *et al.,* 1996). Interestingly, in humans, a pseudogene maps immediately upstream of the serine/threonine kinase gene (Goldmuntz *et al.,* 1997), whereas in mouse the corresponding upstream region codes for a functional kinase gene (Galili *et al.,* 1997). The next transcript identified in this gene-rich region is *DGSI.* It maps only 1.3 kb distal to the 3′ end of the serine/threonine kinase gene (Gong *et al.,* 1996). DGSI, also known as ES2, does not have any significant matches to proteins with known function, but the deduced amino acid sequence has 28% identity and 53% similarity to a *Caenorhabditis elegans* hypothetical protein (Gong *et al.,* 1996, 1997; Rizzu *et al.,* 1996).

A computational approach was used to identify a gene encoding a homeobox protein with highest identity to *goosecoid* (*GSC*) (Gottlieb *et al.,* 1997). Due to this homology, the gene has been named *Goosecoid-like* (*GSCL*). The coding region of *GSCL* is contained within three exons and the first exon maps 2.5 kb distal to *DGSI/ES2.* The proximal deletion boundary of the interstitial deletion in patient G has been positioned immediately distal to *DGSI/ES2* (Rizzu *et al.,* 1996). Thus, *GSCL* is likely to be disrupted or deleted in this patient. Because the related gene *GSC* encodes a transcription factor expressed in developing cephalic neural crest lineages, *GSCL* represents an excellent candidate for contributing to the phenotype associated with DGS/VCFS. However, to date, no mutations have been detected in this gene in nondeleted patients (Gottlieb *et al.,* 1997). The next transcript is located 25 kb distal to *GSCL* (U30597, Gottlieb *et al.,* 1997). It has more than 90% amino acid identity to rat citrate transport protein (CTP) and was isolated using direct cDNA library screening, cDNA selection, and a computational approach (Heisterkamp *et al.,* 1995; Gong *et al.,* 1996; Goldmuntz *et al.,* 1996, respectively). CTP is widely expressed in adult and fetal tissue, and the encoded protein is located in the mitochondrial inner membrane where it participates in electroneutral exchange (Gong *et al.,* 1996; Goldmuntz *et al.,* 1996). It has not been viewed as a likely candidate for the features of DGS/VCFS. A 5.5-kb transcript has been mapped immediately 5′ of *CTP* (Gong *et al.,* 1996). Several laboratories independently characterized this clathrin heavy chain-like gene (*CLTCL*), which was so named because the predicted amino acid sequence demonstrates 85% identity with human clathrin heavy chain (Gong *et al.,* 1996; Sirotkin *et al.,* 1996; Kedra *et al.,* 1996; Lindsay *et al.,*

1996; Long *et al.,* 1996; Holmes *et al.,* 1997). Until the description of *CLTCL* it was generally thought that, like yeast, higher organisms would have a single locus for the clathrin heavy chain. Clathrin is a ubiquitous structural protein that plays a central role in endocytosis and membrane receptor trafficking. The expression of *CLTCL* is much more restricted than clathrin heavy chain, suggesting that *CLTCL* may have a more specialized function. As described in Section IV, the 3′ end of the *CLTCL* gene is disrupted in a patient with a balanced t(21;22) (Holmes *et al.,* 1997). This gene is a candidate for some features of VCFS because the patient manifests several anomalies associated with this disorder.

The next known gene, *TUPLE1/HIRA,* maps approximately 40 kb distal to *CLTCL. TUPLE1/HIRA* (Halford *et al.,* 1993b; Lamour *et al.,* 1995; Lorain *et al.,* 1996) encodes a putative transcriptional regulator with significant similarity to proteins containing WD40 domains. Northern blot analysis and RNA *in situ* hybridization experiments in mouse and chick demonstrate that *TUPLE/HIRA* is widely expressed in the embryo and adult (Halford *et al.,* 1993b; Lamour *et al.,* 1995; Roberts *et al.,* 1997; Wilming *et al.,* 1997). Although it has been shown to be deleted in patients with known 22q11 deletions, there are no reports of mutations in nondeleted patients with DGS, VCFS, or conotruncal cardiac defects. A gene, *UFD1L,* with similarity to yeast ubiquitin fusion degradation 1 protein has been identified 30–40 kb telomeric to *TUPLE1/HIRA* (Pizzuti *et al.,* 1997). *UFD1L* can be considered a possible candidate for features of DGS/VCF because the ubiquitin fusion protein degradation pathway plays a major role in the regulation of protein turnover and has been implicated in development of ectoderm-derived structures. A novel gene, *TMVCF,* has recently been identified in the region between D22S941 and D22S944. *TMVCF* is expressed in several human tissues, with highest levels in adult lung. It is also detectable by Northern blot analysis on Day 9 of mouse development. *TMVCF* is predicted to contain two transmembrane domains and has 56% similarity to RVP.1, a rat protein of unknown function (Sirotkin *et al.,* 1997b).

Glycoprotein Ibβ (GpIbβ), a component of the major platelet receptor for von Willebrand factor, has been mapped approximately 250 kb distal to *TUPLE1/HIRA* (Budarf *et al.,* 1995b). Defects in the GpIb receptor have been associated with a rare autosomal recessive bleeding disorder, Bernard–Soulier syndrome (BSS), which is characterized by excessive bleeding, thrombocytopenia, and very large platelets. Budarf *et al.* (1995b) reported a patient with features of BSS and VCFS with a 22q11 deletion. Haploinsufficiency for this region of 22 unmasked a mutation in the remaining *GpIbβ* allele resulting in manifestations of BSS. This mechanism, unmasking a mutant gene, could also be responsible for some of the phenotypic variability seen among patients with the 22q11 deletions. A longer *GpIbβ* transcript of approximately 3.5 kb was noted in a wide variety of cell types and interestingly this longer transcript appears to be the result of alternative processing of a gene immediately upstream of *GpIbβ*. The gene immediately 5′ of *GpIbβ* has been designated human cell division cycle-related (*hCDCrel-1*) and has similarity to GTP-binding proteins (Zieger *et al.,* 1997). Northern blots using *hCDCrel-1* reveal two transcripts of ~2.5 and ~3.5 kb. The smaller transcript terminates at an imperfect polyadenylation signal and the longer transcript is produced by the alternative use of the polyadenylation of *GpIbβ,* the adjacent 3′ gene.

The human *TBX1* gene has been cloned, characterized, and mapped distal to *GpIbβ* (Chieffo *et al.,* 1997). *TBX1* is a member of a phylogenetically conserved family of genes that share a common DNA-binding domain, the T-box. Mouse *Tbx1* has been previously shown to be expressed during early embryogenesis in the pharyngeal arches, pouches, and otic vesicle. Later developmental expression includes the vertebral column, lung epithelium, and tooth bud. Thus, human *TBX1* is a candidate for some of the features seen in the 22q11 deletion syndrome. *Catechol-O-methyltransferase* (*COMT*) has been mapped to the commonly deleted region (Grossman *et al.,* 1992; Dunham *et al.,* 1992) and is distal to *TBX1.* COMT, important in the metabolism of catecholamines, such as noradrenaline, adrenaline, and dopamine, has been hypothesized to play a role in the development of psychosis associated with the 22q11 deletion syndrome (Dunham *et al.,* 1992). However, a recent study using the polymorphism associated with high/low metabolizing activity of COMT in a cohort of patients with schizophrenia failed to find an excess of either allele among the patient population (Karayiorgou *et al.,* 1996). In contrast, an association between this low-metabolizing polymorphism and the development of bipolar spectrum disorder in deleted VCFS patients has been described (Lachman *et al.,* 1996). Additional studies will be required to determine the significance of this finding. A new member of the catenin family designated *ARVCF* has been isolated and mapped immediately adjacent to *COMT.* Based on the presence of a coiled coil domain and 10 tandem armadillo repeats, it has been proposed that *ARVCF* may play a role in protein–protein interactions at adherens junctions (Sirotkin *et al.,* 1997a). *ARVCF* is expressed in all fetal and adult tissues examined and there are no reports of alterations of this locus in nondeleted DGS/VCFS patients.

Several genes have been mapped more distally in the

commonly deleted region. *N41,* a novel cDNA, was identified by screening fetal liver and brain cDNA libraries with a chromosome specific NotI linking clone (Emanuel *et al.,* 1993). The *N41* cDNA is expressed in adult human tissues (heart, liver, brain, and placenta) and has high sequence identity with a mouse *N41* cDNA clone. Whole mount *in situ* experiments with mouse *N41* show that it is expressed in the developing embryo. Halford *et al.* (1993a) identified a cDNA clone, *T10,* from a mouse embryo library using a cosmid clone, HP500, which had been previously mapped to 22q11.2. The gene was predicted to encode a serine and threonine-rich protein and database searches did not find any similarity to known genes. *T10* has been shown to be expressed during early mouse embryogenesis as well as in human fetal tissue. Aubry *et al.* (1993) screened a chromosome 22-specific cosmid library for zinc finger genes since genes containing this motif have been shown to influence cell proliferation and differentiation. One of the four zinc finger genes they identified, *ZNF74,* was deleted in 23 of 24 DGS patients studied. It was shown by Northern blot analysis to be expressed during mouse embryogenesis as well as in human fetal tissues. Currently, the most distal locus within the commonly deleted region is *LZTR-1. LZTR-1* was isolated from a fetal brain cDNA library and is expressed in a number of fetal tissues (brain, heart, liver, kidney, and lung) (Kurahashi *et al.,* 1995). Sequence analysis demonstrates homology to a basic leucine zipper-like domain and a possible function in regulation of transcription has been suggested. Although the *LZTR-1* gene lies outside the minimal critical region and is not consistently deleted in DGS/VCFS patients, it maps to the region deleted in the patient with only an atypical distal deletion (Kurahashi *et al.,* 1996; see Section IV).

VI. Identification of the Etiologic Gene(s)

Currently, it is still unknown whether haploinsufficiency of a single gene is sufficient to produce the complete DGS/VCF phenotype or whether reduced dosage for several genes is necessary. The identification of numerous genes encoded within this region has only served to highlight the challenge of determining which gene(s) is etiologically related to the features seen in affected individuals. It is likely that the presence of a single copy of a critical gene(s) results in inadequate dosage of a protein(s) at a crucial time in embryonic development. One approach to identify individual genes that play a significant role in DGS/VCFS is to determine whether nondeleted patients have mutations in any of the genes within the DGCR. These studies are in progress and have initially focused on genes in the "SROs." Given the extreme phenotypic variability associated with the 22q11 deletion syndromes, it is likely that other modifying factors play an important role in the phenotypic manifestations. Perhaps there are significant allelic variants that affect the functional status of the nondeleted allele as well as genes upstream or downstream in the complex developmental pathways.

To elucidate the role that haploinsufficiency of these genes plays on the phenotype, expression and functional studies are being undertaken. The expectation is that the important genes will be expressed at the correct time and in the expected place during embryonic development. *In situ* hybridization experiments in the developing mouse embryo have been performed for several of the genes and indicate early embryonic expression (Roberts *et al.,* 1997; Wilming *et al.,* 1997; Taylor *et al.,* 1997; Chapman *et al.,* 1996). To further determine the role of these genes in the etiology of the phenotype, transgenic or "knockout" mice will be required. Toward this end, it has recently been shown that there is at least a 150-kb segment of mouse chromosome 16 with significant homology to the MDGCR (Galili *et al.,* 1997). Southern blot analysis has confirmed the presence, in the proximal portion of mouse chromosome 16, of at least six of the genes described in the transcription map of the MDGCR of Gong *et al.* (1996). Comparative analysis of 38 kb of mouse and human sequence reveals significant conservation of gene content, order, exon composition, and transcriptional direction. These studies provide the framework within which to create a mouse model of the 22q11 deletion. Construction of a mouse model should ultimately permit the experiments that will help to determine whether these disorders are caused by deficiency of a single gene or loss of several genes within 22q11.

References

Amati, F., Mari, F., Diglio, M. C., Mingarelli, R., Marino, B., Gianotti, A., Novelli, G., and Dallapiccola, B. (1995). 22q11 deletions in isolated and syndromic patients with tetralogy of Fallot. *Hum. Genet.* **95,** 472–482.

Aubry, M., Demczuk, S., Desmaze, C., Aikem, M., Aurias, A., Julien, J.-P., and Rouleau, G. A. (1993). Isolation of a zinc finger gene consistently deleted in DiGeorge syndrome. *Hum. Molec. Genet.* **2,** 1583–1587.

Augusseau, S., Jouk, S., Jalbert, P., and Priur, M. (1986). DiGeorge syndrome and 22q11 rearrangements. *Hum. Genet.* **74,** 206.

Bockman, D. E., and Kirby, M. L. (1984). Dependence of thymus development on derivatives of the neural crest. *Science* **223,** 498.

Bockman, D. E., Redmond, M. E., Waldo, K., Davis, H., and Kirby, M. L. (1987). Effect of neural crest ablation on development of the heart and arch arteries in the chick. *Am. J. Anat.* **180,** 332–341.

Bowen, P., Pabst, H., Berry, D., Collins-Nakai, R., and Hoo, J. J. (1986). Thymic deficiency in an infant with a chromosome t(18;22) t(q12.2;p11.2) pat rearrangement. *Clin. Genet.* **29,** 174–177.

Budarf, M. L., Collins, J., Gong, W., Roe, B., Wang, Z., Bailey, L. C., Sellinger, B., Michaud, D., Driscoll, D. A., and Emanuel, B. S. (1995a). Cloning a balanced translocation associated with DiGeorge syndrome and identification of a disrupted candidate gene. *Nat. Genet.* **10,** 269–278.

Budarf, M. L., Konkle, B. A., Ludlow, L. B., Michaud, D., Li, M., Yamashiro, D. J., McDonald-McGinn, D., Zackai, E. H., and Driscoll, D. D. (1995b). Identification of a patient with Bernard-Soulier syndrome and a deletion in the DiGeorge/Velocardiofacial chromosomal region in 22q11.2. *Hum. Molec. Genet.* **4,** 763–766.

Burn, J., and Goodship, J. (1996). Congenital heart disease. In "Emery and Rimoin's Principles and Practice of Medical Genetics" (D. L. Rimoin, J. M. Conner, R. E. Pyeritz, and A. E. H. Emery, eds.), Vol. 1, pp. 767–803. Churchill-Livingston, London.

Burn, J., Takao, A., Wilson, D., Cross, I., Momma, K., Wadey, R., Scambler, P., and Goodship, J. (1993). Conotruncal anomaly face syndrome is associated with a deletion within chromosome 22. *J. Med. Genet.* **30,** 822–824.

Carey, A. H., Kelly, D., Halford, S., Wadey, R., Wilson, D., Goodship, J., Burn, J., Paul, T., Sharkey, A., Dumanski, J., Nordenskjold, M., Williams, R., and Scambler, P. J. (1992). Molecular genetic study of the frequency of monosomy 22q11 in DiGeorge syndrome. *Am. J. Hum. Genet.* **51,** 964–970.

Chapman, D. L., Garvey, N., Hancock, S., Alexiou, M., Agulnik, S. I., Gibson-Brown, J. J., Cebra-Thomas, J., Bollag, R. J., Silver, L. M., and Papaioannou, V. E. (1996). Expression of the T-box family genes, *Tbx1-Tbx5,* during early mouse development. *Dev. Dyn.* **206,** 379–390.

Chieffo, C., Garvey, N., Roe, B., Zhang, G., Silver, L., Emanuel, B. S., and Budarf, M. L. (1997). Isolation and characterization of a gene from the DiGeorge chromosomal region (DGCR) homologous to the mouse Tbx1 gene. *Genomics* **43,** 267–277.

Chisaka, O., and Capecchi, M. R. (1991). Regionally restricted developmental defects resulting from targeted disruption of the mouse homeobox gene hox1.5. *Nature* (*London*) **350,** 473–479.

Conley, M. E., Beckwith, J. B., Mancer, J. F. K., and Tenckhoff, L. (1979). The spectrum of the DiGeorge syndrome. *J. Pediatr.* **94,** 883–890.

Conway, S. J., Henderson, D. J., and Copp, A. J. (1997). Pax3 is required for cardiac neural crest migration in the mouse: Evidence from the splotch (Sp2H) mutant. *Development* (*Cambridge, UK*) **124,** 505–514.

Cormier-Daire, V., Iserin, L., Theophile, D., Sidi, D., Vervel, C., Padovani, J. P., Vekemans, M., Munnich, A., and Lyonnet, S. (1995). Upper limb malformations in DiGeorge syndrome. *Am. J. Med. Genet.* **56,** 39–41.

Cuneo, B. F., Driscoll, D. A., Gidding, S. S., and Langman, C. B. (1997). Evolution of latent hypoparathyroidism in familial 22q11 deletion syndrome. *Am. J. Med. Genet.* **69,** 50–55.

de la Chapelle, A., Herva, R., Koivisto, M., and Aula, P. A. (1981). Deletion in chromosome 22 can cause DiGeorge syndrome. *Hum. Genet.* **57,** 253–256.

Demczuk, S., Aledo, R., Zucman, J., Delattre, O., Desmaze, C., Dauphinot, L., Jalbert, P., Rouleau, G. A., Thomas, G., and Aurias, A. (1995a). Cloning of a balanced translocation breakpoint in the DiGeorge syndrome critical region and isolation of a novel potential adhesion receptor gene in its vicinity. *Hum. Mol. Genet.* **4,** 551–558.

Demczuk, S., Levy, A., Aubry, M., Croquette, M.-F., Philip, N., Prieur, M., Sauer, U., Bouvagnet, P., Rouleau, G. A., Thomas, G., Aurias, A. (1995b). Excess of deletions of maternal origin in the DiGeorge/velo-cardio-facial syndromes: a study of 22 new patients and review of the literature. *Hum. Genet.* **96,** 9–13.

Demczuk, S., Thomas, G., and Aurias, A. (1996). Isolation of a novel gene from the DiGeorge syndrome critical region with homology to *Drosophila gdl* and to human *LAMC1* genes. *Hum. Molec. Genet.* **5,** 633–638.

Devriendt, K., Swillen, A., and Fryns, J.-P. (1996). Renal and urogenital malformations caused by 22q11 deletion. *J. Med. Genet.* **33,** 349.

Devriendt, K., Moerman, P., van Schoubroeck, D., Vandenberghe, K., and Fryns, J.-P. (1997). Chromosome 22q11 deletion presenting as the Potter sequence. *J. Med. Genet.* **34,** 423–425.

DiGeorge, A. M. (1965). Discussion on a new concept of the cellular basis of immunology. *J. Pediatr.* **67,** 907.

Driscoll, D. A., and Emanuel, B. S. (1996). DiGeorge and velocardiofacial syndromes: The 22q11 deletion syndrome. *Ment. Retard. Dev. Disabil. Res. Rev.* **2,** 130–138.

Driscoll, D. A., Budarf, M., McDermid, H., and Emanuel, B. S. (1990). Molecular analysis of DiGeorge Syndrome: 22q11 interstitial deletions. *Am. J. Hum. Genet.* **47,** A215 (abstr.).

Driscoll, D. A., Budarf, M. L., and Emanuel, B. (1992a). A genetic etiology for DiGeorge syndrome: Consistent deletions and microdeletions of 22q11. *Am. J. Hum. Genet.* **50,** 924–933.

Driscoll, D. A., Spinner, N. B., Budarf, M. L., McDonald-McGinn, D. M., Zackai, E. H., Goldberg, R. B., Shprintzen, R. J., Saal, H. M., Zonana, J., Jones, M. C., Mascarello, J. T., and Emanuel, B. S. (1992b). Deletions and microdeletions of 22q11.2 in velo-cardio-facial syndrome. *Am. J. Med. Genet.* **44,** 261–268.

Driscoll, D. A., Salvin, J., Sellinger, B., Budarf, M. L., McDonald-McGinn, D. M., Zackai, E. H., and Emanuel, B. S. (1993). Prevalence of 22q11 microdeletions in DiGeorge and velocardiofacial syndromes: implications for genetic counselling and prenatal diagnosis. *J. Med.Genet.* **30,** 813–817.

Dunham, I., Collins, J., Wadey, R., and Scambler, P. (1992). Possible role for COMT in psychosis associated with velo-cardio-facial sndrome. *Lancet* **340,** 1362.

Du Montcel, S. T., Mendizabal, H., Ayme, S., Levy, A., and Philip, N. (1996). Prevalence of 22q11 microdeletion. *J. Med. Genet.* **33,** 719.

Emanuel, B. S., Driscoll, D., Goldmuntz, E., Baldwin, S., Biegel, J., Zackai, E. H., McDonald-McGinn, D., Sellinger, B., Gorman, N., Williams, S., and Budarf, M. L. (1993). Molecular and phenotypic analysis of the chromosome 22 microdeletion syndromes. *Prog. Clin. Biol. Res.* **384,** 207–224.

Epstein, J. A. (1996). Pax3, neural crest and cardiovascular development. *Trends Cardiovasc. Med.* **6,** 255–261.

Faed, M. J. W., Robertson, J., Swanson Beck, J., Carter, J. I., Bose, B., and Madlon, M. M. (1987). Features of DiGeorge syndrome in a child with 45,XX,−3,−22,+der(3)t(3;22)(p25;q11). *J. Med. Genet.* **24,** 225–234.

Freedom, R. M., Rosen, F. S., and Nadas, A. S. (1972). Congenital cardiovascular disease and anomalies of the third and fourth pharyngeal pouch. *Circulation* **46,** 165–172.

Galili, N., Baldwin, H. S., Lund, J., Reeves, R., Gong, W., Wong, Z., Roe, B., Emanuel, B. S., Nayak, S., Mickanin, C., Budarf, M. L., and Buck, C. A. (1997). A region of mouse chromosome 16 is syntenic to the DiGeorge, velo-cardio-facial syndrome minimal critical region. *Genome Res.* **7,** 17–26.

Giannotti, A., Digilio M. C., Marino, B., Mingarelli, R., and Dallapiccola, B. (1994). Cayler cardiofacial syndrome and del 22q11: Part of the CATCH22 phenotype. *Am. J . Med. Genet.* **53,** 303–304.

Goldberg, R., Marion, R., and Borderon, M. (1985). Phenotypic overlap between velo-cardio-facial syndrome and DiGeorge sequence. *Am. J. Hum. Genet.* **37,** 54A (abstract).

Goldberg, R., Motzkin, B., Marion, R., Scambler, P. J., and Shprintzen, R. (1993). Velo-cardio-facial syndrome. *Am. J. Med. Genet.* **45,** 313–319.

Goldmuntz, E., and Emanuel, B. S. (1997). Genetic disorders of cardiac morphogenesis: The DiGeorge and velocardiofacial syndromes. *Circ. Res.* **80,** 437–443.

Goldmuntz, E., Driscoll, E., Budarf, M. L., Zackai, E. H., McDonald-McGinn, D. M., Biegel, J. A., and Emanuel, B. S. (1993). Microdeletion of chromosomal region 22q11 in patients with congenital conotruncal cardiac defects. *J. Med. Genet.* **30,** 807–812.

Goldmuntz, E., Wang, Z., Roe, B. A., and Budarf, M. L. (1996). Cloning, genomic organization and chromosomal localization of human citrate transport protein to the DiGeorge/velocardialfacial syndrome minimal critical region. *Genomics* **33,** 271–276.

Goldmuntz, E., Fedon, J., Roe, B., and Budarf, M. L. (1997). Molecular characterization of a serine/threonine kinase in the DiGeorge minimal critical region. *Genes* **198,** 379–386.

Goldmuntz, E., Clark, B. J., Mitchell, L. E., Jawad, A. F., Cuneo, B. F., Reed, L., McDonald-McGinn, D., Chien, P., Feuer, J., Zackai, E. H., Emanuel, B. S., and Driscoll, D. A. (1998). Frequency of 22q11 deletions in patients with conotruncal defects. *J. Am. Coll. Cardiol.* **32,** 492–498.

Gong, W., Emanuel, B. S., Collins, J., Kim, D. H., Wang, Z., Chen, F., Zhang, G., Roe, B., and Budarf, M. L. (1996). A transcription map of the DiGeorge and velo-cardio-facial syndrome critical region on 22q11. *Hum. Mol. Genet.* **5,** 789–800.

Gong, W., Emanuel, B. S., Galili, N., Kim, D. H., Roe, B., Driscoll, D., and Budarf, M. L. (1997). Structural and mutational analysis of a conserved gene (DGSI) from the minimal DiGeorge syndrome critical region. *Hum. Mol. Genet.* **6,** 267–276.

Goodship, J., Cross, I., Scambler, P., and Burn, J. (1995). Monozygotic twins with chromosome 22q11 deletion and discordant phenotype. *J. Med. Genet.* **32,** 746–748.

Gottlieb, S., Emanuel, B. S., Driscoll, D. A., Sellinger, B., Wang, Z., Roe, B., and Budarf, M. L. (1997). The DiGeorge syndrome minimal critical region contains a Goosecoid-like (GSCL) homeobox gene, which is expressed early in human development. *Am. J. Hum. Genet.* **60,** 1194–1201.

Greenberg, F., Crowder, W. E., Paschall, V., Colon-Linares, J., Lubianski, B., and Ledbetter, D. H. (1984). Familial DiGeorge syndrome and associated partial monosomy of chromosome 22. *Hum. Genetics.* **65,** 317–319.

Greenberg F., Elder, F. F. B., Haffner, P., Northrup, M., Ledbetter, D. H. (1988). Cytogenetic findings in a prospective series of patients with DiGeorge anomaly. *Am. J. Hum. Genet.* **43,** 605–611.

Gripp, K. W., McDonald-McGinn, D. M., Driscoll, D. A., Reed, L. A., Emanuel, B. S., and Zackai, E. H. (1997). Nasal dimple as part of the 22q11.2 deletion syndrome. *Am. J. Med. Genet.* **69,** 290–292.

Grossman, M. H., Emanuel, B. S., and Budarf, M. L. (1992). Chromosomal mapping of the human catechol-O-methyltransferase gene to 22q11.1->q11.2. *Genomics* **12,** 822–825.

Halford, S., Wilson, D. I., Daw, S. C. M., Roberts, C., Wadey, R., Kamath, S., Wickremasinghe, A., Burn, J., Goodship, J., Mattel, M.-G., Moormon, A. F. M. and Scambler, P. J. (1993a). Isolation of a gene expressed during early embryogenesis from the region of 22q11 commonly deleted in DiGeorge syndrome. *Hum. Mol. Genet.* **2,** 1577–1582.

Halford, S., Wadey, R., Roberts, C., Daw, S. C. M., Whiting, J. A., O'Donnell, H., Dunham, I., Bentley, D., Lindsay, E., Baldini, A., Francis, F., Lehrach, H., Williamson, R., Wilson, D. I., Goodship, J., Cross, I., Burn, J., and Scambler, P. J. (1993b). Isolation of a putative transcriptional regulator from the region of 22q11 deleted in DiGeorge Syndrome, Shprintzen syndrome and familial congenital heart disease. *Hum. Mol. Genet.* **2,** 2099–2107.

Harington, L. H. (1829). Absence of the thymus gland (letter to the Editor). *London Med. Gaz.* **3,** 314.

Heisterkamp, N., Mulder, M. P., Langeveld, A., ten Hoeve, J., Wang, Z., Roe, B. A., and Groffen, J. (1995). Localization of the human mitochondrial citrate transporter protein gene to chromosome 22q11 in the DiGeorge syndrome critical region. *Genomics* **29,** 451–456.

Holder, S. E., Winter, R. M., Kamath, S., and Scambler, P. J. (1993). Velo-cardio-facial syndrome in a mother and daughter -variability of the clinical phenotype. *J. Med. Genet.* **30,** 825–827.

Holmes, S. E., Riazi, M. A., Gong, W., McDermid, H. E., Sellinger, B. T., Hua, A., Chen, F., Wang, Z., Zhang, G., Roe, B., Gonzalez, I., McDonald-MvGinn, D. M., Zackai, E., Emanuel, B. S., and Budarf, M. L. (1997). Disruption of the clathrin heavy chain-like gene (CLTCL) associated with features of DGS/VCFS: A balanced (21;22)(p12;q11) translocation. *Hum. Mol. Genet.* **6,** 357–367.

Hunt, P., and Krumlauf, R. (1991). Deciphering the Hox code: Clues to patterning branchial regions of the head. *Cell* (*Cambridge, Mass.*) **66,** 1075–1078.

Jaquez, M., Driscoll, D. A., Li, M., Emanuel, B. S., Hernandez, I., Jaquez, F., Lembert, N., Ramirez, J., Matalon, R. (1997). Unbalanced 15;22 translocation in a patient with features of both DiGeorge and velocardiofacial syndrome. *Am. J. Med. Genet.* **70,** 6–10.

Junker, A. K., and Driscoll, D. A. (1995). Humoral immunity in DiGeorge syndrome. *J. Pediatr.* **127,** 231–237.

Kaplan, J. C., and Emanuel, B. S. (1989). Report of the committee on the genetic constitution of chromosome 22. *Cytogenet. Cell Genet.* **51,** 372–83.

Karayiorgou, M., Morris, M. A., Morrow, B., Shprintzen, R. J., Goldberg, R., Borrow, J., Gos, A., Nestadt, G., Wolyniec, P. S., Lasseter, V. K., Eisen, H., Childs, B., Kazazian, H. H., Kucherlapati, R., Antonarakis, S. E., Pulver, A. E., and Housman, D. (1995). Schizophrenia susceptibility associated with interstitial deletions of chromosome 22q11. *Proc. Natl. Acad. Sci. USA* **92,** 7612–7616.

Karayiorgou, M., Gogos, J. A., Galke, B. L., Jiang, Q., Wolyniec, P., Nestadt, G., Dombroski, B., Antonarakis, S. E., Housman, D. E., Kazazian, H., Luebbert, H., Budarf, M. L., and Pulver, A. E. (1996). Molecular characterization of the 22q11 schizophrenia susceptibility locus. *Am. J. Hum. Genet.* **59,** A223.

Kedra, D., Peyrard, M., Fransson, I., Collins, J. E., Dunham, I., Roe, B. A., and Dumanski, J. P. (1996). Characterization of a second human clathrin heavy chain polypeptide gene (CLH-22) from chromosome 22q11. *Hum. Molec. Genet.* **5,** 625–631.

Kelley, R. I., Zackai, E. H., Emanuel, B. S., Kistenmacher, M., Greenberg, F., and Punnett, H. H. (1982). The association of the DiGeorge anomalad with partial monosomy of chromosome 22. *J. Pediatr.* **101,** 197–200.

Kelly, D., Goldberg, R., Wilson, D., Lindsay, E., Carey, A., Goodship, J., Burn, J., Cross, I., Shprintzen, R. J., and Scambler, P. J. (1993). Confirmation that the velo-cardiofacial syndrome is associated with haplo-insufficiency of genes at chromosome 22. *Am. J. Med. Genet.* **45,** 308–312.

Kinouchi, A. (1980). A study on specific peculiar facial features of conotruncal anomaly. *J. Tokyo Women's Med. Coll.* (in Japanese) **50,** 396–409.

Kinouchi, A., Mori, K., Ando, M., and Takao, A. (1976). Facial appearance of patients with conotruncal anomalies. *Pediatr. Jpn.* **17,** 84.

Kirby, M. L. (1989). Plasticity and predetermination of mesecephalic and trunk neural crest transplanted into the region of the cardiac neural crest. *Dev. Biol.* **134,** 402–412.

Kirby, M. L., and Waldo, M. L. (1995). Neural crest and cardiovascular patterning. *Circ. Res.* **77,** 211–215.

Kirby, M. L., Gale, T. F., and Stewart, D. E. (1983). Neural crest cells contribute to normal aorticopulmonary septation. *Science* **220,** 1059–1061.

Krumlauf, R. (1994). Hox genes in vertebrate development. *Cell* (*Cambridge, Mass.*) **78,** 191–201.

Kurahashi, H., Akagi, K., Inazawa, J., Ohta, T., Niikawa, N., Kayatani, F., Sano, T., Okada, S., Nishisho, I. (1995). Isolation and characterization of a novel gene deleted in DiGeorge syndrome. *Hum. Mol. Genet.* **4,** 541–549.

Kurahashi, H., Nakayama, T., Osugi, Y., Tsuda, E., Masuno, M.,

Imaizumi, K., Kamiya, T., Sano, T., Okada, S., and Nishisho, I. (1996). Deletion mapping of 22q11 in CATCH22 syndrome: Identification of a second critical region. *Am. J. Hum. Genet.* **58,** 1377–1381.

Lachman, H. M., Morrow, B., Shprintzen, R., Veit, S., Parsia, S. S., Faedda, G., Goldberg, R., Kucherlapati, R., and Papolos, D. F. (1996). Association of codon 108/158 catechol-O-methyltransferase gene polymorphism with the psychiatric manifestations of velocardiofacial syndrome. *Am. J. Med. Genet.* **58,** 468–472.

Lammer, E. J., and Opitz, J. M. (1986). The DiGeorge anomaly as a developmental field defect. *Am. J. Med. Genet.* **29,** 113–127.

Lammer, E. J., Chen, D. T., Hoar, N. D., Agnish, P. J., Benke, J. T., Braun, C. J., Curry, P. M., Fernhoff, A. W., Grix, A. T., Lott, J. M., Richard, J. M., and Sun, S. C. (1986). Retinoic acid embryopathy. A new human teratogen and a mechanistic hypothesis. *N. Engl. J. Med.* **313,** 837–841.

Lamour, V., Lecluse, Y., Desmaze, C., Spector, M., Bodescot, M., Aurias, A., Osley, M. A., and Lipinski, M. (1995). A human homolog of the S. cerevisiae *HIR1* and *HIR2* transcriptional repressors cloned from the DiGeorge syndrome critical region. *Hum. Mol. Genet.* **4,** 791–799.

Leana-Cox, J., Pangkanon, S., Eanet, K. R., Curtin, M. S., and Wulfsberg, E. A. (1996). Familial DiGeorge/velocardiofacial syndrome with deletions of chromosome area 22q11: Report of five families with a review of the literature. *Am. J. Med. Genet.* **65,** 309–316.

Levy, A., Demczuk, S., Aurias, A., Depetris, D., Mattei, M. G., and Philip, N. (1995). Interstitial 22q11 deletion excluding the ADU breakpoint in a patient with DGS. *Hum. Mol. Genet.* **4,** 2417–2418.

Lindsay, E. A., Halford, S., Wadey, R., Scambler, P. J., and Baldini, A. (1993). Molecular cytogenetic characterisation of the DiGeorge syndrome region using fluorescence in situ hybridisation. *Genomics* **17,** 403–407.

Lindsay, E. A., Greenberg, F., Shaffer, L. G., Shapira, S. K., Scambler, P. J., and Baldini, A. (1995). Submicroscopic deletions at 22q11.2: Variability of the clinical picture and delineation of a commonly deleted region. *Am. J. Med. Genet.* **56,** 191–197.

Lindsay, E. A., Rizzu, P., Antonacci, R., Jurecic, V., Delmas-Mata, J., Lee, C.-C., Kim, U.-J., Scambler, P. J. and Baldini, A. (1996). A transcription map in the CATCH22 critical region: identification, mapping, and ordering of four novel transcripts expressed in heart. *Genomics* **32,** 104–112.

Lipson, A. H., Yuille, D., Angel, M., Thompson, P. B., Vandervoord, J. G., Beckenham, E. J. (1991). Velo-cardio-facial (Shprintzen syndrome: An important syndrome for the dysmorphologist to recognize. *J. Med. Genet.* **28,** 596–604.

Lischner, H. W., Huff, D. S., Dacou, C., and DiGeorge, A. M. (1969). Congenital hypoplasia of the thymus and parathyroid glands: Incomplete III-IV pharyngeal pouch syndrome. *Program 79th Annu. Meet. Am. Pediatr. Soc.* Abstract.

Lobdell, D. H. (1959). Congenital absence of the parathyroid glands. *Arch. Pathol.* **68,** 412–415.

Long, K. R., Trofatter, J. A., Ramesh, V., McCormick, M. K., and Buckler, A. J. (1996). Cloning and characterization of a novel human clathrin heavy chain gene (CTCL). *Genomics* **35,** 466–472.

Lorain, S., Demczuk, S., Lamour, V., Toth, S., Aurias, A., Roe, B. A., and Lipinski, M. (1996). Structural organization of the WD repeat protein-encoding gene HIRA in the DiGeorge syndrome critical region of human chromosome 22. *Genome Res.* **6,** 43–50.

Lynch, D. R., McDonald-McGinn, D., Zackai, E. H., Emanuel, B. S., Driscoll, D. A., Whitaker, L. A., and Fischbeck, K. H. (1995). Cerebellar atrophy in a patient with velocardiofacial syndrome. *J. Med. Genet.* **32,** 561–563.

Manley, N., and Capecchi, M. R. (1995). The role of Hoxa-3 in mouse thymus and thyroid development. *Development (Cambridge, UK)* **121,** 1989–2003.

Marmon, L. M., Balsara, R. K., Chen, R., and Dunn, J. M. (1984). Congenital cardiac anomalies associated with the DiGeorge syndrome: A neonatal experience. *Ann. Thorac. Surg.* **38,** 148–150.

Marshall, H., Nonchev, S., Sham, M. H., Muchamore, I., Lumsden, A., and Krumlauf, R. (1992). Retinoic acid alters hindbrain Hox code and induces transformation of rhombomeres 2/3 into a 4/5 identity. *Nature (London)* **360,** 737–741.

Mascarello, J. T., Bastian, J. F., and Jones, M. C. (1989). Interstitial deletion of chromosome 22 in a patient with the DiGeorge malformation sequence. *Am. J. Med. Genet.* **32,** 112–114.

Matsuoka, R., Takao, A., Kimura, M., Imamura, S.-I., Kondo, C., Joh-o, K., Ikeda, K., Nishibatake, M., Ando, M., and Momma, K. (1994). Confirmation that the conotruncal anomaly face syndrome is associated with a deletion within 22q11.2. *Am. J. Med. Genet.* **53,** 285–289.

McDonald-McGinn, D. M., Driscoll, D. A., Bason, L., Christensen, K., Lynch, D., Sullivan, K., Canning, D., Zavod, W., Quinn, N., Weinberg, P., Clark, B. J., Emanuel, B. S., and Zackai, E. H. (1995). Autosomal dominant Opitz syndrome due to a 22q11.2 deletion. *Am. J. Med. Genet.* **5,** 103–113.

McDonald-McGinn, D. M., Driscoll, D. A., Emanuel, B. S., and Zackai, E. H. (1996). The 22q11.2 deletion in African-American patients: an underdiagnosed population? *Am. J. Hum. Genet.* **59,** A20 (abstract).

McDonald-McGinn, D. M., Driscoll, D. A., Emanuel, B. S., Goldmuntz, E., Clark, B. J., Solot, C., Cohen, M., Schultz, P., LaRossa, D., Randall, P., Zackai, E. H. (1997a). Detection of 22q11.2 deletion in cardiac patients suggests a risk for velopharyngeal incompetence. *Pediatrics* **99,** 1–5.

McDonald-McGinn, D. M., LaRossa, D., Goldmuntz, E., Sullivan, K., Eicher, P., Gerdes, M., Moss, E., Wang, P., Solot, C., Schultz, P., Lynch, D., Bingham, P., Keenan, G., Weinzimer, S., Ming, J. E., Driscoll, D., Clark, B. J. III, Markowitz, R., Cohen, A., Moshang, T., Pasquariello, P., Randall, P., Emanuel, B. S., and Zackai, E. H. (1997b). The 22q11.2 deletion: Screening, diagnostic workup, and outcome of results; report on 181 patients. *Genet. Test.* **1,** 99–108.

McLean, S. D., Saal, H. M., Spinner, N. B., Emanuel, B. S., and Driscoll, D. A. (1993). Velo-cardio-facial syndrome: Intra-familial variability of the phenotype. *Am. J. Dis. Child.* **147,** 1212–1216.

Meinecke, A., Beemer, F. A., Schinzel, A., Kushnick, T. (1986). The velo-cardio-facial (Shprintzen) syndrome. *Eur. J. Pediatr.* **145,** 539–544.

Melchionda, S., Diglio, M.C., Mingarelli, R., Novelli, G., Scambler, P., Marino, B. and Dallapiccola, B. (1995). Transposition of the great arteries associated with deletion of chromosome 22q11. *Am. J. Cardiol.* **75,** 95–98.

Mimy, P., Koppers, B., Bogadanova, N., Schulte-Vallentin, M., Horst, J., and Dworniczak, N. (1995). Paternal uniparental disomy 22. *Med. Genet.* **2,** 216(abstr.).

Ming, J. E., McDonald-McGinn, D. M., Megerian, T. E., Driscoll, D. A., Elias, E. R., Russell, B. M., Irons, M., Emanuel, B. S., Markowitz, R. I., and Zackai, E. H. (1997). Skeletal anomalies in patients with deletions of 22q11. *Am. J. Med. Genet.* **72,** 210–215.

Mitnick, R. J., Bello, J. A., and Shprintzen, R. J. (1994). Brain anomalies in velo-cardio-facial syndrome. *Am. J. Med. Genet.* (*Neuropsy Genet.*) **54,** 100–106.

Moerman, P., Goddeeris, P., Lauwerijns, J., and Van der Hauwaert, L. G. (1980). Cardiovascular malformations in DiGeorge syndrome (congenital absence or hypoplasia of the thymus). *Br. Heart J.* **44,** 452–459.

Morrow, B., Goldberg, R., Carlson, C., Das Gupta, R., Sirotkin, H., Collins, J., Dunham, I., O'Donnell, H., Scambler, P., Shprintzen, R., and Kucherlapati, R. (1995). Molecular definition of the 22q11 deletions in velo-cardio-facial syndrome. *Am. J. Hum. Genet.* **56,** 1379–1390.

Moss, E., Wang, P. P., McDonald-McGinn, D. M., Gerdes, M., DaCosta, A. M., Christensen, K. M., Driscoll, D. A., Emanuel, B. S., Batshaw, M. L., and Zackai, E. H. (1995). Characteristic cognitive profile in patients with a 22q11.2 deletion: verbal IQ exceeds nonverbal IQ. *Am. J. Hum. Genet.* **57,** A20 (abstr.).

Moss, E. M., Wang, P. P., McDonald-McGinn, D. M., Gerdes, M., Solot, C., DaCosta, A. M., Keating, T. B., Christensen, K. M., Reed, L., Driscoll, D. A., Emanuel, B. S., Batshaw, M. L., and Zackai, E. H. (1998). A genetic etiology for a nonverbal learning disability? The characteristic cognitive profile of patients with a chromosome 22q11.2 microdeletion. *Pediatrics* (submitted for publication).

Muller, W., Peter, H. H., Wilken, M., Juppner, H., Kallfelz, H. C., Krohn, H. P., Miller, K., and Rieger, C. H. L. (1988). The DiGeorge syndrome I. Clinical evaluation and course of partial and complete forms of the syndrome. *Clin. Immunol. Immunopathol.* **147,** 496–502.

Nishibatake, M., Kirby, M. L., and Van-Mierop, L. H. (1987). Pathogenesis of persistent truncus arteriosus and dextroposed aorta in the chick embryo after neural crest ablation. *Circulation* **75,** 255–264.

O'Donnell, H., McKeown, C., Gould, C., Morrow, B., and Scambler, P. (1997). Detection of a deletion within 22q11 which has no overlap with the DiGeorge syndrome critical region. *Am. J. Hum. Genet.* **60,** 1544–1548.

Opitz, J. M., Summitt, R. L., and Smith, D. W. (1965). Hypertelorism and hypospadias. A newly recognized hereditary malformation syndrome. *J. Pediatr.* **67,** 968.

Papolos, D. F., Faedda, G. L., Veit, S., Goldberg, R., Morrow, B., Kucherlapati, R., and Shprintzen, R. J. (1996). Bipolar spectrum disorders in patients diagnosed with velo-cardio-facial syndrome: Does a hemizygous deletion of chromosome 22q11 result in bipolar affective disorder? *Am. J. Psychiatry* **153,** 1541–1547.

Pizzuti, A., Novelli, G., Ratti, A., Amati, F., Mari, A., Calabrese G., Nicolis, S., Silani, V., Marino, B., Scarlato, G., Ottolenghi, S., and Dallapiccola, B. (1997). UFD1L, a developmentally expressed ubiquitination gene, is deleted in CATCH 22 syndrome. *Hum. Molec. Genet.* **6,** 259–265.

Quaderi, N. A., Scheiger, S., Gaudenz, K., Franco, B., Rugarli, E. I., Berger, W., Feldman, G. J., Volta, M., Andolfi, G., Gilgenkrantz, S., Marion, R. W., Hennekam, R. C. M., Opitz, J. M., Muenke, M., Ropers, H.-H., and Ballabio, A. (1997). Opitz G/BBB syndrome, a defect of midline development, is due to mutations in a new RING finger gene on Xp22. *Nat. Genet.* **17,** 285–291.

Rizzu, P., Lindsay, E. A., Taylor, C., O'Donnell, H., Levy, A., Scambler, P., and Baldini, A. (1996). Cloning and comparative mapping of a gene from the commonly deleted region of DiGeorge and Velocardiofacial syndromes conserved in *C. elegans. Mamm. Genome* **7,** 639–643.

Roberts, C., Daw, S., Halford, S., and Scambler, P. J. (1997). Cloning and developmental expression analysis of chick hira, a candidate gene for DiGeorge syndrome. *Hum. Mol. Genet.* **6,** 237–246.

Robin, N. H., Feldman, G. J., Aronson, A. L., Mitchell, H. F., Weksberg, R., Leonard, C. O., Burton, B. K., Josephson, K. D., Laxova, R., Aleck, K. A., and Allanson, J. E. (1995a). Opitz syndrome is genetically heterogeneous, with one locus on Xp22, and a second locus on 22q11.2. *Nat. Genet.* **11,** 459–461.

Robin, N. H., Sellinger, B., McDonald-McGinn, D., Zackai, E. H., Emanuel, B. S., and Driscoll, D. A. (1995b). Classical Noonan syndrome is not associated with deletions of 22q11. *Am. J. Med. Genet.* **56,** 94–96.

Robinson, H. B., Jr. (1975). DiGeorge's or the III-IV pharyngeal pouch syndrome. Pathology and a theory of pathogenesis. *Perspect. Pediatr. Pathol.* **2,** 173–206.

Rohn, R. D., Leffell, M. S., Leadem, P., Johnson, D., Rubio, T., and Emanuel, B. S. (1984). Familial third and fourth pharyngeal pouch syndrome with apparent autosomal dominant transmission. *J. Pediatr.* **105,** 47–51.

Ryan, A. K., Goodship, J. A., Wilson, D. I., Philip, N., Levy, A., Siedel, H., Schuffenhauer, S., Oechsler, H., Belohradsky, B., Priur, M., Aurias, A., Raymond, F. L., Clayton-Smith, J., Hatchwell, E., McKeown, C., Bemer, F. A., Dallapiccola, B., Novelli, G., Hurst, J., Ignatius, J., Green, A. J., Winter, R. M., Breuton, L., Brondum-Neilsen, K., Stewart, F., Van Essen, T., Patton, M., Patterson, J., and Scambler, P. J. (1997). Spectrum of clinical features associated with interstitial chromosome 22q11 deletions: A European collaborative study. *J. Med. Genet.* **34,** 798–804.

Scambler, P. J., Carey, A. H., Wyse, R. K. H., Roach, S., Dumanski, J. P., Nordenskjold, M., and Williamson, R. (1991). Microdeletions within 22q11 associated with sporadic and familial DiGeorge syndrome. *Genomics* **10,** 201–206.

Schinzel, A. A., Basaran, S., Bernasconi, F., Karaman, B., Yuksel-Apak, M., and Robinson, W. P. (1994). Maternal uniparental disomy 22 has no impact on the phenotype. *Am. J. Hum. Genet.* **54,** 21–24 (abstr.).

Seaver, L. A., Pierpont, J. W., Erickson, R. P., Donnerstein, R. L., Cassidy, S. (1994). Pulmonary atresia associated with maternal 22q11.2 deletion: possible parent of origin effect in the conotruncal anomaly face syndrome. J. *Med. Genet.* **31,** 830–834.

Shimizu, T., Takao, A., Ando, M., and Hirayama, A. (1984). Conotruncal anomaly face syndrome: Its heterogeneity and association with thymus involution. *In* "Congenital Heart Disease: Causes and Processes" (Nora J. J., Takao, A., eds.), pp. 29–41. Futura: Publ. Co., Mt. Kisko, NY.

Shprintzen, R. J., Goldberg, R. B., Lewin, M. L., Sidoti, E. J., Berkman, M. D., Argamaso, R. V., and Young, D. (1978). A new syndrome involving cleft palate, cardiac anomalies, typical facies, and learning disabilities: velo-cardio-facial syndrome. *Cleft Palate J.* **15,** 56–62.

Shprintzen, R. J., Goldberg, R., Golding-Kushner, K. J., and Marion, R. W. (1992). Late onset psychosis in the velo-cardio-facial syndrome. *Am. J. Med. Genet.* **42,** 141–142.

Sirotkin, H., Morow, B., DasGupta, R., Goldberg, R., Patanjali, S. R., Shi, G., Cannizzaro, L., Shprintzen, R., Weissman, S. M., and Kucherlapati, R. (1996). Isolation of a new clathrin heavy gene with muscle-specific expression from the region commonly deleted in velo-cardio-facial syndrome. *Hum. Mol. Genet.* **5,** 617–624.

Sirotkin, H., O'Donnell, H., DasGupta, R., Halford, S., St. Jore, B., Puech, A., Parimoo, S., Morrow, B., Skoultchi, A., Weissman, S. M., Scambler, P., and Kucherlapati, R. (1997a). Identification of a new human catenin gene family member (ARVCF) from the region deleted in velo-cardio-facial syndrome. *Genomics* **41,** 75–83.

Sirotkin, H., Morrow, B., St. Jore, B., Puech, A., DasGupta, R., Patanjali, S. R., Skoultchi, A., Weissman, S. M., and Kucherlapati, R. (1997b). Identification, characterization, and precise mapping of a human gene encoding a novel membrane-spanning protein from the 22q11 regio deleted in velo-cardio-facial syndrome. *Genomics* **42,** 245–251.

Smith, C. A., Driscoll, D. A., Emanuel, B. S., McDonald-McGinn, D. M., Zackai, E. H., and Sullivan, K. E. (1998). Increased prevalance of IgA deficiency in patients with chromosome 22q11.2 deletion syndrome (DiGeorge syndrome/velocardiofacial syndrome). Submitted for publication.

Stevens, C. A., Carey, J. C., and Shigeoka, A. O. (1990). DiGeorge anomaly and velo-cardio-facial syndrome. *Pediatrics* **85,** 526–530.

Strachan, T., and Read, A. P. (1994). PAX genes. *Curr. Opin. Genet. Dev.* **4,** 427–438.

Strong, W. B. (1968). Familial syndrome of right-sided aortic arch, mental deficiency and facial dysmorphism. *J. Pediatr.* **73,** 882–888.

Sullivan, K. E., McDonald-McGinn, D. M., Driscoll, D., Zmijewski,

C. M., Ellabban, A. S., Reed, L., Emanuel, B. S., and Zackai, E. H., Athreya, B. H., and Keenan, G. (1997). JRA-like polyarthritis in chromosome 22q11.2 deletion syndrome (DiGeorge anomald/velocardiofacial syndrome/conotruncal anomaly face syndrome). *Arthritis Rheum.* **40,** 430–436.

Sullivan, K. E., Jawad, A. F., Randall, P., Driscoll, D. A., Emanuel, B. S., McDonald-McGinn, D. M., Zackai, E. H. (1998). Lack of correlation between impaired T cell production, immunodeficiency and other phenotypic features in chromosome 22q11.2 deletion syndromes (DiGeorge syndrome/velocardiofacial syndrome). *Clin. Immunol. and Immunopathol.* **86,** 141–146.

Sutherland, H. F., Wadey, R., McKie, J. M., Tayor, C., Atif, U., Johnstone, K. A., Halford, S., Kim, U.-J., Goodship, J., Baldini, A., and Scambler, P. J. (1996). Identification of a novel transcript disrupted by a balanced translocation associated with DiGeorge syndrome. *Am. J. Hum. Genet.* **59,** 23–31.

Swillen, A., Devriendt, K., Legius, E., Eyskens, B., Dumoulin, M., Gewillig, M., and Fryns, J. P. (1997). Intelligence and psychosocial adjustment in velocardiofacial syndrome: A study of 37 children and adolescents with VCFS. *J. Med. Genet.* **34,** 453–458.

Takahashi, K., Kido, S., Hoshino, K., Ohashi, H., and Fukushima, Y. (1995). Frequency of a 22q11 deletion in patients with conotruncal cardiac malformations: a prospective study. *Eur. J. Pediatr.* **154,** 878–881.

Taylor, C., Wadey, R., O'Donnell, H., Roberts, C., Mattei, M.-G., Kimber, W.L., Wynshaw-Boris, A., and Scambler, P. J. (1997). Cloning and mapping of murine Idd: Conserved synteny of the DiGeorge syndrome critical region. *Mamm. Genome* **8,** 371–375.

Van Mierop, L. H. S. and Kutsche, L. M. (1986). Cardiovascular anomalies in DiGeorge syndrome and importance of neural crest as a possible pathogenic factor. *Am. J. Cardiol.* **58,** 133–137.

Wadey, R., Daw, S., Taylor, C., Atif, U., Kamath, S., Halford, S., O'Donnell, H., Wilson, D., Goodship, J., Burn, J., and Scambler, P. (1995). Isolation of a gene encoding an integral membrane protein from the vicinity of a balanced translocation breakpoint associated with DiGeorge syndrome. *Hum. Mol. Genet.* **4,** 1027–1033.

Wilming, L. G., Snoeren, C. A. S., van Rijswijk, A., Grosveld, F., and Meijers, C. (1997). The murine homologue of HIRA, a DiGeorge syndrome candidate gene, is expressed in embryonic structures affected in CATCH22 patients. *Hum. Mol. Genet.* **6,** 247–258.

Wilson, D. I., Cross, I. E., Goodship, J. A., Coulthard, S., Carey, A. H., Scambler, P. J., Bain, H. H., Hunter, A. S., Carter, P. E., and Burn, J. (1991). DiGeorge syndrome with isolated aortic coarctation and isolated ventricular septal defect in three sibs with a 22q11 deletion of maternal origin. *Br. Heart J.* **66,** 308–312.

Wilson, D. I., Goodship, J. A., Burn, J., Cross, I. E., and Scambler, P. J. (1992). Deletions within chromosome 22q11 in familial congenital heart disease. *Lancet* **340,** 573–575.

Wilson, D. I., Britton, S. B., McKeown, C., Kelly, D., Cross, I. E., Strobel, S., and Scambler, P. J. (1993). Noonan's and DiGeorge syndromes with monosomy 22q11. *Arch. Dis. Child.* **68,** 187–189.

Young, D., Shprintzen, R. J., and Goldberg, R. B. (1980). Cardiac malformations in the velocardiofacial syndrome. *Am. J. Cardiol.* **46,** 643–648.

Zieger, B., Hashimoto, T., and Ware, J. (1997). Alternative expression of platelet glycoprotein Ibβ mRNA from an adjacent 5′ gene with an imperfect polyadenylation signal sequence. *J. Clin. Invest.* **99,** 520–525.

27

Genetics of Human Left–Right Axis Malformations

Brett Casey*[*] and Kenjiro Kosaki[†]

[*]*Department of Pathology, Baylor College of Medicine and Texas Children's Hospital, Houston, Texas, 77030, and*
†Department of Pediatrics, Keio University School of Medicine, Shinjuku, Tokyo 160, Japan

I. Introduction

Development of anatomic left–right (L–R) asymmetry among the thoracic and abdominal organs is an essential component of human embryogenesis. Failure to establish this normal organ position (situs solitus) results in heterotaxy, a term loosely translated from its Greek origins as "other arrangement." Here we describe the clinical genetics of familial heterotaxy, with particular attention to phenotypic variability and to the genetic implications of sporadic cases. Ascertainment of large families with heterotaxy has permitted molecular genetic analysis, which we describe in detail for an X-linked form of the disease.

II. Review of Relevant Embryology and Anatomy

During embryogenesis, anatomic asymmetry emerges from underlying symmetry in three different patterns (Fig. 1). First, all unpaired organs of the chest and abdomen begin development in the midline and subsequently lateralize to their adult positions. The embryonic heart is the first of these structures to show morphologic asymmetry, with the midline heart tube beginning its rightward looping on Embryonic Day 23. During the next 5 weeks, this single tube folds and is remodeled into the asymmetric, four-chambered heart of the newborn. Development of L–R asymmetry in the abdomen begins and ends later than in the chest. At 35

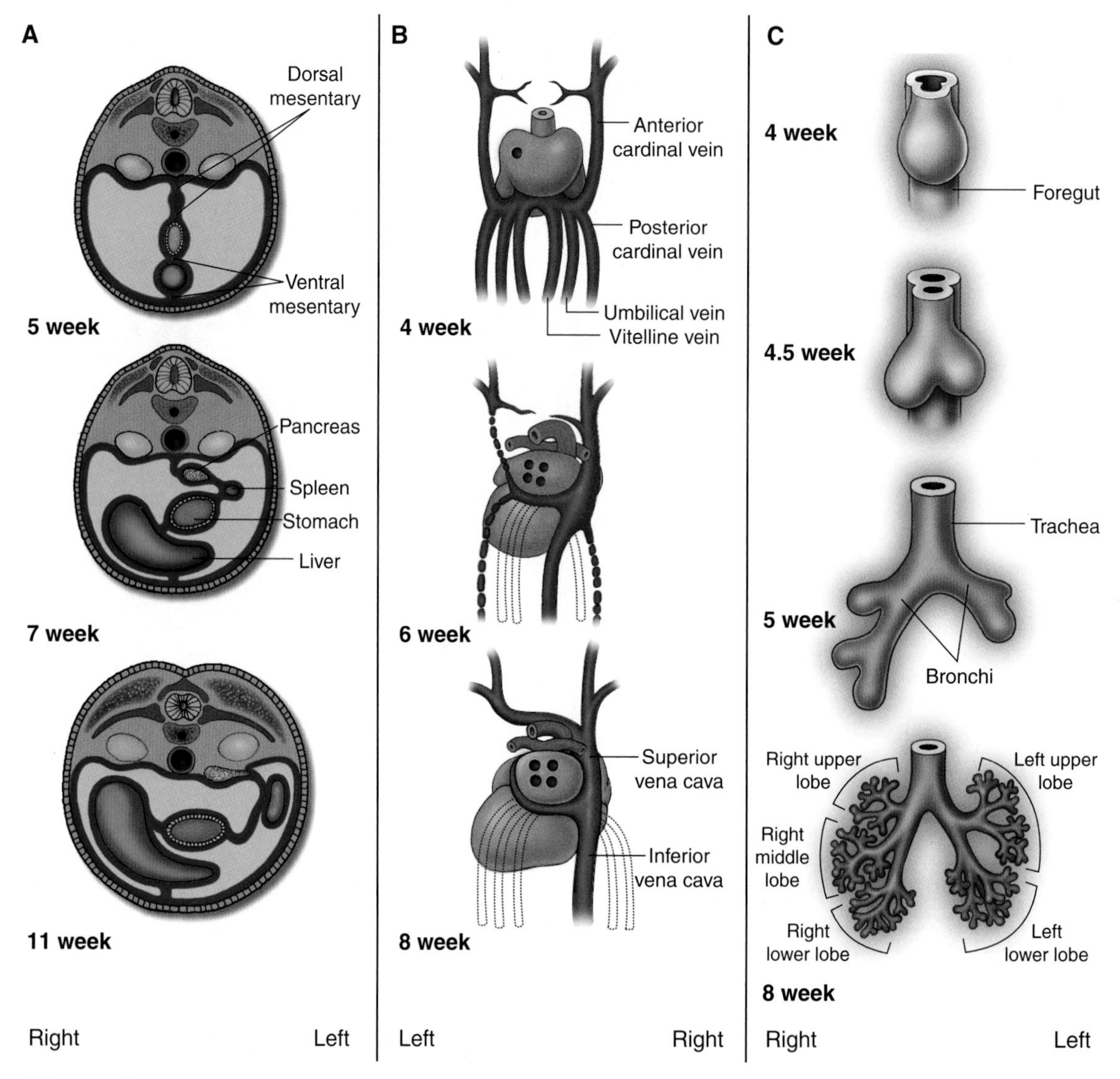

Figure 1 Three different patterns of developing left–right asymmetry; weeks after conception are given for each. (A) Unpaired organs of the abdomen (transverse view) develop within a connective tissue sheath (mesentery) and rotate from the midline into characteristic left–right position [adapted, with permission, from Sadler, T. W. (1985). *Langman's Medical Embryology*, Fig. 11-10, p. 162. Williams & Wilkins, Baltimore]. (B) Paired, bilaterally symmetric veins of the embryo [posterior (dorsal) view] regress differentially, resulting in unpaired, lateralized structures [adapted, with permission, from Larsen, W. (1993). *Human Embryology*, Fig. 7-10, p. 142. Churchill-Livingstone, New York]. (C) Right and left bronchial trees of the lungs [anterior (ventral) view] begin and end development as paired, bilaterally asymmetric structures [adapted, with permission, from Moore, K. (1993). *The Developing Human: Clinically Oriented Embryology,* Fig. 11-6, p. 230. Saunders, Philadelphia].

days the stomach starts its 90° rotation, and by Day 77 the 270° looping of the small and large intestine is complete. During the 1-month period of gut rotation, the liver, gallbladder, pancreas, and spleen also assume their characteristic positions with respect to the L–R axis (Fig. 1A).

Second, unlike the asymmetric, unpaired organs of the chest and abdomen, many of the unpaired lateralized arteries and veins first arise as paired structures with subsequent regression of their mirror-image counterparts. This is exemplified schematically in Fig. 1B for the major veins draining into the developing heart. Asymmetric regression of embryonic veins, accompanied by some remodeling, results in transformation of a paired, bilaterally symmetric system into one of unpaired asymmetry. Third, a review of lung development (Fig. 1C) illustrates that L–R asymmetry can be found in paired organs whose initial development is asymmetric.

The results of these three patterns of asymmetric development are the characteristic L–R differences identified by imaging studies and to a lesser extent by physical examination: lungs trilobed on the right and bilobed on the left, left-pointing cardiac apex and left-sided aortic arch, right-sided liver and gallbladder, left-sided stomach and spleen, and right-to-left course of the large intestine. Any departure from this normal position results by definition in a malformation of L–R axis development.

III. Terminology of Abnormal Left–Right Asymmetry

Anyone approaching the literature of human left–right axis (LRA) malformations finds a confusing array of terms used to describe them. Everyone appears to agree that the normal L–R anatomic arrangement is called situs solitus. Mirror-image reversal of all asymmetric structures has been given a variety of labels, most commonly situs inversus, complete or total situs inversus, and situs inversus totalis. When the entire anatomic LRA is neither normal nor mirror-image reversed, the resulting phenotype has been called situs ambiguus, partial situs inversus, heterotaxy or heterotaxia (sometimes accompanied by the adjective, visceral), laterality or isomerism sequence, and Ivemark, asplenia, or poly(a)splenia syndrome.

Part of the problem with the terminology arises because it is often unclear whether a term refers to an individual structure or to the organism as a whole and whether it describes the underlying disease (abnormal LRA development) or a particular manifestation in any single individual. For simplicity's sake, we favor a somewhat heretical nomenclature. First, we reserve the term situs and its accompanying modifiers to summarize the L–R anatomy of the entire organism: situs solitus for normal, situs inversus for complete, mirror-image reversal of all asymmetric structures, and situs ambiguus for any other abnormality of LRA development. Positional malformations within situs ambiguus individuals are described using the words right, left, and midline, as in "right-sided stomach and spleen" rather than "abdominal situs inversus." An interesting problem of nomenclature arises when families are identified with individuals ranging in phenotype from situs inversus to situs ambiguus and (among obligate carriers) situs solitus. What shall we call the disease running through this family? We favor heterotaxy, a term traditionally used more narrowly to describe individual structures whose L–R position is neither normal nor completely reversed. Heterotaxy implies a primary, underlying disorder of LRA development, usually unaccompanied by other anatomic or physiologic abnormalities. Thus, an individual with a primary abnormality in LRA development has heterotaxy, and his or her overall phenotype is either situs inversus or situs ambiguus.

IV. Syndromic Left–Right Axis Malformations

Sometimes LRA malformations arise as one variable manifestation of a broader spectrum of defects but by themselves are not required features of the overall diagnosis. By far the most common of these so-called syndromic LRA defects is situs inversus occurring as one manifestation of the immotile cilia syndrome (ICS) (Afzelius and Mossberg, 1994). Affected individuals suffer from chronic respiratory tract infections and from a variable combination of infertility (in males), chronic ear infections, and decreased or absent sense of smell. These problems arise as a result of defective cilia and flagella; hence, the diagnosis, ICS. The cilia are functionally abnormal, and electron microscopy usually reveals the absence or abnormalities of the dynein arms connecting the nine pairs of microtubules.

Almost all familial occurrences of ICS are limited to affected offspring of unaffected (and sometimes related) parents with no apparent gender bias, hence the inference of autosomal recessive inheritance. All affected individuals harbor the ciliary and flagellar defects, but only about half are situs inversus, whereas the remainder are situs solitus. To our knowledge situs ambiguus is extremely rare among individuals with ICS (Schidlow *et al.,* 1982; Teichberg *et al.,* 1982; Gershoni-Baruch *et al.,* 1989). Thus, there appears to be a randomization of the overall direction of L–R asymmetry. The relationship between the ciliary defects and LRA formation is unknown.

Albeit much rarer than ICS, agnathia/holoprosencephaly (Stoler and Holmes, 1992) and short-rib/polydactyly (Urioste *et al.,* 1994) syndromes also have situs abnormalities as frequent manifestations. Both situs inversus and situs ambiguus have been described in each of these disorders.

V. Spectrum of Malformations in Situs Ambiguus

Situs ambiguus describes an overall anatomic arrangement that suggests a randomization of L–R position along the superior–inferior axis (anterior–posterior in nonbipedal organisms). Any structure with L–R asymmetry can be normal, completely reversed, or neither. The collections of malformations are often fatal. The heart appears particularly sensitive to perturbation in normal L–R positional information because most children recognized to be situs ambiguus manifest complex cardiac defects. Other organs are affected as well. The anatomy of each lung can be that of either the normal right or normal left. In the abdomen, the spleen often (but not always) is abnormal in position, number, or both. A right-sided stomach and intestinal malrotation are not uncommon. Often the liver is in the midline or reversed in the relative sizes of its right and left lobes.

Malformations among organ systems within a single individual appear in general to arise independently so that, for example, one cannot predict what the heart will look like based on the stomach position or vice versa. There does appear to be a nonrandom association, however, with asplenia or right-sided spleen and a more severe, complex variety of heart malformations (Polhemus and Schafer, 1952; Ivemark, 1955). The heart malformations seen in polysplenia, on the other hand, tend to be less severe, and one report estimates that the heart is unaffected among 5–10% of individuals with polysplenia (Winer-Muram and Tonkin, 1989).

Sometimes the overall anatomic arrangement suggests a retention of symmetry not usually found in the adult organism. The terms bilateral right sidedness (trilobed lungs, an external right atrial appearance to atria, midline liver, and asplenia) and bilateral left sidedness (bilobed lungs, both atria with left atrial appearance externally; also midline liver and polysplenia) have been invoked to describe such an arrangement. While common, these anatomic arrangements are not absolute and do not breed true within families.

Malformations other than those of obvious asymmetric positioning are seen among individuals with situs ambiguus. None of these malformations occurs with enough frequency to be considered part of the usual spectrum of disease, but they do occur more commonly than one would expect by chance. These additional malformations fall into two broad categories: those that may be secondary to altered organ positioning during embryonic development and those that arise in structures that are symmetric about the LRA. Failure to develop, underdevelopment, and regression (aplasia/agenesis, hypoplasia, and atresia) characterize the first group (Fig. 2). Of the aplasias, asplenia is the most common, and absent gallbladder is sometimes seen as well. Sections of bowel can be atretic, the pancreas can be small, and the bile ducts outside the liver can be atretic. These abnormalities may result from mechanical or developmental effects of aberrant positioning (Fig. 2A).

Many of the venous defects commonly observed in situs ambiguus could be interpreted as a failure in the transition from embryologic venous symmetry to adult asymmetry (Fig. 2B) (Freedom *et al.*, 1974). Bilateral superior venae cava (SVC) would result from persistence of the left anterior cardinal vein, whereas regression of the right anterior cardinal vein would lead to left-sided SVC. Regression of the right vitelline vein instead of the left one would lead to absent inferior vena cava (IVC).

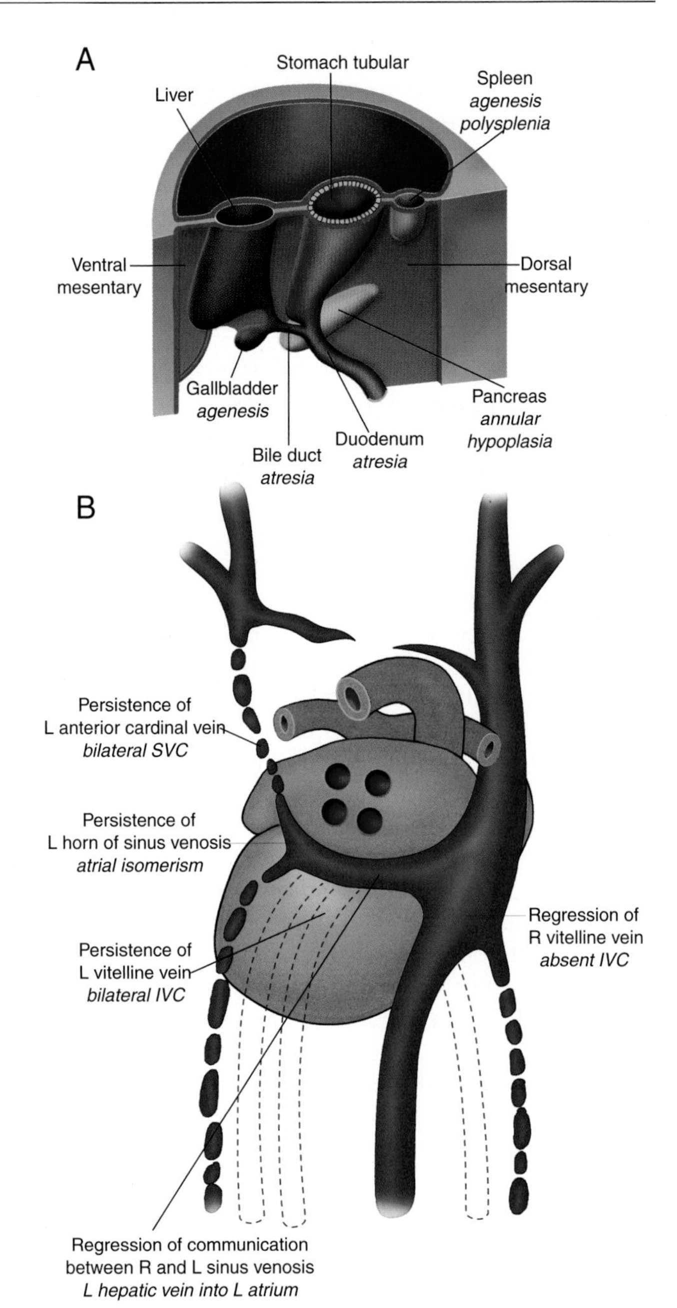

Figure 2 Correlation of abnormal left–right axis formation and some malformations seen in heterotaxy. (A) nonpositional malformations (italics) in the abdomen [adapted, with permission, from Sadler, T. W. (1985). *Langman's Medical Embryology*, Fig. 11-9, p. 162. Williams & Wilkins, Baltimore]. (B) Abnormal regression or persistence of embryonic veins result in characteristic venous malformations of heterotaxy [adapted, with permission, from Larsen W. (1993). *Human Embryology,* Fig. 7-10, p. 142. Churchill-Livingstone, New York].

Malformations of the second group most often occur in the midline along the vertebral column. Defects of neural tube closure are particularly common in this group. Hindgut malformations (e.g., anal atresia or stenosis) are also seen, more often in males and particularly in familial cases with inheritance patterns consistent with X chromosome linkage. Outside of the midline, urinary tract anomalies (renal agenesis and hypoplasia, ureteral malformations) are also seen with some frequency (Phoon and Neill, 1994).

VI. Familial Heterotaxy: Inheritance Patterns and Phenotypic Variability

The literature regarding the spectrum of malformations in heterotaxy has been derived largely from the study of sporadic cases, with particular emphasis on autopsy series. Perforce, this leads to an ascertainment bias in favor of the most severely affected individuals. Familial studies provide the opportunity to appreciate a fuller spectrum of malformations. Furthermore, these families can be used as starting material for the elucidation of the underlying molecular genetics of LRA malformations.

Prior to the mid-1990s, it was thought that autosomal recessive inheritance accounted for most cases of situs abnormalities. Certainly, it has been well documented that ICS with its accompanying situs inversus is inherited as an autosomal recessive disease (Afzelius and Mossberg, 1994). Familial situs inversus outside the setting of ICS has been reported a few times with inheritance patterns suggesting an autosomal recessive trait (Mital *et al.,* 1974; Chib *et al.,* 1977; Cockayne, 1938). In one reported family, four individuals across three generations are situs inversus but do not have ICS. The degree of consanguinity among the relevant matings is quite high, so the inheritance may still be autosomal recessive rather than dominant (Corcos *et al.,* 1989).

Familial situs ambiguus has been reported more frequently than familial situs inversus (Burn, 1991). The usual descriptions are of two or more affected siblings born to normal, often consanguineous parents. Both autosomal dominant and X-linked inheritance were thought to be distinctly uncommon based on the infrequency with which they were reported in the literature and, for the latter, because of relatively equal numbers of affected males and females. Since the malformations of situs ambiguus are quite often fatal, however, dominant inheritance may be difficult to ascertain through clinical family studies.

It has been widely assumed that situs inversus and situs ambiguus do not occur within the same family. As mentioned previously, situs ambiguus in ICS is extremely rare. Reports of situs ambiguus offspring of a situs inversus parent have appeared rarely (Niikawa *et al.,* 1983). Burn and colleagues failed to identify situs inversus among 201 first-degree relatives of 98 unrelated situs ambiguus individuals (Burn, 1991).

Recently, we (Casey *et al.,* 1996) and others (Alonso *et al.,* 1995) have identified several families suggesting that autosomal dominant inheritance may be much more common than previously supposed, and that situs inversus and situs ambiguus can indeed occur in the same family. Figure 3 shows pedigrees of three kindreds that we have studied in some detail. All share the characteristic that situs inversus and situs ambiguus are present among different members of the same family. Furthermore, obligate carriers who are situs solitus appear in each family. Affected individuals of both sexes in

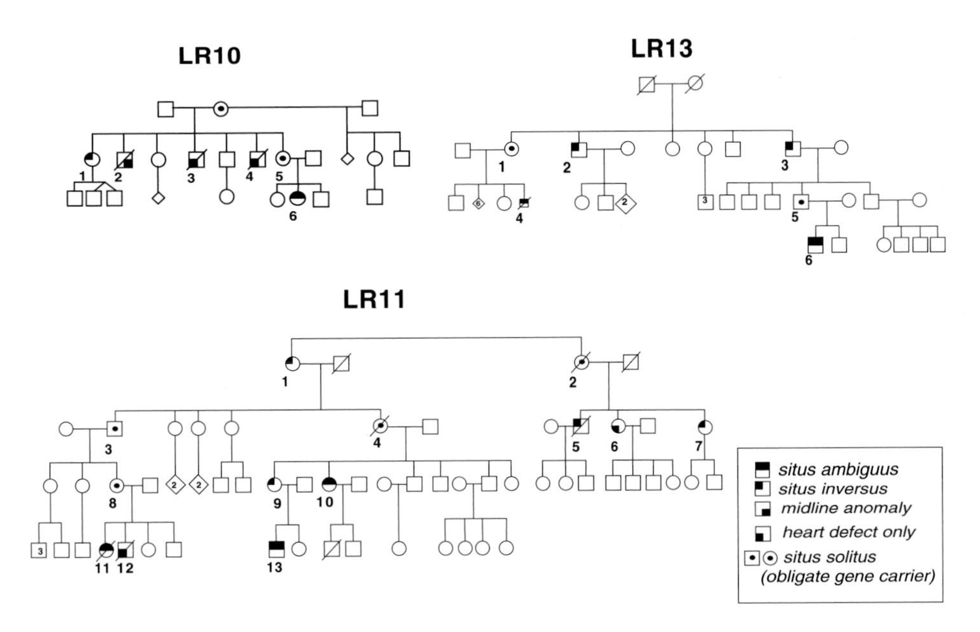

Figure 3 Heterotaxy families with presumed autosomal dominant inheritance.

more than one generation suggests autosomal dominant inheritance, although X-linked transmission cannot be excluded by inspection except in family LR13. Note that in four of the five cases of situs ambiguus, none of the first-degree relatives is situs inversus. Furthermore, if one counts the founders for each pedigree, there are 9 individuals who are obligate disease-gene carriers but are themselves situs solitus.

Ascertainment of these families has important clinical implications. Individuals with asymptomatic situs inversus may be at risk of having offspring with situs ambiguus and its accompanying complications. First and second-degree relatives should be examined carefully for clinically silent manifestations of heterotaxy, including situs inversus or mild manifestations of situs ambiguus (e.g., minor cardiac or venous anomalies and right-sided stomach). These families also suggest that some sporadic cases of heterotaxy may have arisen from a dominant-acting gene, either as a new mutation or inherited from an unaffected carrier. In fact, identification of several unaffected carriers in these families may account in part for the relatively low recurrence risk seen among siblings of situs ambiguus individuals: They may have inherited the disease gene, but for reasons yet unknown, they are not affected.

Notable in two of these families is the appearance of isolated heart malformations in relatives of individuals with heterotaxy. This association has been noted by other observers (Burn, 1991; Alonso *et al.,* 1995) and leads to an intriguing hypothesis: Could some cases of sporadic, isolated heart malformations be manifestations of abnormal LRA development and for some reason the other organs were not affected? Testing of this hypothesis awaits the identification of genes mutated in individuals with indisputable heterotaxy and searching for mutations in these genes among individuals with isolated, heterotaxy-like heart malformations.

VII. Epidemiology of Heterotaxy

It has been estimated that *situs inversus* occurs with an incidence of 1/8000 to 1/25,000 live births, with approximately 1/5 to 1/4 of cases associated with underlying ICS (Afzelius and Mossberg, 1995). These data may underestimate the true incidence of *situs inversus*, since by itself a mirror-image reversal of L–R asymmetry would pose no detriment to the individual. Incidence data for situs ambiguus have been provided through epidemiological studies of congenital cardiovascular disease. The Baltimore–Washington Infant Study estimated an incidence of 1.44/10,000 for all cardiac defects associated with L–R asymmetry malformations (cardiac and/or noncardiac), including corrected (levo) transposition of the great arteries (i.e., with ventricular inversion) (Ferencz *et al.,* 1997). Again, these data may underestimate the true incidence of situs ambiguus because cases of LRA malformations with normal hearts or those with clinically silent cardiac malformations would not have been ascertained in this study.

VIII. Molecular Genetics of Heterotaxy

Advances in human genetics in the past decade have made it possible to begin understanding the molecular pathogenesis of human heterotaxy. The first step involves the identification of genes in which mutations are associated with situs abnormalities. Characterization of candidate genes and positional cloning will provide the most fruitful approaches. The first of these involves isolation of human homologs of genes implicated in LRA development in other organisms. These genes are then analyzed for mutations in unrelated heterotaxy cases. Positional cloning involves the identification of disease genes based solely on their position in the genome. That position can be determined through combinations of linkage analysis, detection of cytogenetic abnormalities, and discovery of small chromosomal rearrangements (e.g., deletions) unseen by routine karyotyping.

The molecular events governing vertebrate L–R asymmetry have begun to emerge from the study of model organisms. The fundamental features of this process will probably be conserved in humans, and one would predict that these genes will be candidates for human heterotaxy. For example, the murine gene *Nodal* is a transforming growth factor-β family member whose normally asymmetric left-sided expression is disrupted in mouse models of LRA development (Collignon *et al.,* 1996; Lowe *et al.,* 1996). Furthermore, some (but not all) progeny mice that are doubly heterozygous for null mutations in *Nodal* and in the transcription factor *HNF-3β* manifest LRA malformations (Collignon *et al.,* 1996), suggesting that heterozygous mutations in *Nodal* could be associated with human situs abnormalities. Demonstrating the point, we have recently isolated human *NODAL* and identified missense amino acid substitutions among a small percentage of heterotaxy cases (M. T. Bassi and B. Casey, unpublished data). One of these individuals also has a mutation in *HNF-3β*. These preliminary results suggest that heterotaxy is likely to be a genetically heterogeneous disease and that in some cases multigenic inheritance may be responsible for the phenotype.

In 1995, putative mutations in a gap junction gene, *connexin 43* (*cx43*), were reported in association with situs ambiguus (Britz-Cunningham *et al.,* 1995). A heterogeneous group of 30 heart transplant patients—including 6 with situs ambiguus—were analyzed by sequencing the 400-base pair portion of *cx43* encoding the cytoplasmic tail of the protein. All 6 situs ambiguus individuals had *cx43* base change. Five of the patients harbored an identical S364P missense substitution, suggesting that mutation occurs frequently among mutant *cx43* chromosomes. It was surprising, however, that all the situs ambiguus individuals in this study had a *cx43* mutation given the genetic heterogeneity of LRA defects suggested by family studies. Subsequent associations of *cx43* mutations with situs ambiguus have not been reported, however, and we and others have failed to detect *cx43* mutations in more than 65 unrelated situs ambiguus patients (Gebbia *et al.,* 1996; Penman-Splitt *et al,* 1997). Furthermore, neither the absence (Reaume *et al.,* 1995) nor misexpression of *cx43* (Ewart *et al.,* 1997) results in situs abnormalities in mice. We conclude that association of putative *cx43* mutations with heterotaxy awaits further clarification.

A small number of chromosomal abnormalities associated with situs ambiguus have been reported (Fig. 4). These may provide clues to the location of genes involved in LRA development. Particularly interesting are those cytogenetic locations that have been implicated in more than one unrelated individual with situs abnormalities. For example, a gene involved in LRA development may be located in 7q22 based on the detection of two independent translocations involving that region in association with situs ambiguus (Figs. 4a and 4b). Furthermore, a situs inversus individual has been identified in which both of his chromosomes 7 have been inherited from his father (Pan *et al.,* 1998). This child has cystic fibrosis (CF) due to the inheritance of a CF gene mutation carried by the father. In this case, uniparental disomy has uncovered one recessive disease (CF) and may be responsible for the appearance of another recessive phenotype (situs inversus). Some of these cytogenetic abnormalities also appear in situs solitus relatives, but recall that we have seen several situs solitus obligate carriers in familial heterotaxy. To our knowledge, none of the candidate LR genes identified in other organisms maps to one of these human cytogenetic breakpoints except one: *Nodal* maps to mouse chromosome 10 in a region syntenic to human chromosome 10q21–q23, lending further support to the hypoth-

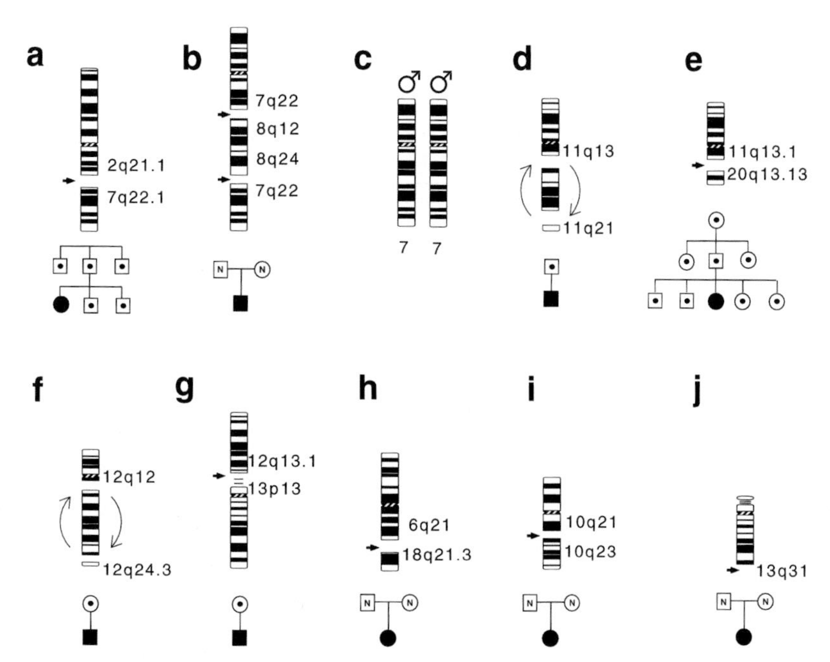

Figure 4 Reported chromosomal anomalies associated with heterotaxy (■, situs ambiguus; ⊡, situs solitus carrier of chromosomal anomaly; N, normal karyotype). a, balanced translocation (bt): 46,XX,t(2;7)(q21.1;q22.1) (Genuardi *et al.,* 1994); b, insertion: 46,XY,ins(7;8)(q22;q12q24) (Koiffmann *et al.,* 1993); c, uniparen-tal disomy: 46,XY, upd(7) (W. Craigen, personal communication); d, inversion: 46,XX,inv(11)(q13q25) (Fukushima *et al.,* 1993); e, *bt:* 46,XX,t(11;20)(q13.1;q13.13) (Freeman *et al.,* 1996); f, inversion: 46,XX,inv(12)(q12q24.3) (R. Sutton and C. Bacino, unpublished data); g, bt: 46,XX,t(12;13)(q13.1;p13) (Wilson *et al.,* 1991); h, bt: 46,XX,t(6;18)(q21;q21.3) (Kato *et al.,* 1996); i, deletion: 46,XX,del(10)(pter $\rightarrow$ q21::q23 $\rightarrow$ qter) (Carmi *et al.,* 1992); j, deletion: 46,XX,del(13)(q31) (Carmi *et al.,* 1992).

esis that *NODAL* mutations account (at least in part) for some cases of human heterotaxy.

Linkage analysis followed by physical mapping and a search for genes has proven effective in the identification of at least one molecular genetic cause of heterotaxy. We have studied several families with possible X-linked heterotaxy (Fig. 5). The largest of these, LR1, contains 9 males in two generations that are documented to be situs ambiguus. Four additional males, all of whom died in infancy or early childhood, are reported by their sisters to have had heart defects. A variety of additional malformations were noted in some of the affected males, including sacral agenesis, posteriorly placed anus, rectal stenosis, cerebellar hypoplasia, and arhinencephaly (the absence of the olfactory nerves). Linkage analysis of this family localized a gene for heterotaxy, *HTX1*, to Xq24–q27.1 (Casey *et al.*, 1993).

A second family, LR14, was also studied by linkage analysis, which yielded a significant statistical likelihood for the same general chromosomal location (M. Gebbia and B. Casey, unpublished data). Although the males in this family are situs ambiguus, some of the females are situs inversus, whereas other obligate heterozygotes are situs solitus. Finding complete, mirror-image reversal in some of these females suggests that the mutated gene functions quite early in LRA specification, at a time when the overall directionality of asymmetry is established.

The original family members from LR1 as well as eight others were analyzed further with additional markers localized to Xq24.1–q27 (Ferrero *et al.*, 1997). Two daughters of obligate carriers inherited disease chromosomes recombinant in this region, which were passed intact to their unaffected sons (Fig. 6a). We hypothesized that *HTX1* could not be in those regions of the disease chromosome inherited by the unaffected males, thus narrowing the search to a 1.2-Mb interval in Xq26.2. In order to provide independent confirmation of *HTX1* localization, a PCR-based search for submicroscopic deletions in this region was performed in unrelated males with sporadic or familial heterotaxy. A cluster of markers failed to amplify in a situs ambiguus individual who also had a deceased, affected brother. Discovery of this deletion supported the linkage analysis results from family LR1 and narrowed the search to a region of <900 kb (Fig. 6b).

A long-range restriction map was generated in order to facilitate further genomic cloning and to identify CG-rich regions (CpG islands) suggestive of nearby genes (Fig. 6b). Overlapping cosmids bridging and extending beyond the CpG islands were isolated. Putative exons were identified within these cosmids. One of these exons was used to isolate a gene with 98% amino acid sim-

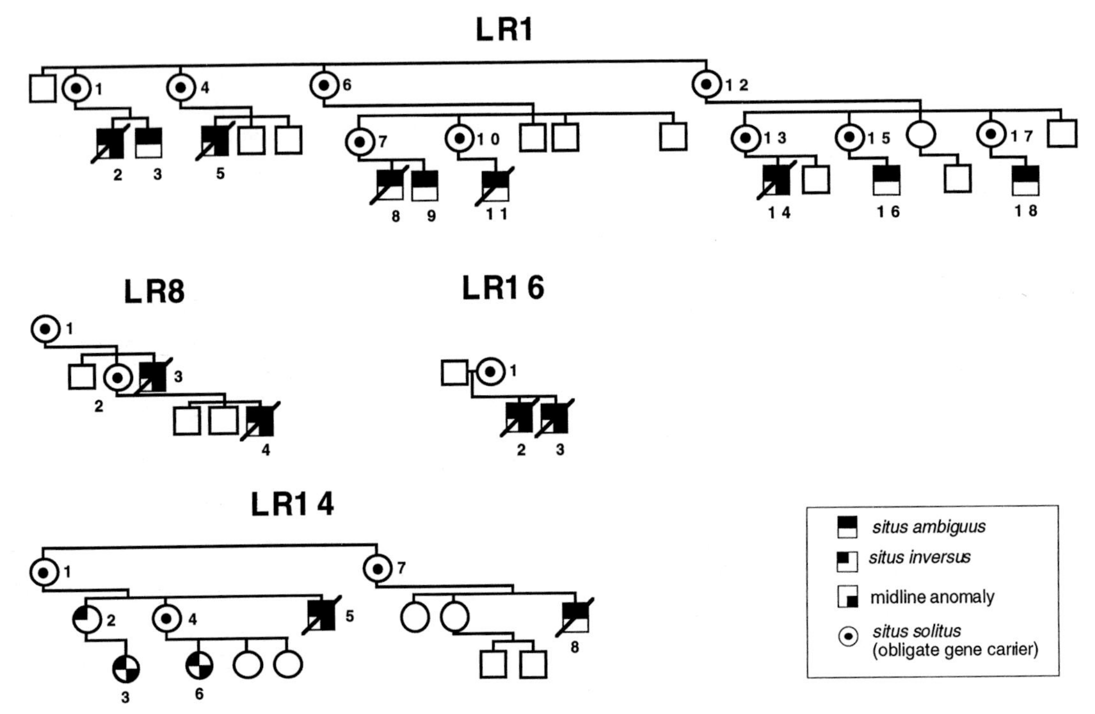

Figure 5 Heterotaxy families with presumed X-linked inheritance. Note frequency of accompanying midline malformations among all families and situs inversus among some females in family LR14.

A

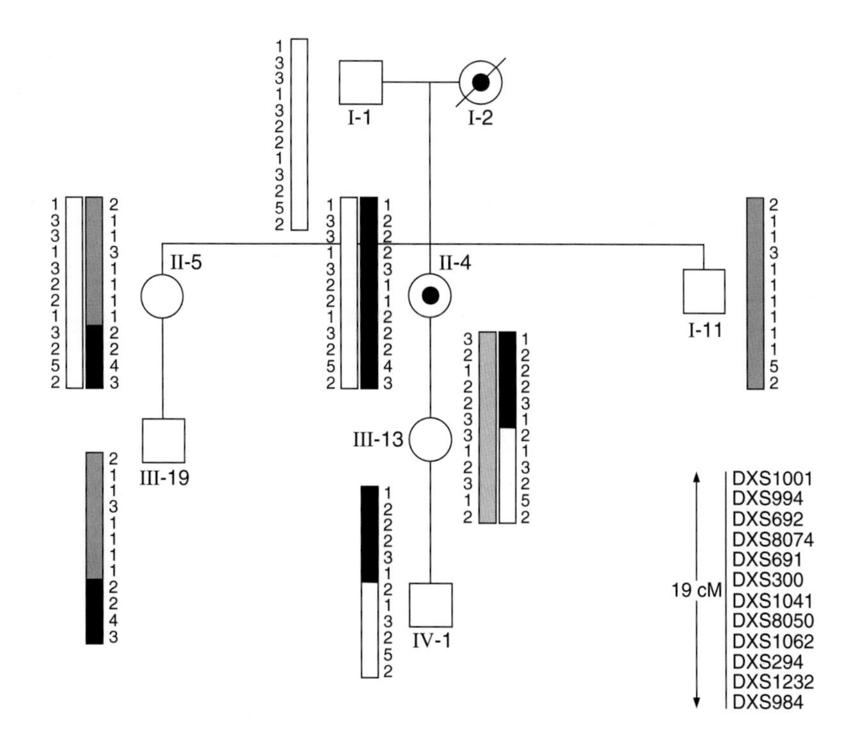

B

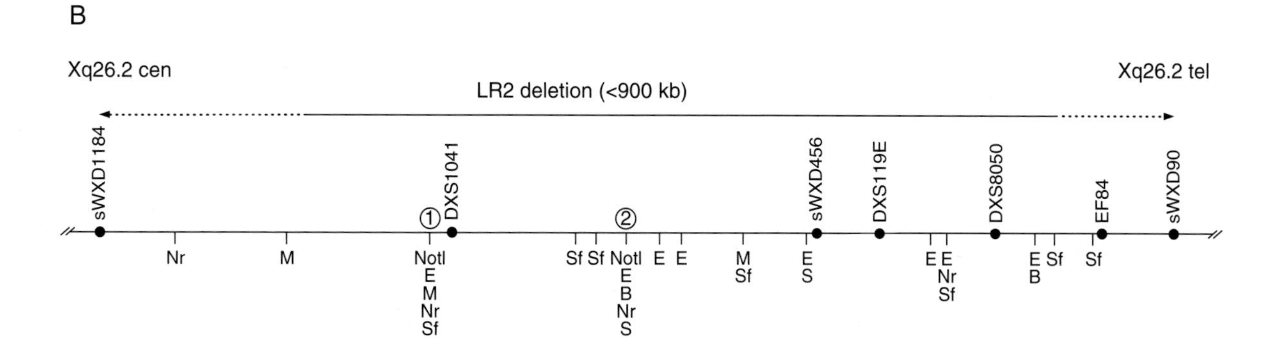

C

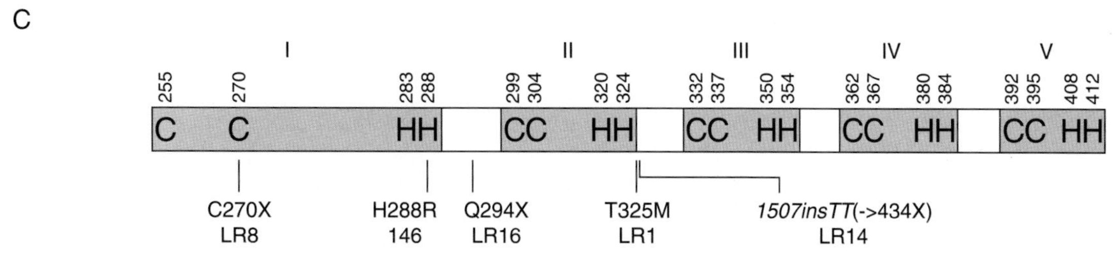

Figure 6 (a) Passage of recombinant disease chromosome to unaffected males in family LR1 [adapted, with permission, from Ferrero, G. B., Gebbia, M., Pilia, G., Witte, D., Peier, A., Hopkin, R. J., Craigen, W. J., Shaffer, L. G., Schlessinger, D., Ballabio, A., and Casey, B. (1997). *Am. J. Hum. Genet.* **61,** Fig. 1. p. 397. Copyright 1997 The University of Chicago Press]; circles with dots, obligate carriers; chromosomal segments between DXS1001 and DXS984 are represented by rectangles; marker order is listed according to the most recently published consensus map of the X chromosome (Nelson *et al.,* 1995); the solid black chromosome section shows the haplotype shared in common by all the obligate carriers and affected males. (b) Extent of deletion detected by PCR-based marker analysis in family LR2; PCR-based markers are indicated above the solid black line (note that DXS1041 and DXS8050 also appear in a), the extent of deletion in family LR2 is shown above the line (dotted lines indicate regions of uncertainty regarding breakpoint locations), and location of CpG islands is indicated by circled numbers. Nr, *Nru*I; M, *Mlu*I; E, *Eag*I; B, *Bss*HII; Sf, *Sfi*I; S, *Sac*I. (c) Amino acid position of mutations in *ZIC3* relative to the five zinc finger domains [adapted, with permission, from Gebbia, M., Ferrero, G. B., Pilia, G., Bassi, M.T., Aylsworth, A. S., Penman-Splitt, M., Bird, L. M., Bamforth, J. S., Burn, J., Schlessinger, D., Nelson, D. L., and Casey, B. (1997) *Nat. Genet.* **17,** Fig. 3, p. 307].

ilarity to *Zic3* (*zinc-finger in cerebellum*), one of four similar genes identified in mouse that encode putative zinc finger transcription factors (Aruga *et al.,* 1994; 1996; 1996 a,b). *Zic3* expression has been reported to occur in the primitive streak as early as Embryonic Day 7.0 and in structures in which malformations occasionally arise among human males with situs ambiguus (e.g., cerebellum, olfactory bulb, and sacrum) (Aruga *et al.,* 1996a; Nagai *et al.,* 1997). All the Zic proteins contain five C2H2 zinc fingers within a domain that shares homology with the *Drosophila* pair-rule gene, *odd-paired* (*opa*) (Benedyk *et al.,* 1994; Cimbora and Sakonju, 1995). Among its functions, *opa* is required for the appropriate expression of *wingless,* a segment-polarity gene that interacts with *hedgehog* to maintain parasegmental identity (Benedyk *et al.,* 1994). Notably, *sonic hedgehog* and a vertebrate homolog of *wingless, cWnt8c,* have been implicated in LRA formation in chick (Levin *et al.,* 1995; Levin, 1997).

We have identified *ZIC3* mutations in all of the families shown in Fig. 5 as well as in one sporadic case (Gebbia *et al.,* 1997). All of the mutations—one frameshift, two missense, and two nonsense—have arisen in the highly conserved domain that encompasses the five zinc fingers (Fig. 6c). Affected males are situs ambiguus, with variable combinations of complex heart malformations, altered lung lobation, splenic abnormalities, and gastrointestinal malrotation. In the sporadic case and in each family there is also at least one affected male with anal malformations. Heterozygous females in three of the families (LR1, -8, and -16) are anatomically normal. In family LR14, however, three females are situs inversus, i.e., they have mirror-image reversal of their normal L–R organ position. Malformations other than anal and LRA (e.g., spinal defects and kidney abnormalities) also appear in some individuals.

ZIC3 has not been implicated previously in vertebrate LR development and is the first gene to be associated unequivocally with human situs abnormalities. The identification of few heterotaxy families with obvious X-linked inheritance and the lack of male excess among sporadic cases suggest that mutations in *ZIC3* will account for only a small percentage of human LRA malformations. Based on our initial studies, however, the presence of anal malformations accompanying a LRA disturbance strongly suggests an underlying *ZIC3* mutation and warrants further molecular studies.

IX. Considerations for the Future

Clinical and molecular studies of familial heterotaxy suggest that human LRA malformations are genetically heterogeneous and quite variable in their manifestations. Several important questions remain unanswered. To what extent will genes implicated in LRA development in model organisms be responsible for human disease? Will multigenic inheritance account for some cases of human heterotaxy? Are some complex, isolated heart malformations actually unrecognized manifestations of aberrant LRA development? Will the positional cloning of additional human disease genes enlarge our general understanding of vertebrate LRA development? Based on recent results, we hope that answers to these questions will soon emerge.

References

Afzelius, B., and Mossberg, B. (1995). Immotile-cilia syndrome (Primary ciliary dyskinesia), including kartagener syndrome. *In* "The Metabolic and Molecular Basis of Inherited Disease" (C. Scriver, A. Beaudet, W. Sly, and D. Valle, eds.), pp. 3943–3954. McGraw-Hill, New York.

Alonso, S., Pierpont, M. E., Radtke, W., Martinez, J., Chen, S. C., Grant, J. W., Dahnert, I., Taviaux, S., Romey, M. C., Demaille, J., and Bouvagnet, P. (1995). Heterotaxia syndrome and autosomal dominant inheritance. *Am. J. Med. Genet.* **56,** 12–5.

Aruga, J., Yokota, N., Hashimoto, M., Furuichi, T., Fukuda, M., and Mikoshiba, K. (1994). A novel zinc finger protein, Zic, is involved in neurogenesis, especially in the cell lineage of cerebellar granule cells. *J. Neurochem.* **63,** 1880–1890.

Aruga, J., Nagai, T., Tokuyama, T., Hayashizaki, Y., Okazaki, Y., Chapman, and V., Mikoshiba, K. (1996a). The mouse *Zic* gene family. *J. Biol. Chem.* **271,** 1043–1047.

Aruga, J., Yozu, A., Hayashizaki, Y., Okazaki, Y., Chapman, V., and Mikoshiba, K. (1996b). Identification and characterization of *Zic4,* a new member of the mouse *Zic* gene family. *Gene* **172,** 291–294.

Benedyk, M., Mullen, J., and DiNardo, S. (1994). *Odd-paired,* a zinc finger pair-rule protein required for the timely activation of engrailed and wingless in *Drosophila* embryos. *Genes Dev.* **8,** 105–117.

Britz-Cunningham, S., Shah, M., Zuppan, C., and Fletcher, W. (1995). Mutations of the connexin43 gap-junction gene in patients with heart malformations and defects of laterality. *N. Eng. J. Med.* **332,** 1323–1329.

Burn, J. (1991). Disturbance of morphological laterality in humans. *In* "Biological Asymmetry and Handedness" (G. R. Bock and J. Marsh, eds.), pp. 282–299. Wiley, Chichester.

Carmi, R., Boughman, J. A., and Rosenbaum, K. R. (1992). Human situs determination is probably controlled by several different genes. *Am. J. Med. Genet.* **44,** 246–249.

Casey, B., Devoto, M., Jones, K., and Ballabio, A. (1993). Mapping a gene for familial situs abnormalities to human chromosome Xq24-q27.1. *Nat. Genet.* **5**(4), 403–407.

Casey, B., Cuneo, B. F., Vitali, C., van Hecke, H., Barrish, J., Hicks, J., Ballabio, A., and Hoo, J. J. (1996). Autosomal dominant transmission of familial laterality defects. *Am. J. Med. Genet.* **61,** 325–328.

Chib, P., Grover, D. N., and Shahi, B. N. (1977). Unusual occurrence of dextrocardia with situs inversus in succeeding generations of a family. *J. Med. Genet.* **14,** 30–32.

Cimbora, D. and Sakonju, S. (1995). *Drosophila* midgut morphogenesis requires the function of the segmentation gene *odd-paired. Dev. Biol.* **169,** 580–595.

Cockayne, E. A. (1938). The genetics of transposition of the viscera. *Q. J. Med.* **7,** 479–493.

Collignon, J., Varlet, I., and Robertson, E. J. (1996). Relationship between asymmetric *nodal* expression and the direction of embryonic turning. *Nature* (*London*) **381,** 155–158.

Corcos, A. P., Tzivoni, D., and Medina, A. (1989). Long QT syndrome and complete situs inversus. Preliminary report of a family. *Cardiology* **76,** 228–233.

Ewart, J. L., Cohen, M. F., Meyer, R. A., Huang, G. Y., Wessels, A., Gourdie, R. G., Chin, A. J., Park, S. M., Lazatin, B. O., Villabon, S., and Lo, C. W. (1997). Heart and neural tube defects in transgenic mice overexpressing the Cx43 gap junction gene. *Development* (*Cambridge, UK*) **124,** 1281–1292.

Ferencz, C., Loffredo, C. A., Correa–Villasenor, A., and Wilson, P. D. (1997). "Genetic & Environmental Risk Factors of Major Cardiovascular Malformations: The Baltimore–Washington Infant Study 1981–1989." Futura Publ. Co., Armonk, NY.

Ferrero, G. B., Gebbia, M., Pilia, G., Witte, D., Peier, A., Hopkin, R. J., Craigen, W. J., Shaffer, L. G., Schlessinger, D., Ballabio, A., and Casey, B. (1997) A submicroscopic deletion in Xq26 associated with familial *situs ambiguus*. *Am. J. Hum. Genet.* **61,** 395–401.

Freedom, R. M., Harrington, D. P., and White, R. I., Jr. (1974). The differential diagnosis of levo–transposed or malposed aorta. An angiocardiographic study. *Circulation* **50,** 1040–1046.

Freeman, S. B., Muralidharan, K., Pettay, D., Blackston, R., and May, K. (1996). Asplenia syndrome in a child with a balanced reciprocal translocation of chromosomes 11 and 20 [46,XX,t(11;20)(q13.1;q13.13)]. *Am. J. Med. Genet.* **61,** 340–344.

Fukushima, Y., Ohashi, H., Wakui, K., Fujiwara, M., Nakamura, Y., and Ogawa, K. (1993). Polysplenia syndrome and paracentric inversion of chromosome 11 [46,XX,inv(11)(q13q25)]. *Am. J. Hum. Genet.* **53** (Suppl.), Abst. No. 1543.

Gebbia, M., Towbin, J., and Casey, B. (1996). Failure to detect *connexin43* mutations in 38 cases of sporadic and familial heterotaxy. *Circulation* **94,** 1909–1912.

Gebbia, M., Ferrero, G. B., Pilia, G., Bassi, M. T., Aylsworth, A. S., Penman–Splitt, M., Bird, L. M., Bamforth, J. S., Burn, J., Schlessinger, D., Nelson, D. L., and Casey, B (1997). X–linked *situs* abnormalities result from mutations in *ZIC3*. *Nat. Genet.* **17,** 305–308.

Genuardi, M., Gurrieri, F., and Neri, G. (1994). Genes for split hand/split foot and laterality defects on 7q22.1 and Xq24–q27.1. *Am. J. Med. Genet.* **50,** 101.

Gershoni–Baruch, R., Gottfried, E., Pery, M., Sahin, A., and Etzioni, A. (1989). Immotile cilia syndrome including polysplenia, situs inversus, and extrahepatic biliary atresia. *Am. J. Med. Genet.* **33,** 390–393.

Ivemark, B. I. (1955). Implications of agenesis of the spleen on the pathogenesis of cono–truncus anomalies in childhood: Analysis of the heart malformations in splenic agenesis syndrome, with fourteen new cases. *Acta Paediatr.* **44,** 1–110.

Kato, R., Yamada, Y., and Niikawa, N. (1996). De novo balanced translocation (6;18)(q21;q21.3) in a patient with heterotaxia. *Am. J. Med. Genet.* **66,** 184–186.

Koiffmann, C., Wajntal, A., de Souza, D. H., Gonzalez, C., and Coates, M. (1993). Human situs determination and chromosome constitution 46,XY, ins(7;8)(q22;q12q24). *Am. J. Med. Genet.* **47,** 568–569.

Larsen, W. (1993). "Human Embryology." Churchill–Livingstone, New York.

Levin, M. (1997). Left–right asymmetry in vertebrate embryogenesis. *BioEssays* **19,** 287–296.

Levin, M., Johnson, R. L., Stern, C. D., Kuehn, M., and Tabin, C. (1995). A molecular pathway determining left–right asymmetry in chick embryogenesis. *Cell* (*Cambridge, Mass.*) **82,** 803–814.

Lowe, L., Supp, D., Sampath, K., Yokoyama, T., Wright, C., Potter, S., Overbeek, P., and Kuehn, M. (1996). Conserved left–right asymmetry of nodal expression and alterations in murine *situs inversus*. *Nature* (*London*) **381,** 158–161.

Mital, O. P., Prasad, R., and Rao, M. S. (1974). *Situs inversus* (totalis) among two brothers: A case report. *Indian J. Chest Dis.* **16,** 188–190.

Moore, K. (1993). "The Developing Human: Clinically–Oriented Embryology." Saunders, Philadelphia.

Nagai, T., Aruga, J., Takada, S., Gunther, T., Sporle, R., Schughart, K., and Mikoshiba, K. (1997). The expression of the mouse *Zicl, Zic2,* and *Zic3* gene suggests an essential role for *Zic* genes in body pattern formation. *Dev. Biol.* **182,** 299–313.

Nelson, D., Ballabio, A., Cremers, F., Monaco, A., and Schlessinger, D. (1995). Report of the sixth international workshop on human X chromosome mapping 1995. *Cytogene. Cell Genet.* **71,** 307–342.

Niikawa, N., Kohsaka, S., Mizumoto, M., Hamada, I., and Kajii, T. (1983). Familial clustering of situs inversus totalis and asplenia and polysplenia syndromes. *Am. J. Med. Genet.* **16,** 43–47.

Pan, Y., McCaskill, C., Thompson, K. H., Hicks, J., Casey, B., Shaffer, L. G., Craigen, W. J. (1998). Paternal uniparental disomy for chromosome 7 associated with complete situs inversus and immotile cilia. *Am. J. Hum. Genet.* **62**(6)**,** 1551–1555.

Penman–Splitt, M., Tsai, M., Burn, J., and Goodship, J. (1997). Absence of mutations in the regulatory domain of the gap junction protein connexin 43 in patients with visceroatrial heterotaxy. *Heart* **77,** 369–370.

Phoon, C. K., and Neill, C. A. (1994). Asplenia syndrome: Insight into embryology through an analysis of cardiac and extracardiac anomalies. *Am. J. Cardiol.* **73,** 581–587.

Polhemus, D. and Schafer, W. (1952). Congenital absence of the spleen: Syndrome with atrioventricularis and situs inversus. *Pediatrics* **9,** 696–708.

Réaume, A., de Sousa, P., Kulkarni, S., Langille, B., Zhu, D., Davies, T., Juneja, S., Kidder, G., and Rossant, J. (1995). Cardiac malformation in neonatal mice lacking Connexin43. *Science* **267,** 1831–1834.

Sadler, T. W. (1985). "Langman's Medical Embryology." Williams & Wilkins, Baltimore, MD.

Schidlow, D. V., Katz, S. M., Tutz, M.G., Donner, R.M., and Capasso S. (1982). Polysplenia and Kartegener syndromes in a sibship: Association with abnormal respiratory cilia. *J. Pediatr.* **100,** 401–403.

Stoler, J. M. and Holmes, L. B. (1992). A case of agnathia, situs inversus, and a normal central nervous system. *Teratology* **46,** 213–216.

Teichberg, S., Markowitz, J., Silverberg, M., Aiges, H., Schneider, K., Kahn, E., and Daum, F. (1982). Abnormal cilia in a child with the polysplenia syndrome and extrahepatic biliary atresia. *J. Pediatr.* **100,** 399–401.

Urioste, M., Martinez–Frias, M. L., Bermejo, E., Jimenez, N., Romero, D., Nieto, C., and Villa, A. (1994). Short rib–polydactyly syndrome and pericentric inversion of chromosome 4. *Am. J. Med. Genet.* **49,** 94–97.

Wilson, G. N., Stout, J. P., Schneider, N. R., Zneimer, S. M., and Gilstrap, L. C. (1991). Balanced translocation 12/13 and situs abnormalities: Homology of early pattern formation in man and lower organisms? *Am. J. Med. Genet.* **38,** 601–607.

Winer–Muram, H. T., and Tonkin, I. L. (1989). The spectrum of heterotaxic syndromes. *Radiol. Clin. North Am.* **27,** 1147–1170.

X

Lessons from Skeletal Muscle

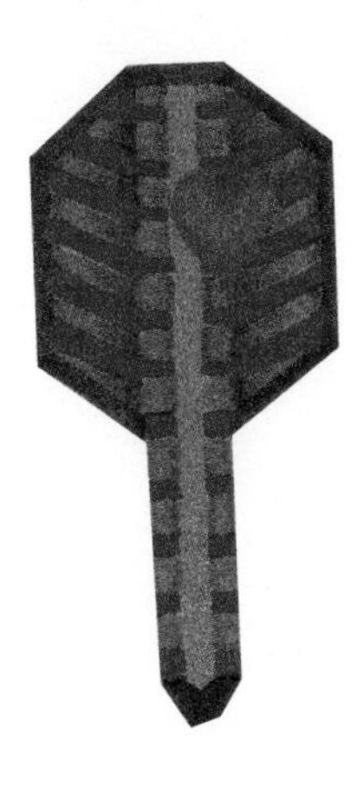

28

Molecular Genetics of Skeletal Muscle Development

Michael J. McGrew,* José Xavier-Neto,[†,‡] Olivier Pourquie,* and Nadia Rosenthal[‡]

*Developmental Biology Institute of Marseille, 13288 Marseille Cedex 9, France
[†]Laboratório de Biologia Molecular, Instituto do Coração, São Paulo SP 05403, Brazil
[‡]Cardiovascular Research Center, Massachusetts General Hospital–East, Boston, Massachusetts 02129

I. Introduction

The objective of this chapter is to explore the similarities and differences between the development of cardiac and skeletal muscle. For the past decade, skeletal muscle differentiation has served as a model system for delineating the gene regulatory pathways responsible for tissue specificity in vertebrate and invertebrate embryos. Our dramatically expanded understanding of heart development, which has emerged from discoveries made in the past few years, has rapidly narrowed the knowledge gap between the two striated muscle systems, such that a comparative analysis may now yield important clues for future research efforts in both fields.

Although vertebrate skeletal and cardiac muscle share a similar sarcomeric organization and function, and express many common protein products, they have different embryonic origins, arising from spatially separated populations of mesodermal progenitors. As described in Sections I and II, the cardiac cell lineage can be traced to a region of anterior lateral plate mesoderm, whereas skeletal muscle cells in the body are derived from cells within the segmented somites which in turn arise from the paraxial mesoderm. The timing of terminal differentiation during embryogenesis is another feature that dramatically distinguishes the two muscle systems. Cardiac precursors express differentiation markers long before they complete their morphogenetic program by migrating to their final locations in the primordial heart tube. Skeletal myoblasts, in contrast, either differentiate *in situ* during somite maturation or undergo differentiation only after they have migrated to their final destination in the embryo. However, the most striking difference between the two muscle cell precursor populations lies in the fact that cardiac precursors differentiate and proliferate at the same time, whereas skeletal myoblasts have to withdraw from the cell cycle in order to differentiate.

In light of these distinctions, many of the similarities between cardiac and skeletal muscle tissues may simply reflect function presupposing structure (Thompson, 1942) rather than a common embryonic developmental program. Nevertheless, it has become increasingly clear that many of the molecular networks regulating the expression of cardiac and skeletal muscle genes, as well as the signaling pathways directing morphogenetic processes in developing heart and skeletal muscle, have been conserved between the two tissue types, albeit with modifications that satisfy their different functional requirements.

The remarkable progress in the field of vertebrate skeletal myogenesis can be traced to the discovery that a muscle phenotype can be induced in cell culture by the controlled expression of specific regulatory genes. When the function of these genes is perturbed in the embryo, muscle cell development is profoundly affected. A confluence of molecular and genetic discoveries during the past decade has established skeletal muscle as a model system for studying transcriptional control of developmental processes.

The complex patterns of contractile skeletal tissue formation in the adult animal depend on early inductive signals that specify particular myogenic fates of primordial precursor cells in the embryonic somites. These cells then migrate to predictable positions in the vertebrate body plan and undergo terminal differentiation at precisely defined developmental stages. Subsequent modulation of the myogenic program by neuronal, hormonal, and mechanical stimuli results in more subtle structural distinctions, which reflect the diversity of functions in the adult skeletal musculature. Molecular analysis of the regulatory pathways underlying the generation of this diversity in skeletal muscle structure and function may provide the necessary tools to uncover equivalent levels of diversity shared by cardiac muscle tissues.

Current research into the molecular biology of both skeletal and cardiac myogenesis has largely concentrated on the fundamental questions of how and when a cell first becomes committed to a muscle phenotype and how tissue-specific gene expression is initiated. Recent progress in this field has emphasized the central role of transcriptional regulation in each step of myogenesis, from the initial selection of skeletal myogenic precursor cell populations from the embryonic mesoderm or of cardiac precursors from the embryonic epiblast to the activation of contractile protein gene expression in either tissue. The availability of skeletal muscle cell lines has enabled the characterization of multiple, muscle-specific DNA control elements which are important for tissue-specific gene expresssion and which constitute binding sites for regulatory proteins. Through transgenic mouse technology, many of these factors have been shown to be critical components of the genetic pathways leading to muscle determination and differentiation in the embryo.

The mechanisms underlying cardiac muscle differentiation and gene regulation are less well understood than those of skeletal muscle, despite the similarities between the two cell types. The more limited understanding in this field can be ascribed in part to the lack of useful cardiac precursor cell lines. In addition, cardiac muscle cell specification begins significantly earlier in development than does skeletal muscle specification. As a result, specified cardiac mesoderm can only be reliably isolated from other nonmuscle tissues when the cardiogenic program is already well under way and most cells are very close to the committed state.

Exploiting sophisticated *in vivo* technology, recent research on heart development has, however, surpassed and in many cases rendered obsolete the kind of molecular analysis in cell culture that originally dominated the skeletal muscle field. Key regulators of cardiac morphogenesis have been shown in the *in vivo* context to choreograph the transformation of seemingly amorphous groups of cardiac progenitor cells into a complex multichambered organ complete with a conduction system and responding to extrinsic signals such as hormones and neurotransmitters. In contrast, we are only just beginning to uncover the equivalent molecular players in the morphogenesis of skeletal muscle tissues.

In general, the factors involved in the initial establishment of skeletal muscle lineages appear to be distinct from those employed in the specification of cardiac muscle lineages described elsewhere in this book. Nevertheless, it is increasingly clear that skeletal and cardiac cell types employ common downstream pathways leading to myogenic cell differentiation and structural gene activation and to functional modulations in response to extrinsic physiologic signals. The two fields are poised to complement each other in unexpected ways. This review will serve to provide a current overview of skeletal muscle development in anticipation of common paradigms emerging from future research.

II. Gene Regulation in Skeletal Muscle

Given the functional similarities of contractile tissues, it is perhaps not surprising to find common aspects of skeletal and cardiac muscle gene regulation. In mammals, both tissues express developmental isoforms of their contractile protein counterparts during fetal development and switch to adult isoforms during the first week after birth, presumably in response to intrinsic programs as well as extrinsic signals, such as thyroid hormone. In the adult, injured or regenerating myocytes in skeletal muscles recapitulate embryonic gene expression patterns. Genes expressed in both cell types, such as those encoding muscle creatine kinase, myoglobin, actins, or β-myosin heavy chain, are likely to be controlled by *cis*-regulatory modules that respond to distinct *trans*-acting signals peculiar to each tissue.

Characterization of the primary genetic components responsible for the regulation of genes transcribed in a tissue-restricted manner has been a necessary prerequisite for building models of the molecular pathways leading to either skeletal or cardiac muscle differentiation. Due to the lack of cardiac tissue culture models, analysis of *cis*-acting DNA elements responsible for cardiac muscle-specific transcription has largely relied on the introduction of mutant or chimeric genes into the germline of transgenic mice (see Chapters 15, 16, and 19).

In contrast, skeletal myoblasts are readily propagated in culture, and myogenic differentiation can be induced by a relatively short exposure to serum-deficient medium, which results in fusion of monocytic myoblasts into multinucleate syncytia. Accompanying these morphological changes, the rapid activation of a battery of genes, including those which encode the components of the muscle contractile apparatus, is coordinated between nuclei in the appropriate stoichiometry to produce a functional contractile cell. From an experimental standpoint, the ability to culture skeletal muscle precursors, to transfect them with expression vectors, and to derive clonal lines from the resulting cultures represents a significant technical advantage. This has resulted in a plethora of information regarding the *cis*-regulatory elements that are employed in response to differentiation.

A. Muscle-Specific DNA Regulatory Elements

Although a common *cis*-acting regulatory element present in all skeletal muscle-specific control regions has not been found, several similar sequence motifs associated with evolutionarily unrelated muscle genes function as binding sites for common nuclear protein complexes. In many cases, deletion or mutation of these motifs impairs the ability of the control region to activate linked gene expression in both cell culture and transgenic animals, underscoring the functional importance of each motif in the muscle transcriptional control circuitry (Arnone and Davidson, 1997). In general, transcriptional regulation of muscle genes appears to reflect the paradigms established for constitutive transcriptional regulatory elements, in which multiple proteins bind to a complex and often overlapping array of recognition sequences. These sequences are often clustered in cassettes, with each driving a different aspect of the gene's expression pattern.

Not surprisingly, skeletal and cardiac muscle genes share *cis*-acting regulatory sequences that form binding sites for the same or closely related transcription factors. For example, two sequence motifs have emerged as critical components of the common regulatory circuitry in striated muscle gene expression: the E box (CANNTG) and an A/T-rich motif (CTA(A/T)4). Together these two sequences appear to form an essential composite for the activation of muscle-specific genes. E boxes were originally characterized by Ephrussi and colleagues (1985) as protein binding sites in the immunoglobulin heavy chain enhancer. In this enhancer, as well as in muscle-specific regulatory elements such as the MCK and MLC1/3 enhancer, E boxes are generally arranged in multiple copies within a given regulatory element (Weintraub *et al.,* 1990; Wentworth *et al.,* 1991) where they act cooperatively in transcriptional activation of linked genes. E boxes are often situated in close proximity to A/T-rich sequences, suggesting a possible synergistic relationship in the functions of the binding proteins that recognize the two motifs. As in other tissue-specific regulatory circuits, individual modules generally constitute necessary but not sufficient genetic information for restricting the expression of a gene to either skeletal or cardiac tissue. Numerous additional elements, such as those described in Section V, are clearly involved in distinguishing skeletal from cardiac gene expression. Ultimately it is the exclusive spectra of transcription factors binding to common sites that presumably determines the unique profile of downstream gene activation peculiar to each tissue type.

B. Myogenic Determination Factors

Identification of a unique group of transcription factors responsible for induction of the myogenic phenotype has represented a significant breakthrough in our understanding of the molecular pathways underlying skeletal muscle differentiation. The four myogenic determination factors, (MDFs), MyoD, myogenin, Myf5, and *MRF4,* were first characterized as nodal points in the myogenic program due to their potent activation of the muscle phenotype when overexpressed in transfected nonmuscle cell cultures (Davis *et al.,* 1987; Edmondson and Olson, 1989; Rhodes and Konieczny, 1989; Wright *et al.,* 1989; Miner and Wold, 1990; Braun *et al.,* 1991).

The four MDF genes have a common three-exon structure and are active almost exclusively in skeletal muscle lineages, with rare exceptions. All share an 80 amino acid basic helix–loop–helix (bHLH) domain which is conserved in many species (Olson, 1990; Venuti *et al.,* 1991; Weintraub *et al.,* 1991; Krause *et al.,* 1992; Muneoka and Sassoon, 1992; Emerson,1993). The basic domain mediates DNA binding to E boxes (Brennan *et al.,* 1991; Davis and Weintraub, *et al.,* 1992; Schwarz *et al.,* 1992; Yutzey and Konieczny, 1992) and the HLH domain enables protein dimerization between different members of the bHLH family (Lassar *et al.,* 1991; Rashbass *et al.,* 1992; Winter *et al.,* 1992).

The MDFs are expressed in specified and committed skeletal muscle precursors during development, and their expression is essential for normal skeletal myogenesis. This skeletal muscle paradigm, in which the same markers of commitment are also able to induce muscle differentiation in different cell types, has not yet found a parallel in cardiac myogenesis. The NK homeobox gene *Nkx2-5* is one of the earliest markers of cardiac commitment; however, it is also expressed in other tissues such as the endoderm and ectoderm flanking the cardiac mesoderm (see Chapters 7 and 16). These studies indicate that *Nkx2-5* expression per se is not sufficient to commit a given cell to a cardiac phenotype, although studies on *Xenopus* and zebrafish have suggested that its overexpression can enlarge the heart field (Cleaver *et al.,* 1996; Chen and Fishman, 1996). Similarly, members of the GATA transcription factor family are expressed in early precardiac mesoderm but are unable to convert uncommitted cells to a cardiogenic phenotype (see Chapter 17). There have been no reports on conversion of other cell types to the cardiac phenotype after forced expression of single or combinations of cardiac restricted transcription factors. Rather, evidence from heterokaryon studies suggests that the cardiac phenotype is recessive to other cell phenotypes (Evans *et al.,* 1994).

None of the MDFs is expressed in the heart. In fact, ectopic expression of at least one of these factors (MyoD) results in embryonic lethality due to severe cardiac defects (Miner *et al.,* 1992). However, MDFs belong to a superfamily of transcription factors, of which some are expressed in cardiac myocytes. Members of this superfamily include cardiac factors dHand and eHand (see Chapters 7 and 9). Differences in the primary structures of skeletal muscle-specific MDFs and other bHLH proteins, such as the Hands, presumably contribute to the different spectra of target genes activated in the two muscle types. The Hand proteins play important roles in both avian and mammalian heart development. Transfection of antisense oligonucleotides for both eHand and dHand mRNAs blocks avian cardiogenesis (Srivastava *et al.,* 1995), whereas targeted disruption of both genes in the mouse results in abnormal cardiac development and embryonic demise (Srivastava *et al.,* 1997; Firulli *et al.,* 1998; Riley *et al.,* 1998). These experiments suggest that both dHand and eHand are necessary for proper heart formation, although alone these bHLH proteins cannot induce the cardiac phenotype in other cell types.

C. Specificity of MDF Function in Muscle Cell Differentiation

All four MDFs can induce skeletal myogenesis when overexpressed in transfected nonmuscle cell cultures. However, they are not equally potent in transactivating muscle gene expression (Chakraborty and Olson, 1991; Fujisawa *et al.,* 1992; Mak *et al.,* 1992), suggesting that they may possess distinct activities. The bHLH proteins function in pairs to activate transcription by binding to their target sites. In muscle cells, the MDFs form heterodimers with other ubiquitously expressed bHLH proteins, known as E proteins, and activate gene transcription by binding to the E boxes in muscle-specific regulatory elements.

Overexpression of MDFs in nonmuscle cell culture also induces transcription of endogenous MDF genes (Braun *et al.,* 1989; Thayer *et al.,* 1989), suggesting that continuous, autoregulatory activation of MDF gene expression may be an important component of the commitment to a myogenic phenotype. The fact that E boxes have been found in the promoters of the MDF genes (Cheng *et al.,* 1993; Yee and Rigby, 1993) suggests that the MDF genes are targets for an autoactivation pathway which may help stabilize or reinforce the cell's commitment to the myogenic pathway.

Although E boxes all share the CANNTG sequence motif, differences in the binding site preferences of various bHLH heterodimers for different versions of the CANNTG motif and its surrounding sequences may play an important role in subtly regulating the expression of different muscle genes (Blackwell and Weintraub, 1990; Chakraborty and Olson, 1991; Schwarz *et al.,* 1992; Yutzey and Konieczny, 1992). It is likely that similar paradigms for E box specificity will be uncovered in the heart, although it is not yet known whether any of the E proteins constitute partners for the Hand proteins in cardiac muscle (see Chapter 9), and it is unclear how the subtle differences between highly conserved bHLH proteins allow discrimination between E box targets associated with skeletal and cardiac muscle-specific genes.

In skeletal muscle the tissue specificity of MDF function is modulated by a number of structural parameters (Brennan *et al.,* 1991; Davis and Weintraub, 1992). The crystal structure of the MyoD–DNA complex (Ma *et al.,*

1994) reveals that a MyoD protein homodimer makes direct contact with the DNA at the E box consensus site where two key amino acids (Ala114 and Thr115) in the basic domain, which are required for muscle-specific regulation by the MDF family, lie buried in the protein–DNA interface (Fig. 1). These two residues do not make direct contact with DNA but are probably important in positioning a third residue (Arg111) in the major groove for transcriptional activation. Perhaps further specificity is conferred by the formation of heterodimers between MDFs and different E proteins (Lassar *et al.*, 1991; Hu *et al.*, 1992; Rashbass *et al.*, 1992), some isoforms of which have yet to be characterized.

Alternatively, the unique amino and carboxyl termini of each MDF may permit specific interactions with different subsets of transcription factors, which may bind other muscle control elements. This possibility is supported by domain-swapping studies in which rearrangements of MDF unique regions confer altered affinities for their target sites (Chakraborty and Olson, 1991; Mak *et al.*, 1992; Schwarz *et al.*, 1992; Winter *et al.*, 1992). Thus, the cell specificity of bHLH action may lie in protein–protein interactions, imparting subtle changes on overall protein conformation. Future studies in this area will undoubtedly focus on detailed structural analysis of mutations in key residues of different bHLH proteins and on interactions with other transcription factors such as the MEF2 family, which is involved in both skeletal and cardiac gene regulation.

D. Regulation of MDF Genes

Although the unique roles played by the four individual MDFs *in vivo* are currently the focus of intense research activity, some insight has been gained from their respective expression patterns in muscle cell cultures. *MyoD* and *Myf5* gene expression is characteristic of proliferating myoblasts (Montarras *et al.*, 1991), although MDF protein function is repressed in these cells. In contrast, expression of the *myogenin* and *MRF4* genes is activated only when myoblasts exit the cell cycle upon growth factor withdrawal (Montarras *et al.*, 1991; Hinterberger *et al.*, 1992). Thus, the mere presence of one myogenic factor is not sufficient to activate expression of all the others. In fact, expression of MyoD or Myf5 is mutually exclusive in established muscle cell lines, and *myf5* expression is reduced in some cells when transfected with a MyoD expression vector (Peterson *et al.*, 1990). The possibility exists that these two particular MDFs may actually inhibit each other's expression, either through direct interaction with the corresponding control regions or through more indirect pathways involving other transcription factors.

General pathways controlling cell proliferation also play an important role in regulating MDF function. For example, the insulin-like growth factors promote myogenic differentiation in muscle cells, at least in part through the stimulation of myogenin gene expression (Florini and Ewton, 1992, Engert *et al.*, 1996). In contrast, the action of MyoD is repressed by fibroblast growth factor, which stimulates protein kinase C to phosphorylate the conserved threonine in the basic region of the MyoD protein and thereby inhibit DNA binding (Li *et al.*, 1992b). Growth factors also serve to maintain high expression of Id, an HLH protein that lacks a basic DNA-binding motif (Benezra *et al.*, 1990a,b). The function of Id *in vivo* may be related to cell proliferation (Wang *et al.*, 1992), and in cell culture it is thought to repress MDF function by binding to selected E proteins, sequestering them and thereby interfering with their interaction with MDFs (Jen *et al.*, 1992).

Other aspects of the cell cycle regulatory circuitry, such as the expression of the dimeric transcription fac-

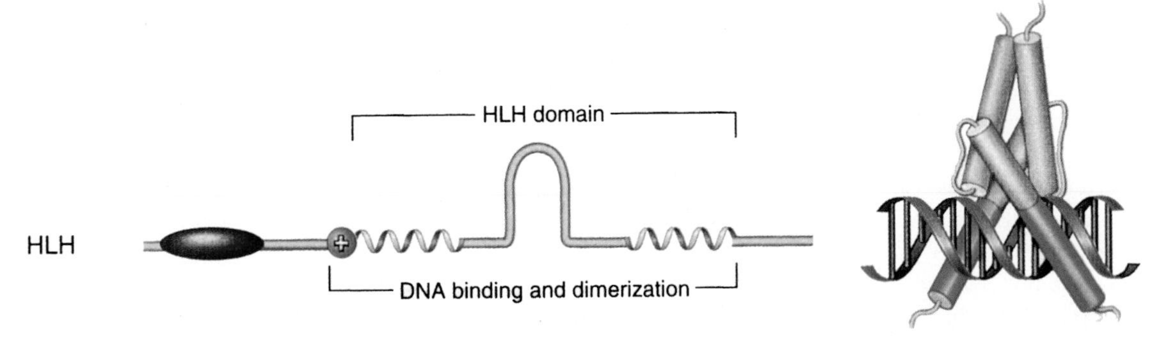

Figure 1 Structure of bHLH transcription factors and their interactions with DNA. The general protein structure of bHLH proteins is shown on the left. A transactivation domain (purple) precedes the areas that contact DNA directly (green) and the helix–loop–helix protein dimerization domain (yellow). The interaction of bHLH dimers with DNA is shown on the right. Cylinders represent α-helical regions.

tors *fos/jun,* also contribute to the repression of MyoD function (Bengal *et al.,* 1992; Li *et al.,* 1992a). Regulatory factors implicated in cell cycle control, such as the adenoviral E1a protein, inhibit myoblast differentiation and the transactivation functions of MDFs (Enkemann *et al.,* 1990; Braun *et al.,* 1992a). Finally, a mechanism for the reciprocal relationship between withdrawal from the cell cycle and the induction of myogenesis is suggested by the interaction of the Rb tumor supressor gene product with MyoD, which maintains Rb in its inactive state and leads to growth arrest (Gu *et al.,* 1993).

It remains to be seen whether parallel regulatory circuitry modulates the function of bHLH factors involved in cardiomyocyte differentiation (see Chapter 23). The studies described previously suggest that the precise mechanisms by which growth factors and signal transduction pathways interact in the reciprocal control of the cell cycle and differentiation may be more readily delineated in the skeletal muscle system, where a direct feedback on the MDF transcriptional regulatory family is likely to be involved.

E. The Role of MEF2 Factors in Skeletal Myogenesis

It is clear from the complex pathways involving the MDFs that they constitute only one part of a network of myogenic regulatory pathways. Although MDFs may control muscle structural gene expression directly, they may also activate expression of intermediate regulatory factors or cooperate with them to control specific aspects of the myogenic phenotype. These include the myocyte enhancer factor-2 (*MEF2*) gene family of transcriptional regulators which can act as intermediates in MDF action in skeletal, cardiac, and smooth muscle cells (see Chapter 8).

MEF2 proteins are encoded by a group of four genes (A–D) and are evolutionarily related to the MADS domain family of proteins which include the serum response factors (Pollock and Treisman, 1991). MEF2 proteins bind an A/T-rich motif present in the *cis*-regulatory regions of numerous muscle-specific genes, which is necessary for high levels of transcriptional activity (Gossett *et al.,* 1989; Braun *et al.,* 1991; Yu *et al.,* 1992). *MEF2* gene expression is induced in non-muscle cells in which MDFs are overexpressed (Martin *et al.,* 1993), indicating that they participate in the myogenic regulatory circuit. Synergistic interactions with the MEF2 group may further refine the target site specificity of MDFs in skeletal muscle. It is likely that certain MEF2 isoforms have distinct roles in the formation of distinct skeletal muscle lineages, perhaps widening the scope of MDF function to include E box-independent gene activation or maintaining the expression of the MDF genes in a positive feedback loop (see Chapter 8).

Increasing evidence points to the MEF2 group of regulators as key modulators of myogenic gene expression in all muscle cell types. This idea is supported by data in *Drosophila* showing that a single *mef2* gene with multiple functions in larval and adult (Ranganayakulu *et al.,* 1995) controls all myogenesis, with mutants lacking *mef2* expression failing to develop skeletal, cardiac, and smooth muscle (Lilly *et al.,* 1995; Bour *et al.,* 1995). In the mouse, selected MEF-2 proteins (MEF2-C and MEF2-B) are first expressed in the mesoderm of the cardiac crescent, concomitantly with expression of *Nkx2-5* and well before sarcomeric genes are activated (Edmondson *et al.,* 1994). This finding suggests that MEF2 proteins are required for terminal cardiac muscle differentiation. Notably, targeted disruption of the mouse *MEF2-C* gene not only decreases expression of differentiation markers such as atrial natriuretic factor, cardiac α-actin and α-myosin heavy chain, but also produces dramatic effects on the architecture of the heart (Lin *et al.,* 1997). *MEF2-C*-null animals fail to develop the right ventricle, a phenotype also shared with dHand knockout mice (see Chapter 9), suggesting that, as in skeletal myogenesis, cardiac myogenesis also involves synergistic interaction between MEF2 proteins and bHLH proteins (Molkentin *et al.,* 1995). Further elucidation of genetic pathways involved in MEF2-mediated functions, as well as in the regulation of the *MEF2* genes, will undoubtedly highlight important conformities between skeletal and cardiac muscle lineages.

III. Skeletal Muscle Development

One of the most dramatic differences between cardiac and skeletal muscle is the manner in which they arise during embryogenesis. Skeletal muscles of the vertebrate body derive from transient metameric structures called somites. These repetitive structures are responsible for generating the segmental pattern of the vertebrate trunk. The majority of classical studies on somite development and differentiation, as with heart development, rely on experiments conducted in the chicken embryo. The stages of somitogenesis in this organism have proven to be comparable with those observed in mammals and probably man. Considerable progress has recently been made in describing somite morphogenesis, with the application of molecular techniques to the classic experimental paradigms. It is now clear that the segregation of different somitic lineages is the result of the local action of inducing factors produced by the adjacent tissues such as the notochord, the neural tube, the ectoderm, and the lateral plate (Lassar and Munsterberg, 1996). These factors act either synergistically or antagonistically on the cells populating the paraxial mesoderm. Among these are inducing molecules implicated in a number of developmental processes from *Drosophila* to mammals. Determination of the molecular mechanisms controlling the development of the somite has revealed many of the influences affecting skeletal muscle precursors at the earliest stages of their inception.

A. Development of the Somite

The paraxial mesoderm in the caudal region of the chicken embryo consists of two distinct bands of mesenchyme situated on either side of the developing neural tube. This region, located between the regressing Hensen's node and the somites formed anteriorly, corresponds to the presomitic mesoderm (psm), also known as the segmental plate. Somites form progressively at the rostral-most part of the psm through the condensation of groups of mesenchymal cells into epithelial spheres surrounded by a basal lamina (Fig. 2). The psm, in turn, maintains its length by incorporating into its caudal end additional cells ingressing from the node or the primitive streak. The uniform periodicity of somite formation results in the repetitive generation of somites as the forming embryo extends caudally. A newly formed somite will steadily assume a more anterior position relative to the point of active somite formation as new somites are formed caudally to it.

The newly formed somite quickly undergoes a series of dynamic morphological changes. The ventral portion of the somite deepithelializes to form the mesenchymal sclerotome. The cells forming this tissue migrate medially and dorsally, surrounding the notochord to supply the progenitors of the axial skeleton: the vertebrae, the intervertebral discs, and part of the ribs. In contrast, the dorsal portion of the somite remains epithelial and constitutes the dermomyotomal compartment of the somite. This structure is the origin of the skeletal musculature of the body as well as the dermis of the back. Thus, the somites represent sequentially formed compartments which give rise to positionally distinct lineages in the vertebrate embryo.

B. Determination of the Epithelial Somite

Several types of experiments have been employed to demonstrate that the cells of the newly formed somite are not predetermined toward a particular cell lineage. Experiments using chick–quail chimeras to surgically replace the last three newly formed somites in chick with their counterparts from a quail embryo reveal that when these somites are rotated 180° along

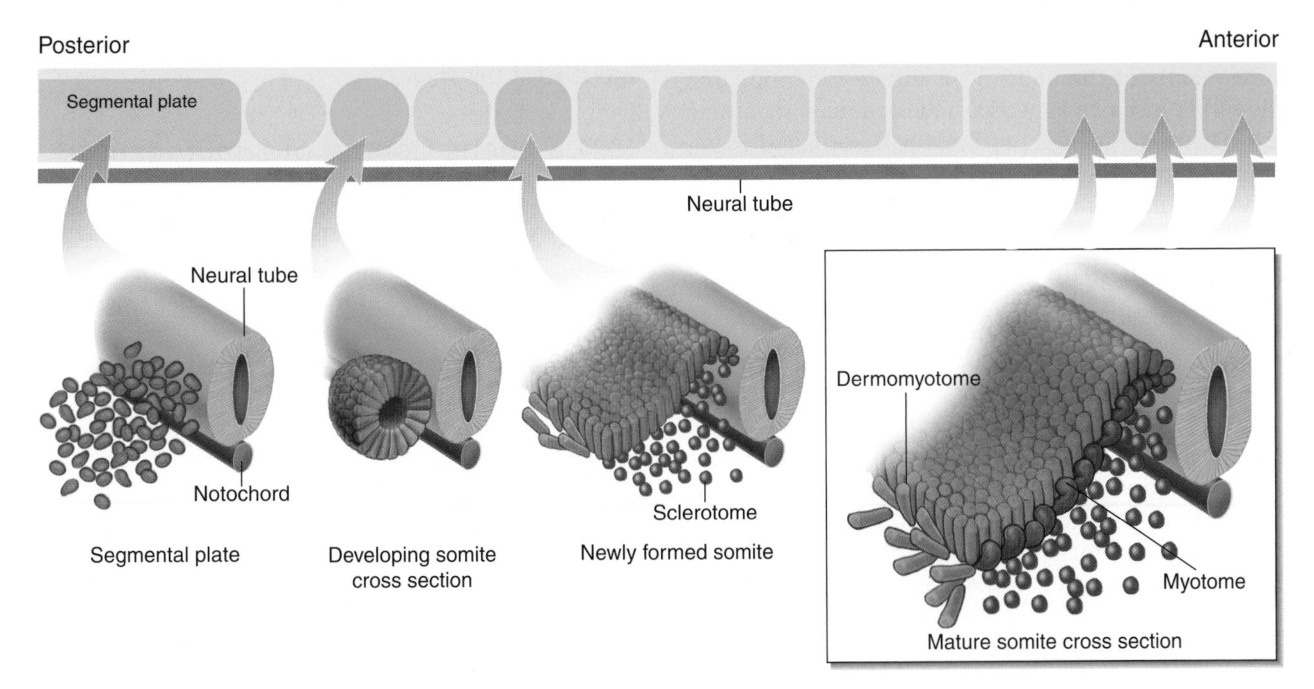

Figure 2 Diagram of vertebrate somitogenesis. The sequential formation and maturation of avian somites from the segmental plate, or presomitic mesoderm, is shown. Somite cross sections represent transitions as an epithelial somite develops to form sclerotome, dermomyotome, and finally the myotome.

the dorsal–ventral axis prior to transplantation, the dermomyotomal and sclerotomal derivatives of the two youngest somites are normal, whereas those of the third, most mature somite are inverted (Aoyama and Asamoto, 1988). This suggests that the dorsal–ventral patterning of the somite is established shortly after its formation but also that the cells of the last formed somite are plastic with regard to their future differentiation. A similar result is observed when the the medial somite half of the most recently formed somite is replaced by the lateral half. The grafted tissue gives rise to derivatives which correspond to its new location in the embryo (Ordahl and Le Douarin, 1992). These experiments confirm the absence of determination of the newly formed somite along the dorsal–ventral and medial–lateral axes. Studies such as these suggest that the regionalization of the somite, and in particular the generation of its different lineages, is likely to be under the control of the immediate environment.

Somites can also be divided into an anterior and a posterior compartment paticularly evident at the sclerotomal level. Experiments rotating the psm along the anterior–posterior axis have demonstrated that at the moment of its formation, the somite is polarized along this axis, as evidenced by distinctive neural crest cell migrations and the position of the axons of motor neurons (Keynes and Stern, 1988). Furthermore, transplantation experiments have established that the psm also possesses positional information along the embryonic anterior–posterior axis. Thus, the transplantation of psm from a thoracic level to a cervical level causes the formation of ribs at the cervical level (Kieny *et al.,* 1972). Consequently, at the moment of somite formation, the cells within it can be viewed as indeterminate regarding their particular lineage but already determined in regard to their anterior–posterior position in the somite and to the embryonic antero–posterior axis.

C. Development and Origins of the Epaxial and Hypaxial Muscles

Two major groups of skeletal muscle are distinguishable in vertebrates. One group is the epaxial musculature which lies within and around the vertebral column. The other is the hypaxial musculature which includes the limb and body wall muscles. These two categories of muscles have separate origins in the somite. Use of chick chimeric embryos in which either the medial or the lateral somitic half was replaced by its equivalent from a quail embryo has demonstrated that the medial portion gives rise to the epaxial muscles, whereas the lateral compartment gives rise to all hypaxial muscles. The formation of the epaxial musculature is coordinated at the level of each somite according to a particular sequence of events. Morphologically, the muscle progenitors are thought to delaminate from the medial dorsal edge of the dermomyotome (dorsal lip) and elongate under this structure in the sagittal plane to form the first mononucleated muscle fibers which constitute the myotome (Fig. 2). These cells are thought to form a scaffolding for later muscle development (Kaehn *et al.,* 1988; Denetclaw *et al.,* 1997). Cells in the lateral part of the dermomyotome differ in their development. They detach from the lateral edge of the dermomyotome and migrate individually through the somatopleura until they reach their site of differentiation (Chevallier *et al.,* 1977; Jacob *et al.,* 1978). Thus, the dermamyotome generates the precursors of two distinct muscle lineages which are spatially separated along the mediolateral axis.

Lineage tracing techniques have permitted the identification of distinct locations at the level of Hensen's node which gives rise to these two populations. More lateral portions of the node will give rise to somitic progeny which are restricted to the medial somite and therefore contribute only epaxial descendants, whereas a region located more caudal in the primitive streak gives rise to descendants distributed preferentially in the lateral somite (Selleck and Stern, 1991; Psychoyos and Stern, 1996). Thus, the precursors of these two lineages show some early spatial segregation before entering the psm, although it is not yet clear whether they are specified at this stage.

D. Genetic Control of Somitogenesis

As the early steps of vertebrate muscle differentiation become more well defined, it is increasingly clear that epaxial and hypaxial muscle precursors are spatially and temporally distinguished by their regulatory gene expression patterns. The earliest of these is *Pax-3* (Fig. 3D). At the moment of somite formation, *Pax-3* is expressed throughout. Its expression quickly becomes restricted to the dorsal somite, the dermomyotome, perhaps through the inhibitory action of *Pax-1,* which is expressed in the underlying sclerotome (Fig. 3B). Later, *Pax-3* remains strongly expressed in the lateral somite and thus by the cells which will give rise to the hypaxial musculature. In the mouse mutant, Splotch, in which the Pax-3 protein is not produced, the epaxial musculature

Figure 3 Expression patterns of regulatory genes in somite development. Whole mount *in situ* hybridizations on chick embryos were sectioned and expression patterns are shown for the following transcripts: A, *Myo D;* B, *Pax-1;* C, *cSim1;* D, *Pax-3;* E, *Noggin;* F, *Wnt1;* G, *BMP-4* (*in situ* hybridizations performed by E. Hirsinger).

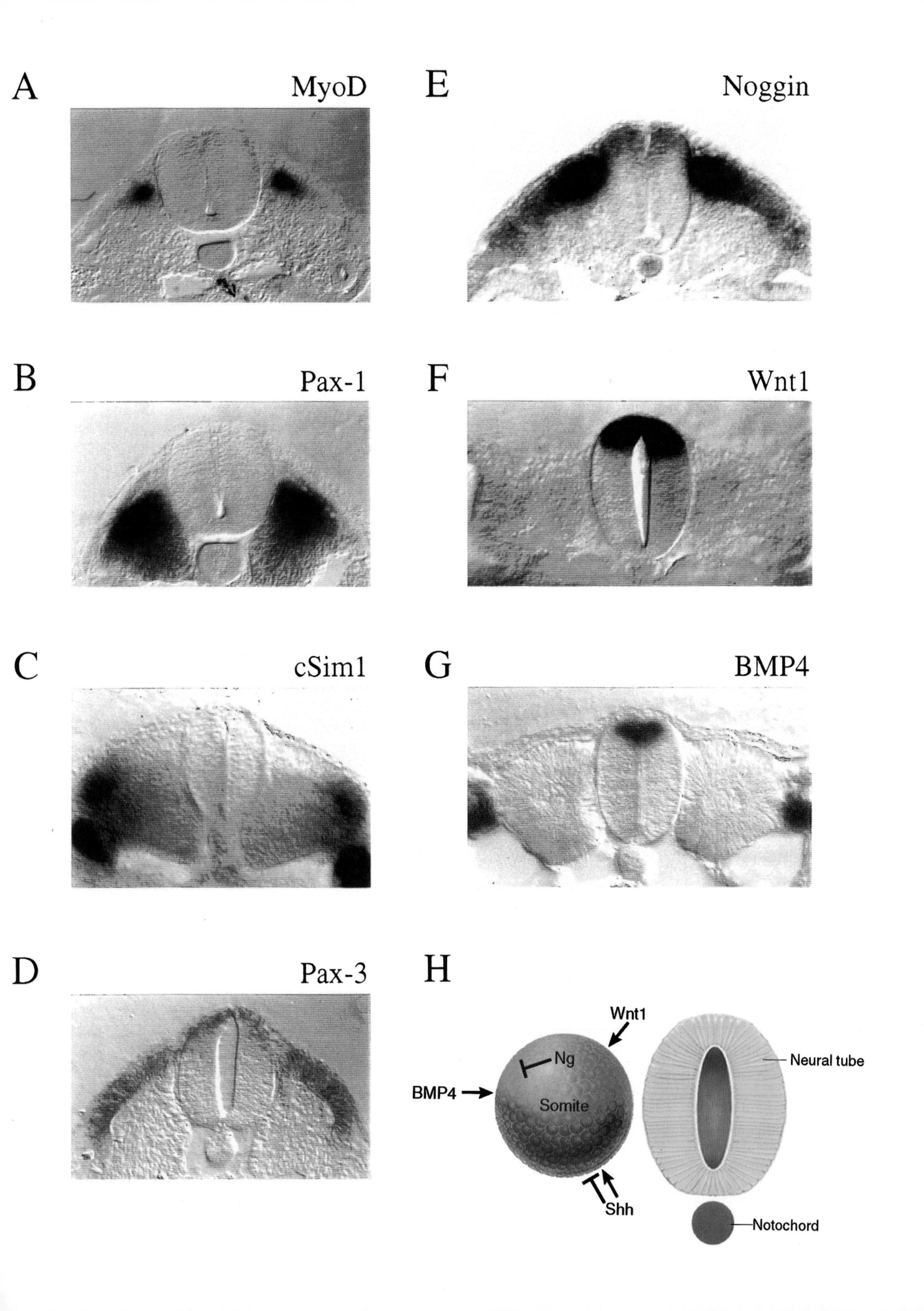
A
MyoD
B
Pax-1
C
cSim1
D
Pax-3
E
Noggin
F
Wnt1
G
BMP4
H
Wnt1
Ng
BMP4
Somite
Shh
Neural tube
Notochord

is relatively normal, whereas the limb musculature is absent (Tajbakhsh *et al.*, 1997). Thus, *Pax-3* is associated with the early stages of muscle development but can also be viewed as essential for the development of the hypaxial musculature.

During somitogenesis, expression of the four MDFs is activated sequentially in the developing myotome, preceding contractile protein gene activation (Fig. 3A). They are expressed from the early steps of myoblast formation and are first detected in the epaxial muscle lineage, appearing in the dorsal medial quadrant of the newly formed somite and later in the myotome (Sassoon *et al.*, 1989). In the hypaxial muscle lineage, expression of MDFs is not detected until several days later in both chick and mouse (Pourquie *et al.*, 1995; Pownall and Emerson, 1992; Smith *et al.*, 1994). The expression of these genes in the somite corresponds to a definitive engagement of the skeletal muscle pathway.

The order of MDF activation during development is species specific. In avian embryos, *MyoD* is activated first, followed by the activation of *myf5* and *myogenin* (Pownall and Emerson, 1992), whereas in mammals, the order is *myf5* > *myogenin* > *MyoD* > *MRF4* (Ott *et al.*, 1991). Expression of *myf5* is transient during mammalian embryogenesis, and expression of both *myogenin* and *MyoD* substantially decreases soon after birth, leaving *MRF4* as the major MDF expressed in adult skeletal muscles. The order of MDF activation in mammalian embryogenesis is in apparent contradiction to the precocious expression of *MyoD* in myoblast cultures. The discrepancy between cell culture and embryos may be explained by a model in which two distinct lineages reside within the myotome: an early lineage that expresses *myf5* gives rise to the first wave of mononucleate myotomal muscles and a later lineage that generates multinucleate axial muscles (Stockdale, 1992; Emerson, 1993). In addition, subdomains within the myotome express different MDFs and combinations of MDFs at distinct developmental stages, suggesting that the order of MDF activation may be discoordinate between different myotomal cells (Smith *et al.*, 1994). This implies that different expression patterns of the four MDF proteins in the embryo may distinguish multiple myogenic cell lineages within the developing somites.

E. Perturbation of MDF Expression during Development

The potential roles of MDFs in myogenesis are further complicated by the results of transgenic mouse studies in which normal MDF expression has been perturbed. That MDFs function in a dominant fashion is graphically illustrated by the effects of overexpressing MyoD in heart muscle, which activates the ectopic expression of skeletal muscle-specific contractile protein genes (Miner *et al.*, 1992). More widespread ectopic expression of MyoD in ectodermal and mesodermal lineages causes embryonic lethalities, although no ectopic myogenic conversion is seen (Faerman *et al.*, 1993).

The effects of targeted disruption of genes encoding each myogenic factor, or combinations thereof, further underscores the complexity of their individual roles *in vivo*. Null mutations in the *myf5* locus result in delayed myotome differentiation, but otherwise muscle development proceeds normally. Notably, rib formation is disrupted in *myf5*-null mice, even though *myf5* is not expressed in the rib precursors (Braun *et al.*, 1992b). This surprising abnormality is presumably due to the perturbation of local interactions in the somite between the developing myotome and the neighboring sclerotome, which gives rise to the ribs and axial cartilage.

Disruption of the *myogenin* gene locus produces a more predictable phenotype in which terminal differentiation of myogenic precursors is largely blocked and muscles fail to form (Hasty *et al.*, 1993; Nabeshima *et al.*, 1993). Embryos lacking myogenin die shortly after birth, presumably due to asphyxiation. Tissues where muscle should form are largely normal in gross morphology yet contain few myofibers and do not express muscle-specific contractile protein genes. Nevertheless, cultured myoblasts from these *myogenin*-null animals are capable of differentiating upon mitogen depletion (Nabeshima *et al.*, 1993), indicating that *myogenin* is not necessary for terminal differentiation per se. This suggests that *myogenin* may overcome a suppressive mechanism provided by the embryonic context, imposed by growth factors or cell–cell interactions, or that cell culture medium supplants essential muscle-specific factors that function in an autoregulatory loop with myogenin. It is also possible that another myogenic factor, *MRF4*, may serve to complement the function of *myogenin* in cell culture by activating the same subset of downstream gene targets necessary for terminal differentiation.

Unexpectedly, mice carrying a null mutation in their *MyoD* loci are viable, although defects in satellite cell regeneration have been noted in these animals (Megeney *et al.*, 1996). Interpretation of the phenotype is further complicated by a compensatory increase in *myf5* expression at later developmental stages, at which time the *myf5* gene would normally be repressed. The extreme phenotype of mouse embryos carrying null mutations in both the *myf5* and *MyoD* loci, generated by mating the two strains together (Rudnicki *et al.*, 1993), suggests that these genes play compensatory roles in normal muscle development. *MyoD/myf5* null animals lack all skeletal muscle and present the skeletal defect in the ribs, typical of *myf5* null animals (Braun *et al.,* 1992b). Unlike the muscle beds of *myogenin*-null mice which contain myoblasts that fail to differentiate, the presumptive muscle beds in the *MyoD/myf5* double-knockout animals are totally devoid of myoblasts.

Finally, the interdependent actions of the MDFs and *Pax-3* were recently demonstrated by the observation that *Splotch* mutant mice carrying a mutated *myf5* gene lack all skeletal body muscle and are unable to activate the *MyoD* gene (Tajbakhsh *et al.,* 1997). Thus, either *Pax-3* or *myf5* is necessary for the formation of the skeletal muscle lineages in the somitic region. One interpretation is that the expression of *Pax-3* is essential to generate the hypaxial musculature, whereas either *Pax-3* or *myf-5* alone is sufficient for formation of the epaxial musculature. These observations also correlate well with the early expression of *Pax-3* in all skeletal muscle cell precursors (Fig. 3D). It remains to be determined whether *Pax-3* and *myf5* activate identical or parallel pathways leading to epaxial muscle formation.

F. Induction of Skeletal Muscle in the Somite

Recent experiments have identified several tissues as inductive sources and candidates for molecular signals involved in somitic patterning, in particular for the generation of myogenic lineages. For example, the ablation of the neural tube and the notochord at the level of the psm in the chicken results in the rapid degeneration of newly formed somites (Rong *et al.,* 1992). Results such as these suggest that the axial organs produce a trophic factor(s) which is necessary for the survival of somitic cells.

In addition to playing a permissive role in the formation of the muscle lineages, numerous *in vitro* and *in vivo* experiments have demonstrated that the neural tube, the notochord, and the dorsal ectoderm play an instructive role in the generation of muscle cells from the psm. Although these results are at times contradictory, they all provide support for the hypothesis that the axial structures synthesize signaling molecules, which include *Sonic hedgehog* (*Shh*) and several *Wnt* molecules, which play a synergistic and sometimes redundant role in the formation of the skeletal muscle lineage (Tajbakhsh and Cossu, 1997). Because the first somitic cells which display committed myogenic markers are located in the medial somite, they may rely on their proximity to axial structures for differentiation signals. Undifferentiated cells in the lateral somite which later form the hypaxial lineage may escape from these inductive influences owing solely to their more distant position.

An alternative hypothesis, based on the premise that dorsal differentiation toward skeletal muscle is a default pathway rather than one requiring instructive cues, is that the medial and the lateral somite compartments possess an equal ability to generate muscle, but that this process is actively inhibited in the lateral somitic compartment by local influences. The default hypothesis of muscle formation is supported by the observation that epiblastic cells when dissociated and cultured *in vitro* form predominately muscle cells, suggesting an early bias toward the muscle lineage (George-Weinstein *et al.,* 1996). Therefore, one of the major roles of the factors regulating somitogenesis would be to repress this early differentiation process.

One observation in support of the default model is the strong expression in the psm of the receptor–ligand pair *Notch* and *Delta,* which have been shown to actively repress myogenesis *in vitro* and *in vivo* (Kopan *et al.,* 1994; Palmeirim *et al.,* 1998). Expression of these two genes is dramatically downregulated upon somite formation. Numerous studies in *Drosophila* have shown that the activation of *Notch* by *Delta* blocks a cells' response to environmental inducers (Artavanis-Tsakonas *et al.,* 1995). Thus, multiple systems functioning actively to repress muscle differentiation seem to exist in the psm, which remains undifferentiated despite the presence of numerous inducing signals emanating from the surrounding tissues.

G. Signals Involved in Medial–Lateral Patterning of the Somite

The differentiation of two muscle lineages formed from the medial and lateral portions of the somite have a very different timing: cells of the medial portion of the somite differentiate approximately 2 days before the lateral cells. The special environment surrounding the lateral somite appears to play an important role in this difference in timing. For example, the isolation of the paraxial mesoderm from the lateral plate results in the immediate expression of *myf5* and *MyoD* in the lateral somite (Pourquie *et al.,* 1995). Activation of the myogenic program is accompanied by the downregulation of Pax-3, which is normally strongly expressed in the lateral somite. This result suggests that tissues lateral to the psm provide a signal which inhibits precocious activation of the muscle differentiation pathway, providing a plausible mechanism for delaying the development of the lateral hypaxial lineage in comparison to the medial epaxial lineage.

Another marker of the lateral somitic compartment is the bHLH protein cSim1, an avian homolog of the *drosophila single-minded* gene (Pourquie *et al.,* 1996). First activated throughout the lateral portion of a newly formed somite, cSim1 expression becomes rapidly restricted to the lateral dermomyotome (Fig. 3C). Several studies established that activation of cSim1 is under the control of a diffusible factor emanating from the lateral plate and that this factor is actively counteracted by a signal produced by the neural tube (Pourquie *et al.*, 1996; Tonegawa *et al.*, 1997).

The protein BMP-4, a member of the transforming growth factor-β superfamily, is produced by the lateral plate and has a temporal expression pattern which correlates well with that of cSim1 (Fig. 3G). When cells producing BMP-4 are grafted between the neural tube and the somite, the medial somite displays lateral characteristics as evidenced by the ectopic expression of cSim1, overexpression of Pax-3, with a concurrent loss of MyoD (Pourquie *et al.*, 1996). BMP-4 is therefore an attractive candidate for the lateralizing factor of the somite. Supporting this model, high levels of BMP-4 induces lateral plate markers in the psm, with lower levels activating the lateral somitic marker, cSim1 (Tonegawa *et al.*, 1997). Thus, an activity gradient of BMP-4 across the somite could establish medial and lateral identity.

The lateralizing role putatively played by BMP-4 is opposed by a signal produced by the axial organs. Several factors which function to counteract the action of BMP-4-related molecules have been identified in other organisms. For example, in *Drosophila* embryos, the specification of different dorsal–ventral regions is established by the opposing interactions of dorsally produced decapentaplegic an analog of BMP-4, and ventrally produced Short gastrulation (SOG) (Ferguson, 1996). Similarly, in *Xenopus*, the specification of mesodermal territories across the dorsal–ventral axis proceeds through the interactions of ventrally produced BMP-4 and dorsaling factors produced by the Spemann organizer such as *Noggin* or *Chordin*, the vertebrate homolog of SOG (Piccolo *et al.*, 1996; Zimmerman *et al.*, 1996).

Specification of medial–lateral territories in chick could be generated in a similar manner since *Noggin* appears to be expressed medially in the somite (Fig. 3E) In fact, cells expressing *Noggin* are able to neutralize the lateraling signals and medialize the lateral somite (Hirsinger *et al.*, 1997; Marcelle *et al.*, 1997; Reshef *et al.*, 1998). Furthermore, *Noggin* expression is under the control of the axial structures and axial signaling molecules such as *Wnt-1* (Fig. 3F) (Hirsinger *et al.*, 1997) and thus could be considered as a potential dorsalizing influence on the somite.

Because of its synthesis in the notochord and floorplate, the diffusible morphogen *Shh* is a candidate antagonist of lateral inhibition, especially in light of recent experiments which have shown that in the caudal neural tube, the specification of cell types along the dorsal–ventral axis results from the antagonizing actions of *Shh* and BMP-4 (Liem *et al.*, 1995). *Shh* has also been shown to be important in establishing dorsal–ventral axis in the somite (Fan and Tessier-Lavigne, 1994; Johnson *et al.*, 1994). Finally, transplantation of *Shh*, expressing cells lateral to the somite has recently been shown to activate the expression of medial somite markers and also activate the expression of *Noggin* (Hirsinger *et al.*, 1997). Thus, *Shh* may also play a role in medial–lateral patterning of the somite. These interactions are summarized in Fig. 3H.

H. Retinoic Acid Signaling in Muscle Development

Retinoic acid (RA) has long been appreciated as a powerful morphogen that affects pattern formation in vertebrate embryos (Summerbell and Maden, 1990). Recent evidence indicates that Hensen's node, a pattern organizing region at the anterior end of the primitive streak in early embryos (Waddington, 1932; Beddington, 1994), may be an endogenous source of RA (Hogan *et al.*, 1992). Ectopic RA administration causes perturbation of segment identity along the anteroposterior body axis, detected by subtle homeotic transformations of vertebral structure (Kessel and Gruss, 1991; Kessel, 1992) and alteration of limb morphogenesis (Tickle *et al.*, 1982).

Given the dramatic phenotypic effects of RA on developing embryos, it has been of interest to investigate the nature of its action on specific cell types, such as skeletal muscle precursors. Both activation and repression of myogenic differentiation in response to RA has been observed in myogenic cells (Langille *et al.*, 1989; Momoi *et al.*, 1992). At high concentrations, RA inhibits myogenesis in chicken embryo craniofacial mesenchyme (Langille *et al.*, 1989) and limb buds (Momoi *et al.*, 1992). By contrast, similar RA concentrations have been reported to induce myogenic differentiation in a rhabdomyosarcoma-derived cell line (Arnold *et al.*, 1992), in C2C12 myoblasts and in primary adult chicken satellite cells (Albagli-Curiel *et al.*, 1993; Halevy and Lerman, 1993). RA has also been reported to induce myogenic differentiation of C2C12 myoblast cells and primary adult chicken satellite cells (Albagli-Curiel *et al.*, 1993; Halevy and Lerman, 1993).

These observed discrepancies in RA action have been at least partially resolved by the discovery that media conditions are as critical as RA concentration in modulating its effects on muscle cultures. Xiao *et al.* (1995) showed that muscle gene expression in C2C12 cells can either be inhibited or induced by the addition or elimination of chick embryo extract in the presence of RA. This involvement of diffusible factors may provide a model for dissecting the pleiotropic functions of RA in different cellular contexts during embryonic development as well.

The distribution of RA receptor (RAR) genes also may provide insight into the mechanism by which RA functions and into its role in the regulation of muscle differentiation and development (de Luca, 1991). RNA *in situ* hybridization studies show that RARs and RXRs are widely expressed throughout the embryo, including the forming heart (Ruberte *et al.*, 1991; Mangelsdorf *et al.*, 1992; Dolle *et al.*, 1994). Among the RARs, RARα is ubiquitously distributed, whereas RARβ and RARγ transcripts are differentially expressed among cell lines and during development, suggesting that they have specific roles in different tissues (de The *et al.*, 1989; Mendelsohn *et al.*, 1992). Congenital abnormalities in mice homozygous for null mutations in single receptor genes are restricted to only a few sites expressing those receptors during embryogenesis making it difficult to assign specific roles to any one receptor in muscle or heart development.

The effects of RA are further modulated by changes in RAR expression in response to RA (de The *et al.*, 1987; Gigure *et al.*, 1987; Petkovich *et al.*, 1987; Benbrook *et al.*, 1988; Brand *et al.*, 1988; Krust *et al.*, 1989; Zelent *et al.*, 1989). The RARβ gene has an RA response element that is involved in a positive autoregulatory feedback loop (Sucov *et al.*, 1990). In most cell lines, the expression of RARβ is induced by RA and impairment of this induction may contribute to neoplastic progression (Hu *et al.*, 1991). In RA-treated midgestation mouse embryos, RARβ is the only significantly induced RAR isoform, especially in limb buds (Harnish *et al.*, 1992). Induction of RARγ transcripts by RA in muscle cultures has been rarely reported in the literature (Wu *et al.*, 1992, Xiao *et al.*, 1995).

The role of RA in both skeletal and cardiac morphogenesis has been traditionally studied utilizing tools such as exposure to exogenous, RA vitamin A deprivation, and inactivation of RA effector molecules by gene targeting (see Chapter 13). These approaches, however, have several shortcomings. The effects of exogenous RA do not necessarily elucidate the physiological role of endogenous RA activity during development. Likewise, the effects of vitamin A deficiency on embryogenesis are more likely to represent the consequences of a protracted RA-deficient state than a critical developmental requirement for RA (Dickman *et al.*, 1997; Smith and Dickman, 1997).

A more informative experimental approach may be to examine the synthetic and inactivating pathways leading to local differences in retinoid concentration in the embryo since restricted production of a ligand with limited diffusion potential may be a key feature of its function. Endogenously produced retinoids are unevenly distributed along the embryonic anteroposterior body axis at early embryonic stages, a result of localized expression of RA synthetic enzymes. The activity of a RA response element (RARE) along the embryonic body axis in RARE transgenic animals (Rossant *et al.*, 1991) is consistent with a concentration-dependent effect on myogenesis. Indeed, in a recent study comparing RA synthetic activity and RA response during mouse heart development (Moss *et al.*, 1998), a striking correlation was observed between the domains of RARE-driven transgene expression and immunoreactivity to a retinaldehyde dehydrogenase (RALDH2) that catalyzes the last step in the RA biosynthetic pathway and is the first isozyme with this activity to appear in the embryo (McCaffrey and Drager, 1995; Zhao *et al.*, 1996) (Fig. 4A). Moreover, patterns of RA synthesis and response correlated inversely with the expression of a gene encoding the RA degrading enzyme P450RA (Moss *et al.*, 1998).

Another hallmark of early embryonic RA synthesis, as inferred by RALDH2 enzyme distribution, is the dynamic aspect of its patterning. This is dramatically illustrated in the distribution of RALDH2 in the heart, in which both RA synthesis and response are limited to the atria at early stages and then expand to include the epicardium and outflow structures (Moss *et al.*, 1998). In quail embryos, RALDH2 enzyme is more abundant in the somites than in the heart but is then downregulated in an anterior to posterior sequence as somitogenesis proceeds. This progression is accompanied by a shifting pattern of RALDH2 distribution within each somite, which is sequentially restricted to the sclerotome during somite maturation (Fig. 4B).

These observations underscore the importance of localized RA concentration in the developing embryo and indicate that morphogenetic actions of RA are implemented through dynamic changes in the expression and activity of its synthetic enzyme rather than by diffusion alone. They may also provide important clues for general molecular mechanisms underlying RA action in myogenesis. In the simplest model, RA may play a role in preventing myogenic cells from differentiating until the appropriate embryonic stage is reached, consistent with its inhibition of myogenic differentiation in cell culture. A more complex interplay of regulatory pathways is likely, however, since embryonic RA controls the expression of certain clustered *Hox* genes, which may influence muscle proliferation and patterning.

The local extracellular environment, involving cell–cell interactions and additional diffusible molecules, is probably as important as RA concentration in defining its effects in the intact embryo. Induction of intermediate regulatory factors such as *Shh,* which is responsive

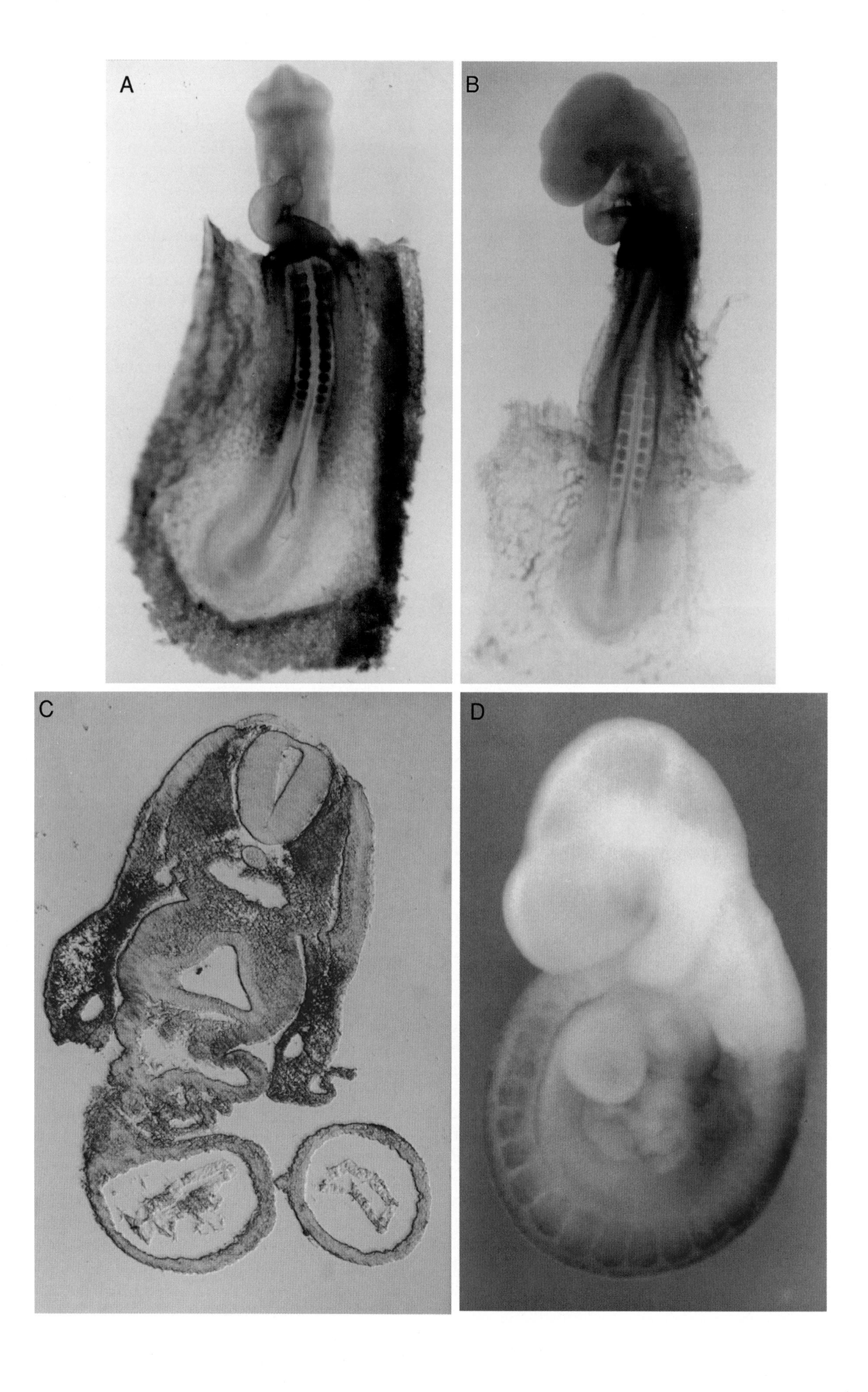
A
B
C
D

to RA and has been implicated in patterning of both axial structures and limb buds in the embryo (Echelard *et al.,* 1993; Riddle *et al.,* 1993), or RA-responsive members of the homeobox gene family, may add to the complexity of the embryo's response to RA. Identification and characterization of the intermediate players which modulate the specific effects of RA on both skeletal and cardiac muscle development may uncover additional similarities in the regulation of myogenesis by this important morphogen.

IV. Positional Information in Striated Muscle

As discussed previously, the induction of skeletal muscle in the vertebrate embryo can be attributed to the combined action of local inducing factors which lead to the expression of MDFs in committed myogenic lineages. Much less is currently known about the molecular cues underlying the correct spatial and temporal control of muscle formation. The complex organization of muscles in the vertebrate body plan is established early in development, guided by positional cues from other cells in the surrounding connective tissue, coordinated with the formation of motor neuronal connections. The positional information prefiguring muscle pattern must ultimately be integrated with regulatory programs for muscle gene transcription. Recent studies have provided increasing evidence that the patterning of certain muscles in vertebrates and invertebrates may be directly determined by the same genetic pathways that control morphogenetic patterning in other cell types, such as connective tissue and the nervous system.

A. Homeobox Protein Function in Muscle Patterning

The establishment of the embryonic body plan is controlled at least in part by the homeodomain gene family, comprising an ancient, structurally related group of transcription factors. These proteins regulate downstream gene expression programs in spatially and temporally restricted patterns to specify the body axes. A common structural motif, the homeobox, consists of a 183-bp stretch of nucleotides encoding the homeodomain, which preferentially binds A/T-rich sequences in DNA (Gehring *et al.,* 1990; Hayashi and Scott, 1990; Treisman *et al.,* 1992).

Homeobox genes can be divided into distinct categories based on their functions in establishing tissue patterns. The first category comprises the clustered *HOM/Hox* genes, which are conserved in both invertebrates (*HOM*) and vertebrates (*Hox*) and expressed in specific patterns or codes to confer regional identities. Perturbation of *HOM/Hox* expression results in characteristic homeotic structural transformations. These morphological changes are considerably more difficult to detect in muscles than in the axial skeleton itself, in which homeotic transformations have traditionally been scored by changes in vertebral morphology.

The second category is represented by many nonclustered homeobox-containing genes (including *Pax-3;* Fig. 3D) whose deletion results in the loss of proper specification of the muscle precursor cells in which they are normally expressed. A final category consists of other nonclustered homeobox genes (including *Pax-1;* Fig. 3B) which exert indirect effects on morphogenetic processes, presumably by controlling the growth or differentiation of myoblasts. Members from each of these categories have been found either in muscle precursors or in the tissues surrounding these precursors and have varying effects on the pattern of cells committed to the myogenic phenotype (Olson and Rosenthal, 1994).

B. Gene Targets for Positional Specification of Skeletal Muscle

The pathways by which homeobox factors control muscle gene transcription *in vivo* remain to be characterized since understanding how positional information and cell identity is conferred by these genes will require

Figure 4 Retinoic acid synthesis in developing muscle. (A–C) Whole mount immunohistochemical analysis of quail embryos, using an antibody to RALDH2, a key enzyme catalyzing the dehydrogenation of retinaldehyde to retinoic acid. (A) Stage 11 Hamilton and Hamburger (HH) quail embryos synthesize high levels of RALDH2 in segmented somites, in the unsegmented paraxial mesoderm, and in the prospective left ventricle, atrium, sinus venosa, and lateral plate mesoderm. Note the anterior–posterior boundary of enzyme expression, located at the level of the prospective left ventricle. (B) Stage 14 HH quail embryos display a more restricted pattern of RALDH2 synthesis. The enzyme recedes in the somites, with immunoreactivity in the anterior somites fading before that of more posterior somites. RALDH2 enzyme is still strong in the prospective left ventricle and in sinoatrial tissues. (C) A transverse section through a stage 16 HH quail embryo indicates that at this stage RALDH enzyme in somites is concentrated in sclerotomal tissues. (D) Whole mount immunohistochemical analysis of an Embryonic Day 9 mouse, using an antibody to RALDH2. The enzyme is abundant in the somites but largely absent from the head and anterior heart at this stage. Note an anterior–posterior boundary of enzyme in the body at approximately the level of the prospective left ventricle.

the identification of downstream muscle gene targets that respond to regional cues during embryogenesis. Some information has been gained by extensive characterization of the *cis*-acting regulatory elements associated with muscle-specific genes. For example, the myosin light chain (*MLC*) 1/3 locus includes distinct transcriptional regulatory elements that are activated in skeletal and cardiac tissues, as established by studies in cell culture and in transgenic mice (McGrew *et al.,* 1996). One subset of these elements directs chamber-specific transgene expression during heart development (see Chapter 19). Another subset directs a gradient of transgene expression in developing somites that increases posteriorly, constituting a positional marker for the epaxial segmented muscles arising from the myotome (Grieshammer *et al.,* 1992). Disappointingly, mutation of an A/T-rich *Hox* binding site in the transgenic MLC regulatory sequences does not change the graded pattern of transgene expression (Rao *et al.,* 1996). However, mutation of an E-box directly adjoining this *Hox* site is sufficient to release the repression of the otherwise unmodified *MLC* transgene in rostral axial muscles, abrogating the rostrocaudal gradient (Ceccarelli and Rosenthal, unpublished observations). It is likely that the complex of factors (including clustered Hox proteins) that bind to this DNA sequence *in vitro* (Gong *et al.,* 1996), may regulate the rostrocaudal position-sensitive activity of this transgene *in vivo*.

A direct role for *HOM* genes in muscle patterning has been elegantly demonstrated in *Drosophila* embryos (Grieg and Akam, 1993; Michelson, 1994). The use of muscle-specific enhancers to target misexpression of selected members of the *HOM BX-C* cluster, normally exclusively in the abdomen to more anterior segments results in transformation of the muscles in these segments to an abdominal identity (Fig. 5). Such studies provide clear evidence for the cell-autonomous role of *HOM* genes in establishing muscle identities during fly development.

Whether positional specification of vertebrate skeletal muscles involves the direct action of clustered *Hox* genes remains unproven. It is clear that although hypaxial muscle pattern can be affected by perturbations in *Hox* gene expression during limb development (Morgan *et al.,* 1992), the action is undoubtedly indirect, arising from altered signals from the surrounding mesenchyme in the embryonic limb. In midline structures, *Hox* genes appear to be targets for positional cues, however, as demonstrated by a recent study showing that the control of rostrocaudal pattern in the developing spinal cord, as judged by *Hox* gene expression patterns, is initiated by signals from the paraxial mesoderm (Ensini *et al.,* 1998). This raises the intriguing possibility that the mechanisms responsible for establishing rostrocaudal distinctions in myogenic cells might also confer upon them positional signaling properties, which in turn could pattern the cell types generated in the adjacent neural tube.

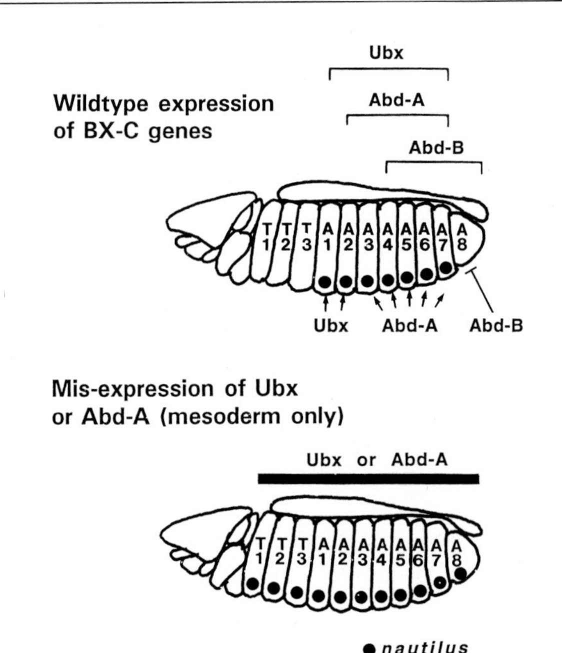

Figure 5 Autonomous determination of muscle patterns by homeotic genes in the *Drosophila* embryo. Analysis of loss-of-function mutations in the BX-C cluster of *HOM* genes has shown that two of the *BX-C* genes, *Ubx* and *Abd-A,* are necessary for proper formation of particular muscle precursors in selected abdominal segments. In contrast, expression of the *BX-C* gene *Abd-B* suppresses the formation of these cells in the last abdominal segment. Because the *BX-C* genes are expressed in the overlying epidermis and in motor neurons as well as muscle precursors, assignment of a cell-autonomous role in muscle patterning to these genes relies on a mesoderm-specific *HOM* ectopic expression scheme. (Top) Diagram of a *Drosophila* embryo, with three thoracic segments (T) and the eight abdominal segments (A). Domains of *BX-C* cluster gene expression (*Ubx, Abd-A,* and *Abd-B*) are indicated above. Wild-type pattern of a muscle marker (nautilus) expression in ventral muscle cells is indicated by black dots. Activation/suppression of nautilus expression by *Ubx, Ab-A,* or *Abd-B* in different segments is indicated below. (Bottom) Effects of mesoderm-restricted misexpression (black bar) of either *Ubx* or *Abd-A* on transformation of anterior muscles to an abdominal identity, as indicated by nautilus expression in all segments. Note that the suppression of nautilus expression in segment 8 by *Abd-B* is also overcome by the ectopic expression of *Ubx* or *Abd-A* in that segment (adapted from Michelson, 1994).

C. *Hox* Gene Action and Cardiac Muscle Patterning

Common paradigms for generating muscle pattern may exist between skeletal muscle cells and those in the myocardium, which are also subject to complex positional cues. During mouse embryogenesis, several clustered Hox proteins are expressed in overlapping domains in the cardiac compartment, suggesting that these proteins may also be involved in cardiovascular patterning. Specifically, knockout of the *Hoxa-3* gene resulted in a mouse with cardiac abnormalities and dysfunction. Conversely, disruption of another member of the *Hox* gene family, *Hoxa-2,* apparently did not affect cardiac function (Gendron-Maguire *et al.,* 1993; Rijli *et al.,* 1993). These knockout experiments reveal that at least some clustered *Hox* genes expressed in the cardiac compartment may play a patterning role in cardiac morphogenesis. The discovery that a nonclustered homeodomain-containing factor, *Nkx-2.5,* is essential for normal vertebrate cardiac morphogenesis (see Chapter 7) suggests that other nonclustered homeodomain genes expressed in developing skeletal muscle may play similar roles in the specification of patterning in this tissue as well.

V. Skeletal Muscle Cell Diversity and Fiber Type Specification

Although skeletal muscle superficially appears to be a relatively homogeneous tissue, it is actually composed of a heterogeneous population of myofibers that together modulate its contractile properties. The multiple types of fibers found in the skeletal muscle of adult vertebrates represent an additional level of complexity in the regulation of muscle-specific gene transcription. The heterogeneous nature of skeletal muscle has been known for over a century (Ranvier, 1874). Fibers were originally classified according to differences in contraction rates. Biochemical, histochemical, and immunological assays have since been developed to allow typing of individual fibers within sections of muscle tissue. These techniques have revealed the molecular diversity among fibers, both at the cell surface and in the energy-generating and contractile systems. A similar diversity undoubtedly exists in the heart, although this is represented largely by chamber-specific gene expression patterns rather than by variation in myocardial cell type within a particular chamber. Changes in fiber composition represent a major component in skeletal muscle degeneration associated with both aging and neuromuscular diseases, and the molecular basis of fiber type diversity is currently an area of active research.

A. Definition of Skeletal Muscle Fiber Types

The contraction rate of a skeletal muscle fiber is primarily regulated by its ATPase activity. This in turn is determined largely by the particular myosin heavy chain (MyHC) isoforms expressed in the fiber. As a result, the phenotypes of specific muscle fibers have been traditionally defined by the MyHC isoforms they express. Each of the major fiber types contains a different MyHC that is expressed by a distinct gene. The MyHCs with the highest ATPase activity are expressed in those fibers that contract most rapidly, whereas MyHC isoforms with low rates of ATP hydrolysis are found in those muscle fibers that contract slowly (Barany, 1967; Guth and Samaha, 1969).

The diversity of MyHC and other fiber-restricted genes expressed throughout the life span of an organism reflects the enormous versatility of skeletal muscle structure and function. At least 10 genes that encode MyHCs expressed in skeletal muscle exist in mice, many of which have been shown to be clustered together in a single locus (Cox *et al.,* 1991). It remains to be seen whether the chromosomal position of these genes within the *MyHC* locus governs the order of developmental expression in a manner similar to that of the *Hox* and *globin* gene loci. Each MyHC isoform has a unique pattern of expression in development and is often expressed in several muscle cell types. These include isoforms that are transiently expressed in skeletal muscle fibers during development (embryonic, perinatal and α-cardiac), two adult slow fiber-specific isoforms (type I or β-cardiac and slow tonic), and five fast-specific isoforms (types IIA, IIB, IIX, IIeom, and IIm). Type I fibers express a MyHC isoform, also present in heart muscle (Fig. 6). Types IIeom and IIm are highly specialized MyHCs found only in the superfast extraocular fibers and in certain jaw muscles of carnivores, respectively. Many adult fibers express more than one isoform at a time and are considered to be hybrid or mixed fiber types. This suggests that MyHC genes expressed together in certain fibers may contain common *cis*-regulatory sequences that are recognized by fiber-specific transcription factors.

In addition to the MyHCs, many other muscle structural proteins are expressed in a fiber-specific manner, including the troponin subunits, myosin light chains, α-tropomyosin, and the sarcoplasmic reticulum Ca^{2+}-ATPase (Kaprielain and Fambrough, 1987; Cox *et al.,* 1991). Each of these proteins is represented by separate isoforms expressed in slow and fast fibers. Together with the MyHCs, contractile proteins are expressed in unique combinations that satisfy the necessary range of muscular function. The signals responsible for generating this diversity are complex and responsive to environmental factors but ultimately rely on gene expression patterns originally established in the developing embryo (Fig. 7).

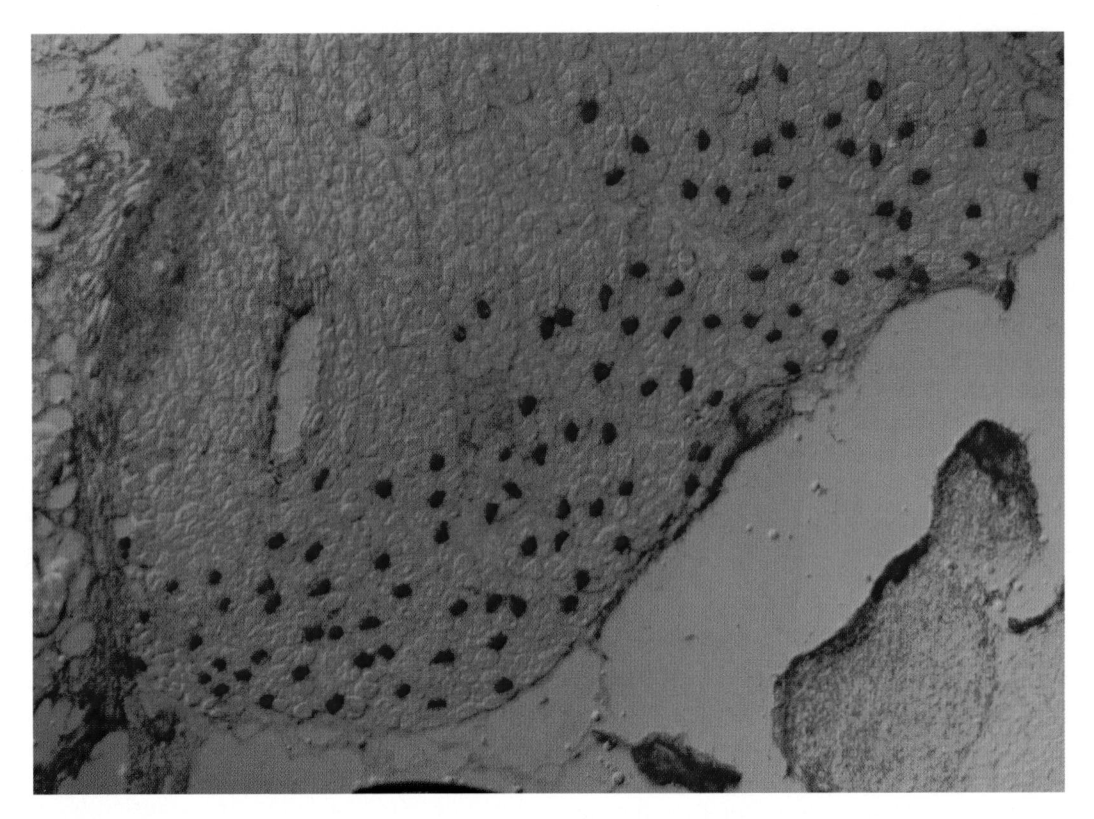

Figure 6 Expression of a cardiac myosin heavy chain in the slow fibers of mouse muscles. Immunohistochemical analysis of a cross section from an adult mouse gastrocnemius muscle, using an antibody directed against the cardiac MyHC, which is also expressed in skeletal muscles, exclusively in slow fibers. Note the characteristic scattered, nonuniform distribution of this fiber type in the muscle bed (courtesy of C. Neville).

B. Establishment of Muscle Fiber Diversity in the Embryo

The mechanisms by which specialized muscle fibers are generated from their less specialized embryonic precursors are of particular importance for understanding the molecular basis of fiber diversity. As previously discussed, myogenic precursor cells originating in the somite give rise to a population of embryonic myoblasts that are heterogenous in character (Stockdale, 1992). Upon differentiation, these myocytes express unique combinations of fiber-specific proteins which continue to shift during embryogenesis and early postnatal life.

There are several alternate models for the generation of fiber diversity, each invoking different pathways of transcriptional control. In the first model, myoblasts are homogeneous and capable of giving rise to all of the different types of fibers in skeletal muscle. Extrinsic influences, such as circulating hormones, growth factors, and innervation, later play a crucial role in determining and maintaining fiber-specific gene expression patterns. An environmental component is certainly evident in the adult, as illustrated by the imposition of fiber type by motor innervation. A classic study (Buller *et al.,* 1960) demonstrated that the fiber-type properties of an adult muscle can be transformed by cross-innervation with a different nerve. The effects of cross-innervation can be duplicated by using an appropriate frequency of electrical stimulation, directly demonstrating the role of electrical activity in modulating gene expression in the mature muscle fiber (Lomo and Rosenthal, 1972). From these results, it was inferred that the nerve also instructed muscle fiber differentiation during development. However, myocyte heterogeneity is evident prior to muscle innervation (Butler *et al.,* 1982; McLennan, 1983; Crow and Stockdale, 1986; Phillips *et al.,* 1986; Condon *et al.,* 1990). Moreover, the initiation of myofiber diversity proceeds normally in embryonic chick and rat limbs that have been denervated *in vivo* by removal of the neural tube or treatment with the drug curare (Fredette and Landmesser, 1991). It thus appears that at least in early development, fiber-specific gene regulation is cell autonomous or determined by influences other than innervation.

In the second model, muscle fiber diversity is the result of distinct lineage directives that are inherited by

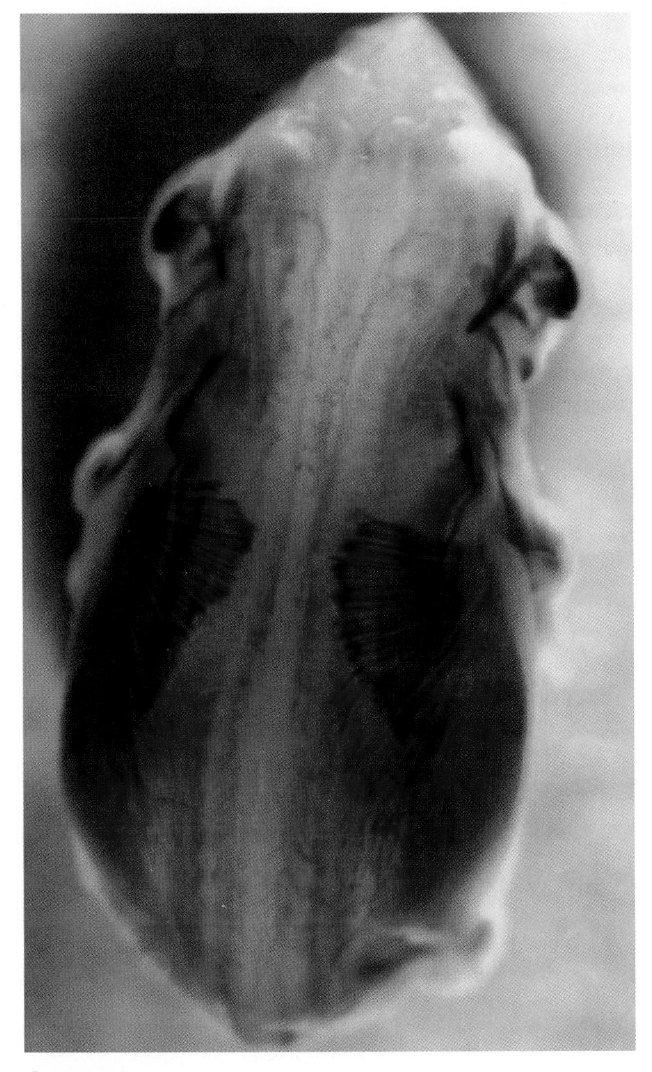

Figure 7 Muscle-specific expression of a transgene during mouse embryogenesis. Regulatory elements from the *myosin light chain* 1/3 locus (Rosenthal *et al.*, 1989) driving a human placental alkaline phosphatase reporter transgene in an Embryonic Day 14.5 mouse embryo. This transgene is destined to be active throughout the life of the animal, exclusively in fast fibers (courtesy of C. Neville).

differentiating cells from their progenitors. Myoblasts committed to the same fate must then fuse with each other, such that different fiber types would be derived from intrinsically distinct types or lineages of myoblasts. This model is supported by experiments involving the fusion of individual primary embryonic chick myoblasts propagated in culture. The resulting myotubes form three classes: those that express only fast isoforms, those that express only slow isoforms, and the remainder express both isoforms. A clonal population of cells is used to derive each set of myotubes and every cell within the set expresses the same MyHC isoforms. The initial myoblasts must therefore be intrinsically different in order to form myotubes with distinct patterns of transcription in the absence of innervation (White *et al.*, 1975).

A third, more complex model invokes an interplay between extrinsic signaling and intrinsic cellular programs. In this multistep model, the range of genes expressed in different lineages becomes increasingly restricted during development (Cox *et al.*, 1991). Transcriptional regulatory pathways which determine a particular fiber type would result in specific patterns of cell surface markers, presented on the surface of muscle precursors destined to respond selectively to extracellular signals, with innervation playing a role in maintaining selected fiber-specific gene expression in the mature myofiber.

Although the molecular details of this hypothetical cascade are not yet known, this last model integrates current information and provides a useful framework for further investigation. The details may differ dramatically in the heart. In skeletal muscle, each myofiber is electrically isolated by a basal lamina, whereas cardiomyocytes are electrically coupled by gap junctions. In addition, it is not clear that the well-documented instructive influence of the motor neuron in mature skeletal muscle holds true for the cardiac system.

In each of the models described previously, differences in skeletal muscle fiber type are ultimately attributed to differences in gene expression, which contribute to the versatility in energy metabolism, fatigue resistance, and contraction rates that satisfy the functional requirements of individual muscles. Identification of the specific mediators that determine fiber type-specific gene expression is an important objective in skeletal muscle research.

C. bHLH Factors and Regulation of Muscle Fiber Type

Since skeletal muscle fiber phenotypes are not faithfully maintained in cell culture, it has been difficult to characterize the *trans*-acting factors that regulate muscle gene transcription in a fiber-specific manner. The MDFs are not activated in fiber-specific patterns during development, although in adult muscle the relative expression levels of myogenin and MyoD have been shown to be higher in slow type I and fast type IIB fibers, respectively (Hughes *et al.*, 1993). A mechanism could be postulated whereby a threshold amount of a particular myogenic factor is required to activate a program of fiber-specific gene expression by preferentially binding to specific E-box targets (Weintraub, 1993). However, this hypothesis is not supported by the phenotypes of MDF-null mouse models, in which animals that survive to adulthood do not display expected perturbations of fiber type.

Alternatively, the involvement of E proteins, ubiquitously expressed bHLH transcription factors, in the establishment of mammalian muscle fiber diversity is suggested by the observation that different E proteins are restricted to distinct subsets of the three adult fast fiber groups (Neville and Rosenthal, unpublished observations). Further clues are provided by the phenotypes of mice carrying homozygous targeted disruptions in each of the three E protein loci *E2A, E2-2,* and *HEB* (Zhuang *et al.,* 1996). In addition to the profound effects on B cell lineage establishment and proliferation, these three lines of E protein null mice each lack one of the three fast fiber types. Fiber-specific gene expression can be rescued by transfection of the appropriate E protein in muscle cell cultures derived from E protein knockout animals. Moreover, forced expression of a transgene encoding an E protein in a different fiber type *in vivo* resulted in the predicted ectopic activation of a fiber-specific target gene (Neville *et al.,* 1998). These results support a model in which heterodimerization of distinct E proteins with myogenic bHLH factors in different fiber types provides sufficient selectivity to recognize the modular genetic targets responsible for patterns of fiber-restricted gene expression.

VI. Future Prospects

The molecular programs governing skeletal muscle myogenesis are likely to be paradigmatic of other developmental processes in the embryonic mesoderm. Delineation of similar genetic pathways in cardiac development may prove to be more challenging, however, since master gene regulators of cardiac differentiation or commitment have yet to be identified in myocardial or vascular precursors. In contrast to the molecular schemes operating in skeletal muscle determination, it is possible that multiple classes of factors may be involved in the initial commitment of cells to a cardiac lineage. Moreover, programs of tissue patterning in the vertebrate heart may have evolved separately using chamber- or structure-specific regulatory pathways. The recent discovery that a specific set of cardiac malformations in man arise from mutations in the human *Nkx-2.5* gene (Schott *et al.,* 1998), a key determinant of heart morphogenesis in *Drosophila* (see Chapter 5) and mice (see Chapter 7), provides a new paradigm to explore these possibilities, and underscores the importance of sharing insights derived from model systems and humans.

It is also not clear to what extent the plasticity of the adult skeletal muscle phenotype extends to the myocardium, altering its repertoire of contractile protein isoforms in response to diverse stimuli. A recent exciting breakthough has delineated at least one molecular mechanism whereby cardiac hypertrophy is induced in response to increased workload or to pathological stimuli, recapitulating fetal gene expression programs through activation of the calcium-mediated calcineurin signaling pathway (Molkentin *et al.,* 1998). Remarkably, the same pathway has been implicated in specifying the slow fiber type in skeletal muscle (Williams *et al.,* 1998). It remains to be seen whether a similar pathway is activated in skeletal muscle hypertrophy, in which case lessons learned from the cardiac system may provide new prospects for the treatment of the devastating atrophy associated with neuromuscular diseases.

References

Albagli-Curiel, O., Carnac, G., Vandromme, M., Vincent, S., Crepieux, P., and Bonnieu, A. (1993). Serum-induced inhibition of myogenesis is differentially relieved by retinoic acid and triiodothyronine in C2 murine muscle cells. *Differentiation* **52,** 201–210.

Aoyama, H., and Asamoto, K. (1988). Determination of somite cells: Independence of cell differentiation and morphogenesis. *Development* **104,** 15–28.

Arnold, H. H., Gerharz, C. D., Gabbert, H. E., and Salminen, A. (1992). Retinoic acid induces myogenin synthesis and myogenic differentiation in the rat rhabdomyosarcoma cell line BA-Han-1C. *J. Cell Biol.* **118**(4), 877–887.

Arnone, M. I., and Davidson, E. H. (1997). The hardwiring of development: Organization and function of genomic regulatory systems. *Development* **124,** 1851–1864.

Artavanis-Tsakonas, S., Matsuno, K., and Fortini, M. E. (1995). Notch signaling. *Science* **268,** 225–232.

Barany, M. (1967). ATPase activity of myosin correlated speed of muscle shortening. *J. Gen. Physiol.* **50,** 197–218.

Beddington, R. S. P. (1994). Induction of a second neural axis by the mouse node. *Development* **120,** 613–620.

Benbrook, D., Lernhardt, E., and Pfahl, M. (1988). A new retinoic acid receptor identified from a hepatocellular carcinoma. *Nature* **333,** 669–672.

Benezra, R., Davis, R. L., Lassar, A., Tapscott, S., Thayer, M., Lockshon, D., and Weintraub, H. (1990a). Id: A negative regulator of helix–loop–helix DNA binding proteins. Control of terminal myogenic differentiation. *Ann. N. Y. Acad. Sci.* **599,** 1–11.

Benezra, R., Davis, R. L., Lockshon, D., Turner, D. L., and Weintraub, H. (1990b). The protein Id: A negative regulator of helix–loop–helix DNA binding proteins. *Cell* **61,** 49–59.

Bengal, E., Ransome, L., Scharfmann, R., Dwarki, V. J., Tapscott, S. J., Weintraub, H., Verma, I. M. (1992). Functional antagonism between c-Jun and MyoD proteins: A direct physical association. *Cell* **68,** 507–519.

Blackwell, T. K., and Weintraub, H. (1990). Differences and similarities in DNA-binding preferences of MyoD and E2A protein complexes revealed by binding site selection. *Science* **250,** 1104–1110.

Bour, B. A., O'Brien, M. A., Lockwood, W. L., Goldstein, E., Bodmer, R., Tagher, P., Abmayr, S., and Nguyen, H. (1995). *Drosophila* MEF2, a transcription factor that is essential for myogenesis. *Genes Dev.* **9,** 730–741.

Brand, N. M., Petkovich, M., Krust, A., Chambon, P., de The, H., Marchio, A., Tiollais, P., and Dejea, A. (1988). Identification of a second human retinoic acid receptor. *Nature* **332,** 850–853.

Braun, T., and Arnold, H. H. (1991). The four human muscle regulatory helix–loop–helix proteins Myf3–Myf6 exhibit similar heterodimerization and DNA binding properties. *Nucleic Acids Res.* **19,** 5645–5651.

Braun, T., Bober, E., Buschhausen Denker, G., Kohtz, S., Grzeschik, K. H., Arnold, H. H., and Kotz, S. (1989). Differential expression of myogenic determination genes in muscle cells: Possible autoactivation by the Myf gene products (published erratum appears in *EMBO J.* 1989, **13,** 4358). *EMBO J.* **8,** 3617–3625.

Braun, T., Tannich, E., Buschhausen Denker, G., Arnold, H. H. (1989). Promoter upstream elements of the chicken cardiac myosin light-chain 2-A gene interact with trans-acting regulatory factors for muscle specific transcription. *Mol. Cell. Biol.* **9,** 2513–2525.

Braun, T., Bober, E., and Arnold, H. H. (1992a). Inhibition of muscle differentiation by the adenovirus E1a protein: Repression of the transcriptional activating function of the HLH protein Myf-5. *Genes Dev.* **6,** 888–902.

Braun, T., Rudnicki, M. A., Arnold, H. H., and Jaenisch, R. (1992b). Targeted inactivation of the muscle regulatory gene Myf-5 results in abnormal rib development and perinatal death. *Cell* **71,** 369–382.

Brennan, T. J., Chakraborty, T., and Olson, E. N. (1991). Mutagenesis of the myogenin basic region identifies an ancient protein motif critical for activation of myogenesis. *Proc. Natl. Acad. Sci. USA* **88,** 5675–5679.

Buller, A. J., Eccles, J. C., and Eccles, R. M. (1960). Differentiation of fast and slow muscles in the cat hind limb. *J. Physiol.* **150,** 399–451.

Butler, J., Cosmos, E., Brierly, J., and Hauschka, S. D. (1982). Differentiation of muscle fiber types in aneurogenic brachial arches of the chick embryo. *J. Exp. Zool.* **224,** 65–80.

Chakraborty, T., and Olson, E. N. (1991). Domains outside of the DNA-binding domain impart target gene specificity to myogenin and MRF4. *Mol. Cell. Biol.* **11,** 6103–6108.

Chen, J. N., and Fishman, M. C. (1996). Zebrafish tinman homolog demarcates the heart field and initiates myocardial differentiation. *Development* **122,** 3809–3816.

Cheng, T. C., Wallace, M. C., Merlie, J. P., and Olson, E. N. (1993). Separable regulatory elements governing myogenin transcription in mouse embryogenesis. *Science* **261,** 215–218.

Chevallier, A., Kieny, M., and Mauger, A. (1977). Limb–somite relationship: Origin of the limb musculature. *J. Embryol. Exp. Morphol.* **41,** 245–258.

Cleaver, O., Patterson, K. D., and Krieg, P. A. (1996). Overexpression of the tinman-related genes XNkx-2.5 and XNkx-2.3 in Xenopus embryos results in myocardial hyperplasia. *Development* **122,** 3549–3556.

Condon, K., Silberstein, L., Blau, H. M., and Thompson, W. J. (1990). Differentiation of fiber types in aneural musculature of the prenatal rat hind limb. *Dev. Biol.* **138,** 275–295.

Conway, S. J., Henderson, D. J., and Copp, A. J. (1997). Pax3 is required for the cardiac neural crest migration in the mouse: Evidence from the splotch (Sp2H) mutant. *Development* **124,** 505–514.

Cox, R. D., Weydert, A., Barlow, D., and Buckingham, M. E. (1991). Three linked myosin heavy chain genes clustered within 370 kb of each other show independent transcriptional and post-transcriptional regulation during differentiation of a mouse mlc line. *Dev. Biol.* **143,** 36–43.

Crow, M. T., and Stockdale, F. E. (1986). Myosin expression and specialization among the earliest muscle fibers of the developing avian limb. *Dev. Biol* **113,** 238–254.

Davis, R. L., and Weintraub, H. (1992). Acquisition of myogenic specificity by replacement of three amino acid residues from MyoD into E12. *Science* **256,** 1027–1030.

Davis, R. L., Weintraub, H., and Lassar, A. B. (1987). Expression of a single transfected cDNA converts fibroblasts to myoblasts, *Cell* **51,** 987–1000.

Denetclaw, W. F., Christ, B., and Ordahl, C. P. (1997). Location and growth of epaxial myotome precursor cells. *Development* **124,** 1601–1610.

de The, H., Marchio, A., Tiollais, P., and Dejean, A. (1987). A novel steroid thyroid hormone receptor-related gene inappropriately expressed in human hepatocellular carcinoma. *Nature* **330,** 667–670.

de The, H., Marchio, A., Tiollais, P., and Dejean, A. (1989). Differential expression and ligand regulation of the retinoic acid receptor, α and β genes. *EMBO J* **8,** 429–433.

Dickman, E. D., Thaller, C., and Smith, S. M. (1997). Temporally-regulated retinoic acid depletion produces specific neural crest, ocular and nervous systems defects. *Development* **124,** 3111–3121.

Dolle, P., Fraulob, V., Kastner, P., and Chambon, P. (1994). Developmental expression of murine retinoid X receptor (RXR) genes. *Mech. Dev.* **45,** 91–104.

Echelard, Y., Epsteon, D. J., St. Jacques, B., Shen, L., Mohler, J., MacMahon, J., and MacMahon, A. P. (1993). Sonic hedgehog, a member of putative signalling molecules, is implicated in the regulation of CNS polarity. *Cell* **75,** 1417–1430.

Edmondson, D. G., and Olson, E. N. (1989). A gene with homology to the myc similarity region of MyoD1 is expressed during myogenesis and is sufficient to activate the muscle differentiation program (published erratum appears in *Genes Dev.* 1990, **4**(8), 1450). *Genes Dev.* **3,** 628–640.

Edmondson, D., Lyons, G., Martin, J., and Olson, E. (1994). Mef2 gene expression marks the cardiac and skeletal muscle lineages during mouse embryogenesis. *Development* **120,** 1251–1263.

Emerson, C. P., Jr. (1993). Skeletal myogenesis: Genetics and embryology to the fore. *Curr. Opin. Genet. Dev.* **3,** 265–274.

Engert, J. C., Berglund, E., and Rosenthal, N. (1996). Proliferation precedes differentiation in IGF-I-stimulated myogenesis. *J. Cell Biol.* **135,** 431–440.

Enkemann, S. A., Konieczny, S. F., and Taparowsky, E. J. (1990). Adenovirus 5 E1A represses muscle-specific enhancers and inhibits expression of the myogenic regulatory factor genes, MyoD1 and myogenin. *Cell Growth Differ.* **1,** 375–382.

Ensini, M., Tsuchida, T. N., Belting, H. G., Jessell, T. M. (1998). The control of rostrocaudal pattern in the developing spinal cord: Specification of motor neuron subtype identity is initiated by signals from paraxial mesoderm. *Development* **125,** 969–982.

Ephrussi, A., Church, G. M., Tonegawa, S., and Gilbert, W. (1985). B lineage-specific interactions of an immunoglobulin enhancer with cellular factors in vivo. *Science* **227,** 134–139.

Evans, S. M., Tai, L., Tan, V., Newton, C., and Chien, K., (1994). Heterokaryons of cardiac myocytes and fibroblasts reveal the lack of dominance of the cardiac muscle phenotype. *Mol. Cell. Biol.* **14,** 4269–4279.

Faerman, A., Pearson White, S., Emerson, C., and Shani, M (1993). Ectopic expression of MyoD1 in mice causes prenatal lethalities. *Dev. Dyn.* **196,** 165–173.

Fan, C. M., and Tessier-Lavigne, M. (1994). Patterning of mammalian somites by surface ectoderm and notochord. Evidence for sclerotome induction by a Hedgehog homolog. *Cell* **79,** 1175–1186.

Ferguson, E. (1996). Conversation of dorsal–ventral patterning in arthropods and chordates. *Curr. Opin. Gen. Dev.* **6,** 424–431.

Firulli, A. B., McFadden, D. G., Lin, Q., Srivastava, D., and Olson, E. (1998). Heart and extra embryonic mesodermal defects in mouse embryos lacking the bHLH transcription factor Hand1. *Nature Genet.* **18,** 266–270.

Florini, J. R., and Ewton, D. Z. (1992). Induction of gene expression in muscle by the IGFs. *Growth Regul.* **2,** 23–29.

Fredette, B. J., and Landmesser, L. T. (1991). A reevaluation of the role of innervation in primary and secondary myogenesis in developing chick muscle. *Dev. Biol.* **143,** 19–35.

Fujisawa Sehara, A., Nabeshima, Y., Komiya, T., Vetsuki, T., and Asakura, A. (1992). Differential trans-activation of muscle-specific regulatory elements including the myosin light chain box by chicken MyoD, myogenin, and MRF4. *J. Biol. Chem.* **267,** 10031–10038.

Gehring, W. J., Muller, M., Affolter, M., Percival-Smith, A., Billeter, M., Qian, Y. Q., Otting, G., and Wuthrich, K. (1990). The structure of the homeodomain and its functional implications. *Trends Genet.* **6,** 323–329.

Gendron-Maguire, M., Mallo, M., Zhang, M., Gridley, T. (1993). Hoxa-2 mutant mice exhibit homeotic transformation of skeletal elements derived from cranial neural crest. *Cell* **75,** 1317–1331.

George-Weinstein, M., Gerhart, J., Reed, R., Flynn, J., Callihan, B., Mattiacci, M., Miehle, C., Foti, G., Lash, J. W., and Weintraub, H. (1996). Skeletal myogenesis: The preferred pathway of chick embryo epiblast cells in vitro. *Dev. Biol.* **173,** 279–291.

Gigure, V., Ong, E. S., Sequi, P., and Evans, R. M. (1987). Identification of receptor for the morphogen retinoic acid. *Nature* (*London*) **330,** 624–629.

Gong, X., Kaushal, S., Ceccarelli, E., Bogdanova, N., Clark, H., Khatib, Z., Valentine, M., Look, T., Rosenthal, N. (1996). Developmental regulation of Zbu1/HIP116, a DNA-binding member of the SWI2/SNF2 family. *Dev. Biol.* **183,** 166–182.

Gossett, L. A., Kelvin, D. J., Sternberg, E. A., and Olson, E. N. (1989). A new myosite-specific enhancer-binding factor that recognizes a conserved element associated with multiple muscle-specific genes. *Mol. Cell. Biol.* **9,** 5022–5033.

Grieg, S. and Akam, M. (1993). Homeotic genes autonomously specify one aspect of pattern in the *Drosophila* mesoderm. *Nature* **362,** 630–632.

Grieshammer, U., Sassoon, D., and Rosenthal, N. (1992). A transgene target for positional regulators marks early rostrocaudal specification of myogenic lineages. *Cell* **69,** 79–93.

Gu, W., Schneider, J. W., Condorelli, G., Kaushal, S., Mahdavi, V., and Nadal, Ginard, B. (1993). Interaction of myogenic factors and the retinoblastoma protein mediates muscle cell commitment and differentiation. *Cell* **72,** 309–324.

Guth, L., and Samaha, F. (1969). Qualitative differences between actomyosin ATPase of slow and fast mammalian muscle. *Exp. Neurol.* **25,** 138–152.

Halavey, O., and Lerman, O. (1993). Retinoic acid induces adult muscle cell differentiation mediated by the retinoic acid receptor-α. *J. Cell. Physiol.* **154,** 566–572.

Harnish, D. C., Jiang, H., Soprano, K. J., Kochhar, D. M., and Soprano, D. R. (1992). Retinoic acid receptor β2 mRNA is elevated by retinoic acid in vivo in susceptible regions of mid-gestation mouse embryos. *Dev. Dyn.* **194,** 239–246.

Hasty, P., Bradley, A., Morris, J. H., Edmondson, D. G., Venuti, J. M., Olson, E. M., and Klein, W. H. (1993). Muscle deficiency and neonatal death in mice with targeted mutation in the myogenin gene (see Comments). *Nature* **364,** 501–506.

Hayashi, S., and Scott, M. P. (1990). What determines the specificity of action of Drosphilia homeodomain proteins? *Cell* **63,** 883–894.

Hinterberger, T. J., Sassoon, D. A., Rhodes, S. J., and Konieczny, S. F. (1991). Expression of the muscle regulatory factor MRF4 during somite and skeletal myofiber development. *Dev. Biol.* **147,** 144–156.

Hinterberger, T. J., Mays, J. L., and Konieczny, S. F. (1992). Structure and myofiber-specific expression of the rat muscle regulatory gene MRF4. *Gene* **117,** 201–207.

Hirsinger, E., Duprez, D., Jouve, C., Malapert, P., Cooke, J., and Pourquie, O. (1997). Noggin acts downstream of Wnt and Sonic Hedgehog to antagonize BMP4 in avian somite patterning. *Development* 124, 4605–4614.

Hu, L., Crowe, D. L., Rheinwald, J. G., Chambon, P., and Gudas, L. J. (1991). Abnormal expression of retinoic acid receptors and keratin 19 by human oral and epidermal squamous cell carcinoma cell lines. *Cancer Res.* **51,** 3972–3981.

Hu, J. S., Olson, E. N., and Kingston, R. E. (1992). HEB, a helix–loop–helix protein related to E2A and ITF2 that can modulate the DNA-binding ability of myogenic regulatory factors. *Mol. Cell. Biol.* **12,** 1031–1042.

Hughes, S. M., Taylor, J. M., Tapscott, S. J., Gurley, C. M., Carter, W. J., and Peterson, C. A. (1993). Selective accumulation of MyoD and myogenin mRNAs in fast and slow adult skeletal muscle is controlled by innervation and hormones. *Development* **118,** 1137–1147.

Jacob, M., Christ, B., and Jacob, H. J. (1978). On the migration of myogenic stem cells into the prospective wing region of chick embryos. A scanning and transmission electron microscope study. *Anat. Embryol.* **153,** 179–193.

Jen, Y., Weintraub, H., and Benezra, R. (1992). Overexpression of Id protein inhibits the muscle differentiation program: In vivo association of Id with E2A proteins. *Genes Dev.* **6,** 1466–1479.

Johnson, R. L., Laufer, E., Riddle, R. D., and Tabin, C. (1994). Ectopic expression of Sonic Hedgehog alters dorsal–ventral patterning of somites. *Cell* **79,** 1165–1173.

Kaehn, K., Jacob, H. J., Christ, B., Hinrichsen, K., and Poelmann, R. E. (1988). The onset of myotome formation in the chick. *Anat. Embryol.* **177,** 191–201.

Kaprielain, Z., and Fambrough, D. M. (1987). Expression of fast and slow isoforms of tre $Ca2^{+}$-ATPase in developing chick skeletal muscle. *Dev. Biol.* **124,** 490–503.

Kessel, M. (1992). Respecification of vertebral identities by retinoic acid. *Development* **115,** 487–501.

Kessel, M., and Gruss, P. (1991). Homeotic transformations of murine vertebrae and concomitant alteration of Hox codes induced by retinoic acid. *Cell* **67,** 89–104.

Keynes, F. J., and Stern, C. D. (1988). Mechanisms of vertebrate segmentation. *Development* **104,** 413–429.

Kieny, M., Mauger, A., and Sengel, P. (1972). Early regionalization of the somitic mesoderm as studied by the development of the axial skeleton of the chick embryo. *Dev. Biol.* **28,** 142–161.

Kopan, R., Nuy, J. S., and Weintraub, H. (1994). The intracellular domain of mouse notch: A constitutively activated repressor of myogenesis directed at the basic helix–loop–helix region of MyoD. *Development* **173,** 2385–2396.

Krause, M., Fire, A., White Harrison, S., Weintraub, H., and Tapscott, S. (1992). Functional conservation of nematode and vertebrate myogenic regulatory factors. *J. Cell Sci. Suppl.* **16,** 111–115.

Krust, A., Kastner, P., Petkovich, M., Zelent, A., and Chambon, P. (1989). A third human retinoic acid receptor hRARy. *Proc. Natl. Acad. Sci. USA* **86,** 5310–5314.

Langille, R. M., Paulsen, D. F., and Solursh, M. (1989). Differential effects of physiological concentrations of retinoic acid in vitro on chondrogenesis and myogenesis in chick craniofacial mesenchyme. *Differentiation* **40,** 84–92.

Lassar, A. B., and Munsterberg, A. E. (1996). The role of positive and negative signals in somite patterning. *Curr. Opin. Neurobiol.* **6,** 57–63.

Lassar, A. B., Davis, R. L., Wright, W. E., Kadesch, T., Murre, C., Voronova, A., Baltimore, D., and Weintraub, H. (1991). Functional activity of myogenic HLH proteins requires hetero-oligomerization with E12/E47-like proteins in vivo. *Cell* **66,** 305–315.

Li, L., Chambard, J. C., Karin, M., and Olson, E. N. (1992a). Fos and Jun repress transcriptional activation by myogenin and MyoD: The amino terminus of Jun can mediate repression. *Genes Dev.* **6,** 676–689.

Li, L., Zhou, J., James, G., Heller Harrison, R., Czech, M. P., and Olson, E. N. (1992b). FGF activates myogenic helix–loop–helix proteins through phosphorylation of a conserved protein kinase C site in their DNA-binding domains. *Cell* **71,** 1181–1194.

Liem, K. F., Tremml, G., Roelink, H., and Jessel, T. M. (1995). Dorsal differentiation of neural plate cells induced by BMP-mediated signals from epidermal ectoderm. *Cell* **82,** 969–979.

Lilly, B., Zhao, B., Ranganayakulu, G., Patterson, B. M., Schulz, R., and Olson, E. (1995). Requirement of MADS domain transcription factor D-MEF2 for muscle formation in Drosophilia. *Science* **267,** 688–693.

Lin, Q., Schwarz, J., Bucana, C., and Olson, E. (1997). Control of mouse cardiac morphogenesis and myogenesis by transcription factor MEF2C. *Science* **276,** 1404–1407.

Lomo, T., and Rosenthal, G. (1972). Control of ACh sensitivity by muscle activity in the rat. *J. Physiol.* **221,** 493–513.

Ma, P. C. M., Rould, M. A., Weintraub, H., and Pabo, C. O. (1994). Crystal structure of MyoD bHLH domain–DNA complex: Perspectives on DNA recognition and implications for transcriptional activation. *Cell* **77,** 451–459.

Mak, K. L., To, R. Q., Kong, Y., and Konieczny, S. F. (1992). The MRF4 activation domain is required to induce muscle-specific gene expression. *Mol. Cell. Biol.* **12,** 4334–4346.

Mangelsdorf, D. J., and Evans, R. M. (1995). The RXR heterodimers and orphan receptors. *Cell* **83,** 841–850.

Marcelle, C., Stark, M. R., and Bronner-Fraser, M. (1997). Coordinate actions of BMPs, Wnts, Shh and noggin mediate patterning of the dorsal somite, *Development* **124,** 3955–3963.

Martin, J. F., Schwarz, J. J., and Olson, E. N. (1993). Myocyte enhancer factor (MEF) 2C: A tissue-restricted member of the MEF-2 family of transcription factors. *Proc. Natl. Acad. Sci. USA* **90,** 5282–5286.

McCaffery, P., and Drager, U. (1995). Retinoic acid synthesizing enzymes in the embryonic and adult vertebrate. *Adv. Exp. Med. Biol.* **372,** 173–183.

McGrew, M., Bogdanova, N., Hasegawa, K., Hughes, S., Kitsis, R., and Rosenthal, N. (1996). Distinct gene expression patterns in skeletal and cardiac muscle are dependent on common regulatory sequences in the MLC1/3 locus. *Mol. Cell. Biol.* **16,** 4524–4534.

McLennan, I. S. (1983). Differentiation of muscle fiber type in the chicken hind limb. *Dev. Biol.* **97,** 222–228.

Means, A. L., and Gudas, L. J. (1995). The roles of retinoids in vertebrate development. *Annu Rev. Biochem.* **64,** 201–233.

Megeney, L. A., Kablar, B., Garrett, K., Anderson, J. E., and Rudnicki, M. (1996). MyoD is required for myogenic stem cell function in adult skeletal muscle. *Genes Dev.* **10,** 1173–1183.

Mendelsohn, C. E., Ruberte, E., and Chamdon, P. (1992). Retinoid receptors in vetebrate limb development. *Dev. Biol.* **152,** 50–61.

Mendelsohn, C., Mark, M., Dolle, P., Dierich, A., Gaub, M.-P., Krust, A., Lampron, C., and Chambon, P. (1994). Retinoic acid receptor (RARβ) null mutant mice appear normal. *Dev. Biol.* **166,** 246–258.

Michelson, A. M. (1994). Muscle pattern diversification in *Drosophila* is determined by the autonomous function of homeotic genes in the embryonic mesoderm. *Development* **120,** 755–768.

Miner, J. H., and Wold, B. (1990). Herculin, a fourth member of the MyoD family of myogenic regulatory genes. *Proc. Natl. Acad. Sci. USA* **87,** 1089–1093.

Miner, J. H., Miller, J. B., and Wold, B. J. (1992). Skeletal muscle phenotypes initiated by ectopic MyoD in transgenic mouse heart. *Development* **114,** 853–860.

Molkentin, J. D., Black, B. L., Martin, J. F., and Olson, E. (1995). Cooperative activation of muscle gene expression by MEF2 and myogenic bHLH proteins. *Cell* **83,** 1125–1136.

Molkentin, J. D., Lu, J.-R., Antos, C. L., Markham, B., Richardson, J., Robbins, J., Grant, S. E., and Olson, E. N. (1998). A calcineurin-dependent transcriptional pathway for cardiac hypertrophy. *Cell* **93,** 215–228.

Momoi, T., Miyagawa-Tomita, S., Nakamura, S., Kimura, I., and Momoi, M. (1992). Retinoic acid ambivalently regulates the expression of MyoD1 in the myogenic cells in the limb buds of the early developmental stages. *Biochem. Biophys. Res. Commun.* **187,** 245–253.

Montarras, D., Chelly, J., Bober, E., Arnold, H., Ott, M. O., Gros, F., and Pinset, C. (1991). Developmental patterns in the expression of Myf5, MyoD, myogenin, and MRF4 during myogenesis. *New Biol.* **3,** 592–600.

Morgan, B. A., Ipzisua-Belmonte, J. C., Deboule, D., and Tabin, C. (1992). Targeted misexpression of HDX 4–6 in the Avian Limb Bud Causes Apparent Homeotic Transformations. *Nature* **358,** 236–239.

Moss, J. B., Xavier-Neto, J., Shapiro, M. D., Nayeem, S. A., McCaffrey, P., Drager, U., and Rosenthal, N. (1998). Dynamic patterns of retinoic acid synthesis and response in the developing mammalian heart. *Dev. Biol.* **199,** 55–71.

Muneoka, K., and Sassoon, D. (1992). Molecular aspects of regeneration in developing vertebrae limbs. *Dev. Biol.* **152,** 37–49.

Nabeshima, Y., Hanaoka, K., Hayasaka, M., Esumi, E., Li, S., and Nonaka, I. (1993). Myogenin gene disruption results in perinatal lethality because of severe muscle defect (see Comments). *Nature* **364,** 532–535.

Niederreither, K., McCaffery, P., Drager, U., Chambon, P., and Dolle, P. (1997). Restricted expression and retinoic-induced downregulation of the retinaldehyde dehydrogenase type 2 (RALDH-2) gene during mouse development. *Mech. Dev.* **62,** 67–78.

Olson, E. N. (1990). MyoD family: A paradigm for development? *Genes Dev.* **4,** 1454–1461.

Olson, E. N., and Rosenthal, N. (1994). Homeobox genes and muscle patterning. *Cell* **79,** 9–12.

Ordahl, C. P., and Le Douarin, N. M. (1992). Two myogenic lineages within the developing somite. *Development* **114,** 339–353.

Ott, M. O., Bober, E., Lyons, G., Arnold, H., and Buckingham, M. (1991). Early expression of the myogenic regulatory gene, myf-5, in precursor cells of skeletal muscle in the mouse embryo. *Development* **111,** 1097–1107.

Palmeirim, I., Dubrulle, J., Henrique, D., Ish-Horowicz, D., and Pourquie, O. (1998). Uncoupling segmentation and somitogenesis in the chick presomitic mesoderm. *Dev. Gen.,* in press.

Peterson, C. A., Gordon, H., Hall, Z. W., Paterson, B. M., and Blau, H. M. (1990). Negative control of the helix–loop–helix family of myogenic regulators in the NFB mutant. *Cell* **62,** 493–502.

Petkovich, M., Brand, N. J., Krust, A., and Chambon, P. (1987). A human retinoic acid receptor which belongs to the family of nuclear receptors. *Nature* (*London*) **330,** 444–450.

Phillips, W. D., Everett, A. W., and Bennett, M. R. (1986). The role of innervation in the establishment of the topographical distribution of primary myotubes during development. *J. Neurocyt.* **15,** 397–405.

Piccolo, S., Sasai, Y., Lu, B., and De Robertis, E. M. (1996). Dorsal-ventral patterning in Xenopus: Inhibition of ventral signals by direct binding of Chordin to BMP-4. *Cell* **86,** 589–598.

Pollock, R., and Treisman, R. (1991). Human SRF-related proteins: DNA-binding properties and potential regulatory targets. *Genes Dev.* **5,** 2327–2341.

Pourquie, O., Coltey, M., Breant, C., Le Douarin, N. M. (1995). Control of somite patterning by signals from the lateral plate. *Proc. Natl. Acad. Sci. USA* **92,** 3219–3223.

Pourquie, O., Fan, C. M., Coltey, M., Hirsinger, E., Watanabe, Y., Breant, C., Grancis-West, P., Brickell, P., Teissier-Lavigne, M., and Le Douarin, N. M. (1996). Lateral and axial signals involved in avian somite patterning: A role for BMP-4. *Cell* **84,** 461–471.

Pownall, M. E., and Emerson, C. P., Jr. (1992). Sequential activation of three myogenic regulatory genes during somite morphogenesis in quail embryos. *Dev. Biol.* **151,** 67–79.

Psychoyos, D., and Stern, C. D. (1996). Fates and migratory routes of primitive streak cells in the chick embryo. *Development* **122,** 1523–1534.

Rao, M. V., Donoghue, M. J., Merlie, J. P., and Sanes, J. R. (1996). Distinct regulatory elements control muscle-specific, fiber-specific and axially graded expression of a myosin light chain gene in transgenic mice. *Mol. Cell. Biol.* **16,** 3909–3922.

Ranganayakulu, G., Zhao, B., Dokodis, A., Molkentin, J. D., Olson, E., and Schulz, R. (1995). A series of mutations in the D-MEF2 transcription factor reveals multiple functions in larval and adult myogenesis in *Drosophila. Dev. Biol.* **171,** 169–181.

Ranvier, L. (1874). Some observations relating to the histology and physiology of striated muscle. *Arch. Physiol. Normal Pathol.* **1,** 5.

Rashbass, J., Taylor, M. V., and Gurdon, J. B. (1992). The DNA-binding protein E12 co-operates with xMyoD in the activation of muscle-specific gene expression in Xenopus embryos. *EMBO J.* **11,** 2981–2990.

Reshef, R., Maroto, M., and Lassar, A. B. (1998). Regulation of dorsal somitic cell fates: BMPs and Noggin control the timing and pattern of myogenic regulator expression. *Genes Dev.* **12,** 290–303.

Rhodes, S. J., and Konieczny, S. F. (1989). Identification of MRF4: A new member of the muscle regulatory factor gene family. *Genes Dev.* **3,** 2050–2061.

Riddle, R., Johnson, R., Laufer, E., and Tabin, C. (1993). Sonic hedgehog mediates the polarizing activity of the ZPA. *Cell* **75,** 1401–1416.

Rijli, F. M., Mark, M., Lakkaraju, S., Dierich, A., Dolle, P., Chambon, P. (1993). A homeotic transformation is generated in the rostral branchial region of the head by disruption of Hoxa-2, which acts as a selector gene. *Cell* **75,** 1333–1349.

Riley, P., Cartwright-Anson, L., and Cross, J. C. (1998). The Hand1 bHALH transcription factor is essential for placentation and cardiac morphogenesis. *Nature Genet.* **18**(3), 271–275.

Rong, P. M., Teillet, M. A., Ziller, C., and Le Douarin, N. M. (1992). The neural tube : notocord complex is necessary for vertebral but not limb and body wall striated muscle differentiation. *Development* **115,** 657–672.

Rosenthal, N., Kornhauser, J. M., Donoghue, M. J., Rosen, K., and Merlie, J. P. (1989). The myosin light chain enhancer activates muscle specific developmentally regulated expression in transgenic mice. *Proc. Natl. Acad. Sci. USA* **86,** 7780–7784.

Rosenthal, N., Berglund, E. B., Wentworth, B. M., Donoghue, M., Winter, B., Bober, E., Braun, T., and Arnold, H. (1990). A highly conserved enhancer downstream of the human MLC1/3 locus is a target for multiple myogenic determination factors. *Nucleic Acid Res.* **18**(21), 6239–6246.

Rossant, J., Zirngibl, R., Cado, D., and Giguere, V. (1991). Expression of a retinoic acid response element-hsplacZ transgene defines specific domains of transcriptional activity during mouse embryogenesis. *Genes Dev.* **5,** 1333–1344.

Ruberte, E., Dolle, P., Krust, A., Zelent, A., Morriss-Kay, G., and Chambon, P. (1990). Specific spatial and temporal distribution of retinoic acid receptor γ transcripts during mouse embryogenesis. *Development* **108**(2), 213–222.

Ruberte, E., Dolle, P., Chambon, P., and Morris-Kay, G. (1991). Retinoic acid receptors and cellular retinoic binding proteins. II. Their differential pattern of transcription during early morphogenesis in mouse embryos. *Development* **111,** 45–60.

Rudnicki, M. A., Schnegelsberg, P. N., Stead, R. H., Braun, T., Arnold, H. H., and Jaenisch, R. (1993). MyoD or Myf-5 is required for the formation of skeletal muscle. *Cell* **75,** 1351–1359.

Sassoon, D. A., Wright, W., Lin, V., Lassar, A. B., Weintraub, H., and Buckingham, M. (1989). Expression of two myogenic regulatory factors, myogenin and MyoD, during mouse embryogenesis. *Nature* **341,** 303–307.

Schott, J.-J., Benson D. W., Basson, C. T., Pease, W., Silberbach, G. M., Moak, J. P., Maron, B. J., Seidman, C. E., and Seidman, J. G. (1998). Congenital heart disease caused by mutations in the transcription factor Nkx 2-5. *Science* **281,** 108–111.

Schwarz, J. J., Chakraborty, T., Martin, J., Zhou, J. M., and Olson, E. N. (1992). The basic region of myogenin cooperates with two transcription activation domains to induce muscle-specific transcription. *Mol. Cell. Biol.* **12,** 266–275.

Selleck, M. A., and Stern, C. D. (1991). Fate mapping and cell lineage analysis of Hensen's node in the chick embryo. *Development* **112,** 615–626.

Smith, S., and Dickman, E. D. (1997). New insights into retinoid signaling in cardiac development and physiology. *Trends Cardiovasc. Med.* **7,** 324–329.

Smith, T. H., Kachinsky, A. M., and Miller, J. B. (1994). Somite subdomains, muscle cell origins, and the four muscle regulatory factor proteins. *J. Cell Biol.* **127,** 95–105.

Srivastava, D., Cserjesi, P., and Olson, E. (1995). A subclass of bHLH proteins required for cardiac morphogenesis. *Science* **270,** 1995–1999.

Srivastava, D., Thomas, T., Lin, Q., Kirby, M., Brown, D., and Olson, E. (1997). Regulation of cardiac mesodermal and neural crest development by the bHLH transcription factor, dHAND. *Nature Genet.* **16,** 154–160.

Stockdale, F. E. (1992). Myogenic cell lineages. *Dev. Biol.* **154,** 284–298.

Sucov, H. M., Murakami, K. K., and Evans, R. M. (1990). Characterization of an autoregulated response element in the mouse retinoic acid receptor type β gene. *Proc. Natl. Acad. Sci. USA* **87,** 5392–5396.

Summerbell, D., and Maden, M. (1990). Retinoic acid, a developmental signaling molecule. *TINS* **13**(4), 142–147.

Tajbakhsh, S., and Cossu, G. (1997). Establishing myogenic identity during somitogenesis. *Curr. Opin. Genes Dev.* **7,** 634–641.

Tajbakhsh, S., Rocancourt, D., Cossu, G., and Buckingham, M. (1997). Redefining the genetic hierarchies controlling skeletal myogenesis: Pax-3 and Myf-5 act upstream of MyoD. *Cell* **89,** 127–138.

Thaller, C., and Eichele, G. (1987). Identification and spatial distribution of retinoids in the developing chick limb bud. *Nature* **327,** 625–628.

Thayer, M. J., Tapscott, S. J., Davis, R. L., Wright, W. E., Lassar, A. B., and Weintraub, H. (1989). Positive autoregulation of the myogenic determination gene MyoD1. *Cell* **58,** 241–248.

Thompson, D'A. (1942). *On Growth and Form.* Cambridge Univ. Press, Cambridge, UK.

Thompson, W. R., Nadal-Ginard, B., and Mahdavi, V. (1991). A MyoD1-independant muscle-specific enhancer controls the expression of the b-myosin heavy chain gene in skeletal and cardiac muscle cells. *J. Biol. Chem.* **266,** 22678–22688.

Tickle, C., Alberts, B., Wolpert, L., and Lee, J. (1982). Local application of retinoic acid to the limb buds mimics the action of the polarizing region. *Nature* **296,** 564–566.

Tonegawa, A., Funayama, N., Ueno, N., and Takahashi, Y. (1997). Mesodermal subdivision along the mediolateral axis in chicken controlled by different concentrations of BMP-4. *Development* **124,** 1975–1984.

Treisman, J., Harris, E., Wilson, D., and Desplan, C. (1992). The homeodomain: A new face for the helix–turn–helix? *Bioessays* **14,** 145–150.

Venuti, J. M., Goldberg, L., Chakraborty, T., Olson, E. N., and Klein, W. H. (1991). A myogenic factor from sea urchin embryos capable of programming muscle differentiation in mammalian cells. *Proc. Natl. Acad. Sci. USA* **88,** 6219–6223.

Waddington, C. H. (1932). Experiments on the development of the chick and the duck embryo cultivated in vitro. *Proc. Trans. R. Soc. London B* **211,** 179–230.

Wang, Y., Benezra, R., and Sassoon, D. A. (1992). Id expression during mouse development: A role in morphogenesis. *Dev. Dyn.* **194,** 222–230.

Weintraub, H. (1993). The MyoD family and myogenesis: redundancy, networks, and thresholds. *Cell* **75,** 1241–1244.

Weintraub, H., Davis, R., Lockshon, D., and Lassar, A. (1990). MyoD binds cooperatively to two sites in a target enhancer sequence: Occupancy of two sites is required for activation. *Proc. Natl. Acad. Sci. USA* **87,** 5623–5627.

Weintraub, H., Davis, R., Tapscott, S., Thayer, M., Krause, M., Benezra, R., Blackwell, T. K., Turner, D., Rupp, R., Hollenberg, S., *et al.* (1991). The MyoD gene family: Nodal point during specification of the muscle cell lineage. *Science* **251,** 761–766.

Wentworth, B. M., Donoghue, M., Engert, J. C., Berglund, E. B., and Rosenthal, N. (1991). Paired MyoD-binding sites regulate myosin light chain gene expression. *Proc. Natl. Acad. Sci. USA* **88,** 1242–1246.

White, N. K., Bonner, P. H., Nelsen, D. R., and Hauschka, S. D. (1975). Clonal analysis of vertebrate myogenesis. IV. Medium dependent classification of colony-forming cells. *Dev. Biol.* **44,** 346–361.

Winter, B., Braun, T., and Arnold, H. H. (1992). Co-operativity of functional domains in the muscle specific transcription factor Myf-5. *EMBO J.* **11,** 1843–1855.

Wright, W. E., Sassoon, D. A., and Lin, V. K. (1989). Myogenin, a factor regulating myogenesis, has a domain homologous to MyoD. *Cell* **56,** 607–617.

Wu, T. C., Wang, L., and Wan, Y. J. (1992). Retinoic acid regulates gene expression of retinoic acid receptor α, β, and γ in F9 teratocarcinoma cells. *Differentiation* **51**(3), 219–224.

Xiao, Y.-H., Grieshammer, U., and Rosenthal, N. (1995). Regulation of a muscle-specific transgene by retinoic acid. *J. Cell Biol.* **129,** 1345–1354.

Yee, S. P., and Rigby, P. W. (1993). The regulation of myogenin gene expression during the embryonic development of the mouse. *Genes Dev.* **7,** 1277–1289.

Yu, Y. T., Breitbart, R. E., Smoot, L. B., Lee, Y., Mahdavi, V., and Nadal Ginard, B. (1992). Human myocyte-specific enhancer factor 2 comprises a group of tissue-restricted MADS box transcription factors. *Genes Dev.* **6,** 1783–1798.

Yutzey, K. E., and Konieczny, S. F. (1992). Different E-box regulatory sequences are functionally distinct when placed within the context of the troponin I enhancer. *Nucleic Acids Res.* **20,** 5105–5113.

Zelent, A., Krust, A., Petkovich, M., Kastner, P., and Chambon, P. (1989). Cloning of murine retinoic acid receptor α and β cDNAs and of a novel third receptor γ predominantly expressed in skin. *Nature (London)* **339,** 714–717.

Zhao, D., McCaffrey, P., Ivins, K. J., Neve, R. L., Hogan, P., Chin, W. W., and Drager, U. C. (1996). Molecular identification of a major retinoic acid synthesizing enzyme, a retinaldehyde-specific dehydrogenase. *Eur. J. Biochem.* **240,** 15–22.

Zhuang, Y., Cheng, P., and Weintraub, H. (1996). β-lymphocyte development is regulated by the combined dosage of three basic helix-loop-helix genes, E2A, E2-2 and HEB. *Mol. Cell. Biol.* **16,** 2898–2905.

Zimmerman, L. B., De Jesus-Escobar, J. M., and Harland, R. M. (1996). The Spemann organizer signal Noggin binds and inactivates bone morphogenetic protein 4. *Cell* **86,** 599–606.

Index

A

B

D

H

I

L

M

N

T

X

Y

Z